# Elements of Metallurgy and Engineering Alloys

Edited by
F.C. Campbell

The Materials Information Society

ASM International®
Materials Park, Ohio 44073-0002
www.asminternational.org

First printing, June 2008
Digital printing, July 2011



Comments, criticisms, and suggestions are invited, and should be forwarded to ASM International.

*Prepared under the direction of the ASM International Technical Book Committee (2007–2008), Lichun L.Chen, Chair.*

*ASM International staff who worked on this project include Scott Henry, Senior Manager of Product and Service Development; Charles Moosbrugger, Technical Editor; Eileen De Guire, Associate Editor; Ann Britton, Editorial Assistant; Bonnie Sanders, Manager of Production; Madrid Tramble, Senior Production Coordinator; Diane Grubbs, Production Coordinator; Rachel Frayser, Production Coordinator; and Kathryn Muldoon, Production Assistant*

Library of Congress Control Number: 2008925434
ISBN-13: 978-0-87170-867-0
ISBN-10: 0-87170-867-1
SAN: 204-7586

ASM International®
Materials Park, OH 44073-0002
www.asminternational.org

Printed in the United States of America

# Contents

# Preface

This book is intended for those wishing to learn more about metallurgy and the metals and engineering alloys used throughout industry. It would be useful to almost anyone who deals with metals, including designers, structural engineers, materials and process engineers, manufacturing engineers, and production personnel. It was written so that both the fundamentals of metallurgy and the specifics of the important engineering alloys could be combined in a single book. Throughout, the emphasis of this book is on practical principles of metallurgy and the engineering alloys.

It should be noted that it is not the objective of this book to replace handbooks on engineering alloys. Rather, the goal is to gather information on engineering alloys in one book and edit it into a uniform format so that the reader is able to compare the metallurgy, properties, and applications of the most important engineering alloy systems. Many of the chapters are adapted from the *ASM Handbook* series. When this was not possible, I wrote chapters to meet this objective.

The book is separated into two parts. Part I covers the fundamentals of physical and mechanical metallurgy, while Part II goes into greater detail for specific metals and their alloys.

The first eleven chapters cover most of the basics of physical metallurgy. Physical metallurgy may be described as the study of how metallic structure and processing affect the final properties of metals and alloys. Chapters on physical metallurgy include metallic structure; crystalline imperfections and plastic deformation; solid solutions; diffusion; phase transformations; phase diagrams; solidification and casting; recovery, recrystallization, and grain growth; precipitation hardening; the iron-carbon system; and the heat treatment of steel. This is followed by five chapters dealing with mechanical metallurgy, which is the study of how loads, either during fabrication or in service, affect metals and their properties. Chapters on mechanical metallurgy include mechanical behavior, fracture, fatigue, creep, and deformation processing. The last two chapters in Part I cover physical properties and corrosion.

Part II deals with specific engineering alloys. The first six chapters cover the important ferrous metal alloys: plain carbon steels, alloy steels, surface hardening of steels, tool materials, stainless steels, and cast irons. The next seven chapters are on the nonferrous metals and their alloys: copper, aluminum, magnesium and zinc, titanium, nickel and cobalt, superalloys, and refractory metals. This is followed by a chapter on several miscellaneous nonferrous metals and alloys. The last chapter deals with metal-matrix composites.

A first course in materials science would be helpful in understanding the material in this book; however, most of the material is self-explanatory and builds on itself as one progresses through the book.

I would like to acknowledge the help and guidance of the ASM International staff. Steve Lampman and Charles Moosbrugger championed the book through the approval process at ASM. Eileen De Guire, with help from Charles Moosbrugger, did the majority of the editorial work. Ann Britton provided help and guidance with the permissions. The efforts of Madrid Tramble, Kathryn Muldoon, and Liz Marquard in converting the manuscript into a book are appreciated. I would also like to thank the ASM peer reviewers for their valuable critiques. Finally, I would like to thank my wife Betty for her support and encouragement during this project.

*F.C. Campbell*
*St. Louis, Missouri*
*December 2007*

# Part I
# Physical and Mechanical Metallurgy

## CHAPTER 1

# Metallic Structure

THE WORD *METAL*, derived from the Greek *metallon*, is believed to have originated as a verb meaning to seek, search after, or inquire about. Today, a metal is defined as any element that tends to lose electrons from the outer shells of its atoms. The resulting positive ions are held together in crystalline structure by the cloud of these free electrons in what is known as the metallic bond. The metallic bond yields three physical characteristics typical of solid metals: (1) metals are good conductors of electricity, (2) metals are good conductors of heat, and (3) metals have a lustrous appearance. In addition, most metals are malleable, ductile, and generally denser than other elemental substances. Those elements that do not display the characteristics of the metallic elements are called nonmetals. However, there are some elements that behave as metals under some circumstances and as nonmetals under different circumstances. These are now called semimetals but have also been called metalloids, meaning like metals. The boundaries separating the regions in the periodic table covered by the different classes of elements are not distinct, except that nonmetals never form positive ions. A simplified periodic table is shown in Fig. 1.1, highlighting the elements that are currently considered to be metals.

## 1.1 Periodic Table

In the periodic table, it is the number of electrons in the outer shell that affects the properties of the elements the most. Those elements that have the same number of electrons

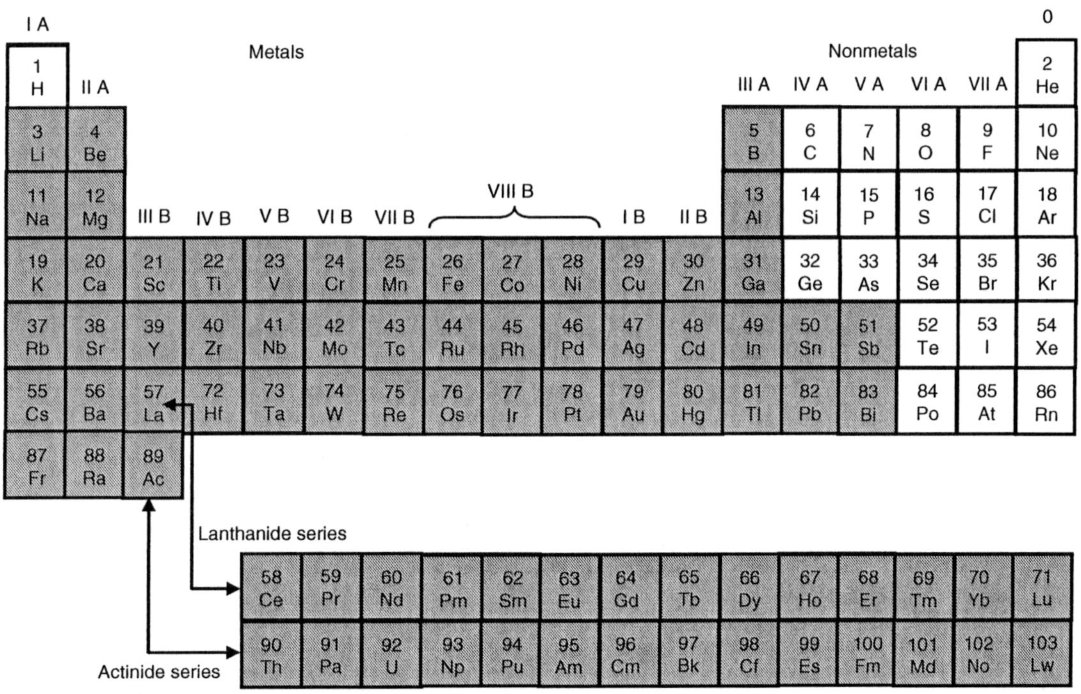


**Fig. 1.1** Periodic table of the elements

in their outermost electron shells, and therefore have similar chemical behavior, are placed in columns. For example, lithium, sodium, and potassium each have a single electron in their outer shells and are chemically very similar. They all oxidize very rapidly and react vigorously with water, liberating hydrogen and forming soluble hydroxides. They are physically very similar, being soft, light metals with a somewhat silver color. At the other end of the periodic table, the gases fluorine and chlorine, with seven electrons in their outer shells, also have similar chemical properties. Both are gases with strong nonmetallic properties. At the far right side of the periodic table, the noble gases helium, neon, and argon contain eight electrons in their outer shells. Since this fills the shell, these gases are nonreactive, or inert, under normal circumstances. Therefore, the chemical interaction between elements is governed by the number of electrons present in the outer shell. When the outer shell is filled, the atom has no further tendency to combine or react with other atoms.

Metallic properties depend on both the nature of their constituent atoms and the way in which they are assembled. Assemblies of atoms can be gases, liquids, or solids. When they are in the solid state, metals are normally arranged in a crystalline structure. The crystalline nature of metals is responsible for their ultimate engineering usefulness, and the crystalline arrangement strongly influences their processing. Although metals can exist as single crystals, they are more commonly polycrystalline solids with crystalline grains of repeating atomic packing sequences. Periodic crystalline order is the equilibrium structure of all solid metals. Crystalline structures are a dominant factor in determining mechanical properties, and crystal structures also play an important role in the magnetic, electrical, and thermal properties. The greatest bonding energy occurs when the atoms are closely packed, and the atoms in a crystalline structure tend to pack as densely as possible. In addition, total metallic bonding energy is increased when each atom has the greatest possible number of nearest neighbor atoms. However, due to a shared bonding arrangement in some metals that is partially metallic and partially covalent, some metals do not crystallize into these close-packed structures. Covalent tendencies appear as one moves closer to the nonmetals on the periodic table. As one moves rightward across the periodic table, progressively greater numbers of metals have looser-packed structures. Most metals bordering the nonmetals possess more complex structures with lower packing densities, because covalent bonding plays a large role in determining their crystal structures. The directionality of covalent bonding dictates fewer nearest neighbors than exist in densely packed metallic crystals. For metals near the nonmetals on the right side of the periodic table, where electronegativities are high, covalency becomes a major part of the bonding.

Properties important to the engineer are strongly influenced by crystal structure. One of the most important properties related to crystal structure is ductility. Densely packed structures usually allow motion on one or more slip planes, permitting the metal to deform plastically without fracturing. Ductility is vital for easy formability and for fracture toughness, two properties that give metals a great advantage over ceramic materials for many engineering uses.

## 1.2 Bonding in Solids

Bonding in solids may be classified as either primary or secondary bonding. Methods of primary bonding include the metallic, ionic, and covalent bonds. Secondary bonds are much weaker bonding mechanisms that are only predominant when one of the primary bonding mechanisms is absent. When two atoms are brought close to each other, there will be a repulsion between the negatively charged electrons of each atom. The repulsion force increases rapidly as the distance of separation decreases. However, when the separation is large, there is attraction between the positive nucleus charge and the negative charge of the electrons. At some equilibrium distance, the attractive and repulsive forces balance each other, and the net force is zero. At this equilibrium distance, the potential energy is at a minimum, as shown in Fig. 1.2. The magnitude of this energy is known as the bond energy, usually expressed in kJ/mol. Primary bond energies range from 100 to 1000 kJ/mol, while the much weaker secondary bonds are on the order of only 1 to 60 kJ/mol. The equilibrium distance, $a_0$, is the bond length. Strong primary bonds have large forces of attraction, with bond lengths of 1 to 2 Å, while the weaker secondary bonds have larger bond lengths of 2 to 5 Å. While it is convenient to discuss the four major types of bonding separately, it should be recognized that although metallic bonding may be predominant, other types of bonding, in particular covalent bonding, may also be present. A comparison of the some of the properties of the different bond types is given in Table 1.1.

### *1.2.1 Metallic Bonding*

Metallic bonding occurs when each of the atoms of the metal contributes its valence electrons to the formation of an electron cloud that surrounds the positively charged metal ions, as illustrated in Fig. 1.3. Hence, the valence electrons are shared by all of the atoms. In this bond, the positively charged ions repel each other uniformly, so they arrange themselves into a regular pattern that is held together by the negatively charged electron cloud. Since the negative electron cloud surrounds each of the positive ions that make up the orderly three-dimensional crystal structure, strong electronic attraction holds the metal together. A characteristic of metallic bonding is the fact that every positive ion is equivalent. Ideally, a symmetrical ion is produced when a valence electron is removed from the metal atom. As a result of this ion symmetry, metals tend to form highly symmetrical, close-packed crystal structures. They also have a large number of nearest neighboring atoms (usually 8 to 12), which helps to explain their high densities and high elastic stiffness.

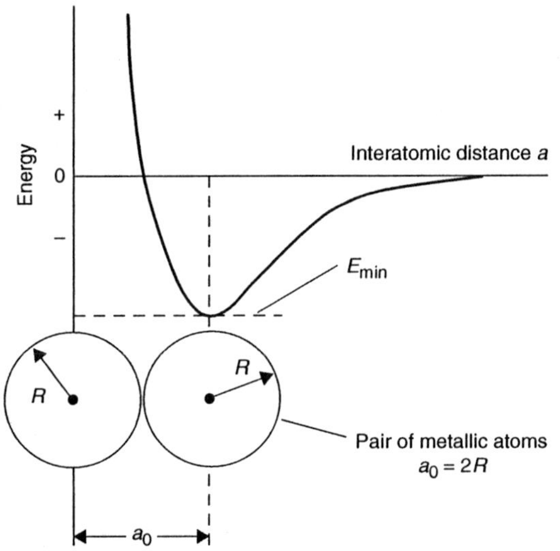


**Fig. 1.2** Bond energy in metallic bond

Since the valence electrons are no longer attached to specific positive ions and are free to travel among the positive ions, metals exhibit high electrical and thermal conductivity. The opaque luster of metals is due to the reflection of light by the free electrons. A light wave striking the surface causes the free electrons to vibrate and absorb all the energy of the wave and prevent transmission. The vibrating electrons then reemit, or reflect, the wave from the surface. The ability of metals to undergo significant amounts of plastic deformation is also due to the metallic bond. Under the action of an applied shearing force, layers of the positive ion cores can slide over each other and reestablish their bonds without drastically altering their relationship with the electron cloud. The ability to alloy, or mix several metals together in the liquid state, is one of the keys to the flexibility of metals. In the liquid state, solubility is often complete, while in the solid state, solubility is generally much more restricted. This change in solubility with temperature forms the basis for heat treatments that can vary the strength and ductility over quite wide ranges.

In general, the fewer the valence electrons and more loosely they are held, the more metallic is

**Table 1.1 General characteristics of bond types**

| Property | Metallic bond | Covalent bond | Ionic bond | Secondary bond |
|---|---|---|---|---|
| Example | Cu, Ni, Fe | Diamond, silicon carbide | NaCl, $CaCl_2$ | Wax, Ar |
| Mechanical | Weaker than ionic or covalent bond<br>Tough and ductile<br>Nondirectional | Very hard and brittle<br>Fails by cleavage<br>Strongly directional | Hardness increases with ionic charge<br>Fails by cleavage<br>Nondirectional | Weak and soft<br>Can be plastically deformed |
| Thermal | Moderately high melting points<br>Good conductors of heat | Very high melting points<br>Thermal insulators | Fairly high melting points<br>Thermal insulators | Low melting points |
| Electrical | Conductors | Insulators | Insulators | Insulators |
| Optical | Opaque and reflecting | Transparent or opaque<br>High refractive index | Transparent<br>Colored by ions | Transparent |

Source: Ref 1

the bonding. Metals such as copper and silver, which have few valence electrons, are very good conductors of electricity and heat, because their few valence electrons are highly mobile. As the number of valence electrons increases and the tightness with which they are held to the nucleus increases, the valence electrons become localized and the bond becomes more covalent. The transition metals, such as iron and nickel, have incomplete *d*-shells and exhibit some covalent bonding, which helps explain their relatively high melting points. Tin is interesting in that it has two crystalline forms, one which is mostly metallic and ductile and another which is mostly covalent and very brittle. Intermetallic compounds can also be formed between two metals in which the bonding is partly metallic and partly ionic. As the electronegativity difference between the two metals increases, the bonding becomes more ionic in nature. For example, both aluminum and vanadium have an electronegativity of 1.5 and the difference is 0, so the compound $Al_3V$ is primarily metallic. On the other hand, aluminum and lithium (electronegativity of 1.0) have an electronegativity difference of 0.5; thus, when they

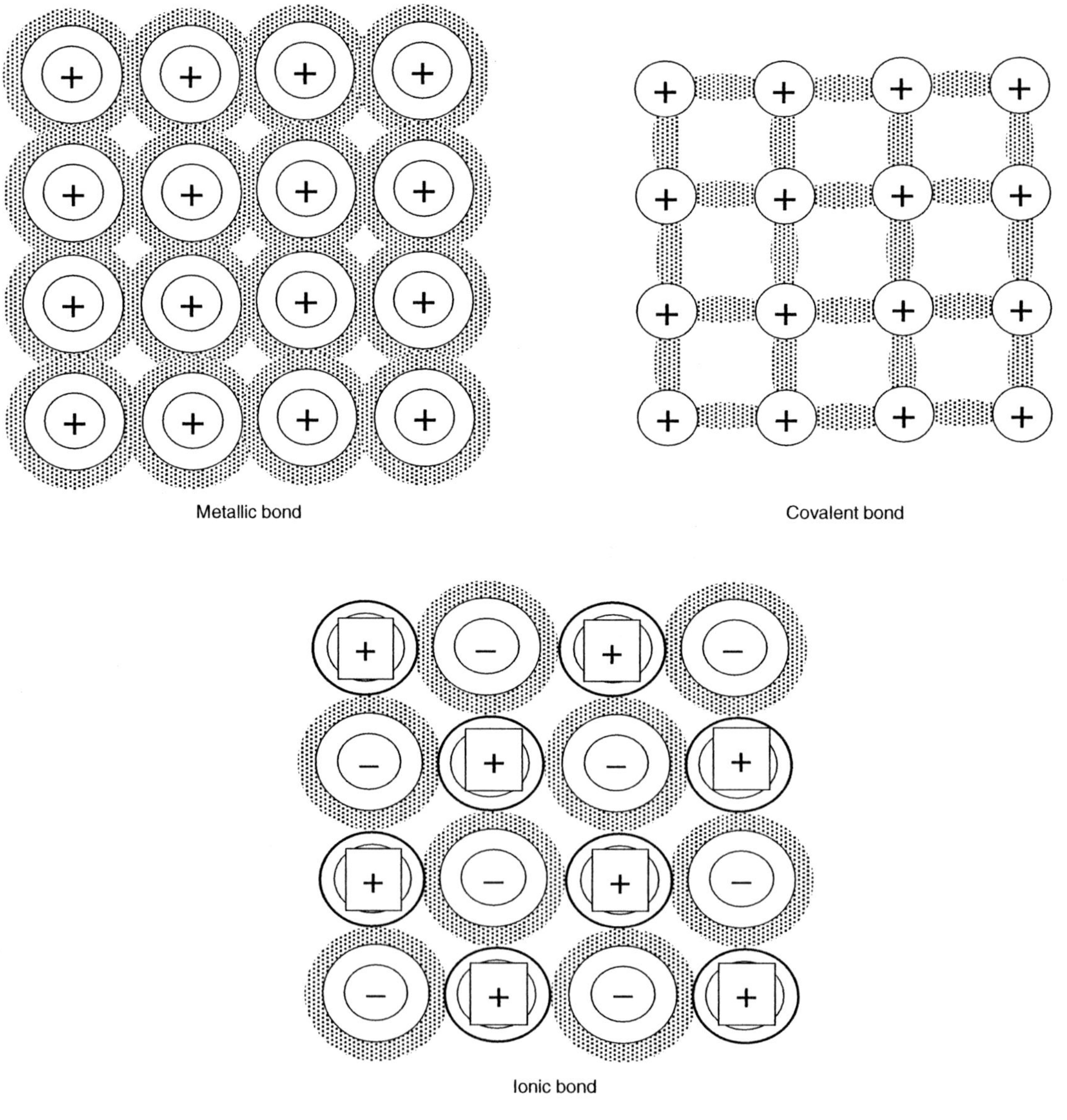


**Fig. 1.3** Primary bonding mechanisms

form the compound AlLi, the bond is a combination of metallic and ionic.

### 1.2.2 Ionic Bonding

Ionic bonding, also shown in Fig. 1.3, is a result of electrical attraction between alternately placed positive and negative ions. In the ionic bond, the electrons are shared by an electropositive ion (cation) and an electronegative ion (anion). The electropositive ion gives up its valence electrons, while the electronegative ion captures them to produce ions having full electron orbitals or suborbitals. As a consequence, there are no free electrons available to conduct electricity. In ionically bonded solids such as salts, there are very few slip systems along which atoms can move. This is a consequence of the electrically charged nature of the ions. For slip in some directions, ions of like charge must be brought into close proximity to each other, and because of electrostatic repulsion, this mode of slip is very restricted. This is not a problem in metals, since all atoms are electronically neutral. No electrical conduction of the kind found in metals is possible in ionic crystals, but weak ionic conduction occurs as a result of the motion of the individual ions. When subjected to stresses, ionic crystals tend to cleave, or break, along certain planes of atoms rather than deform in a ductile fashion as metals do.

Ionic bonds form between electropositive metals and electronegative nonmetals. The further apart the two are on the periodic table, the more likely they are to form ionic bonds. For example, sodium (Na) is on the far left side of the periodic table in group I, while chlorine (Cl) is on the far right side in group VII. Sodium and chlorine combine to form common table salt (NaCl). As shown in Fig. 1.4, the sodium atom gives up its outer valence electron, which is transferred to the outer electron shell of the chlorine atom. Since the outer shell of chlorine now contains eight electrons, similar to the noble gases, it is an extremely stable configuration. In terms of symbols, the sodium ion is written as $Na^+$, and the chlorine ion is written as $Cl^-$. When the two combine to form an ionic bond, the compound (NaCl) is neutral since the charges balance. Since the positively charged cation can attract multiple negatively charged anions, the ionic bond is nondirectional.

### 1.2.3 Covalent Bonding

Many elements that have three or more valence electrons are bound into crystal structures by forces arising from the sharing of electrons. The nature of this covalent bonding is shown schematically in Fig. 1.3. To complete the octet of electrons needed for atomic stability, electrons must be shared with 8-*N* neighboring atoms, where *N* is the number of valence electrons in the given element. High hardness and low electrical conductivity are general characteristics of solids of this type. In covalently bonded ceramics, the bonding between atoms is specific and directional, involving the exchange of electron charge between pairs of atoms. Thus, when covalent crystals are stressed to a sufficient extent, they exhibit brittle fracture due to a separation of electron pair bonds, without subsequent reformation. It should also be noted that ceramics are rarely either all ionically or covalently bonded; they usually consist of a mix of the two types of bonds. For example, silicon nitride ($Si_3N_4$) consists of approximately 70% covalent bonds and 30% ionic bonds.

Covalent bonds also form between electropositive elements and electronegative elements. However, the separation on the periodic table is not great enough to result in electron

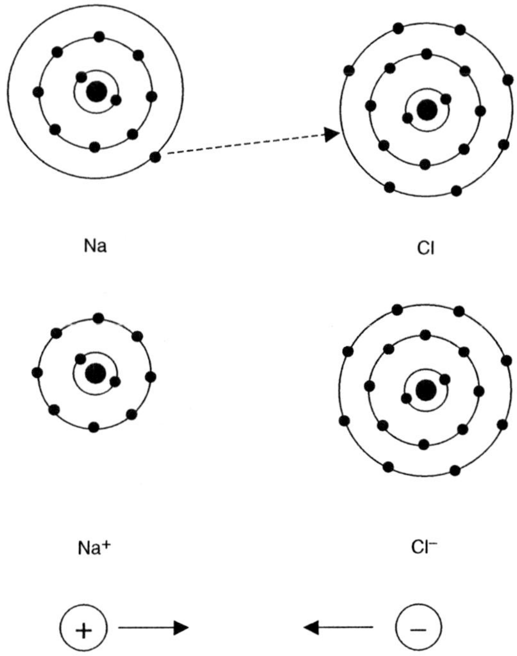


**Fig. 1.4** Ionic bonding in NaCl

transfer, as in the ionic bond. Instead, the valence electrons are shared between the two elements. For example, a molecule of methane ($CH_4$), shown in Fig. 1.5, is held together by covalent bonds. Note that hydrogen, in group I on the periodic table, and carbon, in group IV, are much closer together than sodium and chlorine, which form ionic bonds. In a molecule of methane gas, four hydrogen atoms are combined with one carbon atom. The carbon atom has four electrons in its outer shell, and these are combined with four more electrons, one from each of the four hydrogen atoms, to give a completed stable outer shell of eight electrons held together by covalent bonds. Each shared electron passes from an orbital controlled by one nucleus into an orbital shared by two nuclei. Covalent bonds, since they do not ionize, will not conduct electricity and are nonconductors. Covalent bonds form the basis for many organic compounds, including long-chain polymer molecules. As the molecule size increases, the bond strength of the material also increases. Likewise, the strength of long-chain molecules also increases with increases in chain length.

### *1.2.4 Secondary Bonding*

Secondary, or van der Waals, bonding is weak in comparison to the primary metallic, ionic, and covalent bonds. Bond energies are typically on the order of only 10 kJ/mol (0.1 eV/atom). Although secondary bonding exists between virtually all atoms or molecules, its presence is usually obscured if any of the three primary bonding types is present. While van der Waals forces only play a minor role in metals, they are an important source of bonding for the inert gases that have stable electron structures, some molecular compounds such as water, and thermoplastic polymers where the main chains are covalently bonded but are held to other main chains by secondary bonding. Van der Waals bonding between two dipoles is illustrated in Fig. 1.6.

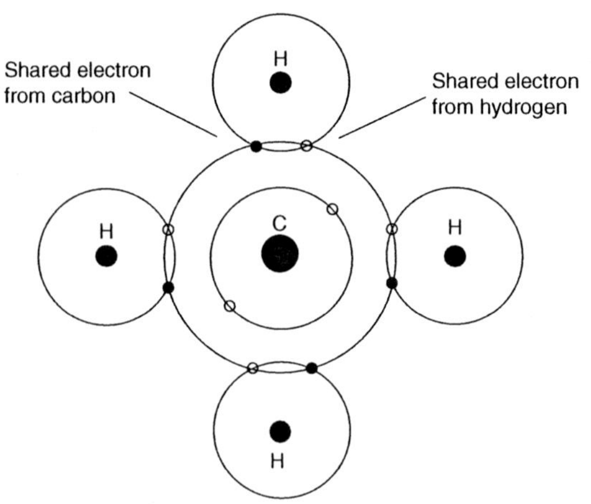


**Fig. 1.5** Covalent bonding in methane

## 1.3 Crystalline Structure

When a substance freezes on cooling from the liquid state, it forms a solid that is either an amorphous or a crystalline structure. An amorphous structure is essentially a random structure. Although there may be what is known as short-range order, in which small groups of atoms are arranged in an orderly manner, it does not contain long-range order, in which all of the atoms are arranged in an orderly manner. Typical amorphous materials include glasses and almost all organic compounds. However, metals, under normal freezing conditions, normally form long-range, orderly crystalline structures. Except for glasses, almost all ceramic materials also form crystalline structures. Therefore, metals and ceramics are, in general, crystalline, while glasses and polymers are mostly amorphous.

### *1.3.1 Space Lattices and Crystal Systems*

A crystalline structure consists of atoms, or molecules, arranged in a pattern that is repetitive in three dimensions. The arrangement of the atoms or molecules in the interior of a crystal is called its crystalline structure. A distribution of points (or atoms) in three dimensions is said to form a space lattice if every point has identical surroundings, as shown in Fig. 1.7. The intersections of the lines, called lattice points, represent locations in space with the same kind of atom or group of atoms of identical composition, arrangement, and orientation. The geometry of a space lattice is completely specified by the lattice constants $a$, $b$, and $c$ and the interaxial angles $\alpha$, $\beta$, and $\gamma$. The unit cell of a

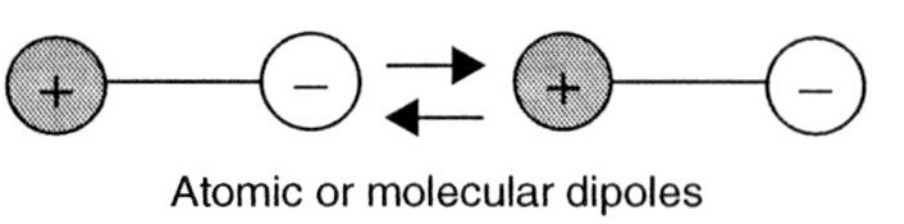


**Fig. 1.6** Van der Waals bonding between two dipoles

crystal is the smallest pattern of arrangement that can be contained in a parallelepiped, the edges of which form the *a*, *b*, and *c* axes of the crystal. Appendix C, "Crystallographic Planes and Directions," in this book describes how the Miller indices, a system for specifying crystallographic planes within the unit cell, are determined.

When discussing crystal structure, it is usually assumed that the space lattice continues to infinity in all directions. In terms of a typical crystal (or grain) of, for example, iron that is 0.2 $cm^3$ (0.01 $in.^3$) in size, this may appear to be a preposterous assumption, but when it is realized that there are $10^{18}$ iron atoms in such a grain, the approximation to infinity seems much more plausible.

All crystal systems can be grouped into one of seven basic systems, as defined in Table 1.2, which can be arranged in 14 different ways,

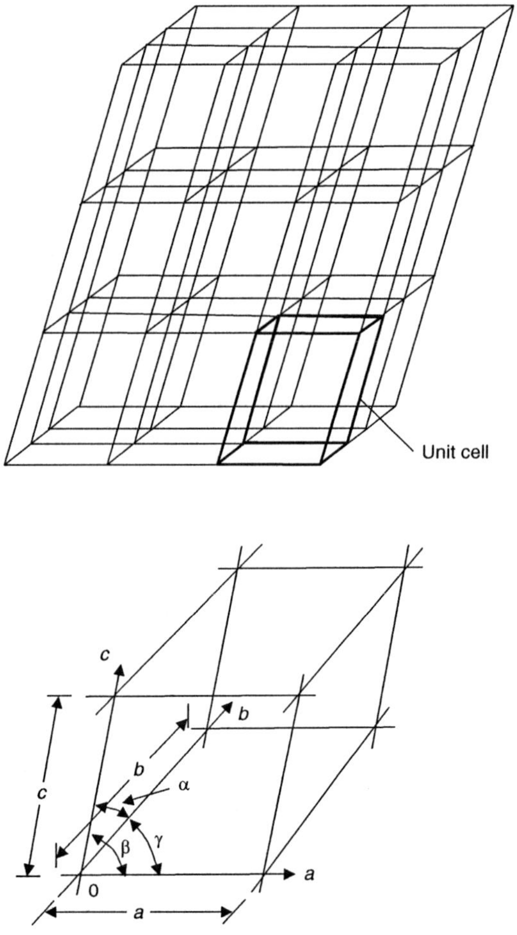


**Fig. 1.7** Space lattice and unit cell

called Bravais lattices, as shown in Fig. 1.8. Almost all structural metals crystallize into one of three crystalline patterns: face-centered cubic, (fcc) hexagonal close-packed, (hcp) or body-centered cubic, (bcc) illustrated in Fig. 1.9 to 1.11. It should be noted that the unit cell edge lengths and axial angles are unique for each crystalline substance. The unique edge lengths are called lattice parameters. Axial angles other than 90° or 120° can also change slightly with changes in composition. When the edges of the unit cell are not equal in all three directions, all unequal lengths must be stated to completely define the crystal. The same is true if all axial angles are not equal.

### *1.3.2 Face-Centered Cubic System*

The face-centered cubic (fcc) system is shown in Fig. 1.9. As the name implies, in addition to the corner atoms, there is an atom centrally located on each face. Since each of the atoms located on the faces belong to two unit cells and the eight corner atoms each belong to eight unit cells, the number of atoms belonging to a unit cell is four. The atomic packing factor (the volume of atoms belonging to the unit cell divided by the volume of the unit cell) is 0.74 for the fcc structure. This is the densest packing that can be obtained. The fcc structure is the most efficient, with 12 nearest atom neighbors (also referred to as the coordination number, or CN); that is, the fcc structure has a CN = 12. Methods of calculating atomic packing factors and coordination numbers are given in Appendix B, "Crystalline System Calculations," in this book. As shown in Fig. 1.12, the stacking sequence for the fcc structure is *ABCABC*. The fcc structure is found in many important metals such as aluminum, copper, and nickel.

### *1.3.3 Hexagonal Close-Packed System*

The atoms in the hexagonal close-packed (hcp) structure (Fig. 1.10) are also packed along close-packed planes. It should also be noted

**Table 1.2 Seven crystal systems**

| Crystal system | Edge length | Interaxial angle |
|---|---|---|
| Triclinic | $a \neq b \neq c$ | $\alpha \neq \beta \neq \gamma \neq 90°$ |
| Monoclinic | $a \neq b \neq c$ | $\alpha = \gamma = 90° \neq \beta$ |
| Orthorhombic | $a \neq b \neq c$ | $\alpha = \beta = \gamma = 90°$ |
| Tetragonal | $a = b \neq c$ | $\alpha = \beta = \gamma = 90°$ |
| Hexagonal | $a = b \neq c$ | $\alpha = \beta = 90°, \gamma = 120°$ |
| Rhombohedral | $a = b = c$ | $\alpha = \beta = \gamma \neq 90°$ |
| Cubic | $a = b = c$ | $\alpha = \beta = \gamma = 90°$ |

that both the fcc and hcp structures are what is known as close-packed structures with crystallographic planes having the same arrangement of atoms; however, the order of stacking the planes is different. Atoms in the hcp planes (called the basal planes) have the same arrangement as those in the fcc close-packed planes. However, in the hcp structure, these

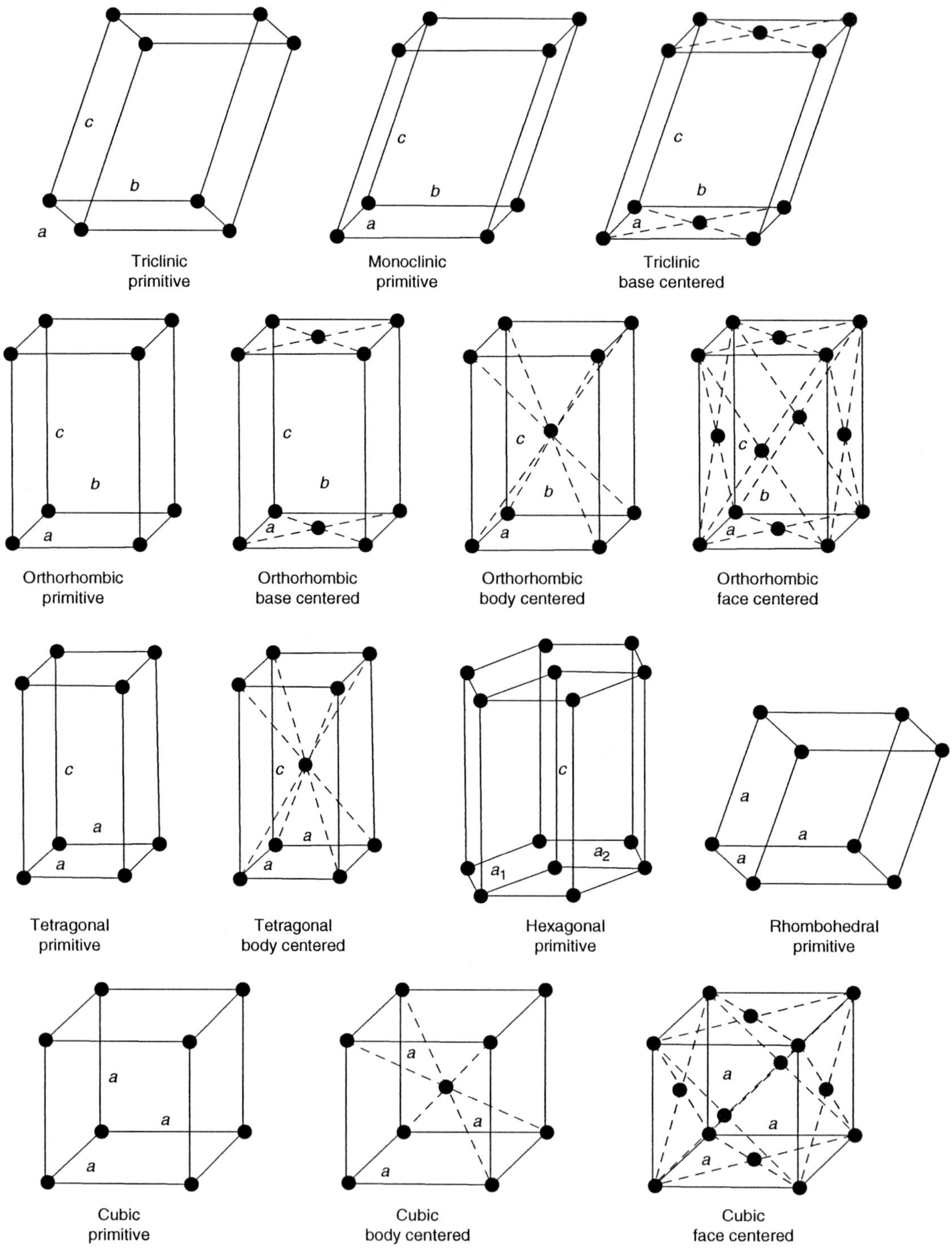


**Fig. 1.8** The fourteen bravais lattices

planes repeat every other layer to give a stacking sequence of . . .*ABA*. . .. The number of atoms belonging to the hcp unit cell is six, and the atomic packing factor is 0.74. Note that this is the same packing factor as was obtained for the fcc structure. Also, the coordination number obtained for the hcp structure (CN = 12) is the same as that for the fcc structure. A basic rule of crystallography is that if the coordination numbers of two different unit cells are the same, then they both will have the same packing factors.

Two lattice parameters, $c$ and $a$, also shown in Fig. 1.10, are needed to completely describe the hcp unit cell. In an ideal hcp structure, the ratio of the lattice constants, $c/a$, is 1.633. In this ideal packing arrangement, the layer between the two basal planes in the center of the structure is located close to the atoms on the upper and lower basal planes. Therefore, any atom in the lattice is in contact with 12 neighboring atoms, and the coordination number is therefore CN = 12. It should be noted that there is often some deviation from the ideal ratio of $c/a = 1.633$. If the ratio is less than 1.633, it means that the atoms are compressed in the $c$-axis direction, and if the ratio is greater than 1.633, the atoms are elongated along the $c$-axis. In these situations, the hcp structure can no longer be viewed as truly being close packed and should be described as just being hexagonal. However, structures deviating from the ideal packing are still normally described as being hcp. For example, beryllium is described as having an hcp structure, but its $c/a$ ratio of 1.57 is unusually low and causes some distortion in the lattice. This distortion and the unusually high elastic modulus of beryllium ($3 \times 10^5$ MPa, or $42 \times 10^3$ ksi) result from a covalent component in its bonding. Contributions from covalent bonding are also present in the hcp metals zinc and cadmium, with $c/a$ ratios greater than 1.85. This lowers their packing density to approximately 65%, considerably less than the 74% of the ideal hcp structure.

### 1.3.4 Body-Centered Cubic System

The body-centered cubic (bcc) system is shown in Fig. 1.11. The bcc system is similar to the simple cubic system except that it has an additional atom located in the center of the structure. Since the center atom belongs completely to the unit cell in question, the number of atoms belonging to the bcc unit cell is two. The coordination number for the bcc structure is eight, since the full center atom is in contact with eight neighboring atoms located at the corner points of the lattice. The atomic packing factor for the bcc structure is 0.68, which is less than that of the fcc and hcp structures. Since the packing is less efficient in the bcc structure, the closest-packed planes are less densely packed (Fig. 1.13).

Even though the bcc crystal is not a densely packed structure, it is the equilibrium structure of 15 metallic elements at room temperature, including many of the important transition elements. This is attributable to two factors: (1) Even though each atom has only eight nearest neighbors, the six second-nearest neighbor atoms are closer in the bcc structure

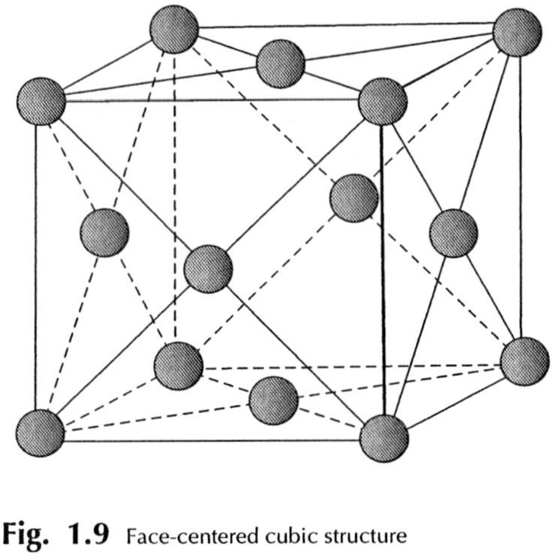

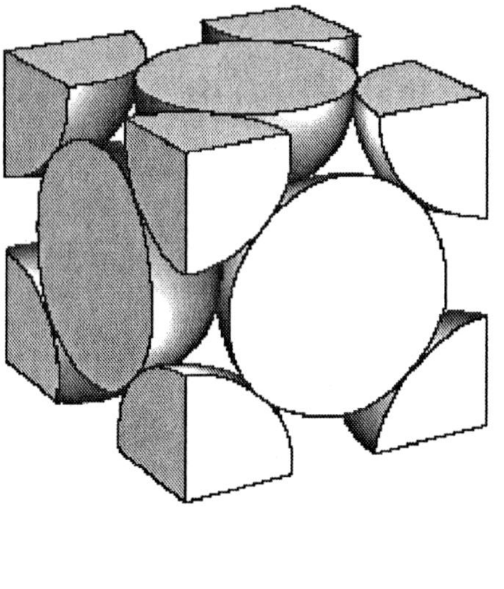

**Fig. 1.9** Face-centered cubic structure

than in the fcc structure. Calculations indicate that these second-nearest neighbor bonds make a significant contribution to the total bonding energy of bcc metals; and, (2) in addition, the greater entropy of the less densely packed bcc structure gives it a stability advantage over the more tightly packed fcc structure at high temperatures. As a consequence, some metals that have the close-packed structures at room temperature transition to bcc structures at higher temperatures.

## 1.4 Slip Systems

Plastic deformation takes place by sliding (slip) of close-packed planes over one another. A reason for this slip plane preference is that

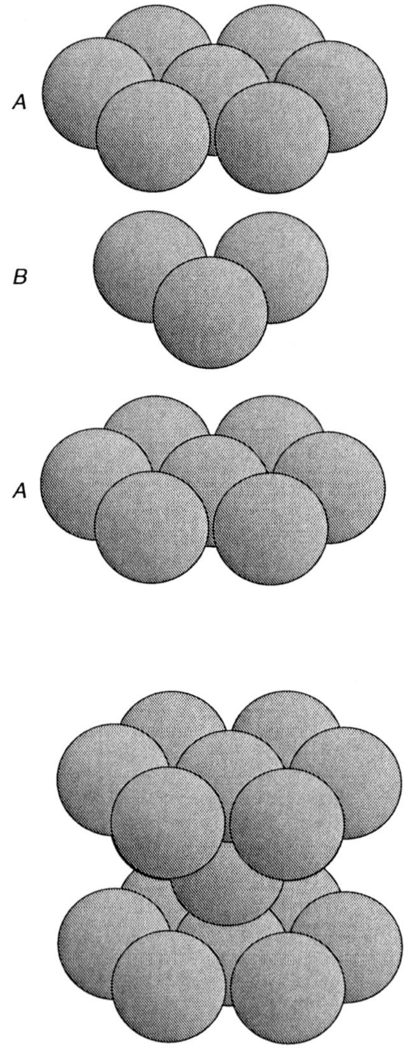


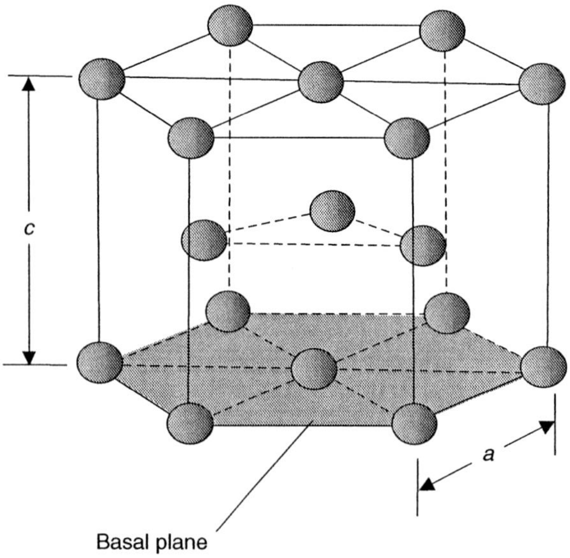


**Fig. 1.10** Hexagonal close-packed structure

the separation between close-packed planes is greater than for other crystal planes, and this makes their relative displacement easier. Furthermore, the slip direction is in a close-packed direction. The combination of planes and directions on which slip takes place constitutes the slip systems of the material. In polycrystalline materials, a certain number of slip systems must be available in order for the material to be capable of plastic deformation. Other things being equal, the greater the number of slip systems, the greater the capacity for deformation. Face-centered cubic metals have a large number of slip systems (12) and are capable of moderate-to-extensive plastic deformation even at temperatures approaching absolute zero. A number of close-packed planes for the fcc, bcc, and hcp structures are illustrated in Fig. 1.14.

Materials having the bcc structure also often display 12 slip systems, although this number comes about differently than it does for the fcc lattice. A closest-packed bcc plane is defined by a unit cell edge and face diagonal. There are only two close-packed directions (the cube diagonals) in the closest-packed bcc plane, but there are six nonparallel planes of this type. Over certain temperature ranges, some bcc

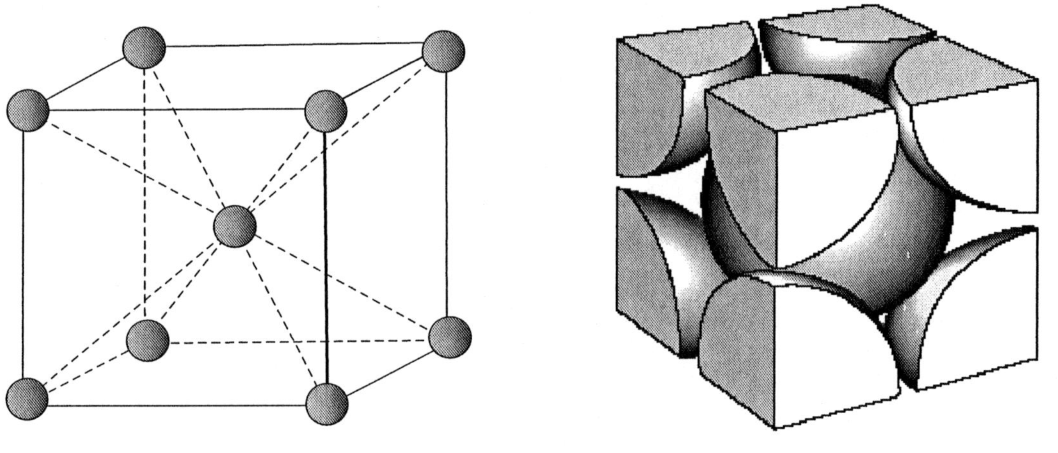

Fig. 1.11 Body-centered cubic structure

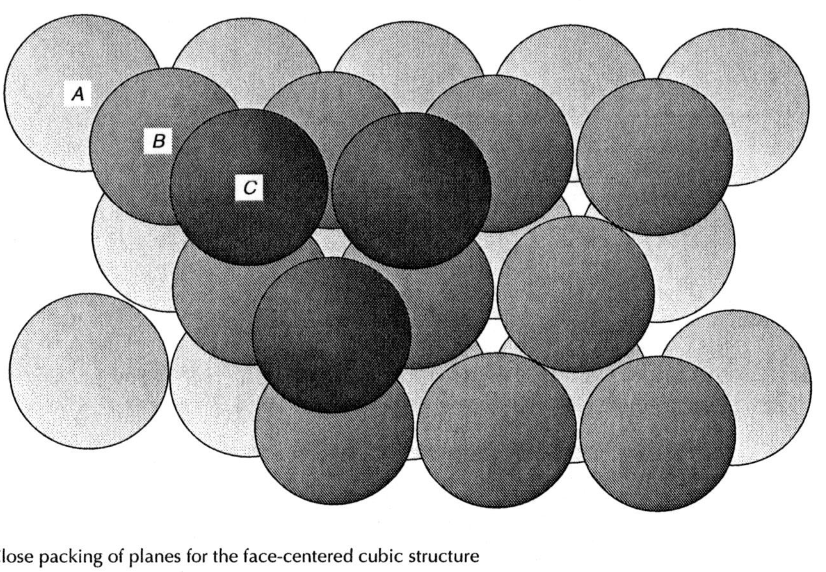


Fig. 1.12 Close packing of planes for the face-centered cubic structure

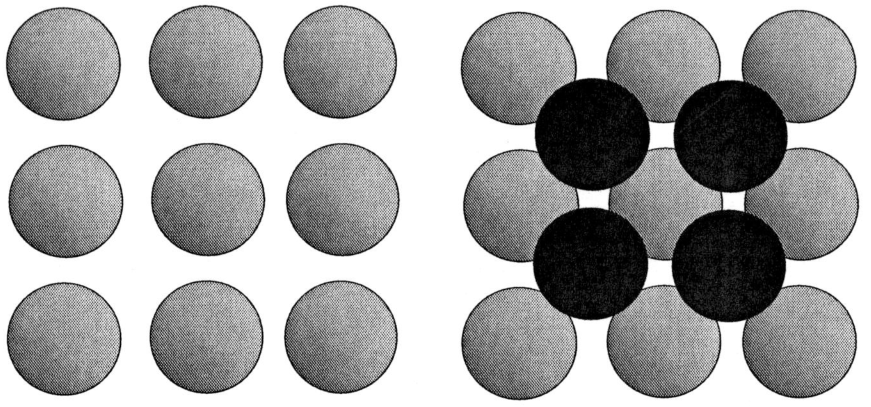

**Fig. 1.13** Loose packing in body-centered cubic structure

metals display slip on other than close-packed planes, although the slip direction remains a close-packed one. Thus, bcc metals have the requisite number of slip systems to allow for plastic deformation. However, bcc metals often become brittle at low temperatures as a result of the strong temperature sensitivity of their yield strength, which causes them to fracture prior to undergoing significant plastic deformation.

Depending on their *c*/*a* ratio, polycrystalline hcp metals may or may not have the necessary number of slip systems to allow for appreciable plastic deformation. The ideal hcp structure has only three slip systems, because there is only one nonparallel close-packed plane in it (the basal plane, which contains three nonparallel close-packed directions). However, three slip systems are insufficient to permit polycrystalline plastic deformation, and so hcp polycrystals for which slip is restricted to the basal plane are not malleable. When *c*/*a* is less than the ideal ratio, basal planes become less widely separated, and other planes compete with them for slip activity. In these instances, the number of slip systems increases, and material ductility is beneficially affected. In addition, polycrystalline hcp metals can also deform by a mechanism called twinning, which is covered in Chapter 2, "Crystalline Imperfections and Plastic Deformation," in this book.

The methods for identifying crystalline planes and crystalline directions can be found in Appendix C, "Crystallographic Planes and Directions."

## 1.5 Allotropy

Depending on pressure and temperature, many metals can exist in more than one crystalline form, a phenomenon known as allotropy. The important metal iron undergoes a series of allotropic transformations during heating and cooling, as shown in Fig. 1.15. Note that an allotropic transformation is a solid-state phase transformation and, as such, occurs at a constant temperature during either heating or cooling.

Under equilibrium cooling conditions, the solidification of pure iron from the liquid occurs at 1540 °C (2800 °F) and forms what is called delta iron ($\delta_{Fe}$), which has a bcc structure. Delta iron is then stable on further cooling until it reaches 1395 °C (2541 °F), where it undergoes a transformation to an fcc structure called gamma iron ($\gamma_{Fe}$). On still further cooling to 900 °C (1648 °F), it undergoes yet another phase transformation, transforming from the fcc structure back to the bcc structure, called ferrite iron ($\alpha_{Fe}$) to distinguish it from the higher-temperature delta iron. This last transformation, $\gamma_{Fe} \rightarrow \alpha_{Fe}$, is extremely important because it forms the basis for the hardening of steel. Note that the $\gamma_{Fe} \rightarrow \alpha_{Fe}$ transformation occurs at 900 °C (1648 °F) on cooling, somewhat lower than the 910 °C (1673 °F) transformation temperature on heating. This temperature differential is known as the temperature hysteresis of allotropic phase transformation, and its magnitude increases with increases in the cooling rate. The temperatures (designated *A*) associated with heating contain the subscript "c," which is

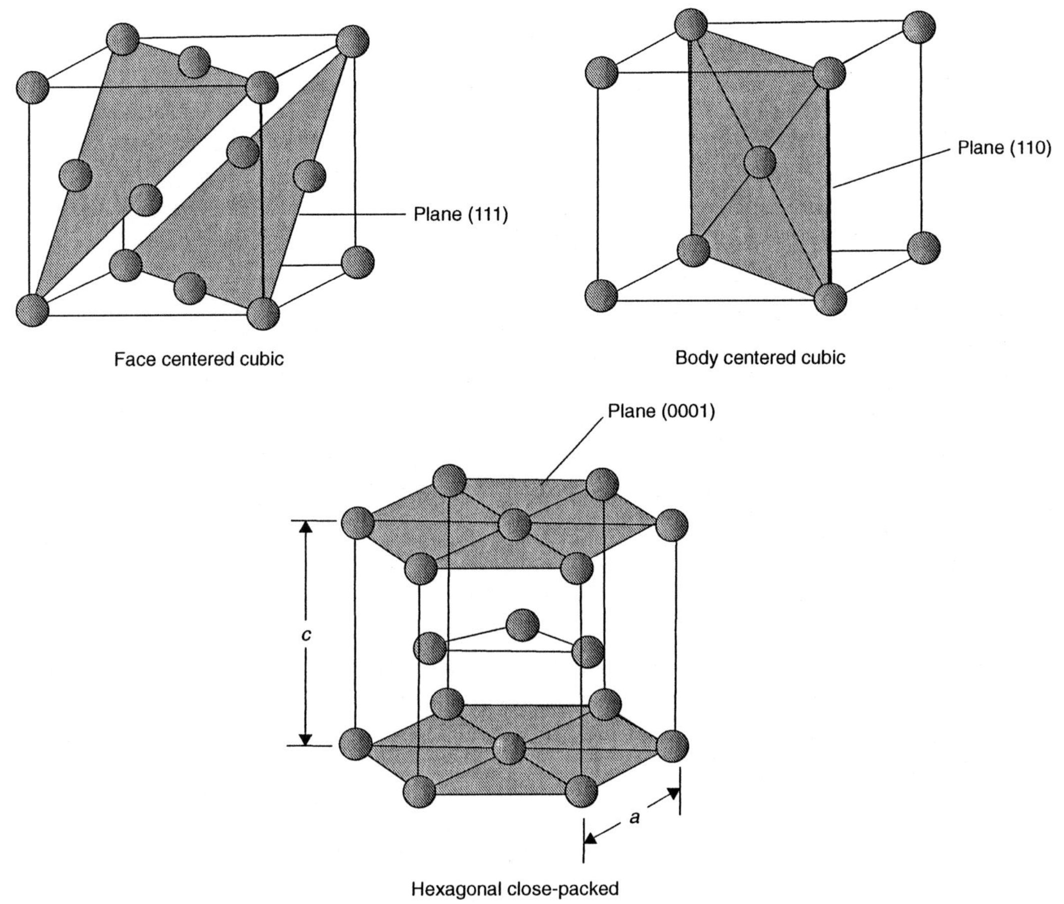


**Fig. 1.14** Close-packed planes

French for *chauffage*, meaning heating, while the ones for cooling have the subscript "r" for the French *refroidissement*, meaning cooling.

Many other metals, as well as some nonmetals, also exhibit allotropic transformations. For example, titanium, zirconium, and hafnium exhibit a transition from an hcp structure to bcc on heating. Note that in each case, a close-packed structure is stable at room temperature, while a looser packing is stable at elevated temperatures. While this is not always the case, it is a trend experienced with many metals.

## REFERENCES

1. V. Singh, *Physical Metallurgy*, Standard Publishers Distributors, 1999
2. M. Tisza, *Physical Metallurgy for Engineers*, ASM International, 2001

## SELECTED REFERENCES

- D.R. Askeland, *The Science and Engineering of Materials*, 2nd ed., PWS-Kent Publishing Co., 1989
- H. Baker, The Chemical Elements, *Metals Handbook Desk Edition*, 2nd ed., ASM International, 1998
- H. Baker, Introduction to Alloy Phase Diagrams, *Metallography and Microstructures*, Vol 9, *ASM Handbook*, ASM International, 2004
- W.D. Callister, *Fundamentals of Materials Science and Engineering*, 5th ed., John Wiley & Sons, Inc., 2001
- T.H. Courtney, Fundamental Structure-Property Relationships in Engineering Materials, *Materials Selection and Design*, Vol 20, *ASM Handbook*, ASM International, 1997

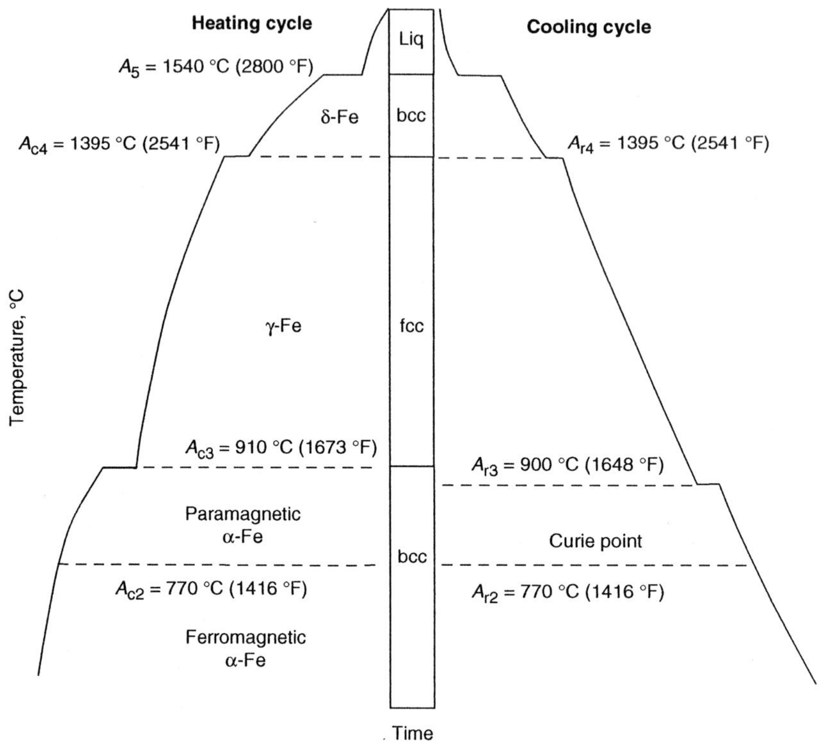


**Fig. 1.15** Allotropic transformations in pure iron. Source: Ref 2


- A.G. Guy, *Elements of Physical Metallurgy*, 2nd ed., Addison-Wesley Publishing Company, 1959
- R.E. Reed-Hill and R. Abbaschian, *Physical Metallurgy Principles*, 3rd ed., PWS Publishing Company, 1991
- A.M. Russell and K.L. Lee, *Structure-Property Relationships in Nonferrous Alloys*, Wiley-Interscience, 2005
- W.F. Smith, *Principles of Materials Science and Engineering*, McGraw-Hill, 1986

# CHAPTER 2

# Crystalline Imperfections and Plastic Deformation

IN A PERFECT crystalline structure, there is an orderly repetition of the lattice in every direction in space. However, real crystals are not perfect, they always contain a considerable number of imperfections, or defects, that affect their physical, chemical, mechanical, and electronic properties. It should be noted that defects do not necessarily have adverse effects on the properties of materials. They play an important role in processes such as deformation, annealing, precipitation, diffusion, and sintering.

All defects and imperfections can be conveniently classified under four main divisions: point defects, line defects, planar defects, and volume defects. Point defects are inherent to the equilibrium state and thus determined by temperature, pressure, and composition. However, the presence and concentration of the other defects depends on the way the metal was originally formed and subsequently processed.

## 2.1 Point Defects

A point defect is an irregularity in the lattice associated with a missing atom (vacancy), an extra atom (interstitial), or an impurity (substitutional) atom. Due to their small size, point defects generally produce only very local distortions in the crystalline lattice. However, their presence can be significant, for example, in aiding diffusion in the crystalline lattice. Vacancies are the simplest defect. A vacancy is simply missing from the crystalline lattice, as illustrated in Fig. 2.1. Vacancies are created during solidification due to imperfect packing. They also occur during processing at elevated temperatures. In an otherwise completely regular lattice, the atoms are constantly being displaced from their ideal locations by thermal vibrations. The frequency of vibration is almost independent of temperature, but the amplitude increases with increasing temperature. For copper, the amplitude near room temperature is approximately one-half its value near the melting point and approximately twice its value near absolute zero. As the temperature is increased, the lattice vibrations become larger, and atoms have a tendency to jump out of their normal positions, leaving a vacant lattice site behind.

The number of vacancies increases exponentially with temperature according to:

$$n_V = Ne^{-E_V/kT} \quad \text{(Eq 2.1)}$$

where $n_V$ is the number of vacancies at temperature, $T$; $N$ is the total number of lattice sites; $E_V$ is the energy necessary to form a vacancy; $k$ is the Boltzmann constant ($1.38 \times 10^{-24}$ J/K); and $T$ is the absolute temperature in degrees Kelvin.

While the number of vacancies would be zero at absolute zero, it is on the order of $10^{-3}$ for metals near their melting point. In Eq 2.1, at any temperature above absolute zero, the equilibrium condition for a metal will contain vacancies; that is, the presence of vacancies is a condition of equilibrium.

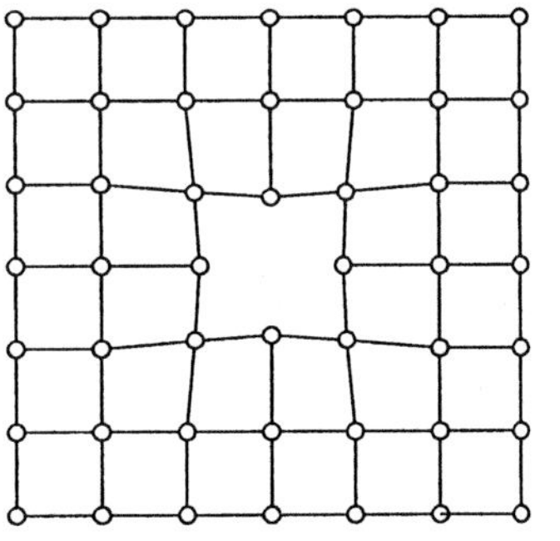

**Fig. 2.1** Vacancy point defect. Source: Ref 1

Vacancies affect the properties of the metal. Density slightly decreases as the number of vacancies increases. The electrical conductivity also decreases as the number of vacancies increases. Vacancies enhance atomic diffusion. Vacancy diffusion is the movement of a vacancy through the lattice, thereby assisting the diffusion of atoms. The number of dislocations is reduced when the vacancies diffuse to grain boundaries or surfaces, which act as sinks. If a metal is heated to a high temperature, the number of vacancies increases. If it is then suddenly quenched to room temperature, the vacancies are trapped in the lattice, because they do not have time to diffuse out.

Vacancies can form by several mechanisms. In the Frenkel mechanism (Fig. 2.2), an atom is displaced from its normal lattice position into an interstitial site. However, this requires quite a bit of energy—the energy to form a vacancy and the energy to form an interstitial. Therefore, the probability is quite low. A more realistic, and lower-energy method, is the Schottky mechanism (Fig. 2.3), in which vacancies originate at free surfaces and move by diffusion into the crystal interior.

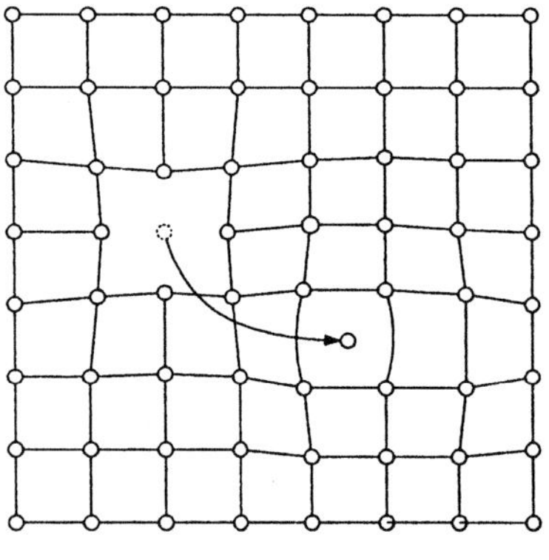

**Fig. 2.2** Frenkel mechanism. Simultaneous formation of vacancy and interstitial atom. Source: Ref 1

Solute atoms of a second metal can be present as impurities or added as intentional alloying elements. These solute atoms can substitute on the crystalline lattice for solvent atoms and form substitutional point defects, or they can be located in the interstitial locations between the atoms of the crystalline lattice to form interstitial defects (Fig. 2.4). If the solute impurities are close to the same diameter as the solvent atoms, they will substitute for solvent atoms to form substitutional defects. Small atoms that can fit in between the larger solvent atoms of a crystalline structure are called interstitials. To form interstitial defects, the atomic diameter of the impurity must be significantly smaller than the solvent atom diameter. Therefore, only atoms with very small diameters, such as carbon, nitrogen, hydrogen, and boron, can form interstitial defects. If the foreign atoms cause harmful or undesirable effects, they are called impurities, while, if they are helpful, they are referred to as alloying elements.

Point defects influence solid-state processes such as diffusion, dislocation motion, phase transformations, and electrical conductivity. Point defects typically strengthen a metal and decrease its ductility by impeding the motion of dislocations. Point defects also decrease electrical conductivity, because they interfere with the flow of electrons through the lattice.

## 2.2 Line Defects

One of the most important defects is the line or edge dislocation. The existence of line defects in crystals, called dislocations, provides the mechanism that allows mechanical deformation. A crystalline metal without dislocations, although extremely strong, would also be extremely brittle and practically useless as an engineering material. Thus, dislocations play a central role in the determination of such important properties as strength and ductility.

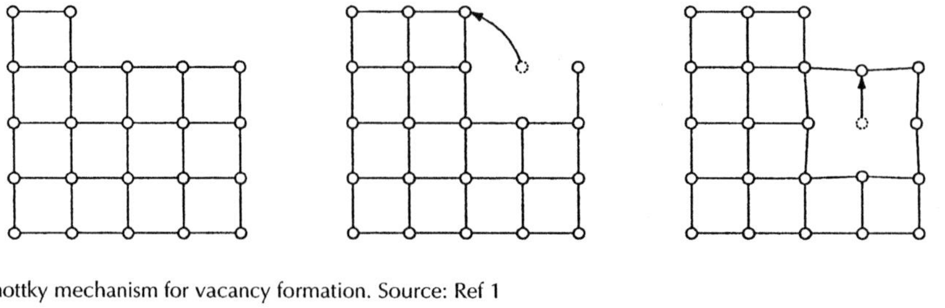

**Fig. 2.3** Schottky mechanism for vacancy formation. Source: Ref 1

In fact, virtually all mechanical properties of metals are, to a significant extent, controlled by the behavior of line imperfections.

As shown in Fig. 2.5, an edge dislocation can be visualized as resulting from the insertion of an extra half-plane of atoms above (or below) the dislocation line. By definition, the dislocation shown in Fig. 2.5 is a positive dislocation. A negative dislocation has the extra half-plane below the dislocation line. An edge dislocation creates a zone of elastic deformation around the dislocation (Fig. 2.6). The lattice below the dislocation is in a state of tension, while above the dislocation, there is a compressive stress field. In the lattice below a dislocation, interstitial atoms usually cluster in regions where the tensile stresses help make more room for them.

A quantitative description of dislocations is given by the Burgers vector, *b,* illustrated in Fig. 2.7. This vector is defined using what is called the Burgers circuit, which is an atom-to-atom path that makes a closed loop in a

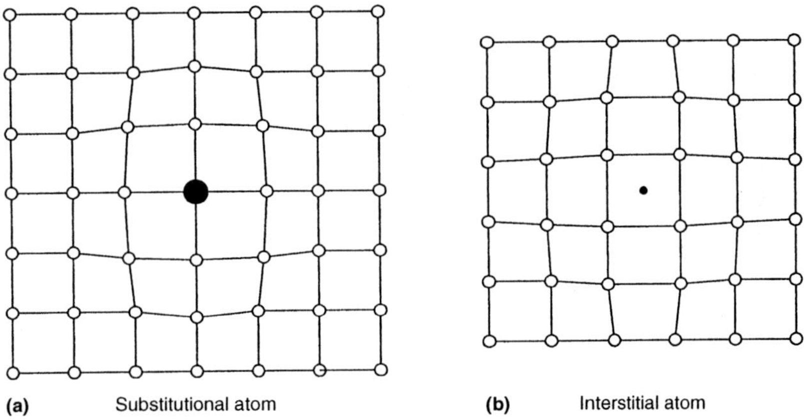


**Fig. 2.4** Foreign atom point defects. Source: Ref 1

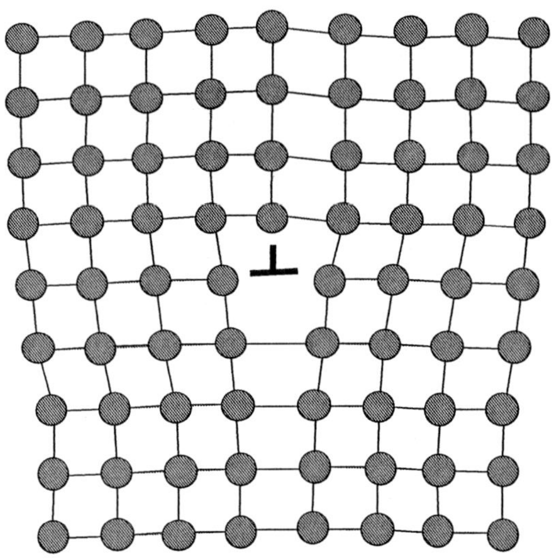

**Fig. 2.5** Line dislocation

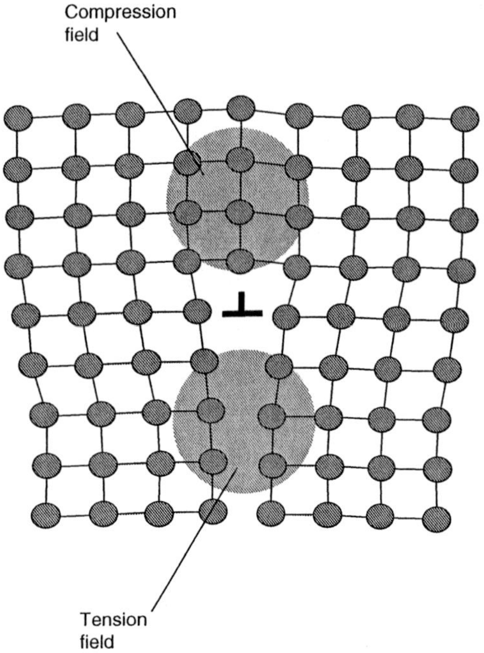


**Fig. 2.6** Stress fields around line dislocations

dislocation-free part of the crystal lattice. Now, if the same Burgers circuit is made to encircle a dislocation, the loop does not close. The vector needed to close the loop (the vector from the end of the Burgers circuit to its starting point) is the Burgers vector, *b,* describing the dislocation. The displacement vector between the two parts of the crystal is denoted by *u,* and the axis of the dislocation is *t.* For an edge dislocation, the Burgers vector, *b,* is perpendicular to the axis of the dislocation, *t* ($b \perp t$), and parallel to the displacement vector, *u* ($b \parallel u$).

The other important type of line dislocation is the screw dislocation (Fig. 2.8). The term *screw dislocation* is used because of the spiral surface formed by the atomic planes around the screw dislocation line. When a Burgers circuit is used to determine the Burgers vector of a screw dislocation, the vector is found to be parallel to the dislocation line rather than perpendicular to it, as in the case of an edge dislocation. A screw dislocation is somewhat like a spiral ramp with an imperfection line running down its axis. In a screw dislocation, the Burgers vector, *b,* is parallel to both the axis of the dislocation, *t,* and the displacement vector, *u*; that is, ($b \parallel t \parallel u$).

An important characteristic of a dislocation is that it cannot end inside the crystal, it must end at a surface such as a grain boundary or at a surface of the crystal. It is possible for a dislocation to change its character inside the crystal, as shown in the mixed dislocation in Fig. 2.9. Here, an edge dislocation is converted to a screw dislocation; in this case, the screw dislocation is shown ending at the surface of the crystal. Dislocations will also form closed loops within a crystal, changing from an edge to a screw and then back to another edge and finally back to a screw to enclose the loop. The material within the loop is visualized as having slipped on the specified slip plane relative to the material around it.

## 2.3 Plastic Deformation

When a mechanical shear load is applied to a metal, it deforms under the applied stress, as shown schematically in Fig. 2.10. If the load is small, the bonds between the atoms will be stretched but will return to their normal lattice positions when the load is removed, this is elastic deformation. However, if a ductile metal is stretched beyond the elastic capability of the bonds, when the load is removed, it will not return to its original shape. It is then said to have undergone plastic deformation. On the other hand, if the material is brittle rather than being ductile, when the elastic stretching of the bonds is exceeded, it immediately fails with little or no evidence of plastic deformation. This is the type of failure that is normally encountered in covalently and ionically bonded solids, such as glasses and crystalline ceramics. The question becomes: Why do metals exhibit moderate-to-large amounts of plastic deformation, while other materials, such as glasses and ceramics, exhibit almost no ductility and fail in a brittle manner?

The nondirectional metallic bond allows metals to deform by shear, as illustrated in Fig. 2.11. For the atoms in the upper plane to slide over those in the lower plane, strong interatomic forces must be overcome by the applied shear stress. When the atoms in the upper plane have been displaced by one-half of their transit distance, the crystal energy is at a maximum and then falls when they reach their

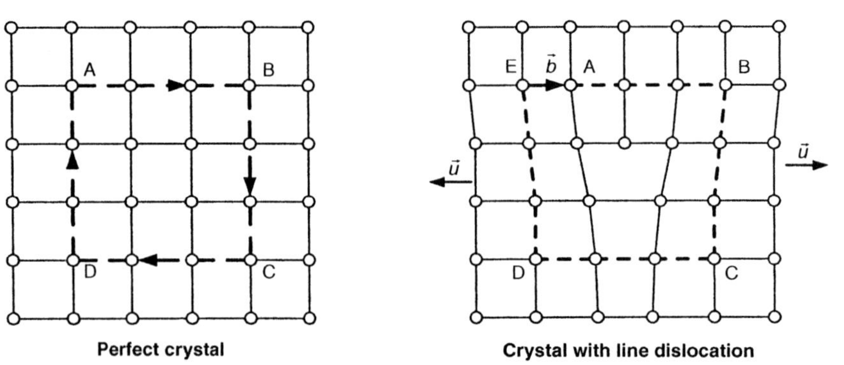


**Fig. 2.7** Burgers circuit and vector for line dislocation. Source: Ref 1

new equilibrium positions. The shear stress required to cause slip is initially zero when the two planes are in coincidence and when the atoms of the top plane are midway between those of the bottom plane, since this is a symmetry position. Between these positions, each atom is attracted to its nearest atom of the other row, so that the shearing stress is a periodic function of the displacement. This shear, or slip, takes the path of least resistance and thus occurs along the close-packed planes in close-packed directions. Atomic bonds are broken and then reestablished as the metallic ions move past one another. This is much more difficult for covalently and ionically bonded solids. In the covalent bond, the bonds between two atoms are well established and do not want to be broken. Remember that covalent bonds are both strong and highly directional, while metallic ions share their valence electrons, allowing freer movement through the electron cloud. The problem with ionic bonds is that one ion is positive and the other is negative. During any type of shear mechanism, when two positively charged ions or two negatively charged ions approach each other too closely, a strong electrical repulsive force will develop between the two and resist plastic movement.

During tensile testing of a single crystal, shown schematically in Fig. 2.12, an applied stress will be reached when the shear stress, resolved onto a slip plane in a slip direction, attains a critical value so that dislocations on that slip plane slip or glide. If the normal, $n$, of the slip plane lies at an angle, $\phi$, to the tensile axis, its area will be $A/\cos\phi$. Similarly, if the slip plane lies at an angle, $\lambda$, to the tensile axis, the component of the axial force, $P$, acting on the

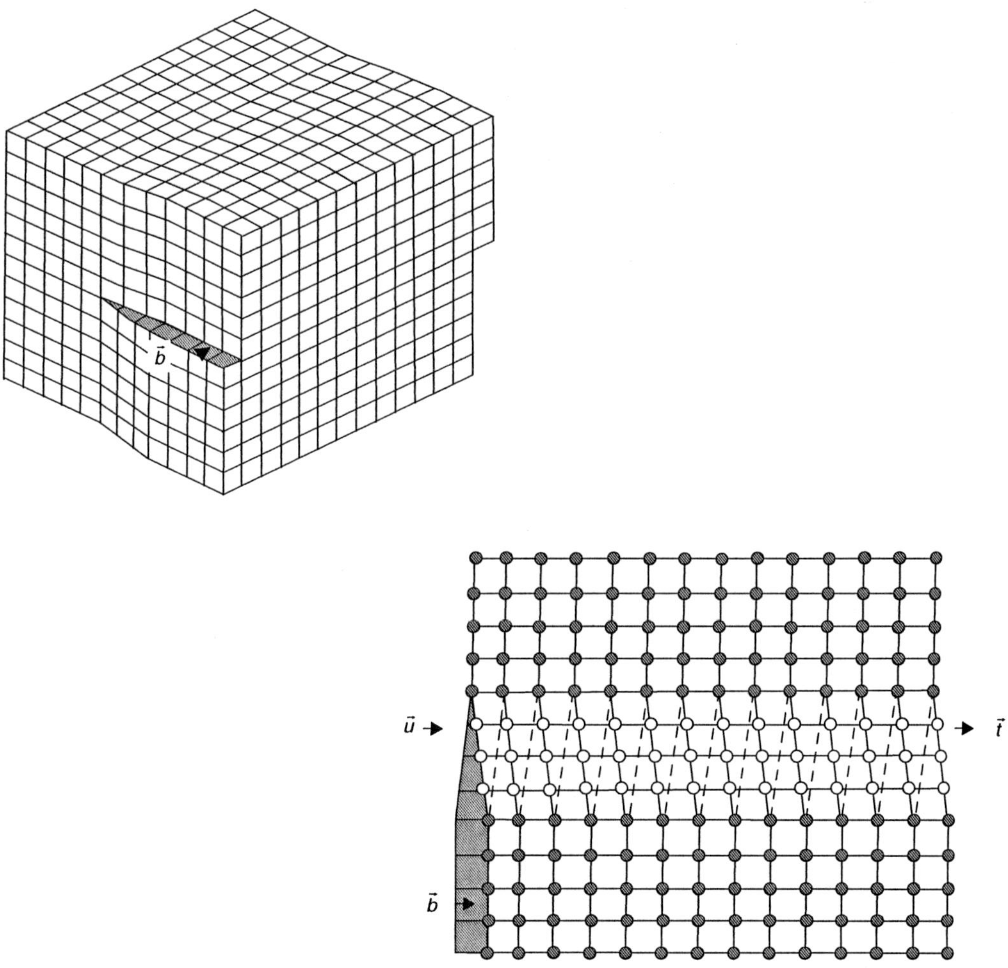


**Fig. 2.8** Screw dislocation

slip direction will be $P \cos \lambda$. The resolved shear stress, $\tau_r$, is then given by:

$$\tau_r = \frac{P \cos \lambda}{(A/\cos \phi)} = \sigma \cos \varphi \cos \lambda \qquad \text{(Eq 2.2)}$$

For a given metal, the value of $\tau$ at which slip occurs is usually found to be a constant, known as the critical resolved shear stress, $\tau_c$. This relationship is known as Schmid's law, and the quantity $\cos \varphi \cos \lambda$ is called the Schmid factor. Since the shear stress at which slip occurs is the yield stress, $\tau_y$, it follows that:

$$\tau_c = \sigma_y \cos \phi \cos \lambda \qquad \text{(Eq 2.3)}$$

However, most metals used in industry are polycrystalline, not single crystals. Under an

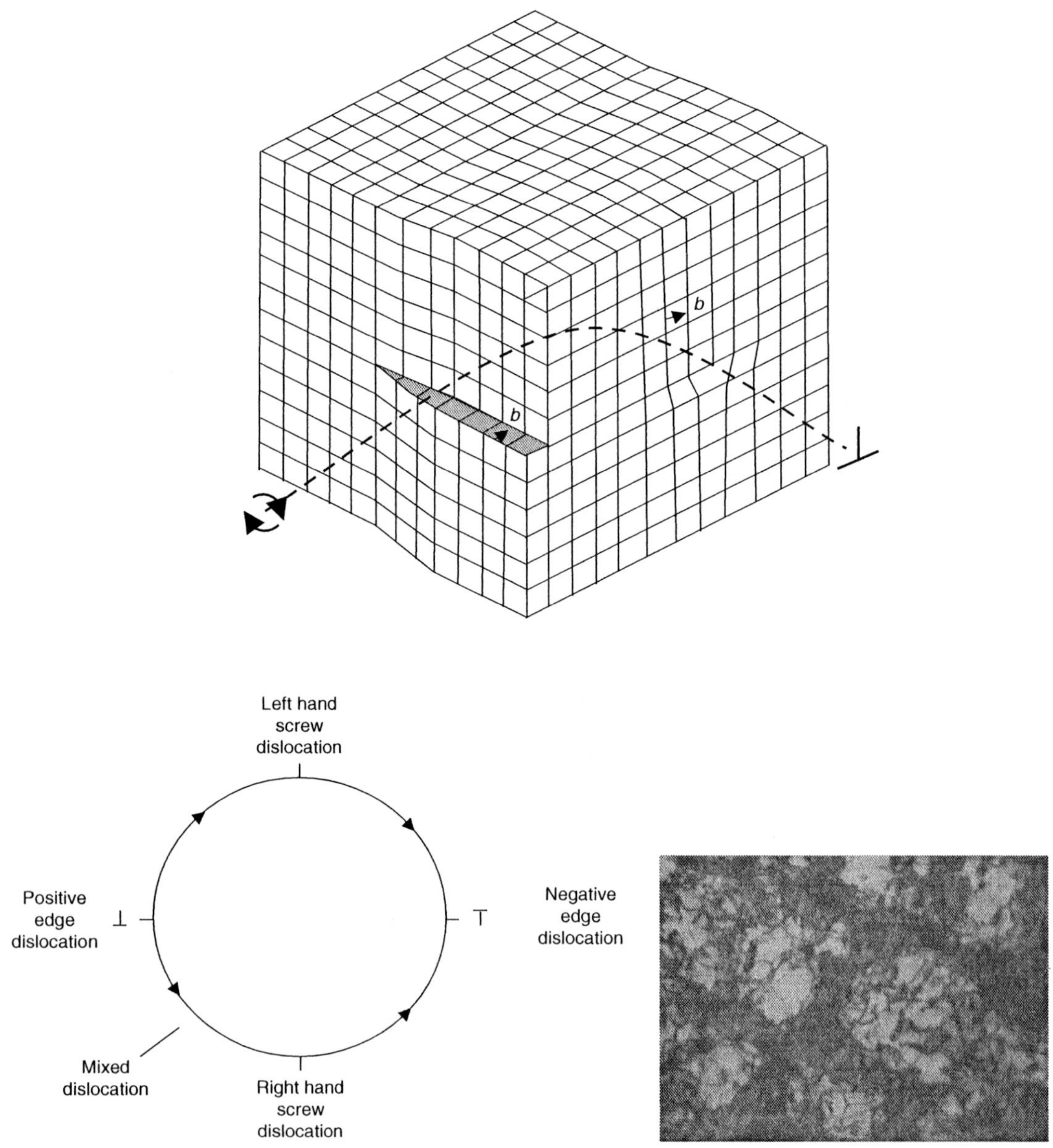


**Fig. 2.9** Combination screw and line dislocations

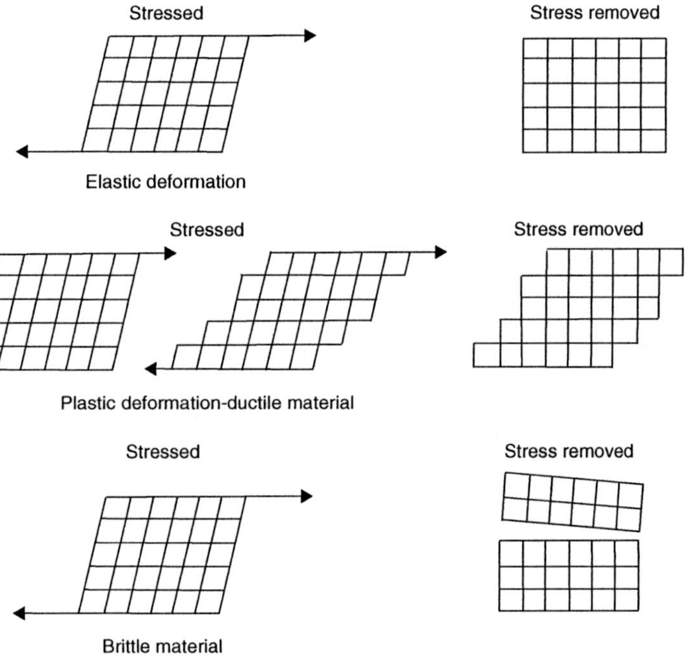


**Fig. 2.10** Material behavior under stress

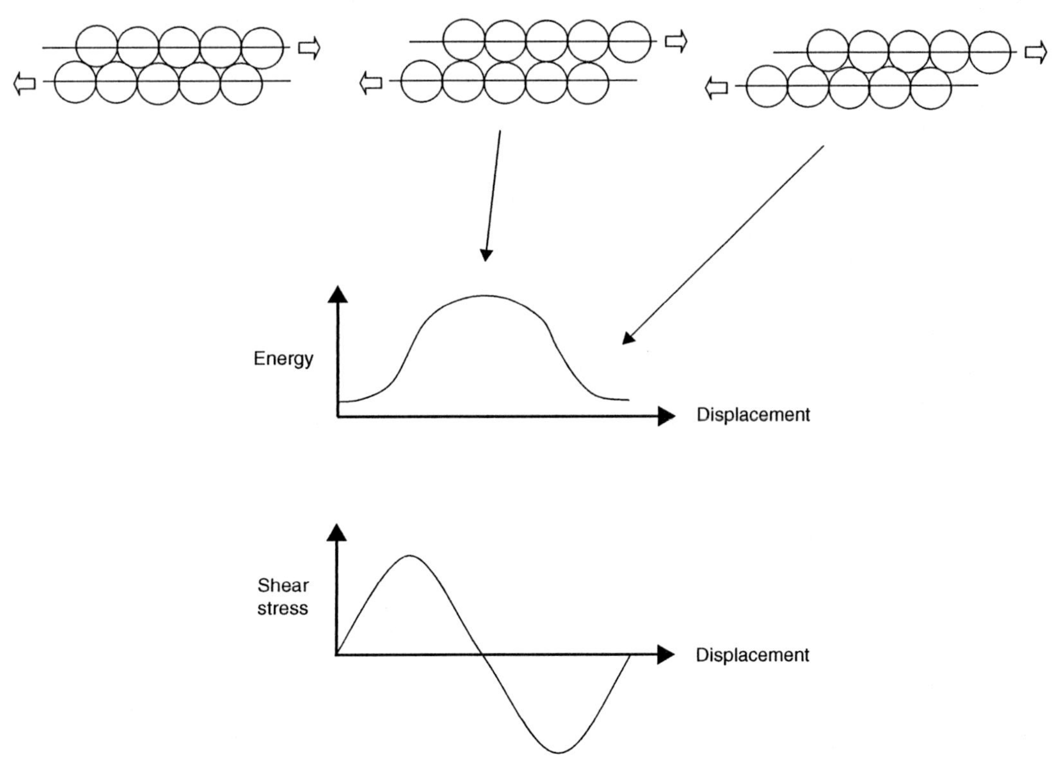


**Fig. 2.11** Planar slip

applied axial load, the Schmid factor will be different for each grain. For randomly oriented grains, the average value of the Schmid factor is ~1/3, which is referred to as the Taylor factor. It then follows that the yield strength should have a value of approximately $3\tau_c$.

### 2.3.1 Dislocations and Plastic Flow

From our knowledge of the metallic bond, it is possible to derive a theoretical value for the stress required to produce slip by the simultaneous movement of atoms along a plane in a metallic crystal. However, the strength actually obtained experimentally on single crystals is only about one-thousandth (1/1000) of the theoretical value, assuming simultaneous slip by all atoms on the plane. Obviously, slip does not occur by the simple simultaneous block movement of one layer of atoms sliding over another, as previously shown in Fig. 2.11. Nor does such a simple interpretation of slip explain work hardening that takes place during mechanical deformation. Earlier theories that sought to explain slip by the simultaneous gliding of a complete block of atoms over another have now been discarded, and the modern concept is that slip occurs by the step-by-step movement of dislocations through the crystal.

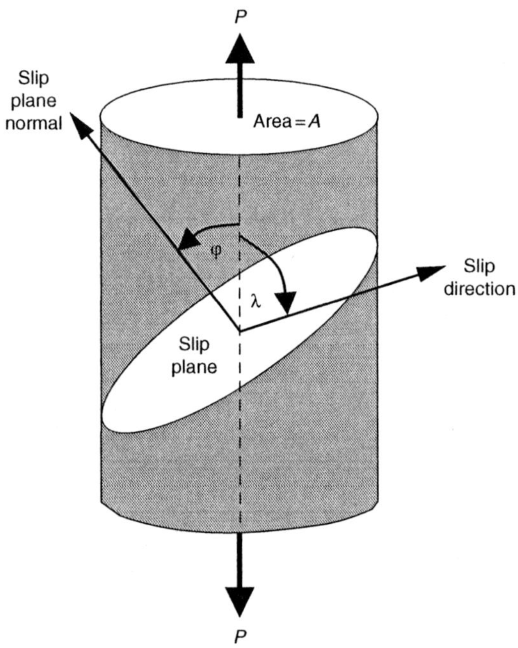


**Fig. 2.12** Tensile test of single crystal

When force is applied such that it shears the upper portion of the crystal to the right, as shown in Fig. 2.13, the plane of atoms above the dislocation can easily establish bonds with the lower plane of atoms to its right, with the result that the dislocation moves one lattice spacing at a time. Note that only single bonds are being broken at any one time, rather than the whole row, as shown in Fig. 2.11. The atomic distribution is again similar to the initial configuration, and so, the slipping of atom planes can be repeated. The movement is much like that of advancing a carpet along a floor by using a wrinkle that is easily propagated down its length. This stress required to cause plastic deformation is orders of magnitude less when dislocations are present than in dislocation-free, perfect crystalline structures. If a large number of dislocations move in succession along the same slip plane, the accumulated deformation becomes visible, resulting in macroscopic plastic deformation. Slip can take place by both edge and screw dislocations, as shown in Fig. 2.14. Note that although the mechanisms are different, the unit slip produced by both is the same.

Dislocations do not move with the same degree of ease on all crystallographic planes nor in all crystallographic directions. Ordinarily, there are preferred planes, and in these planes,

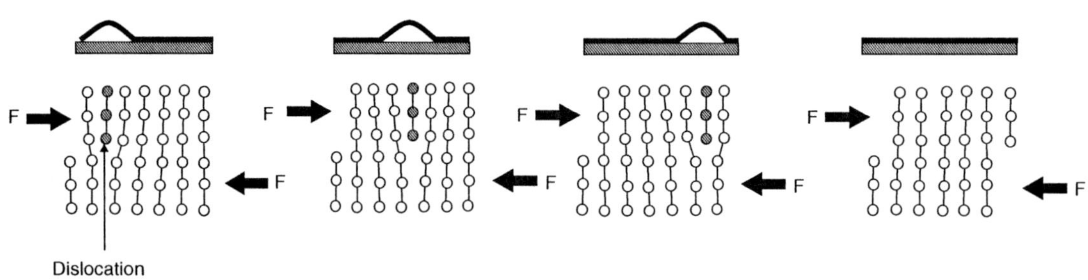


**Fig. 2.13** Line dislocation movement

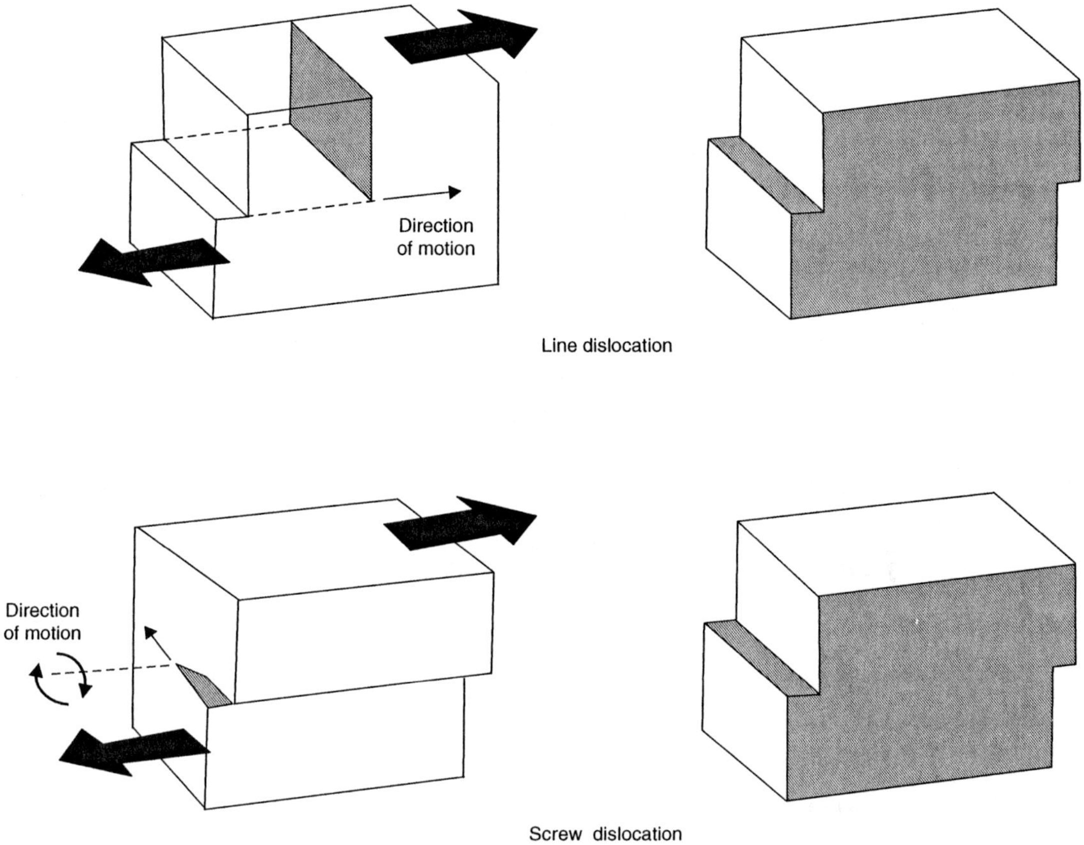


**Fig. 2.14** Displacements caused by line and screw dislocations

there are specific directions along which dislocation motion can occur. These planes are called slip planes, and the direction of movement is known as the slip direction. The combination of a slip plane and a slip direction forms a slip system. For a particular crystal structure, the slip plane is that plane having the most dense atomic packing; that is, it has the greatest planar density. The slip direction corresponds to the direction, in this plane, that is most closely packed with atoms, that is, has the highest linear density. Since plastic deformation takes place by slip, or sliding, on the close-packed planes, the greater the number of slip systems available, the greater the capacity for plastic deformation. The major slip systems for the common metallic crystalline systems are summarized in Table 2.1.

**Table 2.1 Major slip systems for common crystal systems**

| Crystal system(a) | Slip planes | | Slip directions | | No. of slip systems |
|---|---|---|---|---|---|
| Notation | Number | Notation | Number | Notation | Number |
| bcc | {110} | 6 | <111> | 2 | 12 |
| fcc | {111} | 4 | <110> | 3 | 12 |
| hcp | (0001) | 1 | [1120] | 3 | 3 |

(a) bcc, body-centered cubic; fcc, face-centered cubic; hcp, hexagonal close packed. Source: Ref 1

Face-centered cubic (fcc) metals have a large number of slip systems (12) and are therefore capable of moderate-to-extensive plastic deformation. Although body-centered cubic (bcc) systems often have up to 12 slip systems, some of them, like steel, exhibit a ductile-to-brittle transition as the temperature is lowered due to the strong temperature sensitivity of their yield strength, which causes them to fracture prior to reaching their full potential of plastic deformation. In general, the number of slip systems available for hexagonal close-packed (hcp) metals is less than that for either the fcc or bcc metals, and their plastic deformation is much more restricted. The hcp structure normally has only three to six slip systems, only one-fourth to one-half the available slip systems in fcc and bcc structures. Therefore, metals with the hcp structure have poor to only moderate room-temperature ductility. Thus, the hcp metals, such as alloys of magnesium, beryllium, and titanium, often require heating to elevated temperatures, where slip becomes much easier, prior to forming operations.

Dislocations can have two basic types of movement: glide and climb. Glide, or slip, is the type of dislocation movement that has been discussed thus far. It occurs in the plane containing both the dislocation line and the Burgers vector. During each glide step, a single row of atoms changes position with a closest neighbor atom, and the passage of the dislocation displaces the upper part of the grain with respect to the lower part of the grain. The simultaneous glide of many identical dislocations under an applied stress is known as slip and is the typical mechanism of plastic deformation in metals. The second type of dislocation motion is known as climb. As illustrated in Fig. 2.15, climb is directly dependent on vacancies. For an edge dislocation to climb, vacancies must be either created or destroyed. If vacancies are not present in large quantities, climb cannot occur, because it is dependent on diffusion. Although glide can occur at all temperatures, climb is practically nonexistent at temperatures below approximately 0.4 $T_m$, where $T_m$ is the absolute melting point. However, climb becomes an important deformation mechanism when the metal is subjected to stresses at temperatures exceeding approximately 0.4 $T_m$.

The dislocation density for crystals is approximately $10^8$ $cm^{-2}$, corresponding to an average distance between dislocations of a few thousand atoms. If each dislocation produces only one unit of slip, this relatively small number of dislocations could not produce large-scale plastic deformation. Thus, a large number of dislocations must be present to produce macroscopic slip. The Frank-Read spiral mechanism (Fig. 2.16) explains how dislocations can multiply and increase their effectiveness a thousandfold. If a dislocation line becomes immobilized and is pinned at its ends, it will tend to bow out under the influence of an applied shear stress. Eventually, the loop becomes circular and then starts closing in on itself at the ends. This allows the formation of a new loop that again bows out under the influence of a shear stress. This process is repeated over and over again, each time generating a new dislocation loop.

Dislocations are influenced by the presence of other dislocations and interact with each other, as shown for a number of different interactions in Fig. 2.17. Dislocations with Burgers vectors of the same sign will repel each other, while dislocations of opposite signs will attract each other and, if they meet, annihilate each other. If the two dislocations of opposite signs are not on the same slip plane, they will merge to form a row of vacancies. These types of interactions occur because they reduce the internal energy of the system. When a dislocation becomes pinned by an obstacle and is immobile, it is termed a sessile dislocation. A dislocation that is not impeded and can move through the lattice is called a glissile dislocation.

If dislocations could move only by gliding on a single slip plane, they would soon be impeded

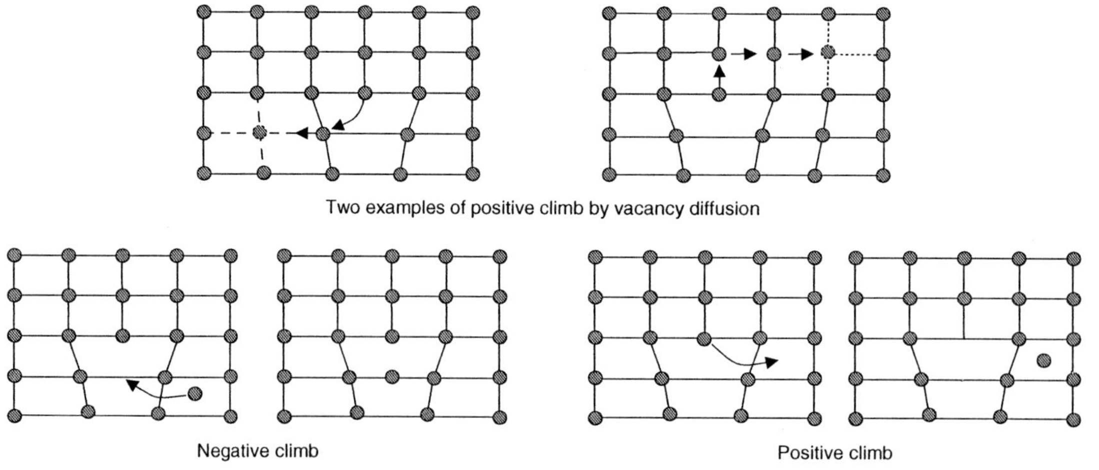


**Fig. 2.15** Dislocation climb associated with interstitial atoms

by obstacles and their motion would be restrained. However, screw dislocations can bypass obstacles on their slip plane by cross slipping onto an alternate plane (Fig. 2.18). While line dislocations cannot cross slip, they can often convert themselves into screw dislocations while they cross slip.

Vacancy diffusion also contributes significantly to high-temperature creep. Specific creep mechanisms are discussed in detail in Chapter 15, "Creep," in this book.

### *2.3.2 Work Hardening*

While slip is required to facilitate plastic deformation and therefore allow a metal to be formed into useful shapes, strengthening metals requires increasing the number of barriers to slip and reducing the ability to plastically deform. Increasing the interference to slip and increasing the strength can be accomplished by methods such as plastic deformation. As a metal is plastically deformed, new dislocations are created, so that the dislocation density becomes higher and higher. In addition to multiplying, the dislocations become entangled and impede each others' motion. The result is increasing resistance to plastic deformation with increasing dislocation density. The number of dislocations is defined by the dislocation density, $\rho$, which is the length of dislocations per unit volume of material. Therefore, the units of $\rho$ are cm/cm$^3$ or cm$^{-2}$. The dislocation density of an annealed metal usually varies between approximately $10^6$ and $10^7$ cm$^{-2}$, while that for a cold-worked metal may run as high as $10^8$ to $10^{11}$ cm$^{-2}$.

This continual increase in resistance to plastic deformation is known as work hardening, cold working, or strain hardening. Work hardening results in a simultaneous increase in strength and a decrease in ductility. Since the work-hardened condition increases the stored energy in the metal and is thermodynamically unstable, the deformed metal will try to return to a state of lower energy. This generally cannot be accomplished at room temperature. Elevated temperatures, in the range of 1/2 to 3/4 of the absolute melting point, are necessary to allow mechanisms, such as diffusion, to restore the lower-energy state. The process of heating a work-hardened metal to restore its original strength and ductility is called annealing. Metals undergoing forming operations often require intermediate anneals to restore enough ductility to continue the forming operation. Approximately 5% of the energy of deformation is retained internally as dislocations when a metal is plastically deformed, while the rest is dissipated as heat.

## 2.4 Surface or Planar Defects

Surface, or planar, defects occur whenever the crystalline structure of a metal is discontinuous across a plane. Surface defects extend in two

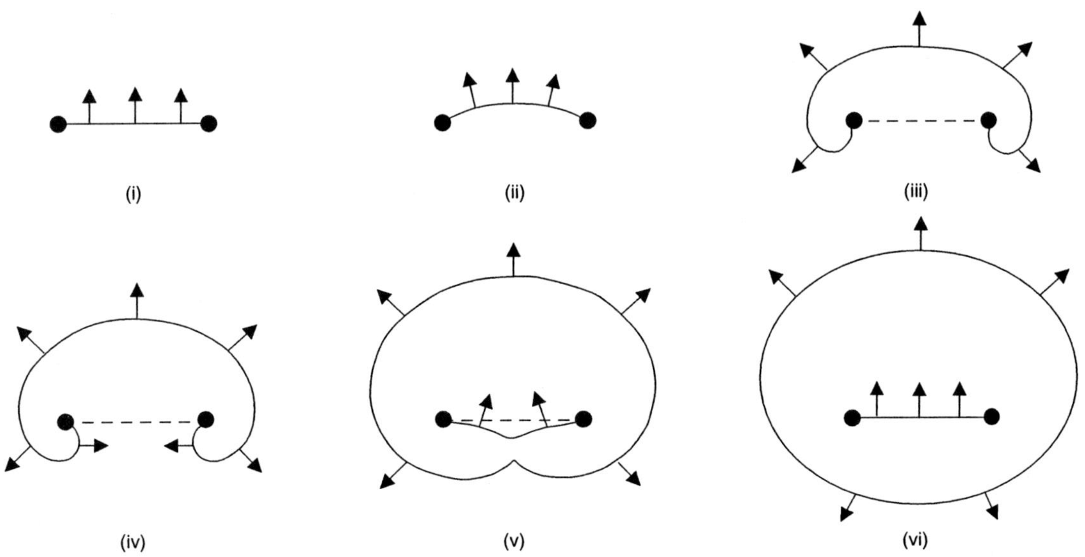


**Fig. 2.16** Frank-Read mechanism for dislocation multiplication

directions over a relatively large surface with a thickness of only one or two lattice parameters. Grain boundaries and phase boundaries are independent of crystal structure, while coherent phase boundaries, twin boundaries, and stacking faults depend on the crystalline structure.

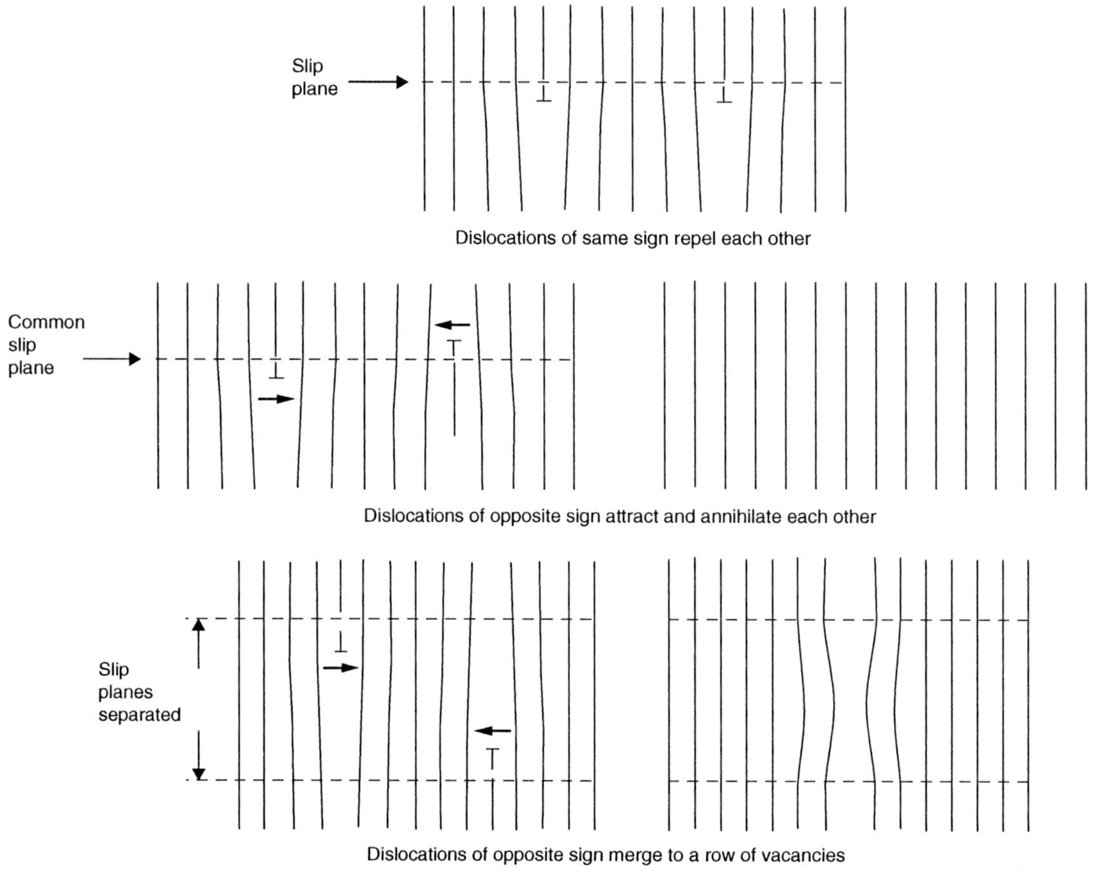


**Fig. 2.17** Examples of dislocation interactions

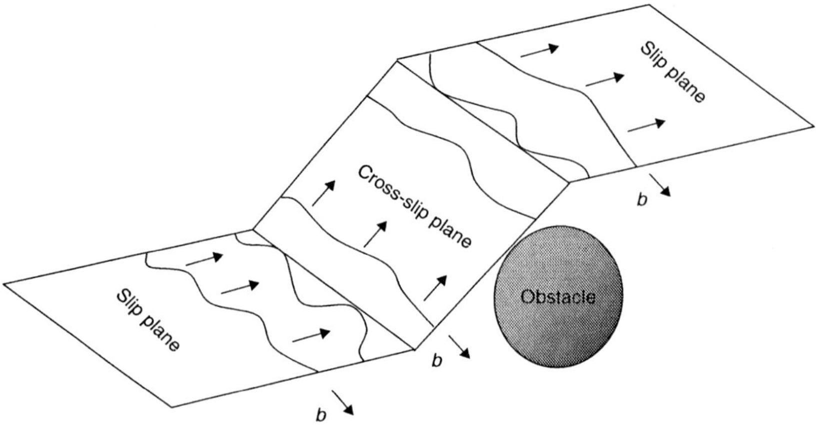


**Fig. 2.18** Cross slip of screw dislocation

### 2.4.1 Grain Boundaries

The most important surface defect is the grain boundary. Most metals are polycrystalline and consist of many small crystallites called grains. The interfaces between these grains are called grain boundaries and are only one or two atoms thick, since the system wants to reduce the free energy as much as possible. Atoms within the grain boundaries are highly strained and distorted; therefore, grain boundaries are high-energy sites. The average diameter of the individual grains within a polycrystalline metal defines the metal grain size. Grain boundaries are a result of the solidification process and occur as a result of the misorientation of the grains as they are frozen into position (Fig. 2.19).

Small-angle grain boundaries occur when the misorientation between grains is small, usually less than 5°. These small-angle misorientations can be represented by a row of somewhat parallel edge dislocations, as shown for the low-angle tilt boundary in Fig. 2.20. The regions between the dislocations consist of an almost perfect fit and have low strain, while regions at the dislocation cores have poor fit and are high-strain regions. The regions surrounded by low-angle grain boundaries are called subgrains

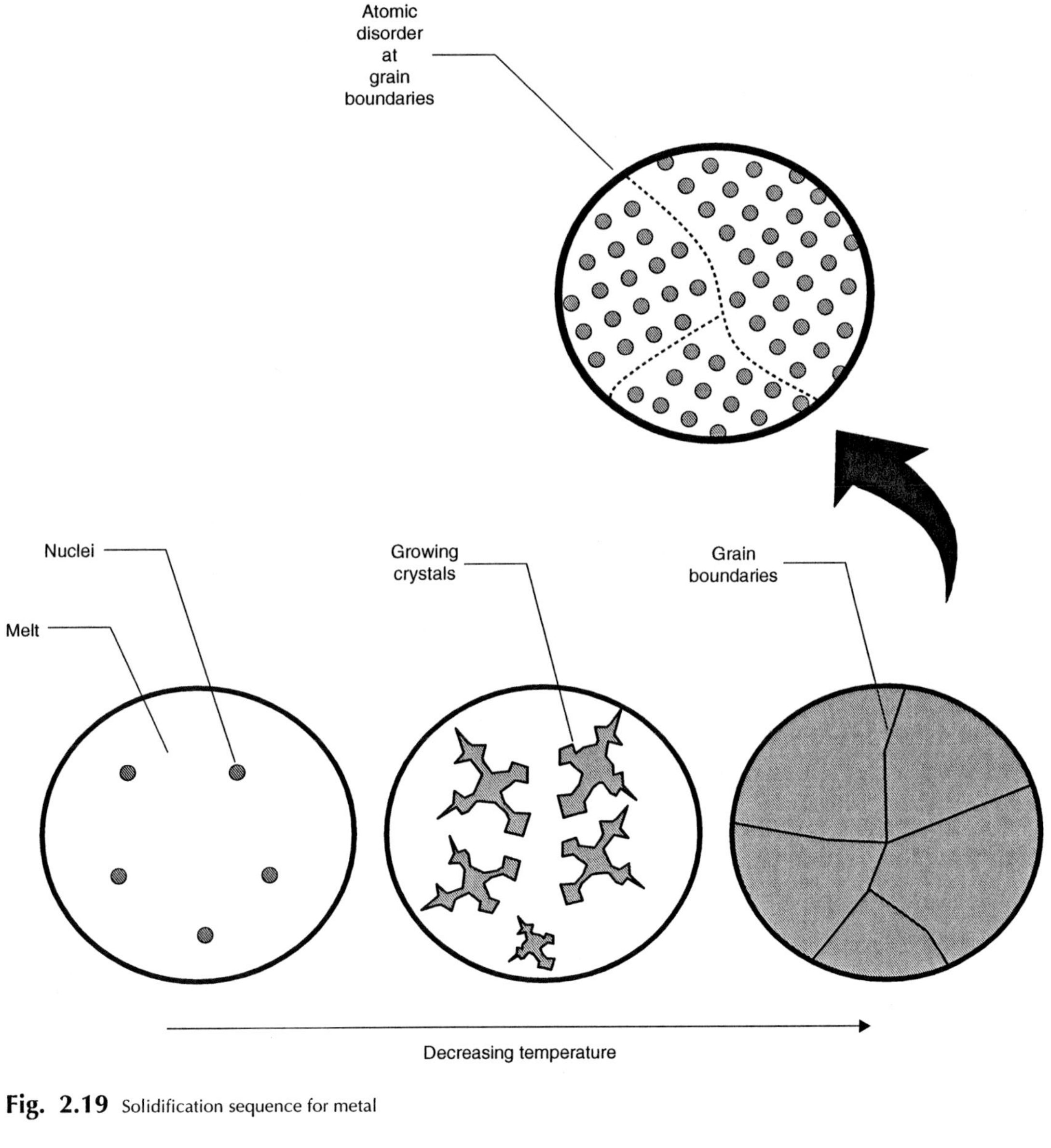


**Fig. 2.19** Solidification sequence for metal

or subcrystals and are essentially free of dislocations. The spacing between dislocations, $D$, is:

$$D = \frac{b}{\sin\theta} \cong \frac{b}{\theta} \qquad \text{(Eq. 2.4)}$$

where $\theta$ is the angular misorientation across the boundary. A gross grid of two screw dislocations can also form a low-angle grain boundary; in this instance, it is called a low-angle twist boundary.

If the misorientation is greater than approximately 10 to 15°, then high-angle grain boundaries will form, as previously illustrated in Fig. 2.19. Since high-angle grain boundaries result in a less ordered arrangement of the atoms with large areas of misfit and a relatively more open structure, the atoms along the grain boundaries have a higher energy than the atoms within the grain interiors. Thus, grain boundaries are regions with many irregularly placed atoms, dislocations, and voids. Compared with high-angle boundaries, low-angle boundaries have less severe defects, obstruct plastic flow less, and are less susceptible to chemical attack and segregation of alloying constituents. In general, mixed types of grain-boundary defects are common. All grain boundaries are sinks into which vacancies and dislocations can disappear.

Grain boundaries, as well as other microstructural features, are often observed by polishing a metal surface, lightly etching it with an acid, and then examining it with a light microscope at magnification. Grain boundaries become visible when the polished surface is etched with the proper acid, creating a microscopically uneven surface that reflects the light slightly differently (Fig. 2.21) than the unetched surface.

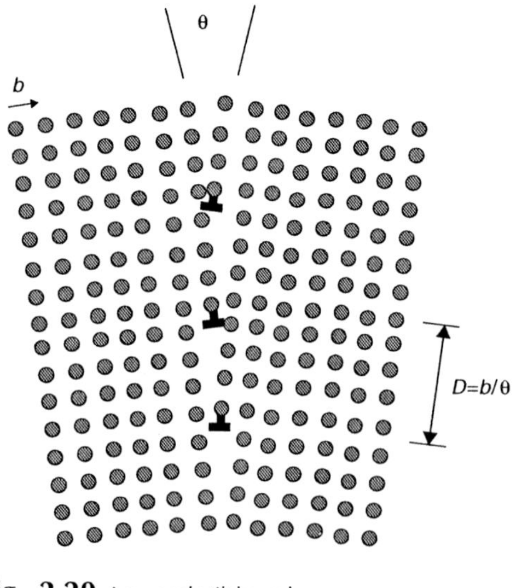


**Fig. 2.20** Low-angle tilt boundary

A grain boundary tends to minimize its area in order to reduce the internal energy of the system. The driving force for this energy reduction is surface tension, which can be reduced by straightening of the irregular-shaped boundaries. If a grain has less than six boundaries, then each boundary will be concave inward and unstable. On the other hand, if a grain has more than six sides, the boundaries will be planar and stable. At high temperatures, for example, during annealing operations at $T > 0.5\,T_m$, there is an exponential increase in the mobility of the atoms. Grains with six or more boundaries will tend to grow, while grains with less than six sides will shrink and be consumed by larger grains. Atoms in the shrinking grains will migrate across the boundary interface to join the larger growing grains. The presence of second-phase particles helps to pin the grain boundaries and impedes grain growth. At equilibrium (Fig. 2.22), all three grain boundaries will have the same surface tension, $\gamma$, and all three will have angles $\theta = 120°$. If all of the boundaries meet at 120°, then the shape of the grain will fill all of the available area and is called a tetrakaidecahedron, which contains 14 faces, 36 edges, and 24 corners. A stack of six tetrakaidecahedra is shown in Fig. 2.23.

Grain boundaries are preferential regions for accumulation and segregation of many types of impurities. Weakening or embrittlement can also occur by preferential phase precipitation or absorption of environmental species, such as hydrogen or oxygen, in the grain boundaries. At room temperature, the grain boundaries are usually stronger than the grain interiors, and failure usually occurs through the grains themselves (transgranular). However, at high temperatures, the grain boundaries typically become the weak link, and failure occurs through the grain boundaries (intergranular). Thus, a coarse grain structure is desirable for high-temperature applications, while fine grains and finely divided phase regions are preferred for most room- and low-temperature applications.

### *2.4.2 Polycrystalline Metals*

In the single-crystal tensile stress, where the critical resolved shear stress was determined

(Fig. 2.12), the slip planes were not restricted by the presence of other grains, and slip occurs as in the left-hand portion of Fig. 2.24. However, in polycrystalline metals, the orientation of the slip planes in adjoining grains is seldom aligned, and the slip plane must change direction when traveling from one grain to another. Reducing the grain size produces more changes in direction of the slip path and also lengthens it, making slip more difficult; therefore, grain boundaries

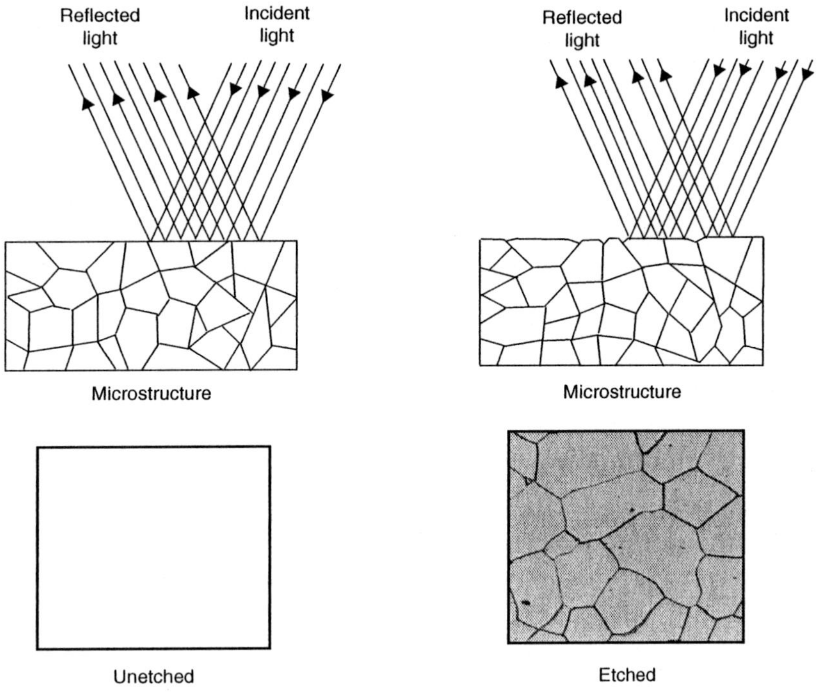


**Fig. 2.21** Metallography of unetched and etched samples

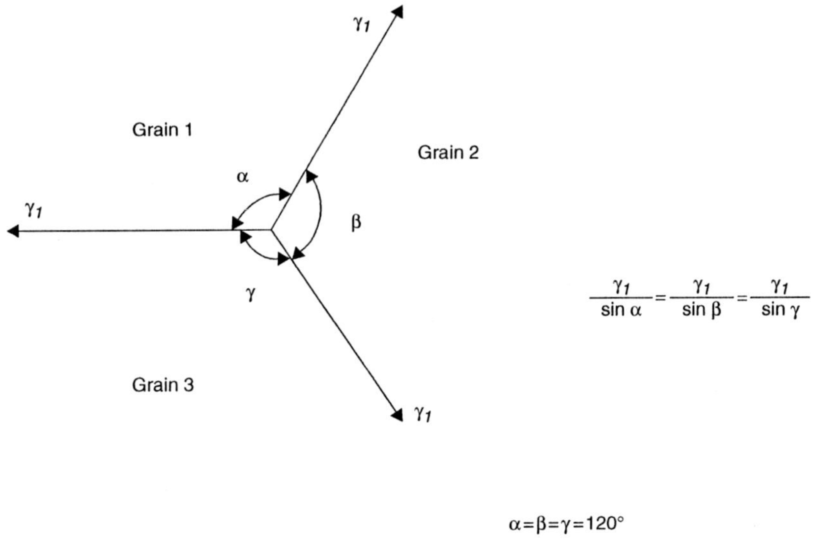


**Fig. 2.22** Grain boundaries in equilibrium. Source: Ref 2

are effective obstacles to slip. In addition, dislocations cannot cross the high-energy grain boundaries; instead, they are blocked and pile up at the boundaries (Fig. 2.25). Decreasing the grain size is effective in both increasing strength and also increasing ductility, and, as such, is one of the most effective strengthening mechanisms. Fracture resistance also generally improves with reductions in grain size, because the cracks formed during deformation, which are the precursors to those causing fracture, are limited in size to the grain diameter. The yield strength of many metals and their alloys has been found to vary with grain size according to the Hall-Petch relationship:

$$\sigma_y = \sigma_0 + k_y d^{-1/2} \quad \text{(Eq 2.5)}$$

where $k_y$ is the Hall-Petch coefficient, a material constant; $d$ is the grain diameter, and $\sigma_0$ is the yield strength of an imaginary polycrystalline metal having an infinite grain size.

The Hall-Petch relationships for a number of metals are shown in Fig. 2.26. The value of the Hall-Petch coefficient varies widely for different metals, and grain size refinement is more efficient for some metals than others. For example, grain size refinement significantly increases the yield strength of low-carbon steels (up to 275 MPa, or 40 ksi), while it provides only

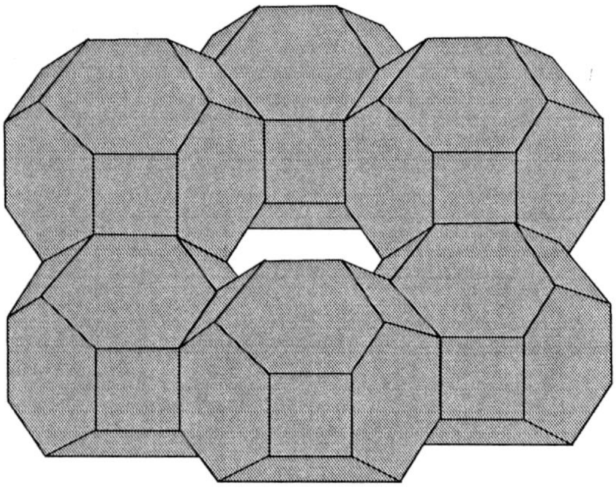

**Fig. 2.23** Stack of tetrakaidecahedra. Source: Ref 2

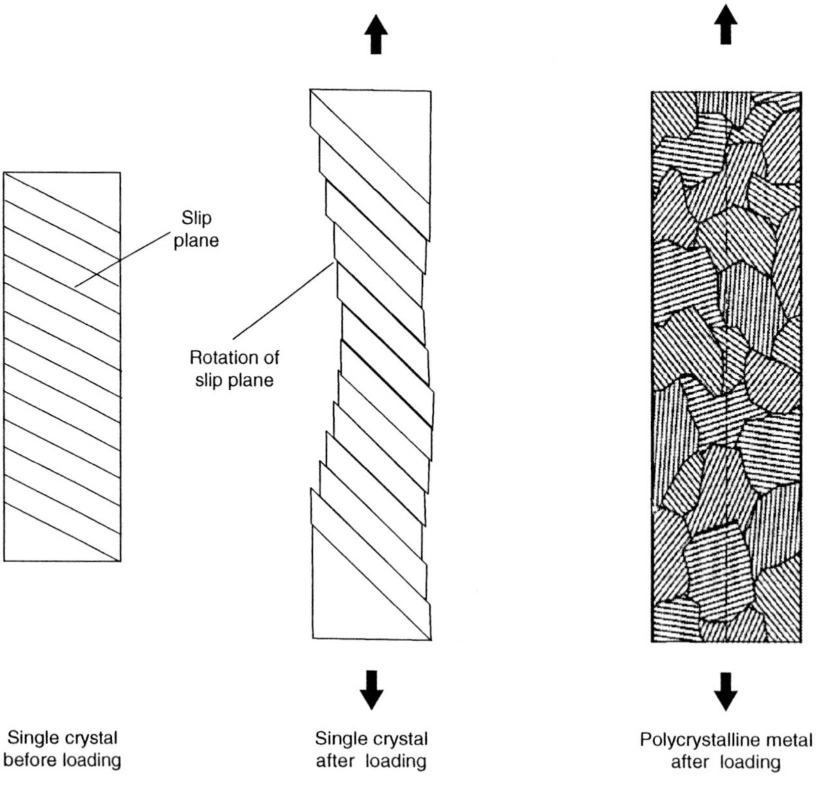


**Fig. 2.24** Tension loading of single and polycrystalline metals

about a 60 MPa (9 ksi) increase for a typical aluminum alloy.

Since the grain size of a metal or alloy has important effects on the structural properties, a number of methods have been developed to measure the grain size of a sample. In all methods, some form of microexamination is used in which a small sample is mounted, polished, and then etched to reveal the grain structure. The most direct method is then to count the number of grains present in a known area of the sample so the grain size can be expressed as the number of grains/area.

ASTM International has developed standard procedures for determining average grain size. The ASTM grain size number provides a convenient method for communicating grain sizes. For materials with a uniform grain size distribution, the ASTM grain size number is derived from the number of grains/in.$^2$ when counted at a magnification of 100 ×. The ASTM index, *N,* is given by:

$$n = 2^{(N-1)} \quad \text{(Eq 2.6)}$$

where $n$ is the number of grains/in.$^2$ at 100 × magnification. To obtain the number of grains per square millimeter at 1×, multiply $n$ by 15.50. This can be rewritten as:

$$\log n = (N-1)\log 2 \quad \text{(Eq 2.7)}$$

or

$$N = \frac{\log n}{0.3010} + 1 \quad \text{(Eq 2.8)}$$

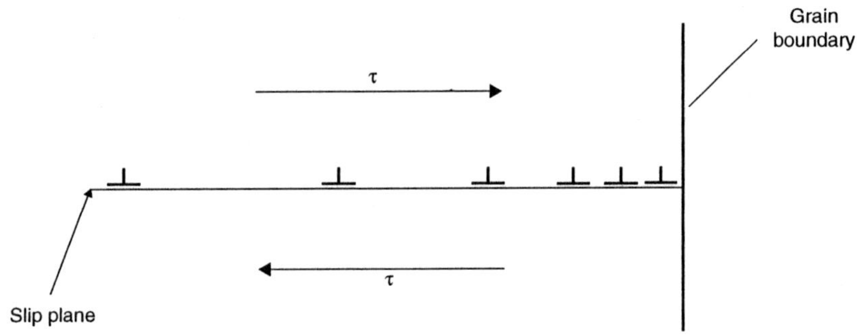


**Fig. 2.25** Dislocation pileup at grain boundary

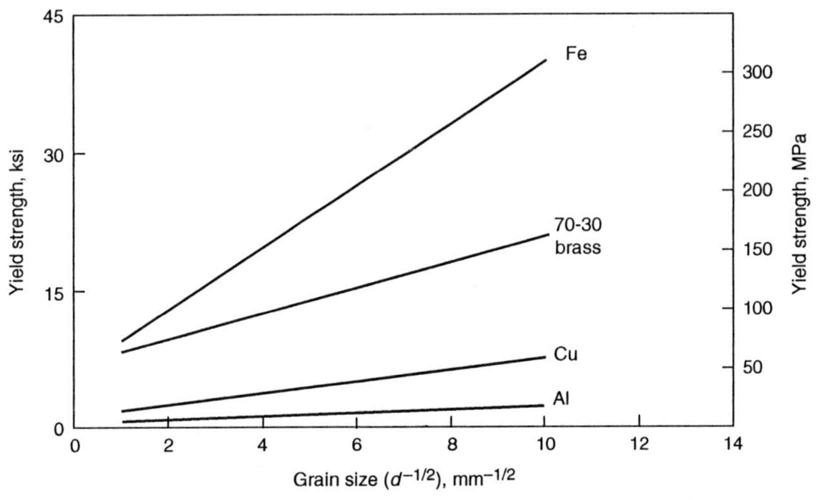


**Fig. 2.26** Hall-Petch relationship

A listing of ASTM grain size numbers and the corresponding grain size is given in Table 2.2. Note that larger ASTM grain size numbers indicate more grains per unit area and finer, or smaller, grain sizes.

### 2.4.3 Phase Boundaries

While a grain boundary is an interface between grains of the same composition and same crystalline structure (α/α interface) with different orientations, a phase boundary is one between two different phases (α/β interface) that can have different crystalline structures and/or different compositions. In two-phase alloys, such as copper-zinc brass alloys containing more than 40% Zn, second phases, such as the one shown in Fig. 2.27, can form due to the limited solid solubility of zinc in copper.

There are three different types of crystalline interfaces that can develop between two phases (Fig. 2.28): coherent, semicoherent, and incoherent. A fully coherent phase boundary (Fig. 2.28a, b) occurs when there is perfect atomic matching and the two lattices are continuous across the interface. The interfacial plane will have the same atomic configuration

**Table 2.2 ASTM grain size, $n = 2^{N-1}$**

| | Average number of grains/unit area | |
|---|---|---|
| Grain size No. ($N$) | No./in.$^2$ at 100× ($n$) | No./mm$^2$ at 1× ($n$) |
| 1 | 1.00 | 15.50 |
| 2 | 2.00 | 31.00 |
| 3 | 4.00 | 62.00 |
| 4 | 8.00 | 124.00 |
| 5 | 16.00 | 496.00 |
| 6 | 32.00 | 992.00 |
| 7 | 64.00 | 1984.0 |
| 8 | 128.00 | 3968.0 |
| 9 | 256.00 | 3968.0 |
| 10 | 512.00 | 7936.0 |

Source: Ref 3

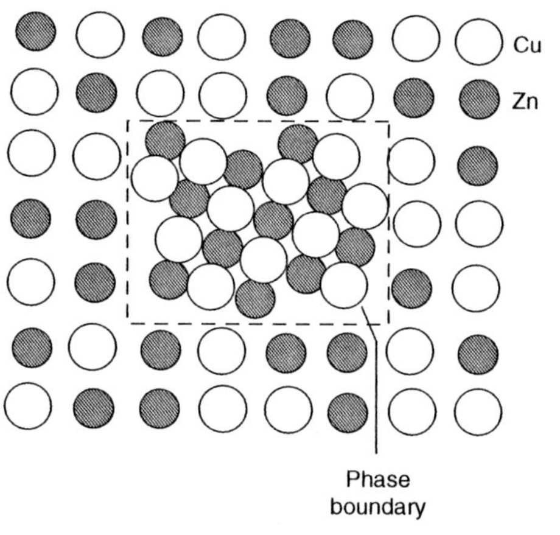


**Fig. 2.27** Phase boundary in copper-zinc system. Source: Ref 4

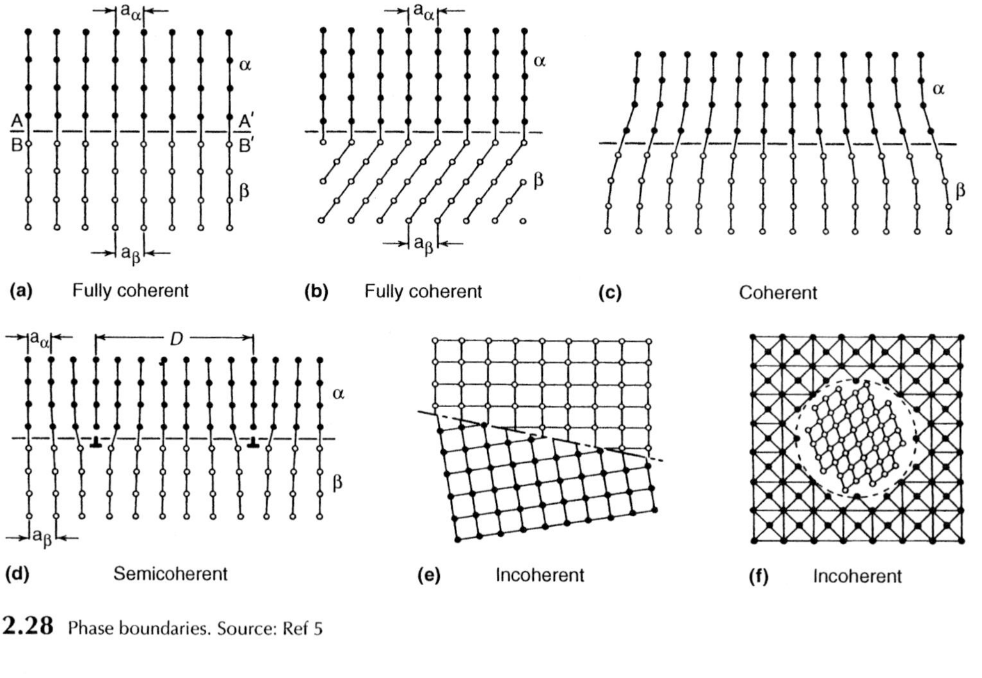


**Fig. 2.28** Phase boundaries. Source: Ref 5

in both planes. Since there is perfect matching at the interface, the interfacial energy is low, typically up to approximately 200 mJ/m$^2$. When the distances between atoms at the interface are not identical (Fig. 2.28c), coherency strains start to develop. However, since there is still perfect atomic matching, it is still a coherent phase boundary; only the interfacial energy will be higher than one with no distortion. When the mismatch becomes sufficiently large, dislocations form to accommodate the growing disregistry. The result is called a semicoherent interface (Fig. 2.28d) that has an interfacial energy of 200 to 500 mJ/m$^2$. Finally, an incoherent interface (Fig. 2.28e, f) is an interphase boundary that results when the matrix and precipitate have very different crystal structures, and little or no atomic matching can occur across the interface. The interfacial energy is even greater, reaching values between 500 and 1000 mJ/m$^2$. An incoherent boundary is essentially the same as a high-angle grain boundary.

In many instances, second phases have a tendency to form at the grain boundaries. This occurs because they reduce their interfacial energy by occupying a grain boundary; that is, by occupying a grain boundary, part of the interfacial energy is eliminated, and the total energy of the system is reduced. Consider the case where two grains of α phase meet with one grain of β phase, as shown in Fig. 2.29. The surface energy, γ, will be in equilibrium if:

$$\gamma_{\alpha\alpha} = 2\gamma_{\alpha\beta} \cos\frac{\theta}{2} \quad \text{(Eq 2.9)}$$

The angle, θ, is called the dihedral angle at the α-to-β interface. If $\gamma_{\alpha\beta} > 1/2\ \gamma_{\alpha\alpha}$, θ will have a finite value; if $\gamma_{\alpha\beta} = \gamma_{\alpha\alpha}$, $\theta = 120°$; and if $\gamma_{\alpha\beta} > \gamma_{\alpha\alpha}$, $\theta > 120°$. However, if $\gamma_{\alpha\alpha} > 2\gamma_{\alpha\beta}$, the previous equation cannot be satisfied, and no

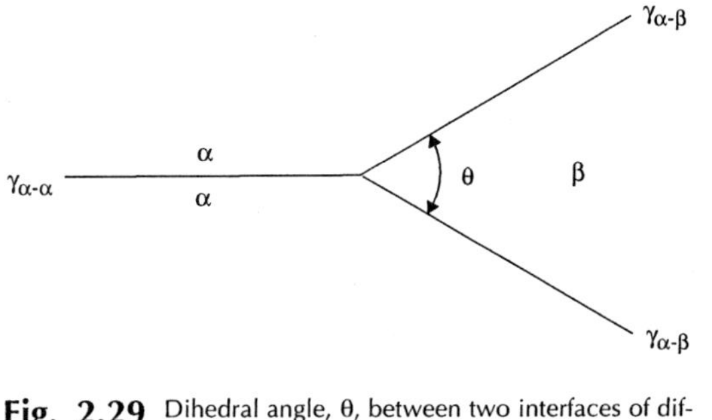


**Fig. 2.29** Dihedral angle, θ, between two interfaces of differing phases. Source: Ref 2

equilibrium will exist. Instead, the β phase will wet the grain boundary and spread out as a thin grain-boundary film. In this case, if the β phase is brittle or has a low melting point, the mechanical properties of the alloy will be impaired even though the α matrix is strong and tough. This potentially disastrous condition, known as grain-boundary embrittlement, is shown in Fig. 2.30.

When the second phase is located at the juncture of three grains, it can form different shapes (Fig. 2.31), depending again on the dihedral angle, θ. At small angles, such as $\theta = 0°$, the second phase can penetrate the grain boundaries and possibly affect alloy properties, while at the other extreme ($\theta = 180°$), it will form round particles that should not inhibit alloy performance. Second-phase particles of lead are often added to alloys to improve machinability by forming cleaner chips. Since they form round particles, such as in the case where $\theta = 180°$, they do not adversely affect strength or ductility. There are even some instances where grain-boundary wetting is desirable, such as during liquid phase sintering of carbide cutting tools. Here, a cobalt matrix wets the tungsten carbide particles and binds them together during sintering.

### 2.4.4 Twinning

Twinning is another mechanism that causes plastic deformation, although it is not nearly as important as dislocation movement. Mechanical twinning is the coordinated movement of large numbers of atoms that deform a portion of the crystal by an abrupt shearing motion. Atoms on each side of the twinning plane, or habit plane, form a mirror image with those on the other side of the plane (Fig. 2.32). Shear stresses along the twin plane cause atoms to move a distance that is proportional to the distance from the twin plane. However, atom motion with respect to one's nearest neighbors is less than one atomic spacing. Twins occur in pairs, such that the change in orientation of the atoms introduced by one twin is restored by the second twin. Twinning occurs on a definite crystallographic plane and in a specific direction that depends on the crystalline structure. Twins can occur as a result of plastic deformation (deformation twins) or during annealing (annealing twins). Mechanical twinning occurs in bcc and hcp metals, while annealing twins are fairly common in fcc metals.

Mechanical twinning increases the strength because it subdivides the crystal, thereby increasing the number of barriers to dislocation movement. Twinning is not a dominant deformation mode in metals with multiple slip systems, such as fcc structures. Mechanical twinning occurs in metals that have bcc and hcp crystalline structures at low temperatures and at high rates of loading, conditions in which the normal slip process is restricted due to few operable slip systems. The amount of bulk plastic deformation in twinning is small compared to slip. The real importance of twinning is that crystallographic planes are reoriented so that additional slip can take place.

Unlike slip, the shear movements in twinning are only a fraction of the interatomic spacing, and the shear is uniformly distributed over volume rather than localized on a number of distinct planes. Also, there is a difference in orientation of the atoms in the twinned region compared to the untwinned region that con-

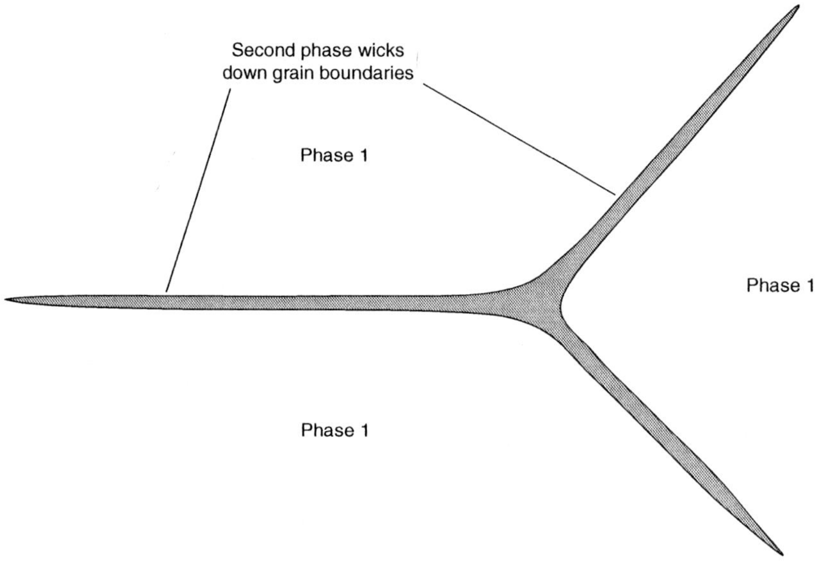


**Fig. 2.30** Grain-boundary embrittlement. Source: Ref 6

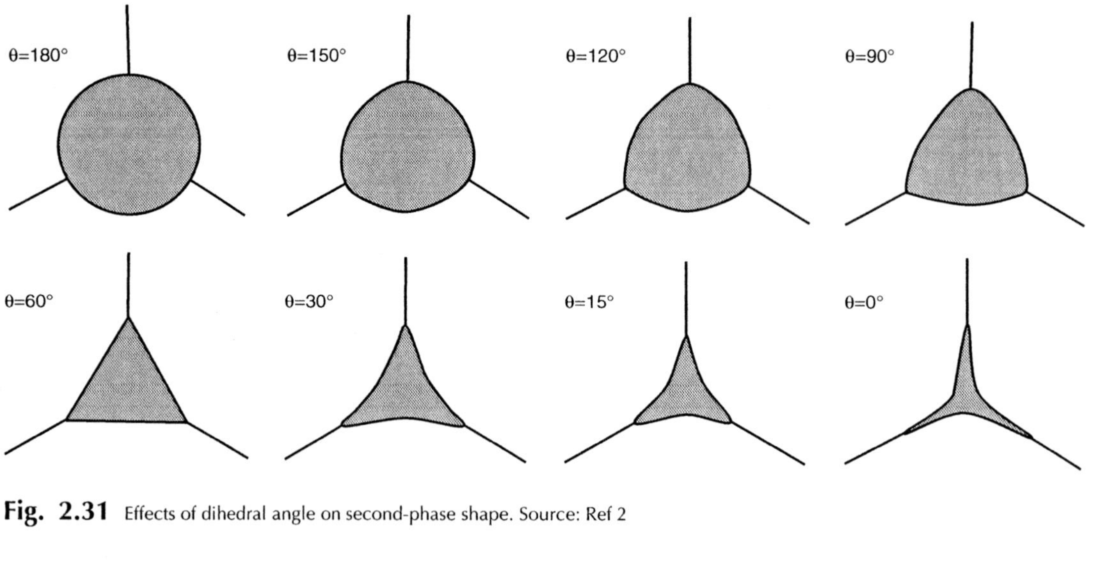


**Fig. 2.31** Effects of dihedral angle on second-phase shape. Source: Ref 2

stitutes a phase boundary. Twins form suddenly, at a rate approaching the speed of sound, and can produce audible sounds, such as "tin cry." Since the amount of atom movement during twinning is small, the resulting plastic deformation is also small.

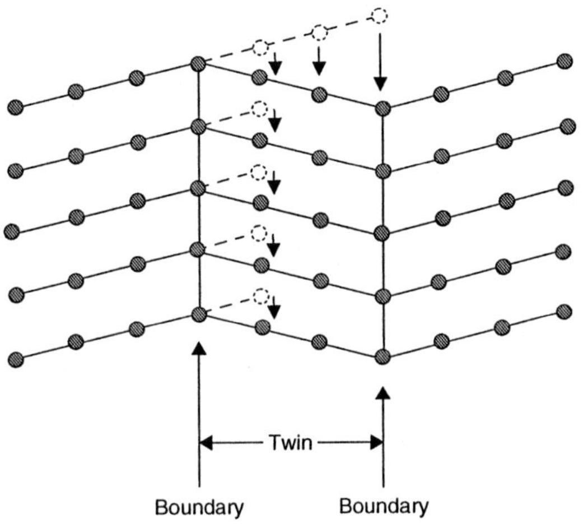


**Fig. 2.32** Deformation by twinning

A comparison of the slip and twinning mechanisms is shown in Fig. 2.33. The differences between the two deformation mechanisms include the following.

**Orientation.** In slip, the orientation above and below the slip plane is the same after slip, while in twinning, there is an orientation change across the twin plane.

**Mirror Image.** Atoms in the twinned portion of the lattice form a mirror image with the untwinned portion. No such relationship exists in slip.

**Deformation.** In slip, the deformation is nonhomogeneous because it is concentrated in bands, while the metal adjacent to the bands is largely undeformed. In twinning, the deformation is homogeneous because all of the atoms move cooperatively at the same time.

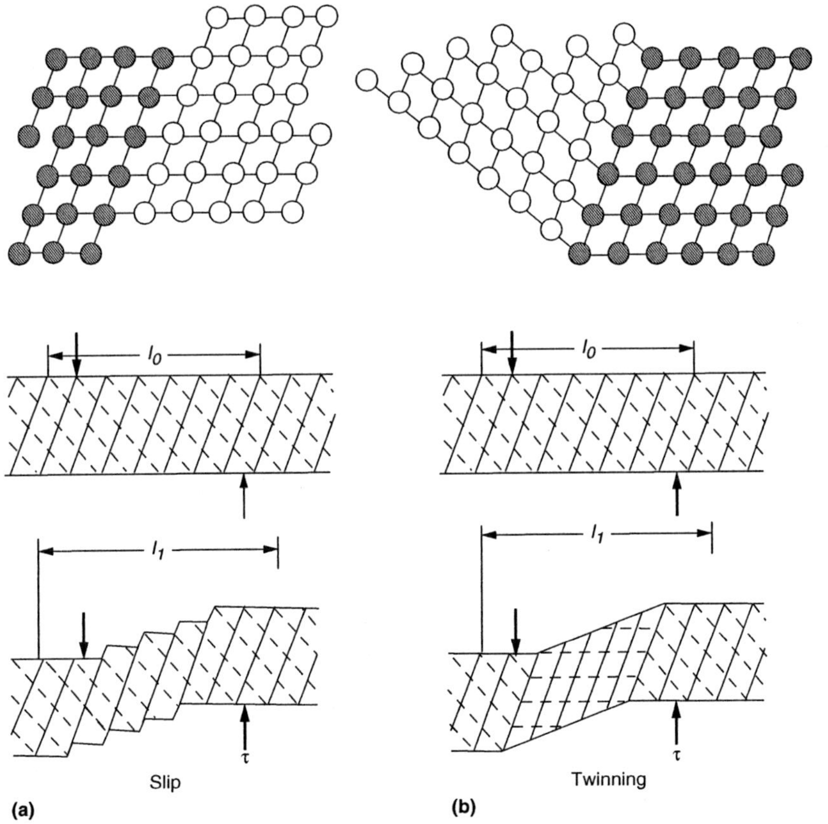


**Fig. 2.33** Comparison of slip and twinning deformation mechanisms occurring over a length, $l$, under a shear stress, $\tau$. Source: Ref 1

**Stress.** In slip, a lower stress is required to initiate it, while a higher stress is required to keep it propagating. In twinning, a high stress is required to initiate, but a very low stress is required for propagation. The shear stress required for twinning is usually higher than that required for slip.

Twin boundaries generally are very flat, appearing as straight lines in micrographs, and are two-dimensional defects of lower energy than high-angle grain boundaries. Therefore, twin boundaries are less effective as sources and sinks of other defects and are less active in deformation and corrosion than ordinary grain boundaries.

Another type of deformation similar to twinning is kink band formation. Kink band formation usually occurs in hcp metals under compressive loading. It occurs when the applied stress is nearly perpendicular to the principal slip plane, normally the basal plane in hcp metals. In this mechanism, the metal shears by the formation of dislocation arrays that produce buckling along the slip plane direction that shears the metal several degrees away from its previous position. As opposed to twinning, the atomic positions do not form a mirror image after shear displacement.

### *2.4.5 Stacking Faults*

The passage of a total dislocation through a crystalline lattice leaves the perfection of the lattice undisturbed. Each atom is shifted from one normal position in the lattice to an adjacent normal position. However, the energy of the system can sometimes be lowered if a total dislocation splits into a partial dislocation (Fig. 2.34). It takes less energy if the total dislocation splits into two partial dislocations that can move in a zigzag path through the valley between atoms, rather than having to climb over an atom. Instead of an atom moving directly from its lattice position to a new position, indicated by the tip of the arrow of the Burgers vector, it can move first to an intermediate vacant site and then again to the final site. Thus, two short jumps are made instead of one longer one, which requires less energy. However, the passage of a partial dislocation leaves behind a planar region of crystalline imperfection.

The planar imperfection produced by the passage of a partial dislocation is called a stacking fault, as illustrated in Fig. 2.35. In an fcc structure, the stacking sequence changes from the normal *ABCABC* to *ABAB,* which is the stacking sequence for the hcp structure. Passage of the second partial dislocation restores the normal *ABCABC* stacking sequence. These partial dislocations are often referred to as Schottky partials. The two partial dislocations that are separated by the faulted area are known as an extended dislocation.

The total energy of a perfect lattice is lower than one with a stacking fault. Thus, a stacking fault has an energy associated with it. The difference in energy between a perfect lattice and one with a stacking fault is known as the stacking fault energy (SFE). Equilibrium occurs between the repulsive energy of the two partials and the surface energy of the fault. The larger the separation between the partial dislocations, the smaller is the repulsive force between them. On the other hand, the surface energy associated with the stacking fault increases with the distance between the two partial dislocations. In general, if the separation between the partial dislocations is small, the metal is said to have a high SFE. If the separation is large, the metal would have a low SFE. For example, the separation in aluminum (high SFE) is on the order of an atomic spacing, while that of copper (low SFE) is approximately 12 atomic spacings.

Stacking fault energy plays a role in determining deformation textures in fcc and hcp

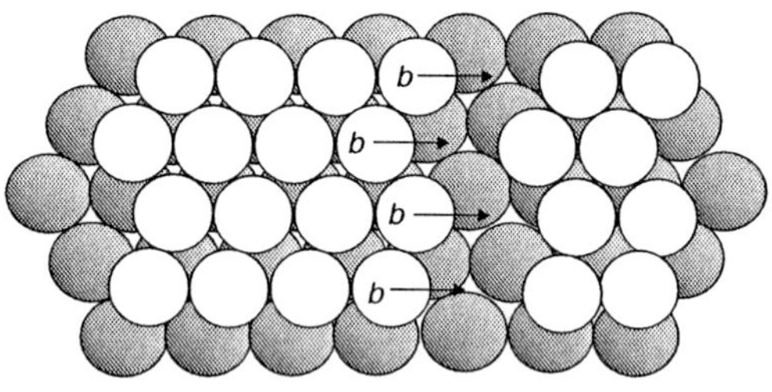


Total dislocation in fcc structure

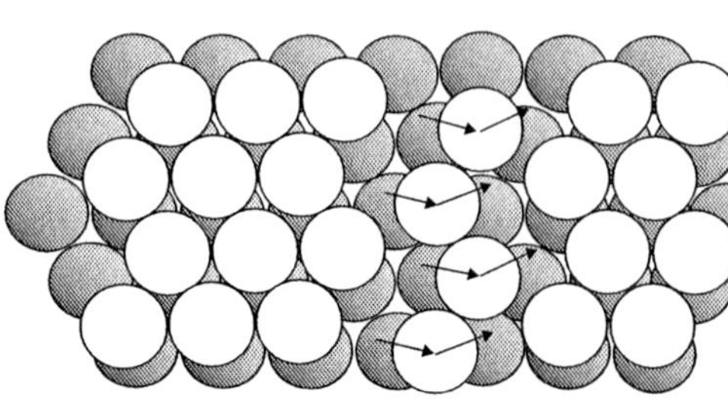

Partial dislocation in fcc structure

**Fig. 2.34** Concept of partial dislocation. Source: Ref 7

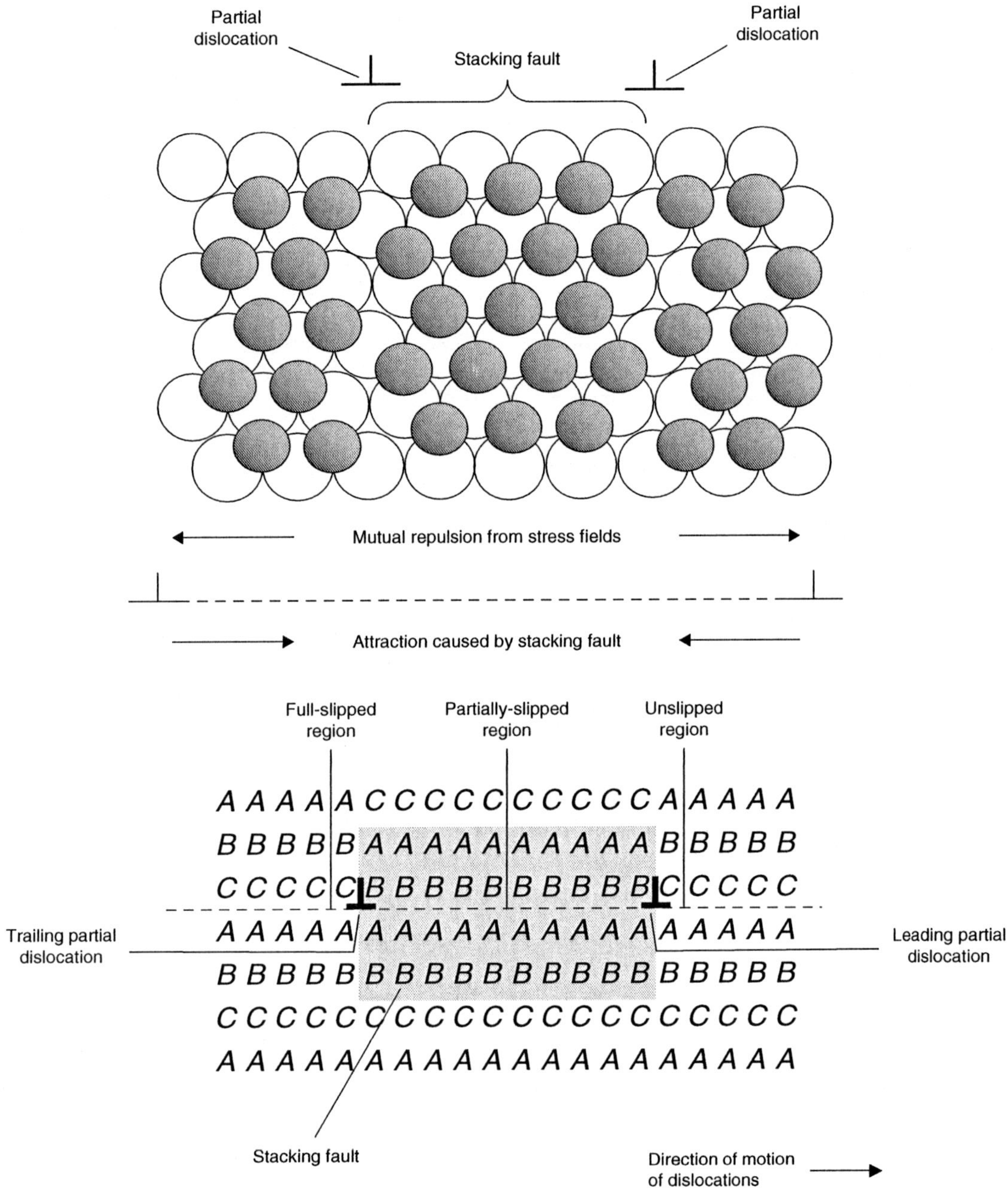


**Fig. 2.35** Stacking fault and extended dislocation. Source: Ref 7, 8

metals. Stacking faults also influence plastic deformation characteristics. Metals with wide stacking faults (low SFE) strain harden more rapidly and twin more readily during annealing than those with narrow stacking faults (high SFE). Some representative SFEs are given in Table 2.3.

## 2.5 Volume Defects

Volume defects, such as porosity and microcracks, almost always reduce strength and fracture resistance. The reductions can be quite substantial, even when the defects constitute only several volume percent. Shrinkage during

**Table 2.3 Approximate stacking fault energies (SFEs)**

| Metal | SFE, mj/m$^2$ |
|---|---|
| Brass | <10 |
| Austenitic stainless steel | <10 |
| Silver | 20–25 |
| Gold | 50–75 |
| Copper | 80–90 |
| Nickel | 130–200 |
| Aluminum | 200–250 |

solidification can result in microporosity, that is, porosity having diameters on the order of micrometers. In metals, porosity is much more likely to be found in castings than in wrought products. The extensive plastic deformation during the production of wrought metals is usually sufficient to heal or close microporosity.

Powder metallurgy products also frequently contain porosity. Powder metallurgy products are usually produced by blending metal powders, pressing them into a shape, and then sintering them at temperatures just below the melting point. Porosity in powder metallurgy products can be reduced if pressure is used during the sintering process by either hot pressing in a press or hot isostatic pressing under gas pressure.

## REFERENCES


1. M. Tisza, *Physical Metallurgy for Engineers,* ASM International, 2001
2. R.M. Brick, A.W. Pense, and R.B. Gordon, *Structure and Properties of Engineering Materials,* 4th ed., McGraw-Hill Book Company, 1977
3. "Determining Average Grain Size," E112-96, *Annual Book of ASTM Standards, Section 3: Metal Test Methods and Analytical Procedures,* ASTM International, 1999, p 237–259
4. D.R. Askeland, *The Science and Engineering of Materials,* 2nd ed., PWS-Kent Publishing Co., 1989
5. M. Epler, Structures by Precipitation from Solid Solution, *Metallography and Microstructures,* Vol 9, *ASM Handbook,* ASM International, 2004
6. V. Singh, *Physical Metallurgy,* Standard Publishers Distributors, 1999
7. R.E. Reed-Hill and R. Abbaschian, *Physical Metallurgy Principles,* 3rd ed., PWS Publishing Company, 1991
8. A.G. Guy, *Elements of Physical Metallurgy,* 2nd ed., Addison-Wesley Publishing Company, 1959


## SELECTED REFERENCES


- M.F. Ashby and D.R.H. Jones, *Engineering Materials 1—An Introduction to Their Properties, and Applications,* 2nd ed., Butterworth Heinemann, 1996
- A.M. Russell and K.L. Lee, *Structure-Property Relationships in Nonferrous Alloys,* Wiley-Interscience, 2005
- D.R. Sadoway, "The Imperfect Solid State," Lecture notes, Introduction to Solid-State Chemistry, Department of Materials Science and Engineering, Massachusetts Institute of Technology, Fall 2006

CHAPTER 3

# Solid Solutions

PURE METALS are rarely used for industrial applications unless high conductivity, high ductility, or good corrosion resistance are required. Since pure metals tend to be much weaker than alloys, alloying elements are added to improve strength and hardness. In addition, alloying element additions can often be accomplished without major reductions in the attributes associated with pure metals, that is, conductivity, ductility, and corrosion resistance. When two metals are mixed in the liquid state to produce a solution, the resulting alloy is called a binary alloy. While a great number of alloys are binary, many more alloys contain a number of alloying elements. In the case of high-temperature superalloys, as many as ten alloying elements may be used to obtain the desired performance.

The metals nickel and copper completely dissolve in each other in the liquid state and then retain their complete solubility in each other on freezing to form a series of alloys. The improvement in strength properties in these alloys is shown in Fig. 3.1. Note that the yield

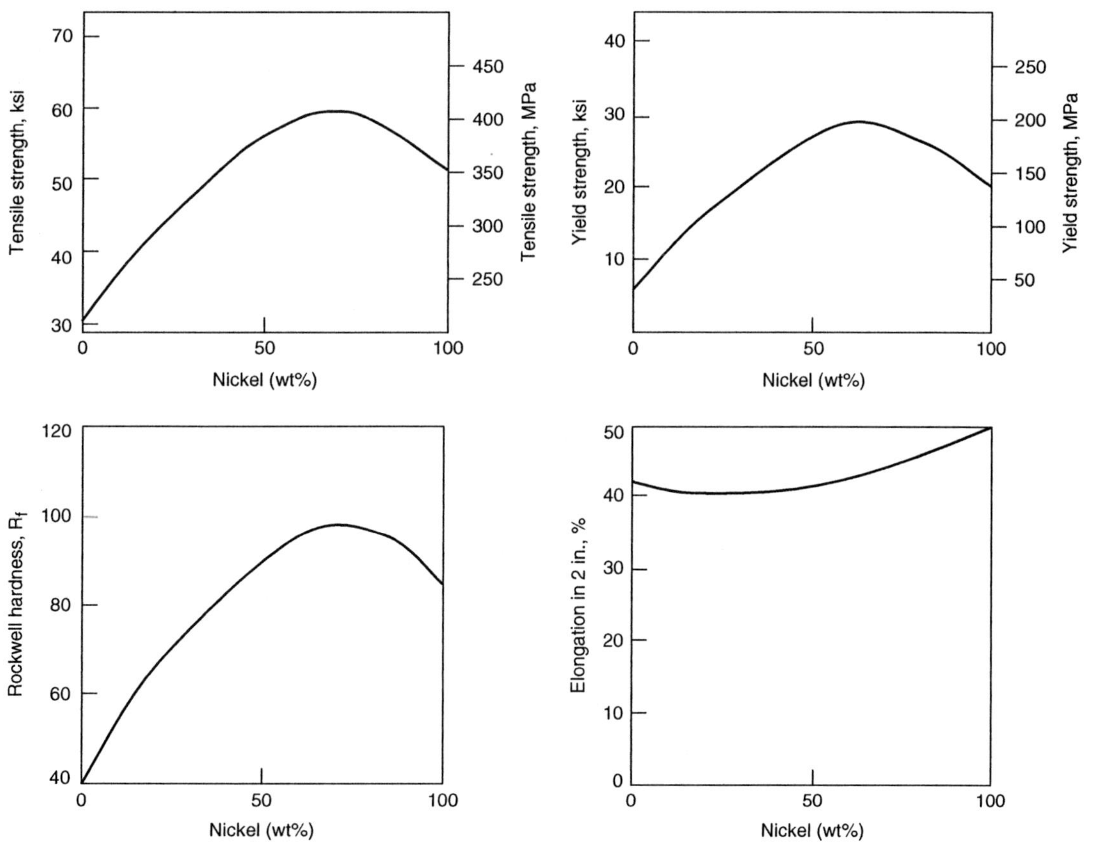


**Fig. 3.1** Solid-solution strengthening for copper-nickel alloys

strength, the tensile strength, and the hardness are improved as a result of alloying. While the ductility is reduced somewhat with alloying, the reduction is very moderate, and these alloys still have very good ductility.

When a metal is alloyed with another metal, either substitutional or interstitial solid solutions (Fig. 3.2) are usually formed. Substitutional solid solutions are those in which the solute and solvent atoms are nearly the same size, and the solute atoms simply substitute for solvent atoms on the crystalline lattice. Interstitial solid solutions are those in which the solute atoms are much smaller and fit within the spaces between the existing solvent atoms on the crystalline structure. However, the only solute atoms small enough to fit into the interstices of metal crystals are hydrogen, nitrogen, carbon, and boron. The other small-diameter atoms, such as oxygen, tend to form compounds with metals rather than dissolve in them. When both small and large solute atoms are present, the solid solution can be both interstitial and substitutional.

The insertion of substitutional and/or interstitial alloying elements strains the crystalline lattice of the host solvent structure (Fig. 3.3). This increase in distortion, or strain, creates barriers to dislocation movement. The distortion energy causes some hardening and strengthening of the alloy and is called solid-solution hardening. The solute atoms have a different size than the host atoms, which alters the crystal lattice. As a result, a moving dislocation is either attracted to or repelled by the solute; however, both situations result in a strength increase. When the dislocation is attracted to a solute, the additional force required to pull the dislocation away from it is the cause of the added strength. If the dislocation is repelled by the solute, an additional force is required to push the dislocation past the solute atom.

Studies of solid-solution hardening indicate that the hardening depends on the differences in elastic stiffness and atomic size between the solvent and solute. In general, larger differences result in greater strengthening, but at the same time, the greater the difference in sizes between the solute and solvent atoms, the more restricted is their mutual solubilities. The solvent phase becomes saturated with the solute atoms and reaches its limit of homogeneity when the distortion energy reaches a critical value determined by the thermodynamics of the system. The effects of several alloying elements on the yield strength of copper are shown in Fig. 3.4. Nickel and zinc atoms are approximately the same size as copper atoms, but beryllium and tin atoms are much different from copper atoms. Increases in the atomic size difference and the amount of alloying result in increases in solid-solution strengthening.

## 3.1 Interstitial Solid Solutions

The four elements carbon, nitrogen, hydrogen, and boron have such small diameters that they can form interstitial solid solutions. In

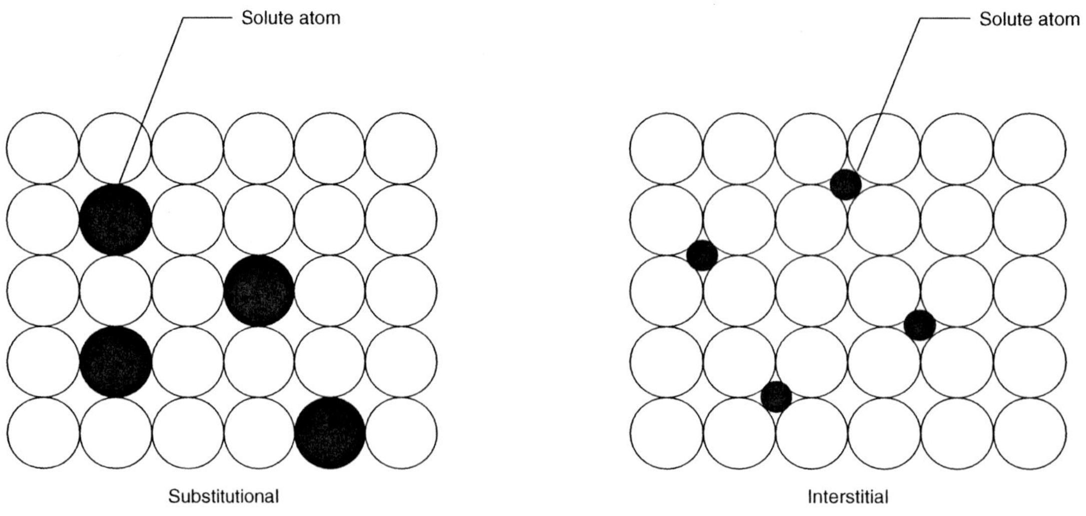


**Fig. 3.2** Solid solutions

general, these interstitial solid solutions have somewhat limited composition ranges. Only the transition metals (e.g., iron, nickel, titanium, and zirconium) have appreciable solubilities for carbon, nitrogen, and boron. Very small atoms, such as carbon, nitrogen, and hydrogen, can fit in the spaces between the larger atoms. These sites are called interstitial sites and can be of either the tetrahedral or octahedral variety (Fig. 3.5). Interstitial atoms generally strengthen a metal more than substitutional atoms do, since the interstitials cause more distortion. Carbon atoms in the body-centered cubic (bcc) form of iron are particularly potent hardeners in this respect.

Carbon, nitrogen, and boron are important alloying elements in steels. Interstitial carbon in iron forms the basis of steel hardening. Indeed, steels are alloys of iron and small amounts of carbon. Although not as important as carbon, nitrogen and boron are also useful alloying elements in certain steels. In addition, carbon and nitrogen are diffused into the surfaces to provide hardness and wear resistance in processes called carburizing and nitriding. On the other hand, hydrogen is almost never a welcome addition to any metal. It usually results in sharp decreases in ductility and produces brittle fracture modes, a mechanism called hydrogen embrittlement.

## 3.2 Substitutional Solid Solutions

The following four rules that give a qualitative estimate of the ability of two metals to form substitutional solid solutions were developed by Hume-Rothery.

**Rule 1—Relative Size Factor.** If the sizes of the solute and solvent atoms differ by less than

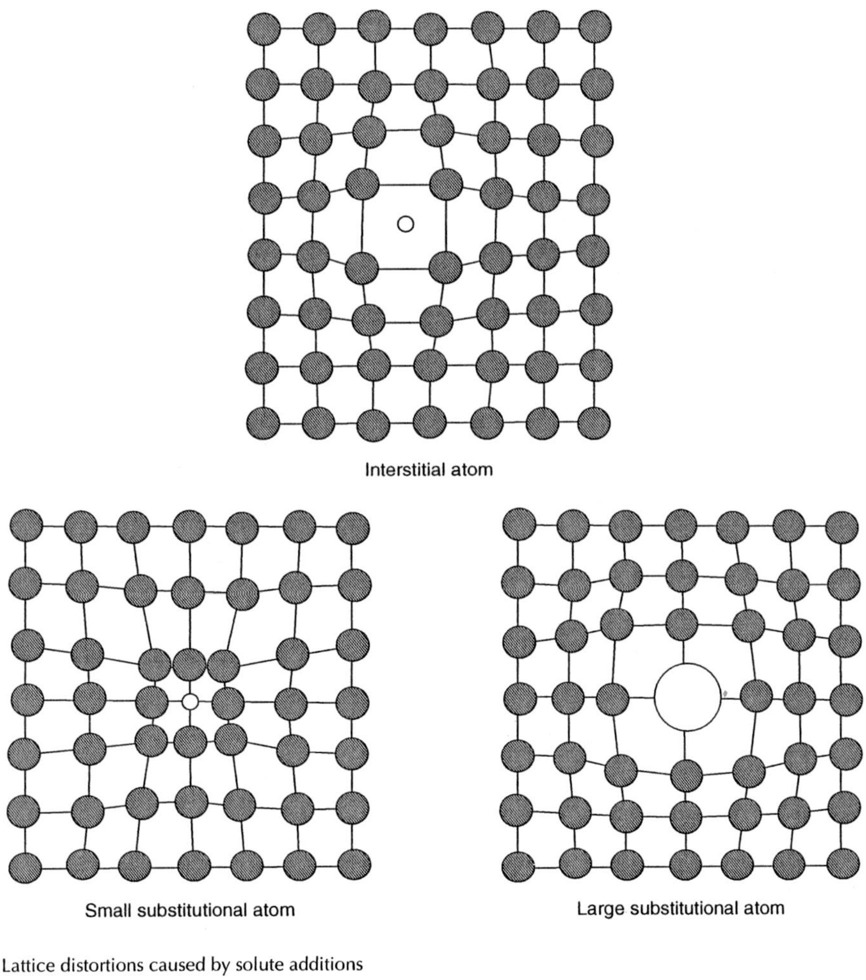



**Fig. 3.3** Lattice distortions caused by solute additions

15%, the metals are said to have a favorable size factor for solid-solution formation. Each of the metals will be able to dissolve an appreciable amount of the other metal, on the order of 10% or more. If the size factor differs by more than 15%, then solid-solution formation tends to be severely restricted.

**Rule 2—Chemical Affinity Factor.** The greater the chemical affinity of two metals, the more restricted is their solid solubility. When their chemical affinity is great, they tend to form compounds rather than a solid solution.

**Rule 3—Relative Valency Factor.** If a solute atom has a different valence from that of the solvent metal, the number of electrons per atom, called the electron ratio (*e*/*a*), will be changed by alloying. Crystal structures are more sensitive to a decrease in the electron ratio than to an increase. Therefore, a metal of high valence can dissolve only a small amount of a lower-valence metal, while a lower-valence metal may have good solubility with a higher-valence metal.

**Rule 4—Lattice Type Factor.** Only metals that have the same type of lattice structure (e.g., face-centered cubic) can form a complete series of solid solutions. Also, for complete solid solubility, the size factor usually must be less than 8%.

There are numerous exceptions to these rules. In general, an unfavorable size factor alone is sufficient to severely limit solid solubility to a minimal level. If the size factor is favorable, then the other three rules should be evaluated to determine the probable degree of solid solubility. Metallic systems that display complete solid solubility are quite rare, with the copper-nickel system being the most important.

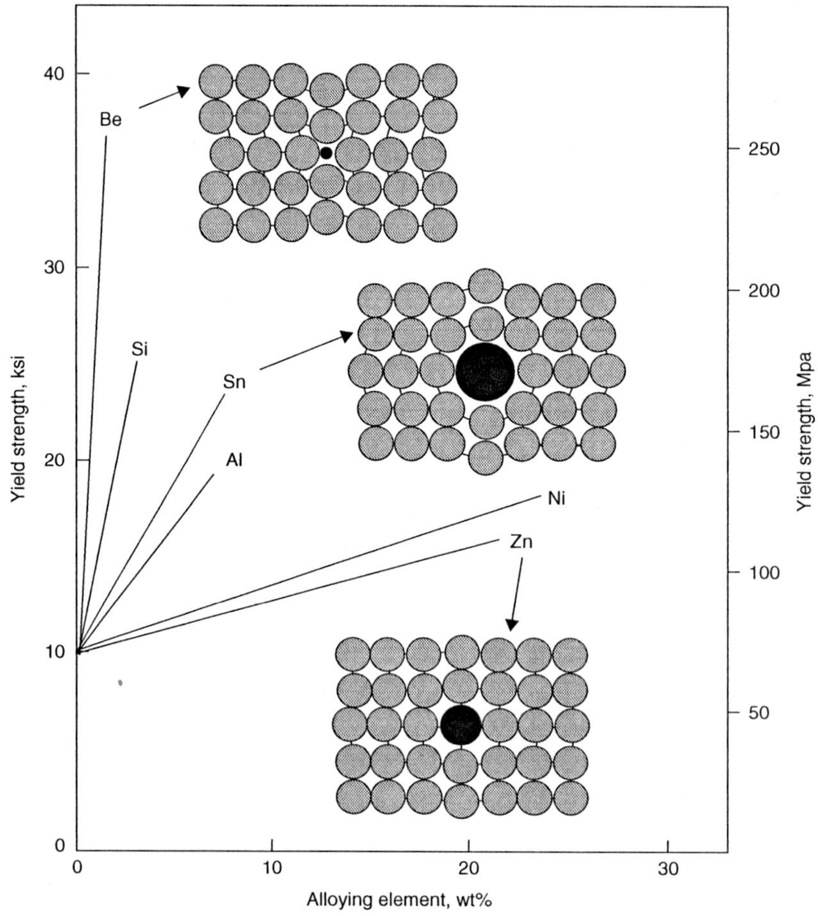


**Fig. 3.4** Effects of several alloying elements on the yield strength of copper. Source: Ref 1

## 3.3 Ordered Structures

Substitutional solid solutions can further be divided into three types: random, clustered, and ordered, as illustrated in Fig. 3.6. Random solid solutions are by far the most common. In a random solution, two types of atoms, *A* and *B,* show no preference in their bonding. In other words, *A-A, A-B,* and *B-B* bonds are equally formed. In clustering, there is a free energy reduction in the system when *A-A* and *B-B* bonds form into *A-A* and *B-B* regions or clusters. Finally, in an ordered solution, the lowest free energy is obtained when *A-B* bonds are preferred. Systems with a simple ratio of *A*:*B* atoms are more inclined to exhibit ordering. Systems that exhibit clustering or ordering at room temperature become random when heated to a high-enough temperature. Essentially, the order decreases with increasing temperature until, at some critical temperature, $T_c$, long-range order breaks down. However, in some systems, the ordering is so strong that it is stable up to the melting point.

An order-disorder transformation typically occurs on cooling from a disordered solid solution to an ordered phase. During this phase transformation, there is a rearrangement of atoms from random site locations in the disordered solution to specific lattice sites in the ordered structure. When atoms periodically arrange themselves into a specific ordered array, they make up what is commonly referred to as a superlattice.

Most alloys that form an ordered structure are disordered at higher temperature, which means that atoms are randomly located on lattice sites. On cooling, small ordered areas will nucleate

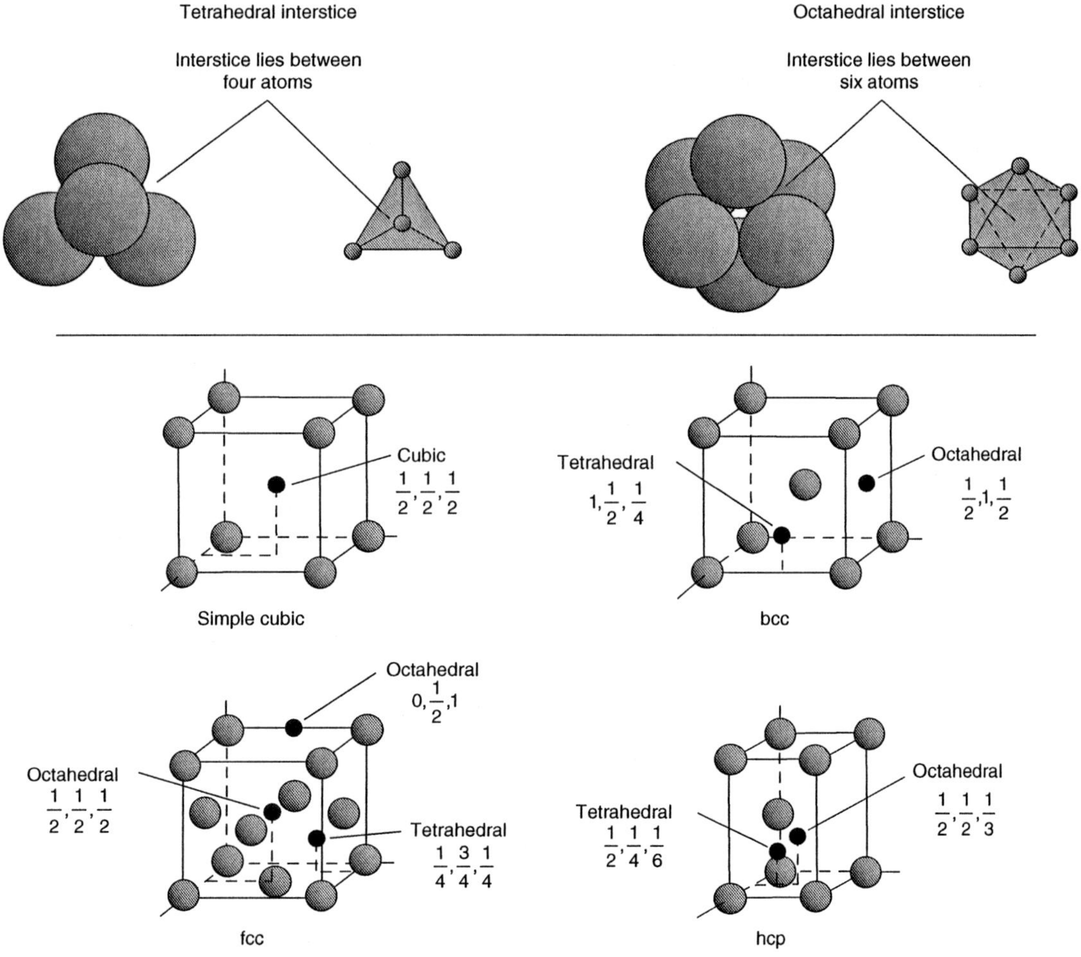


**Fig. 3.5** Representative interstitial sites in unit cells. Source: Ref 1

within the disordered phase and begin to grow into ordered domains. The maximum size of a domain is determined by the grain size in which the domain lies. Usually, there will be a number of ordered domains within a grain. These ordered domains can also form by a continuous ordering mechanism, where local atomic rearrangements occur homogeneously throughout the disordered phase, creating ordered domains. As the temperature is decreased further, the ordered domains will grow until they impinge on or intersect each other and form antiphase boundaries (APBs). Antiphase boundaries are boundaries between two ordered domains where the periodicity of the ordered structure in one domain is out of step with the other. This can be seen in Fig. 3.7, which is a schematic representing the phase transformation from a disordered structure at elevated temperature to the ordered structure, with APBs located where the domains intersect. The APBs are therefore similar to phase boundaries in that they are higher-energy regions within the lattice.

The gold-copper phase diagram (Fig. 3.8) exhibits a series of ordered structures, with the structures forming a maximum number of gold-copper atomic bonds and a minimum number of copper-copper and gold-gold bonds. At higher temperatures, the structure is a random disordered solution due to the thermally induced atomic movements. As the alloys are cooled, they rearrange themselves into ordered structures or superlattices. In the gold-copper system, the transformation from short-range to long-range order produces superlattices with three basic compositional ranges. One range produces the AuCu structure in which there are equal numbers of gold and copper atoms, while the other two have a 1 to 3 ratio of either gold to copper ($AuCu_3$) or copper to gold ($Au_3Cu$). Five different superlattices exist for this system: two corresponding to the AuCu composition, one to the $Au_3Cu$ composition, and two for the $AuCu_3$ composition. When the AuCu composition cools, it initially transforms to the orthorhombic structure AuCu II. On further cooling, it transforms again to the tetragonal AuCu I structure (although only slightly distorted from the cubic structure). In a similar manner, the $AuCu_3$ composition transforms to the orthorhombic

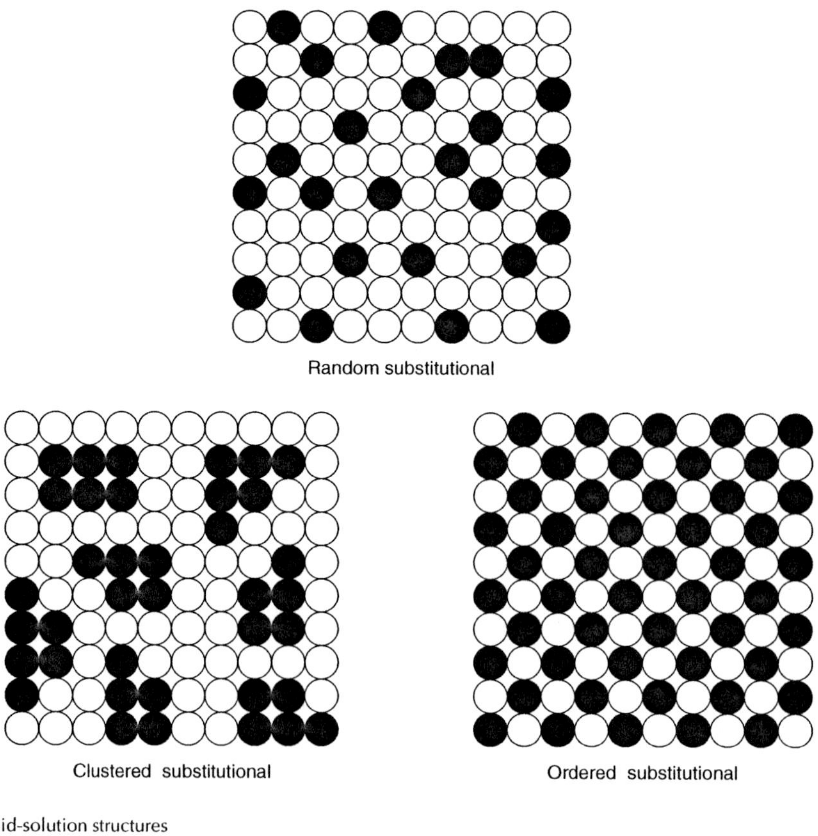


**Fig. 3.6** Solid-solution structures

$AuCu_3$ I structure at high temperatures and then to the face-centered cubic (fcc) $AuCu_3$ structure at lower temperatures. The structure of the $Au_3Cu$ phase is also based on the fcc lattice. A comparison between the high-temperature disordered structure of AuCu and the ordered AuCu I structure is shown in Fig. 3.9.

## 3.4 Intermediate Phases

When the electrochemical properties of the alloying element metals are similar, normal substitutional solid solutions will form during solidification. However, when the metals have widely divergent electrochemical properties, they are more likely to form a chemical compound, often with some degree of covalent or ionic bonding present. For example, strongly electropositive magnesium will combine with weakly electropositive tin to form $Mg_2Sn$, which is often described as being an intermetallic compound.

Between these two extremes of substitutional solid solution on the one hand and intermetallic compound on the other, phases are formed that

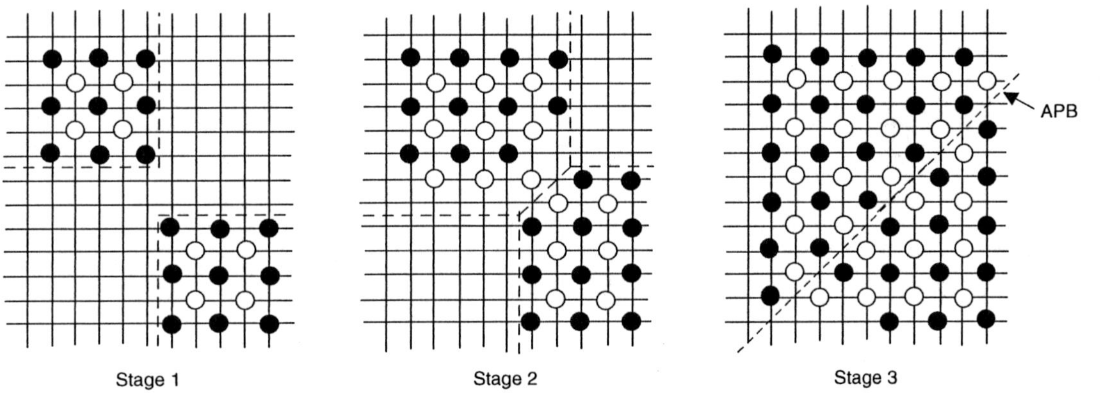


**Fig. 3.7** Formation of antiphase boundaries (APBs). Source: Ref 2

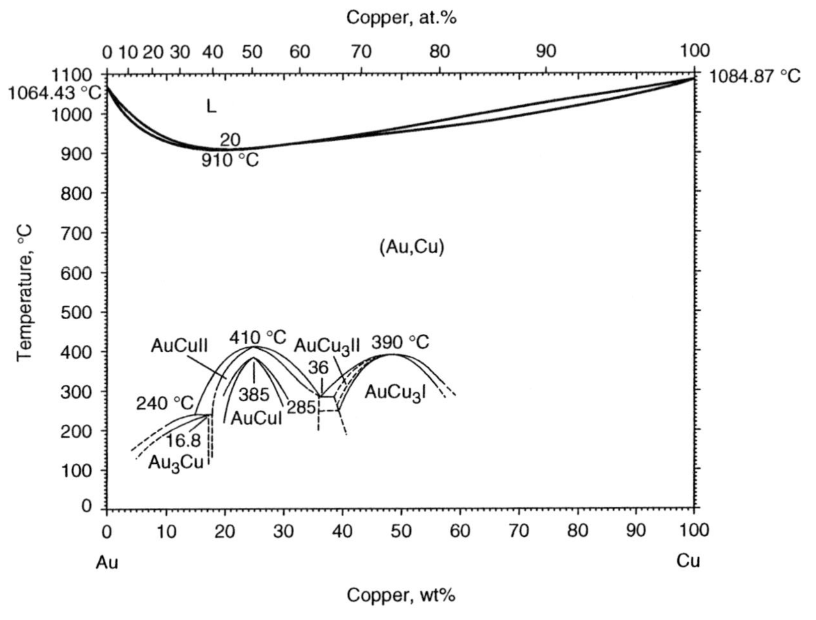


**Fig. 3.8** Gold-copper phase diagram

exhibit a gradation of properties. These phases are collectively termed intermediate phases. At one extreme, there are true intermetallic compounds, while at the other are ordered structures that can be classified more accurately as secondary solid solutions. These intermediate phases are often grouped into categories determined by their structures.

**Intermetallic Compounds.** Two chemically dissimilar metals tend to form compounds with ordinary chemical valences. These compounds have stoichiometric compositions with limited solubility. These compounds are generally formed when one metal has chemical properties that are strongly metallic, such as magnesium, and the other metal has chemical properties that are only weakly metallic, such as tin. Frequently, such a compound has a melting point that is higher than that of either of the parent metals. For example, the intermetallic compound $Mg_2Sn$ melts at 780 °C (1436 °F), whereas pure magnesium and tin melt at 650 and 230 °C (1202 and 450 °F), respectively. This is an indication of the high strength of the bond in $Mg_2Sn$. Since they exhibit either covalent or ionic bonding, they exhibit nonmetallic properties such as brittleness and poor electrical conductivity. Examples include the covalent compounds $Mn_2Sn$, $Fe_3Sn$, and $Cu_6Sn_5$ and the ionic compounds $Mg_2Si$ and $Mg_2Sn$.

**Electron Phases.** These compounds appear at definite compositions and depend on the ratio of electrons to atoms (*e*/*a*) at those compositions. The most important of these are the intermediate phases of the copper-zinc system. The valence of a metal is defined by the number of electrons in the outer shell of the atom. In electron compounds, the normal valence laws are not obeyed, but in many instances, there is a fixed ratio between the total number of valence bonds of all the atoms involved and the total number of atoms in the empirical formula of the compound in question.

There are three such ratios, commonly referred to as Hume-Rothery ratios:

- *Ratio 3/2 (21/14):* β structures, such as CuZn, $Cu_3Al$, $Cu_5Sn$, $Ag_3Al$
- *Ratio 21/13:* γ structures, such as $Cu_5Zn_8$, $Cu_9Al_4$, $Cu_{31}Sn_8$, $Ag_5Zn_8$, $Na_{31}Pb_8$
- *Ratio 7/4 (21/12):* ε structures, such as $CuZn_3$, $Cu_3Sn$, $AgCd_3$, $Ag_5Al_3$

For example, in the β-structure compound CuZn, copper has a valence of 1 and zinc a valence of 2, giving a total of 3 valences and a ratio of 3 valences to 2 atoms. In the compound $Cu_{31}Sn_8$, copper has a valence of 1 and tin a valence of 4. Therefore, 31 valences are donated by the copper atoms and 32 (4 × 8) by the tin atoms, making a total of 63 valences. In all, 39 atoms are present, and the ratio is:

$$\frac{\text{Total number of valences}}{\text{Total number of atoms}} = \frac{63}{39} = \frac{21}{13} \qquad \text{(Eq 3.1)}$$

These phases exist over a range of compositions and are metallic in nature. These Hume-Rothery ratios have been valuable in relating structures that appeared unrelated. However, there are many electron compounds that do not fall into any of these three groups.

**Interstitial Compounds.** These are compounds of transition metals with carbon, nitrogen, hydrogen, or boron; the interstitial atomic radius must be less than ⅔ that of the transition metal atom. These compounds are hard and have very high melting points due to the covalent nature of their bonding. When the solid

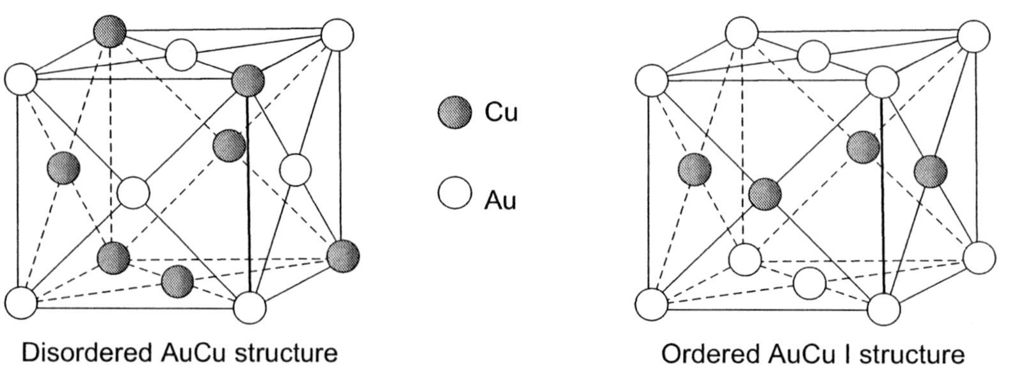


**Fig. 3.9** Comparison of disordered and ordered crystalline structures

solubility of an interstitially dissolved element is exceeded, a compound is precipitated from the solid solution. The small nonmetal atoms still occupy interstitial positions, but the overall crystal structure of the compound is different from that of the original interstitial solid solution. Compounds of this type have some metallic and some nonmetallic properties and comprise carbides, nitrides, hydrides, and borides. Examples include $TiH_2$, TiN, TaC, WC, and $Fe_3C$. All of these compounds are extremely hard, and the carbides find application in tool steels and cemented carbide cutting materials. The compound $Fe_3C$ (cementite) is important in steels.

**Laves Phases.** These compounds have a composition of $AB_2$ that forms because of the dense packing that can be achieved if the ratio of the $B$ atoms to $A$ atoms ($B/A$) is approximately 1.2. In these phases, the $A$ atoms have 12 nearest $B$ neighbors and 4 nearest $A$ neighbors. Each $B$ atom has 6 $A$ and 6 $B$ nearest neighbors. This arrangement produces an average coordination number of $(2 \times 12 + 16)/3 = 13.33$. These hard and brittle phases, such as $NbFe_2$, $TiFe_2$, and $TiCo_2$, cause deterioration in ductility and stress-rupture properties. Undesirable Laves phase formation can be a problem in nickel-, iron-nickel-, and cobalt-base superalloys exposed to elevated temperatures for long periods of time.

**Ordered Phases.** A normal primary solid solution that extends over a large compositional range is sometimes interrupted by the occurrence of an ordered form of that solid solution. The mechanical and physical properties of the localized ordered solid solution are similar to those of the primary solid solution.

Solid-solution strengthening is used for hardening many commercial metals, such as nickel-base superalloys, stainless steels, and brasses. Up to 35 atomic weight percent of zinc, the major alloying element in brasses, can dissolve in solid copper. The most important use of solid-solution strengthening is found in iron-carbon martensites. Carbon does not appreciably dissolve in the low-temperature bcc form of iron. However, appropriate thermal manipulation permits carbon in excess of the equilibrium solubility to be trapped in it. The amount of trapped carbon remains low, less than several weight percent, but significant hardening is provided nonetheless. The reason that carbon is such a potent hardener of iron is that it dissolves interstitially in the bcc lattice, producing a tetragonal distortion. The resulting internal stress interacts strongly with moving dislocations, substantially reducing their mobility. This is discussed in much greater detail in Chapter 11, "Heat Treatment of Steel," in this book.

## 3.5 Dislocation Atmospheres and Strain Aging

Interactions can occur between dislocations and solute atoms. If the diameter of a solute atom is either smaller or larger than the solvent atoms, it creates a lattice strain. A larger solute atom expands the lattice, and a smaller solute atom contracts the matrix. These strains can be reduced if the solute atoms migrate to an edge dislocation. Large solute atoms are attracted to lattice positions below the dislocation where the lattice is expanded, and small solute atoms are attracted to lattice positions above the edge dislocation where the lattice is compressed, as shown in Fig. 3.10.

In the case of interstitial atoms, since they are small, they are attracted to locations below an

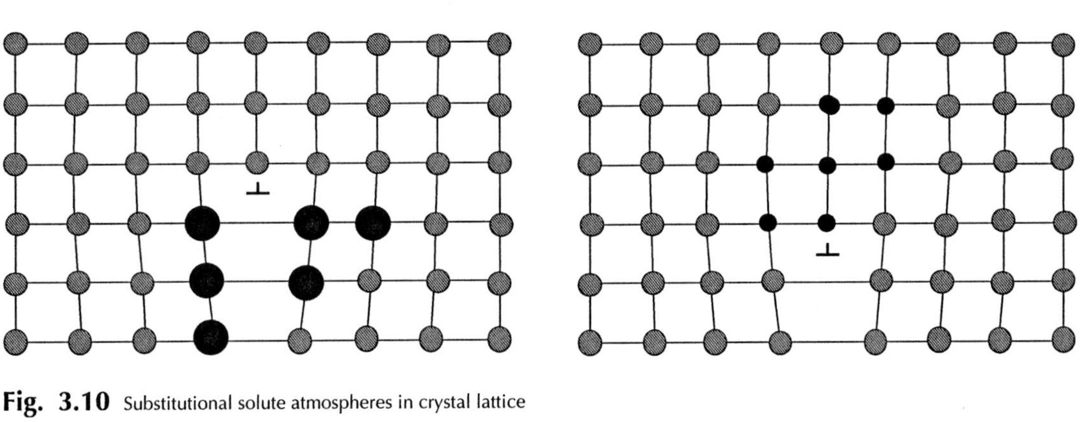

**Fig. 3.10** Substitutional solute atmospheres in crystal lattice

edge dislocation where the lattice is expanded. This excess of interstitial atoms below an edge dislocation is known as a Cottrell atmosphere (Fig. 3.11). As the temperature increases, the increase in entropy causes the atoms to be torn away from the atmospheres, so that at high temperatures they cease to exist.

The interaction between dislocations and solute atoms has practical ramifications for manufacturing processes because of the effect on mechanical properties and workability. When a metal is loaded in tension, two types of stress-strain curves are observed, as shown in Fig. 3.12. The curve on the left, in which the strain increases in a uniform manner with the application of stress, is the most common form of curve. The curve on the right, which has a sharp yield point, is often observed in low-carbon steels. Yield-point reductions on the order of 30% are common. The stress initially rises in a uniform fashion until it reaches point *a,* known as the upper yield point. The material suddenly begins to yield, and the stress drops to point *b,* the lower yield point, where it plastically strains at a constant stress. Eventually, the metal begins to work harden, and the stress-strain curve again rises in a normal manner.

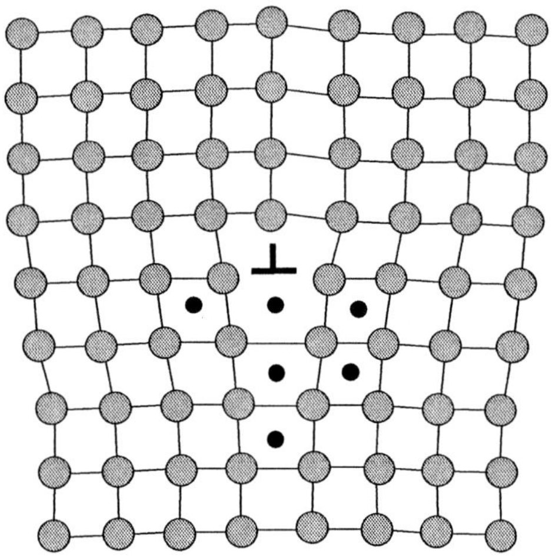

**Fig. 3.11** Interstitial atmospheres in crystal lattice

The presence of a sharp yield point causes problems during sheet metal forming of low-carbon steels, such as stamping or deep drawing of automobile body panels. When plastic deformation starts in a given area, the metal at this point is softer than other areas that have not received as much deformation. The localized deformation forms discrete bands called stretcher strains or Lüders bands, shown in Fig. 3.13. In a tensile test, the fillets are stress concentrations where Lüders bands initiate. They then propagate across the gage length. Since this occurs at a constant stress, it explains the horizontal portion of the curve at the lower yield point. During the forming of low-carbon steel body parts, this nonuniform deformation causes an objectionable rough surface. On the other hand, if a steel with a uniform extension is used, as in the left-hand curve of Fig. 3.12, there is a uniform extension of the material during forming, with no visible stretcher strains.

In the case of low-carbon steel, the sharp yield-point phenomenon is due to the presence of interstitial carbon or nitrogen atoms. One theory is that carbon and nitrogen atoms diffuse to the cores of dislocations and form atmospheres that lock the dislocations in place.

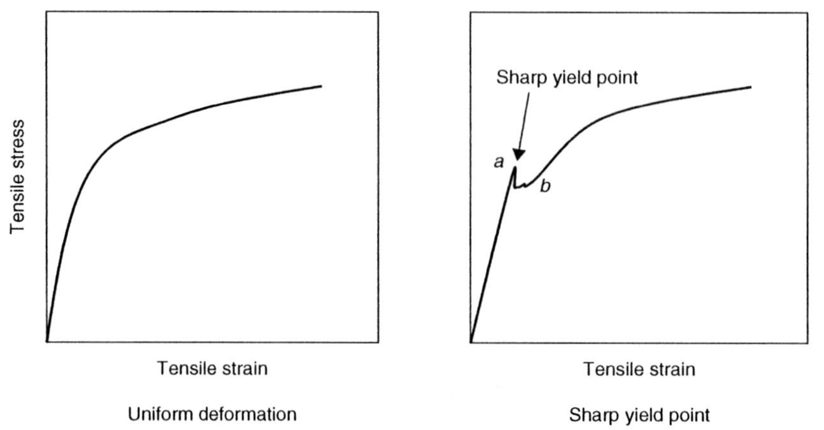


**Fig. 3.12** Tensile stress-strain curves

During the horizontal serrated portion of the stress-strain curve, these dislocations are progressively torn away from their atmospheres. An alternative theory is that when the carbon atmospheres are formed, the dislocations remain locked, and the yield-point phenomenon is caused by the generation and movement of newly formed dislocations. The concentrations of carbon and nitrogen required to produce this effect are very small, carbon concentrations as low as 0.002 wt% and nitrogen concentrations as low as 0.001 to 0.002 wt% are sufficient.

The yield point can be eliminated prior to forming by a small amount (0.5 to 2%) of cold rolling, referred to as temper rolling. This is a sufficient amount of deformation to tear the dislocations away from their interstitial atmospheres. However, since the yield point gradually returns, it is preferable to immediately form the steels after temper rolling. The reappearance of the yield point as a function of room-temperature aging is known as strain aging. Aging raises the stress at which the yield point reappears, and, as a result, the steel is strengthened and hardened, as shown in Fig. 3.14 for a low-carbon steel. Aging at room temperature allows the interstitial atoms to diffuse to the dislocations and again form atmospheres that pin dislocation movement. Because interstitial atoms must diffuse through the lattice in order to accumulate around the dislocations, the reappearance of the yield point is a function of time and temperature. The higher the temperature, the faster the yield point will reappear. However, the yield point is not normally observed above approximately 400 °C (750 °F), because the atmospheres are dispersed by the more intense thermal vibrations at higher temperatures.

At intermediate temperatures, the sharp yield point is replaced by a series of fine serrations in the stress-strain curve, implying that the interstitial atoms are diffusing rapidly enough to keep up with the applied stress. When aging occurs during deformation, it is known as dynamic strain aging. Within the dynamic strain aging temperature range, the plastic flow becomes unstable and can produce a serrated stress-strain

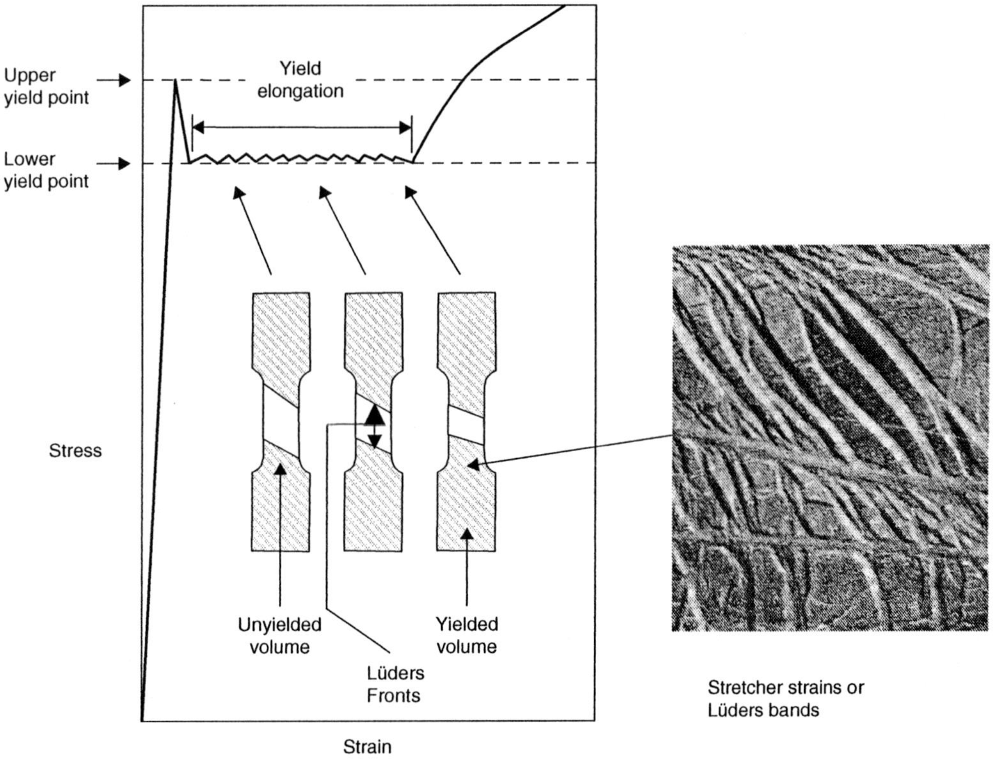


**Fig. 3.13** Discontinuous yielding in plain carbon steels. Source: Ref 3

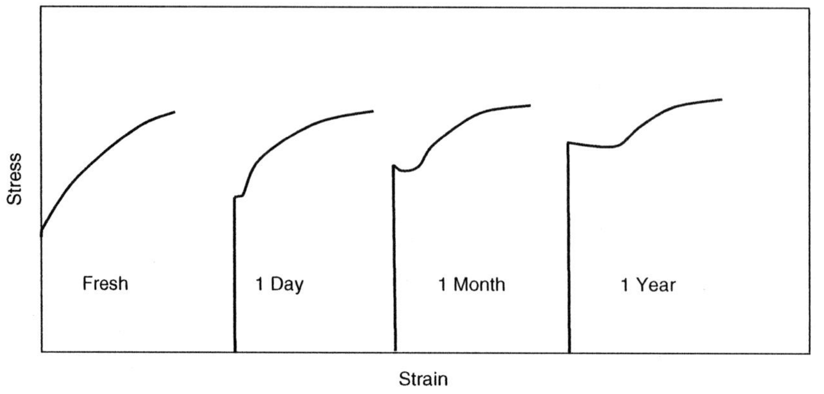


**Fig. 3.14** Strain aging in low-carbon steels

curve (Fig. 3.15). The presence of serrations on a stress-strain curve is referred to as the Portevin-LeChatellier effect.

Although strain aging in low-carbon steels has been emphasized, it can occur in a wide variety of metals to varying degrees. In low-carbon steels, nitrogen has a higher solubility in ferrite than carbon and therefore is actually a bigger problem in steels undergoing severe forming operations. Strong nitride formers, such as aluminum, titanium, and vanadium, can be used to tie up the nitrogen and eliminate the yield-point problem.

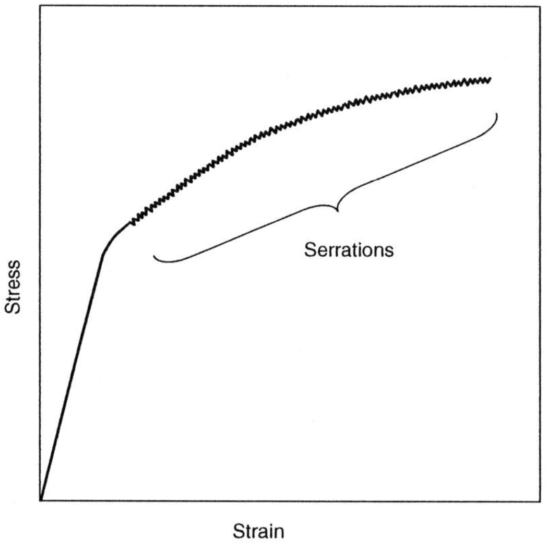


**Fig. 3.15** Stress-strain curve for dynamic strain aging

## REFERENCES

1. D.R. Askeland, *The Science and Engineering of Materials,* 2nd ed., PWS-Kent Publishing Co., 1989
2. D.A. Porter and K.E. Easterling, *Phase Transformations in Metals and Alloys,* Chapman and Hall, 1981
3. B.L. Ferguson, Design for Deformation Processes, *Materials Selection and Design,* Vol 20, *ASM Handbook,* ASM International, 1997

## SELECTED REFERENCES

- M.F. Ashby and D.R.H. Jones, *Engineering Materials 2—An Introduction to Microstructures, Processing, and Design,* 2nd ed., Butterworth Heinemann, 1998
- A. Cottrell, *An Introduction to Metallurgy,* 2nd ed., IOM Communications, 1975
- A.G. Guy, *Elements of Physical Metallurgy,* 2nd ed., Addison-Wesley Publishing Company, 1959
- R.A. Higgins, *Engineering Metallurgy—Applied Physical Metallurgy,* 6th ed., Arnold, 1993
- R.E. Reed-Hill and R. Abbaschian, *Physical Metallurgy Principles,* 3rd ed., PWS Publishing Company, 1991
- J. Regina, Ordered Structures, *Metallography and Microstructures,* Vol 9, *ASM Handbook,* ASM International, 2004
- M. Tisza, *Physical Metallurgy for Engineers,* ASM International, 2001

## CHAPTER 4

# Introduction to Phase Transformations

THIS CHAPTER PROVIDES a short introduction to phase transformations, namely, the liquid-to-solid phase transformations that occur during solidification and the solid-to-solid transformations that are important in processing, such as heat treatment (discussed in detail in subsequent chapters). Important solid-state transformations include annealing, precipitation hardening, and the martensitic transformation that occurs during the hardening of steel. This chapter introduces the concept of free energy that governs whether or not a phase transformation is possible, and then the kinetic considerations that determine the rate at which transformations take place.

## 4.1 Free Energy

Free energy is important because it determines whether or not a phase transformation is thermodynamically possible. Essentially, the change in free energy, either $\Delta F$ or $\Delta G$, must be negative for a reaction to occur; that is, the reaction must result in an overall reduction of the free energy of the system. If the free-energy change is zero, the system is in a state of equilibrium, and no reaction will occur (there is no driving force, or energy, for change). Likewise, if the free-energy change is positive, the reaction will not occur. To summarize:

- If $\Delta F$ or $\Delta G > 0$ or positive, the reaction will not occur.
- If $\Delta F$ or $\Delta G = 0$, the system is in equilibrium, and the reaction will not occur.
- If $\Delta F$ or $\Delta G < 0$ or negative, the reaction will occur.

The more negative the free-energy change (the larger the magnitude of the negative number), the greater is the driving force for a transformation. In other words, if the free-energy change is a small negative number, the driving force is low, and as it becomes more negative, the driving force increases. The following develops these concepts in a little more detail.

The internal energy, $E$, of a system (e.g., an alloy) is made up of two parts: the kinetic energy, which is due to atomic vibrations of the metallic lattice, and the potential energy, which is a function of the bond strengths. The internal energy can also be thought of as the sum of the free energy, $F$, and the bound energy ($TS$):

$$E = F + TS \quad \text{(Eq 4.1)}$$

where $E$ is the internal energy, $F$ is the Hemholtz free energy, $T$ is the absolute temperature in degrees Kelvin (K), and $S$ is entropy, a measure of the randomness of the system.

Solving for $F$, the Helmholtz free energy, can be expressed as:

$$F = E - TS \quad \text{(Eq 4.2)}$$

A related free energy is the Gibbs free energy, which is defined as:

$$G = H - TS \quad \text{(Eq 4.3)}$$

where $H$ is enthalpy.

The enthalpy, $H$, at constant pressure is the heat content:

$$H = E + PV \quad \text{(Eq 4.4)}$$

where $P$ is pressure, and $V$ is volume.

When dealing with liquids and solids, the $PV$ term is usually very small in comparison to $E$; therefore, the enthalpy, $H$, can be considered as being equal to the internal energy, $E$ ($H \sim E$). Therefore, this allows the Gibbs free energy to be expressed as:

$$G = E - TS \quad \text{(Eq 4.5)}$$

which is the same as the Hemholtz free energy of Eq 4.2.

In fact, in older metallurgy texts, one will find extensive references to the Hemholtz free energy, $F$, and in newer texts, the same free-energy curves are now identified with the Gibbs free energy, $G$. However, what is really important in metallurgical processes is not the free energy, $G$, itself, but the change in free energy, $\Delta G$. It can be shown that, at constant temperature and pressure:

$$\Delta G = \Delta H - T\Delta S \qquad \text{(Eq 4.6)}$$

A system is said to be in equilibrium when it attains the state of lowest Gibbs free energy, and the change in $\Delta G$ is then 0:

$$\Delta G = 0 \text{ at constant } T \text{ and } P \qquad \text{(Eq 4.7)}$$

All phase transformations occur to lower the total energy of the system. Any transformation that results in a reduction in free energy, $G$, is possible, that is:

$$\Delta G = G_f - G_i < 0 \qquad \text{(Eq 4.8)}$$

where $G_f$ is the free energy of the final state, and $G_i$ is the free energy of the initial state.

At low temperatures, the solid phase is the most stable state, since there is strong bonding between atoms, and the system has the lowest free energy and entropy. As the temperature increases, the $TS$ term of Eq 4.2 and 4.3 begins to dominate, and the solid phase has more freedom of atomic movement, due to increasing lattice vibrations, until it melts and becomes a liquid phase.

## 4.2 Kinetics

Thermodynamics allow the calculation of the driving force for a phase transformation; however, it tells nothing about how fast the transformation will occur. Kinetics must be used to calculate the speed at which the transformation will occur. In a large number of metallurgical processes, the reaction rate increases exponentially with increasing temperature and can be described by the Arrhenius equation:

$$\text{rate} = Ce^{-Q/RT} \qquad \text{(Eq 4.9)}$$

where $C$ is a temperature-independent pre-exponential constant, $R$ is the universal gas constant, $Q$ is the activation energy, and $T$ is the absolute temperature.

Taking the logarithm of each side allows Eq 4.9 to be rewritten as:

$$\ln(\text{rate}) = \ln C - \frac{Q}{RT} \qquad \text{(Eq 4.10)}$$

A semilogarithmic plot of ln (rate) versus the reciprocal of absolute temperature ($1/T$) gives a straight line, as shown in Fig. 4.1. The slope of this plot is $-Q/R$, and ln $C$ is obtained by extrapolating the plot to $1/T = 0$ ($T = \infty$). Thus, if the rate at two different temperatures is known, the rate at a third temperature can be determined. Likewise, if the rate and activation energy, $Q$, are known at one temperature, then the rate at any other temperature can be determined.

The Arrhenius equation given in Eq 4.9 can also be written as:

$$\text{rate} = Ce^{-q/kT} \qquad \text{(Eq 4.11)}$$

where $q$ is the activation energy per atomic scale unit ($q = Q/N_{AV}$), and $k$ is the Boltzmann constant ($k = R/N_{AV} = 13.8 \times 10^{-24}$ J/K).

The activation energy, $q$, for an atom to move from one stable position to another is shown in Fig. 4.2. In other words, the activation energy is that energy provided by temperature that is necessary to overcome the energy barrier. For an Arrhenius-type reaction, an increase in temperature of approximately 10 °C (20 °F) nearly doubles the reaction rate.

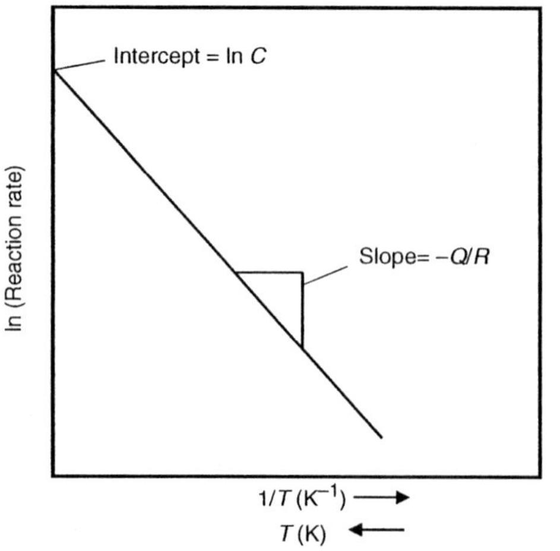


**Fig. 4.1** Typical Arrhenius plot

## 4.3 Liquid-Solid Phase Transformations

At any temperature, the thermodynamically stable state is the one with the lowest free energy. The equilibrium temperature for transition between two states is the temperature at which both have the same free energy. For the liquid-to-solid transition, this is the melting temperature. Compared to a solid metal, a liquid metal has a higher internal energy (equal to the latent heat of fusion) and a higher entropy due to its more random structure. The result is that as the temperature decreases toward the melting point, $T_m$, the liquid phase starts to develop more order, the entropy term decreases, and the free energy for the liquid rises at a faster rate than that of the solid, as illustrated in Fig. 4.3. At $T_m$, the equilibrium melting point, the free energies of both phases are equal. However, solidification does not occur, because the free-energy change is zero ($\Delta G = 0$) and it must be negative. Below $T_m$, the free energy does become negative ($\Delta G < 0$), and the metal solidifies. The free-energy change is thus:

$$\Delta G = G_{solid} - G_{liquid} < 0 \quad \text{(Eq 4.12)}$$

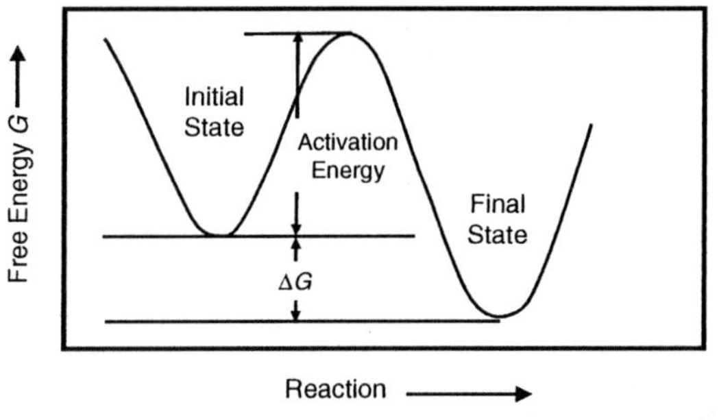


**Fig. 4.2** Activation energy. Initial to final state

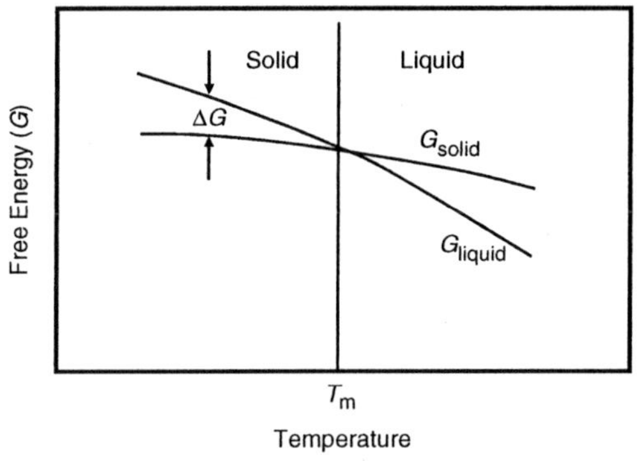


**Fig. 4.3** Gibbs free-energy curves during solidification

Immediately below $T_m$, the free-energy change is very small (small driving force), so solidification occurs slowly, but at larger undercooling or supercooling ($T_m - T$), the free-energy change is greater (larger driving force), and the solidification rate is much more rapid.

As a metal freezes, solidification starts on a small scale, with groups of atoms joining together in clusters (Fig. 4.4). As the temperature falls during the solidification process, the thermal agitation of the atoms in the liquid decreases, allowing small random aggregations of atoms to form into small crystalline regions called embryos. An embryo is a small cluster of atoms that has not yet reached a large enough size to become stable and grow. Therefore, embryos are constantly forming and then remelting. Eventually, as the temperature decreases, some of the embryos will reach a critical size and become nuclei that are stable and capable of growing into crystals. These crystals then continue to grow until they impinge on each other and eventually become grains in the final solidified structure. The crystalline structure within each grain is uniform but changes abruptly at the interfaces (grain boundaries) with adjacent crystals. This process of forming nuclei

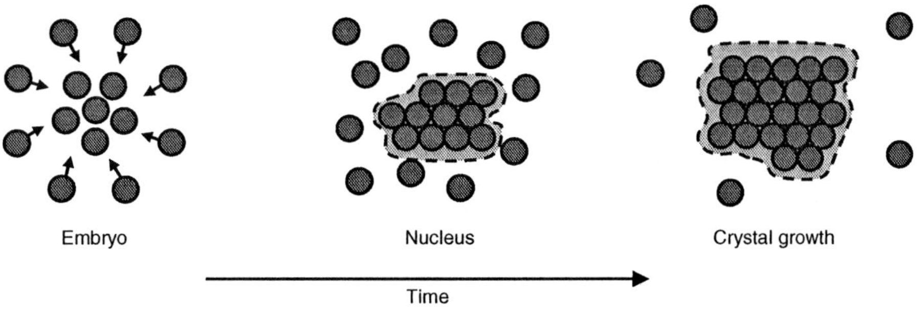


**Fig. 4.4** Nucleation and growth during solidification

in the freezing melt and their subsequent growth is known as a nucleation and growth process.

Nucleation can occur by either homogeneous or heterogeneous nucleation. Homogeneous nucleation occurs when nucleation takes place throughout the bulk of the liquid without preference for any point; that is, the probability of nucleation is the same everywhere within the volume of the liquid. When a solid nucleates preferentially at certain points in the liquid and the probability of nucleation occurring at certain preferred sites is much more than at other sites, it is called heterogeneous nucleation. Heterogeneous nucleation can occur at small solid inclusions within the melt and at surface irregularities on mold walls. While homogeneous nucleation can be made to occur in carefully controlled laboratory conditions, heterogeneous nucleation is the mechanism that is observed in all commercial casting operations.

Consider homogeneous nucleation in which a spherical nuclei of radius $r$ forms within the bulk of the liquid. Since the formation of a nucleus requires the creation of an interface between the solid and liquid, it creates an increase in the free energy of the system. The surface energy required for a sphere of surface area $4\pi r^2$ is $4\pi r^2\sigma$, where $\sigma$ is the surface energy per unit area and is shown as increasing the free energy of the system in Fig. 4.5. As was previously illustrated in Fig. 4.3, there is a free-energy reduction when the metal transforms from a liquid to a solid. This free energy is known as the volume free energy, $G_V$, which, for a spherical nucleus, is $4/3\pi r^3\ G_V$ and contributes to a decrease in free energy in Fig. 4.5. Thus, the total change in free energy, $\Delta G$, is the sum of the decrease in volume free energy and the increase in surface free energy:

$$\Delta G = \frac{4}{3}\pi r^3\ \Delta G_V + 4\pi r^2\sigma \qquad \text{(Eq 4.13)}$$

where $4/3\pi r^3$ is the volume of a spherical embryo of radius $r$, $\Delta G_V$ is the volume free energy, $4\pi r^2$ is the surface area of a spherical embryo, and $\sigma$ is the surface free energy.

The total free-energy, $\Delta G$, curve in Fig. 4.5 shows that there is a critical radius, $r^*$, that the particle must reach before becoming a stable nucleus with continued growth assured. If the embryo is very small, further growth of the embryo would cause the total free energy to increase. Therefore, the embryo remelts and causes a decrease in the free energy. When the particle becomes a nucleus at the critical radius, $r^*$, growth is assured because the total free-energy curve decreases continuously as it becomes larger.

Very large undercoolings and highly polished molds are required to cause homogeneous

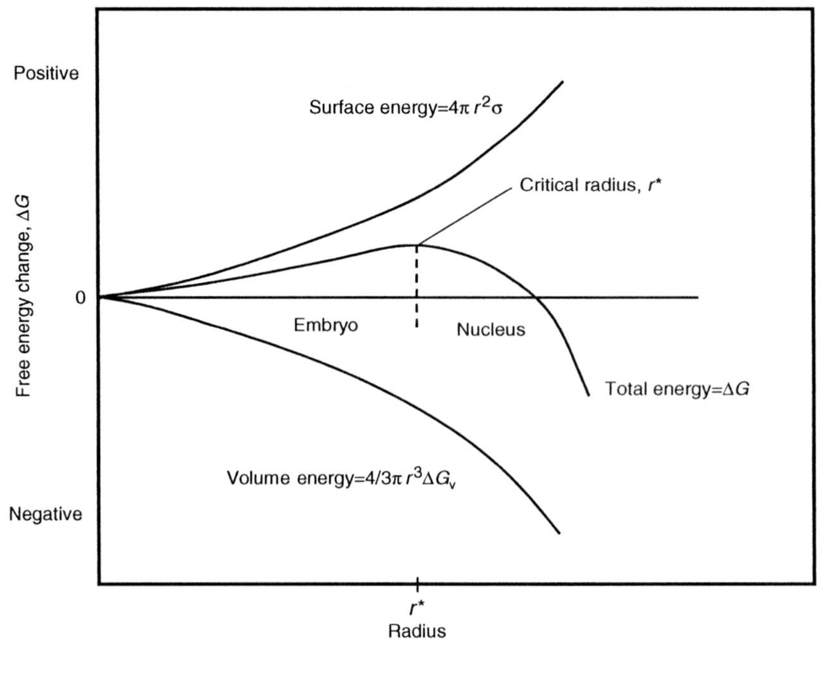


**Fig. 4.5** Free-energy curves for homogeneous nucleation

nucleation, and it is rarely, if ever, observed in commercial practice. In reality, the degree of undercooling is usually very small, often only a degree or two. Whereas homogeneous nucleation assumes that nucleation occurs randomly throughout the liquid, heterogeneous nucleation occurs at preexisting particles in the liquid or at rough areas on the mold, such as cracks and crevices. Thus, heterogeneous nucleation does not incur the free-energy penalty of having to form a new surface; rather, it uses a preexisting surface. However, for an inclusion or a defect on the mold wall to serve as a nucleation site, the liquid metal must be able to wet the surface. In addition, in many castings, a fine grain size is desirable, and nucleation agents are added to the melt to form as many nuclei as possible. Since each of these nuclei eventually grows and forms a grain, a larger number of nuclei results in a finer or smaller grain size. For example, a combination of 0.02 to 0.05% Ti and 0.01 to 0.03% B is added to many aluminum alloys. Solid titanium boride particles form and serve as effective sites for heterogeneous nucleation. Ultrasonic agitation of the mold during casting also promotes finer grain size by breaking up growing particles and helping to distribute them through the melt. When a melt has been populated with nuclei, the solidification rate is controlled almost exclusively by the rate of heat removal.

## 4.4 Solid-State Phase Transformations

Solid-state phase transformations occur when one or more parent phases, usually on cooling, produces a new phase or phases. The most important mechanisms are nucleation, growth, and diffusion. However, not all transformations rely on diffusion. For example, the important martensitic transformation in steels occurs quite suddenly by a combination of shear and heat treatment mechanisms.

Solid-state transformations differ from liquid-to-solid transformations (solidification) in several important ways. In solids, the atoms are bound much tighter than in liquids and diffuse much more slowly. Instead of transforming directly to the equilibrium phase, metastable transition phases can form prior to forming the final equilibrium phase. Like solidification, nucleation in solid-state reactions is almost always heterogeneous. Heterogeneous nucleation occurs at structural defects such as grain boundaries, dislocations, and interstitial atoms.

During a solid-state phase transformation, normally at least one new phase is formed that has different physical/chemical characteristics and/or a different structure than the parent phase. During nucleation, very small particles or nuclei of the new phase, often consisting of only a few hundred atoms, reach a critical size that is capable of further growth. During the growth stage, the nuclei grow in size, resulting in the disappearance of some or all of the parent phase. The transformation is complete when the growth of the new phase particles proceeds until the equilibrium fraction is attained.

Particle growth in a solid is controlled by diffusion (for a detailed discussion, see Chapter 5, "Diffusion," in this book), in which atoms diffuse from the parent phase across the phase boundary and into the growing second-phase particles. Since this is a diffusion-controlled process, the growth rate, $\dot{G}$, is determined by temperature:

$$\dot{G} = Ce^{-Q/RT} \qquad \text{(Eq 4.14)}$$

where $Q$ is the activation energy, and $C$ is the pre-exponential constant.

Both $Q$ and $C$ are independent of temperature. The temperature dependence of the growth rate, $\dot{G}$, and the nucleation rate, $\dot{N}$, is shown in the Fig. 4.6 curves. At a given temperature, the overall transformation rate is given as the product of $\dot{G}$ and $\dot{N}$. For transformations that occur at high temperatures, the nucleation rate will be low and the growth rate will be high. This will result in fewer particles that will grow to large sizes, and the resulting product will be coarse. On the other hand, if the transformation occurs at lower temperatures, where there is a much higher driving force for nucleation, many particles will form where the growth rate is lower, and the resulting product will be much finer.

The rate of transformation and the time required for the transformation to proceed to some degree of completion are inversely proportional to one another. The time to reach a 50% degree of completion in the reaction is frequently used. Therefore, the rate of a transformation is taken as the reciprocal of time required for the transformation to proceed halfway to completion, $t_{0.5}$, or:

$$\text{rate} = \frac{1}{t_{0.5}} \qquad \text{(Eq 4.15)}$$

If the logarithm of the transformation time is plotted versus temperature, the result is a C-shaped curve, like the one shown in Fig. 4.7, which is a mirror image of the transformation-rate curve previously shown in Fig. 4.6.

When the fraction of material transformed is plotted versus the logarithm of time at a constant temperature, an S-shaped curve similar to that in Fig. 4.8 is obtained, which is typical of the kinetic behavior for most solid-state reactions. For solid-state transformations displaying the kinetic behavior in Fig. 4.8, the fraction of transformation, $y$, is a function of time, $t$, and follows the Avrami equation:

$$y = 1 - e^{-kt^n} \quad \text{(Eq 4.16)}$$

where $k$ and $n$ are time-independent constants for the particular reaction. The value for $k$ depends on the temperature and the properties of the initial phase, while the coefficient $n$ has the values listed in Table 4.1.

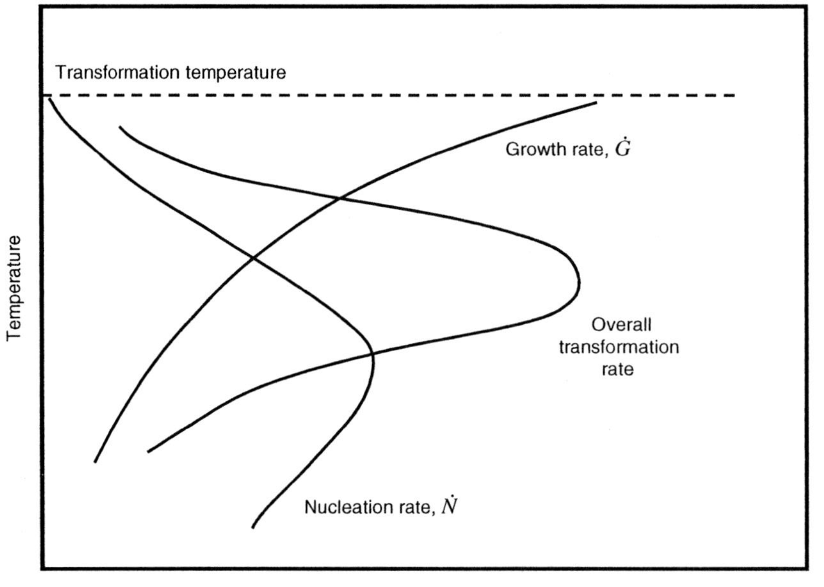


**Fig. 4.6** Solid-state transformation rate as a function of temperature. Source: Ref 1

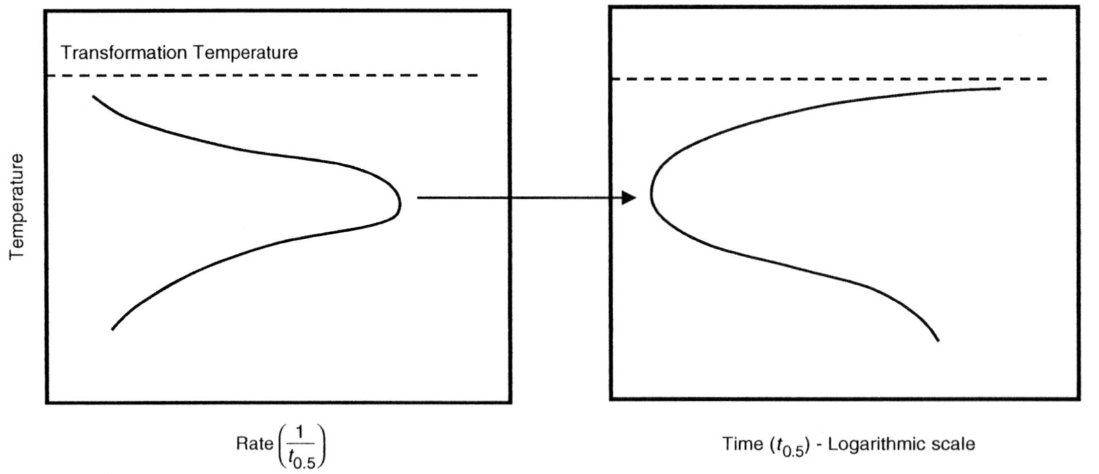


**Fig. 4.7** Transformation rate versus temperature. Source: Ref 1

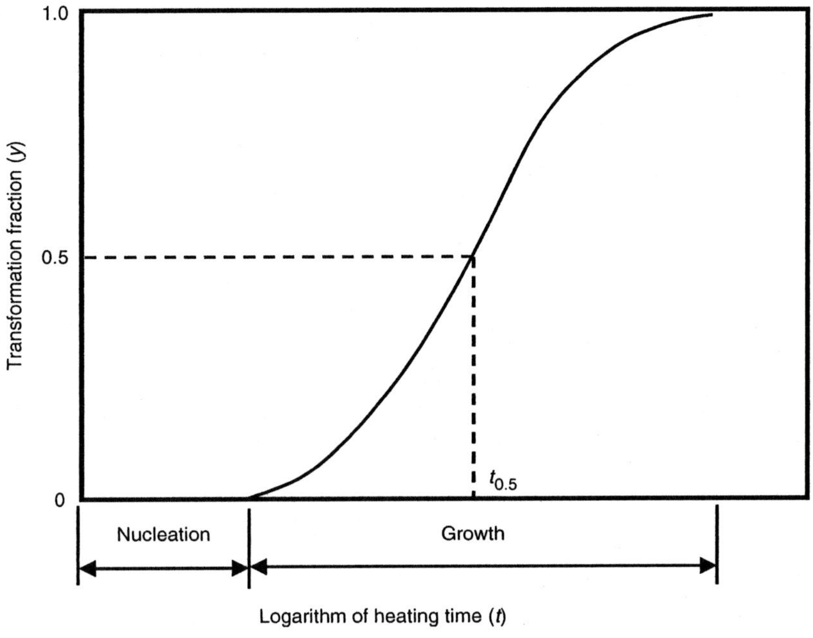


**Fig. 4.8** Fraction reacted as a function of time. Source: Ref 1

The values of $n$ in the Avrami equation distinguish between cellular transformations and precipitation transformations. A cellular transformation is one in which an entirely new phase is formed during the transformation, such as a solid-state transformation in which a higher-temperature $\gamma$ phase transforms in two lower-temperature phases, $\alpha$ and $\beta$. In this reaction, $\gamma$ disappears and is replaced by a mixture of $\alpha$ and $\beta$. It is called a cellular reaction because $\alpha$ and $\beta$ grow as cells or nodules into the $\gamma$ phase.

A precipitation transformation is one in which some of the original phase, $\alpha_{initial}$, remains when it transforms, as in the reaction $\alpha_{initial} \rightarrow \alpha_{final} + \beta$. The transformation of $\alpha_{initial}$ into $\alpha_{final}$ involves a composition change but retains its original crystalline structure. Since the phase forms from $\alpha_{initial}$, atoms of $\alpha_{initial}$ must diffuse to the precipitate $\beta$; the rate of a precipitation reaction is almost always controlled by diffusion. In addition, the initial precipitate phase that forms ($\beta$ in this example) is often not the final equilibrium phase. Metastable phases, which have a higher free energy than the final equilibrium phase, as previously shown in Fig. 4.2, can initially form because their interfacial energy is lower. On further heating, additional activation energy is provided that allows them to eventually transform to the equilibrium phase.

**Table 4.1 Values of $n$ in Avrami equation**

| Type of transformation | | Value of $n$ |
|---|---|---|
| Cellular | Constant rate of nucleation | 4 |
| | Zero rate of nucleation | 3 |
| | Nucleation at grain edges | 2 |
| | Nucleation at grain boundaries | 1 |
| Precipitation | Particle growing from small dimensions | |
| | 1. Constant rate of nucleation | 2.5 |
| | 2. Zero rate of nucleation | 1.5 |
| | Thickening of needles | 1.0 |
| | Thickening of plates | 0.5 |

Source: Ref 2

During a solid-state transformation, the formation of the new phase creates additional strain energy in the alloy. If the new phase forms in an incoherent manner, such as on a grain boundary, then the strain energy of the new particle varies with the shape of the particle (Fig. 4.9). In this figure, the shape of the particle is defined by the $c/a$ ratio, in which $a$ is the equatorial diameter, and $c$ is the polar axis of an ellipsoid. To minimize strain energy, the particle may assume a disc shape, which has a smaller relative amount of boundary interface than a sphere or needle.

Another solid-state transformation is the massive transformation in which a phase $\alpha$ changes crystalline structure as it transforms to phase $\beta$ without a change in composition. A

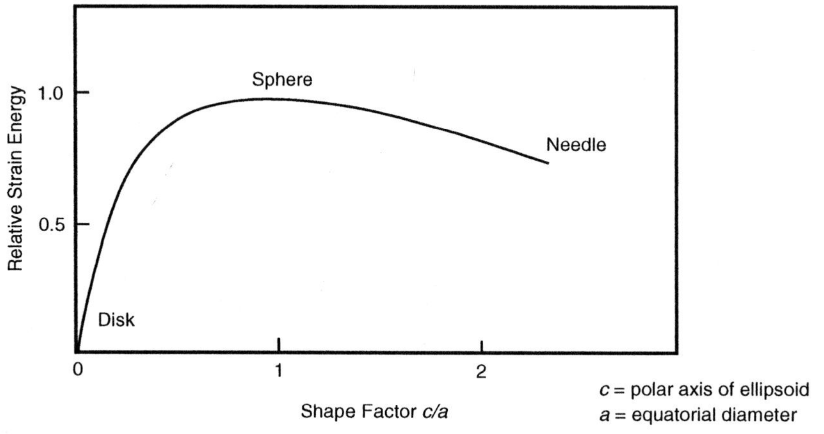


**Fig. 4.9** Strain energy as a function of second-phase particle shape. Source: Ref 2

simple example of a massive transformation is the allotropic (crystal structure) transformations that iron undergoes during heating or cooling; for example, the change from pure γ iron (face-centered cubic) to α iron (body-centered cubic) when it is cooled through the transformation temperature of 900 °C (1648 °F). Massive transformations are occasionally experienced in alloys, normally at high cooling rates that do not allow time for diffusion. Finally, there is the spinodal decomposition. In this reaction, nucleation occurs homogeneously, which is contrary to most nucleation and growth transformations.

## 4.5 Spinodal Decomposition

Spinodal decomposition is a phase separation reaction that does not involve a nucleation step, as in classical nucleation and growth. Instead, spinodal reactions involve spontaneous unmixing, or diffusional clustering of atoms, where a two-phase structure forms spontaneously by growth resulting from small compositional fluctuations. This results in the homogeneous decomposition of a supersaturated single phase into two phases that have essentially the same crystal structure but different compositions than the parent phase.

Consider a phase diagram with a miscibility gap, such as the one shown in Fig. 4.10. If alloy 1 of composition $X_0$ is quickly cooled (quenched) from the α field, it will contain a uniform composition that is initially the same everywhere. However, the alloy is unstable, and small fluctuations in composition will produce *A*-rich and *B*-rich zones. Such segregation lowers the free energy of the system, even though no discrete phase boundary is initially necessary. Because no new boundary area is required, there is no activation energy barrier to be overcome. Thus, there is no nucleation stage in the usual sense of a thermally activated process. The single solid solution can start to separate into two phases without nucleation. However, diffusion is necessary. Diffusion occurs in an "uphill" manner, as shown in Fig. 4.11, until the equilibrium compositions of $X_1$ and $X_2$ are

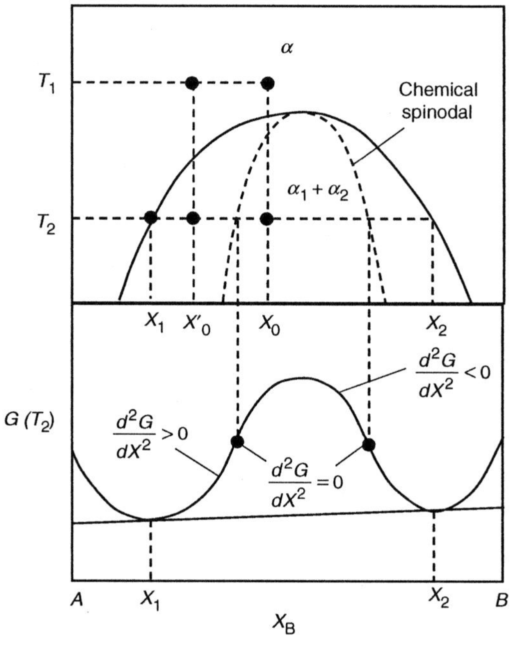


**Fig. 4.10** Spinodal decomposition. Source: Ref 3

obtained. Uphill diffusion refers to the diffusion of atoms up the concentration gradient instead of the normal "downhill" diffusion, where the concentration gradient is reduced. In contrast to classic nucleation and growth, during spinodal decomposition, the boundary between the two phases sharpens rather than moves. For this rather anomalous behavior to occur, the free-energy curve must have regions of negative curvature (i.e., $d^2G/dx^2 < 0$). Therefore, in order for spinodal decomposition to occur, the alloy composition must lie between the two inflection points, as in the alloy $X_0$. If the composition lies outside the inflection points, as in alloy 2 (Fig. 4.11) with a composition $X'_0$ and $d^2G/dx^2 > 0$, normal nucleation and growth will occur by the usual downhill diffusion.

## 4.6 Martensitic Transformation

When a steel is heated to a high enough temperature, the equilibrium phase is a face-centered cubic γ phase called austenite. If the steel is then slowly cooled to room temperature, the austenite transforms to a mixture of body-centered cubic ferrite (α) and iron carbide ($Fe_3C$), called cementite. While metallic phases are usually designated simply with a Greek letter, the phases in steel are so important that they also have names. The transformation that was just described is an example of a cellular reaction. However, if the austenite is rapidly cooled to room temperature (quenched), the carbon atoms in the austenite do not have time to diffuse and form the iron carbide phase. Instead, on rapid cooling, the austenite is distorted into a body-centered tetragonal structure, which is similar to ferrite but distorted into a tetragonal structure by the trapped carbon atoms. Therefore, there is no diffusion in this type of transformation, although there may be a slight rearrangement or "shuffling" of the carbon atoms.

Although the exact transformation depends on the specific alloy and heat treatment, several general features are present. First, the initial

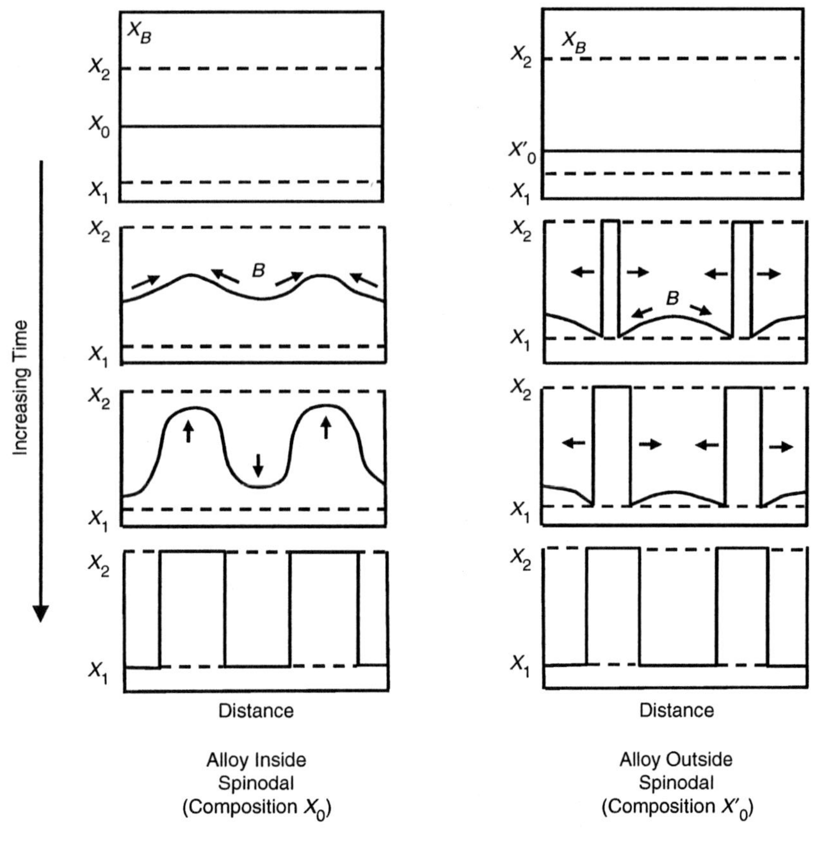


**Fig. 4.11** Composition fluctuations in a spinodal system. Source: Ref 3

phase must be greatly undercooled, usually by quenching past the point where nucleation and growth reactions occur. As a result of undercooling, there is a large driving force for the transformation, and it occurs rapidly, often approaching the speed of sound. Since thermal activation is not required, the reaction is referred to as being an athermal reaction. Since long-range diffusion is not required, and the transformation occurs by the small, coordinated movements of large numbers of atoms, it is also known as a "military" transformation, as opposed to the more random diffusion of atoms in nucleation and growth processes, which are sometimes referred to as "civilian" transformations. Finally, since slip and twinning mechanisms are often involved, they are known as shear mechanisms. The martensitic reaction occurs in alloys other than steels but does not result in the tremendous strengthening that results from the transformation in steels. The martensitic reaction in steel is covered in detail in Chapter 10, "The Iron-Carbon System," and Chapter 11, "Heat Treatment of Steel," in this book.

## REFERENCES

1. W.D. Callister, *Fundamentals of Materials Science and Engineering,* 6th ed., John Wiley & Sons, Inc., 2003
2. J.W. Christian, *The Theory of Transformations in Metals and Alloys,* Pergamon Press, 1965
3. D.A. Porter and K.E. Easterling, *Phase Transformations in Metals and Alloys,* Chapman and Hall, 1981

## SELECTED REFERENCES

- H. Baker, Introduction to Alloy Phase Diagrams, *Metallography and Microstructures,* Vol 9, *ASM Handbook,* ASM International, 2004
- A.G. Guy and J.J. Hren, *Elements of Physical Metallurgy,* 3rd ed., Addison-Wesley Publishing Company, 1974
- S. Para, Spinodal Transformation Structures, *Metallography and Microstructures,* Vol 9, *ASM Handbook,* ASM International, 2004

# CHAPTER 5

# Diffusion

DIFFUSION is the movement of atoms through the crystalline lattice. At all temperatures above absolute zero, the atoms in gases, liquids, and solids are in a constant state of motion. In gases, the atomic movement is very rapid. While the motion in liquids is still rapid, it is much slower than in gases, and in solids, it is even slower. In solids, atomic movements are restricted by the crystal lattice. However, thermal vibrations in solids allow atoms to diffuse, and as the temperature increases, the vibrations of the atoms about their equilibrium positions in the crystal lattice also increase. Thus, atomic vibrations are necessary for diffusion to take place, and as the amplitude of atomic vibrations increases with increasing temperature, the rate of diffusion increases. While random diffusion is constantly occurring in all metals at temperatures above absolute zero, by far the most important case of diffusion occurs when there is a concentration gradient. In this case, the driving force for diffusion is to reduce the Gibbs free energy of the system. For example, if two blocks of the same *A-B* alloy composition are welded together (Fig. 5.1) and then heated for some time at elevated temperature, the free energy of each part of the alloy will initially be $G_1$ and $G_2$, and the free energy of the welded block will be $G_3$. As diffusion occurs to reduce the concentration gradient, the free energies will decrease toward $G_4$, the free energy of a homogeneous *AB* compound. Therefore, a decrease in free energy is produced by the *A* and *B* atoms diffusing away from regions of high concentration to regions of low concentration, or down the concentration gradient. This being said, diffusion can occur up a concentration gradient if the free energy of the system is reduced, such as in the case for spinodal decomposition, which was covered in Chapter 4, "Introduction to Phase Transformations," in this book. What is important is that the free energy of the system is reduced during diffusion.

Diffusion plays an important role in many metallurgical processes, such as phase transformations, annealing, precipitation hardening, diffusion bonding, sintering, carburization of steels, and creep deformation. As an example, diffusion can be used to help eliminate concentration gradients in a casting and produce a more homogeneous composition.

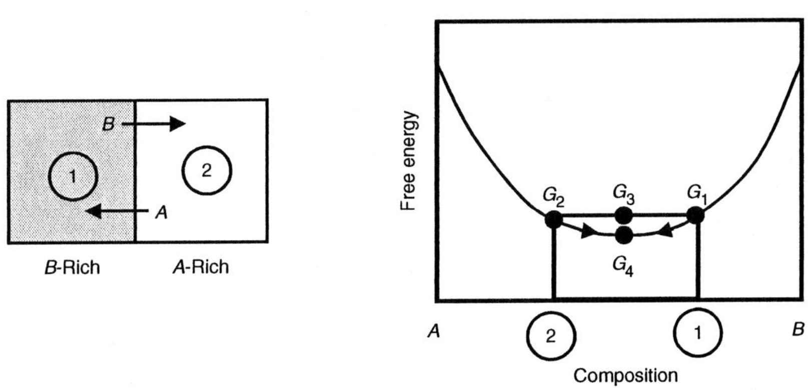


**Fig. 5.1** Free-energy changes during diffusion in a concentration gradient. Source: Ref 1

## 5.1 Mechanisms of Diffusion

There are two main types of diffusion that can occur in solids: interstitial diffusion and substitutional diffusion (Fig. 5.2). Whether diffusion occurs interstitially or substitutionally depends on the relative sizes of the atoms in the solute and solvent. For example, small carbon atoms can diffuse interstitially in iron, while copper atoms, which are approximately the same size as nickel atoms, diffuse substitutionally in nickel.

### *5.1.1 Interstitial Diffusion*

If the solute atom is sufficiently small, it will locate in an interstice between the larger solvent atoms, forming an interstitial solid solution. Diffusion of interstitial atoms occurs by atoms jumping from one interstitial site to another without permanently displacing any other atoms in the crystal lattice. For interstitial diffusion to occur, the size of the diffusing atoms must be very small compared to that of the matrix atoms. Therefore, only small atoms such as carbon, nitrogen, hydrogen, oxygen, and boron can diffuse interstitially. For example, carbon diffuses interstitially in both face-centered cubic (fcc) $\gamma$ iron and body-centered cubic (bcc) $\alpha$ iron.

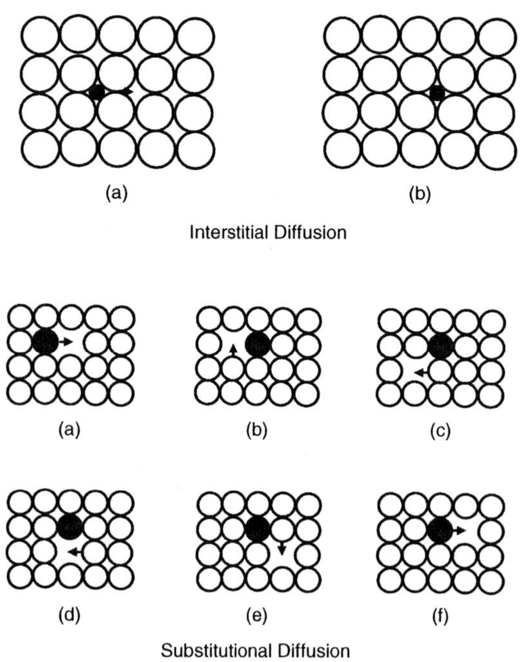


**Fig. 5.2** Interstitial and substitutional diffusion

As the interstitial solute atom increases in size, the activation energy increases, making it more difficult for the atom to move between the solvent atoms to a neighboring interstitial site. For an atom to move from one interstitial position to another, it must pass through a position where the potential energy is a maximum. The difference in the potential energy between this position and that of the normal position is the activation energy for diffusion, which is provided by thermal fluctuations.

The crystalline structure of the solvent metal is important. For example, carbon atoms diffuse faster in bcc iron than in fcc iron under the same carbon concentration gradient, because the atomic packing is not as dense in the bcc structure. The lower atomic packing factor in the bcc structure (0.68) as compared to the fcc structure (0.74) allows easier diffusion in the bcc lattice.

### *5.1.2 Substitutional Diffusion*

Various mechanisms have been proposed to explain substitutional diffusion.

**Direct exchange** diffusion involves two adjacent atoms merely exchanging places.

**Zener ring** diffusion is similar but involves the simultaneous rotation of four atoms.

**Vacancy** diffusion involves the migration of an atom from a lattice position to an adjacent vacancy (vacant lattice site).

The vacancy diffusion mechanism requires less energy than either the direct exchange or Zener ring mechanisms, and therefore, it is now accepted that vacancy diffusion is the major mechanism for substitutional diffusion. However, this is not a restriction, since vacancies are always present in metals and alloys at temperatures above absolute zero. As the temperature increases, both thermal vibrations on the lattice and the number of vacancies increase, greatly facilitating diffusion. Thus, diffusion occurs much more rapidly at higher temperatures. Consider the example of self-diffusion shown in Fig. 5.2. If an atom next to a vacancy has sufficient activation energy, it can move into the vacant site. The activation energy for self-diffusion is equal to the sum of the activation energy to form a vacancy and the activation energy to move the vacancy. As the melting point increases, the activation energy for self-diffusion also increases. This relationship exists because the higher-melting-temperature metals tend to have stronger bonding between their atoms.

It should be recognized that diffusion is constantly occurring, even in pure metals and alloys without concentration gradients. Self-diffusion has been studied using radioactive tracer atoms. For example, when radioactive tracer atoms are introduced onto the surface of nonradioactive gold, the radioactive atoms diffuse into the bulk gold. Eventually, the radioactive atoms become uniformly distributed throughout the entire sample. Although self-diffusion continually occurs in all metals, the effect on the material behavior is insignificant.

When two different metals or alloys are placed in intimate contact, atoms will begin to migrate across the interface. Such diffusion of unlike species under the influence of a chemical (compositional) gradient is called chemical diffusion and is illustrated schematically in Fig. 5.3. The diffusion couple is made up of pure *B* (solid circles) and pure *A* (open circles). As time progresses, mixing on the two sides occurs. At some long time ($t \gg t_0$), complete mixing will be achieved, with the chemical composition being identical on both sides. Chemical diffusion can occur by any of the previously described mechanisms, although vacancy diffusion is the most probable.

## 5.2 Fick's Laws of Diffusion

Adolf Fick, a German physiologist, developed his laws of diffusion in 1855. In addition to describing the diffusion behavior in metals, equations based on Fick's laws have been used to model transport processes in foods, neurons, biopolymers, pharmaceuticals, porous soils, and semiconductor doping processes.

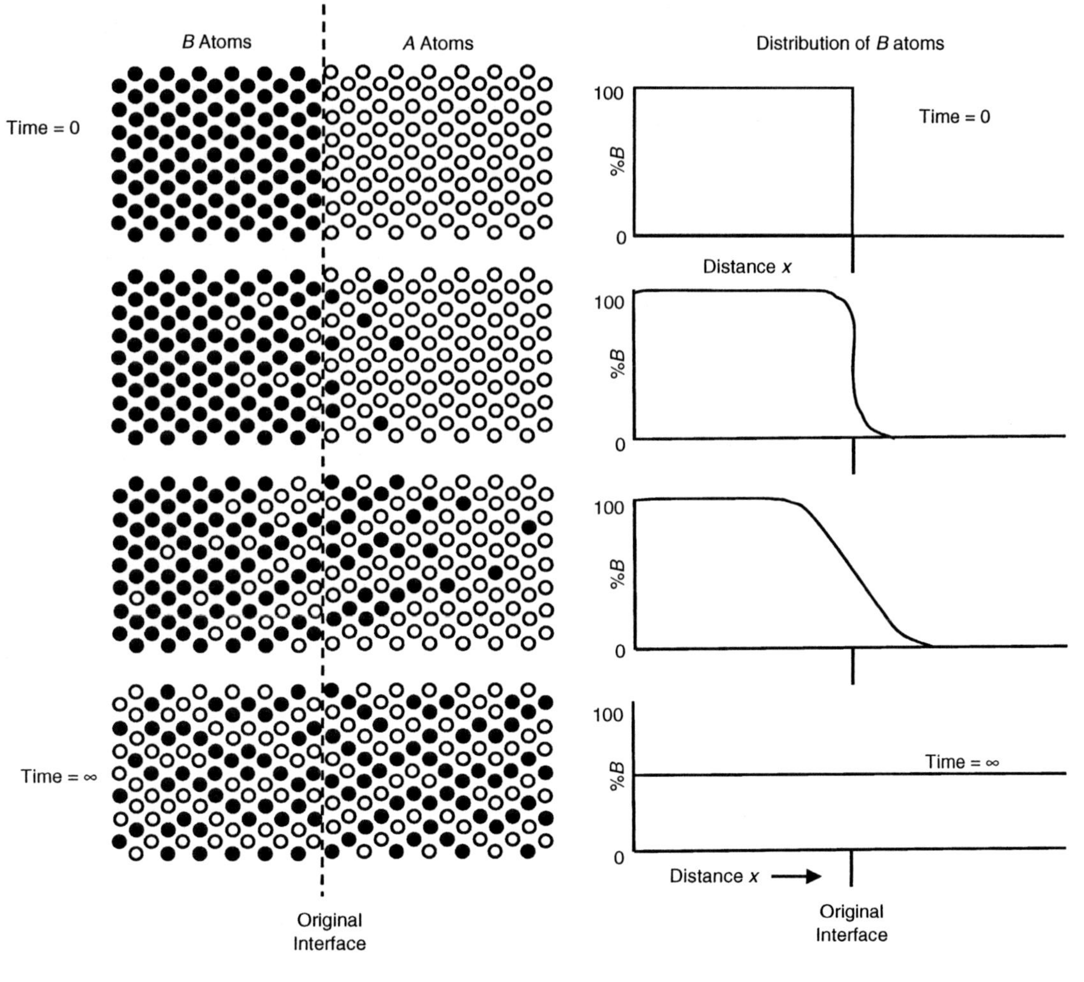


**Fig. 5.3** Diffusion across an interface. Source: Ref 2

### 5.2.1 Fick's First Law of Diffusion

Fick's first law states that in a solution with a concentration gradient, there will be a net flux of atoms from regions of high solute concentration to regions of low solute concentration, and furthermore, the net flux of solute is proportional to the concentration gradient. The flux, $J$, is defined as the number of atoms ($n$) of the diffusing species that pass through a plane of unit area, $A$, which is normal to the diffusion direction, per unit time, $t$. Therefore the flux, $J$, is:

$$J = \frac{dn}{Adt} \quad \text{(Eq 5.1)}$$

The flux of atoms across the plane at $x = x'$ in Fig. 5.4 is the number of atoms crossing the plane area per unit time and is proportional to the gradient $dc/dx$ at that location ($x = x'$). This type of diffusion is governed by Fick's first law of diffusion:

$$J = -D\frac{dc}{dx} \quad \text{(Eq 5.2)}$$

where $J$ is the flux or net flow of solute atoms, $D$ is the diffusion coefficient or diffusivity, $c$ is the solute concentration, and $x$ is the distance.

$$\frac{dc}{dx} = \text{concentration gradient}$$

The proportionality constant, $D$, is the diffusivity or diffusion coefficient. The negative sign is used because the diffusion is from a higher concentration to a lower concentration. Diffusion conditions in which the environment does not change with time are known as steady-state diffusion. An example would be diffusion of a nonreacting gas through a thin metal foil where there is a gas pressure differential on both sides of the foil.

The flux, $J$, may be expressed as solute atoms/($m^2$s), and the concentration, $c$, is then atoms/$m^3$. The diffusion coefficient has the units of $m^2$/s and depends on the solute, the solvent, the concentration, and the temperature:

$$J\left(\frac{\text{atoms}}{\text{m}^2\times\text{s}}\right) = D\left(\frac{\text{m}}{\text{s}}\right)\frac{dc}{dx}\left(\frac{\text{atoms}}{\text{m}^3}\times\frac{1}{\text{m}}\right) \quad \text{(Eq 5.3)}$$

The flux, $J$, and the solute concentration, $c$, may also be expressed in terms of mass of solute. In this case, the units of $J$ are kg solute/$m^2$s, and $c$ is in kg solute/$m^3$.

### 5.2.2 Fick's Second Law of Diffusion

Steady-state diffusion, as defined by Fick's first law, where conditions do not change with time, is not commonly encountered. In most cases, non-steady-state diffusion takes place, where the concentration of solute atoms at any point in the material changes with time. The difference between steady-state and non-steady-state diffusion conditions can readily be visualized by examining Fig. 5.5. In the first case, gas diffuses from an infinite volume ($P_1$ = constant) through a membrane into an infinite volume ($P_2$ = constant). The pressure gradient across the membrane remains constant, as does the diffusive flux. In the second case, gas diffuses from a finite

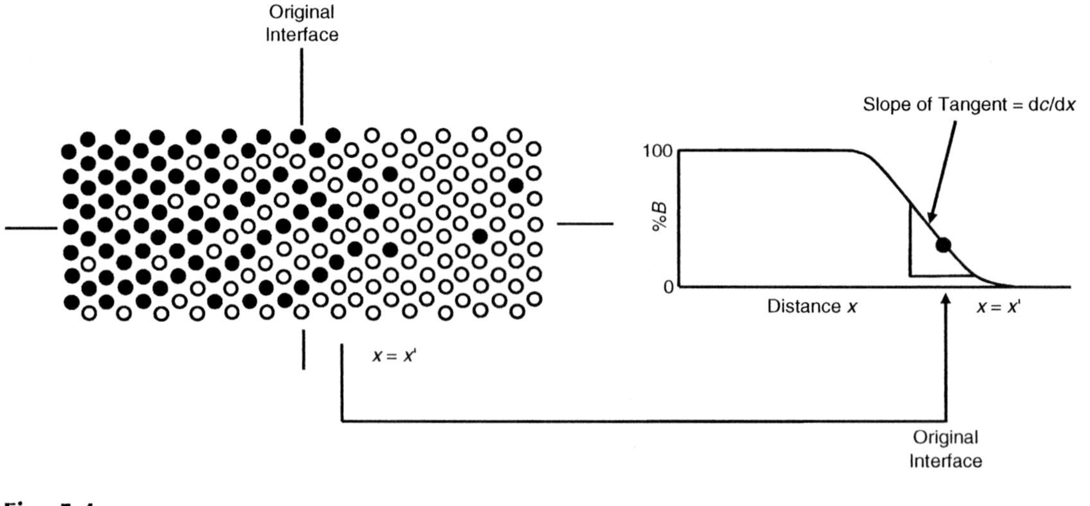


**Fig. 5.4** Flux of atoms across interface. Source: Ref 2

volume through a membrane into another finite volume. The pressures in the reservoirs change with time, creating a pressure gradient across the membrane. For example, to carburize steels, carbon is diffused into the surface layer to harden the surface. The concentration of carbon in the piece at any distance from the surface will change with time as the diffusion process progresses.

Fick's second law (Eq 5.4) takes into account how the concentration at a point changes with time:

$$\frac{dc}{dt} = \frac{d}{dx}\left(D\frac{dc}{dx}\right) \qquad \text{(Eq 5.4)}$$

This law states that the rate of compositional change is equal to the diffusivity times the rate of change of the concentration gradient. In this form, the diffusivity is a function of concentration and distance, $x$. However, it is often appropriate to treat the diffusivity as independent of the concentration. In this case, Fick's second law can be written as:

$$\frac{dc}{dt} = D\frac{d^2c}{dx^2} \qquad \text{(Eq 5.5)}$$

Concentration should be expressed as atoms or mass per volume, but if density changes can be ignored, concentration can be expressed as atomic or weight percent.

### 5.2.3 Several Applications of Fick's Second Law of Diffusion

Fick's laws have already been solved for many problems and are reported in the literature. Also, solutions to Fick's laws have been applied to industrial applications. The following are several examples of these solutions and their applications.

**Carburization.** One application of the second law is calculating the carbon depth obtained during carburization of steel. If the carbon content at the surface is suddenly increased from the initial concentration, $c_o$, to a higher concentration, $c_s$, as shown in Fig. 5.6, then the solution to Eq 5.5 is:

$$\frac{c_s - c_x}{c_s - c_o} = \text{erf}\left[\frac{x}{2\sqrt{Dt}}\right] \qquad \text{(Eq 5.6)}$$

where $c_s$ is the surface concentration of gas diffusing into the surface, $c_o$ is the initial concentration of the element in the solid, $x$ is the distance from the surface, $D$ is the diffusivity of the solute element in the solvent matrix, and $t$ is time.

The error function (erf) is a mathematical function that can be found in standard mathematical tables. An abbreviated table of error functions is given in Table 5.1.

For example, consider the gas carburization of type 1020 steel at 930 °C (1700 °F). Assume that the steel has a nominal carbon content of 0.20%, and the carbon content at the surface is 0.90%. The diffusion coefficient under these conditions is $D_{930\ °C} = 1.28 \times 10^{-11}$ m$^2$/s. The time necessary to increase the carbon content to 0.40% at 0.50mm below the surface can be calculated in the following manner:

$$\frac{c_s - c_x}{c_s - c_o} = \text{erf}\left[\frac{x}{2\sqrt{Dt}}\right]$$

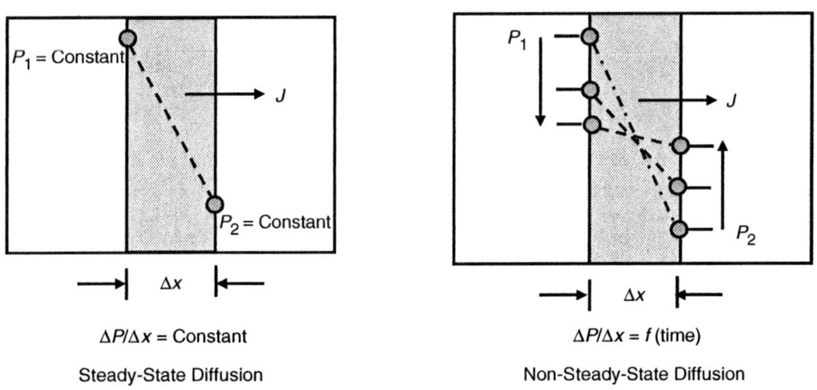


**Fig. 5.5** Steady- and non-steady-state diffusion of gas across a membrane

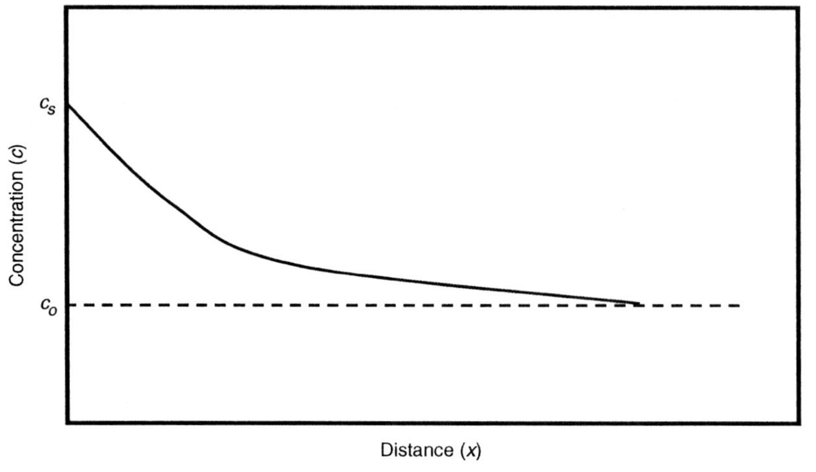


**Fig. 5.6** Fick's second law for constant surface concentration

where

$$c_s = 0.90\%$$

$$c_o = 0.20\%$$

$$c_x = 0.40\%$$

$$x = 0.5\text{mm} = 5 \times 10^{-4}\ \text{m}$$

$$\frac{0.90-0.40}{0.90-0.20} = \text{erf}\left[\frac{5\times10^{-4}}{2\sqrt{1.28\times10^{-11}t}}\right]$$

$$\frac{0.50}{0.70} = \text{erf}\left[\frac{69.88}{\sqrt{t}}\right] = 0.7143$$

Let

$$Z = \frac{69.88}{\sqrt{t}}$$

then

$$\text{erf } Z = 0.7143$$

Since the exact number for $Z$ whose error function is 0.7143 is not available in Table 5.1, interpolation is used to find the value:

$$\frac{0.7143-0.7112}{0.7421-0.7112} = \frac{x-0.75}{0.80-0.75}, \text{ then } x = 0.755$$

$$Z = \frac{69.88}{\sqrt{t}} = 0.755$$

**Table 5.1 Gaussian error function values**

The Gaussian error function is defined as $\text{erf}(Z) = \frac{2}{\sqrt{\pi}} \int_0^z e^{-y^2} dy$.

| Z | erf (Z) | Z | erf (Z) | Z | erf (Z) |
|---|---|---|---|---|---|
| 0 | 0 | 0.55 | 0.5633 | 1.3 | 0.9340 |
| 0.025 | 0.0282 | 0.60 | 0.6039 | 1.4 | 0.9523 |
| 0.05 | 0.0564 | 0.65 | 0.6420 | 1.5 | 0.9661 |
| 0.10 | 0.1125 | 0.70 | 0.6778 | 1.6 | 0.9763 |
| 0.15 | 0.1680 | 0.75 | 0.7112 | 1.7 | 0.9838 |
| 0.20 | 0.2227 | 0.80 | 0.7421 | 1.8 | 0.9891 |
| 0.25 | 0.2763 | 0.85 | 0.7707 | 1.9 | 0.9928 |
| 0.30 | 0.3286 | 0.90 | 0.7970 | 2.0 | 0.9953 |
| 0.35 | 0.3794 | 0.95 | 0.8209 | 2.2 | 0.9981 |
| 0.40 | 0.4284 | 1.0 | 0.8427 | 2.4 | 0.9993 |
| 0.45 | 0.4755 | 1.1 | 0.8802 | 2.6 | 0.9998 |
| 0.50 | 0.5205 | 1.2 | 0.9103 | 2.8 | 0.9999 |

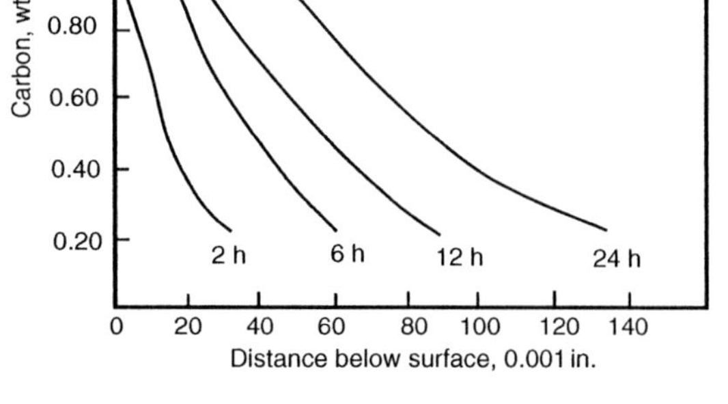


**Fig. 5.7** Carbon concentration during carburizing

$$\sqrt{t} = \frac{69.88}{0.755} = 92.6, \text{ and } t = 143 \text{ min}$$

The carburized steel would be heat treated to convert the carbon-rich area to martensite and then tempered to restore some ductility (discussed in more detail in Chapter 11, "Heat Treatment of Steel," in this book). The resultant product would have an extremely hard surface and a much tougher core with a lower carbon content. A series of curves for 1022 steel carburized at 920 °C (1685 °F) for different times are shown in Fig. 5.7.

**Diffusion Couple.** If two blocks of metal with different compositions, $c_1$ and $c_2$, are joined together, the solution of Fick's second law takes the form:

$$c = \frac{c_1 + c_2}{2} - \left[\frac{c_1 - c_2}{2}\right] \text{erf}\left[\frac{x}{2\sqrt{Dt}}\right] \quad \text{(Eq 5.7)}$$

The solution to this equation is shown in Fig. 5.8. This solution is the same as the one given in Eq 5.6, except that $(c_1 + c_2)/2$ replaces $c_s$, and $(c_1 - c_2)/2$ replaces $c_s - c_o$. An

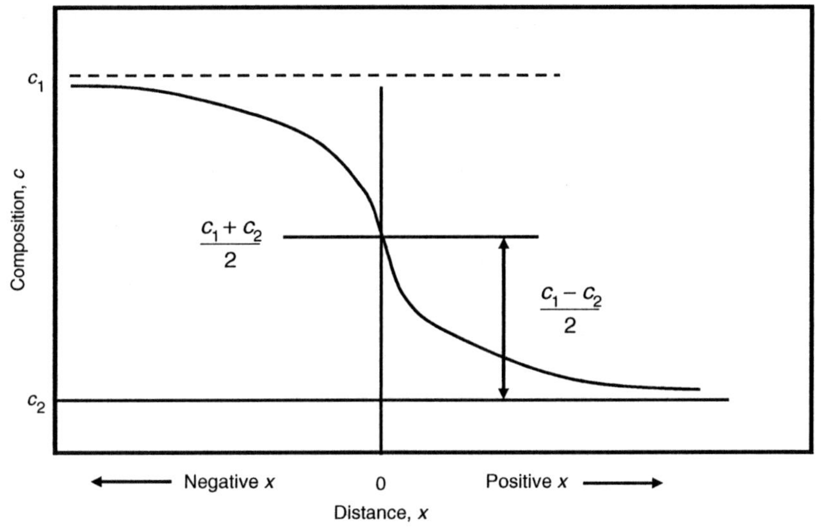


**Fig. 5.8** Fick's second law for diffusion couple. Source: Ref 3

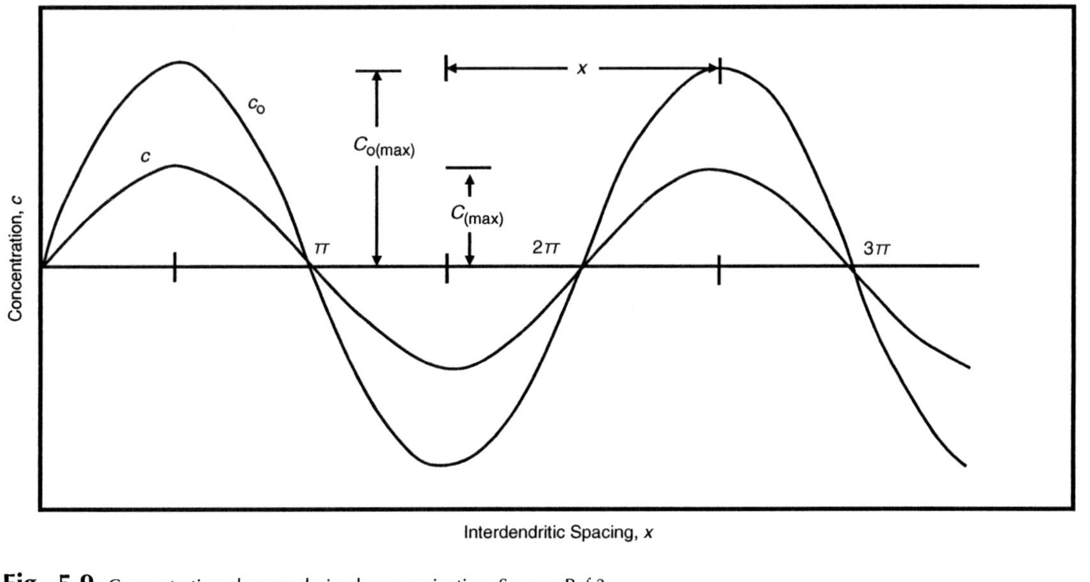


**Fig. 5.9** Concentration changes during homogenization. Source: Ref 3

application of a diffusion couple is when two metal interfaces are joined by diffusion bonding at high pressures and temperatures. For example, superplastically formed and diffusion-bonded titanium structures are used in the aerospace industry. In this application, the parts are heated to 900 °C (1650 °F) and held for several hours under 1000 to 1400 kPa (150 to 200 psig) argon gas pressure to facilitate the forming and bonding operations. This is covered more fully in Chapter 28, "Titanium," in this book.

**Homogenization.** During solidification of an ingot during casting, microsegregation occurs, resulting in local compositional variations that are often removed by thermal aging at high temperatures. The compositional variations can be approximated by a sine wave of wavelength $2x$ and amplitude $c_{omax}$. As shown in Fig. 5.9, homogenization reduces the amplitude $c_{omax}$ to $c_{max}$. If $c$ and $c_o$ are the differences between local concentrations and the average concentration, the extent of homogenization is:

$$\frac{c}{c_o} = e^{-\left(\frac{\pi}{X}\right)^2}$$

where

$$X = \frac{x}{\sqrt{Dt}} \quad \text{(Eq 5.8)}$$

Alloy ingots are normally homogenized by placing them in large soaking pits at elevated temperatures for rather long times. Homogenization is necessary to even out concentration gradients (segregation) that occur during casting. This helps to prevent ingot cracking during initial hot working operations.

In general, all solutions to Fick's second law are of the form:

$$f(\text{concentrations}) = \frac{x}{\sqrt{Dt}} \quad \text{(Eq 5.9)}$$

where $f$(concentrations) is a function of $c$, the concentration at a specific point; $c_o$ is the initial concentration; and $c_s$ is the surface concentration.

In many problems, these concentrations are fixed, so:

$$\frac{x}{\sqrt{Dt}} = \text{constant} \quad \text{(Eq 5.10)}$$

An important practical relation evolves from the solution to Fick's second law, namely, that the time-distance relation for a given concentration, $c$, is $x^2 \approx Dt$. This means, for example, that during a homogenization treatment designed to remove the effects of dendritic segregation (coring), the time is proportional to $x^2$, where $x$ is approximately the dendritic arm spacing.

## 5.3 Temperature Dependence of Diffusion

The temperature at which diffusion occurs greatly affects the value of the diffusion coefficient. As the temperature increases, the diffusion coefficient increases exponentially. The diffusion coefficient, $D$, is given by an Arrhenius equation:

$$D = D_o e^{-Q/RT} \quad \text{(Eq 5.11)}$$

where $D_o$ is the constant for a given diffusion system (cm$^2$/s), $Q$ is the activation energy for the diffusion process (kJ/mol), $R$ is the ideal gas constant (8.31 J/mol · K), and $T$ is the absolute temperature (K).

Typical values for $D_o$ are given in Table 5.2, and the temperature dependence of $D$ is shown for a number of material systems in Fig. 5.10. Since Eq 5.11 is an Arrhenius equation, the plots of logarithmic $D$ versus $1/T$ are straight lines. The change of the diffusion coefficient of carbon, as the concentration of carbon changes in iron at 930 °C (1700 °F), is shown in Fig. 5.11.

**Table 5.2 Representative data for diffusion into selected metals**

| Diffusing species | Solvent metal(a) | $D_o$, m$^2$/s | Activation energy, kJ/mol | Calculated values: Temperature, °C | Calculated values: $D$, m$^2$/s |
|---|---|---|---|---|---|
| Fe | α-Fe (bcc) | $2.8 \times 10^{-4}$ | 251 | 500 | $3.0 \times 10^{-21}$ |
| | | | | 900 | $1.8 \times 10^{-15}$ |
| Fe | γ-Fe (fcc) | $5.0 \times 10^{-5}$ | 284 | 900 | $1.1 \times 10^{-17}$ |
| | | | | 1100 | $7.7 \times 10^{-16}$ |
| C | α-Fe | $6.2 \times 10^{-7}$ | 80 | 500 | $2.4 \times 10^{-12}$ |
| | | | | 900 | $1.7 \times 10^{-10}$ |
| C | γ-Fe | $2.3 \times 10^{-5}$ | 148 | 900 | $5.9 \times 10^{-12}$ |
| | | | | 1100 | $5.3 \times 10^{-11}$ |
| Cu | Cu | $7.8 \times 10^{-5}$ | 211 | 500 | $4.2 \times 10^{-19}$ |
| Zn | Cu | $2.4 \times 10^{-5}$ | 189 | 500 | $4.0 \times 10^{-18}$ |
| Al | Al | $2.3 \times 10^{-4}$ | 144 | 500 | $4.2 \times 10^{-14}$ |
| Cu | Al | $6.5 \times 10^{-5}$ | 136 | 500 | $4.1 \times 10^{-14}$ |
| Mg | Al | $1.2 \times 10^{-4}$ | 131 | 500 | $1.9 \times 10^{-13}$ |
| Cu | Ni | $2.7 \times 10^{-5}$ | 256 | 500 | $1.3 \times 10^{-22}$ |

(a) bcc, body-centered cubic; fcc, face-centered cubic. Source: Ref 4

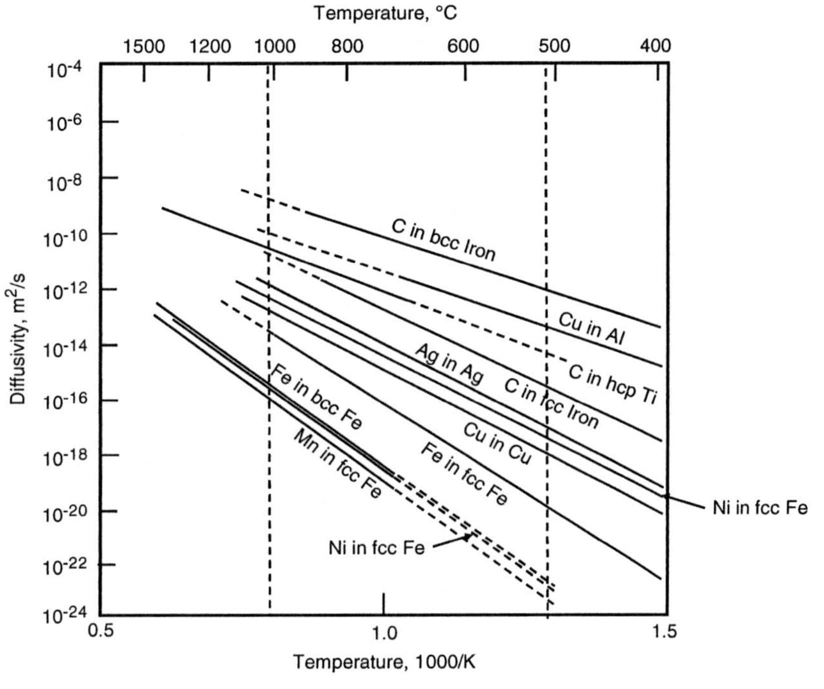


**Fig. 5.10** Arrhenius plot of diffusivity data for a number of metallic systems. Source: Ref 5

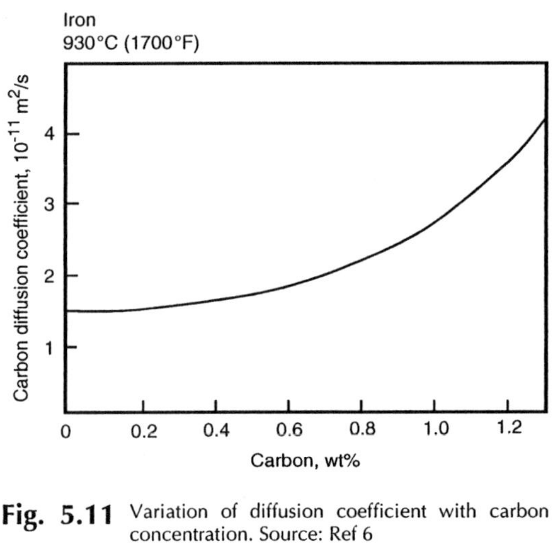


**Fig. 5.11** Variation of diffusion coefficient with carbon concentration. Source: Ref 6

The activation energy, $Q$, reflects the energy required to move an atom over a barrier from one lattice site to another; the barrier is associated with the requirement that the atom must vibrate with a sufficient amplitude to break the nearest neighboring bonds in order to move to a new location.

## 5.4 Intrinsic Diffusion Coefficients (Kirkendall Effect)

At one time, it was assumed that in a binary solid solution, both species of atoms diffused in opposite directions at the same velocity via a direct exchange mechanism. However, Ernest Kirkendall, during a short career as a professor at Wayne State University, showed that this was not the case. He formed a diffusion couple by plating copper onto a bar of brass containing 30% Zn. In order to identify the original interface, inert molybdenum wires were wrapped around the bar prior to plating. After diffusion, the location of the molybdenum wires had moved with respect to their original position (Fig. 5.12). The wires had moved toward the end of the bar at the brass side. The reason for the movement was that the net flux of zinc atoms past the wires in one direction was faster than the net flux of copper atoms in the opposite direction. This observation provided strong evidence that vacancies play an important role in substitutional diffusion.

The plane at the position of the original joint of the brass and copper is called the Matano interface and is important because the two areas,

$A_1$ and $A_2$, in Fig. 5.12 are equal. The areas $A_1$ and $A_2$ represent the amounts of copper and zinc, respectively, that have crossed the interface. Therefore, if the Matano interface is used to describe diffusion behavior, only a single diffusion coefficient, *D,* is needed. Lawrence Darken, a pioneer in chemical metallurgy, subsequently showed that the relation between the measured diffusion coefficient and the intrinsic diffusion diffusivities of the individual atom species for a binary system of atoms *A* and *B* is:

$$D = C_A D_A + C_B D_B \quad \text{(Eq 5.12)}$$

Here, $C_A$ and $C_B$ are the mole fractions of *A* and *B,* respectively, and $D_A$ and $D_B$ are the intrinsic diffusivities of *A* and *B,* respectively. $D_A$ and $D_B$ are concentration dependent.

## 5.5 High Diffusion Paths

Diffusion is much more rapid along free surfaces and grain boundaries than within the interior of the crystalline lattice. In general, the diffusion is highest on free surfaces, next highest at grain boundaries, and slowest within the grains themselves. In other words:

$$D_s > D_b > D_v \quad \text{(Eq 5.13)}$$

where $D_s$ is the diffusivity at a surface, $D_b$ is the diffusivity at grain boundaries, and $D_v$ is the diffusivity within a lattice.

The diffusivities are higher at grain boundaries and free surfaces as a result of the progressively more open structures found at these locations. Although surface diffusion is sometimes important, the total area of the grain boundaries in a metal is usually much greater than the free surface area, and, as such, grain-boundary diffusion is usually more important than surface diffusion.

Typically, the activation energy for grain-boundary diffusion is on the order of only approximately 1/2 that for volume diffusion. Therefore, volume diffusion is more sensitive to temperature changes than grain-boundary diffusion. As the temperature is increased, the rate of volume diffusion increases more rapidly than the rate for grain-boundary diffusion. Conversely, as the temperature is decreased, the rate of grain-boundary diffusion decreases less rapidly. In other words, at very high

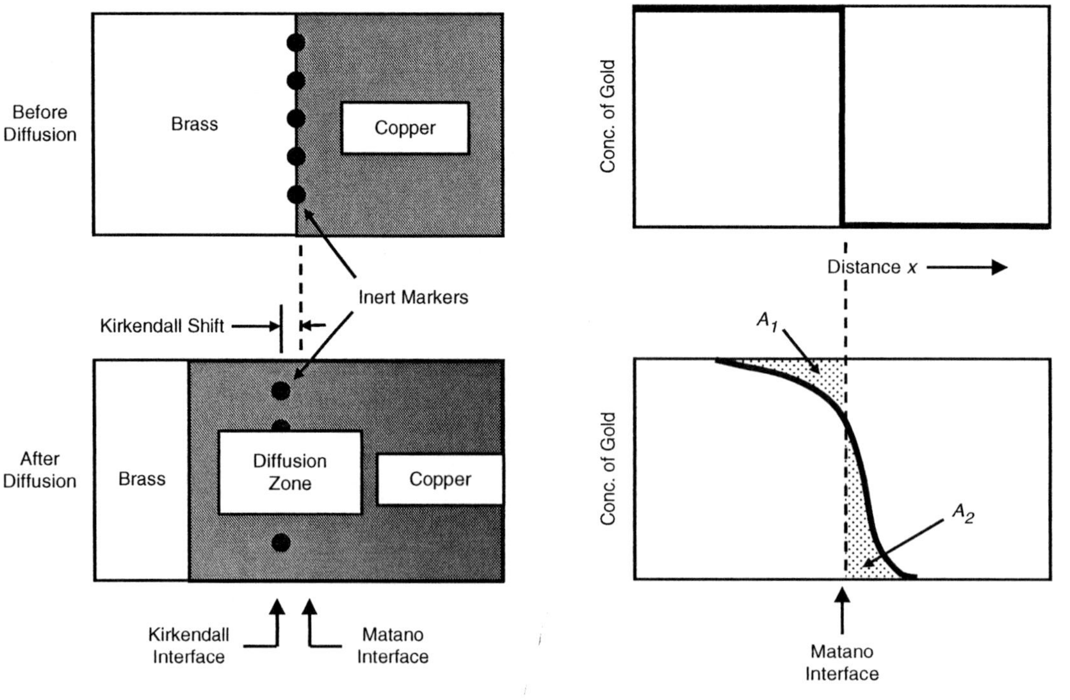


**Fig. 5.12** Kirkendall effect during interdiffusion of brass and copper. Source: Ref 6

temperatures, diffusion through the grains tends to overpower grain-boundary diffusion, but at lower temperatures, grain-boundary diffusion makes a larger contribution to the overall diffusion rate.

## ACKNOWLEDGMENTS

Sections of this chapter were adapted from "Principles of Heat Treating of Nonferrous Alloys" by C.R. Brooks in *Heat Treating,* Volume 4, *ASM Handbook,* ASM International, 1991.

## REFERENCES

1. D.A. Porter and K.E. Easterling, *Phase Transformations in Metals and Alloys,* Chapman and Hall, 1981
2. C.R. Brooks, Principles of Heat Treating of Nonferrous Alloys, *Heat Treating,* Vol 4, *ASM Handbook,* ASM International, 1991
3. W.F. Hosford, *Physical Metallurgy,* Taylor & Francis, 2005
4. E.A. Brandes and G.B. Brook, *Smithell's Metal Reference Book,* 7th ed., Butterworth-Heinemann, 1992
5. L.H. Van Vlack, *Elements of Materials Science and Engineering,* 4th ed., Addison-Wesley Publishing Company, 1980
6. A.G. Guy, *Elements of Physical Metallurgy,* 2nd ed., Addison-Wesley Publishing Company, 1959

## SELECTED REFERENCES

- R.E. Reed-Hill and R. Abbaschian, *Physical Metallurgy Principles,* 3rd ed., PWS Publishing Company, 1991
- W.F. Smith, *Principles of Materials Science and Engineering,* McGraw-Hill Book Company, 1986

# CHAPTER 6

# Phase Diagrams

A PHASE is a homogeneous, physically distinct region of a material that differs in structure and/or composition from another region. A phase may be a single chemical substance or a mixture of two or more substances. Phase diagrams are graphical representations that show the phases present in the material at various compositions, temperatures, and pressures. It should be recognized that phase diagrams represent equilibrium conditions for an alloy, which means that very slow heating and cooling rates are used to generate data for their construction. Since industrial practices almost never approach equilibrium, they should be used with some degree of caution. Nevertheless, phase diagrams are very useful in predicting phase transformations and their resulting microstructures. External pressure also influences the phase structure; however, since pressure is held constant in most applications, phase diagrams are usually constructed at a constant pressure of one atmosphere.

A phase is defined by the type of atomic bonding and arrangement of elements in a material. In pure metals, depending on the conditions of pressure or temperature, the phases may be solid, liquid, or gas. Alloys, which contain two or more elements, also have solid, liquid, and gas phases, but solid alloys may also contain phases with different crystalline structures. Second phases are precipitates that form when a solute alloying element or impurity exceeds its solubility limit in the base solvent metal. For example, cementite ($Fe_3C$) is a second-phase constituent in carbon steel.

## 6.1 Phase Rule

A pure metal can exist in either the solid, liquid, or vapor phases, depending on the conditions of temperature and pressure, as shown in the pressure-temperature diagram in Fig. 6.1. At the triple point, the pressure and temperature are such that the solid, liquid, and vapor phases coexist at the same time. Liquid and vapor phases exist along the vaporization line, and liquid and solid phases along the freezing line.

The phase rule, developed by Willard Gibbs in 1876, allows the determination of the number of phases that can coexist in equilibrium:

$$P+F=C+2 \qquad \text{(Eq 6.1)}$$

where $P$ is the number of phases; $F$ is the number of independent variables, called the degrees of freedom; and $C$ is the number of components in the system.

Usually, a component, $C$, is an alloying element, compound, or solution in the system. The degrees of freedom, $F$, is the number of variables (pressure, temperature, and composition) that can be changed independently without changing the number of phases in equilibrium.

At the triple point in Fig. 6.1, all three phases coexist in equilibrium ($P = 3$). Since there is only one component (the pure metal), $C = 1$. Therefore, the number of degrees of freedom is:

$$P+F=C+2$$

$$3+F=1+2$$

$$F=0$$

Since there are no degrees of freedom, neither the temperature nor the pressure can be changed independently. Thus, the triple point is known as an invariant point.

Next, consider a point along the liquid-solid freezing curve. At any point on the curve, both the liquid and solid phases will coexist. Thus, the number of phases is $P = 2$:

$$P+F=C+2$$

$$2+F=1+2$$

$$F=1$$

In this case, there is one degree of freedom, which means that either the temperature or pressure can be changed independently and still have a system with two phases. Therefore, if the pressure is changed, then the temperature must be adjusted to maintain the two phases.

Finally, consider a point within the solid field. Here, there is only one phase, so $P = 1$:

$$P+F=C+2$$

$$1+F=1+2$$

$$F=2$$

Therefore, with two degrees of freedom, both the temperature and pressure can be changed independently without affecting the phase.

Since pressure is usually fixed as a constant in metallurgical systems, the number of state variables is reduced by one, and the rule becomes:

$$P+F=C+1 \quad \text{(Constant pressure)} \qquad \text{(Eq 6.2)}$$

An easy way to remember this equation is: Police Force = Cops + 1.

## 6.2 Binary Isomorphous System

The phase diagram is, in reality, a chart that shows the relationship between the composition, temperature, and structure of any alloy in a series. Consider briefly how these diagrams are constructed. A pure metal will solidify at a constant temperature, while an alloy will solidify over a temperature range that depends on the alloy composition. Consider the series of cooling curves for the copper-nickel system shown in Fig. 6.2. For increasing amounts of nickel in the alloy, freezing begins at increasing temperatures $A$, $A_1$, $A_2$, $A_3$ up to pure nickel at $A_4$ and finishes at increasing temperatures $B$, $B_1$, $B_2$,

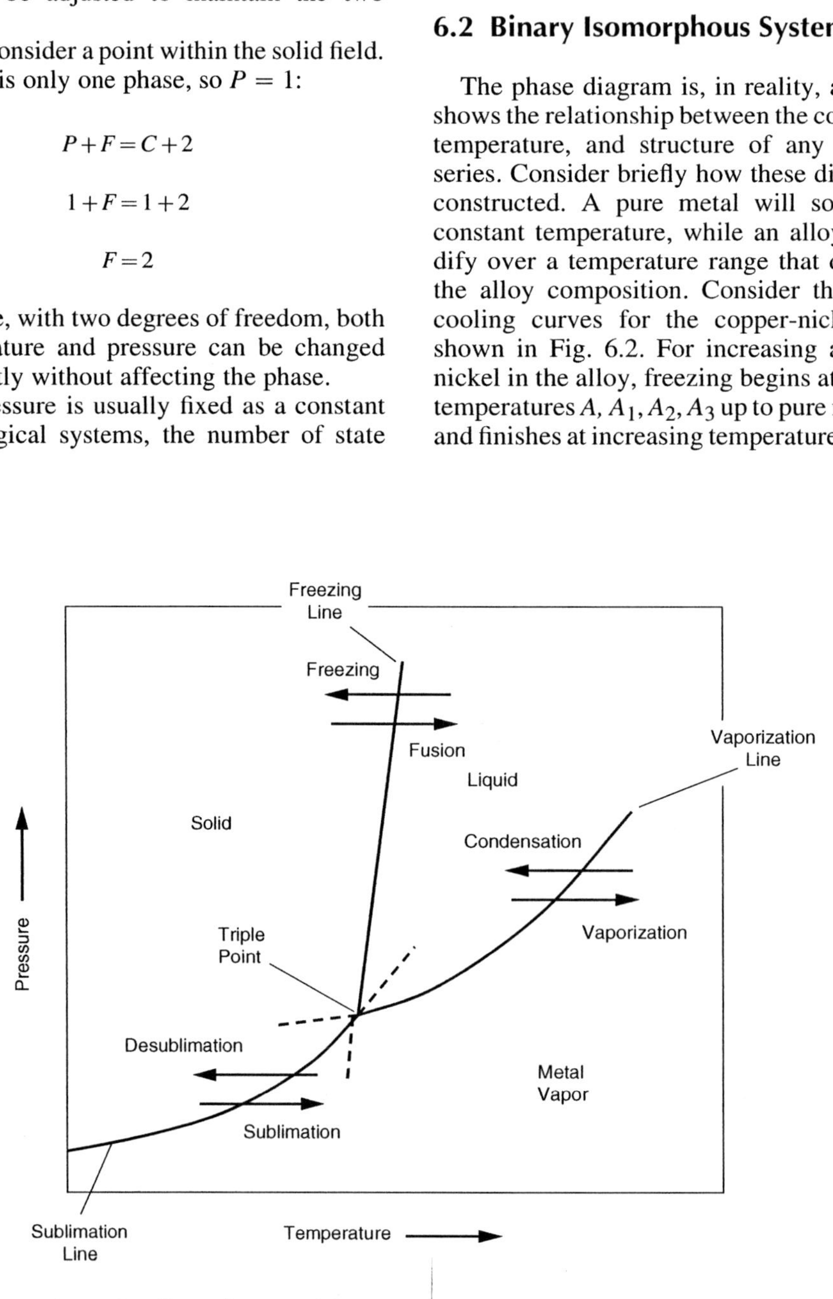


**Fig. 6.1** Pressure-temperature diagram for pure metal

$B_3$ up to pure nickel at $B_4$. If all the points $A$, $A_1$, $A_2$, $A_3$, and $A_4$ are joined, the result is a line known as the liquidus, which indicates the temperature at which any given alloy will begin to solidify. Likewise, by joining the points $B$, $B_1$, $B_2$, $B_3$, and $B_4$, the solidus line, the temperature at which any given alloy will become completely solid, is obtained. In other words, at all temperatures above the liquidus, the alloy will be a liquid, and at all temperatures below the solidus, the alloy will be a solid. At temperatures between the liquidus and the solidus, sometimes referred to as the "mushy zone," both liquid and solid coexist in equilibrium.

Very simple phase diagrams of this type can be constructed by using the appropriate points obtained from time-temperature cooling curves, which indicate where freezing began and where it was complete. However, if the alloy system is one in which further structural changes occur after the alloy has solidified, then the metallurgist must resort to other methods of investigation to determine the phase-boundary lines. These methods include the use of x-rays, electrical conductivity measurements, dilatometry methods that measure changes in volume, and exhaustive microscopic examinations following the quenching of representative specimens from different temperatures. Therefore, the construction of phase diagrams is the result of much tedious experimental work, and since the conditions under which the work is carried out can be so variable, it is not surprising that the values assigned to compositions and temperatures at which phase changes occur are under constant review and amendment by research metallurgists.

A few systems consist of components having the same crystalline structure, and the components of some of these systems are completely soluble, or miscible, in each other in the solid form, thus forming a continuous series of solid solutions. When this occurs in a binary system, the phase diagram usually has the general appearance of the copper-nickel system shown in Fig. 6.3. Temperature is plotted along the ordinate axis, and the alloy composition is shown on the abscissa axis. The composition ranges from 0 wt% Ni (100 wt% Cu) on the extreme left to 100 wt% Ni (0 wt% Cu) on the extreme right. Three different phase regions, or fields, are present on the diagram: a liquid ($L$) field, a two-phase solid plus liquid field ($\alpha + L$), and a solid-solution alpha ($\alpha$) field, where $\alpha$ is a solid solution containing both copper and nickel. Each field is defined by the phase or phases that exist over the range of temperatures and compositions bounded by the phase-boundary lines. At high temperatures, the liquid, $L$, field is a homogeneous liquid solution composed of both copper and nickel. The solid solution, $\alpha$, that exists at lower temperatures is a substitutional solid solution consisting of both copper and nickel atoms with a face-centered cubic (fcc) crystalline structure. When an alloy of any given composition freezes, copper and nickel are mutually soluble in each other and therefore display complete solid solubility. Solid solutions are commonly designated by lowercase Greek letters. The boundaries between the regions are identified as the liquidus and solidus. The upper curve separating the liquid, $L$, and the two-phase, $L + \alpha$, field is termed the liquidus line. The liquidus is the lowest temperature at which

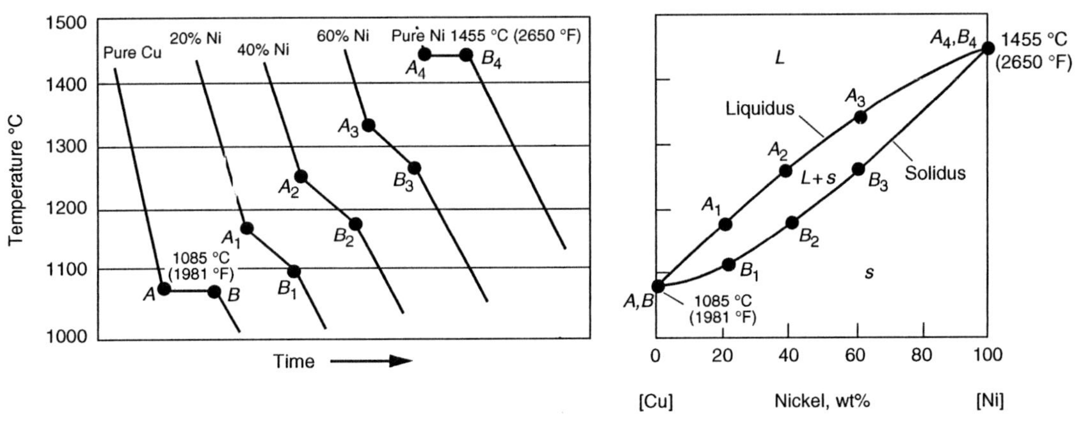


**Fig. 6.2** Phase diagram construction from cooling curves

any given composition can be found in an entirely molten state. The lower curve separating the solid-solution, α, field and the two-phase, $L+\alpha$, field is known as the solidus line. The solidus is the highest temperature when all atoms of a given composition can be found in an entirely solid state. In general, a liquidus is the locus of points in a phase diagram representing the temperatures at which alloys of the various composition of the system begin to freeze on cooling or finish melting on heating. Likewise, a solidus is the locus of points representing the temperatures at which the various alloys finish freezing on cooling or begin melting on heating.

Complete solid solubility is actually the exception rather than the rule. To obtain complete solubility, the system must adhere to the Hume-Rothery rules for solid solutions (Chapter 3, "Solid Solutions," in this book). In this case, both copper and nickel have the fcc crystal structure, have nearly identical atomic radii and electronegativities, and have similar valences. The term *isomorphous* implies complete solubility in both the liquid and solid states. Most alloys do not have such simple phase systems. Typically, alloying elements have significant differences in their atomic size and crystalline structure, and so, the mismatch forces the formation of a new crystal phase that can more

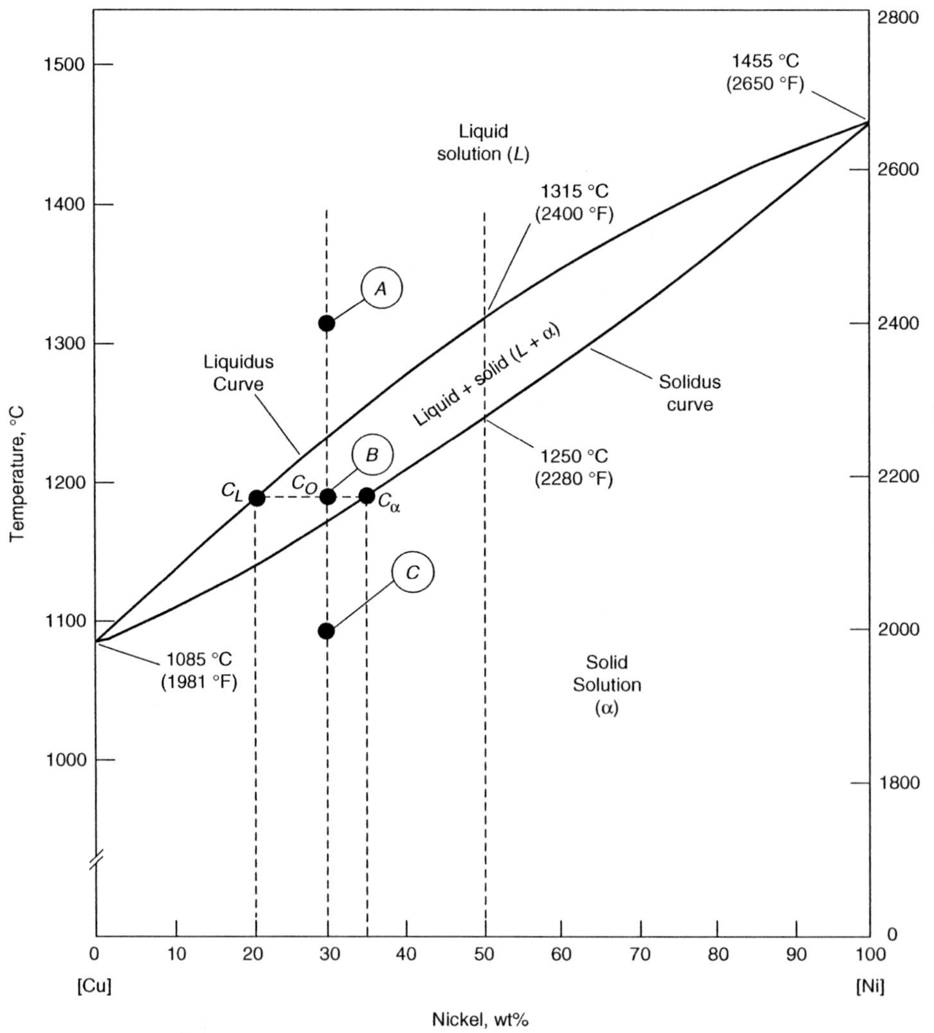


**Fig. 6.3** Copper-nickel phase diagram

easily accommodate alloying elements in the solid state.

The liquidus and solidus lines intersect at the two composition extremities, that is, at the temperatures corresponding to the melting points of pure copper (1085 °C, or 1981 °F) and pure nickel (1455 °C, or 2650 °F). Since pure metals melt at a constant temperature, pure copper remains a solid until its melting point of 1085 °C (1981 °F) is reached on heating. The solid-to-liquid transformation then occurs, and no further heating is possible until the transformation is complete. However, for any composition other than the pure components, melting will occur over a range of temperatures between the solidus and liquidus lines. For example, on heating a composition of 50wt%Cu-50wt%Ni, melting begins at approximately (1250 °C (2280 °F), and the amount of liquid increases until about 1315 °C (2400 °F) is reached, at which point the alloy is completely liquid.

A binary phase diagram can be used to determine three important types of information: (1) the phases that are present, (2) the composition of the phases, and (3) the percentages or fractions of the phases.

**Prediction of Phases.** The phases that are present can be determined by locating the temperature-composition point on the diagram and noting the phase(s) present in the corresponding phase field. For example, an alloy of composition 30wt%Ni-70wt%Cu at 1315 °C (2400 °F) would be located at point *A* in Fig. 6.3. Since this point lies totally within the liquid field, the alloy would be a liquid. The same alloy at 1095 °C (2000 °F), designated point *C*, is within the solid-solution, α, field; only the single α phase would be present. On the other hand, a 30wt%Ni-70wt%Cu alloy at 1190 °C (2170 °F) (point *B*) would consist of a two-phase mixture of solid solution, α, and liquid, *L*.

**Composition of Chemical Compositions of Phases.** To determine the composition of the phases present, locate the point on the phase diagram. If only one phase is present, the composition of the phase is the overall composition of the alloy. For example, for an alloy of 30wt%Ni-70wt%Cu alloy at 1095 °C (2000 °F) (point *C* in Fig. 6.3), only the α phase is present, and the composition is 30wt%Ni-70wt%Cu. For an alloy with composition and temperature coordinates located in a two-phase region, the compositions of the phases can be determined by drawing a horizontal line, referred to as a tie line, between the two phase boundaries at the temperature of interest. Then, drop perpendicular lines from the intersections of each boundary down to the composition axis and read the compositions. For example, again consider the 30wt%Ni-70wt%Cu alloy at 1190 °C (2170 °F), located at point *B* in Fig. 6.3 and lying within the two-phase, $\alpha + L$, field. The perpendicular line from the liquidus boundary to the composition axis is 20%wt%Ni-80wt%Cu, which is the composition, $C_L$, of the liquid phase. In a similar manner, the composition of the solid-solution phase, $C_\alpha$, is read from the perpendicular line from the solidus line down to the composition axis, in this case 35wt%Ni-65wt%Cu.

**Prediction of Amounts of Phases.** The percentages or fractions of the phases present at equilibrium can also be determined with phase diagrams. In a single-phase region, since only one phase is present, the alloy is comprised entirely of that phase; that is, the phase fraction is 1.0 and the percentage is 100%. From the previous example for the 30wt%Ni-70wt%Cu alloy at 1095 °C (2000 °F) (point *C* in Fig. 6.3), only the α phase is present, and the alloy is 100% α.

If the composition and temperature position is located within a two-phase field, a horizontal tie line must be used in conjunction with the lever rule. The lever rule is a mathematical expression (Eq 6.3) based on the principle of conservation of matter. First, a tie line is drawn across the two-phase region at the composition and temperature of the alloy. The fraction of one phase is determined by taking the length of the tie line from the overall alloy composition to the phase boundary for the other phase and dividing by the total tie line length. The fraction of the other phase is then determined in the same manner. If phase percentages are desired, each phase fraction is multiplied by 100. When the composition axis is scaled in weight percent, the phase fractions computed using the lever rule are mass fractions—the mass (or weight) of a specific phase divided by the total alloy mass (or weight). The mass of each phase is computed from the product of each phase fraction and the total alloy mass.

Again, consider the 30wt%Ni-70wt%Cu alloy at 1190 °C (2170 °F) located at point *B* in Fig 6.3 and containing both the solid, α, and the liquid, *L*, phases. The same tie line that was used for determination of the phase compositions can again be used for the lever rule calculation. The overall alloy composition located along the tie line is $C_\alpha - C_L$, or 35−20 wt%. The weight percentage of liquid present is then:

$$\text{wt\% } L = \frac{C_\alpha - C_o}{C_\alpha - C_L} = \frac{35-30}{35-20} = 0.33 \text{ or } 33\% \quad \text{(Eq 6.3)}$$

Likewise, the amount of solid present is:

$$\text{wt\% } S = \frac{C_o - C_L}{C_\alpha - C_L} = \frac{30-20}{35-20} = 0.67 \text{ or } 67\% \quad \text{(Eq 6.4)}$$

The lever rule, or, more appropriately, the inverse lever rule, can be visualized as a scale, such as the one shown in Fig. 6.4. For the scale to balance, the weight percentage of the liquid, which is less, must have the longer lever arm, in this instance 10 units, compared to the solid phase, α, which has a higher weight percentage and thus a shorter lever arm of 5 units.

On occasion, conversion from mass fraction to volume fraction (or vice versa) is desired. Equations that facilitate these conversions are given in Fig. 6.5. In these expressions, $\rho_\alpha$ and $\rho_\beta$ are the densities of the respective phases. When the densities of the phases in a two-phase alloy differ significantly, there will be quite a disparity between mass and volume fractions; conversely, if the phase densities are the same, mass and volume fractions are identical.

It is important to emphasize that equilibrium phase diagrams identify phase changes under conditions of very slow changes in temperature.

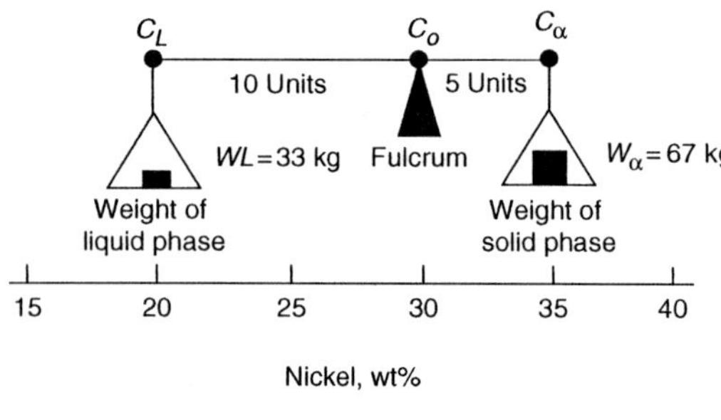


$$\text{Wt\% } L = \frac{C_\alpha - C_o}{C_\alpha - C_L} = \frac{35-30}{35-20} = 0.33 \text{ or } 33\%$$

$$\text{Wt\% } S = \frac{C_o - C_L}{C_\alpha - C_L} = \frac{30-20}{35-20} = 0.67 \text{ or } 67\%$$

**Fig. 6.4** Visual representation of lever rule applied to a 30 wt% Ni-Cu alloy at 1190 °C (2170 °F). Source: Ref 1

In practical situations, where heating and cooling occur more rapidly, the atoms do not have enough time to get into their equilibrium positions, and the transformations may start or end at temperatures different from those shown on the equilibrium phase diagrams. In these practical circumstances, the actual temperature at which the phase transformation occurs will depend on both the rate and direction of temperature change. Nevertheless, phase diagrams provide valuable information for virtually all metal processing operations that involve heating the metal, such as casting, hot working, and all heat treatments.

The copper-nickel system is an example of solid-solution hardening or strengthening. Most of the property changes in a solid-solution system are caused by distortion of the crystalline lattice of the base or solvent metal by additions of the solute metal. The distortion increases with the amount of the solute metal added, and the maximum effect occurs near the center of the diagram, since either metal can be considered as the solvent. As shown in Fig. 6.6, the strength curve passes through a maximum, while the ductility curve, as measured by percent elongation, passes through a minimum. Properties that are almost unaffected by atom interactions vary more linearly with composition. Examples include the lattice constant, thermal expansion, specific heat, and specific volume.

Not all binary isomorphous systems look like the copper-nickel system shown in Fig. 6.3. Some have local temperature maxima or minima like the ones in Fig. 6.7. When the boundaries of a two-phase region intersect, they meet at a maximum or minimum, and the liquidus and solidus curves are tangent to each other and to an isothermal line at the point of intersection. Such points are called congruent points. In Fig. 6.7, they occur at the 50 wt% composition for both

$$V_\alpha = \frac{\dfrac{W_\alpha}{\rho_\alpha}}{\dfrac{W_\alpha}{\rho_\alpha} + \dfrac{W_\beta}{\rho_\beta}} \qquad W_\alpha = \frac{V_\alpha \rho_\alpha}{V_\alpha \rho_\alpha + V_\beta \rho_\beta}$$

$$V_\beta = \frac{\dfrac{W_\beta}{\rho_\beta}}{\dfrac{W_\alpha}{\rho_\alpha} + \dfrac{W_\beta}{\rho_\beta}} \qquad W_\alpha = \frac{V_\beta \rho_\beta}{V_\alpha \rho_\alpha + V_\beta \rho_\beta}$$

**Fig. 6.5** Conversion formulas for weight and volume percentages. Source: Ref 2

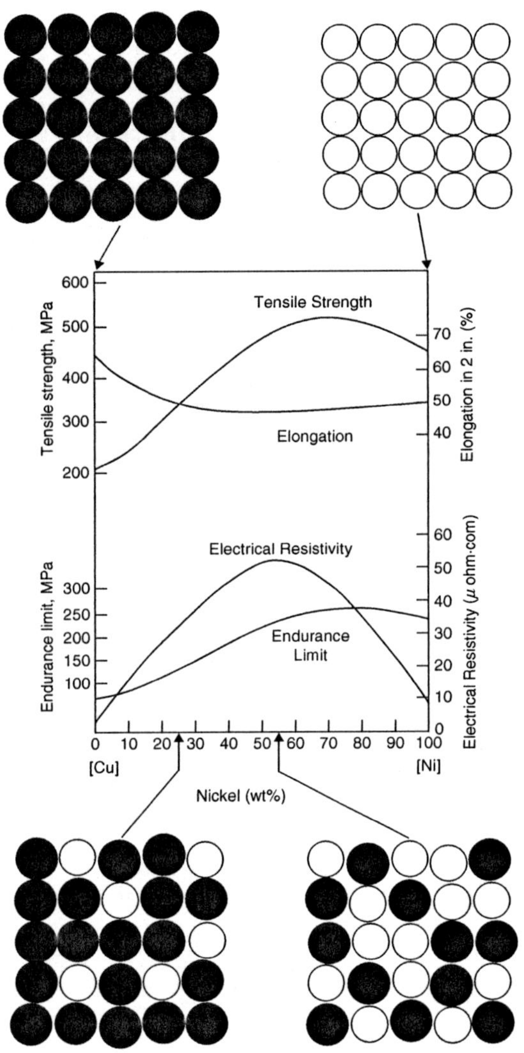


**Fig. 6.6** Typical property variations in copper-nickel system. Source: Ref 1

phase diagrams. It is characteristic of congruent points that solidification can occur with no change in composition or temperature at these points. Thus, the solidification of an alloy at a congruent point is similar to solidification of a pure metal. However, the resulting solid is a solid solution and not a pure component.

## 6.3 Eutectic Alloy Systems

Most alloy systems do not have such simple phase diagrams as the previous completely isomorphous system. One common type of binary phase diagram is a eutectic system, where the term *eutectic* is taken from the Greek for "easy or well melting." A generic eutectic phase diagram is shown in Fig. 6.8. Eutectic systems form when alloying additions cause a lowering of the liquidus lines from both melting points of the pure elements. At a specific composition, there is a minimum melting point, where the mixed solid-liquid phase regions ($L+\alpha$ and $L+\beta$) vanish. This is the eutectic point ($e$), which defines an alloy composition that has the lowest melting point of the $A$–$B$ system. The eutectic composition also solidifies completely at a single temperature that is referred to as an invariant point.

In a eutectic reaction, a liquid freezes to form two solid solutions:

$$\text{Liquid } L \rightarrow \text{Solid } \alpha + \text{Solid } \beta \qquad \text{(Eq 6.5)}$$

An alloy having a composition to the left of the eutectic point is called a hypoeutectic alloy (*hypo* is from the Greek meaning "less than"), while an alloy to the right of the eutectic point is a hypereutectic alloy (*hyper*, also from the Greek, meaning "greater than").

The maximum solid solubility of element $B$ is defined by point $a$ on the $A$-rich side of the diagram, and the maximum solid solubility of element $A$ in the lattice of element $B$ is defined by point $b$. The methods of determining the equilibrium temperature ranges for solidification, fractions of phases, and compositions of phases are similar to those illustrated for the isomorphous systems (Fig. 6.2 through 6.5).

Because eutectic alloys have a single melting/solidification point, eutectic compositions include important types of commercial alloys. Traditional lead-tin solder alloys are based on their eutectic compositions. Casting alloys are often based on eutectic compositions for various reasons, including the minimization of both energy input and coring, or alloy segregation. For example, the iron-carbon system has a eutectic at composition of 4.3 wt% C, which is the basis of cast irons. Many other binary systems have phase diagrams with multiple eutectic transformation compositions.

Note that since the eutectic is the lowest-melting composition in the alloy, it can often cause problems during hot working and heat treating operations. During casting of an ingot, due to nonuniform cooling rates, there is often quite a bit of segregation of the alloying elements. The low-melting eutectic composition

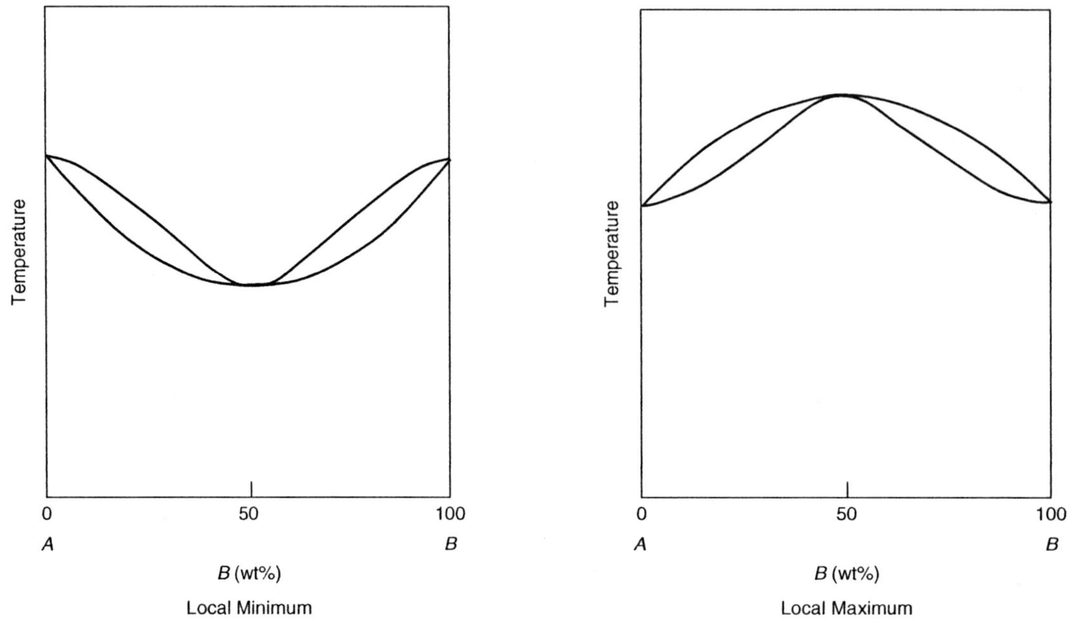


**Fig. 6.7** Binary isomorphous systems with local minima and maxima

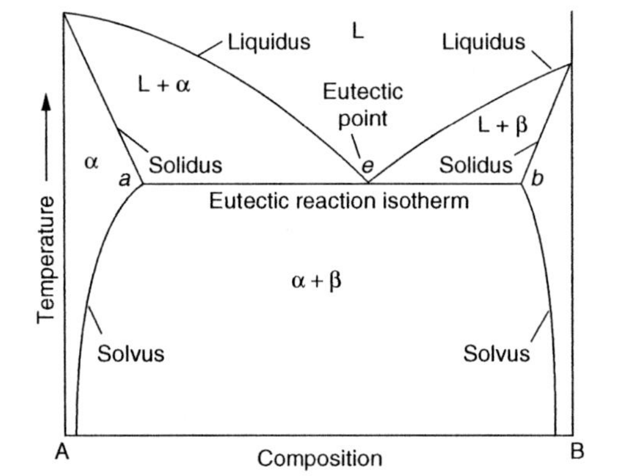


**Fig. 6.8** Phase diagram containing a eutectic reaction. Source: Ref 3

is usually the last portion of the metal to freeze, normally along the grain boundaries. For example, on reheating the metal for hot working operations, if the lowest-melting eutectic temperature is exceeded, melting can occur along the grain boundaries, and the part is frequently ruined. It should also be noted that alloying systems with a number of different alloying elements will often form lower-melting-point eutectics than for a simple binary system. To help prevent some of these problems, as-cast ingots are often reheated to temperatures just below the melting point and soaked for long times (called homogenization) to create more uniform structures prior to hot working or heat treating.

### *6.3.1 Aluminum-Silicon Eutectic System*

Aluminum-silicon alloys are a family of industrially important casting alloys. The phase diagram for the aluminum-silicon system is shown in Fig. 6.9, along with micrographs of representative structures. The microstructure of 99.95 wt% Al has the typical equiaxed structure of a pure metal. The microstructure of the 8 wt% Si alloy shows long dendrites of primary alpha solid solution surrounded by the eutectic microconstituent. In contrast, when the primary solid phase is the silicon-rich beta phase, as in the 20 and 50 wt% Si alloys, the primary crystals have geometric shapes due to the pronounced difference between aluminum, a metal, and silicon, which has predominately nonmetallic properties.

Consider an alloy containing 50 wt% Si. When the alloy is cooled from the liquid state, it starts to solidify at approximately 1080 °C (1975 °F). Solidification begins with the nucleation and growth of primary crystals of beta solid solution. At 870 °C (1600 °F), point *b,* the alloy composition-temperature point lies in the two-phase $L+\beta$ field. The ends of the tie line drawn across the two-phase field determine the chemical composition of the phases; the liquid phase is 34 wt% Si and the beta solid-solution phase is 98 wt% Si. The lever rule can be used to determine the amounts of the two phases:

$$\%\ \beta\ \text{phase} = \frac{ab}{ac} = \frac{50-34}{98-34} \times 100 = 25\ \text{wt\%}$$

Since the remaining portion of the alloy is liquid phase, there must be 75 wt% liquid.

As the 50 wt% Si alloy continues to cool, the amount of primary β phase increases until the eutectic temperature is reached. At this temperature (580 °C, or 1075 °F), the liquid phase reaches the eutectic composition of 12 wt% Si. On cooling through the eutectic temperature, the liquid solidifies at a constant temperature to form the eutectic structure consisting of an intimate mixture of α and β phases. A phase analysis at room temperature is representative of the condition of the alloy in the solid $\alpha+\beta$ region:

$$\%\ \beta\ \text{phase} = \frac{de}{df} = \frac{50-1}{99-1} \times 100 = 50\ \text{wt\%}$$

This amount of β phase (50 wt%) is the sum of the weight fractions of primary and secondary beta. To determine the amount of primary β that

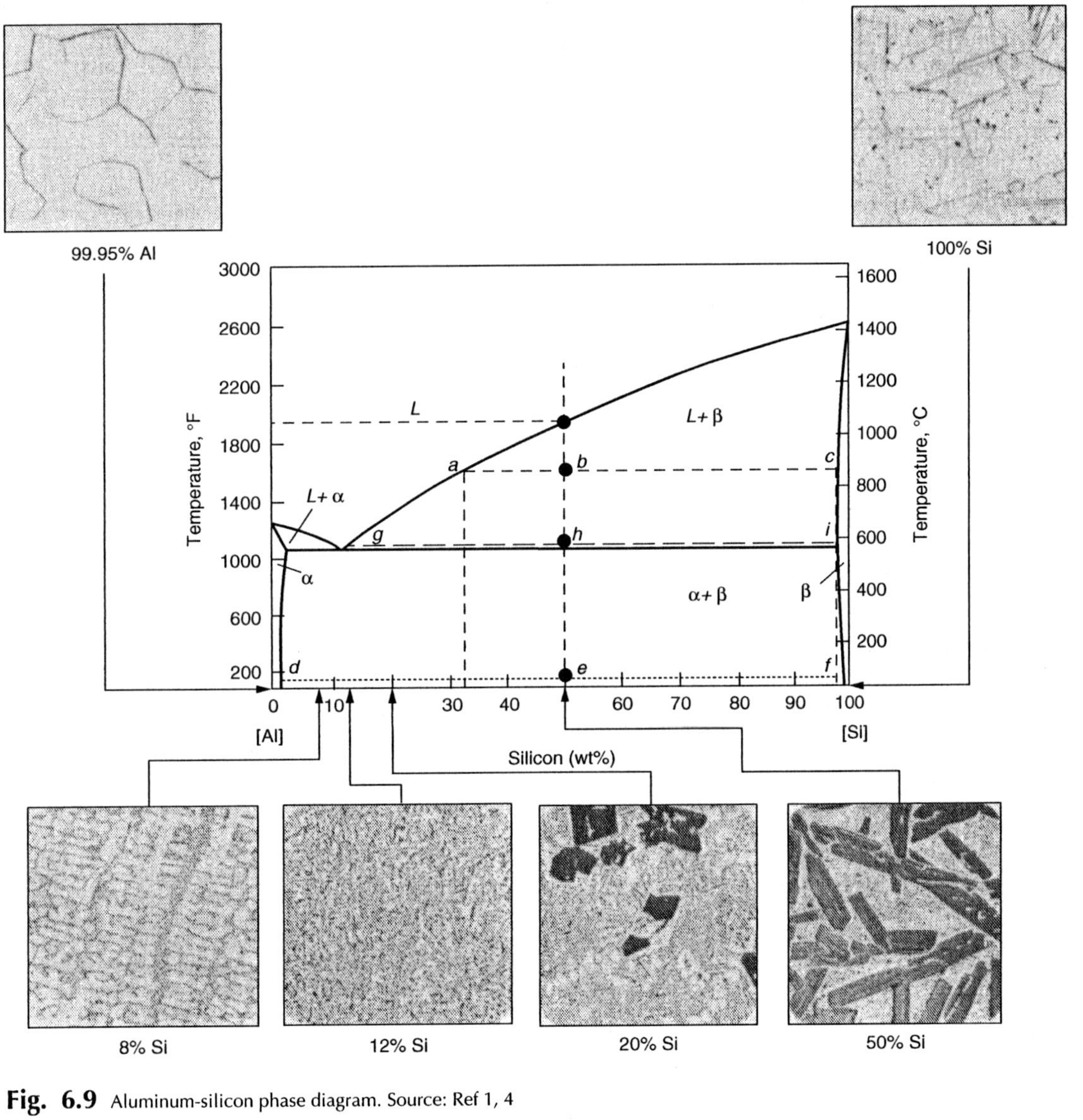


**Fig. 6.9** Aluminum-silicon phase diagram. Source: Ref 1, 4

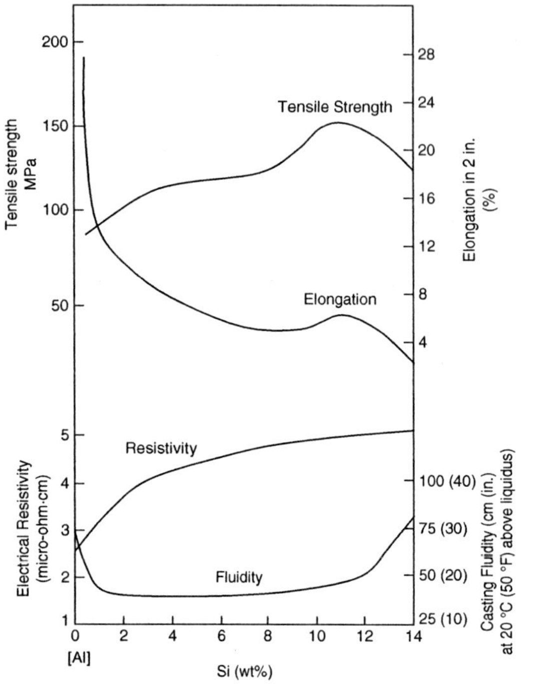


**Fig. 6.10** Selected properties of aluminum-silicon alloys. Source: Ref 1

formed prior to the eutectic reaction (in the $L+\beta$ region of the phase diagram), a phase analysis can be conducted at a temperature immediately above the eutectic temperature. At this temperature, solidification of the primary β is complete, and all the remaining liquid will solidify as part of the eutectic microconstituent:

$$\%\ \text{primary}\ \beta\ \text{phase} = \frac{gh}{gi} = \frac{50-12}{98-12} \times 100 = 44\ \text{wt\%}$$

As previously mentioned, aluminum-silicon alloys are extensively used for castings. The primary reason for their widespread use is that silicon greatly increases the fluidity, or lowers the viscosity, of aluminum during casting, allowing for better mold filling and fewer casting defects. As shown in Fig. 6.10, tensile strength increases with silicon additions, while ductility (elongation) decreases. Slowly cooled aluminum-silicon castings form a eutectic structure in which the eutectic grows as thin, flat plates that appear needlelike. The silicon plates are stress raisers that reduce ductility and toughness. In commercial castings, modifiers such as 0.2% Na or 0.01% Sr are added to the melt to modify the eutectic structure, producing more rounded particles. If the alloy is rapidly cooled, such as in a die casting, modification is not necessary since rapid cooling produces a fine structure. In hypereutectic alloys that are used for engine blocks because of their high-silicon contents and thus high wear resistance, the primary silicon that forms is normally very coarse, causing poor machinability, poor castability, and gravity segregation, where the lighter silicon particles float to the top of the casting. In this case, the addition of 0.05% P encourages the nucleation of primary silicon, refining its size and minimizing its deleterious properties.

### *6.3.2 Lead-Tin Eutectic System*

Although lead-tin alloys are too weak for use as structural materials, they are widely used as solders for everything from joining copper plumbing to soldering electrical circuits. However, there is an international effort to develop lead-free solders because of the health concerns associated with lead. Nevertheless, lead-tin solders are still widely used. The lead-tin phase diagram is shown in Fig. 6.11. Lead-tin alloys containing 0 to 2 wt% Sn solidify exactly like the copper-nickel alloys; a solid-solution α forms during solidification that is stable down to room temperature. Alloys containing 2 to 18.3 wt% Sn also initially solidify to form a solid-solution α. However, on further cooling below the solvus, the solid solubility of β in α is exceeded, and a solid-state reaction occurs in which β precipitates from α.

Note that the lowest melting point of the liquid $L$ occurs at a composition containing 61.9 wt% Sn. At this composition, the alloy undergoes a eutectic reaction in which the liquid freezes to form two solid solutions, α and β:

$$L \rightarrow \alpha + \beta \quad \text{(Eutectic reaction)} \qquad \text{(Eq 6.6)}$$

Since the compositions of the two solid solutions are given by the ends of the eutectic line, liquid containing 61.9 wt% Sn reacts to form solid solutions α of 18.3 wt% Sn and β of 97.8 wt% Sn. Since solidification of the eutectic composition occurs at a single temperature (185 °C, or 361 °F), the cooling curve at the eutectic is similar to that of a pure metal.

A hypoeutectic alloy containing 18.3 to 61.9 wt% Sn forms α particles in the liquid as it freezes through the liquidus line. These particles grow in size as the temperature falls toward the eutectic temperature, increasing the amount of tin in the liquid toward the eutectic composition of 61.9 wt% Sn. When the eutectic temperature of 185 °C (361 °F) is reached, the remaining liquid freezes as a eutectic. Therefore, the final composition contains α particles distributed in a eutectic mixture of α and β. Since the α particles form before the eutectic reaction occurs, the α particles are frequently called primary or proeutectic alpha. The proeutectic alpha does not participate in the eutectic reaction.

Similar to hypoeutectic alloys, hypereutectic alloys form primary or proeutectic β particles when they freeze below the liquidus line, and the liquid mixture becomes enriched with α as the liquid composition moves down the liquidus line. Finally, on reaching the eutectic temperature, the composition of the liquid again contains 61.9 wt% Sn, and the remainder freezes as a eutectic mixture.

Eutectic alloys frequently form a lamellar, or platelike, structure on freezing, as shown in Fig. 6.12. This structure permits the lead and tin atoms to diffuse short distances through the liquid to form the lamellar-type structure. It should be pointed out that although eutectics frequently form a lamellar-type structure on freezing, this is not always the case. As shown in Fig. 6.13, eutectics can form a variety of structures, such as rods, globules, and acicular-type morphologies.

The lead-tin phase diagram is useful for selecting the best solder for a particular application. For soldering delicate electronic components where it is desirable to minimize heat, a low-melting-point solder such as the eutectic solder shown in Fig. 6.11 would be selected. On the other hand, when soldering copper plumbing connections, the plumber's solder would be a logical choice since it stays molten longer and allows the worker to wipe off excess flow from the joint. Finally, for joints that will be exposed to moderately high temperatures, the high-temperature solder may be the best choice.

## 6.4 Free Energy of Alloy Systems

The shapes of liquidus, solidus, and solvus curves in a phase diagram are determined by the

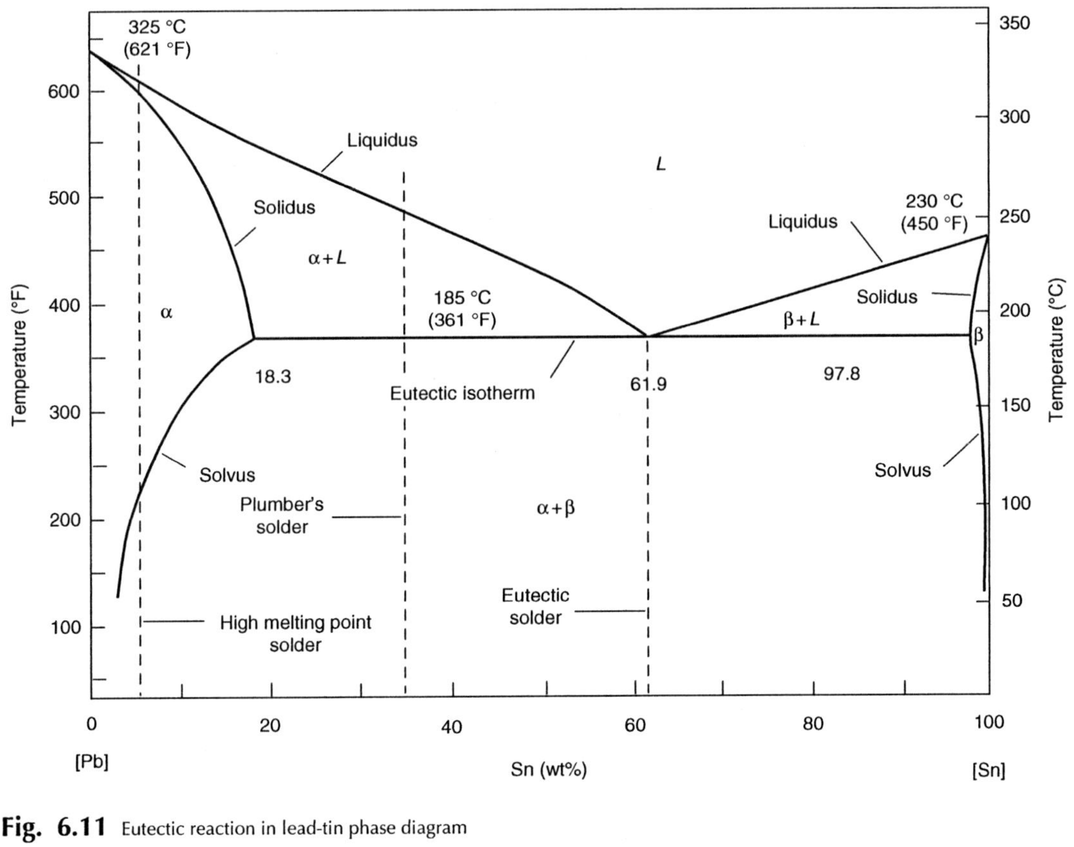


**Fig. 6.11** Eutectic reaction in lead-tin phase diagram

Gibbs free energies of the relevant phases. In this instance, the Gibbs free energy must include not only the energy of the constituent components but also the energy of mixing of these components in the phase. For example, consider the binary isomorphous system shown in Fig. 6.14. The two phases, liquid and solid α, are in stable equilibrium in the two-phase field between the liquidus and solidus lines. The Gibbs free energies at various temperatures are calculated as a function of composition for ideal liquid solutions and for ideal solid solutions of the two components *A* and *B*. The result is a series of plots similar to those shown in Fig. 6.14(a) to (e). At temperature $T_1$, the liquid solution has the lower Gibbs free energy and therefore is the more stable phase. At $T_2$, the melting temperature of *A*, the liquid and solid are equally stable only at a composition of pure *A*. At temperature $T_3$, between the melting temperatures of *A* and *B*, the free energy curves cross. Temperature $T_4$ is the melting temperature of *B*, while $T_5$ is below it.

Construction of the two-phase liquid-plus-solid field of the phase diagram in Fig. 6.14(f) is as follows. According to thermodynamic principles, the compositions of the two phases in equilibrium with each other at temperature $T_3$ can be determined by constructing a straight line that is tangential to both curves in Fig. 6.14(c). This is often referred to as the common tangency rule. The points of tangency, 1 and 2, are then transferred to the phase diagram as points on

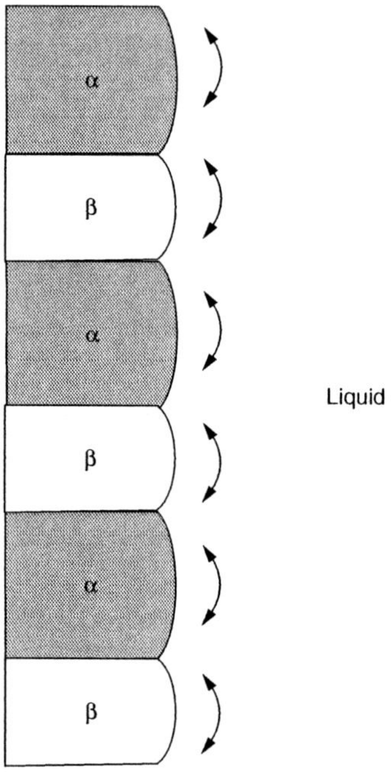


**Fig. 6.12** Lamellar eutectic growth

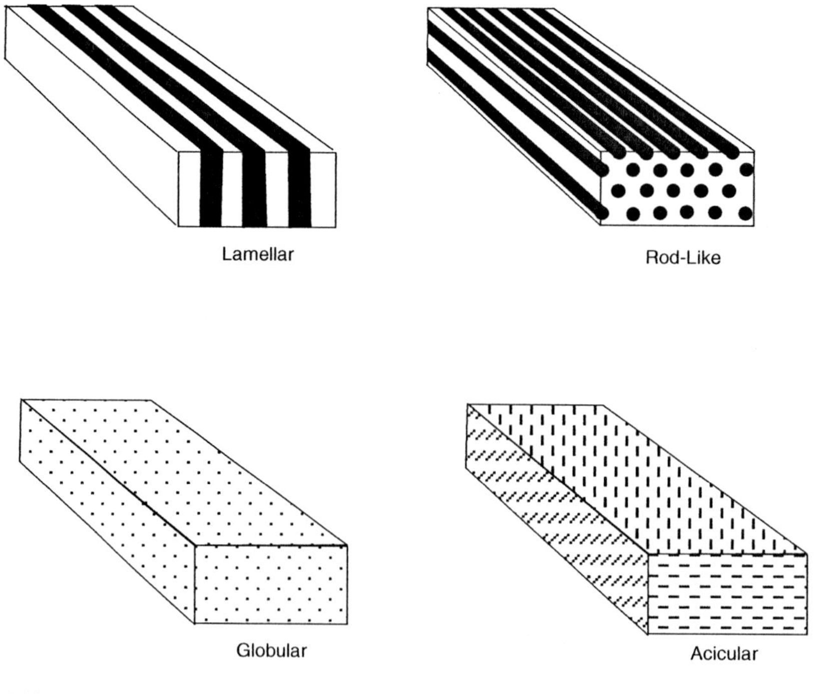


**Fig. 6.13** Solidification structures of eutectics. Source: Ref 5

the solidus and liquidus, respectively. This is repeated at sufficient temperatures to accurately determine the curves. The two-phase field in Fig. 6.14(f) consists of a mixture of liquid and solid phases. As stated previously, the compositions of the two phases in equilibrium at temperature $T_3$ are $C_1$ and $C_2$. The horizontal isothermal line connecting points 1 and 2, where these compositions intersect temperature $T_3$, is the tie line.

Eutectic phase diagrams can also be constructed from free energy curves. Consider the temperatures indicated on the phase diagram in Fig. 6.15(f) and the free energy curves for these temperatures (Fig. 6.15a–e). When the points of tangency on the energy curves are transferred to the phase diagram, the resulting shape forms the typical eutectic system.

## 6.5 Peritectic Reaction

Although not as common as the eutectic reaction, the peritectic reaction is another liquid-solid transformation that at first appears to be similar to the eutectic reaction but is actually very different. The silver-platinum system, shown in Fig. 6.16, contains a peritectic isotherm at 1200 °C (2190 °F) with a peritectic composition of 54 wt% Pt. If the peritectic composition is identified as alloy 1, primary crystals of beta grow in the liquid solution as the alloy is cooled to the peritectic temperature. Just above the peritectic temperature, a phase analysis gives:

Point: 54 wt% Pt alloy at 1200 °C (2190 °F)

Phases: Liquid

Compositions: 31 wt% Pt, 69 wt% Ag

Amounts: $100\% - 8\% = 92\%$

Phases: Beta

Compositions: 86 wt% Pt, 14 wt% Ag

Amounts: $\text{wt\% } \beta = \frac{54-31}{86-31} = \frac{23}{55} = 42 \text{ wt\%}$

Just below the peritectic temperature, a second analysis gives:

Point: 54 wt% Pt alloy at 1150 °C (2100 °F)

Phases: Alpha

Compositions: 54 wt% Pt

Amounts: 100%

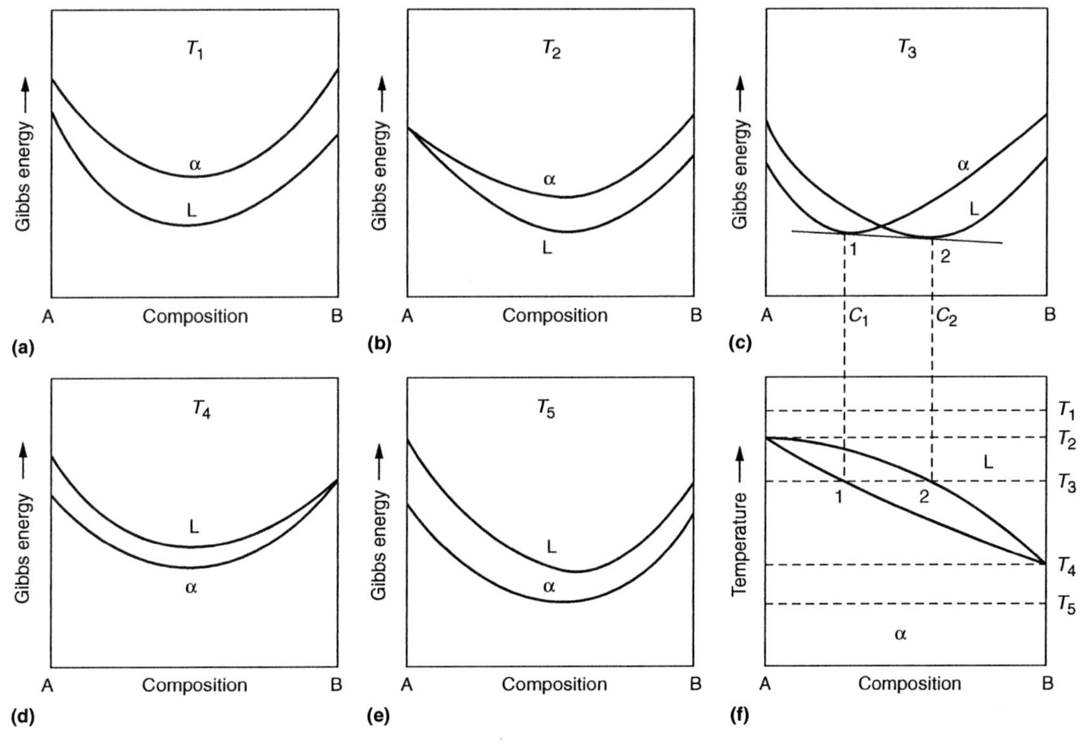


**Fig. 6.14** Gibbs free energy curves and construction of binary phase diagram. Source: Ref 6

This indicates that the liquid and beta phases transformed to the alpha phase on passing through the peritectic.

The generalized form of this peritectic reaction is:

$L+\beta\rightarrow\alpha$ (Peritectic reaction) (Eq 6.7)

Since the α phase surrounds the solid β particles, as shown in Fig. 6.17, β atoms must diffuse through the α crust to reach the liquid for the reaction to continue. However, diffusion through a solid phase is much slower than diffusion through a liquid. As the peritectic reaction continues, the β layer becomes thicker, and the reaction slows down even more. Unless the cooling rate is very slow, a cored segregated structure will result. Thus, equilibrium peritectic reactions are almost never observed in practice.

## 6.6 Monotectic Reaction

Some metals, similar to oil and water, have very limited solubility in each other and are regarded as immiscible liquids. For example, liquid copper and liquid lead are completely soluble in each other at high temperatures. However, as shown in the Fig. 6.18 phase diagram, alloys containing between 36 and 87 wt% Pb separate into two liquids on further cooling. The two liquids coexist in the miscibility gap, or dome, that is typical of all alloys that undergo a monotectic reaction:

$L_1\rightarrow\alpha+L_2$ (Monotectic reaction) (Eq 6.8)

During solidification of a copper-lead alloy containing 20 wt% Pb, the copper-rich α phase forms first. The liquid composition shifts toward the monotectic composition of 36 wt% Pb. Then, the liquid transforms to more solid α and a second liquid containing 87 wt% Pb. The lever rule shows that only a very small amount of the second liquid is present. On further cooling, the second liquid eventually undergoes a eutectic reaction, producing α and β, where the β is almost pure lead. The microstructure of this alloy contains spherical β particles randomly distributed in a matrix of copper-rich α.

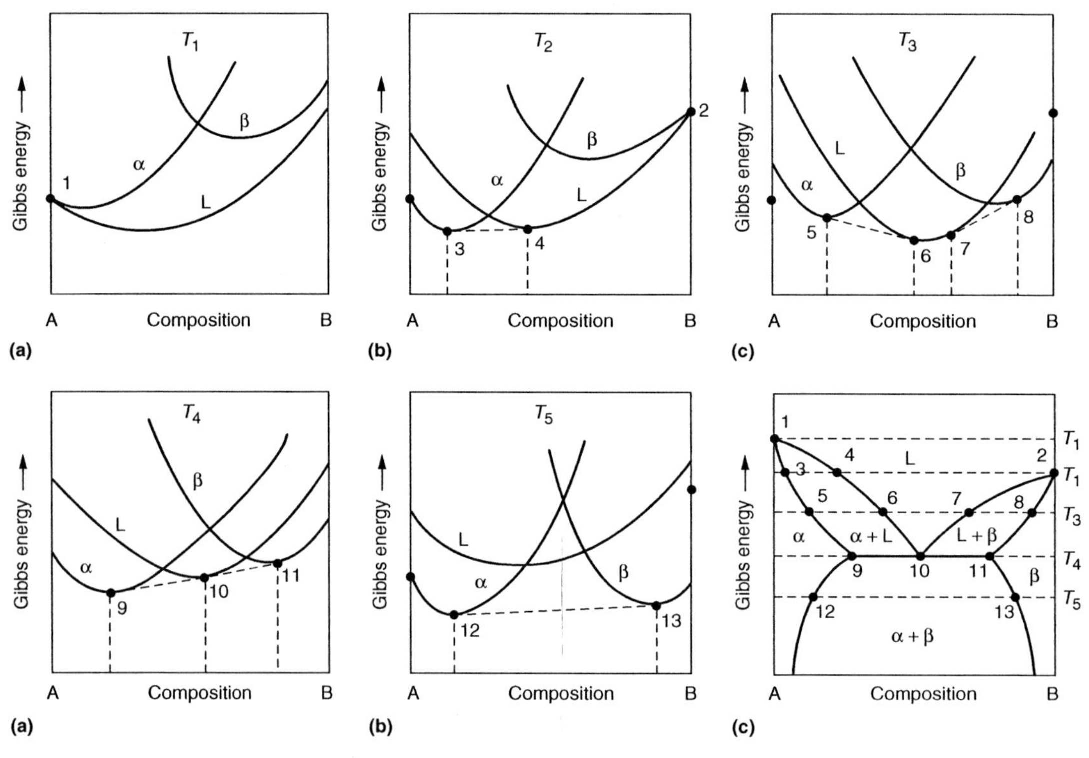


**Fig. 6.15** Gibbs free energy curves and construction of eutectic phase diagram. Source: Ref 6

## 6.7 Intermediate Phases

Phase diagrams are often quite complex, with a number of different reactions occurring at different compositions and temperatures. In most cases, the appearance of several reactions in a binary phase diagram is the result of the presence of intermediate phases. These are phases whose chemical compositions are intermediate between the two pure metals and whose crystalline structures are different from those of the pure metals. Some intermediate phases are intermetallic compounds, when, like $Mg_2Pb$, they have a fixed simple ratio of two kinds of atoms. Note that the intermetallic compound $Mg_2Pb$ divides the lead-magnesium

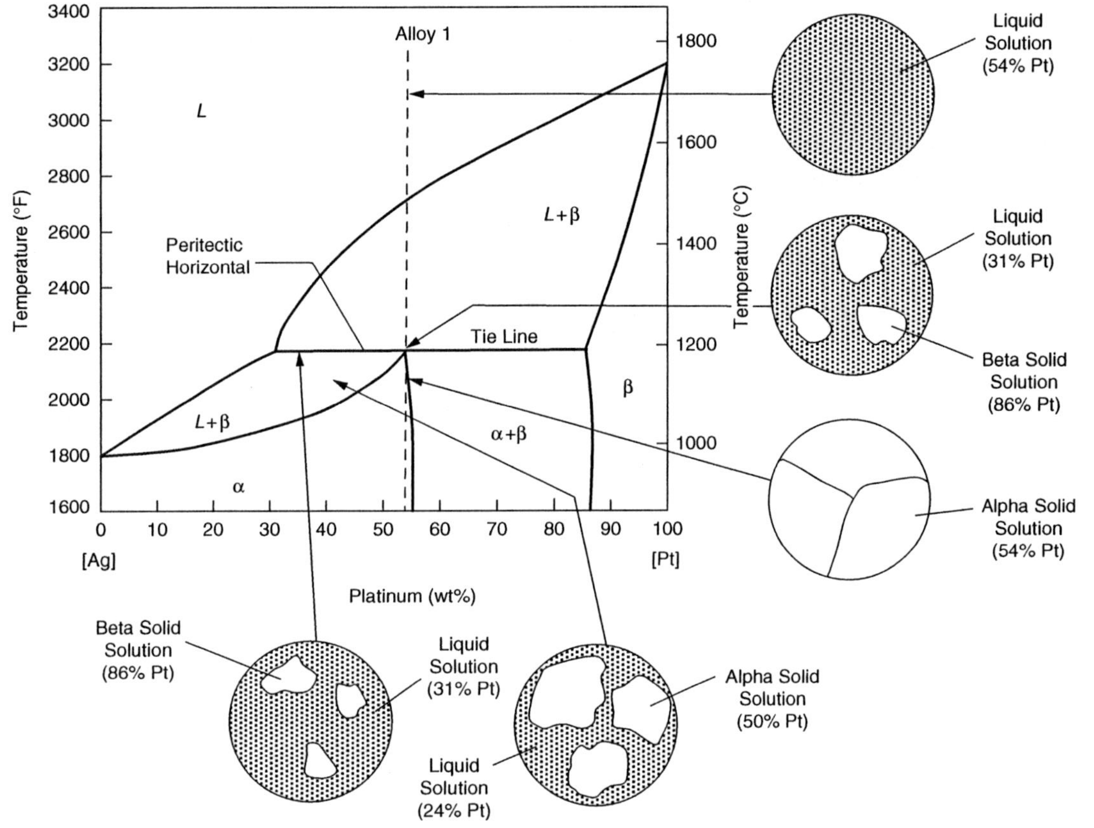


**Fig. 6.16** Peritectic reaction in silver-platinum phase diagram. Adapted from Ref 1

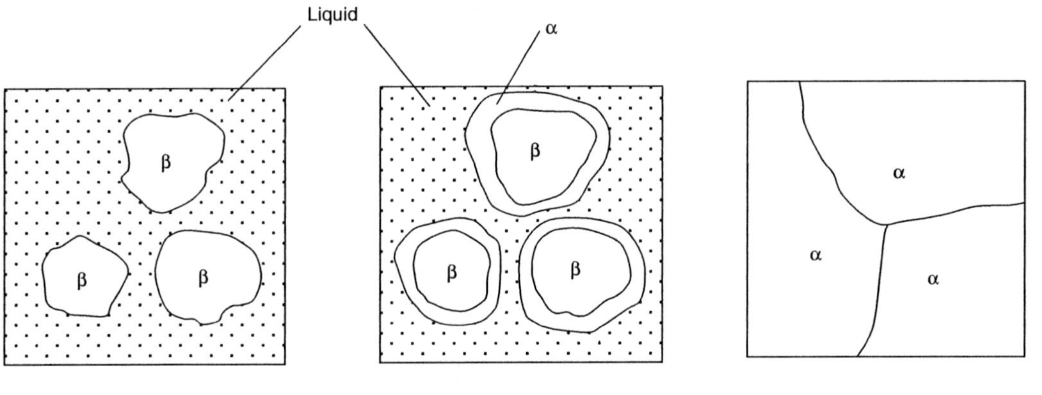


**Fig. 6.17** Schematic of peritectic freezing. Source: Ref 1

diagram (Fig. 6.19) into two separate eutectic reactions that may be analyzed separately. In addition, since $Mg_2Pb$ has a fixed composition, it melts at a constant temperature like a pure metal.

The copper-zinc phase diagram shown in Fig. 6.20 contains four intermediate phases. The copper-rich α solid solution and the zinc-rich η solid solution are the two terminal phases. The intermediate phases are β, γ, δ, and ε.

## 6.8 Solid-State Reactions

Similar to the eutectic and peritectic reactions that describe the decomposition of a liquid into a solid solution, the same types of reactions occur in the solid state. These solid-state reactions are termed eutectoid and peritectoid:

Solid $\gamma \rightarrow$ Solid $\alpha +$ Solid $\beta$ (Eutectoid reaction) (Eq 6.9)

Solid $\alpha +$ Solid $\beta \rightarrow$ Solid $\gamma$ (Peritectoid reaction) (Eq 6.10)

Solid-state reactions differ in two important aspects from liquid reactions in the manner in which they attain the equilibrium conditions predicted by the phase diagram:

- Solid-state reactions occur much more slowly, are subject to greater undercooling, and rarely attain true equilibrium conditions.
- Solid phases consist of atoms arranged in certain crystal structures, and new solid phases forming out of an existing solid phase tend to take definite positions with respect to the existing crystal structure. In other words, the crystal structure of the new phase has a definite orientation relationship to the crystal structure of the phase from which it formed.

The various phase reactions are summarized in Table 6.1.

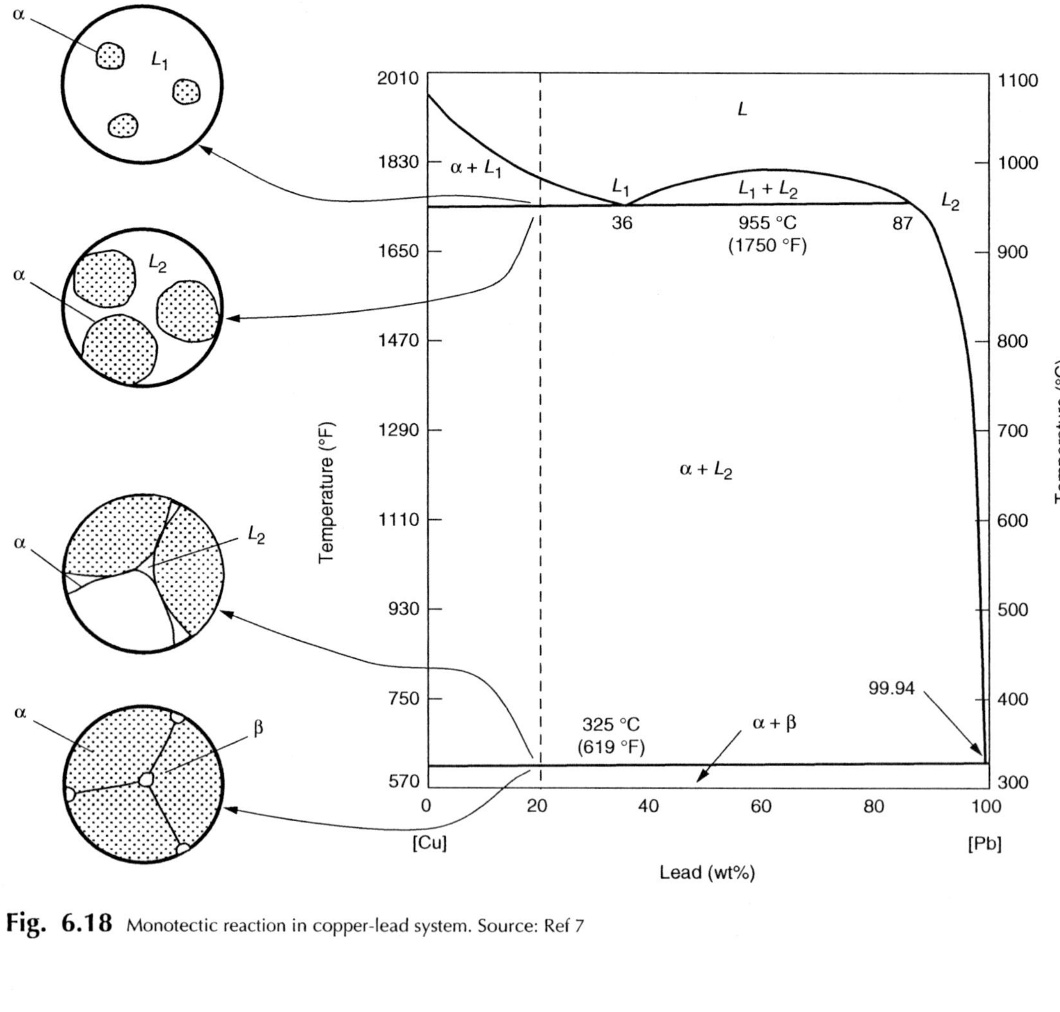


**Fig. 6.18** Monotectic reaction in copper-lead system. Source: Ref 7

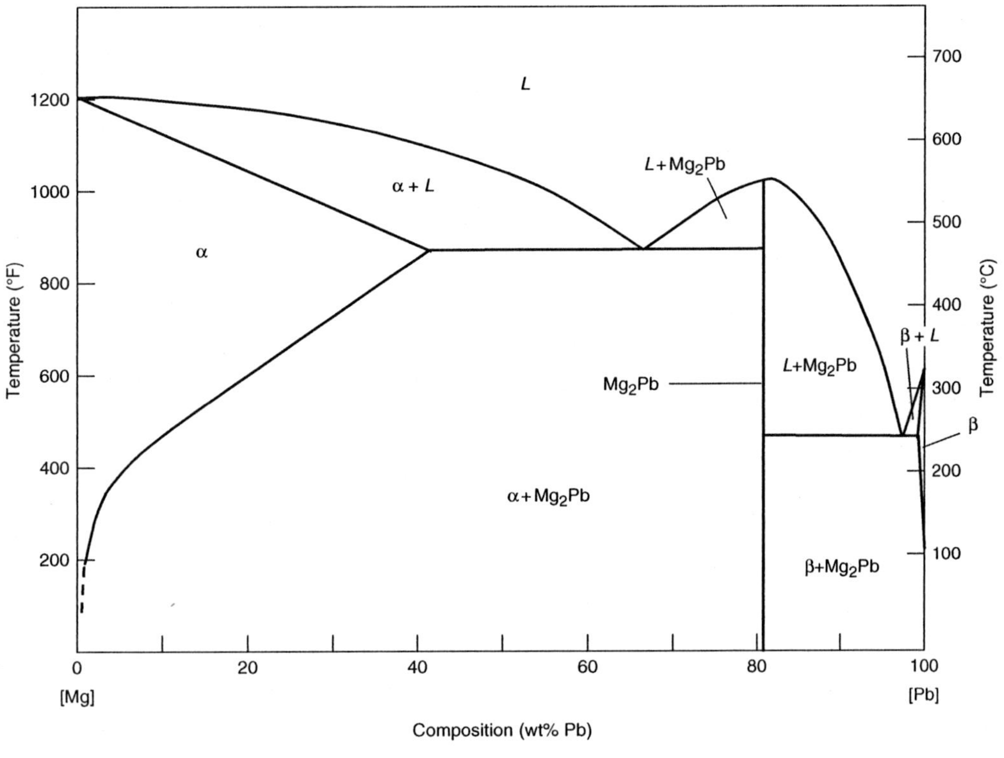


**Fig. 6.19** Compound formation in magnesium-lead system

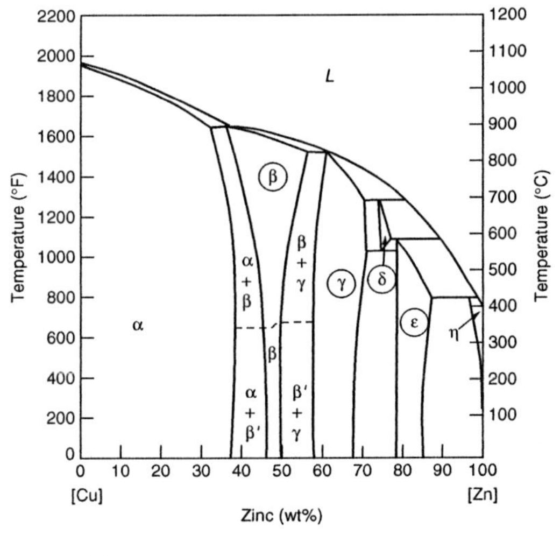


**Fig. 6.20** Copper-zinc phase diagram

### *6.8.1 Eutectoid Reaction*

The fact that steel can be hardened is a direct result of the eutectoid reaction in iron-carbon alloys. Just as a eutectic reaction involves the decomposition of a liquid solution, so does a eutectoid reaction involve the decomposition of solid solution into two other solid phases. In the iron-carbon system (Fig. 6.21), this decomposition involves the solid-solution austenite (γ), which decomposes into a mixture of ferrite (α) and the intermetallic compound cementite ($Fe_3C$):

$$\gamma \rightarrow \alpha + Fe_3C \qquad \text{(Eq 6.11)}$$

At a temperature of 1000 °C (1830 °F), a phase analysis of an alloy containing the eutectoid composition (0.8 wt% C) gives:

Point: 0.8 wt% C alloy at 1000 °C (1830 °F)
Phases: Austenite (γ)
Compositions: 0.8 wt% C
Amounts: 100%

When the alloy is cooled past the eutectoid temperature (720 °C, or 1333 °F), ferrite and cementite form side by side within the austenite to create a nodule of pearlite, the eutectoid microconstituent, which is similar to the morphology shown in Fig. 6.12. As the austenite phase is consumed by the growth of many pearlite nodules, the nodules meet one another to form a lamellar array. A phase analysis just below the eutectoid temperature reveals:

Point: 0.8 wt% C at 700 °C (1290 °F)

Phases: Ferrite (α)

Compositions: 0.3% C

Amounts: 100% − 12% = 88%

Cementite ($Fe_3C$)

6.7 wt% $Fe_3C$

$$\text{wt\% } Fe_3C = \frac{0.80-0.03}{6.7-0.03} = 12 \text{ wt\%}$$

Since the corresponding microstructure is 100% pearlite, it follows that pearlite consists of 88 wt% ferrite and 12 wt% cementite. Due to the importance of the iron-carbon system, a much more thorough analysis of this system is made in Chapter 10, "The Iron-Carbon System," in this book.

## 6.9 Ternary Phase Diagrams

Although binary phase diagrams are invaluable in metallurgy, engineering alloys usually contain more than two alloying elements. This section addresses ternary phase diagrams in which three alloying elements are present. Since it is not possible to represent three alloying elements as a function of temperature in two dimensions, one of the most useful ways of presenting the data is to take isothermal sections. In a typical isothermal section, such as the one shown in Fig. 6.22, each vertex of the triangle represents 100% of one component, such as metal *A*. The base of the triangle opposite vertex *A* represents 0%, and lines parallel to this base represent varying percentages of *A*. Thus, the point representing alloy 1 lies 20% of the distance from the base to vertex *A*. Similar analysis shows that alloy 1 contains 60% *B* and 20% *C*.

An isothermal section of a ternary phase diagram containing a two-phase solid solution field is shown in Fig. 6.23. The composition of

**Table 6.1 Phase reactions**

| Reaction | Equation | Phase diagram characteristic |
|---|---|---|
| Eutectic | $L \rightarrow \alpha + \beta$ | 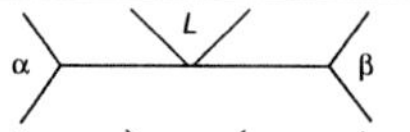 |
| Eutectoid | $\alpha \rightarrow \beta + \gamma$ | 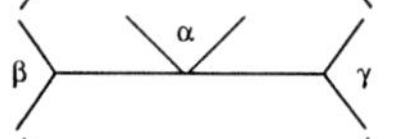 |
| Peritectic | $\alpha + L \rightarrow \beta$ | 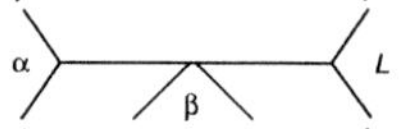 |
| Peritectoid | $\alpha + \beta \rightarrow \gamma$ | 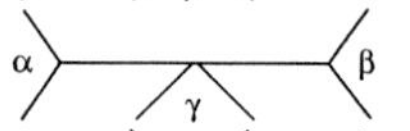 |
| Monotectic | $L_1 \rightarrow \alpha + L_2$ | 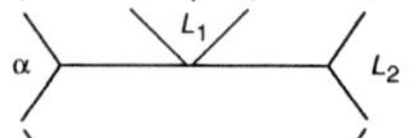 |
| Syntectic | $L_1 + L_2 \rightarrow \alpha$ | 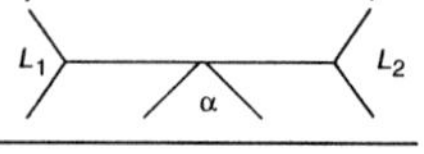 |

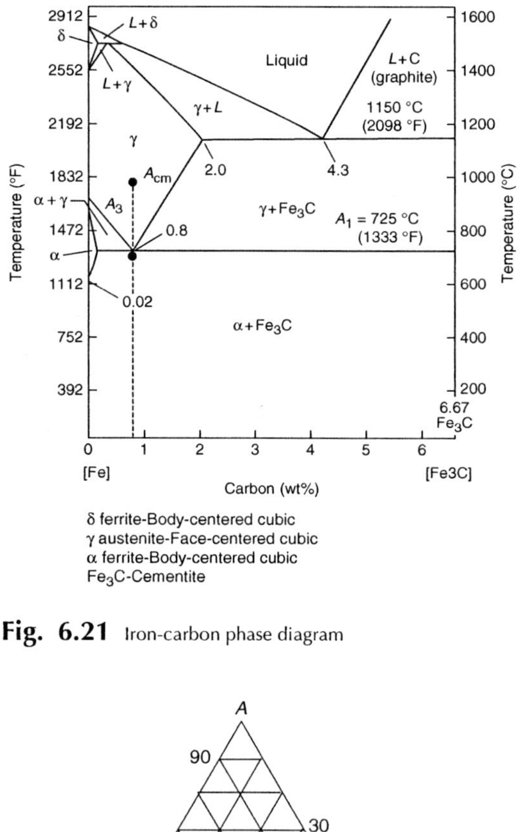


**Fig. 6.21** Iron-carbon phase diagram

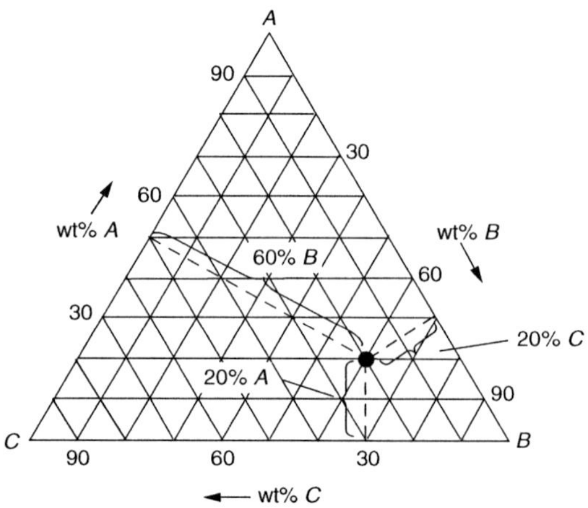


**Fig. 6.22** Method of plotting compositions in ternary alloy system. Source: Ref 1

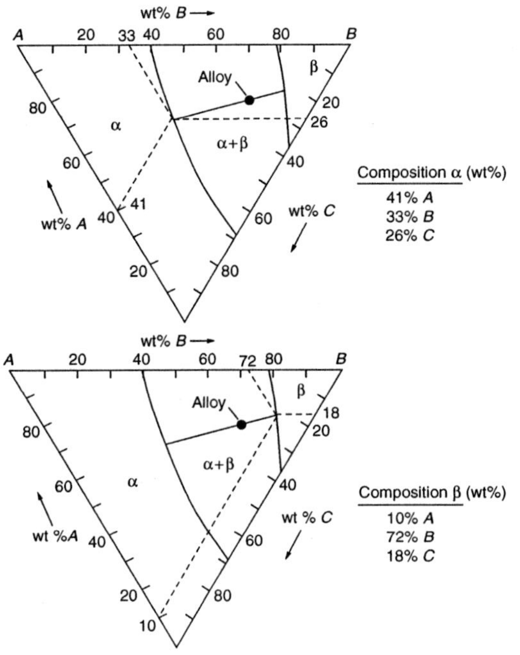


**Fig. 6.23** Determination of phase compositions for ternary phase diagram. Source: Ref 1

the alloy represented by the bullet in the two-phase field is calculated. Construction lines are shown in the upper part of the figure for calculating the phase composition of α and in the lower portion of the figure for calculating the phase composition of β. Note that all but the necessary portions of the phase composition lines have been removed for clarity. The calculations are essentially the same as those developed previously in Fig. 6.22. The phase compositions of α and β can be read from the figures and are:

- α: 41% *A*, 33% *B*, 26% *C*
- β: 10% *A*, 72% *B*, 18% *C*

For the amounts of α and β, the lever law can be used and the lengths measured along the % *B* composition scales, as shown in the construction in Fig. 6.24. Note that the tie lines in isothermal sections are not necessarily parallel to the phase boundaries and must be determined experimentally. The amounts of α and β are:

$$\text{wt\% } \beta = \frac{60-33}{72-33} = \frac{27}{39} = 69 \text{ wt\%}$$

$$\text{wt\% } \alpha = 100\% - 69\% = 31 \text{ wt\%}$$

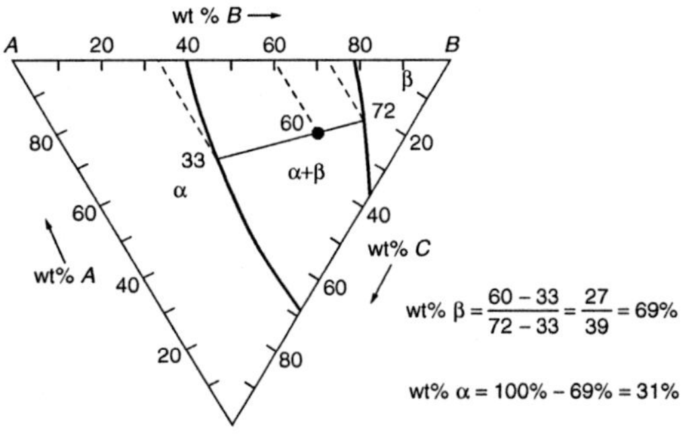


**Fig. 6.24** Determination of phase amounts for ternary phase diagram. Source: Ref 1

## REFERENCES

1. A.G. Guy and J.J. Hren, *Elements of Physical Metallurgy,* 3rd ed., Addison-Wesley Publishing Company, 1974
2. W.D. Callister, *Fundamentals of Materials Science and Engineering,* 6th ed., John Wiley & Sons, Inc., 2003
3. D. Aliya and S. Lampman, Physical Metallurgy Concepts in Interpretation of Microstructures, *Metallography and Microstructures,* Vol 9, *ASM Handbook,* ASM International, 2004
4. E.L. Rooy and A. Kearney, Aluminum Foundry Products, *Properties and Selection: Nonferrous Alloys and Special-Purpose Materials,* Vol 2, *ASM Handbook,* ASM International, 1990
5. W.C. Winegard, *An Introduction to the Solidification of Metals,* Institute of Metals, London, 1964
6. H. Baker, Introduction to Alloy Phase Diagrams, *Alloy Phase Diagrams,* Vol 3, *ASM Handbook,* ASM International, 1992
7. D.R. Askeland, *The Science and Engineering of Materials,* 2nd ed., PWS-Kent Publishing Co., 1989

## SELECTED REFERENCES

- R.A. Higgins, *Engineering Metallurgy—Applied Physical Metallurgy,* 6th ed., Arnold, 1993.
- R.E. Reed-Hill and R. Abbaschian, *Physical Metallurgy Principles,* 3rd ed., PWS Publishing Company, 1991

# CHAPTER 7

# Solidification and Casting

ALMOST ALL METALS AND ALLOYS are produced from liquids by solidification. Sometimes, the liquid metal is poured in a mold with a shape that is close to the desired final shape, a process called casting. The cast part then requires little or no final machining prior to use. On the other hand, simple castings called ingots are initially cast and then shaped by combinations of hot and cold working into wrought product forms, such as plate, sheet, or bar stock. For both castings and wrought products, the solidification process has a major influence on both the microstructure and mechanical properties of the final product.

## 7.1 The Liquid State

The liquid state is intermediate between a gas and a crystalline solid. Liquids possess neither the long-range order of solids nor the complete disorder of gases. Liquids do, however, possess short-range order. In fact, the structure of liquid metals is very close to the solid that freezes from it. For close-packed metals with the face-centered cubic and hexagonal close-packed crystalline structures, melting causes a volume decrease of only 3.5 to 6% and even less in the more loosely packed body-centered cubic metals, which usually exhibit a volume decrease of only 1 to 3% on freezing. Some elements with even looser atomic packing, such as silicon, actually expand on freezing. In addition, the latent heat of fusion is only approximately 3 to 4% of the corresponding latent heat of vaporization. The small amount of volume change and the relatively small latent heats of fusion are evidence that the atomic bonding in liquids and solids is very similar.

X-ray diffraction studies of liquid metals indicate that the atoms in liquids are arranged in an orderly manner over short distances but lack long-range order. This is probably due to the presence of increased amounts of defects, such as vacancies, interstitials, and dislocations. This large population of defects is believed to be responsible for the much higher rates of diffusion in liquid metals. The higher diffusion rates lead to a picture of liquid metals in which the structure is continuously changing. Atoms form clusters with a high degree of short-range order that may exist for a moment and then fall apart to appear again in another location. As the liquid is cooled toward its freezing point, the degree of short-range order increases, and the ordered clusters become larger and more stable.

## 7.2 Solidification Interfaces

The solidifying solid-liquid interface can exhibit one of three types of interfacial growth in the liquid: planar, cellular, or dendritic. As shown in Fig. 7.1, the type of growth is controlled by the manner in which heat is removed from the system. When the liquid ahead of the solid-liquid interface, $x_0$, has a positive temperature gradient, heat is removed from the liquid by conduction through the growing solid. Since the temperature gradient is linear and uniform perpendicular to the interface, a smooth interface is maintained, and the growth is planar into the liquid (Fig. 7.1a). When there is a temperature inversion and the temperature decreases ahead of the solid-liquid interface, then either cellular or dendritic growth will occur (Fig. 7.1b, c). The difference between the two is a matter of degree; small undercoolings will tend to produce cellular growth, while large undercoolings will tend to produce dendritic growth. In pure metals, undercooling can result from thermal supercooling, while in alloys, it can result from a combination of thermal and constitutional supercooling.

**Thermal Supercooling.** Cellular growth, as illustrated in Fig. 7.2, occurs when the advancing planar solid-liquid interface becomes unstable, and a small spike appears on the

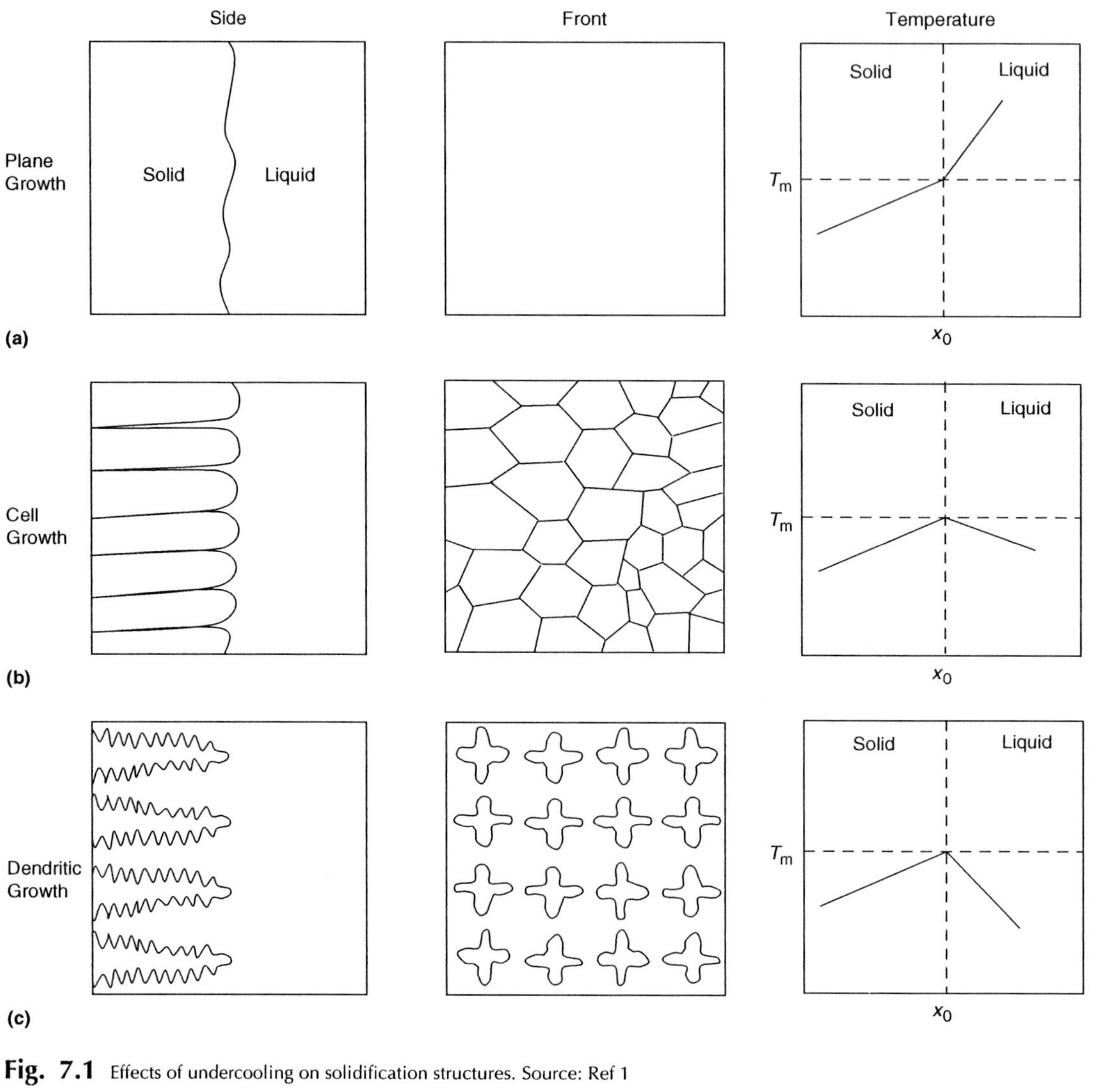


**Fig. 7.1** Effects of undercooling on solidification structures. Source: Ref 1

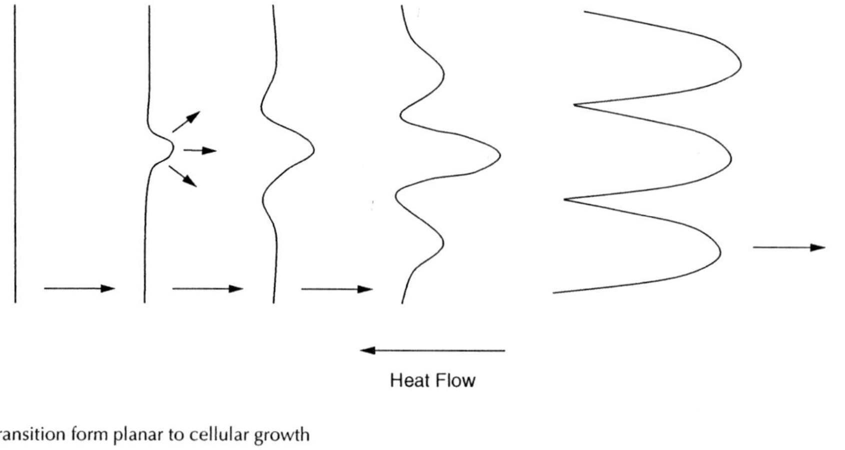


**Fig. 7.2** Transition form planar to cellular growth

interface that then grows into a cellular-type structure. The planar surface becomes unstable, because any part of the interface that grows ahead of the remainder enters a region in the liquid that is at a lower temperature. The initial spikes that form remain isolated at first because, as they grow by solidification, they release their latent heat of fusion into the adjacent liquid, causing a localized increase in the temperature. Consequently, parallel spikes of almost equal spacing advance into the liquid.

Dendritic growth is a further manifestation of cellular growth in which the spikes develop side protrusions. At still higher undercooling and higher growth velocities, the cells grow into rapidly advancing projections, sometimes of complex geometry. Their treelike forms (Fig. 7.3) have given them the name dendrites, after the Greek word *dendros* for tree. The secondary arms of dendrites develop perpendicular to the primary arms because, as the primary arm solidifies and gives off its latent heat of reaction, the temperature immediately adjacent to the primary arm increases. This creates another temperature inversion in the liquid between the primary arms, so secondary arms shoot out in that direction. A similar argument can be made for the formation of the tertiary arms. The spacing of the secondary arms is proportional to the rate at which heat is removed from the casting during solidification, with faster cooling rates producing smaller dendrite arm spacings. Dendrites start as long, thin crystals that grow into the liquid and thicken. The change in interface morphology of a succinonitrile-4% acetone solution with increasing solidification velocity is shown in Fig. 7.4 which demonstrates the evolution of a dendritic structure.

**Constitutional Supercooling.** While dendrites can form to a limited extent in pure metals due to temperature inversions, they are more prevalent in alloys because of the additional undercooling due to constitutional supercooling. Constitutional supercooling arises because of the segregation of alloying elements ahead of the solid-liquid front. The extra concentration of alloying elements reduces the melting point of the liquid. If this reduction is sufficient to reduce the melting point to below the actual temperature at that point, then the liquid is said to be locally constitutionally supercooled; that is, it is

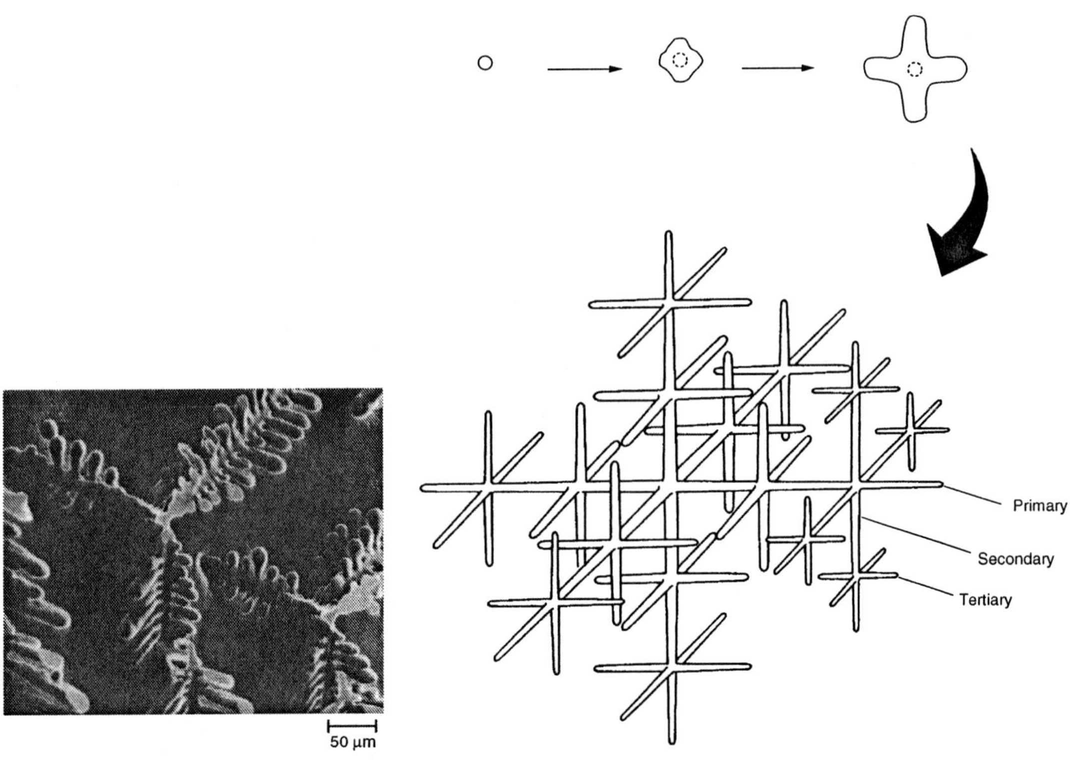


**Fig. 7.3** Dendrite formation and dendritic microstructure in a Cu-10at.%Co alloy. Source Ref 2

effectively undercooled because of a change in the constitution of the liquid.

## 7.3 Solidification Structures

A metal cast into a mold can have up to two or three distinct zones: a chill zone, a zone containing columnar grains, and a center-equiaxed grain zone (Fig. 7.5). It should be noted that all three zones do not always occur. For example, pure metals can exhibit a chill zone and a columnar zone but not contain a center-equiaxed zone.

**Chill Zone.** Because the mold is cooler than the metal, nucleation will occur over the interior surface of the mold. Since the mold wall is much cooler than the liquid metal, the nucleation rate is high, and thus, the average grain size is small. Each nucleation event produces an individual crystal, or grain, that grows dendrictically in a direction roughly perpendicular to the mold wall until it impinges on other grains. As a large amount of latent heat of fusion is released from the solidifying grains and as the superheat of the liquid is dissipated, the rate of growth decreases. The chill zone grains are oriented randomly with respect to the mold; that is, the major axis of each grain is randomly oriented.

**Columnar Zone.** Inside of the chill zone, there are a series of columnar, or column-shaped, grains that are oriented almost parallel to the heat flow direction. Because each metal grows more favorably in one principal crystallographic direction, only those grains favorably oriented with their growth direction most perpendicular to the mold wall will grow into the center of the casting. The axes of the columnar grains are parallel to the direction of heat flow, and they grow along specific crystalline planes.

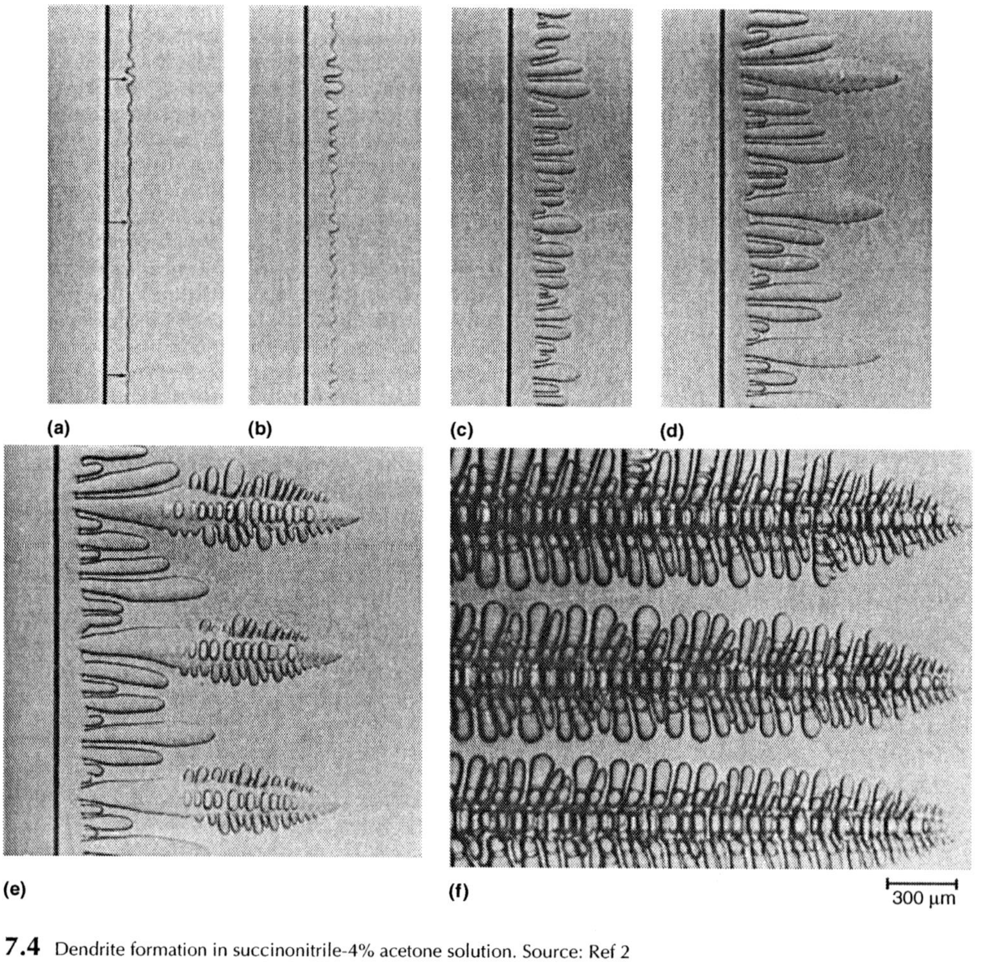


**Fig. 7.4** Dendrite formation in succinonitrile-4% acetone solution. Source: Ref 2

As they grow, the more perfectly oriented grains grow ahead of the less perfectly oriented grains and crowd them out (Fig. 7.6). The fact that the grains most favorably oriented to the mold wall grow the fastest means that the final shape of the grains in a pure metal casting is columnar, with the parallel columns growing progressively from the mold wall into the center of the casting. The final columnar structure results from parallel growth of different colonies of dendrites and the gradual lateral growth, or thickening, between them.

**Center Equiaxed Zone.** A third region at the center of some alloy castings consists of smaller grains that are randomly oriented and nearly equiaxed. As freezing progresses, the thermal gradient decreases, and this causes the dendrites to become very long. Breakdown of columnar growth may occur as a result of fracturing of the very long dendrite grains by convection currents in the melt. These broken arms (Fig. 7.7) can then serve as nuclei for new grains. Another possibility is that new grains nucleate as a result of a low thermal gradient and segregation that is occurring during freezing. At low casting temperatures, the entire casting may solidify with an equiaxed structure.

The amount of the final cast structure that is columnar or equiaxed depends on the alloy composition and on the thermal gradient at the liquid-solid interface during solidification. A thermal gradient is most easily controlled by controlling the rate of heat extraction from the casting, or the cooling rate. Alloys that have a wide spread between the liquidus and the solidus temperatures solidify with a mostly

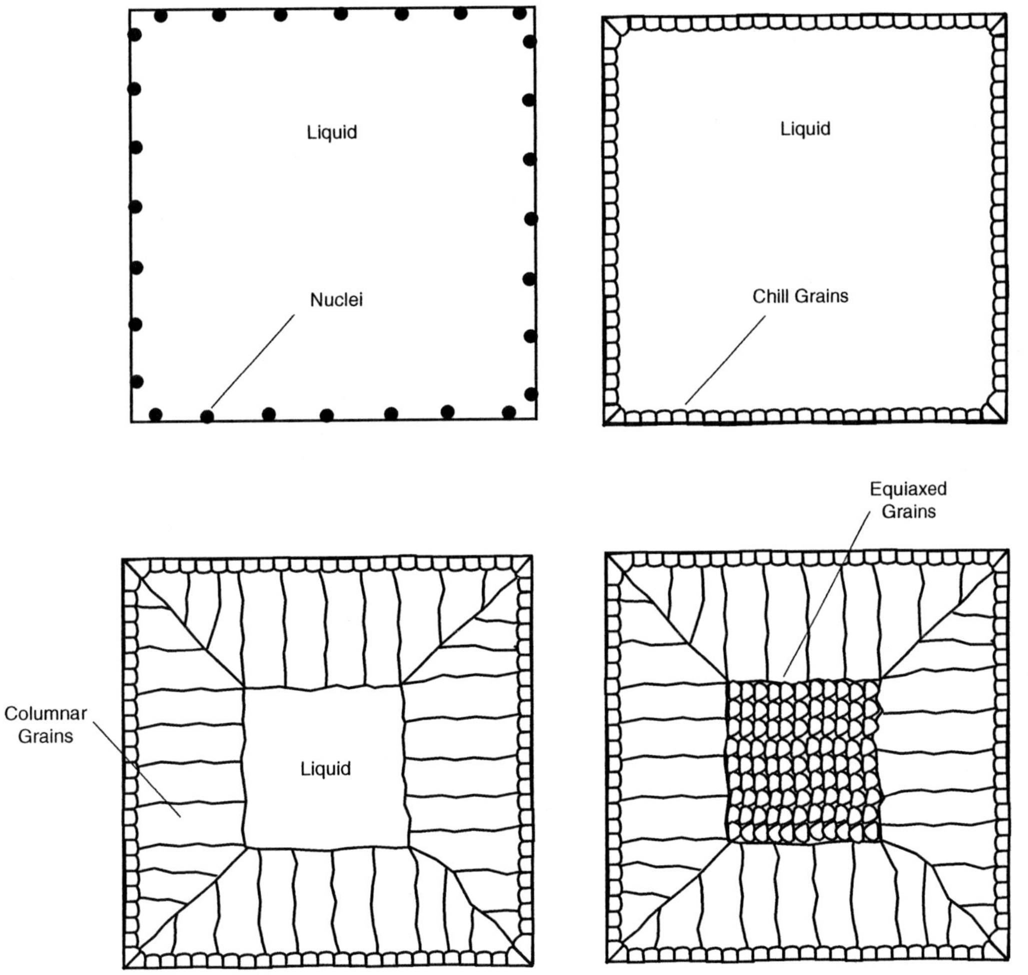


**Fig. 7.5** Freezing sequence for an alloy casting. Source: Ref 3

equiaxed grain structure at normal cooling rates, whereas alloys with small differences in solidus and liquidus temperatures solidify with a mostly columnar structure. High cooling rates encourage columnar solidification because they establish high thermal gradients at the

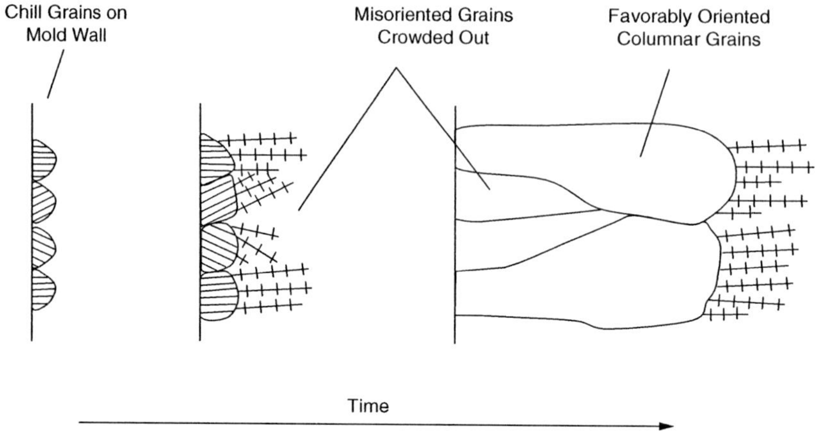


**Fig. 7.6** Columnar growth from mold wall. Source: Ref 4

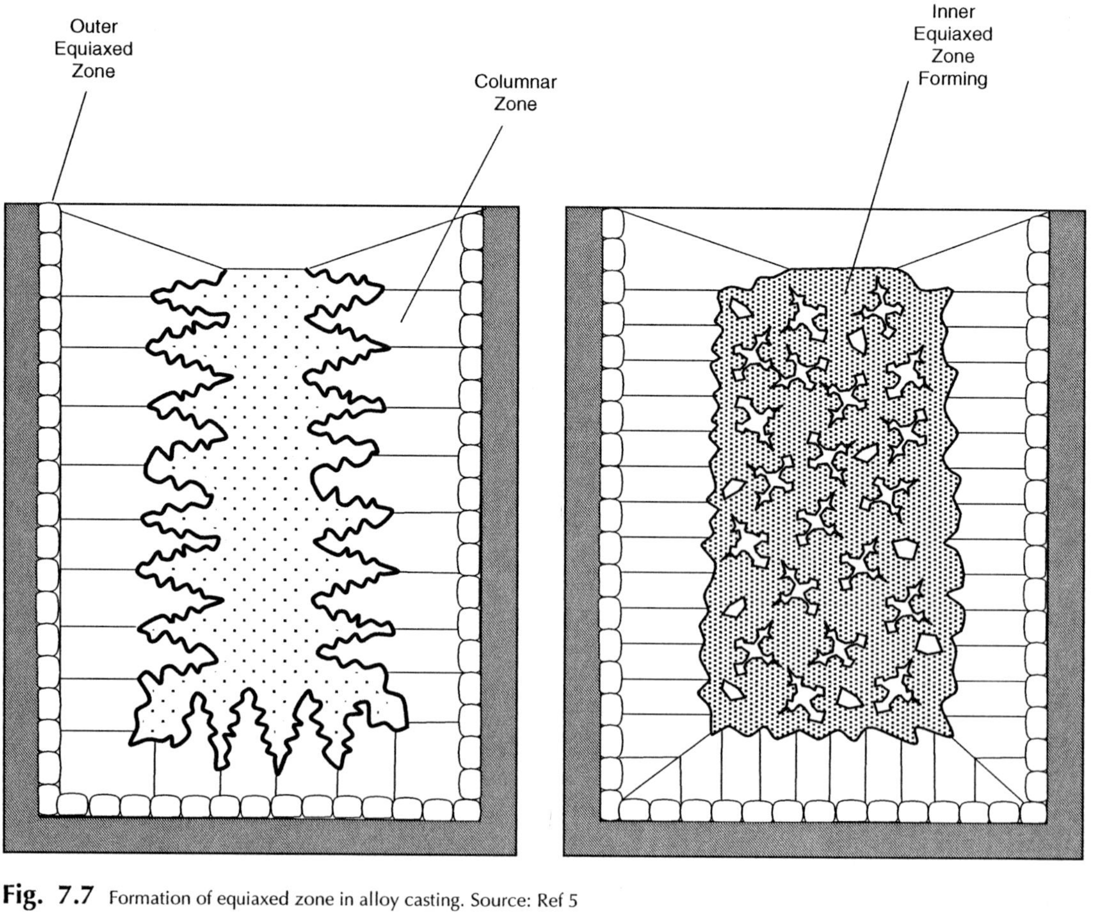


**Fig. 7.7** Formation of equiaxed zone in alloy casting. Source: Ref 5

liquid-solid interface. Low thermal gradients encourage equiaxed solidification.

## 7.4 Segregation

Segregation may be defined as any departure from a uniform distribution of the chemical elements in the alloy. Because of the way in which alloys partition on freezing, it follows that all castings are segregated to different degrees. Alloys can exhibit several types of segregation, and more than one type can be present in any given casting.

**Normal segregation** is the result of rejection of the alloy solute at an advancing interface because it is more soluble in the liquid than the solid. As freezing progresses, there is a buildup of the solute in the liquid that freezes last, such as at the center of the casting. Such long-range variations in composition are called macrosegregation. Normal segregation frequently occurs when the direction of growth is inward, as in columnar growth.

An examination of the relevant portion of a phase diagram (Fig. 7.8) can be used to show the effects of segregation during freezing. The original melt of composition, $C_o$, starts to freeze at the liquidus temperature, $T_L$. The first solid to appear has the composition $kC_o$, where $k$ is known as the partition coefficient. The partition coefficient usually has a value less than 1. For example, for $k = 0.5$, the first solid has only 50% of the concentration of alloying element compared to the original melt; the first metal to appear is therefore usually rather pure. In general, $k$ defines how the solute alloy partitions between the solid and liquid phases. Thus:

$$k = C_S / C_L \qquad \text{(Eq 7.1)}$$

where $C_S$ is the solute fraction by weight in solid, and $C_L$ is the solute fraction by weight in liquid.

For solidus and liquidus lines that are straight, $k$ is constant for all compositions. However, even where they are curved, the relative matching of the curvatures often means that $k$ is still reasonably constant over wide ranges of composition. When $k$ is close to 1, the close spacing of the liquidus and solidus lines indicates little tendency toward segregation. When $k$ is small, then the wide spacing of the liquidus and solidus lines indicates a strongly partitioning alloying element and a tendency to segregate on solidification.

On forming the solid that contains only $kC_o$ amount of alloying element, the alloying element remaining in the liquid must be rejected ahead of the advancing front. Thus, although the liquid ahead of the front was initially of composition $C_o$, after an advance of approximately a millimeter or so, the composition ahead of the front builds up to a peak value of $C_o/k$.

**Gravity segregation** is another type of macrosegregation that occurs due to differences in density in the liquid melt. Denser constituents tend to sink toward the bottom, while lighter ones float toward the top.

**Microsegregation.** As the dendrite grows into the melt and as secondary arms spread from the main dendrite stem, solute is rejected, effectively being pushed aside to concentrate in the tiny regions enclosed by the secondary dendrite arms. This gradation of composition from the inside to the outside of the dendrite is called coring, because on etching a polished section of such dendrites, the progressive change in composition appears as onionlike layers around a central core.

As an example of coring produced by nonequilibrium freezing, consider the 70%Ni-30%Cu alloy in Fig. 7.9 that is rapidly cooled from a temperature $T_0$. The first solid forms at temperature $T_1$ with a composition of $\alpha_1$. On further rapid cooling to $T_2$, additional layers of composition $\alpha_2$ form. The overall composition at $T_2$ lies somewhere between $\alpha_1$ and $\alpha_2$ and is designated as $\alpha'_2$. Since the tie line at $\alpha'_2\ L_2$ is longer than $\alpha_2\ L_2$, there will be more liquid and less solid in the rapidly cooled alloy than if it had been slowly cooled under equilibrium conditions. As rapid cooling continues through $T_3$ and $T_4$, the same processes occur, and the average composition follows the nonequilibrium solidus determined by points $\alpha_1$, $\alpha'_2$, $\alpha'_3$ ... $\alpha'_7$. At $T_7$, freezing is complete, and the average composition of the alloy is 30% Cu. The microstructure consists of regions varying from $\alpha_1$ to $\alpha'_7$, where the core, or center, has a composition of $\alpha_1$.

**Inverse segregation** occurs when the solute content at or near the surface is higher than it is at the center—exactly opposite of what is expected. Under certain conditions, channels form between the dendrites that are growing toward the center of the casting, forming a path in which the last-to-freeze solute-enriched liquid can flow back from the center to the outside. This reverse flow path could be caused by differential contraction between the mold and the ingot, creating a suction effect that pulls the solute-enriched liquid toward the surface. Another possible mechanism is that a large amount of gas evolving at the end of solidification pushes the solute-enriched liquid through the dendritic channels toward the surface.

It should be recognized that this discussion on segregation is rather cursory. An examination of the segregation patterns in the aluminum-killed steel ingot of Fig. 7.10 shows the complexity of segregation that can be observed in real castings. While ingot homogenization treatments can be effective in smoothing out microsegregation, they are not very effective for macrosegregation. Since macrosegregation occurs over distances ranging from 1 to 100 cm (0.4 to 40 in.), the diffusion lengths are just too long, and the times required are prohibitive. Segregation is one of the prime reasons that wrought products are often preferred over castings. The severe plastic deformation that occurs during hot working will heal internal porosity and help break up and eliminate segregated areas.

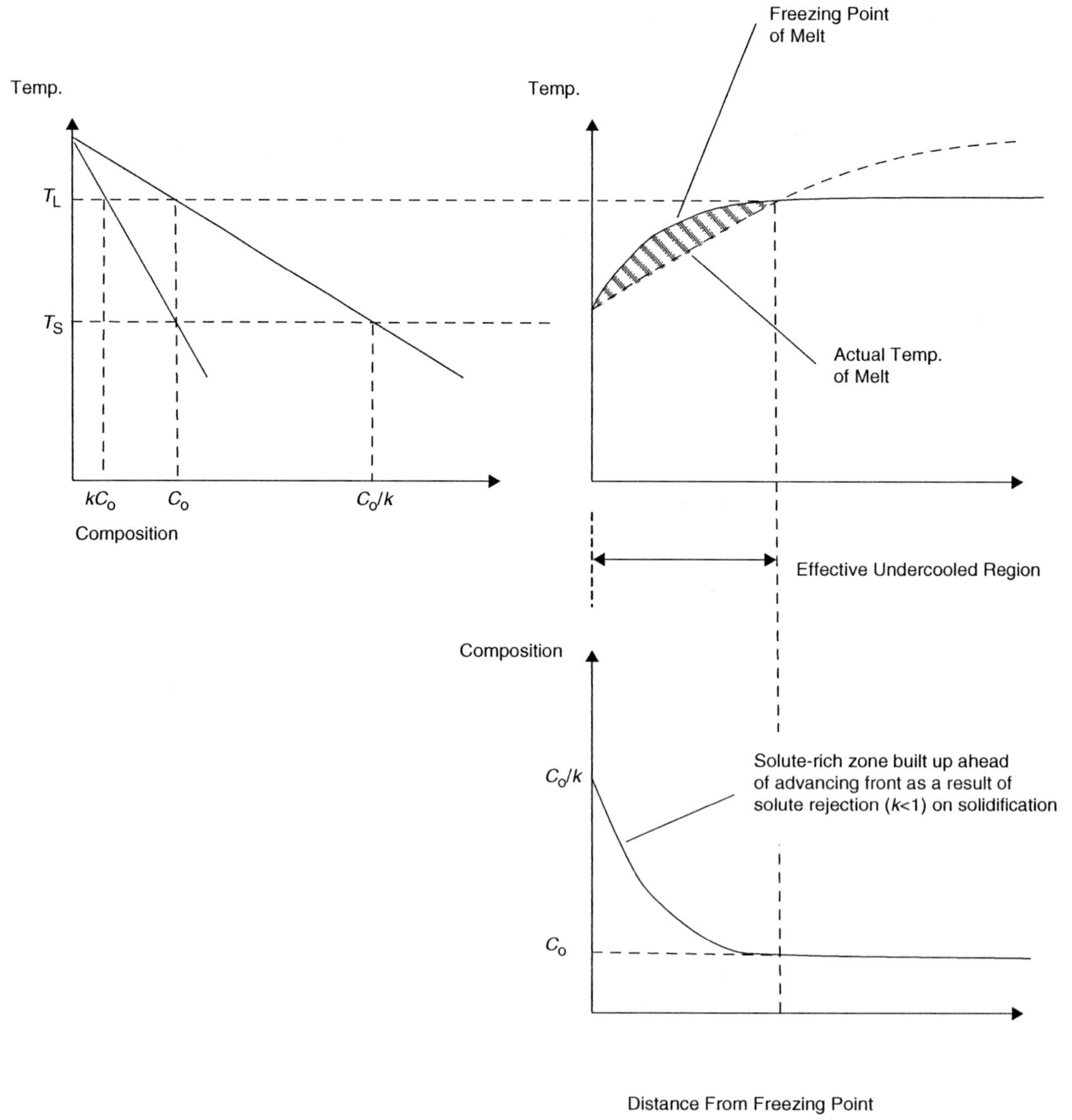


**Fig. 7.8** Constitutional supercooling. Source: Ref 1

## 7.5 Grain Refinement and Secondary Dendrite Arm Spacing

Rapid solidification rates are usually desirable because they tend to produce finer grain sizes and finer dendrites. Grain size control for castings is important because fine grain sizes result in higher strengths and greater ductility. Since the size of porosity in a casting often corresponds somewhat with grain size, finer porosity also goes with finer grain sizes. In addition, shrinkage and hot cracking are more prevalent in castings with a coarse grain size. Grain size is a

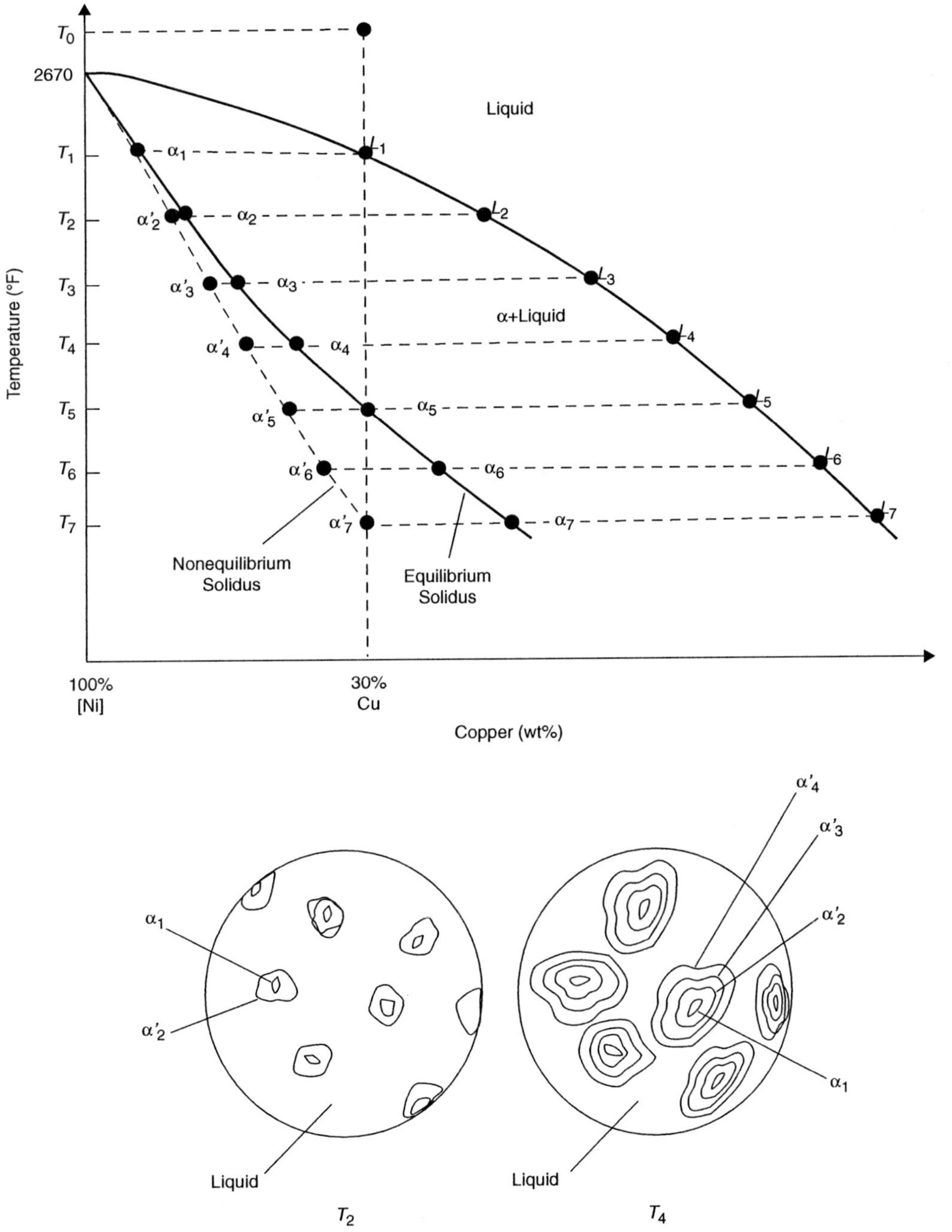


**Fig. 7.9** Coring due to nonequilibrium solidification. Source: Ref 6

function of pouring temperature, solidification rate, and the presence or absence of a grain-refining agent. Low pouring temperatures, faster solidification rates, and grain refiners, such as titanium and boron in aluminum, produce finer grain sizes.

The solidification rate also affects the size of the dendrites that form during freezing. The dendrite size is normally measured as the distance between the secondary dendrite arms, referred to as the secondary dendrite arm spacing (SDAS). As shown in Fig. 7.11, faster solidification rates produce smaller SDAS. The SDAS is related to the solidification time, $t$, by:

$$\text{SDAS} = kt^n \quad \text{(Eq 7.2)}$$

where $k$ and $n$ are constants that depend on the composition of the metal. Small SDASs are associated with higher strengths and improved ductility (Fig. 7.12). Homogenization heat treatments are also dependent on the time, $t$, required to diffuse a solute over a given average distance, $x$. If the coefficient of diffusion in the solid is $D$, then the relationship between time and distance is approximately:

$$x = \sqrt{Dt} \quad \text{(Eq 7.3)}$$

Thus, finer SDAS results in shorter homogenization times or better homogenization in similar times.

## 7.6 Porosity and Shrinkage

Porosity is the presence of pores in the casting. These pores may be connected to the surface, or they may be wholly internal. While mild porosity does not normally degrade static properties, such as tensile strength, it can severely impact dynamic properties, such as fatigue strength. However, a reduction in static properties can occur if the porosity is extensive enough to significantly reduce the cross-sectional area or if it has sharp corners that produce stress

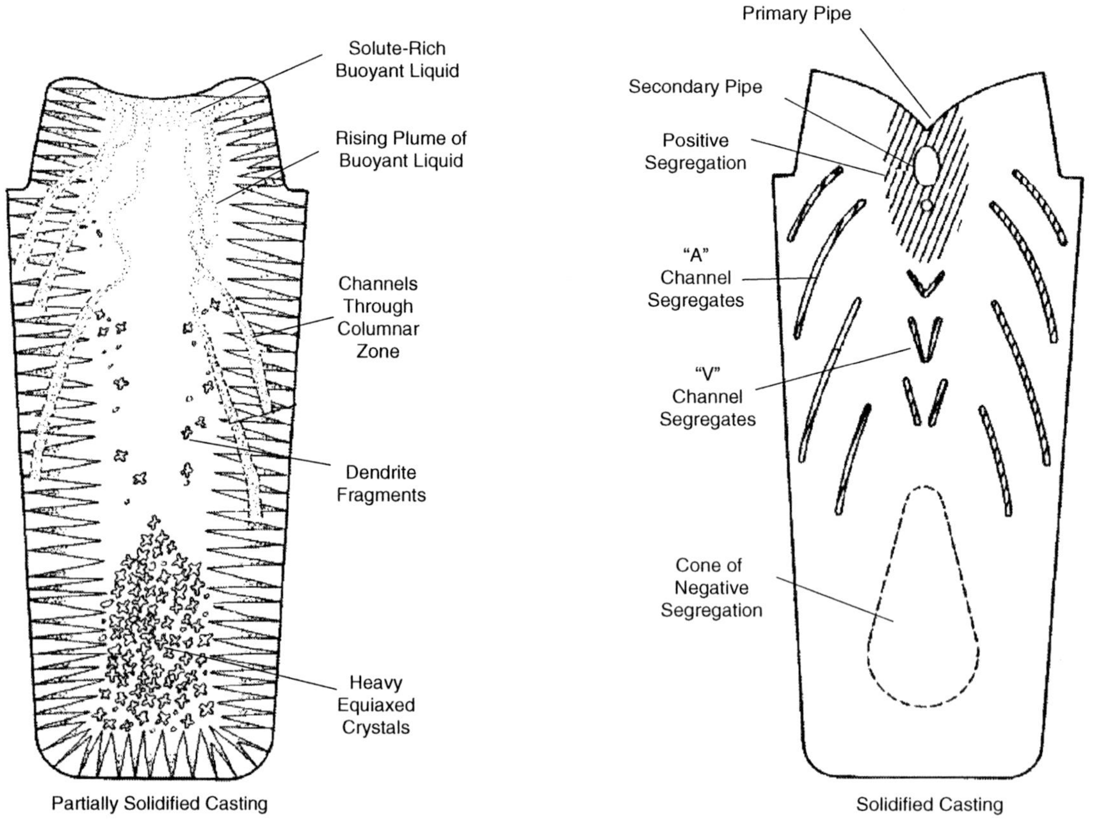


**Fig. 7.10** Segregation morphologies in aluminum-killed steel ingot. Source: Ref 1

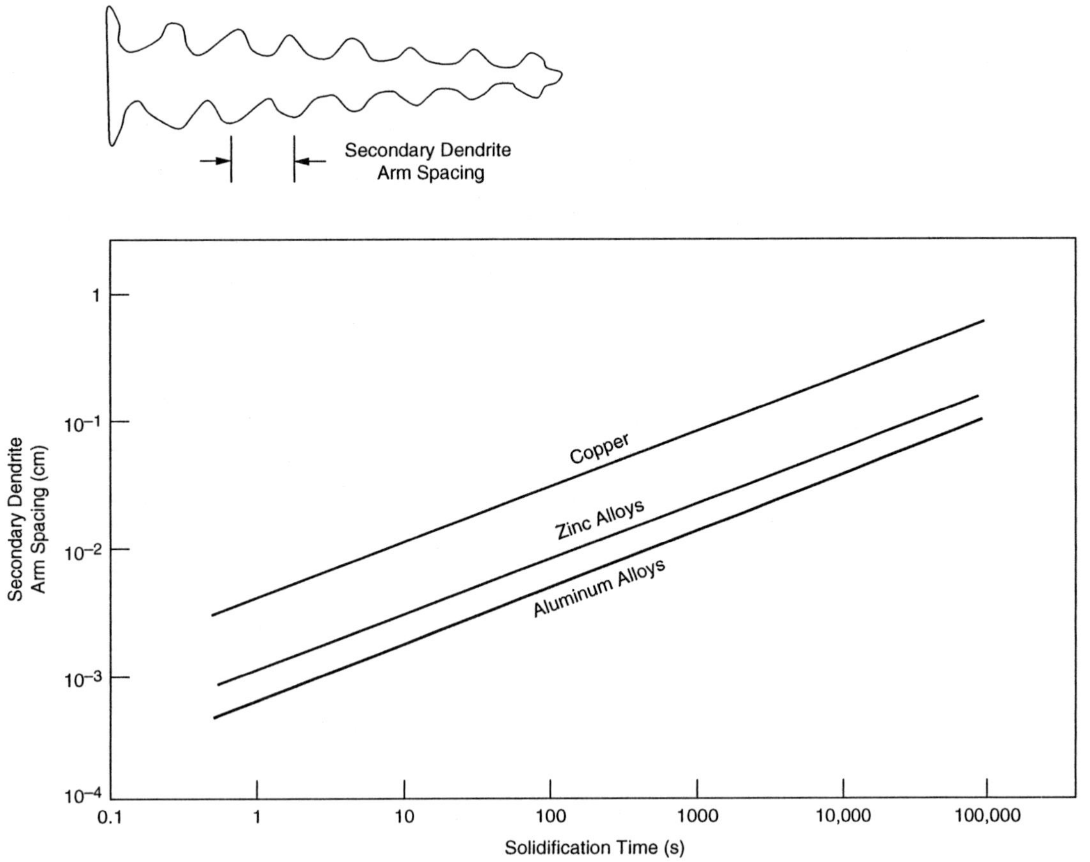


**Fig. 7.11** Effect of solidification time on secondary dendrite arm spacing. Source: Ref 3

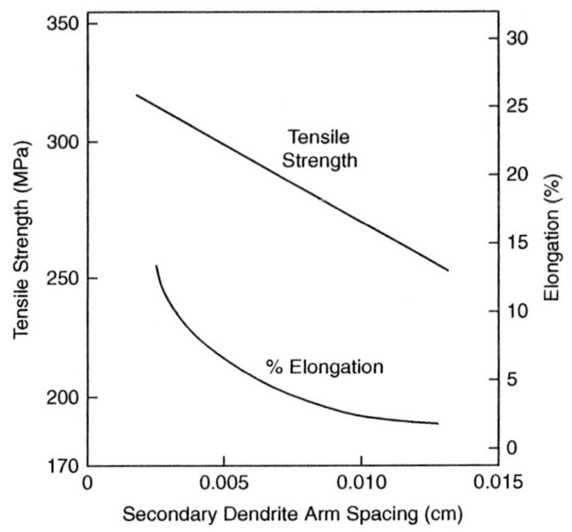


**Fig. 7.12** Effect of secondary dendrite arm spacing on mechanical properties of aluminum casting alloy. Source: Ref 3

concentrations. If the surfaces of small porosity in ingots are not oxidized, they generally will be welded shut during hot working, such as rolling or forging. Small porosity in finished castings must either be accepted or healed by processes such as hot isostatic pressing, where the metal is subjected to significant heat and pressure. Porosity that is open to the surface in finished castings is sometimes ground out and weld-repaired. Both gas evolution and metal shrinkage during casting can produce porosity.

**Gas porosity** is caused by the absorption of gases in the liquid metal prior to casting. The most troublesome gas is hydrogen, which usually comes either from moisture in the air or from burning fossil fuels for heating and melting. Moisture and liquid metal can react to form molecular hydrogen according to:

$$xM + yH_2O \rightarrow M_xO_y + yH_2 \qquad \text{(Eq 7.4)}$$

where M is the metal, and $x$ and $y$ are stoichiometric coefficients balancing the equation. At the high temperatures in the liquid melt, molecular hydrogen decomposes into atomic hydrogen that dissolves in the metal. The solubility of hydrogen in liquid metals increases sharply with temperature, as shown in Fig. 7.13 for nickel and iron. If the hydrogen is not removed prior to solidification, the large decrease in solubility during freezing results in the formation of porosity. One problem with gas porosity is the formation of blisters during subsequent processing, such as heat treating or annealing. The equilibrium gas concentration in a liquid metal can be determined by Sievert's law:

$$G = k\sqrt{p_g} \qquad \text{(Eq 7.5)}$$

where $G$ is the equilibrium gas solubility, $k$ is the equilibrium constant, and $p_g$ is the partial pressure of gas at the metal surface.

Hydrogen is often removed by vacuum degassing prior to casting. For example, premium-quality steels are commonly vacuum degassed. In addition, some metals, such as titanium, are so reactive that all melting and casting operations must be done under vacuum. In open-air casting operations, hydrogen is removed from the melt by using degassing fluxes. The removal of hydrogen is a mechanical, not chemical, process in which the hydrogen attaches itself to the fluxing gas.

The presence of nonmetallic inclusions has an effect on gas porosity formation. It has been shown that nucleation of pores is difficult in the absence of some sort of substrate, such as a nonmetallic inclusion, a grain refiner, or a second-phase particle. Investigations have shown that castings free of inclusions have less porosity than castings that contain inclusions.

As a metal casting cools and solidifies, the volume decreases, contributing to three types of shrinkage: liquid shrinkage, solidification shrinkage, and solid shrinkage. The shrinkage occurring in the liquid metal itself does not cause casting problems. The extra liquid metal required to compensate for this small reduction in volume is provided by a slight extension to the pouring time or by a slight fall in feeder level.

**Solidification shrinkage** occurs at the freezing point when the metal transitions from a liquid to a more highly ordered crystalline solid. The volume change on melting for a number of metals is given in Table 7.1. Solidification shrinkage is usually compensated for by designing the mold with reservoirs for liquid metal, called risers, which allow liquid metal to flow in and fill the areas being vacated by the solidifying metal. Shrinkage porosity often forms in areas that the liquid metal from the risers cannot reach. For example, it is difficult to effectively feed metal into the interdendritic

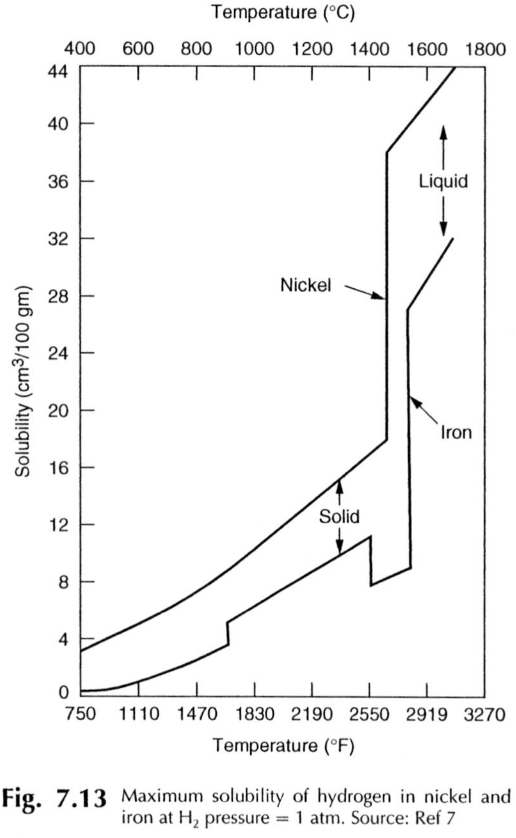


**Fig. 7.13** Maximum solubility of hydrogen in nickel and iron at $H_2$ pressure = 1 atm. Source: Ref 7

**Table 7.1 Volume changes on melting**

| Metal | Crystal structure(a) | Volume change on melting, % |
|---|---|---|
| Al | fcc | 6.0 |
| Au | fcc | 5.1 |
| Cu | fcc | 4.2 |
| Zn | hcp | 4.2 |
| Ag | fcc | 3.8 |
| Pb | fcc | 3.5 |
| Fe | bcc | 3.4 |
| Sn | bct | 2.8 |
| Nb | bcc | 0.9 |
| Sb | Rhombohedral | −0.95 |
| Si | Diamond cubic | −12 |

(a) fcc, face-centered cubic; hcp, hexagonal close-packed; bcc, body-centered cubic; bct, body-centered tetragonal areas where shrinkage is occurring. Because this type of porosity occurs late in solidification, especially in alloys with wide freezing ranges and a large mushy zone, it is particularly difficult to eliminate. A comparison of shrinkage porosity and gas porosity is shown in Fig. 7.14. Shrinkage porosity tends to follow the directions of the dendritic arms and forms along the grain boundaries, while gas porosity occurs throughout the matrix.

Solid shrinkage occurs after the casting has solidified and is cooling down to room temperature. As cooling progresses and the casting attempts to contract, it is constrained to some extent by the mold. This constraint always leads to the casting being somewhat larger than would be expected from free contraction alone. This can lead to difficulties in accurately designing the size of the pattern, since the contraction allowance is not easy to quantify. Mold constraint during cooling can also lead to problems such as hot tearing or cracking of the casting.

## 7.7 Casting Processes

Casting is a major industrial process. This section provides a brief overview of six of the most important of these processes:

- Sand casting
- Plaster mold casting
- Evaporative pattern casting
- Investment casting
- Permanent mold casting
- Die casting

### 7.7.1 Sand Casting

Sand casting is perhaps the oldest known casting process. The molten metal is poured into a cavity shaped inside a body of sand and allowed to solidify. Advantages of sand casting are low equipment costs, design flexibility, ability to cast complex geometries, producibility of large castings, applicability to a large number of casting alloys, and it is economical for low quantities. The biggest disadvantages are that the process does not permit close tolerances and produces rather rough surface finishes. It also requires relatively thick walls for the cast parts (~3 mm, or 0.120 in.), has somewhat lower mechanical properties due to larger grain sizes (resulting from slow cooling rates), and requires a rather high labor cost per part. The steps involved in sand casting are shown in Fig. 7.15 and consist of:

1. Fabricate a pattern, usually wood, of the desired part and split it down the centerline.
2. Place the bottom half of the pattern, called the drag, in a box called a flask.
3. Apply a release coating to the pattern, fill the flask with sand, and then compact the sand by ramming.
4. Turn the drag half of the mold over, and place the top half of the flask on top of it. The top half of the pattern, called the cope, is then placed over the drag half of the pattern and release coated.
5. Risers and a sprue are then installed in the cope half of the flask. The sprue is where liquid metal enters the mold. In a complex casting, the sprue is usually gated to different

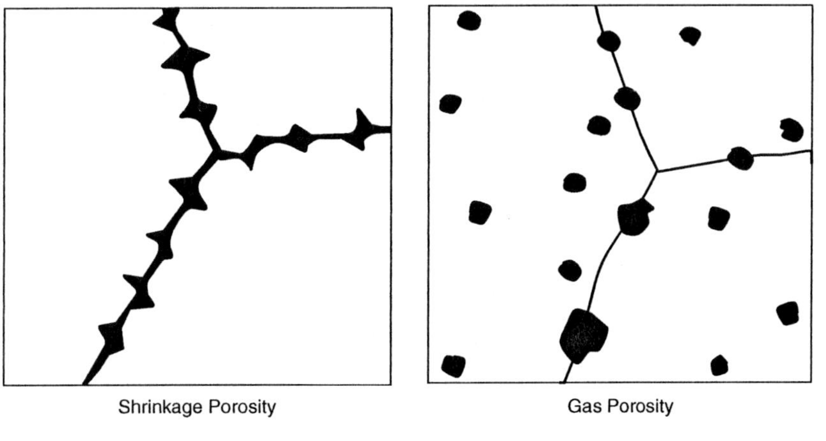


**Fig. 7.14** Porosity in cast metals. Source: Ref 8

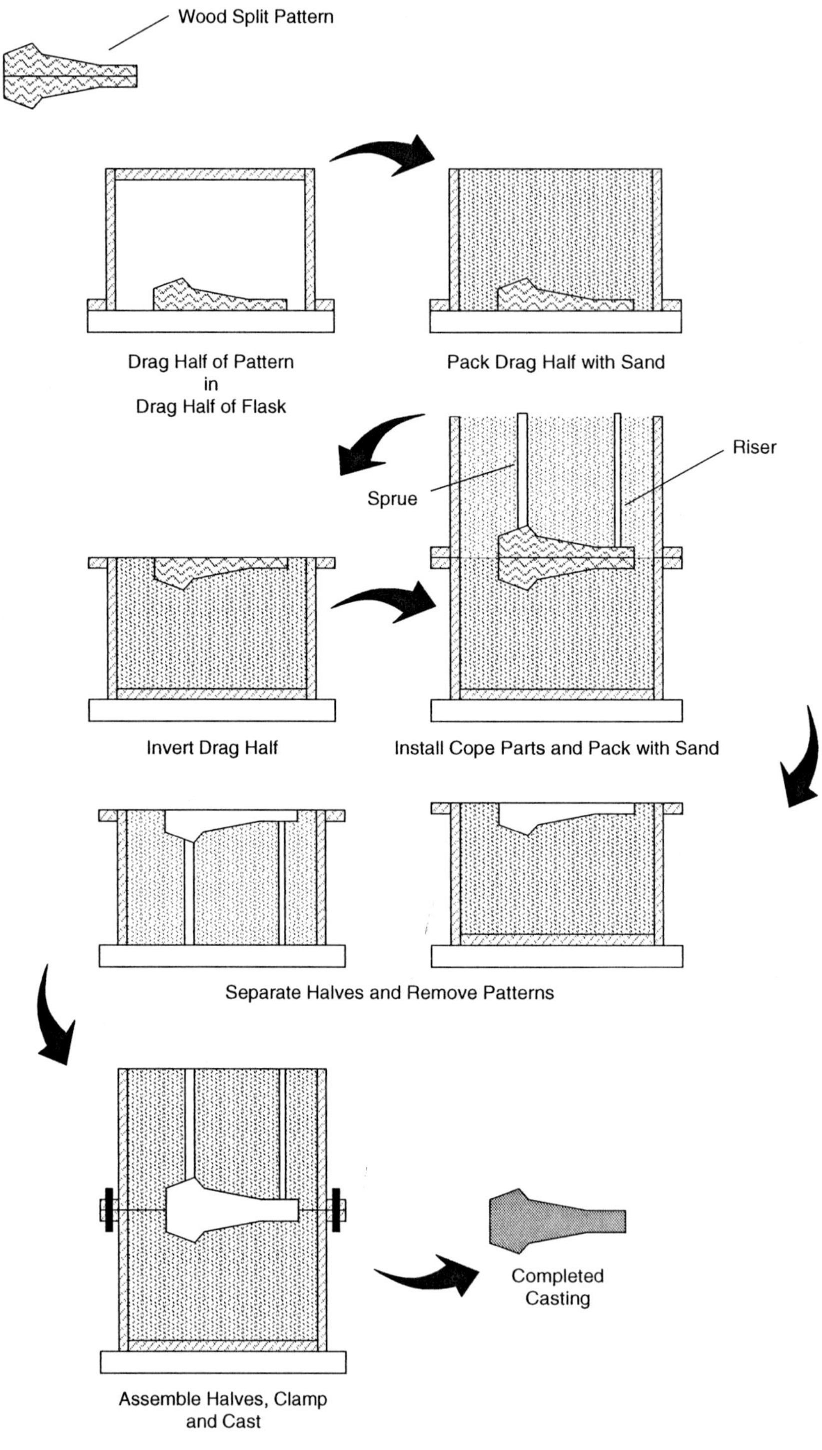


**Fig. 7.15** Sand casting process. Source: Ref 9

positions around the casting. The risers are essentially reservoirs for liquid metal that keep the casting supplied with liquid metal as the metal shrinks and contracts on freezing.
6. The cope half is then packed with sand and rammed.
7. The two halves are separated, and the patterns are removed. If hollow sections are required, a sand core is placed in the drag half of the mold. A gating system is then cut into the sand on the cope half of the mold.
8. The two halves are reassembled and clamped or bolted shut for casting.

Molding sands usually consist of sand grains, a binder, and water. The properties that are important are good flowability or the ability to be easily worked around the pattern, sufficient green strength, and sufficient permeability to allow gas and steam to escape during casting. Sand cores for molded-in inserts can be made using either heat-cured binder systems or no-bake binder systems. No-bake binder systems are usually preferred since they provide greater dimensional accuracy, have higher strengths, are more adaptable to automation, and can be used immediately after fabrication. The no-bake systems typically consist of room-temperature-curing sodium silicate sands, phenolic-urethanes, or furan acids combined with sand.

Gates are used to evenly distribute the metal to the different locations in the casting. The objective is to have progressive solidification from the point most distant from the gate toward the gate; that is, the metal in the casting should solidify before the metal in the gates solidifies. Normally, the area of the gates and runner system connecting the gates should be approximately four times larger than the sprue. When feeding needs to be improved, it is better to increase the number of gates rather than increase the pouring temperature.

### *7.7.2 Plaster and Shell Molding*

Plaster mold casting is basically the same as sand casting except that gypsum plasters are used instead of sand. The advantages are very smooth surfaces, good dimensional tolerances, and microstructural uniformity due to slow uniform cooling. However, as a result of the slow solidification rates (therefore, larger grain sizes), the mechanical properties are not as good as with sand castings. In addition, since plaster can absorb significant moisture from the atmosphere, the mold may require slow drying prior to casting.

Shell molding can also be used in place of sand casting when a better surface finish or tighter dimensional control is required. Since it requires precision metal patterns and more specialized equipment, shell molding should be considered a higher-volume process than sand casting. Shell molding, shown in Fig. 7.16, consists of:

1. A fine silica sand mixed with a phenolic resin is placed in a dump box that can be rotated.
2. A metal pattern is heated to 205 to 260 °C (400 to 500 °F), mold released, and placed in the dump box.
3. The pattern and sand are inverted, allowing the sand to coat the heated pattern. A crust of sand fuses around the part as a result of the heat.
4. The dump box is turned right side up, and the pattern with the shell crust is removed and cured in an oven at 345 to 400 °C (650 to 750 °F).
5. The same process is repeated for the other half of the mold.
6. The two mold halves are clamped together and placed in a flask supported with either sand or metal shot.

### *7.7.3 Evaporative Pattern Casting*

Evaporative pattern casting, also known as the lost foam process, is used quite extensively in the automotive and marine industries. Component patterns of expandable polystyrene are produced in metal dies. The patterns may consist of the entire part, or several patterns may be assembled together. Gating patterns are attached, and the completed pattern is coated with a thin layer of refractory slurry that is allowed to dry. The slurry must be permeable enough to allow mold gases to escape during casting. The slurry-coated pattern is then placed in a flask supported by sand. When the molten metal is poured, it evaporates the polystyrene pattern. A pattern and the resultant casting for an engine block are shown in Fig. 7.17. This process is capable of producing very intricate castings with close tolerances, but the mechanical properties are low because of the large amounts of entrapped porosity caused by offgassing of the polystyrene.

### *7.7.4 Investment Casting*

In the investment casting process (Fig. 7.18), also known as the lost wax process, a pattern of the part is produced from wax. The waxes are formulated to give smooth, defect-free surfaces, have thermal stability, maintain tolerances, and have a relatively long shelf life. The wax

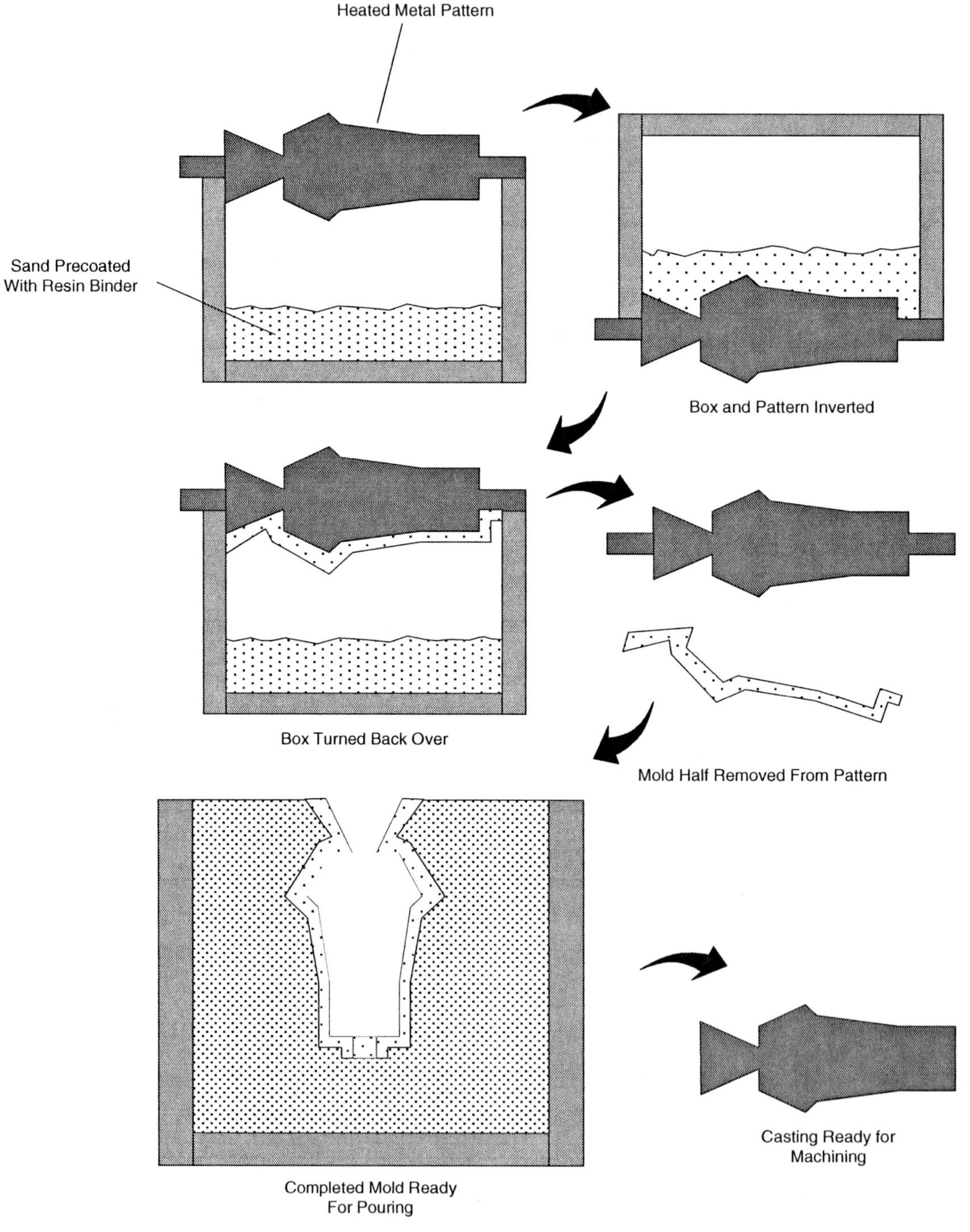


**Fig. 7.16** Shell molding process. Source: Ref 9

patterns are then robotically dipped in a fine, refractory ceramic slurry such as silica or alumina. The coated patterns are then stuccoed with dry, coarse particles of the same material to help the slurry dry faster and ensure adhesion between subsequent layers. The dipping and stuccoing process is repeated until the desired thickness is obtained, usually 6 to 8 times. When the mold is completely dry, it is placed in an oven, and the wax is melted out. The ceramic mold is then fired at approximately 980 °C (1800 °F).

Since the mid-1960s, the hottest parts of jet engines, the blades (rotating) and vanes (nonrotating), have been manufactured by investment casting. As the content of alloying elements in nickel-base superalloys was continually increased to obtain better creep and stress-rupture capability, the alloys became increasingly difficult to forge. To allow even higher contents of alloying elements, it became necessary to change the fabrication process to casting. The original cast blades and vanes were fine-grained polycrystalline structures made using conventional investment casting procedures. These blades were then heat treated to coarsen the grain structures for enhanced creep resistance. Eventually, it became possible to produce directionally solidified (DS) structures

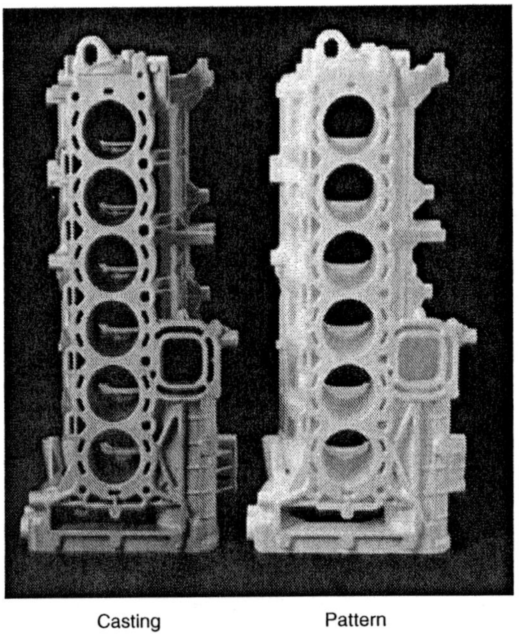


**Fig. 7.17** Evaporative pattern castings

with columnar grains oriented along the longitudinal axis of the blade. The columnar grain structure enhances the elevated-temperature ductility by eliminating the grain boundaries as failure initiation sites. The DS process also creates a preferred low-modulus texture, or orientation, parallel to the solidification direction that helps in preventing thermal fatigue failures. An extension of the DS process was the development of the single-crystal process in which a single crystal grows to form the entire blade.

To develop a DS structure, it is necessary for the dendrites (grains) to grow from one end of the casting to the other. This is accomplished by creating a sharp temperature gradient to remove the majority of the heat from one end of the casting. As shown in Fig. 7.19, a water-chilled mold is slowly withdrawn from the furnace, setting up a strong temperature gradient in the freezing metal. A thin-wall investment casting mold, which is open at the bottom, is placed on a water-cooled copper chill plate. The mold is then heated to the casting temperature. The alloy is heated under vacuum in an upper chamber of the furnace and then poured into the heated mold. After several minutes to allow the grains to nucleate on the chill plate, the mold is slowly withdrawn from the hot zone and moved to the cold zone. The chill plate ensures that there is good nucleation of grains to start the process. Although the grains nucleate with random orientations, those with the preferred growth direction normal to the chill surface grow and crowd out the other grains.

After the development of the DS casting process, it was recognized that if all but one of the growing columnar grains could be suppressed, it would be possible to cast parts with only a single grain, thereby eliminating all of the grain boundaries. Single-crystal castings are produced in a manner similar to DS castings, with one important difference: A method of selecting a single, properly oriented grain is used. In the most prevalent method, a helical section of mold is placed between the chill plate and casting mold. This helix, or spiral grain selector, acts as a filter and allows only a single grain to pass through it. Because superalloys solidify by dendritic growth that can occur in only three mutually orthogonal directions, the continually changing direction of the helix gradually constricts all but one grain, resulting in a single crystal emerging from the helix into the mold.

### *7.7.5 Permanent Mold Casting*

In permanent mold casting, liquid metal is poured into a metal mold and allowed to solidify. Due to the tooling costs, permanent mold casting is usually reserved for high-volume applications. The castings produced are normally small compared to sand casting and rather simple in shape. The process produces fairly uniform wall thicknesses but, unless segmented dies are used, is not capable of undercuts. Compared to sand castings, permanent mold castings are more uniform and have better dimensional tolerances, superior surface finishes, and better mechanical properties due to the faster solidification rates. Mold materials include gray cast iron and hot work die steels, such as H11 and H13. When a disposable sand or plaster core is used with this process, it is referred to as semipermanent mold casting. Another variant of the permanent mold process is low-pressure permanent mold casting. Here, the casting is done inside a pressure vessel, and an inert gas is used to apply 35 to 70 kPa (5 to 10 psi) pressure on the liquid metal. This results in shorter cycle times and excellent mechanical properties.

### *7.7.6 Die Casting*

Die casting is a permanent mold casting process in which the liquid metal is injected into a metal die under high pressure. It is a very high-rate production process using expensive equipment and precision-matched metal dies. Two methods, the hot and cold chamber die casting,

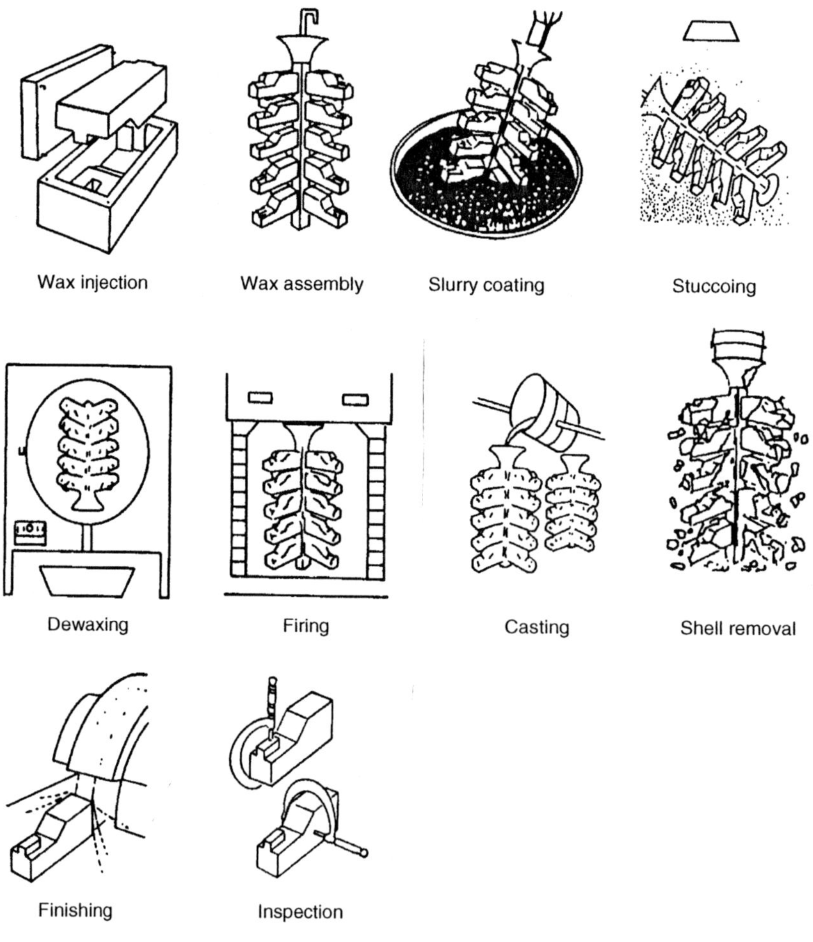


**Fig. 7.18** Investment casting process. Source: Ref 5

are shown in Fig. 7.20. The major die casting metals are zinc, aluminum, magnesium, and, to some extent, copper. Since the solidification rate is high, this process is amenable to high-volume production. It is used to produce very intricate shapes in the small-to-intermediate part size range. Characteristics of the process include extremely good surface finishes and the ability to hold tight dimensions; however, die castings should not be specified where high mechanical

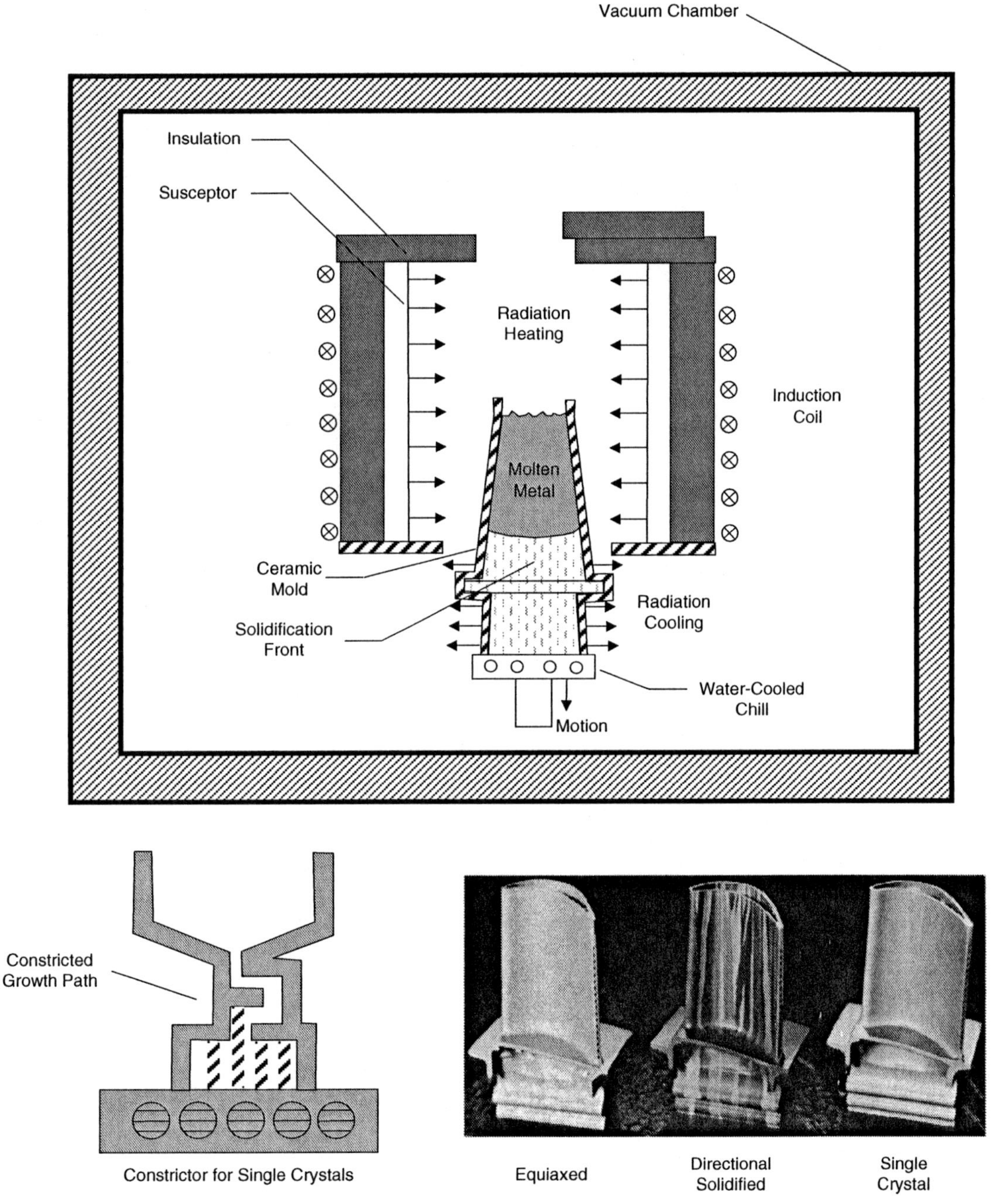


**Fig. 7.19** Directional solidification process for superalloy castings. Source: Ref 10, 11

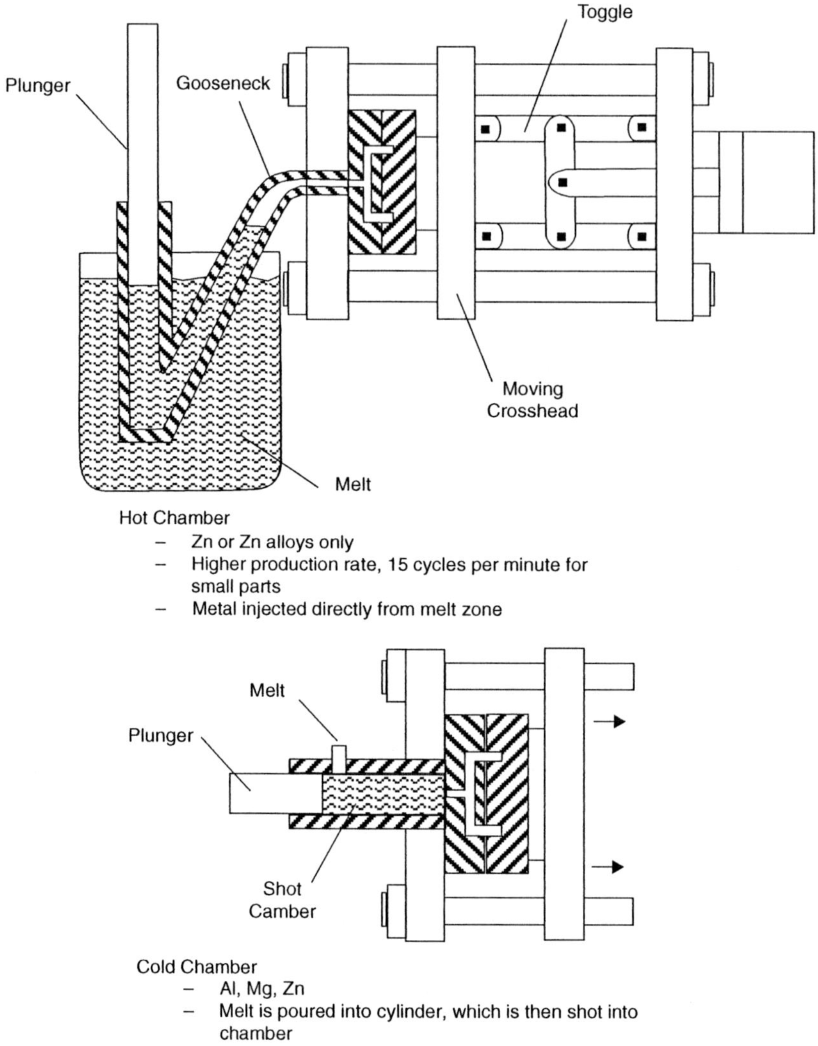


**Fig. 7.20** Hot and cold chamber die casting

properties are important, because of an inherently high porosity level. The high-pressure injection creates turbulence that traps air, resulting in high porosity levels. In fact, die cast parts are not usually heat treated, because the high porosity levels can cause surface blistering. To reduce the porosity level, the process can be done in a vacuum (vacuum die casting).

## REFERENCES

1. J. Campbell, *Castings,* Butterworth Heinemann, 1991
2. D.M. Stefanescu and R. Ruxanda, Fundamentals of Solidification, *Metallography and Microstructure,* Vol 9, *ASM Handbook,* ASM International, 2004
3. D.R. Askeland, *The Science and Engineering of Materials,* 2nd ed., PWS-Kent Publishing Co., 1989
4. M.F. Ashby and D.R.H. Jones, *Engineering Materials 2—An Introduction to Microstructures, Processing and Design,* 2nd ed., Butterworth Heinemann, 1998
5. *Metals Handbook Desk Edition,* 2nd ed., ASM International, 1998
6. V. Singh, *Physical Metallurgy,* Standard Publishers Distributors, 1999

7. A.G. Guy and J.J. Hren, *Elements of Physical Metallurgy,* 3rd ed., Addison-Wesley Publishing Company, 1974
8. R.A. Higgins, *Engineering Metallurgy—Applied Physical Metallurgy,* 6th ed., Arnold, 1993
9. F.C. Campbell, *Manufacturing Technology for Aerospace Structural Materials,* Elsevier Scientific, 2006
10. S.J. Mraz, Birth of an Engine Blade, *Mach. Des.,* June 24, 1997, p 39–44
11. R. Trivedi and W. Kurz, Solidification of Single-Phase Alloys, *Casting,* Vol 15, *ASM Handbook,* ASM International, 1988

## SELECTED REFERENCES

- J. Beddoes and M.J. Bibby, *Principles of Metal Manufacturing Processes,* Arnold, 1999
- P. Haasen, *Physical Metallurgy,* 2nd ed., Cambridge University Press, 1986
- T.S. Piwonka, Design for Casting, *Materials Selection and Design,* Vol 20, *ASM Handbook,* ASM International, 1997
- W.F. Smith, *Principles of Materials Science and Engineering,* McGraw-Hill Book Company, 1986

CHAPTER 8

# Recovery, Recrystallization, and Grain Growth

IF A METAL IS DEFORMED at room temperature, or up to approximately $0.3T_m$ (where $T_m$ is the absolute scale melting point), the microstructure becomes severely deformed due to cold working, also known as work hardening or strain hardening. During cold working, most of the energy is dissipated in the workpiece as heat. However, as much as 10% of the energy is retained as stored energy within the metal in the form of vacancies, dislocations, and stacking faults. The amount of stored energy increases with greater deformation and lower working temperatures. Metals and alloys experience significant changes in their properties as a result of cold working and also as a result of heating the cold-worked structure. These changes must be taken into account during forming operations, especially if the deformation is severe, as in deep drawing. For example, as a metal is deformed by cold working, the strength increases and the ductility decreases. Eventually, a point is reached where it is necessary to anneal the piece to allow further forming operations without the risk of breaking it. In addition, some metals are strengthened primarily by cold working. In this case, it is important that the metal not soften appreciably when placed in service.

Cold working greatly increases the number of dislocations; normal dislocation densities on the order of $10^6$ to $10^7$ cm$^{-2}$ are increased to $10^8$ to $10^{11}$ cm$^{-2}$ during working. The number of dislocations increases with increasing strain; due to their interactions, they become entangled, and the microstructure becomes severely distorted. This increasing dislocation density (dark regions on the micrographs) in a cold-rolled Fe-3%Si alloy is shown in the electron micrographs of Fig. 8.1.

During cold rolling, the grains also become distorted when the grains and their slip planes try to align with the direction of rolling, as shown schematically in Fig. 8.2. With increasing degrees of deformation, the original grains are severely distorted, to the point where they fracture into smaller grains. Heavy cold working also leads to a preferred orientation, or texture, in which the grains become oriented along certain crystallographic planes in a preferred manner with respect to the direction of stress.

Different metals develop different preferred orientations. In Fig. 8.3, cold-rolled copper develops an orientation as shown in the left-hand cube, which has a diagonal plane parallel to the

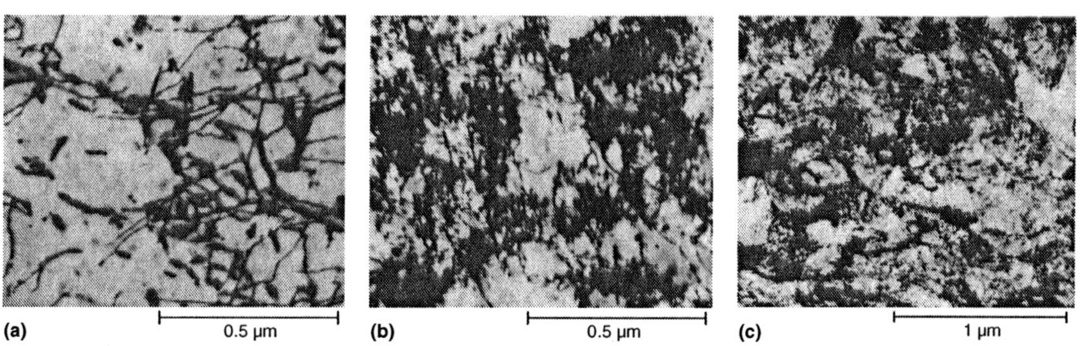


**Fig. 8.1** Dislocation density in Fe-3%Si alloy. (a) Cold rolled 5%. (b) Cold rolled 20%. (c) Cold rolled 90%. Source: Ref 1

rolled surface. After annealing, the orientation changes, as shown in the right-hand cube, where the cube face is now parallel to the rolling direction. The amount of texture increases with the amount of deformation, the type of deformation, the temperature of deformation, and the type of crystalline lattice. Crystallographic texture should not be confused with mechanical fibering or flow lines. Although both will produce directional mechanical properties in deformed metal structures, mechanical fibering is obtained by the alignment of inclusions and second-phase particles in the direction of working and is usually associated with hot working.

Since strain hardening increases with increased deformation, it is not surprising that the strength and hardness increase and the ductility decreases. The changes in properties for a brass alloy are shown in Fig. 8.4. The large number of dislocations that are developed and their entanglements make slip increasingly difficult. In addition, the conductivity decreases as a result of the interference of electron motion with the addition of more and more vacancies. The addition of alloying elements also increases the strength during cold working, as shown for the comparison of copper and brass in Fig. 8.5.

It is apparent that cold-worked materials are in a high energy state and are therefore thermodynamically unstable. Annealing is a heat treatment process where a material is held at an elevated temperature for an extended period of time and then slowly cooled. It is used to soften metals that have been hardened by cold working. Although a reduction in stored energy provides the driving force, annealing usually does not spontaneously occur at room temperature. Since the reduction in stored energy must occur by diffusion, the activation energy needed to start the diffusion process is normally insufficient at room temperature. Therefore, heating is necessary to provide the thermal activation energy needed to transform the material to a lower energy state. At elevated temperatures, the activation energy is provided by an increase in thermal energy. As the internal lattice strains are relieved during annealing, the strength decreases while the ductility increases. The rate of diffusion increases exponentially as the temperature increases. It should be noted that some metals with a low melting point have recrystallization temperatures lower than room temperature and cannot be hardened by cold working. For example, lead, tin, and zinc cannot be hardened by cold working because their recrystallization temperatures are below room temperature.

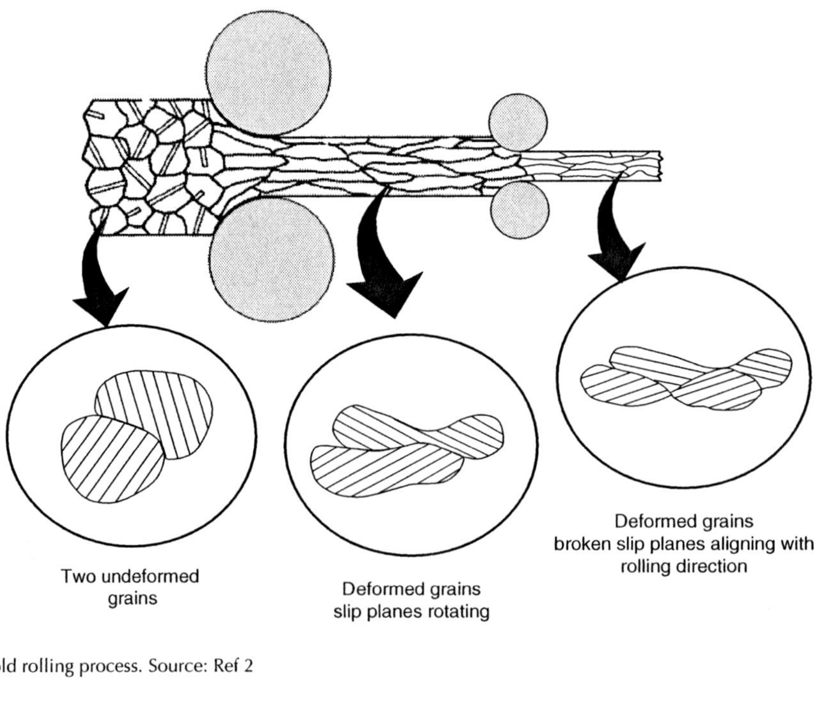


**Fig. 8.2** Cold rolling process. Source: Ref 2

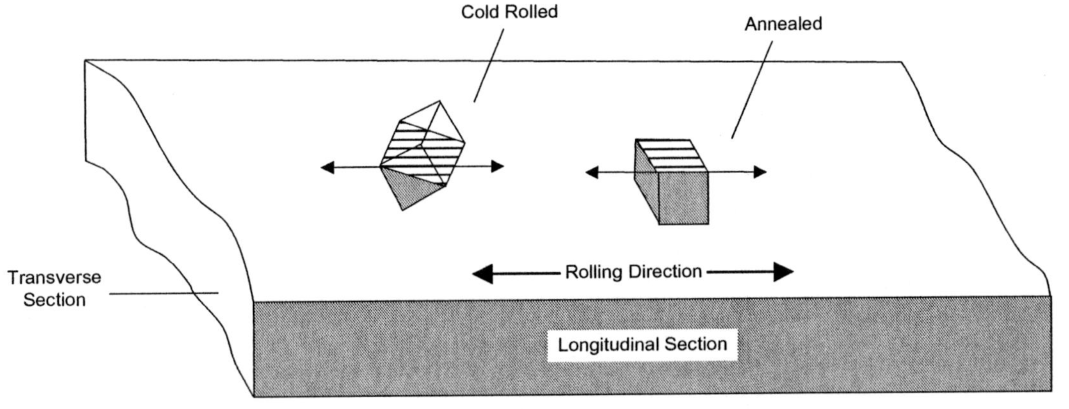


**Fig. 8.3** Preferred orientation in copper during cold rolling. Source: Ref 3

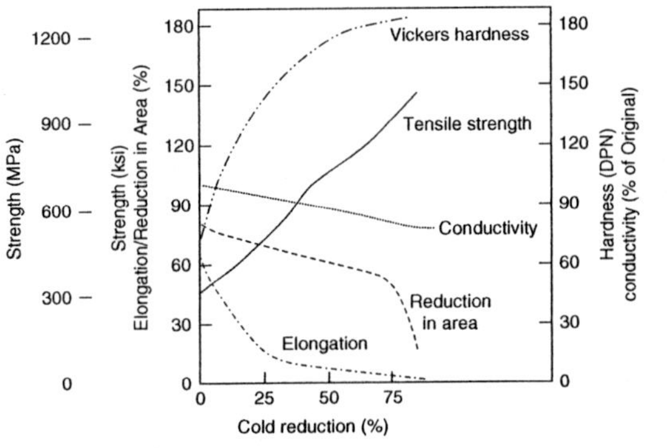


**Fig. 8.4** Effects of cold working on brass (Cu-35%Zn). Source: Ref 3

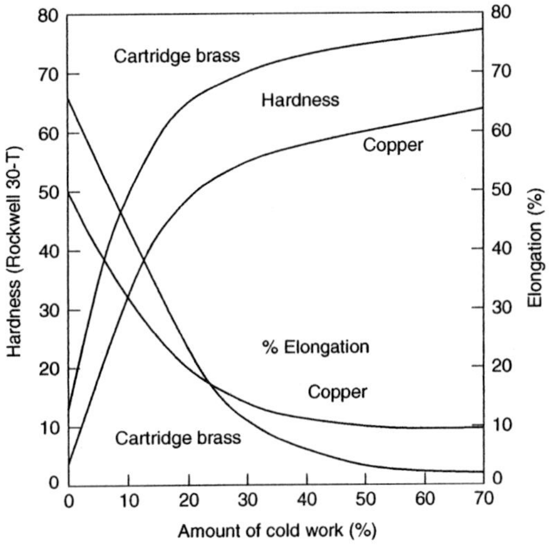


**Fig. 8.5** Effects of alloying elements on cold-worked properties. Source: Ref 3

During annealing, the following three distinct processes can occur.

**Recovery** is the relief of stored internal strain energy by rearrangement of dislocations into lower-energy configurations; however, the grain shape and orientation remain the same. The strength and ductility are largely unaffected. Many physical properties (electrical, thermal, etc.) are "recovered" to their original, pre-cold-working levels.

**Recrystallization** is the replacement of the badly deformed cold-worked grains by new strain-free grains. The strength decreases and the ductility increases, similar to that of the metal before cold working.

**Grain growth** occurs when some of the recrystallized grains grow at the expense of other recrystallized grains. Since fine-grained materials offer the best combination of strength and ductility, in almost all cases, grain growth is an undesirable process.

## 8.1 Recovery

Recovery is the initial stage of the annealing cycle before recrystallization occurs. An anisothermal annealing curve for nickel is shown in Fig. 8.6. This curve was developed by comparing the energy released on annealing for a cold-worked specimen with one that was not deformed. During heating, the cold-worked specimen undergoes reactions that release energy as heat and thus needs less power to heat through the cycle compared to the undeformed specimen. Note that the heat is released in two overlapping stages. During stage I, or recovery, there is a smaller peak followed by a steady release of heat, while during stage II, or recrystallization, there is a large energy release that occurs quickly and then dissipates rapidly. Note that the resistivity is almost completely recovered (approaching 0 Ωcm) before recrystallization, while the hardness (therefore strength) decreases during the recrystallization process.

Recovery can be defined as the process of annihilation and rearrangement of defects within the deformed metal, without the movement or migration of high-angle grain boundaries. During recovery, there are several basic types of processes that occur: (1) the annihilation of excess point defects, particularly vacancies; (2) the rearrangement of dislocations into lower-energy configurations, which also annihilates many of them; and (3) the formation of subgrains that grow and interlock into subboundaries.

At lower temperatures, the vacancies that were generated during cold working are annealed out by migrating to dislocations, grain boundaries, or surfaces. This results in a decrease in the electrical resistivity because the valence electrons that carry the electrical charge are now less scattered. The isothermal changes in electrical resistance of copper during recovery at three different temperatures are shown in Fig. 8.7. Note that recovery is initially very rapid, particularly at higher temperatures, and, with increasing time, becomes slower and eventually levels off. The greater the amount of cold work, the more rapid is this initial state of recovery.

At slightly higher temperatures, the rearrangement of dislocations occurs, and, in the process, the annihilation of dislocations of opposite signs (of the Burger's vector) takes place. The rearrangement of dislocations is assisted by thermal energy, which aids in both climb and slip mechanisms. Since impurities and alloying elements are attracted to both vacancies and dislocations, their presence decreases the mobility of the defects.

One of the key processes of dislocation rearrangement is polygonization. Polygonization is the process of rearranging excess edge dislocations into low-angle tilt boundaries, as shown schematically in Fig. 8.8. The driving force for polygonization is a reduction in the total strain energy. During polygonization, crystal planes that were bent and distorted by plastic deformation become straightened. Dislocations migrate to line up over one another into low-angle tilt boundaries. These low-angle tilt boundaries, which differ in orientation by only a degree or two, form internal subboundaries surrounded by subcrystals, which are essentially

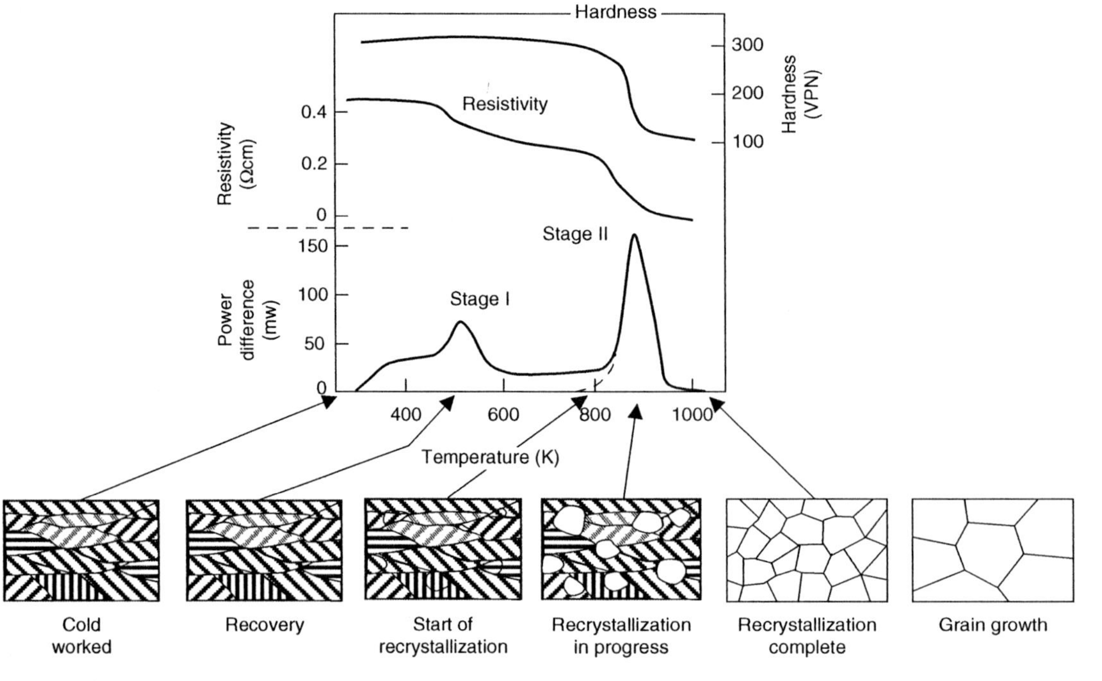


**Fig. 8.6** Anisothermal annealing curve for cold-worked nickel. Source: Ref 2

free of dislocations. As the energy stored as elastic strains is relieved, residual stresses are reduced.

Examine the dislocation distribution in a bent single crystal before and after recovery (Fig. 8.8). When edge dislocations of the same sign accumulate on the same plane, their stress fields are additive; however, when they rearrange themselves vertically, the stress fields have a tendency to partially cancel each other, thereby reducing the overall energy. Dislocation energy is reduced when dislocations of opposite signs (+ and −) meet and partially cancel each other out. Subboundary formation requires that dislocations both climb and slip. Since dislocations cannot climb at low temperatures, thermal energy provides both vacancies, which are required for climb, and makes slip easier. The mechanism for combined climb and slip is shown schematically in Fig. 8.9. After the initial formation of subboundaries, two or more subboundaries will migrate and meet to form a

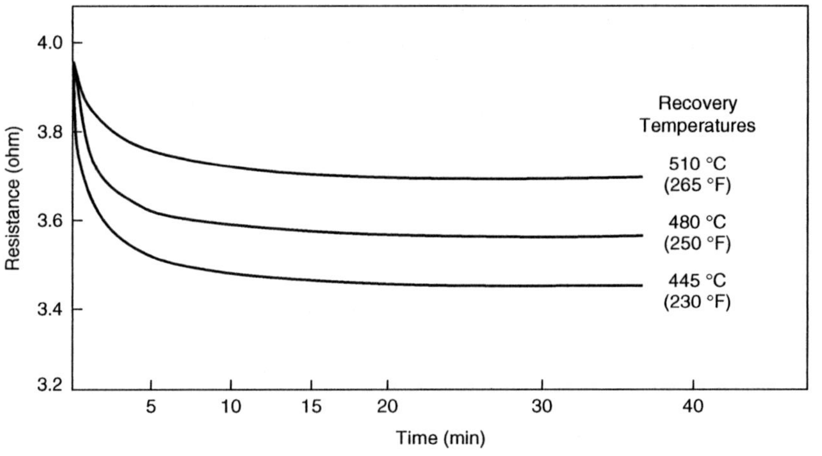


**Fig. 8.7** Electrical resistivity changes of copper during recovery. Source: Ref 2

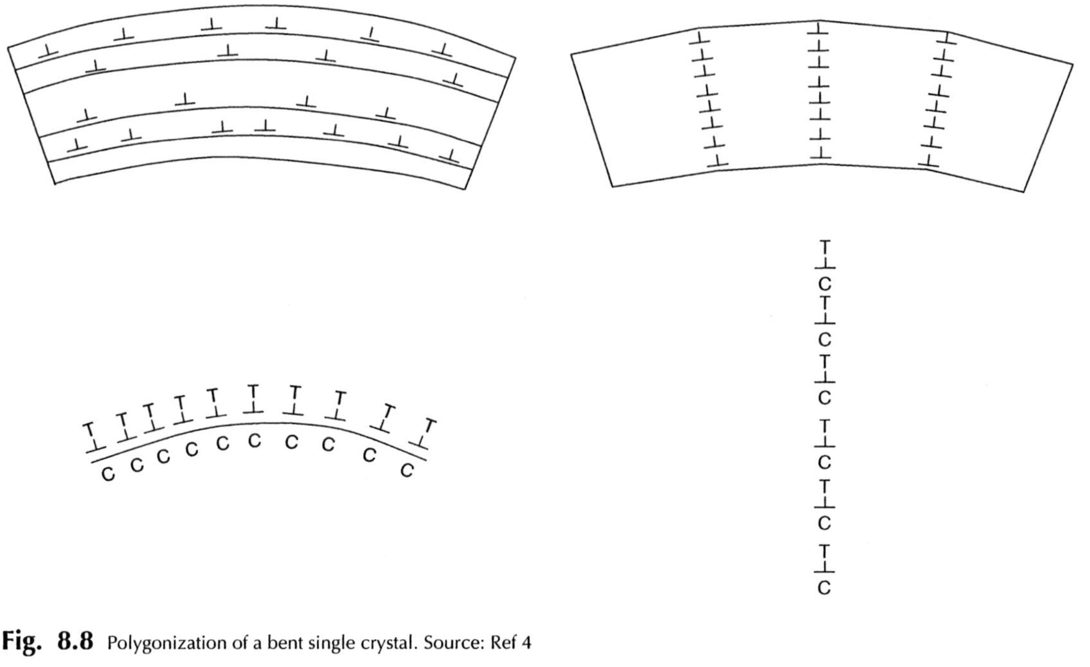

**Fig. 8.8** Polygonization of a bent single crystal. Source: Ref 4

single subboundary, further lowering the internal energy. The size of the subgrains increases with increasing time and temperature but tends to approach a size limit. A greater degree of cold working results in smaller subcrystals, and the hardness decreases as the subcrystals grow. When a steady-state condition is reached, polygonization is complete. Formation of subgrains decreases strength, but the decrease is largely offset by their high dislocation density that restricts defect movement.

Since there is a large decrease in residual stresses during recovery, recovery-type processes are usually conducted to reduce residual stresses, often to prevent stress-corrosion cracking or minimize distortion. During stress-relief operations, the temperature and time are controlled so there is not a major reduction in strength or hardness.

## 8.2 Recrystallization

Recrystallization can be defined as the nucleation and growth of strain-free grains out of the matrix of the cold-worked metal. Cold-worked grains having a high dislocation density are replaced by new grains having a much lower dislocation density. The dislocation density decreases from $10^8$ to $10^{11}$ $cm^{-2}$ to approximately $10^6$ to $10^7$ $cm^{-2}$. New orientations, new grain sizes, and new grain shapes form during recrystallization. The driving force for recrystallization is the remaining stored energy that was not eliminated during recovery. A low-carbon steel sheet that had been severely cold worked (90%) and then annealed at 550 °C (1025 °F) for increasing times is shown in Fig. 8.10. The sample that is only 10% recrystallized still shows highly elongated cold-worked grains. After 40% recrystallization, new strain-free grains are clearly visible. Finally, after 80% recrystallization, the extent of new strain-free grains is much more pervasive.

Recrystallization is considered complete when the mechanical properties of the recrystallized metal approach those of the metal before it was cold worked. Recrystallization and the resulting mechanical softening completely cancel the effects of cold working. An annealing curve for a brass is given in Fig. 8.11, which shows both minimal changes in mechanical properties during recovery and then the large changes in properties that occur during recrystallization. Mechanical properties such as hardness, yield strength, tensile strength, percent elongation, and reduction in area change drastically over a very small temperature range to become typical of the annealed material. Although physical properties, such as electrical conductivity, undergo large increases during recovery, they also continue to increase during recrystallization.

Recovery and recrystallization are two distinctly different processes. At a constant temperature, recovery starts rapidly and then decreases with time. On the other hand, recrystallization, which is a nucleation and growth process, starts slowly and then builds up to a maximum rate before rapidly leveling off. Accordingly, isothermal recrystallization curves are typically sigmoidal in shape (Fig. 8.12) where there is an incubation period during which no visible recrystallization occurs. During incubation, stable nuclei are formed by the coalescence of subgrains, leading to the formation of high-angle grain boundaries. There are two competing forces involved during nucleation. Nucleation will produce new grains that reduce the strain energy in the deformed matrix;

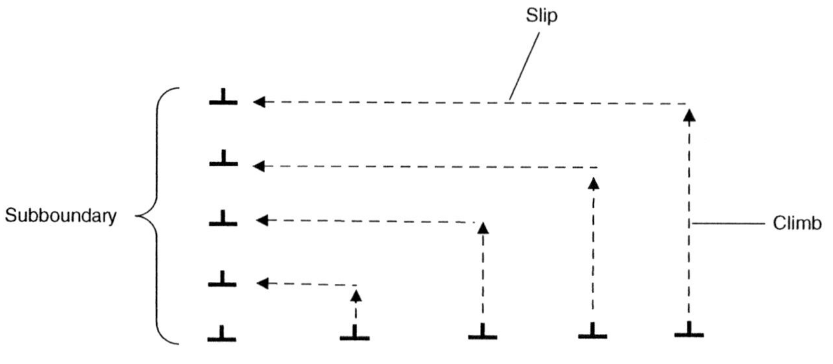


**Fig. 8.9** Polygonization mechanism. Source: Ref 2

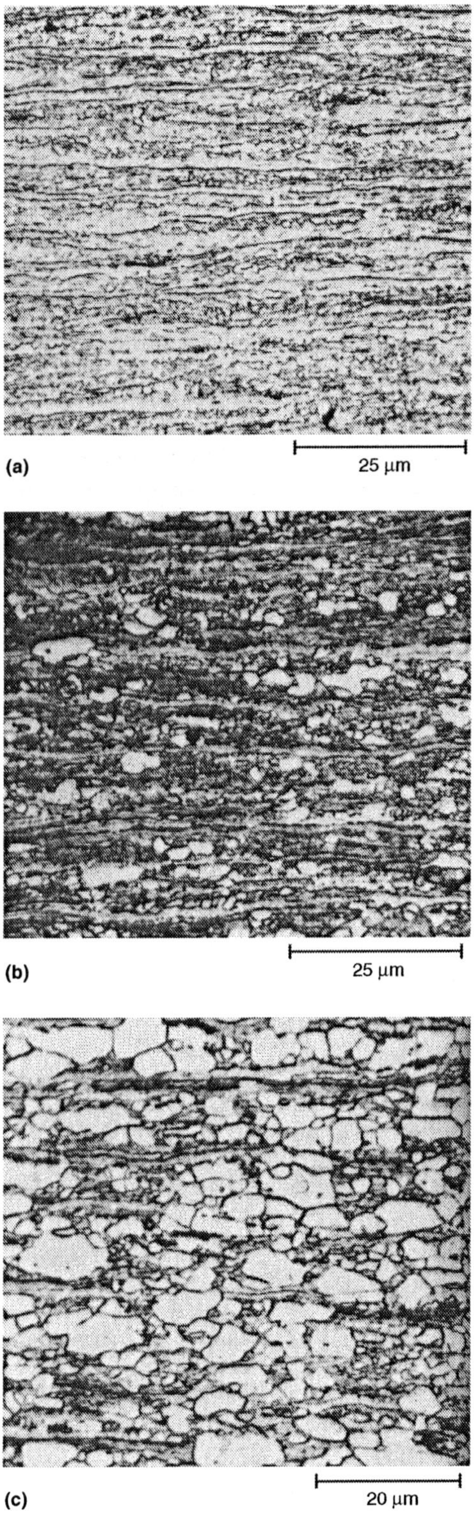


**Fig. 8.10** Recrystallization progression in low-carbon steel. (a) Recrystallized 10%. (b) Recrystallized 40%. (c) Recrystallized 80%. Source: Ref 1

however, their formation is opposed by the energy necessary to form new interfaces. Thus, nuclei will not form until the driving force provided by thermal energy becomes great enough. Since the driving force for recrystallization is the remaining strain energy in the matrix following recovery, nucleation occurs at locations of high dislocation density that have high strain energies, such as at the original grain boundaries, grain-boundary triple points, and at the subgrains formed during polygonization.

Nucleation during recrystallization is a little bit different than during solidification. During solidification, if an embryo or cluster of atoms does not reach a critical size, it redissolves in the melt. However, during annealing, the embryo cannot redissolve, so it just remains as an embryo until there is sufficient energy for it to attract enough atoms and become a nucleus. Thus, the incubation period corresponds to the irreversible growth of embryos.

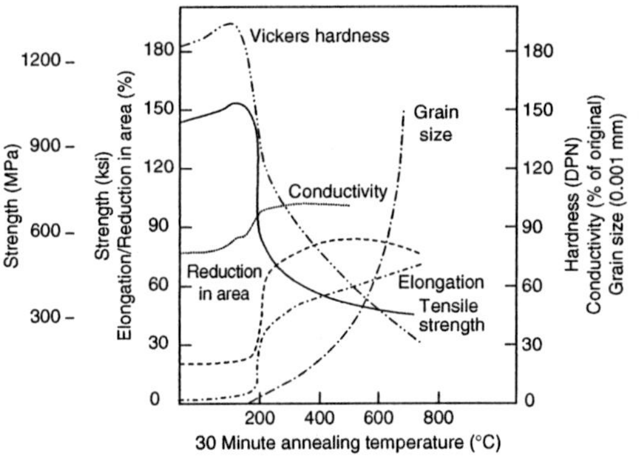


**Fig. 8.11** Effects of annealing brass (Cu-35% Zn). Source: Ref 3

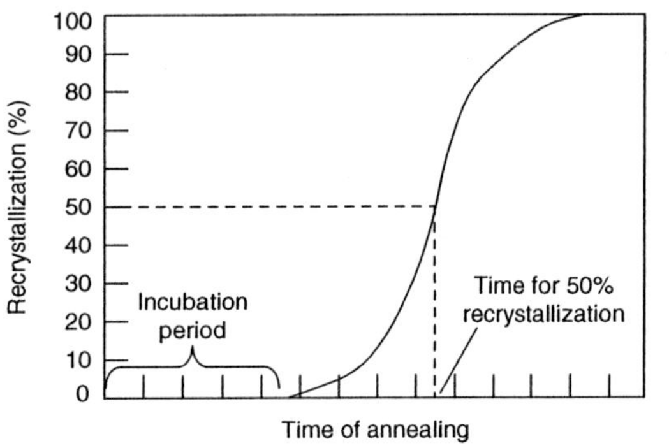


**Fig. 8.12** Typical isothermal recrystallization curve. Source: Ref 5

The nucleation of new grains at grain boundaries and subboundaries is illustrated in Fig. 8.13. The growth of the newly formed strain-free grains at the expense of the polygonized matrix is accomplished by the migration of high-angle grain boundaries. Migration proceeds away from the center of boundary curvature. The strain-free nuclei grow into the deformed metal and gradually absorb the whole of the deformed matrix. The growth of new grains proceeds rapidly because of the high mobility of the new high-angle grain boundaries, which have grain-boundary angles in the range of 30 to 40°. The rate of recrystallization eventually decreases as stored energy in the matrix is depleted and more of the new grains start impinging on each other. The rate of growth is independent of time but increases with deformation and annealing temperature.

If the nucleation rate ($\dot{N}$) is high and the growth rate ($\dot{G}$) is low, then a fine grain size will result; conversely, if the nucleation rate is low and the growth rate is high, the microstructure will have a large grain size. Increasing the amount of cold work increases both the nucleation rate and growth rate, which effectively lowers the recrystallization temperature. However, since the increase in nucleation rate is greater than growth rate, larger amounts of cold work produce finer grain sizes during recrystallization. The smaller the amount of deformation, the higher the temperature required to initiate recrystallization. The ratio $\dot{N}/\dot{G}$, as shown in Fig. 8.14, is often used to interpret

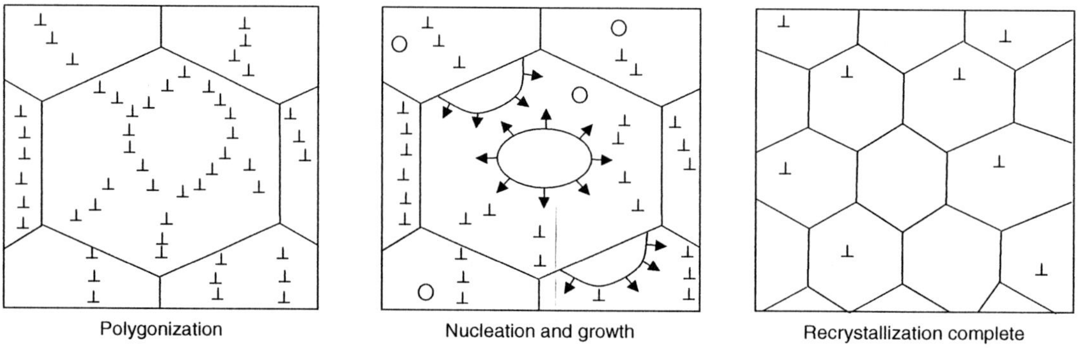


**Fig. 8.13** Nucleation and growth of grains. Source: Ref 6

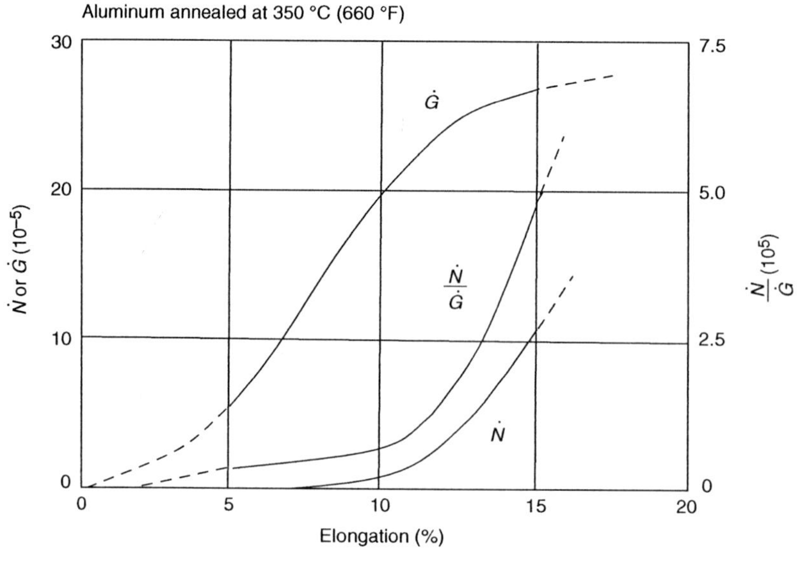


**Fig. 8.14** Nucleation and growth rates during recrystallization. Source: Ref 4

recrystallization data. As the deformation before annealing is reduced to smaller and smaller values, the nucleation rate decreases much faster than the growth rate. Therefore, $\dot{N}/\dot{G}$ decreases with decreasing deformation and approaches zero at several percent elongation. This supports the idea that nuclei form at locations of high strain energy in the cold-worked matrix. The number of these locations should increase with increasing strain and decrease with decreasing strain. A certain minimum amount of cold work or deformation and a certain minimum temperature are necessary to initiate recrystallization. By increasing the degree of cold work, more stored energy is available for the nucleation of new grains with a resultant finer grain size. As the deformation is reduced to smaller and smaller values, the nucleation rate falls much faster than the growth rate. Thus, there is a critical amount of cold work that is required to form the nuclei needed for recrystallization.

The main factors that affect recrystallization are:

- Temperature and time
- Degree of cold work
- Purity of the metal
- Original grain size
- Temperature of deformation

### *8.2.1 Recrystallization—Temperature and Time*

The temperature required for recrystallization is not exact; it depends on the composition of the alloy and, in particular, on the amount of cold work performed. Recrystallization temperatures for metals of commercial purity are in the range of 0.3 to 0.5 of their absolute melting temperatures, as shown for a number of metals in Fig. 8.15. The temperature for recrystallization is often defined as the temperature required for the microstructure to undergo 50% recrystallization in 30 min, or the time for complete recrystallization in approximately 1 h. The approximate recrystallization temperatures for several metals and alloys are given in Table 8.1.

Recovery may affect the recrystallization temperature, since the tendency toward recrystallization is lowered when appreciable recovery has occurred; that is, a higher temperature may then be required to cause recrystallization. Although there is a trade-off between time and temperature, recrystallization occurs more rapidly at higher temperatures and is much more dominant than time. For most kinetic processes, increasing the temperature by approximately 11 °C (20 °F) doubles the reaction rate. When recrystallization is complete, further heating causes grain growth to occur.

A series of isothermal recrystallization curves for high-purity copper (99.999%) are shown in Fig. 8.16. As the curves show, recrystallization occurs more rapidly at higher temperatures. If a horizontal line corresponding to 50% recrystallization is drawn, then its intersection with the various curves gives the time at temperature

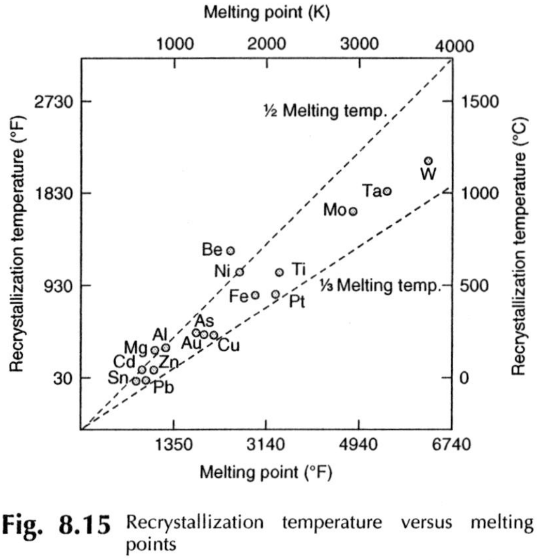


**Fig. 8.15** Recrystallization temperature versus melting points

**Table 8.1 Approximate recrystallization temperatures for several metals and alloys**

| Metal | Recrystallization temperature °C | Recrystallization temperature °F |
|---|---|---|
| Copper (99.999%) | 120 | 250 |
| Copper (OFHC)(a) | 200 | 400 |
| Copper (5% Al) | 290 | 550 |
| Copper (5% Zn) | 320 | 600 |
| Copper (2% Be) | 370 | 700 |
| Aluminum (99.999%) | 80 | 175 |
| Aluminum (99.0% +) | 290 | 550 |
| Aluminum alloys | 320 | 600 |
| Nickel (99.99%) | 370 | 700 |
| Nickel (99.4%) | 590 | 1100 |
| Nickel (30% Cu) | 590 | 1100 |
| Iron (electrolytic) | 400 | 750 |
| Low-carbon steel | 540 | 1000 |
| Magnesium (99.99%) | 65 | 150 |
| Magnesium alloys | 540 | 1000 |
| Zinc | 10 | 50 |
| Tin | −5 | 25 |
| Lead | −5 | 25 |

(a) OFHC, oxygen-free high conductivity. Source: Ref 5

required to recrystallize the structure by 50%. Plotting the logarithm of time for 50% recrystallization versus the reciprocal of absolute temperature yields a straight line. This indicates that recrystallization kinetics follow an Arrhenius behavior:

$$\text{Rate} = \frac{1}{t} = Ae^{-B/T} \qquad \text{(Eq 8.1)}$$

where $A$ and $B$ are constants, and $t$ is the time for 50% recrystallization.

The constant $B$ cannot be considered a true activation energy, $Q$, since it changes with temperature. However, the two constants $A$ and $B$ can be determined if the times for 50% recrystallization at two different temperatures are known by writing the equation for the two different temperatures and solving the two equations for the constants.

Since cold working is a method of strengthening many metals that cannot be strengthened by heat treatment, Eq 8.1 can be used to estimate whether the material will be stable in service for long periods of time at moderate service temperatures. For example, the stability of copper wire at 100 °C (373 K) can be estimated with $A = 10^{12}$ min$^{-1}$ and $B = 15{,}000$. The rate is:

$$\text{Rate} = 10^{12}e^{-15{,}000/373}$$

$$\text{Rate} = 10^{12}e^{-40.2} = 10^{12}10^{-17.45} = 10^{-5.45}$$

$$\text{Rate} = 0.35 \times 10^{-5}\ \text{min}^{-1}$$

therefore:

$$\text{Time for 50\% recrystallization} = 2.9 \times 10^{5}\ \text{min} = 48{,}000\ \text{h} = \sim 5.5\ \text{yr}$$

### *8.2.2 Recrystallization—Degree of Cold Work*

As the amount of cold work increases, recrystallization occurs at lower temperatures and at shorter times. A plot of the recrystallization of aluminum for different amounts of cold work at a constant annealing temperature of 350 °C (660 °F) is shown in Fig. 8.17. As the amount of cold work increases, the time for recrystallization is drastically reduced. Also, the incubation period is shorter for the more highly cold-worked metal. As the amount of cold work increases, there is more strain in the lattice structure, and the dislocation density increases. Thus, higher amounts of cold work produce

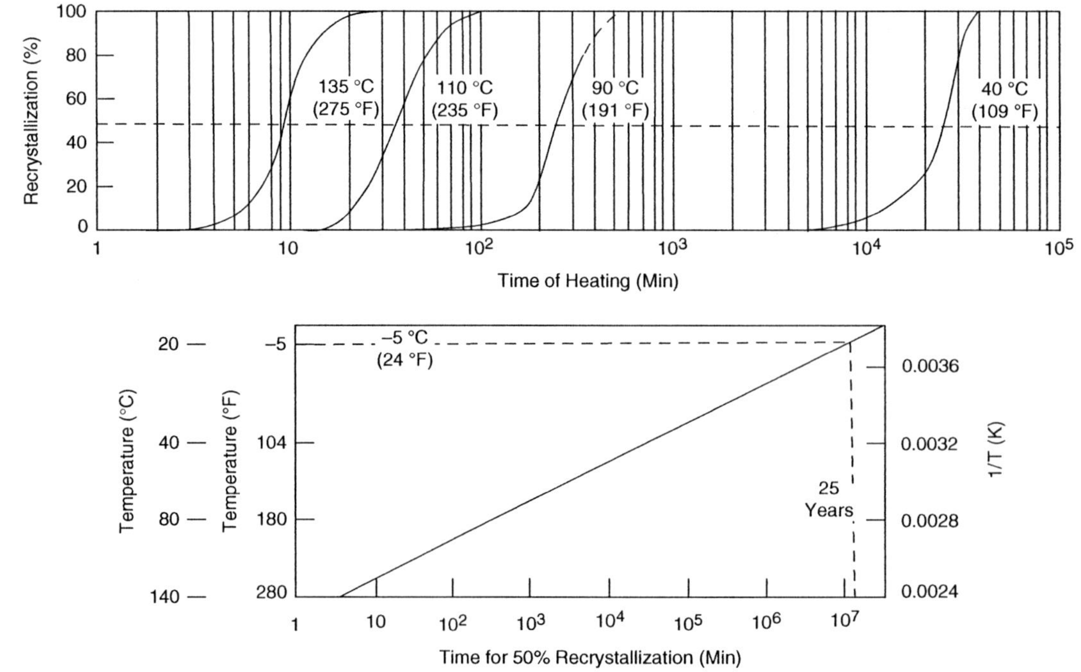


**Fig. 8.16** Isothermal recrystallization of 99.999% pure copper. Source: Ref 5

smaller grain sizes during recrystallization, as shown in Fig. 8.18 for cartridge brass with different degrees of cold working. As a consequence, the growing subgrains will be smaller when they become large-angle grains. In addition, a shorter time is needed at any given temperature for the subgrains to generate active nuclei. At higher temperatures, the nuclei form even quicker.

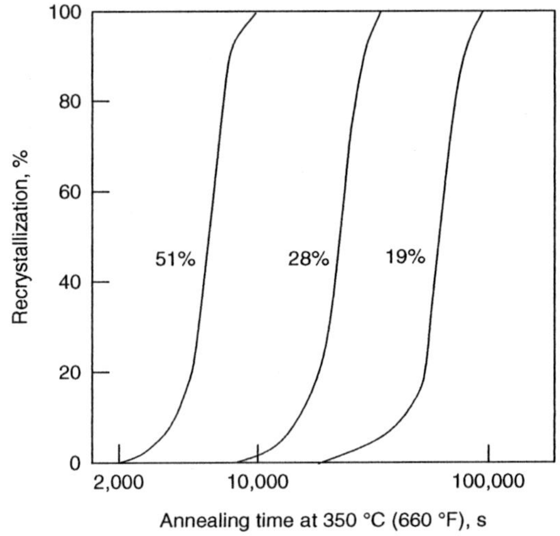


**Fig. 8.17** Effect of cold work on recrystallization of aluminum. Source: Ref 2

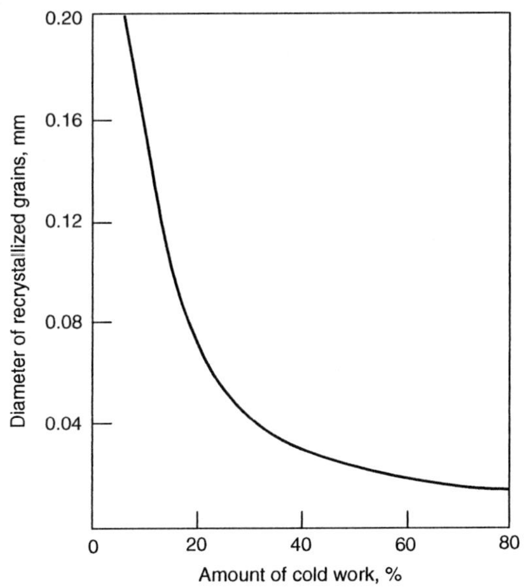


**Fig. 8.18** Effect of cold work on recrystallized grain size in cartridge brass

For recrystallization to occur, a certain amount of deformation, called the recrystallization threshold, is necessary. If the degree of cold work is too low, there will not be a sufficient amount of stored energy to initiate recrystallization (Fig. 8.19). At small deformations, subgrains form on heating and grow, but the boundaries between them remain low-angle boundaries, and recrystallization does not occur. At small deformations, which can range from as low as 2% to as much as 20%, recrystallization does not occur at temperatures even approaching the melting point. The critical degree of deformation can vary with the type of deformation. For example, it would be different for pure tension versus the complex stress state produced during forging. When a polycrystalline specimen is deformed with a very small strain and then annealed at a sufficiently high temperature, recrystallization occurs by strain-induced boundary migration of only a few grains. Near the critical degree of deformation, there is extensive nonuniformity of strain hardening. The boundaries of the severely strained grains migrate rapidly and grow at the expense of the grains with less strain. Grain-boundary nucleation by the bulging out of a section of an initial boundary from the region of low dislocation content into a region of high dislocation content is a consequence of strain-induced boundary migration. These conditions are almost always avoided because extremely large grain sizes can develop, which can result in extremely rough

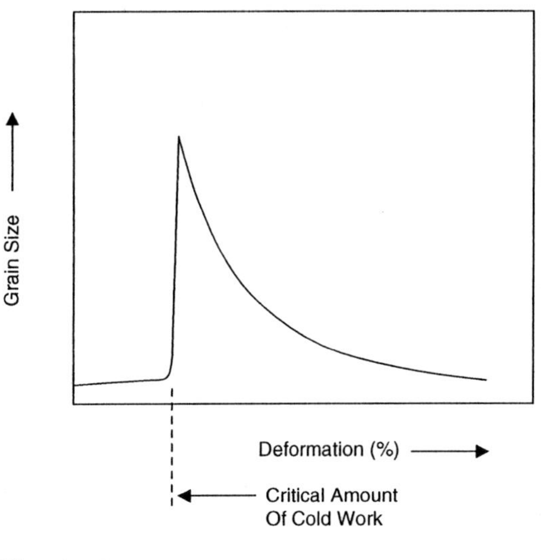


**Fig. 8.19** Critical amount of cold work

and unacceptable surface finishes on parts, requiring further cold working.

### 8.2.3 Recrystallization—Purity of Metal

Very pure metals exhibit lower recrystallization temperatures compared to their alloys. Impurities, solutes, or fine second-phase particles slow down grain-boundary migration; therefore, their presence retards recrystallization. Since impurities and alloying elements slow down grain growth, more time is needed for nucleation and finer recrystallized grain sizes. Solid-solution impurities tend to migrate to dislocations and grain boundaries and slow their motion, thus raising the recrystallization temperature. Second phases also tend to raise the recrystallization temperature. As the amount of solute increases, saturation occurs at these sites, and the increase in recrystallization temperature becomes less. The addition of a small amount of alloying elements usually sharply increases the temperature required for recrystallization. However, with further increases, the recrystallization temperature usually reaches a maximum value and then decreases. Some elements are more effective than others. The optimal addition of magnesium to aluminum is 1%, and for copper the optimal addition is 5% Zn. Zirconium is very effective in aluminum, and molybdenum in steel. The addition of alloying elements is one method that is used in cold-work-hardened alloys to prevent them from softening in service.

### 8.2.4 Recrystallization—Original Grain Size

Smaller initial grain sizes yield smaller recrystallized grain sizes. Since grain boundaries stop dislocation movement during deformation, they become highly strained and smaller, more numerous grain boundaries provide more active sites for nucleation during recrystallization. Final grain size is more dependent on the amount of deformation or cold work than on either the annealing temperature or the time.

The grain size before cold working affects the nucleation rate. For equal amounts of cold work, more strain hardening is produced in fine-grained metal than in coarse-grained metal. Thus, the finer the initial grain size, the lower the recrystallization temperature and the shorter the recrystallization time, as shown in Fig. 8.20 for a low-carbon steel recrystallized at 540 °C (1000 °F). With a smaller grain size, there are more nucleation sites and lower recrystallization temperatures. The larger the original grain size of the material, the greater the amount of cold work required to achieve an equivalent recrystallization for a given temperature and time. For a given amount of work hardening, a higher working temperature is accompanied by a coarser grain size and requires a higher temperature to cause recrystallization. The recrystallized grain size is dependent on both the recrystallization time and temperature, particularly the temperature, with higher temperatures

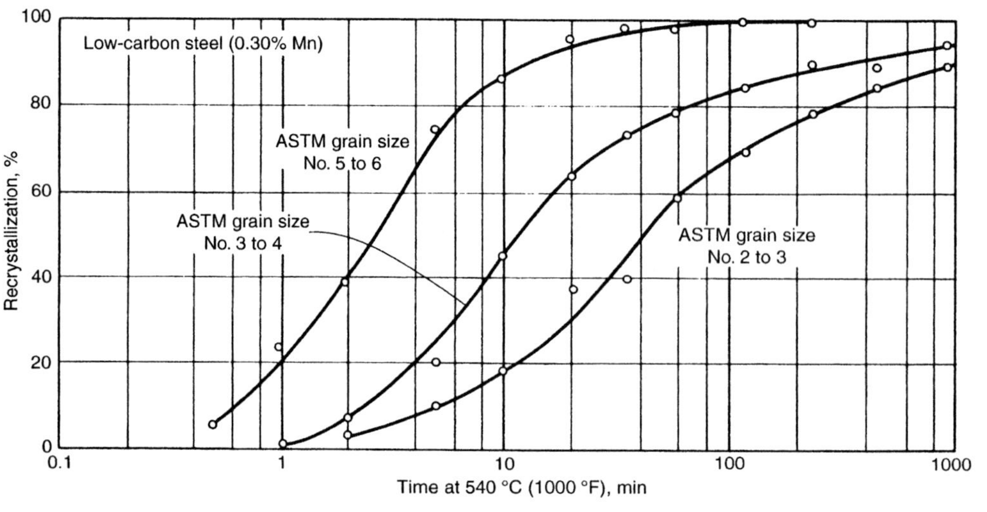


**Fig. 8.20** Effect of grain size on recrystallization kinetics. Source: Ref 1

promoting larger grain sizes, as indicated in Fig. 8.21.

### *8.2.5 Recrystallization—Temperature of Deformation*

As the temperature of cold working is increased above room temperature, a greater amount of cold working is required to obtain an equivalent amount of recrystallization at a given time and temperature. As the cold working temperature increases, dynamic recovery begins to occur during the deformation process. Therefore, there is less stored energy in the metal to serve as a driving force for recrystallization.

## 8.3 Grain Growth

After recrystallization is complete, that is, when the polygonized matrix is replaced by new strain-free grains, further annealing increases the average grain size. This process, known as grain growth, is accomplished by the migration of grain boundaries. Recrystallization consumes the retained energy of deformation, but the grain boundaries still have a finite interfacial energy. The material is still metastable, with thermodynamic stability being achieved only if the structure becomes a single grain or crystal. In contrast to recrystallization, the boundary moves toward its center of curvature. Some of the grains grow, but others become smaller and vanish. Because the volume of the specimen is a constant, the number of the grains decreases during grain growth. As the smaller grains are consumed by larger grains, there will be fewer grains, and their average size increases. The driving force for grain growth is the grain-boundary free energy, which is substantially smaller in magnitude than the driving energy for recrystallization. Grain growth can be further classified into two types: normal or continuous grain growth and abnormal or discontinuous grain growth. The latter has also been termed exaggerated grain growth, coarsening, or secondary recrystallization.

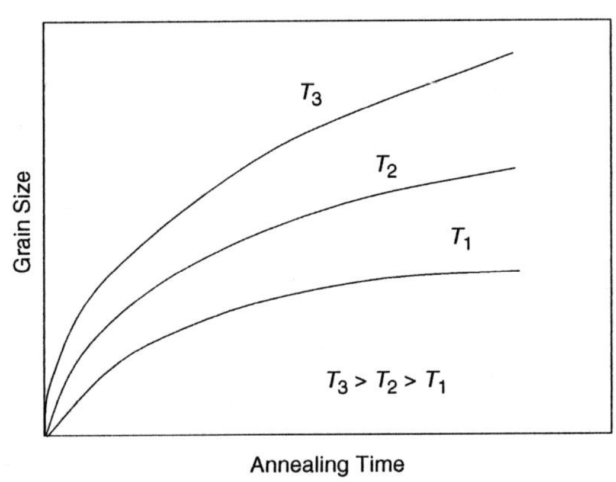


**Fig. 8.21** Grain size versus annealing temperature

### *8.3.1 Normal Grain Growth*

Typically, the smaller the number of sides on the recrystallized grain, the sharper the curvature of the sides and the quicker the grain becomes absorbed by adjacent grains. Grains with less than six sides will have concave curvatures, are unstable, and will tend to shrink in size, while those with more than six sides have convex curvatures and will grow in size (Fig. 8.22). The smallest grains have the fewest sides and are consumed the quickest by the larger grains. Boundaries tend to move toward their center of curvature because, by straightening, they reduce their surface area. The dihedral angle between three grain boundaries moves toward equilibrium at 120°. Grain growth occurs by the movement of grain boundaries resulting from diffusion of atoms. An atom is more stable in the concave-shaped grain rather than on the convex side, where it has more neighboring atoms to form bonds with. Thus, the grain boundary migrates toward its center of curvature. Grain growth occurring with increasing time at a constant temperature is small compared to that occurring at constant time with increasing temperature. Grain-boundary migration rate depends on solute content and orientation. Second-phase particles tend to pin grain boundaries and slow down grain growth.

Empirically, it has been shown that grain growth occurs according to:

$$D = kt^n \quad \text{(Eq 8.2)}$$

where $D$ is the average grain diameter, $t$ is time, $n$ is a constant, and $k = k_0 e^{-Q/2RT}$.

The constant $n$ increases with temperature and approaches a theoretical value of 0.5. It should also be noted that the activation energy, $Q$, also varies with temperature.

When $D_{ave}$ and $t$ are plotted on a logarithmic scale, the result is a straight line, with $k$ as the intercept and $n$ as the slope of the curve. The value of $n$, the time exponent in isothermal grain growth, is usually 0.5 or less. A typical example for isothermal grain growth in zone-refined iron is shown in Fig. 8.23. The deviations from a straight line relationship for very short annealing times at low temperatures are due to recovery, and long annealing times at high temperatures show the limiting effect of the thickness of the sheet specimen.

Coarse grains can result in rough surfaces forming during deformation, a phenomenon called orange peel. Since the rough surface is still apparent after painting, it is undesirable. The effect of grain size on the surface finish for cold-drawn brass is shown in Fig. 8.24, with the finer-grained material yielding a much better surface finish. Orange peel is due to the non-homogeneous deformation of the structure as a result of the anistropic nature of the large grains. However, coarse grains are beneficial for creep resistance in high-temperature alloys because they are more resistant to grain-boundary sliding and rotation.

Recrystallization nuclei are formed from the deformed grains. Therefore, the positions and orientations of the new grains depend on the deformed grain structure. Consequently, the recrystallization microstructure may be equivalent to the deformation microstructure but may also differ from it, depending on the temperature and time of the recrystallization anneal. Recrystallized grains are not usually randomly oriented but have their crystal axes lying near certain preferred directions with respect to the cold-worked grains, as previously shown in Fig. 8.3. Since texture results in anistropic mechanical properties, it is usually objectionable. Texture can be minimized with alloying elements, using a low amount of cold work, and by using a low annealing temperature.

### 8.3.2 Abnormal Grain Growth

Abnormal grain growth, also known as discontinuous grain growth, grain coarsening, and secondary recrystallization, can occur at very high annealing temperatures. Abnormal grain growth occurs when normal growth of the matrix grains is inhibited and when the temperature is high enough to allow a few special grains to overcome the inhibiting force and to grow disproportionately, as illustrated in Fig. 8.25. Initially, the grains grow slowly, followed by rapid growth to sizes on the order of centimeters in some cases.

Abnormal grain growth occurs when the grain boundaries are initially pinned by the presence of a finely precipitated second phase. The presence of inclusions can make a material very

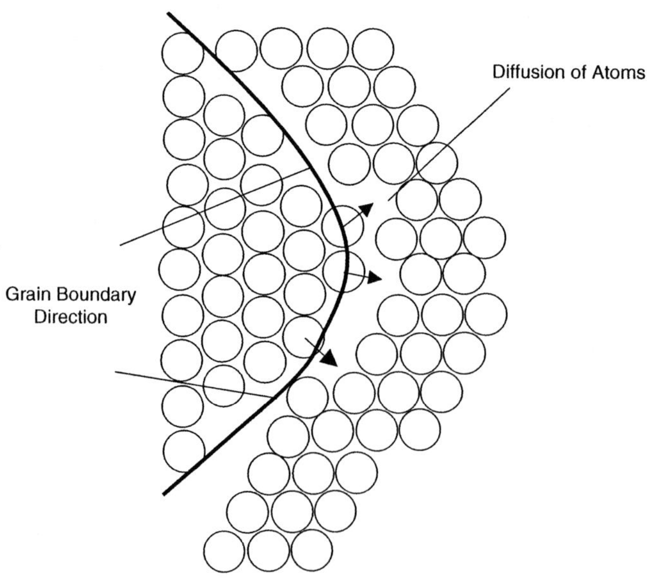


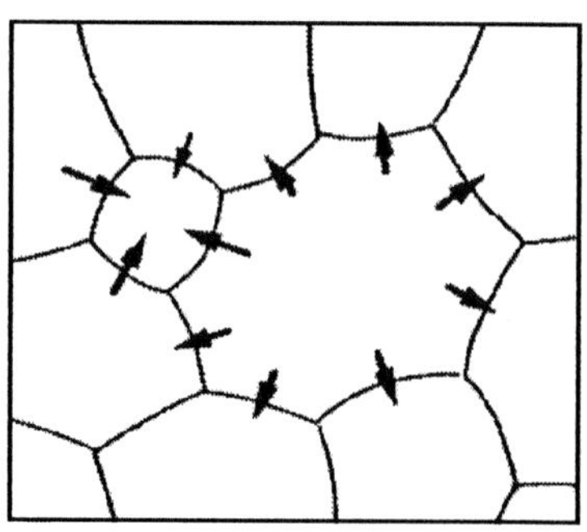

**Fig. 8.22** Diffusion of atoms and grain boundary movement. Source: Ref 2

resistant to grain growth. Such materials are inherently fine grained. However, this resistance can break down at elevated temperatures, resulting in discontinuous or exaggerated grain growth. If the temperature is high enough or the time long enough, the precipitated second-phase particles can coarsen, become larger and fewer in numbers, or even redissolve. Since this is not a uniform process, select boundaries will break away before others and proceed to grow to very large sizes. For example, aluminum-killed steels have a fine dispersion of AlN particles that stabilize the grain size during annealing but redissolve at temperatures above 980 °C (1800 °F). Abnormal grain growth can actually be more severe in initially fine-grained steels as compared to coarse-grained steels (Fig. 8.26). Microstructural features that inhibit grain growth include a fine dispersion of second-phase particles, a discrete grain-boundary precipitate, a strong single-orientation texture, and a stabilized two-dimensional grain structure imposed by sheet thickness.

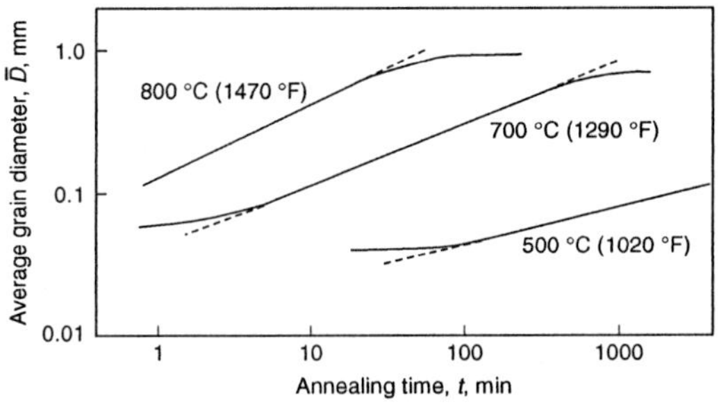


**Fig. 8.23** Grain growth in zone-refined iron during isothermal anneals. Source: Ref 1

Annealing data can be more easily visualized in annealing maps, such as the one shown for

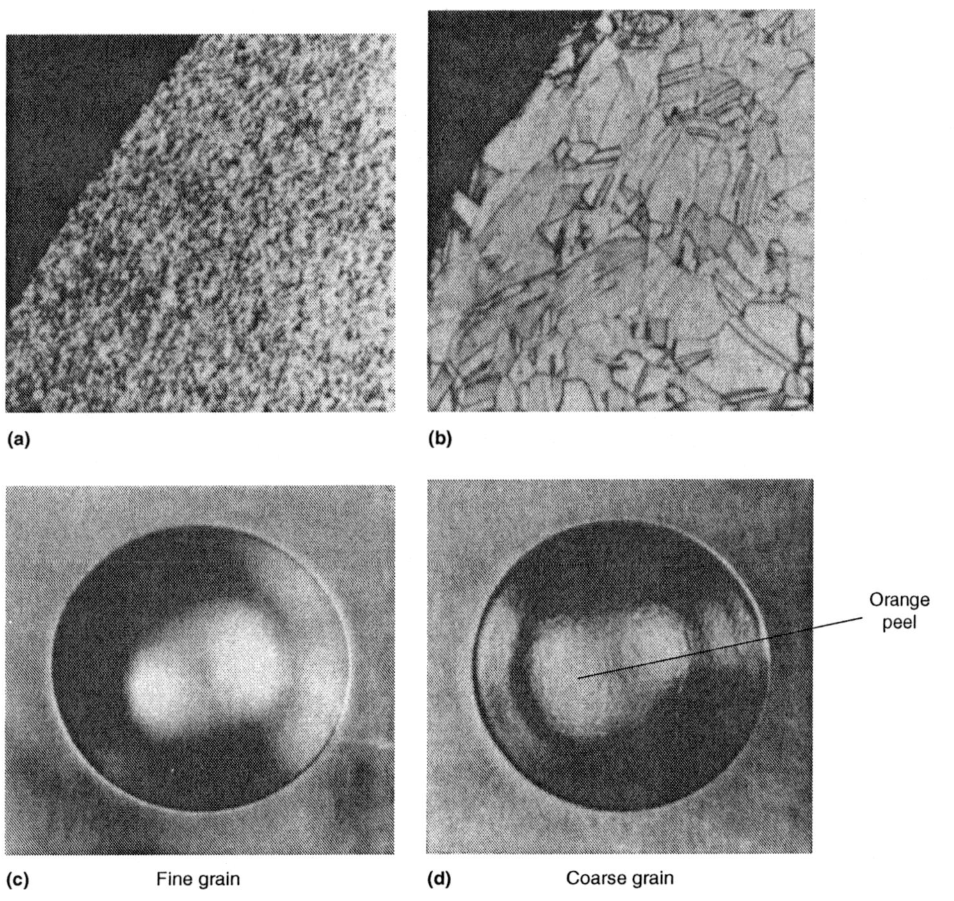


**Fig. 8.24** Effect of grain size on cold-drawn brass sheet. Source: Ref 7

aluminum in Fig. 8.27. Here, the effects of the amount of deformation, the recrystallization temperature, and the recrystallized grain size are presented on one chart. Note the large grain sizes obtained when the amount of deformation is insufficient and when grain growth occurs at high annealing temperatures.

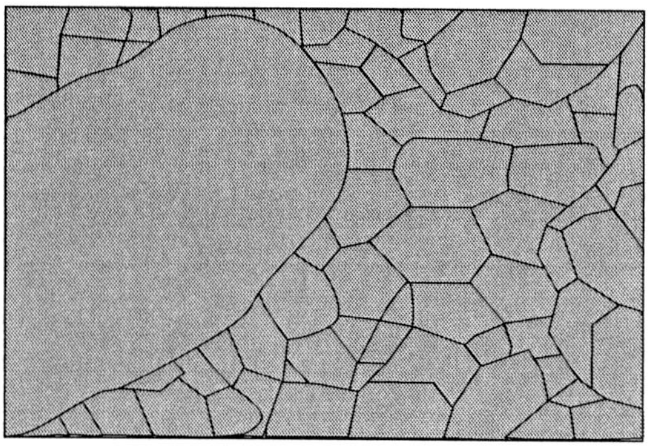

**Fig. 8.25** Abnormal grain growth

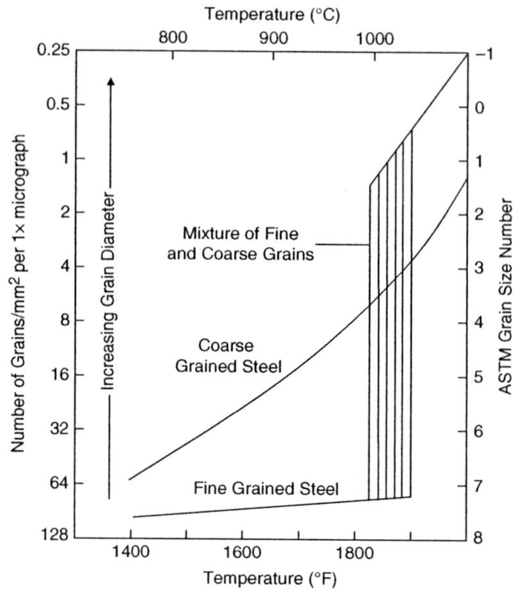


**Fig. 8.26** Grain growth in fine- and coarse-grained steels. Source: Ref 5

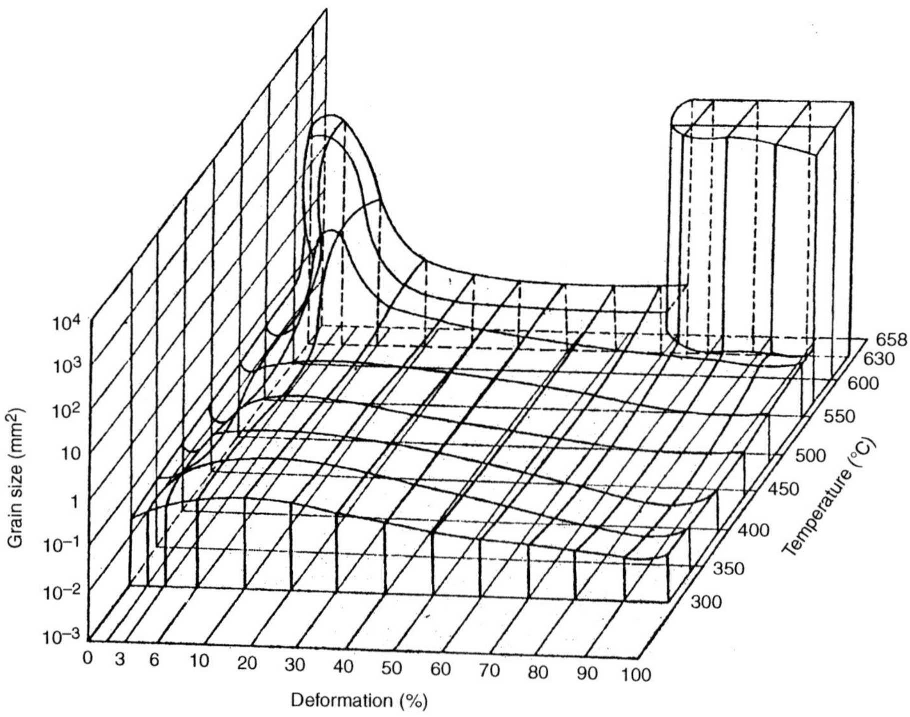


**Fig. 8.27** Annealing map for aluminum. Source: Ref 8

## ACKNOWLEDGMENTS

Sections of this chapter were adapted from "Recovery, Recrystallization, and Grain-Growth Structures" by W.L. Mankins in *Metallography and Microstructures,* Volume 9, *ASM Handbook,* ASM International, 2004.

## REFERENCES

1. W.L. Mankins, Recovery, Recrystallization, and Grain-Growth Structures, *Metallography and Microstructures,* Vol 9, *ASM Handbook,* ASM International, 2004
2. V. Singh, *Physical Metallurgy,* Standard Publishers Distributors, 1999
3. R.M. Brick, A.W. Pense, and R.B. Gordon, *Structure and Properties of Engineering Materials,* 4th ed., McGraw-Hill Book Company, 1977
4. R.E. Reed-Hill and R. Abbaschian, *Physical Metallurgy Principles,* 3rd ed., PWS Publishing Company, 1991
5. A.G. Guy, *Elements of Physical Metallurgy,* 2nd ed., Addison-Wesley Publishing Company, 1959
6. M.F. Ashby and D.R.H. Jones, *Engineering Materials 2—An Introduction to Microstructures, Processing and Design,* 2nd ed., Butterworth Heinemann, 1998
7. Microstructure of Copper and Copper Alloys, *Metallography and Microstructures,* Vol 9, *ASM Handbook,* ASM International, 2004
8. M. Tisza, *Physical Metallurgy for Engineers,* ASM International, 2001

# CHAPTER 9

# Precipitation Hardening

PRECIPITATION HARDENING is used extensively to strengthen aluminum alloys, magnesium alloys, nickel-base superalloys, beryllium-copper alloys, and precipitation-hardening stainless steels. In precipitation hardening, an alloy is heated to a high enough temperature to take a significant amount of an alloying element into solid solution. It is then rapidly cooled (quenched) to room temperature, trapping the alloying elements in solution. On reheating to an intermediate temperature, the host metal rejects the alloying element in the form of fine precipitates that create matrix strains in the lattice. These fine precipitate particles act as barriers to the motion of dislocations and provide resistance to slip, thereby increasing the strength and hardness.

## 9.1 Particle Hardening

Particle, or dispersion, hardening occurs when extremely small particles are dispersed throughout the matrix. When a dislocation encounters a fine particle, it must either cut through the particle or bow (loop) around it, as shown schematically in Fig. 9.1. Two types of particle strengthening are discussed in this chapter: precipitation hardening, which takes place during heat treatment, and true dispersion hardening, which can be achieved by mechanical alloying and powder metallurgy consolidation.

Particles are usually classified as deformable or nondeformable, meaning that the dislocation is able to cut through it (deformable) or the particle is so strong that the dislocation cannot cut through (nondeformable). For effective particle strengthening (Fig. 9.2), the matrix should be soft and ductile, while the particles should be hard and discontinuous. A ductile matrix is better in resisting catastrophic crack propagation. Smaller and more numerous particles are more effective at interfering with dislocation motion than larger and more widely spaced particles. Preferably, the particles should

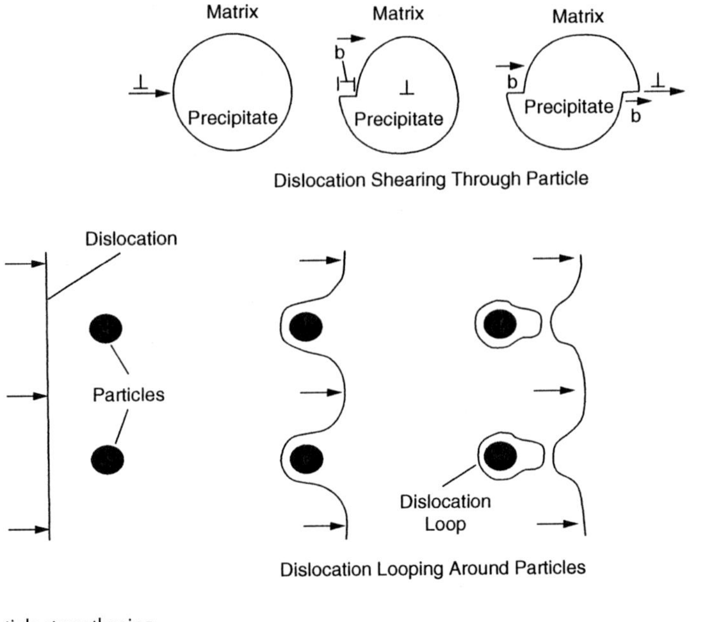


**Fig. 9.1** Particle strengthening

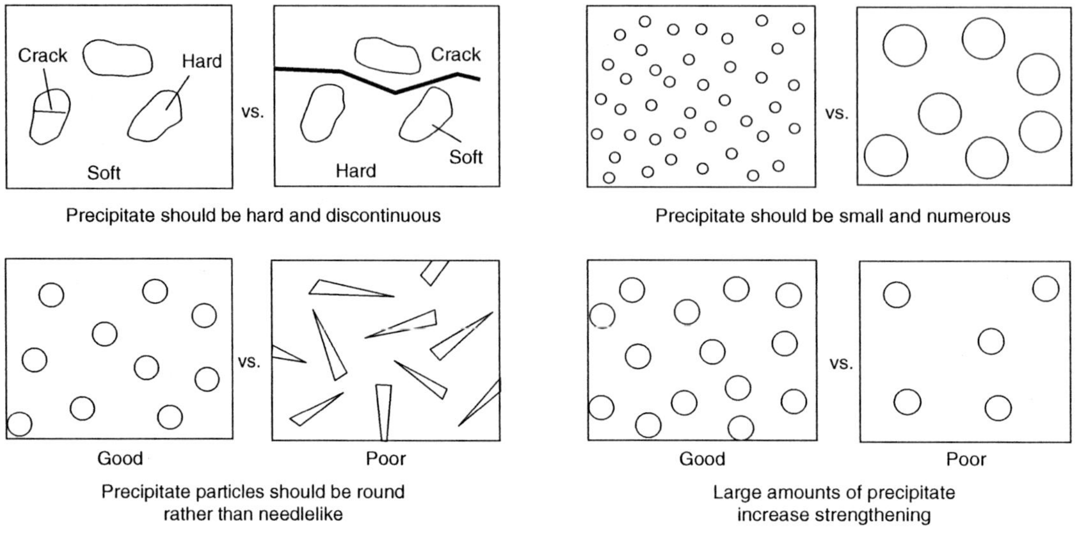


**Fig. 9.2** Particle-hardening considerations. Source: Ref 1

be spherical rather than needlelike to prevent stress-concentration effects. Finally, larger amounts of particles increase strengthening.

## 9.2 Theory of Precipitation Hardening

Precipitation hardening is also known as age hardening or aging, to indicate that the resulting strength increase develops with time. It should be noted that sometimes the term *age hardening* is used to denote alloys that harden on aging at room temperature, while the term *precipitation hardening* is used to denote alloys that must be heated above room temperature for hardening to occur at an appreciable rate. However, in both cases, the hardening mechanism is the precipitation of extremely fine particles, which impedes dislocation movement. A portion of a phase diagram for an alloy system that has the characteristics required for precipitation hardening is shown in Fig. 9.3. Note that the solvent metal at the left-hand edge of the diagram can absorb much more of the solute metal at elevated temperature than it can at room temperature. When the alloy is heated to the solution heat treating temperature and held for a sufficient length of time, the solvent metal absorbs some of the solute metal. Then, when it is rapidly cooled to room temperature, atoms of the solute metal are trapped as a supersaturated solid solution in the solvent metal. On reheating to an intermediate aging temperature, the supersaturated solution precipitates very fine particles that act as barriers to dislocation movement. Note the effects of different aging temperatures, shown in Fig. 9.3. If the metal is aged at too low a temperature ($T_1$), the precipitation process will be incomplete, and the desired strength will not be achieved, a condition known as underaging. On the other hand, aging at too high a temperature ($T_4$) also results in lower-than-desired strength because the precipitate particles coarsen, and the alloy is now said to be overaged. Commercial heat treatments are closer to $T_2$ and $T_3$, in which the optimal strength can be obtained in a reasonable aging time. The alloy used in this example is one that requires artificial or elevated-temperature aging. Some alloys will age satisfactorily at room temperature, a process called natural aging.

Alloys that harden by precipitation hardening do so by forming coherent precipitates within the matrix. The first step in the aging process is the congregation of solute atoms in the matrix lattice. These solute-rich regions are called clusters and are the embryos for nucleation. Solute atoms then diffuse to the clusters from the surrounding matrix and convert some of them to nuclei of a new phase. During the early phases of precipitation, the equilibrium phase does not immediately form, but an intermediate crystal structure related to it grows in close contact with the solid solution. As long as there tends to be atomic matching, or coherency, between the transition phase and the matrix, the transition

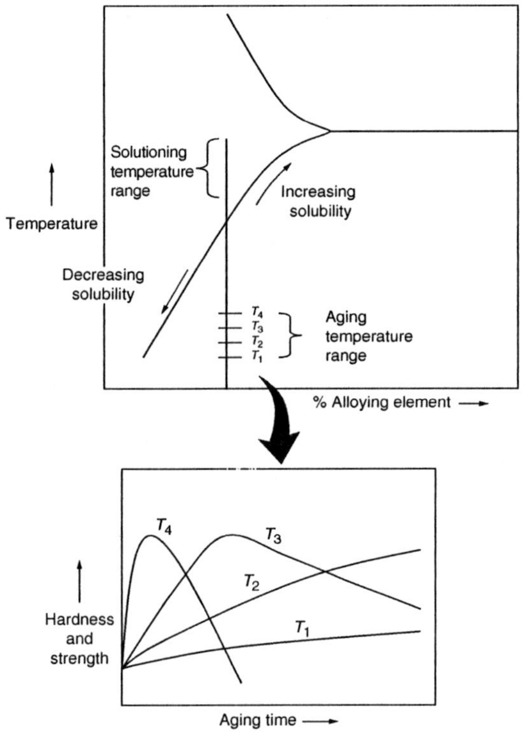


**Fig. 9.3** Partial binary aluminum phase diagram and typical precipitation-hardening heat treatment for aluminum. Source: Ref 2

phase will create a local strain field within the matrix. The initial precipitate particles are often not spherical but can be platelike or rodlike in shape. The combination of a fine precipitate size and the localized strain fields is an effective barrier to dislocation movement.

For effective precipitation hardening, either a coherent or semicoherent interface must be present. The lattice distortion produced by a coherent precipitate, which impedes dislocation motion, is illustrated in Fig. 9.4. When the alloy is overaged, an incoherent interface develops that is accompanied by softening of the alloy. The peak strength is usually attained when the alloy is aged to the point where both particle cutting and particle bowing (looping) of dislocations contribute to the alloy strength (Fig. 9.5) The eventual formation of the equilibrium phase is always energetically favorable. If the alloy is aged at room temperature, equilibrium may never be achieved. However, when the alloy is aged at elevated temperature, either longer aging times or higher temperatures will result in the formation of the final equilibrium precipitate. Since the equilibrium precipitate is larger and more widely spaced, it no longer imposes a strain field on the matrix and is not nearly as effective in blocking dislocation motion, and the strength properties decrease. The equilibrium phase usually nucleates separately from a transition phase(s) and competes with the less stable transition phase for solute atoms, eventually causing the transition phase to redissolve as its particles get smaller. As the volume fraction of equilibrium phase grows, the sizes of the precipitate particles increase, their numbers decrease, and the distance between particles increases, allowing easier passage of dislocations. It is important to remember that precipitation-hardened alloys are metastable; equilibrium is always lurking around the corner. All precipitation-hardened alloys will eventually soften if heated to high enough temperatures or if they are exposed to somewhat lower temperatures for long periods of time. This being said, some are extremely stable at elevated temperatures, such as the ordered $Ni_3(Al,Ti)$ precipitate in precipitation-hardened nickel-base superalloys that are used in the hottest portions of jet engines.

Coarsening of the particles occurs because the microstructure of a two-phase alloy is not stable unless the interfacial energy is at a minimum. A lower density of larger particles has less total interfacial energy than a higher density of small particles, which provides the driving force for particle coarsening that occurs by diffusion of solute atoms. Thus, the rate of coarsening increases with temperature. Ostwald ripening is the mechanism by which smaller precipitates dissolve, and the solute is redistributed to larger, stable precipitates. Smaller particles have a higher free energy due to increased pressure as a result of their high surface curvature. Reversion occurs when an alloy containing a coherent precipitate, or an intermediate semicoherent precipitate, is heated above its solvus temperature, allowing the particles to redissolve into the matrix.

Precipitate formation is not always uniform. Remember that grain boundaries are high-energy sites, and precipitates often form along the grain boundaries. Some alloys will form discontinuous precipitates at the grain boundaries, in which lamellae of the second phase are interspersed with the solute-depleted matrix. Small additions of nickel or cobalt are used in beryllium-copper alloys to minimize this effect, since it adversely affects mechanical properties.

A number of alloy systems that can be precipitation hardened are given in Table 9.1. Aluminum alloys are one of the most important series of alloys that can be precipitation hardened, including the 2*xxx* (aluminum-copper), 6*xxx* (aluminum-magnesium-silicon), 7*xxx* (aluminum-zinc), and some of the 8*xxx* (aluminum-lithium) alloys. Some of the copper alloys, in particular beryllium-copper, can be precipitation hardened. The iron- and nickel-base superalloys are another particularly important class of precipitation-hardening alloys. In the nickel-base superalloys, the precipitate $Ni_3(Al,Ti)$ has very little lattice misfit with the nickel matrix (<2%), which produces very low strain energies (~10 to 30 mJ/m$^2$) and provides resistance to overaging for prolonged periods at high temperatures.

## 9.3 Precipitation Hardening of Aluminum Alloys

The process of strengthening by precipitation hardening plays a critical role in high-strength aluminum alloys. Precipitation hardening consists of three steps:

1. Solution heat treating
2. Rapid quenching
3. Aging

In solution heat treating, the alloy is heated to a temperature that is high enough to put the soluble alloying elements in solution. After holding at the solution-treating temperature long enough for diffusion of solute atoms into the solvent matrix to occur, it is quenched to a lower temperature (e.g., room temperature) to keep the

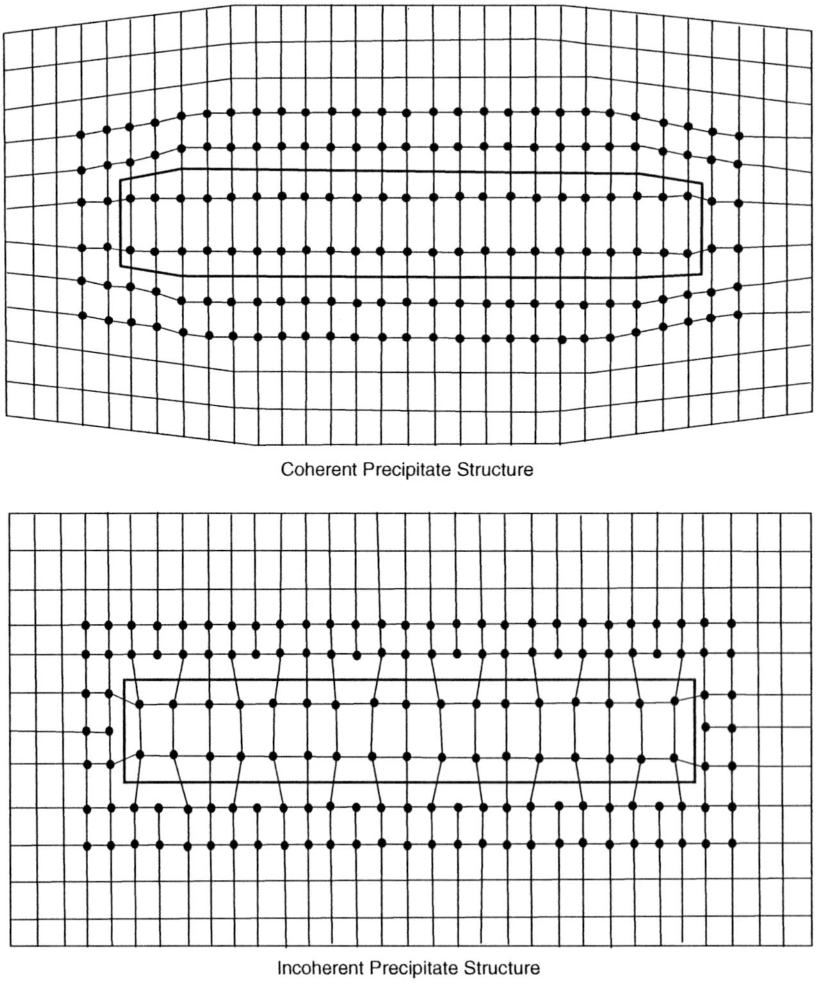


**Fig. 9.4** Coherent and incoherent precipitates

alloying elements trapped in solution. During aging, the alloying elements trapped in solution precipitate to form a uniform distribution of very fine particles. Some aluminum alloys will harden after a few days at room temperature, a process called natural aging, while others are artificially aged by heating to an intermediate temperature.

Consider the aluminum-copper system shown in Fig. 9.6. The equilibrium solid solubility of copper in aluminum increases from approximately 0.20% at 250 °C (480 °F) to a maximum of 5.65% at the eutectic melting temperature of 550 °C (1018 °F). It is even lower than 0.20% at temperatures below 250 °C (480 °F). For aluminum-copper alloys containing from 0.2 to 5.6% Cu, two equilibrium solid states are possible. At temperatures above the solvus curve, copper is completely soluble, and when the alloy is held at such temperatures for sufficient time to permit diffusion, copper will be taken completely into the α solid solution. At temperatures below the solvus, the equilibrium state consists of two solid phases: solid solution, α, plus the equilibrium intermetallic compound, θ ($CuAl_2$).

If an alloy of aluminum containing 4% Cu is heated to 500 °C (940 °F) and held for 1 h, the copper will go into solution in the aluminum. After solution heat treating, the alloy is quenched in cold water to room temperature to keep the copper in solution. The alloy is then artificially aged at 170 °C (340 °F) for 10 h. During the aging process, very fine particles of aluminum-copper are precipitated, and the strength and hardness increase dramatically. Precipitation heat treatments generally are low-temperature, long-term processes. Aging temperatures for aluminum range from 115 to 190 °C (240 to 375 °F), with times between 5 and 48 h.

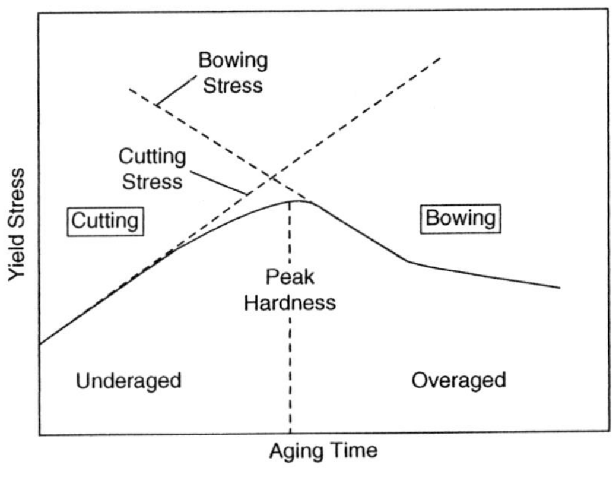


**Fig. 9.5** Relative contributions of particle cutting and dislocation bowing

On quenching, the copper is trapped as a supersaturated solution in the matrix. There is a strong driving force to precipitate the copper as the equilibrium precipitate, θ ($CuAl_2$). However, in aluminum alloys, precipitation occurs by one or more metastable transition precipitates appearing before the appearance of the final equilibrium phase, with each successive stage lowering the free energy of the system. The transition precipitates are often crystallographically similar to the matrix, allowing the formation of a low-energy coherent interface during the nucleation process. Often, the precipitation sequence begins with the nucleation of small, fully coherent phases known as Guinier-Preston (GP) zones. The GP zones are ordered, solute-rich clusters resulting from phase separation that nucleate and grow.

The GP zones are the first to nucleate, because of their small size and coherency with the matrix. The interfacial energy term is extremely low, providing a low barrier to nucleation. The GP zones are extremely fine, with sizes in the range of tens of angstroms. The exact shape, size, and distribution of the GP zones depend on the specific alloy and on the thermal and mechanical history of the product. The GP zones typically take the shape of small, spherical

**Table 9.1 Some common precipitation-hardening systems**

| Matrix | Solute | | Transition structures(a) | Equilibrium precipitate |
|---|---|---|---|---|
| Al | Cu | (i) | Platelike solute-rich GP I zones | θ-$CuAl_2$ |
| | | (ii) | Ordered GP II zones | |
| | | (iii) | θ″-phase | |
| | | (iv) | θ′-phase | |
| Al | Mg, Si | (i) | GP zones rich in Mg and Si atoms | β″-$Mg_2Si$ |
| | | (ii) | Ordered zones of β″ | |
| Al | Mg, Cu | (i) | GP zones rich in Mg and Cu atoms | S-$CuAl_2Mg$ |
| | | (ii) | S′-platelets | |
| Al | Mg, Zn | (i) | Spherical zones rich in Mg and Zn | η-$MgZn_2$ |
| | | (ii) | Platelets of η′-phase | |
| Cu | Be | (i) | Be-rich regions | γ-CuBe |
| | | (ii) | γ′ spherical GP zones | |
| Cu | Co | (i) | Spherical GP zones | β-Co |
| Ni | Al, Ti | (i) | γ′ cubes | γ-$Ni_3$(AlTi) |
| Fe | C | (i) | α′ martensite | $Fe_3C$ |
| | | (ii) | α″ martensite | |
| | | (iii) | ε carbide | |
| Fe | N | (i) | α′ nitrogen martensite | |
| | | (ii) | α″ nitrogen martensite | $Fe_4N$ |

(a) GP, Guinier-Preston. Source: Ref 3

particles or disk-shaped particles that are approximately two atomic layers thick, 10 nm in diameter, with spacings on the order of 10 nm. Their formation requires movements of atoms over very short distances, and their densities can approach $10^{17}$ to $10^{18}$ $cm^{-3}$. The GP precipitates then grow into more stable transition phases and, eventually, the equilibrium phase.

The phases that nucleate and grow from the GP zones are termed transition phases. Transition phases form because they have lower activation energies of formation than the equilibrium phase, θ, although the system eventually wants to move toward the formation of θ because, as the equilibrium phase, it has the overall lowest free energy. As shown in Fig. 9.7, the transition phases have a crystal structure intermediate between the matrix and the equilibrium phase. This minimizes the strain energy between the precipitate and the matrix, making it more favorable in the nucleation sequence than the equilibrium phase, which is incompatible with the matrix and has a high interfacial energy. A typical reaction sequence for aluminum-copper systems is:

$$\text{Super saturated solid solution} \rightarrow \text{Clustering} \rightarrow \text{GP zones} \rightarrow \theta'' \rightarrow \theta' \rightarrow \theta$$

where $\theta''$ and $\theta'$ are transition precipitates, and θ is the final equilibrium precipitate. While the GP zones are totally coherent with the matrix, the much larger transition precipitates are only semicoherent. As each new precipitate forms, the matrix (α) becomes more and more depleted in copper. The GP zones and $\theta''$ precipitates, also known as GP II zones, can be resolved by a transmission electron microscope because of the lattice coherency strains, as shown in Fig. 9.8. The zones themselves are too small to be resolved, but the resulting strain fields can be resolved. Each step results in the previously precipitated phase being replaced with a new, more stable phase with a lower free energy.

During heating, the GP zones develop an intermediate precipitate ($\theta''$), which has a tetragonal structure that forms as plates and maintains coherency with the matrix and further increases the strain in the matrix, providing peak strength levels. On still further heating, $\theta''$ is replaced by a second intermediate precipitate, $\theta'$, which is not coherent with the matrix, and the strength starts to decrease; the alloy is now termed overaged. However, in the highest strength condition, both $\theta''$ and $\theta'$ are generally present. Both the precipitate particles themselves and the strains they produce in the lattice structure inhibit dislocation motion, and thus, both contribute to strengthening. To maximize strengthening, aging is typically carried out at temperatures between those where precipitation of $\theta''$ and $\theta'$ occurs, because this spacing and lattice strain are ideal for hindering dislocation motion. Further heating of the alloy causes $\theta'$ to transform to the equilibrium precipitate, θ, which is stoichiometric $CuAl_2$.

The solvus lines for the GP zones and the transition phases can be shown as lines on the phase diagram (Fig. 9.9). This series of lines defines the upper temperature limit of the various transition phases for different concentrations. For example, if aging is carried out above the $\theta''$ solvus but below the $\theta'$ solvus, then the first precipitate to form will be $\theta'$.

The GP zones will normally develop on aging at room temperature. The fact that this will happen at room temperature is somewhat surprising and can be attributed to a high vacancy concentration. When the alloy is heated to the solution heat treating temperature, the equilibrium number of vacancies increases as the temperature increases. On quenching, the vacancies become trapped in solution. These excess vacancies are then available to help accelerate the nucleation and growth process.

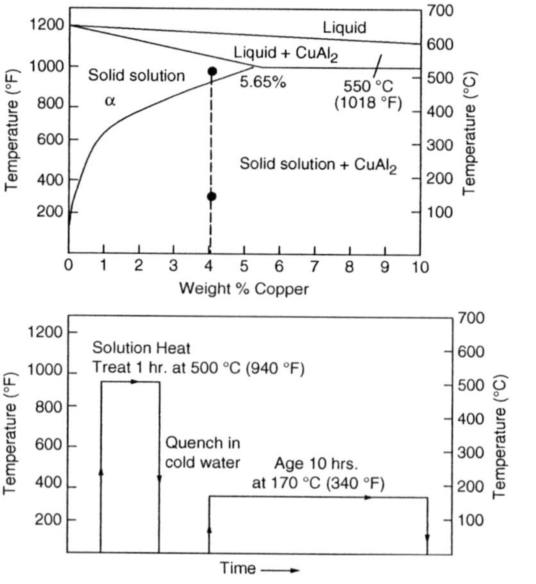


**Fig. 9.6** Precipitation hardening of an aluminum-copper alloy

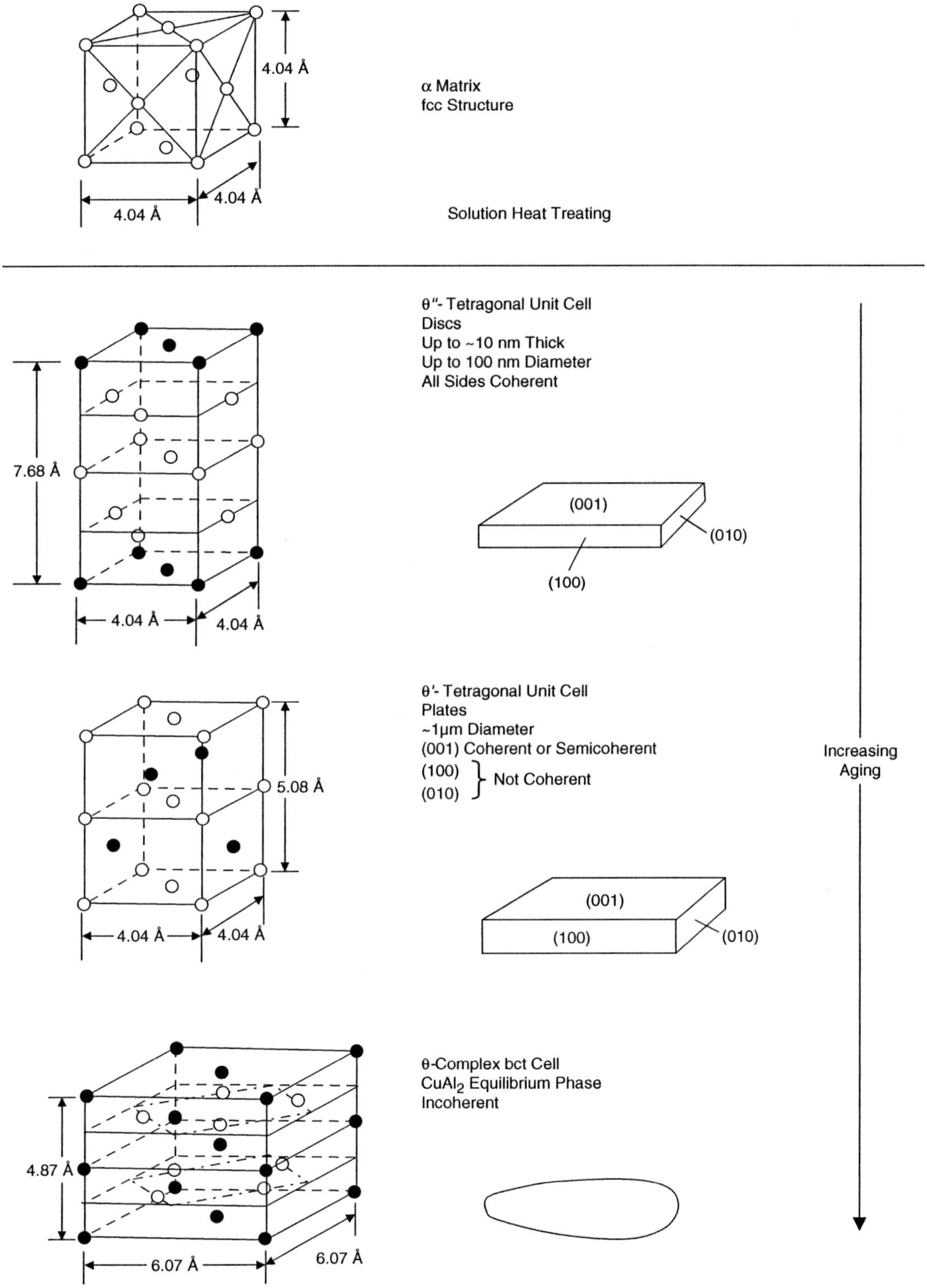


**Fig. 9.7** Aluminum-copper precipitation sequence. fcc, face-centered cubic; bct, body-centered tetragonal. Source: Ref 4

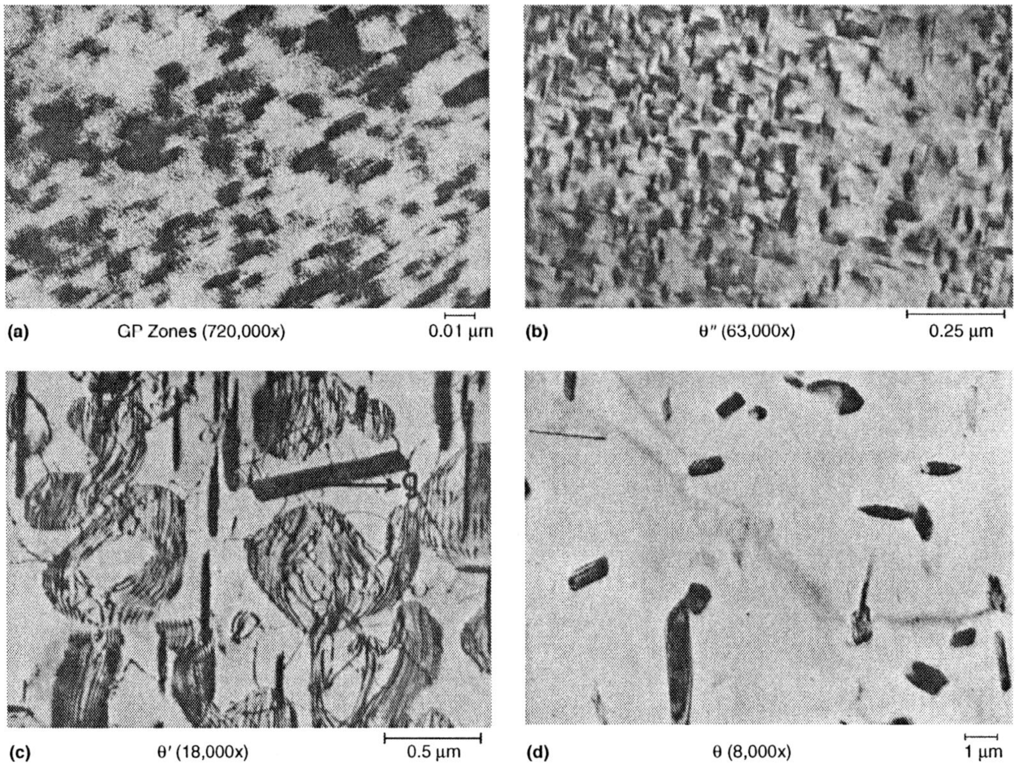


**Fig. 9.8** Transmission electron micrographs of aluminum-copper precipitation sequence. GP, Guinier-Preston. Source: Ref 5

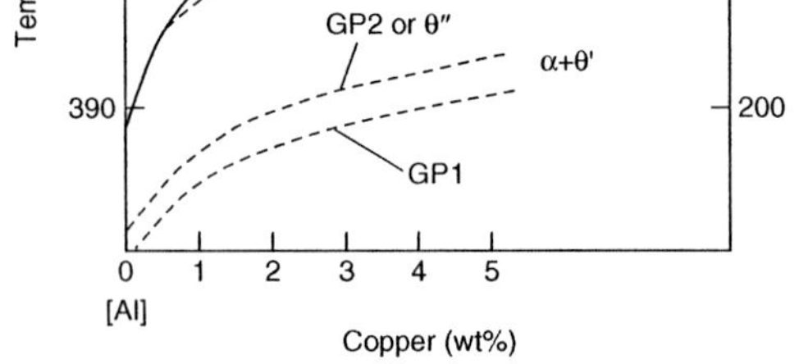


**Fig. 9.9** Aluminum-copper binary diagram with Guinier-Preston (GP) 1, θ″, and θ′ solvus lines. Source: Ref 4

The other heat treatable aluminum alloys behave similarly, but the precipitates are, of course, different. The progression of particle growth during the aging process is shown schematically in Fig. 9.10, and a series of aging curves for an Al-4%Cu alloy is shown in Fig. 9.11. Reactions carried out beyond maximum strengthening are overaged, because the beneficial effects of precipitation strengthening are lost as the precipitates grow larger in size and spacing. Both the underaged and overaged conditions have lower strengths and hardness levels than the peak-aged condition.

In aluminum alloys, precipitate-free zones (PFZs) can occur adjacent to the grain boundaries (Fig. 9.12). The grain boundaries themselves and the interior of the grains contain precipitate particles, but there is a zone adjacent to the boundaries with very few particles. These differences in chemical composition can set up galvanic effects that lead to intergranular corrosion. There are two plausible explanations for PFZs. The first is due to vacancy migration. Grain boundaries are a major sink for vacancies that migrate to the grain boundaries and deplete the areas adjacent to the boundaries of vacancies. The lack of vacancies in these depleted areas inhibits the nucleation and growth of precipitate particles, even though the matrix in these areas contains sufficient solute. The second explanation is that precipitation occurs directly on the higher-energy grain boundaries, and the adjacent areas become depleted of solute. There is direct experimental evidence for both of these mechanisms. Special heat treatments are used to minimize the formation of PFZs. Precipitate-free zones can be eliminated by a two-stage heat treatment where nucleation is induced homogeneously at a low temperature, and the precipitates are then allowed to grow during the second higher-temperature aging treatment. In addition, higher solution-treating temperatures and faster quenching rates also reduce the PFZ widths. The beneficial effect of a faster quenching rate is shown in Fig. 9.13.

### *9.3.1 Solution Heat Treating*

The first step is to produce a supersaturated solid solution. Solution heat treating is accomplished by heating the alloy to a high enough temperature and soaking for a sufficient time to achieve a nearly homogeneous solid solution. Nominal commercial solution heat treating temperatures are determined by the composition limits of the alloy, with an allowance for unintentional temperature variations. The time at temperature required for solution treating should be long enough for diffusion to occur to produce a supersaturated solid solution. The time required can be less than a minute for thin sheet to as much as 20 h for large inhomogeneous castings.

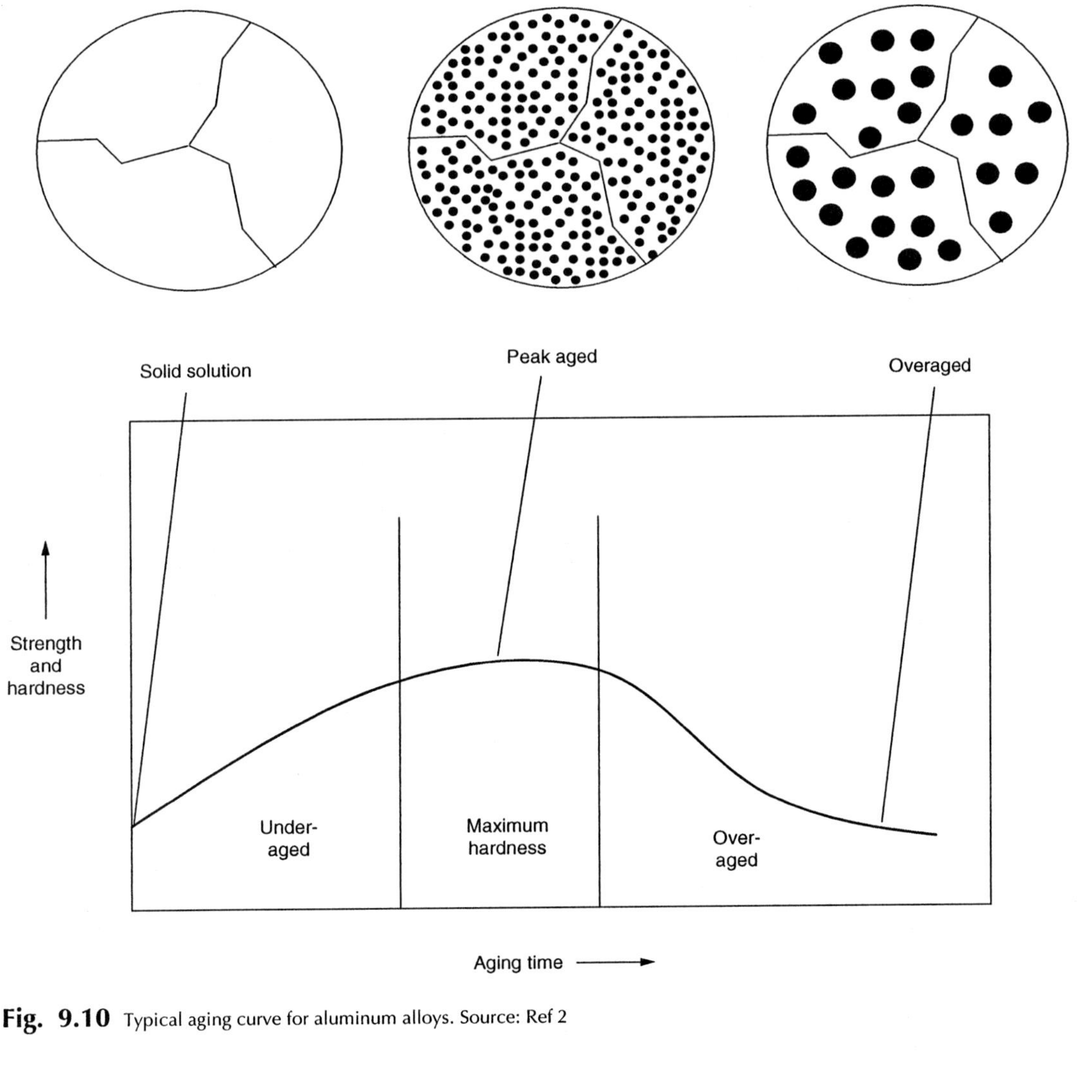


**Fig. 9.10** Typical aging curve for aluminum alloys. Source: Ref 2

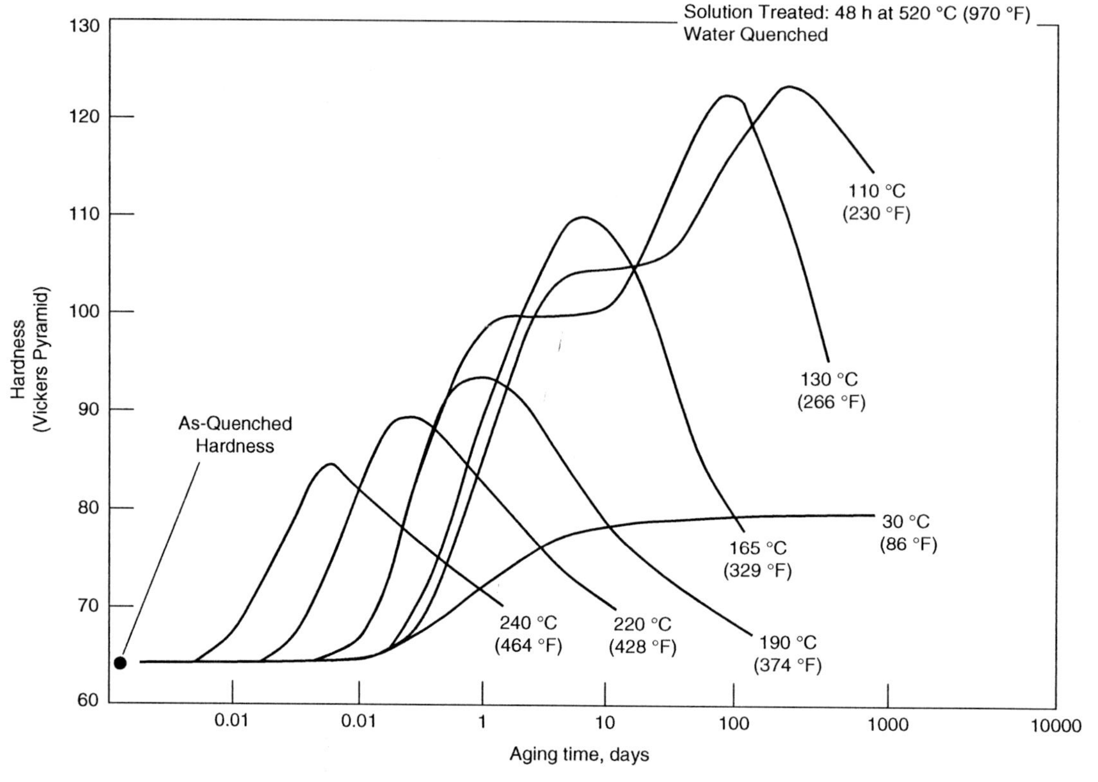


**Fig. 9.11** Aging curves for Al-4%Cu alloy

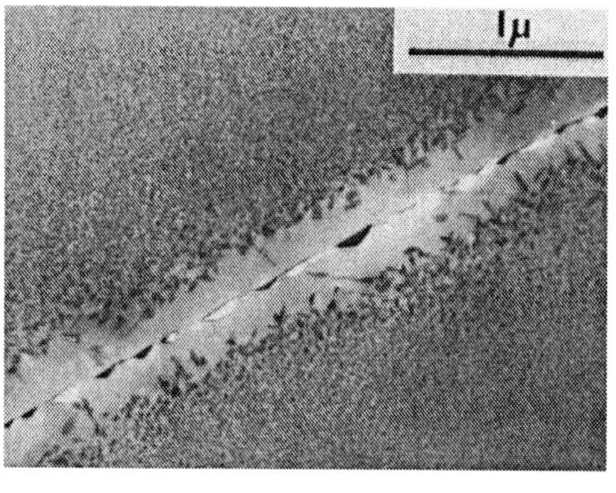


**Fig. 9.12** Precipitate-free zone in an Al-Zn-Mg alloy. Source: Ref 6

Castings, which normally contain microsegregation from the solidification process, usually require homogenization before solution treating. Homogenization consists of heating to a high temperature, but below the eutectic temperature, and then soaking for times long enough to allow diffusion to eliminate a majority of the microsegregation. If appreciable segregation is present when the alloy is heated to the solution-treating temperature, low-melting microconstituents, usually located at the grain boundaries, can melt and ruin the part, a process frequently called burning. Even in a fairly homogeneous wrought product, close control of the temperature during solution treating, usually within 3 to 6 °C (5 to 10 °F) is necessary, because the objective is to heat the alloy to as high a temperature as possible to dissolve the maximum amount of solute, but not to heat it so high that grain-boundary melting occurs. When the temperatures attained during solution treating are appreciably below the normal range, solution will be incomplete, and the resultant strength will be somewhat lower than normal when the alloy is aged.

### *9.3.2 Quenching*

During quenching, it is important to cool the alloy quickly enough so that the supersaturated solid solution is maintained on cooling to room temperature. On the other hand, rapid quenching often results in high residual stresses and part distortion, so less drastic quenches frequently are used with some sacrifice in properties to minimize distortion.

To avoid appreciable precipitation during the quenching process, two requirements must be met: (1) The time required to transfer the part from the furnace to the quench medium must be short enough to preclude slow cooling in the temperature range where very rapid precipitation can occur; and (2) the cooling-rate capacity of the quench medium must be such that little or no precipitation occurs during cooling. For example, if a quench tank is used, the tank must be large enough that there is not a significant rise in the tank temperature during quenching. Aluminum alloy parts are usually tank quenched in cold water to provide the optimum strength and corrosion resistance. For plate stock that is heat treated at the mill, quenching is conducted with high-velocity sprays of cold water. For parts with complex shapes that are subject to distortion, milder quenches are often used, such as aqueous solutions of polyalkaline glycol or hot water maintained at 65 to 80 °C (150 to 180 °F).

Immediately after quenching, most aluminum alloys are nearly as ductile as they are in the annealed condition. Thus, straightening and forming operations are often conducted in the as-quenched temper. Because precipitation hardening will occur at room temperature, forming or straightening is usually conducted as soon as possible after quenching. Forming and straightening operations vary in degree from minor corrections of warpage to complete forming of complex parts from solution-treated flat blanks. In either case, the plastic deformation must be completed before an appreciable amount of precipitation hardening takes place.

### 9.3.3 Aging

After solution treatment and quenching, hardening is achieved by aging at room temperature (natural aging) or by reheating to an intermediate temperature (artificial aging). In some alloys, sufficient precipitation occurs in a few days at room temperature to yield stable products with properties that are adequate for many applications. These alloys are sometimes artificially aged to provide increased strength and hardness in wrought or cast products. Other

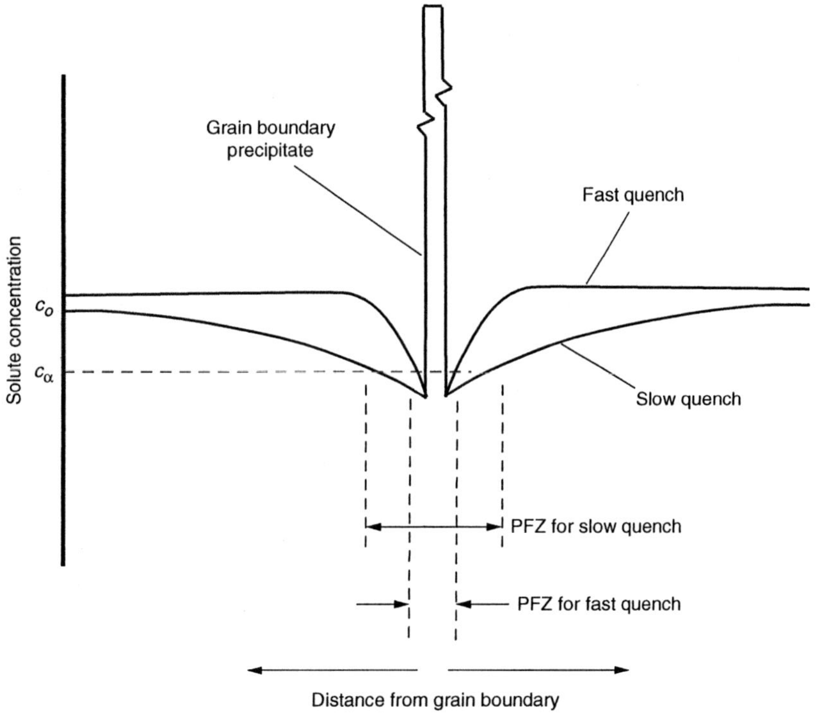


**Fig. 9.13** Precipitate-free zone (PFZ) width as a function of quench rate. Source: Ref 4

alloys with slow precipitation reactions at room temperature are always artificially aged.

**Natural Aging.** The wrought heat treatable 2*xxx* alloys generally contain magnesium in addition to copper as an alloying element, which greatly improves their response to natural aging. Other significant alloying additions include titanium to refine the grain structure during ingot casting, and transition element additions (manganese, chromium, and/or zirconium) that form dispersoid particles ($Al_{20}Cu_2Mn_3$, $Al_{18}Mg_3Cr_2$, and $Al_3Zr$), which help control the wrought grain structure. Iron and silicon are considered impurities and are held to an absolute minimum because they form intermetallic compounds ($Al_7Cu_2Fe$ and $Mg_2Si$) that are detrimental to both fatigue and fracture toughness.

For aluminum alloys, solution treating, quenching, and natural aging is designated as the T4 temper, a heat treatment used for many of the aluminum-copper 2*xxx* alloys. For these alloys, the relatively high supersaturation of atoms and vacancies retained by rapid quenching causes rapid formation of GP zones, and strength increases rapidly, attaining nearly maximum stable values in four to five days. In some of the 2*xxx* alloys, cold working of the freshly quenched alloy greatly increases its response to later natural aging by increasing the density of dislocations that serve as sites for nucleation of additional precipitate particles. This heat treatment scheme of solution treating, quenching, cold working, and natural aging is designated as the T3 temper. A typical thermal cycle for the T3 and T4 tempers is shown in Fig. 9.14, with the temperatures being the same but with the T3 incorporating a cold working sequence prior to

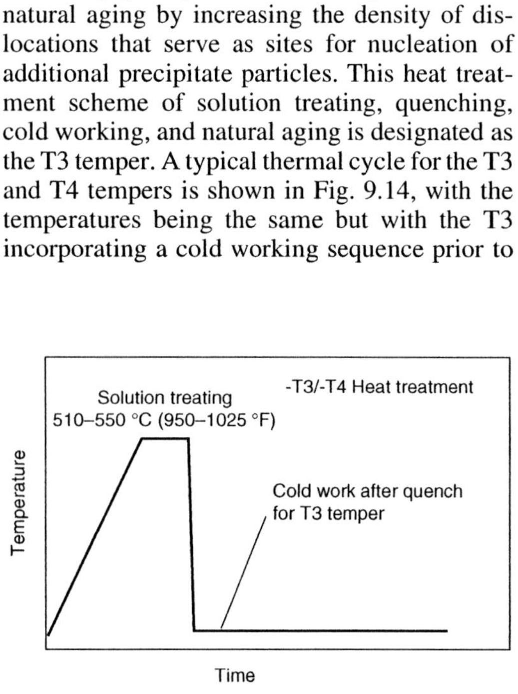


**Fig. 9.14** Time-temperature profile for solution-treated and naturally aged alloys

natural aging. At the mill, roller or stretcher leveling to achieve flatness or straightness produces modest strain (~1 to 4%) that also increases strength properties. Cold rolling and additional stretching operations result in larger strains and greater strength increases. Both the T3 and T4 tempers are characterized by high ratios of tensile to yield strength, high fracture toughness, and good fatigue resistance. A comparison of properties for 2024 Al heat treated to the T3 and T4 conditions in Fig. 9.15 shows the dramatic improvement in properties as compared to the annealed (O) condition. Note that even though the strength is significantly increased, the elongation to failure is little affected. In the 2*xxx* alloys for which T3 or T4 tempers are standard, changes that occur on further natural aging are relatively minor, and these tempers are regarded as essentially stable after approximately one week.

In some alloys of the 2*xxx* series, strain produced by cold working after solution heat treatment and quenching but before artificial aging nucleates a finer, denser precipitate dispersion that considerably increases strength. This is the basis for the higher-strength T8 tempers, which are produced by applying controlled amounts of cold rolling, stretching, or combinations of these operations. Since the 7*xxx* alloys do not respond favorably to cold working before aging, the T8 temper is not used for this series of alloys.

Natural aging can often be suppressed by refrigeration for several days at −18 °C (0 °F) or lower. Thus, it is normal practice to complete straightening and forming operations before aging changes the mechanical properties appreciably. When scheduling makes this impractical, aging may be avoided in some alloys by refrigerating prior to forming. It is also conventional practice to refrigerate alloy 2024-T4 rivets to maintain good driving characteristics.

**Artificial Aging.** The wrought heat treatable 7*xxx* alloys are more responsive to precipitation hardening than the 2*xxx* alloys and can obtain even higher strength levels. These alloys are based on the Al-Zn-Mg(-Cu) system. The 7*xxx* alloys can be naturally aged but are not, because they are unstable when aged at room temperature; that is, their strength will gradually increase with increasing time and can continue to do so for years. Therefore, all 7*xxx* alloys are artificially aged to produce stable alloys. For aluminum alloys, solution treating, quenching,

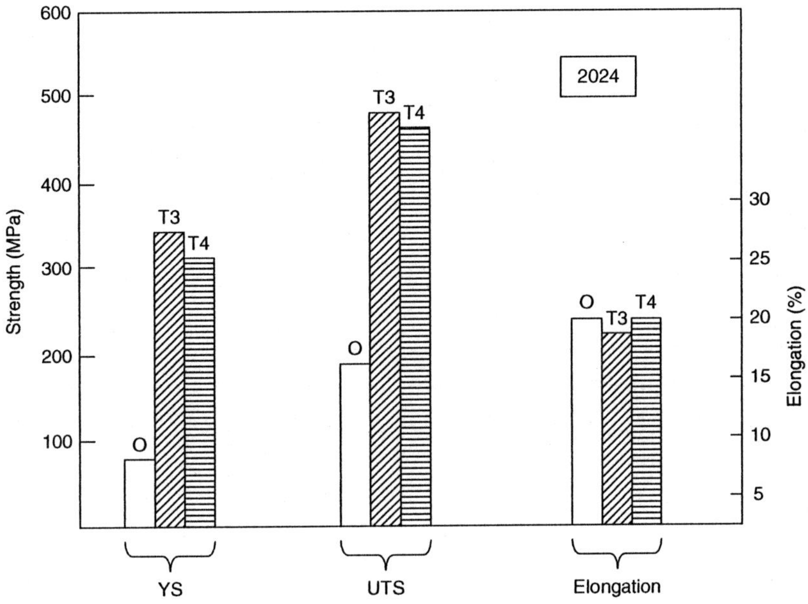


**Fig. 9.15** Effects of T3 and T4 heat treatments on 2024 Al. YS, yield strength; UTS, ultimate tensile strength. Source: Ref 2

and artificial aging is designated as the T6 temper, as shown in the temperature-time profile of Fig. 9.16. This temper provides the highest strength levels that can be obtained, as illustrated for 7075-T6 aluminum in Fig. 9.17.

In the peak-aged T6 condition, thick plate, forgings, and extrusions of many of the 7*xxx* alloys are susceptible to stress-corrosion cracking (SCC), particularly when stressed through the thickness. For example, SCC of 7075-T6 has frequently occurred in service. To combat the SCC problem, a number of overaged T7 tempers have been developed for the 7*xxx* alloys. Although there is usually some sacrifice in strength, these tempers have dramatically reduced the susceptibility to both SCC and exfoliation corrosion. An additional benefit is that they improve the fracture toughness, especially the through-the-thickness (i.e., short transverse) fracture toughness of thick plate. For example, the overaged T73 temper, which was originally developed to reduce the susceptibility to SCC, reduces the yield strength of 7075 by approximately 15% but increases the threshold stress for SCC for 7075 by a factor of 6 (6×). Other overaged tempers, such as the T74, T75, and the T77 tempers, were then developed to provide trade-offs in strength and SCC resistance between the T6 and T73 tempers.

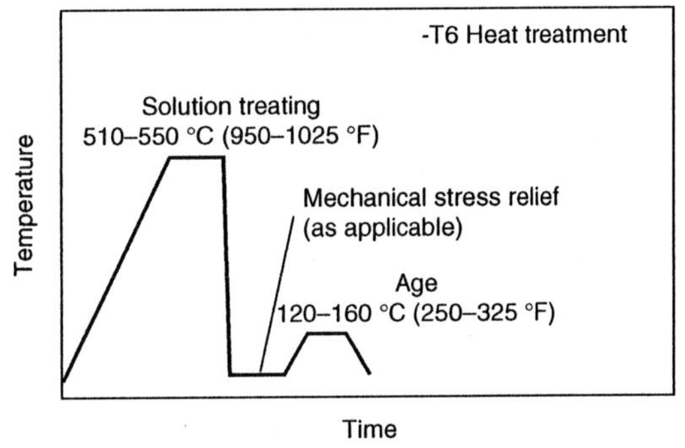


**Fig. 9.16** Time-temperature profile for solution-treated and peak-aged alloys

The precipitation heat treatments used to produce the T73 and T76 tempers consist of either a two-stage isothermal precipitation heat treatment (Fig. 9.18) or a single-stage precipitation heat treatment by heating at a controlled rate to the aging temperature. The desired corrosion-resistance properties can be achieved by using only a single-stage precipitation heat treatment at a temperature above approximately 150 °C (300 °F), but higher strength is obtained by preceding this with a lower-temperature treatment or with a slow,

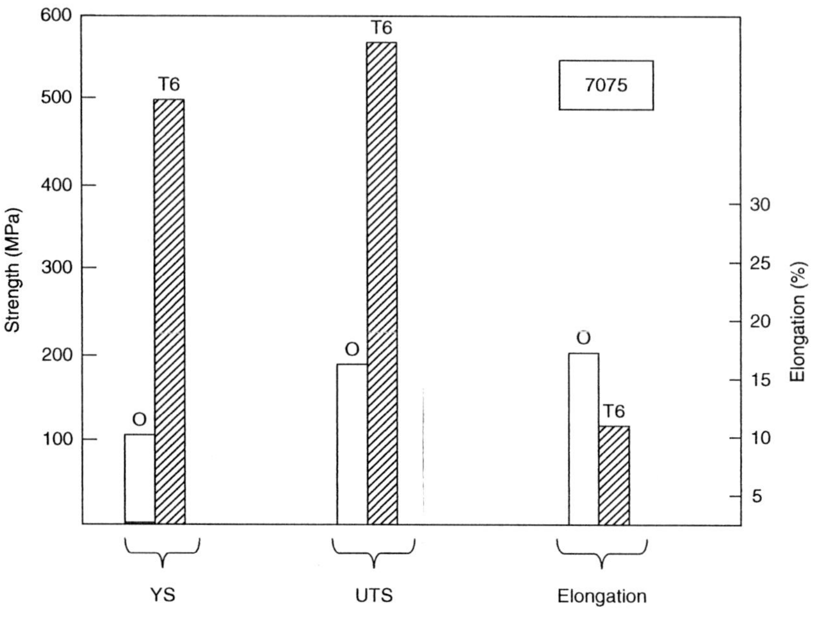


**Fig. 9.17** Effects of T6 heat treatment on 7075 Al. YS, yield strength; UTS, ultimate tensile strength. Source: Ref 2

controlled heatup. Either during the first-stage treatment or during slow heatup, a fine, high-density dispersion of GP zones is nucleated. Both the time and temperature of the first treatment, or the rate of heating, must be controlled to produce GP zones that will not dissolve but will transform to the $\eta'$ precipitate when heated to an aging temperature above 150 °C (300 °F). The aging treatment that produces the results in the shortest time depends on the GP zone solvus temperature. This temperature, in turn, depends on the vacancy concentration, a factor influenced by solution heat treating temperature, quench rate, and composition. If the first-step aging time is too short, the aging temperature is too far below the GP zone solvus, or the heating rates are too high, the GP zones will dissolve above 150 °C (300 °F), and the resultant coarse and widely distributed precipitate will provide lower strength. The T76 treatments use the same sequence but employ second-stage heating that is only long enough to develop exfoliation corrosion resistance that is higher than that provided by the T6 tempers. Materials in the T73 temper also have high resistance to exfoliation corrosion. The best combination of strength and corrosion resistance is produced by the T77 temper, which is a form of what is known as a retrogression and overaging treatment.

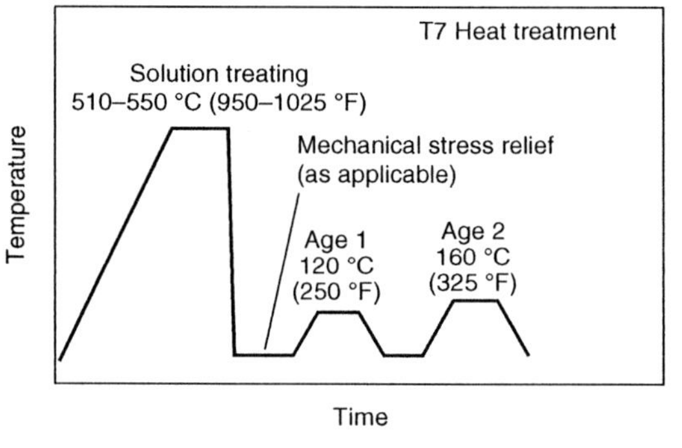


**Fig. 9.18** Typical time-temperature profile for precipitation and overaged alloys

A summary of the effects of different tempers on 7*xxx* alloys is shown in Fig. 9.19.

## 9.4 Dispersion Hardening

Mechanical alloying (MA) is a process used to make oxide-dispersion-strengthened (ODS) alloys, such as the nickel-base superalloy MA 6000. Some of these alloys combine precipitation hardening along with oxide dispersion so that their stress-rupture properties at 1000 °C

(1830 °F) (and higher) are superior to precipitation strengthening alone.

The elevated-temperature 1000 h rupture life as a function of temperature of alloy MA 6000 is compared with those of the directionally solidified (DS) superalloy DS MAR-M 200 and a thoriated nickel alloy bar in Fig. 9.20. This diagram shows the effect of the two strengthening mechanisms in MA 6000. At intermediate temperatures, approximately 815 °C (1500 °F), the strength of MA 6000 approaches that of the complex, highly alloyed DS MAR-M 200 containing hafnium and is almost four times that of an unalloyed ODS metal such as thoriated nickel (TD nickel). At high temperatures (~1095 °C, or 2000 °F), where the alloy DS MAR-M 200 containing hafnium has lost most of its strength due to growth and dissolution of its precipitate, MA 6000 still has useful strength due to the presence of the oxide dispersion. At temperatures between these extremes, the strength of MA 6000 is superior to both the cast nickel-base superalloy and the ODS metal, because the two strengthening mechanisms supplement one another.

Mechanical alloying is a high-energy dry ball milling process used to make composite metallic powders with a controlled, fine microstructure. As shown in Fig. 9.21, powders of the desired composition are blended and then placed in a high-energy ball mill. The intensive milling process repeatedly fractures and then rewelds the powder particles. During each collision with the grinding balls, the particles are plastically deformed to the extent that the surface oxides are broken, exposing clean metal surfaces. On subsequent impacts, the clean surfaces are welded together. This cold welding process increases the size of the particles, while at the same time, additional impacts are fracturing particles and reducing their size. As the process continues, the microstructure of the particles is continually being refined. Very fine yttria (25 nm) powder, which is added to the mixture, becomes entrapped between fragments of the composite metal powders. Milling is continued until the powder particles contain the same composition as the starting mix.

Powders for mechanical alloying include pure metals, master alloys, and refractory compounds. Pure metals include nickel, iron, chromium, cobalt, tungsten, molybdenum, and niobium, while master alloys are nickel-base alloys with large concentrations of the more reactive metals aluminum, titanium, zirconium, and hafnium. Approximately 2% volume of fine yttria ($Y_2O_3$) is added to form the dispersoid. After milling, a uniform interparticle spacing of approximately 0.5 μm is achieved.

After mechanical alloying, the powder is packed into steel tubes and extruded at 1065 to 1120 °C (1950 to 2050 °F), with a reduction ratio of 13 to 1. The consolidated extruded stock is then hot rolled into mill shapes at 955 to 1065 °C (1750 to 1950 °F) to produce product forms such as bar, plate, sheet, and wire.

**Solution Heat Treated & Aged Tempers**

- **T6*xx* (~0.080 in. max.)**
  - Highest tensile strength
  - Possible exfoliation corrosion
- **T76*xx* (~0.250 in. max.)**
  - Medium tensile strength
  - Best exfoliation resistance
- **T73*xx*/T74*xx* (~0.250+ in.)**
  - Lower tensile strength
  - Best stress corrosion resistance
- **Stress relieved tempers**
  - Eliminates residual stresses
  - T*xx*51 - Stress relieved by stretching
  - T*xx*52 - Stress relieved by compression

| | |
|---|---|
| -T6 | Peak Aged |
| -T77 | Retrogression and Overaging |
| -T79 | Very Limited Overaging |
| -T76 | Overaged to Achieve Enhanced Exfoliation Resistance |
| -T74 | Overaged to Achieve Enhanced Exfoliation and SCC Resistance |
| -T73 | Overaged to Achieve Optimum Exfoliation and SCC Resistance |

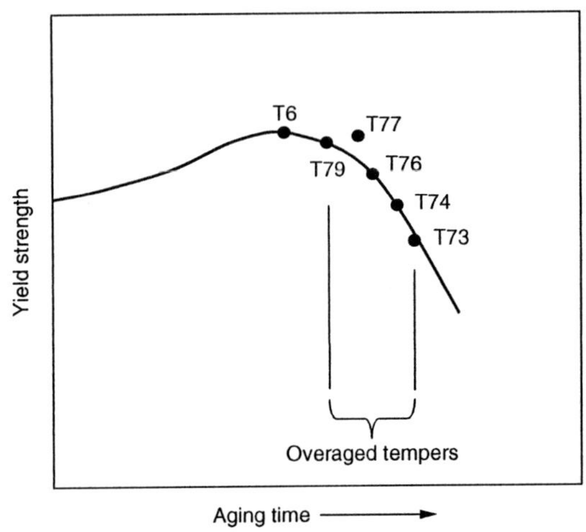


**Fig. 9.19** Effects of different tempers on 7*xxx* alloys. SCC, stress-corrosion cracking

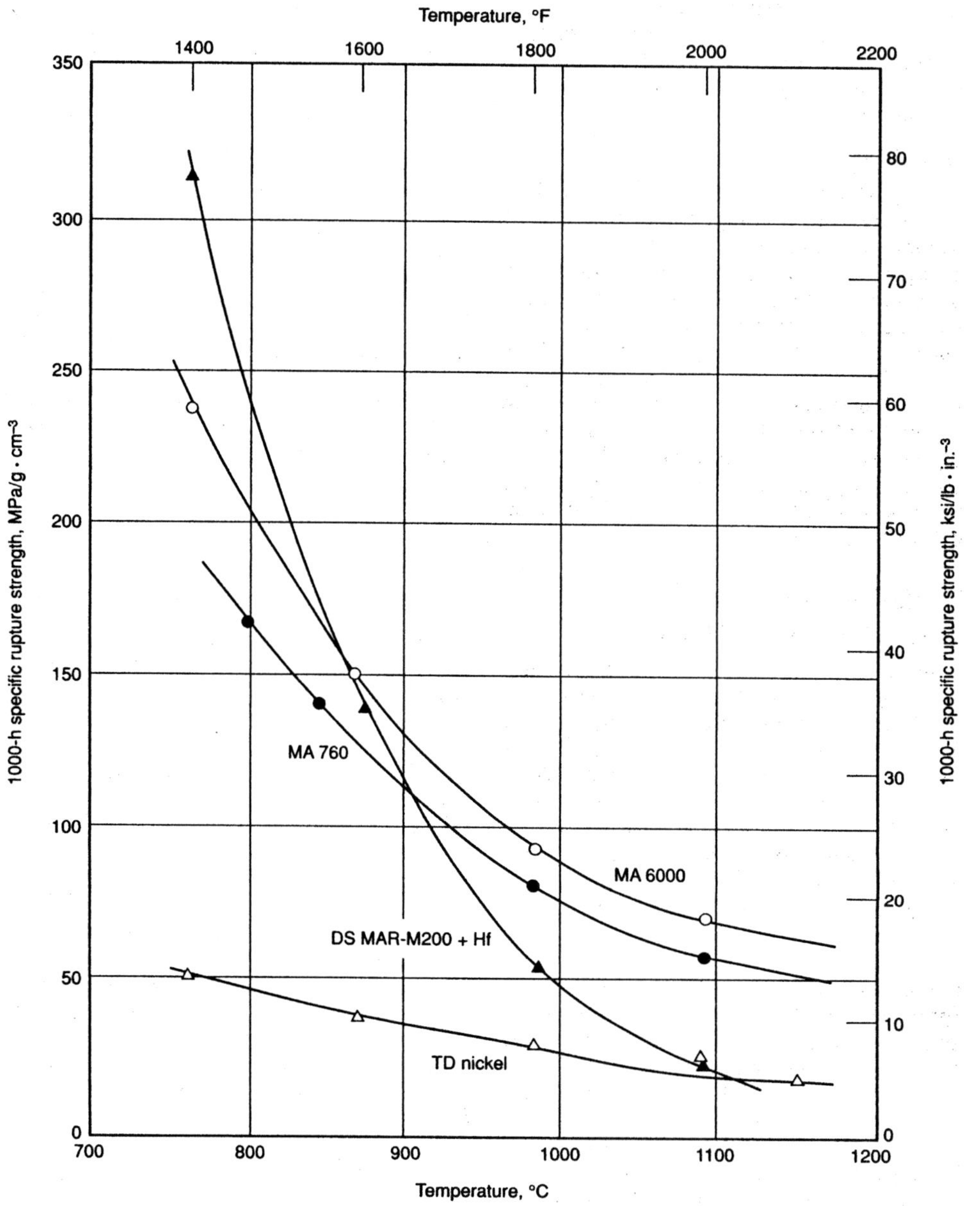


**Fig. 9.20** Stress-rupture strength of MA 6000. Source: Ref 7


## REFERENCES

1. D.R. Askeland, *The Science and Engineering of Materials,* 2nd ed., PWS-Kent Publishing Co., 1989
2. F.C. Campbell, *Manufacturing Technology for Aerospace Structural Materials,* Elsevier Scientific, 2006
3. R.E. Smallman and R.J. Bishop, *Modern Physical Metallurgy and Materials Engineering,* Butterworth Heinemann, 1999
4. D.A. Porter and K.E. Easterling, *Phase Transformations in Metals and Alloys,* Chapman and Hall, 1981
5. M. Epler, Structures by Precipitation from Solid Solution, *Metallography and*

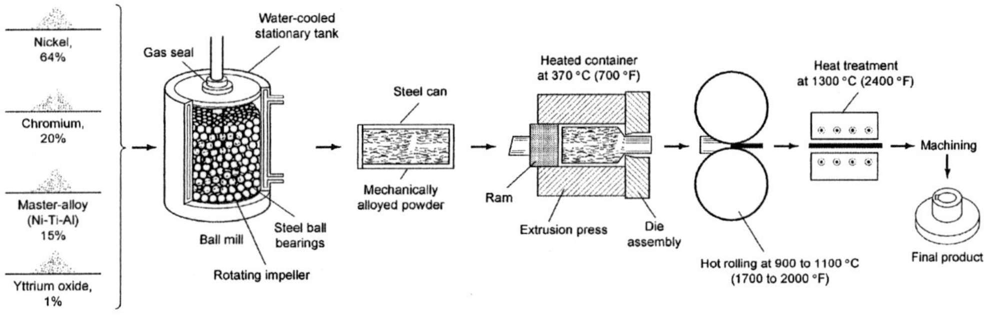


**Fig. 9.21** Mechanical alloying powder metallurgy process. Source: Ref 8

*Microstructures,* Vol 9, *ASM Handbook,* ASM International, 2004

6. W.T. Becker and S. Lampman, Fracture Appearance and Mechanisms of Deformation and Fracture, *Failure Analysis and Prevention,* Vol 11, *ASM Handbook,* ASM International, 2002
7. J.J. deBarbadillo and J.J. Fischer, Dispersion-Strengthened Nickel-Base and Iron-Base Alloys, *Properties and Selection: Nonferrous Alloys and Special-Purpose Materials,* Vol 2, *ASM Handbook,* ASM International, 1990
8. J.R. Davis, *Alloying: Understanding the Basics,* ASM International, 2001

## SELECTED REFERENCES

- A.G. Guy, *Elements of Physical Metallurgy,* 2nd ed., Addison-Wesley Publishing Company, 1959
- P. Haasen, *Physical Metallurgy,* 2nd ed., Cambridge University Press, 1986
- Heat Treating of Nonferrous Alloys, *Metals Handbook Desk Edition,* 2nd ed., ASM International, 1998

## CHAPTER 10

# The Iron-Carbon System

SINCE OVER 80% by weight of all metals in use are iron and steel alloys, it is appropriate to discuss the iron-carbon system in some detail. Most carbon steels contain up to 1.5% C, while cast irons normally contain 2 to 4% C. As discussed in Chapter 1, "Metallic Structure," iron is allotropic and changes its crystalline structure on heating or cooling. When pure iron solidifies on cooling from its melting point of 1540 °C (2800 °F), it assumes a body-centered cubic (bcc) structure designated as δ-ferrite. On further cooling between 1395 and 912 °C (2541 and 1674 °F), it has a face-centered cubic (fcc) structure called austenite, designated as γ. Below 912 °C (1674 °F), it again has a bcc crystal structure called ferrite, designated as α.

The iron-carbide phase diagram is shown in Fig. 10.1. The composition only extends to 6.70% C; at this composition the intermediate compound iron carbide, or cementite ($Fe_3C$), is formed, represented by the vertical line on the right-hand side of the diagram. Since all steels and cast irons have carbon contents less than 6.70%, the remainder of the diagram with 6.7 to 100% C is of no engineering interest. Cementite is a hard and brittle carbide

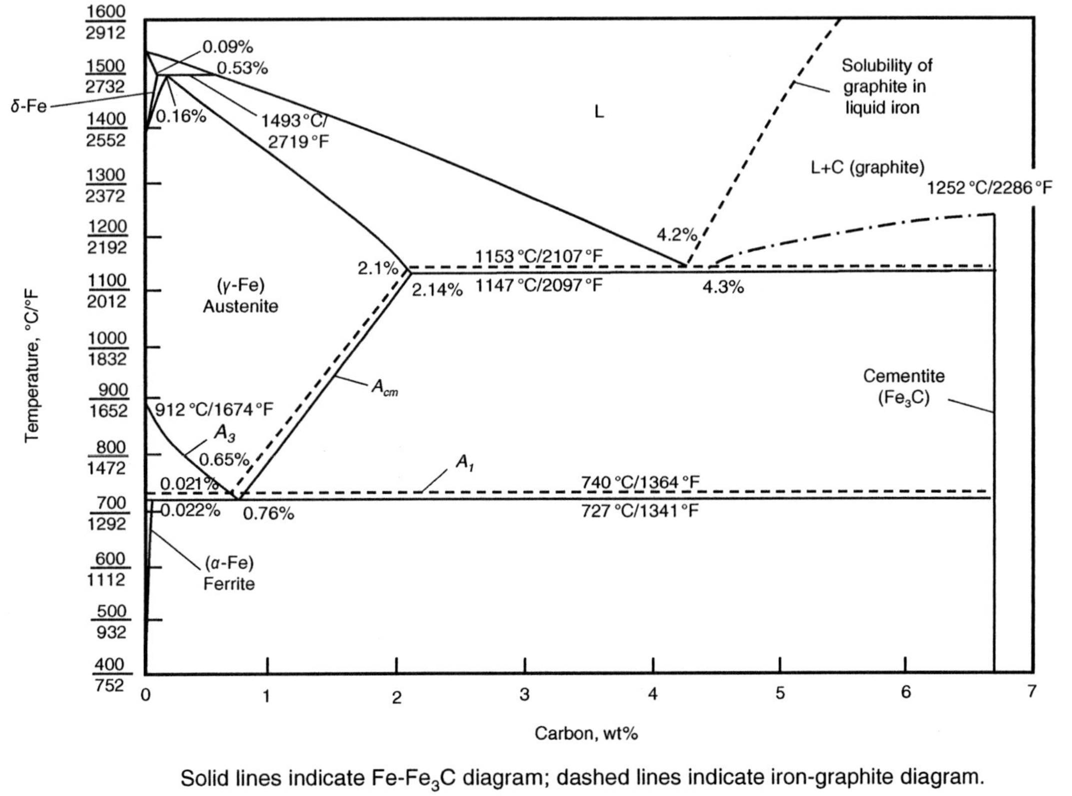


Solid lines indicate $Fe$-$Fe_3C$ diagram; dashed lines indicate iron-graphite diagram.

**Fig. 10.1** The $Fe$-$Fe_3C$ diagram. Source: Ref 1

compound with an orthorhombic crystalline structure.

Carbon forms a series of solid solutions with iron, as indicated by the δ, γ, and α fields. However, carbon has very limited solubility in iron. The maximum solubility of carbon in ferrite is only approximately 0.022 wt% at 727 °C (1341 °F), which increases to 2.14 wt% C at 1147 °C (2097 °F). At room temperature, it is even less, with the solubility of carbon in iron only 0.005%. Thus, the solubility of carbon in fcc iron is approximately 100 times greater than in bcc ferrite. This is a result of the fcc interstitial positions being larger, and therefore, the strains imposed on the surrounding iron lattice are much lower. Even though present in very small amounts, carbon significantly influences the mechanical properties of ferrite.

Steels contain carbon in the form of cementite, while cast irons can contain either cementite or carbon in the form of graphite. Graphite is a more stable, carbon-rich phase than cementite. Graphite formation is promoted by a higher carbon content and the presence of large amounts of certain alloying additions, in particular silicon. Therefore, graphite is an important phase in cast irons but is rarely found in steels. Cementite is metastable and will remain a compound at room temperature indefinitely. However, if it is heated to 650 to 705 °C (1200 to 1300 °F) for several years, it will gradually transform into iron and graphite. When graphite does form, the solubility limits and temperature ranges of phase stability are changed slightly, as indicated by the dashed lines in Fig. 10.1.

In ferrous alloys, iron is the prime alloying element, but carbon and other alloying elements are frequently used. Based on carbon content, there are three types of ferrous alloys: irons, steels, and cast irons. Commercially pure iron contains less than 0.008 wt% C and is composed almost exclusively of ferrite when it cools to room temperature. In steels, which have carbon contents in the range of 0.008 to 2 wt%, the microstructure consists of ferrite and cementite when they are slowly cooled to room temperature. Although a steel may contain as much as 2 wt% C, carbon contents are usually restricted to 1.5 wt% or less because of excessive brittleness. Generally, the carbon content is kept low in steels that require high ductility, high toughness, and good weldability but is used at higher levels in steels that require high strength, high hardness, fatigue resistance, and wear resistance. Cast irons are classified as those alloys that contain between 2 and 6.70 wt% C; however, most cast irons contain 2 to 4 wt% C.

## 10.1 Ferrite

Ferrite is usually present in steels as a solid solution of iron containing carbon or one or more alloying elements such as silicon, chromium, manganese, and nickel. Carbon is an interstitial element, while the larger elements are substitutional. Carbon occupies specific interstitial sites in the bcc iron crystalline lattice, while the larger substitutional elements replace or substitute for iron atoms. The two types of solid solutions impart different characteristics. For example, interstitial carbon can easily diffuse through the open bcc lattice, whereas substitutional elements diffuse much more slowly. Therefore, carbon responds quickly during heat treatment, whereas substitutional alloying elements behave more sluggishly. Interstitial alloying elements have a much stronger influence on the properties of iron than substitutional alloying elements. However, even a small addition of carbon, less than its solubility limit, into pure iron substantially increases the room-temperature yield strength of the material (Fig. 10.2).

The influence of alloying elements on the yield strength of ferrite is shown in Fig. 10.3. The strong effects of the interstitial elements carbon and nitrogen, as well as phosphorus, are clearly evident. On the other hand, the substitutional solid-solution elements silicon, copper, manganese, molybdenum, nickel, aluminum, and chromium are far less effective as ferrite strengtheners. The strength of a ferritic steel is also determined by its grain size according to the Hall-Petch relationship, as shown in Fig. 10.4 and previously discussed

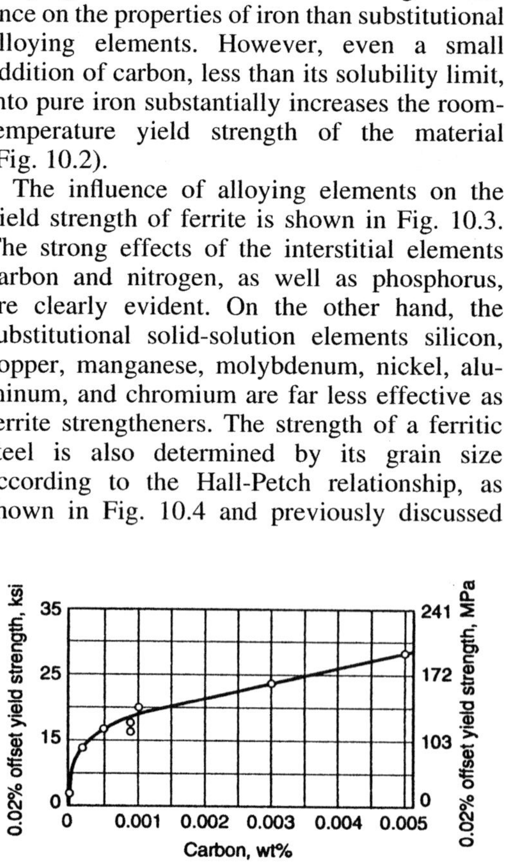


**Fig. 10.2** Effect of small carbon additions on room-temperature yield strength of iron. Source: Ref 2

in Chapter 2, "Crystalline Imperfections and Plastic Deformation"; that is, a finer grain size results in a higher yield strength. Control of the grain size through thermomechanical treatments, heat treatments, and/or alloying is vital to the control of strength and toughness of most steels. While a wide variety of steels contain ferrite, only a few commercial steels are totally ferritic. An example of the microstructure of a fully ferritic ultra-low-carbon steel is shown in Fig. 10.5.

## 10.2 Eutectoid Structures

When a steel with the eutectoid composition (0.76 wt% C) is slowly cooled through the eutectoid temperature of 727 °C (1341 °F), austenite decomposes into ferrite and cementite in the manner shown in Fig. 10.6:

$$\gamma\text{-austenite (0.76 wt\% C)} \rightarrow \alpha\text{-ferrite (0.022 wt\% C)} + Fe_3C\text{-cementite (6.7 wt\% C)}$$

The eutectoid mixture of ferrite and cementite that forms is called pearlite, a lamellar construction consisting of alternating layers of ferrite and cementite. As shown in Fig. 10.7, either ferrite or cementite nuclei form at

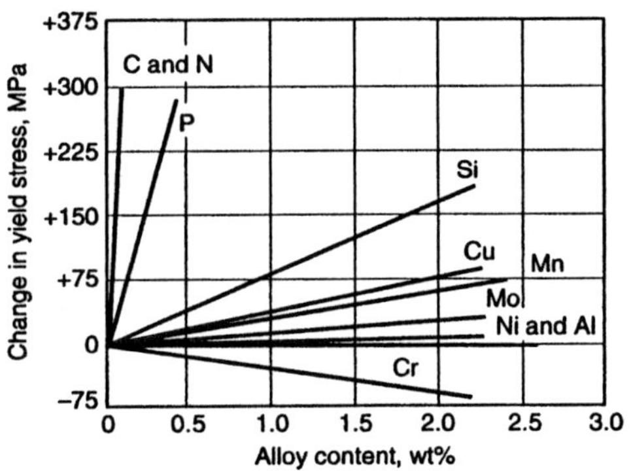


**Fig. 10.3** Effect of alloying elements on yield strength. Source: Ref 2

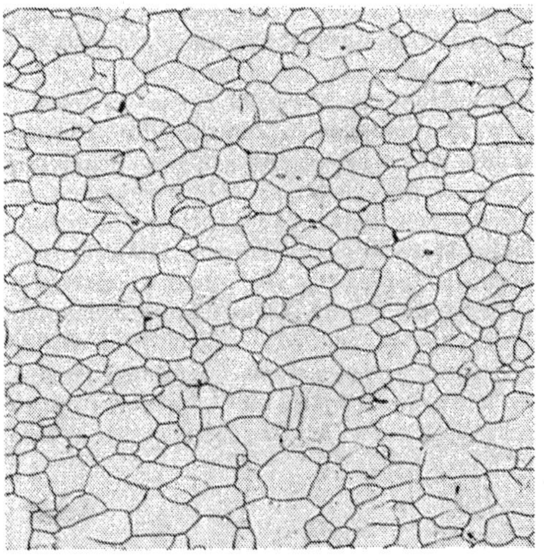

**Fig. 10.5** Microstructure of ultra-low-carbon ferritic steel. Source: Ref 2

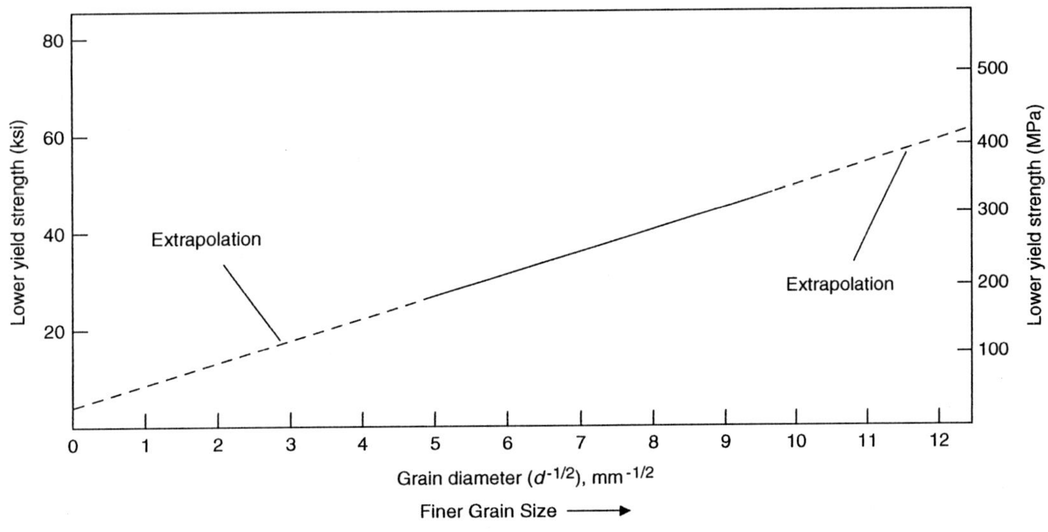


**Fig. 10.4** Hall-Petch relationship for low-carbon ferritic steel

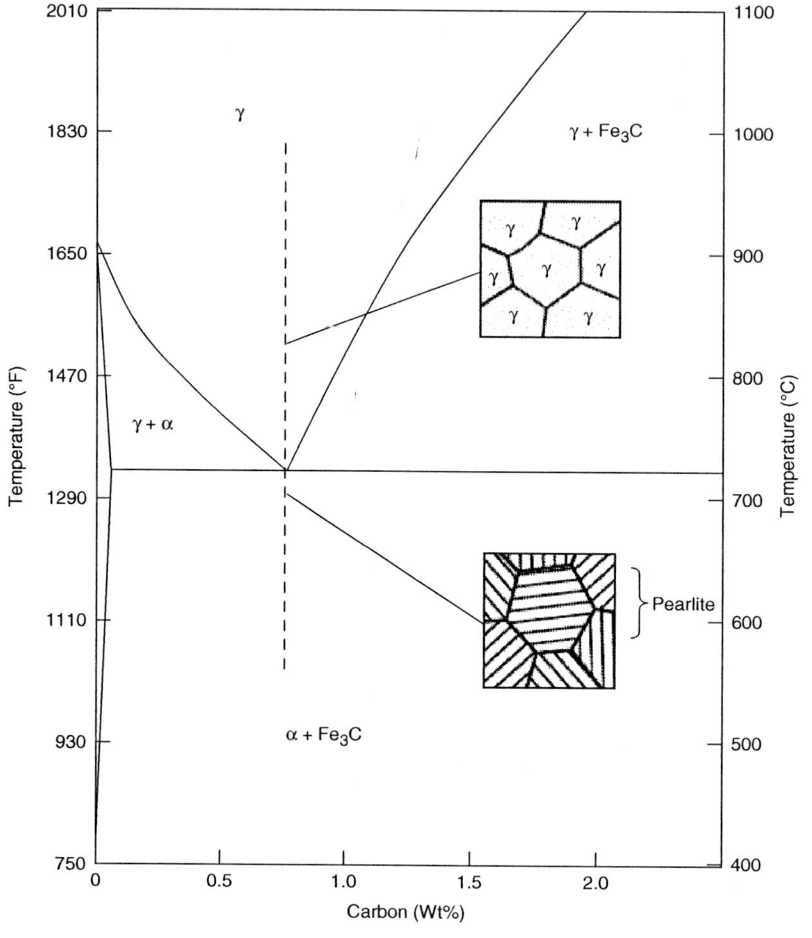


**Fig. 10.6** Equilibrium cooling of a eutectoid steel

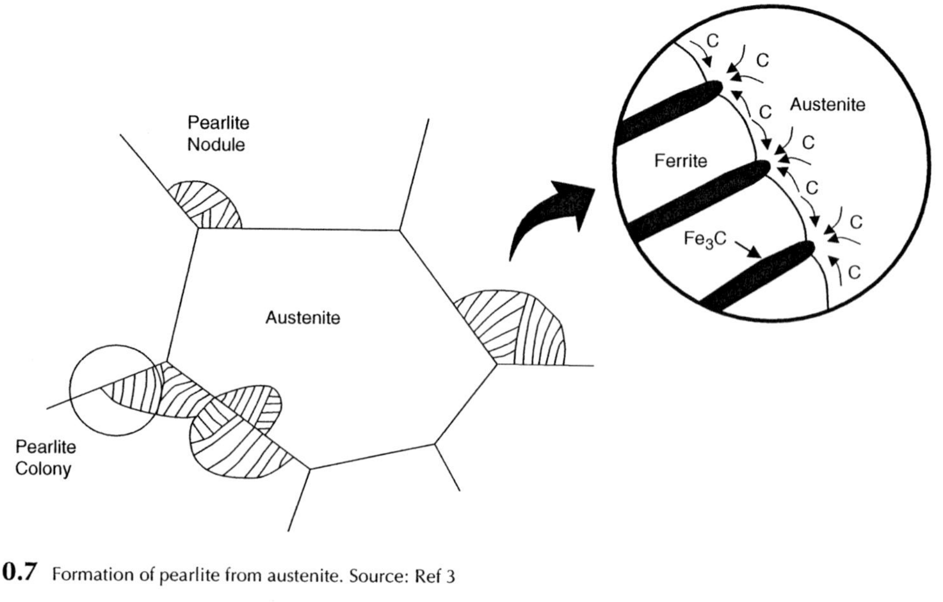


**Fig. 10.7** Formation of pearlite from austenite. Source: Ref 3

austenite grain boundaries and then grow as colonies into the austenite grains, with the layers oriented in essentially the same direction within the colony. The concurrent growth of several colonies leads to the formation of a larger nodule. The interface between the growing pearlite and austenite is incoherent. Provided that the temperature is constant, pearlite advances at a constant velocity and tends to produce a hemispherical nodule. The mechanism of pearlite growth involves the rejection of carbon by the growing ferrite plates and its incorporation into the cementite. This process requires diffusion of carbon by one or more of three paths: through the austenite, through the ferrite, or along the interface. The lower the transformation temperature, the greater is the chemical energy driving the reaction, resulting in a finer pearlite size. Pearlite gets its name because its appearance at low magnification under a light microscope (Fig. 10.8) resembles mother of pearl.

The alternating α-ferrite and $Fe_3C$ layers in pearlite form because the carbon content of the austenite parent phase (0.76 wt% C) is different from that of both ferrite (0.022 wt% C) and cementite (6.7 wt% C), and the phase transformation requires that there be a redistribution of the carbon by diffusion. Carbon atoms diffuse away from the ferrite regions to the cementite layers as the pearlite extends from the grain boundary into the austenite grains. The relative layer thickness of the eutectoid ferrite-to-cementite layers is approximately 8 to 1, meaning that the ferrite layers are a lot thicker than the cementite layers. When the rate of cooling is slow enough for the transformation to begin and end just below the eutectoid isotherm, the product will be coarse pearlite. As the cooling rate increases, pearlite with increasingly fine structure will develop.

Although energy minimization would normally predict a structure of dispersed spheroids, the alternating platelet structure forms because the carbon diffuses a shorter distance than is required to form a dispersed spheroid microstructure. Diffusion is possible because the small interstitial carbon atoms can diffuse quite rapidly through the iron lattice at elevated temperatures. In addition, diffusion is faster in the more loosely packed bcc ferrite structure than the closer packed fcc austenite structure. However, diffusion in solids during a eutectoid reaction is much slower than in liquids during the eutectic reaction. Therefore, nonequilibrium transformations are even more important in eutectoid solid-state reactions than in the liquid-to-solid eutectic reactions. Subsequent cooling of pearlite from the eutectoid to room temperature produces only insignificant changes in the microstructure.

The properties of fully pearlitic steels are determined by the spacing between the ferrite-cementite lamellae, a dimension called the interlamellar spacing. A relationship similar to the Hall-Petch relationship for grain size also exists for the interlaminar spacing in pearlite, with a smaller interlaminar spacing producing a higher yield strength. Thus, strength is related to interlamellar spacing, pearlite colony size, and prior-austenite grain size, with finer sizes giving better properties. The thickness of the cementite lamellae can also influence the properties of pearlite. Fine cementite lamellae can be deformed, as compared with coarse lamellae, which tend to crack during deformation.

Although fully pearlitic steels with a lamellar microstructure have high strength, high hardness, and good wear resistance, they also have rather poor ductility and toughness. For example, a low-carbon, fully ferritic steel will typically have a total elongation of more than 50%, whereas a fully pearlitic steel will typically have

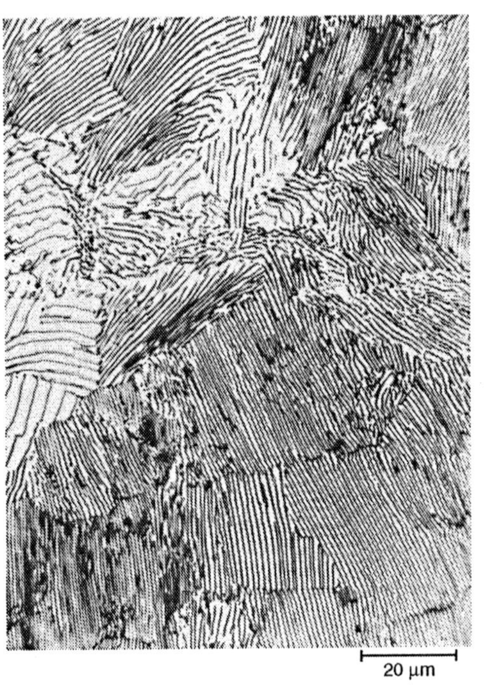


**Fig. 10.8** Pearlitic structure in eutectoid steel. Source: Ref 4

a total elongation of only approximately 10%. A low-carbon, fully ferritic steel will have a room-temperature Charpy V-notch impact energy of approximately 200 J (150 ft · lbf), whereas a fully pearlitic steel will have room-temperature impact energy of under 10 J (7 ft · lbf). In addition, the transition temperature of a fully pearlitic steel is always well above room temperature. This means that at room temperature, the general fracture mode is the brittle fracture mode cleavage. Therefore, fully pearlitic steels with lamellar microstructure should not be used in applications where toughness is important. Also, pearlitic steels with carbon contents slightly or moderately higher than the eutectoid composition (i.e., hypereutectoid steels) have even lower toughness. Because of their low ductility/toughness, there are only a few applications for fully pearlitic steels, examples being railroad rails and wheels and high-strength wire. A fully pearlitic rail steel provides excellent wear resistance during railroad wheel/rail contact.

## 10.3 Hypoeutectoid and Hypereutectoid Structures

Hypoeutectoid, meaning compositions with carbon content less than the eutectoid, alloys contain 0.022 to 0.76 wt% C. Cooling of a hypoeutectoid alloy is represented by the vertical line in the Fig. 10.9 phase diagram. After solidification, at high temperatures, such as point *a,* the alloy is in the single-phase austenite field and is entirely austenite. When the alloy is cooled to approximately 800 °C (1475 °F) (point *b*), the alloy enters the two-phase $\alpha + \gamma$ phase field. Small particles of ferrite start forming along the austenite grain boundaries. This ferrite is called proeutectoid ferrite because it forms prior to cooling through the eutectoid

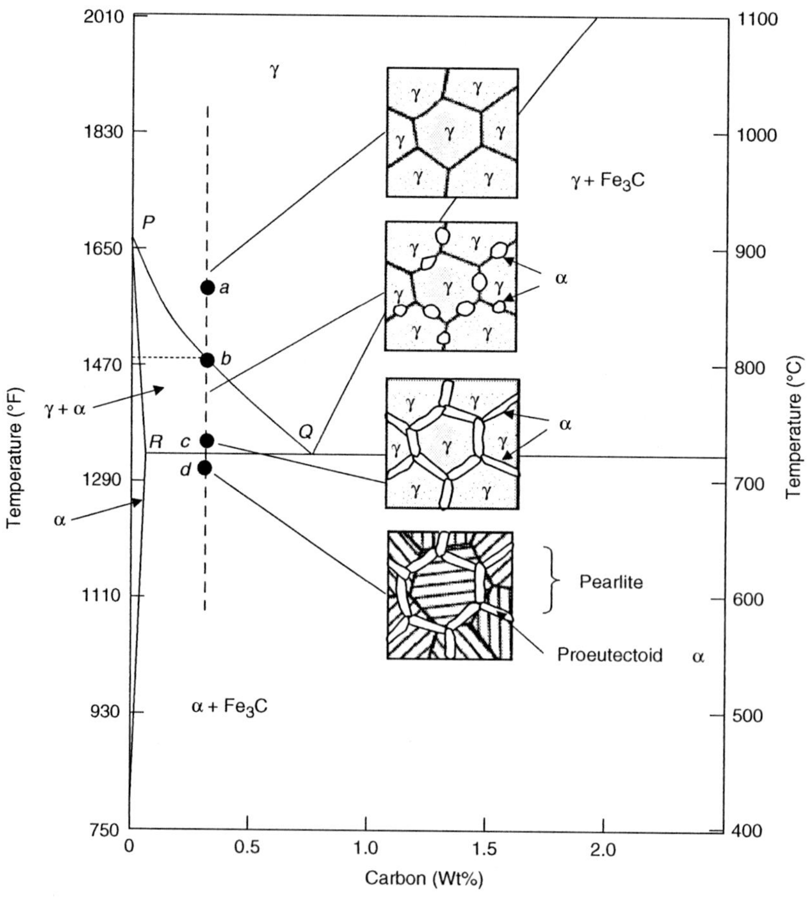


**Fig. 10.9** Equilibrium cooling of a hypoeutectoid steel

temperature. This is in contrast to the eutectoid ferrite that forms at the eutectoid temperature as part of the constituent pearlite. While cooling the alloy through the $\alpha + \gamma$ phase region, the composition of the ferrite phase changes with temperature along the $\alpha$-($\alpha + \gamma$) phase boundary, line *PR,* and becomes slightly richer in carbon. The change in composition of the austenite is more drastic, proceeding along the ($\alpha + \gamma$)-$\gamma$ boundary, line *PQ,* as the temperature is reduced. As the alloy further cools within the $\alpha + \gamma$ field (point *c*), the ferrite particles become larger in size and thicken along the austenite grain boundaries. When the alloy is finally cooled through the eutectoid temperature (point *d*), all of the remaining austenite transforms to pearlite. Since this reaction does not really affect the proeutectoid ferrite that has already formed at the austenite grain boundaries, the final microstructure is one of pearlite within the grain interiors surrounded by proeutectoid ferrite at the prior-austenite grain boundaries.

The microstructure of a hypoeutectoid steel depends on how rapidly it is cooled from the austenitic phase region. Very slow cooling, such as furnace cooling, will result in the formation of proeutectoid ferrite at the austenite grain boundaries. The remaining austenite grains will then convert to pearlite as the steel cools through the eutectoid temperature. As the temperature decreases, the morphology of proeutectoid ferrite changes. Growth along the grain boundaries becomes markedly easier than growth normal to them. Therefore, films of proeutectoid ferrite form at the sites of the former austenite boundaries. If the cooling rate is fast, ferrite side-plates protrude from the proeutectoid ferrite and project into the austenite grains. These side-plates, called Widmanstätten plates, have a wedgelike shape, and their broad faces comprise flat areas separated by ledges, as shown in Fig. 10.10. The growth process now involves the nucleation and sideways migration of the ledges that separate the flats. Thus, the microstructure of hypoeutectoid steel that is cooled rapidly will contain Widmanstätten ferrite and fine pearlite.

The transformation of hypereutectoid steels, those containing 0.077 to 2 wt% C, is exactly analogous to hypoeutectoid steels, except that in this instance, proeutectoid cementite forms on the grain boundaries in the manner shown in the Fig. 10.11 phase diagram. This phase diagram also illustrates why hypereutectoid steels are only used where extreme hardness is required, such as in cutting tools. Note that the cementite forms a continuous network along the prior-austenite grain boundaries. Since cementite is a hard and brittle compound, these steels have very little ductility and can be prone to sudden brittle failures. When hypereutectoid steels are fractured, it is common to find a region of the fracture that appears to be intergranular where the crack propagated along the proeutectoid cementite grain-boundary films.

In most ferrite-pearlite steels, the carbon content and the grain size determine the microstructure and resulting properties. For example, the effect of carbon on tensile and impact properties is shown in Fig. 10.12. The ultimate tensile strength steadily increases with increasing carbon content. This is caused by the increase in the volume fraction of pearlite in the microstructure, which has much higher strength than that of ferrite. Thus, increasing the volume fraction of pearlite has a profound effect on increasing tensile strength (Fig. 10.13). However, the yield strength is relatively unaffected by carbon content, rising from approximately 280 MPa (40 ksi) to about 420 MPa (60 ksi) over the range of carbon content shown. This occurs because yielding in a ferrite-pearlite steel is controlled by the ferrite matrix, which is generally the continuous matrix phase in the microstructure. Therefore, pearlite plays only a minor role in yielding behavior. Ductility, as represented by reduction in area, steadily decreases with increasing carbon content. For example, a steel with 0.10 wt% C has a reduction in area of approximately 75%, whereas a steel with 0.70 wt% C has a reduction in area of only 25%.

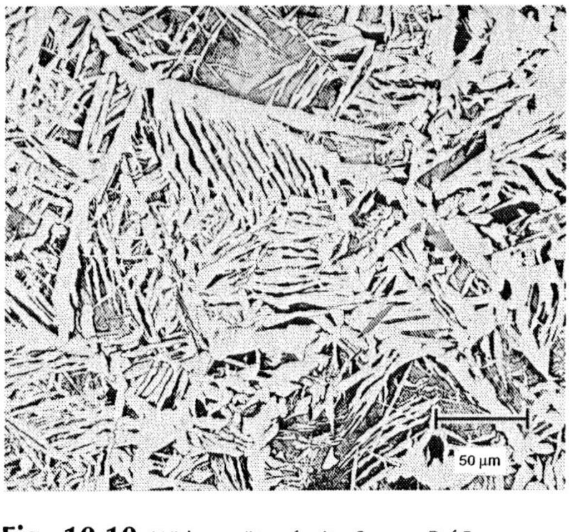


**Fig. 10.10** Widmanstätten ferrite. Source: Ref 5

During plastic deformation in pearlite lamellar microstructures, stress concentrations are created at the ends of pileups where the dislocations in the ferrite meet the ferrite-carbide interfaces, leading to carbide cracking. Cracking occurs more rapidly when the ferrite plates are thicker, because more dislocations can participate in an individual pileup for a given applied stress. The generation of cracks in cementite plates can initiate general fracture, which implies that coarse pearlite should fail before fine pearlite, because the cementite plates are expected to crack sooner when the ferrite lamellae are thicker. Also, the cracks will be longer and therefore more able to propagate. In other words, fine pearlite is both stronger and more ductile than coarse pearlite.

The toughness of ferrite-pearlite steels is another important consideration. This is true for other bcc metals, almost all of which, unlike fcc metals, have a ductile-to-brittle transition temperature under impact loading. This is often measured by the Charpy V-notch impact test (Chapter 13, "Fracture," in this book) in which a notched square-section bar is broken by a swinging pendulum. Measuring the height of the pendulum at the beginning of the test and the height to which it rises after breaking the specimen enables the energy consumed in breaking the bar to be calculated. When the bar

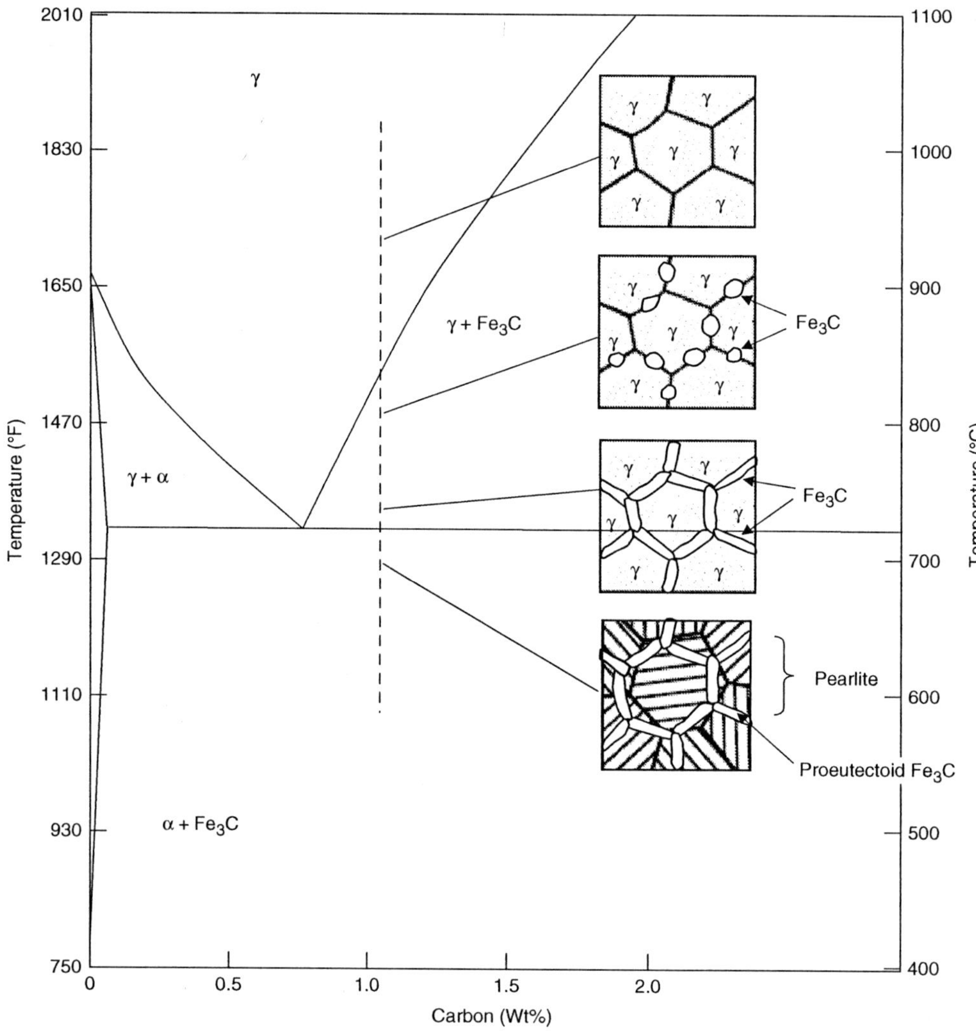


**Fig. 10.11** Equilibrium cooling of a hypereutectoid steel

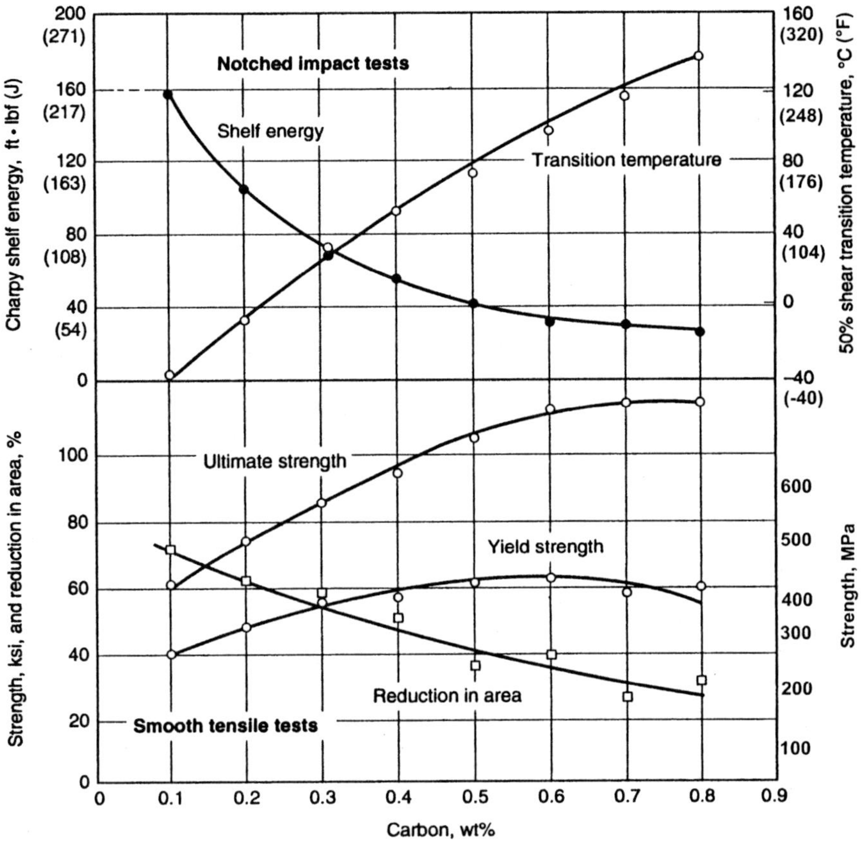


**Fig. 10.12** Effect of carbon content on mechanical properties. Source: Ref 2

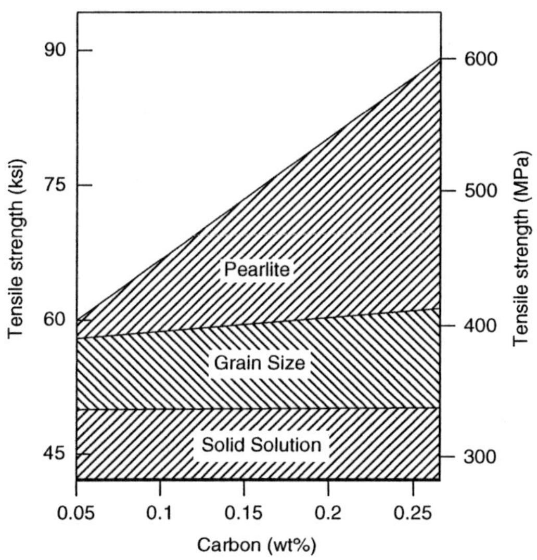


**Fig. 10.13** Effect of pearlite content on mechanical properties. Source: Ref 6

breaks in a brittle mode, the energy consumed is small. In this case, the bar breaks into two pieces, and the pendulum swings onward and upward. When fracture occurs in a ductile fashion, a large amount of energy is expended by the deformation processes that precede fracture, as well as those accompanying fracture. Then, the distance traveled by the pendulum after the fracture is smaller. Extremely ductile samples may not even break, stopping the pendulum. The nature of the fracture and the amount of energy expended in an individual test can vary with the testing temperature; therefore, it is standard practice to perform Charpy tests over a range of temperatures.

It has long been known that the absorbed energy in a Charpy V-notch test decreases as the carbon content increases (Fig. 10.14). In this graph showing impact energy versus test temperature, the upper-shelf energy decreases from approximately 200 J (150 ft·lbf) for a

0.11 wt% C steel to approximately 35 J (25 ft·lbf) for a 0.80 wt% C steel. Also, the transition temperature increases from approximately −50 to 150 °C (−60 to 300 °F) over this range of carbon content. The effect of carbon is due mainly to its effect on the percentage of pearlite in the microstructure. The eutectoid steel, which is entirely pearlitic, shows a gradual transition of the energy consumed as a function of temperature. Presumably, the gradual transition at the high pearlite levels occurs because, even though the cementite plates crack, the total fracture path is tortuous in comparison to cleavage in the ferrite. Cleavage fracture in ferrite is a brittle fracture process in which the fracture path follows particular crystallographic planes in each grain, forming large, relatively flat facets.

Until approximately 20 years ago, hypoeutectoid steels with mixed ferrite-pearlite microstructures were commonly selected for structural applications because of their useful strength levels. Applications included beams for bridges and high-rise buildings, plates for ships, and reinforcing bars for roadways. These steels are relatively inexpensive and are produced in large tonnages. They also have the advantage of being able to be produced with a wide range of properties. The microstructures of typical ferrite-pearlite steels are shown in Fig. 10.15. Unfortunately, medium-to-high-carbon steels are not very ameable to welding and are therefore usually bolted or riveted together. In addition, their low toughness and high ductile-to-brittle transition temperatures can cause catastrophic failures. Today, high-strength low-alloy (HSLA) steels are replacing conventional hypoeutectoid steels in many structural applications. The HSLA steels (Chapter 20, "Alloy Steels," in this book), which have very low carbon contents, have an attractive combination of properties, including high strength, good toughness, a low ductile-to-brittle transition temperature, and are more readily weldable.

## 10.4 Nonequilibrium Cooling—TTT Diagrams

The cooling rates considered so far are equilibrium cooling rates, such as may be approximated by a very large piece of steel undergoing very slow furnace cooling. Under normal industrial processing conditions, the cooling rates are almost always much faster, and new phases that are not on the equilibrium phase diagram can appear. When the cooling rate is extremely rapid, such as when quenching steel from temperatures in the austenite phase field (>740 °C, or 1364 °F) into cold water, the

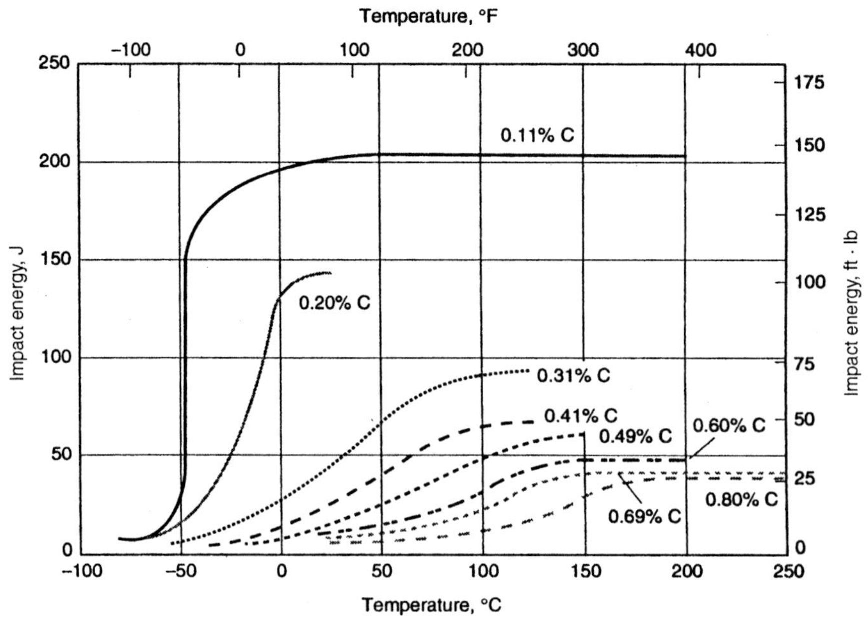


**Fig. 10.14** Effect of carbon content (wt%) on impact properties. Source: Ref 2

iron-carbon phase diagram can no longer be used, because quenching is such a radical departure from equilibrium. At rapid cooling rates, it is necessary to use a time-temperature-transformation (TTT) diagram. The TTT diagrams are so named because they depict Transformations as a function of Time and Temperature.

To construct a TTT diagram, very small samples, approximately the size of a dime, are first heated into the single-phase austenite field, also referred to as austenitization. Samples are then withdrawn from the furnace and immediately quenched into salt baths maintained at a given intermediate temperature. The specimen is held in the salt bath for a precise time and then quenched in a water bath maintained at room temperature. This procedure is shown in Fig. 10.16. At each intermediate temperature, the sample is held for a precise period of time, as shown in the example in Fig. 10.17. After quenching, the samples are examined by metallography to determine the microstructure. In the example, note that the amount of pearlite in the microstructure increases as the length of time in the intermediate salt bath increases. After a series of samples has been evaluated at one temperature, other temperatures are evaluated, for example, 700, 600, 500, 400, 300 °C (1200, 1100, 1000, 900, 800 °F), and so on down toward room temperature. When the data are collected, they are used to plot the extent of transformation product in the microstructure on a temperature-time plot in the manner shown in Fig. 10.18. A fully developed TTT curve with the transformation products marked on the diagram is shown in Fig. 10.19 for the eutectoid carbon steel 1080, which nominally contains 0.80 wt% C.

A TTT diagram is essentially a visual map that allows one to determine which constituents form as a function of temperature and time. For example, a sample rapidly cooled from the austenitic phase field to a temperature of 600 °C (1110 °F) will begin to transform to pearlite if held at the intermediate temperature for approximately 2 to 3 s. On further holding at 600 °C (1110 °F), it will completely transform to pearlite in approximately 10 s. In this particular steel, pearlite forms at all temperatures along the start-of-transformation curve from the $A_1$ temperature (725 °C, or 1340 °F) to the nose of the diagram (540 °C, or 1000 °F). At the higher transformation temperatures, the pearlite interlamellar spacing is very coarse. The interlamellar spacing is increasingly fine as the temperature decreases toward the nose of the diagram. This is significant since a steel with a coarse pearlite interlamellar spacing is softer and of lower strength than a steel with a fine pearlite interlamellar spacing.

At small amounts of undercooling below the $A_1$ isotherm, the transformation to pearlite takes a relatively long time to nucleate. As the degree of undercooling increases, the incubation time decreases rapidly, even though the transformation is occurring at lower temperatures where diffusion is less rapid. Eventually, the incubation time reaches a minimum and then begins to increase again. This minimum

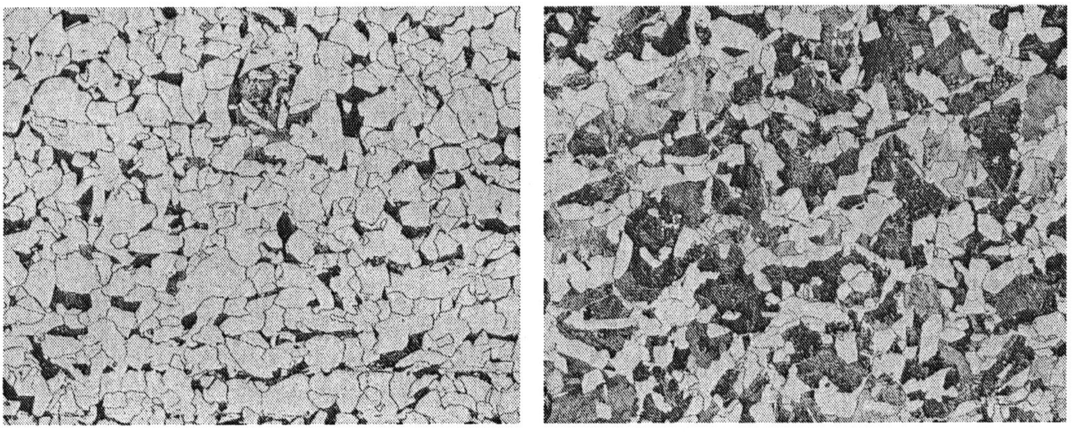

(a) (b)

**Fig. 10.15** Microstructures of hypoeutectoid steels. (a) 0.10% C, (b) 0.25% C. Source: Ref 2 incubation time occurs at the nose of the TTT diagram. The shape of the TTT diagram is a result of changing reaction kinetics. At high temperatures, the driving force for reaction is lower due to less undercooling, but the diffusion rates are faster, while at lower temperatures, the driving for reaction is increasing due to greater degrees of undercooling, while the diffusion rates are decreasing due to lower temperatures.

If the steel sample is quenched to 400 °C (750 °F) and held for various times, pearlite does not form; instead, a totally new constituent, called bainite, forms. Like pearlite, bainite consists of ferrite and cementite, but the morphology is different. Also, note that there are two types of bainite shown on the diagram: upper bainite, which forms at higher temperatures, and lower bainite, which forms at lower temperatures.

Finally, if the specimens are quenched in a salt bath to 150 °C (300 °F), another new constituent, called martensite, forms. Martensite is quite different from both pearlite and bainite. It is not a mixture of ferrite and cementite. Instead, martensite is a form of ferrite that is supersaturated with carbon. Because of the very fast cooling rate, the carbon atoms do not have time to diffuse from their interstitial positions in the bcc lattice to form cementite particles. Since martensite forms suddenly, there are no start-of-transformation or finish-of-transformation curves, as for pearlite and bainite. Instead, there are martensite start ($M_s$) and martensite finish ($M_f$) temperatures.

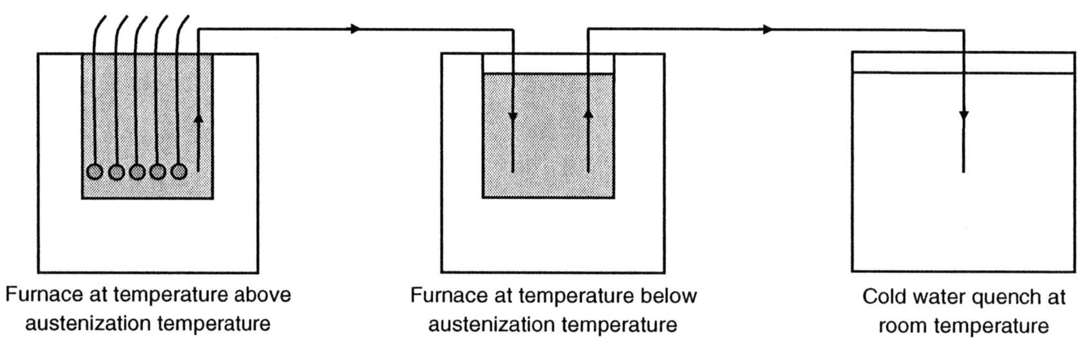


**Fig. 10.16** Method for determining isothermal transformations. Source: Ref 7

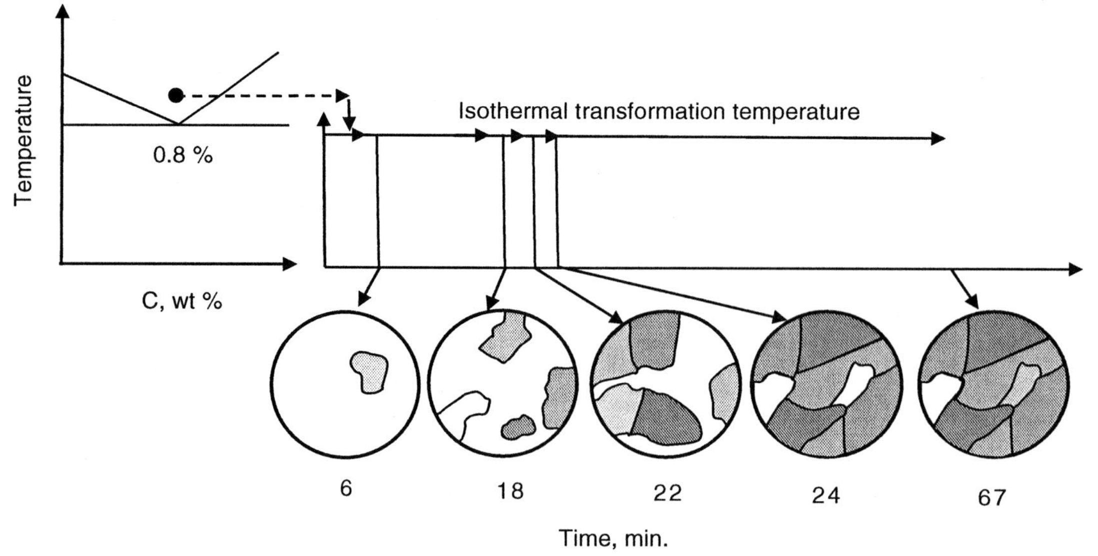


**Fig. 10.17** Isothermal transformation of plain carbon steel. Source: Ref 7

The hardness of the various transformation products is indicated on the right axis of the diagram and shows that the hardness increases with decreasing transformation temperature. Steel products produced with an as-quenched martensitic microstructure are very hard and brittle. Therefore, most martensitic products are tempered by heating to temperatures between approximately 345 and 650 °C (650 and 1200 °F). The tempering process allows some of the carbon to diffuse from the supersaturated iron lattice and form a carbide phase, softening the steel and restoring some ductility. The degree of softening is determined by the tempering temperature and the time at the tempering temperature. The higher the temperature and the longer the time of tempering, the softer the steel becomes. Thus, most steels with martensite are used in the quenched and tempered condition.

The balance of this chapter examines the bainite and martensite transformations in greater detail. In Chapter 11, "Heat Treatment of Steel," control and optimization of steel properties by heat treatment is covered in detail.

## 10.5 Bainite

Bainite is an isothermal decomposition product that forms on cooling in the temperature range below that of pearlite but above that of martensite. Like pearlite, bainite forms when austenite (γ) transforms into ferrite (α) and cementite ($Fe_3C$). Bainite has some of the characteristics of the diffusion-controlled growth of pearlite and some of the athermal shear transformation characteristics observed in martensite; that is, bainite is somewhat of a cross between pearlite and martensite. Somewhat like pearlite, bainite is a two-phase structure that contains ferrite-cementite regions that form from the eutectoid decomposition of austenite.

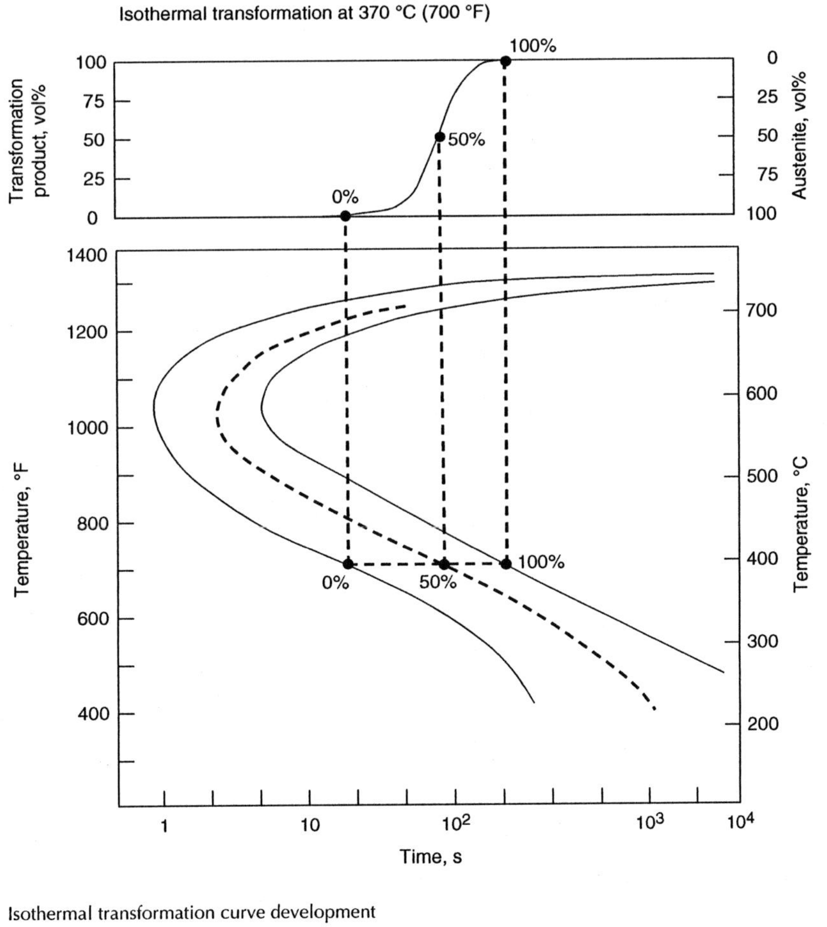


**Fig. 10.18** Isothermal transformation curve development

However, the carbide particles in bainite are small and chunky, similar to tempered martensite rather than cementite platelets characteristic of pearlite. Because of these structural differences, bainite has different property characteristics than pearlite. While pearlitic steels have high strength but relatively poor toughness, bainitic steels, in particular lower-bainite steels, display both high strength and good toughness.

The lamellar microstructure of pearlite has a strong effect on its toughness, with the toughness increasing as the cementite spacing decreases. Bainite does not develop a lamellar structure. Instead, bainite forms as an array of ferrite sheaves of very fine laths, needles, or platelets in hypoeutectoid steels, and the fine carbides are discrete particles. As a consequence, bainitic microstructures are generally tougher than pearlitic microstructures. Because of the fine scale of its microstructure, bainite is also harder than pearlite in steels of the same carbon contents.

The bainite transformation occurs below a bainite start temperature, and the fraction transformed increases with decreasing temperature. During isothermal transformation, the amount of bainite transformed increases as a function of time, eventually approaching a limit that does not change with prolonged isothermal holding. Bainitic transformations can also occur during continuous cooling, provided the cooling path travels through the transformation zone. In plain carbon steels, there is a significant overlap between the bainite and pearlite transformation temperature ranges. However, in many alloy steels, the separation between the pearlite and bainite transformation temperature ranges is distinct, resulting in the formation of a bay between the two transformation curves on the TTT curve, as shown in Fig. 10.20.

Bainite is grouped into two types, upper and lower bainite, depending on the temperature range at which the transformation occurs. Upper bainite forms isothermally in the temperature range of 400 to 550 °C (750 to 1020 °F), while lower bainite forms isothermally in the temperature range of 250 to 400 °C (480 to 750 °F). Both types of bainite have an acicular morphology, with upper bainite being coarser than lower bainite. At high temperatures, the carbon-supersaturated plates of bainitic ferrite reject carbon into the surrounding austenite by

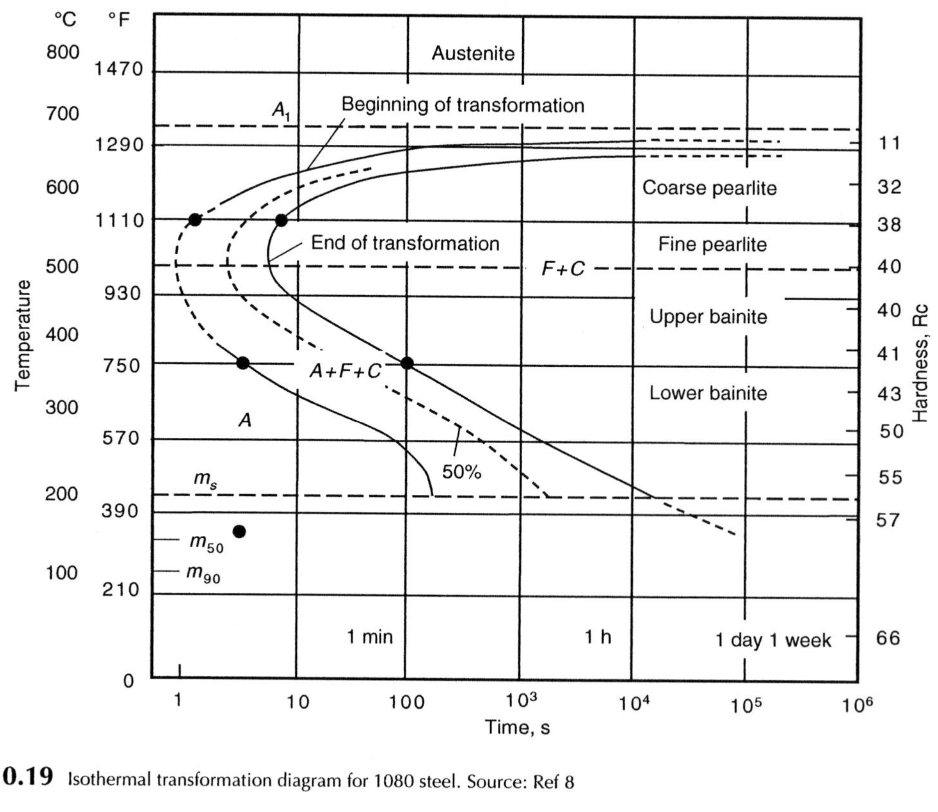


**Fig. 10.19** Isothermal transformation diagram for 1080 steel. Source: Ref 8

diffusion, leaving upper-bainitic ferrite free of internal carbides. Thus, the increased carbon content of the austenite creates a driving force for cementite precipitation in the interlath regions of the microstructure. At low temperatures, carbon diffusion out of bainitic ferrite is slower and incomplete, leading to the precipitation of carbides in both the interlath region and the ferrite lath interior. These two growth processes are shown schematically in Fig. 10.21. Because of these differences in morphology, upper and lower bainite have different mechanical properties. Lower bainite, with a fine acicular structure and carbides within the laths, has higher strength and higher toughness than upper bainite with its coarser structure.

### 10.5.1 Upper Bainite

Upper bainite grows as stacks (parallel laths) or needles of ferrite. Nucleation of ferrite typically starts at the grain boundaries of austenite, and cementite grows as elongated particles that are generally oriented parallel to the direction of growth of the ferrite laths. Cementite forms at the interfaces of newly formed ferrite/austenite, because the ability of the cementite to nucleate at those locations is much greater than that at the lath interfaces within the newly formed ferrite through the diffusion of excess carbon in the ferrite in this temperature range. The interparticle spacing of cementite particles in upper bainite is determined primarily by the size of the ferrite laths, and this spacing is generally quite large relative to the lamellar spacing of pearlite. As the carbon-supersaturated plates of bainitic ferrite reject carbon into the surrounding austenite by diffusion, the ferrite is left free of internal carbides. Lower transformation temperatures promote more ferrite plates in close proximity to each other and increase the extent of cementite precipitation. Upper bainite in carbon steel becomes progressively coarser and more feathery with lower carbon contents, and particles of cementite can be resolved when the carbon content is less than approximately 0.25 wt%.

### 10.5.2 Lower Bainite

Lower bainite is significantly different from upper bainite in terms of its kinetics and the carbide phase. At lower transformation temperatures, closer to the $M_s$ temperature, ferrite tends to form martensite needles with a distinctly more lenticular shape, as compared to upper bainite. Lower bainite also consists of a carbide that is crystallographically distinct from that of cementite. The carbide in lower bainite is epsilon carbide ($Fe_{2.4}C$) or Hägg's carbide.

Diffusion plays a role in tempering the needles of lower bainite, and the appearance of lower bainite is very similar to tempered martensite. However, lower bainite does not exhibit transformation twins that characterize martensite in high-carbon steels. Lower bainite, with its fine acicular structure and carbides within the laths, has higher strength and higher toughness than the coarser-structure upper bainite. At lower temperatures, carbon diffusion out of bainitic ferrite is slower and incomplete, leading to the precipitation of carbides in both the interlath region and the ferrite lath interior.

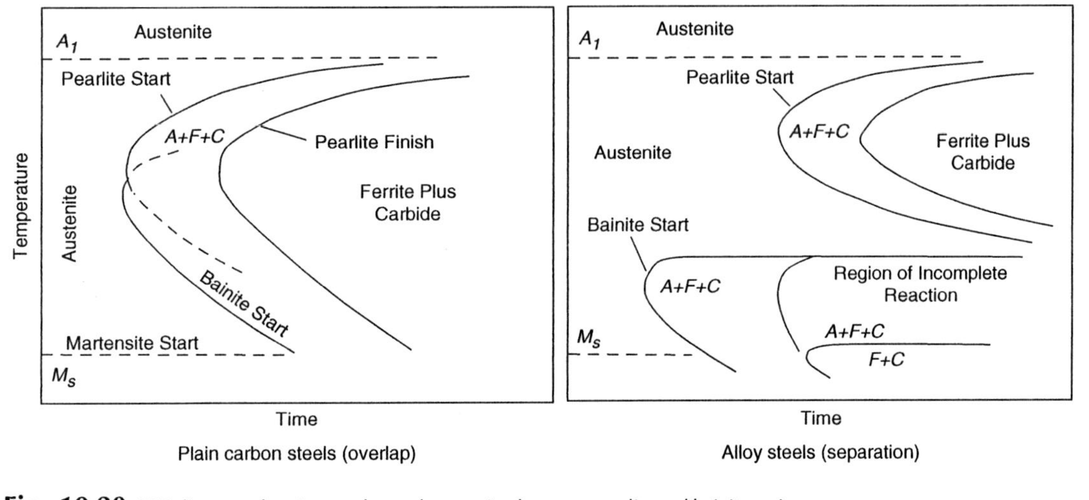


**Fig. 10.20** TTT diagrams showing overlap and separation between pearlite and bainite regions

Bainitic steels have a wide range of mechanical properties, depending on their morphology and composition. For example, the yield strength can range from 450 to 965 MPa (65 to 140 ksi), and the tensile strength from 520 to 1205 MPa (75 to 175 ksi). Bainitic steels are used in many applications, including pressure vessels, backup rolls, turbine rotors, die blocks, die casting molds, nuclear reactor components, and earthmoving equipment. One major advantage of bainitic steels is that an optimal strength/toughness combination can be produced without an expensive heat treatment such as quenching and tempering, which is used for martensitic steels. In designing a bainitic steel with a wide transformation region, it becomes critical that the pearlite and ferrite regions are pushed as far to the right as possible on the TTT diagram. Alloying elements used for this purpose include nickel, chromium, molybdenum, and manganese.

For low-carbon bainitic steels, the relationship between transformation temperature and tensile strength is shown in Fig. 10.22. Note the rapid increase in tensile strength as the transformation temperature decreases. With low-carbon bainitic steels, upper bainite has inferior toughness to lower bainite. In both cases, strength increases as the transition temperature decreases. In upper bainite, the carbides are much coarser than in lower bainite and have a tendency to crack and initiate brittle fracture. With lower bainite, the small carbides have less tendency to fracture.

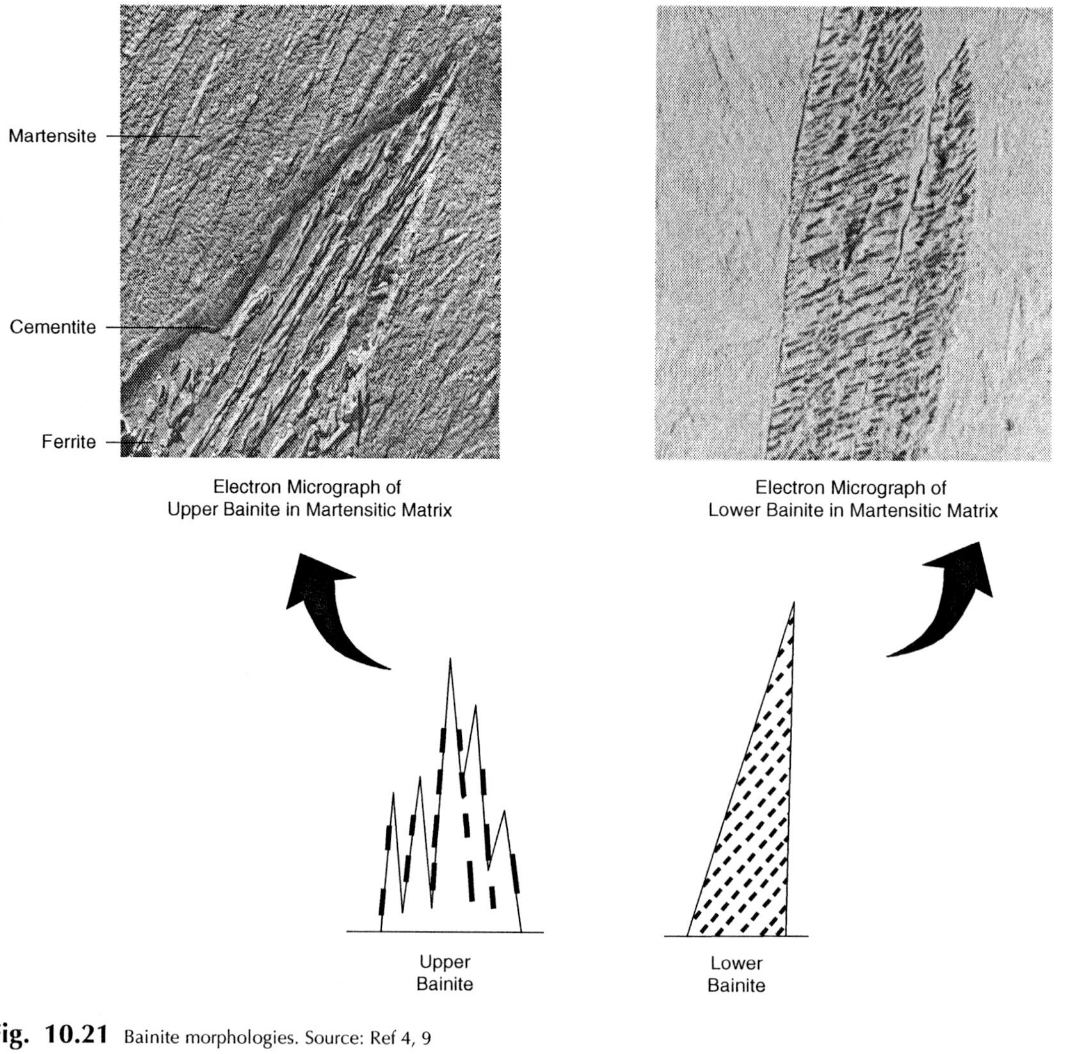


**Fig. 10.21** Bainite morphologies. Source: Ref 4, 9

## 10.6 Martensite

Martensite is formed in steels when they are heated into the austenitic field and then rapidly quenched to room temperature. The martensite transformation differs from other phase transformations in several ways: the reaction is a diffusionless phase transformation, no chemical composition change occurs between martensite and the parent phase, there is a coordinated crystal lattice structural change ("military mode") during transformation, the movement of atoms in the crystal is no more than the distance between two atoms in the crystal lattice, and there are certain crystallographic relationships between martensite and the parent phase. While still a nucleation and growth mechanism, martensite forms through the diffusionless transformation of austenite. In general, a martensitic reaction can be defined as a mechanism for forming a new crystallographic structure that does not require atomic diffusion.

To form martensite, the cooling rate must be great enough to avoid the nose of the TTT diagram. Martensite starts forming when it is cooled below the $M_s$ temperature and is completely transformed when it reaches the $M_f$ temperature. Above the $M_s$ temperature, martensite does not form, even at high cooling rates. The transformation only occurs at temperatures between the $M_s$ and $M_f$. Because of matrix constraints, the width of the martensitic units is limited, and the transformation proceeds primarily by the successive nucleation of new crystals. This process occurs only on cooling to lower temperatures and is therefore independent of time. The transformation ceases when the steel falls below the $M_f$ temperature. Some remaining austenite may still be present when the transformation stops and is referred to as retained austenite. Refer to Fig. 10.19 to note the locations of the $M_s$ and the 50 and 90% martensite transformation temperatures, $M_{50}$ and $M_{90}$.

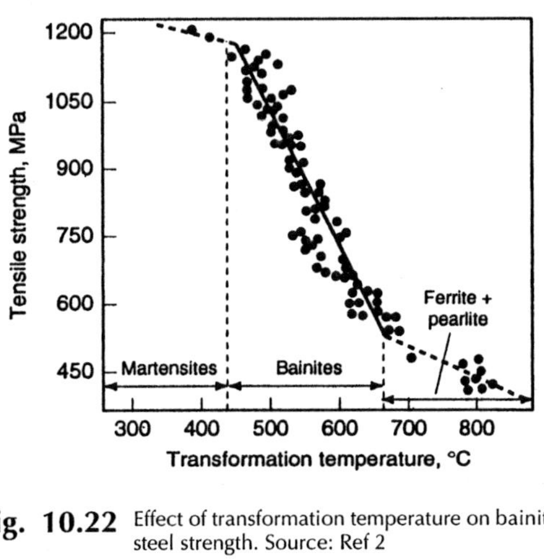


**Fig. 10.22** Effect of transformation temperature on bainitic steel strength. Source: Ref 2

The martensitic transformation is effectively athermal because it occurs during cooling and begins when the sample is cooled below a particular temperature; the amount of the new phase that forms depends on the temperature to which the sample is cooled rather than the time it spends at that temperature. If cooling is done quickly enough, structural rearrangement of atoms occurs by shear displacement over a small distance, on the order of approximately an interatomic spacing. This shear movement of atoms and phase growth occurs rapidly. It is not an instantaneous transformation, but it is quicker and distinct from the slower, diffusion-controlled process in isothermal transformations.

The $M_s$ temperature can be raised by the application of stress during transformation. This occurs because a crystal that forms during a martensitic transformation has a different shape and volume than that of the austenite from which it is forming. Therefore, if there is an applied stress available to achieve this shape change, the transformation will be accomplished more easily. If it can occur more easily, then it will require less thermodynamic driving force for it to occur, which means that the necessary degree of undercooling will be smaller. Thus, if an applied stress is locally oriented so that it helps the transformation to produce the shape change, martensite will be able to form at a higher temperature than for the unstressed condition. In general, tensile stresses are found to be more effective than compressive stresses, but rolling can have appreciable effects. The highest temperature to which the $M_s$ temperature can be raised by applied stresses is defined as the $M_d$ temperature, the martensite deformation temperature. When the $M_d$ temperature is above room temperature and the $M_s$ is below room temperature, it is possible to retain the austenite at room temperature and then to form some martensite by working the metastable austenite at room temperature.

As-quenched martensite has a very high strength but a very low fracture resistance or toughness. Therefore, almost all steels that are quenched to martensite are also tempered, or reheated, to some temperature below $A_1$ in order to increase toughness. In tempered martensite, the carbide consists of small, chunky particles rather than platelets. These finely divided particles of cementite in tempered martensite provide strengthening and restore ductility, depending on the specific tempering temperature. In general, higher tempering temperatures result in lower strength but greater ductility.

The hardness of heat treated martensite is determined by its carbon content, as shown in Fig. 10.23. Martensite attains a maximum hardness of HRC 66 at carbon contents of 0.8 to 1.0 wt%. The reason that the hardness does not monotonically increase with carbon is that retained austenite starts forming when the carbon content is above approximately 0.4 wt%, and retained austenite is much softer than martensite.

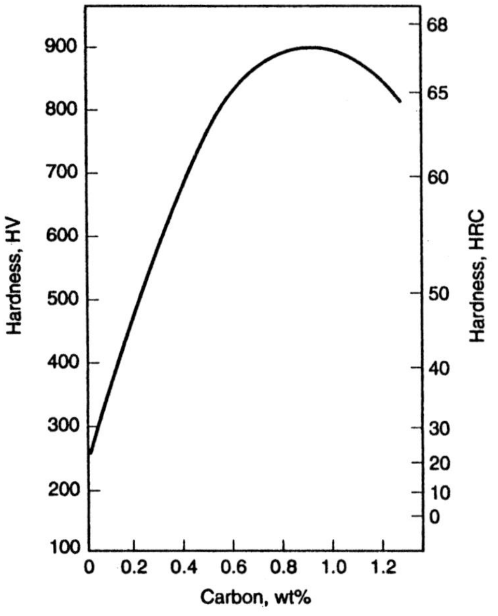


**Fig. 10.23** Hardness (Rockwell C) versus carbon content for quenched steels. Source: Ref 2

### *10.6.1 Formation of Martensite in Steels*

In a martensitic transformation process, large numbers of atoms experience cooperative movements with only a slight displacement of each atom relative to its neighbors. The fcc austenite transforms to a body-centered tetragonal (bct) structure (Fig. 10.24). This structure is essentially the same as the ferrite bcc structure, except that it has been distorted into a tetragonal structure due to the entrapment of carbon that did not have time to diffuse out and form cementite. Since the martensitic transformation does not involve diffusion, it occurs

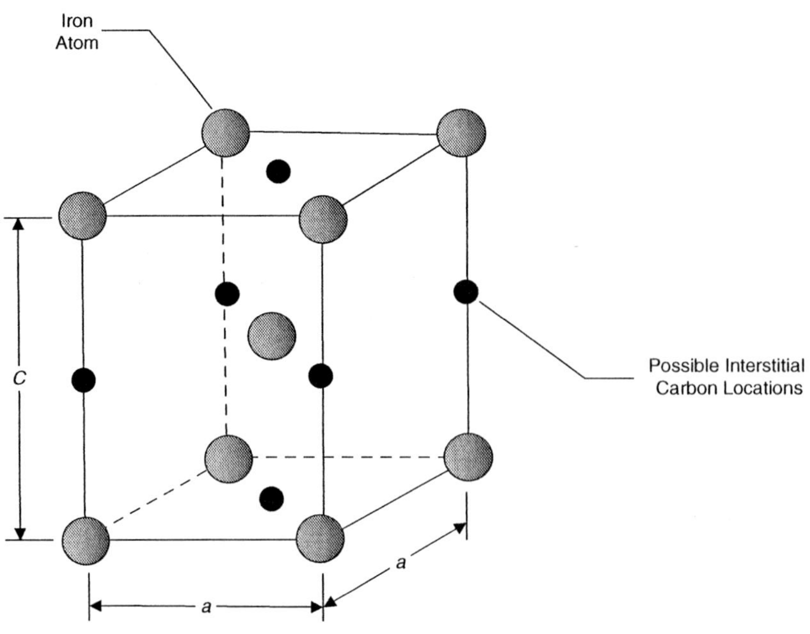


**Fig. 10.24** Body-centered tetragonal structure of martensite

almost instantaneously, approaching the speed of sound within the austenite matrix. All the carbon atoms remain as interstitial impurities in martensite, and as such, they constitute a supersaturated solid solution that is capable of rapidly transforming to other structures if heated to temperatures at which diffusion rates become appreciable. Since the bct crystal structure is less densely packed than the fcc structure of austenite, the transformation results in a volumetric expansion and a hardening of steel. As shown in Fig. 10.25, the *c*/*a* ratio of the tetragonal unit cell increases as the carbon content increases. To a good approximation, the variations in *c* and *a* are linear, with the *c* value increasing at a greater rate with increasing carbon contents than the rate the *a* parameter decreases.

Two adjacent fcc unit cells of austenite are shown in Fig. 10.26(a), in which a bct unit cell has been identified. The atoms identified in Fig. 10.26(a) have been isolated in the left-hand portion of Fig. 10.26(b). At this stage, the dimensions of the bct cell are still those derived from the austenite lattice parameter. The unit cell in the right-hand portion of Fig. 10.26(b)

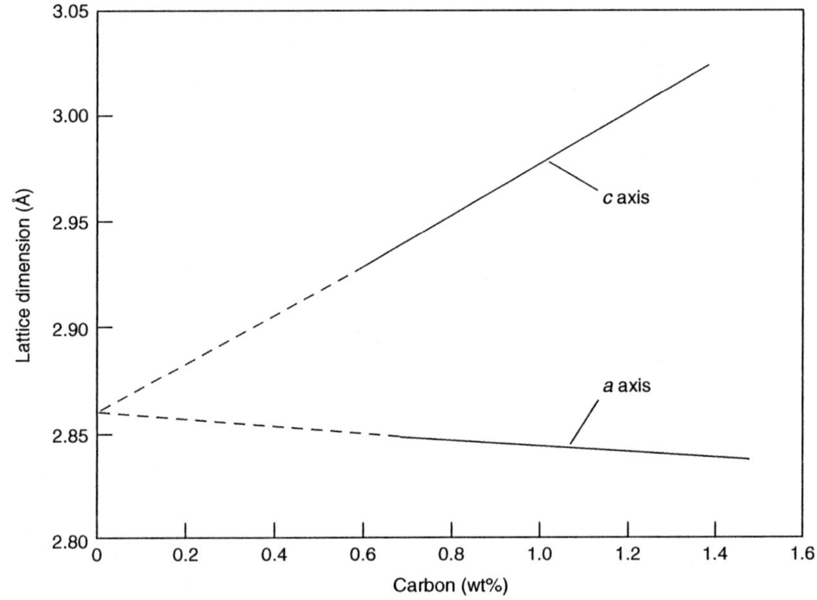


**Fig. 10.25** Variation of *c*/*a* parameters with carbon content

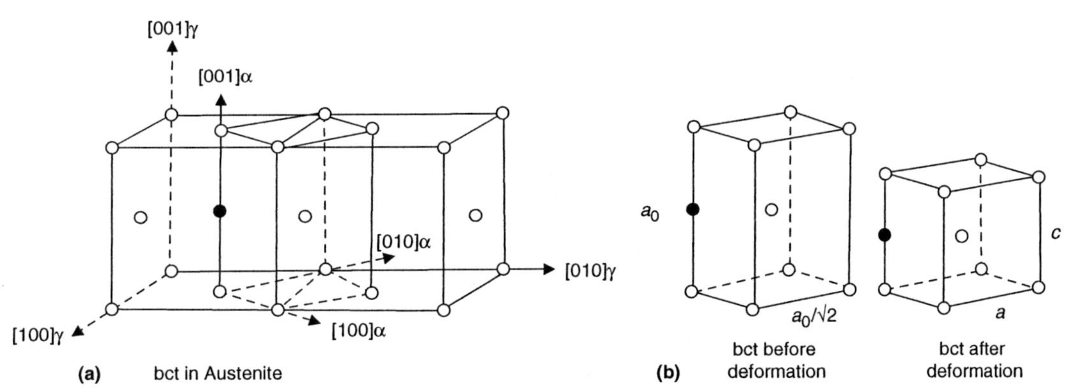


**Fig. 10.26** Transformation from austenite to martensite

is that of martensite with lattice parameters $a$ and $c$, corresponding to a given carbon content shown in Fig. 10.25. Note that a lattice deformation is required to produce martensite from austenite. This deformation was first identified by Bain and is referred to as the Bain strain. As previously shown in Fig. 10.25, the Bain strain produces an expansion along the $c$ axis and a contraction along the $a$ axis.

The formation of a martensitic crystal is shown schematically in Fig. 10.27. Shearing occurs parallel to a fixed crystallographic plane, termed the habit plane, and produces a uniformly tilted surface relief on a free surface. This mechanism involves shear displacement of iron atoms on specific planes, namely the {111} planes. In addition to the lattice distortion caused by the formation of the bct structure, the resulting martensite is simultaneously deformed because of the constraints created by maintaining an unrotated and undistorted habit plane within the bulk austenite. The deformation of martensite is referred to as a lattice invariant deformation, and it produces a high density of dislocations or twins in martensite. The result is a bct crystal structure that is less densely packed than the fcc structure of austenite. Hence, the martensitic transformation results in a volumetric expansion and a hardening of steel. This fine structure, together with the carbon atoms trapped within the octahedral interstitial sites of the bct structure, produces the very high strength of as-quenched martensite.

### *10.6.2 Morphology of Martensite*

In steels, there are two distinctly different microstructures that are found, depending on the carbon content of the steel (Fig. 10.28). For alloys containing less than approximately 0.6 wt% C, the martensite grains form as laths (i.e., long and thin plates) that form side by side and are aligned parallel to one another. Furthermore, these laths are grouped into larger structural entities, called blocks. The morphology of this lath or massive martensite is shown in Fig. 10.29. Lenticular, or plate, martensite is typically found in alloys containing greater than approximately 0.6 wt% C. With this structure, the martensite grains take on a needlelike (i.e., lenticular) or platelike appearance, which grows across the complete austenite grain, as shown in Fig. 10.30. Generally, plate martensite can be distinguished from lath martensite by its plate morphology with a central midrib. As the carbon content increases, twins begin to replace dislocations within the plates, so that high-carbon martensite is composed mainly of twinned plates. The transformation is also associated with an appreciable volume increase, because it replaces the close-packed fcc structure with the more loosely packed bct structure. This

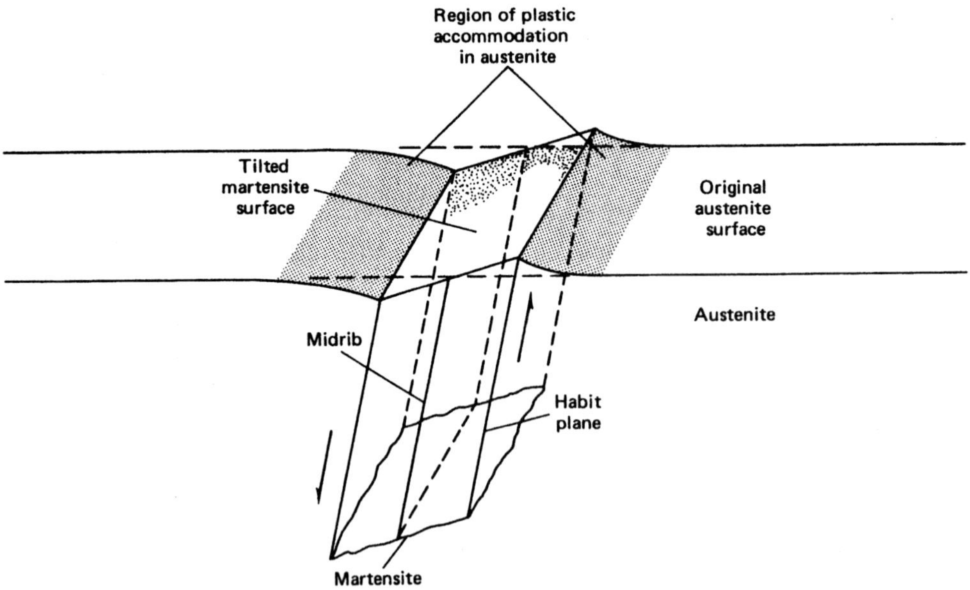


**Fig. 10.27** Shear and surface tilting during martensite formation. Source: Ref 10

transformation creates residual stresses that are related to the specific volume change, in addition to the strains due to the misfit of the interstitial solute atoms. At high carbon levels, these stresses can become so severe that the material cracks as the martensite forms. These cracks can range from small microcracks that require microscopy to be detected, to large cracks easily visible to the unaided eye. The cracks form during transformation when a growing plate impinges on an existing plate. Because of these microcracks, plate martensite is generally avoided in most applications. With carbon contents between 0.6 and 1.0 wt%, the martensite is a mixture of lath and plate morphologies.

## 10.7 Retained Austenite

Austenite starts transforming to martensite when it reaches the $M_s$ temperature and continues to transform until the $M_f$ temperature is reached. When the transformation ceases, some remaining austenite may still be present. When retained austenite is present, it is usually as small island grains surrounded by martensite (Fig. 10.31). Because martensite expands when it is formed, the remaining austenite is effectively pressurized from all sides. This makes it more difficult for it to expand, which it needs to do in order to transform into martensite. Because elastic deformation results in a volume expansion, impact-related stresses or other service loading conditions, as well as low-temperature exposure, can allow retained austenite to transform to martensite in service. Unless the service temperature is somewhat elevated, the new martensite will remain in an essentially untempered condition.

## 10.8 Carbon Content

The addition of carbon and other alloying elements causes significant changes in the isothermal transformation curves. As shown in Fig. 10.32, increasing amounts of carbon push the nose of the TTT diagram to the right. The practical consequence is that steels with higher carbon contents can be hardened to form fully martensitic structures. For example, the steels containing 0.06 and 0.21 wt% C cannot be hardened to fully martensitic structures because there is not a quench fast enough to miss the nose of the TTT diagram. Thus, even with fast cooling, some pearlite will form. On the other hand, the steels containing 0.50 and 0.80 wt% C could, at least in thin sections, be hardened to fully martensitic structures. As is discussed in Chapter 11, "Heat Treatment of Steel," the

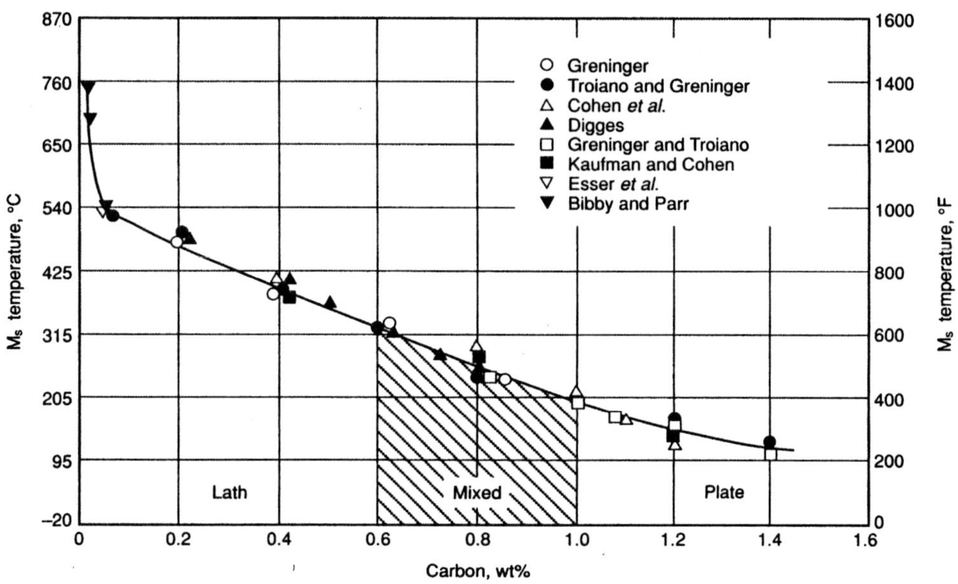


**Fig. 10.28** Martensite morphology versus carbon content. Source: Ref 10

main advantage of using substitutional alloying elements, rather than increasing carbon contents, is that the steel can be hardened to a greater depth without causing the brittleness associated with high carbon contents.

The $M_s$ and $M_f$ temperatures also depend on the carbon content of steel. When the carbon content is increased, carbide formation ($Fe_3C$) becomes more dominant under both isothermal and athermal conditions, and thus, the $M_s$ and $M_f$ temperatures are lowered. Almost all other alloying elements also lower $M_s$. The reduction of the $M_s$ and $M_f$ temperatures, as a function of carbon content and the alloying element manganese, is shown in Fig. 10.33. It is important to recognize that the martensitic transformation rarely goes to 100% completion. A small amount of the microstructure in a eutectoid plain carbon steel can be retained austenite. The amount of retained austenite varies with composition, being higher in steels containing high carbon contents and other alloying elements.

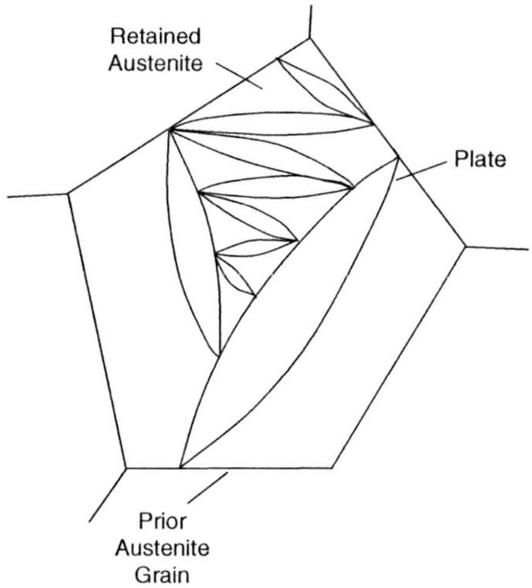


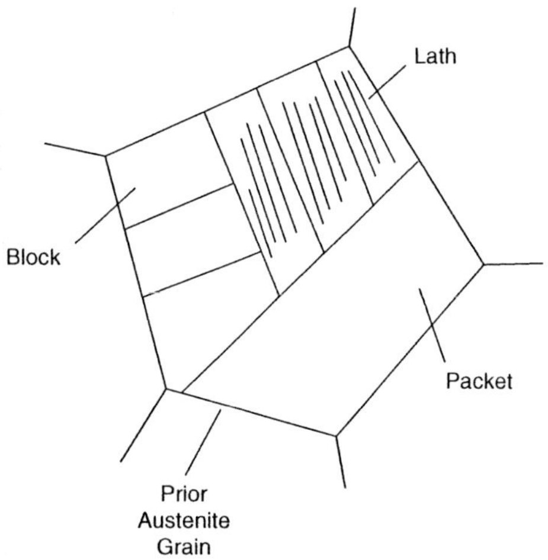


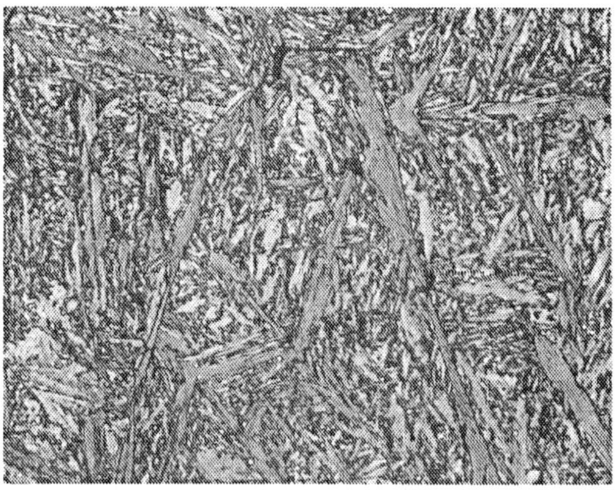

**Fig. 10.30** Plate martensite. Source: Ref 2

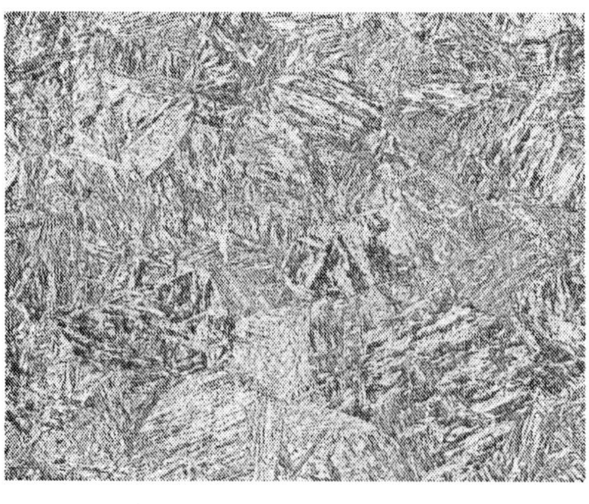

**Fig. 10.29** Lath martensite. Source: Ref 2

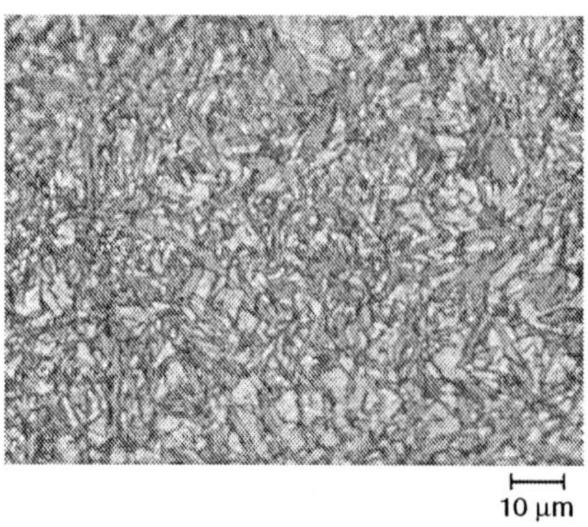


**Fig. 10.31** Martensite microstructure with retained austenite (light areas). Source: Ref 4

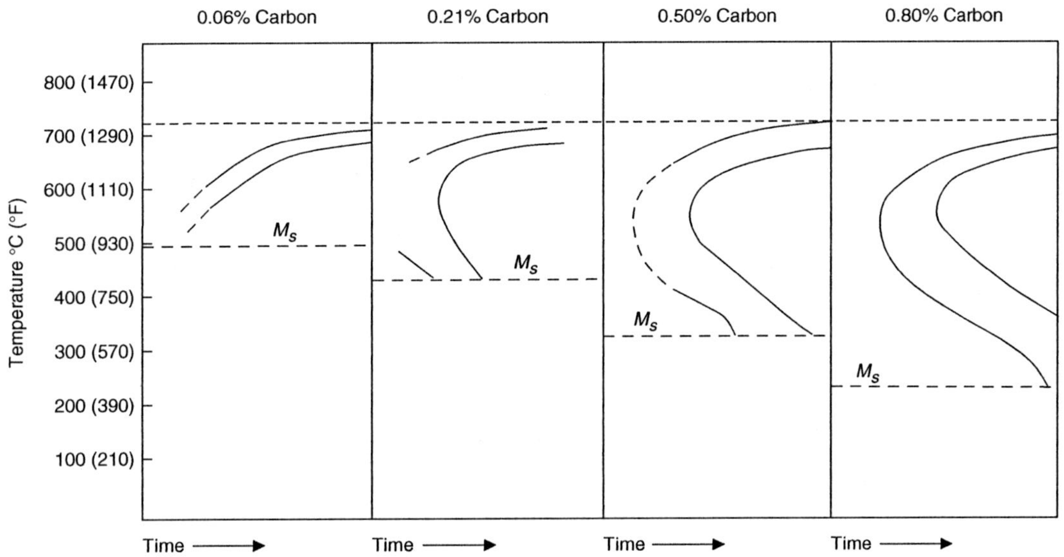


**Fig. 10.32** Effect of carbon content on TTT diagram

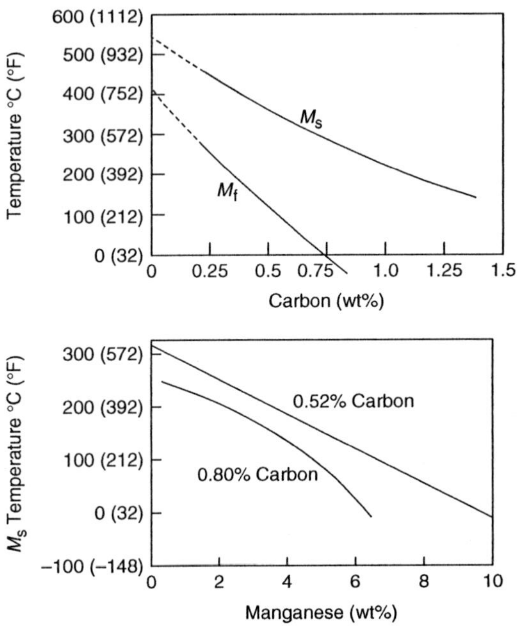


**Fig. 10.33** Increasing carbon and alloy content lowers the $M_s$ temperature.

## ACKNOWLEDGMENTS

Sections of this chapter were adapted from "Microstructures, Processing, and Properties of Steels" by G. Krauss in *Properties and Selections: Irons, Steels, and High-Performance Alloys,* Volume 1, *ASM Handbook,* ASM International, 1990 and "Effects of Composition, Processing, and Structure on Properties of Irons and Steels" by B.L. Bramfitt in *Materials Selection and Design,* Volume 20, *ASM Handbook,* ASM International, 1997.

## REFERENCES

1. T.B. Massalski, H. Okamoto, P.R. Subramanian, and L. Kacprzak, *Binary Alloy Phase Diagrams,* 2nd ed., ASM International, 1990
2. B.L. Bramfitt, Effects of Composition, Processing, and Structure on Properties of Irons and Steels, *Materials Selection and Design,* Vol 20, *ASM Handbook,* ASM International, 1997
3. A.G. Guy and J.R. Hren, *Elements of Physical Metallurgy,* 3rd ed., Addison-Wesley Publishing Company, 1974
4. D. Aliya and S. Lampman, Physical Metallurgy Concepts in Interpretation of Microstructures, *Metallography and Microstructures,* Vol 9, *ASM Handbook,* ASM International, 2004
5. B.L. Bramfitt and S.J. Lawrence, Metallography and Microstructures of Carbon and Low-Alloy Steels, *Metallography and Microstructures,* Vol 9, *ASM Handbook,* ASM International, 2004
6. R.A. Higgins, *Engineering Metallurgy,* 6th ed., Arnold, 1993

7. W.F. Smith, *Principles of Materials Science and Engineering,* McGraw-Hill, 1986
8. H.E. Boyer, Ed., *Atlas of Isothermal Transformation and Cooling Transformation Diagrams,* American Society for Metals, 1977
9. R.F. Hehemann, Ferrous and Nonferrous Bainitic Structures, *Metallography, Structures and Phase Diagrams,* Vol 8, *Metals Handbook,* 8th ed., American Society for Metals, 1973
10. G. Krauss, Microstructures, Processing, and Properties of Steels, *Properties and Selections: Irons, Steels, and High-Performance Alloys,* Vol 1, *ASM Handbook,* ASM International, 1990

## SELECTED REFERENCE

- E.C. Bain and H.W. Paxton, *Alloying Elements in Steel,* 2nd ed., American Society for Metals, 1966

# CHAPTER 11

# Heat Treatment of Steel

ONE OF THE PRIMARY ADVANTAGES of steels is their ability to attain high strengths through heat treatment while still retaining some degree of ductility. This ability of steels to be strengthened is a direct consequence of the amount of carbon present (Fig. 11.1). As the carbon content is increased, higher strength levels are obtainable. Although the ductility decreases with increasing strength, it is still high enough to satisfy most engineering applications. The type of heat treatment used also influences the properties. As shown in Fig. 11.2, a steel hardened by heating into the austenite field, followed by quenching and then tempering, has a much higher strength than one subjected to either a normalizing or spheroidizing heat treatment. Heat treatments can be used to not only harden steels but also to provide other useful combinations of properties, such as ductility, formability, and machinability. The various heat treatment processes covered in this chapter include annealing, stress relieving, normalizing, spheroidizing, and hardening by quenching and tempering.

In all of these processes, the steel is heated fairly slowly to some predetermined temperature and then cooled. It is the temperature and the cooling rate that determine the resultant structure of the steel and hence the mechanical properties, as illustrated in the time-temperature transformation (TTT) diagram in Fig. 11.3. The final structure is independent of the heating rate, provided it has been slow enough for the steel to reach structural equilibrium at the maximum temperature. However, the subsequent cooling rate, which determines the nature of the final structure, is critical and may vary between slow furnace cooling to rapid cooling by quenching in water.

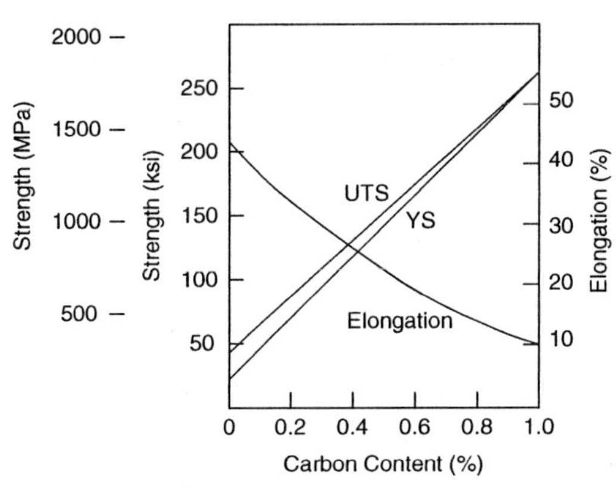


**Fig. 11.1** Effect of carbon content on steel strength. UTS, ultimate tensile strength; YS, yield strength

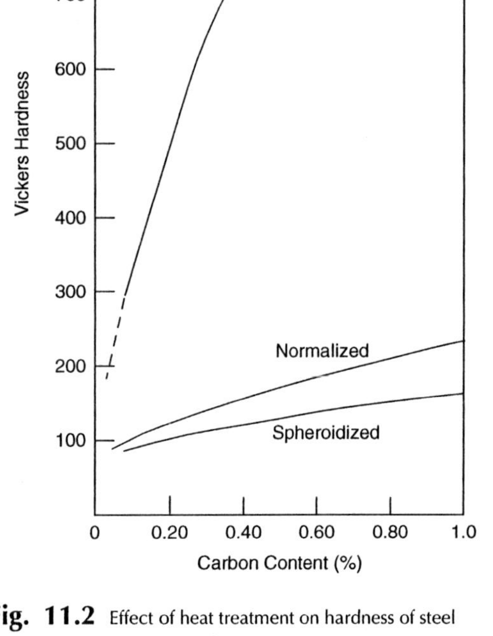


**Fig. 11.2** Effect of heat treatment on hardness of steel

## 11.1 Annealing

The term *annealing* is a heat treatment in which a metal or alloy is heated to a certain high temperature and held for a certain period of time, followed by slow cooling. In steels, annealing usually has the meaning of a heat treatment with furnace cooling from the austenitizing range shown in Fig. 11.4. Annealing is used to reduce hardness, obtain a relatively near-stable microstructure, refine grain size, improve machinability, and facilitate cold working. For hypoeutectoid steels, full annealing consists of heating to 10 to 40 °C (50 to 100 °F) above the $A_3$ temperature, and for hypereutectoid steels, heating above the $A_1$ temperature and followed, in both cases, by very slow cooling. As a result of the very slow cooling rates, the microstructure consists of coarse ferrite or coarse ferrite plus pearlite, depending on the carbon and alloy content of the steel.

Isothermal annealing is often more efficient for small parts because, instead of furnace cooling, the parts can be transferred between constant-temperature molten salt baths. The steel is also protected from scaling and oxidation by the molten bath. These salts are various mixtures of compounds such as alkali metal hydroxides and nitrates that are chosen for particular temperature ranges. Baths of molten lead were commonly used before concerns about the dangers of lead to human health and the environment became widespread.

## 11.2 Process Annealing and Stress Relief

As the hardness of steel increases during cold working, ductility decreases, and additional cold reduction becomes so difficult that the material must be annealed to restore its ductility. Such annealing between processing steps is referred to as in-process or simply process annealing. It may consist of any appropriate treatment; however, in most instances, a subcritical treatment is adequate and the least costly. The term *process annealing* without further qualification usually refers to an in-process subcritical anneal. Process annealing usually consists of heating steel to a temperature just below the $A_1$ eutectoid

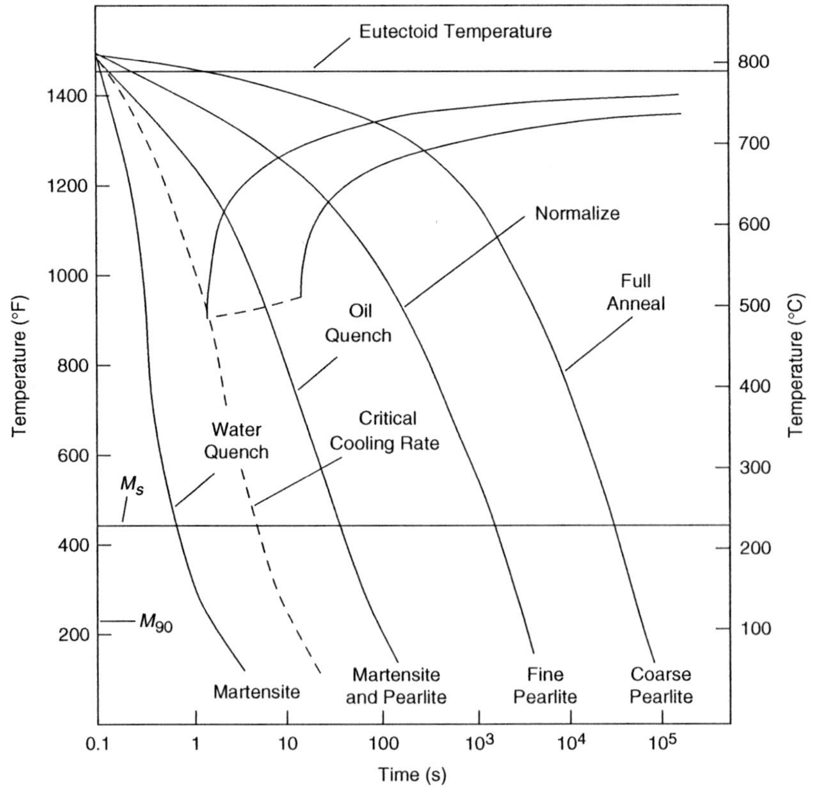


**Fig. 11.3** Effect of cooling rate on microstructure. Source: Ref 1

temperature for a short time. This provides stress relief, makes the steel easier to form, and is applied to low-carbon cold-rolled sheet steels to restore ductility. The temperatures used range from 550 to 650 °C (1020 to 1200 °F). Slow cooling is not essential for process annealing, since any cooling rate from temperature below $A_1$ will not affect the microstructure or hardness. Although recrystallization can occur due to the stored energy from cold working, there are no phase changes, and the ferrite and cementite constituents remain the same throughout the process. Process annealing is generally carried out in either batch-type or continuous furnaces.

## 11.3 Normalizing

Steel is normalized by heating 30 to 45 °C (90 to 110 °F) into the austenite phase field above the $A_3$ temperature, somewhat higher than those used by annealing, followed by cooling at a medium rate. For carbon steels and low-alloy steels, normalizing means air cooling. Many steels are normalized to establish a uniform ferrite plus pearlite microstructure along with a uniform grain size. The faster cooling rate during normalizing results in a much finer pearlitic structure, which is harder and stronger than the coarse pearlite produced by full annealing. Steel is normalized to refine grain size, make its structure more uniform, make it more responsive to hardening, and to improve machinability. When steel is heated to a high temperature, the carbon can readily diffuse, resulting in a reasonably uniform composition from one area to the next. The steel is then more homogeneous and will respond to the heat treatment more uniformly. The properties of normalized steels depend on their chemical composition and on the cooling rate, with the cooling rate being a function of the size of the part. Although there can be a considerable variation in the hardness and strengths of normalized steels, the structure usually contains fine pearlite.

## 11.4 Spheroidizing

To produce a steel in its softest possible condition with minimum hardness and maximum ductility, it can be spheroidized by heating just above or just below the $A_1$ eutectoid temperature and then holding at that temperature for

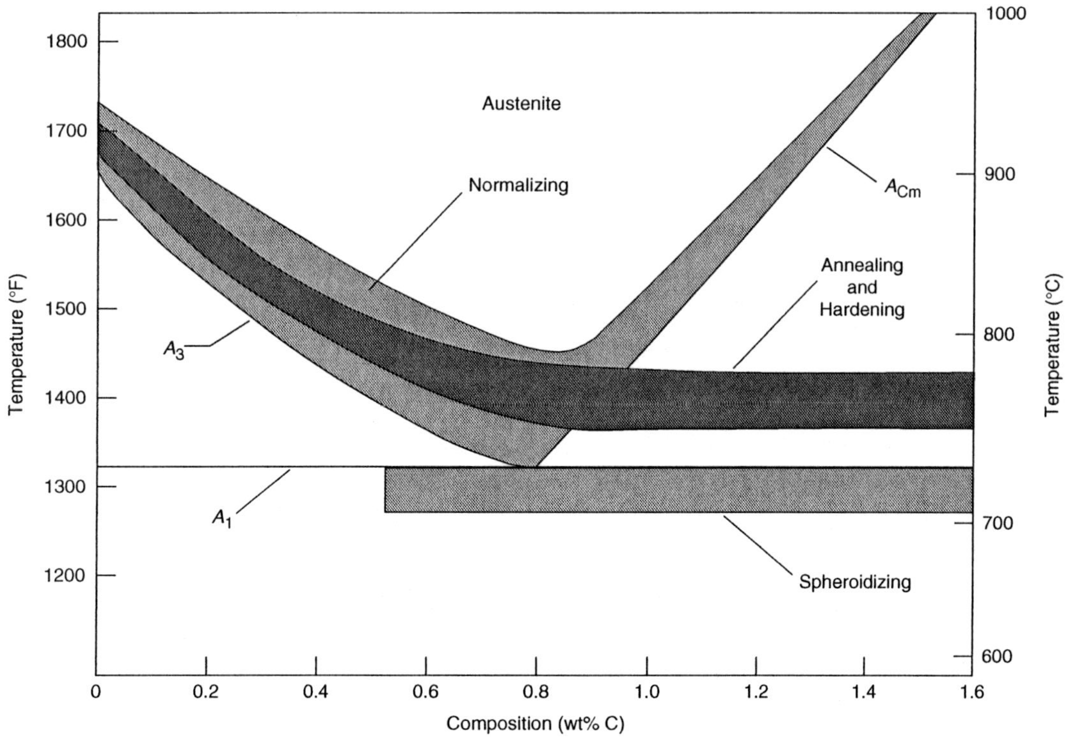


**Fig. 11.4** Steel heat treating ranges. Source: Ref 2

an extended period of time. Spheroidizing can also be conducted by cyclic processing, in which the temperature of the steel is cycled above and below the $A_1$ line. This process breaks down lamellar pearlite into small pieces and forms small cementite spheroids through diffusion in a continuous matrix of ferrite (Fig. 11.5). Surface tension causes the carbide particles to develop a spherical shape. Since a fine initial carbide size accelerates spheroidization, the steel is often normalized prior to spheroidizing. Another option is to start with a martensitic structure that produces a very uniform dispersion of cementite spheroids, because carbon is more uniformly distributed in martensite than in lamellar pearlite. The cementite lamellae must first dissolve and then redistribute the carbon as spheroids, whereas cementite spheroids can form directly from martensite.

## 11.5 Hardening

Steels are hardened by austenitizing, quenching, and then tempering to their final hardness. Since the composition of steels varies quite a bit, it is important to: (1) understand the maximum section thickness that can be hardened in a specific quench medium (e.g., water or oil), and (2) realize the large variations in final strength and ductility that can be obtained by tempering at different temperatures. The generalized procedure for quenching and tempering is illustrated in Fig. 11.6. It should be noted that tempering is not used to harden steel. Steel is hardened by austenitizing and quenching. Tempering is conducted to restore a portion of the ductility that was lost during hardening, and often, appreciable softening is produced by tempering.

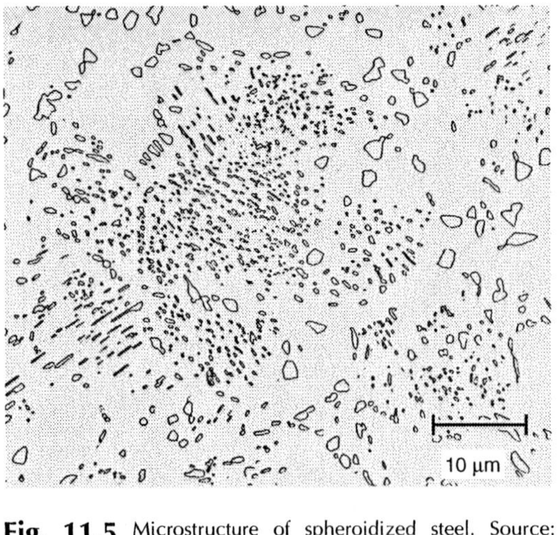


**Fig. 11.5** Microstructure of spheroidized steel. Source: Ref 3

The carbon content is critical to the ability to harden steel. Since ductility decreases with increasing carbon content, the carbon content is held to approximately 0.45 wt% in many engineering steels. However, when wear resistance is required, for example, in tool and die steels, it may be increased to over 1.0 wt%. The addition of alloying elements shifts the nose of the TTT diagram to the right, thus allowing thicker sections to be hardened or allowing less drastic quenches. The effect of alloying elements and section size on hardenability is illustrated in Fig. 11.7. In this example, both the 1040 and 4140 steels contain nominal carbon contents of 0.40%, and yet, due to the alloying elements in 4140, 4140 hardens to a much greater depth. However, as the diameter of a bar of 4140 is increased from 5 to 10 cm (2 to 4 in.), the depth of hardening (hardenability) decreases. Some more highly alloyed steels have TTT diagrams shifted so far to the right that they will form fully martensitic structures in fairly thick sections by air cooling to room temperature. Therefore, while some alloying elements may not directly increase the hardness of martensite significantly, they do increase the hardenability, which is the depth from the surface of the martensite microstructure that can be produced in steel of a given carbon content during quenching.

### *11.5.1 Continuous Cooling Transformation Diagrams*

Isothermal transformation diagrams (TTT curves) provide valuable information about the kinetics of the transformation of austenite and are useful tools for developing heat treatments. However, in practical heat treatments, where the size of the part being treated is usually significant, the temperature and the rate of change of temperature vary with position within the part being heat treated. Also, there are heat treatments in which the steel is continuously cooled. Therefore, for industrial applications, continuous cooling transformation (CCT) diagrams are more useful than TTT diagrams. During continuous cooling, part of the time is spent at higher temperatures where the transformation rates are slower. Therefore, the time for transformation is longer for continuous cooling than for isothermal cooling. The lines representing the start of pearlite and bainite formation are shifted to longer times and lower temperatures. Since alloying elements tend to delay the formation of pearlite more than they delay bainite formation, bainite formation can occur during continuous cooling.

The TTT and CCT diagrams for 4340 are compared in Fig. 11.8 and 11.9, respectively. There are two main types of CCT diagrams.

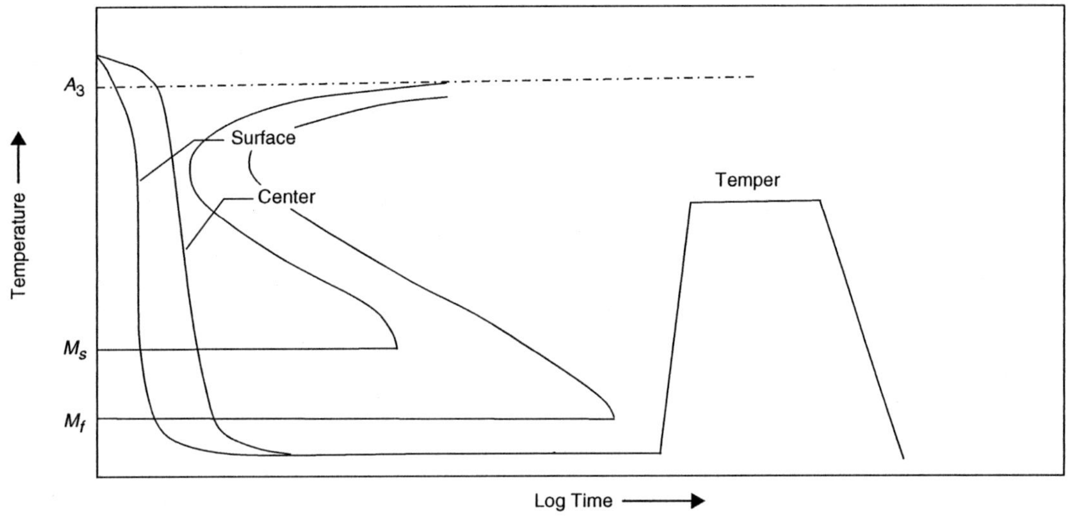


**Fig. 11.6** Hardening heat treatment for steels

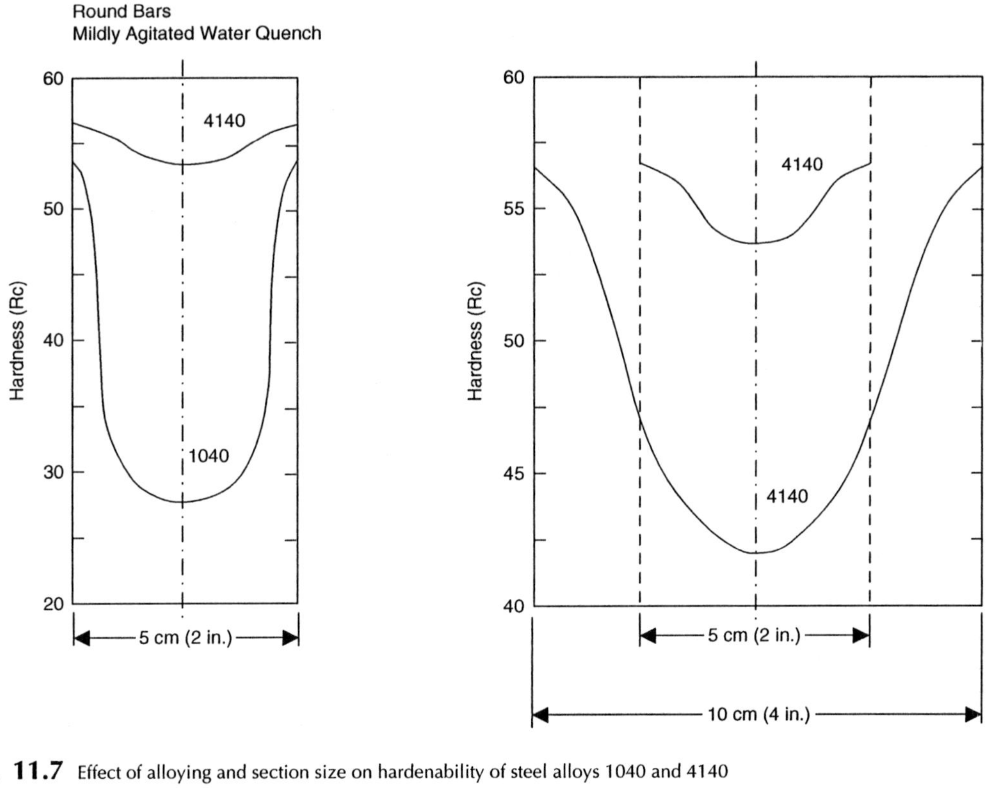


**Fig. 11.7** Effect of alloying and section size on hardenability of steel alloys 1040 and 4140

The first type, shown in Fig. 11.9, has axes that show the temperature of the steel on the ordinate axis and time in seconds plotted on a logarithmic scale on the abscissa. Lines are drawn on the diagram to show the cooling path in terms of the variation of temperature with time. By considering such a cooling path and traveling along it downward from the austenite, the start of a transformation is encountered. The location of this point is marked on the diagram and joined to similar points on other cooling paths to create curved lines that denote the initiation of the decomposition of the austenite. Similar lines that show the progress of the transformation in terms of the percent completed are often placed between the start and finish curves. Diagrams of this type are determined experimentally by dilatometry.

Most commercial martensitic steels contain alloying additions intended to suppress the formation of other constituents, namely ferrite, pearlite, and bainite, during continuous cooling. This means that these constituents form at slower cooling rates, allowing martensite to form at the faster cooling rates, for example, during oil and water quenching. Most of the conventional alloying elements in steel promote hardenability. For example, the 4340 steel shown has significant levels of carbon, manganese, nickel, copper, and molybdenum to promote hardenability.

The second type of CCT diagram retains temperature as the ordinate axis but defines the abscissa in terms of diameters of steel bars cooled in air, oil, and water, respectively, as shown in Fig. 11.10. These quenching media are commonly used in industrial practice and extract heat from the surface of a bar at different rates, with the water quench being the most rapid. Since different oil quenchants do not extract heat at the same rates, the diagrams provide only qualitative guidelines. There are also variations in chemical composition among steels of the same designation, which can affect the transformations.

### *11.5.2 Austenitizing*

During austenitization, the steel is heated into the austenite (γ) field and held for a sufficient

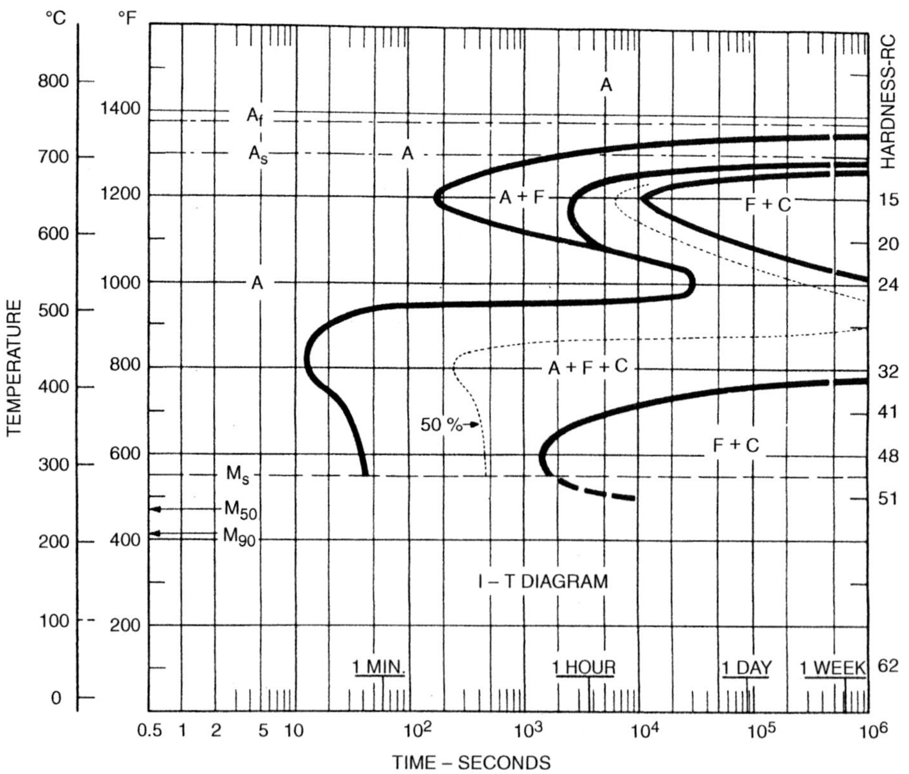


**Fig. 11.8** Isothermal transformation diagram for 4340 steel. Austenitized at 870 °C (1600 °F), ASTM grain size 6–7. Source: Ref 4

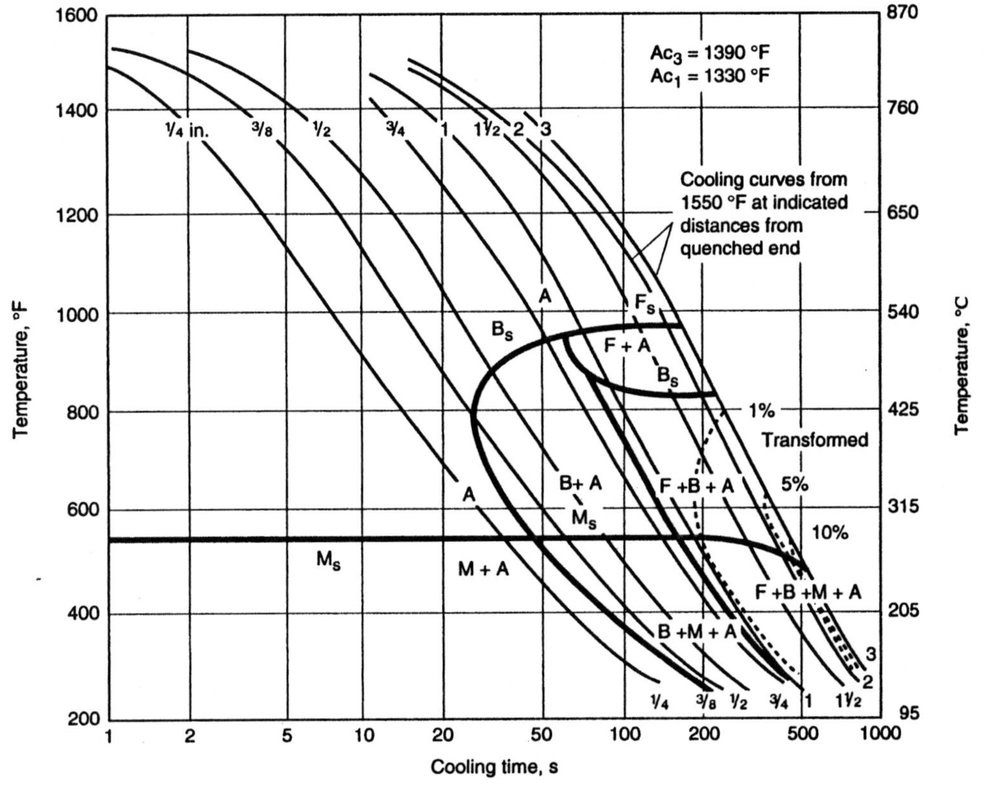


Fig. 11.9 Continuous cooling transformation diagram for 4340 steel. Source: Ref 5

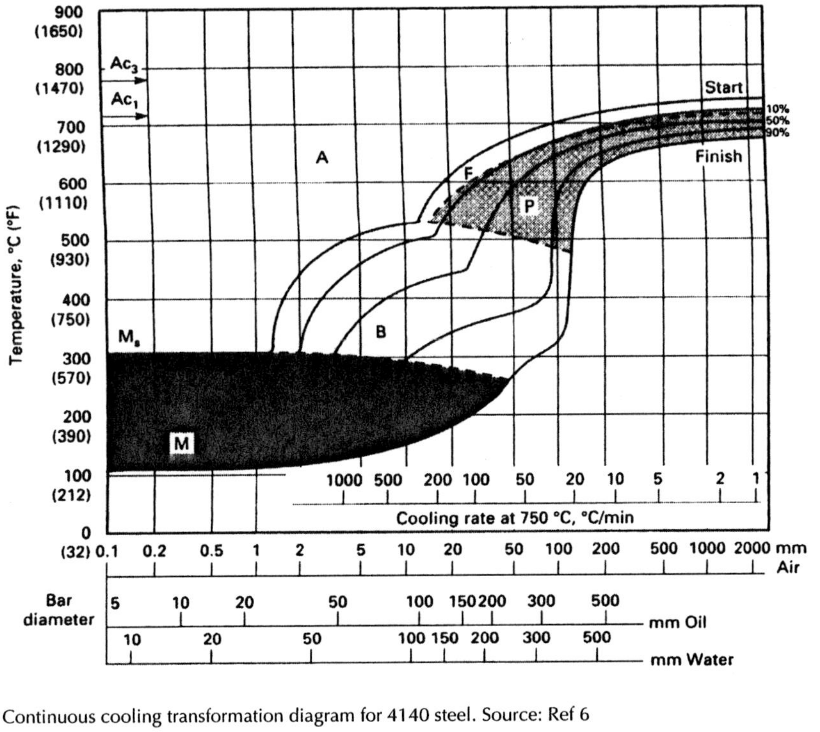


Fig. 11.10 Continuous cooling transformation diagram for 4140 steel. Source: Ref 6

period of time to dissolve most of the carbides and put them into solid solution. As previously shown in Fig. 11.4, the temperature required for austenitization is a function of the carbon content in carbon steel. With increasing carbon contents, the temperature decreases along the $A_3$ line to a minimum value at $A_1$, the eutectoid composition (0.8% C), and then increases again along the $A_{Cm}$ line. The first stage in the formation of austenite is the nucleation and growth of austenite from pearlite (ferrite + $Fe_3C$). Even after the complete disappearance of pearlite, some carbides will remain in the austenite. The formation of homogeneous austenite is accelerated by increasing the temperature and increasing the fineness of the initial carbide particles. To minimize the time for austenitization, the temperature used is approximately 55 °C (100 °F) above the minimum temperature for 100% austenite, and the time is approximately 1 h per 2.5 cm (1 in.) of thickness. However, it is also important to keep the austenitization temperature as low as possible to reduce the tendency toward cracking and distortion, minimize oxidation and decarburization, and minimize grain growth.

The temperatures needed for obtaining 100% austenite in hypereutectoid steels can be quite high; however, austenite suitable for hardening in these steels can be obtained at approximately 770 °C (1425 °F). The small amount of undissolved carbides dispersed in the austenite has little effect on the final mechanical properties. High temperatures are also required for very low carbon steels, but carbon steels containing less than 0.2% C respond poorly to quenching and are seldom used in the quenched and tempered condition.

### *11.5.3 Quenching*

While the face-centered cubic (fcc) austenite that forms during austenitization is capable of dissolving as much as 2 wt% C, only a small fraction of carbon can be retained in the lower-temperature body-centered cubic (bcc) ferrite. If the steel is slowly cooled from the austenitization temperature, carbon atoms are rejected as the fcc austenite transforms to the bcc ferrite, and alternating layers of ferrite and cementite form pearlite through a nucleation and growth process. However, if the steel is rapidly cooled from the austenitization temperature (quenched), the carbon does not have time to diffuse out of the austenite microstructure when it transforms into the body-centered tetragonal (bct) structure called martensite. Thus, the objective of the quenching process is to cool at a sufficient rate to form martensite. The distortion of the bct structure results in high strength and hardness of the quenched steel. As previously shown in Fig. 11.6, the steel must be cooled past the nose of the isothermal transformation diagram to form 100% martensite. Martensite does not form until it reaches the martensite start temperature ($M_s$) and is complete after it is cooled below the martensite finish temperature ($M_f$). The addition of alloying elements increases the hardenability of steels by moving the nose of the isothermal transformation diagram to the right, allowing slower cooling rates for alloy steels to form martensite. However, alloying elements also depress the $M_s$ and $M_f$ temperatures, so that some highly alloyed steels must be cooled to below room temperature to obtain fully martensitic structures. Many steels are quenched in either water or oil to produce adequate cooling rates. While water quenching produces the fastest cooling rates, it also produces the highest residual stresses and can often cause warpage and distortion; therefore, higher alloy grades are often used so that a milder oil quench can be applied.

The rate at which a given portion of a steel bar cools from the austenitizing temperature depends on two factors: (1) the temperature to which the surface of the bar is cooled by the quenching medium, and (2) the rate of heat flow in the bar itself. Since heat flow in the bar itself is relatively low, the center of a steel bar cools much more slowly than the surface. Different quench media are denoted by $H$, which represents the cooling power of the quenching medium. An ideal quench is one in which there is no resistance to heat transfer from the bar to the quench media ($H = \infty$), so the surface immediately achieves the temperature of the quench bath. Representative quench values are shown in Table 11.1. While water quenching produces a fairly rapid cooling rate, it also produces high residual stresses that can lead to cracking and distortion. Since large temperature gradients can cause distortion, it is often necessary to use a milder quench, such as oil or an air blast.

There are three stages of heat removal during quenching in liquids, as illustrated in Fig. 11.11: (1) vapor blanket stage, (2) nucleate boiling stage, and (3) liquid cooling stage. The vapor blanket stage is characterized by the formation of a uniform vapor blanket around the part. The vapor blanket is maintained while the supply of heat from the interior of the part to the surface exceeds the amount of heat needed to evaporate the quenchant and maintain the vapor phase. The highest cooling rates occur during the nucleate boiling stage. During this period, the vapor envelope collapses, and high heat extraction rates are achieved that are associated with nucleate boiling of the quenchant on the metal surface. Heat is rapidly removed from the surface as liquid quenchant contacts the metal surface and is vaporized. The liquid cooling stage begins when the temperature of the metal surface is reduced below the boiling point of the quenching liquid. Below this temperature, boiling stops, and cooling takes place by conduction and convection into the quenchant.

Agitation refers to liquid quenchant movement relative to the part. Agitation is usually obtained by stirring the liquid, but in some cases, it is obtained by moving the part in the liquid. Agitation causes mechanical disruption of the vapor blanket and a faster transition to the nucleate boiling stage cooling. However, increasing agitation usually produces faster cooling rates in all three regions. Nonuniform quenching can result if agitation is not used, due to localized hot spots resulting from uneven heat removal from the metal surface. This can lead to spotty hardness, increased surface cracking, distortion, and higher residual stresses.

Many different media have been used for quenching, such as water, brine solutions, oils, and synthetic polymer solutions. A relative comparison of several different quench media is shown in Fig. 11.12. The ideal quenchant would have a high initial quenching effect during the vapor phase and boiling range periods but would cool slowly through the final convection range (liquid cooling phase). Cold water, and especially brine solutions, have the highest initial quenching speeds, but they also quench very fast at the end of the quenching process, that is, during the convection phase. Thus, they are restricted to quenching simple shapes and steels of comparatively low hardenability. For other workpieces, they would cause either intolerable degrees of distortion or warpage and possibly cracking. All quenching oils have considerably lower quenching rates than water or salt solutions. Moreover, their heat extraction is more uniform and particularly slow at the end of the cooling cycle, that is, during the convection range. As a result, the dangers of distortion or cracking are greatly reduced.

**Table 11.1 Quench values for several quenching media**

| Agitation | Quench severity, $H$ | | |
|---|---|---|---|
| | Oil | Water | Brine |
| None | 0.25–0.30 | 0.9–1.0 | 2 |
| Mild | 0.30–0.35 | 1.0–1.1 | 2.0–2.0 |
| Moderate | 0.35–0.40 | 1.2–1.3 | ... |
| Good | 0.40–0.50 | 1.4–1.5 | ... |
| Strong | 0.50–0.80 | 1.6–2.0 | ... |
| Violent | 0.80–1.1 | 4.0 | 5.0 |

Source: Ref 2

### 11.5.4 Hardenability

Hardenability is the depth from the steel surface to which martensite can be produced by a given heat treatment. Hardenability can be quantitatively defined by several methods. The Jominy end-quench test is one of the simplest and most widely used. In this test, shown in Fig. 11.13, a 10 cm (4 in.) long by 2.5 cm (1 in.) diameter normalized bar is first heated to the appropriate austenitization temperature, rapidly moved to a fixture, and then quenched at one end with a water spray. After quenching, parallel flats 180° apart are ground 0.4 mm (0.015 in.) deep on the cylindrical surface. Starting from the water-quenched end, Rockwell C hardness measurements are taken at intervals of 1.5 mm (1/16 in.) for alloy steels and 0.8 mm (1/32 in.) for carbon steels.

The cooling rate at the quenched end is fast enough to produce 100% martensite, and the hardness at this end is a function of the carbon content, with higher carbon contents producing greater hardness. The depth of hardening, measured as the distance from the quenched end,

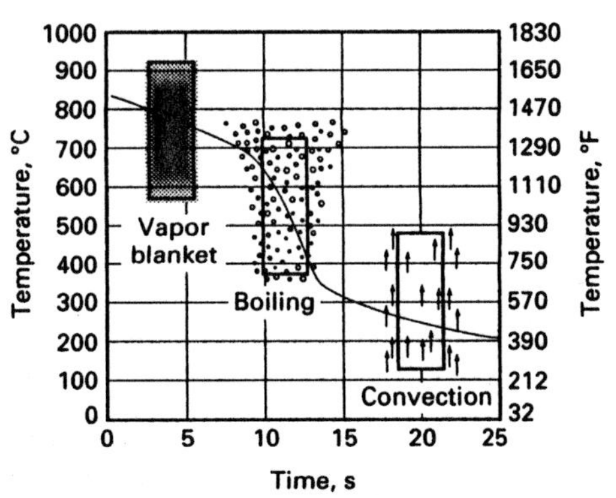


**Fig. 11.11** Three stages of quenching. Source: Ref 7

is a function of the type and amount of alloying elements present. There are large quantities of Jominy hardenability curves available in the literature, such as the ones shown in Fig. 11.14.

### *11.5.5 Prediction of Hardenability*

The hardenability of a steel can also be predicted from its composition and grain size. The ideal diameter ($D_I$) is defined as the diameter of a steel bar that will harden to 50 vol% martensite when quenched in an ideal quench medium. The significance of this value is that a bar with a diameter larger than $D_I$ cannot be hardened all the way through its cross section, even in an infinitely rapid quench. Grossman originally developed a method for calculating hardenability. Using the carbon content and the grain size, the base diameter ($D_B$) can be read from Fig. 11.15. Each of the alloying elements in the steel has what is known as a multiplying factor that increases the depth of hardening. Multiplying factors are given in Table 11.2.

Consider a steel containing 0.45 wt% C with an ASTM grain size of 6 and an alloy content of 0.5 wt% Mn, 0.2 wt% Si, and 0.35 wt% Cr. Using Fig. 11.15, the base diameter ($D_B$) of a steel with 0.45 wt% C and a grain size of 6 is 6.5 mm (0.25 in.). The multiplying factors from Table 11.2 are as follows:

$$0.5\text{ wt\% Mn} = 2.667$$
$$0.2\text{ wt\% Si} = 1.140$$
$$0.35\text{ wt\% Cr} = 1.756$$

The ideal diameter ($D_I$) is:

$$\begin{aligned} D_I &= D_B \times (M_f\text{Mn}) \times (M_f\text{Si}) \times (M_f\text{Cr}) \qquad \text{(Eq 11.1)} \\ &= (6.5\text{ mm, or } 0.25\text{ in.})(2.667)(1.140)(1.756) \\ &= 34\text{ mm, or } 1.33\text{ in.} \end{aligned}$$

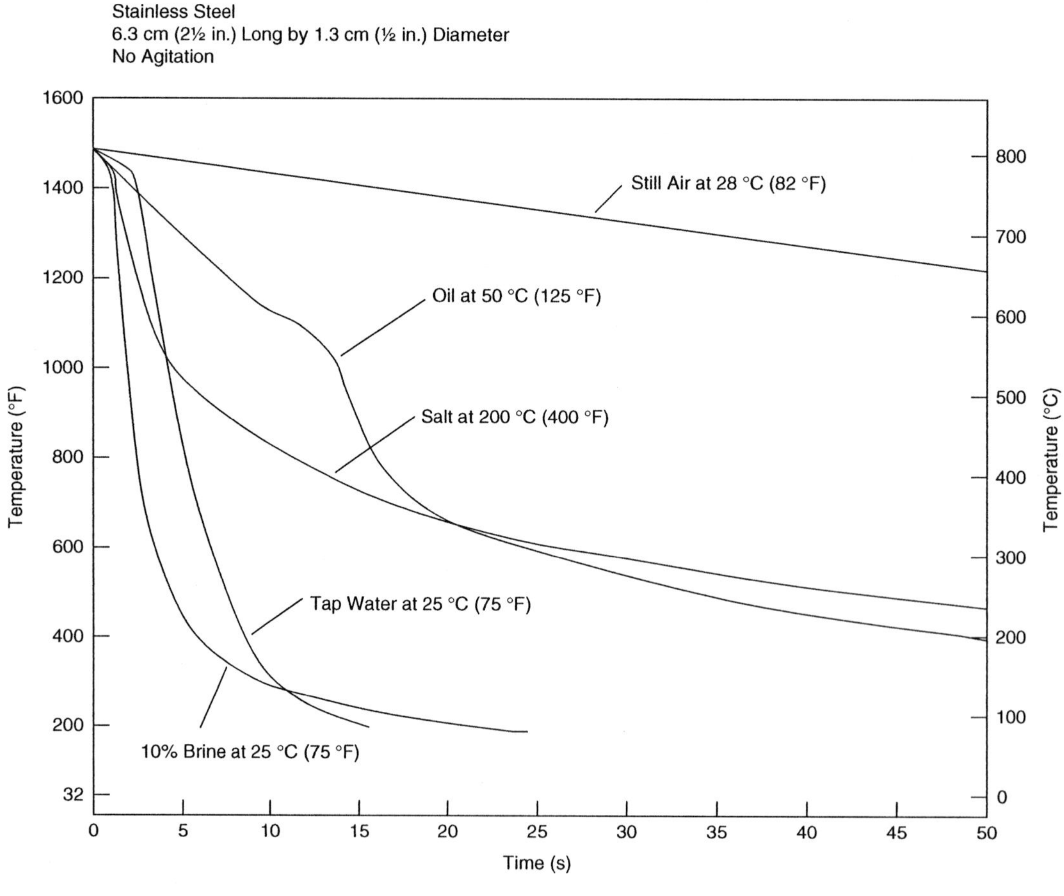


**Fig. 11.12** Comparison of different quench media

However, the critical diameter depends on the quench severity. The relationship between ideal diameter ($D_I$) and critical diameter ($D_C$) for various quench severities is given in Fig. 11.16. If this steel were quenched in moderately agitated oil, an $H = 0.37$ can be determined from Table 11.1. Since the curves in Fig. 11.16 are somewhat difficult to read, the lower portion of the curves shown in Fig. 11.16 is reproduced as Fig. 11.17, with all the curves removed except for the one needed for this calculation. As shown, the critical diameter is determined to be

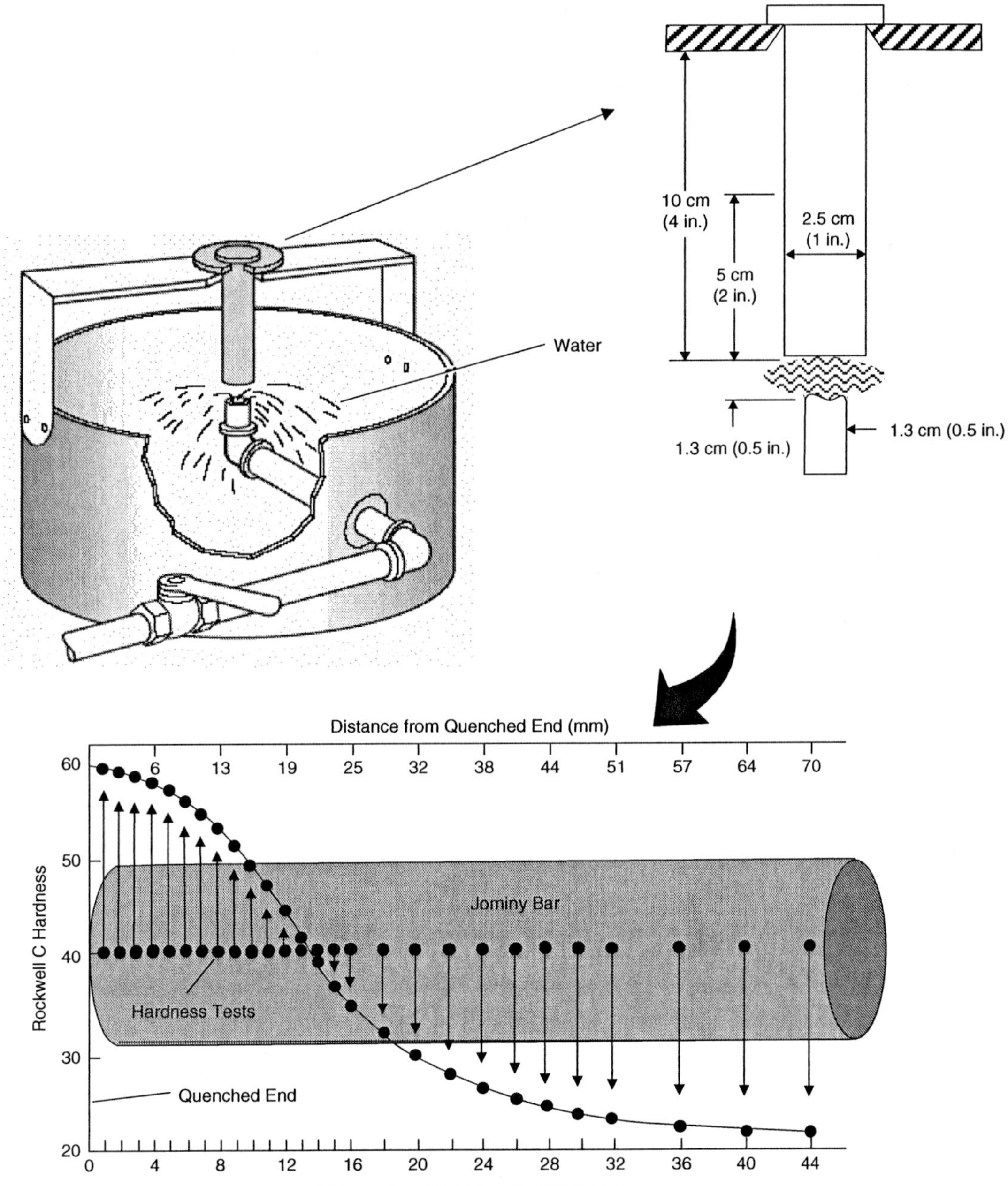


**Fig. 11.13** Jominy end-quench test

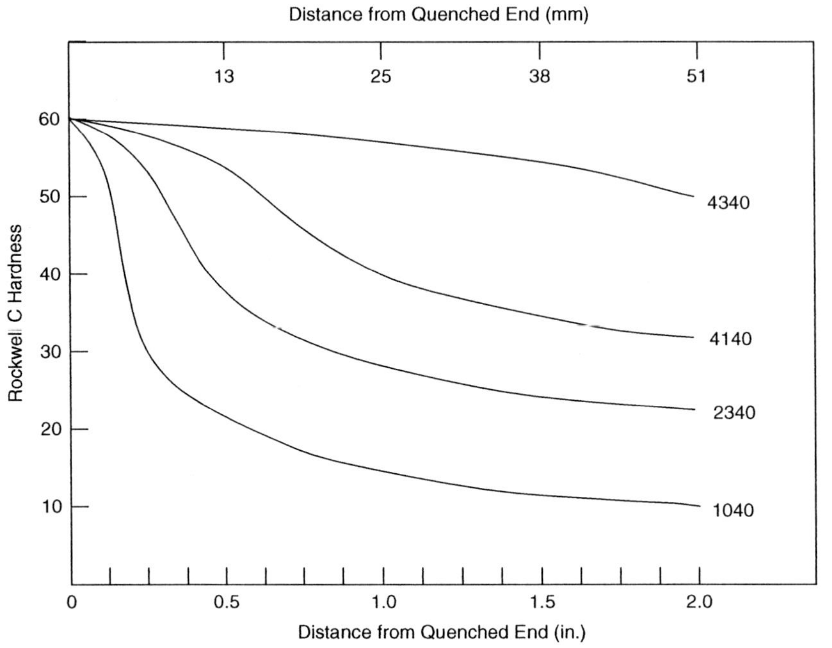


**Fig. 11.14** Typical Jominy hardenability curves for medium-carbon steels austenitized at 845 °C (1550 °F) from initial normalized condition

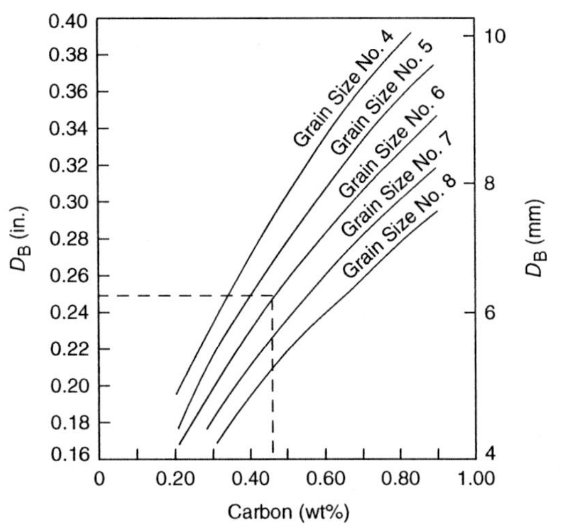


**Fig. 11.15** Plot of base diameter, $D_B$, used for calculating ideal diameter. Source: Ref 2

$D_C$ = 7.5 mm (0.30 in.). The section size of bars with square cross sections and plates is compared with round bars in Fig. 11.18. There is also an empirical relationship between the diameter of steel and the Jominy distance at which there will be 50% martensite at the center, as shown in Fig. 11.19.

**Table 11.2 Multiplying factors for alloying elements**

| | Alloying element | | | | |
|---|---|---|---|---|---|
| Percent | Mn | Si | Ni | Cr | Mo |
| 0.00 | 1.00 | 1.00 | 1.00 | 1.00 | 1.00 |
| 0.05 | 1.167 | 1.035 | 1.018 | 1.108 | 1.15 |
| 0.10 | 1.333 | 1.070 | 1.036 | 1.216 | 1.30 |
| 0.15 | 1.500 | 1.105 | 1.055 | 1.324 | 1.45 |
| 0.20 | 1.667 | 1.140 | 1.073 | 1.432 | 1.60 |
| 0.25 | 1.833 | 1.175 | 1.091 | 1.540 | 1.75 |
| 0.30 | 2.000 | 1.210 | 1.109 | 1.648 | 1.90 |
| 0.35 | 2.167 | 1.245 | 1.128 | 1.756 | 2.05 |
| 0.40 | 2.333 | 1.280 | 1.146 | 1.864 | 2.20 |
| 0.45 | 2.500 | 1.315 | 1.164 | 1.972 | 2.35 |
| 0.50 | 2.667 | 1.350 | 1.182 | 2.080 | 2.50 |
| 0.55 | 2.833 | 1.385 | 1.201 | 2.188 | 2.65 |
| 0.60 | 3.000 | 1.420 | 1.219 | 2.296 | 2.80 |
| 0.65 | 3.167 | 1.455 | 1.237 | 2.404 | 2.95 |
| 0.70 | 3.333 | 1.490 | 1.255 | 2.512 | 3.10 |
| 0.75 | 3.500 | 1.525 | 1.273 | 2.620 | 3.25 |
| 0.80 | 3.677 | 1.560 | 1.291 | 2.728 | 3.55 |
| 0.85 | 3.833 | 1.595 | 1.309 | 2.836 | 3.70 |
| 0.90 | 4.000 | 1.630 | 1.321 | 2.944 | |
| 0.95 | 4.167 | 1.665 | 1.345 | 3.052 | |
| 1.00 | 4.333 | 1.700 | 1.364 | 3.160 | |

Source: Ref 8

### 11.5.6 Effect of Grain Size

The hardenability of a carbon steel can increase by as much as 50% with an increase in ASTM austenite grain size of the base material from 8 to 3. This effect becomes more

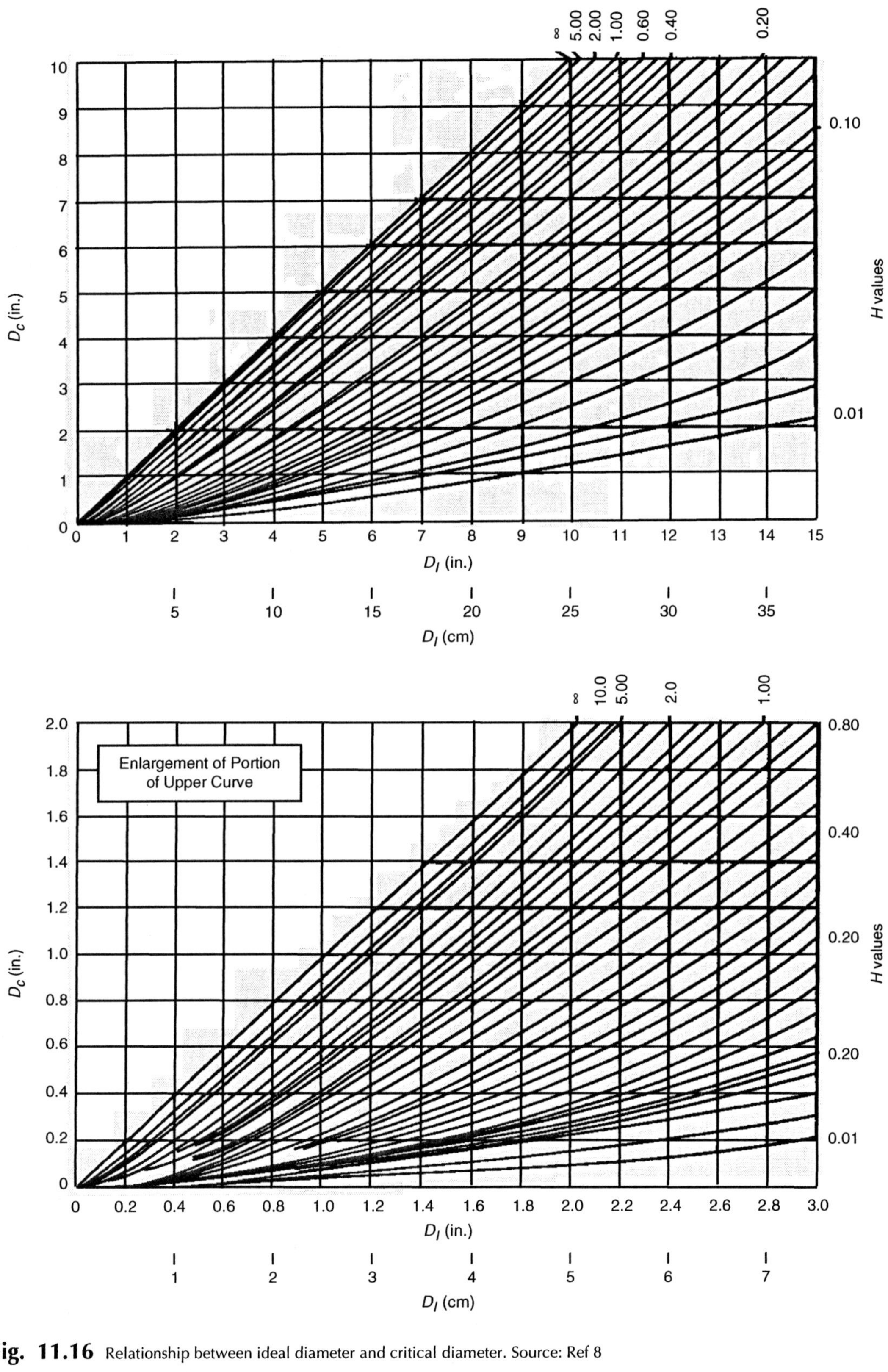


**Fig. 11.16** Relationship between ideal diameter and critical diameter. Source: Ref 8

pronounced if the carbon content is increased at the same time. The hardenability increases with larger austenite grain sizes, because larger grain sizes means there are fewer nucleation sites for diffusion-controlled phase transformations (that is, pearlite, ferrite, or bainite). When the danger of quench cracking is remote (i.e., no abrupt changes in section thickness) and engineering considerations permit, it may sometimes appear to be more practical to use a coarse-grained steel than a fine-grained steel to obtain hardenability. However, this is not recommended, because the use of coarser-grained steels usually involves a serious sacrifice in notch toughness.

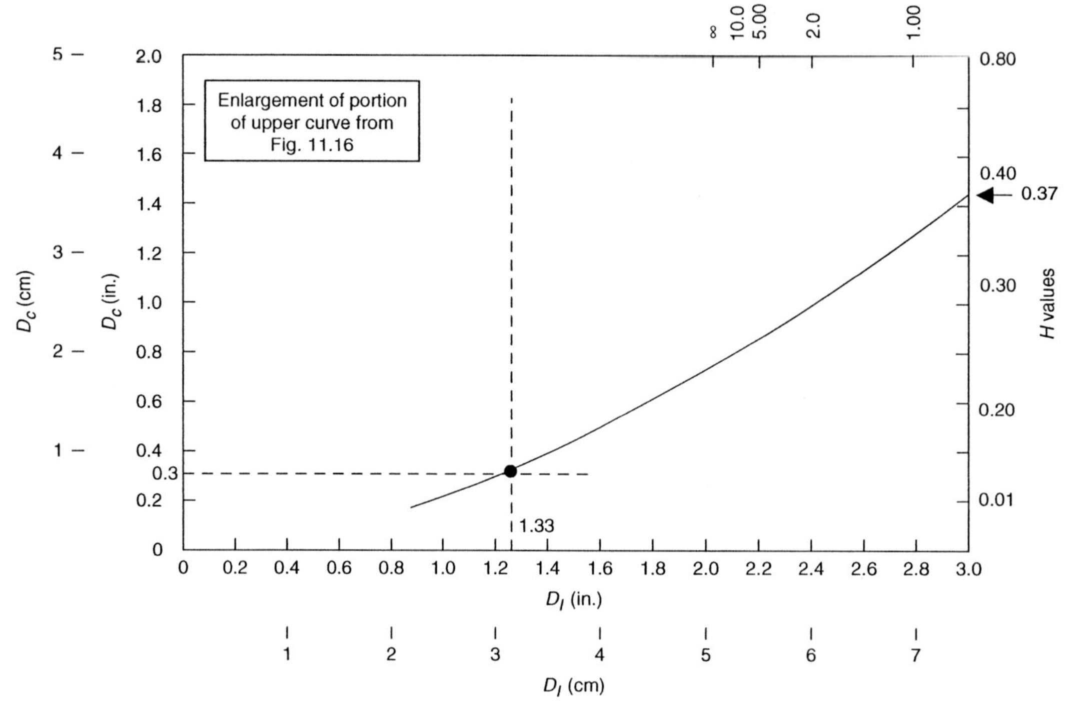


Fig. 11.17 Determination of critical diameter ($D_C$)

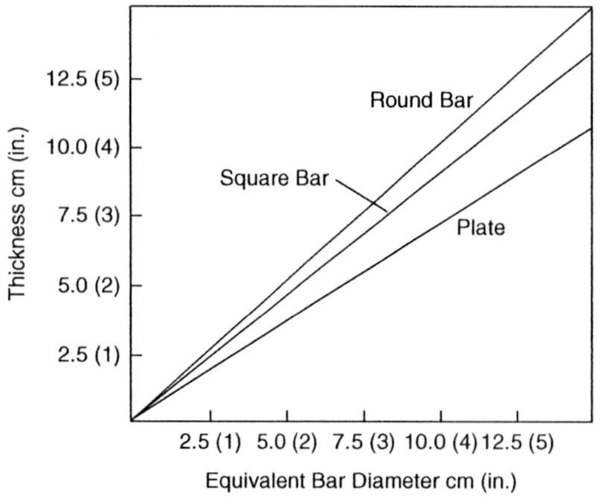


Fig. 11.18 Comparison of different geometries with the same cooling rate at the center. Source: Ref 8

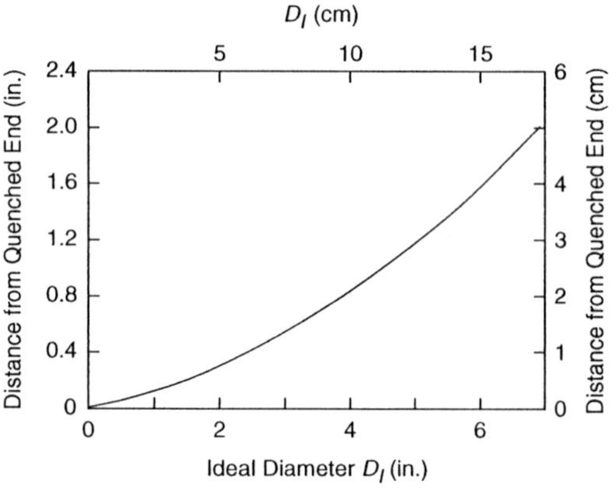


Fig. 11.19 Relationship between ideal diameter and Jominy position. Source: Ref 8

### *11.5.7 Effect of Alloying Elements*

Alloying elements are used to increase the hardenability of steel. While the carbon content determines the hardness achievable in the steel, alloying elements allow greater depths of hardness to be developed during heat treatment. This is accomplished by shifting the nose of the TTT or CCT diagram to the right, thus allowing slower quenching rates, or greater thicknesses, to be quenched without forming pearlite.

Alloying elements also permit use of a lower carbon content for a given application. The decrease in hardenability accompanying a decrease in carbon content can be offset by the hardenability effect of added alloying elements, and the lower-carbon steel will exhibit a much lower susceptibility to quench cracking. This lower susceptibility results from greater plasticity of the low-carbon martensite and from the generally higher temperature range at which martensite is formed in the lower-carbon materials. Quench cracking is seldom encountered in steels containing 0.25 wt% C or less. The susceptibility to cracking increases progressively with increasing carbon content.

### *11.5.8 Tempering*

While as-quenched steel is extremely hard and strong, it is also very brittle. Tempering, a process where the steel is reheated to an intermediate temperature, is used to increase the ductility and toughness, with some loss of strength and hardness. During tempering, the highly strained bct structure starts losing carbon to transformation products, which reduces lattice strains, producing an increase in ductility and a reduction in strength.

Tempering involves reheating the steel to some temperature below $A_1$ and holding for some period of time, usually 1 to 2 h. As in most heat treatment processes, the tempering temperature is much more important than tempering time. As the tempering temperature is increased, the hardness decreases (Fig. 11.20). Plain carbon and low-alloy steels can be tempered at lower or higher temperature ranges, depending on the balance of properties required. Tempering between 150 and 200 °C (300 and 400 °F) will retain much of the hardness and strength of the quenched martensite and provide a small improvement in ductility and toughness. This treatment can be used for bearings and gears that are subjected to compression loading. Tempering above 430 °C (800 °F) significantly improves ductility and toughness but at the expense of hardness and strength. The effect of tempering temperature on the tensile properties of oil-quenched 4340 medium-carbon low-alloy steel is shown in Fig. 11.21. The tensile strength

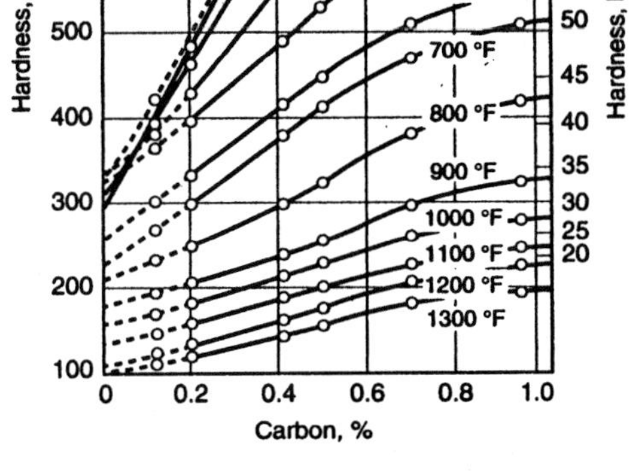


**Fig. 11.20** Effect of carbon content on tempering of plain carbon steels. Source: Ref 5

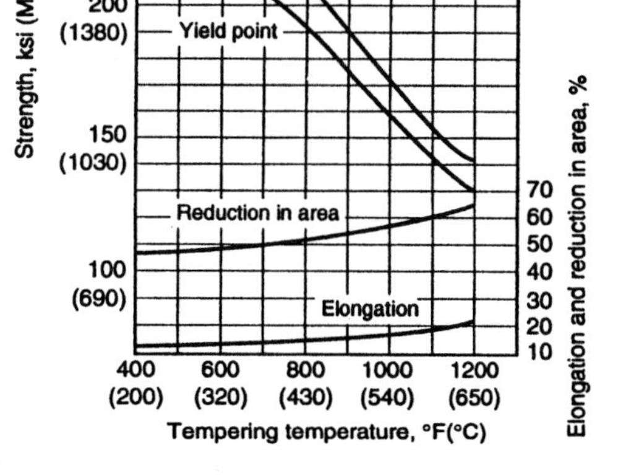


**Fig. 11.21** Effects of tempering temperature on 4340 steel. Source: Ref 5

has decreased from 1900 MPa (275 ksi) at a 200 °C (400 °F) tempering temperature to approximately 965 MPa (140 ksi) at a 650 °C (1200 °F) tempering temperature. However, the ductility, as measured by total elongation and reduction in area, increases dramatically. The tempering range of 260 to 370 °C (500 to 700 °F) is often avoided because it can result in low impact toughness (Fig. 11.22) due to temper embrittlement, which is discussed in section 11.7.1, "Tempered Martensite Embrittlement," in this chapter.

During tempering, carbon atoms dispersed in martensite form carbide precipitates of increasing size. Tempering occurs in stages as the steel is heated to higher and higher temperatures. However, it should be noted that the stages of tempering are somewhat arbitrary and that they overlap considerably, with reactions continually occurring as the part is heated to higher and higher temperatures. The three stages of tempering involve: stage 1, the formation of a transition carbide and the reduction of the carbon content in the martensitic matrix; stage 2, the transformation of any retained austenite to ferrite and cementite; and stage 3, the replacement of the transition carbide and low-carbon martensite with ferrite and cementite.

**Stage 1.** The first stage of tempering occurs at temperatures between 95 and 260 °C (200 and 500 °F). At the beginning of stage 1 tempering of low-carbon steels, the carbon atoms redistribute themselves to lower-energy sites such as dislocations. Actually, a large percentage of the redistribution of carbon atoms takes place during quenching through the temperature range in which martensite forms. For carbon contents of less than approximately 0.2 wt%, up to 90% of the carbon segregation takes place during quenching. Since the carbon atoms can reduce their energies more by segregating to dislocation sites than by forming transition carbides, no transition carbides form in steels with less than approximately 0.2 wt% C. In steels containing more than 0.2 wt% C, the initial carbon segregation occurs by precipitation clustering. Very fine particles of transition carbide then nucleate and grow within the martensite. The carbon content of the martensitic matrix is reduced by the formation of transition carbides.

The transition carbides can be ε carbide (epsilon carbide), which has the approximate composition formula $Fe_{2.4}C$ and a hexagonal crystal structure, and/or η carbide (eta carbide), with the formula $Fe_2C$ and an orthorhombic crystal structure. Both epsilon carbide and eta carbide have substantially higher carbon contents than the cementite $Fe_3C$ that forms later when tempering is conducted at higher temperatures. When the transition carbides form, the martensite retains some degree of tetragonality, because it still contains more carbon in solid

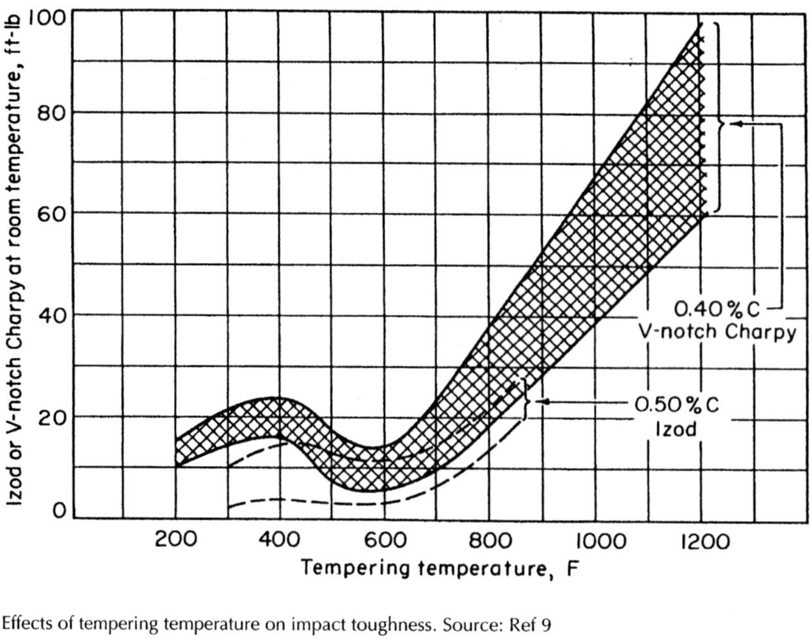


**Fig. 11.22** Effects of tempering temperature on impact toughness. Source: Ref 9

solution than ferrite does. Therefore, when the total carbon content is high enough, the first stage of tempering involves the segregation of carbon to various defects in the microstructure and the conversion of martensite to low-carbon martensite and a transition carbide. During stage 1 tempering, there are also changes in the physical properties, such as electrical resistivity, which can be used to monitor the progress of the changes. However, there is not much of a reduction in the hardness; in fact, it can increase slightly for steels of medium-to-high-carbon contents.

**Stage 2.** The second stage of tempering occurs between 205 and 315 °C (400 and 600 °F). It is not very important in low-carbon steels, because it involves the conversion of retained austenite to ferrite and cementite. The quantity of retained austenite in low-carbon steel is not large. However, it can be large in alloy steels, especially those in which the $M_f$ temperature is below room temperature. Thus, stage 2 is important when there are significant quantities of retained austenite in the martensite. The transformations of retained austenite in 4130 and 4340 steels, which contain approximately 2 and 4 vol% of retained austenite, respectively, are shown in Fig. 11.23. For tempering times of 1 h, the transformation of retained austenite begins above 205 °C (400 °F) and is complete by 315 °C (600 °F). Note that the percentage of cementite increases as the amount of retained austenite decreases.

**Stage 3.** The third stage of tempering, which begins at approximately 260 °C (500 °F), results in the formation of ferrite and cementite. An additional transition carbide, known as Hägg or chi (χ) carbide, with the formula $Fe_5C_2$ and a monoclinic structure, has been identified in high-carbon steels during tempering between 205 and 315 °C (400 and 600 °F). It is a metastable carbide with a composition intermediate between epsilon carbide and cementite. As the tempering temperature is increased in the range of 260 to 705 °C (500 to 1300 °F), the transition carbides dissolve and are replaced by cementite, $Fe_3C$. The diffusion of carbon to cementite causes the martensite matrix to lose tetragonality and become bcc ferrite as its carbon content drops. The initial shape of cementite is needle-like when tempered between 205 and 315 °C (400 and 600 °F). In high-carbon martensite, $Fe_3C$ nucleates at twin boundaries, and in low-carbon lath martensite, it nucleates at the martensite lath boundaries as well as at dislocations. From 400 to 595 °C (750 to 1100 °F), the lathlike carbides coalesce to form spheroidite, which reduces the overall surface energy. From 595 to 704 °C (1100 to 1300 °F), spheroidite coarsens even more, with the smaller particles dissolving. The driving force for coalescence is the reduction of the overall surface energy of the cementite in the ferrite matrix. The hardness after tempering decreases, an effect that becomes more and more significant as the tempering temperature becomes higher. Eventually,

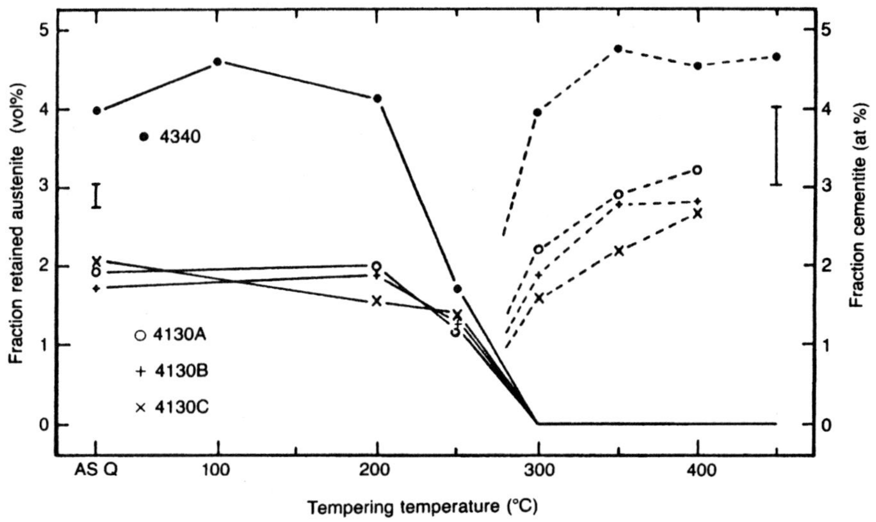


**Fig. 11.23** Transformation of retained austenite in 4130 and 4340 steels. Source: Ref 10

the temperature becomes high enough to cause the dislocation content to decrease in a manner similar to the changes in the dislocation content of cold-worked metals during recovery processes. At approximately 595 °C (1100 °F), the dislocation density is substantially reduced. The lath morphology in lath martensite and the twin boundaries in plate martensite disappear, and the ferrite grains grow and become equiaxed. These morphological changes in the ferrite grains are similar to those that occur during recrystallization in cold-worked ferrite. Secondary hardening of certain alloy steels can also occur during stage 3 tempering, although secondary hardening is sometimes referred to as stage 4 tempering.

Except for stage 1, the hardness and strength of the steel decreases during tempering, while the ductility and toughness increase. Although these various stages offer a convenient way to classify the different tempering reactions, it is important to recognize that the temperatures are merely guidelines. The stages overlap because the dominant type of transformation and the reaction kinetics depend on temperature and time, both of which change continuously.

Alloying elements slow the rate of tempering and lessen the decrease in hardness. The tempering process can be retarded by the addition of alloying elements such as vanadium, molybdenum, manganese, chromium, and silicon. These larger substitutional alloying elements diffuse slower than the interstitial carbon, and some of them form carbides that are harder than cementite. In some high-carbon steels with significant alloying additions, some retained austenite will convert to martensite on cooling from the tempering operation. Since this adversely affects the toughness, such steels are often double or even triple tempered.

Some alloy steels containing molybdenum, chromium, and tungsten exhibit secondary hardening during tempering, as shown in Fig. 11.24. These elements lead to precipitation hardening during tempering. The cementite-rich carbides are replaced with a new fine array of alloy carbides. Rapidly diffusing carbon enables relatively pure cementite to form first because the nucleation and growth of the alloy carbides is paced by substitutional diffusion. When the dense array of fine alloy carbides forms, gliding dislocations meet a higher density of obstacles to slip, and the hardness increases. The particular temperature at which secondary hardening occurs depends on the specific carbide that forms and increases as the stability of the carbide increases. Secondary hardening is quite important in certain tool steels, where it helps to retain hardness at elevated temperatures.

Full hardening, followed by tempering, usually results in the best combination of properties. However, in practice, some pearlite and/or bainite may be mixed with the martensite if the local cooling rate is less than the critical rate necessary to avoid the nose of the CCT diagram. Retained austenite may also be present if the $M_f$ temperature is below room temperature. In both cases, the properties will not be as good as those with a fully hardened and tempered martensite matrix.

## 11.6 Interrupted Quenching

When a steel is quenched to room temperature, the exterior cools much faster than the interior, setting up compressive residual stresses in the steel. Then, when the austenite transforms to martensite, there is a volume expansion that causes large residual tensile stresses on the surface. The combined effect of these contraction and expansion stresses can cause distortion and even cracking. Some special steels have such a high alloy content that pushes the nose of the TTT diagram far enough to the right that they can be fully hardened by air cooling, thus avoiding development of these stresses. However, due to their high alloy contents, they are more expensive than normal alloy steels. The lower-alloy steels can be heat treated to avoid

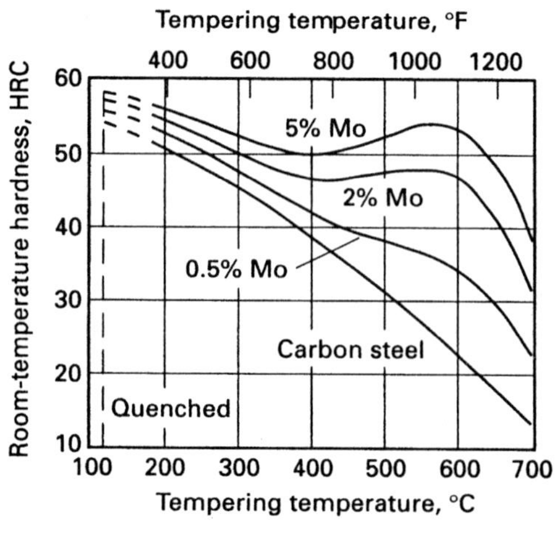


Fig. 11.24 Secondary hardening of molybdenum alloy steels. Source: Ref 11

most of these stresses if the surface and interior can be made to transform to martensite at the same time. Interrupted quenching is a two-step procedure where the workpiece is first placed in a quench that is at a temperature higher than the Ms temperature. After an appropriate holding period, the workpiece is removed and then slowly cooled past the Ms temperature. In this way, the steel can be rapidly cooled past the nose of the TTT diagram; it can then be cooled slowly through the martensite reaction or be allowed to transform to lower bainite.

### 11.6.1 Martempering

The most frequent type of interrupted quenching process is called martempering, shown in Fig. 11.25. In martempering, the part is quenched into a molten salt bath at a temperature just above the $M_S$ temperature. When the part reaches thermal equilibrium, but before bainite starts to form, it is removed from the salt bath and allowed to air cool past the $M_S$ temperature, thus forming martensite. The objective is to delay the cooling just above the martensitic transformation for long enough to equalize the temperature throughout the piece and minimize distortion, cracking, and residual stresses. It should be noted that even though this process is called martempering, it is not a tempering process; instead, it is a hardening process. The martensite produced by martempering still requires a subsequent tempering process. In fact, this process is sometimes more correctly referred to as marquenching. The time lapse before

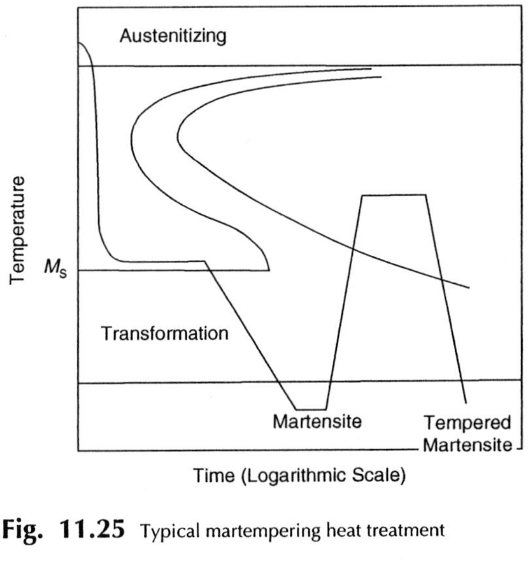


**Fig. 11.25** Typical martempering heat treatment

tempering is not as critical because the stress is greatly reduced. Formation of martensite occurs fairly uniformly throughout the workpiece during cooling to room temperature, thereby avoiding formation of excessive amounts of residual stress. Straightening or forming is also easily accomplished upon removal from the martempering bath while the part is still hot but has not yet transformed to martensite.

The advantage of martempering is the reduced thermal gradient between the surface and center as the part is quenched to the isothermal temperature and then air cooled to room temperature. Residual stresses developed during martempering are lower than those developed during conventional quenching, because the greatest thermal variations occur while the steel is in the relatively plastic austenitic condition, and because final transformation and thermal changes occur throughout the part at approximately the same time. Martempering also reduces or eliminates susceptibility to cracking.

Martempering can be accomplished in a variety of baths, including hot oil, molten salt, molten metal, or a fluidized particle bed. Molten salt and hot oil are both widely used for martempering. Oils are used for martempering at temperatures up to 205 °C (400 °F), while molten salt is used in the range of 160 to 400 °C (320 to 750 °F).

Martempering is conducted on certain alloy, cast, tool, and stainless steels. Alloy steels generally are more adaptable than carbon steels to martempering. In general, any steel that is normally quenched in oil can be martempered. Some carbon steels that are normally water quenched can be martempered at 205 °C (400 °F) in sections thinner than 5 mm (0.2 in.) using vigorous agitation of the martempering medium. In addition, thousands of gray cast iron parts are martempered on a routine basis. Any steel that is to be martempered successfully must contain sufficient carbon or alloying additions to move the nose of the TTT curve to the right, thus permitting sufficient time for quenching of workpieces past the nose of the TTT curve. Martempering is especially appropriate for carburized parts, such as splined shafts, cams, and gears, because these parts generally are more difficult to grind, more costly to fabricate, and are manufactured with closer tolerances than are parts made of through-hardening steels.

The limitation of section thickness or mass must be considered when martempering. With a given severity of quench, there is a limit to bar size beyond which the center of the bar will not cool fast enough to transform entirely to martensite. Agitation of the martempering salt or oil considerably increases the hardness obtainable for a given section thickness. In some instances, the rapid cooling produced by the most vigorous agitation increases distortion; thus, mild agitation is often used.

### 11.6.2 Austempering

Austempering is an interrupted quenching process in which the objective is to form a bainitic structure. Like martempering, the part is quenched into a salt bath above the $M_s$ temperature. However, it is allowed to remain at that temperature until the transformation to bainite is complete (Fig. 11.26). Since the salt bath used for austempering is usually at a higher temperature than one used for martempering, a more hardenable steel is required for austempering than for martempering.

Austempering offers several potential advantages, including increased ductility, toughness, and strength at a given hardness, reduced distortion, and the ability to heat treat steels to a hardness of HRC 35 to 55 without having to temper. One of the outstanding attributes of austempered steels is their high impact resistance. For example, an austempered steel heat treated to the same hardness as a quenched and tempered steel may have a Charpy V-notch impact energy of 54 J (40 ft·lbf) compared to only 20 J (15 ft·lbf) for a conventionally hardened steel. Since austempering is an interrupted quenching process that produces bainite rather than martensite, parts can usually be produced with less dimensional change than is typical through conventional quenching and tempering. In addition, austempered parts do not normally require a tempering operation.

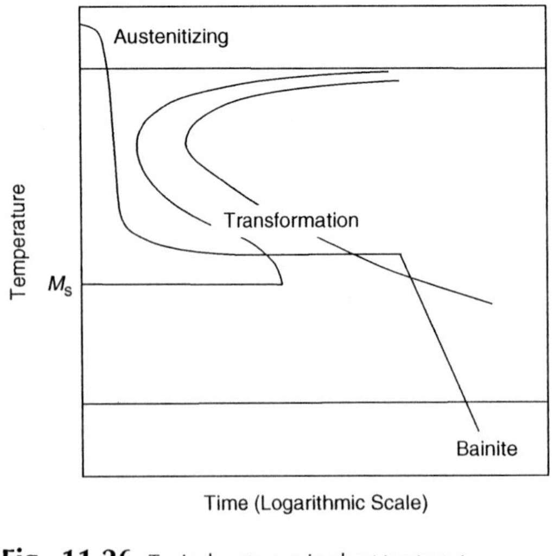


**Fig. 11.26** Typical austempering heat treatment

A typical austempering cycle would include heating to a temperature within the austenitizing range (790 to 910 °C, or 1450 to 1675 °F); quenching in a salt bath maintained at a constant temperature, usually in the range of 260 to 400 °C (500 to 750 °F); allowing time for the austenite to isothermally transform to bainite; and then air cooling to room temperature. For true austempering, the metal must be cooled from the austenitizing temperature to the temperature of the austempering bath fast enough so that no transformation of austenite occurs during cooling, and then held at bath temperature long enough to ensure complete transformation of austenite to bainite.

The selection of steel for austempering must be based on transformation characteristics as indicated in the TTT diagram. Important considerations are the location of the nose of the TTT curve, cooling rate of the quench being used, and the time required for complete transformation of austenite to bainite at the austempering temperature. Maximum thickness of section is also important in determining whether or not a part can be successfully austempered. For a 0.80 wt% C steel, a section thickness of approximately 5 mm (0.2 in.) is the maximum that can be austempered to a fully bainitic structure. Carbon steels of lower carbon content are restricted to a proportionately smaller thickness. However, lower-carbon steels containing boron can be successfully austempered in thicker sections. In some alloy steels, section thicknesses up to approximately 2.5 cm (1 in.) can be austempered to fully bainitic structures. However, in industrial austempering practice, a sizable percentage of applications are successful with less than 100% bainite. Approximately an 85 vol% bainitic structure has been found to be satisfactory for some applications. The range of austempering applications generally encompasses parts fabricated from bars of small diameter or from sheet or strip of small cross section. Austempering is particularly applicable to thin-section carbon steel parts requiring exceptional toughness at a hardness of HRC 40 to 50.

## 11.7 Temper Embrittlement

During tempering, a decrease in hardness is usually accompanied by an increase in toughness. However, embrittlement can occur during tempering. This reduction in toughness raises the ductile-to-brittle transition temperature and can reduce the energy absorbed during fracture at and above room temperature. There are two types of embrittlement. One is called tempered martensite embrittlement. It is also referred to as 500 °F embrittlement and one-step embrittlement. The other is simply called temper embrittlement or, perhaps preferably, two-step embrittlement. Commercial heat treatments and compositional controls to eliminate unwanted impurities are designed to avoid these embrittlement mechanisms.

### *11.7.1 Tempered Martensite Embrittlement*

This form of embrittlement can occur in standard low-alloy steels (e.g., 4140) that are heat treated to high strength levels. It can also occur in plain carbon steels that contain phosphorus. A typical treatment that produces it is tempering in the region of 230 to 290 °C (450 to 550 °F). The embrittlement manifests itself by raising the ductile-to-brittle transition temperature and by reducing the amount of energy absorbed during impact in comparison to steel tempered immediately above and below this range. The loss in impact energy for several alloy steels is shown in Fig. 11.27. This range of tempering temperatures corresponds to the stage of tempering in which sheets of cementite needles and platelets precipitate heterogeneously on interlath boundaries. Sheets of cementite particles can also be deposited from retained austenite when the austenite transforms during the tempering treatment. Since the properties of the whole sample are dominated by those of the most continuous phase, one may expect a deterioration of toughness with these microstructural changes. This seems to be true when the steels are relatively free of phosphorus and nitrogen, because the fracture paths follow the sheets of cementite. Furthermore, embrittlement is affected by the tempering time required to produce these microstructures. These times are typically about an hour. The fracture paths are transgranular, and the failure mode is often a mixture of cleavage and microvoid coalescence. However, when the steel contains high amounts of phosphorus (e.g., 0.03 wt% P), embrittlement is associated with intergranular failure along prior-austenite grain boundaries. This suggests that the embrittlement is also associated with a loss of cohesive strength across prior-austenite grain boundaries due to the segregation of impurities within the steel as well as with precipitates of cementite. It is thought to be due to the impurity elements phosphorus, nitrogen, and possibly sulfur. Manganese may play an indirect role by helping the impurity elements segregate to the grain boundaries.

### *11.7.2 Temper Embrittlement*

This type of embrittlement causes a decrease in notch toughness when tempered alloy steels are isothermally aged in the temperature range of 370 to 560 °C (700 to 1040 °F) or are slowly cooled after tempering. It is also known as two-step embrittlement because it develops on aging or slow cooling through this critical temperature range but can be removed by heating to a high temperature (600 °C, or 1110 °F). The embrittlement occurs again on slow cooling through the critical temperature range. Temper embrittlement results in intergranular failure modes and is also attributed to the presence of impurities that segregate to the grain boundaries. The ductile-to-brittle transition is directly dependent on the grain-boundary concentration of impurities. The relative effect of these impurities has been found to be tin > antimony > phosphorus. Alloying elements can also cosegregate to the grain boundaries along with the impurities; for example, nickel cosegregates with antimony. The rate and amount of impurity segregation, and hence the resultant intergranular embrittlement, depends on the total composition of the system. Nickel, chromium, silicon, and manganese increase two-step temper embrittlement caused by antimony, tin, phosphorus or arsenic. However, additions of molybdenum retard temper embrittlement, since molybdenum inhibits the segregation of impurities by readily precipitating as phosphides in the matrix and inhibiting segregation. Heating to over 600 °C (1110 °F), followed by rapid cooling, can reverse this type of embrittlement.

#### ACKNOWLEDGMENTS

Sections of this chapter were adapted from "Microstructures, Processing, and Properties of

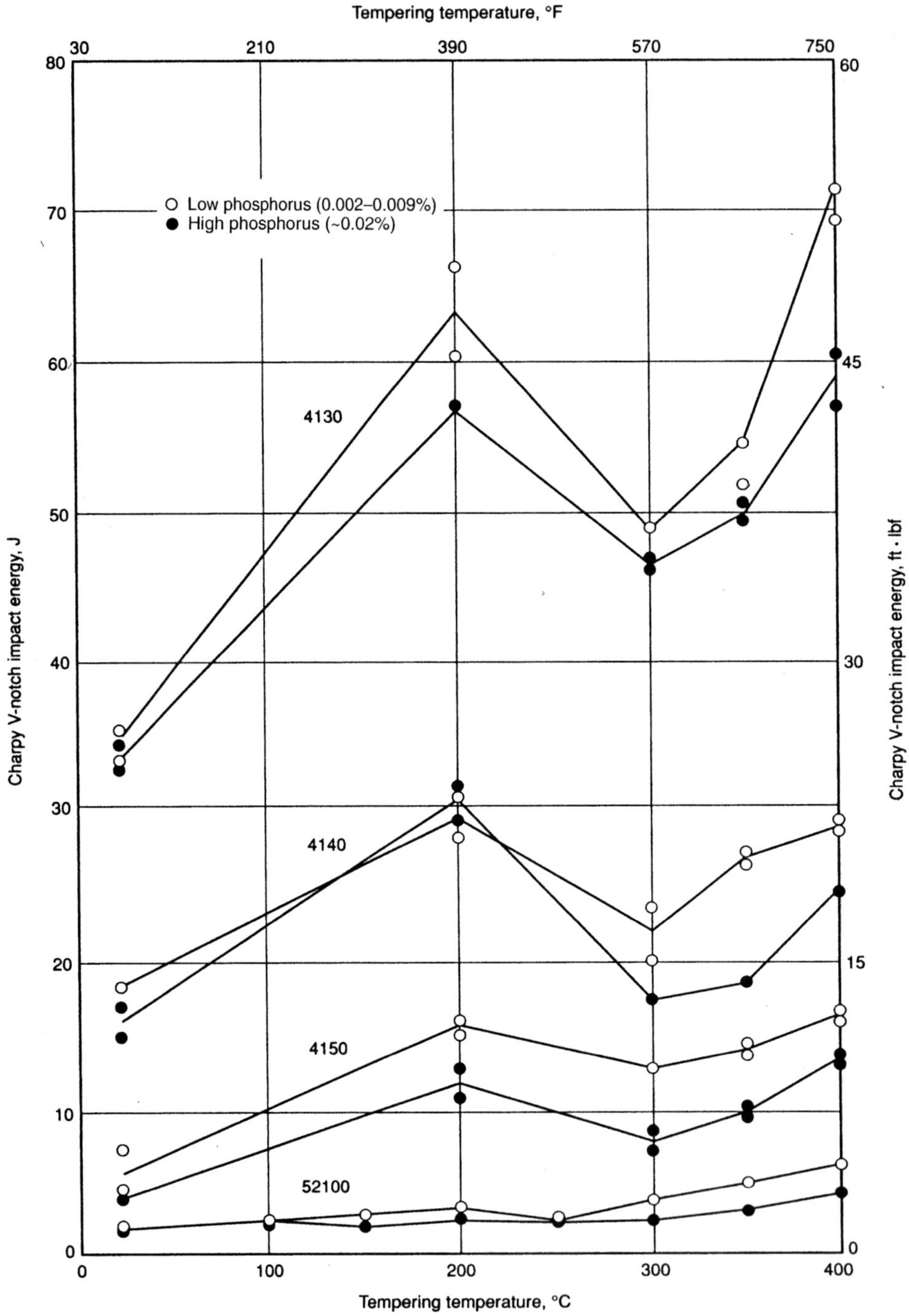


**Fig. 11.27** Tempered martensite embrittlement. Source: Ref 10

Steels" by G. Krauss in *Properties and Selections: Irons, Steels, and High-Performance Alloys,* Volume 1, *ASM Handbook,* ASM International, 1990 and "Effects of Composition, Processing, and Structure on Properties of Irons and Steels" by B.L. Bramfitt in *Materials Selection and Design,* Volume 20, *ASM Handbook,* ASM International, 1997.

## REFERENCES


1. R.E. Reed-Hill and R. Abbaschian, *Physical Metallurgy Principles,* 3rd ed., PWS Publishing Company, 1991
2. A.G. Guy, *Elements of Physical Metallurgy,* 2nd ed., Addison-Wesley Publishing Company, 1959
3. B.L. Bramfitt and S.J. Lawrence, Metallography and Microstructures of Carbon and Low-Alloy Steels, *Metallography and Microstructures,* Vol 9, *ASM Handbook,* ASM International, 2004
4. *Atlas of Isothermal and Continuous Cooling Diagrams,* American Society for Metals, 1977
5. B.L. Bramfitt, Effects of Composition, Processing, and Structure on Properties of Irons and Steels, *Materials Selection and Design,* Vol 20, *ASM Handbook,* ASM International, 1997
6. Basic Principles and Design Guidelines for Heat Treating of Steel, *Metals Handbook Desk Edition,* 2nd ed., ASM International, 1998
7. T. Ericsson, Principles of Heat Treating of Steels, *Heat Treating,* Vol 4, *ASM Handbook,* ASM International, 1991
8. W.F. Hosford, *Physical Metallurgy,* Taylor & Francis, 2005
9. M.A. Grossmann and E.G. Bain, *Principles of Heat Treatment,* American Society for Metals, 1964
10. G. Krauss, Microstructures, Processing, and Properties of Steels, *Properties and Selections: Irons, Steels, and High-Performance Alloys,* Vol 1, *ASM Handbook,* ASM International, 1990
11. M. Wisti and M. Hingwe, Tempering of Steel, *Heat Treating,* Vol 4, *ASM Handbook,* ASM International, 1991


## SELECTED REFERENCES


- D. Aliya and S. Lampman, Physical Metallurgy Concepts in Interpretation of Microstructures, *Metallography and Microstructures,* Vol 9, *ASM Handbook,* ASM International, 2004
- R. Honeycombe and H.K.D.H. Bhadeshia, *Steels—Microstructure and Properties,* 2nd ed., Arnold, 1995
- J.R. Keough, W.J. Laird, and A.D. Godding, Austempering of Steel, *Heat Treating,* Vol 4, *ASM Handbook,* ASM International, 1991
- G. Kraus, *Steels: Processing, Structure, and Performance,* 3rd ed., ASM International, 2005
- Heat Treating of Steel, *Metals Handbook Desk Edition,* 2nd ed., ASM International, 1998
- H. Webster and W.J. Laird, Martempering of Steel, *Heat Treating,* Vol 4, *ASM Handbook,* ASM International, 1991

# CHAPTER 12

# Mechanical Behavior

MANY MATERIALS, both during initial fabrication and later when placed in service, are subjected to forces or loads. So that excessive deformation or failure does not occur, it is important to know what effects these loads have on the part. The mechanical behavior of a material is its response to an applied load or force. Important mechanical properties are strength, hardness, stiffness, and ductility. There are three principal ways in which a load may be applied: tension, compression, and shear, as illustrated in Fig. 12.1. A large number of mechanical property tests have been developed to determine a material response to applied loads or forces. Since the results of these tests are often used by regulatory agencies, it is important that these test methods are consistent and standardized. In the United States, ASTM International is the most widely recognized source of standardized testing procedures.

## 12.1 Tension

The tensile test is the most commonly used mechanical property test. Its chief use is to determine the properties related to the elastic design of structures. In addition, the tensile test gives information on the plasticity and fracture of a material. The main advantages of the tensile test are that the stress state is well established, the test has been carefully standardized, and the test is relatively easy and inexpensive to perform. Important tensile properties include yield strength, ultimate tensile strength, ductility, resilience, and toughness.

### *12.1.1 Engineering Stress-Strain Curve*

The tensile properties of a material are determined by applying a tensile load to a specimen and measuring the elongation or extension in a load frame, such as the one shown in Fig. 12.2. The load can be converted to engineering stress, *s*, by dividing the load by the original cross-sectional area of the specimen:

$$s=\frac{P}{A_o} \text{ in pascals (Pa) or pounds per square inch (psi)} \quad \text{(Eq 12.1)}$$

where $P$ is the load in Newtons (N) or pounds (lb) and $A_o$ is the original cross-sectional area in $m^2$ or $in.^2$.

The engineering strain ($e$) can be calculated by dividing the change in gage length by the original gage length:

$$e=\frac{l-l_o}{l_o} \quad \text{(Eq 12.2)}$$

where $l$ is the gage length, and $l_o$ is the original gage length.

Since strain has the units of length/length, it is dimensionless.

A typical stress-strain curve for a metal is shown in Fig. 12.3. The shape and magnitude of the stress-strain curve of a metal depends on its composition, heat treatment, prior history of plastic deformation, and the strain rate, temperature, and state of stress imposed during the testing. The different stress-strain behaviors for a brittle metal, one with limited ductility, and a fully ductile metal are shown in Fig. 12.4, and some typical stress-strain curves for a number of engineering alloys are shown in Fig. 12.5. The parameters used to describe the stress-strain curve of a metal are the tensile strength, yield strength or yield point, percent elongation, and reduction in area. The first two are strength parameters, and the last two are indications of ductility.

As long as the specimen is loaded within the elastic region, the strain is totally recoverable; that is, the specimen will return to its original length when the load is removed. However, when the load exceeds a value corresponding to the yield stress, the specimen undergoes plastic deformation and is permanently deformed when the load is removed. The stress to produce continued plastic deformation increases with increasing strain as the metal strains or work hardens. To a good engineering approximation, the volume remains constant during plastic deformation ($Al = A_o l_o$), and as the specimen elongates, it decreases uniformly in cross-sectional area along its gage length. Initially, strain hardening more than compensates for this decrease in area, and the engineering stress continues to rise with increasing strain. However, eventually, a point is reached where the decrease in area is greater than the increase in deformation load from strain hardening. This condition will be reached first at some point in the specimen that is slightly weaker

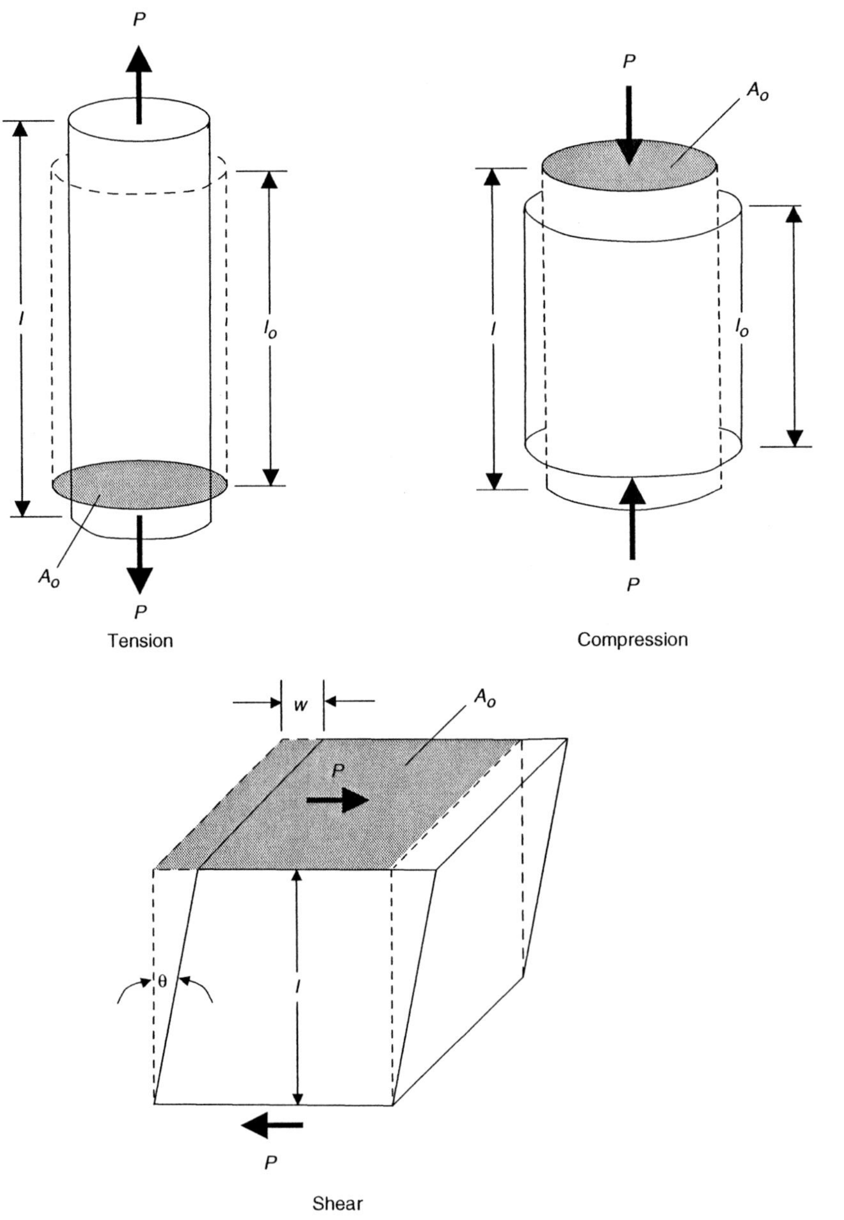


**Fig. 12.1** States of stress. Source: Ref 1

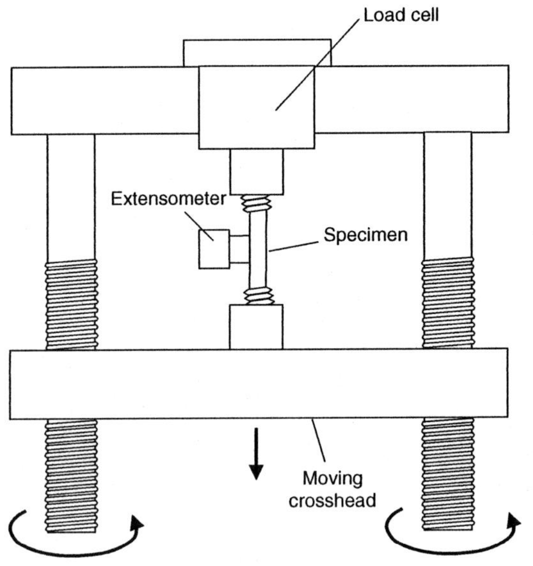


Fig. 12.2 Typical tensile test setup

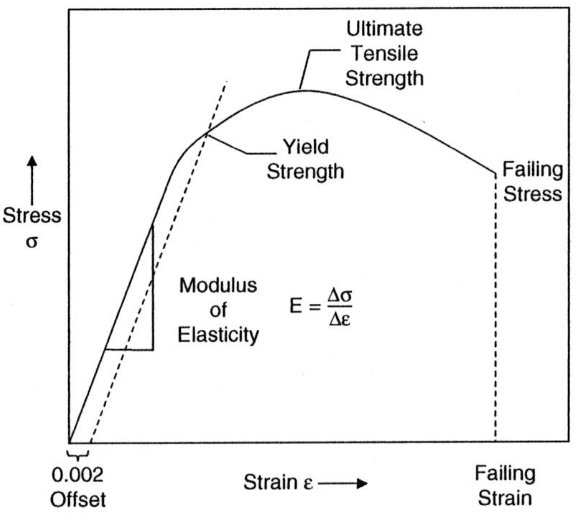


Fig. 12.3 Typical stress-strain curve

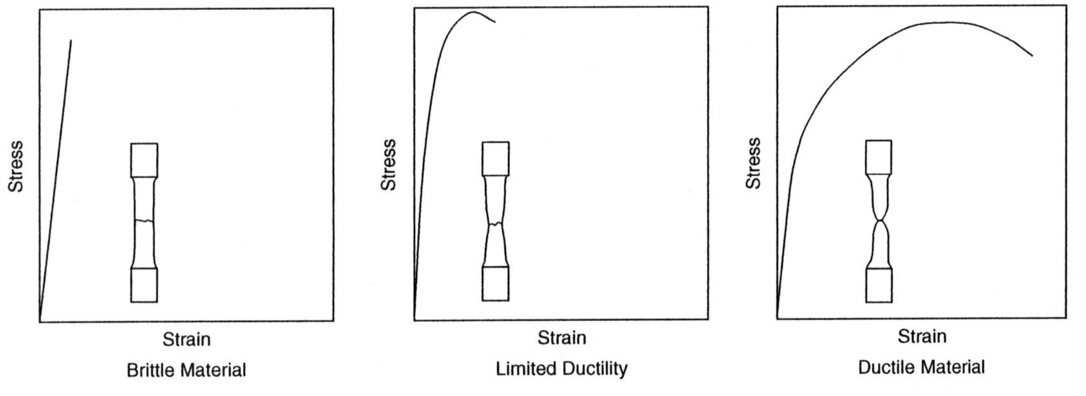


Fig. 12.4 Different stress-strain responses

than the rest. All further plastic deformation is then concentrated in this region, and the specimen begins to neck or thin locally. Because the cross-sectional area is now decreasing far more rapidly than the deformation load is being increased by strain hardening, the engineering stress continues to decrease until fracture occurs.

As shown in Fig. 12.3, the ultimate tensile strength is the maximum stress that occurs during the test. Although the tensile strength is the value most often listed from the results of tensile testing, it is not generally the value that is used in design. Static design of ductile metals is usually based on the yield strength, since most designs do not allow any plastic deformation. However, for brittle metals that do not display any appreciable plastic deformation, tensile strength is a valid design criterion.

With most metals, there is a gradual transition from elastic to plastic behavior, and the point at which plastic deformation actually begins is difficult to define with precision. The transition from elastic to plastic deformation is illustrated in Fig. 12.6. When a specimen is loaded into the plastic range and then unloaded, the elastic strain immediately disappears. With the passage of time, the anelastic strain recovers. The remaining strain is the plastic strain. In uniaxial loading, three criteria for the initiation of yielding have been used: the elastic limit, the proportional limit, and the yield strength.

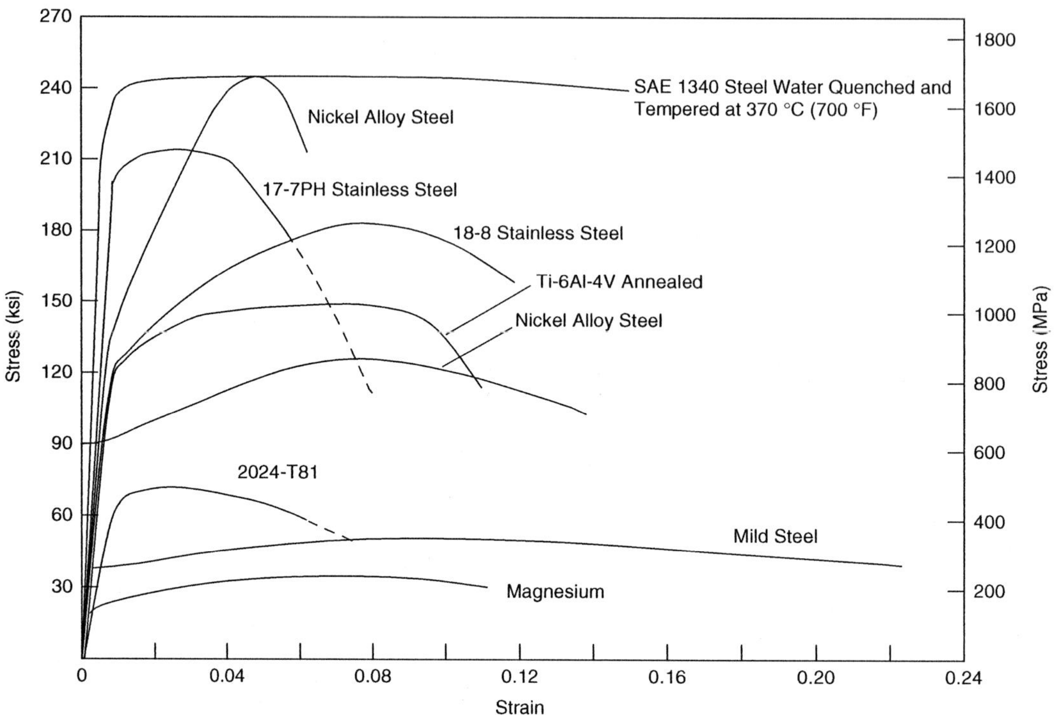


**Fig. 12.5** Typical engineering stress-strain curves

The elastic limit shown in Fig. 12.7 is the greatest stress the material can withstand without any measurable permanent strain remaining after the complete release of load. However, the determination of the elastic limit requires a tedious incremental loading-unloading test procedure. For this reason, it is often replaced by the proportional limit. The proportional limit is the highest stress at which stress is directly proportional to strain. It is obtained by observing the deviation from the straight line portion of the stress-strain curve. Finally, the yield strength (sometimes referred to as the yield point) is the stress required to produce a small specified amount of plastic deformation. The usual definition of this property is the offset yield strength determined by the stress corresponding to the intersection of the stress-strain curve offset by a specified strain. For metals without a definite yield point, the yield strength is determined by drawing a straight line parallel to the initial straight line portion of the stress-strain curve. The line is normally offset by a strain of 0.2% (0.002). The offset yield strength is often referred to in Great Britain as the proof stress, where offset values are either 0.1 or 0.5%. The yield strength obtained by an offset method is commonly used for design and specification purposes, because it avoids the practical difficulties of measuring the elastic or proportional limit. As previously stated, yield strength is generally a more important design parameter than tensile strength, since the possibility of plastic yielding (permanent deformation) is unacceptable for almost all structures.

For some materials, for example, soft copper and gray cast iron, there is essentially no linear portion in the stress-strain curve. For these materials, the offset method cannot be used, and the usual practice is to define the yield strength as the stress to produce a specified total strain, for example, $e = 0.005$. On the other hand, some metals, particularly annealed low-carbon steel, show a localized heterogeneous type of transition from elastic to plastic deformation that produces a definite yield point in the stress-strain curve. Metals that yield discontinuously have a stress-strain curve with a gradual transition from elastic to plastic behavior and produce a load-elongation diagram similar to that shown in Fig. 12.8. Instead of a discrete yield strength (yield point), the yield strength ranges from the lower yield point, to the upper yield point, as shown in Fig. 12.8. The load increases steadily

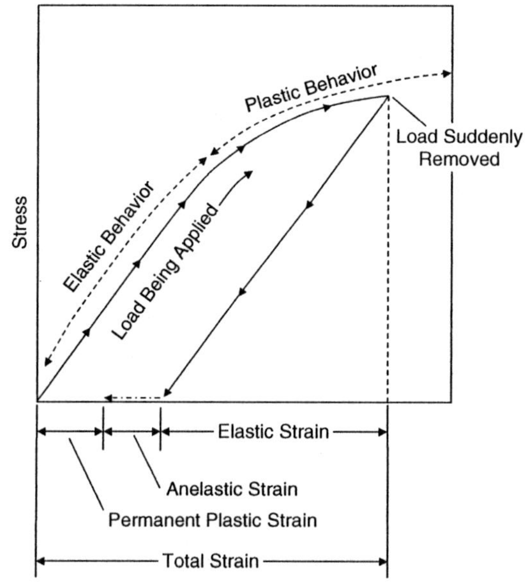


**Fig. 12.6** Elastic and plastic behavior during tensile loading. Source: Ref 2

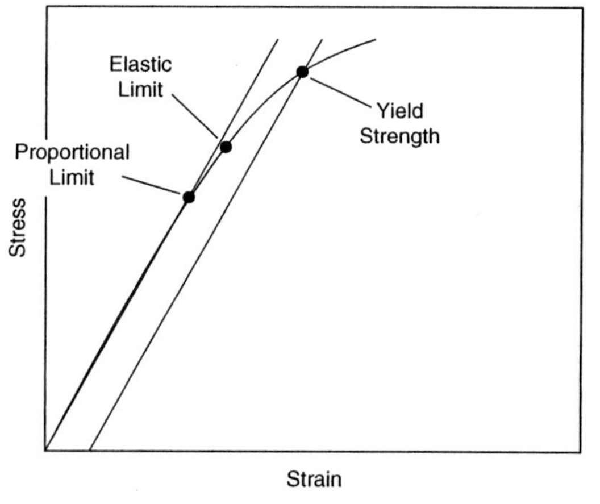


**Fig. 12.7** Transition from elastic to plastic behavior

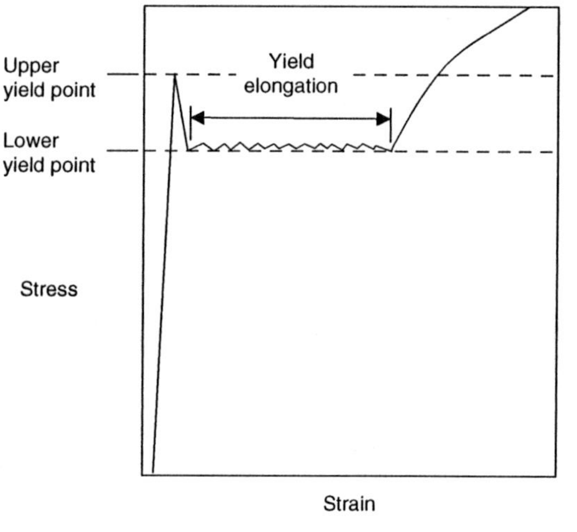


**Fig. 12.8** Discontinuous yielding in plain carbon steels. Source: Ref 3

with elastic strain, drops suddenly, fluctuates about some approximately constant value of load, and then rises with further strain. This type of stress-strain behavior was previously discussed in Chapter 3, "Solid Solutions," in this book.

### *12.1.2 Ductility*

Measures of ductility that are obtained from the tension test are the engineering strain at fracture ($e_f$) and the reduction of area at fracture ($q$). Both are usually expressed as percentages, with the engineering strain at failure often reported as the percent elongation. Both of these properties are obtained after fracture by putting the specimen back together and taking measurements of the final length, $l_f$, and final specimen cross section at fracture, $A_f$. Percent elongation can be determined by:

$$\%\ \text{elongation}\ (e_f) = \frac{l_f - l_o}{l_o} \times 100 \qquad \text{(Eq 12.3)}$$

where $l_o$ is the original gage length, and $l_f$ is the final length of the gage section.

Likewise, reduction in area can be determined by:

$$\%\ \text{reduction in area}\ (q) = \frac{A_o - A_f}{A_o} \times 100 \qquad \text{(Eq 12.4)}$$

where $A_o$ is the original area of the gage section, and $A_f$ is the final area of the gage section at fracture.

Because an appreciable fraction of the deformation is concentrated in the necked region of the specimen, the value of $e_f$ will depend on the gage length, $l_o$, over which the measurement was taken. The smaller the gage length of the specimen, the greater will be the contribution from the neck, and the higher the value of $e_f$. Since percent elongation is sensitive to the gage length in the manner shown in Fig. 12.9, it is important to record the gage length used when reporting percent elongation. To eliminate this difficulty and to provide a measure of ductility that correlates with forming operations in which the gage length is very short, it is possible to

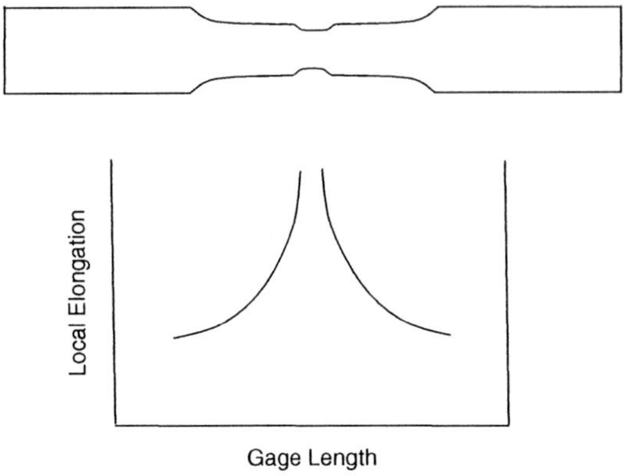


Fig. 12.9 Variation of local elongation along gage length. Source: Ref 4

determine the zero-gage-length elongation, $e_o$. From the constancy-of-volume relationship for plastic deformation ($Al = A_o l_o$):

$$\frac{l}{l_o} = \frac{A_o}{A} = \frac{1}{1-q} \quad \text{(Eq 12.5)}$$

$$e_o = \frac{l - l_o}{l} = \frac{A_o}{A} - 1 = \frac{1}{1-q} - 1 = \frac{q}{1-q} \quad \text{(Eq 12.6)}$$

This represents the elongation based on a very short gage length near the fracture.

When a tensile specimen is loaded, it elongates in the direction of the load and contracts in the transverse or lateral direction. The ratio of the lateral strain ($e_t$) to the tensile or longitudinal strain ($e_l$) is known as Poisson's ratio, ν:

$$\nu = -\frac{\text{Lateral strain}}{\text{Tensile strain}} = -\frac{e_t}{e_l} \quad \text{(Eq 12.7)}$$

Values of Poisson's ratio for most materials range from 0.3 to 0.5.

The slope of the initial linear portion of the stress-strain curve is the modulus of elasticity, or Young's modulus, as shown in Fig. 12.3. The modulus of elasticity ($E$) is a measure of the stiffness of the material. The greater the modulus of elasticity, the smaller will be the elastic strain resulting from a given stress. In other words, metals with higher moduli of elasticity are stiffer than those with lower moduli. Because the modulus of elasticity is needed for computing deflections of beams and other members, it is another important design value.

The modulus of elasticity is determined by the binding forces between atoms. Because these forces cannot be changed without changing the basic nature of the material, the modulus of elasticity is one of the most microstructure-insensitive of the mechanical properties. Generally, it is only slightly affected by alloying additions, heat treatment, or cold work:

$$E = \frac{\Delta s}{\Delta e} \quad \text{(Eq 12.8)}$$

The units of $E$ are megapascals (MPa) or millions of pounds/in.$^2$ (msi).

### 12.1.3 Resilience

Resilience is the ability of a metal to absorb energy when elastically deformed and then to return it when it is unloaded. It is usually measured by the modulus of resilience, which is the strain energy per unit volume required to stress the material from zero to the yield stress. In uniaxial tension, the modulus of resilience, $U_R$, is:

$$U_R = \frac{s_o^2}{2E} \quad \text{(Eq 12.9)}$$

where $s_o$ is the yield strength, and $E$ is the modulus of elasticity. For applications where the metal must not undergo permanent plastic deformation but resist energy loads, the best materials will have a high yield stress and a low modulus of elasticity. An example is golf club driver heads that use heat treated titanium on the face to give a higher rebound than the older steel-faced heads.

### 12.1.4 Toughness

Toughness is the ability of a metal to absorb energy in the plastic range. Although there are a number of approaches to defining toughness, one of the oldest is to consider it as the total area under the stress-strain curve. This area is an indication of the amount of work per unit volume that can be done on a material without causing it to fail. As shown in Fig. 12.10, this definition of toughness implies that toughness is a function of both strength and ductility.

### 12.1.5 True Stress-Strain Curve

The engineering stress-strain curve, based on the original dimensions of the specimen, is used in engineering design because almost all structural designs do not permit yielding or plastic deformation. However, for analysis of metal forming operations, the true stress-strain curve, also known as the flow curve, should be used. The true stress-strain curve is based on the actual dimensions of the test specimen as it undergoes plastic deformation and has the shape shown in Fig. 12.11. The true stress, $\sigma$, is defined as:

$$\sigma = \frac{P}{A} \text{ in pascals (Pa) or psi} \quad \text{(Eq 12.10)}$$

where $P$ is the load in Newtons (N) or pounds (lb), and $A$ is the instantaneous cross-sectional area in $m^2$ or $in.^2$.

The true stress is related to the engineering stress as follows:

$$\sigma = \frac{P}{A} = \frac{P}{A_o}\frac{A_o}{A} = s\frac{A_o}{A} \quad \text{(Eq 12.11)}$$

Using the constancy-of-volume relationship:

$$\frac{A_o}{A} = \frac{l}{l_o} = e + 1 \quad \text{(Eq 12.12)}$$

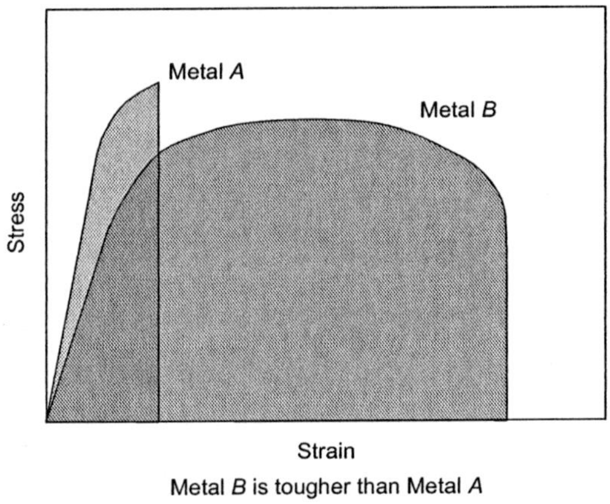


**Fig. 12.10** Area under stress-strain curve as a measure of toughness

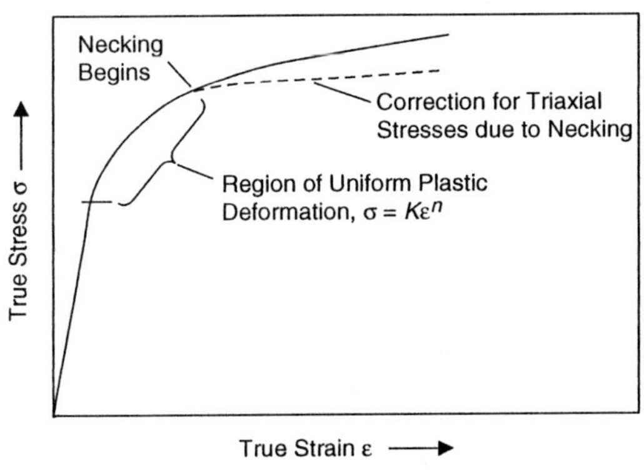


**Fig. 12.11** True stress-true strain behavior

therefore:

$$\sigma = \frac{P}{A_o}(e+1) = s(e+1) \quad \text{(Eq 12.13)}$$

which is applicable up to the onset of necking.

The true strain is defined as:

$$\varepsilon = \int_{l_o}^{l} \frac{dl}{l} = \ln\frac{l}{l_o} \quad \text{(Eq 12.14)}$$

where $l$ is the gage length, and $l_o$ is the original gage length.

The relationship between the true strain and the engineering strain is:

$$\varepsilon = \ln(1+e) \quad \text{(Eq 12.15)}$$

The two values of strain give almost identical values for strains up to 0.1. This relationship is also only valid up to the onset of necking.

Beyond the onset of necking, the true strain is based on the actual area or diameter measurements. Because of the constancy of volume ($A_o l_o = Al$), the true strain can be written in terms of either length or area:

$$\varepsilon = \ln\frac{l}{l_o} = \ln\frac{A_o}{A} = \ln\frac{(\pi/4)D_o^2}{(\pi/4)D^2} = 2\ln\frac{D_o}{D} \quad \text{(Eq 12.16)}$$

For cylindrical test specimens, the true strain at failure, $\varepsilon_f$, can be determined from the reduction of area, $q$, by:

$$\varepsilon_f = \ln\frac{1}{1-q} \quad \text{(Eq 12.17)}$$

The flow curve for many metals in the region of uniform plastic deformation can be expressed by the Holloman power curve relation:

$$\sigma = K\varepsilon^n \quad \text{(Eq 12.18)}$$

where $n$ is the strain-hardening exponent, and $K$ is the strength coefficient. Equation 12.18 can be rewritten as:

$$\log\sigma = \log K + n \log\varepsilon \quad \text{(Eq 12.19)}$$

As shown in Fig. 12.12, a log-log plot of true stress versus true strain up to the maximum load (onset of necking) will give a straight line if the

Holloman relationship holds for the metal in question. The slope of the straight line is $n$, and $K$ is equal to the true stress at a true strain $\varepsilon = 1.0$. The strain-hardening exponent can have values from $n = 0$ for a perfectly plastic solid up to $n = 1.0$ for a perfectly elastic solid. Metals typically have values of $n$ between 0.10 and 0.50. It should be noted that $n$ is equal to:

$$n = \frac{d(\log \sigma)}{d(\log \varepsilon)} = \frac{d(\ln \sigma)}{d(\ln \varepsilon)} = \frac{\varepsilon d\sigma}{\sigma d\varepsilon} \qquad \text{(Eq 12.20)}$$

Necking begins at the maximum load of the engineering stress-strain curve. The engineering stress-strain curve continues upward as the metal strain hardens, and this counteracts the reduction in cross-sectional area due to deformation. Eventually, a point is reached where the increase in strain hardening can no longer keep up with the reducing area, and the stress begins to decrease until fracture. On the other hand, since the true stress-strain curve is based on the instantaneous cross-sectional area, the curve continues to rise until failure.

The strain rate $\dot{\varepsilon} = d\varepsilon/dt$ can also influence the yield or flow stress, particularly at elevated temperatures. The strain-rate sensitivity of metals is low at room temperature but increases with temperature, especially at temperatures above one-half of the absolute melting point. A series of curves for 6063-O aluminum are shown in Fig. 12.13. Note that the strain-rate sensitivity is much greater at the higher temperatures. For deformation occurring at constant volume, the relationship between flow stress and strain rate, at a constant strain and temperature, is given by:

$$\sigma = C\dot{\varepsilon}^m \qquad \text{(Eq 12.21)}$$

where $m$ is the strain-rate sensitivity, which is the slope of a plot of log $\sigma$ versus log $\dot{\varepsilon}$, as shown in Fig. 12.14.

The strain-rate sensitivity, $m$, can also be determined by measuring the change in flow stress, $\sigma$, due to a change in $\dot{\varepsilon}$ at a constant strain, $\varepsilon$, and temperature, $T$:

$$m = \frac{\Delta \log \sigma}{\Delta \log \dot{\varepsilon}} = \frac{\log(\sigma_2/\sigma_1)}{\log(\dot{\varepsilon}_2/\dot{\varepsilon}_1)} \qquad \text{(Eq 12.22)}$$

In hot working operations, $m$ is typically in the range of 0.1 to 0.2.

The strain-rate sensitivity describes the ability of a material to resist plastic instability or necking. Superplasticity is the ability of certain polycrystalline metallic materials to extend plastically to large strains when deformed in tension. For superplasticity, $m$ is usually greater than 0.5, with the majority of superplastic materials having an $m$ value in the range of 0.4 to 0.8, where a value of 1.0 would indicate a perfectly superplastic material. The presence of a neck in a material undergoing a tensile strain results in a locally high strain rate and, for a high value of $m$, leads to a sharp increase in the yield stress within the necked region; that is, the neck undergoes strain hardening, which restricts its further development. Therefore, a high strain-rate sensitivity resists neck formation and leads

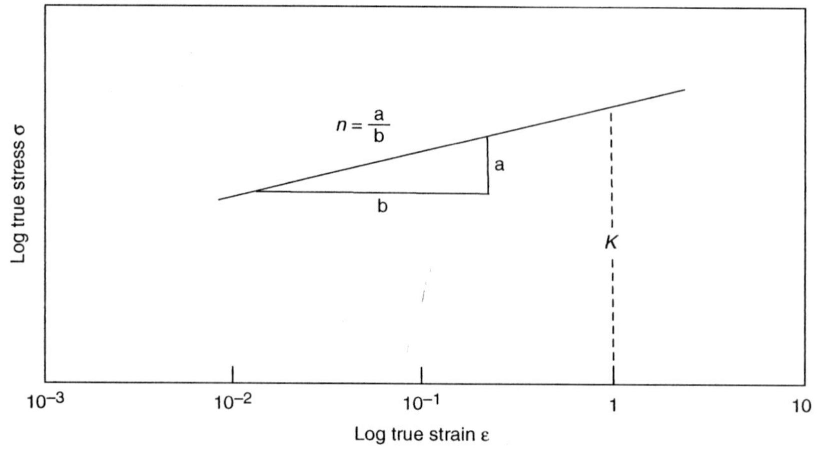


**Fig. 12.12** Plot to determine strain-hardening exponent, $n$, and strength coefficient, $K$. Source: Ref 5

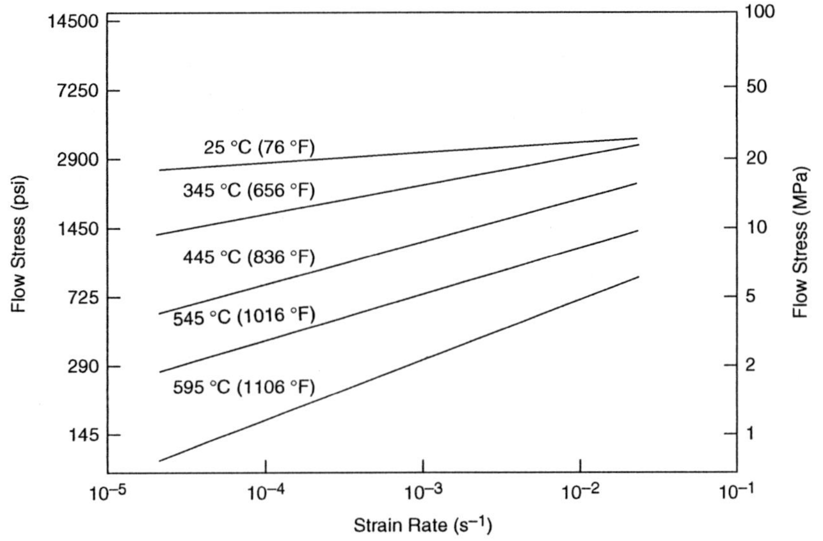


**Fig. 12.13** Flow stress versus strain rate for 6063-O aluminum. Source: Ref 6

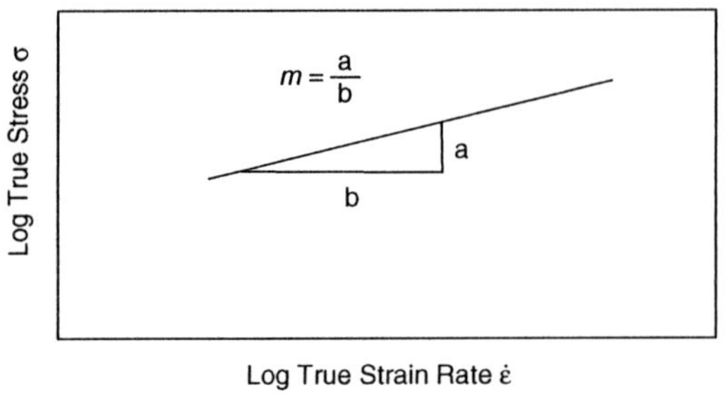


**Fig. 12.14** Plot to determine strain-rate sensitivity, *m*

to the high tensile elongations observed in superplastic materials. The yield stress decreases and the strain-rate sensitivity increases with increasing temperature and decreasing grain size. The elongation to failure tends to increase with increasing *m*.

## 12.2 Stress Concentrations

Geometrical features, such as holes, fillets, and radii, produce higher local stresses than encountered in the body of the material. For example, the tensile stresses at the top and bottom of the hole shown in Fig. 12.15 are three times greater than they are in the body of the material. These higher stresses are a result of the inability of the stresses to pass through the hole. Since the hole wants to widen in the direction of the load, compressive stresses are produced at the sides of the hole (the load is making the circular cross section of the hole tend toward an ellipse). Similar stress perturbations occur around notches, as shown in Fig. 12.16. The regions adjoining the notch must carry this portion of the load in addition to their normal share, creating a concentration of force lines at the notch.

Such stress increases are described by the stress-concentration factor, *K*:

$$K = \frac{\text{Maximum actual stress}}{\text{Nominal stress}} \quad \text{(Eq 12.23)}$$

While the stress-concentration factor *K* for a round hole is 3, much higher stress-concentration factors occur when there is a sharp notch in the material. The stress-concentration factors for three different geometries are shown in Fig. 12.17. Note that the value of the stress-concentration factor is strongly dependent on the actual geometry, and that values significantly higher than the value for a round hole are achieved for some geometries. Values for a wide range of geometries can be found in handbooks dealing with stress analysis.

Because most metals undergo plastic deformation at sufficiently high stresses, the stress values corresponding to the theoretical stress concentrations are often not developed. Ductile metals deform locally in the highly stressed regions and thus partially relieve the stress.

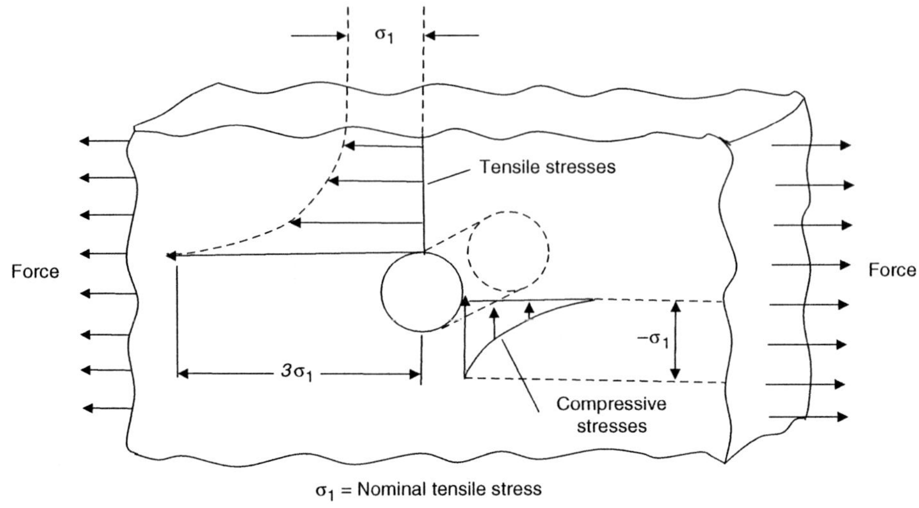


**Fig. 12.15** Stress-concentration effect around a hole. Source: Ref 2

Often, stress concentrations are not dangerous in ductile metals under static loads. However, under cyclic loads (fatigue), even ductile metals are generally affected by stress concentrations. Most brittle metals are sensitive to stress-concentration effects under both static and dynamic loading.

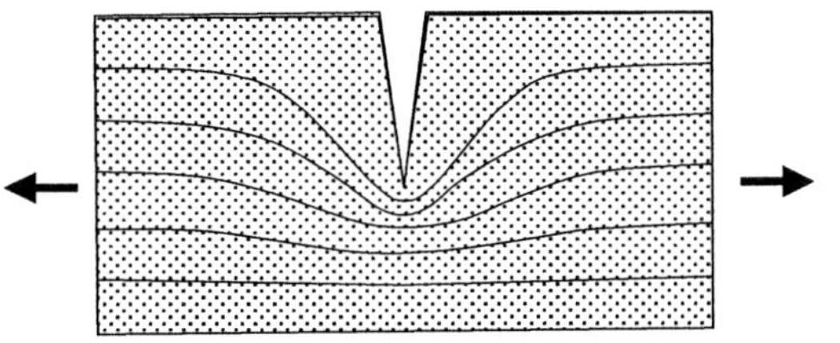

**Fig. 12.16** Lines of force around notch. Source: Ref 7

## 12.3 Notched Tensile Test

Ductility measurements on standard smooth tensile specimens do not always reveal metallurgical or environmental changes that can lead to reduced local ductility. The tendency for reduced ductility in the presence of a triaxial stress field caused by a notch is called notch sensitivity. A common way of evaluating notch sensitivity is a tension test using a notched specimen. The notched tensile test has been used extensively for investigating the properties of high-strength steels, for studying hydrogen embrittlement in steels and titanium, and for investigating the notch sensitivity of high-temperature alloys.

The most common notched tensile specimen uses a 60° notch with a root radius of 0.03 mm (0.001 in.) or less introduced into a round (circumferential notch) or flat (double-edge notch) tensile specimen. Usually, the depth of the notch is such that the cross-sectional area at the root of the notch is one-half of the area in the unnotched section. The specimen is loaded in tension until fracture occurs. The notch strength is defined as the maximum load divided by the original cross-sectional area at the notch. Because of the plastic constraint at the notch, this value will be higher than the tensile strength of an unnotched specimen if the material possesses some ductility. Therefore, the common way of detecting notch brittleness, or high notch sensitivity, is by determining the notch strength ratio (NSR):

$$\text{NSR} = \frac{\text{Net tensile strength for notched specimen}}{\text{Tensile strength for unnotched specimen}} \tag{Eq 12.24}$$

If the NSR is less than 1, the material is considered to be notch brittle.

As strength, hardness, or some metallurgical variable restricting plastic flow increases, the metal at the root of the notch is less able to deform, and fracture becomes more likely. Notch brittleness begins at the strength level where the notch strength begins to fall or, more conventionally, at the strength level where the

NSR becomes less than unity. The sensitivity of notch strength for detecting metallurgical embrittlement in an alloy steel is illustrated in Fig. 12.18. Note that the elongation measured on a conventional smooth specimen was unable to detect the fall in notch strength produced by tempering in the 315 to 480 °C (600 to 900 °F) range.

## 12.4 Compression

Many metalworking operations, such as rolling, extrusion, and forging, are performed with the workpiece under compressive loads. A compression test is conducted in a manner similar to the tensile test, except that the force is compressive, and the specimen contracts along the direction of the stress. However, the deformation shown in Fig. 12.1 is idealized; in reality, frictional forces develop at the two ends and cause barreling, as shown in Fig. 12.19. During deformation, as the metal spreads over the compression anvils to increase its diameter, frictional forces oppose the outward flow of metal. The metal at midheight in the specimen can flow outward, leading to the barreled shape. Internally, a region of undeformed metal is created

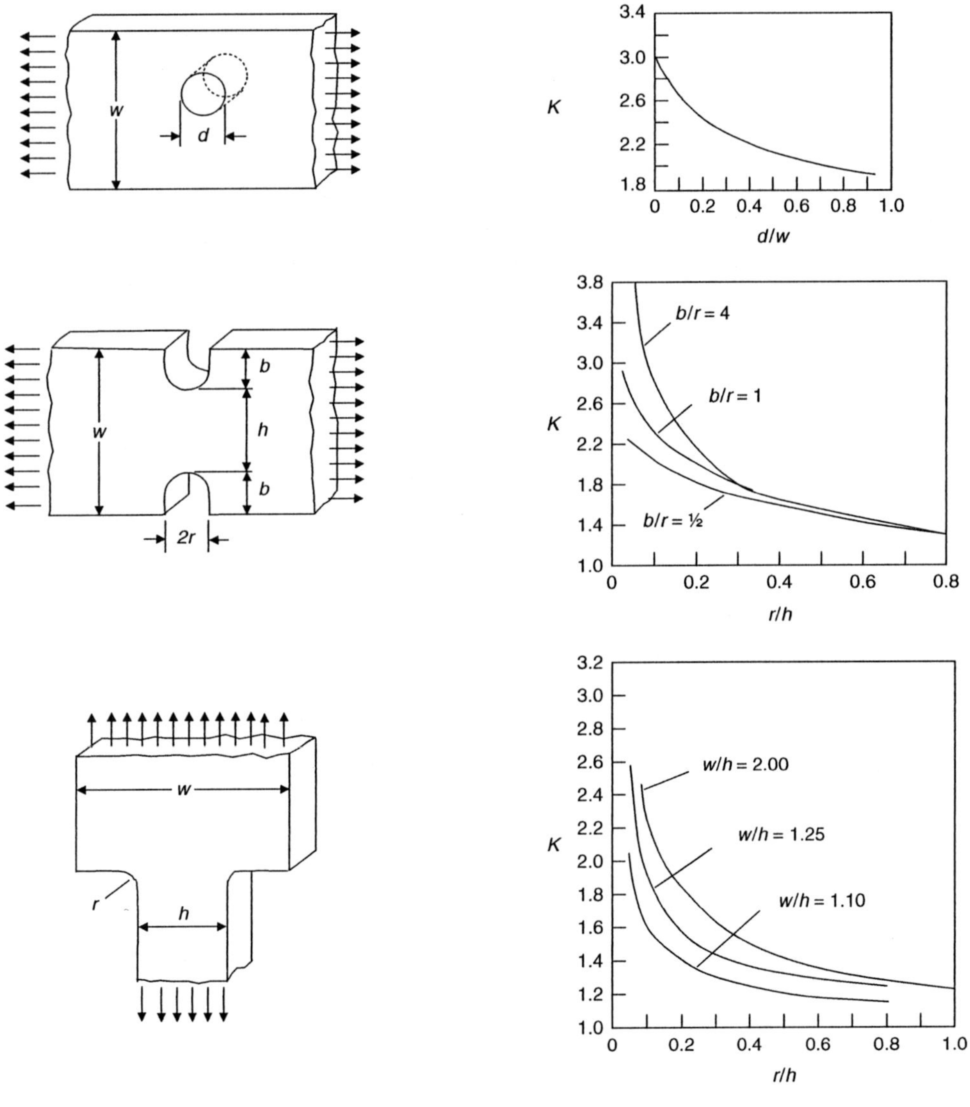


**Fig. 12.17** Stress-concentration factors, $K$, for three geometries. Source: Ref 8

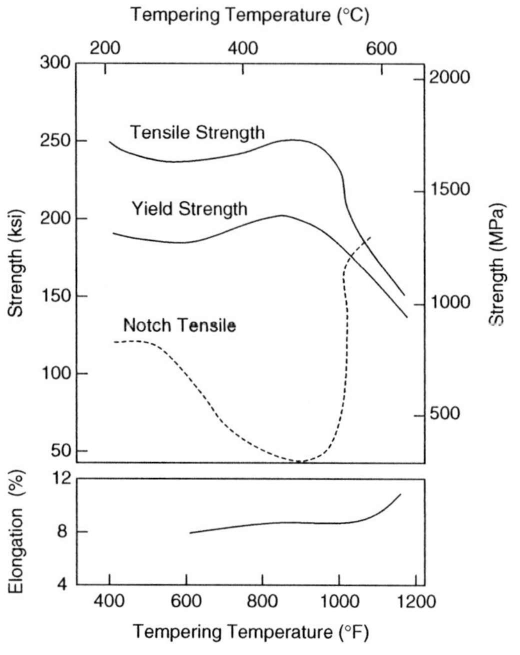


**Fig. 12.18** Unnotched and notched properties of an alloy steel. Source: Ref 9

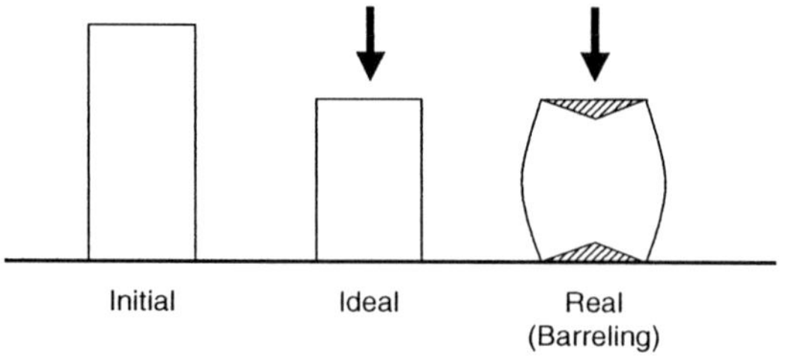


**Fig. 12.19** Barreling during compression test

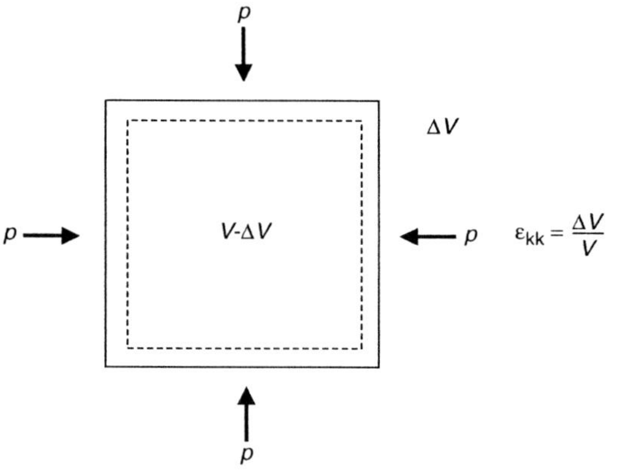


**Fig. 12.20** Dilatation under hydrostatic pressure

near the anvil surfaces. The addition of lubrication and/or Teflon (E.I. DuPont de Nemours & Co., Inc.) strips helps in reducing the frictional forces.

A solid under hydrostatic pressure (Fig. 12.20) undergoes a volume change called dilatation. If the volume change is $\Delta V$, the dilatation, $\varepsilon_{kk}$, or volume strain, is defined as:

$$\varepsilon_{kk} = \frac{\Delta V}{V} \qquad \text{(Eq 12.25)}$$

Since positive pressure causes a volume reduction, the negative of dilatation is proportional to the pressure, so that:

$$p = -K\varepsilon_{kk} \qquad \text{(Eq 12.26)}$$

where $K$ is known as the bulk modulus.

Annealed metals have approximately the same tensile and compressive yield strengths. When a metal is plastically deformed in tension, the familiar work-hardening effect is observed. However, the same tensile deformation that increases the tensile yield strength in tension reduces the compressive yield strength. In addition, plastic deformation in compression decreases the yield strength in tension. This phenomenon by which plastic deformation increases yield strength in the direction of plastic flow and decreases it in other directions is known as the Bauschinger effect. This directionality of strain hardening occurs in all polycrystalline metals but is especially pronounced in some metals, such as titanium and zirconium. For example, the titanium alloy Ti-6Al-4V develops more than a 50% difference between its tensile and compressive yield strengths after only a 2% tensile elongation. The Bauschinger effect is relieved at elevated temperatures and can be totally eliminated by hot working or a stress-relief anneal after cold working.

The development of the Bauschinger effect is demonstrated in Fig. 12.21. When a tensile specimen is loaded to a plastic strain (point 2), it strain hardens since the yield strength (point 1) has been exceeded. When the load is removed, the specimen contracts elastically along path 2–3. If the metal is then loaded in compression, its compressive yield strength (point 4) would be considerably less than its tensile yield strength (point 2). The Bauschinger effect is caused by the tendency of dislocations to pile up and become entangled during the initial tensile loading, producing strain hardening. However, when the metal is loaded in compression, it is

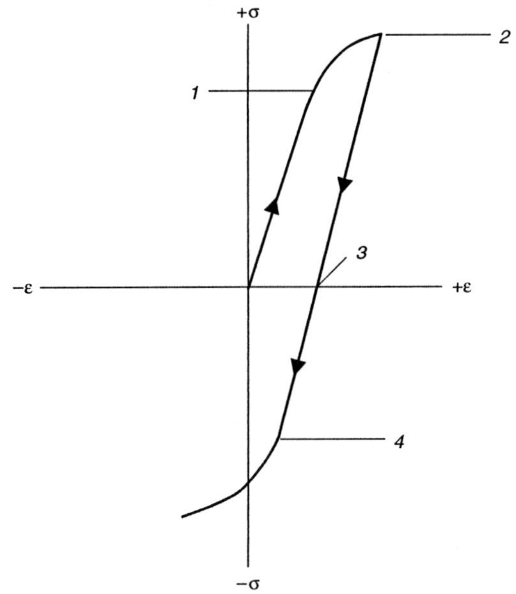


**Fig. 12.21** Bauschinger effect

easier for the dislocations to reverse their path and travel backward in a direction that now contains fewer dislocations. Compressive loading from point 3 also produces Frank-Read dislocation loops of the opposite sign of those generated during tension loading, and these annihilate each other when they meet.

## 12.5 Shear

For tests performed using a pure shear force, as shown in Fig. 12.1, the shear stress, $\tau$, is computed according to:

$$\tau = \frac{P}{A_o} \quad \text{(Eq 12.27)}$$

where $P$ is the load or force imposed parallel to the upper and lower faces, each of which has an area of $A_o$. The shear strain, $\gamma$, is defined as the tangent of the strain angle, $\theta$:

$$\gamma = \frac{w}{l} = \tan\theta \quad \text{(Eq 12.28)}$$

but for small elastic strains, $\gamma = \theta$.

The units for shea· stress and strain are the same as for their tensile counterparts. The shear modulus ($G$) is the slope of the linear portion of the shear stress-strain curve:

$$G = \frac{\tau}{\gamma} \quad \text{(Eq 12.29)}$$

Torsion is a variation of pure shear where a structural member is twisted. Torsional forces produce a rotational motion about the longitudinal axis of one end of the member relative to the other end. Torsional tests are normally performed on cylindrical solid shafts or tubes. The shear stress, $\tau$, is a function of the applied torque, $T$, while the shear strain, $\gamma$, is related to the angle of twist, $\varphi$.

## 12.6 Stress-Strain Relationships

When the state of stress is multiaxial, it is necessary to calculate the elastic strains induced by the multiaxial state of stress in three mutually perpendicular directions by using the generalized Hooke's law:

$$e_x = \frac{1}{E}[\sigma_x - \nu(\sigma_y + \sigma_z)] \quad \text{(Eq 12.30)}$$

$$e_y = \frac{1}{E}[\sigma_y - \nu(\sigma_x + \sigma_z)] \quad \text{(Eq 12.31)}$$

$$e_z = \frac{1}{E}[\sigma_z - \nu(\sigma_x + \sigma_y)] \quad \text{(Eq 12.32)}$$

where $\sigma_x$, $\sigma_y$, and $\sigma_z$ are the normal stresses in three directions, and $\varepsilon_x$, $\varepsilon_y$, and $\varepsilon_z$ are the elastic strains in the three coordinate directions.

The modulus of elasticity, $E$, the shear modulus, $G$, the bulk modulus, $K$, and Poisson's ratio, $\nu$, are related to each other according to:

$$E = 2G(1+\nu) \quad G = \frac{E}{2(1+\nu)} \quad \nu = \frac{E-2G}{2G}$$

$$K = \frac{E}{3(1-2\nu)} \quad E = \frac{9KG}{3K+G} \quad \nu = \frac{3K-E}{6K}$$

and for many metals and other materials:

$$K \sim E, \quad G \sim \frac{3}{8}E, \quad \nu \sim 0.33$$

## 12.7 Combined Stresses

For combined stress states that have a combination of tensile, compression, and shear stresses, the stress state can be resolved into principal stresses that are the maximum and minimum values of $\sigma$ for a set of stresses $\sigma_1$, $\sigma_2$, and $\tau_1$. Principal stresses act on axes that differ from the axes along which the stresses are acting and represent the maximum ($\sigma_1$) and minimum ($\sigma_2$) values of the normal stresses for the particular point under consideration. In two dimensions, principal stresses along the $x$- and $y$-axes are:

$$\sigma_1 = \frac{\sigma_x + \sigma_y}{2} + \sqrt{\left(\frac{\sigma_x - \sigma_y}{2}\right)^2 + \tau_{xy}^2} \quad \text{(Eq 12.33)}$$

$$\sigma_2 = \frac{\sigma_x + \sigma_y}{2} - \sqrt{\left(\frac{\sigma_x - \sigma_y}{2}\right)^2 + \tau_{xy}^2} \quad \text{(Eq 12.34)}$$

$$\tau_1 = \pm\sqrt{\left(\frac{\sigma_x - \sigma_y}{2}\right)^2 + \tau_{xy}^2} \quad \text{(Eq 12.35)}$$

where $\sigma_1$, $\sigma_2$, $\tau_1$ are the principal stress components, and $\sigma_x$, $\sigma_y$, $\tau_{xy}$ are the calculated stress components shown in Fig. 12.22, all of which are determined at any particular point. Principal stresses can also be determined graphically, using the Mohr's circle construction shown in Fig. 12.23. More detailed descriptions of Mohr's circle can be found in almost any book dealing with mechanics or strength of materials. Similar equations and Mohr's circle constructions can be done for three-dimensional principal stress analyses.

## 12.8 Yield Criteria

In many instances, for example, metalworking operations, the material is not subjected to simple tensile and compression stresses; instead, a triaxial (three-dimensional) stress state will exist. For the case of simple uniaxial loading, the onset of yielding occurs at the yield strength of the material as determined by a simple tensile test. If the loading is more complicated, and a multiaxial state of stress is produced by the loads, the onset of yielding can no longer be predicted by comparing any one of the simple stress components with uniaxial material yield strength, not even the maximum principal stress. The onset of yielding for multiaxially stressed parts is more accurately predicated through the use of a combined stress theory of failure. The two most widely accepted theories for predicting the onset of yielding are the distortion energy theory (von Mises criterion) and the maximum shearing stress theory (Tresca criterion). The distortion energy theory is usually somewhat more accurate, while the maximum shearing stress theory may be slightly easier to use.

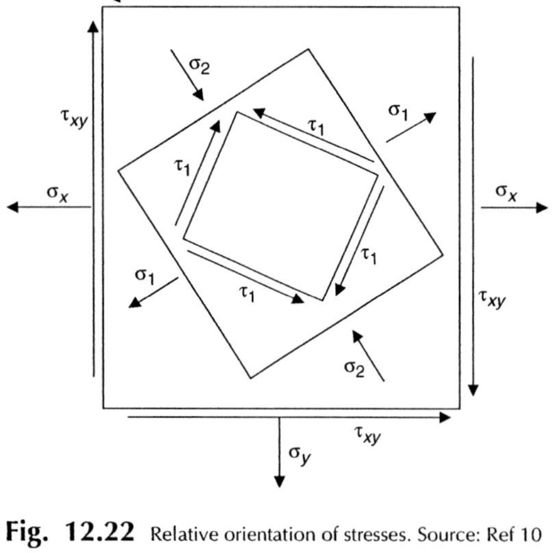


**Fig. 12.22** Relative orientation of stresses. Source: Ref 10

In the distortion energy theory, yielding is predicted to occur in a multiaxial state of stress when the distortion energy per unit volume becomes equal to or exceeds the distortion energy per unit volume at the time of yielding in a simple uniaxial stress test using a specimen of the same material. Therefore, yielding is predicted by the distortion energy theory if:

$$\frac{1}{\sqrt{2}}[(\sigma_1 - \sigma_2)^2 + (\sigma_2 - \sigma_3)^2 + (\sigma_3 - \sigma_1)^2]^{1/2} \geq \sigma_y \quad \text{(Eq 12.36)}$$

In the maximum shearing stress theory, yielding is predicted to occur in a multiaxial state of stress when the maximum shearing stress magnitude becomes equal to or exceeds the maximum shearing stress magnitude at the time of yielding in a simple uniaxial stress test using a specimen of the same material. In this instance, yielding is predicted by the maximum shearing stress theory if:

$$\sigma_1 - \sigma_3 \geq \sigma_y \quad \text{(Eq 12.37)}$$

where $\sigma_1$, $\sigma_2$, and $\sigma_3$ are the principal stresses at a point, ordered such that $\sigma_1 > \sigma_2 > \sigma_3$, and $\sigma_y$ is the uniaxial yield strength in tension.

A comparison of the two criteria is shown in Fig. 12.24. In the simpler Tresca criterion shown in Fig. 12.24(a), stress combinations lying within the solid line envelope do not result in plastic flow, while those lying outside the envelope develop plastic flow. The situation is the same for the von Mises criterion shown in Fig. 12.24(b), which also shows the Tresca criterion for comparison and shows that the Tresca criterion is the more conservative of the two. However, experimental data shown in Fig. 12.24(c) fit the von Mises criterion somewhat better, but the difference between the two is not great.

## 12.9 Residual Stresses

Residual stresses are the stresses existing in a body at rest, in equilibrium, at uniform temperature, and not subjected to external forces. One source of residual stresses are manufacturing processes. They may be mild and unnoticeable; moderate, as in the slight warpage of a piece of sheet metal; or severe enough to cause cracking and splitting apart of a freshly quenched piece of steel.

Plastic deformation by cold working of a metal always produces residual stresses. However, the resulting stress pattern varies greatly, depending on the type of working employed. If the metal is only lightly worked, so that plastic deformation is limited to regions near the surface, the surface layers will be left in a state of residual compression. An example of this type of stress pattern is shown in Fig. 12.25(a), which may be produced by processes such as shot peening, light cold drawing, or surface rolling. Compressive residual stress states on the surface are desirable in fatigue loading applications, since the cyclic tensile stresses have to overcome the residual compressive surface stresses before the bulk metal starts being loaded in tension. However, if the working is severe, so that deformation occurs throughout the cross section of the metal, there will be tensile residual stresses on the surface, as shown in Fig. 12.25(b). Using the same logic as for compression residual stresses, tensile residual

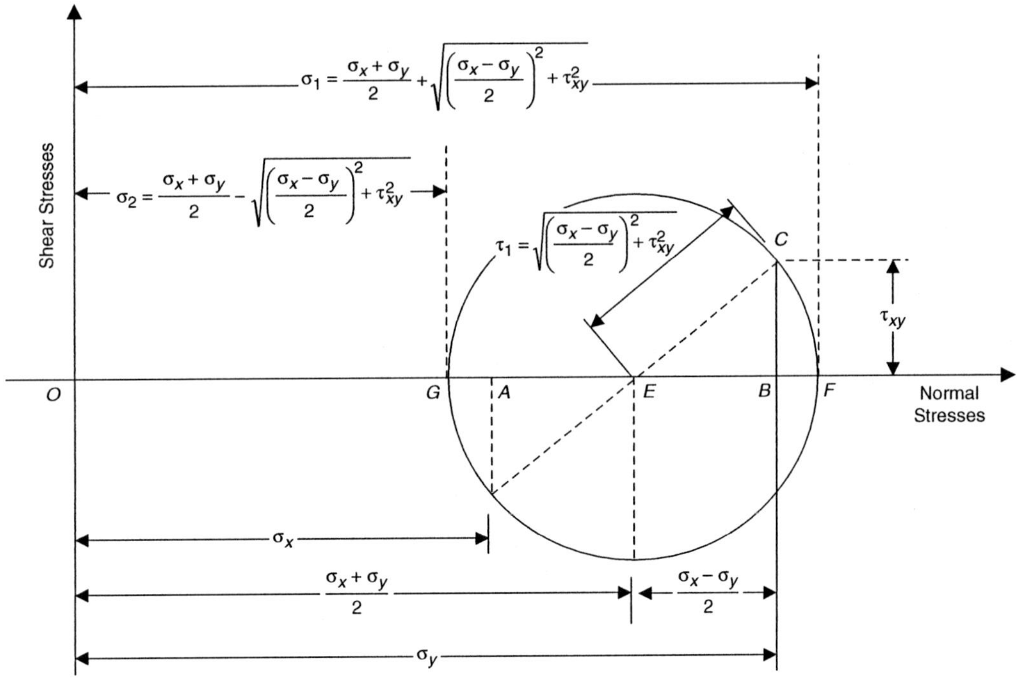


1. Let the axes $x$ and $y$ be chosen to represent the directions of the applied normal and shearing stresses, respectively.
2. Construct to suitable scale distances $OA = \sigma_x$, $OB = \sigma_x$, and $BC = AD = \tau_{xy}$.
3. With point $E$ as a center, construct the circle DFC.
4. Then $OF$ and $OG$ are the principal stresses $\sigma_1$ and $\sigma_2$, respectively, and $EC$ is the maximum shear stress $\tau_1$.

**Fig. 12.23** Mohr's circle for determination of principal stresses in two dimensions. Source: Ref 10 stresses can drastically lower the fatigue strength, since the residual stresses are added to the applied tensile stresses. Therefore, heavily worked sections are frequently given a final light deformation to convert the residual surface tensile stresses to compressive stresses, as shown in Fig. 12.25(c), for the example of a light surface rolling of heavily worked cold-drawn steel.

Another major source of residual stresses is the nonuniform change in volume induced by thermal effects. In many cases, such as fusion welding, several thermal factors may act simultaneously to produce a final, complex pattern of residual stresses. These factors include thermal expansion/contraction on heating and cooling, a change in volume, and phase changes, such as from the liquid to the solid state. When a hot solid metal is quenched in water, the surface cools more rapidly than the center and thus shrinks faster. This nonuniform shrinkage

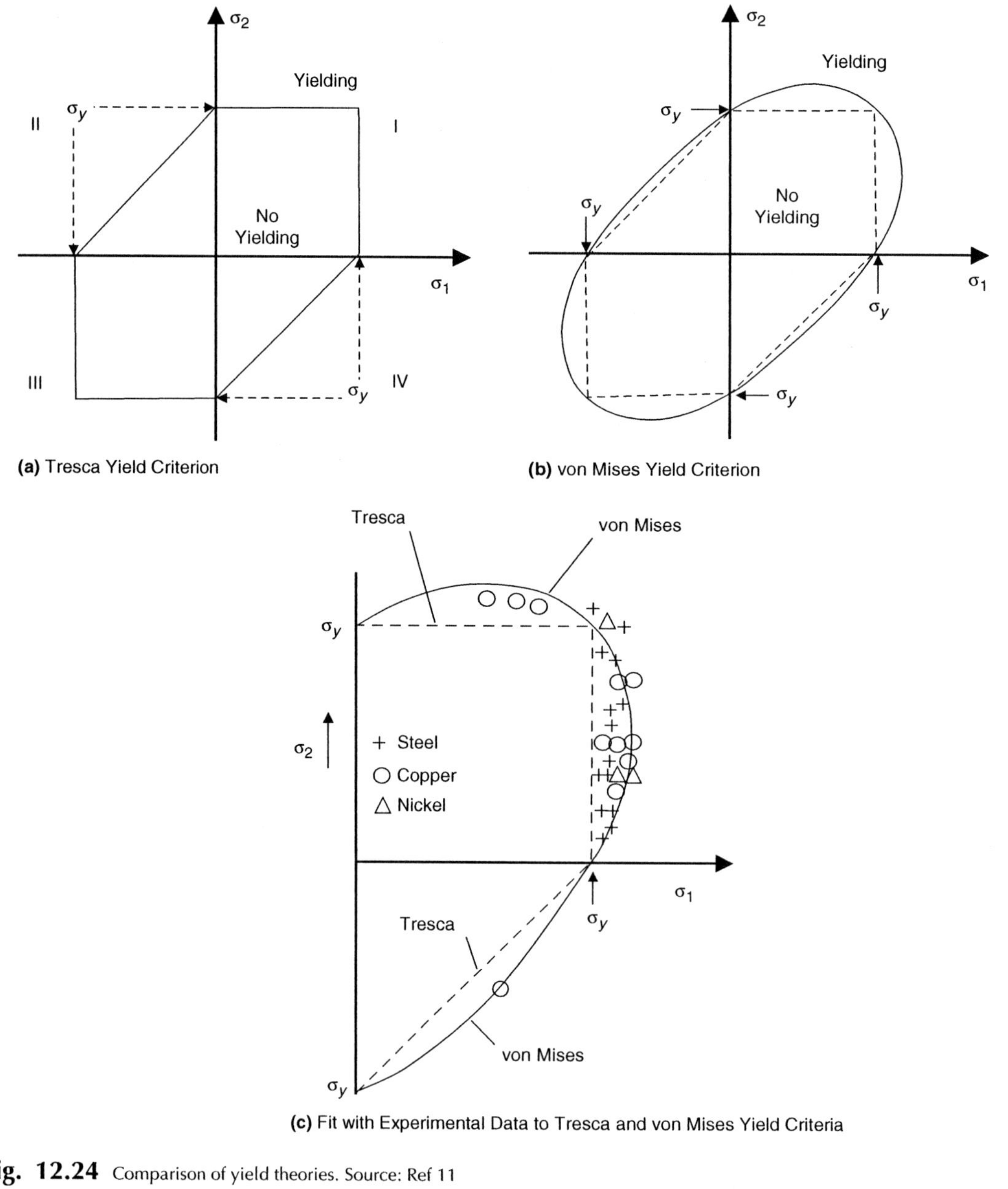


**Fig. 12.24** Comparison of yield theories. Source: Ref 11

causes plastic deformation of the hotter, softer interior. After the surface has cooled, the interior continues to contract as its temperature continues to fall. When both the surface and interior are able to support large stresses, the contraction of the interior produces compressive stresses at the surface, similar to the pattern shown in Fig. 12.25(a). These stresses can be quite large, approximately −205 MPa (−30 ksi) in steel and approximately −105 MPa (−15 ksi) in softer metals such as aluminum. The use of a milder quench will reduce the residual-stress level.

In steels, the formation of martensite from austenite during quenching results in an expansion. This expansion during martensite formation can result in tensile residual stresses at the surface and compressive residual stresses in the interior, similar to that shown in Fig. 12.25(b), with maximum residual tensile stresses at the surface approaching +550 MPa (+80 ksi). In the actual hardening of steel, the stress patterns are usually much more complex than this, since the final pattern is due to a combination of thermal contraction and martensitic expansion. However, this does help explain why freshly quenched steels can often crack and spall. Thus, it is common practice to initiate tempering as soon as possible after quenching. Combine these residual stresses with stress concentrations, such as sharp corners, and it is easy to see how problems can develop. Typical causes of distortion and cracking are sharp corners, re-entrant angles, and abrupt changes in cross section. Milder quenches and immediate tempering are helpful in reducing problems.

Stresses induced during casting can result from the shrinkage of the metal during solidification. These stresses can be great enough to cause the formation of cracks while the casting is solidifying, commonly known as hot tearing. When approximately 80% of solidification is complete, the dendrites become interlocked and hinder further contraction of the casting. Hot tearing can occur during the last 20% of solidification and is related to the length of the temperature interval over which the remaining solidification occurs.

## 12.10 Hardness

Hardness is the resistance to penetration, and the majority of hardness testers force a small sphere, pyramid, or cone into a specimen by means of an applied load. A number is obtained, and the hardness can often be correlated to the tensile strength of the metal. For example, the tensile strength of a steel in pounds per square inch is roughly equal to 500 times its Brinell hardness number. The correlations between tensile strength and hardness for steels, brass, and nodular cast iron are shown in Fig. 12.26.

A hardness test is easy to execute and provides information on the yield strength of a material. In brief, a hard material (the indentor) is placed above the sample surface. A load is

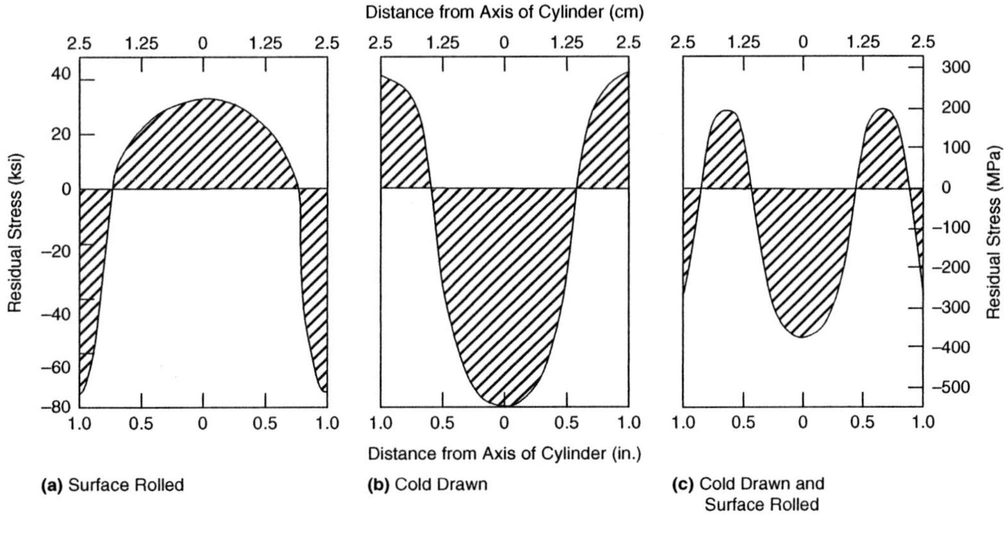


**Fig. 12.25** Residual stress patterns in 5 cm (2 in.) diameter steel bars. Source: Ref 2

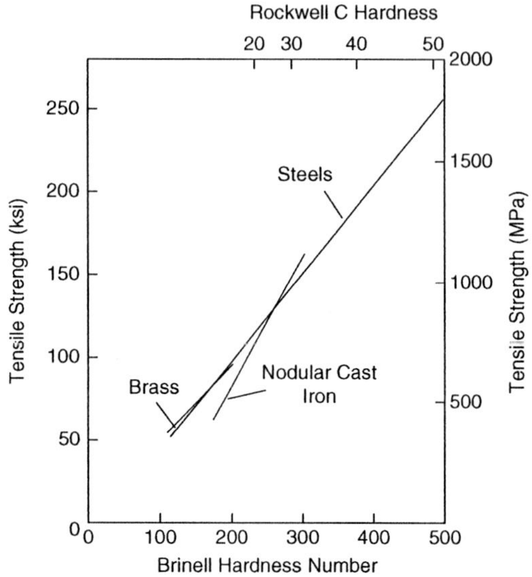


**Fig. 12.26** Correlation of hardness with tensile strength

then placed on the indentor, which then penetrates the surface of the specimen. The size of the resulting indentation, a measure of the permanent deformation that the material undergoes during the test, relates to the material hardness; a small indentation indicates high hardness and vice versa. When the load is divided by the indentation area (sometimes the projected area of the indentation), a hardness number, $H$, is defined. Hardness has units of stress, and, since it relates to permanent deformation, $H$ correlates with $\sigma_y$. The stress state arising during a hardness test differs from that in a tensile test, however. This results in $H$ being greater than $\sigma_y$. To a good approximation, $H \sim (2.5 \text{ to } 3.0)\sigma_y$; the approximation $H = 3\,\sigma_y$ is often used.

A number of important hardness tests are summarized in Fig. 12.27. One of the most widely used is the Rockwell hardness tester, shown in Fig. 12.28. For determining the

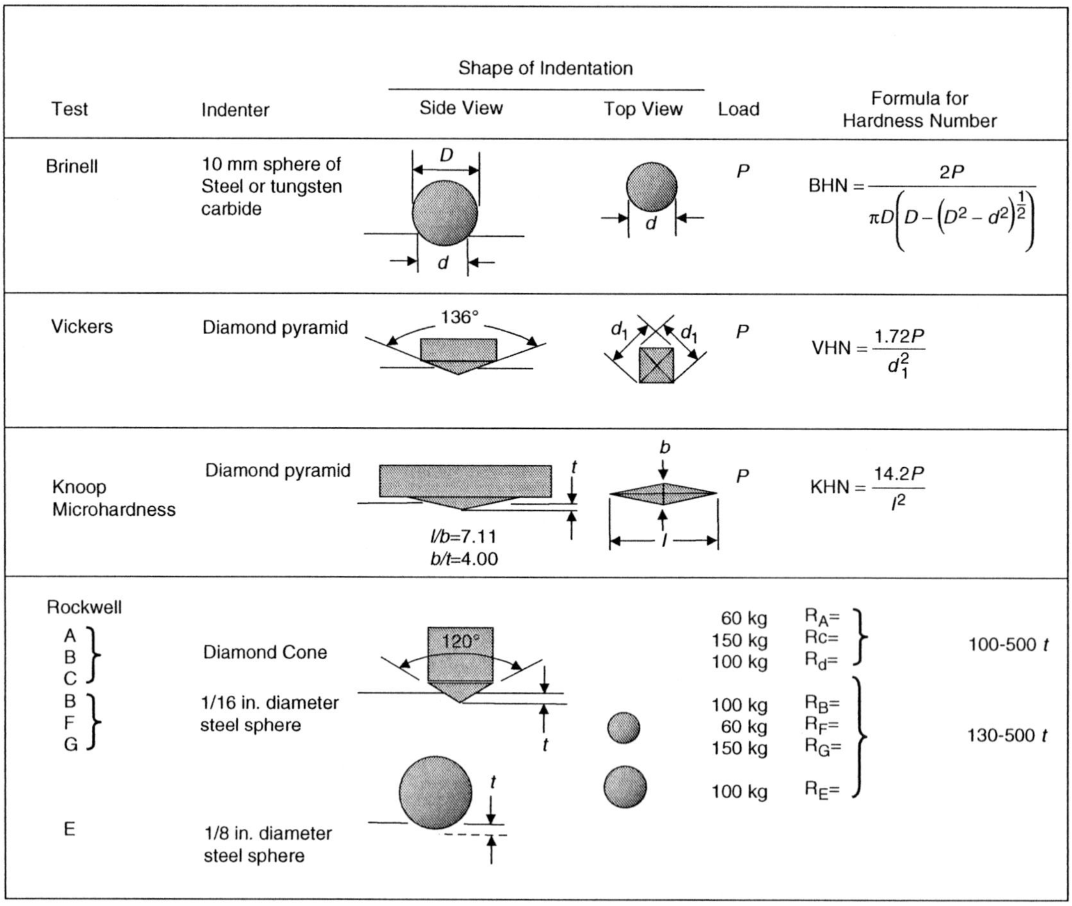


**Fig. 12.27** Hardness tests. Source: Ref 12

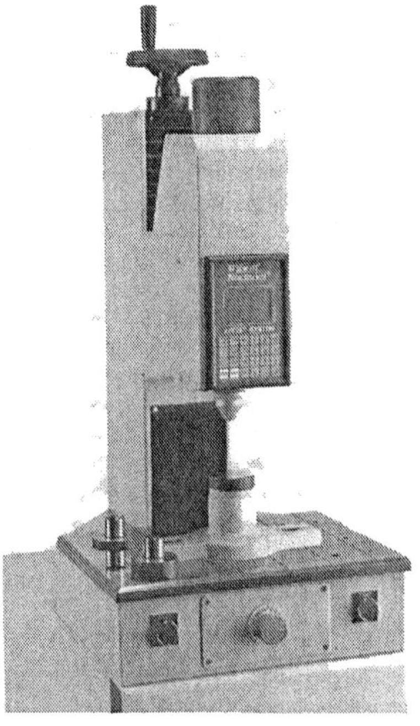

(a)

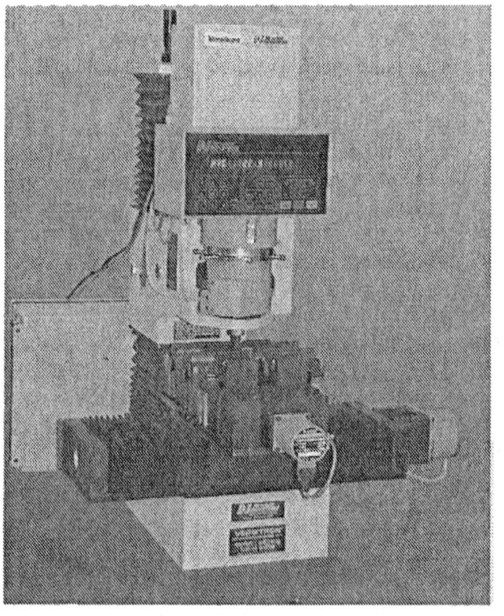

(b)

**Fig. 12.28** (a) High-speed Rockwell tester. (b) Automated Rockwell tester for high-rate testing

Rockwell hardness, a small load is initially applied, followed by a larger load. The penetrator is forced into the surface, and when the load is released, a dial indicator indicates the hardness, which is a function of the depth reached by the penetrator. The Brinell hardness tester is another widely used test for large steel parts, such as crankshafts, where a minimum of depth of penetration is needed. Microhardness tests (Vickers and Knoop) can be made on individual constituents of a microstructure, and it is possible to determine the hardness of particles as small as 0.2 mm (0.007 in.).

## ACKNOWLEDGMENTS

Sections of this chapter were adapted from "Mechanical Behavior Under Tensile and Compressive Loads" by G.E. Dieter in *Mechanical Testing and Evaluation,* Volume 8, *ASM Handbook,* ASM International, 2000.

## REFERENCES

1. W.D. Callister, *Fundamentals of Materials Science and Engineering,* 5th ed., John Wiley & Sons, Inc., 2001
2. A.G. Guy and J.J. Hren, *Elements of Physical Metallurgy,* 3rd ed., Addison-Wesley Publishing, 1974
3. B.L. Ferguson, Design for Deformation Processes, *Materials Selection and Design,* Vol 20, *ASM Handbook,* ASM International, 1997
4. G.E. Dieter, *Mechanical Metallurgy,* 3rd ed., McGraw-Hill Book Co., 1986
5. G.E. Dieter, Mechanical Behavior Under Tensile and Compressive Loads, *Mechanical Testing and Evaluation,* Vol 8, *ASM Handbook,* ASM International, 2000
6. R. D'Antonia, C. Caciag, and K. Mukherjee, *Metall. Trans.,* Vol 1, 1970, p 1771
7. M.A. Meyers and K.K Chawla, *Mechanical Metallurgy—Principles and Applications,* Prentice-Hall Inc., 1984
8. G.H. Neugebauer, *Prod. Eng.,* Vol 14, 1943, p 82–87
9. G.B. Espey, M.H. Jones, and W.F. Brown, *ASTM Proceedings,* Vol 59, American Society for Testing Materials, 1959, p 837
10. J. Marin, *Mechanical Properties of Materials and Design,* McGraw-Hill, 1942

11. T.H. Courtney, *Mechanical Behavior of Materials,* 2nd ed., McGraw-Hill Book Co., 2000
12. H.W. Hayden, W.G. Moffatt, and J. Wulff, *The Structure and Properties of Materials,* Vol III, Wiley, 1965

## SELECTED REFERENCES

- M.F. Ashby and D.R.H. Jones, *Engineering Materials 1—An Introduction to Their Properties and Applications,* 2nd ed., Butterworth Heinemann, 1996
- D.R. Askeland, *The Science and Engineering of Materials,* 2nd ed., PWS-Kent Publishing Co., 1989
- J. Collins and S. Daniewicz, Failure Considerations, Chap. 18, *Mechanical Engineer's Handbook,* 12th ed., John Wiley & Sons, Inc., 1998
- F.E. Fisher, Stress Analysis, Chap. 10, *Mechanical Engineer's Handbook,* 12th ed., John Wiley & Sons, Inc., 1998
- R.A. Higgins, *Engineering Metallurgy—Applied Physical Metallurgy,* 6th ed., Arnold, 1993
- *Metals Handbook Desk Edition,* 2nd ed., ASM International, 1998
- O.D. Sherby, Appendix: Superplasticity in Iron-Base Alloys, *Forming and Forging,* Vol 14, *ASM Handbook,* ASM International, 1988
- E.L. Tobolski, Macroindentation Hardness Testing, *Mechanical Testing and Evaluation,* Vol 8, *ASM Handbook,* ASM International, 2000

# CHAPTER 13

# Fracture

FRACTURE is the separation of a solid body into two or more pieces under the action of stress. Fracture can be classified into two broad categories: ductile fracture and brittle fracture. As shown in the Fig. 13.1 comparison, ductile fractures are characterized by extensive plastic deformation prior to and during crack propagation. Brittle fractures occur with little or no gross plastic deformation and usually occur suddenly, without warning. The tendency for brittle fracture increases with decreasing temperature, increasing strain rate, and under triaxial stress conditions, usually due to a notch or other stress concentration.

## 13.1 The Brittle Fracture Problem

Brittle fracture can be defined as fracture that takes place at stresses below the net section yield strength, with very little observable plastic deformation and minimal absorption of energy. Brittle fracture is very dangerous because it occurs abruptly, with little or no warning. Since

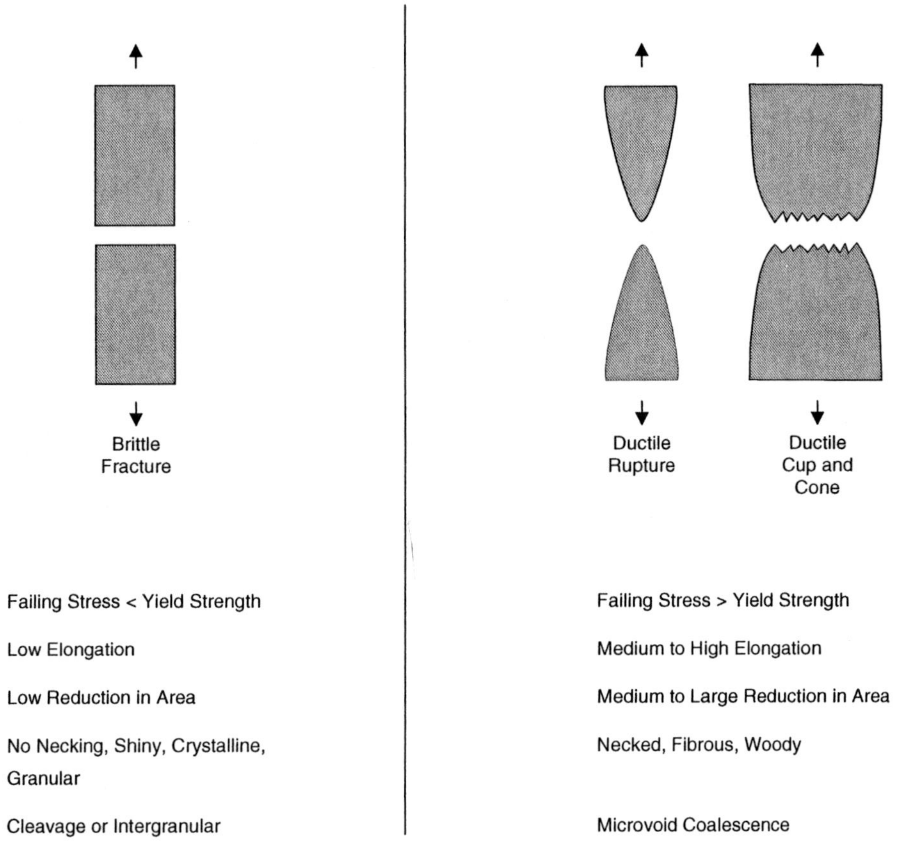


**Fig. 13.1** Comparison of brittle and ductile fracture modes

brittle fracture can result in serious economic losses and loss of life, a great deal of effort has been expended in developing methodologies to avoid brittle fractures.

The catastrophic nature of brittle fracture was dramatically exemplified with the failure of Liberty ships during World War II. An example of one of these failures is shown in Fig. 13.2, where the USS *Schenectady* broke in half while just sitting at dock. To produce ships quickly for the war effort, all-welded construction was used for the hulls instead of the traditional riveted design. Of the 2700 Liberty ships built, approximately 400 sustained fractures, 90 of which were considered serious. In 20 of these, the hull fractured in half. The failures were caused by three factors: (1) defective welds containing cracklike flaws, (2) stress concentrations at square hatch covers (the sites of most fracture initiations), and (3) the steel used in their construction was relatively brittle by today's (2007) standards. When the hatch covers were retrofitted with rounded reinforcements and riveted structures replaced some of the welded structures, the incidence of fracture was greatly reduced. Even if a crack initiated at a defective weld, it would be arrested at a rivet hole before it reached catastrophic dimensions.

After the war, G.R. Irwin and his staff at the Naval Research Laboratory laid the foundation for what is known today as fracture mechanics. Liberty ships are just one example of catastrophic brittle fracture. It has also been a recurring problem in aircraft, bridges, train wheels, and other heavy equipment. Although brittle fracture does not occur today with the frequency it once did, it can still be a problem if proper design and manufacturing practices are not used.

## 13.2 Brittle and Ductile Fracture

In a typical scenario for a brittle fracture, a small flaw is introduced during fabrication or when the part is placed in service. Unfortunately, the initial flaw size is so small that it often goes undetected. During service, the flaw initially propagates in a stable manner under static or repeated loads, often aided by corrosion. When the flaw reaches a critical size, final fracture is sudden, proceeding at velocities approaching the speed of sound.

Brittle fractures are generally flat, with little or no evidence of localized necking. Glasses

**Fig. 13.2** Fracture of the USS *Schenectady* at San Diego, CA, pier. Source: Ref 1

and crystalline ceramics, when fractured at room temperature, fracture in a brittle manner, with no appreciable evidence of plastic deformation. Brittle fractures can also occur in body-centered cubic (bcc) and hexagonal close-packed (hcp) metals but not in face-centered cubic (fcc) metals unless grain-boundary embrittlement has occurred. However, even the most brittle metal will exhibit some slight evidence of plastic deformation. Crack initiation normally occurs at a small flaw, such as a defect, notch, or discontinuity, which acts as a stress concentration, and rapidly propagates through the metal. Cracks resulting from machining, quenching, hydrogen embrittlement, or stress corrosion can cause brittle failures. Even normally ductile metals can fail in a brittle manner at low temperatures, in thick sections, at high strain rates such as impact loading, or when there are preexisting flaws. Brittle failures normally initiate as a result of cleavage that occurs by breaking of the atomic bonds. Brittle failures are characterized by rapid crack propagation, with less energy expenditure than in ductile fractures. Factors promoting brittle failures are:

- High yield strengths that allow storage of high elastic energy levels
- Low temperatures that cause a ductile-to-brittle transition in bcc metals
- Large grain sizes that build up stress from dislocation pileups
- High strain rates that do not allow time for stress redistribution
- Coarse carbides or other inclusions that are themselves susceptible to cracking
- Deep notches that create constraint at the crack tip
- Thick sections that cause plane-strain loading

Cleavage fracture is a brittle fracture mode. Cleavage fractures are characterized by a planar crack that changes planes by the formation of discrete steps. River patterns are formed at grain boundaries (Fig. 13.3) where the cleavage plane in one grain is not parallel to the plane in the adjacent grain, the difference being accommodated by a series of steps. The river patterns eventually diminish as the crack propagates and adopts the cleavage plane of the new grain before being reformed at the next grain boundary.

As opposed to brittle fractures, ductile failures are associated with large amounts of plastic deformation. As a result of plastic deformation, localized necking or distortion is often present. Ductile failures occur by tearing of the metal, with a large expenditure of energy. Ductile fractures can take several forms. In a very ductile polycrystalline metal, such as gold or lead, they may actually be drawn down to a point before failure, generally referred to as ductile rupture. In most ductile metals, failure occurs by microvoid nucleation and growth. Microvoids form at stress concentrations and are most frequently initiated by second-phase particles, followed by void formation and growth around the particles, or by particle cracking (Fig. 13.4). The microvoids then coalesce and grow to produce larger voids until the remaining area becomes too small to support the load, and final failure occurs. The ligaments between the voids fail in shear on the plane of highest shear stress at 45° to the tensile axis. Shear lips, due to slip mechanisms, often occur at angles approaching 45° to the applied tensile stress, to form the well-known cup-and-cone fracture appearance.

Ductile fractures proceed only as long as the material is being strained; that is, when the deformation ceases, the crack stops propagating. At the other extreme, when a brittle crack is initiated, it propagates through the material at velocities approaching the speed of sound, with no possibility of arresting it. There is insufficient plastic deformation to blunt the crack. This makes brittle fractures extremely dangerous; that is, there is usually no warning of impending fracture.

As shown in Fig. 13.5, some bcc and hcp metals, and steels in particular, exhibit a ductile-to-brittle transition when loaded under impact. At high temperatures, the impact energy is high and the failure modes are ductile, while at low temperatures, the impact energy absorbed is low and the failure mode changes to a brittle fracture. The transition temperature is sensitive to both alloy composition and microstructure. For example, reducing the grain size of steels lowers the transition temperature. Not all metals display a ductile-to-brittle transition. Those having an fcc structure, such as aluminum, remain ductile down to even cryogenic temperatures.

As shown in Fig. 13.6, crack propagation is either transgranular (i.e., through the grains) or intergranular (i.e., along the grains). At room temperature, the grain boundaries are usually stronger than the grains themselves, and thus, fracture normally occurs in a transgranular manner. Intergranular failure at room temperature often implies some embrittling behavior, such the formation of brittle grain-boundary films or the segregation of impurities or inclusions at the grain boundaries. However, at temperatures high enough for creep to dominate, the grain boundaries become weaker than the grains themselves, and intergranular failure modes are common.

## 13.3 Ductile-to-Brittle Transition Testing

As discussed in Chapter 12, "Mechanical Behavior," the area under the stress-strain curve is an indication of material toughness. Thus, metals that have a good combination of strength and ductility should exhibit good toughness. For metals that do not contain flaws and are loaded in uniaxial tension, this approach is valid. However, in many engineering applications, a notch is present that introduces a triaxial stress state. Since the triaxial stress state reduces the ability of the metal to deform plastically at the notch, the toughness is reduced. In addition, impact loading reduces the thermal energy available for plastic deformation, further reducing the toughness.

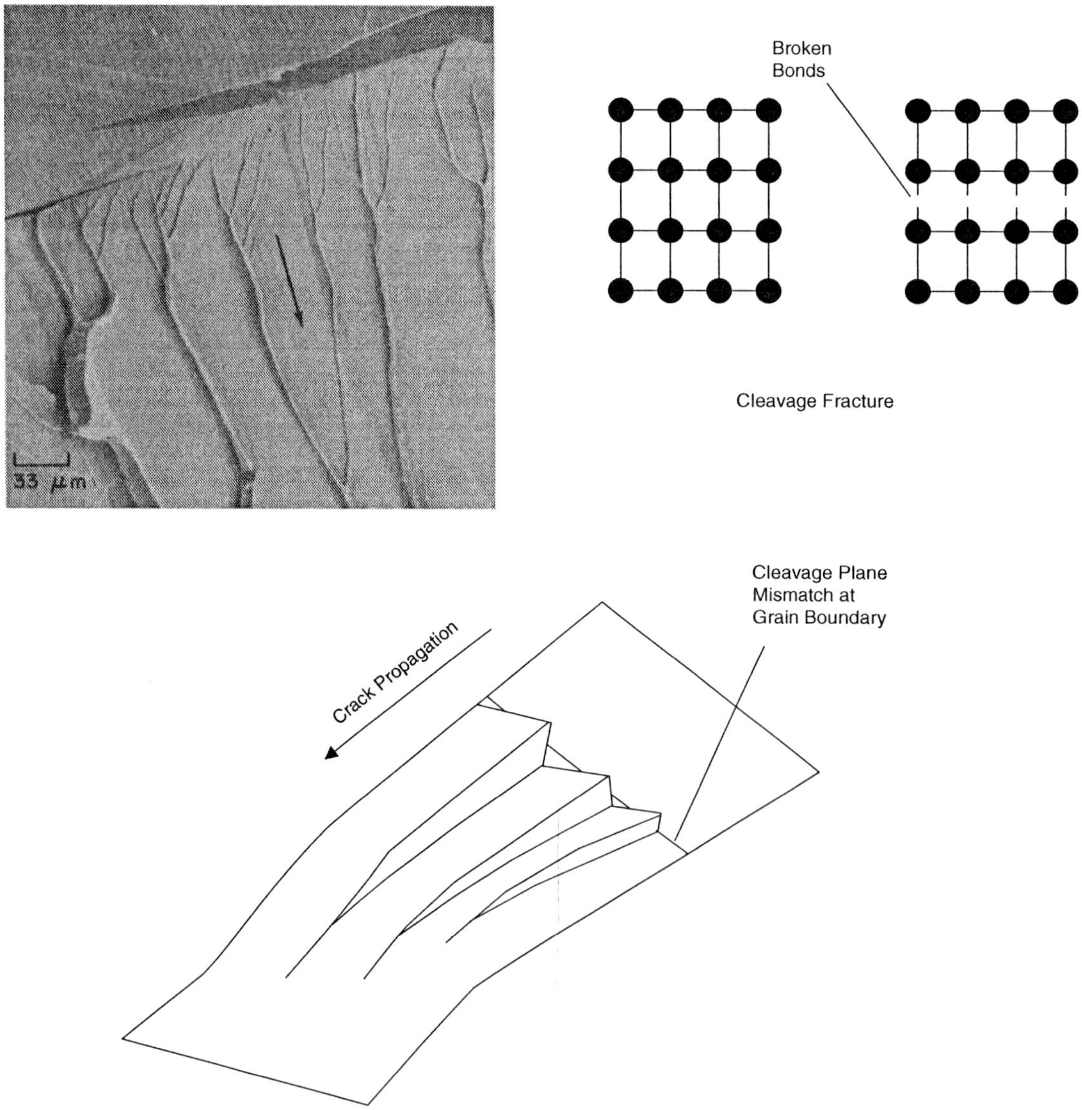


**Fig. 13.3** Facetted brittle failure with river lines. Source: Ref 2

Prior to the formulation and maturity of linear elastic fracture mechanics, the notch toughness characteristics of low- and intermediate-strength steels were determined primarily by notched bar impact testing. The purpose of these tests is to determine the temperature at which a normally ductile failure transitions to a brittle failure. Therefore, tests are conducted at a series of different temperatures, and the impact loads (J or ft · lbf) are plotted versus test temperature. As the test temperature is lowered, there is a transition from ductile failures at high temperatures to brittle failures at lower temperatures (Fig. 13.7). This type of test is particularly applicable to steels. Because of their bcc crystalline structure, steels undergo a change in their

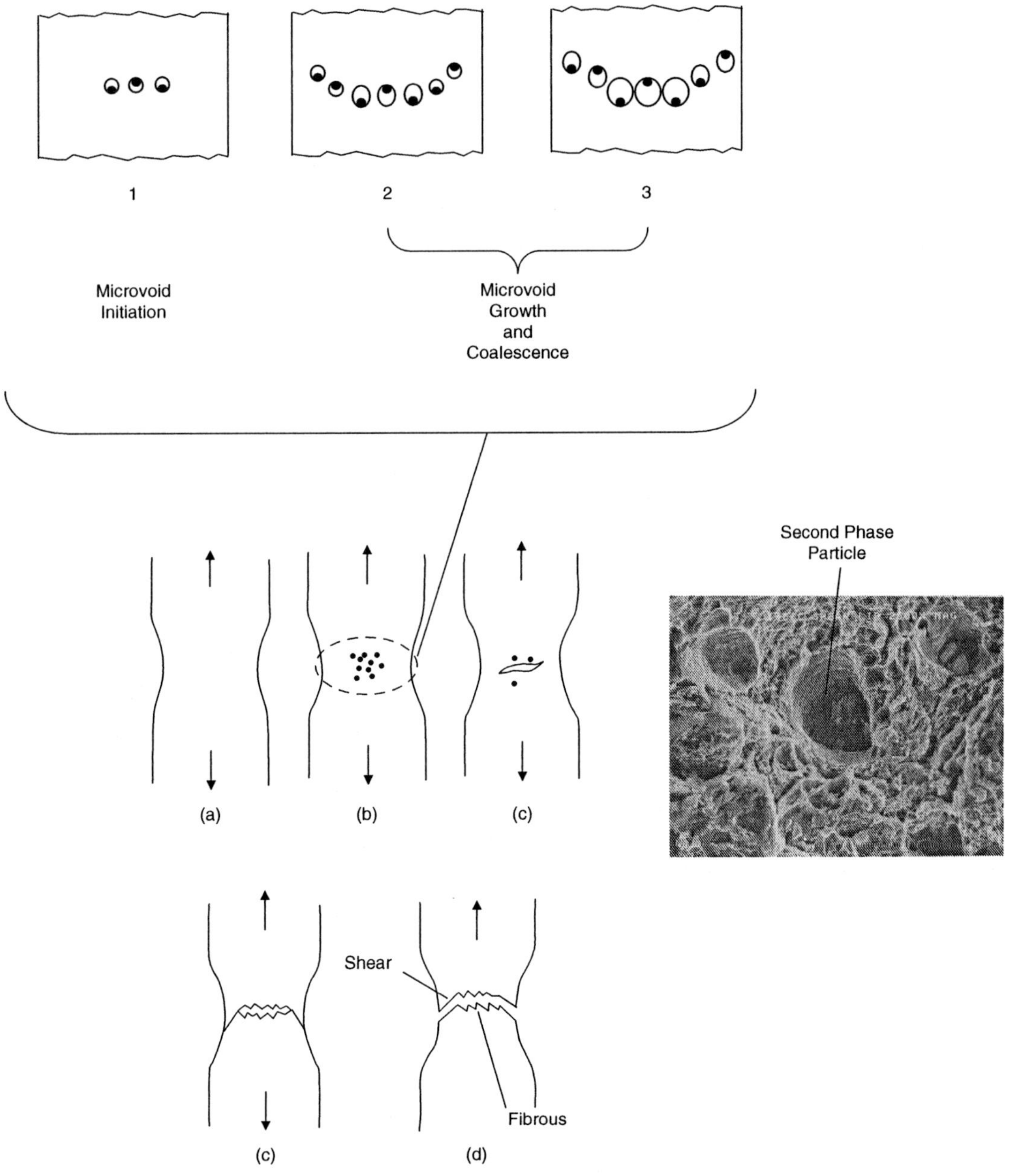


**Fig. 13.4** Microvoid coalescence during ductile failure. Source: Ref 3

fracture behavior with changes in temperature. At high temperatures, fracture occurs in a ductile manner, with large amounts of plastic deformation, while at low temperatures, fracture occurs in a brittle manner, with little or no evidence of plastic deformation.

The most widely used specimen for characterizing the ductile-to-brittle transition behavior of steels is the Charpy V-notch impact specimen, which is described in ASTM E 23. The Charpy specimen (Fig. 13.8) is a three-point bend specimen containing a machined notch in the center of the face facing away from the impacting device. A cantilever bend test, the Izod test, is similar to the Charpy V-notch test and is preferred in Great Britain. The Izod specimen has a V-notch that is located toward one end of the specimen instead of in the middle, as in the Charpy V-notch specimen. The same testing machine can be used for both tests; however, the specimen geometry and computation of the impact toughness is different. The impact tester consists of a heavy pendulum that strikes the specimen at the bottom of its arch, where both the velocity and kinetic energy are at a maximum. As the specimen deforms and fractures, a portion of the kinetic energy of the pendulum is transferred to the specimen. The specimen is broken into two pieces as the pendulum continues its swing to a somewhat lower position than it was released from. The differences in these heights and the mass of the pendulum determine how much energy was absorbed by the specimen. Most impact testers have a gage that reports the energy level directly in ft · lbf. Impact tests such as the Charpy and Izod tests are severe tests, because they occur at high strain rates and, because of the notches, experience a high degree of triaxial loading that constrains plastic deformation at the notch.

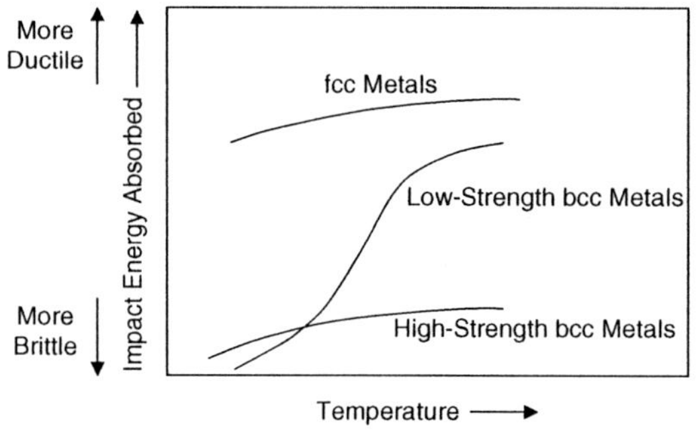


**Fig. 13.5** Ductile-to-brittle temperature comparison for different metals: fcc, face-centred cubic; bcc, body-centered cubic

Specimens are tested at different temperatures, and the impact notch toughness at each test temperature is determined from the energy absorbed during fracture. In addition, the percent shear (fibrous) fracture on the fracture surface and the change in the width of the specimen (lateral expansion) are frequently examined. An example of the ductile-to-brittle transition as a function of temperature for each of these parameters is presented in Fig. 13.9. The actual values for each parameter and the locations of the ductile-brittle transition along the temperature axis are usually different for different steels and even for a steel of a given composition. The rate of change from ductile to brittle behavior depends on many factors, including the strength

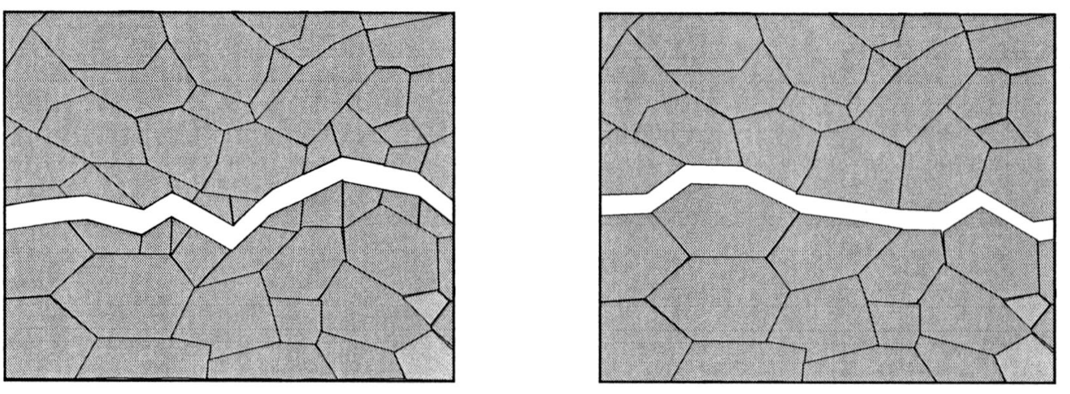


**Fig. 13.6** Transgranular and intergranular failures. Source: Ref 4

and composition of the material. Because the transition occurs over a range of temperatures, it is customary to define a single temperature within the transition range that reflects the behavior of the steel under consideration. Several equally useful definitions are in use, including the 15 ft·lbf (20 J) temperature, the 15 mil (0.4 mm) lateral expansion temperature, and the 50% shear or ductile failure temperature.

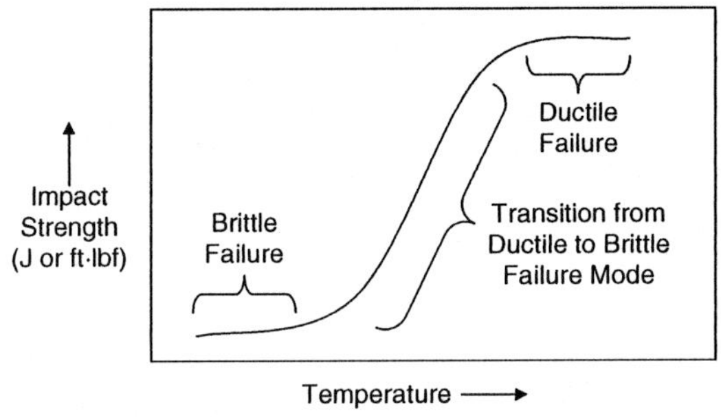


Fig. 13.7 Ductile-brittle transition typical of steels

Several variations on the basic Charpy test specimen are shown in Fig. 13.10. The most common Charpy test uses a V-notched specimen. Three others are the Charpy keyhole test, Charpy U-notch test, and the precracked Charpy test. The Charpy keyhole and U-notch tests are similar to the Charpy V-notch test. Only the geometry of the machined notch is different. However, since the notch is not as severe, the keyhole and U-notch tests generally yield higher values of toughness and lower transition

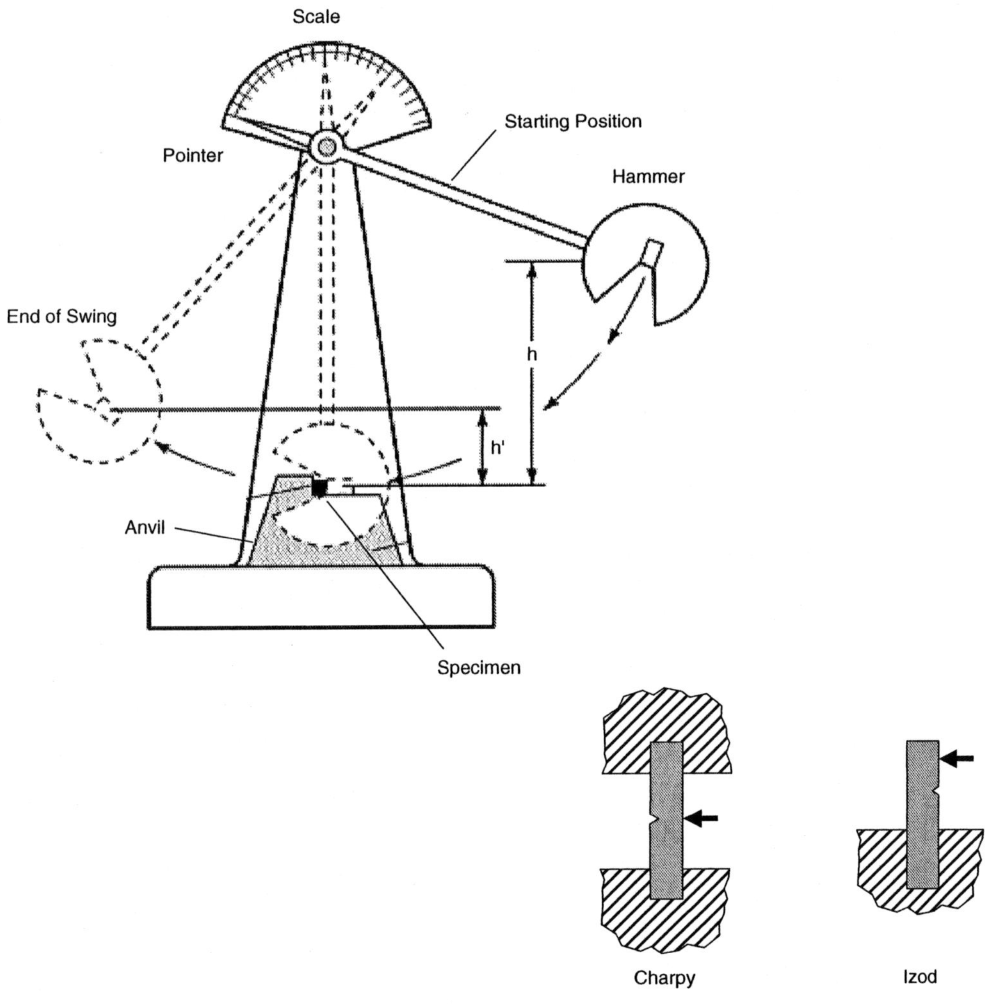


Fig. 13.8 Notched bar impact testing. Source: Ref 5

temperatures. They are more appropriate when testing materials that are less ductile or have a high notch sensitivity.

Applications of the notched bar impact test include comparisons of batch variations in steels, evaluation of material behavior during either intentional or accidental high rates of loading, evaluation of irradiation embrittlement of steels, evaluation of the effects of microstructure and fabrication on toughness, and studies of the fundamental aspects of deformation in bcc materials. It should be noted that the notched bar impact tests do not provide a direct correlation to how an alloy will perform in service. However, they are useful indicators of how the material may behave in service. In spite of the limitations of notched bar impact testing, these tests are relatively simple, quick, and inexpensive methods for testing the dynamic fracture behavior of materials.

One of the major drawbacks of the Charpy test is that it does not provide much information about the fracture process itself. Therefore, instrumented Charpy tests have been developed. A strain gage is mounted on the arm of the pendulum, and a fast triggered data-acquisition

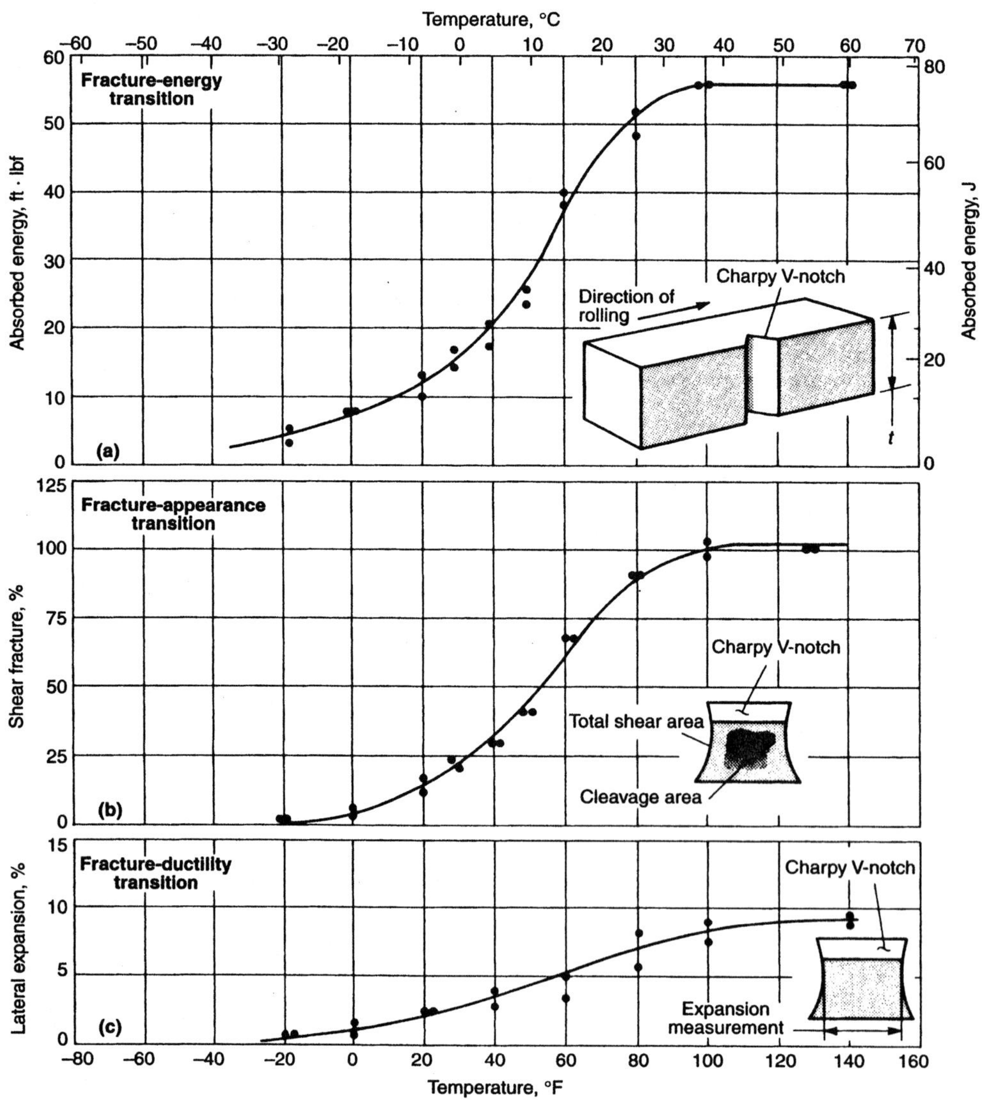


**Fig. 13.9** Charpy impact ductile-to-brittle temperature transition criteria. Source: Ref 6

system records the impact. The data provide load-time profiles (Fig. 13.11) that show the different stages of deformation and fracture: general yield, maximum load, fast fracture, and arrest load after fast fracture. In addition, the actual energy absorbed can be obtained by accounting for the decrease in velocity of the pendulum as it fractures the specimen. The ability to separate the total energy absorbed into its different components makes the instrumented test a more effective analytical tool.

## 13.4 Griffith Theory of Brittle Fracture

The fracture strength of a solid should be a function of the cohesive forces that hold the atoms together. Using this criterion, the theoretical cohesive strength of a brittle elastic solid can be estimated to be in the range of *E*/10, where *E* is the modulus of elasticity. However, the true fracture strengths of real materials are much lower, normally 10 to as much as 1000 times below their theoretical values. In the 1920s, A.A. Griffith, while testing glass rods, observed that the longer the rod, the lower the strength. This led to the idea that the strength variation was due to defects, primarily surface defects, in the glass rods. As the rods became longer, there was a higher probability of encountering a flaw large enough to cause failure. These flaws lower the fracture strength because they amplify the stress at the crack tip. This led to an instability criterion that considered the energy released in a solid at the time a flaw grew catastrophically under an applied stress.

Consider the internal crack shown in the Fig. 13.12 plate. As shown in the adjacent stress profile, the localized stress is high at the crack and then diminishes to the nominal applied stress ($\sigma_o$) at distances far removed from the crack. If it is assumed that the crack is similar to an elliptical hole through a plate and is oriented perpendicular to the applied stress, the maximum stress, $\sigma_m$, at the crack tip is given by:

$$\sigma_m = \sigma_o\left[1 + \frac{2a}{b}\right] = \sigma_o\left[1 + 2\left(\frac{a}{\rho_t}\right)^{1/2}\right] \qquad \text{(Eq 13.1)}$$

where $\rho_t = b^2/a$ is the radius tangential at the tip. Note that for a round hole ($a = b$), the quantity $[1 + 2a/b]$ reduces to 3, which is the same stress-concentration factor that was introduced in section 12.2, "Stress Concentration," in Chapter 12 in this book.

Here, $\sigma_o$ is the magnitude of the applied nominal tensile stress, $\rho_t$ is the radius of the crack tip, and $a$ is the length of a surface crack or one-half the length of an internal crack. For a relatively long microcrack that has a small tip radius (sharp crack), the quantity $(a/\rho_t)^{1/2}$ is large, and Eq 13.1 reduces to:

$$\sigma_m = 2\sigma_o\left(\frac{a}{\rho_t}\right)^{1/2} \qquad \text{(Eq 13.2)}$$

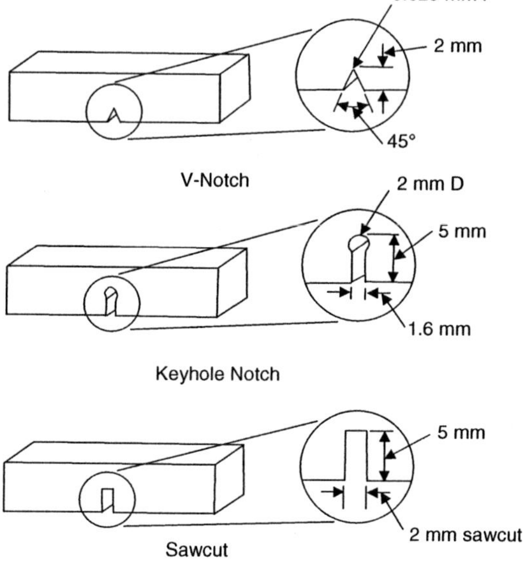


**Fig. 13.10** Charpy impact specimen configurations

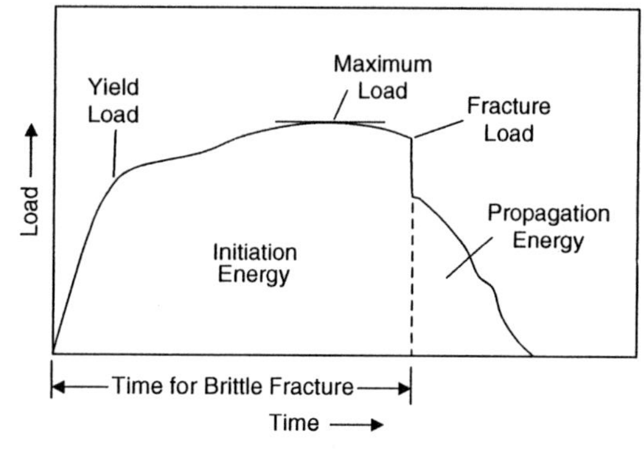


**Fig. 13.11** Load-time curve for instrumented charpy impact test. Source: Ref 7

The ratio $\sigma_m/\sigma_o$ is called the stress-concentration factor, $K_t$:

$$K_t = \frac{\sigma_m}{\sigma_o} = 2\left(\frac{a}{\rho_t}\right)^{1/2} \quad \text{(Eq 13.3)}$$

which is the amount the applied nominal stress is magnified at the crack tip. Note that the stress-concentration factor, $K_t$, increases with increasing crack length, $a$, and decreasing crack radius, $\rho_t$. Therefore, all cracks, if present, should be kept as small as possible. However, the effect of a stress raiser is much more significant in a brittle material than a ductile one. In a ductile material, when the concentrated stress exceeds the yield strength, the material will locally yield at the crack tip to relieve the stress. Since brittle materials do not have the capability to yield, the crack propagates through the material until it reaches a critical length, and the material fails. At the first part of this section, it was stated that the theoretical strength of an elastic solid should be close to the cohesive strength of the material. This has been demonstrated with defect-free single-crystal whiskers that do indeed approach their theoretical strength.

Griffith developed a criterion for the elliptical crack in a plate using an energy balance approach. He equated the elastic strain energy that is stored in the material as it is elastically deformed to the surface energy created when two new free surfaces form during crack propagation. He concluded that the crack will propagate when the elastic energy released as a result of crack propagation exceeds the energy required to propagate the crack.

His analysis showed that the critical stress required to propagate a crack in a brittle material is:

$$\sigma_c = \sqrt{\frac{2E\gamma_s}{\pi a}} \quad \text{(Eq 13.4)}$$

where $E$ is the modulus of elasticity, $\gamma_s$ is the surface energy, and $a$ is one-half the length of an internal crack. Although this equation does not contain the crack tip radius, $\rho_t$, Griffith's analysis was for brittle materials containing sharp cracks.

The Griffith equation is valid only for brittle materials that deform elastically, such as glasses and most ceramics. Since metals deform plastically, Orowan later modified the Griffith equation, replacing $\gamma_s$ with $\gamma_s + \gamma_p$, where $\gamma_p$ is the plastic deformation associated with crack extension. Griffith's equation can then be rewritten as:

$$\sigma_c = \sqrt{\left[\frac{2E(\gamma_s + \gamma_p)}{\pi a}\right]} \quad \text{(Eq 13.5)}$$

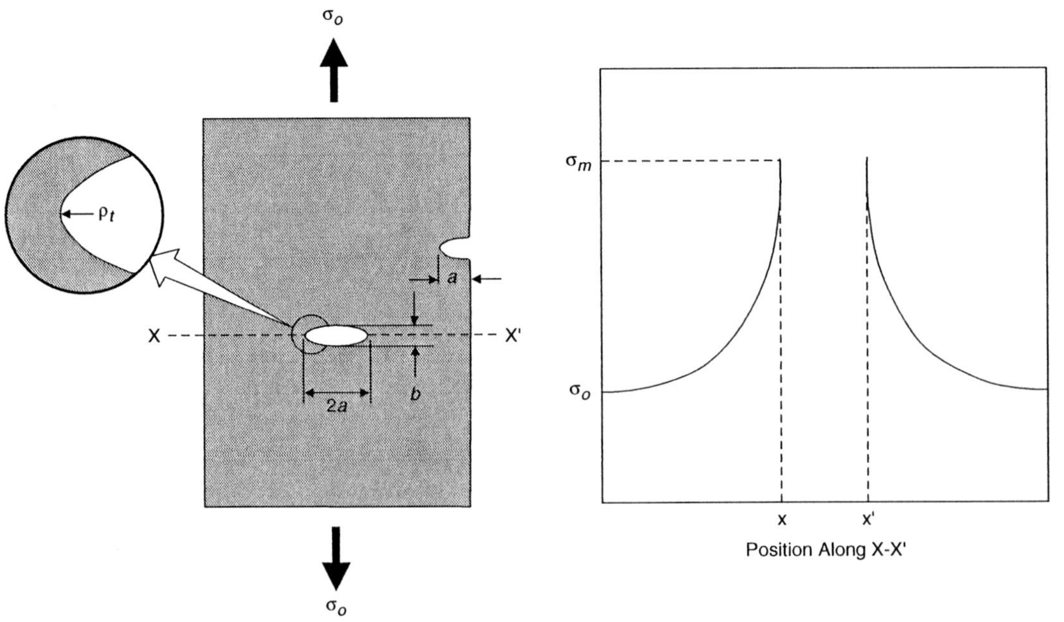


**Fig. 13.12** Analysis of a crack in a wide plate. Source: Ref 8

If the material is highly ductile, then $\gamma_p > \gamma_s$ and:

$$\sigma_c = \sqrt{\frac{2E\gamma_p}{\pi a}} \quad \text{(Eq 13.6)}$$

In the 1950s, G.R. Irwin incorporated $\gamma_s$ and $\gamma_p$ into a single term, $G_c$, known as the critical strain-energy release rate:

$$G_c = 2(\gamma_s + \gamma_p) \quad \text{(Eq 13.7)}$$

The strain-energy release rate or the crack extension force, $G$, is the change in potential energy, $U$, of the system per unit increase in crack area ($G = dU/da$).

The original Griffith criterion can now be written for both brittle and ductile materials as:

$$G_c = \frac{\pi\sigma^2 a}{E} \quad \text{(Eq 13.8)}$$

Therefore, crack extension occurs when $\pi\sigma^2 a/E$ exceeds the value of $G_c$ for the material in question.

## 13.5 Fracture Mechanics

Fracture mechanics is the science of predicting the load-carrying capabilities of structures and components containing cracks. Fracture mechanics is based on a mathematical description of the stress field that surrounds a crack in a loaded body. Two categories of fracture mechanics are linear elastic fracture mechanics (LEFM) and elastic-plastic fracture mechanics (EPFM). LEFM is used when the crack tip is sharp and there is only a small amount of plastic deformation at or near the crack tip. LEFM is used for high-strength metals, such as high-strength steels, titanium, and aluminum alloys. EPFM is used when the crack tip is not sharp and there is some crack tip plasticity (blunting). EPFM is used in the design of materials, such as lower-strength, higher-toughness steels.

LEFM assumes that the component contains a crack or other flaw, the crack is a flat surface in a linear elastic stress field, and the energy released during rapid crack propagation is a basic material property that is not influenced by part size. As shown in Fig. 13.13, there are three modes of crack tip opening displacement. Mode I, tensile opening, is the most important. The other two, modes II and III, are sliding and tearing, respectively. It should be noted that mode I is by far the most important of the three because it almost always turns out to be the limiting or critical case.

For mode I, the stresses acting on an element of material are shown in Fig. 13.14. The tensile ($\sigma_x$ and $\sigma_y$) and shear ($\tau_{xy}$) stresses are functions of both the radial distance, $r$, and the angle, $\theta$, as follows:

$$\sigma_x = \frac{K}{\sqrt{2\pi r}}\cos\frac{\theta}{2}\left(1 - \sin\frac{\theta}{2}\sin\frac{3\theta}{2}\right) \quad \text{(Eq 13.9)}$$

$$\sigma_y = \frac{K}{\sqrt{2\pi r}}\cos\frac{\theta}{2}\left(1 + \sin\frac{\theta}{2}\sin\frac{3\theta}{2}\right) \quad \text{(Eq 13.10)}$$

$$\tau_{xy} = \frac{K}{\sqrt{2\pi r}}\sin\frac{\theta}{2}\cos\frac{\theta}{2}\cos\frac{3\theta}{2} \quad \text{(Eq 13.11)}$$

If the plate is thin, then $\sigma_z = 0$, and a condition of plane stress exists. On the other hand, if the plate is thick and $\varepsilon_z = 0$ and $\sigma_z = \nu(\sigma_x + \sigma_y)$, where $\nu$ is Poisson's ratio, then a condition of plane strain exists. In these equations, the parameter $K$ is known as the stress-intensity factor. Note that the stress-intensity factor, $K$, and the stress-concentration factor, $K_t$, although similar, are not equivalent. In general, the stress-intensity factor is related to the applied stress and crack length by:

$$K = Y\sigma\sqrt{\pi a} \quad \text{(Eq 13.12)}$$

where $Y$, or $Y(a/w)$, is a dimensionless parameter that depends on the specific crack, specimen geometry, and type of loading. The stress-intensity factor, $K$, has the units $\text{MPa}\sqrt{\text{m}}$ ($\text{ksi}\sqrt{\text{in.}}$). The stress-intensity factor, $K$, represents a single parameter that includes both the effect of the applied stress on a sample and the effect of a crack of a given size. The stress-intensity factor can have a simple relation to applied stress and crack length, or the relation can involve complex geometry factors for complex loading, various configurations of structural components, and variations in crack shapes. Examples of several stress-intensity factors for different geometries are shown in Fig. 13.15.

When the applied stress, $\sigma$, exceeds some critical value, $\sigma_c$, the crack propagates and failure occurs. Therefore, there exists some critical value of $K$ that corresponds to the critical value of applied stress, $\sigma_c$. This critical value of $K$ is called the fracture toughness, $K_c$, and is simply:

$$K_c = Y\sigma_c\sqrt{\pi a} \qquad \text{(Eq 13.13)}$$

Fracture toughness, $K_c$, is a measure of a material resistance to brittle fracture when it contains a crack. As the ratio ($a/w$) approaches 0 for the case of a very wide plate with a short crack, the function $Y(a/w)$ approaches 1. The function $Y(a/w)$ for different geometries and different types of cracks can be found in handbooks on fracture mechanics.

For relatively thin specimens (plane stress), the value of $K_c$ will depend on the specimen thickness, $B$, as indicated in Fig. 13.16. As the specimen gets thicker and thicker, the plane-stress condition transitions into plane strain. When the fracture toughness, $K_c$, finally reaches a constant and lower value, this is known as the plane-strain fracture toughness, $K_{Ic}$:

$$K_{Ic} = Y\sigma_c\sqrt{\pi a} \qquad \text{(Eq 13.14)}$$

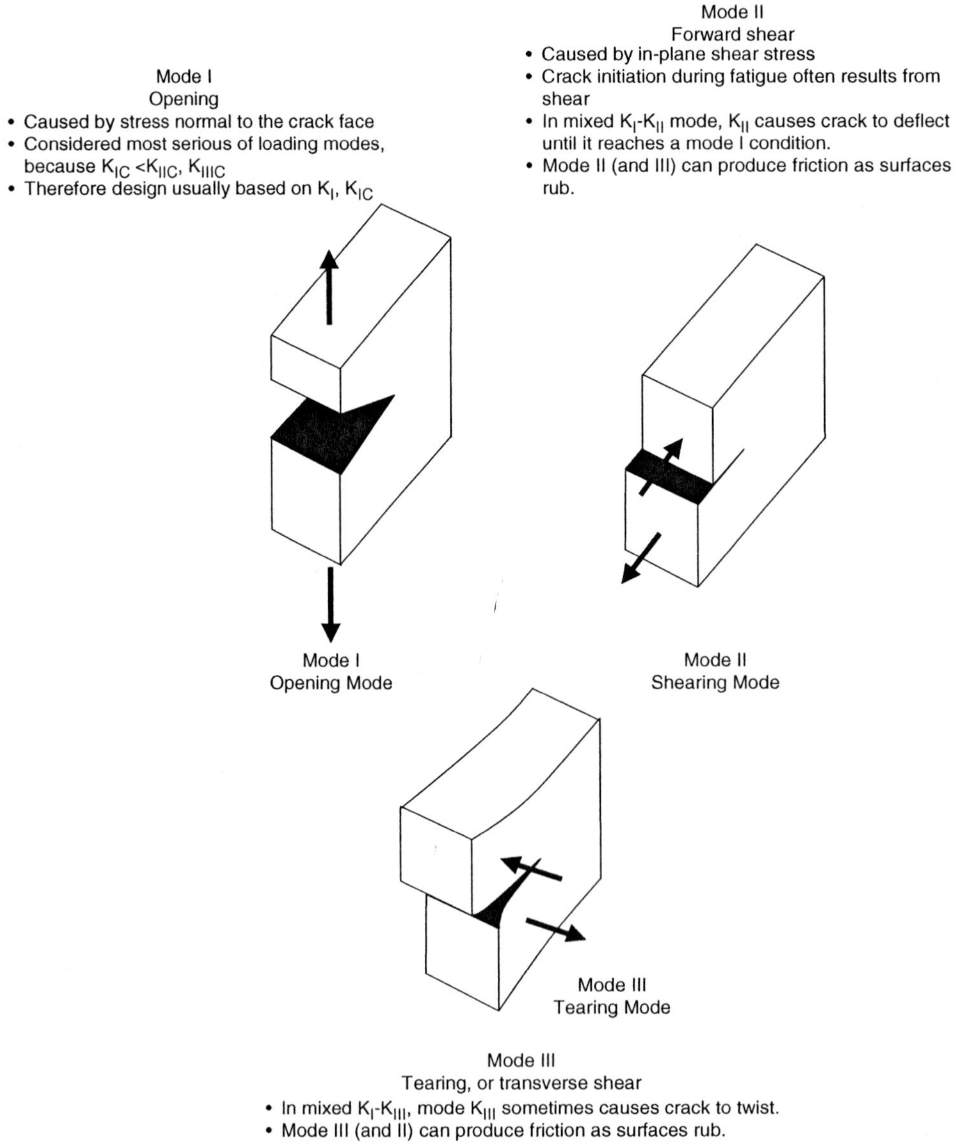


**Fig. 13.13** Crack opening displacement modes

The subscript "I" refers to the mode I crack-opening mode, since it is almost always the limiting value in fracture analysis. Since plastic deformation, which can blunt a sharp crack, is not possible in brittle elastic solid materials, these materials will have a low plane-strain fracture toughness and are subject to sudden and catastrophic failures. More ductile materials, which can plastically deform and blunt cracks, will naturally have much higher values of $K_{Ic}$ and typically will exhibit more graceful failure modes. It should be noted that the plane-strain fracture toughness, $K_{Ic}$, of a material is a material property, much as the yield strength or the elongation. This is not true for the plane-stress condition, where the fracture toughness, $K_c$, varies with the specimen thickness. The plane-strain fracture toughness, $K_{Ic}$, depends on a number of factors, such as temperature, strain rate, material processing, and the final microstructure. The value of $K$ decreases with increasing strain rate, decreasing temperature, and increasing yield strength. It generally increases with finer grain sizes. The value of $K_{Ic}$ determined for a given material is unaffected by specimen dimensions or the type of loading, provided that the specimen dimensions are large enough relative to the plastic zone to ensure plane-strain conditions at the crack tip. Therefore, plane-strain fracture toughness, $K_{Ic}$, is particularly pertinent in materials selection because, unlike other measures of toughness, it is independent of specimen configuration.

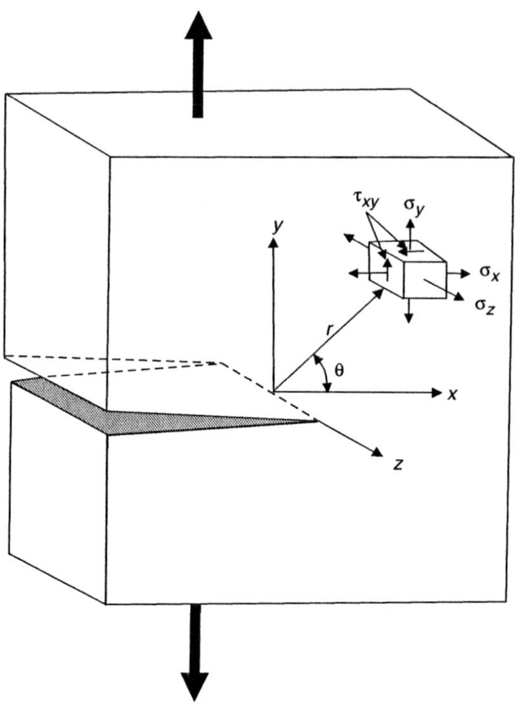


**Fig. 13.14** Stresses at front of crack tip in mode I loading

## 13.6 Plasticity Corrections

When the stress at the crack tip exceeds the yield strength, the material in the vicinity of the crack tip begins to deform plastically and absorb energy. As shown in Fig. 13.17, instead of rising to the tensile stress, $\sigma_c$, predicted by linear elastic analysis, the peak stress is reduced to the yield strength, $\sigma_y$. This explains why metals are so much tougher than brittle materials, such as ceramics, that do not yield.

Irwin proposed that the existence of a plastic zone makes the crack act as if it were longer than its actual size. In other words, as a result of crack tip plasticity, the displacements are longer and the stiffness is lower than predicted by elastic analysis. He proposed that the effective length of the crack is equal to the actual length plus the radius of the plastic zone:

$$a_{eff} = a + r_p \quad \text{(Eq 13.15)}$$

where

$$r_p = \frac{1}{2\pi}\frac{K^2}{\sigma_{ys}} \text{ for plane stress} \quad \text{(Eq 13.16)}$$

and

$$r_p = \frac{1}{6\pi}\frac{K^2}{\sigma_{ys}} \text{ for plane strain} \quad \text{(Eq 13.17)}$$

As a result of the constraint imposed by the triaxial stress state in plane strain, the plastic zone, $r_p$, is less for plane strain than for plane stress. Thus, a thick specimen will have a larger plastic zone or enclave at the edges, due to plane-stress conditions, than in the middle, which is under pure plane-strain conditions (Fig. 13.18).

## 13.7 Plane-Strain Fracture Toughness Testing

ASTM International has developed detailed procedures for determination of the plane-strain fracture toughness ($K_{Ic}$) as described in ASTM E

399. There are a number of acceptable specimen configurations for this test, but the single edge-notched bend specimen and compact tension specimens (Fig. 13.19) are probably the most widely used. The first step is to grow a fatigue starter crack at the base of the machined notch. Such cracks are used to ensure reproducibility and to simulate actual service conditions. The specimen is then loaded to failure at a specified load rate while the load-versus-displacement curve is recorded. The displacement is measured using a strain gage clip gage that is positioned over the mouth of the crack in the specimen. Typical load-versus-displacement curves are shown in Fig. 13.20. For the test to be valid, the load-displacement curve must be virtually linear, and the crack front must be essentially straight.

The data are then analyzed to determine a provisional $K_{Ic}$ value, labeled $K_Q$. This provisional value is determined from a provisional load ($P_Q$) and the crack length. The $P_Q$ value is determined from a secant slope on the load-displacement curve. The $P_Q$ value is determined by drawing the original loading slope of the load-displacement curve. A slope

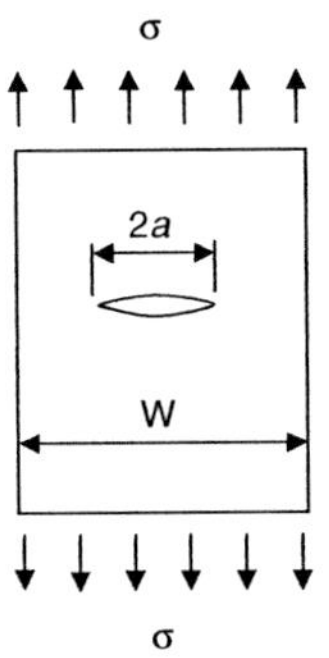


Center Crack: $K_I = Y\sigma\sqrt{\pi a}$

$$Y = 1 + 0.256\left(\frac{a}{W}\right) - 1.152\left(\frac{a}{W}\right)^2 + 12.200\left(\frac{a}{W}\right)^3$$

or $$Y = \sqrt{\sec\left(\frac{\pi a}{W}\right)}$$

or $$Y = \frac{1}{\sqrt{1-\left(\frac{2a}{W}\right)^2}}$$

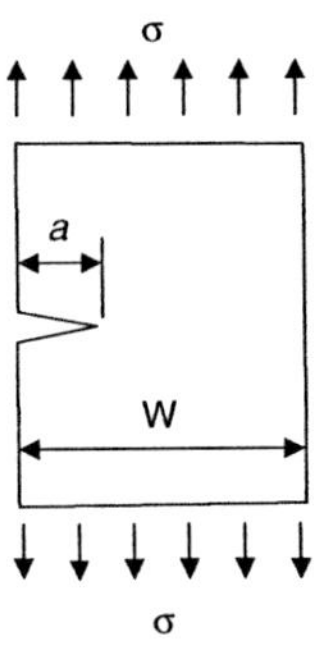


Single Edge Crack: $K_I = Y\sigma\sqrt{\pi a}$

$Y = 1.12$ for small cracks

or $$Y = 1.12 - 0.231\left(\frac{a}{W}\right) + 10.55\left(\frac{a}{W}\right)^2 - 21.72\left(\frac{a}{W}\right)^3 + 30.39\left(\frac{a}{W}\right)^4$$

up to $a/W$=0.6

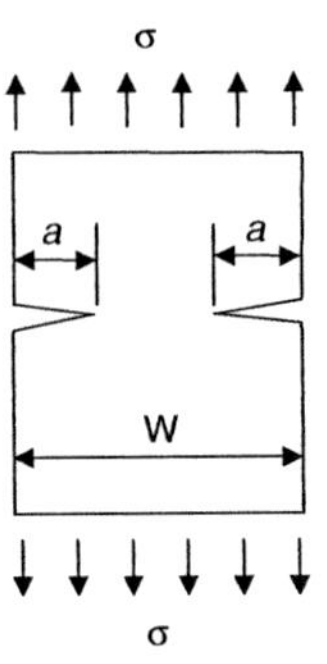


Double Edge Crack:

$Y = 1.12$ for small cracks

or $$Y = \frac{1.222 - 0.561\left(\frac{a}{W}\right) - 0.205\left(\frac{a}{W}\right)^2 + 0.471\left(\frac{a}{W}\right)^3 - 0.190\left(\frac{a}{W}\right)^4}{\sqrt{1-\frac{a}{W}}}$$

**Fig. 13.15** Examples of crack configurations and stress-intensity factors of 5% less than the original slope is then drawn from the origin. For a monotonically increasing load, $P_Q$ is taken where the 5% secant intersects the load-displacement curve, as illustrated as type I in Fig. 13.20. For other curves in which an instability or other maximum load is reached before the 5% secant, the maximum load reached up to and including the possible intersection of the 5% secant is taken as $P_Q$, as illustrated by type II in Fig. 13.20.

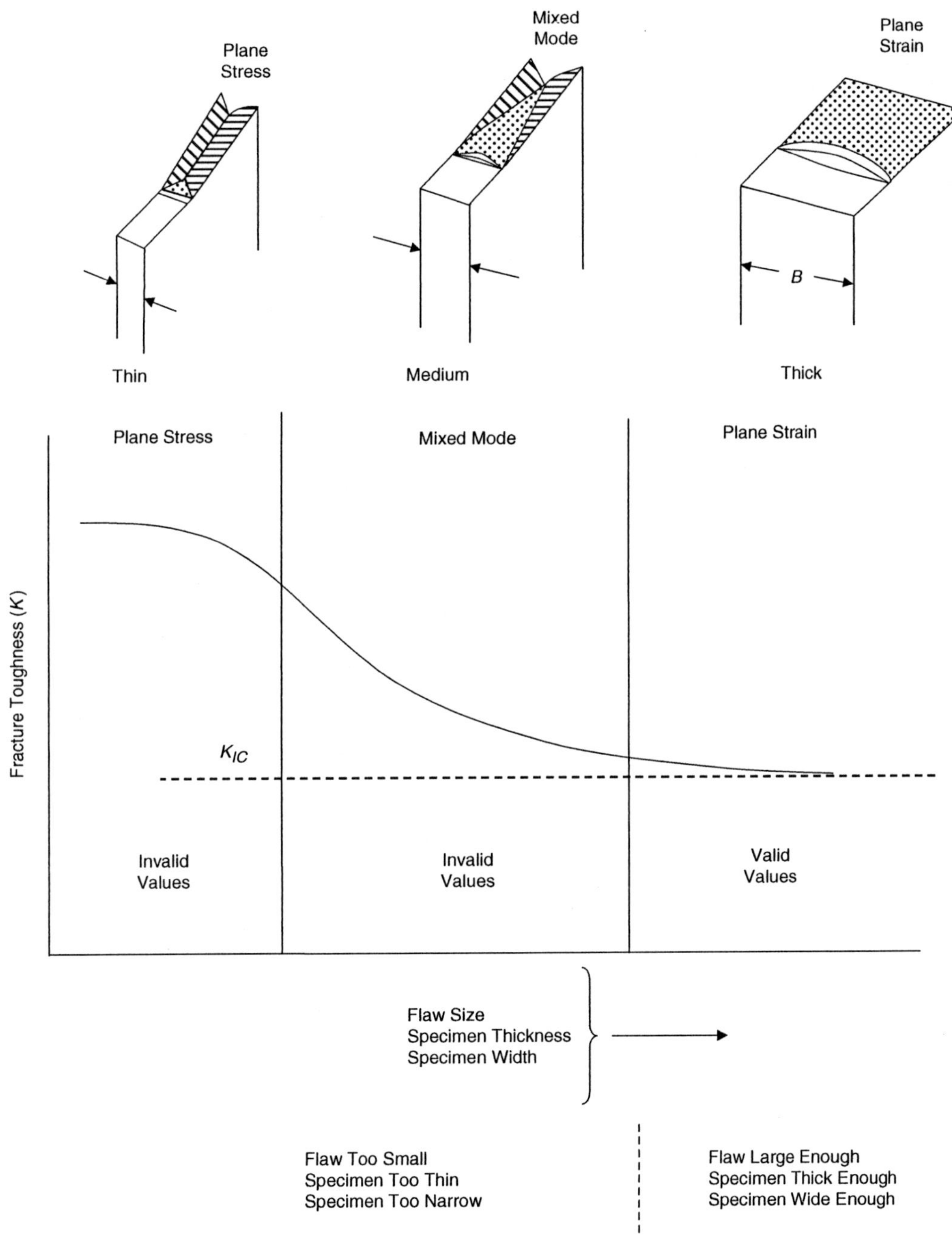


**Fig. 13.16** Variation of $K_c$ with specimen dimensions

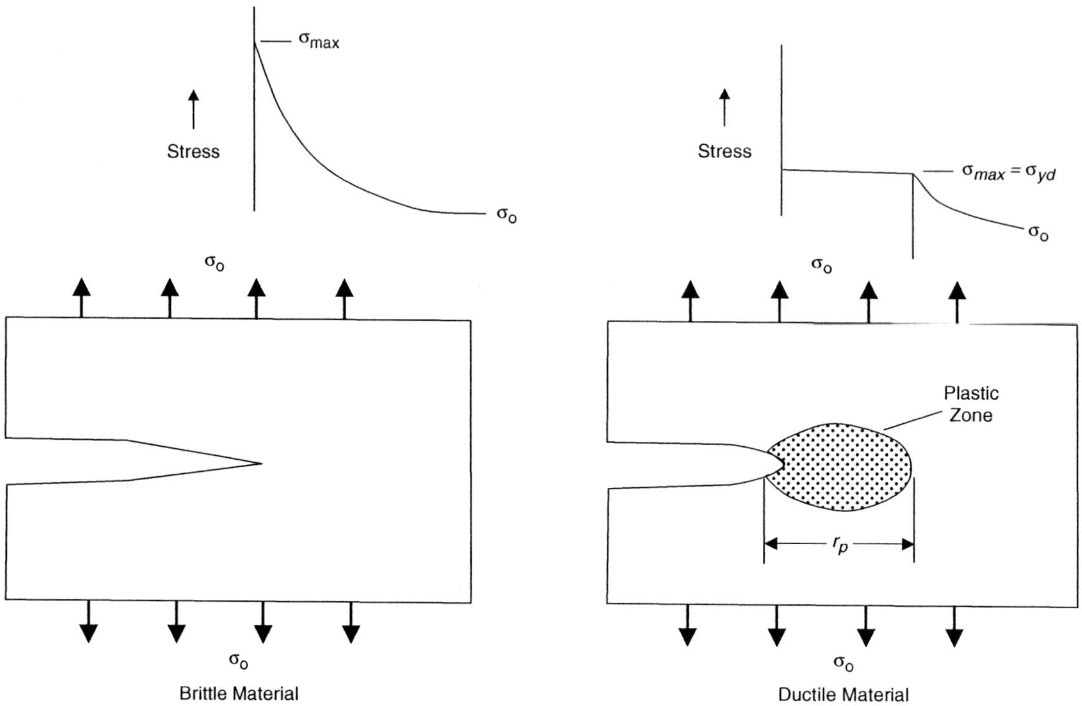


**Fig. 13.17** Comparison of stress at crack tip for brittle and ductile materials

The $P_Q$ value is then used to determine the corresponding $K_Q$ value. This is calculated from the equation:

$$K_{Ic} = \frac{P}{B\sqrt{W}} f\left(\frac{a}{W}\right) \quad \text{(Eq 13.20)}$$

where $K_{Ic}$ is the plane-strain fracture toughness, $P$ is the fracture load, $B$ is the specimen thickness, $W$ is the specimen width, and $f(a/W)$ is the calibration function.

For the calculation of $K$, the crack length value, $a$, is required. This comes from a physical measurement on the fracture surface of the broken specimen. The specimen must be fractured into halves if it did not fracture during the test. The crack length is measured to the tip of the precrack using an averaging formula in the ASTM standard. This value of crack length, normalized by the width ($W$), is used in the calibration function $f(a/W)$ to determine the $K_Q$ value.

The $K_Q$ is a provisional $K$ value that may be the $K_{Ic}$ if it passes several validity requirements. The two major validity requirements are to ensure that crack resistance does not increase significantly with crack growth and that linear elastic loading and plane-strain thickness are achieved. The first of these two requirements is:

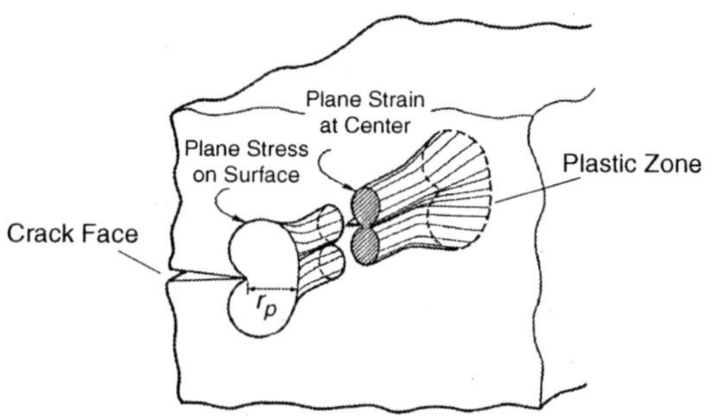


**Fig. 13.18** Plastic zone formation at crack tip

$$\frac{P_{max}}{P_Q} \leq 1.10 \quad \text{(Eq 13.21)}$$

which limits the crack extension behavior to an essentially flat trend and ensures some physical crack extension. The second requirement is:

$$a, B \geq 2.5\left(\frac{K_Q}{\sigma_{ys}}\right) \quad \text{(Eq 13.22)}$$

which guarantees linear elastic loading and plane-strain thickness. $P_{max}$ is the maximum value of load reached during the test, and $\sigma_{ys}$ is the 0.2% offset yield strength. Values of $K_Q$ that pass these validity requirements are labeled as valid $K_{Ic}$ values.

Originally, the technology of fracture mechanics was limited to relatively high-strength materials that could be tested in sizes that met certain requirements for linear elastic displacement. More recently, tests such as the crack tip opening displacement and the *J*-integral tests have extended the use of fracture mechanics to elastic-plastic conditions, which are associated with lower-strength materials and smaller cross-sectional sizes. However, when stresses approach or exceed the yield strength, the elastic stress field surrounding the crack departs from plane strain. With increasing load, slow, stable crack extension (tearing) may accompany the increasing plastic zone size. Onset of rapid fracture occurs when the increase in crack tip stress field, as measured by *K*, equals or exceeds resistance to crack extension due to an increase in plastic zone size, crack tip blunting, and a change from flat to slant fracture. This behavior is most clearly seen in fracture of relatively tough, thin plate and sheet alloys. Unstable fracture under these conditions cannot be described as a material property, since the events leading to rapid fracture are dependent on specimen configuration and size.

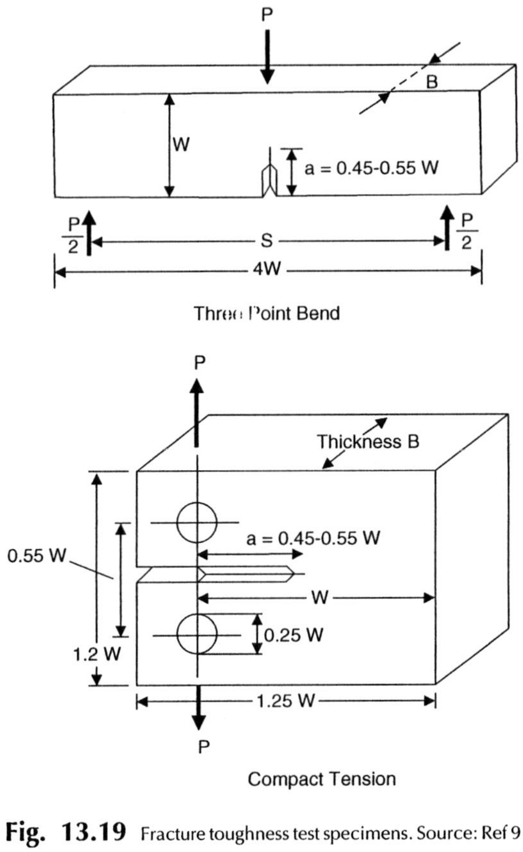


**Fig. 13.19** Fracture toughness test specimens. Source: Ref 9

## 13.8 Fracture Toughness of Engineering Alloys

Processing of engineering alloys usually involves mechanical working at intermediate and high temperatures. Hot working can produce highly directional grain structures (e.g., aluminum alloys) and significant degrees of texturing (e.g., titanium alloys). Fracture toughness is affected by both of these directional effects, and it is customary to report fracture toughness values in terms of the orientation relative to the principal direction of working. In the nomenclature shown in Fig. 13.21, the first letter is the load direction, and the second letter is crack direction. Generally, the fracture toughness will be high for orientations in which the grains are elongated normal to the crack plane (e.g., L-S, L-T) and proportionally lower for crack planes parallel to the elongated grain direction (e.g., S-T, S-L). The data in Table 13.1, which gives typical fracture toughness values for a number of engineering alloys, illustrate this variation in fracture toughness for different orientations. When the crack plane is parallel to the rolling

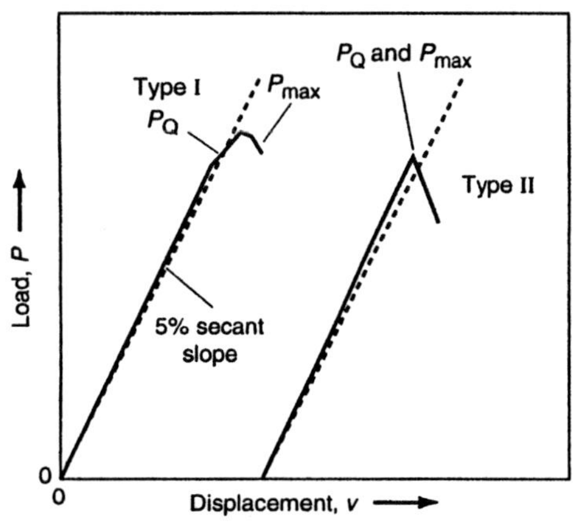


**Fig. 13.20** Typical load-displacement curves for plane-strain fracture toughness test. Source: Ref 10

direction, segregated impurities and intermetallics that lie in these planes represent easy fracture paths, and the fracture toughness is low. However, when the crack plane is perpendicular to these weak planes, crack tip blunting and stress reduction take place, thus effectively toughening the material.

The fracture toughness of high-strength materials is almost always inversely proportional to their yield strength, as shown in the data for steel alloys in Fig. 13.22. The high-fracture-toughness steels normally have more ductile low-carbon martensite and retained metastable austenite as dominant phases in their microstructures, while steels that contain large amounts of ferrite and pearlite have lower fracture toughness. Impurities, inclusions, and coarse microconstituents also lower the toughness. Decreasing grain size produces greater fracture toughness and strength in virtually all metals and is the only method of simultaneously increasing both strength and fracture toughness.

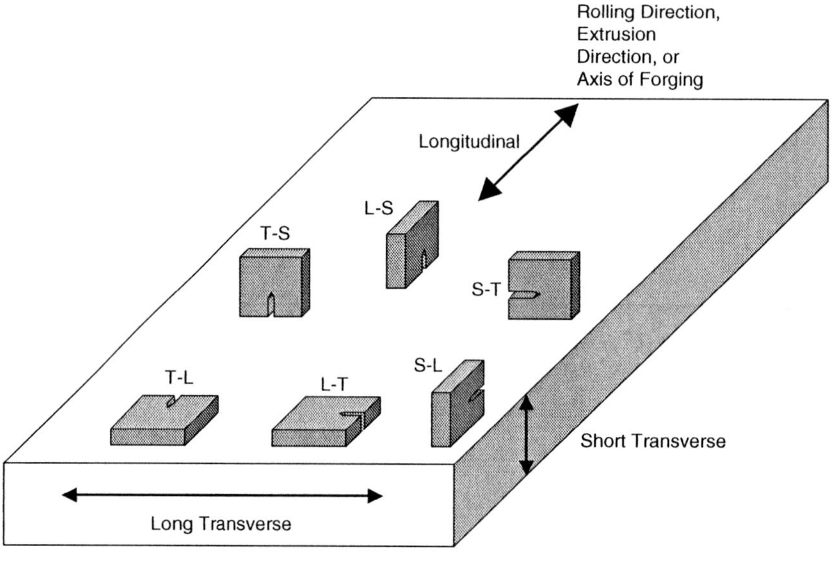


**Fig. 13.21** Fracture toughness orientation nomenclature

**Table 13.1 Selected fracture toughness values for some engineering alloys**

| | Yield strength | | | Temperature | | Plane-strain fracture toughness(b) | |
|---|---|---|---|---|---|---|---|
| Alloy | Mpa | Ksi | Orientation(a) | °C | °F | Mpa$\sqrt{m}$ | ksi$\sqrt{in.}$ |
| 2042-T351 | 385 | 56 | L-T | 29 | 84 | 31 | 28 |
| 2024-T351 | 292 | 42 | S-L | 32 | 90 | 21 | 19 |
| 7075-T651 | 530 | 77 | L-T | 28 | 82 | 32 | 29 |
| 7075-T651 | 446 | 64.5 | S-L | 29 | 84 | 21 | 19 |
| 4140 | 1379 | 200 | L-T | 24 | 75 | 65 | 59 |
| 4140 | 1586 | 230 | L-T | 24 | 75 | 55 | 50 |
| 4340 | 1455 | 211 | L-T | 21 | 70 | 83 | 75.5 |
| D6AC | 1496 | 217 | L-T | 21 | 70 | 102 | 93 |
| HP9-4-.20 | 1282 | 186 | L-T | 26 | 79 | 151 | 137 |
| HP9-4-.20 | 1310 | 190 | T-L | 26 | 79 | 138 | 125.5 |
| 250 Maraging | 1607 | 233 | L-T | 24 | 75 | 86 | 78 |
| 250 Maraging | 1600 | 232 | T-L | 24 | 75 | 86 | 78 |
| Ti-6Al-4V | 889 | 129 | L-T | 24 | 75 | 64 | 58 |
| Ti-6Al-4V | 910 | 132 | T-L | 24 | 75 | 68 | 62 |
| Ti-6Al-4V | 883 | 128 | S-L | 24 | 75 | 75 | 68 |
| Inconel 718 | 1041 | 151 | T-L | 24 | 75 | 87 | 79 |
| Inconel 718 | 986 | 143 | S-L | 24 | 75 | 73 | 66 |

(a) The first letter gives the direction normal to the crack plane while the second letter gives the direction of crack propagation. In the L-T orientation, the crack is normal to the rolling direction (L) and propagates in the transverse direction (T). (b) All numbers are rounded off so as not to imply a greater precision than can be justified by the experimental procedure. Source: Ref 1

A summary of the effects of microstructure on the toughness of steels is shown in Table 13.2.

The fracture resistance of aluminum alloys is strongly sensitive to purity, aging, the presence of intermetallic compounds, thermomechanical treatment, grain size, and orientation or texture. A list of variables and the nature of their effect on the fracture toughness of aluminum alloys is given in Table 13.3. The *7xxx* alloys have the highest combination of strength and toughness of any family of aluminum alloys (Fig. 13.23). In the *7xxx* alloys, the highest strength is associated with the T6 peak-aged temper. However, higher fracture toughness with only a minimal reduction in strength can be obtained by overaging to one of the T7 tempers. In addition, the overaged tempers provide better resistance to stress-corrosion cracking (SCC) and exfoliation corrosion. The *2xxx* series are used in both the naturally aged and artificially aged conditions. Naturally aged *2xxx* alloys (i.e., the T3 and T4 tempers) provide good combinations of toughness and strength. Artificial aging to the precipitation-hardened T8 tempers produces higher strength with some reduction in toughness but offers greater stability of mechanical properties at high temperatures and higher resistance to exfoliation and SCC.

One of the primary contributors to improving the fracture toughness of high-strength aluminum alloys has been to improve the purity of the alloys, primarily by reducing the iron and silicon contents. This minimizes the formation of relatively large, insoluble constituents (>1 μm). Constituent particles participate in the fracture process through void formation at particle/matrix interfaces or by fracturing during primary processing. A comparison of the fracture toughness of two conventional alloys (2024-T851 and

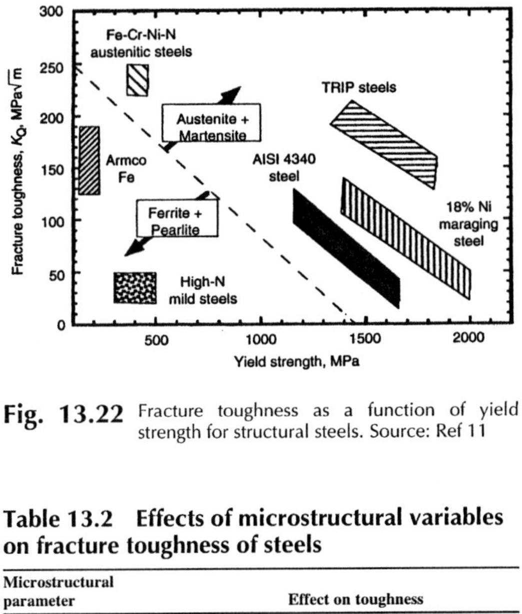


**Fig. 13.22** Fracture toughness as a function of yield strength for structural steels. Source: Ref 11

**Table 13.2 Effects of microstructural variables on fracture toughness of steels**

| Microstructural parameter | Effect on toughness |
|---|---|
| Grain size | Increase in grain size increases $K_{Ic}$ in austenitic and ferritic steels |
| Unalloyed retained austenite | Marginal increase in $K_{Ic}$ by crack blunting |
| Alloyed retained austenite | Significant increase in $K_{Ic}$ by transformation-induced toughening |
| Interlath and intralath carbides | Decrease $K_{Ic}$ by increasing the tendency to cleave |
| Impurities (P, S, As, Sn) | Decrease $K_{Ic}$ by temper embrittlement |
| Sulfide inclusions and coarse carbides | Decrease $K_{Ic}$ by promoting crack or void nucleation |
| High carbon content (>0.25%) | Decrease $K_{Ic}$ by easily nucleating cleavage |
| Twinned martensite | Decrease $K_{Ic}$ due to brittleness |
| Martensite content in quenched steels | Increase $K_{Ic}$ |
| Ferrite and pearlite in quenched steels | Decrease $K_{Ic}$ of martensitic steels |

Source: Ref 11

**Table 13.3 Effects of microstructural variables on fracture toughness of aluminum alloys**

| Variable | Effect on fracture toughness |
|---|---|
| Quench rate | Decrease in $K_{Ic}$ at low quench rates |
| Impurities (Fe, Si, Mn, Cr) | Decrease in $K_{Ic}$ with high levels of these elements |
| Grain size | Decrease in $K_{Ic}$ at large grain sizes due to coarse grain-boundary precipitation |
| Grain-boundary precipitates | Increase in size and area fraction decrease $K_{Ic}$ |
| Underaging | Increases toughness |
| Peak aging | Decreases fracture toughness |
| Overaging | Increases fracture toughness |

Source: Ref 11

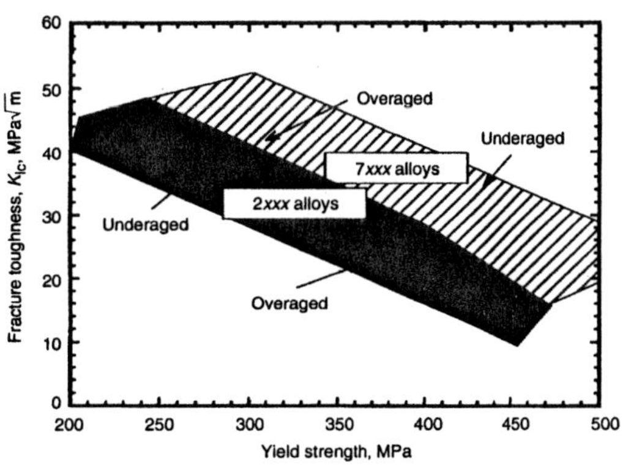


**Fig. 13.23** Effect of alloy type and aging on strength fracture toughness of aluminum alloys. Source: Ref 11

7075-T7351) with their lower-impurity counterparts (2124-T851 and 7475-T7351) in Fig. 13.24 illustrates the marked improvement in toughness for the higher-purity alloys.

Titanium alloys are used in aerospace applications because of their good combination of specific strength, ductility, and fracture toughness. As in steels and aluminum alloys, this combination is achieved by control of microstructure. Among the two phases (α and β) in titanium alloys, β is more ductile and is preferable in increasing the fracture toughness of titanium alloys. The three broad classes of titanium alloys are near-α, α + β, and β alloys, grouped according to the levels of α or β stabilizing elements. Typically, the β content by volume in near-α alloys is less than 10%, between 10 and 25% in α + β alloys, and greater than 25% in β alloys. The fracture toughness/strength relationships for different titanium alloys are shown in Fig. 13.25. The β alloys possess the highest combination of strength and toughness due to a large volume fraction of β phase and fine aged-α precipitates. The dominant variables that influence fracture toughness in titanium alloys are the interstitial elements, grain size, microstructural morphology, and relative proportions of α and β phases. These variables and their effects on fracture toughness are listed in Table 13.4.

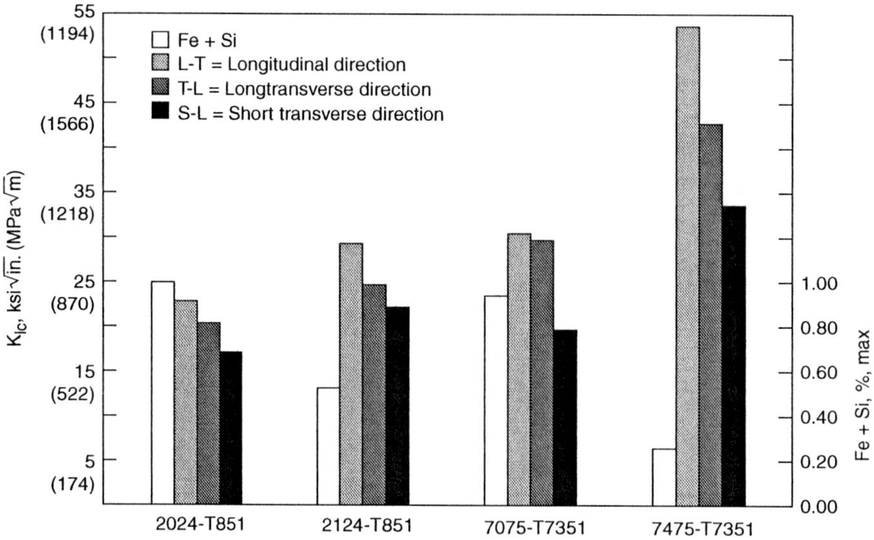


**Fig. 13.24** Effect of alloy purity on fracture toughness of aluminum alloys. Source: Ref 12

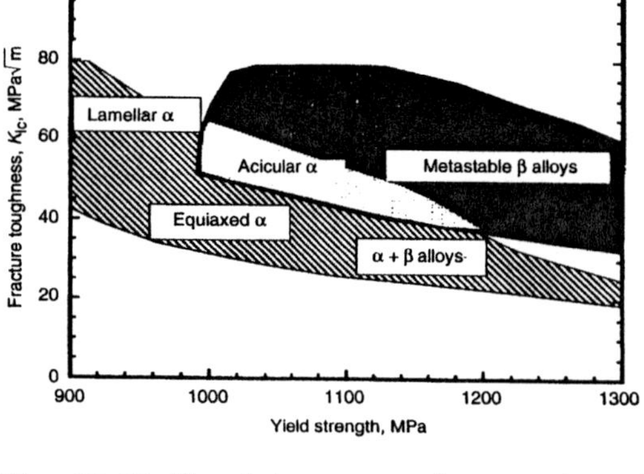


**Fig. 13.25** Effect of microstructure on fracture toughness of titanium alloys. Source: Ref 11

**Table 13.4 Effects of microstructural variables on fracture toughness of titanium alloys**

| Variable | Effect on fracture toughness |
|---|---|
| Interstitials (O, H, C, N) | Decrease in $K_{Ic}$ |
| Grain size | Increase in grain size decreases $K_{Ic}$ |
| Lamellar colony size | Increase in colony size increases $K_{Ic}$ |
| β phase | Increases in β volume fraction and continuity increase $K_{Ic}$ |
| Grain-boundary α phase | Increases in thickness and continuity increase $K_{Ic}$ |
| Shape of α phase | Increase in aspect ratio of α phase increases $K_{Ic}$ |
| Orientation | Crack oriented for easy cleavage along basal planes gives low $K_{Ic}$ |

Source: Ref 11

## ACKNOWLEDGMENTS

Sections of this chapter were adapted from "An Introduction to Fracture Mechanics" by S.D. Antolovich and B.F. Antolovich, and "Fracture Resistance of Structural Alloys" by K.S. Ravichandran and A.K. Vasudevan, both in *Fatigue and Fracture*, Volume 19, *ASM Handbook*, ASM International, 1996.

## REFERENCES

1. S.D. Antolovich and B.F. Antolovich, An Introduction to Fracture Mechanics, *Fatigue and Fracture*, Vol 19, *ASM Handbook*, ASM International, 1996
2. W.T. Becker and S. Lampman, Fracture Appearance and Mechanisms of Deformation and Fracture, *Failure Analysis and Prevention*, Vol 11, *ASM Handbook*, ASM International, 2002
3. C.Q. Bowles, Fracture and Structure, *Fatigue and Fracture*, Vol 19, *ASM Handbook*, ASM International, 1996
4. F.C. Campbell, *Manufacturing Technology for Aerospace Structural Materials*, Elsevier Scientific, 2006
5. H.W. Hayden, W.G. Moffatt, and J. Wulff, *The Structure and Properties of Materials*, Vol III, John Wiley & Sons, 1965
6. Fracture Toughness and Fracture Mechanics, *Mechanical Testing and Evaluation*, Vol 8, *ASM Handbook*, ASM International, 2000
7. G.E. Dieter, *Mechanical Metallurgy*, 3rd ed., McGraw-Hill Book Co., 1986
8. W.D. Callister, *Fundamentals of Materials Science and Engineering*, 5th ed., John Wiley & Sons, Inc., 2001
9. M.A. Meyers and K.K. Chawla, *Mechanical Metallurgy—Principles and Applications*, Prentice-Hall Inc., 1984
10. J.D. Landes, Fracture Toughness Testing, *Fatigue and Fracture*, Vol 19, *ASM Handbook*, ASM International, 1996
11. K.S. Ravichandran and A.K. Vasudevan, Fracture Resistance of Structural Alloys, *Fatigue and Fracture*, Vol 19, *ASM Handbook*, ASM International, 1996
12. R.J. Bucci, G. Nordmark, and E.A. Starke, Selecting Aluminum Alloys to Resist Failure by Fracture Mechanisms, *Fatigue and Fracture*, Vol 19, *ASM Handbook*, ASM International, 1996

## SELECTED REFERENCES

- T.L. Anderson, *Fracture Mechanics: Fundamentals and Applications*, 3rd ed., Taylor & Francis, 2005
- T.H. Courtney, *Mechanical Behavior of Materials*, 2nd ed., McGraw-Hill Book Co., 2000
- *Fractography*, Vol 12, *ASM Handbook*, ASM International, 1987
- R.W. Hertzberg, *Deformation and Fracture Mechanics of Engineering Materials*, 3rd ed., John Wiley & Sons, Inc., 1983

# CHAPTER 14

# Fatigue

FATIGUE FAILURES OCCUR due to the application of fluctuating stresses that are much lower than the stress required to cause failure during a single application of stress. It has been estimated that fatigue contributes to approximately 90% of all mechanical service failures. Fatigue is a problem that can affect any part or component that moves. Automobiles on roads, aircraft wings and fuselages, ships at sea, nuclear reactors, jet engines, and land-based turbines are all subject to fatigue failures. Fatigue was initially recognized as a problem in the early 1800s when investigators in Europe observed that bridge and railroad components were cracking when subjected to repeated loading. As the century progressed and the use of metals expanded with the increasing use of machines, more and more failures of components subjected to repeated loads were recorded. Today, structural fatigue has assumed an even greater importance as a result of the ever-increasing use of high-strength materials and the desire for higher performance from these materials.

## 14.1 Stress Cycles

There are three basic factors necessary to cause fatigue: (1) a maximum tensile stress of sufficiently high value, (2) a large enough variation or fluctuation in the applied stress, and (3) a sufficiently large number of cycles of the applied stress. There are many types of fluctuating stresses. Several of the more common types encountered are shown in Fig. 14.1. A fully reversed stress cycle, shown in the top graph of Fig. 14.1, where the maximum and minimum stresses are equal, is commonly used in testing. This is the type of stress produced, for example, by the R.R. Moore rotating-beam fatigue machine shown in Fig. 14.2, which is similar to what a shaft may encounter during service. Since this was the original type of machine used to generate fatigue data, quite a bit of the data in the literature is for fully reversed bending with no mean stress applied on top of it. Another common stress cycle is the repeated stress cycle, in which there is a mean stress ($\sigma_m$) applied on top of the maximum and minimum

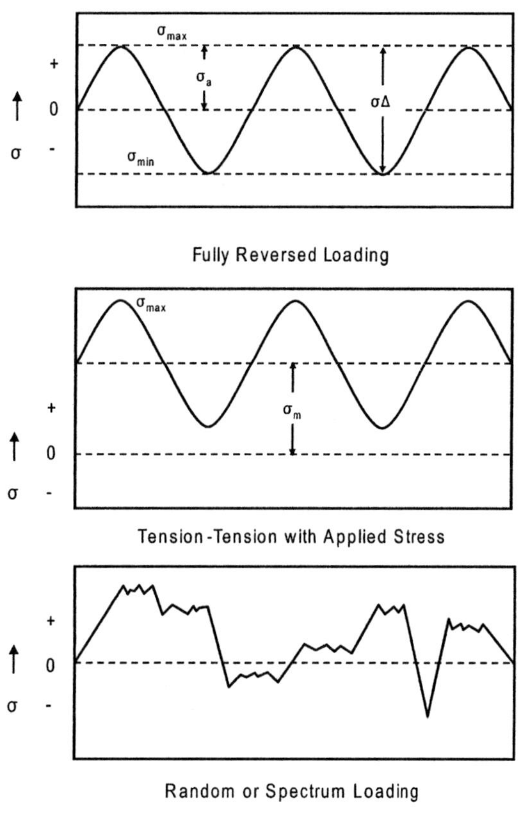


Cyclic Stress Range: $\sigma\Delta = \sigma_{max} - \sigma_{min}$

Cyclic Stress Amplitude: $\sigma_a = \dfrac{\sigma_{max} - \sigma_{min}}{2}$

Mean Stress: $\sigma_m = \dfrac{\sigma_{max} + \sigma_{min}}{2}$

Stress Ratio: $R = \dfrac{\sigma_{min}}{\sigma_{max}}$

**Fig. 14.1** Typical loading cycles

stresses. The middle graph in Fig. 14.1 shows the condition where both stresses (cyclic and applied) are tensile (greater than zero), but it is also possible to test with both stresses in compression. In addition, the maximum and minimum stresses in the cycle do not necessarily have to be equal in value. The last type of loading cycle is the random or irregular stress cycle, in which the part is subjected to random loads during service, as shown in the bottom graph of Fig. 14.1. Although a majority of the fatigue data in the literature is for fully reversed bending, there are also axial test machines (Fig. 14.3) that are capable of tension and compression loading in both the high- and low-cycle fatigue ranges. These modern test frames are closed-loop servohydraulically controlled and can be programmed with almost any desired fatigue spectrum.

A fluctuating stress is made up of two components: a mean or steady stress, $\sigma_m$, and an alternating or variable stress, $\sigma_a$. The stress range, $\sigma_r$, is the difference between the maximum and minimum stress in a cycle:

$$\sigma_r = \sigma_{max} - \sigma_{min} \quad \text{(Eq 14.1)}$$

The alternating stress is one-half the stress range:

$$\sigma_a = \frac{\sigma_r}{2} = \frac{\sigma_{max} - \sigma_{min}}{2} \quad \text{(Eq 14.2)}$$

The mean stress is the algebraic average of the maximum and minimum stress in the cycle:

$$\sigma_m = \frac{\sigma_{max} + \sigma_{min}}{2} \quad \text{(Eq 14.3)}$$

Two ratios frequently used in presenting fatigue data are:

$$\text{Stress ratio} \quad R = \frac{\sigma_{min}}{\sigma_{max}} \quad \text{(Eq 14.4)}$$

$$\text{Amplitude ratio} \quad A = \frac{\sigma_a}{\sigma_m} = \frac{1-R}{1+R} \quad \text{(Eq 14.5)}$$

## 14.2 High-Cycle Fatigue

High-cycle fatigue involves a large number of cycles ($N > 10^5$ cycles) and an elastically applied stress. High-cycle fatigue tests are usually carried out for $10^7$ cycles and sometimes $5 \times 10^8$ cycles for nonferrous metals. Although the applied stress is low enough to be elastic, plastic deformation can take place at the crack tip. High-cycle fatigue data are usually presented as a plot of stress, *S*, versus the number of cycles to failure, *N*. A log scale is used for the number of cycles. The value of stress, *S*, can be the maximum stress, $\sigma_{max}$, the minimum stress, $\sigma_{min}$, or the stress amplitude, $\sigma_a$. The *S*-*N* relationship is usually determined for a specified value of the mean stress, $\sigma_m$, or one of the two ratios, *R* or *A*.

The fatigue life is the number of cycles to failure at a specified stress level, while the fatigue strength (also referred to as the endurance limit) is the stress below which failure does not occur. As the applied stress level is decreased, the number of cycles to failure increases. Normally, the fatigue strength increases as the static tensile strength increases. For example,

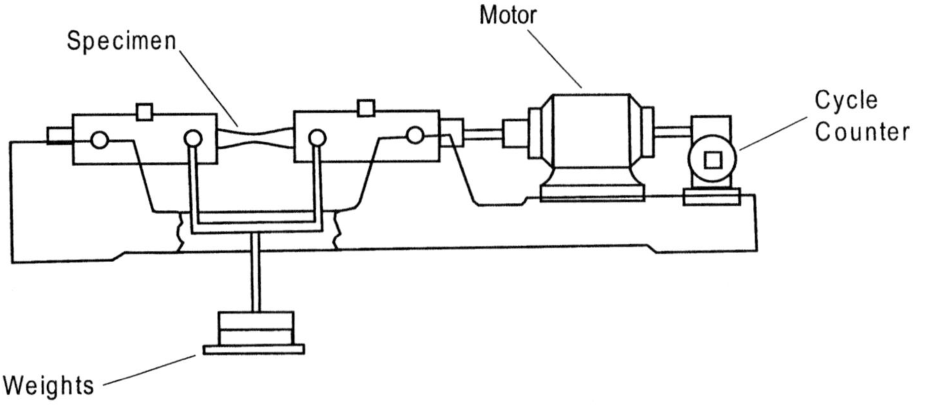


**Fig. 14.2** Schematic of R.R. Moore reversed-bending fatigue machine. Source: Ref 1

high-strength steels heat treated to over 1400 MPa (200 ksi) yield strengths have much higher fatigue strengths than aluminum alloys with 480 MPa (70 ksi) yield strengths. A comparison of the *S-N* curves for steel and aluminum is shown in Fig. 14.4. Note that steel not only has a higher fatigue strength than aluminum, but it also has an endurance limit. Below a certain stress level, the steel alloy will never fail due to cyclic loading alone. On the other hand, aluminum does not have a true endurance limit. It will always fail if tested to a sufficient number of cycles. Therefore, the fatigue strength of aluminum is usually reported as the stress level it can survive at a large total number of cycles, usually $5 \times 10^8$ cycles. It should be noted that there is a considerable amount of scatter in fatigue test results. It is therefore important to test a sufficient number of specimens to obtain statistically meaningful results.

For a large number of steels, there is a direct correlation between tensile strength and fatigue strength; higher-tensile-strength steels have higher endurance limits. The endurance limit is normally in the range of 0.35 to 0.60 of the tensile strength. This relationship holds up to a hardness of approximately 40 HRC (~1200 MPa, or 180 ksi tensile strength), and then the scatter becomes too great to be reliable (Fig. 14.5). This does not necessarily mean it is wise to use as high a strength steel as possible to maximize fatigue life because, as the tensile strength increases, the fracture toughness decreases and the environmental sensitivity increases. The endurance limit of high-strength steels is extremely sensitive to surface condition, residual-stress state, and the presence of inclusions that act as stress concentrations.

Fatigue cracking can occur quite early in the service life of the component by the formation of a small crack, generally at some point on the external surface. The crack then propagates slowly through the material in a direction roughly perpendicular to the main tensile axis (Fig. 14.6). Ultimately, the cross-sectional area of the member is reduced to the point that it can no longer carry the load, and the member fails in tension. The fracture surface of a fatigued high-strength part is shown in Fig. 14.7. The portion of the fracture surface due to fatigue crack growth and the portion finally cracked due to overload are clearly evident.

As previously mentioned, much of the fatigue data in the literature has been determined for completely reversed bending with $\sigma_m = 0$. However, the effects of mean stress are important,

**Fig. 14.3** Modern fatigue test frame

and an increase in mean stress will always cause a reduction in the fatigue life (Fig. 14.8). A number of mathematical models have been developed that allow the effects of mean stress on stress amplitude to be predicted from fully reversed-bending data. Goodman developed a linear model, while Gerber used a parabolic model (Fig. 14.9). Test data for ductile metals usually fall closer to the Gerber parabolic curve; however, because of the scatter in fatigue data and the fact that notched data fall closer to the Goodman line, the more conservative Goodman relationship is often used in practice. If the component design is based on yield rather than ultimate strength, as most are, then the even more conservative Soderberg relationship can be used. Mathematically, the three relationships can be expressed by:

$$\sigma_a = \sigma_e \left[1 - \left(\frac{\sigma_m}{\sigma_u}\right)^x\right] \quad \text{(Eq 14.6)}$$

where $x = 1$ for the Goodman line, $x = 2$ for the Gerber curve, $\sigma_u = \sigma_y$ for the Soderberg curve, and $\sigma_e$ is the fatigue limit for completely reversed bending.

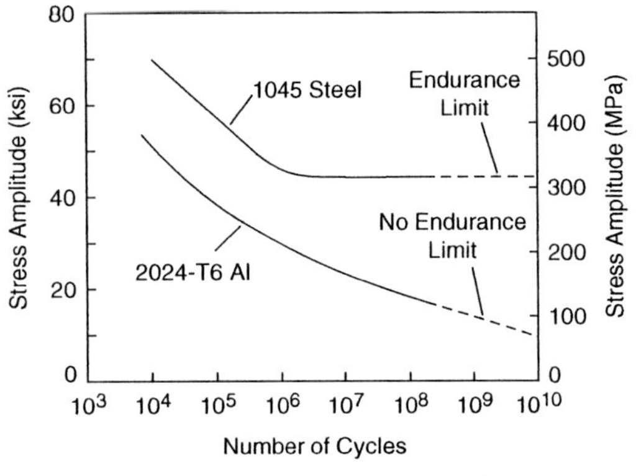


**Fig. 14.4** Comparison of steel and aluminum fatigue behavior. Source: Ref 2

## 14.3 Low-Cycle Fatigue

During cyclic loading within the elastic regime, stress and strain are directly related through the elastic modulus. However, for cyclic

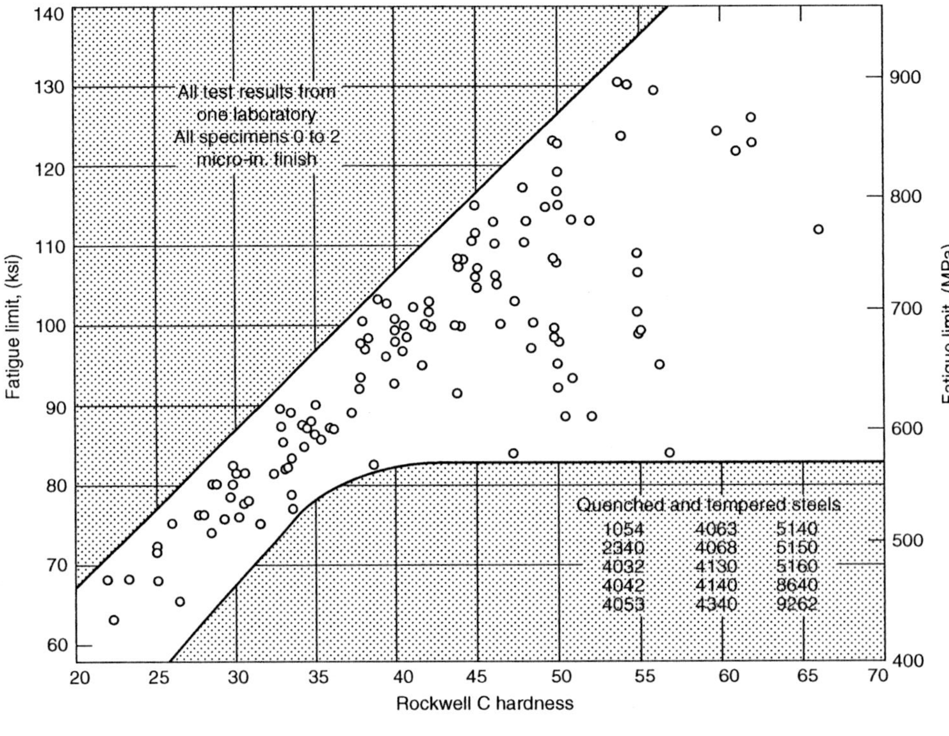


**Fig. 14.5** Endurance limit versus hardness for steels. Source: Ref 3

loading that produces plastic strains, the responses are more complex and form a hysteresis loop (Fig. 14.10). From point *O* up to point *A*, the component is in tension. On unloading, the strain response of the specimen follows the curve from *A* to *D*. At *D*, the component is under no stress. As the component is subjected to compressive stress, the strain response follows the curve from *D* to *B*. Releasing the compressive stress from *B* and reapplying tensile stress, the component stress-strain condition returns to point *A* along the curve defined by *B*, *C*, and *A*. Points *A* and *B* represent the cyclic stress and strain limits. The total strain, $\Delta\varepsilon$, consists of both elastic and plastic components:

$$\Delta\varepsilon = \Delta\varepsilon_e + \Delta\varepsilon_p \qquad \text{(Eq 14.7)}$$

where $\Delta\varepsilon_e$ is the elastic strain and $\Delta\varepsilon_e = \Delta\sigma/E$, and $\Delta\varepsilon_p$ is the plastic strain and $\Delta\varepsilon_p$ is the width of the loop at its center, that is, the distance *CD* in Fig. 14.10. The area of the hysteresis loop is equal to the work done or the energy loss per cycle.

In cyclic strain-controlled fatigue, the strain amplitude is held constant during cycling. Since plastic deformation is not completely

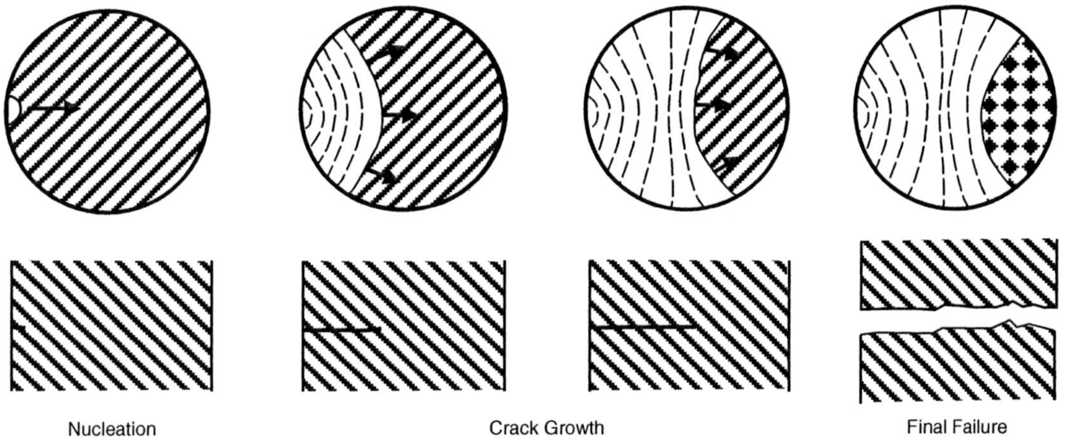


**Fig. 14.6** Typical propagation of a fatigue crack. Source: Ref 4

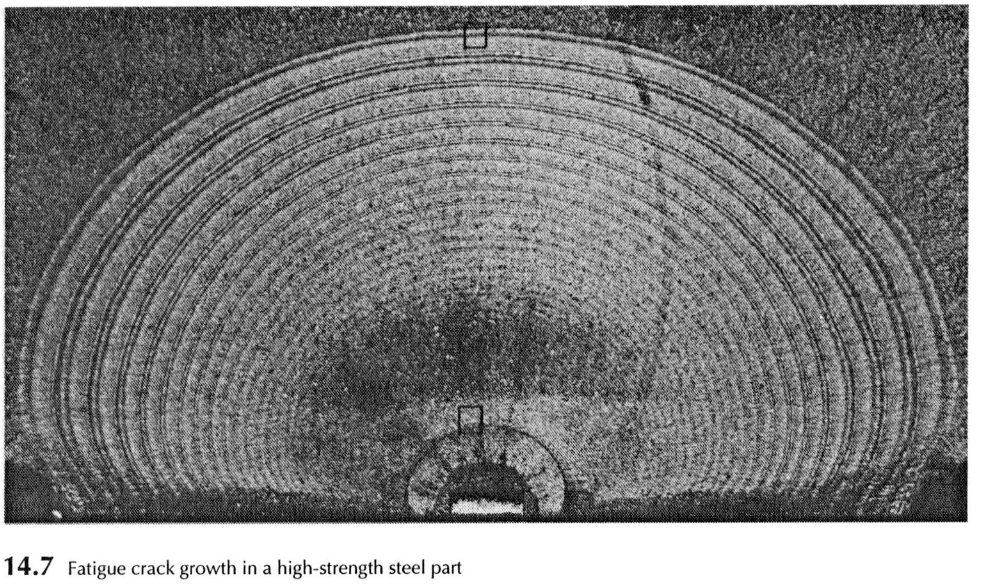

**Fig. 14.7** Fatigue crack growth in a high-strength steel part

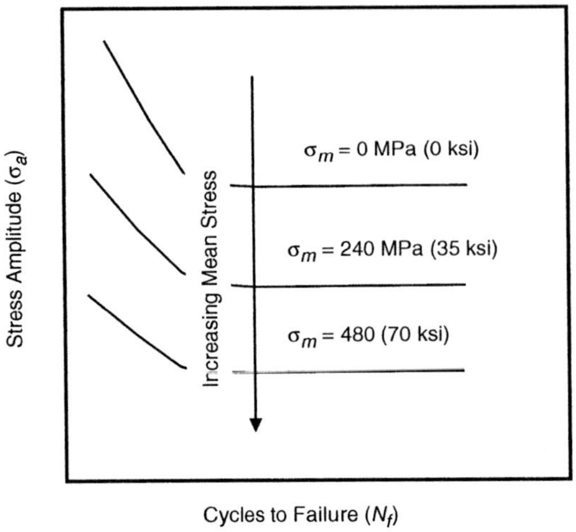


**Fig. 14.8** Effect of mean stress on fatigue life

reversible, the stress-strain response during cycling can change, largely depending on the initial condition of the metal. The metal can either undergo cyclic strain hardening, cyclic strain softening, or remain stable. Cyclic strain hardening and softening are illustrated schematically in Fig. 14.11. In both cases, the hysteresis loops change with successive cycles. Cyclic hardening leads to increasing peak strain with increasing cycles, while cyclic softening results in decreasing strain levels with increasing cycles. In general, strong metals tend to cyclically soften, and low-strength metals tend to cyclically harden. However, the hysteresis loop tends to stabilize after a few hundred cycles, when the material attains a stable condition for the imposed strain level.

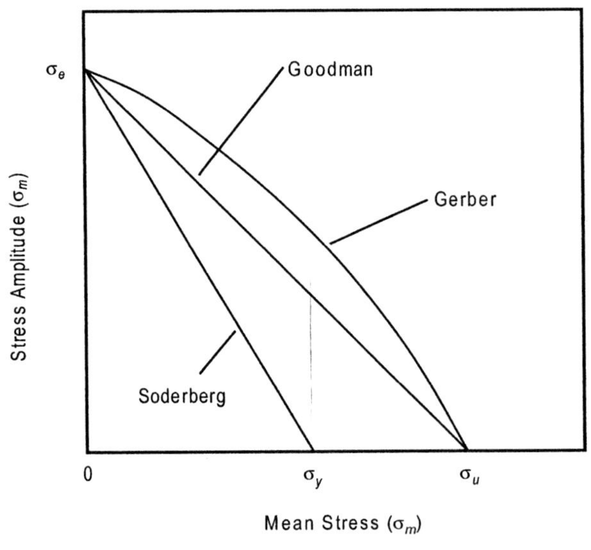


Goodman $$\sigma_a = \sigma_e\left[1-\left(\frac{\sigma_m}{\sigma_u}\right)\right]$$

Soderberg $$\sigma_a = \sigma_e\left[1-\left(\frac{\sigma_m}{\sigma_y}\right)\right]$$

Gerber $$\sigma_a = \sigma_e\left[1-\left(\frac{\sigma_m}{\sigma_u}\right)^2\right]$$

where

$\sigma_e$ = Fatigue strength for N cycles under zero mean stress
$\sigma_a$ = Fatigue strength for N cycles under mean stress of $\sigma_m$
$\sigma_u$ = Ultimate tensile strength
$\sigma_y$ = Yield strength

**Fig. 14.9** Comparison of Goodman, Gerber, and Soderberg models for relating mean stress to stress amplitude

Typically, metals harden if $\sigma_u/\sigma_y \geq 1.4$ and soften if $\sigma_u/\sigma_y \leq 1.2$.

The reason for hardening or softening is related to the dislocation microstructure of the metal. When the metal is highly work hardened and the dislocation density is high, cyclic strain allows the rearrangement of dislocations into more stable networks, thereby reducing the stress at which plastic deformation occurs. Conversely, when the initial dislocation density is low, the cyclic strain increases the dislocation density, increasing the amount of elastic strain and stress on the material.

Low-cycle fatigue test data are often presented as a plot of the plastic strain range, $\Delta\varepsilon_p$, versus cycles to failure, $N$. When plotted on log-log coordinates, a straight line is obtained that is described by the Coffin-Manson relation:

$$\frac{\Delta\varepsilon_p}{2} = \varepsilon_f'(2N)^c \qquad \text{(Eq 14.8)}$$

where $\Delta\varepsilon_p/2$ is the plastic strain amplitude, and $\varepsilon_f'$ is the fatigue ductility coefficient defined by the strain intercept at $2N = 1$. For many metals, $\varepsilon_f'$ is approximately equal to the true fracture strain, $\varepsilon_f$. $2N$ is the number of strain reversals to failure, where one cycle is two reversals, and $c$ is the fatigue ductility exponent, which usually varies between $-0.5$ and $-0.7$. A smaller value of $c$ results in longer fatigue lives.

The Basquin equation, which describes the high-cycle, low-strain regime where the nominal strains are elastic, is:

$$\sigma_a = \frac{\Delta\varepsilon_e}{2} E = \sigma_f'(2N)^b \qquad \text{(Eq 14.9)}$$

where $\sigma_a$ is the alternating stress amplitude, $\Delta\varepsilon_e/2$ is the elastic strain amplitude, $E$ is the modulus of elasticity, and $\sigma_f'$ is the fatigue strength coefficient, defined as the stress intercept at $2N = 1$. $\sigma_f'$ is approximately equal to the true fracture stress, $\sigma_f$. $2N$ is the number of load reversals to failure, and $b$ is the fatigue strength exponent, which varies for most metals between $-0.05$ and $-0.12$. A smaller $b$ results in a longer fatigue life.

Combining Basquin's equation and the Coffin-Manson equation gives an equation

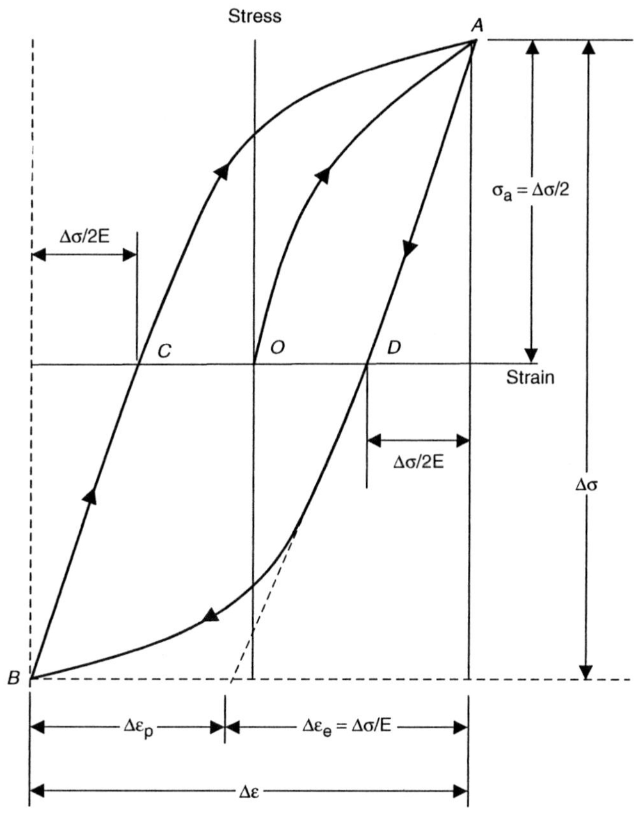


**Fig. 14.10** Stress-strain hysteresis loop for cyclic loading

that can be used to estimate the entire range of fatigue lives:

$$\frac{\Delta\varepsilon}{2} = \frac{\Delta\varepsilon_e}{2} + \frac{\Delta\varepsilon_p}{2} \quad \text{(Eq 14.10)}$$

$$\frac{\Delta\varepsilon}{2} = \frac{\sigma_f'}{E}(2N)^b + \varepsilon_f'(2N)^c \quad \text{(Eq 14.11)}$$

The fatigue strain-life curve tends toward the plastic curve at large strain amplitudes and

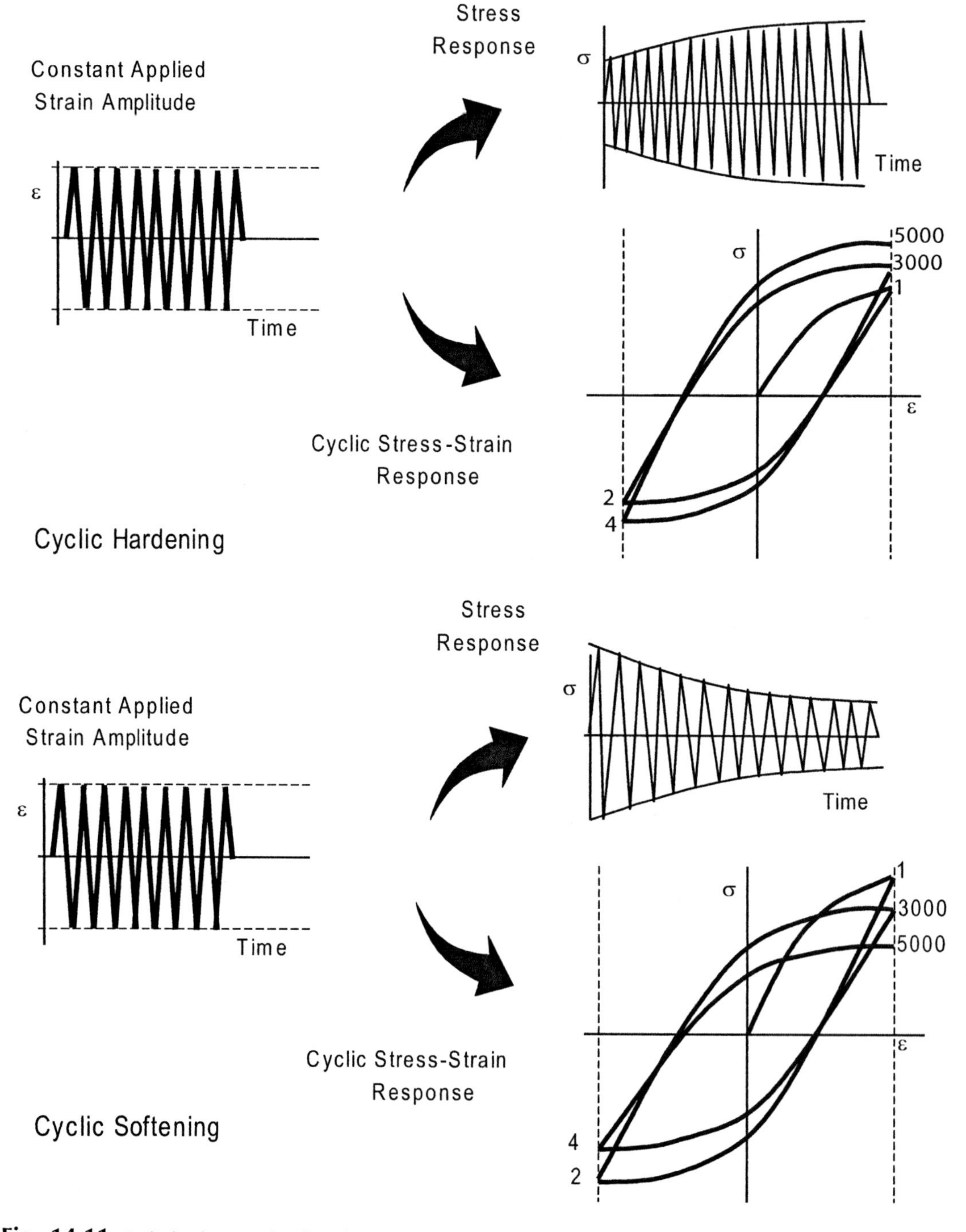


**Fig. 14.11** Cyclic hardening and cyclic softening

toward the elastic curve at small strain amplitudes (Fig. 14.12). For high-cycle strain conditions, ductile metals have the longest lives, while at low-cycle strain conditions, strong metals have the longest lives.

## 14.4 Cumulative Damage

Fatigue tests are often conducted under simple conditions, such as constant amplitude and constant frequency. However, in real structures, the loading conditions are rarely simple. Many structures are subjected to a range of fluctuating loads, mean stress levels, and variable frequencies. Thus, it is important to be able to predict the life of a component subjected to variable amplitude loading using data generated in constant amplitude laboratory tests. Cumulative damage theories consider the fatigue process to be one of damage accumulation until the life of the component is exhausted. Consider the fatigue life diagram in Fig. 14.13. At a constant stress, $\sigma_1$, the life is 200 cycles, while at $\sigma_2$, the life is 400 cycles. According to cumulative damage theory, fatigue life is gradually exhausted. That is, at points *A* and *C*, 100% of the life is available, while at points *B* and *D*, the respective lives are completely exhausted. If fatigue damage accumulates in a linear manner, each cycle contributes the same amount of damage at a given stress level. For example, at $\sigma_1$, on cycling the metal from *A* to *E*, one-fourth of the fatigue life is consumed (50 of 200 cycles). If the stress level is reduced to $\sigma_2$, the percentage of life exhausted at $\sigma_2$ is equivalent to the percentage of life exhausted at $\sigma_1$. That is, if one-fourth of the fatigue life is consumed at $\sigma_1$, the component still will have lost one-fourth of its life even if the stress is reduced to $\sigma_2$. For example, at point *E*, 50 cycles of 200 are consumed. If, when the component reaches the number of cycles defined by point *E*, the stress is reduced to $\sigma_2$, the component will be in the condition defined by point *F*, with 100 of its 400 cycles consumed. The same kind of change can be described for a low-to-high stress traverse.

Cumulative damage during fatigue is often determined using the Palmgren-Miner rule, which assumes that the total life of a part can be estimated by adding up the percentage of life consumed by each stress level:

$$\frac{n_1}{N_1}+\frac{n_2}{N_2}+\cdots\cdots+\frac{n_k}{N_k}=1 \text{ or } \sum_{j=1}^{j=k}\frac{n_j}{N_j}=1 \qquad \text{(Eq 14.12)}$$

where $n_1$, $n_2$ · · · · $n_k$ represent the number of cycles at a specific stress level, and $N_1$, $N_2$ · · · · · $N_k$ represent the fatigue life in cycles at the same stress level.

This rule should be used with caution. It is a linear relationship that does not consider the effects of understressing or overstressing. Understressing, in which relatively low stress levels are initially applied, can result in longer fatigue lives at a higher stress level, due to localized strain hardening. On the other hand, overstressing, in which high stresses are initially applied, can result in shorter fatigue lives at a low stress level, due to damage induced by the higher stress levels.

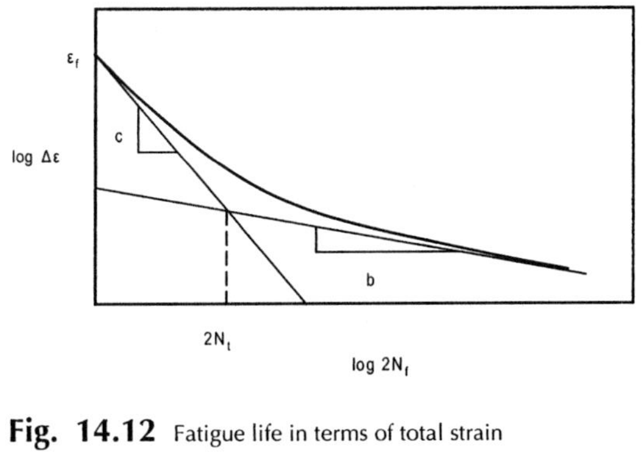


**Fig. 14.12** Fatigue life in terms of total strain

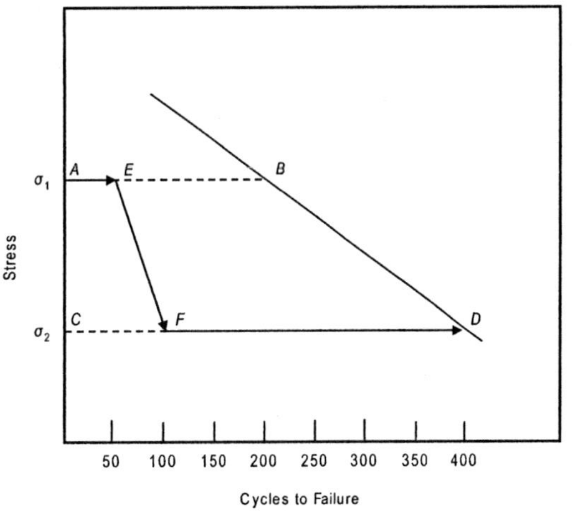


**Fig. 14.13** Cumulative damage during high-to-low loading sequence. Source: Ref 5

## 14.5 Fatigue Crack Nucleation and Growth

Fatigue cracks almost always initiate at free surfaces, usually external surfaces but also internal surfaces if the metal contains defects such as voids and cracked second-phase particles. Common external surface defects include geometric notches and surface roughness.

Fatigue crack nucleation and growth occurs in the following stages.

**Stage I.** Crack initiation usually starts at a notch or other surface discontinuity. Even in the absence of a surface defect, crack initiation will eventually occur due to the formation of persistent slip bands (PSBs), so called because traces of the bands persist even when the surface damage is polished away. Slip bands are a result of the systematic buildup of fine slip movements on the order of only 1 nm. However, the plastic strain within the PSB can be as much as 100 times greater than that in the surrounding material. The back-and-forth movement of the slip bands leads to the formation of intrusions and extrusions at the surface, eventually leading to the formation of a crack (Fig. 14.14). The initial crack propagates parallel to the slip bands. The crack propagation rate during stage I is very low, on the order of 1 nm per cycle, and produces a practically featureless fracture surface. The crack initially follows the slip bands at approximately 45° to the principal stress direction. When the crack length becomes sufficient for the stress field at the tip to become dominant, the overall crack plane changes and becomes perpendicular to the principal stress, and the crack enters stage II.

**Stage II** crack growth occurs when the stage I crack changes direction and propagates in a direction normal to the applied stress. Crack growth proceeds by a continual process of crack sharpening followed by blunting, as shown in the Fig. 14.15 illustration. Crack propagation during crack growth often produces a pattern of fatigue striations (Fig. 14.16), with each striation representing one cycle of fatigue. Although striations are indicative of fatigue, fatigue failures can occur without the formation of striations. Striations are microstructural details that are best examined with a scanning electron microscope and are not visible to the naked eye. Frequently, visible examination of a fatigued surface will reveal a series of concentric markings on the surface, referred to as beach marks (Fig. 14.6). These are present as a result of stress changes during fatigue, for example, the starting and stopping of a rotating shaft. Each of the beach marks can contain thousands or even tens of thousands of fatigue cycles.

**Stage III.** Ultimate failure occurs when the fatigue crack becomes long enough that the remaining cross section can no longer support the applied load.

## 14.6 Fatigue Crack Propagation

Linear elastic fracture mechanics assumes that all structures contain flaws. Cracks grow from an initial size, $a_0$, to a critical size, $a_c$, corresponding to failure as a function of the number of load cycles (Fig. 14.17). The crack growth rate, *da/dn,* can be determined from the slope of the curve. Initially, the crack growth rate is slow but increases with increasing crack length. Of course, the crack growth rate is also higher for higher applied stresses. If one can characterize the crack growth, it is then possible to estimate

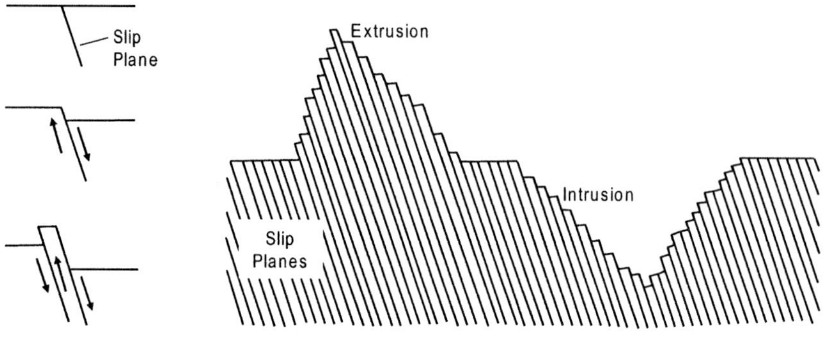


**Fig. 14.14** Development of extrusions and intrusions during fatigue. Source: Ref 4

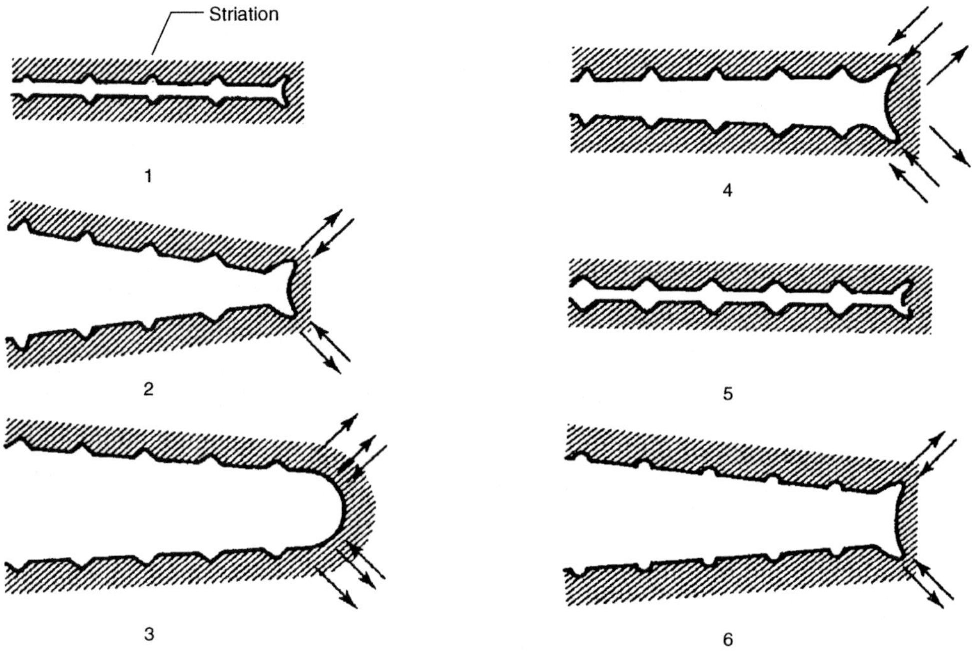


**Fig. 14.15** Fatigue crack propagation. Source: Ref 6

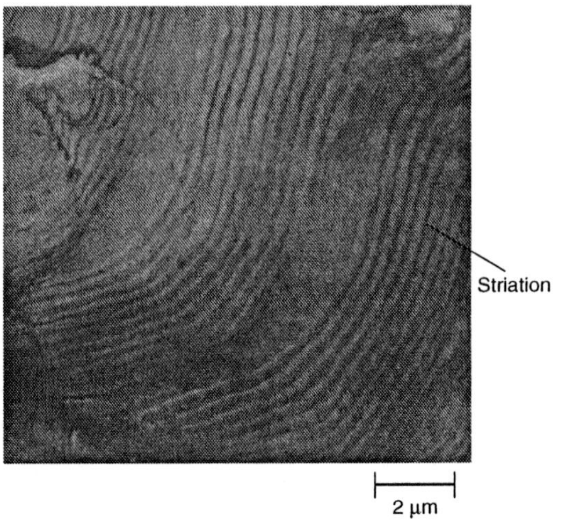


**Fig. 14.16** SEM image showing fatigue striations

the service life or inspection intervals required under specific loading conditions and service environment. In the fracture mechanics approach to fatigue crack growth, the crack growth rate, or the amount of crack extension per loading cycle, is correlated with the stress-intensity parameter, $K$. This approach makes it possible to estimate the useful safe lifetime and to establish inspection intervals. An idealized $da/dn$ versus $\Delta K$ curve is shown in Fig. 14.18. In region I, $\Delta K_{th}$ is the fatigue crack growth threshold, which is at the lower end of the $\Delta K$ range, where crack growth rates approach zero. In region II, the crack growth rate is stable and essentially linear and can be modeled by power law equations, such as the Paris equation:

$$da/dn = C(\Delta K)^m \quad \text{(Eq 14.13)}$$

where: $a$ is the flaw or crack size in inches; $n$ is the number of cycles; $C$ and $m$ are constant parameters and are related to material variables, environment, temperature, and fatigue stress conditions; and $\Delta K = \Delta K_{max} - \Delta K_{min}$ is the stress-intensity parameter range. The constants $C$ and $m$ are material parameters that must be determined experimentally. Typically, $m$ is in the range of 2 to 4 for metals and 4 to 100 for ceramics and polymers.

During stage II growth in the linear crack growth region, the Paris law can be used to determine the number of cycles to failure. $\Delta K$ can be expressed in terms of $\Delta\sigma$:

$$\Delta K = \Delta K_{max} - \Delta K_{min} = Y\sigma_{max} \times \sqrt{\pi a} - Y\sigma_{min}\sqrt{\pi a} = Y\Delta\sigma\sqrt{\pi a} \quad \text{(Eq 14.14)}$$

where $Y$ depends on the specific specimen geometry. Thus, the Paris law becomes:

$$\frac{da}{dn} = C(Y\Delta\sigma\sqrt{\pi a})^m \quad \text{(Eq 14.15)}$$

One of the goals of fatigue analysis is to be able to predict the fatigue life of structures. The fatigue life, $n$, can be solved for by rearranging Eq 14.13:

$$dn = \frac{da}{C(\Delta K)^m} \quad \text{(Eq 14.16)}$$

which may then be integrated to give:

$$n_f = \int_0^{n_f} dn = \int_{a_o}^{a_c} \frac{da}{C(\Delta K)^m} \quad \text{(Eq 14.17)}$$

Substitution of the expression for $\Delta K$ (Eq 14.14) gives:

$$n_f = \int_{a_o}^{a_c} \frac{da}{C(Y\Delta\sigma\sqrt{\pi a})^m} = \frac{1}{C\pi^{m/2}(\Delta\sigma)^m} \int_{a_o}^{a_c} \frac{da}{Y^m a^{m/2}} \quad \text{(Eq 14.18)}$$

It is assumed that $\Delta\sigma$ (or $\sigma_{max} - \sigma_{min}$) is constant and that $Y$ depends on the crack length and therefore cannot be removed from within the integral. For cases where $Y$ depends on crack length, these integrations will generally be performed using numerical techniques. During stage II crack growth, some metals are sensitive to the load ratio, $R$, as shown in Fig. 14.19 for 7075-T651 aluminum plate. Since the Paris equation does not account for the load ratio, there are other expressions that do account for the sensitivity of crack growth rate to the load ratio. One of the most utilized is the one developed by Foreman and his associates:

$$\frac{da}{dn} = \frac{C(\Delta K)^m}{(1-R)K_c - \Delta K} \quad \text{(Eq 14.19)}$$

In these expressions, it is necessary to determine the initial crack length, $a_o$, and the final or critical crack length, $a_c$. Crack lengths can be detected using a number of nondestructive testing techniques. If no cracks are detected, it must be assumed that a crack exists below the resolution of the detection system being used. A subcritical crack will eventually grow to a length at which the metal immediately fails, that is:

$$K_{max} \rightarrow K_c \quad \text{(Eq 14.20)}$$

or

$$Y\sigma_{max}\sqrt{\pi a_f} \rightarrow K_c \quad \text{(Eq 14.21)}$$

Solving for $a_c$ gives:

$$a_c = \frac{K_c^2}{\pi Y^2 \sigma_{max}^2} \quad \text{(Eq 14.22)}$$

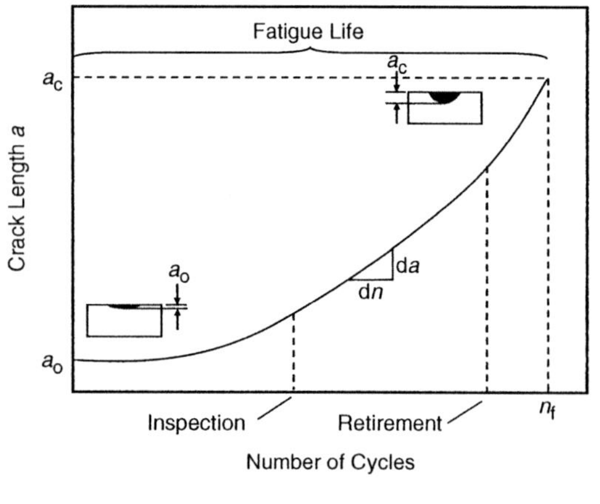


**Fig. 14.17** Crack length as a function of cycles. Source: Ref 5

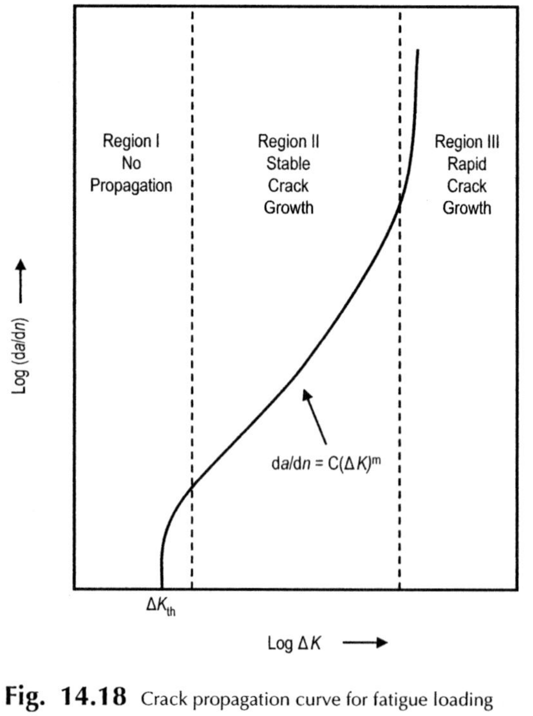


**Fig. 14.18** Crack propagation curve for fatigue loading

Note that even if a component contains a detectable crack, it can remain in service, provided that it is periodically inspected. This philosophy forms the basis for what is known as the damage-tolerance design approach.

Finally, in region III, the crack growth rate accelerates, since the fracture toughness of the material is approached, and there is a local tensile overload failure.

## 14.7 Crack Closure

During fatigue cycling, the crack growth rate can slow due to crack closure. For example, if the structure is subjected to a series of high-tensile overloads that are followed by loads at lower tensile stress levels, the crack may temporarily stop propagating, leading to a decrease in the crack growth rate, *da*/*dn,* and consequently have a longer fatigue life than would normally be expected. Crack closure mechanisms effectively reduce the $\Delta K$ during fatigue cycling, thus retarding the crack growth. Since $\Delta K$ is equal to $K_{max} - K_{min}$ and the actual $\Delta K$ is reduced because crack closure is hindered, $\Delta K$ is effectively smaller, and crack growth rate is then reduced.

Four mechanisms that can cause crack closure are shown in Fig. 14.20. Plasticity-induced closure results from compressive residual stresses developing in the plastic wake. Roughness-induced closure is due to crack meandering, in which the crack deviates from mode I displacements and is subject to mode II shear displacements. These displacements cause a mismatch between the upper and lower crack faces, which results in contact between the crack faces during cycling. Oxide debris-induced closure results from corrosion products becoming wedged in the crack interface. Finally, fluid

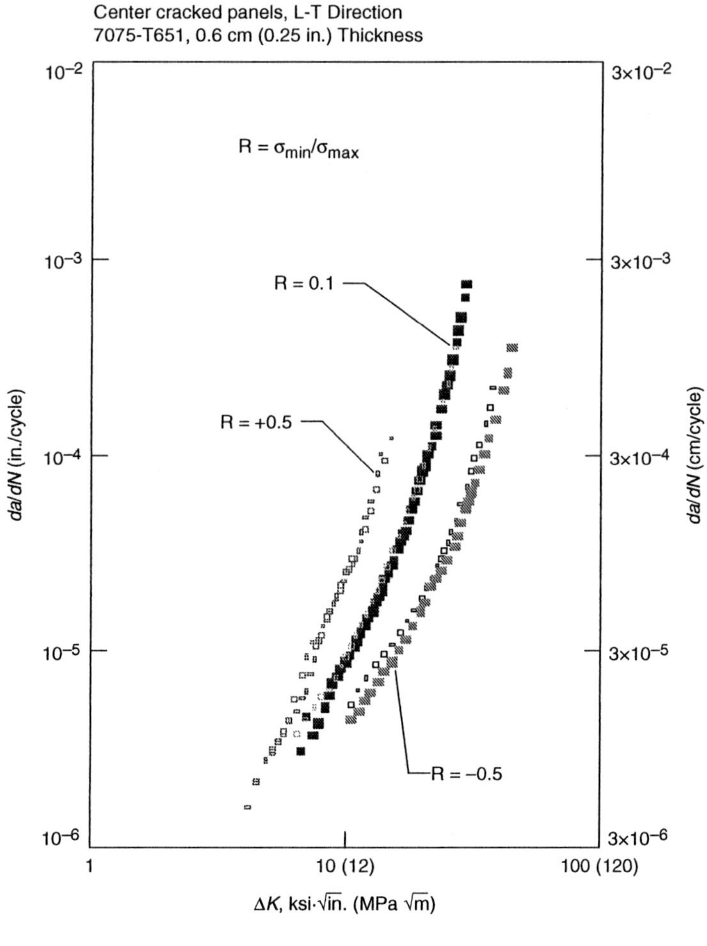


**Fig. 14.19** Stress ratio effects on an aluminum alloy

pressure can act as a wedge, preventing total crack closure.

## 14.8 Geometrical Stress Concentrations

In most structures, fatigue cracking usually initiates at a stress concentration. The stress concentration may by inherent in the design, such as a fillet, hole, thread, or other geometrical feature, or the stress concentration can result from a manufacturing process, such as a rough surface finish or residual tensile stresses introduced by heat treatment.

The effect of geometrical stress concentrations on fatigue is often studied by testing notched specimens. When a notch is present in a specimen under uniaxial loading, three effects are present: (1) there is an increase or concentration of stress at the root of the notch, (2) there is a stress gradient from the notch toward the center of the specimen, and (3) a triaxial state of stress exists. The dramatic reduction in fatigue life of normalized 4340 steel sheet containing different types of stress concentrations is shown in Fig. 14.21.

The effect of notches on fatigue strength is determined by comparing the *S-N* curves of notched and unnotched specimens. The data for the notched specimens are usually plotted in terms of nominal stress based on the net cross section of the specimen. The effect of the notch in decreasing the fatigue strength is reported as the fatigue strength reduction factor, or the fatigue notch factor, $K_f$:

$$K_f = \frac{\text{Fatigue limit unnotched}}{\text{Fatigue limit notched}} \quad \text{(Eq 14.23)}$$

For metals that do not have a definite fatigue limit, the fatigue notch factor is based on the fatigue strength at a specified number of cycles. Values of the fatigue notch factor vary with the severity of the notch, the type of notch, the material, the type of loading, and the applied stress level.

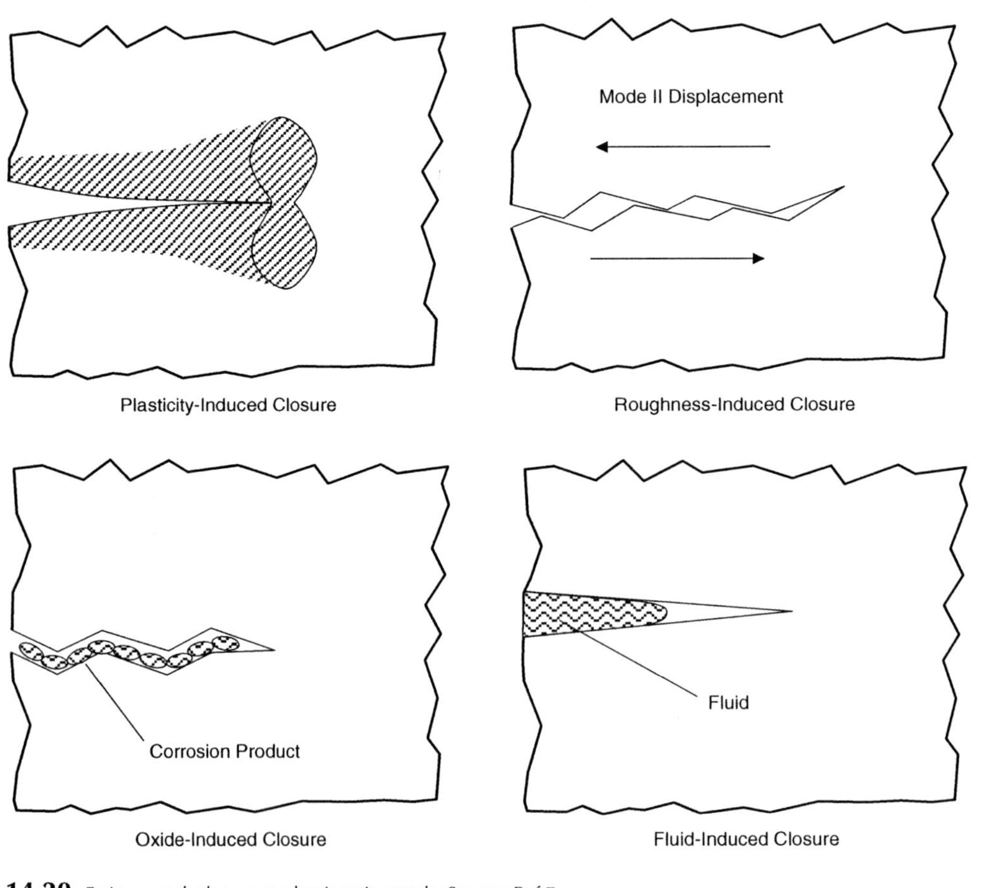


**Fig. 14.20** Fatigue crack closure mechanisms in metals. Source: Ref 7

Notched fatigue data are also reported using a notch sensitivity factor, $q$:

$$q = \frac{K_f - 1}{K_t - 1} \quad \text{(Eq 14.24)}$$

This relationship compares the theoretical stress-concentration factor, $K_t$, to the fatigue notch factor, $K_f$. In this relationship, a material that experiences no reduction in fatigue due to a notch will have a notch sensitivity factor of $q = 0$, while one that experiences a reduction in fatigue up to the full theoretical value will have a notch sensitivity factor of $q = 1$. The value of $q$ is dependent on the material and the radius of the notch root, as illustrated by the plots shown in Fig. 14.22 for steels. The notch sensitivity factor, $q$, significantly decreases with smaller notch radii. Although this seems counterintuitive, it occurs because $K_f$ increases more slowly than $K_t$ with decreasing notch radius. For notches with large radii, $K_f$ is almost equal to $K_t$. For materials with small notches, $K_f$ is less than $K_t$. In addition, lower-strength metals are less affected than high-strength metals by geometric discontinuities that reduce the fatigue resistance, because high-strength metals have a limited capacity for deformation and crack tip blunting.

## 14.9 Manufacturing Stress Concentrations

Because almost all fatigue failures start on a surface, the surface finish and residual-stress state near the surface can have a profound effect on the fatigue strength. Since the early days of fatigue investigations, it was recognized that the fatigue life of a component is very dependent on the surface finish produced by machining or grinding operations. As shown in Fig. 14.23, highly polished steel specimens perform much better in fatigue than even carefully machined surfaces. Further degradation in fatigue strength occurs for hot rolled and forged surfaces. Finally, when corrosion is introduced, the fatigue life can be seriously degraded.

Parts that are formed at room temperature will contain residual stresses. For example, the surface of a part that was formed in tension will contain residual compressive stresses, and a surface that was in compression during forming will contain residual tensile stresses. Since fatigue always occurs under tensile loading, the surface with the residual tensile stresses will be the most prone to fatigue cracking. Similar to forming, some quenching operations during heat treatment can result in a tensile residual-stress pattern on the surface that will adversely affect fatigue strength.

The fatigue data shown in Table 14.1 for 4340 steel forgings tested in fully reversed bending illustrate two important points. First, vacuum melting, which removes inclusions, improves both the longitudinal and transverse endurance limits. Second, since forging tends to elongate inclusions in the longitudinal or working direction, the fatigue endurance limit is higher in the longitudinal direction. The lower transverse endurance limit in steels containing inclusions is attributed to stress concentrations at the inclusions, which can be quite high when an elongated stringer is oriented transverse to the tensile stress. For the case of the vacuum-melted steel, the transverse limit is still somewhat lower than

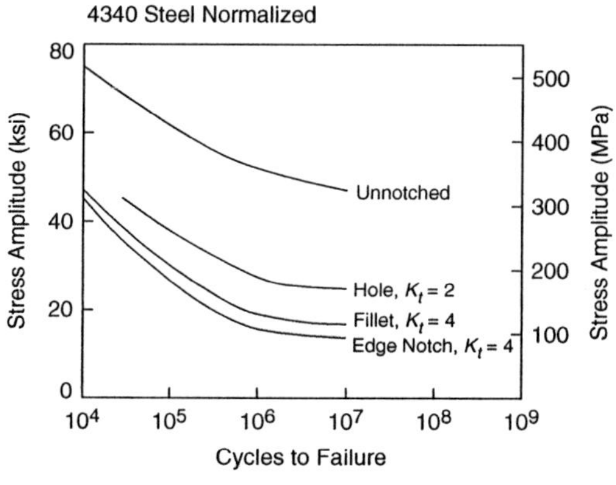


**Fig. 14.21** Effect of geometrical stress concentrations on fatigue life. Source: Ref 8

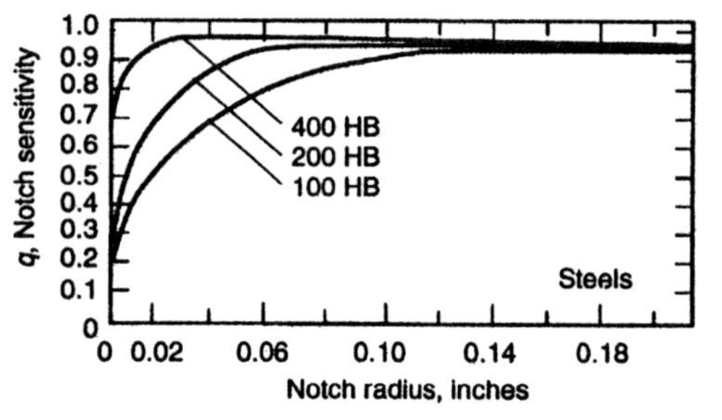


**Fig. 14.22** Notch sensitivity versus notch radius for steels. Source: Ref 9

for the longitudinal direction, indicating the extreme sensitivity of transverse fatigue properties to microstructure.

Due to their extreme strength and hardness, high-strength steels are often finished by grinding to provide precise dimensions, remove any nicks or scratches, and provide smooth surfaces. However, great care must be taken when grinding these ultrahigh-strength steels. Improper or abusive grinding can result in grinding burns, in which the surface is heated above the austenitizing temperature and the austenite formed converts to untempered martensite on cooling (Fig. 14.24). This untempered martensitic surface layer is brittle and susceptible to forming a network of fine cracks that can reduce the fatigue strength by as much as 30%. Even if the grinding temperature is not high enough to produce austenite, it may cause the original tempered martensite microstructure on the surface to become overtempered, thereby reducing the strength and hardness.

## 14.10 Environmental Effects

**Corrosion Fatigue.** The simultaneous action of fatigue and corrosive attack is known as corrosion fatigue. In the absence of fatigue loading, corrosive attack can often cause pitting of a metal surface. These pits can then act as stress concentrations that will initiate fatigue cracking. When corrosion and fatigue occur simultaneously, the chemical attack greatly accelerates fatigue crack growth. Materials such as steels, which show a definite endurance limit when tested in air, do not exhibit a definite endurance limit when tested in a corrosive environment (Fig. 14.25). Frequently, fatigue will corrupt the protective films that would normally slow the corrosive attack, allowing continued and accelerated attack. Since corrosion is time dependent, slow cycling rates will result in greater life reductions than accelerated fatigue cycling. Another possible effect of a corrosive environment on fatigue life is the

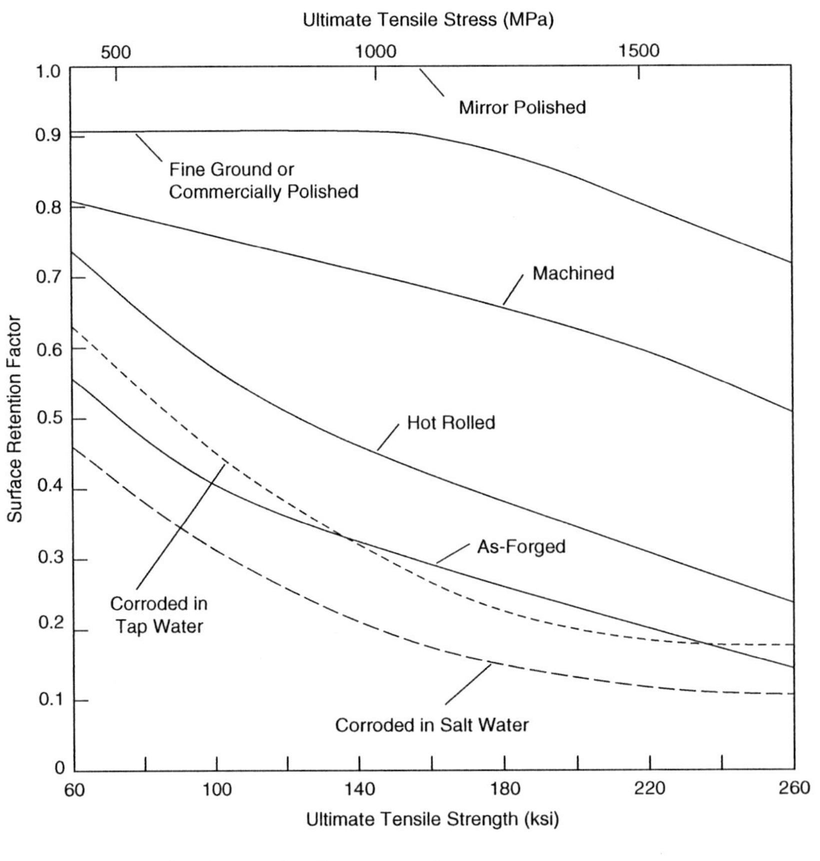


**Fig. 14.23** Effect of surface finish on fatigue of steels. Source: Ref 10

**Table 14.1 Effect of inclusions in 4340 forged steel**

| Property | Electric furnace melted | Vacuum melted |
|---|---|---|
| Longitudinal fatigue limit MPa (ksi) | 800 (116) | 960 (139) |
| Transverse fatigue limit MPa (ksi) | 545 (79) | 830 (120) |
| Ratio transverse/longitudinal | 0.68 | 0.86 |
| Hardness (HRC) | 27 | 29 |

Source: Ref 11

absence of the stable stage II crack growth that occurs according to the Paris law. A reduction in fatigue life of 50% or more is not unusual.

There are a number of methods used to combat corrosion fatigue. First, the choice of a material for an aggressive environment should probably be based on its corrosion resistance rather than its fatigue properties. For example, stainless steel, bronze, or beryllium-copper would probably yield better service than heat treated steels. Corrosion-resistant coatings that isolate the surface from the environment are successful as long as the coating does not rupture during cycling. Zinc and cadmium coatings on steel and aluminum alclad coatings on aluminum generally produce better fatigue lives in aggressive environments, even though the coatings may reduce fatigue lives somewhat when tests are conducted in air. The introduction of surface compressive residual-stress patterns by shot peening or nitriding can be very useful.

**Low-Temperature Fatigue.** Fatigue life normally increases with decreasing temperatures. Although steels become more notch sensitive at low temperatures, there is no experimental evidence indicating that any sudden change in fatigue properties occurs at temperatures below the ductile-to-brittle transition temperature.

**High-Temperature Fatigue.** In general, the fatigue strength of metals decreases with increasing temperatures. An exception is mild steel, which exhibits a maximum in fatigue strength between 205 and 300 °C (400 and 575 °F) due to strain aging. As the temperature exceeds approximately half the melting point, $T_m$, creep can become the dominant cause of failure. Ferrous alloys, which usually exhibit an endurance limit at room temperature, will no longer have an endurance limit when the temperature exceeds approximately 425 °C (800 °F). In general, the higher the creep strength of an alloy, the higher will be its high-temperature fatigue strength. However, there are

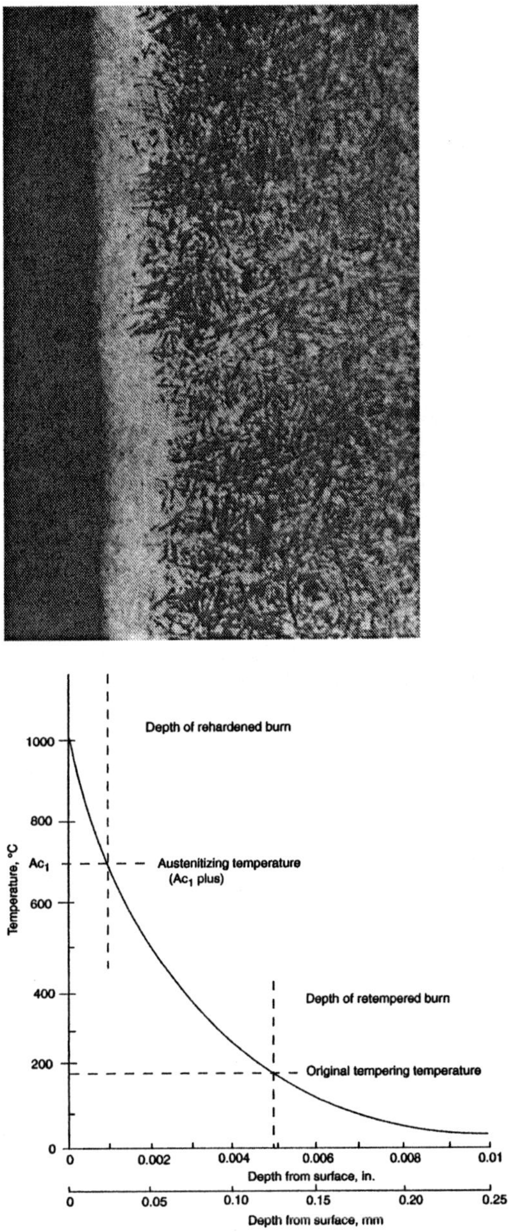


**Fig. 14.24** Effects of grinding burns (untempered martensite) on high-strength steel. Source: Ref 12

conflicting requirements, with fine grain sizes being beneficial to fatigue properties, while coarse grain sizes are preferred in certain creep regimes. In addition, the beneficial effects of fatigue improvement processes such as shot peening, which are effective at room temperature, may be annealed out at elevated temperatures.

**Thermal Fatigue.** Thermal stresses result when dimensional changes in response to temperature changes are restricted by some type of constraint imposed on the component. For the case of a simple bar constrained rigidly at both ends, the thermal stress, $\sigma$, developed by a temperature change, $\Delta T$, is:

$$\sigma = \alpha E \Delta T \qquad \text{(Eq 14.25)}$$

where $\alpha$ is the linear coefficient of thermal expansion, and $E$ is the modulus of elasticity.

If failure occurs during one cycle of thermal stress because of a rapid temperature change, the phenomenon is known as thermal shock, while failures occurring after multiple thermal stress cycles caused by a cyclic temperature change are referred to as thermal fatigue. An example of thermal fatigue occurs in ceramic thermal barrier coatings that are applied to nickel-base superalloys used in jet engines. The large differences in the coefficients of thermal expansion of the ceramic coating and the underlying nickel blade result in cracking and spalling of the coating. Another example of thermal fatigue occurs in electronics. Heat generated during operation of electronic devices can cause fatigue of the soldered joints.

The tendency for thermal fatigue has been related to the parameter:

$$\sigma_f k / E\alpha \qquad \text{(Eq 14.26)}$$

where $\sigma_f$ is the mean fatigue strength, and $k$ is the thermal conductivity. A high value of this parameter indicates good resistance to thermal fatigue.

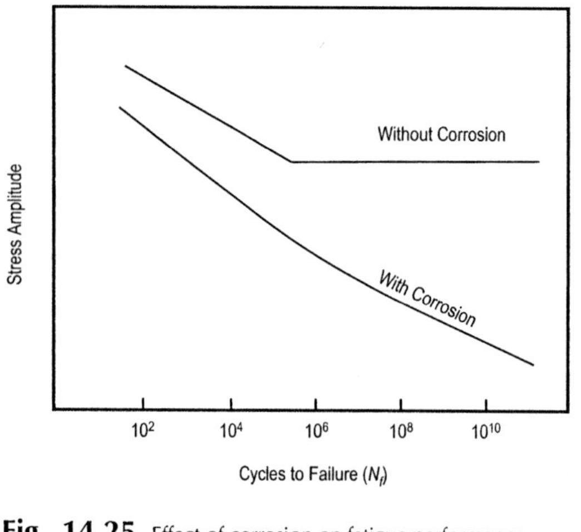


**Fig. 14.25** Effect of corrosion on fatigue performance

## 14.11 Fatigue Life Improvement

One of the most effective methods of improving fatigue life is to induce residual compressive stresses on the surface of the part. This is often accomplished by shot peening or by surface rolling with contoured rollers. Shot peening involves propelling fine steel or cast iron shot into the surface at high velocities. The severity of the stress produced by shot peening is monitored by measuring the residual deformation of shot-peened specimens called Almen strips. In addition to inducing compressive stress on the surface, the surface layers are also strengthened by the cold working that occurs during shot peening. Important variables in shot peening include the hardness of the shot, the size, the shape, and the velocity. Shot peening must be carefully controlled so as not to introduce surface damage to the part. Shot peening

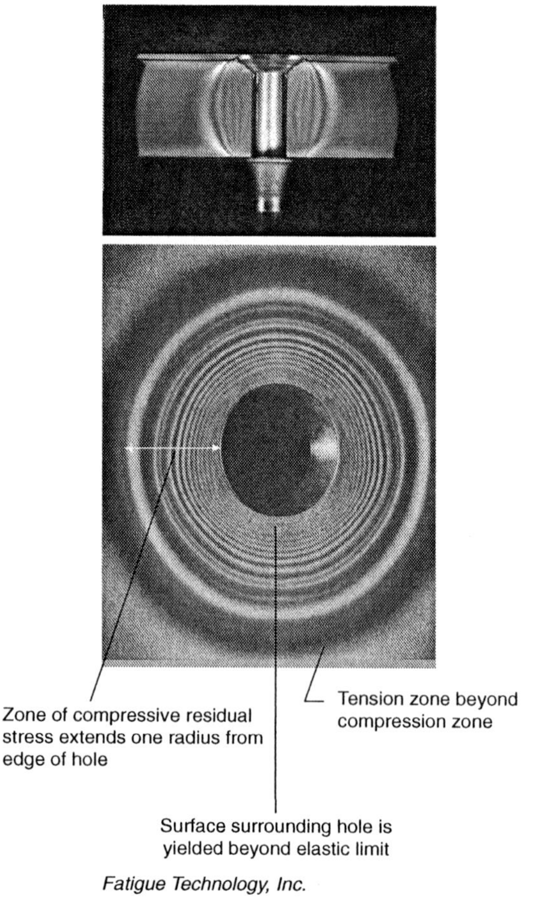


**Fig. 14.26** Residual-stress state around cold-worked hole

imparts greater fatigue life during high-cycle fatigue than it does for higher-stress low-cycle fatigue. In low-cycle fatigue, large stresses in the plastic range cause "fading" of the residual-stress pattern.

Steel parts subjected to wear conditions are often carburized or nitrided to harden the surface for greater wear resistance. Since both of these processes also introduce residual compressive stresses on the surface, there is an improvement in fatigue life. The greatest improvement occurs where a high stress gradient occurs, as in bending or torsion, with less improvement in axial loading. Nitriding is extremely effective in improving the fatigue strength of notched specimens. Other surface-hardening methods, such as flame or induction hardening, produce similar effects. Case hardening is more effective for components that have a stress gradient, as in bending, or where there is a notch. For axially loaded components, fatigue cracks can form under the case in the softer, weaker core at the boundary between the surface and bulk.

Many large structures are assembled with mechanical fasteners. In the aerospace industry, fatigue life improvement of aluminum structures is often accomplished with cold working of fastener holes and/or installing interference fit fasteners. Since fatigue cracks often initiate at fastener holes in metallic structures, methods such as cold working fastener holes and interference fit fasteners have been developed to improve fatigue life. Both cold working and interference fit fasteners set up residual compressive stress fields in the metal immediately adjacent to the hole (Fig. 14.26). The applied tension stress during fatigue loading must then overcome the residual compressive stress field before the hole becomes loaded in tension. The fatigue improvement due to cold working in 2024-T851 aluminum is shown in Fig. 14.27.

Cold working of holes is usually conducted using either the split-sleeve or split-mandrel method. Both methods involve pulling a mandrel through the hole that expands the hole diameter, creating plastic deformation of material around the hole and a resulting residual compressive stress field. The residual-stress field, depending on the material and the amount of expansion, will extend approximately one radius from the edge of the hole. In the split-sleeve process (Fig. 14.28), a stainless steel split sleeve is placed over a tapered mandrel and inserted into the hole. The hole is cold worked when the largest part of the mandrel is drawn back through the sleeve. After cold working, the sleeve is removed and discarded. In the split-mandrel process, a collapsible mandrel is placed in the

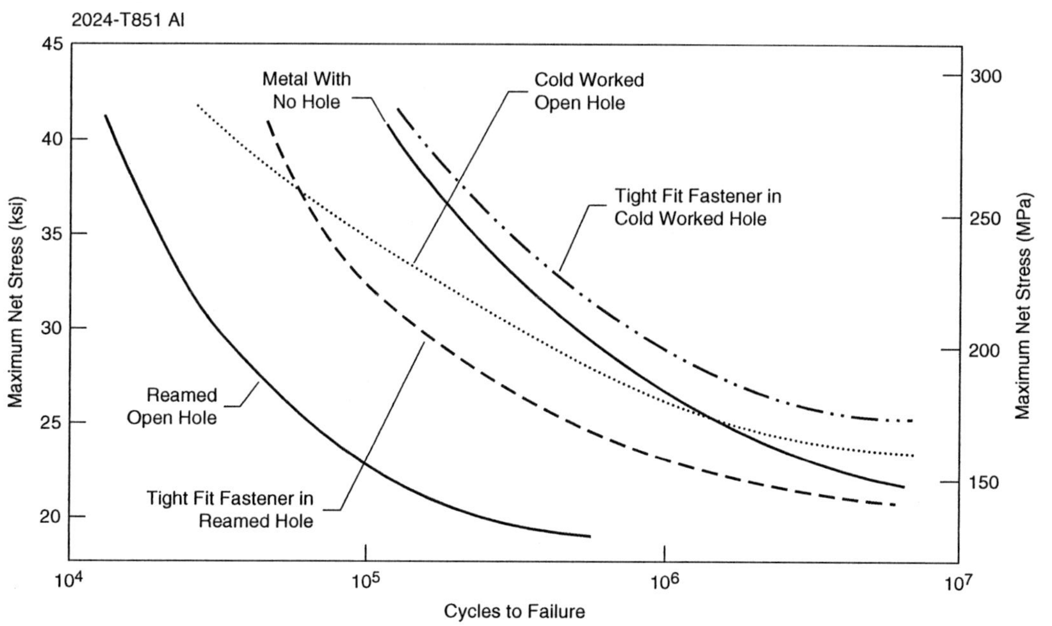


**Fig. 14.27** Fatigue life improvement with cold working. Source: Ref 13

hole, and as the mandrel is withdrawn, it expands to cold work the hole.

Interference fit fasteners can also be used in metallic structures to improve fatigue life. When an interference fit fastener is installed in metal, it also plastically deforms a small zone around the hole, setting up a compressive stress field, which again is beneficial when fatigue loading is primarily tensile. The amount of interference can vary, depending on structural requirements, but it is usually in the range of 0.07 to 0.10 mm (0.003 to 0.004 in.). In some highly loaded holes, both cold working and interference fit fasteners are used. While cold working and the use of interference fit fasteners are proven methods of improving fatigue resistance, both increase assembly costs and should only be specified when they are really needed.

Castings can be hot isostatic pressed (HIP) to help reduce internal porosity. The HIP of aluminum castings is usually conducted using argon pressure at 105 MPa (15 ksi) and temperatures in the range of 480 to 525 °C (900 to 980 °F). The use of HIP usually results in improved mechanical properties, especially fatigue strength, but, of course, it adds to the cost and cycle time. The improvement in fatigue life for the aluminum casting alloy A201.0-T7 as a result of HIP is shown in Fig. 14.29.

## 14.12 Fatigue Design Methodologies

Since the 1800s, a number of design philosophies or methodologies have evolved to deal with design against fatigue failures.

**Infinite-Life Design.** This is the oldest of the design philosophies and is based on maintaining the stresses at some fraction below the fatigue strength of the metal. The initial material is assumed to be free of flaws. This methodology is based on the classic *S-N* curve. It is most

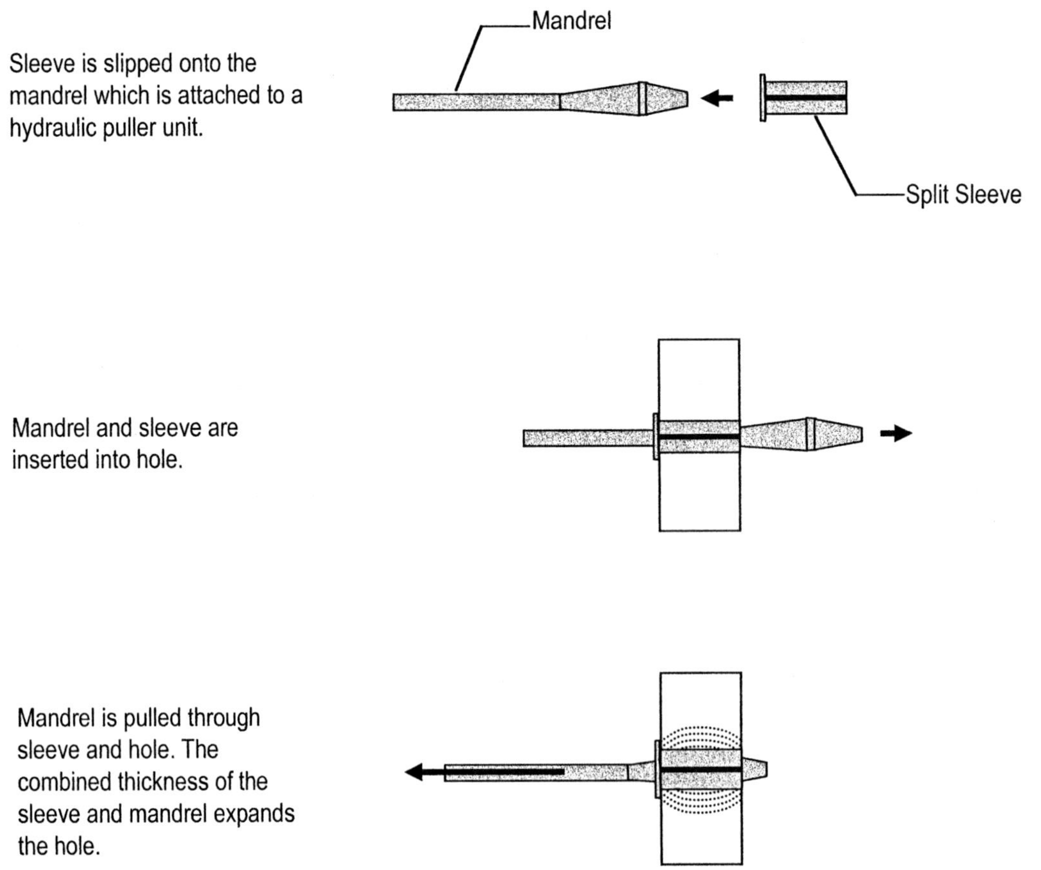


**Fig. 14.28** Split-sleeve cold working process. Reprinted with permission from SAE Paper # 982145 © 1998 SAE International. Source: Ref 14

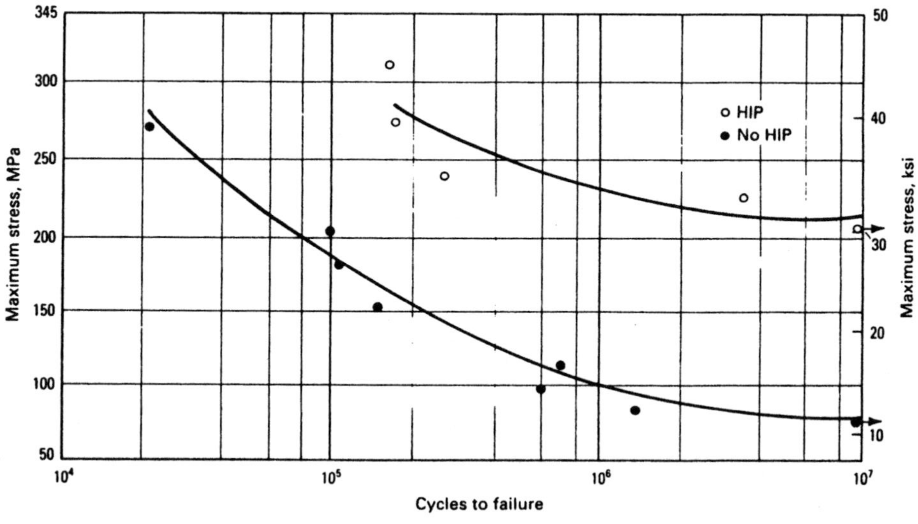


**Fig. 14.29** Effect of hot isostatic pressing (HIP) on fatigue life of A201.0-T7 aluminum casting. Source: Ref 15

applicable when the component is only stressed in the elastic range and displays a distinct endurance limit, such as for steels. Although it has largely been superseded by later methodologies, it is simple to use and can be useful where periodic inspection is either impractical or not warranted. However, since it is a conservative methodology, it usually results in a heavier design than some of the later methodologies.

**Safe-life design** is based on the assumption that the part is initially flaw-free and has a fine life in which to develop a critical crack size. Like the infinite-life methodology, safe-life assumes an initial flaw-free component. This methodology was developed to account for parts that were subjected to higher loads that produced plastic strains. Under these conditions, the description of local events in terms of strain made more sense and resulted in the development of assessment techniques that used strain as a determining quantity, that is, log strain (ε) versus log number of cycles (*N*). The failure criterion is usually the detection of a small crack or some equivalent measure related to a substantial change in load-deflection response, although failure may also be defined as fracture.

**Fail-Safe Design.** The fail-safe design philosophy assumes that fatigue cracks will be detected and repaired before they lead to failure. This methodology was developed in the aircraft industry where large safety margins were weight-prohibitive. Fail-safe designs incorporate multiple load paths and crack stoppers in the structure. In other words, if a primary load path fails, the load will be picked up by an alternate load path to prevent the structure from failing. A major part of this methodology is rigid certification criteria along with the capability to detect and inspect cracks.

**Damage-tolerant design** is the latest design methodology and is an extension of fail-safe design. The refinement of the damage-tolerant methodology is the assumption that structures do contain cracks and that fracture mechanics approaches can be used to determine crack growth rates. If the crack growth rate can be calculated, it is then possible to inspect the structure at various intervals and either repair or retire it prior to the crack reaching critical dimensions.

Although much progress has been made in the design against fatigue failures, they still occur with disturbing frequency. For any design, it is imperative that part or component testing be conducted prior to placing the part in service. In addition, it is important that the test conditions are representative of the actual environment that the component will experience in service.

## REFERENCES

1. Fatigue Failures, *Failure Analysis and Prevention,* Vol 11, *ASM Handbook,* ASM International, 2002

2. S. Kalpakjian, *Manufacturing Engineering and Technology,* 3rd ed. Addison-Wesley Publishing Co., 1995
3. *Properties and Selection of Metals,* Vol 1, *Metals Handbook,* 8th ed., American Society for Metals, 1961
4. R.A. Higgins, *Engineering Metallurgy—Applied Physical Metallurgy,* 6th ed., Arnold, 1993
5. M.A. Meyers and K.K. Chawla, *Mechanical Metallurgy—Principles and Applications,* Prentice-Hall Inc., 1984
6. C.M. Laird, STP 415, American Society for Testing and Materials, 1966, p 130
7. T.L. Anderson, *Fracture Mechanics: Fundamentals and Applications,* 3rd ed., Taylor & Francis, 2005
8. J. Collins and S. Daniewicz, Failure Considerations, *Mechanical Engineers Handbook,* John Wiley & Sons, Inc., 1998
9. M.R. Mitchell, Fundamentals of Modern Fatigue Analysis for Design, *Fatigue and Fracture,* Vol 19, *ASM Handbook,* ASM International, 1996
10. R.C. Juvinall, *Stress, Strain, and Strength,* McGraw-Hill Book Co., 1967
11. G.E. Dieter, *Mechanical Metallurgy,* 3rd ed., McGraw-Hill Book Co., 1986, p. 375–431
12. G.E. Toten, et al., Factors Relating to Heat Treating Operations, *Failure Analysis and Prevention,* Vol 11, *ASM Handbook,* ASM International, 2002
13. F.C. Campbell, *Manufacturing Technology for Aerospace Structural Materials,* Elsevier Scientific, 2006
14. A. Leon, "Developments in Advanced Coldworking," SAE 982145, Society of Automotive Engineers, 1998
15. J.R. Davis, *ASM Specialty Handbook: Aluminum and Aluminum Alloys,* ASM International, 1993

## SELECTED REFERENCES

- M.F. Ashby and D.R.H. Jones, *Engineering Materials 1 — An Introduction to Their Properties and Applications,* 2nd ed., Butterworth Heinemann, 1996
- T.H. Courtney, *Mechanical Behavior of Materials,* 2nd ed., McGraw-Hill Book Co., 2000, p 566–629
- H.W. Hayden, W.G. Moffatt, and J. Wulf, *The Structure and Properties of Materials,* Vol III, John Wiley, 1965
- M.P. Kaplan and T.A. Wolff, Fatigue-Life Assessment, *Failure Analysis and Prevention,* Vol 11, *ASM Handbook,* ASM International, 2002
- E. Krempl, Design for Fatigue Resistance, *Materials Selection and Design,* Vol 20, *ASM Handbook,* ASM International, 1997
- R.A. Lund and S. Sheybany, Fatigue Fracture Appearances, *Failure Analysis and Prevention,* Vol 11, *ASM Handbook,* ASM International, 2002

CHAPTER 15

# Creep

FOR METALS AT LOW TEMPERATURES, yield strength is usually the limiting design factor. However, at high temperatures, permanent deformation can occur over a period of time at stresses well below the yield strength. This time-dependent deformation is known as creep and occurs at temperatures greater than approximately 0.3 to 0.5 of the absolute melting point. In creep, thermal activation enables plastic deformation at stresses below those needed to deform the lattice without thermal activation. When a metal is placed under a constant load, it stretches elastically but also gradually extends plastically. Thus, a metal subjected to a constant tensile load at elevated temperature will creep and undergo a time-dependent increase in length.

Since the mobility of atoms increases with increasing temperature, diffusion-controlled mechanisms become active. Dislocation mobility increases, slip becomes easier, new slip systems become available, and dislocation climb is aided by both increases in temperature and the presence of a greater number of vacancies. Deformation at grain boundaries also becomes a possibility. The metallurgical stability of the alloy can become an issue. For example, precipitation-hardened alloys may undergo overaging, with a resulting strength loss. Oxidation and intergranular attack can occur.

Creep occurs in any metal or alloy at a temperature where atoms become sufficiently mobile to allow the time-dependent rearrangement of structure. Since the elevated-temperature strength of metals is closely related to their melting points, it is normal practice to specify the homologous temperature, which is the ratio of the exposure temperature to the melting point ($T/T_m$), on an absolute scale. Creep behavior of a polycrystalline metal or alloy often is considered to begin at approximately 1/3 to 1/2 of its melting point (~0.3 to 0.5 $T_m$). However, creep becomes important when the mechanical strength of a metal becomes limited by creep rather than by its yield strength. Since this transition in engineering design is not directly related to melting temperature, the temperature at which the mechanical strength of a metal becomes creep limited must be determined individually for each specific metal or alloy.

Approximate temperatures at which creep behavior begins for several metals and alloys are listed in Table 15.1. Low-melting-point metals, such as lead and tin, may deform by creep at or slightly above room temperature. In contrast, refractory and nickel-base superalloys require temperatures exceeding 980 °C (1800 °F) to activate the onset of creep deformation. Creep is a critical consideration in a number of high-temperature applications, such as power and chemical plants, and turbine components. Often, the main concern is dimensional stability, although prolonged creep will eventually lead to rupture.

## 15.1 The Creep Curve

Two types of elevated-temperature, time-dependent tests are conducted to evaluate metals for long-term elevated-temperature service. In a creep test, the time-dependent strain, or extension, is measured under long-term elevated

**Table 15.1 Approximate temperatures for onset of creep**

| Metal or alloy | °C | °F | Temperature as ratio of melting point ($T_m$) K |
|---|---|---|---|
| Aluminum alloys | 150–200 | 300–400 | 0.48–0.54 $T_m$ |
| Titanium alloys | 315 | 600 | 0.3 $T_m$ |
| Low-alloy steels | 370 | 700 | 0.36 $T_m$ |
| Austenitic iron-base heat-resisting alloys | 540 | 1000 | 0.49 $T_m$ |
| Nickel- and cobalt-base heat-resisting alloys | 650 | 1200 | 0.56 $T_m$ |
| Refractory metals and alloys | 980–1540 | 1800–2800 | 0.4–0.45 $T_m$ |

Source: Ref 1

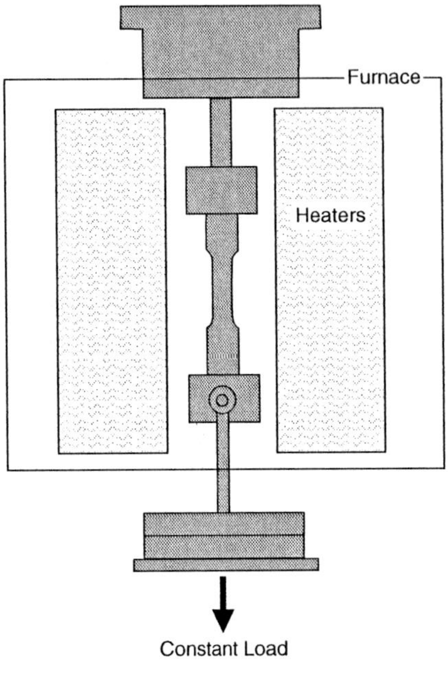


**Fig. 15.1** Typical creep test fixture

temperature. A shorter test, the stress-rupture test, measures the time it takes for a metal to fail at a given stress at elevated temperature.

Engineering creep is measured by applying a constant load on a tensile specimen at a constant temperature and measuring the strain, or extension, of the specimen as a function of time. A schematic of a test setup for a creep test is shown in Fig. 15.1. Creep tests may be run for as short as several months to as long as 10 years. Creep tests are usually conducted at constant load rather than constant stress, and, as the specimen elongates, the cross-sectional area decreases, so the applied stress increases with time. Constant load creep curves typically, but not always, consist of three distinct stages, as shown in Fig. 15.2.

**Primary Creep.** During primary creep, the specimen undergoes an initial elongation, $\varepsilon_0$, and then the creep rate ($\dot{\varepsilon} = d\varepsilon/dt$) rapidly decreases with time. Primary creep, also known as transient creep, represents a stage of adjustment in the metal during which rapid, thermally

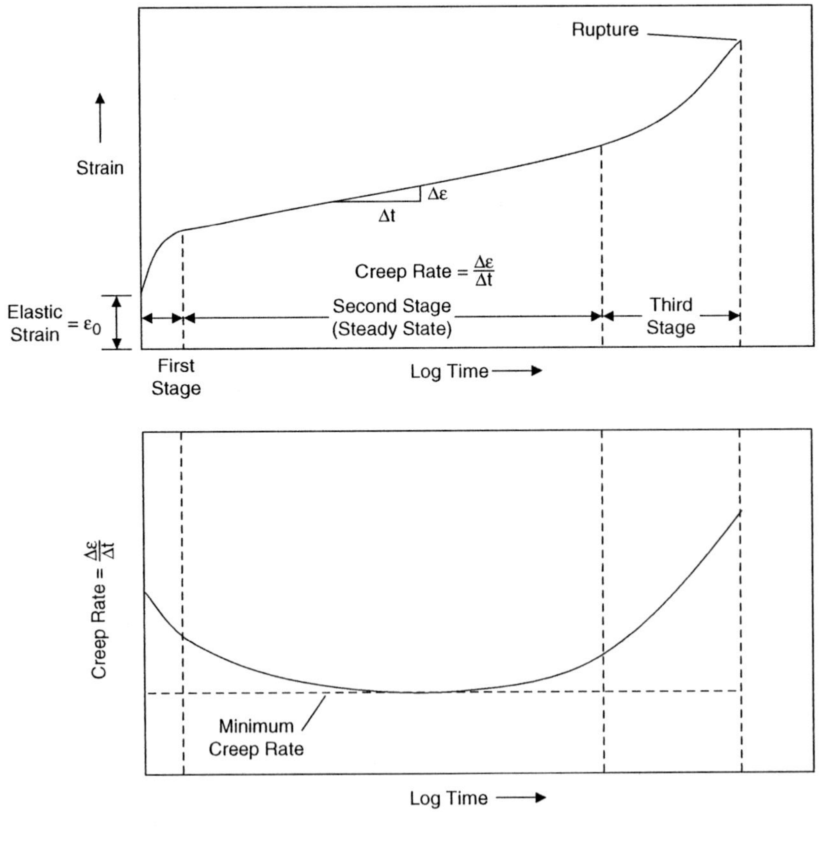


**Fig. 15.2** Typical creep curve

activated plastic strain occurs. The competing processes of strain hardening and recovery eventually lead to a somewhat stable dislocation configuration. Primary creep occurs in the first few moments after initial strain and decreases in rate as crystallographic imperfections within the metal undergo realignment. This realignment leads to secondary creep.

**Secondary Creep.** Following primary creep is the region of secondary creep, where the creep rate is nominally constant at a minimum rate, generally known as the minimum creep rate, shown in Fig. 15.2. Secondary creep, also known as steady-state creep, occurs when there is a balance between the competing processes of strain hardening and recovery. Secondary creep often occupies the major portion of the duration of the creep test, and the strain rate in this region for many creep-resistant materials is sufficiently constant to be considered as a steady-state creep rate. For these materials, the minimum creep rate is a steady-state value that can be empirically related to creep-rupture life and is widely used in engineering analyses.

During creep, significant microstructural changes occur on all levels. On the atomic scale, dislocations are created and forced to move through the material. This leads to work hardening as the dislocation density increases and the dislocations encounter barriers to their motion. At low temperatures, an ever-diminishing creep rate results. However, if the temperature is sufficiently high, dislocations can rearrange and annihilate themselves through recovery. During creep deformation, the material also is progressively degraded or damaged as the amount of creep strain increases over time.

Most engineering applications for high-temperature components fall within the range of 0.4 to 0.6 $T_m$. Since creep is a thermally activated process, it is not surprising that steady-state creep obeys an Arrhenius-type rate equation. The steady-state creep rate, $\dot{\varepsilon}_s$, in this intermediate temperature range (0.4 $T_m < T < 0.6\ T_m$) can be expressed as a power law function:

$$\dot{\varepsilon}_s = A\sigma^n e^{-Q/RT} \qquad \text{(Eq 15.1)}$$

where $Q$ is the activation energy for creep; $A$ is the pre-exponential constant; $n$ is a constant, usually between 3 and 10; $R$ is the universal gas constant; and $T$ is the absolute temperature.

This equation describes the dependence of creep rate on the key variables, temperature and stress. Specific values for $n$ and $Q$ are associated with specific creep mechanisms. The parameters $A$, $n$, and $Q$ depend on the particular material condition, stress level, and temperature. A correlation of creep and diffusion data for pure metals shows that the activation energy for creep, $Q$, is equal to the activation energy for self-diffusion (Fig. 15.3). At high stress levels, where creep rates begin to increase more rapidly with stress, this relationship no longer holds and is commonly referred to as power law breakdown.

The minimum creep rate during steady-state creep is usually the most important design parameter obtained during a creep test. Two criteria are common: (1) the stress to produce 1% creep in 10,000 h, or (2) the stress to produce 1% creep in 100,000 h. The first criterion is often used for jet engine components and the second for steam turbines.

**Tertiary creep** is a region of drastically increasing strain rate with rapid extension to fracture. Tertiary creep is dominated by a number of weakening metallurgical instabilities, such as localized necking, corrosion, intercrystalline fracture, microvoid formation, precipitation of brittle second-phase particles, and dissolution of second phases that originally contributed to strengthening of the alloy. In addition, recrystallization of the strain-hardened grains can destroy the balance between the material hardening and softening processes. Unlike constant load tests, constant stress tests often do not show tertiary behavior.

During service or during creep testing, tertiary creep may be accelerated by a reduction in cross-sectional area resulting from cracking or

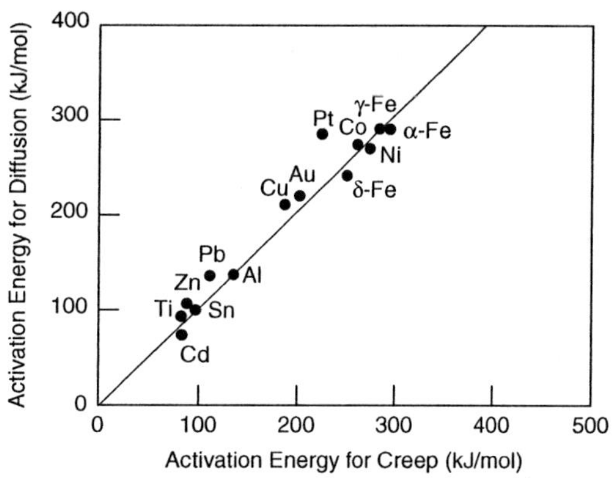


**Fig. 15.3** Correlation between diffusion and creep activation energies

necking. Environmental effects, such as oxidation, that reduce the cross section may also initiate tertiary creep or increase the tertiary creep rate. In many commercial creep-resistant alloys, tertiary creep is apparently caused by inherent deformation processes and occurs at creep strains of 0.5% or less. When designing components for service at elevated temperatures, the steady-state creep rate is usually the significant design parameter. However, the duration of tertiary creep is also important, because it constitutes a safety factor that may allow detection of a failing component before catastrophic fracture.

Not all creep curves will show these three distinct regions, because they are strongly dependent on the applied stress and temperature. As depicted in Fig. 15.4, creep curves with three well-defined regions will be found at only certain combinations of stress and temperature. The higher the temperature and/or the greater the stress, the greater will be the creep rate.

It is important to note that the region of initial elastic strain under load ($\varepsilon_0$) is not usually identified as a stage of creep deformation, and it is common practice to ignore this contribution to total strain when plotting creep curves. Therefore, creep curves generally show only the time-dependent plastic strain that follows this initial elastic strain. Although this procedure is acceptable for research, the initial strain may amount to a substantial fraction of the total strain and should not be omitted from design analysis. In addition, while creep tests are normally conducted at constant temperatures and loads, components in service hardly ever operate under constant conditions. Start-stop cycles, reduced power operation, thermal gradients, and other factors result in variations in stresses and temperatures.

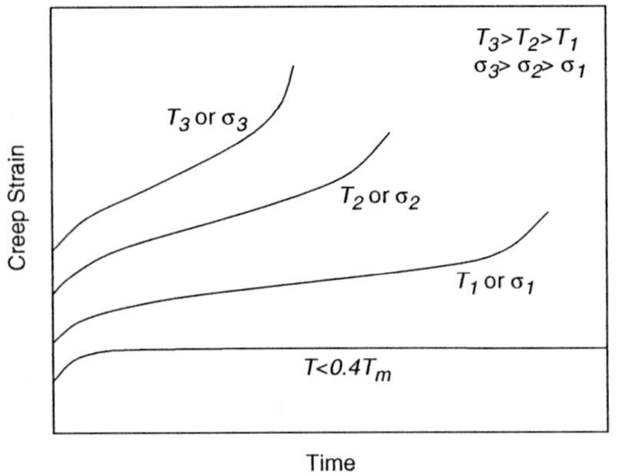


**Fig. 15.4** Effects of temperature and stress on creep curve

## 15.2 Stress-Rupture Test

During the stress-rupture test, the specimen is normally loaded at higher stresses than for the creep test and is complete upon failure of the specimen. In the creep test, the total strain is often less than 0.5%, while in the stress-rupture test, the strain can approach 50%. An advantage of the stress-rupture test is time; a stress-rupture test can often be concluded at approximately 1000 h. As shown in Fig. 15.5, the stresses at different temperatures are plotted against the time to failure. Changes in the slope of the curve often indicate that some type of metallurgical instability has taken place, such as phase changes, internal oxidation, or the change from transgranular to intergranular failure.

Stress-rupture ductility is an important consideration when stress concentrations and localized defects, such as notches, are a factor in design. Most engineering alloys can lose ductility during high-temperature service, because the diffusion of impurities to the grain boundaries becomes more pronounced. On creep curves, there are two meaningful measures of elongation: (1) true elongation, which is defined as the elongation at the end of the second stage of creep, and (2) total elongation, which is the elongation at fracture. The common practice is to plot both elongations versus rupture life. As an example, a plot of both total and true elongation at two different testing temperatures for the high-temperature alloy S-590 is shown in Fig. 15.6. At both temperatures, total elongation shows appreciable data scatter. Since true

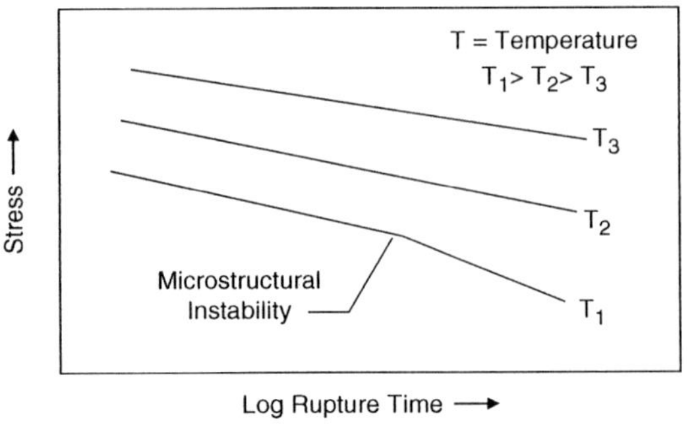


**Fig. 15.5** Stress-rupture curves

elongation is not influenced by the localized creep mechanisms (e.g., necking) that can occur during tertiary creep, it provides a more accurate representation of ductility in metals at elevated temperatures. The large differences between total and true elongation are a function of crack volume and distribution during tertiary creep. Stress-rupture ductility is an important factor in alloy selection, because premature failures have resulted from lack of ductility during tertiary creep. Since component designs are frequently based on 1% creep, a low stress-rupture ductility may prevent using the alloy to its full strength potential. As shown by the schematic creep curves in Fig. 15.7, a higher rupture ductility for the same load and temperature conditions provides a higher safety margin.

Since the creep and stress-rupture tests are similar, there are a number of empirical relationships that relate steady-state creep rate, $\dot{\varepsilon}_s$, to rupture life, $t_R$. Monkman and Grant (Ref 2) developed one that implies an inverse relationship between the two:

$$\log t_R + m \log \dot{\varepsilon}_s = B \qquad \text{(Eq 15.2)}$$

where $t_R$ is the rupture life, $\dot{\varepsilon}_s$ is the steady-state creep rate, and $m$ and $B$ are constants.

For a number of aluminum-, copper-, titanium-, iron-, and nickel-base alloys, Monkman and Grant found that $0.77 < m < 0.93$ and $0.48 < B < 1.3$. As long as the validity of the relationship can be established along with the values of the constants, the rupture life can be estimated from the steady-state creep rate.

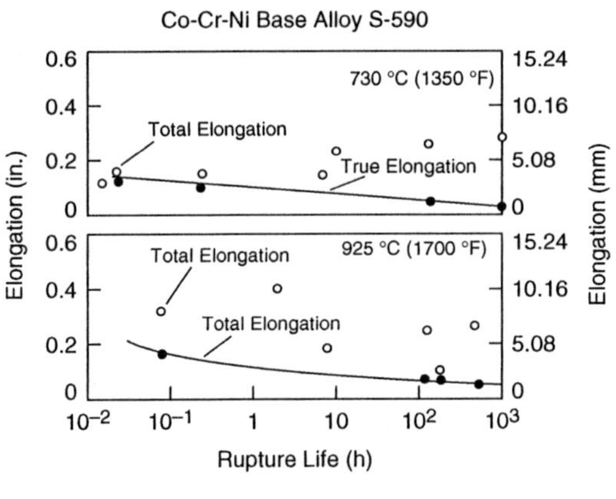


**Fig. 15.6** Comparison of true and total elongation. Source: Ref 1

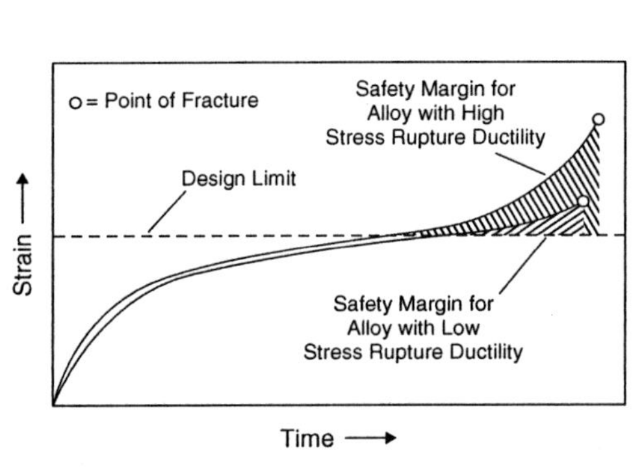


**Fig. 15.7** Stress-rupture curves with high and low rupture ductility. Source: Ref 1

## 15.3 Creep Deformation Mechanisms

The major classes of creep mechanisms are those that are governed by dislocation motion and those that are diffusion controlled. The dominating mechanism is determined by both the stress and temperature; however, several mechanisms may be active at the same time. In general, the ones governed by dislocation motion are more prevalent at lower temperatures and higher stresses, while those controlled by diffusion occur at higher temperatures and lower stresses.

Dislocation creep is the result of dislocations, in combination with vacancies and thermal activation, climbing over obstacles that would normally impede their motion at lower temperatures. The relevant dislocation mechanisms are dislocation glide and climb. As the temperature is increased, slip systems that were not available at room temperature become active, promoting dislocation glide. When dislocations encounter an obstacle, they are blocked and tend to pile up against the obstacle (Fig. 15.8). At low stress levels, the applied stress is insufficient to enable the dislocations to bow around or cut through the obstacle. However, at elevated temperature, a dislocation may climb by diffusion to a parallel slip plane. Having climbed, the

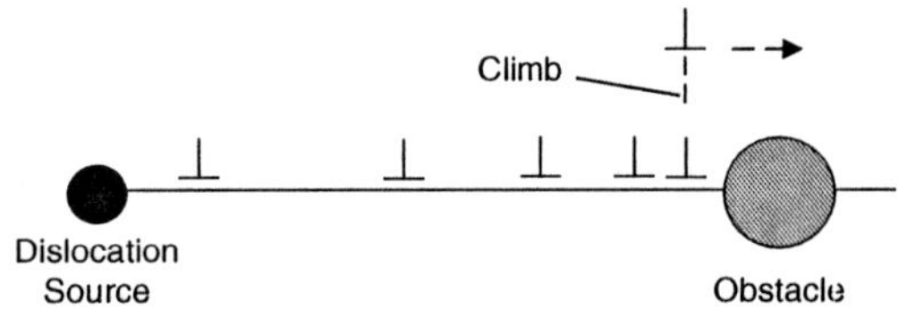


**Fig. 15.8** Thermally activated dislocation climb

dislocation proceeds along the new slip plane until it encounters another resistant obstacle, whereupon it climbs (or descends) to another parallel plane and the process repeats. Since dislocation motion depends on both dislocation glide and climb, the term *climb-glide creep* is used to describe this form of creep. Climb-glide creep depends more strongly on stress than does diffusion creep. At temperatures in excess of 0.5 $T_m$, dislocations can escape from the obstacles by climbing out of the slip plane and thus can continue to glide. The rate-determining step is the climb process, but the strain is produced during the glide to the next obstacle. In creep-resistant alloys, precipitates or dispersions are added to prevent dislocation glide.

Diffusion creep is often the dominating mechanism at high temperatures and low stresses. Two types of diffusion creep occur: Nabarro-Herring creep and Coble creep. Under the driving force of an applied stress, atoms diffuse from the sides of the grains to the tops and bottoms in the manner shown in Fig. 15.9. The grain becomes longer as the applied stress does work, and the process will be faster at high temperatures because there are more vacancies. It should be noted that atomic diffusion in one direction is the same as vacancy diffusion in the opposite direction.

**Nabarro-Herring creep** results from diffusion of vacancies in the grains from grain boundaries that are undergoing tensile stress to areas that are in compression. At the same time, atoms flow in the opposite direction, resulting in grain elongation in the direction of axial load. It is intrinsically a very slow process but becomes significant at stresses too low for dislocation motion to be activated and at relatively high temperatures, where the diffusion is fast enough to produce a measurable creep rate.

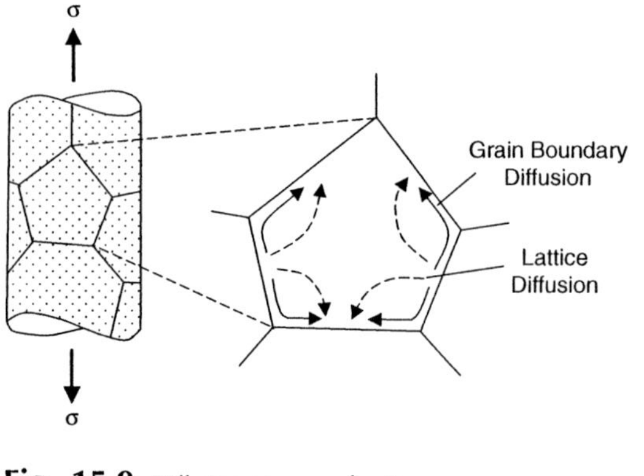


**Fig. 15.9** Diffusion creep mechanisms

**Coble creep** is similar, except that it occurs at lower temperatures, and all of the flow, vacancies and atoms occurs within the grain boundaries themselves. Because diffusion is very sensitive to temperature, at lower temperatures the main diffusion path is along the grain boundaries, since the activation energy for grain-boundary diffusion is considerably less than that for bulk diffusion.

Grain-boundary sliding is often observed in the final stages of creep just prior to failure. As the grains change shape, relative movement of the grain centers is necessary to maintain continuity at the grain boundaries. The grains actually start rotating and elongating in the direction of load. This often leads to intergranular failure modes. Grain-boundary precipitates are often used to inhibit grain-boundary sliding.

Grain size plays a role in creep of metals. The finer the grain size, the more rapid the mass transport, causing permanent deformation. Thus, under conditions where creep is solely due to diffusion, the creep resistance is improved by increasing the grain size. Note that this is different from low-temperature behavior, where a fine grain size is almost always beneficial to strength and ductility.

The dominant creep mechanism at a specific stress-temperature combination is found by calculating the creep rate for the several mechanisms and determining the maximum rate. Although the net creep rate is the sum of the several rates, the maximum one is usually much greater than the others. Deformation mechanism diagrams having axes of stress and temperature can be used to display the results of such calculations. Regions in the diagram essentially show stress-temperature combinations in which a given creep mechanism dominates. Ashby and his coworkers (Ref 3) have developed deformation mechanism maps that are useful in determining the dominant creep mechanism at different combinations of stress and temperature. Two deformation mechanism maps for the turbine blade alloy MAR-M 200 are shown in Fig. 15.10. In these maps, the shear stress (τ) normalized by the shear modulus ($G$) is plotted against the homologous temperature ($T/T_m$). Also, it is seen that diffusion creep mechanisms operate at low stresses and high temperatures, while at intermediate stresses, dislocation creep or power law creep is dominant. By comparing

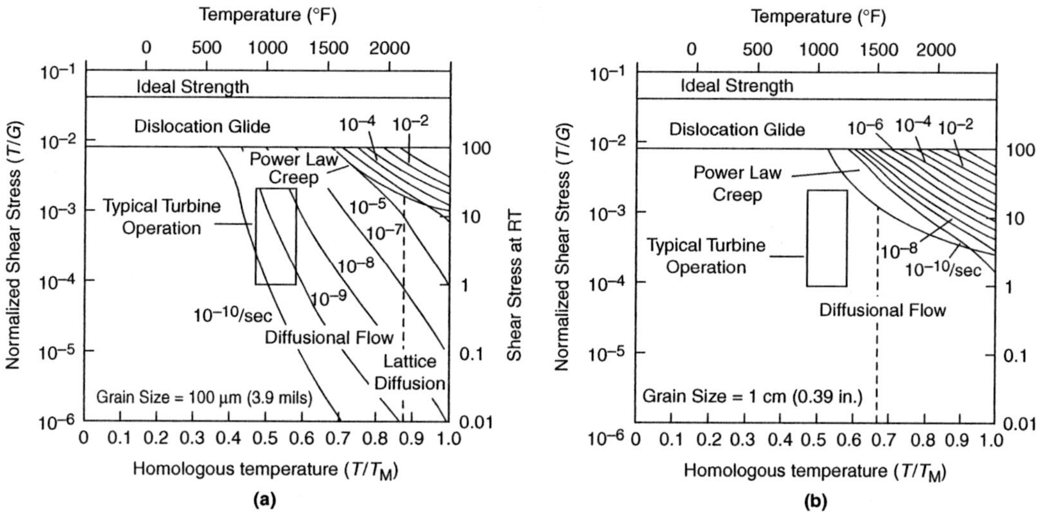


**Fig. 15.10** Deformation mechanism maps for MAR-M 200 superalloy. Source: Ref 4

Fig. 15.10(a) and (b), it can also be seen that increasing the grain size from 100 μm to 1 cm (3.9 mils to 0.39 in.) expands the power law creep regime and appreciably decreases the creep rate. These two maps for the same material with different grain sizes illustrate the wide variation in creep rates at a given combination of stress and temperature and help explain why large grain sizes are favorable for many creep-limited applications.

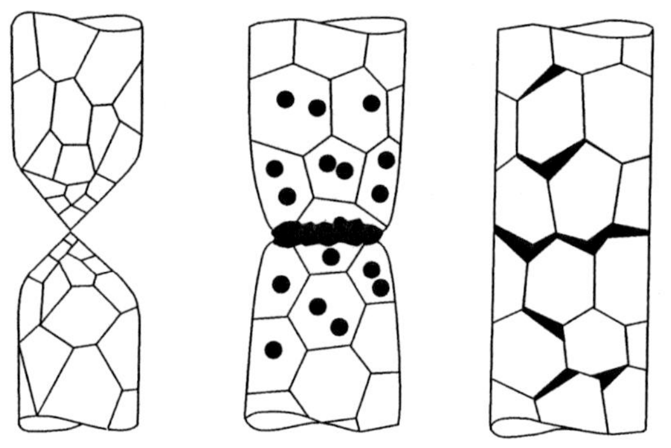

**Fig. 15.11** High-temperature failure modes. Source: Ref 5

## 15.4 Elevated-Temperature Fracture

The three primary modes of elevated fracture are rupture, transgranular fracture, and intergranular fracture, as illustrated in Fig. 15.11. Rupture, which is characterized by a reduction of area approaching 100%, occurs at high stresses and high temperatures. This type of failure can occur during hot working at high strain rates. Dynamic recovery and recrystallization are usually associated with rupture failures. Since it is not really a creep-dominated failure mode, it will not be considered further in this section. Transgranular fractures can occur during creep when the stress levels and strain levels are fairly high. Voids nucleate, usually around inclusions, and then grow and coalesce until fracture occurs. This type of fracture is very similar to ductile fracture modes experienced at room temperature, except that at elevated temperature, void formation and growth is aided by diffusion. Intergranular fracture is a creep fracture mode that occurs at lower stresses and longer times. Again, voids nucleate, grow, and coalesce, but their formation and growth is primarily restricted to the grain boundaries themselves. There is very little macroscopic plastic flow, and the failure appears brittle in nature.

As temperature increases, the failure mechanism of metals undergoes a transition from transgranular fracture to intergranular fracture. In other words, at low temperatures, the grain boundaries are usually stronger than the grains themselves, while at high temperatures, the grains are stronger than the grain boundaries. An example of an intergranular fracture is shown in Fig. 15.12 for Inconel 751 that was stress-rupture tested at 730 °C (1350 °F) and 380 MPa (55 ksi) for 125 h. The equicohesive temperature (ECT) is the temperature at which the strengths of the grains and the grain boundaries are the same. The ECT is not a fixed temperature. In addition to the obvious effects of stress and temperature, the strain rate is important. Decreasing the strain rate lowers the ECT and increases the tendency for intergranular failure. Since the total grain-boundary area decreases with increasing grain size, a material with a larger grain size will have a higher strength above the ECT but a lower strength below the ECT.

Depending on the alloy, temperature, and strain rate, stress-rupture fractures may be macroscopically either ductile or brittle. A brittle fracture is usually intergranular and occurs with little elongation or necking. In general, lower creep rates, longer rupture times, and higher temperatures promote intergranular fractures. Ductile fractures are transgranular and are typically accompanied by more pronounced elongation and necking. Transgranular creep ruptures, which generally result from high applied stresses, fail by a void-forming process similar to that of microvoid coalescence in room-temperature dimple rupture. Some stress-rupture fractures exhibit both transgranular and intergranular fracture paths. In such instances, it is usually found that the transgranular fractures were initiated by prior intergranular cracking that decreased the cross-sectional area, causing an increase in the stress level.

Intergranular (IG) fracture is the more common fracture path for most stress-rupture failures, depending on stress and temperature. Low stresses and/or high temperatures increase the likelihood of IG fracture. Sometimes, however, IG cracking may not be readily discernible on the surface of a brittle stress-rupture fracture, due to a buildup of surface oxides. Intergranular creep fracture depends on the nucleation, growth, and subsequent linking of voids on grain boundaries to form two types of cavities: wedge-type (w-type) cavities or isolated, rounded-type (r-type) cavities (Fig. 15.13). Cavitation damage, or the nucleation and growth of voids, is the most common microstructural change during creep and stress-rupture testing. Stress-rupture fractures are characterized by multiple creep voids adjacent to the main fracture. Wedge-shaped (w-type) cracks initiate along the grain boundaries that aligned in shear, while round or elliptical cavities (r-type) form along the boundaries that are aligned in tension. Under low stress conditions, IG fractures occur by void formation at the grain boundaries. These cavities form along grain edges rather than at grain corners. Because they appear to be round or spherical on metallographic cross sections, these voids are sometimes referred to as r-type cavities.

Wedge or w-type are usually associated with cracking at grain-boundary triple points and are also referred to as triple-point or grain-corner cracks. Wedge-type cracks form at triple points due to grain-boundary sliding and may be promoted by decohesion at interfaces between grain-boundary precipitates and the matrix.

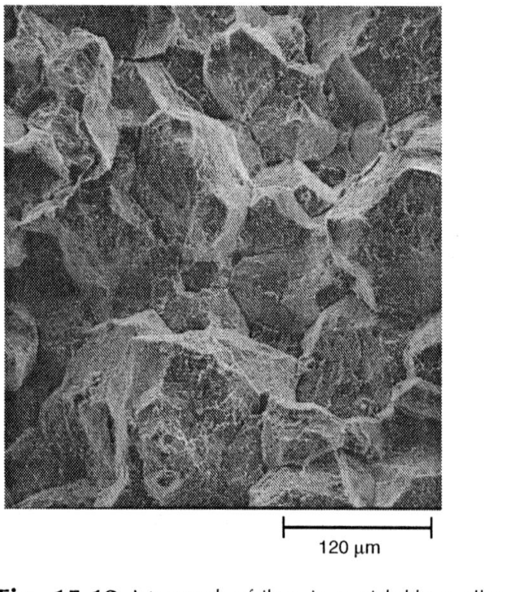


**Fig. 15.12** Intergranular failure in a nickel-base alloy. Source: Ref 6

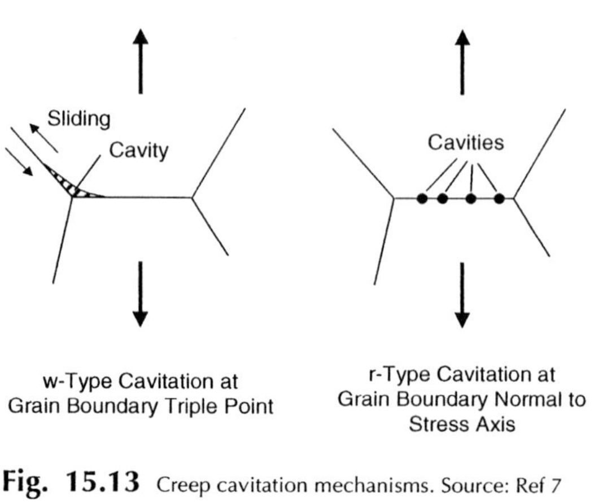


**Fig. 15.13** Creep cavitation mechanisms. Source: Ref 7

High stresses and lower temperatures promote wedge crack formation. Wedge cracking at triple points produces a rough fracture surface with identifiable grain-boundary precipitates.

Other significant microstructural features also are associated with stress-rupture fracture. When IG cracking occurs, not only do the crack paths follow grain boundaries at and beneath the fracture surfaces, but also the grains appear equiaxed even after considerable plastic deformation and total elongation, which implies dynamic recrystallization. In contrast, transgranular fracture, frequently is characterized by severely elongated grains in the vicinity of the fracture, with no recrystallization.

## 15.5 Metallurgical Instabilities

Stress, time, temperature, and the environment may act to change the metallurgical structure during service and hence change the strength. A sharp change of slope in a rupture life curve can occur from metallurgical instabilities (Fig. 15.5), such as transgranular-intergranular fracture transition, recrystallization, aging or overaging, intermetallic phase precipitation, oxidation, IG corrosion, or contamination by trace elements.

**Overaging.** Precipitation-hardening alloys, which include certain stainless steels, heat-resisting alloys, and aluminum alloys, are characteristically unstable in the heat treated condition. Overaging can occur when they are exposed to elevated operating temperatures without being loaded. Under creep conditions, temperature-induced and stress-induced atomic migration usually causes aging to continue, resulting in reduced strength. The extent and nature of this change depends on several factors, including the condition of the alloy prior to creep and the temperature, stress, and time of exposure.

**Intermetallic Phase Precipitation.** Topologically close-packed phases, such as sigma, mu, and Laves phases, can form at elevated temperatures in austenitic high-temperature alloys. The precipitate shape determines the effect they have on creep strength. For example, needlelike precipitates reduce the toughness and creep strength. The effect of sigma phase, a hard, brittle intermediate phase, on the nickel-base superalloy U-700 creep strength at 815 °C (1500 °F) is shown in Fig. 15.14, where the pronounced break in the stress-rupture curve at 1000 h is due to sigma-phase embrittlement. However, it has been found that sigma does not have a similar effect on certain other nickel-base alloys. The amount, location, and shape of sigma-phase precipitation determine whether sigma strengthens or weakens an alloy, or has no effect. Sigma and other intermetallics can severely reduce ductility and toughness on subsequent cooling to room temperature.

**Carbide Reactions.** Several types of carbides are found in steels and nickel- and cobalt-base superalloys. Carbides are normally present both within the grains and along the grain boundaries; however, it is the grain-boundary carbides that can have the most significant influence on the creep behavior. If the grain-boundary carbides are present as discrete particles, they are beneficial in slowing down creep, because they tend to pin the grain boundaries. However, they are detrimental if they form a brittle, continuous grain-boundary film. Continuous carbide films formed at grain boundaries almost always decrease rupture life. An electron micrograph of grain-boundary carbide films in a Waspaloy forging is shown in Fig. 15.15.

## 15.6 Creep Life Prediction

A frequent problem encountered in materials selection for component design is that it may be desirable to use a new alloy in a high-temperature structure; however, the alloy is so new that there are not sufficient data to substantiate its long-term creep resistance. Therefore, it is necessary to extrapolate higher-temperature creep data to much longer times at lower temperatures. The creep test is accelerated

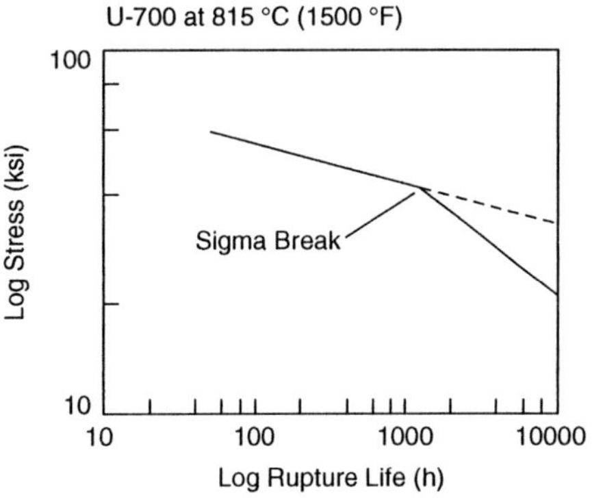


**Fig. 15.14** Sigma break in stress-rupture curve. Source: Ref 1

by raising the temperature, and the time difference is accounted for by using the activation energy, assuming that it is the same at all temperatures. Data are required at a range of temperatures and stresses. In order to extrapolate the data to the required times, a considerable amount of effort has been expended in developing empirical prediction methods. One of the most widely used is the Larson-Miller parameter, $P$:

$$P = T(\ln t + C) = \frac{Q}{R} \quad \text{(Eq 15.3)}$$

where $T$ is the absolute temperature, $t$ is the time to rupture in hours, $Q$ is the activation energy for creep, $R$ is the universal gas constant, and $C$ is the Larson-Miller constant, which typically has a value in the range of 30 to 65. A value of 46 is often used for metals when a specific value is not determined.

If $P$ is evaluated for pairs of $t$ and $T$ for a number of different stress levels, a single master curve can be plotted for the material. An actual set of stress-rupture data is shown for the nickel-base alloy Inconel 718 as log stress versus log time to rupture in Fig. 15.16(a). The data are then replotted as constant-time curves on coordinates of stress versus temperature in Fig. 15.16(b). Dashed horizontal lines have been added at stress levels of 550, 620, 760, and 830 MPa (80, 90, 110, and 120 ksi). Values for $T$ at the intercepts of these dashed lines and the constant-time curves are then determined and plotted in Fig. 15.16(c) on coordinates of log $t$ versus $10^4/T_A$. By extending the data in Fig. 15.16(c), a set of converging isostress lines meeting on the ordinate at a value of log $t = -25$ can be obtained. The Larson-Miller equation for this set of data is:

$$P = T_A\ (\log t_r + 25) \quad \text{(Eq 15.4)}$$

where absolute temperature, $T_A$, is in units of the Kelvin or Rankine temperature scales. The final master Larson-Miller curve is shown in Fig. 15.16(d).

One of the dangers of any of the extrapolation methods is an unanticipated structural instability in the material that may go undetected. Since the instabilities usually occur at higher temperatures, several tests should be conducted at substantially higher temperatures than the actual part will see in service. It should also be pointed out that the Larson-Miller parameter is only one of many methods for extrapolating stress-rupture data.

**Fig. 15.15** Continuous grain-boundary carbide film in Waspaloy. Source: Ref 1

## 15.7 Creep-Fatigue Interaction

In elevated-temperature structures, it is quite common to have both creep and fatigue conditions operating at the same time. A possible creep-fatigue interaction is shown in the elevated-temperature constant strain fatigue curves of Fig. 15.17, where dwelling at both the high and low stress levels produces creep that reduces the stress. While either creep or fatigue is complicated enough when they are the sole effect, combining creep and fatigue produces a truly complex situation.

There are a number of approaches to addressing this combined problem, but one of the simplest is the damage-accumulation approach. This approach is essentially an extension of the Palmgren-Miner rule for predicting cumulative fatigue damage. In other words, the independent damage from fatigue and creep are summed and compared to the limit of damage the material can withstand:

$$\sum_{j=1}^{j=k} \left(\frac{n}{N_d}\right)_j + \sum_{l=1}^{l=m} \left(\frac{t}{t_d}\right)_l \leq D \quad \text{(Eq 15.5)}$$

where $n$ is the number of fatigue cycles applied at loading condition $j$, $N_d$ is the number of

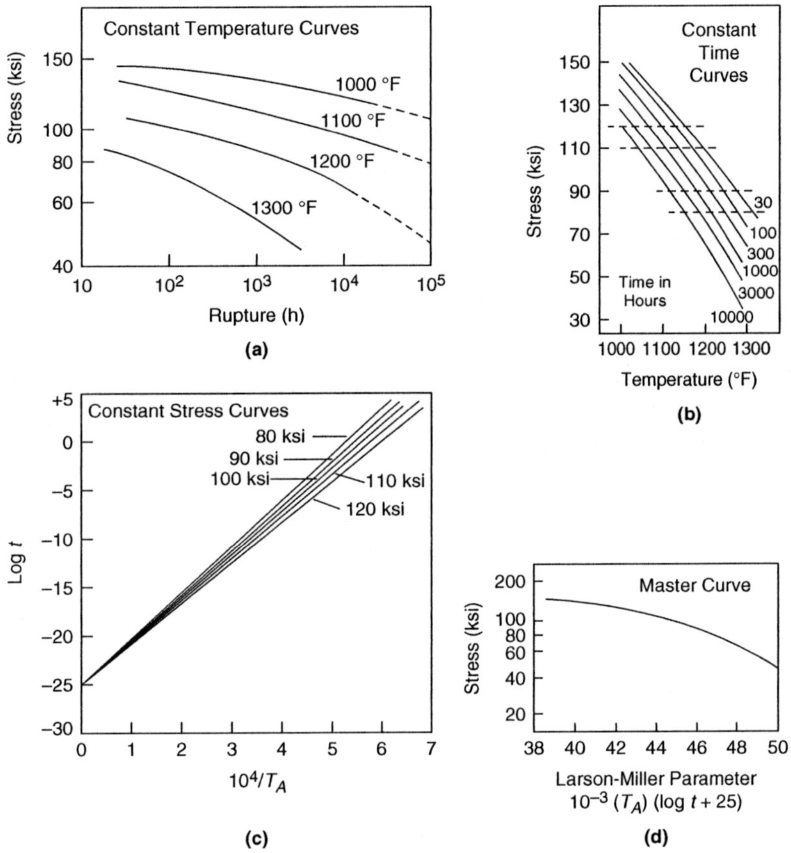


**Fig. 15.16** Larson-Miller curve generation for Inconel 718. Source: Ref 8

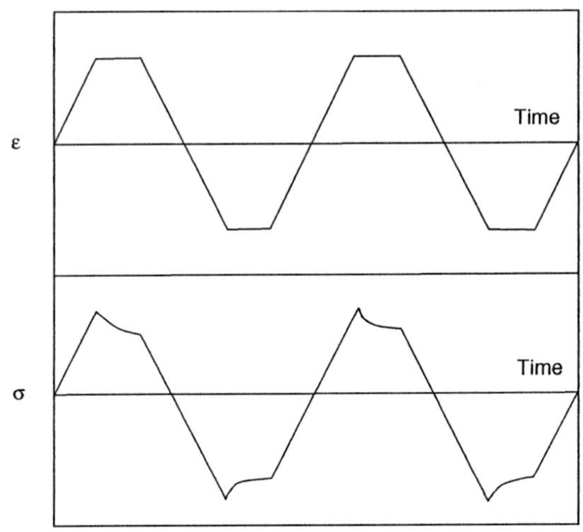


**Fig. 15.17** Example of fatigue-creep interaction

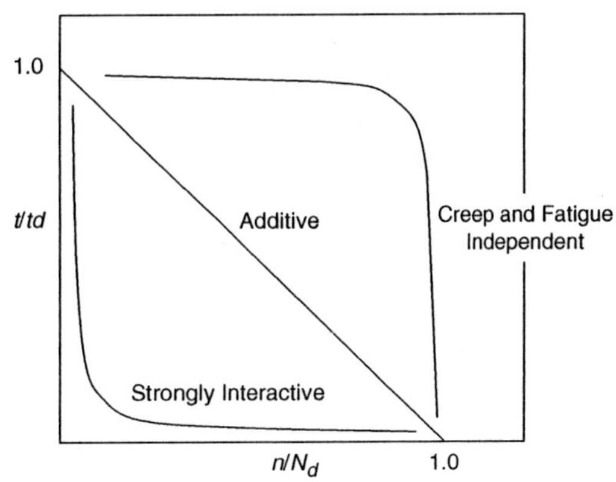


**Fig. 15.18** Example of fatigue-creep interaction

design-allowable fatigue cycles at loading condition $j$, $t$ is the time under the applied creep load condition $l$, $t_d$ is the allowable time at load condition $l$, $D$ is the total allowable creep-fatigue damage, which is usually taken as 1.0 but sometimes less than 1.0.

A plot of the combined effects of creep and fatigue is shown in Fig. 15.18. Three types of interaction are shown. When the contributions of creep and fatigue are the same, Eq 15.5 is obeyed, and $D$ is equal to 1.0 and a straight-line interaction is obtained. In essence, this results in a reduction of lifetime by half in a case where creep and fatigue would be expected to result in failure at the same number of cycles. For the case where there is no interaction, creep and fatigue have no effect on each other, and their respective lifetimes are unchanged by damage from the other mechanism. Finally, if there is a strong interaction, the lifetimes for both creep and fatigue are much shorter as a result of the interaction of the two. In this case, if lifetime is being predicted on the basis of the creep and fatigue data separately, there will be a serious problem.

## 15.8 Design Against Creep

Design against creep can take several approaches. First, it makes sense to use a material with a high melting temperature in high-temperature applications since, to a good approximation, diffusion coefficients on which the creep rate depends are proportional to the homologous temperature ($T/T_m$). Face-centered cubic (fcc) metals generally have superior creep resistance to body-centered cubic (bcc) metals at equivalent homologous temperatures, because the slightly more open bcc structure results in greater diffusivities.

Design against creep may also involve first determining the dominant creep mechanism. If diffusion creep dominates, then increasing the grain size is a beneficial microstructural alteration. Further, for diffusion creep, improved creep resistance can sometimes be obtained by placing inert particles, such as carbides, on grain boundaries. This helps to pin the grain boundaries, enhancing creep resistance. Creep-resistant alloys include carbon steels, chromium-molybdenum steels, chromium-molybdenum-vanadium steels, stainless steels, nickel and cobalt alloys, and superalloys. A comparison of the stress-rupture properties for a number of different metallic alloy families is shown in Fig. 15.19.

Depending on the in-service stress, carbon steels can be used at operating temperatures between 400 and 480 °C (750 and 900 °F). Above approximately 480 °C (900 °F), the pearlite in carbon steels begins to transform to graphite, which leads to embrittlement. Chromium-molybdenum steels contain 0.5 to 9.0 wt% Cr and either 0.5 or 1.0 wt% Mo. The chromium content increases oxidation resistance, while molybdenum increases elevated-temperature strength and forms carbides

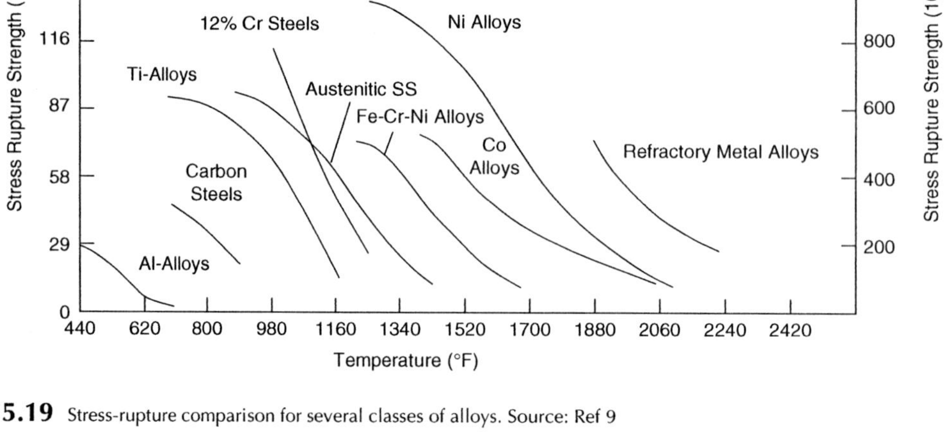


**Fig. 15.19** Stress-rupture comparison for several classes of alloys. Source: Ref 9

that help prevent graphitization. Chromium-molybdenum-vanadium steels contain up to 0.5 wt% C, 5 wt% Cr, 1 wt% Mo, and 0.9 wt% V. The vanadium addition increases the softening point of the steel. Both the chromium-molybdenum and the chromium-molybdenum-vanadium steels extend the operating range up to 540 °C (1000 °F).

Stainless steels are also used for elevated-temperature applications. Martensitic stainless steels are used at temperatures up to 540 to 650 °C (1000 to 1200 °F) but must be tempered at approximately 55 °C (100 °F) higher than the operating temperature to prevent softening in service. Ferritic stainless steels, due to their high chromium contents, have the best scaling resistance but are limited to service temperatures of approximately 370 °C (700 °F) due to the precipitation of embrittling phases at higher temperatures. Austenitic stainless steels have the best creep resistance of the stainless steels and are used at temperatures up to 870 °C (1600 °F). However, a major problem with the austenitic grades is their high coefficients of thermal expansion, which must be compensated for during design.

Nickel-base alloys are frequently used for high-temperature service in corrosive environments. These alloys contain alloying elements such as chromium, molybdenum, and iron. The addition of chromium and molybdenum to nickel produces alloys with resistance to both oxidizing and reducing corrosives. Resistance to oxidizing environments is provided by the chromium, while molybdenum gives the resistance to reducing environments. These alloys are the most corrosion resistant of the nickel alloys. The Ni-Cr-Mo alloys are used in the chemical processing, pollution control, and waste treatment industries to utilize their excellent combination of heat and corrosion resistance.

Superalloys are heat-resistant alloys of nickel, iron-nickel, and cobalt that frequently are used at temperatures exceeding 540 °C (1000 °F). However, some superalloys are capable of being used in load-bearing applications in excess of 85% of their incipient melting temperatures. They exhibit the best combination of high strength, good fatigue and creep resistance, good corrosion resistance, and the ability to operate at elevated temperatures for extended periods of time (i.e., metallurgical stability). Their combination of elevated-temperature strength and resistance to surface degradation is unmatched by other metallic materials.

Nickel-base superalloys have the best creep resistance of the superalloys and are used as turbine blades in engines operating at temperatures as high as 1290 °C (2350 °F) ($T/T_m = 0.9$). The microstructure consists of fine $Ni_3Al$ particles dispersed in an fcc nickel-rich solid solution. These alloys are remarkably resistant to particle coarsening. Solid-solution hardening is also used to strengthen the matrix, and when used in polycrystalline form, carbides are dispersed on the grain boundaries. Increases in engine operating temperatures over the last 50 years have been gradual but significant, from approximately 400 °C (750 °F) to the previously noted 1290 °C (2350 °F). This increase in operating temperature translates into improved engine efficiency.

Improved high-temperature material properties are largely responsible for the increase in the engine operating temperature. Some improved material performance has come about from minor alterations in alloy chemistry, and some has resulted from processing changes. For example, the major stress axis in a turbine blade is parallel to the blade axis. In a polycrystalline blade, this stress is normal to some grain boundaries, and this causes voids, precursors to fracture, to initiate on the boundaries. Directional solidification (i.e., solidifying the material sequentially from the blade bottom to its top) results in columnar grains having boundaries aligned along the blade axis. This significantly reduces the cavitation problem. Further improvements came with the casting of single-crystal blades. Increasing the superalloy grain size from 100 μm to 10 mm (0.004 to 0.4 in.) (the order of the thickness of a turbine blade) reduces the creep rate by approximately 6 orders of magnitude when Coble creep is the dominant creep mechanism, as it often is under typical blade operating conditions.

## ACKNOWLEDGMENTS

Sections of this chapter were adapted from "Assessment and Use of Creep-Ruptures Properties" by H.R. Voorhess and M. Prager in *Mechanical Testing and Evaluation*, Volume 8, *ASM Handbook,* ASM International, 2000, and "Creep and Strees-Rupture Failures" in *Failure Analysis and Prevention,* Volume 11, *ASM Handbook,* ASM International, 2002.

## REFERENCES


1. Creep and Stress-Rupture Failures, *Failure Analysis and Prevention,* Vol 11, *ASM Handbook,* ASM International, 2002
2. F.C. Monkman and N.J. Grant, *Proceedings ASTM 56,* American Society for Testing and Materials, 1953, p 593
3. M.F. Ashby, The Microstructure and Design of Alloys, *Proceedings, Third International Conference on Strength of Metals and Alloys,* Vol 2 (Cambridge, England), Institute of Metals, Iron and Steel Institute, and U.S. Army European Research Office, 1973
4. Creep and Creep-Rupture Testing, *Mechanical Testing and Evaluation,* Vol 8, *ASM Handbook,* ASM International, 2000
5. T.H. Courtney, *Mechanical Behavior of Materials,* 2nd ed., McGraw-Hill Book Co., 2000
6. S. Lampman, Intergranular Fracture, *Failure Analysis and Prevention,* Vol 11, *ASM Handbook,* ASM International, 2002
7. M.A. Meyers and K.K. Chawla, *Mechanical Metallurgy—Principles and Applications,* Prentice-Hall Inc., 1984
8. H.R. Voorhees and M. Prager, Assessment and Use of Creep-Rupture Properties, *Mechanical Testing* and *Evaluation,* Vol 8, *ASM Handbook,* ASM International, 2000
9. A. Weronski and T. Hejwowski, *Thermal Fatigue of Metals,* Marcel Dekker, Inc., 1991, p 83


## SELECTED REFERENCES


- M.F. Ashby and D.R.H. Jones, *Engineering Materials 1—An Introduction to Their Properties and Applications,* 2nd ed., Butterworth Heinemann, 1996
- G.E. Dieter, *Mechanical Metallurgy,* 3rd ed., McGraw-Hill Book Co., 1986
- J.C. Gibeling, Creep Deformation of Metals, Polymers, Ceramics, and Composites, *Mechanical Testing and Evaluation,* Vol 8, *ASM Handbook,* ASM International, 2000

CHAPTER 16

# Deformation Processing

METALWORKING PROCESSES are commonly classified as either hot working or cold working operations. Primary metalworking processes, such as the bulk deformation processes used to conduct the initial breakdown of cast ingots, are always conducted hot. The term *bulk deformation* implies large amounts of material movement, such as in hot rolling or forging. Secondary processes, which are used to produce the final product shape, are also conducted either hot or cold. Some secondary processes, such as sheet forming, do not involve large amounts of deformation.

Hot working processes are conducted at temperatures above the recrystallization temperature, which is approximately 0.5 $T_m$. Cold working processes are conducted at or near room temperature, while warm working processes are conducted at intermediate temperatures. Hot working produces a recrystallized grain structure, while the grain structure due to cold working is unrecrystallized and retains the effects of the working operation.

Bulk deformation changes the shape of a workpiece by plastic deformation through the application of compressive forces, as for the typical bulk deformation processes in Fig. 16.1. In addition to shaping the metal, bulk deformation is used to refine the structure that results from solidification. To refine the

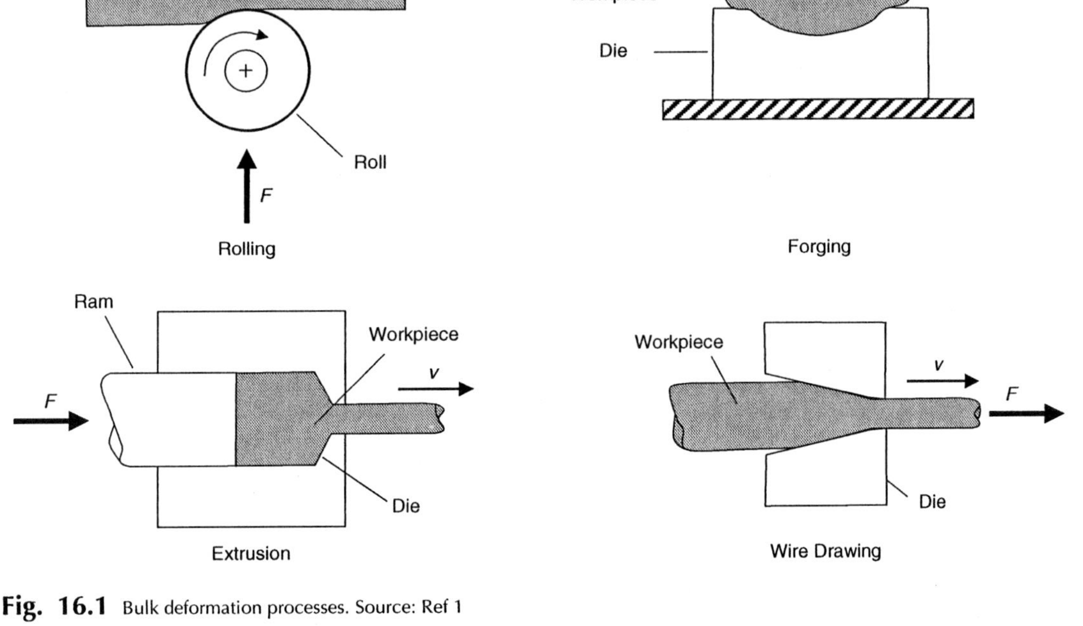


**Fig. 16.1** Bulk deformation processes. Source: Ref 1

inhomogeneous structure resulting from solidification, cast ingots and continuously cast slabs and blooms are typically hot worked into intermediate product forms, such as plates, bars, and sheet. Large plastic deformation in combination with heat is very effective in refining the microstructure, breaking up macrosegregations, collapsing and sealing porosity, and refining the grain size. This product may be suitable for its intended application, but in many cases, it provides the starting material for secondary deformation processes such as drawing, hot or cold forging, and sheet metalworking.

Because most processes involve sliding contact between the workpiece and a tool or die, friction affects material flow, die pressures, and the force and energy requirements. In many instances, lubricants are used to minimize friction. Lubricants also reduce die wear, help in providing temperature control, and minimize high-temperature oxidation.

## 16.1 Hot Working

Hot working normally takes place at approximately 70 to 80% of the absolute melting temperature. During hot working, the strain-hardened and distorted grain structure produced by deformation is rapidly eliminated by the formation of new strain-free grains as a result of recrystallization. Dynamic recrystallization occurs during deformation, while static recrystallization occurs after deformation is complete but while the workpiece is still hot. For processes such as hot rolling, hot extrusion, and hot forging, the time within the deformation zone is usually short, and grain refinement is usually accomplished by static recrystallization after hot working. A high level of hot deformation followed by holding the workpiece at an elevated temperature causes static recovery and recrystallization, resulting in a fine grain size. This may occur during hot rolling (Fig. 16.2) where there is time between roll passes, or after hot forging, where the workpiece slowly cools to room temperature.

Very large deformations are possible in hot working, and since recovery processes keep pace with the deformation, hot working occurs at essentially a constant flow stress (the shear stress required to cause plastic deformation of the metal). Since the flow stress decreases with increasing temperature, metals become more malleable, and less energy is needed to produce a given amount of deformation. However, as the temperature increases, the strength also decreases. Therefore, hot working processes usually involve the use of compressive forces to prevent cracking or failure. Because recovery processes take time, flow stress, $\sigma_f$, is a function of strain rate, $\dot{\varepsilon}$:

$$\sigma_f = C\dot{\varepsilon}^m \quad \text{(Eq 16.1)}$$

where $C$ is the strength coefficient (decreasing with increasing temperature), and $m$ is the strain-rate sensitivity.

A high value of $m$ means that any incipient neck that develops becomes stronger and spreads to neighboring material, allowing more deformation in tension. In some very fine-grained metals, the value of $m$ may reach 0.4 to 0.5 but only at very low strain rates and within a limited temperature range.

Because hot working processes may hold a workpiece at a high temperature for a long time, grain growth can occur. In fact, in an extended hot working process such as ingot breakdown during rolling, a cyclic history of grain

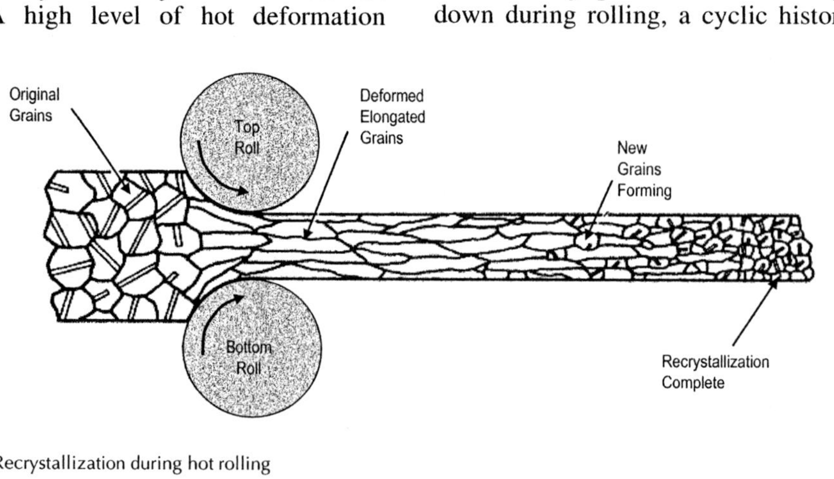


**Fig. 16.2** Recrystallization during hot rolling

deformation, recrystallization, and grain growth occurs for each deformation step. The ability to put work into the grains at a level sufficient to cause recrystallization is the reason that fine grains can be developed from a coarse-grained structure by hot working. Thus, hot working processes must balance recovery and recrystallization against grain growth.

At temperatures above the equicohesive temperature, the grain interiors are more resistant to deformation than the grain boundaries, and the grain boundaries can sustain substantial deformation. If a very fine grain size can be achieved and maintained during deformation at a low strain rate, grain-boundary sliding can occur. Creep forming, hot die forging, isothermal forging, and isothermal rolling are processes that rely in part on grain-boundary sliding and other thermally activated deformation mechanisms.

The workability, or the ease with which a metal is shaped by plastic deformation, is lower for cast structures than for wrought structures (Fig. 16.3). Workability is also dependent on grain size and grain structure. When the grain size is relatively large, as in cast structures, workability is lower because cracks can initiate and propagate along grain boundaries. Also, with cast structures, impurities frequently segregate to the center, to the top, or to the surface of the ingot, creating areas of low workability. Because alloying elements are not distributed uniformly, the temperature range over which an ingot structure can be worked is somewhat limited.

The upper limit for hot working is determined by the temperature at which either melting or excessive oxidation occurs. Generally, the maximum working temperature is limited to approximately 40 °C (100 °F) below the

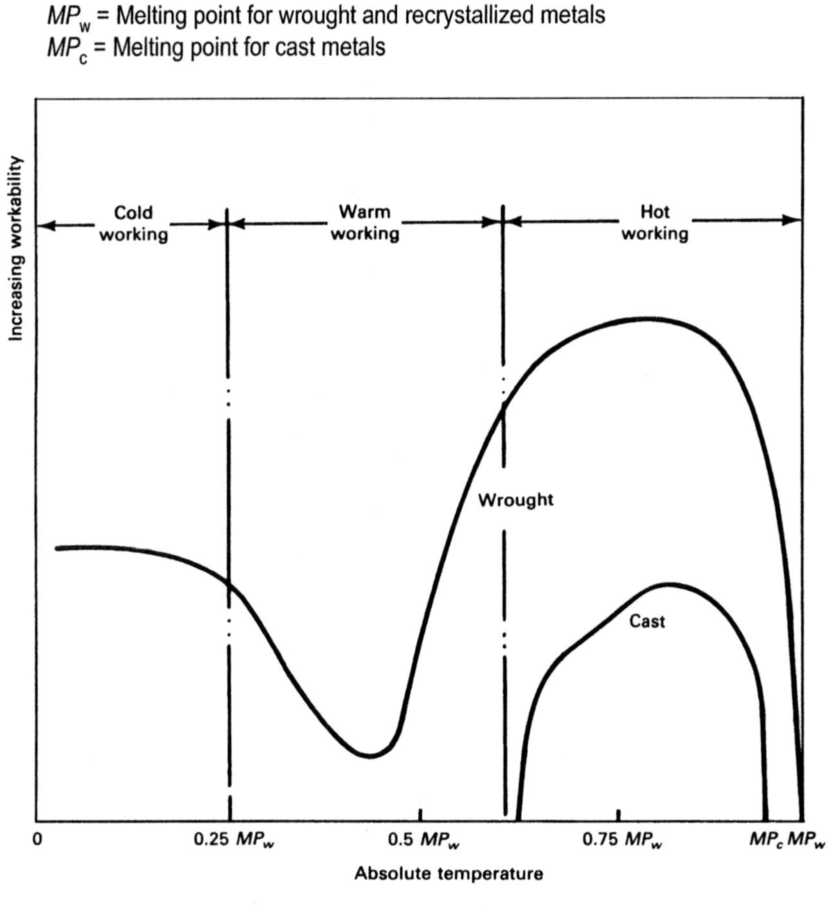


**Fig. 16.3** Relative workabilities of wrought and cast metals. Source: Ref 2

melting temperature to prevent the occurrence of hot shortness. The melting point of an alloy in the as-cast condition is usually lower than that of the same alloy in the fine-grain, recrystallized condition, because of chemical inhomogeneities and the presence of low-melting-point compounds that frequently occur at grain boundaries. Deformation at temperatures too close to the melting point of these compounds may lead to grain-boundary cracking when the heat developed by deformation increases the workpiece temperature and produces local melting, a fracture mode called hot shortness. Hot shortness can be prevented by using a sufficiently low deformation rate that allows the heat developed by deformation to be dissipated by the tooling, by using lower working temperatures, or by subjecting the workpiece to a homogenization heat treatment prior to hot working. The intermediate temperature region of low ductility shown in Fig. 16.3 can occur at temperatures that are sufficiently high for grain-boundary sliding to cause grain-boundary cracking but are not high enough for the cracks to be healed by dynamic recrystallization.

The advantages of hot working include:

- Flow stresses are low; hence, forces and power requirements are relatively low. Even very large workpieces can be deformed with equipment of reasonable size.
- Ductility is high, allowing for large deformations.
- Complex part shapes can be produced.

## 16.2 Cold Working

Cold working produces better surface finishes, can impart higher strengths, and allows thinner products than hot working but requires higher forces. Cold working results in a deformed, unrecrystallized grain structure with the grains elongated in the direction of metal flow. An empirical relationship between flow stress and plastic strain during cold working is:

$$\sigma_f = K\varepsilon^n \quad \text{(Eq 16.2)}$$

where $\sigma_f$ is the flow stress, $K$ is the strength coefficient (stress when $\varepsilon = 1.0$), $\varepsilon$ is the plastic strain, and $n$ is the work-hardening exponent.

A high strength coefficient indicates a high initial resistance to plastic flow, and metals with a high $K$ require large forces for deformation. Work hardening is a measure of how the resistance to plastic flow increases as the metal is deformed. Typically, $n$ has values of 0.1 to 0.5 for cold working, with 0 being a perfectly plastic metal that will not work harden. A high $n$ value allows more tensile deformation before a localized neck develops. A metal with a high work-hardening exponent but a low strength coefficient will achieve a high strength level after a large amount of deformation. Examples include coppers, brasses, and low-carbon steels that are cold worked to produce higher hardness and strength in the formed part. The value of $K$ increases and $n$ decreases in the course of cold working; thus, ductility is reduced and the forces increase. Intermediate process anneals can restore ductility and reduce forces but at an additional cost. Warm working in an intermediate temperature range can combine some of the benefits of hot and cold working. Some values of $K$ and $n$ are given in Table 16.1. For steels, $K$ increases with carbon content, while $n$ generally decreases. Both copper and brass have a much higher work-hardening exponent than steels. Over the range of strain rates at which cold deformation processes are conducted (0.1 to 100/s), the sensitivity to strain rate for most metals is low. Rather, the strain level and the strength coefficient, $K$, control the flow stress.

During cold working, the grains distort, the grain boundaries align, and a well-defined, fiberlike grain pattern develops. Nonmetallic inclusions can further produce a definite directionality in the microstructure and mechanical properties due to the deformation-induced fibrous microstructure. Extremely deformed microstructures, such as in cold-rolled sheet products, can also develop into aligned crystallographic planes or texture. This results in anisotropic behavior of the deformed material, both in service or during subsequent deformation steps.

Although cold rolling is usually conducted at room temperature, the work of deformation can increase the temperature by as much as 95 to 205 °C (200 to 400 °F). A material subjected to cold rolling strain hardens considerably, and grains become elongated in the direction of major deformation. Dislocation density increases, and when a tension test is performed on the strain-hardened material, a higher stress will be needed to initiate and maintain plastic deformation; that is, the yield stress increases. However, the ductility of the material, as expressed by total elongation and reduction of

**Table 16.1 Work-hardening exponents and strength coefficients**

| Metal | Condition | Work-hardening exponent ($n$) | Strength coefficient ($K$) MPa | Strength coefficient ($K$) (ksi) |
|---|---|---|---|---|
| 1100 aluminum | Annealed | 0.20 | 180 | 26 |
| 2024 aluminum | Solution treated and naturally aged | 0.16 | 690 | 100 |
| 6061 aluminum | Annealed | 0.20 | 895 | 30 |
| 6061 aluminum | Solution treated and artificially aged | 0.05 | 405 | 59 |
| 7075 aluminum | Annealed | 0.17 | 400 | 58 |
| Copper | Annealed | 0.54 | 315 | 46 |
| 70/30 brass | Annealed | 0.49 | 895 | 130 |
| 85/15 brass | Cold rolled | 0.34 | 580 | 84 |
| 0.05% C steel | Annealed | 0.26 | 530 | 77 |
| 4135 steel | Annealed | 0.17 | 1015 | 147 |
| 4135 steel | Cold rolled | 0.14 | 1105 | 160 |
| 4340 steel | Annealed | 0.15 | 640 | 93 |
| 0.6% C steel | Quenched and tempered at 540 °C (1000 °F) | 0.10 | 1570 | 228 |
| 0.6% C steel | Quenched and tempered at 705 °C (1300 °F) | 0.19 | 1225 | 178 |
| 304 stainless steel | Annealed | 0.45 | 1275 | 185 |
| 410 stainless steel | Annealed | 0.10 | 960 | 139 |
| Cobalt alloy | Heat treated | 0.50 | 2070 | 300 |

Source: Ref 3

area, drops because of the higher initial dislocation density.

Advantages of cold working include:

- In the absence of cooling and oxidation, tighter tolerances and better surface finishes are obtained.
- Thinner walls are possible.
- The final properties of the workpiece can be closely controlled, and, if desired, the high strength obtained during cold rolling can be retained, or, if high ductility is needed, grain size can be controlled by annealing.
- Lubrication is, in general, easier.

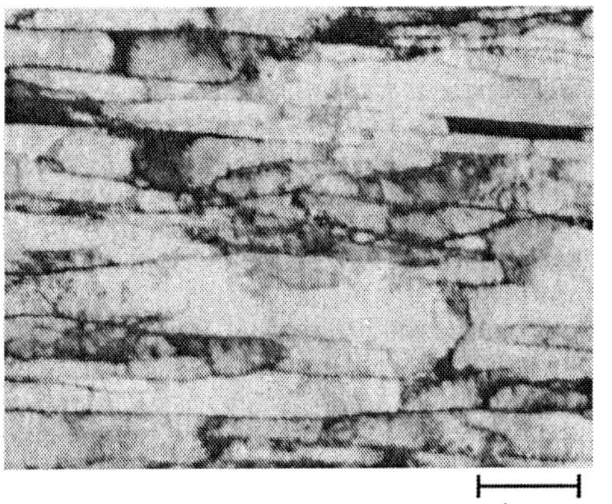


**Fig. 16.4** Grain directionality due to rolling of electrolytic iron single crystal. See also Fig. 26.7. Source: Ref 4

## 16.3 Rolling

Rolling is reduction of the cross-sectional area of the metal stock, or the general shaping of the metal products, through the use of the rotating rolls. The rolls rotate to pull and simultaneously squeeze the work between them. After ingot casting, rolling is perhaps the most important metalworking process. More than 90% of all wrought steel, aluminum, and copper produced go through at least one rolling process. Hot rolling helps to break up the as-cast structure and provides a more uniform grain size and a better distribution and size of constituent particles. During hot rolling, the grain structure becomes elongated in the rolling direction, as shown in Fig. 16.4. This grain directionally can have a substantial effect on some of the mechanical properties, especially fracture toughness and corrosion resistance, in which the properties are lowest in the through-the-thickness or short-transverse direction.

The primary objectives of the rolling process are to reduce the cross section of the incoming material while improving its properties and to obtain the desired shape. The process can be carried out hot, warm, or cold, depending on the application and the material involved. During rolling, compression deformation is accomplished by using two work rolls. Because the rolls rotate with a surface velocity exceeding the speed of the incoming metal, friction along the contact interface grabs the metal and propels it forward. The amount of deformation that can be achieved in a single pass between a given pair of rolls depends on the frictional conditions along the interface. If too much deformation is attempted, the rolls will simply skid over stationery metal, while too little deformation per pass results in excessive cost.

The main advantage of rolling lies in its ability to produce desired shapes from relatively large pieces of metals at very high speeds in a somewhat continuous manner. Many engineering metals are often cast into ingots and then processed by hot rolling into blooms, slabs, and billets (Fig. 16.5), which are then rolled into other products, such as plate, sheet, tube, rod, bar, and structural shapes. Rolling of blooms, slabs, billets, plates, and structural shapes is usually done at temperatures above the recrystallization temperature, where large reductions in height or thickness are possible with moderate forming pressures. Sheet and strip are often cold rolled in order to maintain closer thickness tolerances and provide superior surface finishes.

The three principal types of rolling mills are the two-high, three-high, and four-high mills shown in Fig. 16.6. As their names imply, this classification is based on the way the rolls are arranged in the housings. Two-high mills may be either pull-over mills or reversing mills. In pull-over mills, the rolls run in only one direction. The workpiece must be returned over the top of the mill for further rolling. Reversing mills employ rolls in which the direction of rotation can be reversed. Rolling then takes place alternately in two opposite directions. Reversing mills are among the most widely used in industry and can be used to produce slabs, blooms, plates, billets, rounds, and partially formed sections suitable for rolling into finished shapes on other mills. In three-high mills, the top and bottom rolls rotate in the same direction, while the middle roll rotates in the opposite direction. This allows the workpiece to be passed back and forth alternately through the top and middle rolls, and then through the bottom and middle rolls without reversing the direction of roll rotation. Since the roll-work contact length is reduced with a smaller roll radius, leading to lower forces, torque, and power, the four-high rolling mill uses two smaller-diameter rolls to contact the work, with two larger backing rolls for support. The larger backup rolls reinforce the smaller work rolls, thus allowing fairly large reductions without excessive amounts of roll deflection. Four-high mills are used to produce wide plates and hot-rolled or cold-rolled sheet, as well as strip of uniform thickness.

Another important type of mill that uses smaller working rolls against the workpiece is the cluster mill, the most common being the Sendzimir mill. In a typical Sendzimir mill design (Fig. 16.7), each work roll is supported through its entire length by two rolls, which in

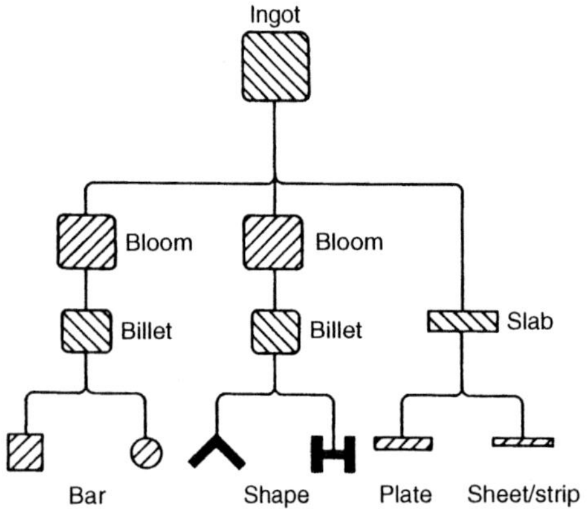


**Fig. 16.5** A bloom has a nearly square cross section with an area larger than 32 in.$^2$ (206 cm$^2$); the minimum cross section of a billet is about 1.5 × 1.5 in. (3.8 × 3.8 cm), and a slab is a hot-rolled ingot with a cross-sectional area greater than 16 in.$^2$ (103 cm$^2$) and a section width of at least twice the section thickness. Plates are generally thicker than 0.25 in. (0.64 cm), whereas sheets are thinner-gage materials with very large width-to-thickness ratios. Sheet material with a thickness of a few thousandths of an inch is referred to as foil.

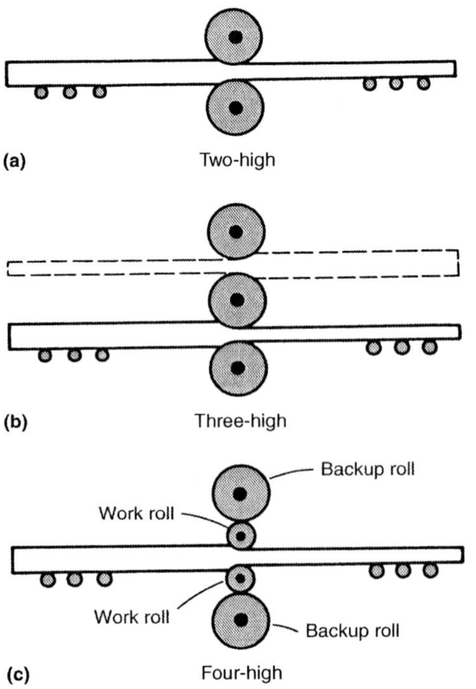


**Fig. 16.6** Common types of rolling mills. Source: Ref 5

turn are supported by three rolls. These rolls transfer roll separating forces through four large backup rolls to a rigid cast steel housing. Sendzimir mills are used for the cold rolling of sheet and foil to precise thicknesses. To achieve higher throughput rates in standard products, a tandem rolling mill is often used, consisting of a series of rolling stands. However, with each rolling step, work velocity increases so the rolling speeds at each stand must be synchronized. Hot rolling is also used to produce structural shapes. As shown in the example for an I-beam in Fig. 16.8, multiple stands are used to gradually produce the desired shape.

After hot rolling, many products undergo further processing by cold rolling. Prior to cold rolling, coils of steel are usually annealed to remove any cold work introduced during hot rolling. Cold rolling, usually performed at room temperature, permits the thickness of the strip to be closely controlled while also improving the surface quality. Since cold rolling work hardens the metal, sheet metal that is going to be subsequently used for forming or pressing operations is annealed to soften it. Depending on the particular equipment and cold rolling parameters, different patterns of residual stresses can be produced in cold-rolled sheet (Fig. 16.9). If the product is going to be used in the cold-rolled state, it is better to use smaller rollers and/or smaller reductions per pass to produce compressive residual stresses on the surface.

Another form of cold processing is cold drawing (Fig. 16.1). Metal in rod form is drawn through a series of dies that progressively reduce the rod circumference to produce wire. The drawing process substantially increases the material tensile strength. Drawn wires can be spun into huge cables that have very high strengths.

## 16.4 Forging

During forging, large compressive forces are used to deform the metal into a new shape. There are two main reasons for forging: (1) forging improves the properties and homogeneity of the microstructure; therefore, forging is often used to prepare cast ingots for other bulk deformation processes, such as hot rolling; and (2) forging is also a major method of producing semifinished or near-net shapes. Forging processes can be described as open-die forging or closed-die forging (Fig. 16.10). Open-die forging is carried out between two flat dies or dies with very simple configurations, such as V-shaped or semicircular cavities. In closed-die forging, the metal to be forged is placed between two dies that have the upper and lower impressions of the desired shape of the forging. Closed-die forging can be carried out by using a single pair of dies or with multiple impression dies. Hammers or presses are used to apply the forces to the

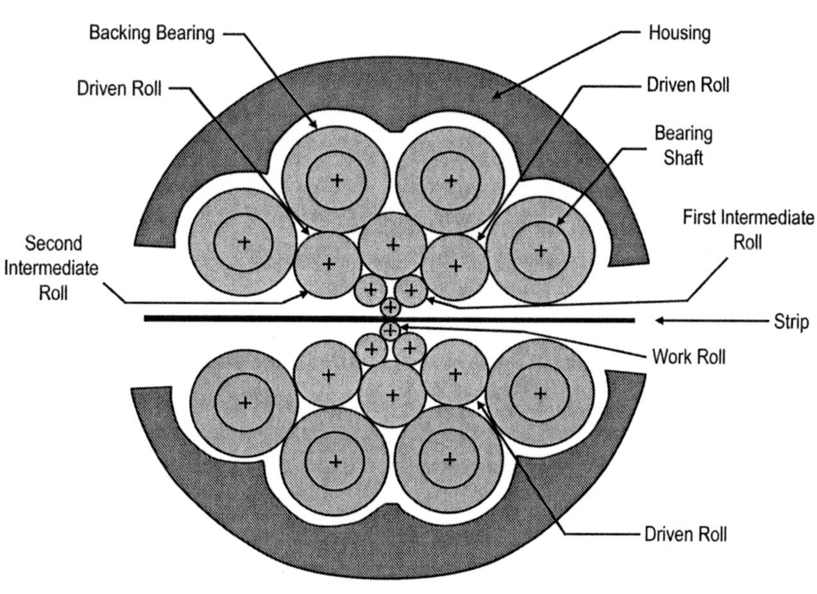


**Fig. 16.7** Sendzimir mill. Source: Ref 5

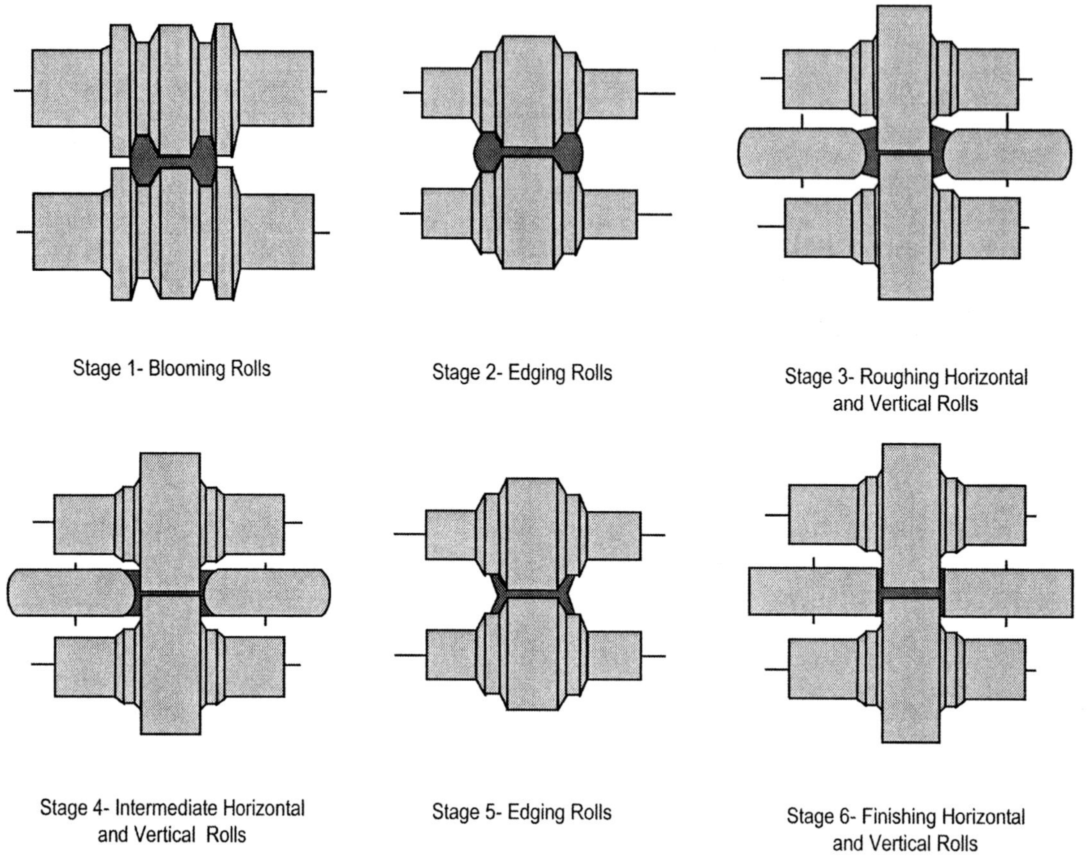


**Fig. 16.8** Rolling of an I-beam. Source: Ref 6

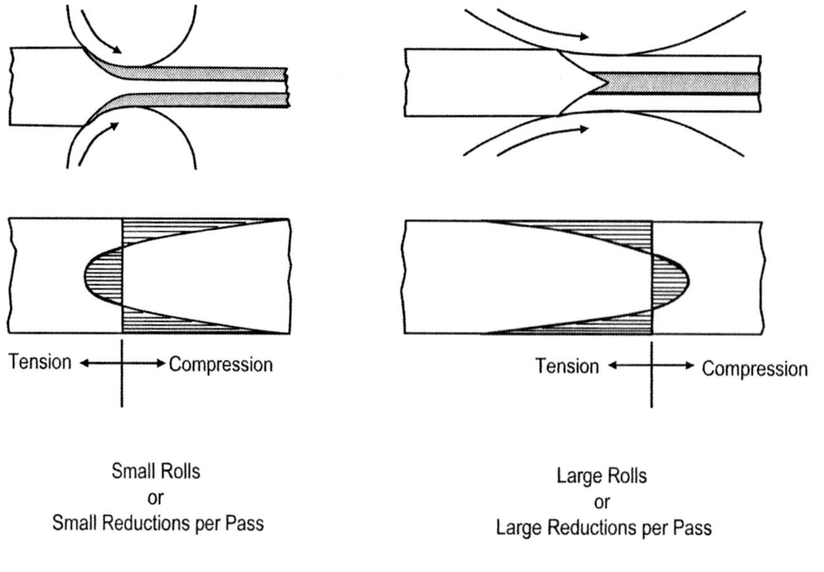


**Fig. 16.9** Residual stress patterns produced by rolling. Source: Ref 6

workpiece, which is usually heated to above its recrystallization temperature, although cold forging is also used to produce finished shapes.

During the breakdown of cast ingots and working by forging, nonuniformities in alloy chemistry, second-phase particles, inclusions, and the crystalline grains are broken up and redistributed, improving the homogeneity of the microstructure. This is important in cast structures that are going to be processed by hot rolling. Without the refinement of the microstructure due to forging, the ingot will often crack under the forces of the rolling operation. When the deformation produced by forging becomes large, such as during the forging of a structural shape, the microstructure becomes aligned in the direction of greatest metal flow. This directional pattern of grains and second-phase particles is known as the grain flow pattern. This pattern is responsible for the familiar fiber structure of forgings (Fig. 16.11). It also produces a directional variation in strength, ductility, fracture toughness, and fatigue strength. This anisotropy in properties is greatest between the working (longitudinal) direction and the transverse direction. In a properly designed forging, the largest stress should be parallel to the fibrous structure, and the parting line of the dies should be located so as to minimize disruption to the grain flow lines.

Metals can be forged using hammers, mechanical presses, or hydraulic presses. Hammer forging operations can be conducted with

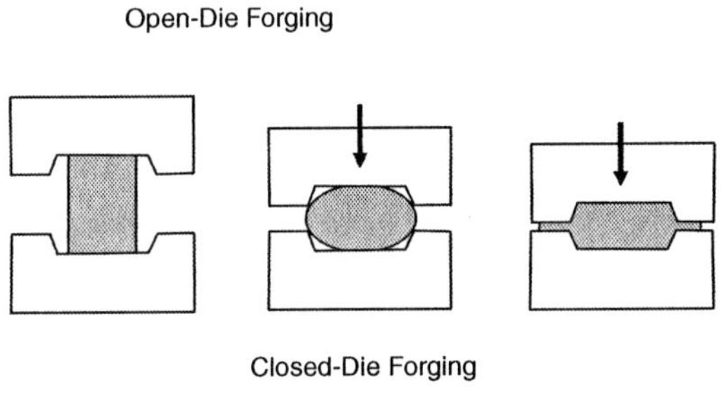


**Fig. 16.10** Open- and closed-die forging

either gravity or power drop hammers and are used for both open- and closed-die forgings. Hammers deform the metal with high deformation speed; therefore, it is necessary to control the length of the stroke, the speed of the blows, and the force being exerted. Hammer operations are frequently used to conduct preliminary shaping prior to closed-die forging. Both mechanical and screw presses are used for forging moderate-sized parts of modest shapes and are often used for high-volume production runs. Mechanical and screw presses combine impact with a squeezing action that is more compatible with the flow characteristics of some metals than are hammer processes. Hydraulic presses are the best method for producing large and thick forgings, because the deformation rate is slower and more controlled than with hammers or mechanical/screw presses. The deformation or strain rate can be very fast ($>10\ s^{-1}$) for processes such as hammer forging or very slow ($<0.1\ s^{-1}$) for hydraulic presses. Since higher strain rates increase the flow stress (decrease forgeability), hydraulic presses are usually preferred for forging alloys with low forgeability. Hydraulic presses are available in the range of 225 to 34,000 kg (500 to 75,000 lb) and can produce forgings up to approximately 1360 kg (3000 lb). A large hammer forge and hydraulic press are shown in Fig. 16.12.

Open-die forgings, also know as hand forgings, are produced in dies that do not provide lateral restraint during the forging operation. In this process, the metal is forged between either flat or simply-shaped dies. This process is used to produce small quantities where the small quantities do not justify the expense of matched dies. Although open-die forgings somewhat improve the grain flow of the material, they offer minimal economic benefit in reduced machining costs. Open-die forging is often used to produce preforms for closed-die forging.

Flow localization, caused by the formation of a dead-metal zone between the workpiece and the tooling, is a common problem with large reductions. This can arise from poor lubrication at the workpiece/tool interface. Upsetting is the working of a metal such that the cross-sectional area of all or part of the workpiece is increased. Upsetting of a cylinder with poorly lubricated platens is shown in Fig. 16.13. When the workpiece is constrained from sliding at the interface, it takes on a barrellike shape, and a friction-hill pressure distribution is created over the interface. The inhomogeneity of deformation throughout the cross section leads to a dead-metal zone at the tool interface and a region of intense shear deformation. If the shear deformations become too high, the forging can fracture.

Closed-die forging shapes the part between two die halves; thus, productivity is increased, albeit at the expense of higher die costs. Excess metal is allowed to escape in the flash; therefore, pressure is kept within safe limits, while die filling is ensured. More complex shapes, thinner walls, and thinner webs may necessitate forging in a sequence of die cavities. At high width-to-thickness ratios, friction sets a limit to minimum web thickness that decreases with effective lubrication. During true closed-die forging, the material is trapped in the die cavity. An important consideration when closed-die forging is for the metal to flow during forging to fill the cavities of the impression die. The forgeability of the metal is a way of expressing this capability. Poor forgeability can be caused by rupture before the die is filled or by a high flow strength, which causes the metal to flow past recesses without filling them, or causes underfilling with the maximum loads available.

Closed-die forgings are produced by forging ingots, plates, or extrusions between a matched set of dies. Closed-die forging uses progressive sets of dies to gradually shape the part to near-net dimensions. Die forgings can be subdivided into four categories from the lowest cost, least intricate to the highest cost, most intricate. A comparison of the relative amount of part definition for these different forging processes is shown in Fig. 16.14.

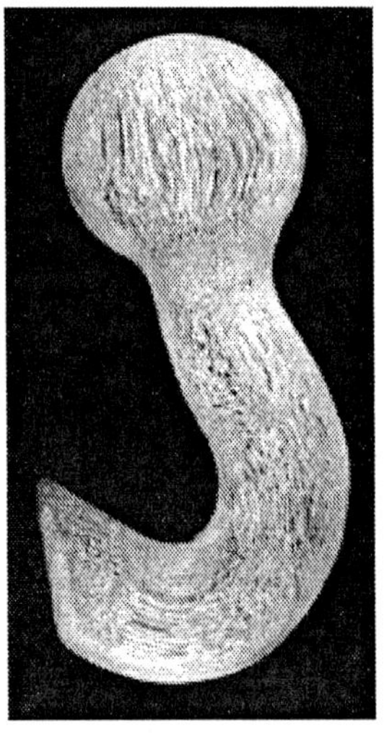

Etched 4140 Forging Showing Grain Flow

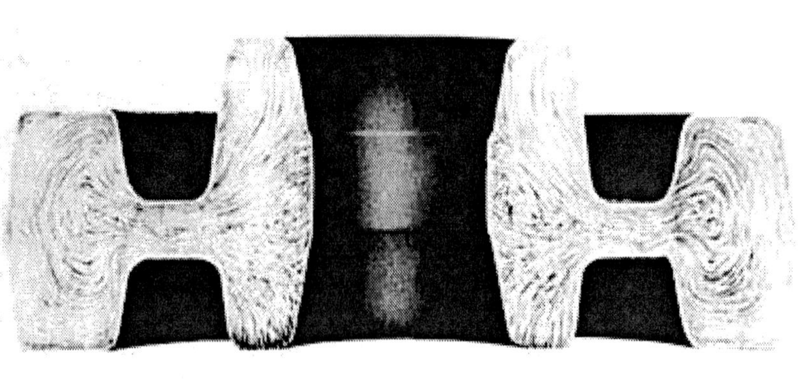

**Fig. 16.11** Grain flow from forging. Source: Ref 7

**Blocker forgings** may be chosen if the total quantities are small (e.g., 200). Since they have large fillet and corner radii, extensive machining is required to produce a finished part. The fillets are approximately 2 times the radius, and the corner radii approximately 1.5 times the radii of conventional forgings. Therefore, a blocker forging costs less than a conventional forging but requires more machining. Finish-only forgings are similar to blocker forgings in that only one set of dies is used; however, since they have one more squeeze applied, they have somewhat better part definition. Fillets are approximately 1.5 times the radius of conventional forgings, with corner radii about the same as conventional forgings. A quantity of approximately 500 may justify the use of finish-only forgings.

**Conventional forgings** require two to four sets of dies, with the first set producing a blocker-type forging that is subsequently finished in the other sets. This is the most common type of forging and is usually specified for quantities of 500 or more. Conventional forgings have more definition and require less machining than blocker forgings, but the die cost is higher.

**High-definition forgings** contain even better definition and tolerance control than conventional forgings, with less machining costs. These forgings are near-net shape forgings produced on multiple die sets. For some applications, some of the forged surfaces may not require machining.

**Precision forgings** produce the best component definition and highest quality but are, of course, the most expensive. These forgings have tighter tolerances than those produced by even high-definition forgings, with better grain flow. Minimal or no machining is required to finish these forgings.

Other common forging methods include roll forging, orbital or rotary forging, spin forging, ring rolling, and mandrel forging. The choice of a particular forging method depends on the shape required and the economics of the number of pieces required traded off against higher quality and lower machining costs.

Isothermal forging, shown in comparison with conventional forging in Fig. 16.15, is a process where the dies are maintained at the same temperature as the workpiece. Since this temperature is often on the order of 980 to

Large Forge Hammer

(a)

15,000 Ton Hydraulic Press

(b) *Ladish*

**Fig. 16.12** Forging equipment

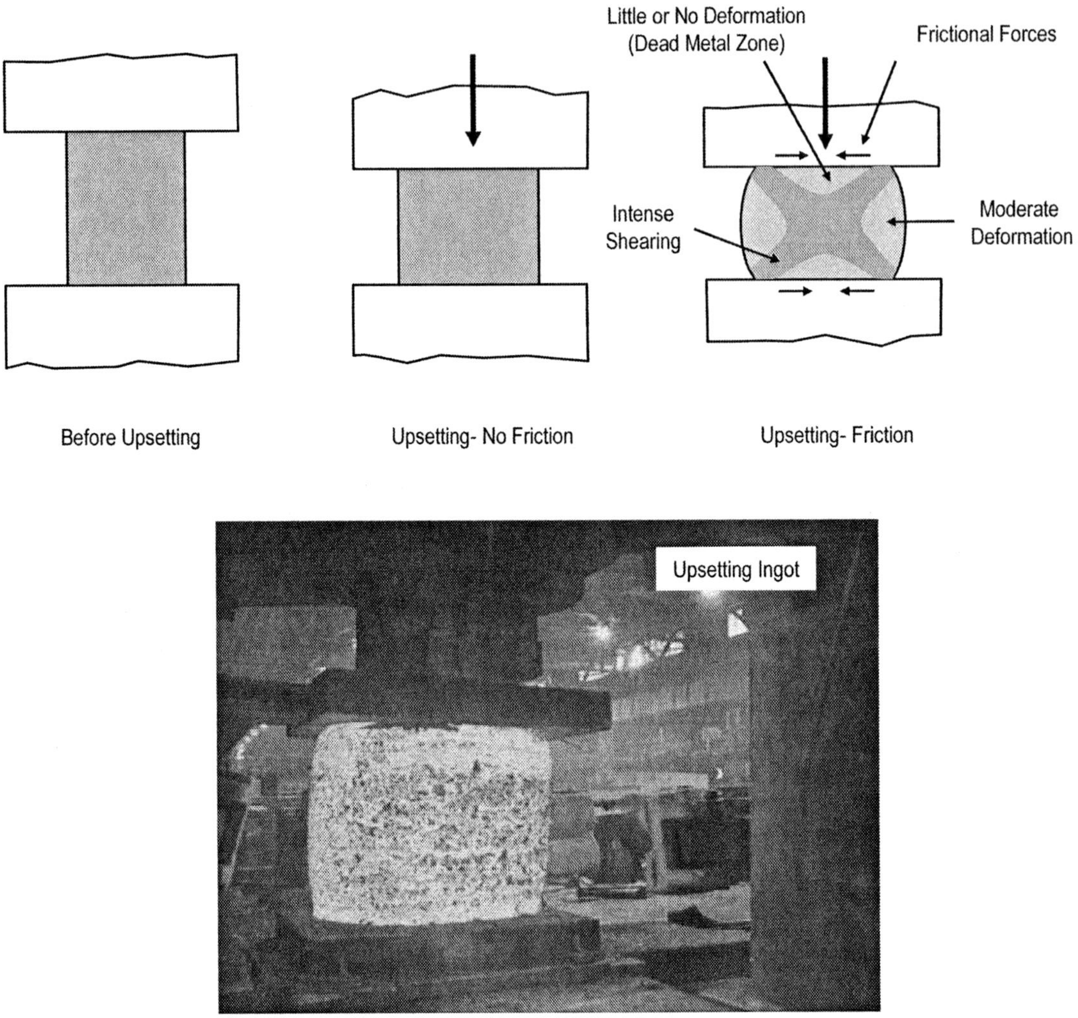


**Fig. 16.13** Upsetting

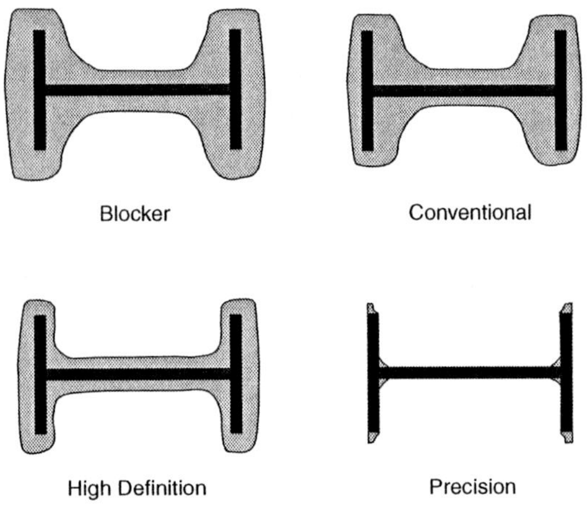


**Fig. 16.14** Part definition produced by different die forging methods. Source: Ref 8

1095 °C (1800 to 2000 °F), the dies are usually made of titanium-zirconium-molybdenum for elevated-temperature strength. In addition, isothermal forging is conducted in a vacuum or inert atmosphere in order to protect the part and die material from oxidation. Compared to conventional forging, isothermal forging deformation rates are slow; hydraulic press speeds of approximately 0.3 cm/min (0.1 in./min) are typical. The slow strain rate helps to control both the shape of the forged part and the microstructure. Accurate temperature and strain rate can be controlled by sensors in conjunction with the press equipment to give a closed-loop control system. The slower production rate is largely offset by the ability to forge complex shapes to closer tolerances, which leads to less machining

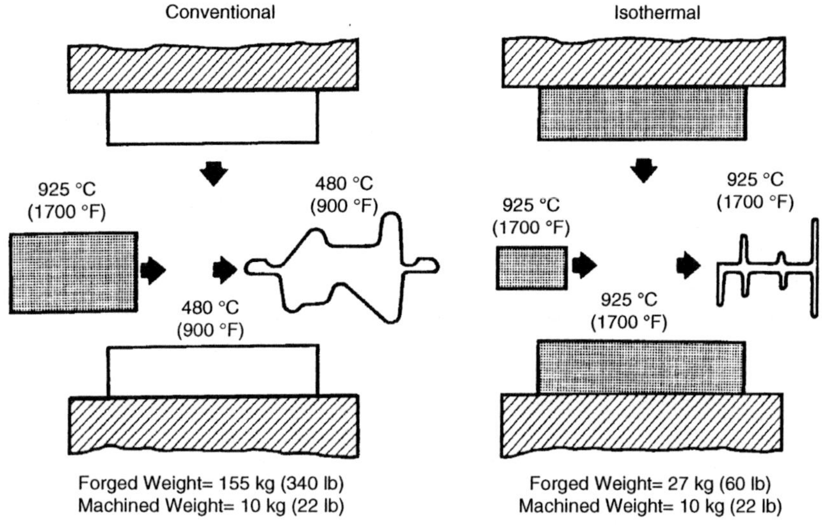


**Fig. 16.15** Comparison of conventional and isothermal forging. Source: Ref 8

and substantial material savings. In addition, a large amount of deformation is accomplished in one operation. Pressures are low, and uniform microstructures are achieved. For example, a finish-machined 68 kg (150 lb) Astroloy nickel-base superalloy disk started with an as-forged weight of 111 kg (245 lb) for conventional forging versus 73 kg (160 lb) for a corresponding isothermal forging.

Cold forging, when conducted so that a complex shape is formed in a single step, requires special lubricants, often with a conversion coating. Alternatively, the shape can be developed by moving the bar or slug through a sequence of cavities using a liquid lubricant. Cold forging is often combined with cold extrusion. It is the preferred process for mass producing near-net shape parts such as bolts, nuts, rivets, and many automotive and appliance components.

**Fig. 16.16** Aluminum extrusions. Source: Ref 9

## 16.5 Extrusion

Extrusion is a bulk deformation process in which a metal under high pressure is reduced in cross section by forcing it through a die. During extrusion, the metal billet is forced by a hydraulic ram through a die so that the metal is continuously deformed into a long length of metal with the desired cross section. Most metals are hot extruded, since their resistance to deformation is lower than in the cold state. The extrusion process is used primarily for producing bar shapes, tubes, and irregular shapes, such as those shown in Fig. 16.16.

In hot extrusion, a heated billet is forced through the die opening. The temperature used for hot extrusion depends on the alloy and can range from as low as 95 °C (200 °F) for lead to

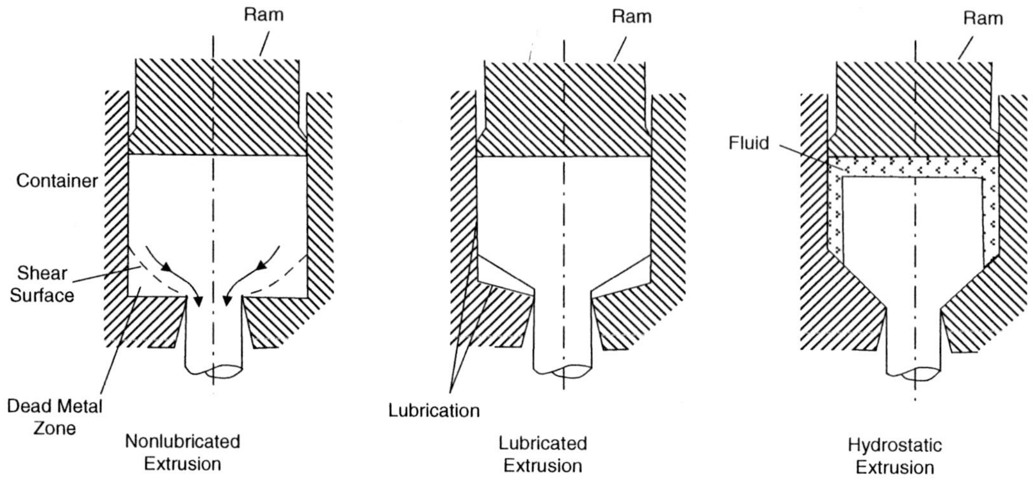


**Fig. 16.17** Extrusion processes. Source: Ref 10

1260 °C (2300 °F) for steel and nickel alloys. Hot extrusion is used to produce long, straight metal products of constant cross section, such as bars, solid and hollow sections, tubes, wires, and strips from materials that cannot be formed by cold extrusion. The three basic types of hot extrusion are nonlubricated, lubricated, and hydrostatic (Fig. 16.17). In nonlubricated hot extrusion, the material flows by internal shear, and a dead-metal zone is formed in front of the extrusion die. Lubricated extrusion, as the name implies, uses a suitable lubricant, usually glass powder or grease, between the extruded billet and the die. With hydrostatic extrusion, a fluid film present between the billet and the die exerts pressure on the deforming billet. Hydrostatic extrusion is used primarily when conventional lubrication is inadequate, for example, in the extrusion of special alloys or clad materials. For all practical purposes, hydrostatic extrusion can be considered an extension of the lubricated hot extrusion process.

Nonlubricated hot extrusion uses no lubrication on the billet, container, or die. It can produce very complex sections with mirror-surface finishes and close dimensional tolerances that are considered to be net extrusions. There are basically two methods of hot extruding materials without lubrication: (1) forward, or direct, extrusion; and (2) backward, or indirect, extrusion (Fig. 16.18). In direct extrusion, the metal billet is placed in the container of an extrusion press and forced directly through the die by the hydraulic ram. During the indirect extrusion process, a hollow ram holds the die, which is

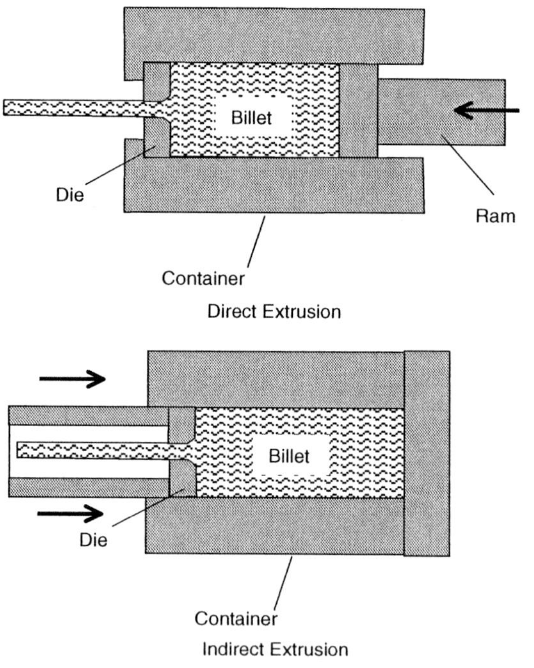


**Fig. 16.18** Direct and indirect extrusion

forced onto the metal billet. The billet is held in the extrusion container during this process. The advantage of indirect extrusion is that there is approximately a 25 to 30% reduction in maximum load relative to direct extrusion, and since no friction is produced between the billet and container, the temperatures are lower, leading to longer die lives. The disadvantage of indirect extrusion is that impurities or defects on the billet surface affect the surface finish of the extrusion and are not automatically retained as a shell or discard in the container. As a result, machined billets are used in many cases. In addition, the cross-sectional area of the extrusion is limited by the size of the hollow stem.

Generally, aluminum alloys are extruded without lubrication, but copper alloys, titanium alloys, alloy steels, stainless steels, and tool steels are extruded with a variety of graphite and glass-based lubricants. The choice between grease and glass lubricants is based mainly on the extrusion temperature. At low temperatures, since lubrication is used only to reduce friction, greases and graphite are commonly used. At temperatures above 980 °C (1800 °F), glasses are used because, in addition to providing lubrication, they help to thermally insulate the tooling from overheating. The lubricating film can also impede oxidation.

In hydrostatic extrusion, the billet in the container is extruded through the die by the action of a liquid pressure medium rather than by direct application of the load with a ram. The process of pure hydrostatic extrusion differs from conventional extrusion processes in that the billet is completely surrounded by a fluid, which is sealed off and pressurized sufficiently to extrude the billet through the die. Hydrostatic extrusion can be done hot, warm, or cold and can be used to extrude brittle materials that cannot be processed by conventional extrusion. Hydrostatic extrusion also allows greater reductions in area than either cold or conventional hot extrusion.

In cold extrusion, the slug or preform enters the extrusion die at room temperature. Any subsequent increase in temperature, which may amount to several hundred degrees, is caused by the conversion of the work of deformation into heat. Cold extrusion can be conducted by direct, indirect, or combined direct-indirect processes. Aluminum and aluminum alloys, copper and copper alloys, low- and medium-carbon steels, modified carbon steels, low-alloy steels, and stainless steels are the metals that are most commonly cold extruded. Cold extrusion competes with alternative metal forming processes such as cold heading, hot forging, hot extrusion, machining, and sometimes casting. Cold extrusion is used when the process is economically attractive because of savings in material; reduction or elimination of machining and grinding operations, because of the good surface finish and dimensional accuracy of cold-extruded parts; and elimination of heat treating operations because of the increase in the mechanical properties of cold-extruded parts. Impact extrusion is a term employed for the cold extrusion of thin-walled products such as toothpaste tubes.

## 16.6 Sheet Metal Forming Processes

Sheet metal forming processes usually employ hot- or cold-rolled sheet or strip material that is normally cold formed into the desired shape. Deformation is primarily by tension or combined tension-compression, and the limits are set by the formability of the material and only rarely by force or die pressure. Many sheet metal parts are fairly simple in shape, as shown in Fig. 16.19.

## 16.7 Blanking and Piercing

Blanking is a process in which a shape is sheared from a larger piece of sheet, while piercing produces a hole in the sheet by punching out a slug of metal (Fig. 16.20). Both blanking and piercing operations are usually performed in a punch press. The clearance between the punch and die must be controlled to obtain a uniform shearing action. Clearance is the distance between the mating surfaces of the punch and die, usually expressed as a percentage of sheet thickness. The walls of the die opening are tapered to minimize sticking, and the use of lubricants, such as mineral oil mixed with small quantities of fatty oils, also reduces sticking tendencies. Dull cutting edges on punches and dies have effects similar to excessive clearance, with burrs becoming excessive. With sharp tools and proper clearance, the cuts are clean without evidence of secondary shearing or excessive burring. When the clearance is too small, secondary shearing can occur, and if the clearance is too large, the sheared edge will have a large radius and a stringy burr.

## 16.8 Bending

In bending (Fig. 16.21), the sheet is placed over a die and pressed down by a punch that is actuated by the hydraulic ram of a press brake. The material is stressed beyond the yield strength but below the ultimate tensile strength. The surface area of the material does not change much. Bending usually refers to deformation about one axis and is a flexible process by which many different shapes can be produced. Standard die sets are used to produce a wide variety of shapes. The material is placed on the die, positioned in the machine with stops and/or gages, and is held in place with holddowns. The upper part of the press, the ram with the appropriately shaped punch, descends and forms the bend. Press brakes normally have a capacity of

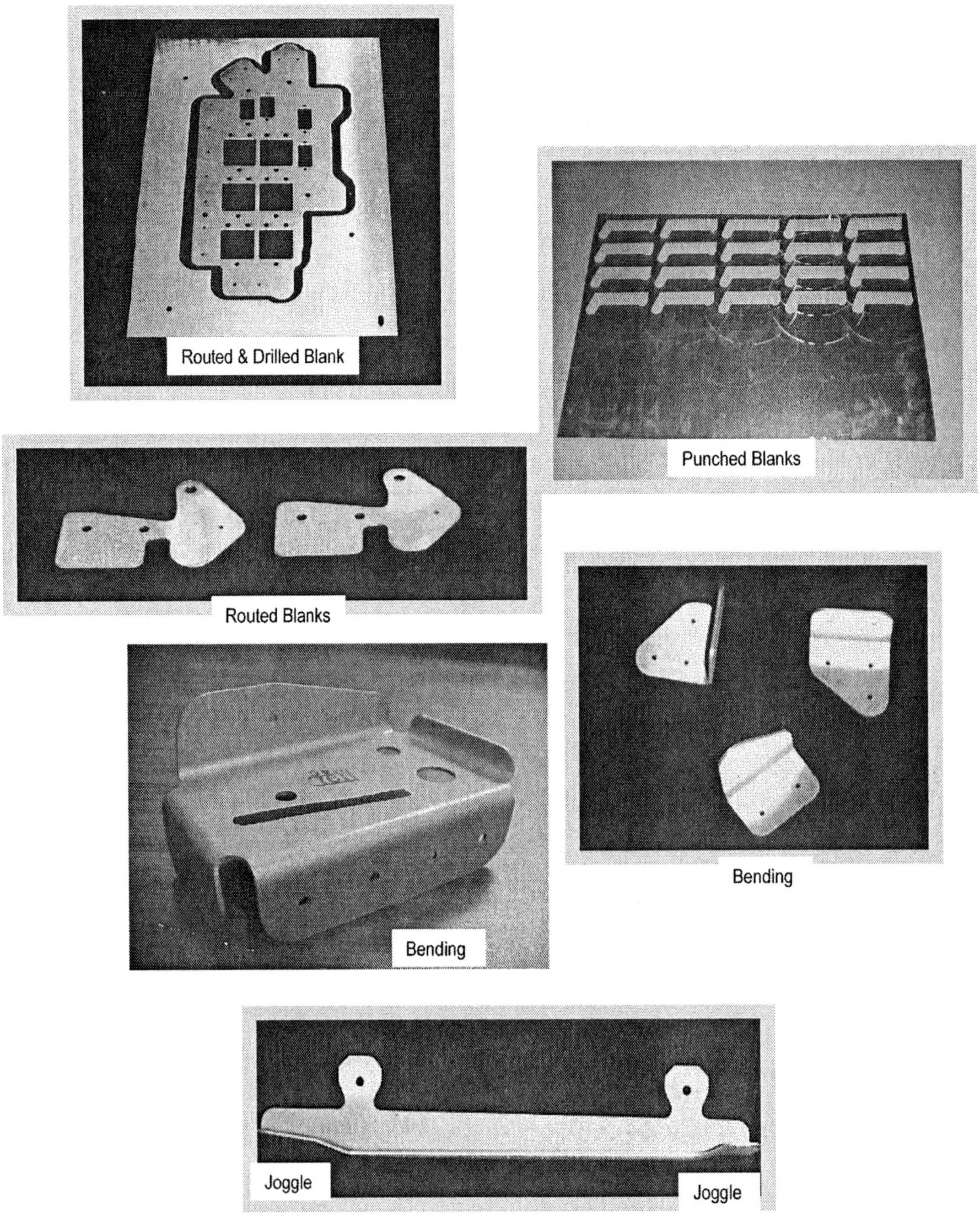


**Fig. 16.19** Simple sheet metal parts

18 to 181 Mg (20 to 200 tons) to accommodate stock from 1 to 5 m (3 to 15 ft). Larger and smaller presses are used for specialized applications.

Springback is the partial return of the part to its original shape after forming. The amount of springback is a function of the yield strength of the material being formed, the bend radius, and the sheet thickness. Springback is compensated by overbending the material beyond the final angle so that it springbacks to the desired angle. The springback allowance (i.e., the amount of overbend) increases with increasing yield strength and bend radius but varies inversely with sheet thickness. The smallest angle that can be safely bent, called the minimum bend radius, depends on the yield strength and on the design, dimensions, and conditions of the tooling. The most severe bends can be made across the rolling direction. If similar bends are to be made in two or more directions, it is best to make all bends at an angle to the direction of rolling.

Bending a sheet along a curved line is termed flanging. Circular or other close-shaped flanges (collars) are mass produced in preparation for joining tubes and fasteners to sheet, as in heat exchangers. Limits are set by fracture in a stretch flange and by buckling in a shrink flange. Related processes are flanging and necking of tubes and cans, as for beverage cans.

## 16.9 Stretch Forming

In stretch forming (Fig. 16.22), the material is stretched over a tool to produce the desired shape. The blank is firmly clamped at much or all of its circumference, and the shape is developed by penetration of a punch, at the expense of thickness. Small quantities can be produced with only a punch, although at higher material cost, as in the aircraft industry. Mating dies with carefully designed blankholders are more economical for the large quantities typical of the

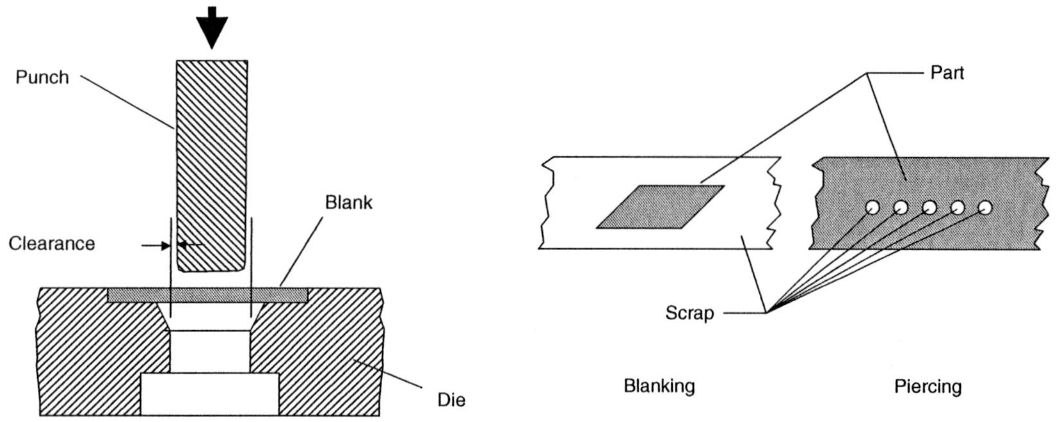


**Fig. 16.20** Blanking and piercing. Source: Ref 8

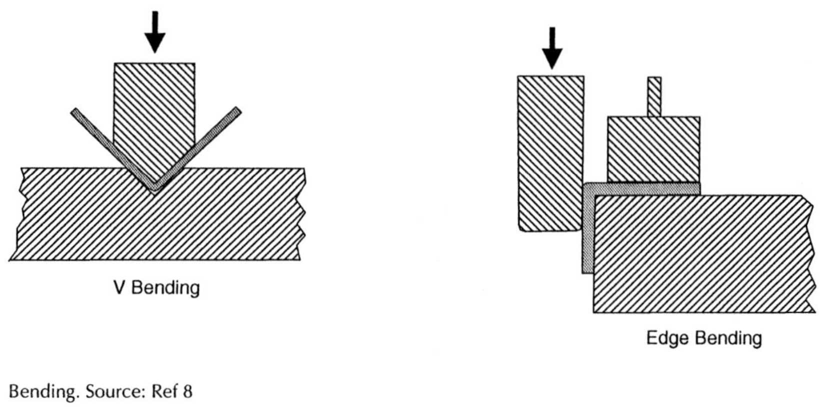


**Fig. 16.21** Bending. Source: Ref 8

automotive and appliance industries. An important aspect is control of or compensation for elastic recovery (springback), which distorts the shape. Large compound shapes can be formed by stretching the sheet both longitudinally and transversely. In addition, extrusions are frequently stretch formed to final curvature. Variants of stretch forming include stretch draw forming, stretch wrapping, and radial draw forming. Forming lubricants are recommended except when self-lubricating, smooth-faced plastic dies are used; however, the use of too much lubricant can result in workpiece bucking. Material properties that help in stretch forming are a high elongation, a large spread between the yield and ultimate strengths (called the forming range), toughness, and a fine grain structure. Alloys with a narrow spread between the yield and ultimate strengths are more susceptible to local necking and failure. For example, when the aluminum alloy 7075 has been quenched but not aged (W condition), it has a yield strength of 140 MPa (20 ksi), an ultimate strength of 330 MPa (48 ksi), a forming range of 190 MPa (28 ksi) (330 − 140 MPa), and a stretchability rating of 100. In contrast, 7075 that has been aged to peak strength (T6) has a yield strength of 460 MPa (67 ksi), an ultimate strength of 525 MPa (76 ksi), a forming range of only 65 MPa (9 ksi), and a stretchability rating of only 10.

## 16.10 Drawing

In drawing, a blank of sheet metal is restrained at the edges, and the middle section is forced by a punch into a die to stretch the metal into a cup-shaped drawn part, as shown for deep drawing in Fig. 16.23. Drawn components can be just about any cross section but are usually circular or rectangular. Drawing can be either shallow or deep, depending on the amount of deformation. For shallow drawing, the depth of draw is less than the smallest dimension of the opening; otherwise, it is deep drawing. Drawing leads to wrinkling and puckering at the edge where the sheet metal is clamped. Ears (Fig. 16.24) are usually removed by a separate trimming operation.

In drawing, the shape is developed by drawing material into the die, and the average thickness is approximately preserved. The flange is in circumferential compression and forms wrinkles, unless the blank is thick relative to the blank diameter or is adequately restrained with a blankholder. Fracture occurs at the base or in the wall if stresses exceed the strength of the partly formed cup; therefore, a limit is set to the attainable reduction in diameter or, as more frequently expressed, a limiting draw ratio

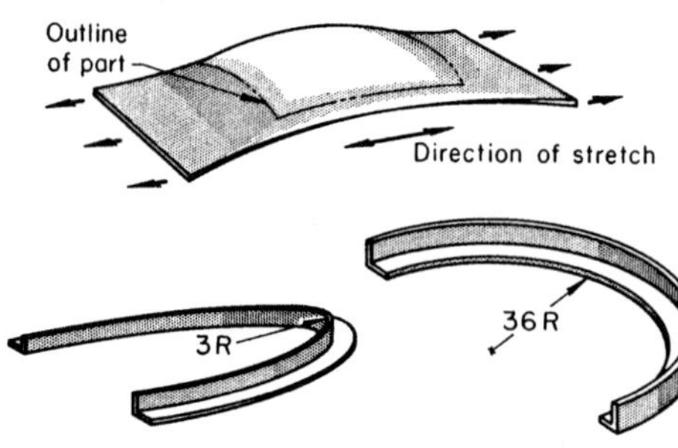


Fig. 16.22 Stretch forming. Source: Ref 11

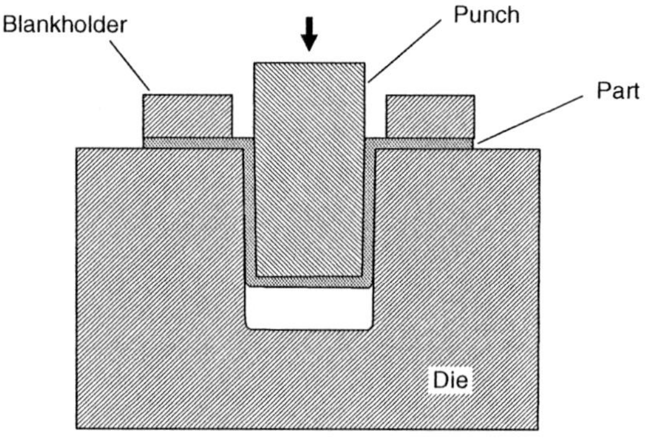


Fig. 16.23 Deep drawing. Source: Ref 8

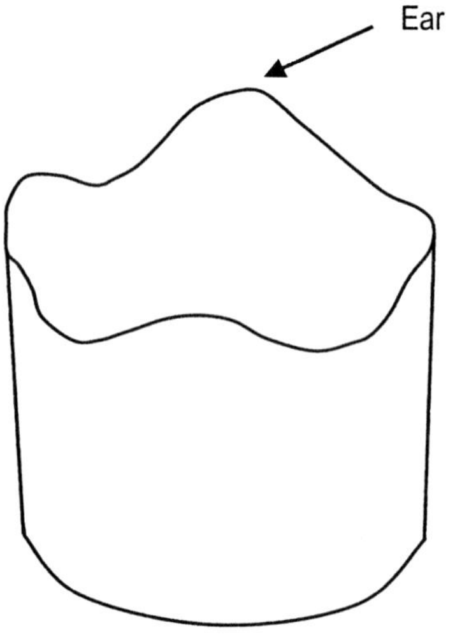


Fig. 16.24 Earing in deep-drawn cup

(LDR = diameter of blank/diameter of cup) is reached. The LDR is a system property, affected by material, especially the *r*-value, die design, and lubricant. (An explanation of the significance of the *r*-value can be found in section 19.9.1. in Chapter 19, "Plain Carbon Steels," in this book.) Cups deeper than allowed by the LDR are produced by redrawing, or reverse redrawing, or by thinning the wall by ironing (increasing length by reducing wall thickness while maintaining a constant inner diameter).

Punch presses are used for most deep-drawing operations. In a typical deep-drawing operation, a punch or male die pushes the sheet into the die cavity while it is supported around the periphery by a blankholder. Single-action presses can be operated at 27 to 43 m/min (90 to 140 ft/min), while double-action presses operate at 12 to 30 m/min (40 to 100 ft/min) for mild draws and at less than 15 m/min (50 ft/min) for deep draws. Clearances between the punch and die are important. Excessive clearance can result in wrinkling of the sidewalls of the drawn shell, while insufficient clearance increases the force required for drawing and tends to burnish the part surfaces. If the punch radius is too large, wrinkling can result, and if the radius is too small, sheet fracture is a possibility. Draw punches and dies should have a surface finish of 0.4 μm (16 μin.) or less for most applications. Tools are often chrome plated to minimize friction and dirt that can damage the part finish. Lubricants for deep drawing must allow the blank to slip readily and uniformly between the blankholder and die. Stretching and galling during drawing must be avoided. During blank preparation, excessive stock at the corners must be avoided because it obstructs the uniform flow of metal under the blankholder, leading to wrinkles or cracks. Severe forming operations of relatively thick or large blanks of high-strength alloys generally must be conducted at elevated temperature, where the lower strength and partial recrystallization aids in forming, but the time at temperature should be minimized to limit grain growth.

## 16.11 Rubber Pad Forming

Rubber pad forming uses a rubber pad to exert nearly equal pressure over the part as it is formed down over a form block. Rubber pad forming, and a closely related process, fluid cell forming, are shown in Fig. 16.25. The rubber pad acts somewhat like a hydraulic fluid, spreading the force over the surface of the part. The pad can either consist of a solid piece or may be several pieces laminated together. The pad is usually in the range of 15 to 30 cm (6 to 12 in.) thick and must be held in a sturdy retainer, because the pressures generated can be as high as 140 MPa (20 ksi). Rubber pad forming can often be used to form tighter radii and more severe contours than other forming methods because of the multidirectional nature of the force exerted on the workpiece. The rubber acts somewhat like a blankholder, helping to eliminate the tendency for wrinkling. This process is very good for making sheet metal parts with integral stiffening beads. Most rubber pad forming is conducted on sheet 1.6 mm (0.063 in.) or less in thickness; however, material as thick as 16 mm (0.625 in.) has been successfully formed. Fluid cell forming, which uses a fluid cell to apply pressure through an elastomeric membrane, can form even more severe contours than rubber pad forming. Due to the high pressures employed in this process, as high as 100 to 140 MPa (15 to 20 ksi), many parts can be formed in one operation with minimal or no springback. However, fluid cell forming presses are usually expensive.

## 16.12 Superplastic Forming

Superplasticity is a property that allows sheet to elongate to quite large strains without localized necking and rupture. In uniaxial tensile testing, elongations to failure in excess of 200% are usually indicative of superplasticity. Micrograin superplasticity occurs in some materials with a fine grain size, usually less than 10 μm, when they are deformed in the strain range of 0.00005 to 0.01/s at temperatures greater than 0.5 $T_m$, where $T_m$ is the absolute melting point. Although superplastic behavior can produce strains in excess of 1000% (Fig. 16.26), superplastic forming (SPF) processes are generally limited to approximately 100 to 300%. The advantages of SPF include the ability to make part shapes not possible with conventional forming, reduced forming stresses, improved formability with essentially no springback, and reduced machining costs. The disadvantages are that the process is rather slow, and the equipment and tooling can be relatively expensive.

The main requirement for superplasticity is a high strain-rate sensitivity. In other words, the strain-rate sensitivity, $m$, should be high where $m$ is defined as:

$$m = \frac{d(\ln \sigma)}{d(\ln \dot{\varepsilon})} \qquad \text{(Eq 16.3)}$$

where $m$ is the strain-rate sensitivity, $\sigma$ is the flow stress, and $\dot{\varepsilon}$ is the strain rate.

The strain-rate sensitivity describes the ability of a material to resist plastic instability or necking. For superplasticity, $m$ is usually greater than 0.5, with the majority of superplastic materials having an $m$ value in the range of 0.4 to 0.8, where a value of 1.0 would indicate a perfectly superplastic material. The presence of a neck in a material undergoing a tensile strain results in a locally high strain rate and, for a high value of $m$, to a sharp increase in the yield stress within the necked region; that is, the neck undergoes strain hardening, which restricts its further development. Therefore, a high strain-rate sensitivity resists neck formation and leads to the high tensile elongations observed in superplastic materials. The yield stress decreases and the strain-rate sensitivity increases with increasing temperature and decreasing grain size. The elongation to failure tends to increase with increasing $m$.

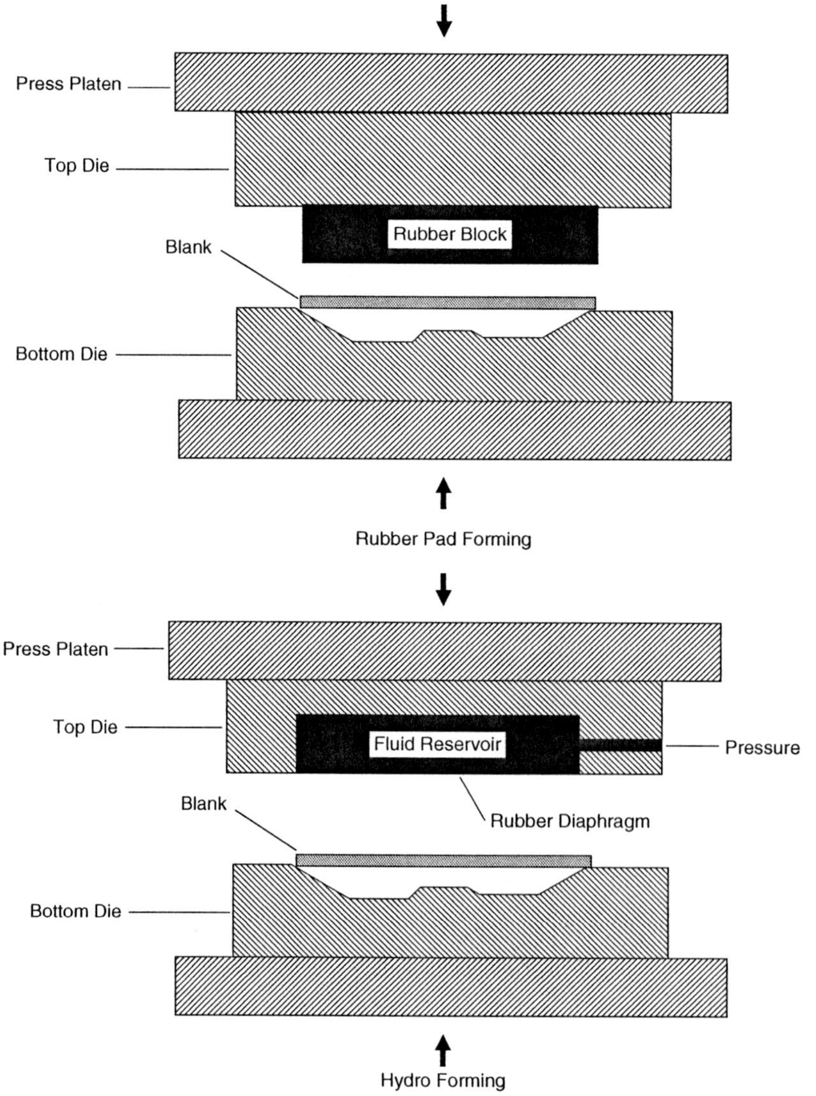


**Fig. 16.25** Rubber pad and hydro forming. Source: Ref 8

Superplasticity depends on microstructure and exists only over certain temperature and strain-rate ranges. A fine grain structure is a prerequisite since superplasticity results from grain rotation and grain-boundary sliding, and increasing grain size results in increases in yield stress. Equiaxed grains are desirable because they contribute to grain-boundary sliding and grain rotation. A duplex structure also contributes to superplasticity by inhibiting grain growth at elevated temperature. Grain growth inhibits superplasticity by increasing the flow stress and decreasing $m$. The Ashby and Verrall model for superplasticity, based on grain-boundary sliding with diffusional accommodation, is shown in Fig. 16.27, in which grains switch places with their neighbors to facilitate elongation. Since the grains are not all the same size, some rotation must also take place. Slow strain rates are necessary to allow the

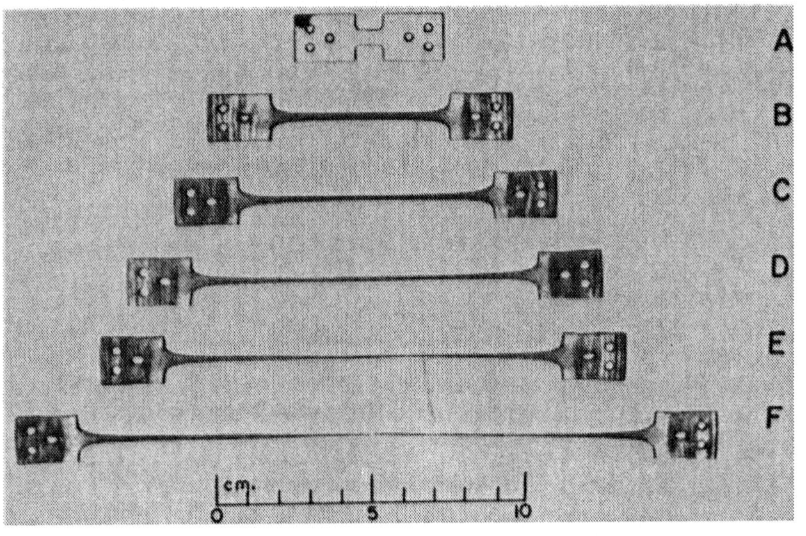


**Fig. 16.26** Superplastic elongation. Source: Ref 12

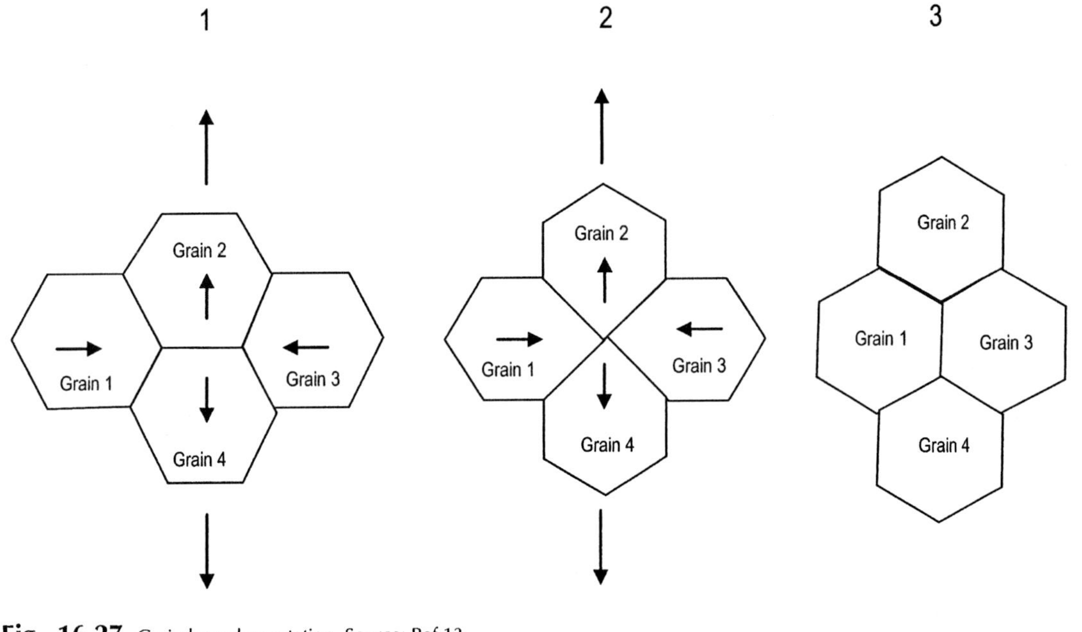


**Fig. 16.27** Grain-boundary rotation. Source: Ref 13

diffusion mechanisms time to allow this rearrangement.

In the single-sheet SPF process (Fig. 16.28), a single sheet of metal is sealed around its periphery between an upper and lower die. The lower die is either machined to the desired shape of the final part or a die inset is placed in the lower die box. The dies and sheet are maintained at the SPF temperature, and gas pressure is used to form the sheet down over the tool. The lower cavity is maintained under vacuum or can be vented to the atmosphere. After the sheet is heated to its superplastic temperature range, gas pressure is injected through inlets in the upper die. This pressurizes the cavity above the metal sheet, forcing it to superplastically form to the shape of the lower die. Gas pressurization is applied slowly so that the strains in the sheet are maintained in the superplastic range, and the pressure is varied during the forming process to maintain the required slow strain rate. Typical forming cycles for aluminum alloys are 5 to 6 MPa (700 to 900 psi) at 450 to 520 °C (840 to 975 °F) and 0.7 to 1.4 MPa (100 to 200 psi) and 900 °C (1650 °F) for titanium alloys.

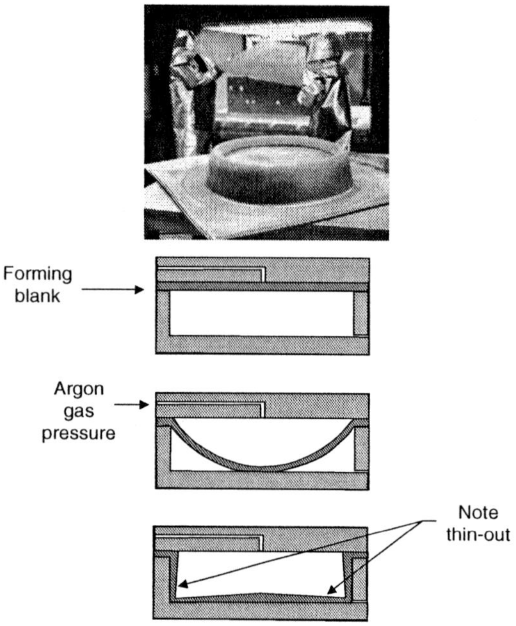


**Fig. 16.28** Single-sheet superplastic forming. Source: Ref 8

In aluminum alloys, the hard particles at the grain boundaries that help control grain growth may contribute to the formation of voids, a process called cavitation. Cavitation on the order of 3% can occur after approximately 200% of superplastic deformation. Cavitation can be minimized, or eliminated, by applying a hydrostatic back pressure to the sheet during forming, as shown schematically in Fig. 16.29. Back pressures of 0.7 to 3.5 MPa (100 to 500 psi) are normally sufficient to suppress cavitation.

For titanium alloys, superplastic forming can be combined with diffusion bonding (SPF/DB) to form a one-piece unitized structure (Fig. 16.30). Titanium is very amenable to diffusion bonding because the thin, protective oxide layer ($TiO_2$) dissolves into the titanium above 620 °C (1150 °F), leaving a clean surface. Internal sheets of the multilayer preform are formed into integral stiffening members between the outer sheets, with geometry of the stiffening core determined by the welding patterns. Truss core, sinusoidal, egg-crate, and other internal stiffening geometries can be produced in a single forming step using automated welding patterns. Low forming pressures and single-step processing significantly reduce

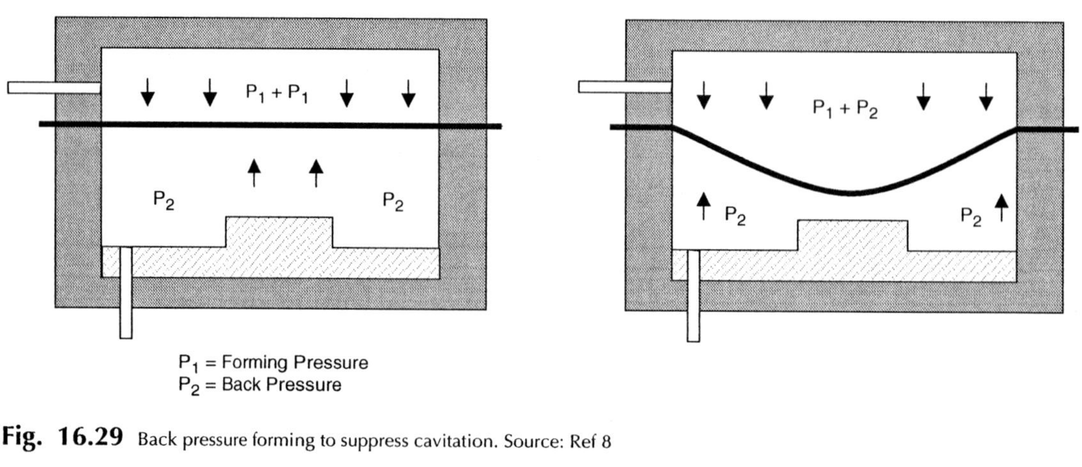


**Fig. 16.29** Back pressure forming to suppress cavitation. Source: Ref 8

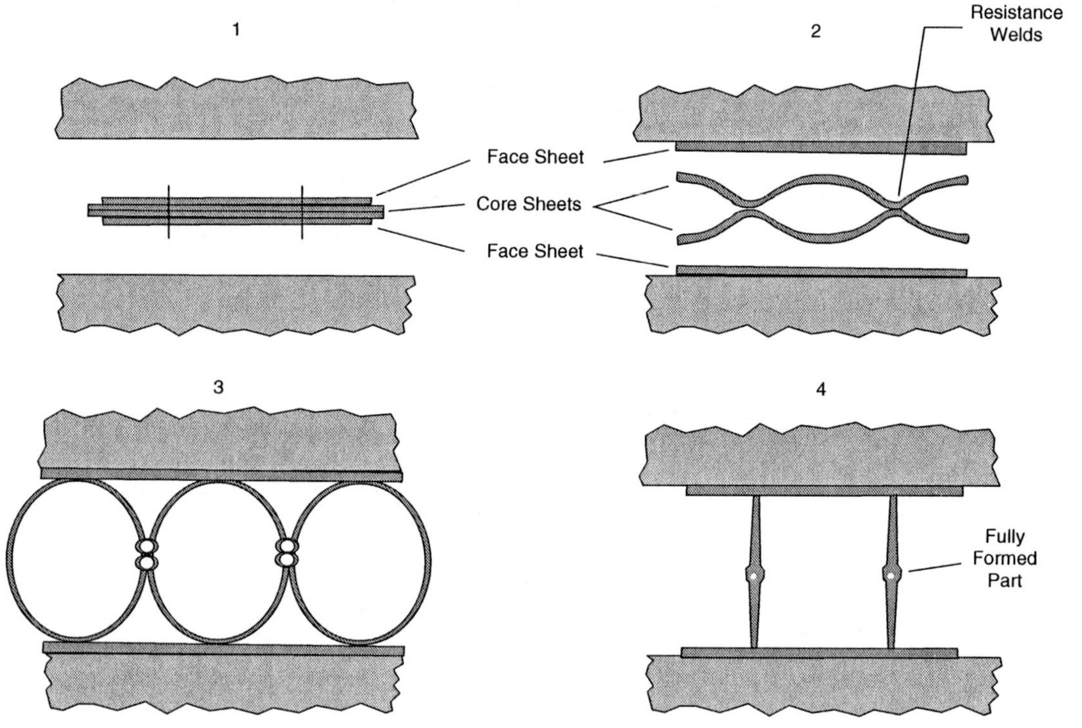


**Fig. 16.30** Fabrication of part by subjecting four metal sheets to superplastic forming and diffusion bonding. Source: Ref 8

tooling costs compared to conventional methods. However, the protective aluminum oxide ($Al_2O_3$) coating on aluminum does not dissolve and must either be removed or ruptured to promote diffusion bonding. Although diffusion bonding of aluminum alloys has successfully been demonstrated in the laboratory, SPF/DB of aluminum alloys is not yet a commercial process.

## ACKNOWLEDGMENTS

Sections of this chapter were adapted from "Flat, Bar, and Shape Rolling" by G.D. Lahoti, "Conventional Hot Extrusion," "Hydrostatic Extrusion," and "Cold Extrusion" by P.S. Raghupathi, W.C. Setzer, and M. Baxi, all in *Forging and Forming,* Volume 14, *ASM Handbook,* ASM International, 1988.

## REFERENCES

1. J. Beddoes and M.J. Bibby, *Principles of Metal Manufacturing Processes,* Arnold, 1999
2. G.E. Dieter, Introduction to Workability, *Forming and Forging,* Vol 14, *ASM Handbook,* ASM International, 1988
3. S. Kalpakjian, Mechanical Testing for Metalworking Processes, *Mechanical Testing and Evaluation,* Vol 8, *ASM Handbook,* ASM International, 2000
4. W.L. Mankins, Recovery, Recrystallization, and Grain-Growth Structures, *Metallography and Microstructures,* Vol 9, *ASM Handbook,* ASM International, 2004
5. G.D. Lahoti, Flat, Bar, and Shape Rolling, *Forming and Forging,* Vol 14, *ASM Handbook,* ASM International, 1988
6. S. Kalpakjian, *Manufacturing Processes for Engineering Materials,* 2nd ed., Addison-Wesley Publishing Company, 1991
7. Forging of Carbon and Alloy Steels, *Forming and Forging,* Vol 14, *ASM Handbook,* ASM International, 1988
8. F.C. Campbell, *Manufacturing Technology for Aerospace Structural Materials,* Elsevier Scientific, 2006

9. J.G. Kaufman, *Introduction to Aluminum Alloys and Tempers,* ASM International, 2000
10. Conventional Hot Extrusion, *Forming and Forging,* Vol 14, *ASM Handbook,* ASM International, 1988
11. Forming of Aluminum Alloys, *Forging and Forming,* Vol 14, *ASM Handbook,* ASM International, 1988
12. F.A. Mohamed, Superplastic Deformation at Elevated Temperatures, *Mechanical Testing and Evaluation,* Vol 8, *ASM Handbook,* ASM International, 2000
13. M.F. Ashby and R.A. Verrall, Diffusion Accommodated Flow and Superplasticity, *Acta Metall.,* Vol 21, 1973, p 149

## SELECTED REFERENCES

- B.L. Ferguson, Design for Deformation Processes, *Materials Selection and Design,* Vol 20, *ASM Handbook,* ASM International, 1997
- Hydrostatic Extrusion, *Forming and Forging,* Vol 14, *ASM Handbook,* ASM International, 1988
- P.S. Raghupathi, W.C. Setzer, and M. Baxi, Cold Extrusion, *Forging and Forming,* Vol 14, *ASM Handbook,* ASM International, 1988
- J.A. Schey, Manufacturing Processes and Their Selection, *Materials Selection and Design,* Vol 20, *ASM Handbook,* ASM International, 1997

# CHAPTER 17

# Physical Properties of Metals

THE PHYSICAL PROPERTIES of a material are those properties that can be measured or characterized without application of force and without changing the material identity. They are relatively insensitive to structure. Common physical properties include density, lattice parameter, electrical properties such as conductivity and dielectric permittivity, thermal properties such as thermal expansion and specific heat capacity, magnetic properties such as magnetic permeability, and optical properties such as refractivity. Some physical properties for a number of metals are given in Table 17.1.

## 17.1 Density

Mass properties include atomic weight and density. The atomic weight, or relative atomic mass, is the ratio of the average mass per atom of an element to $^1/_{12}$ of the mass of the atom of carbon 12. Density, ρ, is the mass per unit volume and, for solids, is usually expressed in g/cm$^3$ or lb/ft$^3$. The reciprocal of density is the specific volume, $v$:

$$\text{Density, } \rho = \frac{\text{Mass}}{\text{Volume}} \qquad \text{(Eq 17.1)}$$

$$\text{Specific volume, } v = \frac{1}{\rho} \qquad \text{(Eq 17.2)}$$

Density increases regularly with increasing atomic numbers in each subgroup of the periodic table. On alloying, the density of a metal changes, because the mass of the solute atoms differs from that of the solvent, and also because the lattice parameter usually changes on alloying. The parameter change may often be deduced from Vegard's law, which assumes that the

**Table 17.1 Physical properties of some metals at room temperature**

| Metal | Density, g/cm$^3$ | Specific heat capacity ($C_p$), kJ/kg · K | Coefficient of linear thermal expansion, μm/m · K | Thermal conductivity, W/m · K | Electrical resistivity, nΩ · m |
|---|---|---|---|---|---|
| Aluminum | 2.6989 | 0.900 | 22.8 | 247 | 26.2 |
| Beryllium | 1.848 | 1.886 | 9 | 210 | 40 |
| Chromium | 7.19 | 0.4598 | 6.2 | 67 | 130 |
| Cobalt | 8.832 | 0.414 | 13.8 | 69.04 | 52.5 |
| Copper | 8.93 | 0.3846 | 16.5 | 398 | 16.730 |
| Gold | 19.302 | 0.128 | 14.2 | 317.9 | 23.5 |
| Hafnium | 13.31 | 0.147 | 519 | 23 | 351 |
| Iron | 7.870 | 0.4473 | 11.8 | 80.4 | 97.1 |
| Lead | 11.34 | 0.1287 | 26.5 | 33.6 | 206.43 |
| Magnesium | 1.738 | 1.025 | 25.2 | 155 | 44.5 |
| Manganese | 7.43 | 0.475 | 21.7 | 7.79 | 1440 |
| Molybdenum | 10.22 | 0.276 | 5 | 142 | 52 |
| Nickel | 8.902 | 0.471 | 13.3 | 82.9 | 68.44 |
| Niobium | 8.57 | 0.27 | 7.1 | 52.3 | 146 |
| Platinum | 21.45 | 0.132 | 9.1 | 71.1 | 106 |
| Rhenium | 21.02 | 0.138 | 6.6 | 71.2 | 193 |
| Silver | 10.49 | 0.235 | 19.0 | 428 | 14.7 |
| Tantalum | 16.6 | 0.1391 | 6.5 | 54.4 | 135.0 |
| Tin | 5.765 | 0.205 | 21 | 62.8 | 110 |
| Titanium | 4.507 | 0.5223 | 8.41 | 11.4 | 420 |
| Tungsten | 19.254 | 0.128 | 3.01–8.87 | 160 | 53 |
| Uranium | 19.05 | 0.117 | 12 | 27.6 | 300 |
| Vanadium | 6.16 | 0.498 | 8.3 | 31.0 | 248 |
| Zinc | 7.133 | 0.382 | 39.7 | 113 | 59.16 |
| Zirconium | 6.505 | 0.30 | 5.85 | 21.1 | 450 |

Source: Ref 1

lattice parameter of a solid solution varies linearly with atomic concentration; however, numerous deviations from this ideal behavior exist. Density depends on the mass of the atoms, their size, and their packing arrangement. Metals are dense because their atoms are heavy and tightly packed. Ceramics have lower densities because they contain the light atoms carbon, nitrogen, and oxygen; and polymers have even lower densities because their chains consist almost entirely of light atoms.

## 17.2 Thermal Properties

Thermal properties are the response of a material to heat energy. As a solid absorbs heat energy, its temperature rises and its dimensions increase. If a temperature gradient exists, energy will be transported from the hotter to the cooler regions of the metal. If enough thermal energy is supplied, the metal will melt. Important thermal properties include melting and boiling points, thermal expansion, specific heat capacity, thermal conductivity, and thermal diffusivity.

### *17.2.1 Melting and Boiling Points*

The melting point is the temperature at which the solid and liquid phases of a pure material are in equilibrium. The boiling point is the temperature at which the vapor pressure of a liquid equals the pressure of the surroundings, normally taken as 1 atm (14.7 psia).

### *17.2.2 Thermal Expansion*

Metals expand on heating and contract when cooled. The change in length with temperature is expressed as the linear coefficient of thermal expansion, α, during a temperature change of $T_o$ to $T_f$:

$$\alpha = \frac{l_f - l_o/l_o}{T_f - T_o}; \quad \frac{\Delta l/l_o}{\Delta T} \qquad \text{(Eq 17.3)}$$

where $l_o$ is the original length, and $l_f$ is the final length. The linear coefficient of thermal expansion is a material property that indicates the extent to which a material expands on heating and has units of reciprocal temperature, such as $K^{-1}$ or μm/m · K. The coefficient of thermal expansion increases as the temperature increases. In general, coefficient of thermal expansion values for metals fall between those of ceramics, which have lower values, and polymers, which have higher values.

Heating or cooling affects all the dimensions of the material, with a resultant change in volume. The volume coefficient of thermal expansion, $\alpha_V$, is:

$$\alpha_V = \frac{\Delta V/V_o}{\Delta T} \qquad \text{(Eq 17.4)}$$

where $V_o$ is the original volume, and $\Delta V$ is the change in volume. In many materials, the value of $\alpha_V$ is anisotropic; that is, it varies depending on the crystallographic direction along which it is measured. For materials in which the thermal expansion is isotropic, $\alpha_V$ is approximately 3α.

### *17.2.3 Heat Capacity and Specific Heat Capacity*

When a metal is heated, it absorbs some energy, and the temperature increases. Heat capacity is the ability of a material to absorb heat from external surroundings and is the amount of energy required to produce a unit temperature rise. Heat capacity is proportional to the amount of material; for example, a bathtub of water has a larger heat capacity than a bucket of water. Heat capacity, *C*, is expressed as follows:

$$C = \frac{dQ}{dT} \qquad \text{(Eq 17.5)}$$

where $dQ$ is the energy required to produce a $dT$ temperature change. Heat capacity is ordinarily specified per mole of material (e.g., J/mol · K). The specific heat capacity, or specific heat (*c*), is used in many applications. Specific heat capacity is the quantity of heat per unit mass required to change the temperature of a unit of material by one degree, expressed as J/kg · K. Lower and higher specific heat capacity values correspond to lower and higher temperatures, respectively.

There are two ways in which the heat capacity can be measured, according to the environmental conditions under which heat is transferred. One is the heat capacity at constant volume ($C_V$), while the other is at constant external pressure ($C_p$). The magnitude of $C_p$ is always greater than $C_V$; however, the difference is very slight for most solid materials at room temperature and below.

In solid metals, thermal energy causes an increase in the vibrational energy of the atoms within the crystalline lattice. The atoms are constantly vibrating at very high frequencies and with relatively small amplitudes. However, the atoms do not vibrate independently; rather, the vibrations of adjacent atoms are coupled together as a result of their atomic bonding. These vibrations are coordinated such that traveling lattice waves are produced that have short wavelengths and high frequencies and propagate through the crystalline lattice at the speed of sound. Vibrational thermal energy consists of a series of these waves, which have a range of distributions and frequencies. Only certain energy levels are allowed, and a single quantum of vibrational energy is called a phonon. Vibrational waves are responsible for the thermal scattering of free electrons during electrical conduction.

The vibrational contribution to the heat capacity at constant volume ($C_V$) as a function of temperature is shown in Fig. 17.1. At absolute zero, the heat capacity, $C_V$, is zero but increases rapidly with temperature and increases as a result of an increase in the vibrational energy in the lattice waves. At low temperatures, the relationship between $C_V$ and absolute temperature, $T$, is:

$$C_V = AT^3 \qquad \text{(Eq 17.6)}$$

where A is a temperature-independent constant. Above the Debye temperature, $\theta_D$, the heat capacity, $C_V$, levels off at a value of $3R$ ($R$ is the universal gas constant) and becomes essentially independent of temperature. Therefore, even though the total energy of the metal increases with temperature, the quantity of energy required to produce a one-degree temperature change is constant. The value of $\theta_D$ is below room temperature for metals, with 25 J/mol·K (6 cal/mol·K) being a reasonable room-temperature approximation for $C_V$.

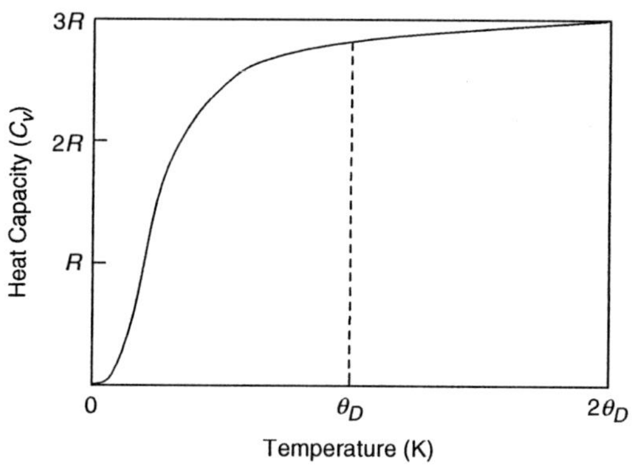


**Fig. 17.1** Variation of heat capacity at constant volume with temperature

### 17.2.4 Thermal Conductivity

Thermal conduction is the process by which heat is transported, or transferred, from high to low temperature regions of a material. The property that characterizes the ability of a material to transfer heat is the thermal conductivity, $q$, which is defined as:

$$q = -k\left(\frac{dT}{dx}\right) \qquad \text{(Eq 17.7)}$$

where $q$ is the heat flux, or heat flow, per unit time per unit area (W/m$^2$), $k$ is the thermal conductivity (W/m·K), and $dT/dx$ is the temperature gradient through the conducting medium.

This equation is only valid for steady-state heat flow, that is, for situations in which the heat flux $q$ does not change with time. The minus sign in the equation indicates that the direction of heat flow is from hot to cold, or down the temperature gradient.

Heat is transported in solid metals by a combination of lattice vibrational waves (phonons) and free electrons. The total thermal conductivity is the sum of the lattice vibrations and free electron thermal conductivities. The phonon contribution is in the direction form high to low temperature regions or down the temperature gradient. Free electrons in the hotter region also migrate to colder areas, where some of their additional kinetic energy is transferred to the colder atoms as vibrational energy, due to collisions with phonons or other imperfections in the crystalline lattice. As the free electron concentration in a metal increases, the relative contribution of the free electrons to total thermal conductivity increases, since more electrons are available to participate in the heat-transfer process. In general, metals are better conductors of heat than ceramics or polymers, because they have large numbers of free electrons that can participate in thermal conduction. In high-purity metals, the free electron mechanism of heat transport is much more efficient than the phonon contribution, because electrons are not as easily scattered as phonons, and they have higher velocities.

Since free electrons are the dominant mechanism for both electrical and heat conduction in pure metals, the two conductivities can be related according to the Wiedemann-Franz law:

$$L = \frac{k}{\sigma T} \quad \text{(Eq 17.8)}$$

where $L$ is the constant called the Lorentz number, σ is the electrical conductivity, and $T$ is the absolute temperature.

The theoretical value for the Lorentz number, $L$, is $2.44 \times 10^8\ \Omega \cdot W/(K)^2$. If heat energy is transported entirely by free electrons, $L$ should be independent of temperature and have the same value for all metals. However, the addition of alloying elements reduces the thermal conductivity, since alloying elements act as scattering centers and reduce the efficiency of electron motion. Because of the difficulty involved in measuring thermal conductivity, as compared to electrical resistivity, this relation is useful in estimating the thermal conductivities of metals and alloys. It should be noted that because both electrical and thermal conductivities rely on the movement of free electrons, temperature is not the only factor that affects these properties. Anything that distorts the crystal structure, such as thermal vibrations of the ions, crystal vacancies and dislocations, grain boundaries, interstitial and solute ions, and elastic and plastic strains, will increase the resistance to conduction.

### *17.2.5 Thermal Diffusivity*

Thermal diffusivity is a measure of the speed at which heat spreads throughout a metal. Thermal diffusivity can be calculated by dividing the thermal conductivity by the product of the specific heat capacity and the density of the metal; that is, thermal diffusivity is equal to $k/C\rho$. The units of thermal diffusivity are $m^2/s$.

### *17.2.6 Thermal Stresses*

Thermal stresses are stresses induced in a body as a result of temperature changes. Thermal stresses are important because they can lead to undesirable plastic deformation or even fracture. If a metal bar is heated or cooled uniformly, there are no temperature gradients, and the bar will be stress free. However, if the bar is rigidly constrained by end supports, the magnitude of thermal stresses will be:

$$\sigma = \alpha E \Delta T \quad \text{(Eq 17.9)}$$

where σ is the linear coefficient of thermal expansion, $E$ is the modulus of elasticity, and $\Delta T$ is the change in temperature.

On heating, the stress is compressive, since expansion of the bar is constrained, and on cooling below the initial stress-free temperature, the bar will be in tension. The stress is the same as that which would be required to elastically compress (or elongate) a rod specimen back to its original length after it had been allowed to freely expand (or contract) with the $\Delta T$ temperature change.

When a metal is heated or cooled, the internal temperature distribution depends on its size and shape, the thermal conductivity, and the rate of temperature change. Thermal stresses are the result of temperature gradients across the body, which are frequently caused by rapid heating or cooling, in that the surface temperature changes more rapidly than the interior. Differential dimensional changes restrain the free expansion or contraction of adjacent volume elements within the piece. For example, on heating, the exterior of a specimen is hotter and will therefore expand more than the interior regions. Thus, compressive surface stresses are induced and are balanced by tensile interior stresses. On rapid cooling, the interior-exterior stress state is reversed, so that the surface is in a state of tension with the interior in compression. A more extensive discussion of residual stresses can be found in section 12.10 in Chapter 12, "Mechanical Behavior," in this book.

## 17.3 Band Theory of Metals

The original theory of an electron gas (Drude-Lorentz theory) neglected all of the electrons except the valence electrons. The total energy of the electrons was determined as the sum of their kinetic and potential energies relative to a free electron in space outside the metal. While this model was a good start, it led to a serious overestimation of specific heat. Later, the free electron theory retained the concept of uniform potential within the metal but used quantum mechanics to calculate the total energy, $E$. For calculating the energies of the valence electrons in a given metal, the density of states was used, which is the number of states per unit energy level, *N(E)*. The free electron theory correctly predicted the specific heat of metals, a value of only approximately 1/200 that was yielded by the Drude-Lorentz theory. While the free electron theory satisfactorily accounted for electrical conductivity in most metals, it failed to explain why other substances that also contain mobile electrons have virtually no conductivity and are insulators. The solution to this problem led to the zone or band theory.

A solid may be thought of as consisting of a large number of atoms (*N*), initially separated from one another, that are subsequently brought together and bonded to form the ordered atomic arrangement found in a crystalline matrix. At relatively large separation distances, each atom is independent of all others and will have an atomic energy level and electron configuration as if it were isolated from all other atoms. However, as the atoms come within close proximity of each other, their electrons are acted upon, or perturbed, by the electrons and nuclei of adjacent atoms. The result is that each distinct atomic state may split into a series of closely spaced electron states to form what is termed an electron energy band. The broadening of the atomic electron levels as the atoms are brought together is shown in the example in Fig. 17.2. The extent of splitting depends on the interatomic separation and begins with the outermost electron shells, since they are the first to be perturbed as the atoms are brought closer together. Within each band, the energy states are discrete, yet the difference between adjacent states within a band is exceedingly small. At the equilibrium spacing, band formation may not occur for the electron subshells nearest the nucleus. Further, gaps may exist between adjacent bands, and normally, energies lying within these band gaps are not available for electron occupancy; they are called forbidden zones. It is the extent of the reduction in the mean energy of the outer electrons that determines the stability of a metal structure. The equilibrium spacing between the atoms in a metal is attained when any further decrease in the atomic spacing would lead to an increase in the repulsive interaction of the positive ions as they are forced closer into contact with each other.

In a metallic structure, the free electrons can be thought of as occupying a series of discrete energy levels at very close intervals. Each atomic level that splits into a band contains the

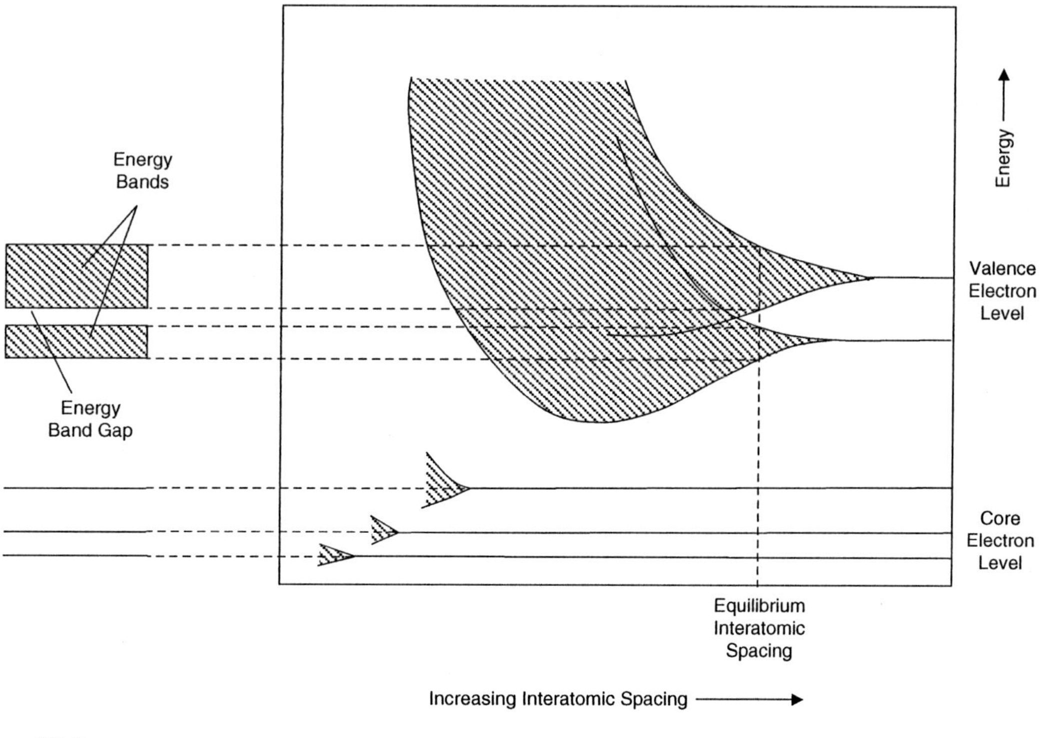


**Fig. 17.2** Broadening of atomic energy levels

same number of energy levels as the number, $N$, of atoms in the metal. In order to comply with the Pauli exclusion principle, only two electrons of opposite spin can occupy any one level, so that a band can contain a maximum of $2N$ electrons. In the lowest energy state, all the lower energy levels are occupied. The energy gap between successive levels is not constant but decreases as the energy of the levels increases. This is usually expressed in terms of the density of electronic states, $N(E)$, as a function of the energy, $E$. The quantity $N(E)dE$ gives the number of energy levels in a small energy interval, $dE$, and for free electrons forms a parabolic energy function of the form shown in Fig. 17.3. The energy corresponding to the highest filled state at absolute zero is called the Fermi energy, $E_f$. At temperatures above absolute zero, some of the electrons are thermally excited to higher energy levels that correspond to $E_{max}$, as indicated in the dashed curve.

Because only two electrons can occupy each level, the energy of an electron occupying a low energy level cannot be increased unless it has sufficient energy to allow it to jump to an empty level at the top of the band. The energy width of these bands is normally approximately 5 to 6 eV; therefore, considerable energy would have to be put into the metal to excite an inner-shell electron. Since these energy levels do not occur at normal temperatures, only those free electrons with energies close to that of the top of the band can take part in thermal and electrical processes.

The electrons can be thought of as moving continuously through the structure with an energy that depends on the energy level of the band it occupies. For electrons in the regular array of a metallic lattice, the lattice acts like a three-dimensional diffraction grating, since the atom cores are positively charged and interact with the negatively charged moving electrons. At certain wavelengths, determined by the spacing of the atoms on the metallic lattice, the electrons will experience strong diffraction effects. The result is that electrons having energies corresponding to certain wavelengths will not be able to move freely through the lattice. Therefore, electron bands at certain energy levels cannot be occupied, and there will be energy gaps in the otherwise continuous energy spectrum within a band.

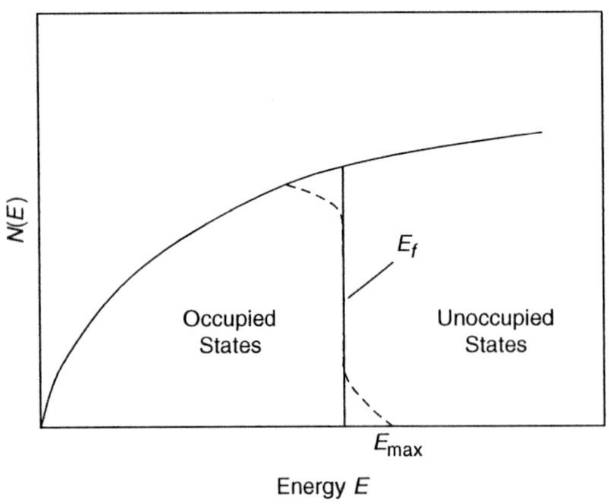


**Fig. 17.3** Density-of-states parabola. Source: Ref 2

For electrical conduction to occur, it is necessary that the electrons at the top of a band have their energy increased when an electric field is applied, so that a net flow of electrons in the direction of the applied potential, which produces an electric current, can take place. Three different types of band structures are shown in Fig. 17.4. Two representations are shown for each type of structure, a density of states versus energy curve and a conventional band structure representation, as previously shown in Fig. 17.2. Metallic conduction is due to the fact that the number of electrons per atom is insufficient to fill the band up to the point where an energy gap occurs. For example, in copper, a monovalent metal, as illustrated in Fig. 17.4(a), the 4S valence electrons fill only one-half of the outer $s$-band. In other metals, such as the divalent metal magnesium (Fig. 17.4b), the valence band overlaps a higher energy band, and the electrons near the Fermi level are free to move into the empty states of a higher band. When the valence band is completely filled and the next higher band, separated by an energy gap, is completely empty (Fig. 17.4c), the material is either an insulator or a semiconductor. If the gap is several electron volts wide, such as in diamond, where it is 7 eV, extremely high electric fields would be necessary to raise electrons to the higher band, and the material is an insulator. If the gap is small enough, such as 1 to 2 eV, as in silicon, then thermal energy may be sufficient to excite some electrons into the higher band, and the material is then a semiconductor.

Although all metals are relatively good conductors of electricity, they exhibit a range of resistivities. There are a number of reasons for this variability. The resistivity of a metal depends on the density of states of the most energetic electrons at the top of the band, and the shape of the *N(E)-E* curve at this point. In the transition metals, apart from producing strong magnetic properties, great strengths, and high melting points, the unfilled *d*-band is also responsible for the poor electrical conductivity and high electronic specific heat. When an electron is scattered by a lattice irregularity, it jumps into a different quantum state. The more vacant quantum states there are available in the same energy range, the more likely will be the electron to deflect at the irregularity. Thus, the high resistivities of the transition metals can be explained by the ease with which electrons can be deflected into vacant *d*-states.

Phonon-assisted *s-d* scattering gives rise to the nonlinear variation of electrical resistance with temperature observed at high temperatures. The high electronic specific heat is also due to

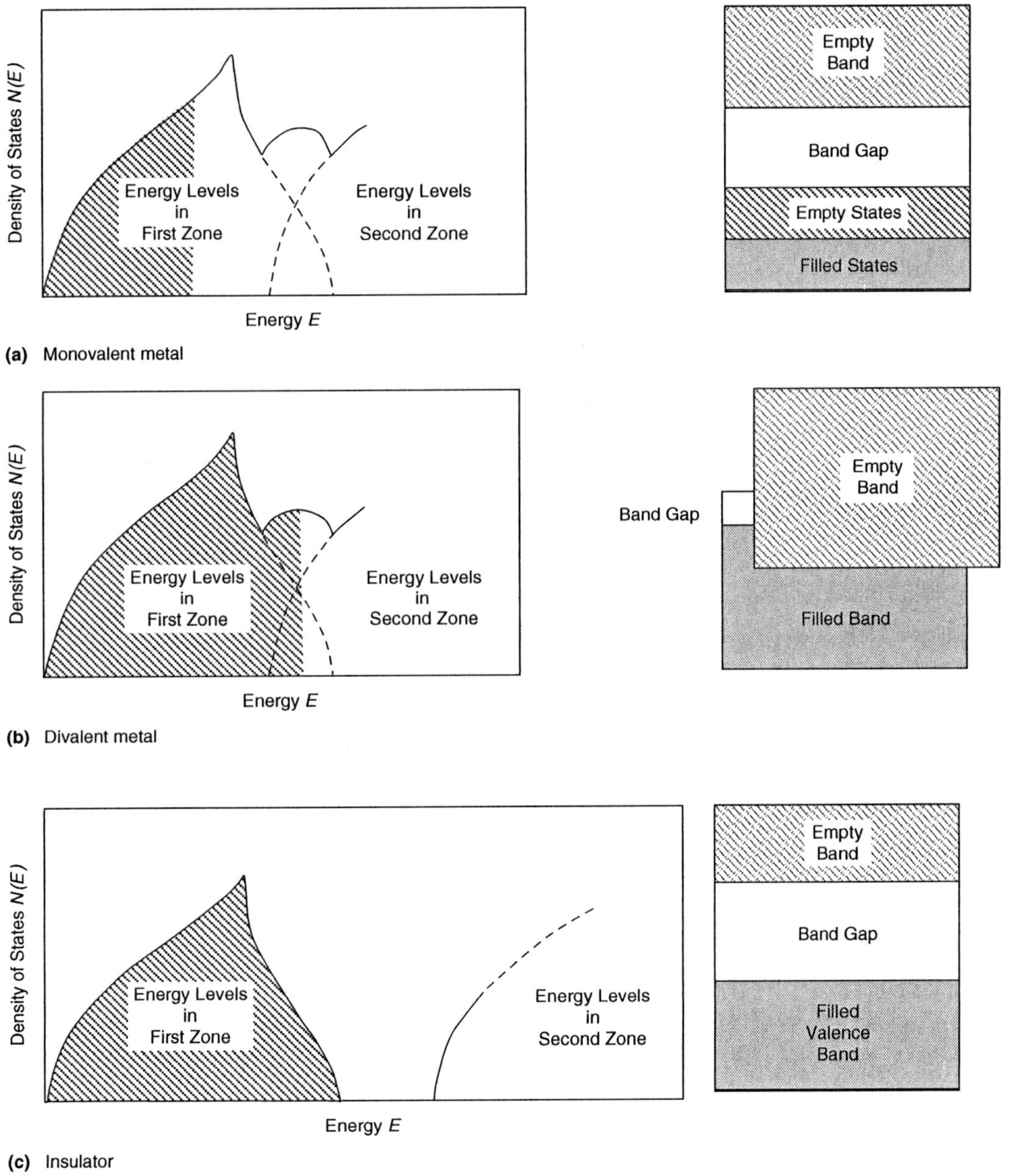


**Fig. 17.4** Density of state-energy curves and band structures

the high density of states in the unfilled *d*-band, since this gives rise to a considerable number of electrons at the top of the Fermi distribution that can be excited by thermal activation. In copper, there are no unfilled levels at the top of the *d*-band into which electrons can go, and consequently, both the electronic specific heat and electrical resistance are low. Conductivity also depends on the degree to which the electrons are scattered by the ions of the metal that are thermally vibrating, and on impurity atoms or other defects present in the metal.

Insulators can also be modified either by the application of high temperatures or by the addition of impurities. Clearly, insulators may become conductors at elevated temperatures if the thermal agitation is sufficient to enable electrons to jump the energy gap into the unfilled zone above.

## 17.4 Electrical Properties

One of the most important electrical characteristics of a solid material is the ease with which it transits an electric current. Ohm's law relates the current ($I$) to the applied voltage ($V$) as follows:

$$V = IR \quad \text{(Eq 17.10)}$$

where $R$ is the resistance of the material through which the current is passing. The units for $V$, $I$, and $R$ are volts (V), amperes (A), and ohms ($\Omega = \text{V/A}$), respectively. The value of $R$ is influenced by specimen configuration, and for many materials is independent of current.

The resistivity, designated by $\rho$, is independent of specimen shape but is related to $R$ through the expression:

$$\rho = \frac{RA}{l} \quad \text{(Eq 17.11)}$$

where $l$ is the distance between the two points at which the voltage is measured, and $A$ is the cross-sectional area perpendicular to the direction of the current. The units for $\rho$ are $\Omega \cdot \text{m}$ (often given as $\mu\text{m}\Omega \cdot \text{m}$ or $\text{n}\Omega \cdot \text{m}$). From the expression for Ohm's law and the previous equation:

$$\rho = \frac{VA}{Il} \quad \text{(Eq 17.12)}$$

Electrical conductivity ($\sigma$) is also used to specify the electrical characteristics of a material. It is simply the reciprocal of the resistivity, or $\sigma = 1/\rho$.

Metals exhibit a wide range of resistivities, from the low resistivity (high conductivity) of silver and copper to the high resistivity (low conductivity) of manganese and bismuth. Resistivity values of metals increase with increasing temperature. The semimetals boron, carbon, silicon, white phosphorus, sulfur, germanium, and tellurium have electrical resistivities that are intermediate between those of metal conductors and nonmetal insulators. Therefore, they are called semiconductors. Their resistivities range from approximately $10^3$ to $10^{13}$ $\Omega \cdot \text{m}$, and they decrease with increasing temperature.

### 17.4.1 Electron Mobility

When an electric field is applied, a force is brought to bear on the free electrons, and by virtue of their negative charge, they all migrate in a direction opposite to that of the field. According to quantum mechanics, there is no interaction between an accelerating electron and atoms in a perfect crystal lattice. All the free electrons should accelerate as long as the electric field is applied, which would give rise to a continuously increasing electric current with time. However, a current reaches a constant value the instant that a field is applied, indicating that there exists what may be termed frictional forces, which counter the acceleration from the external field.

Frictional forces result from the scattering of electrons by imperfections in the crystal lattice, including impurity atoms, vacancies, interstitial atoms, dislocations, and even the thermal vibrations of the atoms themselves. Each scattering event causes an electron to lose kinetic energy and to change its direction. However, there is a net electron motion in the direction opposite to the field, producing an electric current.

Scattering phenomenon can be considered to be a resistance to the passage of an electric current. Several parameters are used to describe the extent of scattering, namely the drift velocity and the mobility of an electron. The drift velocity, $V_d$, represents the average electron velocity in the direction of the force imposed by the applied field. It is directly proportional to the electric field, $\varepsilon$, as follows:

$$V_d = \mu_e \varepsilon \quad \text{(Eq 17.13)}$$

The constant $\mu_e$ is called the electron mobility, which is an indication of the frequency of scattering events. Its units are square meters per volt · second ($m^2/V \cdot s$).

The conductivity, $\sigma$, of most materials may be expressed as:

$$\sigma = n|e|\mu_e \quad \text{(Eq 17.14)}$$

where $n$ is the number of free or conducting electrons per unit volume (e.g., per cubic meter), and $|e|$ is the absolute magnitude of the electrical charge on an electron ($1.6 \times 10^{-19}$ coulombs). Thus, the electrical conductivity is proportional to both the number of free electrons and the electron mobility.

### 17.4.2 Electrical Resistivity

Metals are extremely good conductors of electricity because of the large numbers of free electrons that have been excited into empty states above the Fermi energy. Since crystalline defects serve as scattering centers for conduction electrons, increasing their number increases the resistivity. The concentration of these imperfections depends on temperature, composition, and the degree of cold work. Plastic deformation raises the electrical resistivity as a result of increased numbers of electron scattering dislocations. Matthiessen's rule states that the total resistivity is the sum of the independent contributions from thermal vibrations, impurities, and plastic deformation:

$$\rho_{total} = \rho_t + \rho_i + \rho_d \quad \text{(Eq 17.15)}$$

where $\rho_t$, $\rho_i$, and $\rho_d$ represent the individual thermal, impurity, and deformation resistivity contributions, respectively.

For the pure metal and all the copper-nickel alloys, the resistivity increases linearly with temperatures above approximately −185 °C (−300 °F). Thus:

$$\rho_t = \rho_o + aT \quad \text{(Eq 17.16)}$$

where $\rho_o$ and $a$ are constants for each particular metal. This dependence of the thermal resistivity component on temperature is due to the increase in thermal vibrations and other lattice irregularities with temperature.

For additions of a single impurity that forms a solid solution, the impurity resistivity, $\rho_i$, is related to the impurity concentration, $c_i$, in terms of the atom fraction (at.%/100) as follows:

$$\rho_i = Ac_i(1 - c_i) \quad \text{(Eq 17.17)}$$

where A is a composition-independent constant that is a function of both the impurity and host metals.

For a two-phase alloy consisting of $\alpha$ and $\beta$ phases, a rule-of-mixtures expression may be used to approximate the resistivity as follows:

$$\rho_t = \rho_\alpha V_\alpha + \rho_\beta V_\beta \quad \text{(Eq 17.18)}$$

where the $V$'s and $\rho$'s represent volume fractions and individual resistivities for the respective phases.

### 17.4.3 Electrical Conductor Alloys

Electrical and other properties of copper make it the most widely used metallic conductor. Oxygen-free high-conductivity copper, having extremely low oxygen and other impurity contents, is produced for many electrical applications. Even though aluminum has a lower electrical conductivity (only approximately one-half that of copper), it is also frequently used as an electrical conductor. Silver has a higher conductivity than either copper or aluminum; however, its use is restricted because of its high cost. On occasion, it is necessary to improve the mechanical strength of a metal alloy without significantly impairing its electrical conductivity. Both solid-solution alloying and cold working improve strength at the expense of conductivity, and thus a trade-off must be made. However, alloying with certain elements can be used to improve strength with only minor reductions in electrical conductivity. For example, up to 1 wt% Cd is added to copper telephone wires to provide greater strength. This alloy, when cold worked by drawing, has a tensile strength of approximately 462 MPa (67 ksi) compared to 338 MPa (49 ksi) for cold-worked pure copper, and the electrical conductivity is still over 90% of that for soft pure copper.

For some applications, such as furnace heating elements, a high electrical resistivity is desirable. The energy loss by electrons that are scattered is dissipated as heat energy. These materials must have not only a high resistivity but also resistance to oxidation at elevated temperatures and a high melting temperature. Nickel-chromium alloys, such as Nichrome, are commonly employed as heating elements.

## 17.5 Magnetic Properties

Magnetic materials are important industrially and are necessary for many engineering applications, especially in the field of electrical engineering. In general, there are two types of magnetic materials: soft and hard magnets. Soft magnets are used for applications where the material must be easily magnetized and demagnetized, such as for cores of transformers and stator and rotor materials for motors and generators. Hard magnets retain their magnetization and are not easily demagnetized. They are used for applications such as loudspeakers, telephone receivers, and synchronous and brushless motors.

### *17.5.1 Magnetic Fields*

The metals iron, nickel, and cobalt are the only three common metals that, when magnetized at room temperature, can produce a strong magnetic field around themselves and are said to be ferromagnetic. As shown in Fig. 17.5, a bar magnet has two magnetic poles in which magnet field lines leave one pole and enter the other. In fact, every magnet has two magnetic poles that are separated by a definite distance, a characteristic that extends down to even the small magnetic dipoles found in atoms.

Magnetic fields are also produced by current-carrying conductors, as shown in Fig. 17.6(a), which illustrates the formation of a magnetic field around a long coil of copper wire, called a solenoid, whose length is long with respect to its

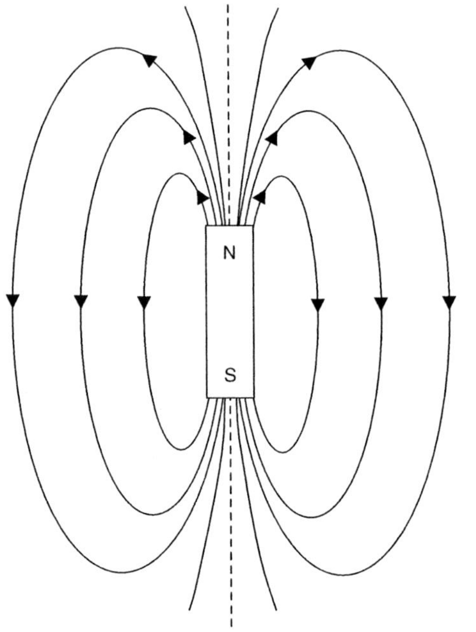


**Fig. 17.5** Magnetic field lines around a bar magnet

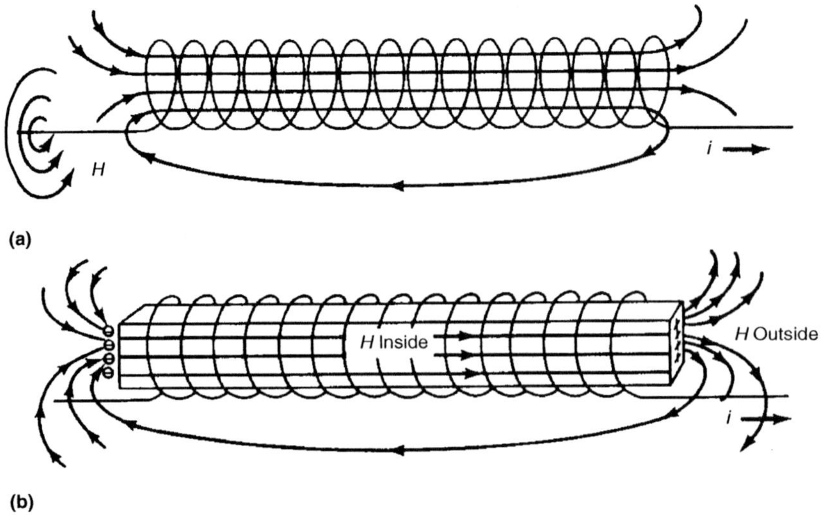


**Fig. 17.6** Magnetic field created by solenoid. Source: Ref 3

radius. For a solenoid of $n$ turns and length $l$, the magnetic field strength, $H$, is:

$$H = \frac{0.4\pi ni}{l} \qquad \text{(Eq 17.19)}$$

where $i$ is the current. The magnetic field strength, $H$, has units of amperes per meter (A/m) or oersteds (Oe).

### 17.5.2 Magnetic Induction

If a demagnetized iron bar is placed inside a solenoid and a current is applied to the solenoid, the magnetic field outside the solenoid is then stronger due to the magnetized bar inside the solenoid (Fig. 17.6b). The enhanced magnetic field outside the solenoid is due to the sum of the solenoid field itself and the external magnetic field of the magnetized bar. This additive magnetic field is called magnetic induction, or flux density, or simply induction and is expressed by the symbol $B$. Therefore, the magnetic induction $B$ is the sum of the applied field $H$ and the external field that arises from the bar inside the solenoid. The induced magnetic moment per unit volume due to the bar is called the intensity of magnetization, or simply the magnetization ($M$). Therefore:

$$B = \mu_o H + \mu_o M = \mu_o (H + M) \qquad \text{(Eq 17.20)}$$

where $\mu_o$ is the permeability of free space and is equal to $4\pi \times 10^{-7}$ tesla·meters per ampere. However, for ferromagnetic materials, the magnetization $\mu_o M$ is much greater than the applied field $\mu_o H$, so the relationship $B = \mu_o M$ is used. Thus, for ferromagnetic materials, the quantities $B$ (magnetic induction) and $M$ (magnetization) can be used interchangeably.

### 17.5.3 Magnetic Permeability

When a ferromagnetic material is placed in an applied magnetic field, the intensity of the field is increased. This increase in magnetization is measured by the magnetic permeability, $\mu$:

$$\mu = \frac{B}{H} \qquad \text{(Eq 17.21)}$$

If there is only a vacuum in the applied magnet field, then:

$$\mu_o = \frac{B}{H} \qquad \text{(Eq 17.22)}$$

An alternative method for defining magnetic permeability is to use the quantity relative permeability, $\mu_r$:

$$\mu_r = \frac{\mu}{\mu_o} \qquad \text{(Eq 17.23)}$$

then

$$B = \mu_o \mu_r H \qquad \text{(Eq 17.24)}$$

Relative permeability, $\mu_r$, is a dimensionless quantity. The relative permeability is a measure of the degree to which a material can be magnetized. For a ferromagnetic material, it changes as the material is magnetized and is measured from the slope of the $B$-$H$ magnetizing curve (Fig. 17.7). Both initial permeability, $\mu_i$, and maximum permeability, $\mu_{max}$, are commonly measured.

### 17.5.4 Magnetic Susceptibility

Since the magnetization, $M$, of a magnetic material is proportional to the applied field, $H$, a proportionality factor, called the magnetic susceptibility, $\chi_m$, is defined as:

$$\chi_m = \frac{M}{H} \qquad \text{(Eq 17.25)}$$

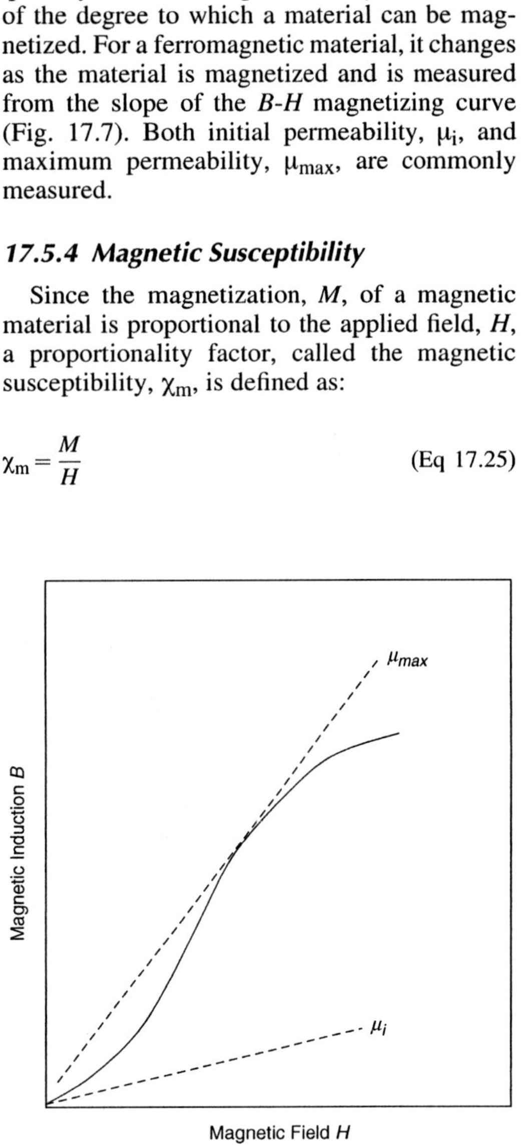


**Fig. 17.7** Measurement of magnetic permeability. Source: Ref 3

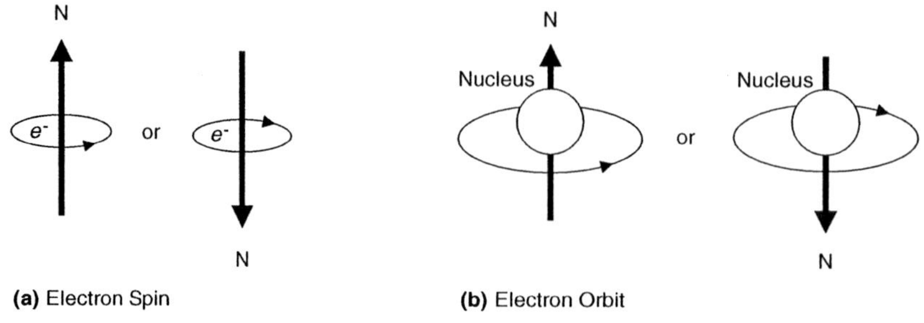


**Fig. 17.8** Electron spin and orbit. Source: Ref 4

which is a dimensionless quantity. Weak magnetic responses of materials are often measured in terms of magnetic susceptibility.

### *17.5.5 Types of Magnetism*

Magnetization occurs when induced or permanent magnetic dipoles are oriented by the interaction between a magnetic material and a magnetic field. Each electron has two magnetic moments. A magnetic moment, called the Bohr magnetron, is the strength of the magnetic field associated with the electron:

$$\text{Bohr magnetron} = \frac{qh}{4\pi m_e} = 9.27 \times 10^{-24}\ \text{A} \cdot \text{m}^2 \quad \text{(Eq 17.26)}$$

where $q$ is the charge on the electron, $h$ is the Plank's constant, and $m_e$ is the mass of the electron.

Magnetic moments are due to the orbital motion of the electron around its nucleus and the spin of the electron about its own axis. The spin of an electron produces a magnetic field (shown by vector **N** in Fig. 17.8a), and electrons orbiting about the nucleus create a magnetic field about the atom (shown by vector **N** in Fig. 17.8b). Depending on the electronic structure of the material, these moments can either cancel with no resultant magnetism or be additive to various degrees, producing magnetic responses in varying degrees. It is the amount that the moments are additive that determines the degree and therefore the type of resultant magnetism.

Each discrete energy level can contain two electrons, but the two electrons must have an opposite spin to comply with the Pauli exclusion principle. Therefore, whenever an energy band is completely full, the magnetic moments of each electron are opposed, and consequently, there is no net magnetic moment. An unpaired valence electron is also not effective, because the valence electrons from each atom interact and, on average, cancel each other, so that no net magnetic moment is produced. However, the transition elements contain an unfilled 3*d* inner shell. Most of the transition elements react in such a way that the sum of their magnetic moments is zero. However, the atoms in iron, nickel, and cobalt undergo what is called an exchange mechanism, in which the orientation of the dipole in one atom influences the surrounding atoms to have the same dipole orientation, thus producing net magnetic moments.

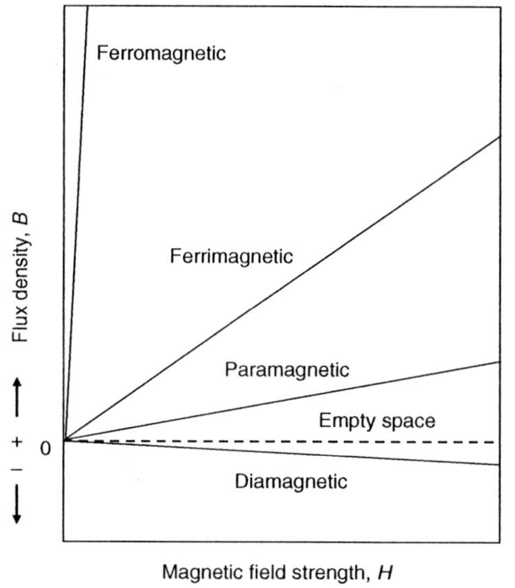


**Fig. 17.9** Magnetic response of several types of magnetic materials. Source: Ref 1

The response of ferromagnetic materials to a magnetic field is very strong, while it is much weaker in ferrimagnetic, paramagnetic, and diamagnetic materials, as shown in Fig. 17.9. As a result, ferromagnetic materials are by far the strongest and thus most important of the magnetic materials.

**Diamagnetic Behavior.** Diamagnetic materials are those in which the magnetic induction is slightly less than the applied magnetic field. For diamagnetic materials, the permeability is slightly less than that of empty space, and the magnetic susceptibility is negative and small. Diamagnetic behavior has no important applications for magnetic materials or devices and, in important magnetic materials, is overwhelmed by the other types of magnetism.

**Paramagnetism.** Paramagnetic materials are those in which the permanent magnetic dipole moments are only partially aligned, and the magnetic induction is only slightly greater than the applied magnetic field strength. Paramagnetic materials exhibit a small positive magnetic susceptibility in the presence of a magnetic field, but the paramagnetic effect disappears when the applied field is removed. While the permeability of a ferromagnetic metal can reach values on the order of $10^6$ times that of empty space, the permeability of a paramagnetic metal is only slightly greater than that of empty space. Most metals are paramagnetic.

**Ferrimagnetism.** Ferrimagnetic materials are those in which the magnetic moments of atoms or ions tend to assume an ordered but nonparallel arrangement. In ferrimagnetic ceramic materials, called ferrites, the various constituent ions have different magnetic moments. In a magnetic field, the dipoles of ion *A* may line up with the field, while dipoles of ion *B* oppose the field. Because the strengths of the two dipoles are not equal, a net magnetization results.

**Antiferromagnetism.** In antiferromagnetic materials, dipoles of equal strength align themselves in opposite directions, the dipoles cancel each other, and there is zero magnetism. Examples of antiferromagnetic materials are manganese and chromium.

**Ferromagnetism.** Ferromagnetic materials are those in which the magnetic moments are aligned parallel and reinforce each other even after the applied field is removed, thereby producing a permanent magnet. The positive interaction that characterizes ferromagnetic metals gradually decreases with increases in temperature and thermal energy and falls to zero at the Curie temperature, as shown in Fig. 17.10 for iron. Above 770 °C (1418 °F), iron switches from being a strongly magnetic material (ferromagnetic) to a very weak magnetic material (paramagnetic). Nickel and cobalt, the other two important ferromagnetic metals, show a similar behavior, with nickel having a Curie temperature of 355 °C (670 °F) and cobalt at 1115 °C (2039 °F).

### 17.5.6 Magnetic Domains

When a ferromagnetic material becomes magnetized in a magnetic field, such as one generated by an electric current in a coil, there are differences in the degree of magnetism that is attained when it is removed. Many materials are easily demagnetized after the field is removed and are referred to as being magnetically soft. On the other hand, some materials are magnetically hard and strongly retain their magnetism after the field is removed. A comparison of a magnetically soft and hard material is shown in Fig. 17.11. In this figure, the small bars represent magnetic domains. A domain is a region in which the magnetic moments are aligned and all point in the same direction. A magnetic material consists of many domains, instead of one large one. While the interaction energy would favor one large domain, that would result in magnetic poles on the surface of the metal, and the associated magnetostatic energy would be very high. Therefore, small, closed-loop domains are formed (Fig. 17.12). The boundary between

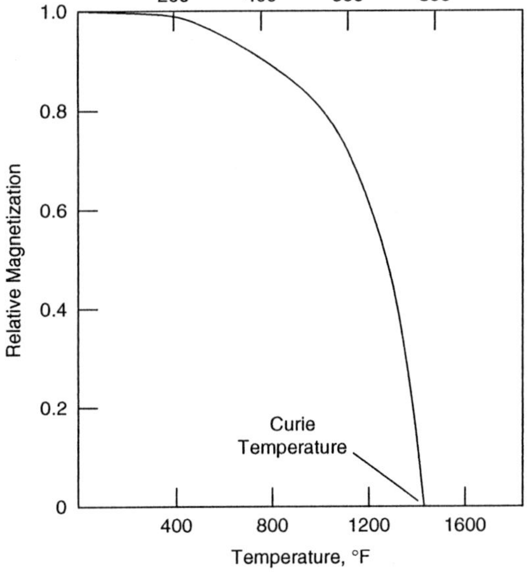


**Fig. 17.10** Decrease in magnetization of iron with temperature

magnetic domains is called the Bloch wall. Since the direction of magnetization must turn out of the direction of easiest magnetization when it passes through a Bloch wall, the anisotropy energy of the domains is increased. However, the increase in energy is a minimum if the transition is gradual, such as occurs over a region of approximately 1000 atoms wide. As a result of the relatively high energy of domain boundaries, if the dimensions of a ferromagnetic substance are sufficiently small, each particle will exist as a single domain and will have exceptional resistance to demagnetization.

The ease of magnetization varies with the crystallographic direction, as shown in Fig. 17.13 for a single crystal of iron. In iron, the [100] direction is the easiest direction of magnetization. In nickel, the easiest direction is [111], and in cobalt, it is the hexagonal axis [0001]. If magnetization is to occur in a direction other than the easiest direction, then additional energy, called the anisotropy energy, is required.

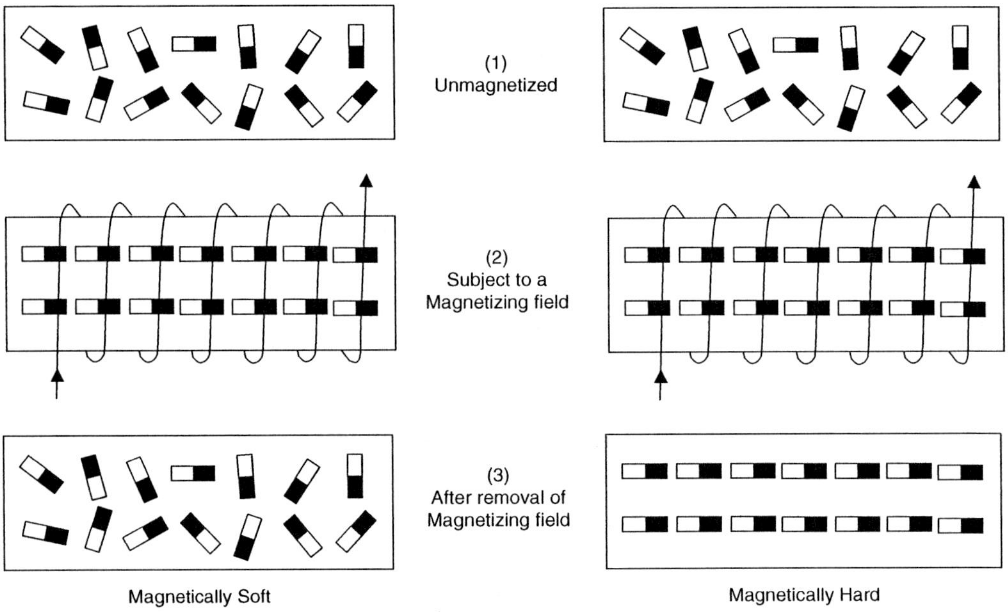


**Fig. 17.11** Domain alignment of magnetically hard and soft materials. Source: Ref 5

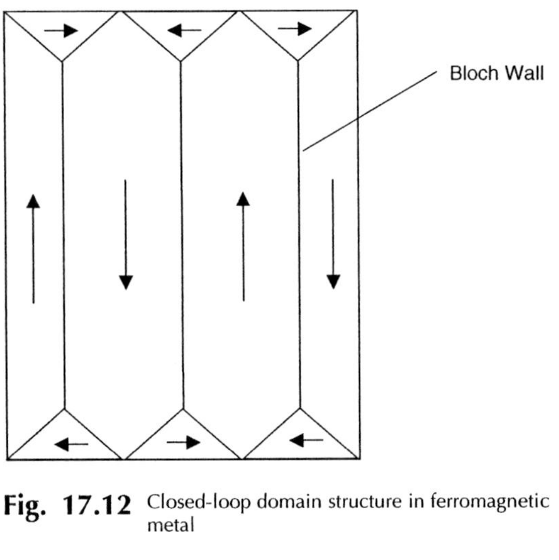


**Fig. 17.12** Closed-loop domain structure in ferromagnetic metal

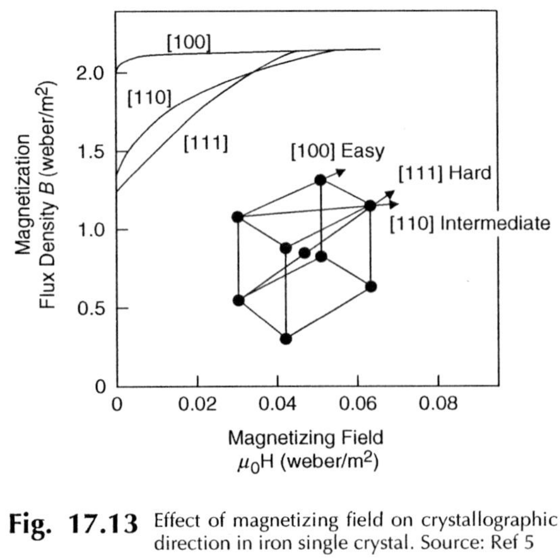


**Fig. 17.13** Effect of magnetizing field on crystallographic direction in iron single crystal. Source: Ref 5

### *17.5.7 Magnetically Soft Materials*

The magnetic behavior of a magnetically soft metal can be described using the *B-H* diagram in Fig. 17.14. If the metal is initially unmagnetized, then the magnetization, *B*, will increase along the dashed curve as the magnetizing force, *H*, increases. The degree of magnetization eventually reaches a maximum, called the saturation magnetization, $B_S$, where all of the dipoles are aligned in the direction of the field. When the magnetizing force is reduced back toward zero, *B* decreases along the solid curve and reaches the residual magnetization, or remanence, $B_r$, when $H=0$. As the magnetizing force is decreased, it becomes zero at the coercive force, $H_c$. As the reversing magnetizing force becomes even greater, the magnetization becomes reversed and eventually reaches $-B_S$. The remaining half of the solid curve is obtained when the magnetizing force is again reversed and then increases. When a magnetic field is applied, the material changes dimensions, either expanding or contracting in the direction of the field. This phenomenon is called magnetostriction and is related to attraction or repulsion between dipoles.

The change in the domain structure during magnetization is shown in Fig. 17.15. In the unmagnetized structure, the magnetic effects of the domains cancel out, and the metal is nonmagnetic. When a small magnetic field, $H>0$, is applied in the direction indicated by the arrow, domains with orientations close to this direction have lower energies and are more stable. The more stable domains increase in size by movement of the Bloch wall separating them from the less stable domains. The final stage of magnetization occurs when a relatively large magnetic field, $H \gg 0$, is applied, and the domains rotate from their original easy directions toward the direction of the applied field. If the magnetic field is applied cyclically, the frequency at which the material responds is important. If the

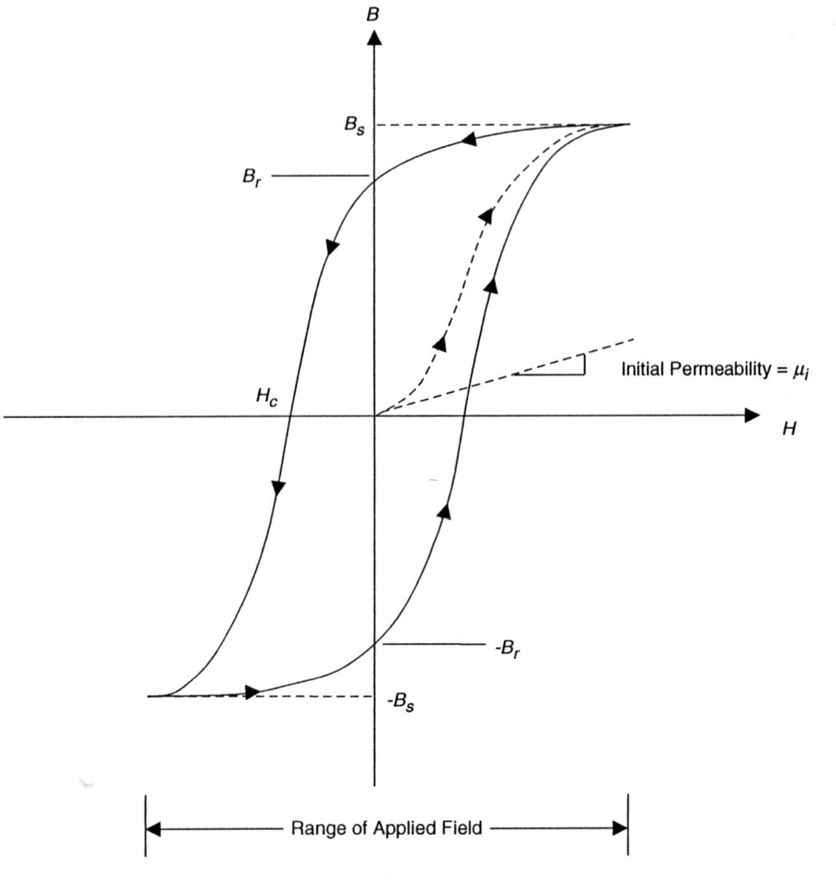


**Fig. 17.14** Hysteresis loop for a magnetically soft material

frequency is so high that the domains do not have time to realign during each cycle, dipole friction can produce heat. Another source of energy loss is through electrical eddy currents that can be induced and consequent Joule heating ($I^2R$). Eddy current losses can be minimized by using magnetic materials with a high resistivity.

The following magnetic properties are usually of importance for soft magnets.

**Saturation flux density, $B_s$,** determines the degree of magnetization that can be developed in a magnetic material, and therefore, a high value of $B_s$ is almost always desirable.

**Relative Permeability, $\mu_r$.** The larger the value of $\mu_r$, the smaller is the magnetizing force, $\mu_0 H$, necessary to produce a given magnetization, $B$. A large initial relative permeability is important when only a weak magnetizing force is available.

**Coercive Force, $\mu_0 H$.** Since the coercive force is the reverse magnetizing force necessary to eliminate the residual flux density, $B_r$, the coercive force should be small in magnetically soft materials.

**Total core loss** is the total power loss. It is the sum of the hysteresis and eddy current losses. The total core loss in an iron-silicon steel can be reduced by approximately 50% by increasing the silicon content from 1.0 to 4.25 wt%.

The hysteresis loops of magnetically soft and hard materials are compared in Fig. 17.16. Soft magnets have narrow hysteresis loops with low coercive forces, while hard magnets have wide hysteresis loops with high coercive forces.

The properties of several magnetically soft materials are given in Table 17.2. The most extensively used soft magnetic materials are the iron-3 to 4 wt% Si alloys. Silicon produces several benefits: Silicon increases the electrical resistivity of low-carbon steel and thus reduces stray current (eddy current) losses; silicon decreases the magnetoanistropy energy of iron and increases magnetic permeability, thus

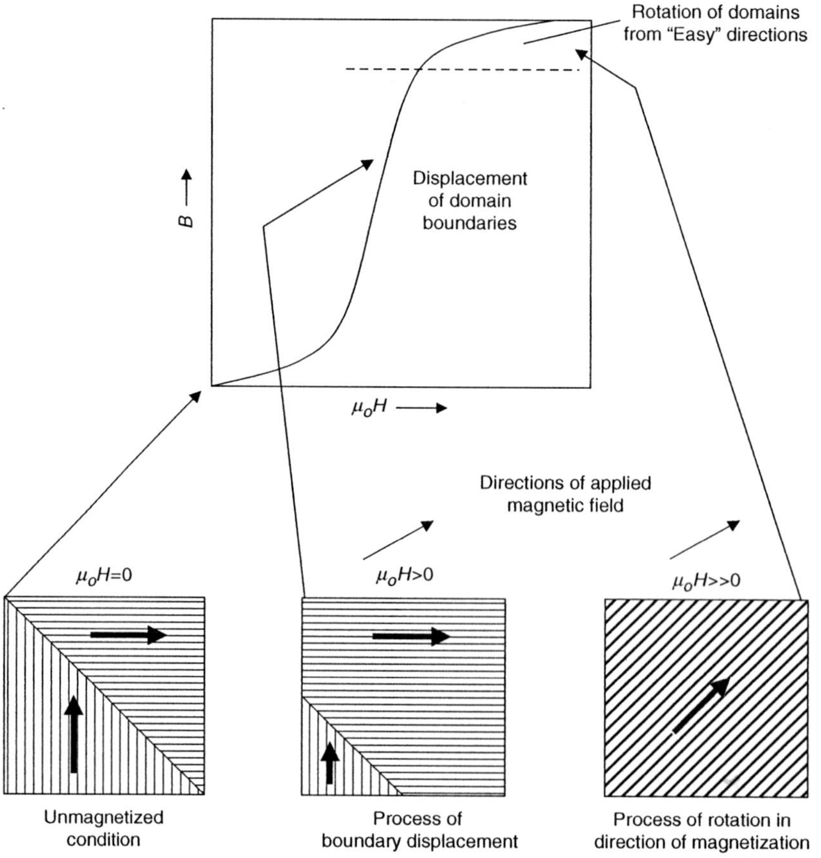


**Fig. 17.15** Process of magnetization for a ferromagnetic material. Source: Ref 5

decreasing core losses; and silicon additions reduce hysteresis losses and transformer noise or "hum." However, silicon also reduces the ductility of iron, and the practical limit is approximately 4 to 5 wt%. Silicon also decreases the saturation induction and the Curie temperature.

Design and processing are also used to improve the performance of iron-silicon magnetic alloys. For example, laminated designs can be used to reduce eddy current losses. Thin sheets (0.3 to 3.6 mm, or 0.010 to 0.14 in.) of alloy are stacked between layers of insulation to make a laminated construction. The insulation prevents stray eddy currents from flowing perpendicularly through the sheets. Another method to reduce core losses is to produce grain-oriented sheet by a combination of cold working and recrystallization treatments. Essentially, the grains are oriented so that the domains are oriented for easy magnetization when a magnetic field is applied parallel to the rolling direction of the sheet. This produces a material with a higher permeability and lower hysteresis losses than a random texture.

The magnetic permeabilities of pure iron and iron-silicon alloys are relatively low at low applied fields. Low initial permeability is not a problem for power applications such as transformer cores, since this equipment

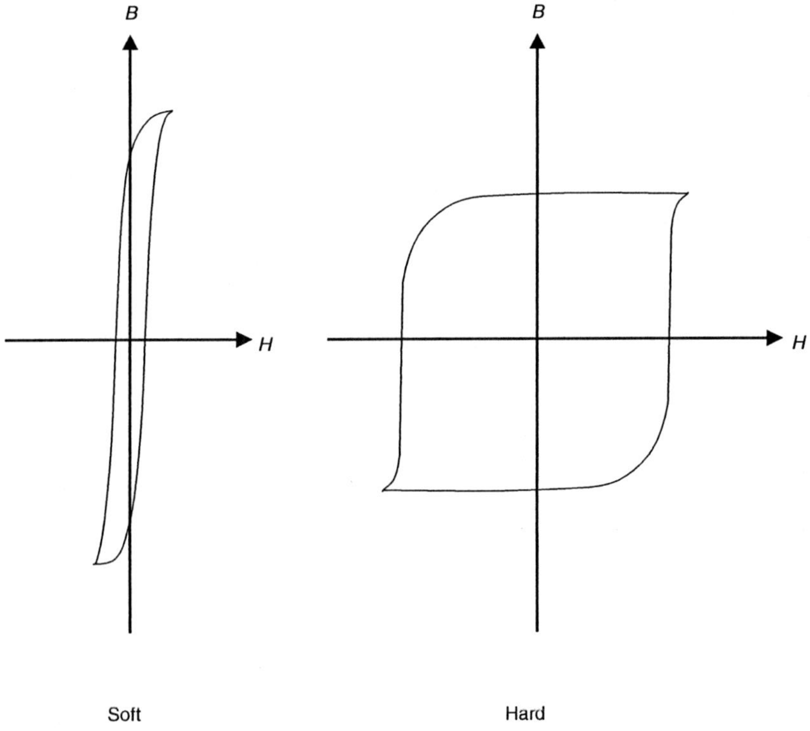


**Fig. 17.16** *B-H* curves for soft and hard magnets

**Table 17.2 Typical properties for several soft magnetic materials**

| Material | Composition, wt% | Initial relative permeability ($\mu_i$) | Saturation flux density ($B_s$), tesla (gauss) | Hysteresis loss/cycle, J/m$^3$ (erg/cm$^3$) | Resistivity ($\rho$), ohm · m |
|---|---|---|---|---|---|
| Commercial ingot iron | 99.95Fe | 150 | 2.14 (21,400) | 270 (2700) | $1.0 \times 10^{-7}$ |
| Silicon-iron (oriented) | 97Fe, 3Si | 1400 | 2.01 (20,100) | 40 (400) | $4.7 \times 10^{-7}$ |
| 45 Permalloy | 55Fe, 45Ni | 2500 | 1.60 (16,000) | 120 (400) | $4.5 \times 10^{-7}$ |
| Supermalloy | 79Ni, 15Fe, 5Mo, 0.5Mn | 75,000 | 0.80 (8000) | ... | $6.0 \times 10^{-7}$ |
| Ferroxcube A | $48MnFe_2O_4$, $52ZnFe_2O_4$ | 1400 | 0.33 (3300) | ~40 (~400) | 2000 |
| Ferroxcube B | $36NiFe_2O_4$, $64ZnFe_2O_4$ | 650 | 0.36 (3600) | ~35 (~350) | $10^{-7}$ |

Source: Ref 6

is used at high magnetizations. However, for high-sensitivity communication equipment used to detect or transmit small signals, nickel-iron alloys are commonly used because they have much higher permeabilities in weak magnetic fields. There are two broad classes of these alloys: those with approximately 50 wt% Ni (e.g., 45 Permalloy) and those with approximately 79 wt% Ni (Supermalloy). The alloy 45 Permalloy has a rather moderate permeability and a high saturation induction, while Supermalloy has high permeability and a lower saturation induction. However, the initial permeability of the iron-nickel alloys containing approximately 56 to 58 wt% Ni, such 45 Permalloy, can be increased by three to four times by annealing the alloy in the presence of a magnetic field. The magnetic anneal causes directional ordering of the atoms in the crystalline lattice. These alloys are used for audio and instrument transformers, instrument relays, and for rotor and stator laminations.

### 17.5.8 Magnetically Hard Materials

Although the most powerful magnets are the electromagnets made of magnetically soft materials, permanent or hard magnets are useful because they retain their magnetic properties when the electrical field is removed. Permanent magnets require high remanence, high permeability, high coercive fields, and high power.

A useful design property for permanent magnets is the maximum external energy or power, $(BH)_{max}$, which is related to the size of the hysteresis loop, or the maximum product of $B$ and $H$. The area of the largest rectangle that can be drawn in the second or fourth quadrant of the $B$-$H$ curve is related to the energy required to demagnetize the magnet (Fig. 17.17). Maximum

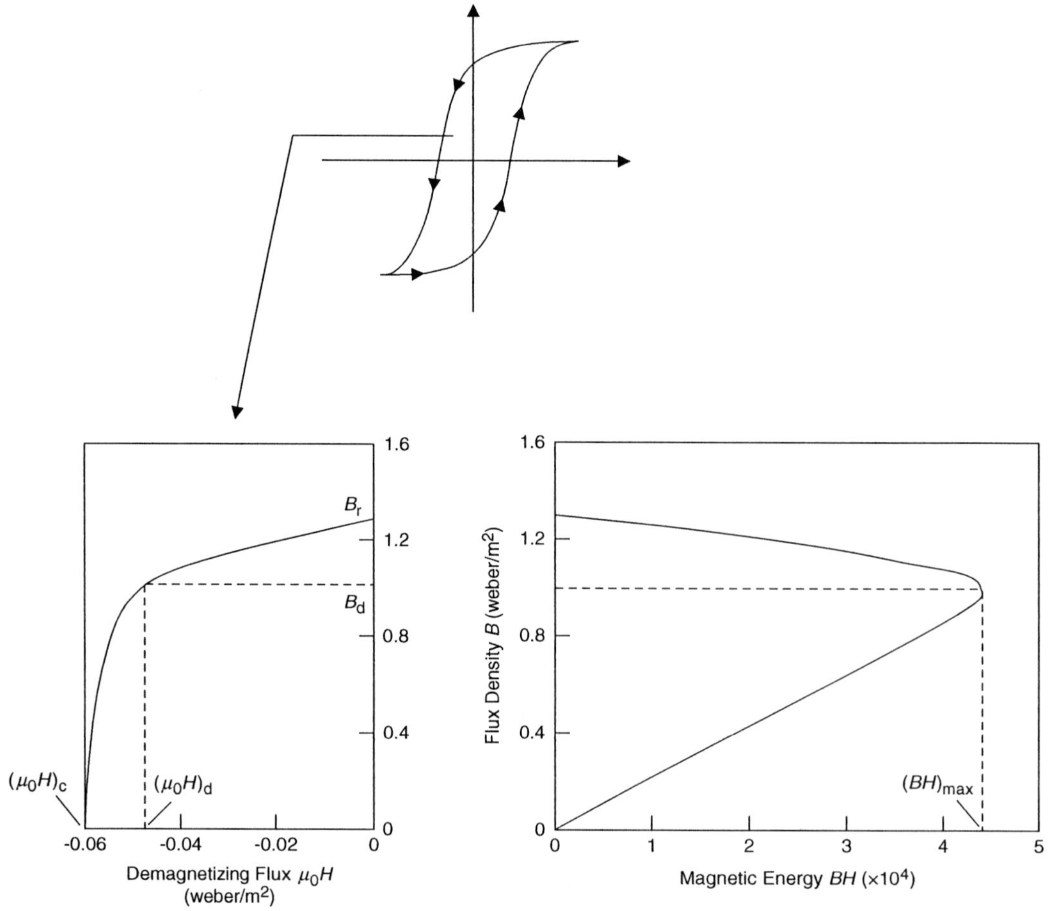


**Fig. 17.17** Method of determining maximum external energy, $(BH)_{max}$. Source: Ref 5

external energy is derived from the *B-H* curve by finding the value of magnetization, $B_d$, at which the magnetic energy required for demagnetization reaches a maximum value. The product of $B_d$ and $H_d$ is the demagnetizing field, $(BH)_{max}$. The properties of several magnetically hard materials are shown in Table 17.3.

Important classes of permanent magnets include tungsten steels, iron-nickel-copper alloys, aluminum-nickel-cobalt (Alnico) alloys, rare earth alloys, iron-chromium-cobalt alloys, and ferrites. Good permanent magnets can be produced by making the grain size so small that only one domain is present in each grain. When the boundaries are grain boundaries rather than domain boundaries (Bloch walls), the domains can change their orientation only by rotating, which takes more energy than domain growth. Two techniques are used to produce these materials: phase transformations and powder metallurgy. Alnico alloys have a single-phase body-centered cubic (bcc) structure at high temperatures. When Alnico alloys are cooled below 800 °C (1470 °F), a second bcc phase, rich in iron and cobalt, precipitates. The second phase is so fine that each precipitate particle is a single domain, producing a very high remanence, coercive force, and power. The alloys are also cooled and transformed in a magnetic field to align the domains as they cool. The Alnico alloys are the most important magnetic alloys in use today and account for approximately 30% of the permanent magnet market in the United States.

Powder metallurgy is used for a group of rare earth metal alloys, such as the samarium-cobalt alloys. The composition $SmCo_5$, an intermetallic compound, is crushed and ground to produce a fine powder in which each particle is a single magnetic domain. The powder is compacted under the influence of a magnetic field and then carefully sintered to avoid grain growth.

Ferrites, magnetic ceramic materials, are produced as both soft and hard magnetic materials. They are ferrimagnetic due to a net magnetic moment produced by their ionic structures. Most magnetically soft ferrites have the basic composition $MO \cdot Fe_2O_3$, where "M" is a divalent ion such as $Fe^{++}$, $Mn^{++}$, or $Ni^{++}$. These materials are used for low-signal, memory core, audiovisual, and recording head applications. Since they are insulators, they can be used for high-frequency applications where eddy currents are a problem with alternating fields. Magnetically hard ferrites have the general formula $MO \cdot 6Fe_2O_3$, where "M" is usually a barium or strontium ion, and are used for applications requiring low-cost, low-density permanent magnets. Applications include loud speakers, telephone ringers and receivers, and holding devices for doors, seals, and latches.

## 17.6 Optical Properties

When light strikes a metallic surface, it is either absorbed, reflected, or transmitted. Absorptivity is the fraction of light absorbed, reflectivity is the fraction of light reflected, and transmissivity is the fraction of light transmitted. Because metals are opaque, transmissivity is very low except in extremely thin films of metal (approximately 0.1 μm or less), and the absorptivity and reflectivity are relatively high. Incident radiation of visible light over a wide range of frequencies excites electrons to unoccupied states of higher energy and thus is absorbed. The excited electrons soon decay back to lower energy levels, and light is re-emitted from the surface in the form of photons, giving rise to reflectance.

All frequencies of visible light are absorbed by metals because of the continuously available empty electron states. Metals are opaque to all electromagnetic radiation on the low end of the

**Table 17.3 Typical properties for several hard magnetic materials**

| Material | Composition, wt% | Remanence ($B_r$), tesla (gauss) | Coercivity ($H_c$), amp · turn/m (Oe) | $(BH)_{max}$, kJ/m³ (MGO$_e$) | Curie temp. ($T_c$)°C (°F) | Resistivity (ρ), ohm · m |
|---|---|---|---|---|---|---|
| Tungsten steel | 92.8Fe, 6W, 0.5Cr, 0.7C | 0.95 (9500) | 5900 (74) | 2.6 (0.33) | 760 (1400) | $3.0 \times 10^{-7}$ |
| Cunife | 20Fe, 20Ni, 60Cu | 0.54 (5400) | 44,000 (550) | 12 (1.5) | 410 (770) | $1.8 \times 10^{-7}$ |
| Sintered Alnico 8 | 34Fe, 7Al, 15Ni, 35Co, 4Cu, 5Ti | 0.76 (7600) | 125,000 (1550) | 36 (4.5) | 860 (1580) | ... |
| Sintered ferrite 3 | $BaO \cdot 6Fe_2O_3$ | 0.32 (3200) | 240,000 (3000) | 20 (2.5) | 450 (840) | $10^4$ |
| Cobalt rare earth 1 | $SmCo_5$ | 0.92 (9200) | 720,000 (9000) | 170 (21) | 725 (1340) | $5.0 \times 10^{-7}$ |
| Sintered neodymium-iron-boron | $Nd_2Fe_{14}B$ | 1.16 (11,600) | 848,000 (10,600) | 255 (32) | 310 (590) | $1.6 \times 10^{-6}$ |

Source: Ref 6

frequency spectrum, from radio waves through infrared, the visible, and into approximately the middle of the ultraviolet radiation. However, metals are transparent to high-frequency x- and γ-ray radiation. Most of the absorbed radiation is re-emitted from the surface in the form of visible light of the same wavelength, which appears as reflected light. The reflectivity for most metals is between 0.90 and 0.95, with some small fraction of the energy from electron decay processes being dissipated as heat.

Since metals are opaque and highly reflective, their perceived color is determined by the wavelength distribution of the radiation that is reflected and not absorbed. A bright, silvery appearance when exposed to white light indicates that the metal is highly reflective over the entire range of the visible spectrum. In other words, for the reflected beam, the composition of these re-emitted photons, in terms of frequency and number, is approximately the same as for the incident beam. The surface of clean, bare, unoxidized metals is lustrous. Because the metal silver is highly reflective over the entire visible range of light, its surface color is white metallic. Copper and gold exhibit red-orange and yellow colors, respectively, because incident light of short wavelengths excites electrons in filled *d*-bands to empty levels in the *s*-bands. The electrons decay by a different path, resulting in absorption of green, blue, and violet light, whereas yellow, orange, and red light are reflected. Many metals, such as iron and nickel, absorb light of various wavelengths and therefore have grayish or dull colors, or luster, due to relatively low reflectivity.

Because of their chemical activities, thin oxides rapidly form on the fresh exposed surfaces of most metals, and these oxides can affect their appearance. Rough surfaces reflect light in a variety of directions. The presence of oxides, hydroxides, or other foreign materials greatly increases the absorption of light and therefore decreases the reflectivity of metallic surfaces. Ruthenium, rhodium, palladium, silver, osmium, iridium, platinum, and gold are the exceptions. These eight metals have such low chemical activity that they are called noble metals.

When a metal is heated to a very high temperature, such as a tungsten filament in an incandescent light bulb, its electrons are thermally excited to high energy levels, and, because many energy levels are involved, light of many wavelengths is emitted when the electrons decay back to lower energy levels. As the temperature of the metal increases, the energies absorbed by the electrons increase, causing white light to be emitted, in contrast with the red light emitted from metals at relatively low temperatures (595 to 980 °C, or 1100 to 1800 °F).

## ACKNOWLEDGMENTS

Sections of this chapter were adapted from "Properties of Metals" by H. Baker in *Metals Handbook Desk Edition*, 2nd ed., ASM International, 1998.

## REFERENCES

1. H. Barker, Properties of Metals, *Metals Handbook Desk Edition,* 2nd ed., ASM International, 1998
2. R.E. Smallman and R.J. Bishop, *Modern Physical Metallurgy and Materials Engineering,* Butterworth Heinemann, 1999
3. W.F. Smith, *Principles of Materials Science and Engineering,* McGraw-Hill, 1986
4. D.R. Askeland, *The Science and Engineering of Materials,* 2nd ed., PWS-Kent Publishing Co., 1989
5. A.G. Guy and J.J. Hren, *Elements of Physical Metallurgy,* 3rd ed., Addison-Wesley Publishing Company, 1974
6. W.D. Callister, *Fundamentals of Materials Science and Engineering,* 5th ed., John Wiley & Sons, Inc., 2001

## SELECTED REFERENCES

- G.F. Carter, *Principles of Physical and Mechanical Metallurgy,* American Society for Metals, 1979
- A. Cottrell, *An Introduction of Metallurgy,* 2nd ed., IOM Communications, 1975
- J.F. Schakelford, *Introduction to Materials Science for Engineers,* 5th ed., Prentice Hall, 2000
- L.H. Van Vlack, *Elements of Materials Science and Engineering,* 4th ed., Addison-Wesley Publishing, 1980

# CHAPTER 18

# Corrosion

CORROSION is the gradual degradation of a material due to the environment. In fact, the word corrode is derived from the Latin *corrodere,* which means "to gnaw to pieces." Metallic corrosion is a chemical or electrochemical process in which surface atoms of a solid metal react with a substance in contact with the exposed surface and produce a deterioration of the material and its properties. The corroding medium is usually a liquid substance, but gases and even solids can also act as corroding media. Corrosion can manifest itself in numerous ways and may or may not be obvious. For example, while the rusting of a steel surface is fairly obvious, the intergranular corrosion of stainless steel is less obvious but just as damaging.

Although the principles of corrosion are fairly well understood, corrosion continues to cost billions of dollars a year in the United States alone. As shown in Table 18.1, the cost in the United States exceeds $300 billion per year. Fortunately or unfortunately, depending on how one looks at it, approximately 35% of the total cost could be avoided by proper prevention methods.

This chapter first covers some basic principles of electrochemical corrosion and then some of the various types of corrosion. This is followed by a short section on corrosion control. The last section deals with high-temperature oxidation that usually occurs in the absence of moisture.

**Table 18.1 Cost of metallic corrosion in the United States**

| Industry | Billions of U.S. dollars | |
|---|---|---|
| | 1975 | 1995 |
| All industries | | |
| Total | 82 | 296 |
| Avoidable | 33 | 104 |
| Motor vehicles | | |
| Total | 31 | 94 |
| Avoidable | 23 | 65 |
| Aircraft | | |
| Total | 3 | 13 |
| Avoidable | 0.5 | 3 |
| Other industries | | |
| Total | 48 | 189 |
| Avoidable | 9 | 36 |

Source: Ref 1

## 18.1 Basics of Electrochemical Corrosion

Electrochemical corrosion is a process resulting in part or all of the metal being transformed from the metallic to the ionic state. Electrochemical corrosion in metals is caused by a flow of electricity from one metal to another metal or from one part of a metal surface to another part of the same surface. For a current to flow, a complete electrical circuit is required. In a corroding system (Fig. 18.1), this circuit is made up of four components:

- The anode is the electrode of an electrolytic cell at which oxidation is the principal reaction. Electrons flow away from the anode in the external circuit. It is the electrode at which corrosion occurs and metal ions enter solution.
- The electrolyte is an electrical conducting solution that contains ions, which are atomic

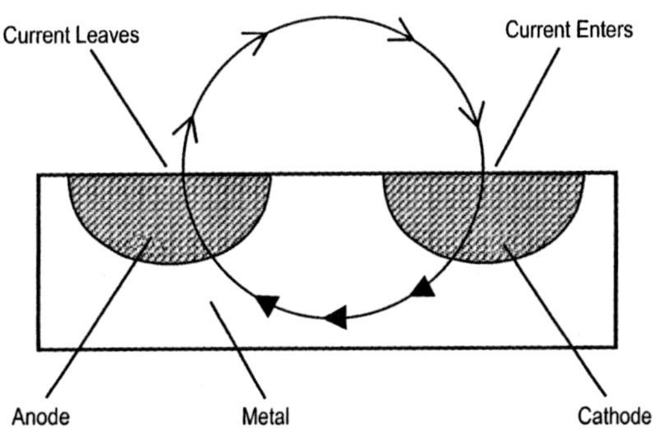


**Fig. 18.1** A corroding system. Source: Ref 2

particles or radicals possessing an electrical charge. The electrolyte is a conductive liquid through which the current is carried by positively charged ions (cations) to the cathode. Negatively charged ions (anions) are simultaneously attracted to the anode. Charged ions are present in solutions of acids, alkalis, and salts. Water, especially saltwater, is an excellent electrolyte. In pure water, there are positively charged hydrogen ions ($H^+$) and negatively charged hydroxyl ions ($OH^-$) in equal amounts. The metal undergoing electrochemical corrosion need not be immersed in a liquid but may be in contact with moist soil or may just have moist areas on the metal surface.

- The cathode is the electrode of an electrolytic cell at which reduction is the principal reaction. Electrons flow toward the cathode in the external circuit. The cathode does not corrode.
- A metallic path is an external circuit to complete the connection between anode and cathode.

If any one of these four conditions is absent, the electric circuit is incomplete and corrosion cannot occur. In essence, this is the strategy of electrochemical corrosion-prevention methods.

It is important to note that separate anode and cathode metals are not required for corrosion. As shown in Fig. 18.2, distinct anode and cathode areas can be caused by inhomogeneities within a single piece of metal. Second phases in alloys are regions with an electrode potential different from that of the bulk metal and can therefore cause localized areas of corrosion. Other inhomogeneities include grain boundaries, segregation of impurities, cold-worked areas, and nonuniformly heat treated areas. Discrete anodes and cathodes can also be caused by variations in the electrolyte, such as temperature differences or concentration gradients in the solution, of ions in the solution, or of dissolved gases such as oxygen.

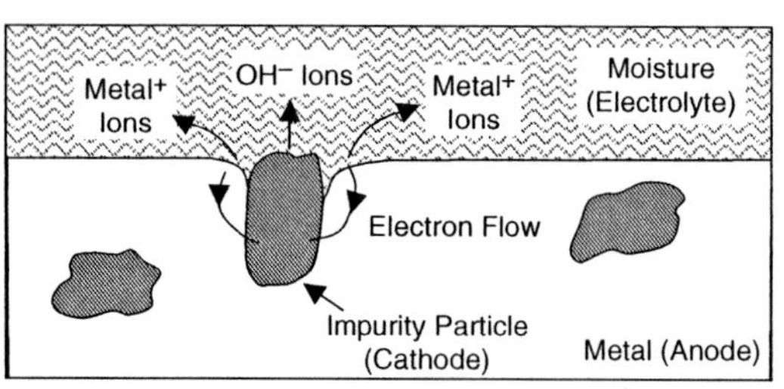


**Fig. 18.2** Electrolytic corrosion of steel. Source: Ref 3

A simple corrosion cell is shown in Fig. 18.3. The cell contains an anode, a cathode, and an electrolyte. If a voltmeter is connected to the circuit, a potential difference between the anode and cathode would be measured, indicating that a direct current is flowing between the two. Anodic reactions are always oxidation reactions and therefore tend to destroy the anode metal by causing it to dissolve as an ion or to revert to a combined state such as an oxide. Therefore, in a corrosion cell, corrosion always takes place at the anode, which is the metal ($M_a$) that undergoes an oxidation reaction and gives up electrons to the circuit:

$$M_a \rightarrow M^{n+} + n^{e-} \qquad \text{(Eq 18.1)}$$

A reduction reaction, which is the reverse of the anode reaction, occurs at the cathode:

$$M^{n+} + n^{e-} \rightarrow M_c \qquad \text{(Eq 18.2)}$$

Cathodic reactions are always reduction reactions and usually do not affect the cathode metal. During metallic corrosion, the rate of oxidation equals the rate of reduction.

As an example, iron immersed in water corrodes according to the mechanism shown in Fig. 18.4. The iron contains discrete areas that are anodic to the rest of the metal surface. At these areas, iron is oxidized according to the equation:

$$Fe \rightarrow Fe^{2+} + 2e^- \qquad \text{(Eq 18.3)}$$

If the water is pure and contains no dissolved oxygen, the cathodic reaction is the reduction of ionic hydrogen:

$$2H^+ + 2e^- \rightarrow H_2 \qquad \text{(Eq 18.4)}$$

To maintain overall electrical neutrality, these reactions must proceed in balance. Therefore, two hydrogen ions must be reduced for every iron atom that corrodes. In pure water, only one water molecule in approximately 10 million dissociates to produce hydrogen and hydroxide ions, so that the supply of hydrogen ions is quite limited. For this reason, the cathodic reaction is quite slow, and corrosion rates are very low. Cases in which the rate of attack is limited by the speed of the cathodic reaction are known as cathodically controlled reactions.

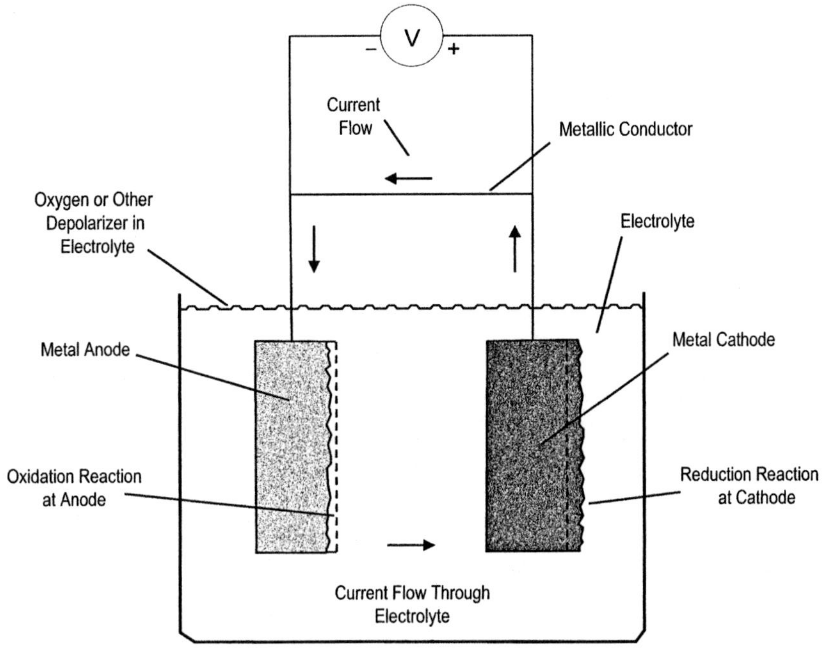


**Fig. 18.3** Basic electrochemical cell

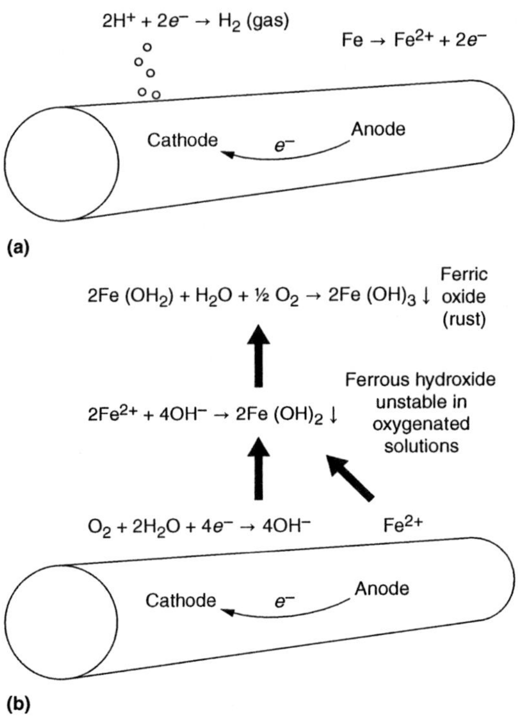


**Fig. 18.4** Corrosion of iron in (a) unaerated and (b) aerated water

If, as is usually the case, dissolved oxygen is present in the water, another cathodic reaction can occur, the reduction of oxygen:

$$1/2 O_2 + H_2O + 2e^- \rightarrow 2OH^- \qquad \text{(Eq 18.5)}$$

This reaction will support a more rapid rate of attack since it depends only on the diffusion of oxygen to cathodic areas of the iron. When iron corrodes in water, the hydroxyl ions from the cathodic reaction react with the ferrous ions from the anodic reaction to form ferrous hydroxide [$Fe(OH)_2$]. The ferrous hydroxide further reacts with oxygen to form ferric hydroxide [$Fe(OH)_3$], which is familiar as reddish-brown rust. Unfortunately, this corrosion product is porous and not very adherent and therefore does not prevent further corrosion.

In practice, a metal often exhibits a high initial rate of corrosion. However, the rate can often diminish with time, an effect known as polarization. Polarization can result from reactions at either the anode or the cathode. In some corrosion reactions (e.g., iron in aerated water) that produce anodic polarization, the corrosion rate diminishes due to an accumulation of insoluble corrosion products that become somewhat protective of the iron anode. Conversely, cathodic polarization can result from reactions at the cathode. For example, when iron is immersed in nonaerated neutral water, the absence of dissolved oxygen allows the development of an adsorbed film of hydrogen that quickly covers the surface:

$$Fe + 2H^{+} \rightarrow Fe^{2+} + 2H \quad \text{(Eq 18.6)}$$

Since oxygen is not present, it is difficult for the hydrogen atoms to combine to form hydrogen gas and escape from the surface. When oxygen is introduced and the hydrogen atoms can escape the surface of the iron, depolarization occurs and corrosion resumes. Thus, in this case, the corrosion rate of iron is not controlled by the primary reaction with hydrogen ions but by the depolarization reaction involving oxygen.

Another characteristic of oxygen and oxidizing media is their ability to make certain metals and alloys passive by forming complex oxide films on the surface. A metal is less reactive and corrosion-prone in the passive condition than in the normal or active condition. As an example, notice in the galvanic series (Table 18.2) that stainless steels are listed in both the passive and active conditions, with the passive conditions being more cathodic than the active conditions. Chromium in stainless steels oxidizes and forms a thin, tightly adherent layer of oxide ($Cr_2O_3$) on the surface that normally prevents corrosion in saltwater. However, if a pit develops on the surface and destroys the oxide layer, it forms a local anodic area, and corrosion is accelerated.

One of the most important factors influencing corrosion is the difference in electrical potential of two different metals when they are coupled together and immersed in an electrolyte. The electrode potential, measured in volts, provides an electrical measure of a metal to give up electrons. Electrode potentials are measured with two half-cells: one for the corroding metal and the other for a standard hydrogen half-cell consisting of gaseous hydrogen ($H^{+}$) at unit concentration with a specially prepared platinum electrode. This half-cell is called the standard hydrogen electrode. The electromotive series is shown in Table 18.3. However, the electromotive series is determined under ideal laboratory conditions that may not reflect reality. Therefore, the galvanic series for seawater (Table 18.2) is often more useful. Again, the more anodic metal will corrode. For example, when iron and copper are coupled together in an electrolyte, iron will corrode because it is more anodic than copper. However, if iron is coupled to zinc, the iron is now protected, since it is cathodic compared to zinc. The further away from each other two metals are on the galvanic series, the greater will be the tendency for corrosion. On the other hand, a metal coupled with another close to it on the series will usually

**Table 18.2 Galvanic series in seawater**

| | |
|---|---|
| Cathodic (protected) | Platinum |
| | Gold |
| | Graphite |
| | Titanium |
| | Silver |
| | Stainless steels (passive) |
| | Nickel-base alloys (passive) |
| | Cu-35%Zn brass |
| | Nickel-base alloys (active) |
| | Manganese bronze |
| | Cu-40%Zn |
| | Tin |
| | Lead |
| | 316 stainless steel (active) |
| | 50%Pb-50%Sn solder |
| | 410 stainless steel (active) |
| | Cast iron |
| | Low-carbon steel |
| | 2024 aluminum |
| | 2017 aluminum |
| | Cadmium |
| | Alclad |
| | 6053 aluminum |
| | 1100 aluminum |
| | 3003 aluminum |
| | 5052 aluminum |
| | Zinc |
| Anodic (corrodes) | Magnesium alloys |
| | Magnesium |

**Table 18.3 Electromotive series**

| | Electrode potential | Standard electrode potential ($E^0$), V |
|---|---|---|
| ↑ | $Au^{3+} + 3e^- \rightarrow Au$ | +1.420 |
| | $O_2 + 4H^+ + 4e^- \rightarrow H_2O$ | +1.229 |
| | $Pt^{2+} + 2e^- \rightarrow Pt$ | ~+1.2 |
| | $Ag^+ + e^- \rightarrow Ag$ | +0.800 |
| | $Fe^{3+} + 3e^- \rightarrow Fe^{2+}$ | +0.771 |
| Increasingly cathodic (inert) | $O_2 + H_2O + 4e^- \rightarrow 4(OH^-)$ | +0.401 |
| | $Cu^{2+} + 2e^- \rightarrow Cu$ | +0.340 |
| | $2H^+ + 2e^- \rightarrow H_2$ | 0.000 (reference) |
| | $Pb^{2+} + 2e^- \rightarrow Pb$ | −0.126 |
| | $Sn^{2+} + 2e^- \rightarrow Sn$ | −0.136 |
| | $Ni^{2+} + 2e^- \rightarrow Ni$ | −0.250 |
| | $Co^{2+} + 2e^- \rightarrow Co$ | −0.277 |
| | $Cd^{2+} + 2e^- \rightarrow Cd$ | −0.403 |
| | $Fe^{2+} + 2e^- \rightarrow Fe$ | −0.440 |
| Increasingly anodic (active) | $Cr^{3+} + 3e^- \rightarrow Cr$ | −0.744 |
| | $Zr^{2+} + 2e^- \rightarrow Zn$ | −0.763 |
| | $Al^{3+} + 3e^- \rightarrow Al$ | −1.662 |
| | $Mg^{2+} + 2e^- \rightarrow Mg$ | −2.363 |
| | $Na^+ + e^- \rightarrow Na$ | −2.714 |
| ↓ | $K^+ + e^- \rightarrow K$ | −2.924 |

corrode more slowly than when coupled with one further away from it.

## 18.2 Forms of Corrosion

There are many different types of corrosion, and more than one corrosion mechanism may be operating at the same time. Some types of corrosion depend only on the environment, while others need mechanical or microbiological assistance. Some are unique to certain metals and alloys, while others attack many, if not most, metals and alloys. In this section, some of the more common types of corrosion are discussed. A number of these are illustrated in Fig. 18.5.

### *18.2.1 Uniform Corrosion*

Uniform corrosion is the fairly uniform attack of the entire metal surface, resulting in the gradual thinning of the metal. It is by far the most common form of attack, accounting for the greatest corrosion loss of metal. Since the attack is relatively linear with time, the life of equipment can be predicted with reasonable accuracy. Uniform corrosion is uniform because it results from the formation and dissolution of multiple anodic and cathodic areas that move around the surface with time. Uniform corrosion is often caused by exposure to the atmosphere but can be aggravated by industrial pollution, brackish and salt waters, and soils and chemicals.

Uniform corrosion rates are measured as the average metal thickness loss with time, in mils per year. A convenient rating for metals subject to uniform attack based on corrosion rates is as follows:

- *Excellent:* Rate of less than 2 mils/yr. Metals suitable for making critical parts
- *Satisfactory:* Rate of 2 to 20 mils/yr. Metals generally suitable for noncritical parts where a higher rate of attack can be tolerated
- *Acceptable:* Rate of 20 to 50 mils/yr. A rate tolerable for massive equipment with a generous corrosion allowance
- *Unsatisfactory:* Rates of over 50 mils/yr. Metals usually not acceptable in the environment

Good judgment must be used when comparing corrosion rates determined by tests. Although the rates are relatively linear in time, test results may be based on short-duration tests under controlled simulated conditions, which will not accurately predict much longer-term performance. Other tests may be based on in situ long-term exposure but with uncontrolled conditions.

Some metals form protective passive films on their surface when exposed to air that inhibits further attack. For example, stainless steel forms a protective oxide of $Cr_2O_3$, while aluminum forms a protective film of alumina ($Al_2O_3$). Unfortunately, although steel also produces a surface film, the film is porous and does not adhere very well, so the corrosion continues.

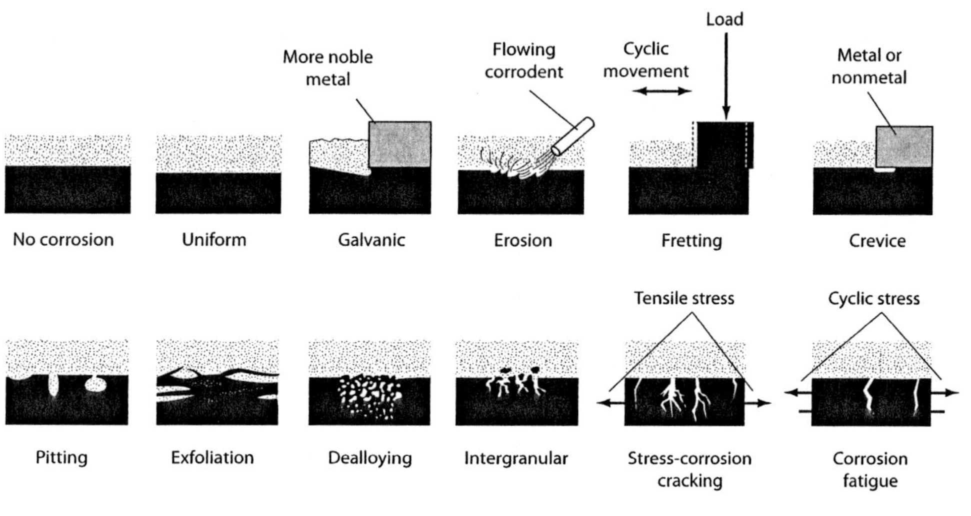


**Fig. 18.5** Various forms of corrosion

Uniform corrosion is normally prevented by selecting a metal that forms a protective film, by applying coatings that isolate the metal from the environment, or by cathodic protection.

In cathodic protection, the metal to be protected is electrically connected to a more anodic metal, so that the sacrificial anodic metal corrodes instead of the metal being protected. For example, steel is often coated with zinc in a process called galvanizing. The zinc coating, being more anodic than steel, corrodes preferentially to the underlying steel. This is the passive type of cathodic protection. The active (impressed-current) type is discussed in section 18.3.3, "Electrochemical Control," in this chapter.

### *18.2.2 Galvanic Corrosion*

When two different metals (or a metal and a conducting nonmetal) are placed in electrical contact in the presence of an electrolyte, a potential or voltage difference is established. This potential difference causes a current to flow, and the less noble, or more anodic, metal corrodes while the more noble, or cathodic metal, is unaffected. The rate of attack depends on the relative voltage difference between the two metals, the relative areas of each metal that are exposed, and the particular corrosive environment. An example of galvanic corrosion between a metal and a conducting nonmetal is carbon-fiber composites in direct contact with aluminum. If the carbon fiber contacts the aluminum alloy in the presence of an electrolyte such as water, the aluminum will corrode.

The relative tendency for galvanic corrosion is given by the galvanic series shown in Table 18.3. Metals close to each other in the series generally do not have a tendency to react. The further the two metals are separated on the series, the greater is the tendency for the more anodic metal to corrode. It should be noted that it is possible for some metals to reverse their positions in certain environments; however, their normal positions are maintained in natural water and normal atmospheres.

The ratio of the anodic area ($S_a$) to the cathodic area ($S_c$) is important in galvanic corrosion. Galvanic corrosion is accelerated when the anodic area is small in relation to the cathodic area; that is, the surface area ratio ($S_a/S_c$) is small. This results in a high current density on the anode and causes severe corrosion of the anode. For example, a large area of stainless steel in contact with a small surface area of carbon steel, such as when the wrong mechanical fastener is used, is undesirable. The potential difference will tend to corrode the carbon steel, and the very large area of stainless steel will make that corrosion occur more quickly. The reverse condition is preferred. That is, a small area of stainless steel (or more noble metal) may be coupled with a much larger area of carbon steel (anodic) with a much slower rate of attack.

Although the galvanic series indicates the potential for corrosion, actual corrosion is difficult to predict. Electrolytes may be poor conductors, long distances may increase the resistance to the point that corrosion does not occur, or the reaction rate may be very sluggish. Corrosion products can also form a partially insulating layer over the anode. A cathode having a layer of adsorbed gas bubbles resulting from a corrosion reaction can become polarized and reduce the corrosion rate.

The passivity of stainless steels is a result of either the presence of a corrosion-resistant oxide film ($Cr_2O_3$) or an oxygen-caused polarizing effect, durable only as long as there is sufficient oxygen to maintain the effect. In most natural environments, stainless steels will remain in a passive state and thus tend to be cathodic to ordinary iron and steel. A change to an active state usually occurs only where chloride concentrations are high, as in seawater, or in reducing solutions. Oxygen starvation also produces a change in the active state, leading to accelerated corrosion. This occurs where the oxygen supply is limited, as in crevices and beneath contamination on partially biologically fouled surfaces.

Galvanic corrosion can be prevented or reduced by proper materials selection, that is, selection of combinations of metals as close together as possible in the galvanic series, insulating dissimilar metals, applying a barrier coating to both the anodic and cathodic metal, applying a sacrificial coating (such as zinc on steel), applying or building nonmetallic films (e.g., anodizing aluminum alloys), and by providing cathodic protection.

### *18.2.3 Pitting*

Pitting is a form of highly localized attack characterized by the formation of small pits on the surface. Several pits in an austenitic stainless steel thin-walled bellows are shown in Fig. 18.6. Pitting occurs on alloys with passive films and is considered to be more dangerous than uniform corrosion damage because it is more difficult to detect, predict, design against, and can lead to a sudden loss of function of the material. Pitting occurs when discrete areas of a material undergo rapid attack while most of the adjacent surface remains virtually unaffected. Although the total metal loss may be small, the part may be rendered useless due to perforation. In addition to the localized loss of thickness, corrosion pits can also act as stress raisers, leading to fatigue or stress-corrosion cracking.

Pitting occurs when the anodic or corroding area is small in relation to the cathodic or protected area. Pitting can occur in protected metals when there are small breaks in the continuity of the metal coating. Pitting can also occur on bare, clean metal surfaces as a result of irregularities in their physical or chemical structure. The rate of penetration into the metal by pitting can be 10 to 100 times that caused by general (uniform) corrosion. Pitting can cause structural failure from localized weakening while considerable sound metal still remains.

Pitting usually requires a rather long initiation period before attack becomes visible. However, once a pit has started, the attack continues at an accelerating rate. Pits tend to grow in a manner that undermines or undercuts the surface. Typically, a very small hole is seen on the surface. Poking at the hole with a sharp instrument may reveal a much larger hole under what had looked like solid metal. Pitting can cause visible pits, or they may be covered with a semipermeable membrane of corrosion products. Pitting corrosion may assume different shapes. Pits can be either hemispherical or cup-shaped. In some cases, they are flat-walled, revealing the crystal structure of the metal, or they may have a completely irregular shape.

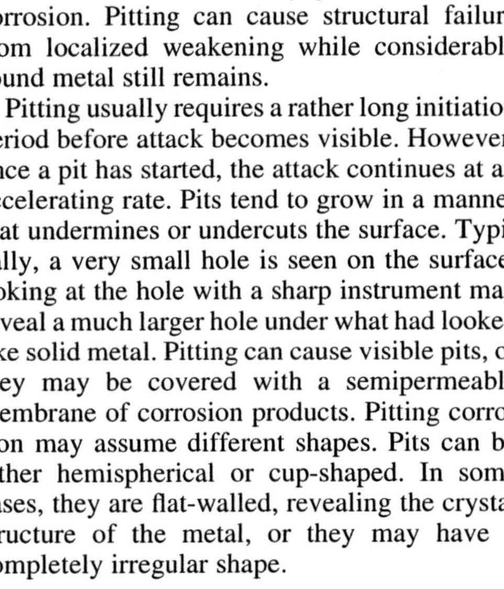

**Fig. 18.6** Corrosion pits in thin-walled austenitic stainless steel sheet approximately 0.5 mm (0.02 in.). Source: Ref 4, courtesy of M.D. Chaudhari

Pitting normally occurs in a stagnant environment. Concentration cells can accelerate pitting. Concentration cells are areas on the metal surface where the oxygen or conductive salt concentrations in water differ. As a pit becomes deeper, an oxygen concentration cell is started by depletion of oxygen in the pit. The rate of penetration of such pits is accelerated proportionately as the bottom of the pit becomes more anodic. Pitting attack increases with temperature. Variations in soil conditions can also trigger pitting.

The depth of pitting can be expressed by the pitting factor ($p/d$), as defined in Fig. 18.7. A value of 1 would represent uniform corrosion. The maximum depth of penetration ($p$) can be measured by several methods, including metallographic examination, machining, use of a micrometer, or a microscope. The average penetration depth ($d$) is calculated from the weight lost by the sample. The maximum penetration depth is extremely significant if the metal is part of a barrier or tank or is part of a pressurized system. For a mechanical component, the density of pits (number per unit surface area) and size may be a more critical characteristic than the maximum depth. The loss of effective cross section can decrease the strength of the component, and pits can become sites of stress concentrations, leading to either static overload or fatigue failures.

Pitting occurs in most commonly used metals and alloys. While iron buried in the soil corrodes with the formation of shallow pits, carbon steels in contact with hydrochloric acid or stainless steels immersed in seawater develop deep pits. Mill scale is cathodic to steel, and discontinuities in it are a common cause of pitting. The potential difference between steel and mill scale often amounts to 0.2 to 0.3 V, enough to cause serious attack. When the anodic area is relatively small, the metal loss is concentrated and may be very serious. As the size of the

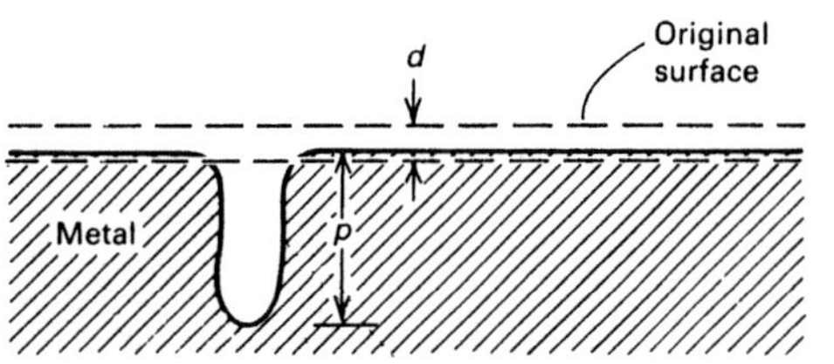


**Fig. 18.7** Pitting factor, p/d. Source: Ref 4

anodic area decreases, the degree of penetration increases. A pit can form wherever there is a break in the mill scale. Aluminum tends to pit in waters containing chloride ions, particularly in stagnant water.

Despite their good resistance to general corrosion, stainless steels are susceptible to pitting. High-alloy stainless steels containing chromium, nickel, and molybdenum are more resistant to pitting but are not immune under all service conditions. The pitting resistance equivalent number (PREN), or pitting index, can be used to quantify and compare the resistance of stainless steel alloys to pitting:

$$PREN = \%Cr + 3.3(\%Mo) + 16(\%N) \qquad \text{(Eq 18.7)}$$

The higher the number, the more resistant the alloy is to pitting and crevice corrosion. A number of different PREN equations have been developed for specific alloy and weld metal groups. Equations include the effects of nickel, manganese, nitrogen, tungsten, and carbon.

Typical approaches to alleviating or minimizing pitting corrosion include the following: using defect-free barrier coatings; reducing the aggressiveness of the environment, for example, chloride ion concentrations, temperature, acidity, and oxidizing agents; using more resistant metals, for example, using molybdenum-containing (4 to 6% Mo) stainless steels or Mo-W-Ni-base alloys; using corrosion-resistant alloy linings; modifying the design of the system, for example, avoiding crevices and the formation of deposits; circulating/stirring to eliminate stagnant solutions; and by ensuring proper drainage. When contact between dissimilar metals is unavoidable and the surface is painted, it is advisable to paint both metals. If only one surface is painted, it should be the cathode. If only the anode is coated, any weak points, such as pinholes in the coating, can result in intense pitting.

### 18.2.4 Crevice Corrosion

Crevice corrosion is a form of localized attack that occurs at narrow openings, spaces, or gaps between metal-to-metal or nonmetal-to-metal components (Fig. 18.8). Attack results from a concentration cell formed between the electrolyte within the crevice, which is oxygen starved, and the electrolyte outside the crevice, where oxygen is more plentiful. The material within the oxygen-starved crevice becomes the anode, while the exterior material becomes the cathode. The attack usually occurs in small volumes of stagnant solution under gasket surfaces, lap joints, marine fouling, solid deposits, in the crevices under bolt heads, and the mating surfaces of male and female threads. Crevice corrosion can progress very rapidly, on the order of tens to hundreds of times faster than the normal rate of general corrosion in the same given solution. Susceptibility to crevice corrosion increases rapidly with increases in temperature.

Crevices are produced by both design and accident. Examples of crevices created by design include gaskets, flanges, washers, bolt holes, threaded joints, riveted seams, lap joints, or anywhere close-fitting surfaces are present.

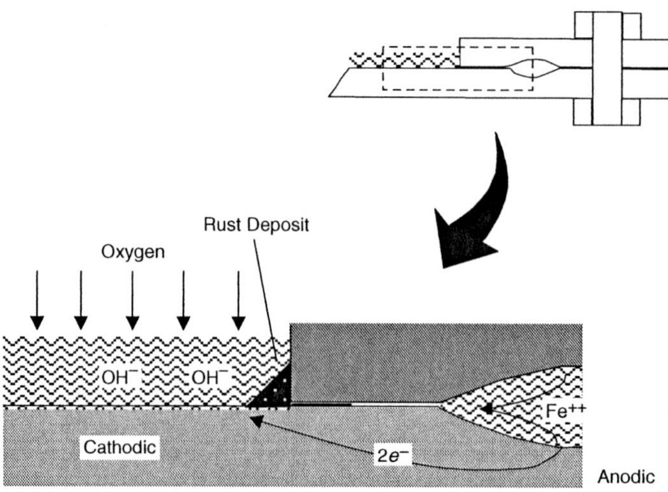


**Fig. 18.8** Crevice corrosion of steel plates. Source: Adapted from Ref 3

Unintentional crevices include cracks, seams, and metallurgical defects. Although crevice corrosion affects both active and passive metals, the attack is often more severe for passive alloys, particularly those in the stainless steel group. Breakdown of the passive film within a restricted area leads to rapid metal loss and penetration in that area. Any layer of solid material on the surface of a metal that can exclude oxygen from the surface or allows the accumulation of metal ions beneath the deposit because of restricted diffusion is a candidate site for crevice corrosion.

Riveted and bolted joints are prime sites for crevice corrosion; therefore, they require careful attention in design and assembly to minimize crevices, as well as provisions to ensure uniform aeration and moderate but not excessive flow rates at the joints. Not only does the geometry of the joint affect crevice corrosion, but corners and cracks that collect debris will increase the potential for corrosion.

Crevice corrosion can be prevented or reduced through improved design to avoid crevices, regular cleaning to remove deposits, by selecting a more corrosion-resistant material, and by coating carbon steel or cast iron components with organic barrier coatings. Replacement of mechanical joints with welded joints can eliminate crevice corrosion, provided special care is taken in welding and subsequent finishing of the welds to provide smooth, defect-free joints. Weld splatter is another source of crevice corrosion.

### *18.2.5 Erosion-Corrosion*

Erosion-corrosion is the acceleration of corrosive attack due to the simultaneous action of corrosion and erosion. The erosion-corrosion of mild steel in flowing water is illustrated in Fig. 18.9. The attack is more severe than if just corrosion or erosion alone were acting. The erosive action removes metal from the surface as dissolved ions, as particles of solid corrosion products, or as elemental metal. Erosion-corrosion can be primarily erosive attack or primarily chemical attack, or somewhere in between. Both gases and liquids can cause attack. It is encountered when particles in a liquid or gas impinge on a metal surface, causing the removal of protective surface films, such as protective oxide films or adherent corrosion products, thus exposing new reactive surfaces that are anodic to noneroded neighboring areas on the surface. This results in rapid localized corrosion of the exposed areas in the form of smooth-bottomed, shallow recesses. As temperature increases, the protective film may become more soluble and/or less resistant to abrasion. Hence, the same flow rates, but at higher temperature, can cause an increase in corrosion rates. Hot gases may oxidize a metal and then, at high velocities, blow the protective film off of the metal. Slurries, present in solids as liquid suspensions, are particularly aggressive media.

Nearly all turbulent corrosive media can cause erosion-corrosion. The attack may exhibit a directional pattern related to the path taken by the corrosive as it moves over the surface of the metal. Erosion-corrosion is characterized in appearance by grooves, waves, rounded holes, and/or horseshoe-shaped grooves. Affected areas are usually free of deposits and corrosion products, although corrosion products can sometimes be found if erosion-corrosion occurs intermittently and/or the liquid flow rate is relatively low.

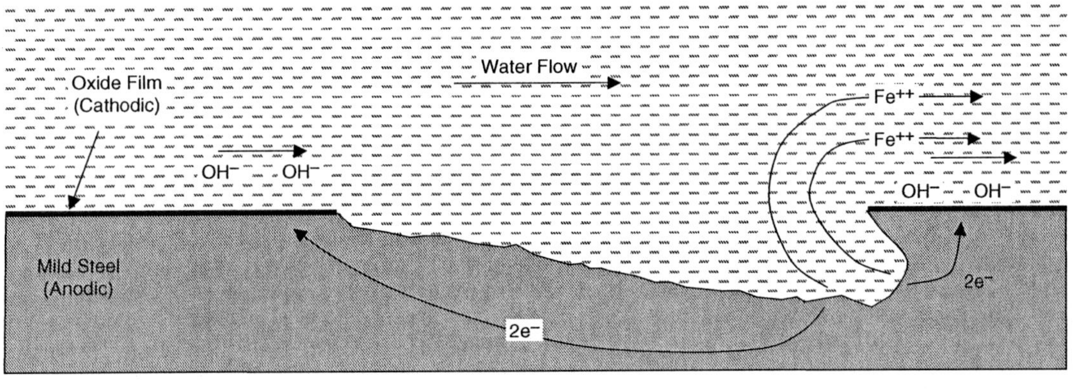


**Fig. 18.9** Erosion-corrosion of mild steel. Source: Ref 3

Most metals are susceptible to erosion-corrosion under specific conditions. Metals that develop thick protective coatings of corrosion product are often susceptible because thick coatings frequently exhibit poorer adhesion than thin coatings. Thin-film oxide coatings, such as those that form in stainless steel and titanium alloys, are relatively more immune to erosion-corrosion under similar conditions. On the other hand, soft metals, such as copper and lead, are quite susceptible to attack.

Impingement corrosion is a severe form of erosion-corrosion. It occurs frequently in turns or elbows of tubes and pipes and on surfaces of impellers or turbines. It occurs as deep, clean, horseshoe-shaped pits with the deep, or undercut, end pointing in the direction of flow. Impingement corrosion attack can also occur as the result of partial blockage of a tube. A stone, a piece of wood, or some other object can cause the main flow to deflect against the wall of the tube. The impinging stream can rapidly perforate tube walls. Water that carries sand, silt, or mud will have an additional severely erosive effect on tubes. The energy transfer is a function of the rate of change of the momentum of the flowing medium. As energy transfer takes place over a smaller and smaller area per unit time, the energy or power density of the process becomes damaging to the substrate. Steam erosion is another form of impingement corrosion, occurring when high-velocity wet steam impacts a metal surface. The resulting attack usually produces a roughened surface showing a large number of small cones with the points facing in the direction of flow.

Erosion-corrosion can be prevented, or reduced, through improved design, such as increasing the diameter and using streamlined bends in pipes, by altering the environment by deaeration or the addition of inhibitors, and by using hard and tough protective coatings.

### 18.2.6 Cavitation

Cavitation is a form of erosion-corrosion caused by the formation and collapse of vapor bubbles of a liquid against a metal surface. Cavitation frequently occurs in hydraulic turbines, pump impellers, ship propellers, and on any surfaces in contact with high-velocity liquids subject to changes in pressure. Cavitation pits are similar to those in conventional pitting except that the surfaces of the pits are much rougher. The high liquid pressures experienced during cavitation usually remove any corrosion products from the surface of the pits.

When a liquid is subjected to sudden differential pressures, vapor bubbles form in the liquid due to the reduced pressure. The bubbles condense and collapse due to a rise in pressure in a process that occurs in milliseconds, and water is ejected from the collapsing bubbles at velocities in the range of 90 to 500 m/s (300 to 1650 ft/s). This creates large pressures that can dislodge protective surface films and even plastically deform the metal surface. When the bubbles collapse, they impose hammerlike blows, which produce stresses on the order of 415 Mpa (60 ksi), simultaneously with the initiation of tearing action that appears to pull away portions of the surface. The tearing action can remove protective oxide films on the surface of a metal, exposing active metal to the corrosive influence of the liquid environment. Damage occurs as this cycle is repeated over and over again.

Cavitation can be controlled, or minimized, by improving the design to minimize hydrodynamic pressure differences, using harder and more corrosion-resistant materials, specifying a smooth finish on all critical metal surfaces, and coating with tough, resilient materials such as rubber and some plastics. For turbine blades, aeration of water serves to cushion the damage caused by the collapse of bubbles. Neoprene or similar elastomeric coatings on metals are somewhat resistant to damage.

### 18.2.7 Fretting Corrosion

Fretting corrosion is a combination of wear and corrosion. It is caused by slight movements between two materials in which fine metal particles are removed from the surface and act as an abrasive. Oxidation is the most common form of fretting corrosion. The fine particles removed by wear oxidize and, in ferrous alloys, produce a red material that oozes from the joint. This type of corrosion is quite common in mechanically fastened joints that become loose with age, allowing movement between the two members. Fretting often appears as rather deep pits on the interface members.

Fretting can be controlled by lubricating the faying surfaces, designing the joint to prevent or restrict movement during service, carburizing or nitriding steel to produce hard, wear-resistant surfaces, or by applying protective coatings.

### *18.2.8 Intergranular Corrosion*

Intergranular corrosion is the selective attack of grain boundaries or the immediate adjacent regions, usually with only slight or negligible attack of the grains themselves. In essence, the grain-boundary area becomes anodic to that of the grain interiors. These differences can occur during manufacturing or in-service exposure. When the grain boundaries become anodic, the metal is said to be sensitized and is susceptible to intergranular attack in a corrosive environment. The classic example of intergranular corrosion occurs when stainless steels become sensitized by the diffusion of chromium and trace amounts of carbon to the grain boundaries during elevated temperatures, resulting in the precipitation of chromium carbides (Fig. 18.10) when the steel is slowly cooled. Some aluminum alloys also exhibit a similar behavior, in particular those containing copper that precipitate $CuAl_2$, resulting in precipitate-free zones.

When the attack is severe, entire grains can be dislodged due to the complete deterioration of the grain boundaries. When an alloy is undergoing intergranular corrosion, its rate of weight loss can accelerate with time. As the grain-boundary area dissolves, the unaffected grains are undermined and fall out, thus increasing the weight loss. Intergranular corrosion can occur in a number of alloy systems under specific circumstances. The susceptibility depends on the corrosive environment and on alloy composition, fabrication, and heat treatment parameters. Susceptibility of a component to intergranular corrosion can be corrected by proper heat treatment to distribute alloying elements more uniformly, by modification of the alloy, or by the use of a completely different alloy.

Although stainless steels provide resistance against general corrosion, the 300 and 400 series of stainless steels can be susceptible to intergranular corrosion by sensitization. Susceptible stainless steels are those that have normal carbon contents (generally > 0.04%) and do not contain titanium and niobium carbide stabilizing elements. Sensitization is caused by the precipitation of chromium carbides at grain boundaries during exposure to temperatures from 450 to 870 °C (850 to 1600 °F), with the maximum effect occurring near 680 °C (1250 °F). As shown in Fig. 18.11, the resulting depletion in chromium adjacent to the chromium-rich carbides provides a selective path for intergranular corrosion. Precipitation can occur from the heat of welding, from slow cooling after annealing, or from prolonged exposure to intermediate temperatures in service. For exposures at very long times or at the high end of the temperature range, diffusion of chromium back into the depleted zone can restore the corrosion resistance. An effective means of combating intergranular corrosion in stainless steels is to restrict the carbon content of the alloy. In the stainless L-grades, limiting the carbon content to a maximum of 0.03% is often sufficient. High chromium and molybdenum additions also reduce the chance of intergranular attack. However, even better performance can be obtained from the stabilized types, which contain sufficient titanium and niobium that combine preferentially with carbon to form titanium and niobium carbides.

The typical appearance of intergranular corrosion of stainless steels is shown in Fig. 18.12 for sensitized type 304 stainless steel attacked by a water solution containing a low concentration of fluorides at 80 °C (180 °F). Intergranular corrosion of this type is more or less randomly

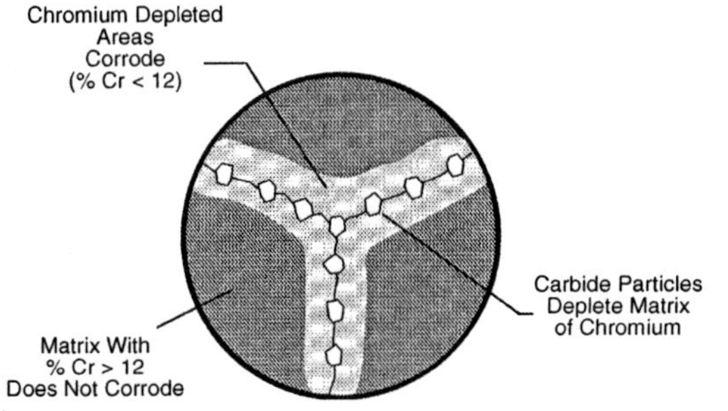


**Fig. 18.10** Intergranular corrosion of stainless steel. Source: Ref 5

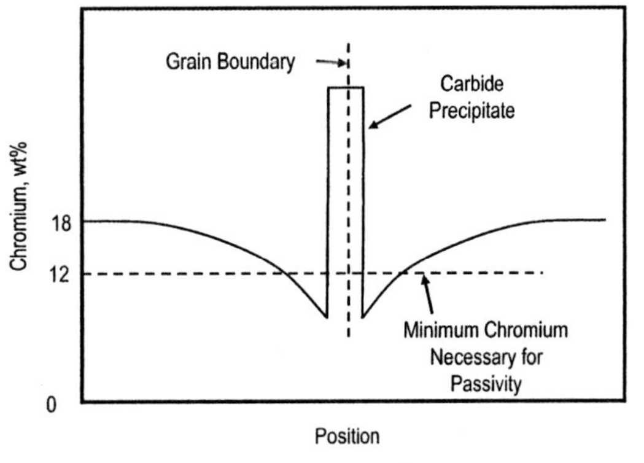


**Fig. 18.11** Precipitation of chromium carbide at grain boundaries

oriented and does not have highly localized propagation, as does intergranular stress corrosion in which cracking progresses in a direction normal to applied or residual stresses. For austenitic stainless steels, the susceptibility to intergranular corrosion is mitigated by solution heat treating at 1070 to 1120 °C (1950 to 2050 °F) followed by water quenching. In this treatment, the chromium carbides are redissolved in solid solution and then retained in solid solution as a result of the quench. However, a solution heat treatment may be difficult on many welded assemblies and is generally impracticable on large equipment or when making repairs.

Under severe conditions, such as multipass welding, even the stabilized alloys will sensitize. They are also susceptible to a highly localized form of intergranular corrosion known as knife-line attack, which occurs in the base metal immediately adjacent to the weld fusion line. In some cases, these alloys are given stabilizing heat treatments after solution heat treatment for maximum resistance in the as-welded condition. For example, type 321 stainless steel is given a stabilize anneal at 900 °C (1650 °F) for 2 h before fabrication to avoid knife-line attack. However, type 321 may still be susceptible because titanium has a tendency to form an oxide during welding; therefore, its role as a carbide stabilizer may be diminished. For this reason, type 321 is always welded with a niobium-stabilized weld filler metal, such as type 347 stainless.

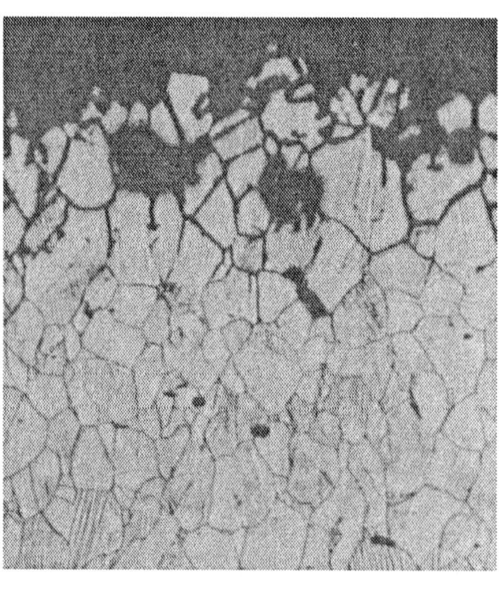

**Fig. 18.12** Intergranular corrosion in sensitized type 304 stainless steel. Original magnification 100×

Sensitization and intergranular corrosion can also occur in ferritic stainless steels. A wider range of corrosive environments can produce intergranular attack in ferritic grades than for austenitic grades. The thermal processes causing intergranular corrosion in ferritic stainless steels are also different from those in austenitic stainless steels. In the case of welds, the attacked region is usually larger for ferritic grades than for austenitic grades because temperatures above 930 °C (1700 °F) are involved in causing sensitization. However, ferritic grades with less than 15% Cr are not susceptible. New grades of ferritic stainless steels are also alternatives to the more common 300- and 400-series stainless steels.

All the 2*xxx* aluminum alloys and most of the 7*xxx* alloys contain copper. The presence of copper can contribute to intergranular corrosion, depending on material processing. These alloys can precipitate $CuAl_2$ at the grain boundaries, which are more anodic than the grains themselves. During exposure to chloride solutions, in particular saltwater, galvanic couples form between the precipitate-rich grain boundaries and the grain interiors. One approach to mitigating this type of intergranular attack is to employ special heat treatments that control the distribution of the precipitate particles.

Although many types of intergranular corrosion are not associated with a potential difference between the grain-boundary region and the adjacent grains, intergranular corrosion of aluminum alloys can occur due to potential differences. In 2*xxx*-series alloys, a narrow band on either side of the boundary is depleted of copper; in 5*xxx*-series alloys, it is the anodic constituent $Mg_2Al_3$ that forms a continuous path along a grain boundary; in copper-free 7*xxx*-series alloys, it is generally considered to be the anodic zinc- and magnesium-bearing constituents at the grain boundary; and in the copper-bearing 7*xxx*-series alloys, it appears to be the copper-depleted bands along the grain boundaries. The 6*xxx*-series alloys generally resist this type of corrosion, although slight intergranular attack has been observed in aggressive environments.

### 18.2.9 Exfoliation

Exfoliation is a form of intergranular corrosion that primarily affects aluminum alloys in industrial or marine environments. Corrosion proceeds laterally from initiation sites on the surface and generally proceeds intergranularly along planes parallel to the surface, as shown in the Fig. 18.13 example. The corrosion products that form in the grain boundaries force metal away from the underlying base material, resulting in a layered or flakelike appearance.

Resistance to exfoliation corrosion is attained through proper alloy and temper selection. Again, the most susceptible alloys are the high-strength, heat treatable 2*xxx* and 7*xxx* alloys. Exfoliation corrosion in these alloys is usually confined to relatively thin sections of highly worked products.

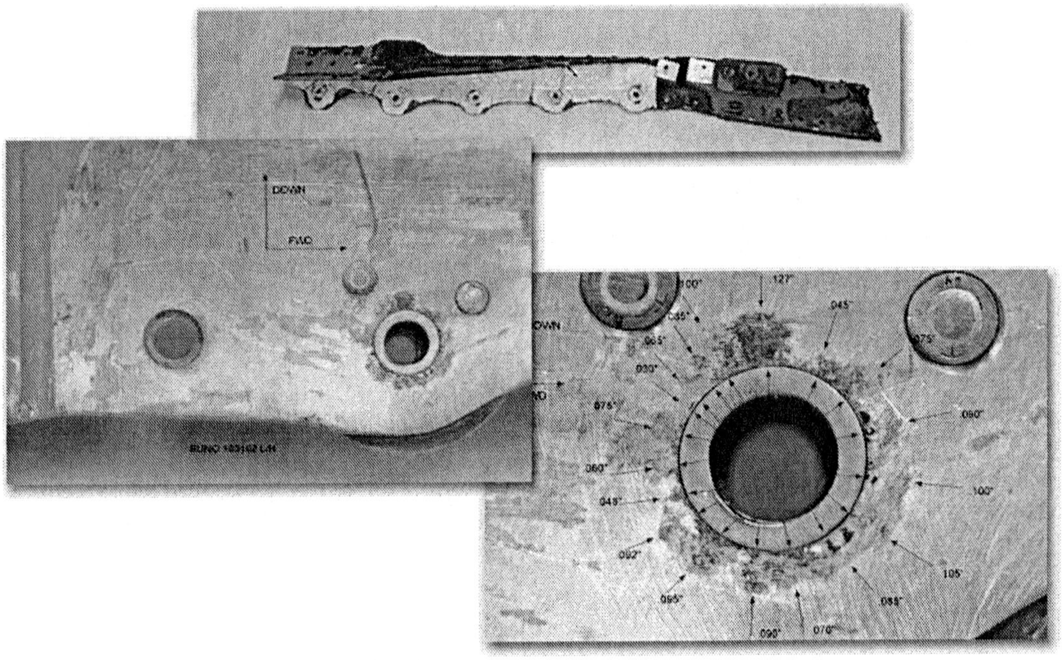

(a)

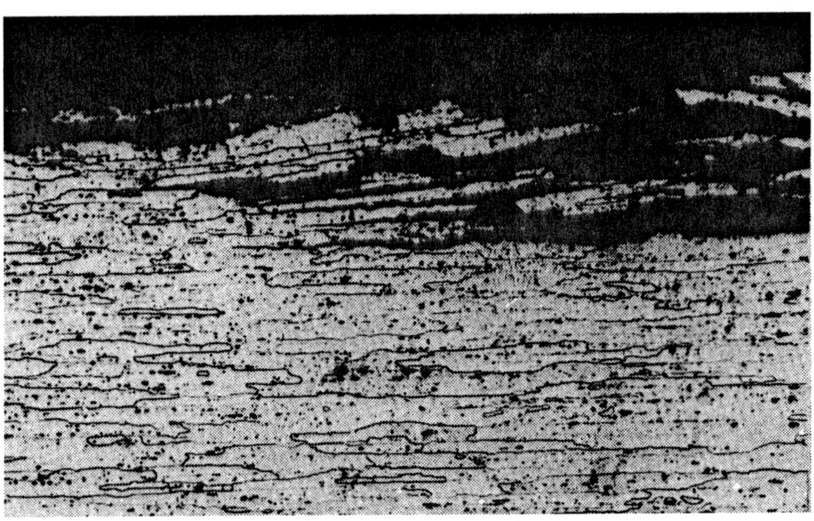

(b)

**Fig. 18.13** Exfoliation corrosion (a) around a fastener hole in a 7049-T73 aluminum alloy longeron. (Radial lines indicate measurements taken to assess damage). (b) Aluminum plate, 7178-T651, exposed to maritime environment. Source: (a) Ref 6

### *18.2.10 Dealloying Corrosion*

Dealloying, also known as selective leaching or parting corrosion, is a destructive process in which the more anodic alloying element is selectively removed from the alloy, leaving behind a porous, spongy mass. Specific categories of dealloying are often known by the name of the dissolved element. The preferential leaching of zinc from brass is known as dezincification, while the loss of iron from gray iron is called graphitic corrosion. During dealloying, typically one of two processes occurs: alloy dissolution and replating of the more cathodic element, or selective leaching of the more anodic alloying element.

Dezincification occurs in copper-zinc brasses containing more than 15% Zn. Dezincification leaves behind a porous and weak layer of copper and copper oxide. High potentials and low pH values favor selective removal of zinc. At negative potentials and acidic conditions, copper and zinc dissolve. At zero to slightly positive potentials and acidic conditions, a region exists in which copper is expected to redeposit. Both dezincification mechanisms can occur independently or in conjunction, depending on the given environmental conditions.

Two types of damage can be characterized: one type of dezincification is uniform, and the second is plug type. Uniform or layer-type dezincification results in a relatively uniform zone of dezincified material, with the underlying material remaining unaffected. Brasses with high zinc content in an acidic environment are prone to uniform dezincification. Plug-type dezincification, shown in Fig. 18.14, results in localized penetrations of dezincified areas that progress through the wall thickness of the material. The overall dimensions of the material do not change. The dezincified areas are weakened or, in some cases, perforated. Plug-type corrosion is most likely to occur in basic or neutral environments and at elevated temperatures.

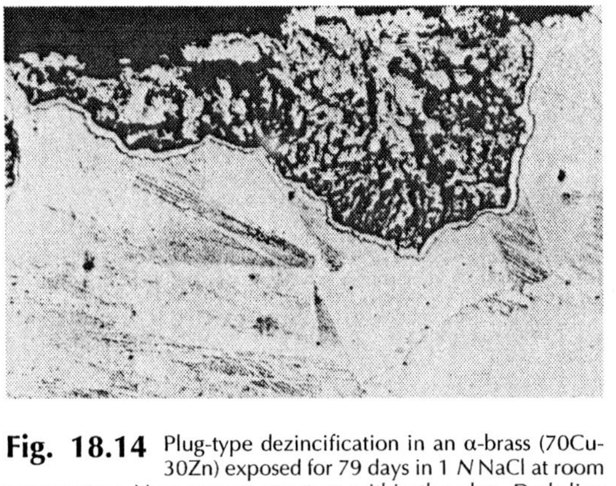

**Fig. 18.14** Plug-type dezincification in an α-brass (70Cu-30Zn) exposed for 79 days in 1 *N* NaCl at room temperature. Note porous structure within the plug. Dark line surrounding the plug is an etching artifact. Total width shown is 0.56 mm (22 mils).

Prevention of dezincification can be achieved most readily by proper alloy selection. Alloys containing greater than 85% Cu are considered resistant to dezincification. Tin is added to act as an inhibitor. In addition, inhibited copper-zinc alloys containing 0.020 to 0.6% arsenic, antimony, or phosphorus are also considered resistant to dezincification. This is likely due to the formation of a redeposited film of protective elements. Other possible remedies for dezincification include the use of cathodic protection, liners, or coatings.

Graphitic corrosion is a form of dealloying that occurs in cast irons. This corrosion mechanism is usually found in gray cast irons and is associated with the presence of graphite flakes. The graphite flakes are cathodic to the iron matrix. Exposure to an electrolyte results in selective leaching of the iron matrix, leaving behind a porous mass of graphite flakes. Graphitic corrosion is generally a long-term mechanism, resulting from exposures of 50 years or more. Pipelines made of cast iron, especially those buried in clay-based soils and soils containing sulfates, are susceptible. In cases where graphitic corrosion has caused extensive wall loss, a reduction in component strength will occur. Thus, it is not unusual for cracking to accompany graphitic corrosion. In some cases, graphitic corrosion has been observed on the fracture surfaces, indicating the long-term existence of the crack. Graphitic corrosion is reduced by alloy substitution, for example, by the use of a ductile or alloyed iron rather than gray iron, by raising the water pH to neutral or slightly alkaline levels, by the addition of inhibitors, and by avoiding stagnant water conditions.

Other forms of dealloying include dealuminification of aluminum bronze and nickel-aluminum bronze alloys; denickelification involving the removal of nickel from copper-nickel alloys; destanification, or the removal of tin, in cast tin bronzes; desiliconification of silicon bronzes; and dealloying in copper-gold and silver-gold alloys.

### *18.2.11 Stress-Corrosion Cracking*

Stress-corrosion cracking (SCC) is a failure process that occurs because of the simultaneous presence of a tensile stress, a specific environment, and a susceptible material. Removal of or changes in any one of these three factors will often eliminate or reduce the susceptibility to SCC. Stress-corrosion cracking occurs by subcritical crack growth involving crack initiation at selected sites, crack propagation, and finally overload fracture of the remaining section. Failure by SCC is frequently encountered in seemingly mild chemical environments at tensile stresses well below the yield strength. Failures often take the form of fine cracks that penetrate deeply into the metal with little or no evidence of corrosion. Therefore, during casual inspection, no macroscopic evidence of impending failure is seen. Stress-corrosion cracking continues to be a cause of significant service failures. It is very likely that for every alloy there is an environment that will cause SCC, but most of the ones of industrial significance are known and avoidable. Materials selection is the first line of defense. Lowering of the applied stresses and elimination of residual stresses can also go a long way toward eliminating problems. Sometimes, minor changes or additions to the environment can help.

Stress-corrosion cracks ordinarily undergo extensive branching and proceed in a general direction perpendicular to the stresses contributing to their initiation and propagation, as shown in Fig. 18.15 for intergranular SCC of 316L stainless steel. The surfaces of some stress-corrosion cracks resemble those of brittle mechanical fractures, although they are actually the result of local corrosion in combination with tensile stress. In some metals, cracking propagates intergranularly and in others transgranularly. In certain metals, such as high-nickel alloys, iron-chromium alloys, and brasses, either type of cracking can occur, depending on the specific metal-environment combination.

A simplified mechanism for SCC is shown in Fig. 18.16. Stress causes rupture of the oxide film at the crack tip, which exposes fresh metal that corrodes and forms another thin oxide film. The oxide ruptures again, allowing more corrosion, and the crack slowly propagates through the alloy until the crack reaches a critical length and failure occurs. Since the cathodic reaction during corrosion can often produce hydrogen, hydrogen can contribute to SCC, often making it difficult to distinguish between SCC and hydrogen embrittlement.

A characteristic of SCC is the existence of a minimum stress for failure, or threshold stress, for smooth components and a threshold stress intensity for crack propagation for precracked components. The threshold stress is that stress below which the probability for cracking is extremely low, and it depends on the temperature, composition, metallurgical structure of the alloy, and the environment. In some tests, cracking has occurred at an applied stress as low as approximately 10% of the yield strength, and

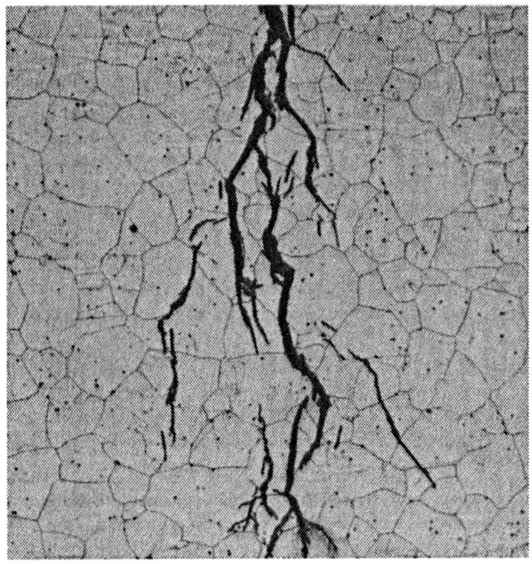

**Fig. 18.15** Chloride-induced stress-corrosion cracking of type 316 stainless steel pipe. Source: Ref 7

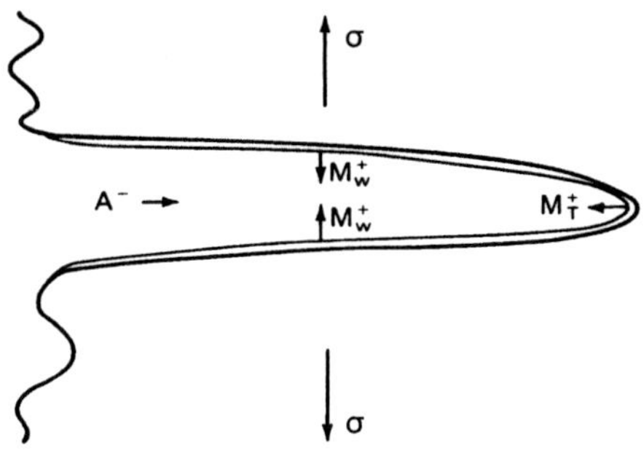


**Fig. 18.16** Schematic of stress-corrosion crack showing important transport and corrosion reactions. $A^-$ represents negatively charged anions migrating to the crack tip, $M_w^+$ represents metal ions entering the crack solution from the crack walls, and $M_T^+$ indicates metal ions entering the crack solution from the crack tip. Source: Ref 8

for other metal-environment combinations, the threshold stress is as high as 70% of yield strength. The effect of alloy composition on threshold stress is shown in Fig. 18.17, which illustrates the relationship between applied stress and average time-to-fracture in boiling 42% magnesium chloride solution for two 18–8 stainless steels (types 304 and 304L) and two more highly alloyed stainless steels (types 310 and 314). As indicated by the nearly level portions of the curves, the threshold stress in this environment is approximately 240 MPa (35 ksi) for high-alloy stainless steels and only 83 Mpa (12 ksi) for the 18-8 stainless steels. For most metal-environment combinations that are susceptible to SCC, there appears to be a threshold stress intensity, $K_{ISCC}$, below which SCC does not occur. The level of $K_{ISCC}$ with respect to the critical fracture stress intensity ($K_{Ic}$) of a material gives a measure of its susceptibility to SCC, shown schematically in Fig. 18.18.

If the right combination of stress and environment is present, almost every metal can be prone to SCC. However, only specific combinations of alloys and environments result in SCC. Susceptibility of a given metal to SCC in a specific environment depends on its condition, that is, its overall and local chemical composition and its metallurgical structure, as determined by thermal processing and cold working. Some aspects of the metallurgical condition that are significant include phase distribution, grain size and shape, grain-boundary precipitation, grain-boundary segregation, cold work, and inclusion type and distribution. Stress-corrosion cracking frequently occurs in commercial alloys, such as low-carbon steels, high-strength steels, austenitic stainless steels and other austenitic alloys (especially in the sensitized condition), high-strength aluminum alloys, and brasses and certain other copper alloys. A partial list of alloy and environmental combinations exhibiting SCC is given in Table 18.4.

### 18.2.12 Corrosion Fatigue

Metals are subject to fatigue degradation by the application of cyclic stress, and all metals have their fatigue strengths further reduced by a corrosive environment. As opposed to SCC, the cracks from corrosion fatigue exhibit little or no branching and are typically full of corrosion debris. The phenomenon can occur as a single crack but more frequently occurs as a multiple series of parallel cracks. The cracks are frequently associated with pits, grooves, or some other type of stress concentration. Transgranular failure modes are more prevalent than intergranular failures.

Making the surface more resistant to either fatigue or corrosion will obviously help. Surface enhancements can include shot peening and

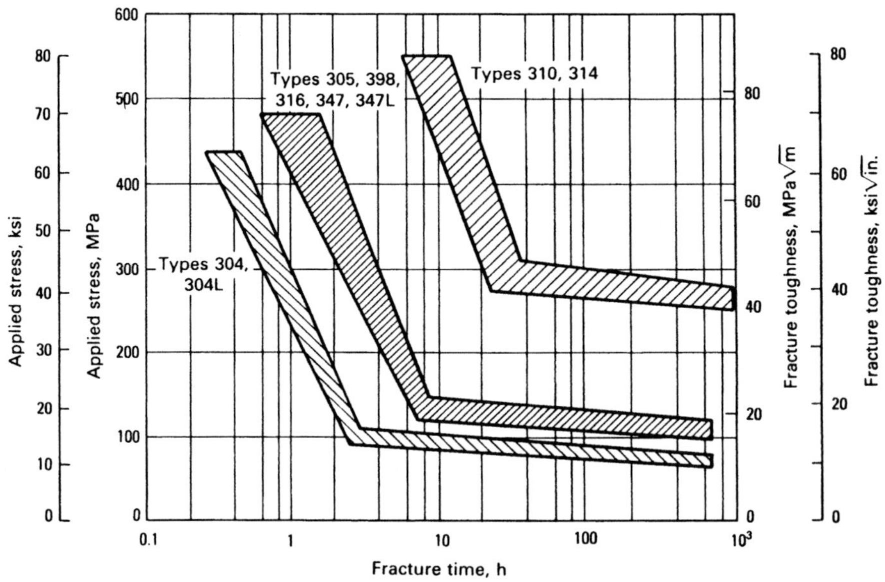


**Fig. 18.17** Relative stress-corrosion cracking behavior of austenitic stainless steels in boiling magnesium chloride. Source: Ref 9

nitriding of steels, as well as protective coatings that reduce access to corrosive agents. Corrosion fatigue is also briefly covered in section 14.10 in Chapter 14, "Fatigue," in this book.

### *18.2.13 Hydrogen Damage*

Hydrogen damage, like SCC, is another insidious type of attack that can strike without warning. In addition, there are a number of different mechanisms associated with hydrogen damage. The problem with hydrogen is that it is an extremely small molecule that can quite easily penetrate into the metallic lattice. As opposed to stress corrosion, hydrogen damage can occur without an applied or residual tensile stress, although the presence of tensile stresses can often make a bad situation worse. Different mechanisms operate in different metals. Specific types of hydrogen damage include:

- *Hydrogen embrittlement:* Occurs most often in high-strength steels, primarily quenched-and-tempered and precipitation-hardened steels with tensile strengths greater than approximately 1035 MPa (150 ksi)
- *Hydrogen-induced blistering:* Also commonly referred to as hydrogen-induced cracking; occurs in unhardened lower-strength steels, typically with tensile strengths less than approximately 550 MPa (80 ksi)
- *Cracking from precipitation of internal hydrogen:* During cooling from the melt, hydrogen diffuses and precipitates in voids and discontinuities. Examples include shatter cracks, flakes, and fish eyes found in steel forgings, weldments, and castings.
- *Hydrogen attack:* A high-pressure, high-temperature form of hydrogen damage commonly experienced in steels used in petrochemical plant equipment that often handles hydrogen and hydrogen-hydrocarbon streams at pressures as high as 21 MPa (3 ksi) and temperatures up to 540 °C (1000 °F). Damage can occur at temperatures as low as 205 °C (400 °F).
- *Hydride formation:* Occurs when excess hydrogen is picked up during melting or welding of titanium, tantalum, zirconium, uranium, and thorium. Brittle, needlelike hydride particles cause a significant loss in strength and large losses in ductility and toughness.

**Characteristics of Hydrogen Embrittlement.** This mechanism has particularly been a problem in heat treated high-strength steels. In general, the higher the strength level of the steel, the greater is the susceptibility to hydrogen embrittlement. Hydrogen embrittlement occurs primarily in body-centered cubic and hexagonal close-packed metals, while face-centered cubic metals are generally not susceptible.

Hydrogen embrittlement results in sudden failures at stress levels below the yield strength. It is normally a delayed failure, in which an appreciable amount of time passes between the time hydrogen is introduced into the metal and failure occurs. Hydrogen embrittlement is a complex process, and different mechanisms may operate in different metals under different environments and operating stresses. However, hydrogen is a small molecule that can dissociate into monatomic hydrogen that readily diffuses into the crystalline structure. Very small amounts of hydrogen can cause damage; for example, as little as 0.0001% hydrogen can cause cracking in steel. Typical sources of hydrogen include a hydrogen-rich environment,

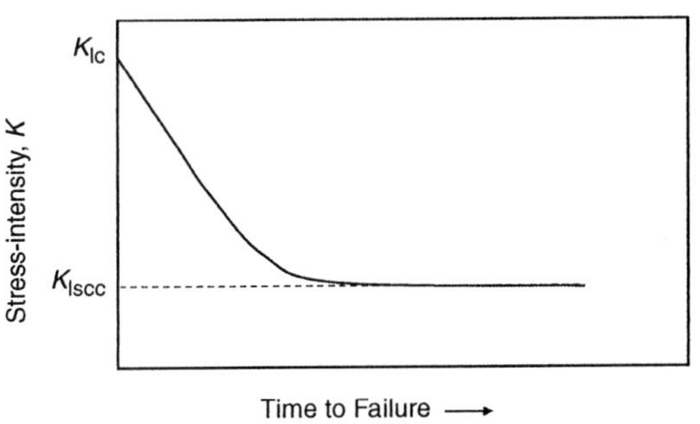


**Fig. 18.18** Concept of threshold stress intensity ($K_{Iscc}$). $K_{Ic}$, fracture stress intensity

**Table 18.4 Typical environment-metal combinations subject to stress-corrosion cracking**

| Alloy | Environment |
|---|---|
| Carbon steel | Hot nitrate, hydroxide, and carbonate/bicarbonate solutions |
| High-strength steels | Aqueous electrolytes, particularly when containing $H_2S$ |
| Austenitic stainless steels | Hot, concentrated chloride solutions; chloride-contaminated steam |
| High-nickel alloys | High-purity steam |
| α-brass | Ammonia solutions |
| Aluminum alloys | Aqueous $Cl^-$, $Br^-$, and $I^-$ solutions |
| Titanium alloys | Aqueous $Cl^-$, $Br^-$, and $I^-$ solutions; organic liquids; $N_2O_4$ |
| Magnesium alloys | Aqueous $Cl^-$ solutions |
| Zirconium alloys | Aqueous $Cl^-$ solutions; organic liquids; $I_2$ at 350 °C (660 °F) |

melting operations, heat treatments, welding, pickling, and plating. In addition, the cathodic reaction during in-service corrosion can also produce hydrogen.

Characteristics of hydrogen embrittlement include strain-rate sensitivity, a temperature dependence, and delayed fracture. As opposed to many forms of brittle fracture, hydrogen embrittlement is enhanced by slow strain rates. In addition, it does not occur at low or high temperatures but occurs at intermediate temperature ranges. For steels, the most susceptible temperature is near room temperature.

A comparison between hydrogen-free notched tensile specimens and ones charged with hydrogen in a static tensile test is shown in Fig. 18.19. Note that there is a time delay before failure occurs, hence the term *static fatigue*. Also, below a certain stress level, failure does not occur. The higher the hydrogen content, the lower the stress level that can be endured before failure. There is also a large reduction in ductility associated with embrittlement (Fig. 18.20). There is no single fracture mode associated with hydrogen embrittlement. Fracture can be transgranular, intergranular (Fig. 18.21), and can exhibit characteristics of both brittle and ductile failure modes. If a steel part is not under stress when it contains hydrogen, then hydrogen can usually be safely removed without damage to the part by baking the part at elevated temperature. The use of a vacuum during baking is even more effective. High-strength steels are usually baked at 185 to 195 °C (365–385 °F) for at least 8 to 24 h to remove any hydrogen after chromium or cadmium plating operations.

The primary factors controlling hydrogen damage are material, stress, and environment. Hydrogen damage can often be prevented by using a more resistant material, changing the manufacturing processes, modifying the design to lower stresses, or changing the environment. Inhibitors and postprocessing bake-out treatments can also be used. Baking of electroplated high-strength steel parts reduces the possibility of hydrogen embrittlement by removing the hydrogen from the metal.

## 18.3 Corrosion Prevention

The first line of defense against corrosion is good design practices. During the previous discussion on the various forms of corrosion, a number of good design practices were mentioned for the different types of corrosion. In addition, by retarding either the anodic or cathodic reactions, the rate of corrosion can be reduced. This can be achieved in several ways: (1) conditioning of the metal, (2) conditioning the environment, and (3) by electrochemical control (removing or altering the components of at least one of the electrochemical cells).

### *18.3.1 Conditioning the Metal*

Conditioning the metal can be subdivided into two main groups: coating the metal or by using a more corrosion-resistant alloy. Coating the metal provides a more corrosion-resistant barrier between the metal and the environment. The

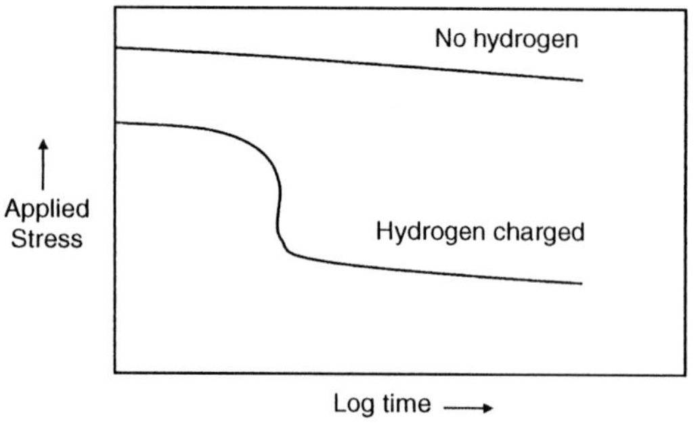


**Fig. 18.19** Hydrogen effect on static tensile strength

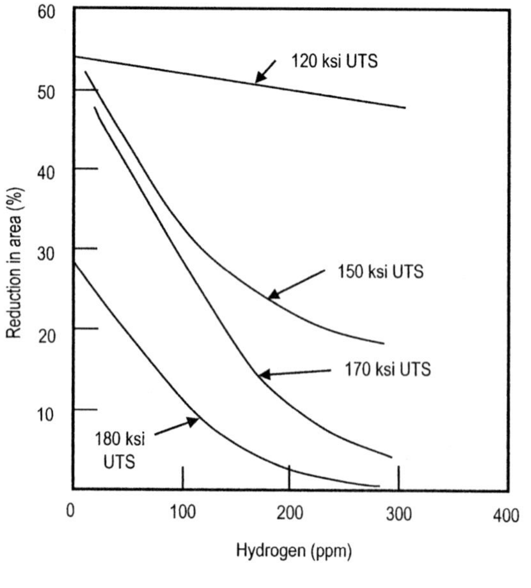


**Fig. 18.20** Effect of hydrogen on ductility of steels grouped by ultimate tensile strength (UTS)

coating may consist of another metal; a protective coating derived from the metal itself, such as a protective oxide; or an inorganic or organic coating, such as resins, plastics, glasses, elastomers, paints, and enamels.

**Protective coatings** are the most widely used to control corrosion. Protective coatings include organic coatings such as epoxy, polyesters, polyurethanes, vinyl, or chlorinated rubber. The coating system must be considered, with surface preparation an important step. Special primers are used to provide passivation, galvanic protection, corrosion inhibition, or mechanical or electrical barriers to corrosive action. A slightly soluble inhibitor incorporated into the primer coat can have a considerable protective influence. Most paint primers contain a partially soluble inhibitive pigment such as zinc chromate, which reacts with the steel substrate to form iron salts that slow corrosion of steel. Chromates, phosphates, molybdates, borates, and silicates are commonly used for this purpose. Some pigments add alkalinity, slowing chemical attack on steel. Alkaline pigments are effective, provided that the environment is not too aggressive. In addition, many new pigments have been introduced to the paint industry, such as zinc phosphosilicate and zinc flake.

The compatible topcoat is chosen based on factors such as the design environment, application method, cost, aesthetics, and regulations. Multiple coatings are preferred to a single thick coating, because the chance of a break in the barrier (holiday) is lessened.

**Metallic coatings** include tin-plated and zinc-plated (galvanized) steel. A continuous coating of either isolates the steel from the electrolyte. However, when the coating is scratched or otherwise penetrated, exposing the underlying steel, the two coatings behave differently (Fig. 18.22). Because zinc is anodic to steel, it continues to be effective in protecting the steel. Since the area of the exposed steel cathode is small, the zinc coating corrodes at a very slow rate, and the underlying steel is protected. On the other hand, steel is anodic to tin in air, so that a small steel anode is created when the tin plate is breached, leading to rapid corrosion of the underlying steel. (Tin is anodic to steel in anaerobic conditions, so tin coating performs well in unopened tin cans but corrodes after opening.)

**Alloying** is often used to produce a more corrosion-resistant alloy. Not all metals exhibit passivity, but the ones that do are among the most widely used corrosion-resisting materials. Nickel, chromium, titanium, and zirconium spontaneously react with the oxygen in air to form protective films. In the case of steels, chromium (at least 10.5%) and nickel are the major alloying components added to iron-base alloys to make stainless steel. The chromium in stainless steel forms a thin oxide of $Cr_2O_3$ that protects the metal. Chromium additions to

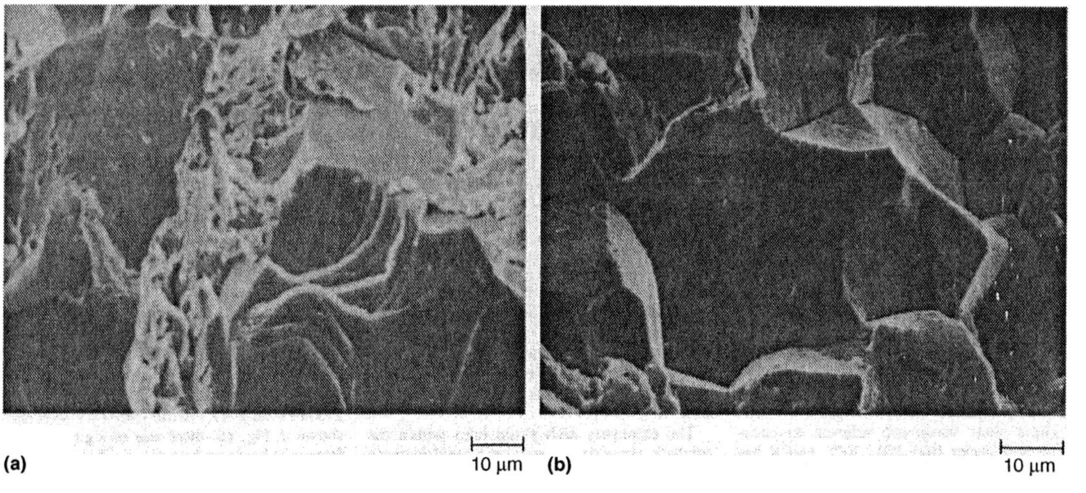


**Fig. 18.21** Hydrogen-embrittled steels. (a) Transgranular cleavage fracture in a hydrogen-embrittled annealed type 301 austenitic stainless steel. (b) Intergranular decohesive fracture in 4130 steel heat treated to 1280 MPa (185 ksi) and stessed at 980 MPa (142 ksi) while being charged with hydrogen. Ref 10

nickel-base alloys also generally enhance resistance.

Other forms of surface conditioning are phosphate, chromate, and chromate-free conversion coatings, anodizing of aluminum, chemical and physical vapor deposition, metallic electroplated coatings, and metallic and nonmetallic thermal spray coatings.

### *18.3.2 Conditioning the Corrosive Environment*

A corrosion inhibitor is a chemical additive that, when added to a corrosive environment in small concentrations, reduces the corrosion rate. Corrosion inhibitors can be liquid or vapor and consist of anodic inhibitors, cathodic inhibitors, adsorption-type, or mixed inhibitors.

**Anodic Inhibitors.** As their name implies, anodic inhibitors interfere with the anodic reaction:

$$Fe \rightarrow Fe^{2+} + 2e^- \qquad \text{(Eq 18.8)}$$

However, if an anodic inhibitor is not present at a concentration level sufficient to block all the anodic sites, localized attack, such as pitting, can become a serious problem due to the oxidizing nature of the inhibitor, which raises the metal potential and encourages the anodic reaction. Thus, anodic inhibitors are often classified as "dangerous inhibitors."

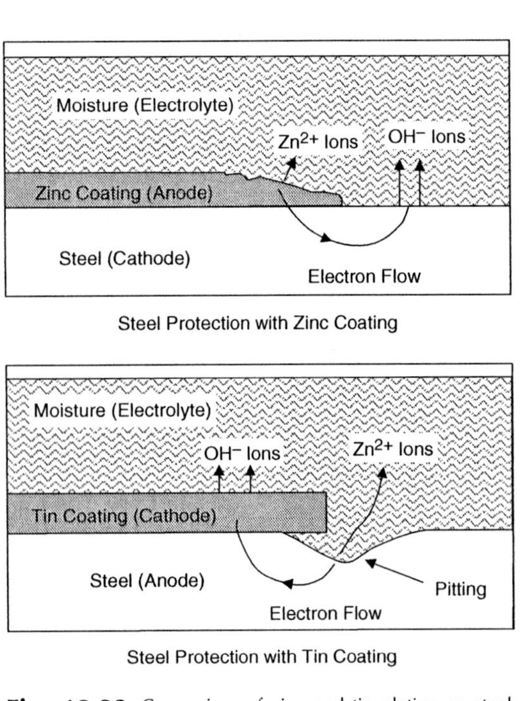


**Fig. 18.22** Comparison of zinc and tin plating on steel. Source: Ref 3

**Cathodic Inhibitors.** The major cathodic reaction in cooling systems is the reduction of oxygen:

$$\tfrac{1}{2}O_2 + H_2O + 2e^- \rightarrow 2OH^- \qquad \text{(Eq 18.9)}$$

Cathodic inhibitors include other additives that suppress the cathodic reaction. They function by reducing the available area for the cathodic reaction. This is often achieved by precipitating an insoluble species onto the cathodic sites. For example, zinc ions are used as cathodic inhibitors because of the precipitation of $Zn(OH)_2$ at cathodic sites as a consequence of the localized high pH. Cathodic inhibitors are classified as safe because they do not cause localized corrosion.

**Adsorption-Type Inhibitors.** Many organic inhibitors work by an adsorption mechanism. The resultant film of chemisorbed inhibitor is then responsible for protection either by physically blocking the surface from the corrosion environment or by retarding the electrochemical processes. The main functional groups capable of forming chemisorbed bonds with metal surfaces are aminos ($-NH_2$), carboxyls (-COOH), and phosphonates ($-PO_3H_2$), although other functional groups or atoms can also form coordinate bonds with metal surfaces.

**Mixed Inhibitors.** Because of the danger of pitting when using anodic inhibitors alone, a common practice incorporates a cathodic inhibitor along with an anodic inhibitor. A common formulation would consist of a mix of zinc and chromate ions. In general, inhibitors added to the environment work best in closed systems that can be monitored, such as boilers and closed cooling systems. For example, the antifreeze that is added to water in a car radiator contains corrosion inhibitors. It is important to change the antifreeze according to the manufacturer's instructions because the inhibitors become depleted over time.

### *18.3.3 Electrochemical Control*

Steel pipes, pipelines, and tanks buried in moist soils and structures and vessels immersed in water are often protected by cathodic protection. Cathodic protection can be achieved by using a direct-current power supply (impressed-current active system) or by applying cathodic current from the anodic dissolution of a metal lower in the galvanic series, such as aluminum, zinc, or magnesium sacrificial anodes (passive system). Both of these approaches are shown schematically in Fig. 18.23.

Protective coatings are normally used in conjunction with cathodic protection. Because the cathodic protection current must protect only the bare or poorly insulated areas of the surface, coatings that are highly insulating, very durable, and free of discontinuities lower the current requirements and system costs. A good coating also enables a single impressed-current installation to protect many miles of piping. Coal tar enamel, epoxy powder coatings, and vinyl resin are examples of coatings that are most suitable for use with cathodic protection.

## 18.4 High-Temperature Oxidation and Corrosion

High-temperature oxidation or gaseous corrosion refers to the reaction of metals with oxygen at high temperatures, usually in the absence of moisture. A surface film or scale forms that acts as the electrolyte in the absence of a liquid electrolyte. The nature of the oxide film usually takes one of three forms: (1) the oxide is unstable, such as in gold, and oxidation does not occur; (2) the oxide is volatile, such as in the case of refractory metals, and oxidation occurs at a constant, relatively high rate; or (3) more commonly, one or more oxides form a layer or layers on the metal surface.

When the thickness is less than approximately 300 nm, surface oxide layers are called films, and when thicker, they are referred to as scales. Thick oxide or scale layers are divided into two categories, protective and nonprotective, on the basis of the Pilling-Bedworth (P-B) ratio, which is the volume of oxide produced to the volume of metal consumed. According the P-B ratio, the oxide is protective if the volume of the oxide is at least as great as the volume of metal from which it formed. If the volume of oxide is less than this amount, the scale is not continuous and is not effective in blocking oxygen from the surface. Although there are many exceptions to the P-B ratio, it is a useful guide when the oxidation characteristics of a metal are unknown.

Three common types of oxide growth, shown in Fig. 18.24, are linear, parabolic, and logarithmic. Metals that have nonprotective oxides tend to increase the weight, $W$, of their scale at a linear rate according to:

$$W = At \quad \text{(Eq 18.10)}$$

where $A$ is a constant, and $t$ is time. In linear growth, oxygen usually passes right through pores or fissures in the oxide layer. When a protective oxide forms on the surface, diffusion must occur for additional growth (Fig. 18.25). The metal ionizes at the surface, and then both the metal ions $M^{2+}$ and the electrons diffuse through the oxide layer to the oxygen surface. Electrons aid in forming the oxygen ion. The ions react near the oxygen surface to form the MO oxide. In this case, the rate of oxidation is slower than that for linear growth and occurs by

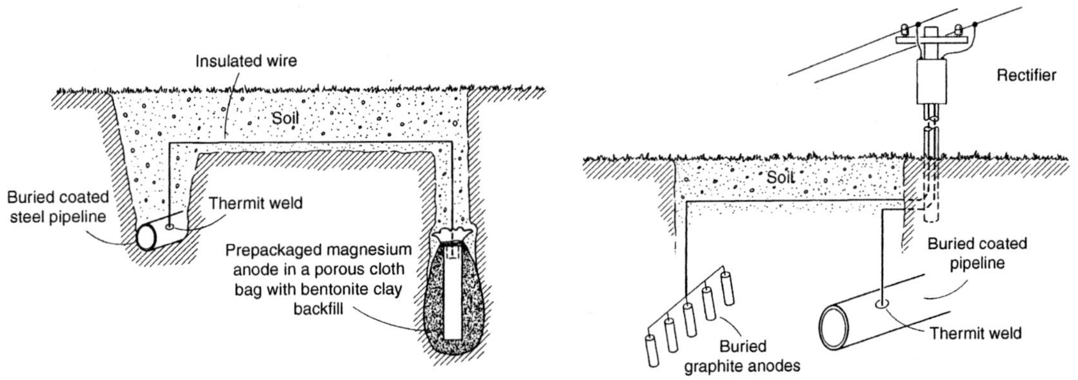


(a) (b)

**Fig. 18.23** Cathodic protection of a buried pipeline. (a) Using a buried magnesium anode. (b) Impressed-current cathodic protection of a buried pipeline using graphite anodes parabolic growth rate:

$$W^2 = Bt \quad \text{(Eq 18.11)}$$

where the constant $B$ depends on the temperature. Some metals that form protective oxides have reaction rates that diminish more rapidly with time than a parabolic relationship predicts. A logarithmic increase in the weight of oxide follows a relationship of the form:

$$W = C \log (Dt + E) \quad \text{(Eq 18.12)}$$

where the constants $C$ and $D$ depend on temperature.

When an alloy, rather than a pure metal, undergoes oxidation, additional factors must be considered. The separate oxides may form a solid solution, a multiphase scale may appear, or only a single component of the alloy may undergo oxidation, a process called selective oxidation. In the case of selective oxidation, benefit can be derived from oxide films if they are adherent, have poor electrical conductivity, and possess a complex crystal structure that hinders diffusion through them. Under certain conditions, oxidation of an alloying element can occur below the surface of the base metal. This internal oxidation can be a problem with some copper and silver alloys that are given high-temperature treatments in mildly oxidizing environments. Compared with ordinary oxide films, the subscales are difficult to remove with conventional cleaning methods. Oxidation rarely occurs in only one of these ways described; instead, a combination of two or three types of reactions may occur simultaneously in different parts of the metal. The stress condition and the orientation of the oxide layer may vary with time or thickness, and discontinuous cracking or spalling of the oxide layer can cause sudden changes in the reaction rate. If more than one oxide layer is stable under oxidation conditions, a series of oxide layers may be formed on the metal surface.

The two major environmental effects on superalloys are oxidation and hot corrosion. At temperatures of approximately 870 °C (1600 °F) and lower, oxidation of superalloys is not a major problem; however, at higher temperatures, oxidation can rapidly occur. Since $Cr_2O_3$ forms as a protective oxide, the level of oxidation resistance at temperatures below 980 °C (1800 °F) is a function of the chromium content. These high-temperature nickel alloys contain 8 to 48% Cr. At temperatures above 980 °C (1800 °F), the aluminum content becomes an important component as $Al_2O_3$ becomes the dominant oxide protector, especially when thermal cycling is involved. Chromium and aluminum can contribute in an interactive manner to provide oxidation protection. For example, the higher the chromium content, the less aluminum that may be required. However, the alloy content of many superalloys is insufficient to provide long-term protection, and protective coatings are usually required to provide satisfactory life.

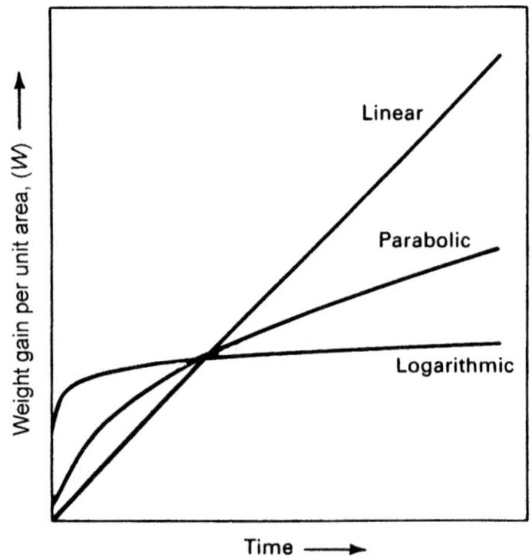


**Fig. 18.24** Oxidation growth rate curves

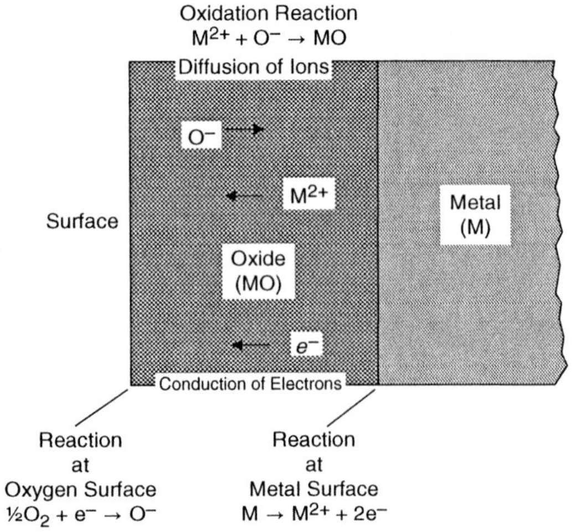


**Fig. 18.25** Oxidation of metal through an oxide layer. Source: Ref 11

Hot corrosion of superalloys, often referred to as sulfidation, is classified as either type I or II, depending on the temperature. Type I occurs at higher temperatures (900 to 1050 °C, or 1650 to 1920 °F), while type II occurs at lower temperatures (680 to 750 °C, or 1255 to 1380 °F). Both are triggered by the presence of sulfur in fuels combining with salt from the environment. Hot corrosion is an accelerated, often catastrophic, surface attack of parts in the hot gas path. It is believed that the presence of alkali metal salts (i.e., $Na_2SO_4$) is a prerequisite for hot corrosion.

## ACKNOWLEDGMENTS

Sections of this chapter were adapted from *Surface Engineering* edited by J.R. Davis, ASM International, 2001, and "Forms of Corrosion" in *Failure Analysis and Prevention,* Volume 11, *ASM Handbook,* ASM International, 2002.

## REFERENCES

1. "Economic Effects of Metallic Corrosion in the United States," Battelle Columbus Laboratories and the National Institute of Standards and Technology, 1978 and Battelle updates in 1995
2. "High-Performance Alloys for Resistance to Aqueous Corrosion," Special Metals Corporation
3. R.A. Higgins, *Engineering Metallurgy—Applied Physical Metallurgy,* 6th ed., Arnold, 1993
4. Forms of Corrosion, *Failure Analysis and Prevention,* Vol 11, *ASM Handbook,* ASM International, 2002
5. D.R. Askeland, *The Science and Engineering of Materials,* 2nd ed., PWS-Kent Publishing Co., 1989
6. K.K. Sankaran, R. Perez, and H. Smith, Military Aircraft Corrosion Fatigue, *Corrosion: Environments and Industries,* Vol 13C, *ASM Handbook,* ASM International, 2006, p 195–204
7. Atlas of Fractographs, Austenitic Stainless Steels, *Fractography,* Vol 12, *ASM Handbook,* ASM International, 1987
8. R.H. Jones, Stress-Corrosion Cracking, *Corrosion: Fundamentals, Testing, and Protection,* Vol 13A, *ASM Handbook,* ASM International, 2003, p 346–366
9. E. Denhard, Effect of Composition and Heat Treatment on the Stress Corrosion Cracking of Austenitic Stainless Steels, *Corrosion,* Vol 16 (No. 7), 1960, p 131–141
10. V. Kerlins and A. Phillips, Effect of Environment on Failure Modes, *Fractography,* Vol 12, *ASM Handbook,* ASM International, 1987
11. A.G. Guy and J.J. Hren, *Elements of Physical Metallurgy,* 3rd ed., Addison-Wesley Publishing, 1974

# Part II
# Engineering Alloys

# CHAPTER 19

# Plain Carbon Steels

PLAIN CARBON STEELS are the most important group of engineering alloys and account for the vast majority of steel produced. Their relatively low cost and wide range of useful properties make them attractive as engineering materials. Applications for plain carbon steel are countless, with product forms consisting of sheet, strip, plate, bar, wire, and tubular products. Plain carbon steels are members of the family of ferrous alloys (Fig. 19.1), which also includes alloy steels, stainless steels, tool steels, and cast irons. In the past, steel has been described as an alloy of iron and carbon. Today, this description is no longer applicable since in some very important steels, such as interstitial-free steels and some stainless steels, carbon is considered an impurity and is present in quantities of only a few parts per million. Steels are iron-base alloys containing one or more alloying elements, generally including carbon, manganese, silicon, nickel, chromium, molybdenum, vanadium, titanium, niobium, and aluminum.

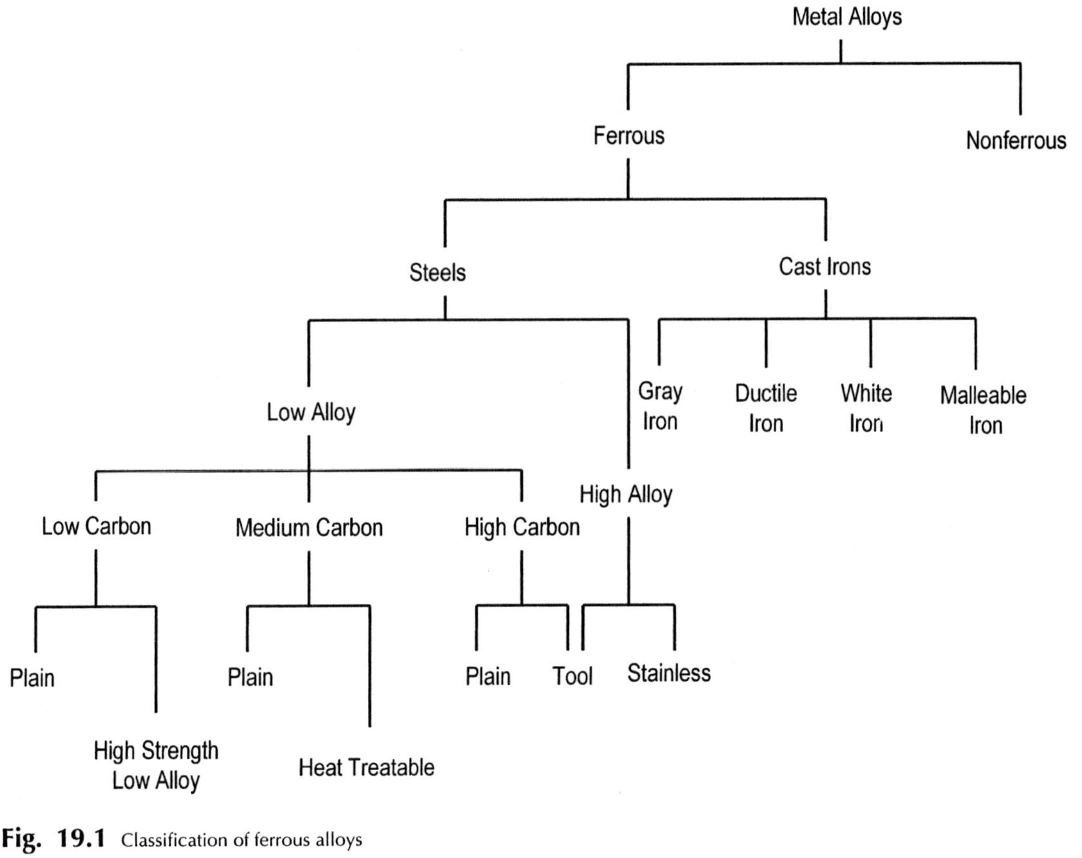


**Fig. 19.1** Classification of ferrous alloys

## 19.1 Brief History of Steel

Steel started to replace bronze in approximately 1200 B.C. Cast iron alloys predate steel because cast iron, with its higher carbon content, melts at lower temperatures. Early steel alloys were produced by solid-state smelting that produced iron with a low carbon content and high density of entrapped slag inclusions. Heavy hammering or forging was used to fragment and disperse the inclusions. In approximately 350 B.C., wootz steel was produced in India by adding carbon to wrought iron and then carburizing it in crucibles with charcoal. During this period, similar processes evolved in other parts of the world. Since these early processes provided both an economic and military advantage, they were closely guarded secrets. Despite the drawbacks of these early processes, early blacksmiths produced remarkable objects, such as the Damascus and Japanese swords that had sharp cutting edges, high hardness and strength, good fracture resistance, and were also objects of great beauty. Damascus swords, first produced in approximately 500 A.D., were forged from blocks of high-carbon wootz steel and were known for their highly decorative surface patterns caused by fine bands of dispersed alloy carbides. Japanese swords, which evolved about the same time, were made by welding alternating layers of low- and high-carbon steel together in multiple forging steps. However, it was not until the middle of the 19th century that a large-scale process emerged for making steel, when, in 1856, Bessemer patented a process in which hot air was blown through molten pig iron to reduce the carbon and silicon contents. In 1858, Siemens first successfully operated an open-hearth furnace in which liquid pig iron and scrap were melted with a hot gas flame. The key factor in both the Bessemer and Siemens processes was the oxidation and removal of carbon, silicon, and other impurities by oxides, such as CO in the case of carbon. There are two main processes

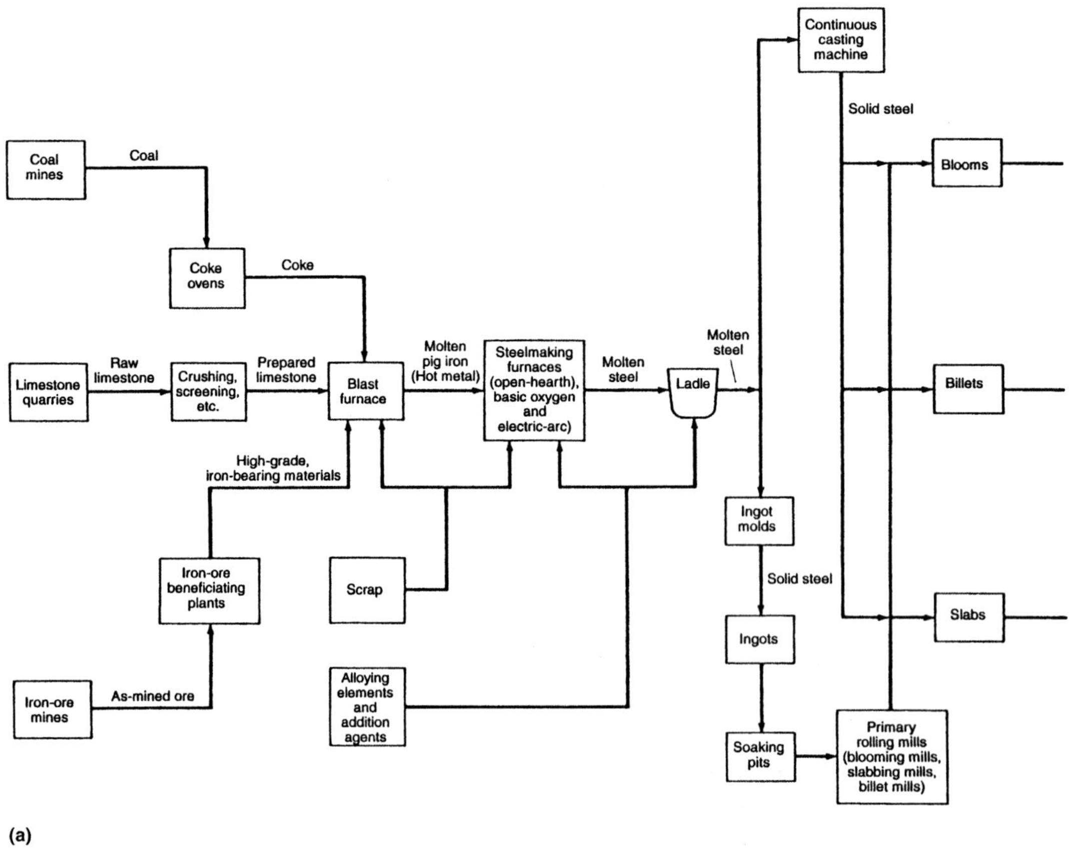


**Fig. 19.2** Principal steps in steelmaking. Source: Ref 1

used for making steel today: the electric arc furnace and the basic oxygen furnace (BOF). While the electric arc furnace mainly uses scrap steel, the BOF requires a charge of molten iron, which is produced in blast furnaces.

## 19.2 Steel Production

A general diagram for the production of steel from raw materials to finished mill products is shown in Fig. 19.2. Steel production starts with the reduction of ore in a blast furnace into pig iron. Since pig iron is rather impure and contains carbon in the range of 3 to 4.5 wt%, it must be further refined in either a basic oxygen or an electric arc furnace to produce steel that usually has a carbon content of less than 1 wt%. After the pig iron has been reduced to steel, it is cast into ingots or continuously cast into billets. Cast steels are hot worked to improve homogeneity, refine the as-cast microstructure, and fabricate desired product shapes. After initial hot rolling operations, semifinished products are worked by hot rolling, cold rolling, forging, extruding, or drawing. Some steels are used in the hot rolled condition, while others are heat treated to obtain specific properties. However, the great majority of plain carbon steel products are low-(<0.30 wt% C) carbon steels that are used in the annealed condition. Medium (0.30 to 0.60 wt% C) and high (0.60 to 1.00 wt% C) plain carbon steels are often quenched and tempered to provide higher strengths and hardness.

## 19.3 Ironmaking

The first step in making steel from iron ore is to make iron by chemically reducing the ore (iron oxide) with carbon, in the form of coke, according to the general equation:

$$Fe_2O_3 + 3CO \rightarrow 2Fe + 3CO_2 \quad \text{(Eq 19.1)}$$

The ironmaking reaction takes place in a blast furnace, shown schematically in Fig. 19.3,

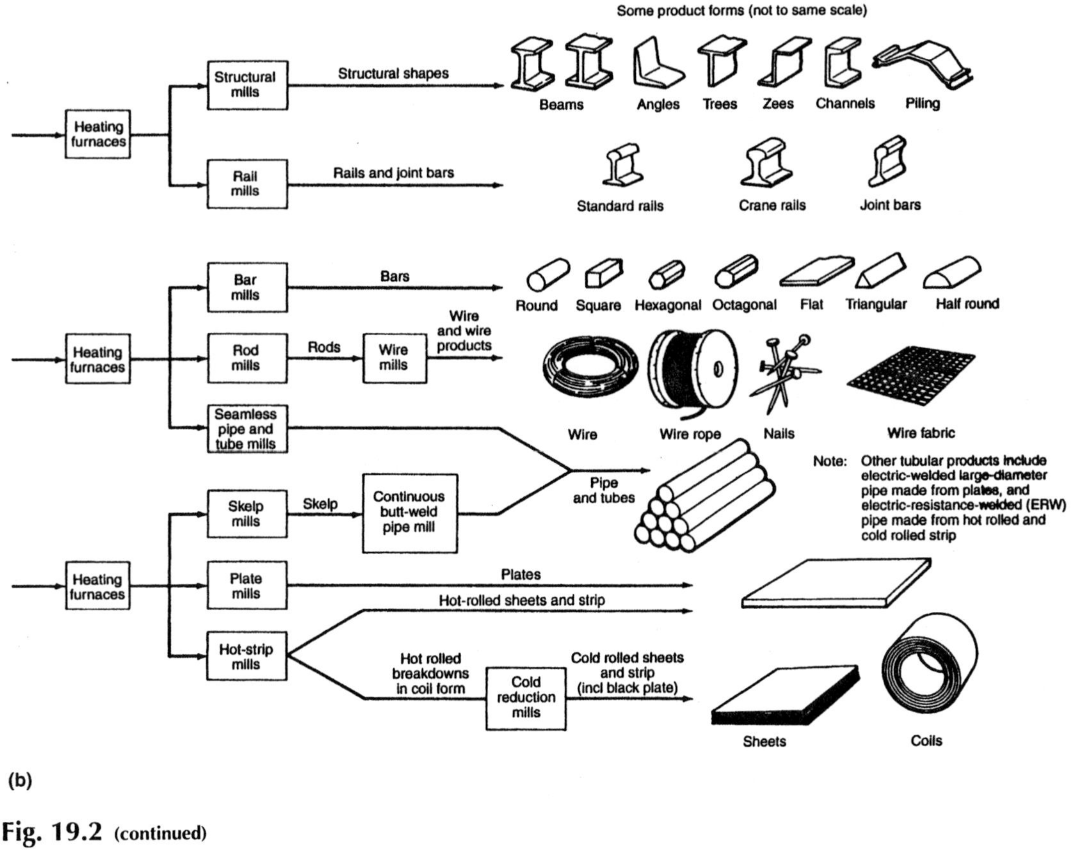


**Fig. 19.2** (continued)

(a)

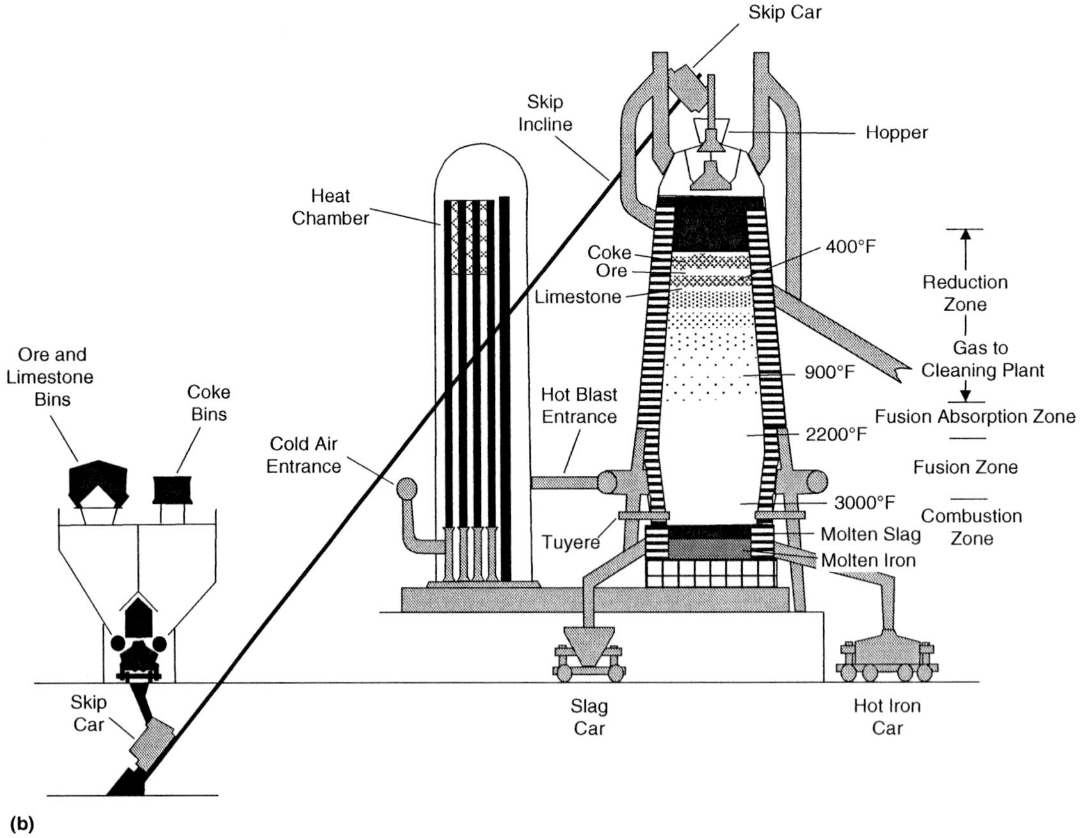


(b)

**Fig. 19.3** The iron blast furnace and general view of 1000 ton furnace. Source: Ref 4, 5 which is essentially a tall, hollow, cylindrical structure with a steel outer shell lined on the inside with refractory brick. The raw materials for a blast furnace charge are iron ore, coking coal, and fluxes, mainly limestone. Coke is a spongelike carbon mass that is produced from coal by heating coal to expel the organic matter and gasses. In a process called carbonization, blended coal is first heated in ovens to produce coke. The gas produced during carbonization is extracted and used for fuel elsewhere in the steelworks. Other by-products, such as tar, are also extracted for further refining and sale. After carbonization, the coke is pushed out of the ovens and cooled. Limestone, mainly calcium carbonate, is added as a flux for easier melting and slag formation. (Slag floats on top of the molten iron and absorbs many of the unwanted impurities.) Fine ore is mixed with the coke and fluxes and heated in a sinter plant on a continuous moving belt on which the coke is ignited. The high temperatures generated fuse the ore particles and fluxes together to form a porous clinker called sinter. The use of sinter in the blast furnace helps make the ironmaking process more efficient.

Iron ore lumps and pellets, coke, sinter, and possibly extra flux are carried to the top of the blast furnace on a conveyor and charged into the furnace. The crushed or pelletized ore, coke, and limestone are added as layers through an opening at the top of the furnace. Hot air at approximately 900 °C (1650 °F) is blasted into the bottom of the furnace, an area called the bosh, through water-cooled copper nozzles called tuyeres. The oxygen in the air reacts with the coke to form carbon monoxide gas according to Eq 19.1 and, at the same time, generates a great deal of heat.

Frequently, oil or coal is injected with the air, which allows less expensive coke to be used. The carbon monoxide gas flows up through the blast furnace, removing oxygen from the iron ore and leaving iron. The iron in the ore reduces to metallic iron from iron because the free energies of CO and $CO_2$ are both lower than that of iron oxide. This greatly increases the temperature and provides the required carbon for steelmaking.

The resulting liquid iron is tapped at regular intervals by opening a hole in the bottom of the furnace, and the hot metal flows into specially constructed railway containers that transport the liquid iron to the BOF, where it is made into steel. The molten slag, which floats on the iron, is removed by tapping at regular intervals. A successful steelmaking furnace campaign can last for ten continuous years or more. If the furnace were allowed to cool, damage to the refractory lining bricks could result from their contraction as they cooled. Eventually, the refractory brick linings are eroded away, the steelmaking campaign is stopped, and the furnace is relined with new bricks.

Iron produced in a blast furnace, called pig iron, has a carbon content of 3 to 4.5 wt% as well as a number of other impurities, which makes it extremely brittle. The product of the furnace is called pig iron because, in the early days, the molten iron was drawn from the furnace and cast directly into branched mold configurations on the cast house floor. The central branch of iron leading from the furnace was called the sow, and the side branches were called pigs.

## 19.4 Steelmaking

The two dominant steelmaking methods during the 20th century were the Bessemer and open-hearth processes. In the Bessemer process, developed in 1856, air was blown through molten pig iron to reduce the carbon and silicon contents to tolerable levels. In the open-hearth processes, developed shortly after in 1858, steel was made in a very large, shallow furnace in which carbon reduction was achieved by an oxidizing slag. Although the open-hearth furnace required a longer time (8 h versus 30 min for the Bessemer process), it was more widely used because much larger amounts of steel could be produced. Both of these processes are now obsolete, and the BOF has largely replaced both of these older processes.

### *19.4.1 Basic Oxygen Furnace*

Most modern bulk steels are made in the BOF according to the process shown in Fig. 19.4. Up to 30% of the BOF is charged with scrap steel, followed by liquid pig iron from the blast furnace. A water-cooled lance is then lowered into the vessel, through which very pure oxygen is blown at high pressure. The oxygen interacts with the molten pig iron to oxidize undesirable elements, including excess carbon, manganese, and silicon from the ore, limestone, and other impurities such as sulfur and phosphorus.

Carbon in the steel reacts with iron oxide to form iron and carbon monoxide:

$$FeO + C \rightarrow Fe + CO \qquad \text{(Eq 19.2)}$$

A careful balance between the relative amounts of pig iron and scrap charged into the converter is maintained as a means of controlling the temperature and to ensure that steel of the required specification is produced. After a sample has been taken to verify the chemical composition of the steel, the vessel is tilted to allow the molten steel to flow out. The steel is tapped into a ladle where further composition adjustments are made. During tapping, small quantities of other metals and fluxes are often added to control the state of oxidation and to meet requirements for particular grades of steel. Finally, the vessel is turned upside down to remove the remaining slag. A modern BOF vessel can make up to 318,000 kg (700,000 lb) of steel in approximately 40 min with the desired carbon content and a low level of impurities such as sulfur and phosphorus.

While in the ladle, certain alloying elements can be added to the steel to control the state of oxidation and produce the desired chemical composition. The ladle furnace is maintained at a specified temperature by external heat from electrodes in the lid that covers the ladle. After the desired chemical composition is achieved, the ladle can be placed in a vacuum chamber to remove undesirable gases such as hydrogen and oxygen. Degassing is used for higher-quality steel products, such as railroad rail, sheet, plate, bar, and forged products.

### *19.4.2 Electric Arc Furnace*

Unlike the BOF, the electric arc furnace (Fig. 19.5) does not use molten pig iron but uses steel scrap. Steel scrap is charged into the furnace from an overhead crane, and a lid is swung into position over the furnace. The lid holds

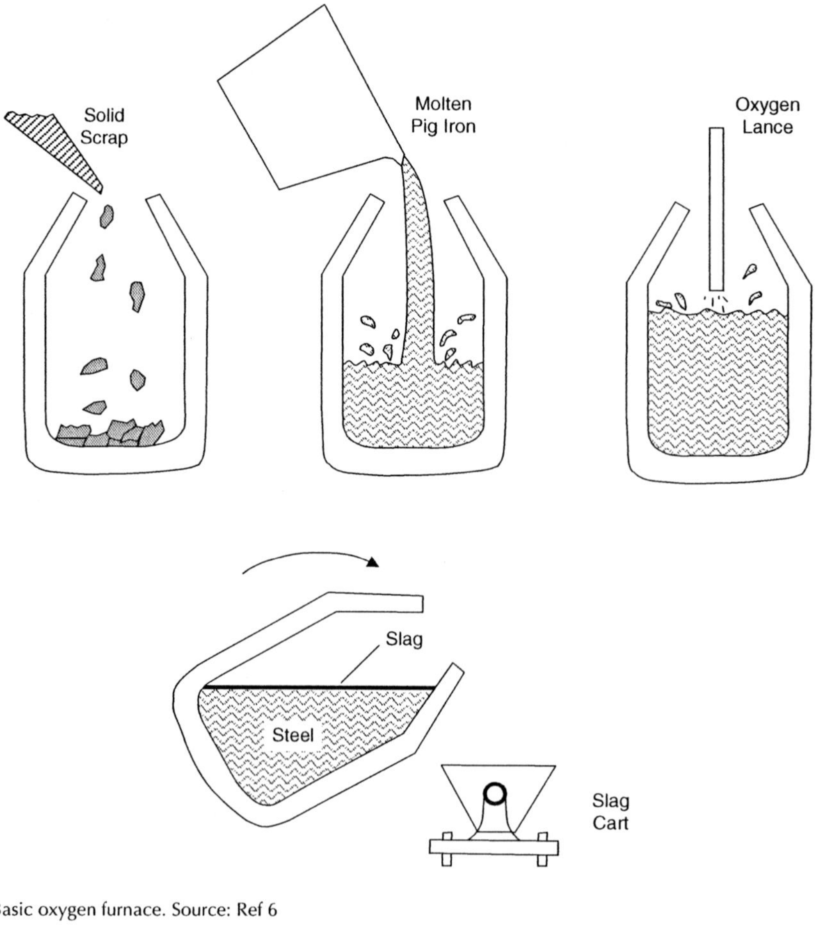


**Fig. 19.4** Basic oxygen furnace. Source: Ref 6

graphite electrodes that are lowered into the furnace. An electric current is passed through the electrodes to form an arc, which generates the heat necessary to melt the scrap. During melting, alloying elements are added to the steel to give it the required chemical composition. After samples have been taken to check the chemical composition, the furnace is tilted to allow the floating slag to be poured off. The furnace is then tilted in the other direction, and the molten steel is tapped into a ladle, where it either undergoes secondary steelmaking or is transported to the caster. The modern electric arc furnace typically makes approximately 136,000 kg (300,000 lb) of steel in about 90 min.

Since the electric arc furnace has a relatively low capital equipment cost and uses steel scrap, this process is used where local supplies of steel scrap are available and has given rise to what are known as "mini" mills. The electric arc furnace is also used for producing alloy steels that contain appreciable amounts of easily oxidized alloying elements, such as chromium, tungsten, and molybdenum. It can also be used to make steels requiring very low sulfur and phosphorus contents. Special slags are used to lower the sulfur and phosphorus levels and to protect against oxidation of alloying elements. An additional benefit is that temperature control with the electric arc process is very good.

### 19.4.3 Ladle Metallurgy

The demand for higher-quality and cleaner steels has led to refining operations after the steel is made by either the basic oxygen or electric arc processes. These refining processes are conducted in the liquid steel transfer ladle into which the steel has been poured after the basic oxygen or electric arc processes are complete. By conducting these refining processes outside the steelmaking furnace, valuable steelmaking resources are freed up. In addition, reducing atmospheres are more easily applied for desulfurization. Vacuum degassing is also possible with the steel in a ladle, and argon lances can be used to stir the steel to make the composition more homogeneous. Vacuum degassing produces ultralow-carbon steels, with carbon contents as low as 0.002 wt%. This allows these ultralow-carbon steels to be continuously annealed and still have the high formability required for deep-drawing applications. Vacuum degassing also removes hydrogen that can result in hydrogen flaking and porosity.

### 19.4.4 Residual Elements and Cleanliness

Various manufacturing practices can affect the oxygen, nitrogen, and sulfur contents and hence the cleanliness of the product. The term *cleanliness* usually refers to the nonsteel phases,

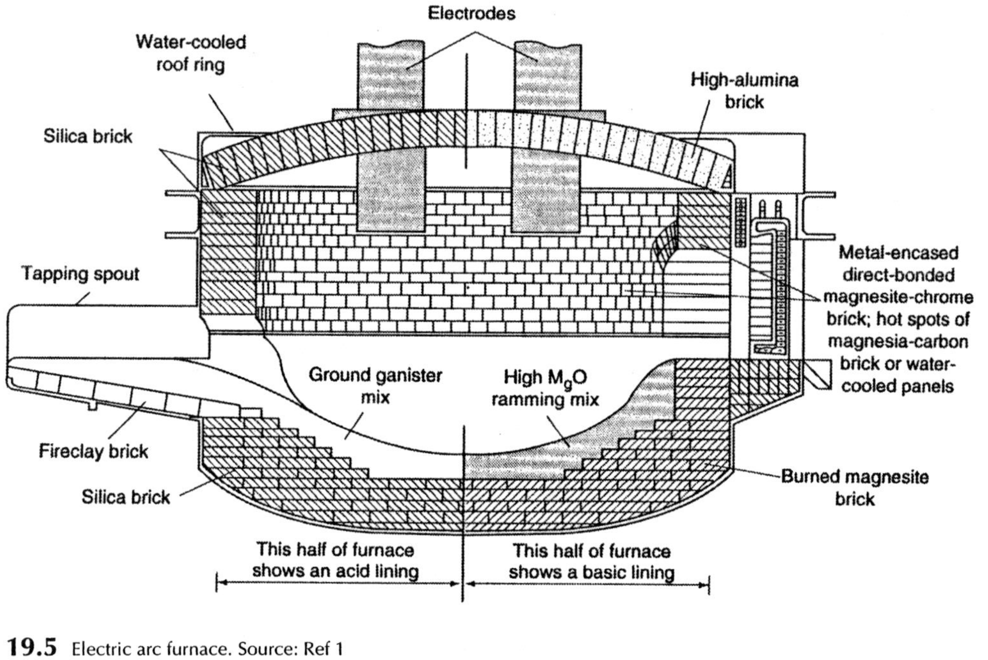


**Fig. 19.5** Electric arc furnace. Source: Ref 1

such as oxides, sulfides, and silicates, that can be present in steel. The smaller the amount of these phases, the cleaner is the steel. They are present in the form of inclusions, which can have significant undesirable effects on the properties of steel. Tin, antimony, arsenic, and copper are considered residual or tramp elements in steel, although copper is added as an alloying element to improve the corrosion resistance of some steels. Tramp elements remain in steel because they are difficult to remove during steelmaking and refining. Steels made by electric furnace melting employing scrap as a raw material contain higher levels of residual elements than steels made in an integrated steelmaking facility using the blast furnace-BOF route. Some electric furnace melting shops use direct reduced-iron pellets to dilute the effect of these residuals.

Hydrogen gas is also a residual element that can be very deleterious. Hydrogen is soluble in liquid steel and somewhat soluble in austenite. However, it is insoluble in ferrite and is rejected as atomic hydrogen ($H^+$). If trapped inside the steel, usually in products such as thick plate, heavy forgings, or railroad rail, hydrogen will accumulate on the surfaces of manganese sulfide inclusions. Eventually, enough molecular hydrogen ($H_2$) accumulates and sufficient pressure develops to create internal cracks. As the cracks grow, they form what are termed hydrogen flakes, and the product must be scrapped. However, if the product is slow cooled from the rolling temperature, atomic hydrogen has sufficient time to diffuse out of the product, thus avoiding hydrogen damage. Vacuum degassing is used to remove hydrogen from liquid steel.

There have been major advances in the production of steel during the last 20 years, and continuous casting, in which great attention is being paid to the cleanliness of the steel, has become the dominant production method. Vacuum deoxidization is also being used to eliminate oxygen, and the steel is protected by argon atmospheres in covered tundishes that yield cleaner steel with a lower inclusion content. This is beneficial to the mechanical properties and uniformity of the final product. Continuous casting also produces a product that is much closer to a shape that is amenable to hot rolling.

More and more steel sheet is now being produced by minimills. These mills, employing electric arc furnaces, continuously cast steel into slabs several inches thick and a few feet wide. The slab is immediately fed through a long furnace to the hot rolling mill. However, since steel scrap is the primary raw material, controlling the residual elements in the composition can be problematic. Copper is particularly troublesome because it is not easily removed from liquid steel, and, as its concentration increases, it can produce cracks due to hot shortness by penetrating the grain boundaries and causing grain-boundary cracking during hot rolling. Hot shortness is the tendency for alloys to separate along grain boundaries when stressed or deformed at temperatures near the melting point. Hot shortness is caused by a low-melting constituent, often present only in minute amounts, that is segregated at grain boundaries.

### *19.4.5 Ingot Casting*

After the ladle refining operations are complete, the liquid steel is cast in molds to produce ingots or continuously cast in a continuous casting machine. Although ingot casting has been the traditional method, continuous casting has rapidly evolved as the method of choice because of cost and quality advantages. During ingot casting, the ladle is moved by an overhead crane so that is can be tapped or teemed into individual upright-standing molds on rail cars. Ingot molds are slightly tapered, as shown in Fig. 19.6, for easier removal of the ingots after solidification. After stripping from the molds, the hot ingots are transferred to soaking pits where they are reheated for hot rolling.

During solidification, excess gases are expelled from the solidifying metal. Oxygen in the form of FeO reacts with carbon to produce carbon monoxide according to Eq 19.2. Since steel solidifies over a range of temperatures,

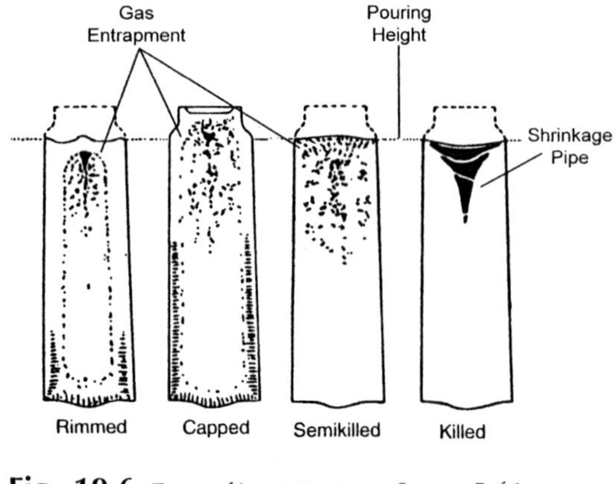


**Fig. 19.6** Types of ingot structures. Source: Ref 1

gases evolving from the solidifying metal are trapped at the solid-liquid interfaces, producing porosity known as blowholes. The amount of oxygen dissolved in the liquid steel just before casting can be controlled by adding deoxidizing agents such as aluminum or ferrosilicon. Depending on the amount of gases (mainly oxygen) remaining in the liquid steel during the solidification process, the resulting ingot structure is either rimmed, capped, semikilled, or killed, as shown in Fig. 19.6. The deoxidation method used is based on economics and end use of the steel.

**Rimmed steels** are only slightly deoxidized, so that a large evolution of gas occurs as the metal begins to solidify. The gas is a product of the reaction between the carbon and oxygen in the molten steel, which occurs at the boundary between the solidified metal and the remaining molten metal. As a result, the outer "rim" of the ingot is practically free of carbon. Rimming may be stopped mechanically after a desired period, or it may continue until the rimming action subsides and the top freezes over, thereby ending all gas evolution. The center portion of the ingot, which solidifies after the rimming ceases, has a carbon composition somewhat above that of the original molten metal as a result of the segregation tendencies. The low-carbon surface layer of rimmed steel is very ductile. Proper control of the rimming action will result in a very sound surface for subsequent rolling. Consequently, rimmed grades are particularly adaptable to applications involving cold forming and where the surface finish is of prime importance. The presence of an appreciable percentage of carbon or manganese decreases the oxygen available for the rimming action. If the carbon content is greater than 0.25 wt% and the manganese is greater than 0.60 wt%, the action will be very sluggish or nonexistent. If a rim is formed, it will be quite thin and porous. As a result, the cold forming properties and surface quality are seriously impaired. It is therefore standard practice to specify rimmed steel only for grades with low percentages of carbon and manganese. Rimmed steel is cheaper to produce since the top portion of the ingot does not have a large gas cavity that must be scrapped, which means the yield is higher.

**Capped steels** are much the same as rimmed steels except that the duration of the rimming action is curtailed. A deoxidizer is usually added during pouring of the ingot, with the result that a sufficient amount of gas is entrapped in the solidifying steel to cause the metal to occupy a larger volume in the mold. The rising metal contacts the cap, thereby stopping the action. A similar effect can be obtained by adding ferrosilicon or aluminum to the ingot top after the ingot has cooled for the desired time. Rimming times of 1 to 3 min prior to capping are most common. Capped steels have a thin, low-carbon rim that imparts the surface and cold forming characteristics of rimmed steel. The remainder of the cross section approaches the uniformity typical of semikilled steels.

**Semikilled steels** are intermediate in deoxidation between the rimmed and killed grades. Sufficient oxygen is retained so that its evolution counteracts the shrinkage on solidification, but there is no rimming action. Consequently, the composition is more uniform than rimmed steel, but there is also a greater possibility of segregation than in killed steels. Semikilled steels are used where neither the surface nor the cold forming characteristics of rimmed steels nor the greater uniformity of killed steels are essential requirements.

**Killed steels** are strongly deoxidized and characterized by a relatively high degree of uniformity in composition and properties. The metal shrinks during solidification, forming a cavity or pipe in the extreme upper portion of the ingot. Generally, these grades are poured in big-end-up molds. A "hot top" brick is placed on top of the mold before pouring and filled with metal after the ingot is poured. The pipe formed will be confined to the hot-top section of the ingot, which is removed by cropping before rolling. The most severe segregation of the ingot will also be removed by cropping. While killed steels are more uniform in composition and properties than any other type, they are nevertheless susceptible to some degree of segregation. As in the other grades, the top center portion of the ingot will exhibit greater segregation than the balance of the ingot. The uniformity of killed steel makes it most suitable for applications involving operations such as forging, piercing, carburizing, and heat treatment. Aluminum-killed steels are widely used for cold rolled sheet that will be used for severe forming or deep drawing. These steels exhibit a minimum of strain aging and have a fine grain size.

### *19.4.6 Continuous Casting*

Today, most steel is cast into solid form in a continuous casting or strand casting machine.

In the continuous casting process (Fig. 19.7), the ladle of molten steel is transported to an elevated casting platform above the casting machine. The molten steel is poured into a rectangular trough, called a tundish, which acts as a reservoir for the steel. From a spout in the bottom of the tundish, the molten steel is poured into a water-cooled mold with a movable temporary bottom. As the molten steel enters the mold, the metal at the surface of the mold solidifies, forming a thin skin. The skin thickens as the metal passes through the mold, and the temporary bottom is slowly lowered to allow metal to be continuously poured into the mold. The remaining metal in the center of the ingot is solidified by spraying cold water onto the ingot as it leaves the mold. The solid metal billet is pulled by rollers so that a long, continuous steel slab is produced. At the end of the machine, it is straightened and cut to the required length. Fully formed slabs, blooms, and billets emerge from the end of this continuous process. The continuous casting process runs for days or weeks as ladle after ladle of molten steel feeds the casting machine. The advantages of the continuous casting process include reduced costs, improved quality, increased yield, lower energy costs, and less pollution. It is now the process of choice for high-volume, low-cost plain carbon steels.

Quality improvements include less variability in chemical composition, both through the thickness and along the length of the continuously cast slab. The surface quality of the slab is also higher than for an ingot, having fewer surface defects such as seams and scabs. The yield for continuous casting is also higher, since it is not necessary to crop the ends of continuously cast slabs. Energy savings are achieved, since the continuously cast slabs are sent directly to rolling mills and do not require soaking pits for reheating. In addition, since the thickness of continuously cast slabs is approximately half the thickness of individual ingots, much less hot rolling is required.

## 19.5 Hot Rolling

Semifinished blooms, billets, and slabs are transported from the steelmaking plant to the

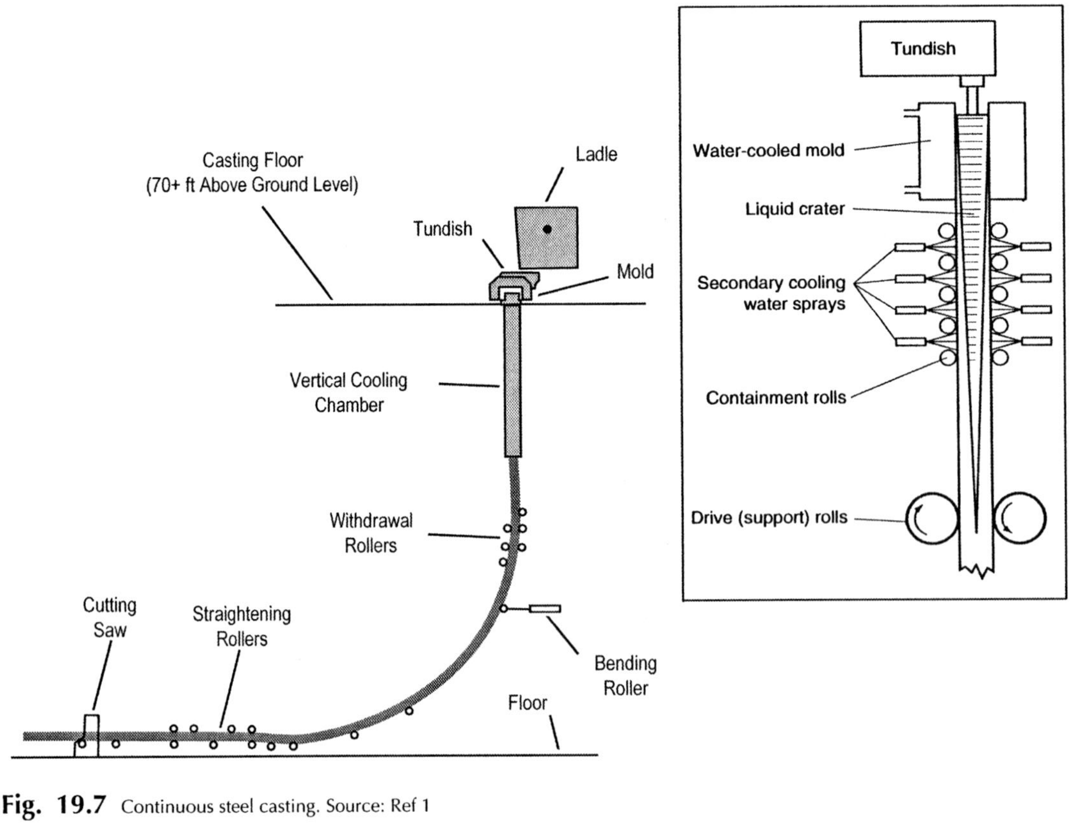


**Fig. 19.7** Continuous steel casting. Source: Ref 1

rolling mills. Steel products are classified into two basic types according to their shape: flat products and long products. Slabs are used to roll flat products, while blooms and billets are mostly used to roll long products. Billets are smaller than blooms and therefore are used for smaller long products. Semifinished products are heated until they are red hot (~1200 °C, or 2200 °F) and passed through a roughing stand, a collection of steel rolls that squeeze the hot steel into the required shape. The steel is passed backward and forward through the roughing stands several times, with each pass approaching the shape and dimension of the finished product.

### 19.5.1 Plate Mills

Slabs are used to make plate—large, flat pieces of steel in the range of 6 to 51 mm (0.25 to 2.00 in.) thick and up to 5 m (15 ft) wide. After leaving the plate mill roughing stand, they are passed through a finishing stand, a reversing mill where the steel is passed backward and forward through the mill. It is usually rotated 90° and rolled sideways at one stage during the process.

### 19.5.2 Strip Mills

Slabs are also used to make steel strip, called hot rolled coil. After leaving the roughing stand, the slab passes continuously through a series of finishing stands that progressively reduce the thickness. As the steel becomes thinner, it also becomes longer and moves through the rolls faster. Since different parts of the same piece of steel are traveling at different speeds through the different rolls, this process requires very close computer control of the speeds at each individual stand of rolls. By the time it reaches the end of the mill, the steel can be traveling at speeds approaching 64 km/h (40 mph). As the long strip of steel comes off the strip mill, it is coiled and allowed to cool. Hot rolled strip frequently goes through further stages of processing, such as cold rolling.

### 19.5.3 Long Product Mills

Long products are so called because they come off the mill as long bars of steel. They are produced in a vast range of shapes and sizes, with cross sections shaped like an "H," "I," "U," or a "T." Bars can have cross sections in the shape of squares, rectangles, circles, hexagons, and angles. Rod is coiled and drawn into wire or fabricated into products used to reinforce concrete buildings. Other types of long product include railway rails and piling.

Blooms and billets are used to make long products. After leaving the roughing stand, the steel passes through a succession of stands that reduces their size and changes their shape. In a universal mill, all faces are rolled at the same time. In other mills, only two sides of the steel are rolled at any one time, the piece of steel being turned over to allow the other two sides to be rolled.

## 19.6 Cold Rolling and Drawing

After hot rolling, many steel products are cold rolled. Although cold rolling does not alter the shape, it does reduce the thickness and work hardens the steel. During cold rolling, the previously hot rolled strip is uncoiled and passed through a series of rolling mill stands that progressively reduce its thickness. The cold rolled strip is then recoiled. Because the cold rolling process produces a hard sheet with little ductility, it is usually annealed by either continuous or batch annealing.

Batch annealing has traditionally been used to anneal coiled rolls of cold-worked steel. However, much faster continuous annealing lines are now being used. A comparison of heating profiles for the two processes is shown in Fig. 19.8. The most common method of annealing cold rolled sheet is by box annealing, in which coils of steel are stacked three or four high and placed under a cover (Fig. 19.9). They are then heated to approximately 590 to 760 °C (1100 to 1400 °F) in a reducing atmosphere to prevent decarburization. Very slow heat-up and cool-down rates are associated with batch annealing, resulting in a coarse pearlitic structure. Since almost all of the carbon in the steel is precipitated as pearlite, steel produced in this manner normally has a low susceptibility to strain aging. However, batch annealing requires several days.

A much faster process that is increasingly used is continuous annealing. The steel sheet is uncoiled and rapidly passed through high-temperature zones in continuous annealing furnaces. An integrated continuous cold rolling and annealing line is depicted in Fig. 19.10. Continuous annealing requires only minutes to recrystallize a section of sheet as it passes through the hot zone of a furnace. However, since the heat-up and cool-down rates are much faster, various cooling profiles, called overaging, must be used to remove carbon and nitrogen from solution and reduce the susceptibility to strain aging. Generally, the rapid anneal cycle of a continuous anneal process results in material properties that are less ductile than those resulting from a box anneal cycle. However, continuous annealing results in more uniformity of properties throughout the length of a coil.

Most cold rolled steels are subcritically annealed; that is, they are annealed below the $A_1$ temperature (Fig. 19.8). However, continuous annealing lines have made possible intercritical heating into the ferrite-austenite field, with cooling that is rapid enough to cause the austenite to transform to martensite. The martensite formation introduces a dislocation density exceeding that which can be pinned by the available carbon. As a result, early yielding is continuous and occurs with high rates of strain hardening. Intercritically annealed steels with ferrite-martensite microstructures are referred to as dual phase steels that have strength

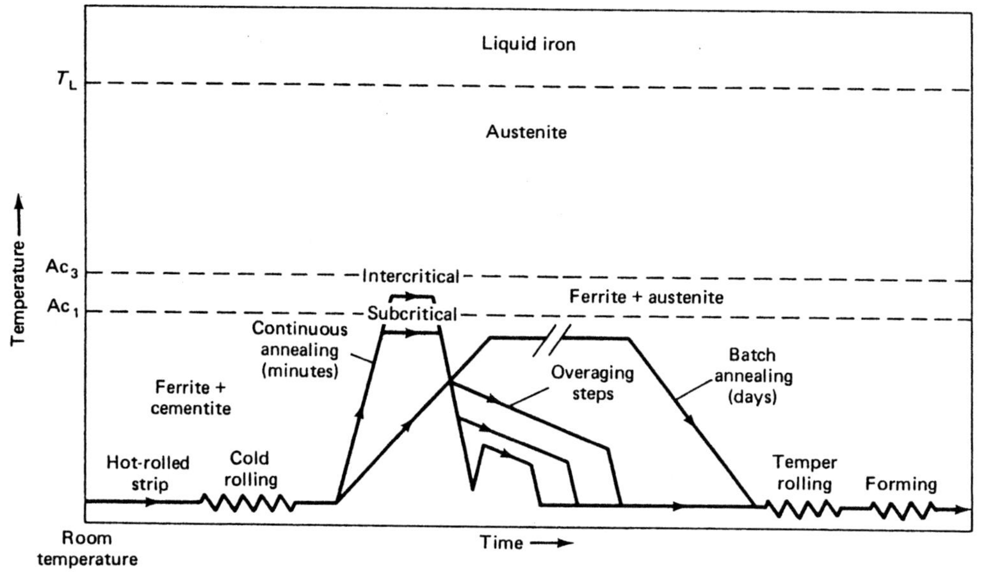


**Fig. 19.8** Batch and continuous annealing cycles. Source: Ref 7

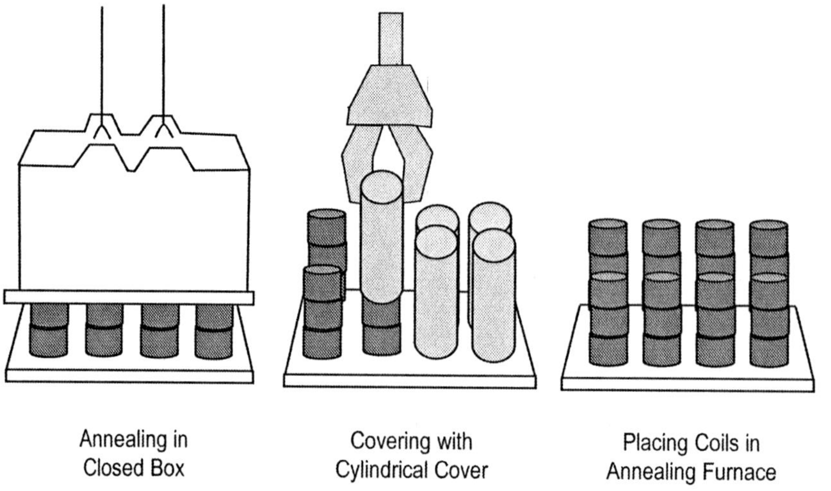


**Fig. 19.9** Three methods of batch annealing

levels in the 345 to 550 MPa (50 to 80 ksi) range.

Rimmed or capped ingot cast steel traditionally has been used because of its lower price. However, rimmed and capped steels are susceptible to quench and strain aging. Strain aging and quench aging can produce localized deformation and discontinuous yielding. This causes stretcher strains during forming and unacceptable surface finishes on formed parts. To eliminate stretcher strains, cold rolled and annealed sheet steels are temper rolled. Temper rolling introduces just enough strain so that the deformed sheet will strain uniformly or continuously. However, as explained in section 3.5 of Chapter 3, "Solid Solutions," the yield point will reappear with time. Where strain aging is to be avoided and/or when exceptional formability is required, steel killed with aluminum is preferred, because nitrogen, a primary contributor to strain aging, is tied up as inert AlN particles. In addition, all continuously cast steel must be fully killed to allow casting.

Another form of cold working is cold drawing. Steel rod is drawn through a series of dies that progressively reduce the circumference to produce wire. The drawing process substantially increases the tensile strength. Steel wires can be spun into huge ropes strong enough to support large suspension bridges.

## 19.7 Classification and Specifications for Steels

Plain carbon steels are classified by several different systems, depending on the type of steel and its application. In addition, there are literally thousands of different types of steels, making it difficult to classify them in a simple, straightforward manner. A classification system, originally developed by the Society of Automotive Engineers (SAE) and later refined in conjunction with the American Iron and Steel Institute (AISI), is frequently used. Recently, a unified numbering system was established that incorporates the SAE/AISI number.

Many steel products are purchased by specifications describing specific compositional, dimensional, and property requirements. Specification organizations, such as ASTM Internationl, have developed numerous specifications for steel products and the testing of steel products. In addition to specifying chemical composition, ASTM standards also set mechanical property limits and often specify fabrication procedures and heat treatments. Some specific product-user groups in the United States have developed their own specifications: the American Bureau of Ships for ship plate and other marine products, Aerospace Materials Specifications for aerospace applications, the American Railway Engineering and Maintenance of Way Association for rail and rail products, the Society of Automotive Engineers for automotive applications, and the American Society of Mechanical Engineers for steels produced for pressure vessels.

## 19.8 Plain Carbon Steels

A plain carbon steel is essentially an alloy of iron and carbon that also contains manganese

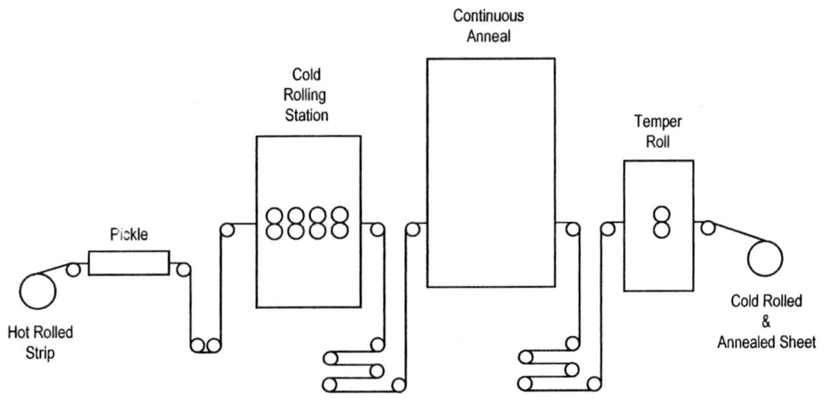


**Fig. 19.10** Integrated cold rolling and annealing line

and a variety of residual elements, in particular sulfur, phosphorus, and silicon. The residual elements were either present in the raw materials (e.g., iron ore and scrap steel additions) or they were added during the production process for a specific purpose, such as silicon additions for deoxidization. Hence, they are called residual elements to distinguish them from alloying elements that are deliberately added according to specified minimum amounts. The AISI has defined a plain carbon steel as an alloy of iron and carbon that contains 1.65 wt% maximum manganese, less than 0.6 wt% Si, less than 0.6 wt% Cu, and which does not have any other deliberately added alloying elements. It is also usual for maximum amounts (i.e., 0.05 wt%) of sulfur and phosphorus to be specified.

In the SAE/AISI system, the carbon steels are classified as follows:

- Nonresulfurized carbon steels, 10*xx* series
- Resulfurized steels, 11*xx* series
- Rephosphorized and resulfurized steels, 12*xx* series
- High manganese carbon steels, 15*xx* series

The combined SAE/AISI numbering system uses a numerical code beginning with either 10*xx* or 10*xxx* for plain carbon steels. The newer unified numbering system uses the designation G 10*xxx*. In both systems, the 10 indicates that the steel is a plain carbon steel, while the subsequent *xxx*'s specify the nominal carbon content in points of carbon (1 point = 0.01% C). For example, 1045 steel would nominally contain 0.45% C, which actually varies between 0.43 and 0.50%. Typically, compositions run from 1005 to 1095, but the designation 10125 would indicate a plain carbon steel containing 1.25% C. Because of the wide range in carbon content, the 10*xx* carbon steels are the most commonly used steels. The mechanical properties of 10*xx* hot rolled and cold drawn steels are given in Table 19.1.

The principal factors affecting the properties of the plain carbon steels are carbon content and microstructure, with the microstructure being determined largely by the composition and the final rolling, forging, or heat treating operations. As the carbon content increases, the strength and hardness increase while the ductility and toughness decrease. The influence of carbon content on these properties is shown in Fig. 19.11, along with some of the usages of steels with different carbon contents. Unfortunately, as the carbon content increases, the ductile-to-brittle transition temperature also increases (Fig. 19.12), which means that brittle fractures will occur at higher temperatures. As shown in Fig. 19.13, the form of carbon in the steel is also important: fine pearlite caused by more rapid cooling rates

**Table 19.1 Typical properties of plain carbon steels**

| Steel | Condition | Ultimate tensile strength | | Yield strength | | Elongation in 2 in., % | Reduction in area, % | Hardness, HB |
|---|---|---|---|---|---|---|---|---|
| | | MPa | ksi | MPa | ksi | | | |
| 1010 | Hot rolled | 325 | 47 | 180 | 26 | 28 | 50 | 95 |
| | Cold drawn | 365 | 53 | 305 | 44 | 20 | 40 | 105 |
| 1020 | Hot rolled | 380 | 55 | 205 | 30 | 25 | 50 | 111 |
| | Cold drawn | 420 | 61 | 350 | 51 | 15 | 40 | 121 |
| 1025 | Hot rolled | 400 | 58 | 220 | 32 | 25 | 50 | 116 |
| | Cold drawn | 440 | 64 | 370 | 54 | 15 | 40 | 126 |
| 1030 | Hot rolled | 470 | 68 | 260 | 37.5 | 20 | 42 | 137 |
| | Cold drawn | 525 | 76 | 440 | 64 | 12 | 35 | 149 |
| 1035 | Hot rolled | 495 | 72 | 270 | 39.5 | 18 | 40 | 143 |
| | Cold drawn | 550 | 80 | 460 | 67 | 12 | 35 | 163 |
| 1040 | Hot rolled | 525 | 76 | 290 | 42 | 18 | 40 | 149 |
| | Cold drawn | 585 | 85 | 490 | 71 | 12 | 35 | 170 |
| 1050 | Hot rolled | 620 | 90 | 340 | 49.5 | 15 | 35 | 179 |
| | Cold drawn | 690 | 100 | 580 | 84 | 10 | 30 | 197 |
| | Annealed, cold drawn | 655 | 95 | 550 | 80 | 10 | 40 | 189 |
| 1060 | Hot rolled | 675 | 98 | 370 | 54 | 12 | 30 | 201 |
| | Spheroidized annealed, cold drawn | 620 | 90 | 485 | 70 | 10 | 45 | 183 |
| 1070 | Hot rolled | 705 | 102 | 385 | 56 | 12 | 30 | 212 |
| | Spheroidized annealed, cold drawn | 640 | 93 | 495 | 72 | 10 | 45 | 192 |
| 1080 | Hot rolled | 770 | 112 | 425 | 61.5 | 10 | 25 | 229 |
| | Spheroidized annealed, cold drawn | 675 | 98 | 515 | 75 | 10 | 40 | 192 |
| 1090 | Hot rolled | 840 | 122 | 460 | 67 | 10 | 25 | 248 |
| | Spheroidized annealed, cold drawn | 695 | 101 | 540 | 78 | 10 | 40 | 197 |
| 1095 | Hot rolled | 825 | 120 | 455 | 66 | 10 | 25 | 248 |
| | Spheroidized annealed, cold drawn | 680 | 99 | 525 | 76 | 10 | 40 | 197 |

Source: Ref 8

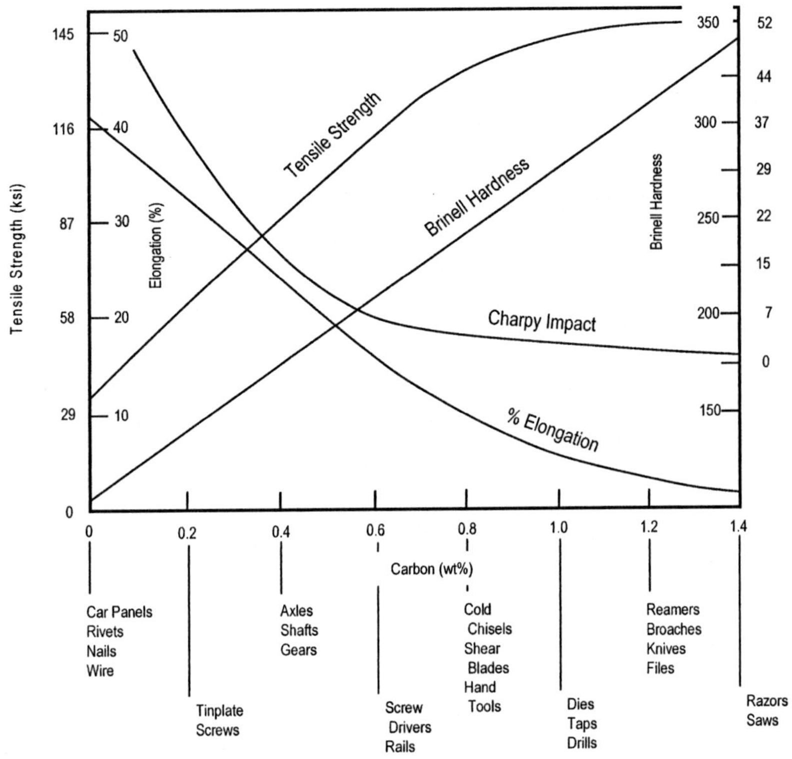


**Fig. 19.11** Properties of cold rolled plain carbon steel

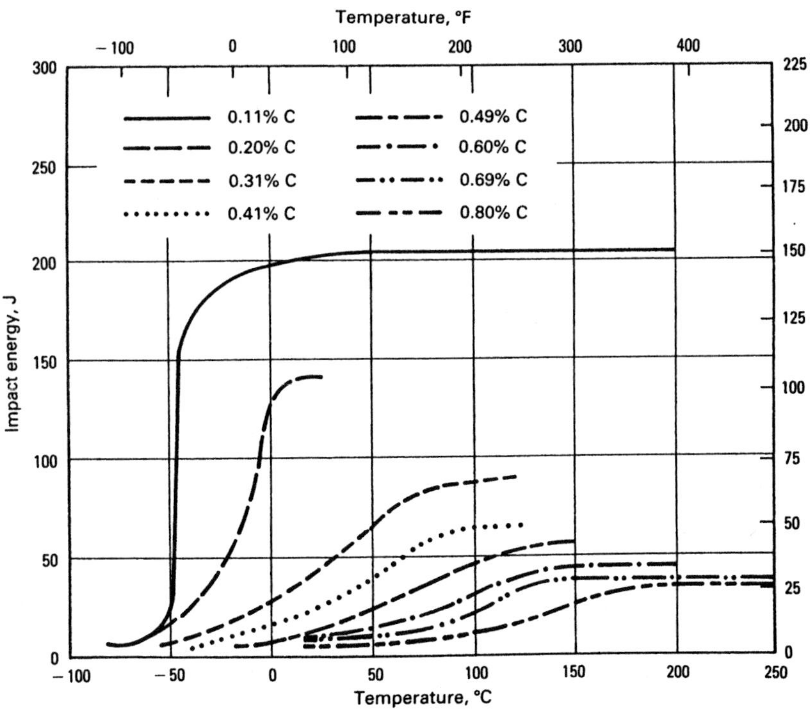


**Fig. 19.12** Ductile-to-brittle transitions for carbon steels. Source: Ref 8

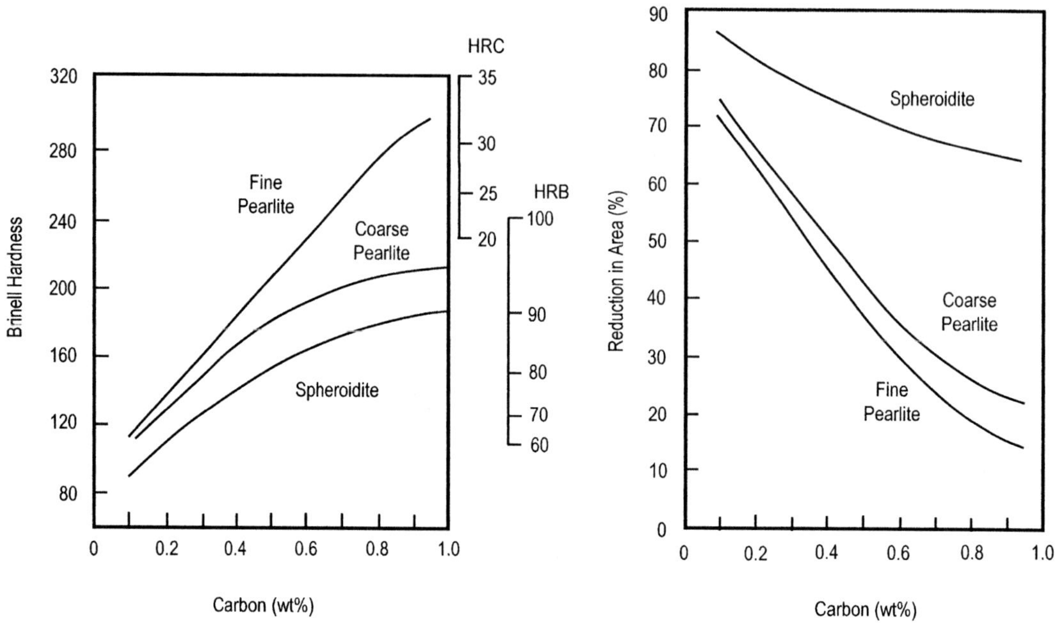


**Fig. 19.13** Hardness and ductility as a function of carbon content. Source: Ref 9

(e.g., normalizing) will have higher strength but lower ductility than slowly cooled coarse pearlite (annealing). A typical microstructure of a low-carbon steel is shown in Fig. 19.14, with pearlite (dark) being dispersed in a matrix of ferrite (light). Of course, as the carbon content increases toward the eutectoid composition (0.77 wt% C), the amount of pearlite in the microstructure becomes greater. The softest structures are spheroidized structures that contain discrete carbide particles embedded in a ferrite matrix.

All 10*xx*-series carbon steels contain manganese at levels between 0.25 and 1.00 wt%. Manganese is added to counteract the effects of sulfur, which is always present as an impurity. Sulfur reacts with iron to form a continuous network of iron sulfide along the grain boundaries. Since iron sulfide melts at a relatively low temperature, hot shortness or grain-boundary melting can occur during hot working of a cast ingot, causing cracking. If manganese is present, it reacts preferentially with the sulfur to form much higher-melting-point manganese sulfide (MnS) particles (Fig. 19.15), thereby avoiding cracking during hot working. However, the MnS particles tend to elongate and form a banded structure during hot working that reduces the fracture resistance in the through-the-thickness direction. Splitting of steels parallel to the rolling direction can sometimes result from elongated MnS particles. The ductility in the transverse and through-the-thickness directions is much less than in the longitudinal direction. Cerium can be substituted for manganese to form sulfides that are much more resistant to elongation during rolling. However, with the development of steelmaking practices that

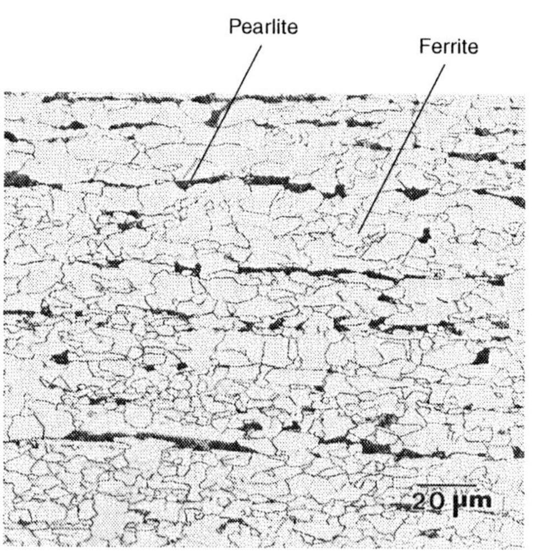


**Fig. 19.14** Microstructure of low-carbon steel. Source: Ref 10

produce very low-sulfur steel, manganese is becoming less important in this role.

The 11*xx* series of resulfurized steels contain between 0.08 and 0.33 wt% S. Sulfur is usually considered an undesirable impurity and is restricted to less than 0.05 wt%. However, in the 11*xx* and 12*xx* series of steels, sulfur is intentionally added to form excess manganese sulfide inclusions. These are the free-machining steels that have improved machinability over lower-sulfur steels due to enhanced chip breaking and lubrication created by the MnS inclusions.

The 12*xx* series are also free-machining steels and contain both sulfur (0.16 to 0.35 wt%) and phosphorus (0.04 to 0.12 wt%). The 13*xx* steels contain more than 1wt% Mn and also have a minimum silicon content of 0.15 to 0.35 wt%. The 15*xx* series contain higher manganese levels (up to 1.65 wt%) than the 10*xx* series of carbon steels. Finally, some steels also have lead additions for improved machinability, and an "L" is included in the designation (e.g., 12L14). Typically, lead is added in amounts between 0.15 and 0.35 wt%.

## 19.9 Low-Carbon Steels

Low-carbon steels, those containing less than 0.30 wt% C, constitute the highest tonnage of steels produced. Uses range from structural shapes and beams for buildings and bridges, and plate for pipelines, to automotive sheet applications. Requirements are good formability, weldability, strength, and fracture resistance. The recent trend in steel metallurgy is toward higher-strength steels with lower carbon contents for improved formability, weldability, and fracture resistance.

A typical microstructure of a hot rolled low-carbon steel contains ferrite as the major microconstituent. Since ferrite is relatively low in strength, alloying elements and different processing approaches are used to improve the strength. Some steels are designed for maximum formability with only moderate strength, while others have higher strength but less formability. Alloying and processing can be used to refine the ferrite grain size, which improves the strength, formability, and fracture resistance. For improved fracture resistance, the carbon contents, and hence the amount of pearlite, along with reduced impurity levels are used to improve the fracture resistance.

Low-carbon steels can be further subdivided into two groups: (1) steels containing less than approximately 0.15 wt% C that are used primarily for cold forming and drawing applications, and (2) structural steels containing carbon in the range of 0.15 to 0.30 wt%.

### *19.9.1 Low-Carbon Sheet Steels*

Low-carbon steels are used for sheet forming applications, such as in the automotive and appliance industries. Low-carbon sheet steels are supplied in both hot rolled and cold rolled conditions. Since the surfaces of hot rolled steels are rougher than cold rolled steels, hot rolled steels are limited to applications where surface appearance is not important. For surface-appearance-critical applications, cold rolled steel that has been annealed is normally used so that it can be more easily formed. If the steel is

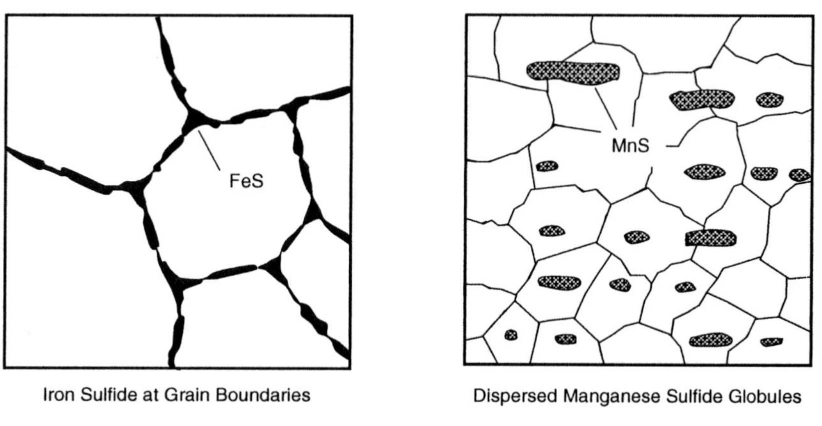


**Fig. 19.15** Manganese sulfides in steel. Source: Ref 6

going to be used for deep drawing or forming, coils of cold rolled sheet are softened by annealing.

The important characteristics of sheet forming steels are surface finish, strain hardening, strain-rate sensitivity, anisotropy, freedom from yield point, and yield strength.

Strain hardening is expressed by the Holloman equation:

$$\sigma = K\varepsilon^n \qquad \text{(Eq 19.3)}$$

where $\sigma$ is the stress, $K$ is the strength coefficient, $\varepsilon$ is the true strain, and $n$ is the strain-hardening coefficient as determined during a tensile test. A high strain-hardening coefficient, $n$, is desirable since it indicates high stretchability during forming.

Strain-rate sensitivity is expressed by the equation:

$$\sigma = C\dot{\varepsilon}^m \qquad \text{(Eq 19.4)}$$

where $C$ is the strength coefficient, $\dot{\varepsilon}$ is the strain rate, and $m$ is the strain-rate sensitivity. A high strain-rate sensitivity, $m$, is also desirable. In steels with a high $m$, necking appears less rapidly during forming.

Anisotropy can be expressed by the plastic strain ratio, $r$:

$$r = \varepsilon_w / \varepsilon_t \qquad \text{(Eq 19.5)}$$

where $\varepsilon_w$ is the tensile strain in the width direction, and $\varepsilon_t$ is the tensile strain in the thickness direction.

Since plastic strain ratios are a function of orientation in the sheet, the average plastic strain ratio, $r_m$, is often used:

$$r_m = \frac{1}{4}(r_0 + 2r_{45} + r_{90}) \qquad \text{(Eq 19.6)}$$

where $r_0$ is the $r$-value parallel to the rolling direction, $r_{45}$ is the $r$-value at 45° to the rolling direction, and $r_{90}$ is the $r$-value transverse to the rolling direction. In general, high plastic strain ratios $r$ and $r_m$ imply increased drawability and less wrinkling. High $r$-value steels have excellent deep-drawing ability and can form difficult parts.

The final important characteristic is yield strength. The trend in industry is to use high-yield-strength steels so that the sheet gage can be reduced.

**Interstitial-Free Steels.** There are a number of steels that are produced with very low carbon levels (less than 0.002% C), and all the remaining free carbon in the steel is tied up as carbides. A typical ferrite microstructure is shown in Fig. 19.16. These steels are known as interstitial-free (IF) steels, which means that the interstitial elements of carbon and nitrogen are no longer present in elemental form in the iron lattice but are combined with elements such as titanium or niobium as carbides and nitrides (carbonitrides). These steels have excellent formability, as evidenced by the high $r_m$-values shown in Fig. 19.17. However, since the carbon contents are extremely low, they are not very strong, with yield strengths of only 138 to 179 MPa (20 to 26 ksi) and tensile strengths of 290 to 338 MPa (42 to 49 ksi). Small amounts of phosphorus can be added to strengthen the ferrite, but there is the problem of phosphorus segregating to the grain boundaries and causing brittleness. To reduce this segregation problem, small amounts of boron are sometimes added.

**Deep-Quality Special-Killed Steels.** Another type of low-carbon steel sheet forming steel is a special class called deep-quality special-killed (DQSK) steel. This type of aluminum-treated steel also has a preferred orientation and high $r$-value. The preferred orientation is produced by hot rolling the steel on a hot strip mill, followed by rapid cooling. The rapid cooling keeps the aluminum and interstitial elements from forming aluminum nitride (AlN) particles (i.e., the aluminum and nitrogen atoms are in solid solution in the iron lattice).

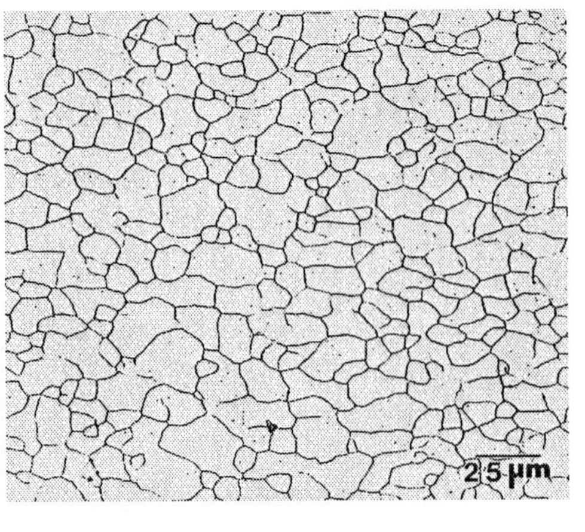


**Fig. 19.16** Microstructure of interstitial-free steel. Source: Ref 10

After rolling, the steel is annealed to allow aluminum nitride to precipitate. The aluminum nitride plays an important role in the development of the optimal crystallographic texture. The DQSK steel is used in deep-drawing applications that are not as demanding as those requiring IF steel.

**Bake Hardening and Enameling Steels.** Bake hardening steels also have a low but controlled carbon content. These steels gain strength during the paint bake cycle used during automotive production. Controlled amounts of both carbon and nitrogen combine with carbonitride-forming elements such as titanium and niobium during the baking cycle (generally 175 °C, or 350 °F, for 30 min). The precipitation of these carbonitrides during the paint bake cycle strengthens the steel by an aging process.

**Enameling steels** are produced with as little carbon as possible because, during the enameling process, carbon in the form of carbides can react with the frit (the particles of glasslike material that melt to produce the enamel coating) to cause defects in the coating. Therefore, steels used for enameling are generally decarburized in a special reducing atmosphere during batch annealing. After decarburization, the sheet steel is essentially pure iron. Enamel coatings are used for many household appliances, such as washers and dryers, stovetops, ovens, and refrigerators. Also, steel tanks in most hot water heaters have a glass (or enameled) interior coating.

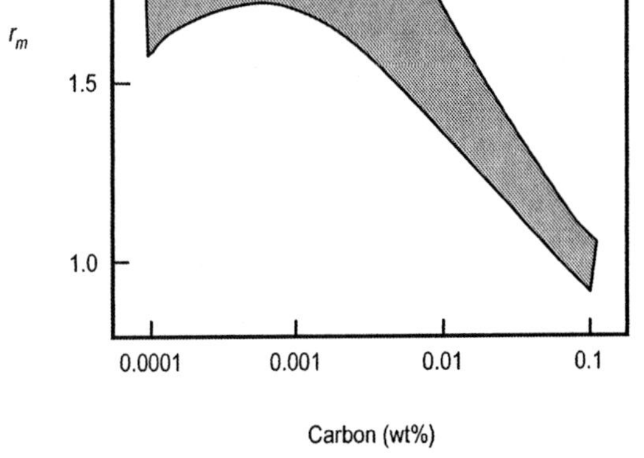


**Fig. 19.17** Effect of carbon content on plastic strain ratio ($r_m$) values. Source: Ref 11

### *19.9.2 Low-Carbon Structural Steels*

Steels containing 0.15 to 0.30 wt% C and less than 0.75 wt% Mn are commonly referred to as mild steels. These steels have increased strength and hardness and reduced formability compared to the lower-carbon sheet forming steels. They are not deliberately strengthened by alloying elements other than carbon and contain some manganese for sulfur stabilization and silicon for deoxidation. Mild steels are mostly used in the as-rolled, forged, or annealed condition and are seldom quenched and tempered. A number of these steels are used for carburizing. An increase in carbon content results in a higher core hardness when carburized, while an increase in the manganese content results in a hardness increase in both the case and core.

A relatively new class of steels with very small amounts of alloying elements, referred to as high-strength low-alloy (HSLA), has replaced low-carbon structural steels in many applications. Since these are technically alloy steels, they are covered in Chapter 20, "Alloy Steels," in this book. Before the advent of HSLA steels, mild steels were commonly used for the structural parts of automobiles, bridges, and buildings. However, as lighter-weight automobiles became desirable during the energy crisis, there was a trend to reduce weight by using higher-strength steels with suitable ductility for forming operations. The trend for structural steels used in the construction of bridges and buildings has also been away from mild steels and toward HSLA steels. The HSLA steels are a superior substitute for mild steels because HSLA steels provide higher yield strengths without adverse effects on weldability. Weathering HSLA steels also provide better atmospheric corrosion resistance than carbon steel.

## 19.10 Medium-Carbon Plain Carbon Steels

Steels with carbon contents of 0.30 to 0.60 wt% are selected for applications where higher mechanical properties are required. They are frequently hardened by quench and temper heat treatments or by cold working. The properties of several medium- and high-carbon steels hardened by quenching and tempering are given in Table 19.2. Selection of a specific carbon and manganese content is governed by a number of factors. Increases in mechanical properties, section thickness, or depth of hardening normally require higher carbon, higher manganese, or both. These steels can also be hardened by flame or induction heat treatment.

Excellent properties can be obtained by quenching and tempering heat treatments. However, plain carbon steels have serious limitations compared to alloy steels. The time-temperature transformation diagrams for plain carbon steels show that the formation of martensite requires very rapid cooling rates. These cooling rates are so fast that it is not possible to transform large parts into martensite. Thus, the excellent properties provided by tempered martensite cannot be obtained in thick components made of plain carbon steel simply because they cannot be cooled fast enough. Very severe quenching rates are necessary, and even then, only small sections can be converted to martensite throughout. Thus, plain carbon steels have low hardenability. Such rapid quench rates are undesirable because there is no opportunity for stress relief during cooling; that is, warping and distortion are more likely to occur during rapid quenching. These stresses, superimposed on the transformation stresses caused by the volume expansion that occurs when martensite forms, can cause quench cracking.

Steels in this group with lower carbon and manganese contents can be cold formed, provided that the bends are not too severe. After cold forming, the steel can be annealed, normalized, or quenched and tempered, depending on the application. Higher-carbon grades are cold drawn to specific strength levels with no further heat treatment. They can be used with or without heat treatment, depending on the strength level required. All of these steels can be used for forgings and also for parts machined from bar stock. Although it is possible to weld these steels, they are not as weldable as the lower-carbon grades, and pre- and postheating is often required to prevent weld cracking.

## 19.11 High-Carbon Plain Carbon Steels

Steels with carbon contents in the range of 0.60 to 1.00 wt% are used for applications where either higher strengths or greater wear resistances are required. In general, cold forming is not used, and weldability is poor. Practically all high-carbon steels are hardened by quenching and tempering. The ultimate tensile strengths for these steels range from

**Table 19.2 Mechanical properties of quenched and tempered plain carbon steels**

| AISI No. | Tempering temperature | | Ultimate tensile strength | | Yield strength | | Elongation, % | Reduction in area, % | Hardness, HB |
|---|---|---|---|---|---|---|---|---|---|
| | °C | °F | MPa | ksi | MPa | ksi | | | |
| 1040 | 205 | 400 | 779 | 113 | 593 | 86 | 19 | 48 | 262 |
| | 315 | 600 | 779 | 113 | 593 | 86 | 20 | 53 | 255 |
| | 425 | 800 | 758 | 110 | 552 | 80 | 21 | 54 | 241 |
| | 540 | 1000 | 717 | 104 | 490 | 71 | 26 | 57 | 212 |
| | 650 | 1200 | 634 | 92 | 434 | 63 | 29 | 65 | 192 |
| 1050 | 205 | 400 | ... | ... | ... | ... | ... | ... | ... |
| | 315 | 600 | 979 | 142 | 724 | 105 | 14 | 47 | 321 |
| | 425 | 800 | 938 | 136 | 655 | 95 | 20 | 50 | 277 |
| | 540 | 1000 | 876 | 127 | 579 | 84 | 23 | 53 | 262 |
| | 650 | 1200 | 738 | 107 | 469 | 68 | 29 | 60 | 223 |
| 1060 | 205 | 400 | 1103 | 160 | 779 | 113 | 13 | 40 | 321 |
| | 315 | 600 | 1103 | 160 | 779 | 113 | 13 | 40 | 321 |
| | 425 | 800 | 1076 | 156 | 765 | 111 | 14 | 41 | 311 |
| | 540 | 1000 | 965 | 140 | 669 | 97 | 17 | 45 | 277 |
| | 650 | 1200 | 800 | 116 | 524 | 76 | 23 | 54 | 229 |
| 1080 | 205 | 400 | 1310 | 190 | 979 | 142 | 12 | 35 | 388 |
| | 315 | 600 | 1303 | 189 | 979 | 142 | 12 | 35 | 388 |
| | 425 | 800 | 1289 | 187 | 951 | 138 | 13 | 36 | 375 |
| | 540 | 1000 | 1131 | 164 | 807 | 117 | 16 | 40 | 321 |
| | 650 | 1200 | 889 | 129 | 600 | 87 | 21 | 50 | 255 |
| 1095 | 205 | 400 | 1289 | 187 | 827 | 120 | 10 | 30 | 401 |
| | 315 | 600 | 1262 | 183 | 813 | 118 | 10 | 30 | 375 |
| | 425 | 800 | 1213 | 176 | 772 | 112 | 12 | 32 | 363 |
| | 540 | 1000 | 1089 | 158 | 676 | 98 | 15 | 37 | 321 |
| | 650 | 1200 | 896 | 130 | 552 | 80 | 21 | 47 | 269 |

Oil quench. Source: Ref 12

620 to 869 MPa (90 to 126 ksi), while their elongations range from 9 to 25%. In most cases, high-carbon steels are heat treated by oil quenching and tempering. Water quenching is used for heavier sections or when cutting edges are required. Because of their good wear resistance when heat treated, they are often used in applications requiring wear and abrasion resistance.

## 19.12 Corrosion of Iron and Steel

Irons and steels corrode in the presence of moisture or water. Factors that accelerate attack include increases in the velocity or the acidity of the water, increases in the relative motion or agitation between the metal and the water, increases in temperature or aeration, and the presence of certain forms of bacteria. Note that water must contain oxygen to be corrosive, unless it is acidic or contains anaerobic bacteria. If oxygen-free water is maintained at a neutral pH or at a slight alkalinity, it is practically noncorrosive to steels. Steam boilers and water supply systems can be protected by deaerating the water.

Soils containing dispersed metallic particles or pockets of bacteria can provide a natural electrical pathway for buried steel. If the soil has an electrolyte such as moisture, and the soil has a negative charge, an electrical circuit exists and corrosion can occur. Differences in soil conditions, such as moisture content, are commonly responsible for creating anodic and cathodic areas. If the oxygen concentration of the soil is different in different areas, cathodic areas develop at locations of relatively high oxygen concentration and anodes at areas of low concentration.

In an acidic environment, even without the presence of oxygen, the metal becomes the anode and is rapidly attacked. Hydrogen gas is released at the cathode. If a salt forms on the surface during corrosion, it can actually slow the rate of attack because the salt, rather than the underlying metal, is preferentially attacked. Environments that contain chlorine or other halogens are particularly aggressive.

The oxide scale formed on steel during hot rolling varies with the operation performed and the rolling temperature. The dissimilarity of steel and the mill scale formed during hot rolling can cause corrosion to occur. Unfortunately, mill scale is cathodic to steel, and an electric current can easily be produced between the steel, and the mill scale. This electrochemical action will corrode the steel without affecting the mill scale.

## 19.13 Corrosion-Resistant Coatings

Since plain carbon steels exposed to air will rust, they must be coated, for example, with paint, for protection. Steel makers often improve the corrosion resistance by coating at the factory prior to delivery. A wide range of different coatings is available.

**Zinc coating,** or galvanizing, is when the zinc is applied either electrolytically, which provides a thinner coating, or by hot dipping the steel in a bath of molten zinc. An advantage of zinc coating is that small breaks in the coating do not result in corrosion. Much of the sheet used to produce car bodies is zinc coated, which allows thinner steels to be used, thus saving weight and improving fuel consumption. Wire is also frequently galvanized for corrosion protection.

**Tinplate** is thin steel sheet with a thin coating of tin applied. A disadvantage of tin plating is that any breaks in the coating will allow the steel to rust.

**Organic coatings,** such as paint and baked-on enamels, are applied for corrosion protection, while, at the same time, they provide an attractive appearance. Frequently, a combination of galvanizing and organic coatings is used.

### ACKNOWLEDGMENTS

Sections of this chapter were adapted from "Steel Making Practices and Their Influence on Properties" by B. Mishra and "Classifications and Designations of Carbon and Alloy Steels" in *Metals Handbook Desk Edition,* 2nd ed., ASM International, 1998, and "Sheet Formability of Steels" by W.G. Granzow in *Properties and Selections: Irons, Steels, and High-Performance Alloys,* Volume 1, *ASM Handbook,* ASM International, 1990.

### REFERENCES

1. B. Mishra, Steel Making Practices and Their Influence on Properties, *Metals Handbook Desk Edition,* 2nd ed., ASM International, 1998

2. Classifications and Designations of Carbon and Alloy Steels, *Metals Handbook Desk Edition,* 2nd ed., ASM International, 1998
3. W.G. Granzow, Sheet Formability of Steels, *Properties and Selection: Irons, Steels, and High-Performance Alloys,* Vol 1, *ASM Handbook,* ASM International, 1990
4. H.E. McGannon, *The Making, Shaping, and Treating of Steel,* 9th ed., United States Steel Corporation, 1971
5. E.E. Thum, Ed., *Modern Steels: Manufacture, Inspection, Treatment, and Uses,* American Society for Metals, 1939
6. R.A. Higgins, *Engineering Metallurgy—Applied Physical Metallurgy,* 6th ed., Arnold, 1993
7. G. Krauss, Microstructures, Processing, and Properties of Steels, *Properties and Selection: Irons, Steels, and High-Performance Alloys,* Vol 1, *ASM Handbook,* ASM International, 1990
8. B.L. Bramfitt, Structure/Property Relationships in Irons and Steels, *Metals Handbook Desk Edition,* 2nd ed., ASM International, 1998
9. W.D. Callister, *Fundamentals of Materials Science and Engineering,* 5th ed., John Wiley & Sons, Inc., 2001
10. A.O. Benscoter and B.L. Bramfitt, Metallography and Microstructures of Low-Carbon and Coated Steels, *Metallography and Microstructures,* Vol 9, *ASM Handbook,* ASM International, 2004
11. W.B. Hutchinson, K.I. Nilsson, and J. Hirsch, Annealing Textures in Ultra-Low Carbon Steels, *Metallurgy of Vacuum-Degassed Steel Products,* TMS, 1990, p 109–125
12. Mechanical Properties of Carbon and Alloy Steels, *Metals Handbook Desk Edition,* 2nd ed., ASM International, 1998


## SELECTED REFERENCES


- B.L. Bramfitt, Carbon and Alloy Steels, *Handbook of Materials Selection,* John Wiley & Sons, Inc., 2002
- B.L. Bramfitt and S.J. Lawrence, Metallography and Microstructures of Carbon and Low-Alloy Steels, *Metallography and Microstructures,* Vol 9, *ASM Handbook,* ASM International, 2004
- R.M. Brick, A.W. Pense, and R.B. Gordon, *Structure and Properties of Engineering Materials,* 4th ed., McGraw-Hill Book Company, 1977
- High-Strength Structural and High-Strength Low-Alloy Steels, *Properties and Selection: Irons, Steels, and High-Performance Alloys,* Vol 1, *ASM Handbook,* ASM International, 1990
- W.F. Hosford, *Physical Metallurgy,* Taylor & Francis, 2005
- G. Krauss, *Steels: Processing, Structure, and Performance,* 3rd ed., ASM International, 2005
- W.F. Smith, *Principles of Materials Science and Engineering,* McGraw-Hill Book Company, 1986
- W.F. Smith, *Structure and Properties of Engineering Alloys,* 2nd ed., McGraw-Hill Inc., 1993

## CHAPTER 20

# Alloy Steels

ALTHOUGH PLAIN CARBON STEELS are widely used, they are not adequate for all engineering applications because of the following limitations:

- They cannot be strengthened beyond approximately 690 MPa (100 ksi) without a significant loss in toughness and ductility.
- They are not hardenable to great depths, thus limiting the maximum cross section that can be through hardened.
- Severe quenches, such as brine or water, are often required to produce the desired hardness, which greatly increases the susceptibility to distortion and cracking.
- Plain carbon steels have poor corrosion resistance, and they readily oxidize at elevated temperatures.

Alloy steels are alloys of iron with the addition of carbon and one or more of the following elements: manganese, chromium, nickel, molybdenum, niobium, titanium, tungsten, cobalt, copper, vanadium, silicon, aluminum, and boron. Alloy steels exhibit superior mechanical properties compared to plain carbon steels as a result of alloying additions. For information on the effects of alloying elements on steel microstructure, see Chapter 10, "The Iron-Carbon System," in this book. The details of alloy steel heat treatment and hardenability are covered in Chapter 11, "Heat Treatment of Steel."

## 20.1 Effects of Alloying Elements

Alloying elements are added to steels for a variety of reasons, including:

- Deeper hardening of quenched and tempered steels for a given carbon content, resulting in higher strength and improved mechanical properties. The deeper hardening is possible because the alloying elements shift the nose of the time-temperature transformation (TTT) diagram to the right.
- Higher tempering temperatures are possible while maintaining high strength and good ductility.
- Improved mechanical properties at high and low temperatures.
- Greater corrosion and elevated-temperature oxidation resistance.
- Formation of hard carbides to provide greater wear resistance. Alloying elements that are strong carbide formers, such as titanium or vanadium, help in preventing grain growth at elevated temperatures by pinning the grain boundaries.

A list of important alloying elements and their effects are given in Table 20.1. The most important reason for adding alloying elements to steel is to increase hardenability. The most effective alloying elements for increasing hardenability are manganese, chromium, molybdenum, and nickel. Generally, the higher the alloy content, the greater the hardenability, and the higher the carbon content, the greater the strength. By increasing the depth of hardening, larger sections can be through hardened, providing the strength and toughness advantage of tempered martensite. Also, by increasing the depth of hardening, a slower quench rate can be used, which reduces cooling stresses. Oil or air quenching reduces thermal gradients that can lead to distortion or cracking.

Carbide formers, such as molybdenum and chromium, also retard tempering. By increasing the resistance to softening during tempering, alloy steels are able to resist softening at elevated temperatures. Thus, a lower-carbon-content alloy steel can be used to obtain the same tempered hardness as a higher-carbon-content plain carbon steel. A steel with a lower carbon content is usually tougher than one with a higher carbon content. In addition, the ability to temper at higher temperatures allows greater relaxation of stresses while maintaining the same hardness.

Other elements that are added to standard alloy steels include silicon, aluminum, vanadium, lead, and boron. As for plain carbon steels, silicon and aluminum are used for deoxidation during the steelmaking process. Vanadium is a very strong carbide former and forms fine carbides that, along with the aluminum compounds in killed steels, inhibit grain growth during austenitization. Vanadium carbides are effective because they are stable at temperatures up to approximately 1095 °C (2000 °F), which is much higher than typical austenitization temperatures used for standard low-alloy steels. Leaded steels contain 0.15 to 0.35 wt% Pb for improved machinability; however, lead is no longer favored as an alloying addition because of health concerns. Very small amounts (0.0005 to 0.003 wt%) of boron can dramatically increase hardenability. When boron is present in a standard alloy steel, the code designation is amended by inserting the letter "B" between the second and third digit, for example, 50B44.

The distribution of alloying elements can be rather complex, depending on the amount of various alloying elements present and how they interact with each other. However, there are some general trends in the way in which alloying elements are distributed in the steel (Table 20.2).

**Table 20.1 Effects of alloying elements in steel**

| Alloying element | Purpose |
|---|---|
| Manganese | Used in all carbon and alloy steels. Combines with embrittling sulfur to form manganese sulfide (MnS) at a minimum manganese-to-sulfur ratio of 20 : 1. Provides substitutional hardening. Manganese contributes markedly to hardenability, especially in amounts greater than 0.8 wt%. |
| Nickel | Provides substitutional hardening. Strong austenite stabilizer and forms basis of austenitic stainless steel. Improves toughness in low-alloy steels |
| Molybdenum | Very potent substitutional hardener. Forms carbides for good wear resistance. Carbides delay softening during tempering. Added to minimize temper embrittlement. Enhances creep resistance of low-alloy steels at elevated temperatures |
| Chromium | Substitutional hardener. Forms carbides for good wear resistance. Retards softening during tempering. Enhances corrosion resistance and forms the basis for stainless steels |
| Silicon | Primary purpose is as a deoxidizer. Stabilizes ferrite. Retards formation of cementite during tempering. Used in transformer steels that have high magnetic permeability and low core loss |
| Copper | Copper is usually added to alloy steels for its contribution to atmospheric corrosion resistance and at higher levels for precipitation hardening. |
| Aluminum | Deoxidizer that removes oxygen and reduces porosity in castings. Forms AlN precipitates that provide optimal crystallographic texture in deep-drawing steels |
| Tungsten | Used primarily in high-speed tool steels where it forms hard, wear-resistant carbides |
| Vanadium | Forms extremely hard VC and VN carbides. Used in tool steels for wear resistance. High-strength low-alloy (HSLA) steels are hardened by the precipitation of vanadium carbides. |
| Titanium | Forms TiC and TiN precipitates. Helps to refine grain structure. Used in HSLA steels. Carbide stabilizer in stainless steel |
| Niobium | Forms precipitation hardening in HSLA steels. Carbide stabilizer in stainless steels |
| Boron | On a weight percent basis, boron is the most powerful hardenability element in steel. Only 0.003 wt% B provides hardenability in low-alloy steels. |
| Calcium | Used to replace manganese to tie up sulfur as calcium sulfide particles that do not elongate on hot rolling, eliminating anistropy |
| Lead and selenium | Added to improve machinability by providing lubricity and chip-breaking ability |

**Table 20.2 Approximate distribution of alloying elements in steels**

| Element | Dissolved in ferrite | Combined in carbide | Typical carbides | Typical compounds and inclusions | Elemental |
|---|---|---|---|---|---|
| Nickel | XXXXXX | ... | | $Ni_3Al$, NiSi | |
| Silicon | XXXXXX | ... | | $M_xO_y \cdot SiO_2$ | |
| Manganese | XXXXX | X | $Mn_{15}C_4$, $Mn_{23}C_6$ | MnS, $MnO \cdot SiO_2$ | |
| Chromium | XXXX | XX | $Cr_7C_3$, $Cr_{23}C_6$ | | |
| Molybdenum | XXX | XXX | $Mo_2C$, $Mo_6C$ | | |
| Tungsten | XXX | XXX | WC, $W_2C$ | | |
| Vanadium | XX | XXXX | CC, $V_4C_3$ | | |
| Titanium | XX | XXXX | TiC | | |
| Niobium | X | XXXXX | NbC | | |
| Aluminum | XXXXXX | ... | | $Al_2O_3$; $Al_xN_y$ | |
| Copper | XXXX | ... | | | Cu (>0.8 wt%) |
| Lead | ... | ... | | | Pb |

The X's indicate the relative tendencies of the elements to dissolve in ferrite or combine in carbides. Adapted from Ref 4

Nickel and silicon dissolve primarily in ferrite, although some of the silicon will combine with oxygen during the deoxidation process. Most of the manganese dissolves in ferrite, with some entering cementite as $(Fe,Mn)_3C$. Chromium also partitions between the ferrite and carbide phases, with the distribution depending on the amount of carbon and other carbide-forming elements. Tungsten and molybdenum combine with carbon to form carbides if sufficient carbon is present and if other stronger carbide-forming elements, such as titanium and niobium, are absent. Vanadium, titanium, and niobium are all strong carbide formers and will be present mainly as carbides. Aluminum combines with oxygen and nitrogen to form $Al_2O_3$ and AlN, respectively.

By examining the iron carbide binary phase diagram, it can be seen that adding carbon to iron favors the existence of austenite over ferrite. This occurs because carbon decreases the temperature of the austenite to α-ferrite transition (the $A_3$ point) and also increases the temperature of the austenite to δ-ferrite transition. Thus, carbon expands the gamma field and is therefore an austenite stabilizer. Those elements that behave similarly are also austenite stabilizers. Austenite stabilizers can be subdivided into two groups (Fig. 20.1). In the first subgroup, which includes carbon, the austenite phase field is limited by the appearance of the carbide cementite. Nitrogen and copper act similarly in that they expand the γ-field. Nitrogen forms a nitride phase, while the phases in the iron-copper system are solid solutions. In the second subgroup, the addition of the alloying elements depresses the face-centered cubic (fcc) to body-centered cubic (bcc) transition temperature toward room temperature. Examples of these elements are nickel and manganese. The elements that stabilize austenite generally have an fcc crystal structure like that of austenite. They dissolve substitutionally in austenite and retard the transformation of austenite to ferrite.

Another group of alloying elements favor ferrite formation over austenite. Again, these can be divided into two subgroups (Fig. 20.2). In the first subgroup, the fcc field is restricted and separated from the bcc field, in which α- and δ-ferrite fields merge together by a two-phase area. The fcc area is called a gamma (γ) loop. This group of alloying elements includes chromium, molybdenum, aluminum, silicon, titanium, and vanadium. The second subgroup restricts the γ-field by lowering the δ-ferrite transition temperature and raising that of

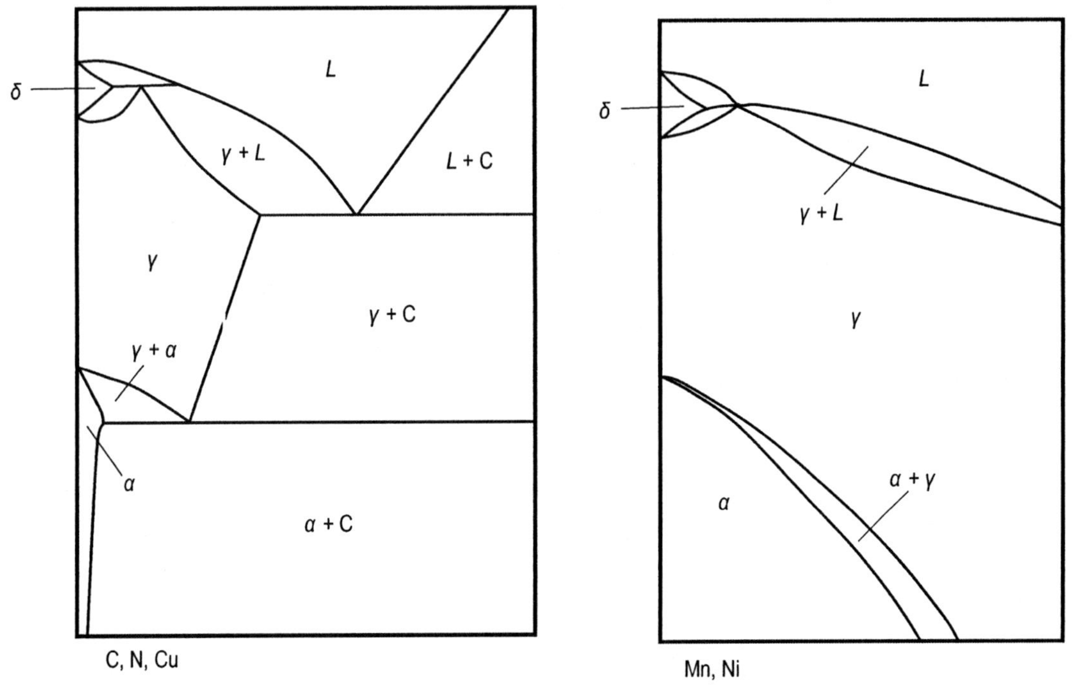


**Fig. 20.1** Austenite stabilizers

α-ferrite while also producing other phases. These elements are boron, niobium, tantalum, and zirconium. The ferrite-stabilizing elements are principally those that, like α-iron, have a bcc crystal structure. They therefore dissolve substitutionally more readily in α-iron than in γ-iron, stabilizing ferrite over a wider temperature range.

Some alloying elements have a strong tendency to combine with carbon to form hard carbides. Carbide formers include chromium, tungsten, vanadium, molybdenum, titanium, and niobium. Examples of carbides are $Cr_7C_3$, $W_2C$, $Mo_2C$, and VC, while double or complex carbides containing iron and one or more other metals are also formed. Carbide formers are often ranked according to their stability in the iron matrix: moderate carbide formers include chromium and molybdenum; strong carbide formers include tungsten, tantalum, and niobium; and very strong carbide formers are vanadium, titanium, and zirconium. When significant amounts of carbide formers are present, the microstructural constituents can become complicated. Steel microstructures can consist of cementite, mixed carbides (i.e., cementite containing dissolved alloying elements), and a variety of alloy carbides in which there is little iron. Nitrogen tends to prefer to reside in carbides, giving rise to carbonitride compounds. It can also combine with aluminum to form nitride precipitates.

Some alloying elements dissolve in ferrite and thermodynamically oppose the formation of cementite. This encourages the formation of graphite, and so these elements are called graphite stabilizers. Examples are silicon and nickel. In fact, silicon is so effective that it enables the formation of graphite from the melt in cast irons and is therefore a particularly important alloying element in cast irons.

Alloying additions reduce the solubility of carbon in austenite and displace the eutectoid point toward the left on the phase diagram (Fig. 20.3). Thus, an alloy steel can be completely pearlitic even though it contains less than 0.8 wt% C. Therefore, low-alloy steels contain less carbon than plain carbon steels for equivalent properties. At the same time, the eutectoid temperature ($A_1$) is altered by alloying (Fig. 20.4). The ferrite stabilizers raise the eutectoid temperature in the same way that they

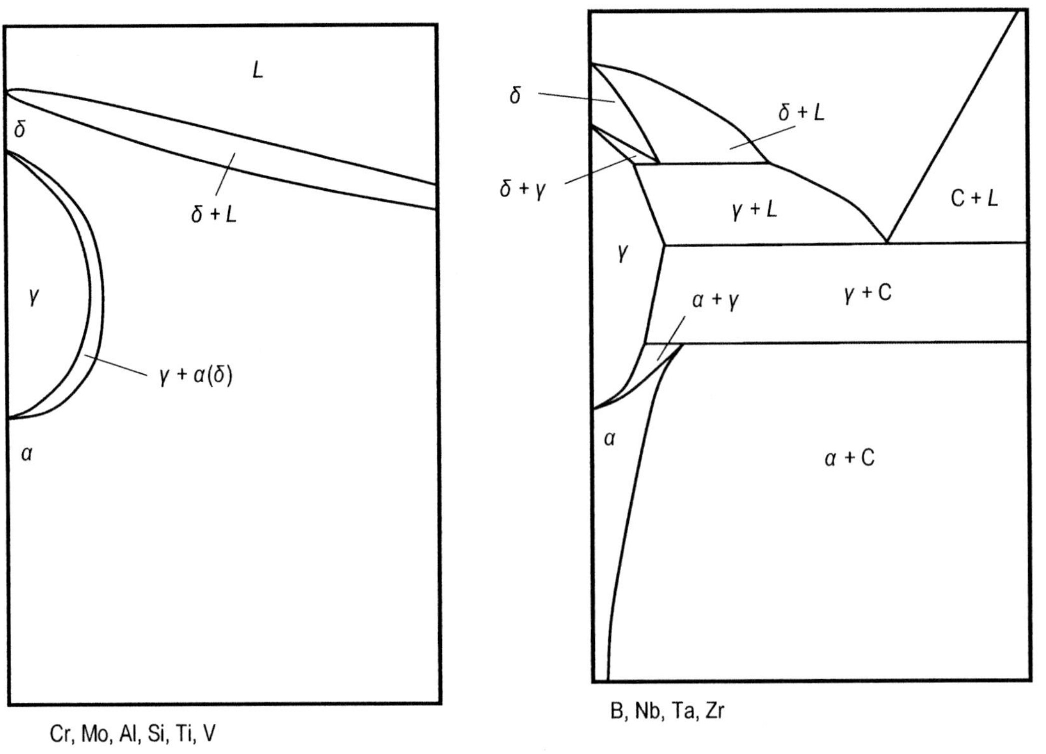


**Fig. 20.2** Ferrite stabilizers

raise $A_3$, while the austenite stabilizers lower the eutectoid temperature.

## 20.2 Low-Alloy Structural Steels

Hot rolled and heat treated carbon-manganese steels are used for many structural applications, and many are covered by ASTM International specifications. Comparative properties of several hot rolled, normalized, and quenched and tempered structural steels are listed in Table 20.3.

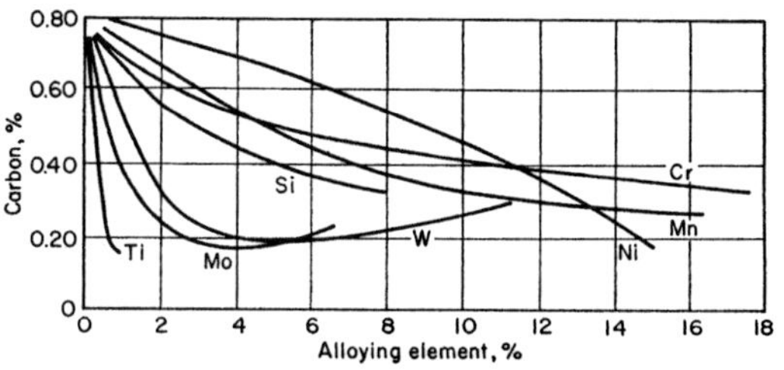


**Fig. 20.3** Effect of alloying elements on eutectoid composition. Source: Ref 4

### 20.2.1 *Hot Rolled Carbon-Manganese Structural Steels*

One method of achieving higher strengths in rolled structural plate and sections is to increase the manganese content. Manganese is a mild solid-solution strengthener in ferrite and is the principal strengthening element when it is

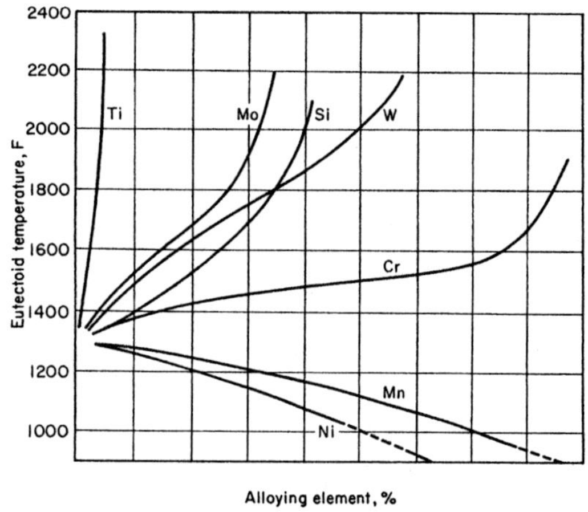


**Fig. 20.4** Effect of alloying elements on eutectoid temperature. Source: Ref 4

**Table 20.3 Examples of high-strength structural steels**

| Specification and grade or class | Product form | Thickness mm | Thickness in. | Composition, % C | Composition, % Mn | Composition, % Si | Composition, % Cu | Yield strength MPa | Yield strength ksi | Ultimate tensile strength MPa | Ultimate tensile strength ksi | Elongation, % |
|---|---|---|---|---|---|---|---|---|---|---|---|---|
| **As-hot-rolled carbon-manganese steels** | | | | | | | | | | | | |
| ASTM A 529 | Bar, plate, and shapes | 13 | 0.50 | 0.27 | 1.20 | ... | 0.20 | 290 | 42 | 415–585 | 60–85 | 19 |
| ASTM A 570 grades 45, 50, 55 | Sheet | 6 | 0.229 | 0.25 | 1.35 | ... | 0.20 | 310–380 | 45–55 | 415–480 | 60–70 | 14–10 |
| ASTM A 662, grade B | Plate | 40 | 1.5 | 0.19 | 0.85–1.50 | 0.15–0.40 | ... | 275 | 40 | 450–585 | 65–85 | 20 |
| ASTM A 662, grade C | Plate | 40 | 1.5 | 0.20 | 1.00–1.60 | 0.15–0.50 | ... | 295 | 43 | 485–620 | 70–90 | 18 |
| **Normalized structural carbon-manganese steels** | | | | | | | | | | | | |
| ASTM A 537, class 1 | Plate | 40 | 1.5 | 0.24 | 0.70–1.35 | 0.15–0.50 | 0.35 | 345 | 50 | 485–620 | 70–90 | 18 |
| | Plate | 40–65 | 1.5–2.5 | 0.24 | 1.00–1.60 | 0.15–0.50 | 0.35 | 345 | 50 | 485–620 | 70–90 | 18 |
| | Plate | 65–100 | 2.5–4 | 0.24 | 1.0–1.60 | 0.15–0.50 | 0.35 | 310 | 45 | 450–585 | 65–85 | 18 |
| ASTM A 633, grade A | Plate | 100 | 4 | 0.18 | 1.00–1.35 | 0.15–0.50 | ... | 290 | 42 | 430–570 | 63–83 | 18 |
| ASTM A 662, grade A | Plate | 40–50 | 1.5–2 | 0.14 | 0.90–1.35 | 0.15–0.40 | ... | 275 | 40 | 400–540 | 58–78 | 20 |
| ASTM A 662, grade B | Plate | 40–50 | 1.5–2 | 0.19 | 0.85–1.50 | 0.15–0.40 | ... | 275 | 40 | 450–585 | 65–85 | 20 |
| ASTM A 662, grade C | Plate | 40–50 | 1.5–2 | 0.20 | 1.00–1.60 | 0.15–0.50 | ... | 295 | 43 | 485–620 | 70–90 | 18 |
| ASTM A 738, grade A | Plate | 65 | 2.5 | 0.24 | 1.50 | 0.15–0.50 | 0.35 | 295 | 45 | 515–655 | 75–95 | 20 |
| **Quenched and tempered structural carbon-manganese steels** | | | | | | | | | | | | |
| ASTM A 678, grade A | Plate | 40 | 1.5 | 0.16 | 0.90–1.50 | 0.15–0.50 | 0.20 | 345 | 50 | 485–620 | 70–90 | 22 |
| ASTM A 678, grade B | Plate | 40 | 1.5 | 0.20 | 0.70–1.35 | 0.15–0.50 | 0.20 | 415 | 60 | 550–690 | 80–100 | 22 |
| | | 40–65 | 1.5–2.5 | 0.20 | 1.00–1.60 | 0.15–0.50 | 0.20 | 415 | 60 | 550–690 | 80–100 | 22 |
| ASTM A 678, grade C | Plate | 20 | 0.75 | 0.22 | 1.00–1.60 | 0.20–0.50 | 0.20 | 515 | 75 | 655–790 | 95–115 | 19 |
| | | 20–40 | 0.75–1.5 | 0.22 | 1.00–1.60 | 0.20–0.50 | 0.20 | 485 | 70 | 620–760 | 90–110 | 19 |
| | | 40–50 | 1.5–2 | 0.22 | 1.00–1.60 | 0.20–0.50 | 0.20 | 450 | 65 | 585–720 | 85–105 | 19 |
| ASTM A 738, grade B | Plate | 65 | 2.5 | 0.20 | 0.90–1.50 | 0.15–0.50 | 0.20 | 415 | 60 | 585–705 | 85–102 | 20 |

Source: Ref 5

present in amounts over 1 wt% in hot rolled low-carbon (<0.20 wt% C) steels. Manganese also improves the toughness of low-carbon steels and lowers the ductile-to-brittle transition temperature (Fig. 20.5).

Hot rolled structural steels with 0.4 wt% C and 1.5 wt% Mn produce yield strengths in the range of 345 to 400 MPa (50 to 58 ksi). As the strength and amount of pearlite increase, the notch toughness and weldability decrease. However, this increase in tensile strength is not accompanied by an increase in yield strength, which is normally the main design criterion. In applications requiring greater toughness, small additions of aluminum can be added for grain refinement. In spite of these limitations, carbon-manganese hot rolled steels are used in a variety of applications, such as stampings, forgings, tubing, and boiler plates.

### *20.2.2 Heat Treated Carbon-Manganese Structural Steels*

Heat treatment of carbon-manganese steels consists of either normalizing or quenching and tempering. Both of these heat treatments improve the mechanical properties of structural plate, bar, and occasionally structural shapes. Structural shapes, such as I-beams, channels, and other shapes, are primarily used in the hot rolled condition because warpage is difficult to prevent during heat treatment. Nevertheless, some normalized or quenched and tempered structural sections can be produced in a limited number of section sizes.

Normalizing produces essentially the same ferrite-pearlite microstructure as that of hot rolled carbon steel, except that normalizing results in a finer grain size. The grain refinement makes the steel stronger, tougher, and more uniform throughout. Quenching and tempering are conducted by heating to approximately 900 °C (1650 °F), water quenching, and tempering at temperatures of 480 to 595 °C (900 to 1100 °F) or higher. Quenching and tempering produces a tempered martensitic or bainitic microstructure that has a better combination of strength and toughness. An increase in the carbon content to approximately 0.5 wt%, usually accompanied by an increase in manganese, allows these steels to be used in the quenched and tempered condition.

The yield strength of quenched and tempered carbon-manganese steel plate ranges from 317 to 552 MPa (46 to 80 ksi), depending on section thickness. Minimum Charpy V-notch impact toughness may be as high as 27 to 34 J (20 to 25 ft·lbf) at temperatures as low as −68 °C (−90 °F) for steel having a yield strength of 345 MPa (50 ksi). However, for quenched and tempered steel with a 690 MPa (100 ksi) yield strength, the impact values are lower, normally about 20 J (15 ft·lbf) at −59 °C (−75 °F). All grades can be grain refined with aluminum to improve toughness.

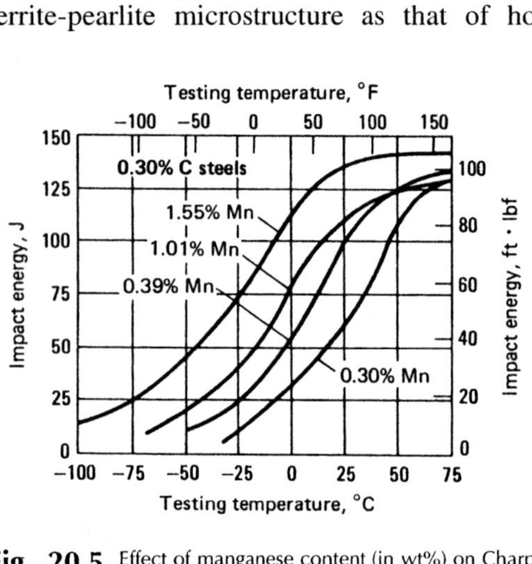


**Fig. 20.5** Effect of manganese content (in wt%) on Charpy V-notch impact strength. Source: Ref 5

### *20.2.3 High-Nickel Steels for Low-Temperature Service*

For applications involving exposure to temperatures from 0 to −200 °C (32 to −320 °F), ferritic steels with low carbon and high nickel contents are typically used. Such applications include storage tanks for liquefied hydrocarbon gases and structures and machinery designed for use in cold regions. The compositions for a number of these steels are given in Table 20.4. These steels use the effect of nickel content in reducing the ductile-to-brittle transition temperature, thereby improving toughness at low temperatures.

These steels are made to ASTM International specifications that have specific requirements for Charpy V-notch impact results. For example, for tests conducted on 9 wt% Ni steel at −195 °C (−320 °F), the transverse Charpy V-notch values must not be less than 27 J (20 ft·lbf), and the longitudinal values must be at least 34 J (25 ft·lbf). Typical tensile properties of 5 and 9 wt% Ni steels at room temperature and at subzero temperatures are presented in Table 20.5. Yield and tensile strengths increase as the testing temperature is decreased. These steels remain ductile at the lowest testing temperatures. The 5 wt% Ni steel retains relatively high toughness at −162 °C (−260 °F), while the 9 wt% Ni steel retains relatively high toughness at −196 °C (−320 °F), which are the approximate minimum temperatures at which these steels should be used.

## 20.3 SAE/AISI Alloy Steels

The SAE/AISI four-digit classification system for low-alloy steels is summarized in Table 20.6. As in the carbon steels, the first two digits are for the alloy class, and the last two or three digits designate the carbon content. Because of the various combinations of elements, the system is broader and more complex than that used for the carbon steels. While some specifications are based on compositions, such as the SAE/AISI specifications, others, such as the ASTM International specifications, specify the mode of manufacture and permissible amounts of sulfur and phosphorus but often leave it up to the steelmaker to choose the carbon level necessary to achieve specified tensile properties. The nominal compositions of a number of SAE/AISI steels are given in Table 20.7.

Alloy steels are supplied in the form of bar, plate, and forged products and are usually heat treated to obtain specific mechanical properties, especially high strength and toughness. Quenched and tempered low-alloy steels are used in a large number of applications requiring high strength and good toughness. When the hardenability is of prime importance, it is possible to obtain steels that are manufactured with closely controlled compositions so as to ensure that a specified hardenability is obtained. Such steels are identified by adding the letter "H" to the numerical code.

### *20.3.1 Manganese Steels (13xx)*

The 13*xx* manganese steels contain 0.30 to 0.45 wt% C and 1.6 to 1.9 wt% Mn. This is a higher manganese content than the 0.25 to 1.0 wt% that is used in plain carbon steels. As a result of the increased manganese levels, the 13*xx* steels have somewhat better strengths and hardenability than their plain carbon counterparts. Manganese shifts the nose of the TTT diagram slightly to the right, which improves hardenability and also refines and strengthens

**Table 20.4 High-nickel steels for low-temperature service**

| ASTM specification | Compositions of plates < 50 mm (2 in.) thick, % | | | | | | | |
|---|---|---|---|---|---|---|---|---|
| | C | Mn | P | S | Si | Ni | Mo | Others |
| A 645 | 0.13 | 0.30–0.60 | 0.025 | 0.025 | 0.20–0.35 | 4.75–5.25 | 0.20–0.35 | 0.02–0.12 Al, 0.020 N |
| A 353 | 0.13 | 0.90 | 0.035 | 0.040 | 0.15–0.30 | 8.5–9.5 | ... | ... |
| A 553 I | 0.13 | 0.90 | 0.035 | 0.040 | 0.15–0.30 | 8.5–9.5 | ... | ... |

Source: Ref 5

**Table 20.5 Properties of high-nickel steels for low-temperature service**

| Temperature | | Ultimate tensile strength | | Yield strength | | Elongation, % | Reduction in area, % |
|---|---|---|---|---|---|---|---|
| °C | °F | MPa | ksi | MPa | ksi | | |
| **A 645 plate (longitudinal orientation)** | | | | | | | |
| 24 | 75 | 715 | 104 | 530 | 76.8 | 32 | 72 |
| −168 | −270 | 930 | 135 | 570 | 82.9 | 28 | 68 |
| −196 | −320 | 1130 | 164 | 765 | 111 | 30 | 62 |
| **A 353 plate (longitudinal orientation)** | | | | | | | |
| 24 | 75 | 780 | 113 | 680 | 98.6 | 28 | 70 |
| −151 | −240 | 1030 | 149 | 850 | 123 | 17 | 61 |
| −196 | −320 | 1190 | 172 | 950 | 138 | 25 | 58 |
| −253 | −423 | 1430 | 208 | 1320 | 192 | 18 | 43 |
| −269 | −452 | 1590 | 231 | 1430 | 208 | 21 | 59 |
| **A 553-I plate (longitudinal orientation)** | | | | | | | |
| 24 | 75 | 770 | 112 | 695 | 101 | 27 | 69 |
| −151 | −240 | 995 | 144 | 885 | 128 | 18 | 42 |
| −196 | −320 | 1150 | 167 | 960 | 139 | 27 | 38 |

Source: Ref 5

**Table 20.6 Summary of AISI/SAE designations for carbon and low-alloy steels**

| Designation(a) | Approximate alloy content, wt% |
|---|---|
| **Carbon steels** | |
| 10*xx* | Plain carbon |
| 11*xx* | Resulfurized |
| 12*xx* | Resulfurized and rephosphorized |
| 15*xx* | 1.00–1.65 Mn |
| **Manganese steels** | |
| 13*xx* | 1.75 Mn |
| **Molybdenum steels** | |
| 40*xx* | 0.25 Mo |
| 44*xx* | 0.40–0.52 Mo |
| **Molybdenum-chromium steels** | |
| 41*xx* | 0.12–0.30 Mo, 0.50–0.95 Cr |
| **Molybdenum-chromium-nickel steels** | |
| 43*xx* | 0.25 Mo, 0.50 or 0.80 Cr, 1.82 Ni |
| 47*xx* | 0.25 or 0.35 Mo, 0.45 Cr, 1.45 Ni |
| 81*xx* | 0.12 Mo, 0.40 Cr, 0.30 Ni |
| 86*xx* | 0.20 Mo, 0.50 Cr, 0.55 Ni |
| 87*xx* | 0.25 Mo, 0.50 Cr, 0.55 Ni |
| 94*xx* | 0.12 Mo, 0.40 Cr, 0.45 Ni |
| **Molybdenum-nickel steels** | |
| 46*xx* | 0.25 Mo, 0.85 or 1.82 Ni |
| 48*xx* | 0.25 Mo, 3.50 Ni |
| **Chromium steels** | |
| 50*xx*(*x*) | 0.27 to 0.65 Cr |
| 51*xx*(*x*) | 0.80 to 1.05 Cr |
| 52*xx* | 1.45 Cr |
| **Chromium-vanadium steels** | |
| 61*xx* | 0.6 to 0.95 Cr, 0.15 V |
| **Silicon-manganese steels** | |
| 92*xx* | 1.40 or 2.00 Si, 0.70–0.87 Mn, 0 or 0.70 Cr |
| **Boron steels** | |
| *yy*B*xx* | 0.00005 to 0.003B |

(a) Replace *xx* or *xxx* with carbon content in hundredths of a percent. Replace *yy* with any two digits from earlier in the table to indicate additional alloying content. Source: Ref 6

pearlite. The mechanical properties of 1330 and 1340 are given in Table 20.8. These steels are used when slightly better properties are required than obtainable with mild steels and the application does not warrant a more expensive alloy steel.

### *20.3.2 Chromium Steels (5xxx)*

Alloy steels in the 5*xxx* series contain from 0.20 to 0.60 wt% C and from 0.8 to 0.9 wt% Cr. Chromium improves hardenability, strength, and wear resistance. Since chromium has a bcc structure, it is a strong ferrite stabilizer. Chromium also combines with carbon to form carbides. Since the chromium content is less than 2 wt%, chromium replaces iron in $Fe_3C$ to form the carbide $(Fe,Cr)_3C$. When these steels are quenched and tempered, they develop high strengths and hardness, but their ductility is rather low (Table 20.9). Under some conditions, they are also susceptible to temper brittleness. Some of the higher-carbon grades are used for ball bearings.

### *20.3.3 Molybdenum Steels (40xx)*

Molybdenum, added in small amounts, increases both the strength and hardenability of steels. The 40*xx* series of alloys contain approximately 0.25 wt% Mo, since this amount has been found to produce the optimal balance of strength, hardenability, and toughness. Steels containing molybdenum are less susceptible to temper brittleness. The low-carbon grades of the 40*xx* alloys are often used as carburized parts. The properties of a typical molybdenum steel (4042) are given in Table 20.10.

### *20.3.4 Chromium-Molybdenum Steels (41xx)*

The 41*xx* series of alloys are classified as chromium-molybdenum steels and contain 0.5 to 0.95 wt% Cr and 0.13 to 0.20 wt% Mo. Chromium is added to increase hardenability and strength; however, the addition of chromium can also make this series susceptible to temper embrittlement. Due to its low-to-intermediate hardenability, 4130 must be water quenched. It has good tensile, fatigue, and impact properties up to approximately 370 °C (700 °F); however, the impact properties at cryogenic temperatures are low. The alloy 4140 is similar to 4130 except for a higher carbon content, which results in higher strengths with some sacrifice in formability and weldability. When heat treated to high strength levels, 4140 is susceptible to hydrogen embrittlement that can result from acid pickling or from electroplating with cadmium or chromium. After pickling or electroplating, it is baked for 2 to 4 h at 190 °C (375 °F) to remove any absorbed hydrogen. Typical properties are shown in Table 20.11.

### *20.3.5 Nickel-Chromium-Molybdenum Steels (43xx and 8xxx)*

Nickel is added along with chromium and molybdenum to form the 43*xx* class of alloys. Their composition (by weight) is approximately 0.5 to 0.8% Cr, 0.20% Mo, and 1.8% Ni. In the

## Table 20.7 Nominal compositions of select SAE/AISI steels

| Steel | | Composition, % | | | | | | | |
|---|---|---|---|---|---|---|---|---|---|
| Type No. | UNS No. | C | Mn | P max | S Max | Si | Ni | Cr | Mo |
| **Manganese steels** | | | | | | | | | |
| 1330 | G13300 | 0.28–0.33 | 1.60–1.90 | 0.035 | 0.040 | 0.15–0.35 | ... | ... | ... |
| 1340 | G13400 | 0.38–0.43 | 1.60–1.90 | 0.035 | 0.040 | 0.15–0.35 | ... | ... | ... |
| **Chromium steels** | | | | | | | | | |
| 5120 | G51200 | 0.17–0.22 | 0.70–0.90 | 0.035 | 0.040 | 0.15–0.35 | ... | 0.70–0.90 | ... |
| 5130 | G51300 | 0.28–0.33 | 0.70–0.90 | 0.035 | 0.040 | 0.15–0.35 | ... | 0.80–1.10 | ... |
| 5140 | G51400 | 0.38–0.43 | 0.70–0.90 | 0.035 | 0.040 | 0.15–0.35 | ... | 0.70–0.90 | ... |
| 5160 | G51600 | 0.56–0.64 | 0.75–1.00 | 0.035 | 0.040 | 0.15–0.35 | ... | 0.70–0.90 | ... |
| E52100 | G52986 | 0.98–1.10 | 0.25–0.45 | 0.025 | 0.025 | 0.15–0.35 | ... | 1.30–1.60 | ... |
| **Molybdenum steels** | | | | | | | | | |
| 4023 | G40230 | 0.20–0.25 | 0.70–0.90 | 0.035 | 0.040 | 0.15–0.35 | ... | ... | 0.20–0.30 |
| 4037 | G40370 | 0.35–0.40 | 0.70–0.90 | 0.035 | 0.040 | 0.15–0.35 | ... | ... | 0.20–0.30 |
| 4047 | G40470 | 0.45–0.50 | 0.70–0.90 | 0.035 | 0.040 | 0.15–0.35 | ... | ... | 0.20–0.30 |
| **Chromium-molybdenum steels** | | | | | | | | | |
| 4118 | G41180 | 0.18–0.23 | 0.70–0.90 | 0.035 | 0.040 | 0.15–0.35 | ... | 0.40–0.60 | 0.08–0.15 |
| 4130 | G41300 | 0.28–0.33 | 0.40–0.60 | 0.035 | 0.040 | 0.15–0.35 | ... | 0.80–1.10 | 0.15–0.25 |
| 4140 | G41400 | 0.38–0.43 | 0.75–1.00 | 0.035 | 0.040 | 0.15–0.35 | ... | 0.80–1.10 | 0.15–0.25 |
| **Nickel-molybdenum steels** | | | | | | | | | |
| 4620 | G46200 | 0.17–0.22 | 0.45–0.65 | 0.035 | 0.040 | 0.15–0.35 | 1.65–2.00 | ... | 0.20–0.30 |
| 4820 | G48200 | 0.18–0.32 | 0.50–0.70 | 0.035 | 0.040 | 0.15–0.35 | 3.25–3.75 | ... | 0.20–0.30 |
| **Nickel (1.83%)-chromium-molybdenum steels** | | | | | | | | | |
| 4320 | G43200 | 0.17–0.22 | 0.45–0.65 | 0.035 | 0.040 | 0.15–0.35 | 1.65–2.00 | 0.40–0.60 | 0.20–0.30 |
| 4340 | G43400 | 0.38–0.43 | 0.60–0.80 | 0.035 | 0.040 | 0.15–0.35 | 1.65–2.00 | 0.70–0.90 | 0.20–0.30 |
| **Nickel (0.55%)-chromium-molybdenum steels** | | | | | | | | | |
| 8620 | G86200 | 0.18–0.23 | 0.70–0.90 | 0.035 | 0.040 | 0.15–0.35 | 0.40–0.70 | 0.40–0.60 | 0.15–0.25 |
| 8640 | G86400 | 0.38–0.43 | 0.75–1.00 | 0.035 | 0.040 | 0.15–0.35 | 0.40–0.70 | 0.40–0.60 | 0.15–0.25 |
| 8655 | G86550 | 0.51–0.59 | 0.75–1.00 | 0.035 | 0.040 | 0.15–0.35 | 0.40–0.70 | 0.40–0.60 | 0.15–0.25 |
| **Silicon steels** | | | | | | | | | |
| 9260 | G92600 | 0.56–0.64 | 0.75–1.00 | 0.035 | 0.040 | 1.80–2.20 | ... | ... | ... |

Source: Ref 7

## Table 20.8 Properties of 1330 and 1340 manganese steels

| | Tempering temperature | | Ultimate tensile strength | | Yield strength | | | | |
|---|---|---|---|---|---|---|---|---|---|
| AISI No. | °C | °F | MPa | ksi | MPa | ksi | Elongation, % | Reduction in area, % | Hardness, HB |
| 1330 | 205 | 400 | 1600 | 232 | 1455 | 211 | 9 | 39 | 459 |
| | 315 | 600 | 1427 | 207 | 1282 | 186 | 9 | 44 | 402 |
| | 425 | 800 | 1158 | 168 | 1034 | 150 | 15 | 53 | 335 |
| | 540 | 1000 | 876 | 127 | 772 | 112 | 18 | 60 | 263 |
| | 650 | 1200 | 731 | 106 | 572 | 83 | 23 | 63 | 216 |
| 1340 | 205 | 400 | 1806 | 262 | 1593 | 231 | 11 | 35 | 505 |
| | 315 | 600 | 1586 | 230 | 1420 | 206 | 12 | 43 | 453 |
| | 425 | 800 | 1262 | 183 | 1151 | 167 | 14 | 51 | 375 |
| | 540 | 1000 | 965 | 140 | 827 | 120 | 17 | 58 | 295 |
| | 650 | 1200 | 800 | 116 | 621 | 90 | 22 | 66 | 252 |

Source: Ref 8

8*xxx* series, the nickel content is reduced to 0.55%. Nickel in combination with chromium improves strength and provides greater hardenability, higher impact strength, and better fatigue resistance. The addition of 0.2 wt% Mo further increases hardenability and minimizes the susceptibility to temper embrittlement. The properties of several alloys are given in Table 20.12.

Alloy 4340 is the benchmark by which other high-strength steels are judged. It combines deep hardenability with high strength, ductility, and toughness. It also has good fatigue and creep resistance. The effects of different tempering temperatures on the properties of 4340 are shown in Fig. 20.6. It is often used where high strength in thick sections is required. Alloy 4340 can be oil quenched to full hardness in sections

**Table 20.9 Properties of select chromium steels**

| AISI No. | Tempering temperature °C | Tempering temperature °F | Ultimate tensile strength MPa | Ultimate tensile strength ksi | Yield strength MPa | Yield strength ksi | Elongation, % | Reduction in area, % | Hardness, HB |
|---|---|---|---|---|---|---|---|---|---|
| 5130 | 205 | 400 | 1613 | 234 | 1517 | 220 | 10 | 40 | 475 |
| | 315 | 600 | 1496 | 217 | 1407 | 204 | 10 | 46 | 440 |
| | 425 | 800 | 1276 | 185 | 1207 | 175 | 12 | 51 | 379 |
| | 540 | 1000 | 1034 | 150 | 938 | 136 | 15 | 56 | 305 |
| | 650 | 1200 | 793 | 115 | 689 | 100 | 20 | 63 | 245 |
| 5140 | 205 | 400 | 1793 | 260 | 1641 | 238 | 9 | 38 | 490 |
| | 315 | 600 | 1579 | 229 | 1448 | 210 | 10 | 43 | 450 |
| | 425 | 800 | 1310 | 190 | 1172 | 170 | 13 | 50 | 365 |
| | 540 | 1000 | 1000 | 145 | 862 | 125 | 17 | 58 | 280 |
| | 650 | 1200 | 758 | 110 | 662 | 96 | 25 | 66 | 235 |
| 5150 | 205 | 400 | 1944 | 282 | 1731 | 251 | 5 | 37 | 525 |
| | 315 | 600 | 1737 | 252 | 1586 | 230 | 6 | 40 | 475 |
| | 425 | 800 | 1448 | 210 | 1310 | 190 | 9 | 47 | 410 |
| | 540 | 1000 | 1124 | 163 | 1034 | 150 | 15 | 54 | 340 |
| | 650 | 1200 | 807 | 117 | 814 | 118 | 20 | 60 | 270 |
| 5160 | 205 | 400 | 2220 | 322 | 1793 | 260 | 4 | 10 | 627 |
| | 315 | 600 | 1999 | 290 | 1772 | 257 | 9 | 30 | 555 |
| | 425 | 800 | 1606 | 233 | 1462 | 212 | 10 | 37 | 461 |
| | 540 | 1000 | 1665 | 169 | 1041 | 151 | 12 | 47 | 341 |
| | 650 | 1200 | 896 | 130 | 800 | 116 | 20 | 56 | 269 |

Source: Ref 8

**Table 20.10 Properties of 4042 molybdenum steel**

| AISI No. | Tempering temperature °C | Tempering temperature °F | Ultimate tensile strength MPa | Ultimate tensile strength ksi | Yield strength MPa | Yield strength ksi | Elongation, % | Reduction in area, % | Hardness, HB |
|---|---|---|---|---|---|---|---|---|---|
| 4042 | 205 | 400 | 1800 | 261 | 1662 | 241 | 12 | 37 | 516 |
| | 315 | 600 | 1613 | 234 | 1455 | 211 | 13 | 42 | 455 |
| | 425 | 800 | 1289 | 187 | 1172 | 170 | 15 | 51 | 380 |
| | 540 | 1000 | 986 | 143 | 883 | 128 | 20 | 59 | 300 |
| | 650 | 1200 | 793 | 115 | 689 | 100 | 28 | 66 | 238 |

Source: Ref 8

with up to 76 mm (3 in.) diameter. Thicker sections require water quenching; however, water quenching significantly increases the danger of quench cracking. It is immune to temper embrittlement. Parts exposed to hydrogen during pickling or plating operations are baked to remove any hydrogen. The effects of various baking cycles on the notched-bar strength of hydrogen-charged 4340 are shown in Fig. 20.7. Unfortunately, 4340 is very susceptible to stress-corrosion cracking when heat treated to high strength levels of 1520 to 1930 MPa, or (220 to 280 ksi). Alloy 4340 can be modified with vanadium (4340 V), which forms a stable, high-melting-point carbide that helps in pinning grain boundaries and preventing grain growth during hot working operations. The vanadium addition also serves as a grain refiner that increases toughness.

When the silicon content of 4340 is increased to approximately 2%, the strength and toughness increase in the manner shown in Fig. 20.8. The increased silicon content provides deeper hardenability, increases solid-solution strengthening, and provides better high-temperature resistance. The increase in toughness is attributed to silicon retarding the precipitation of cementite from retained austenite during tempering and to the stabilization of carbides. Silicon added to the basic 4340 composition forms the alloy 300 M, which nominally contains 1.6 wt% Si. Vanadium is added for grain refinement, and the sulfur and phosphorus levels are kept very low to reduce temper embrittlement and increase toughness and transverse ductility. Alloy 300 M is also vacuum arc remelted to lower the hydrogen and oxygen contents. The lower oxygen content minimizes the formation of oxide inclusions and increases toughness. However, due to the high silicon and molybdenum contents of 300 M, it is extremely prone to decarburization

**Table 20.11 Properties of select chromium-molybdenum steels**

| AISI No. | Tempering temperature °C | Tempering temperature °F | Ultimate tensile strength MPa | Ultimate tensile strength ksi | Yield strength MPa | Yield strength ksi | Elongation, % | Reduction in area, % | Hardness, HB |
|---|---|---|---|---|---|---|---|---|---|
| 4130 | 205 | 400 | 1627 | 236 | 1462 | 212 | 10 | 41 | 467 |
| | 315 | 600 | 1496 | 217 | 1379 | 200 | 11 | 43 | 435 |
| | 425 | 800 | 1282 | 186 | 1193 | 173 | 13 | 49 | 380 |
| | 540 | 1000 | 1034 | 150 | 910 | 132 | 17 | 57 | 315 |
| | 650 | 1200 | 814 | 118 | 703 | 102 | 22 | 64 | 245 |
| 4140 | 205 | 400 | 1772 | 257 | 1641 | 238 | 8 | 38 | 510 |
| | 315 | 600 | 1551 | 225 | 1434 | 208 | 9 | 43 | 445 |
| | 425 | 800 | 1248 | 181 | 1138 | 165 | 13 | 49 | 370 |
| | 540 | 1000 | 951 | 138 | 834 | 121 | 18 | 58 | 285 |
| | 650 | 1200 | 758 | 110 | 655 | 95 | 22 | 63 | 230 |
| 4150 | 205 | 400 | 1931 | 280 | 1724 | 250 | 10 | 39 | 530 |
| | 315 | 600 | 1765 | 256 | 1593 | 231 | 10 | 40 | 495 |
| | 425 | 800 | 1517 | 220 | 1379 | 200 | 12 | 45 | 440 |
| | 540 | 1000 | 1207 | 175 | 1103 | 160 | 15 | 52 | 370 |
| | 650 | 1200 | 958 | 139 | 841 | 122 | 19 | 60 | 290 |

Source: Ref 8

**Table 20.12 Properties of select nickel-chromium-molybdenum steels**

| AISI No. | Tempering temperature °C | Tempering temperature °F | Ultimate tensile strength MPa | Ultimate tensile strength ksi | Yield strength MPa | Yield strength ksi | Elongation, % | Reduction in area, % | Hardness, HB |
|---|---|---|---|---|---|---|---|---|---|
| 4340 | 205 | 400 | 1875 | 272 | 1675 | 243 | 10 | 38 | 520 |
| | 315 | 600 | 1724 | 250 | 1586 | 230 | 10 | 40 | 486 |
| | 425 | 800 | 1469 | 213 | 1365 | 198 | 10 | 44 | 430 |
| | 540 | 1000 | 1172 | 170 | 1076 | 156 | 13 | 51 | 360 |
| | 650 | 1200 | 965 | 140 | 855 | 124 | 19 | 60 | ... |
| 300M(a) | 95 | 200 | 2344 | 340 | 1931 | 280 | 6 | 10 | ... |
| | 205 | 400 | 2137 | 310 | 1655 | 240 | 8 | 27 | ... |
| | 315 | 600 | 1993 | 289 | 1689 | 245 | 10 | 34 | ... |
| | 425 | 800 | 1793 | 260 | 1482 | 215 | 9 | 23 | ... |
| 8630 | 205 | 400 | 1641 | 238 | 1503 | 218 | 9 | 38 | 465 |
| | 315 | 600 | 1482 | 215 | 1393 | 202 | 10 | 42 | 430 |
| | 425 | 800 | 1276 | 185 | 1172 | 170 | 13 | 47 | 375 |
| | 540 | 1000 | 1034 | 150 | 896 | 130 | 17 | 54 | 310 |
| | 650 | 1200 | 772 | 112 | 689 | 100 | 23 | 63 | 240 |
| 8640 | 205 | 400 | 1862 | 270 | 1669 | 242 | 10 | 40 | 505 |
| | 315 | 600 | 1655 | 240 | 1517 | 220 | 10 | 41 | 460 |
| | 425 | 800 | 1379 | 200 | 1296 | 188 | 12 | 45 | 400 |
| | 540 | 1000 | 1103 | 160 | 1034 | 150 | 16 | 54 | 340 |
| | 650 | 1200 | 896 | 130 | 800 | 116 | 20 | 62 | 280 |

(a) 300M is not an AISI designation. Source: Ref 8

during heat treatment, and when heat treated to strength levels above 1380 MPa (200 ksi), it is also susceptible to hydrogen embrittlement.

High-strength steels are available in a variety of quality levels, depending on the type of melting practice used. While many of these steels were originally air melted, the trend has been to move to more advanced melting techniques such as vacuum degassing, electroslag remelting (ESR), vacuum arc remelting (VAR), and double vacuum melting (vacuum induction melting followed by vacuum arc remelting, or VIM-VAR) for improved cleanliness and higher quality. These methods reduce both the quantity of dissolved gases (hydrogen, oxygen, and nitrogen) and nonmetallic inclusions. As the high-strength steels have evolved since the mid-1970s, improvements in melting process control and inspection have steadily increased fracture toughness, ductility, and fatigue resistance. A comparison of air- and vacuum-melted 300 M, shown in Fig. 20.9, illustrates the property advantages imparted by vacuum processing. Both VAR and ESR are acceptable melting methods, since the mechanical properties are essentially equivalent for both methods.

## 20.4 High-Fracture-Toughness Steels

The three high-fracture-toughness steels, HP-9-4-30, AF1410, and AerMet 100, have somewhat lower carbon contents than the medium-carbon low-alloy steels 4340 and 300 M. The lower carbon content significantly contributes to their better ductilities and higher fracture toughness. In addition, these alloys have high nickel contents that provide deep hardening and toughness and cobalt that helps to prevent retained austenite. To obtain the desired fracture toughness, all of these steels are vacuum melted. The mechanical properties for HP-9-4-30, AF1410, and AerMet 100 are given in Table 20.13. These alloys are not corrosion resistant, and parts must be protected with a corrosion-resistant coating.

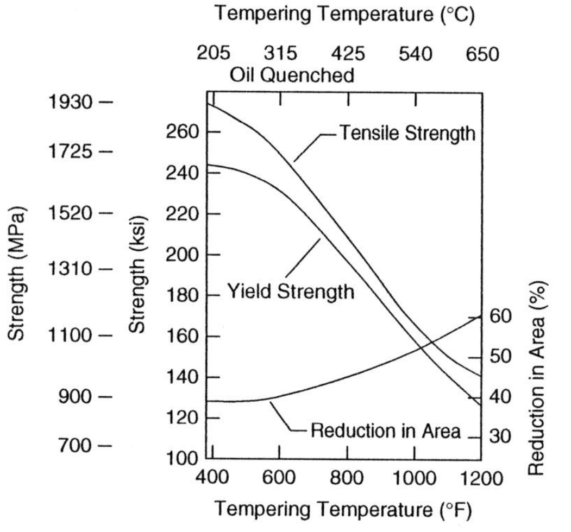


**Fig. 20.6** Effects of tempering temperature on 4340 steel. Source: Ref 9

The 9Ni-4Co family of steels was developed as high-fracture-toughness steels capable of being heat treated to high strength levels in thick sections. The highest strength of these alloys, HP-9-4-30, nominally contains 0.30 wt% C, 9 wt% Ni, and 4 wt% Co. It is capable of being hardened in sections up to 15 cm (6 in.) thick to an ultimate tensile strength level of 1520 to 1655 MPa (220 to 240 ksi) while maintaining a fracture toughness, $K_{Ic}$, of 110 MPa$\sqrt{m}$ (100 ksi$\sqrt{in.}$). Double tempering is normally employed to prevent retained austenite. Alloy HP-9-4-30 is available as billet, bar, rod, plate, sheet, and strip. It can be formed by bending, rolling, or shear spinning. Heat treated HP-9-4-30 can be welded using gas tungsten arc welding (GTAW) without preheating or postheating. Welded parts should be stress relieved at 540 °C (1000 °F) for 24 h.

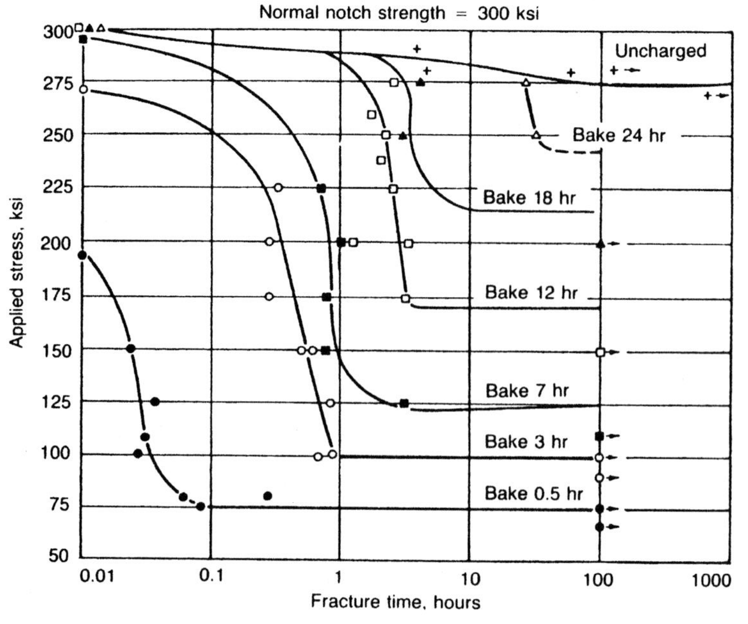


**Fig. 20.7** Static fatigue of 4340 notched bars. Baked at 300 °F (150 °C) Source: Ref 10

Alloy AF1410 was developed specifically to have high strength, excellent fracture toughness, and excellent weldability when heat treated to 1620 to 1760 MPa (235 to 255 ksi) ultimate tensile strength. The nominal composition is 14 wt% Co, 10 wt% Ni, 2 wt% Cr, 1 wt% Mo and 0.15 wt% C. Alloy AF1410 maintains good toughness at cryogenic temperatures and has high strength and stability at temperatures up to 425 °C (800 °F). The general corrosion resistance is similar to the maraging steels. The alloy is highly resistant to stress-corrosion cracking compared to other high-strength steels. Alloy AF1410 is produced by VIM followed by VAR. It is available as billet, bar, plate, and die forgings. Alloy AF1410 has good weldability using GTAW, provided high-purity welding wire is used and oxygen contamination is avoided. Preheating prior to welding is not required.

AerMet 100 is a nickel-cobalt high-strength steel (0.23 wt% C, 3.1 wt% Cr, 1.2 wt% Mo, 11.1 wt% Ni, 3.4 wt% Co) that can be heat treated to 1930 to 2070 or 2000 to 2140 MPa (280 to 300 or 290 to 310 ksi) tensile strength while exhibiting excellent fracture toughness and high resistance to stress-corrosion cracking. AerMet 100 is replacing older steels such as 4340, 300 M, HP-9-4-30, and AF1410 in many applications due to its good combination of strength (ultimate tensile strength is 1965 MPa, or 285 ksi) and toughness ($K_{Ic}$ is 110 MPa$\sqrt{m}$, or 100 ksi$\sqrt{in.}$). Other advantages include good toughness at cryogenic temperatures, a critical flaw length of nearly 6.3 mm (0.25 in.), and an operating temperature up to 400 °C (750 °F). It is highly resistant to stress-corrosion cracking compared to other high-strength steels of the same strength level. AerMet 100, produced by VIM followed by VAR, is available as billet, bar, sheet, strip, plate, wire, and die forgings. Impurity concentrations and inclusions are kept to a minimum by double vacuum melt processing. Unlike conventional steels, the manganese and silicon concentrations are also kept close to zero, because both reduce austenite grain-boundary cohesion. AerMet 100 has good weldability and does not require preheating prior to welding.

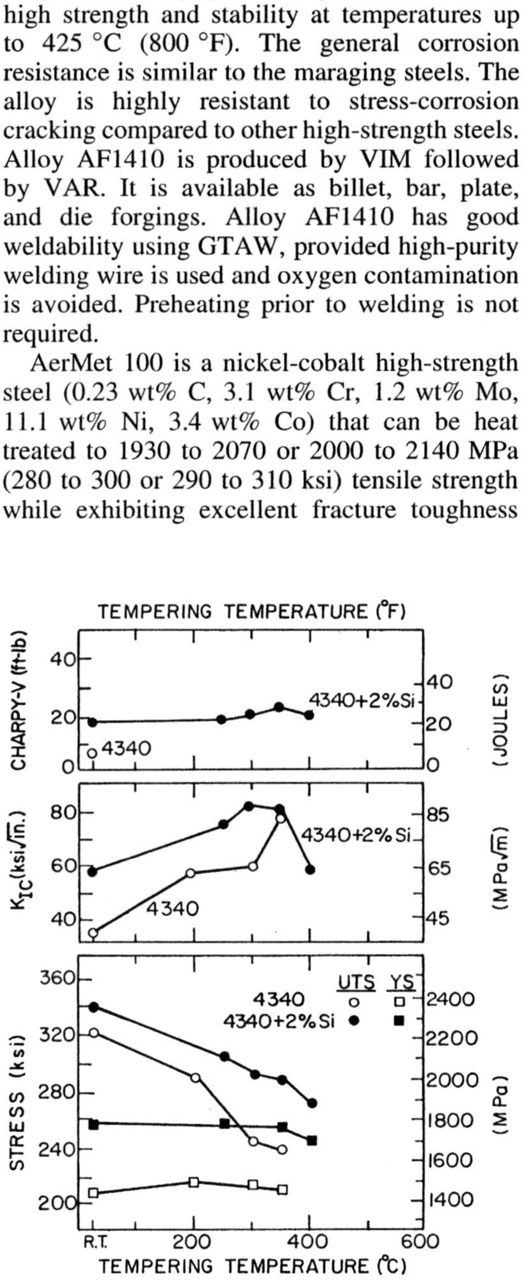


**Fig. 20.8** Effects of silicon content on 4340 steels. YS, yield strength; UTS, ultimate tensile strength Source: Ref 11

## 20.5 Maraging Steels

Maraging steels are a class of high-strength steels with very low carbon contents (0.030 wt% maximum) and additions of substitutional alloying elements that produce age hardening of iron-nickel martensites. The term *maraging* was derived from the combination of the words "martensite" and "age hardening." Maraging steels have high hardenability and high strength combined with high toughness. The maraging steels have a nominal composition by weight of 18% Ni, 7 to 9% Co, 3 to 5% Mo, less than 1% Ti, and very low carbon contents. Carbon is considered an impurity and kept to as low a level as possible to minimize the formation of titanium carbide (TiC), which can adversely affect ductility and toughness. During air cooling from the annealing or hot working temperature, maraging steels transform to a relatively soft martensite (30 to 35 HRC), which can be easily machined or formed. They are then aged to high strength levels at 455 to 510 °C (850 to 950 °F) for times ranging from 3 to 9 h.

The commercial maraging steels, 18Ni(200), 18Ni(250), 18Ni(300), and 18Ni(350), have nominal yield strengths after heat treatment of 1380, 1725, 2070 and 2415 MPa (200, 250, 300, and 350 ksi), respectively. Typical properties of maraging steels are shown in Table 20.14.

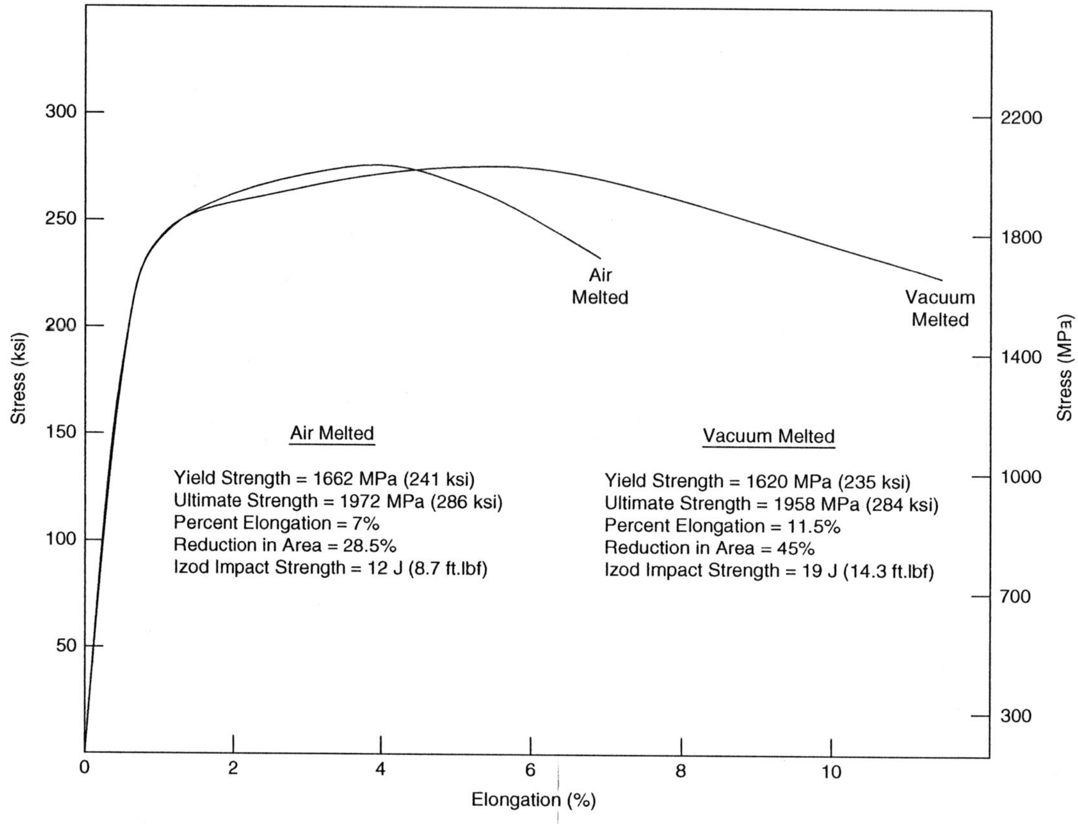


**Fig. 20.9** Comparison of air- and vacuum-melted 300M steel. Source: Ref 12

## Table 20.13 Typical properties of high-fracture-toughness steels

| Alloy | Heat treatment | Ultimate tensile strength, MPa | Ultimate tensile strength, ksi | Yield strength, MPa | Yield strength, ksi | Elongation (4D gage), % | Reduction in area, % | Fracture toughness ($K_{Ic}$), MPa $\sqrt{m}$ | Fracture toughness ($K_{Ic}$), (ksi $\sqrt{in.}$) |
|---|---|---|---|---|---|---|---|---|---|
| HP-9-4-30 | 845 °C (1550 °F), oil quench, cool to −75 °C (−100 °F), double temper at 205 °C (400 °F) | 1650–1790 | 240–260 | 1380–1450 | 200–210 | 8–12 | 25–35 | 66–99 | 60–90 |
| | 845 °C (1550 °F), oil quench, cool to −75 °C (−100 °F), double temper at 550 °C (1025 °F) | 1520–1650 | 220–240 | 1310–1380 | 190–200 | 12–16 | 35–50 | 99–115 | 90–105 |
| AF1410 | 900 °C (1650 °F), air cool, 830 °C (1525 °F), air cool, cool to −75 °C (−100 °F), temper at 510 °C (950 °F) | 1680 | 244 | 1475 | 214 | 16 | 69 | 174 | 158 |
| AerMet 100 | 885 °C (1625 °F), air cool, cool to −75 °C (−100 °F) aged at 480 °C (900 °F) | 1965 | 285 | 1724 | 250 | 13–14 | 55–65 | 100–115 | 110–126 |

Source: Ref 3

**Table 20.14 Typical properties of maraging steels**

| Alloy | Heat treatment | Ultimate tensile strength, MPa | Ultimate tensile strength, ksi | Yield strength, MPa | Yield strength, ksi | Elongation in 2 in., % | Reduction in area, % | Fracture toughness ($K_{Ic}$), MPa $\sqrt{m}$ | Fracture toughness ($K_{Ic}$), ksi $\sqrt{in.}$ |
|---|---|---|---|---|---|---|---|---|---|
| 18Ni(200) | 815 °C (1500 °F), air cool, age 3 h at 480 °C (900 °F) | 1503 | 218 | 1400 | 203 | 10 | 60 | 154–242 | 140–220 |
| 18Ni(250) | | 1793 | 260 | 1703 | 247 | 8 | 55 | 121 | 110 |
| 18Ni(300) | | 2048 | 297 | 2000 | 290 | 7 | 40 | 80 | 73 |
| 18Ni(350) | 815 °C (1500 °F), air cool, age 12 h at 480 °C (900 °F) | 2468 | 358 | 2448 | 355 | 6 | 25 | 35–49 | 32–45 |

Source: Ref 13

Because of their extremely low carbon contents, the fracture toughness of maraging steels is considerably higher than that of conventional high-strength steels. Maraging steels can be used for prolonged service at temperatures up to 400 °C (750 °F). Maraging steels are also more resistant to hydrogen embrittlement than the medium-carbon low-alloy steels. Although they are susceptible to stress-corrosion cracking, they are more resistant than the medium-carbon low-alloy steels. Processing techniques that improve the fracture toughness, such as vacuum melting, proper hot working, and keeping residual impurities low, also improve the resistance to stress-corrosion cracking. The alloys are available in the form of sheet, plate, bar, and die forgings; however, most applications use bar or forgings.

Maraging steels are either air melted followed by VAR or VIM followed by VAR. Aerospace grades are triple melted using air, VIM, and VAR to minimize the residual elements carbon, manganese, sulfur, and phosphorus and the gases oxygen, nitrogen, and hydrogen. Carbon and sulfur are the most deleterious impurities because they tend to form brittle carbide, sulfide, carbonitride, and carbosulfide inclusions that can crack when the metal is strained, lowering the fracture toughness and ductility.

Heat treatment consists of solution annealing, air cooling, and then aging. Solution annealing is usually conducted at 815 °C (1500 °F) for 1 h. Since the nickel content is so high, austenite transforms to martensite on cooling from the austenitic temperature. The martensite start temperature ($M_S$) is approximately 155 °C (310 °F), and the martensite finish temperature ($M_f$) is approximately 100 °C (210 °F). The formation of martensite is not affected by cooling rate, and thick sections can be air cooled and still be fully martensitic. Since the martensitic transformation involves only an austenite-to-martensite transformation of iron-nickel and does not involve carbon to any considerable extent, the martensite formed is relatively ductile. Before aging, maraging steels have yield strengths in the range of 655 to 830 MPa (95 to 120 ksi). The effect of aging temperature on 18Ni(250) is shown in Fig. 20.10.

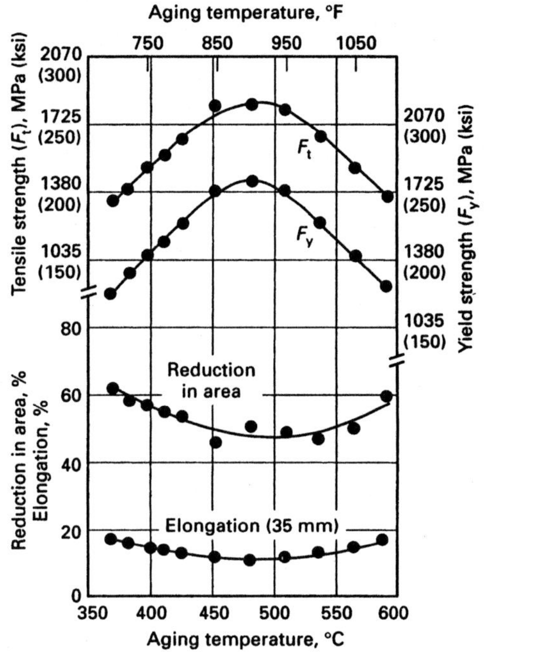


**Fig. 20.10** Effect of aging temperature on 18Ni(250) maraging steel. Source: Ref 13

## 20.6 Austenitic Manganese Steels

The first austenitic manganese steel was invented by Sir Robert Hadfield in 1882. This

**Table 20.15 Composition ranges of austenitic manganese steel casting alloys**

| ASTM A 128 grade | Composition, % | | | | | | |
|---|---|---|---|---|---|---|---|
| | C | Mn | Cr | Mo | Ni | Si (max) | P (max) |
| A | 1.05–1.35 | 11.0 min | ... | ... | ... | 1.00 | 0.07 |
| B-1 | 0.9–1.05 | 11.5–14.0 | ... | ... | ... | 1.00 | 0.07 |
| B-2 | 1.05–1.2 | 11.5–14.0 | ... | ... | ... | 1.00 | 0.07 |
| B-3 | 1.12–1.28 | 11.5–14.0 | ... | ... | ... | 1.00 | 0.07 |
| B-4 | 1.2–1.35 | 11.5–14.0 | ... | ... | ... | 1.00 | 0.07 |
| C | 1.05–1.35 | 11.5–14.0 | 1.5–2.5 | ... | ... | 1.00 | 0.07 |
| D | 0.7–1.3 | 11.5–14.0 | ... | ... | 3.0–4.0 | 1.00 | 0.07 |
| E-1 | 0.7–1.3 | 11.5–14.0 | ... | 0.9–1.2 | ... | 1.00 | 0.07 |
| E-2 | 1.05–1.45 | 11.5–14.0 | ... | 1.8–2.1 | ... | 1.00 | 0.07 |
| F | 1.05–1.35 | 6.0–8.0 | ... | 0.9–1.2 | ... | 1.00 | 0.07 |

Source: Ref 1

steel quickly became known for its outstanding wear resistance and toughness. These steels have high carbon contents (1 wt%) and very high manganese contents (11 to 14 wt%). Since manganese is a strong austenite stabilizer, they retain their austenitic structure on cooling to room temperature. As a result of their fcc crystalline structure, these steels have a combination of high toughness and ductility, a high work-hardening capacity, and good abrasion and wear resistance. Because these alloys work harden during use, they are used in applications involving earthmoving, mining and quarrying, and railway track. Since they are difficult to hot work, they are usually cast into the final product form. The most common alloys are castings that are covered in ASTM A 128, with the compositions given in Table 20.15.

As shown in Fig. 20.11, the mechanical properties of austenitic manganese steels increase with increasing carbon contents up to 1.05 to 1.35 wt%. As the carbon content increases, it becomes increasingly difficult to retain all of the carbon in solid solution, which may account for the reductions in tensile strength and ductility. However, because abrasion resistance tends to increase with carbon content, carbon contents higher than 1.20 wt% are used, even though the ductility is lower. Carbon contents above 1.4 wt% are seldom used because of the difficulty of obtaining an austenitic structure free of grain-boundary carbides, which are detrimental to strength and ductility.

Manganese content has little effect on yield strength; however, ultimate tensile strength and ductility increase fairly rapidly with increasing manganese content up to approximately 12 wt% and then tend to level off, although small improvements normally continue up to approximately 13 wt% Mn. The most common

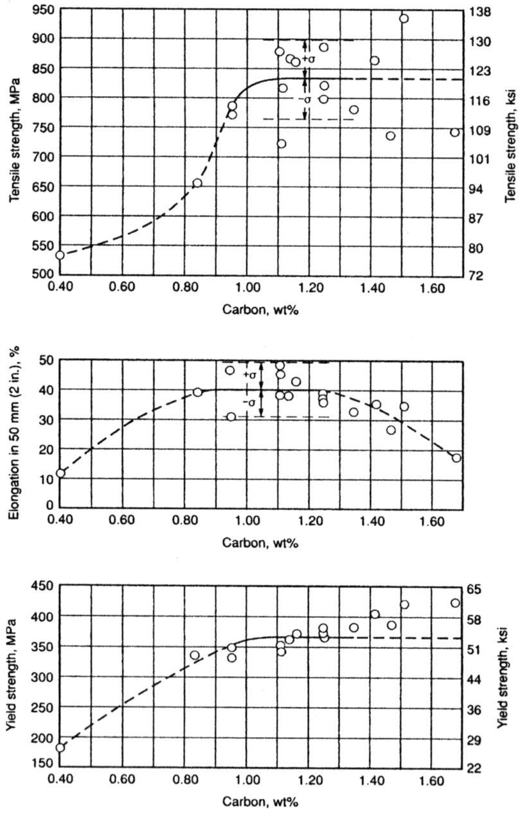


**Fig. 20.11** Effect of carbon content on properties of austenitic manganese steel. Source: Ref 1

alloying elements are chromium, molybdenum, and nickel. Both chromium and molybdenum increase the yield strength and flow resistance under impact when added to steels with the normal carbon content of 1.15 wt%. Nickel additions increase ductility, decrease yield strength slightly, and lower the abrasion resistance of manganese steel. However, nickel is particularly effective for suppressing precipitates of carbides, which can form between approximately 300 and 550 °C (570 and 1020 °F).

Heat treatment is used to put carbon in solution, since carbide precipitates along grain boundaries decrease strength and ductility. Heat treatment consists of slowly heating to 1010 to 1095 °C (1850 to 2000 °F), soaking for 1 to 2 h per inch of thickness to dissolve the carbon in solution, and then quenching in agitated water to prevent carbide precipitation. The microstructures of a 76 mm (3 in.) section of austenitic manganese steel in the as-cast condition and after solution annealing and quenching are shown in Fig. 20.12. A fully austenitic structure, essentially free of carbides and reasonably homogeneous with respect to carbon and manganese, is desired in the as-quenched condition, although this is not always attainable in heavy sections or in steels containing carbide-forming elements such as chromium, molybdenum, vanadium, and titanium. If carbides exist in the as-quenched structure, it is desirable for them to be present as relatively innocuous particles or nodules within the austenite grains rather than as continuous envelopes at grain boundaries.

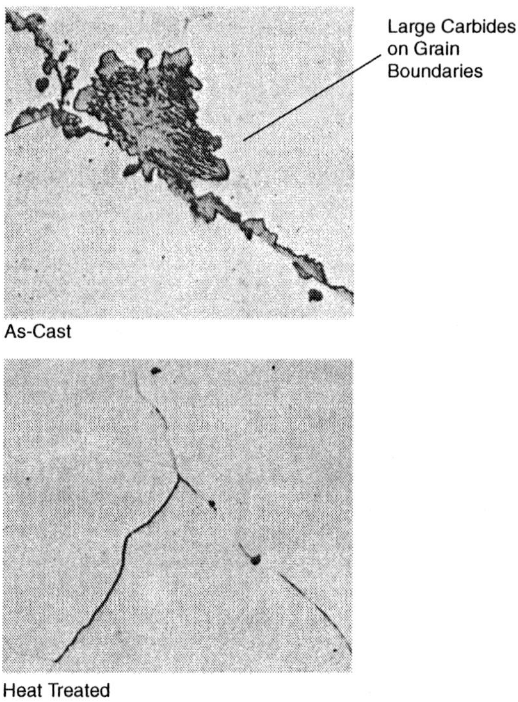


**Fig. 20.12** Effect of heat treatment on microstructure of austenitic manganese casting. Original magnification: 500 × . Source: Ref 1

Manganese steels are unequaled in their ability to work harden, exceeding even the metastable austenitic stainless steels. For example, a standard grade of manganese steel containing 1.0 to 1.4 wt% C and 10 to 14 wt% Mn can work harden from an initial level of 220 HV to a maximum of more than 900 HV. Maximum attainable hardness depends on many factors, including specified composition, service limitations, method of work hardening, and preservice hardening procedures. It appears that rubbing under heavy pressure can produce higher values of maximum attainable hardness than can be produced by simple impact.

Austenitic manganese steels have certain properties that tend to restrict their use. They are difficult to machine and usually have yield strengths of only 345 to 415 MPa (50 to 60 ksi). Consequently, they are not well suited for parts that require close-tolerance machining or parts that must resist plastic deformation when highly stressed in service.

## 20.7 High-Strength Low-Alloy Steels

Conventional hot rolled mild steels have rather low strengths but are readily weldable. As the carbon content is increased to increase strength, the amount of lamellar pearlite increases, reducing weldability and toughness and increasing the ductile-to-brittle transition temperature (DBTT). The development of high-strength low-alloy (HSLA) steels in the early-1980s provided an answer to this dilemma. High-strength low-alloy steels are a hybrid between plain carbon steels and alloy steels and are often considered to be a separate class of steel. They contain alloying elements; however, the alloy content is usually on the order of only 0.1 wt% and is referred to as microalloying. The HSLA steels are essentially low-carbon steels (0.03 to 0.1 wt% C) containing approximately 1.5 wt% Mn and less than 0.1 wt% of niobium, titanium, and/or vanadium, which have been hot rolled under controlled conditions to produce ultrafine ferrite grain sizes of less than 5 to 10 μm. They attain yield strengths of 275 to 550 MPa (40 to 80 ksi) and tensile strengths of 415 to 690 MPa (60 to 100 ksi), with a DBTT of approximately −75 °C (−100 °F). These steels have better mechanical properties and sometimes better corrosion resistance than hot rolled plain carbon steels. For example, a plain carbon hot rolled steel containing 0.2 wt% C will have a ferrite grain size in the range of 20 to 30 μm, a yield strength of approximately 205 MPa (30 ksi), and a tensile strength of approximately 380 MPa (55 ksi). Due to their low carbon and interstitial contents, HSLA steels have good toughness and weldability. Since the higher strength of HSLA steels can be obtained at lower carbon contents, the weldability of many HSLA steels is comparable to or better than that of mild steel. In addition, since HSLA steels do not contain a large percentage of alloying elements, they can be competitively priced against plain carbon steels.

The main factors responsible for the increased strength of HSLA steels are a fine ferrite grain size, precipitation hardening, and solid-solution hardening, with a fine ferrite size being the most important. A comparison of the grain sizes of a conventional and an HSLA steel is shown in Fig. 20.13. Note the much finer grain size for the HSLA steel. Microalloying, on the order of 0.1 wt%, with niobium, vanadium, and/or titanium is used along with controlled rolling. The very fine grain sizes in HSLA steels result from the control of the austenite grain size by the formation of carbides, carbonitrides, and nitrides during hot rolling as the temperature of the steel falls. These fine precipitates pin the austenite grain boundaries, hindering grain growth. At still lower rolling temperatures, they inhibit recrystallization of the severely deformed austenite grains. The elongated and pancaked grains can then rapidly transform to fine ferrite. These fine precipitates also provide additional locations for ferrite nuclei to form during cooling, resulting in an even finer ferrite grain size. In addition, as shown in Fig. 20.14, nuclei locations are also formed within the austenite grains at locations of deformed shear bands.

The microalloying additions of niobium, vanadium, and/or titanium perform similar, although somewhat different, functions during steel processing. The most stable of the precipitates is TiN. It forms either during solidification or during soaking at relatively high temperatures (1205 to 1315 °C, or 2200 to 2400 °F). High soaking temperatures are used to dissolve as much niobium, vanadium, and/or titanium as possible so that they can precipitate later during hot rolling. TiN precipitates at the grain boundaries to restrict austenite grain growth during soaking and during dynamic recrystallization of austenite at the high initial rolling temperatures. Niobium is the most effective element in modifying the recrystallization behavior of austenite during hot rolling in the range of 1040 to 815 °C (1900 to 1500 °F). The presence of niobium precipitates does not allow recrystallization at temperatures as high as 925 °C (1700 °F). Since the solubility of niobium in less than vanadium or titanium, grain refinement can also be obtained with smaller additions.

The effect of microalloying elements on preventing recrystallization is shown in Fig. 20.15. Each of the four steels in the figure was subjected to hot rolling deformation of 50% at

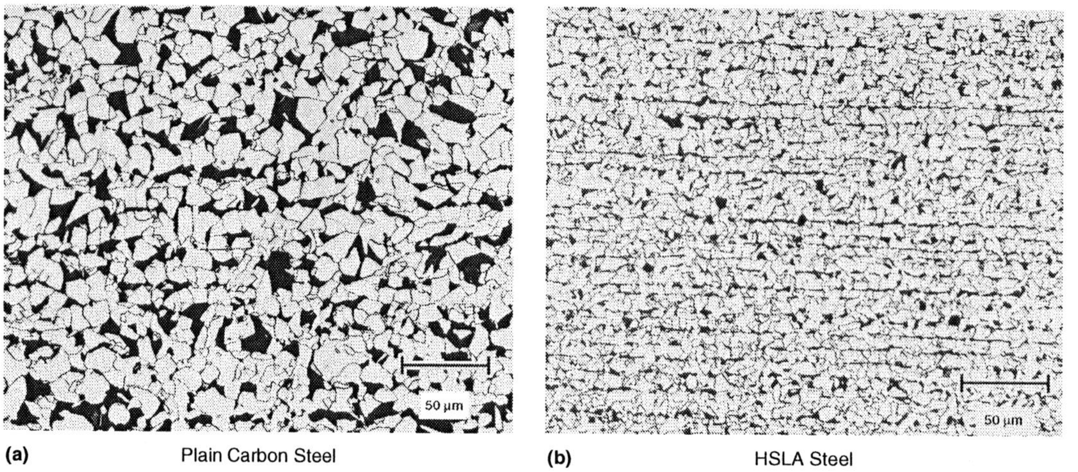


(a) Plain Carbon Steel (b) HSLA Steel

**Fig. 20.13** Fine grain size in high-strength low-alloy (HSLA) steel. Source: Ref 14

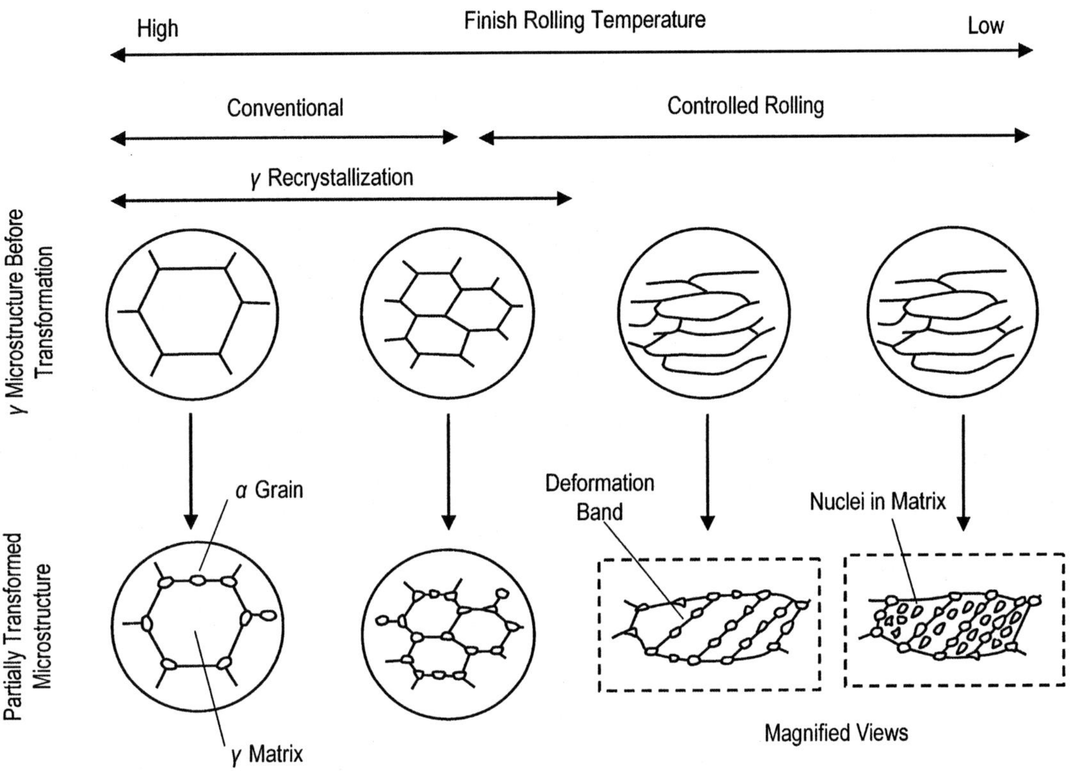


**Fig. 20.14** Microstructure development in low-carbon steels. Source: Ref 15

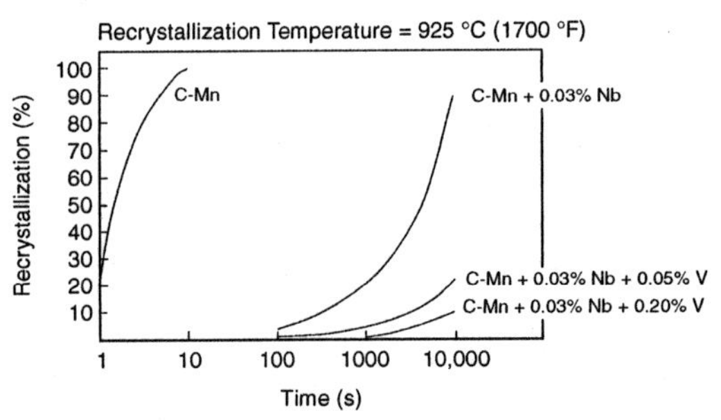


**Fig. 20.15** Recrystallization kinetics of microalloyed steels. Source: Ref 16

955 °C (1750 °F) before recrystallization. The plain carbon steel, containing 0.09 wt% C and 1.9 wt% Mn, recrystallized in approximately 10 s. The same steel, with an addition of only 0.03 wt% Nb, did not recrystallize until 10,000 s, where 90% recrystallization was achieved. When small amounts of vanadium were added to the steel containing 0.03 wt% Nb, the recrystallization process was further retarded. This illustrates the extremely strong effects of even very small amounts of niobium in retarding the recrystallization kinetics of austenite. In addition, when small amounts of vanadium are added along with niobium, recrystallization is further retarded. The ability of these elements to slow down recrystallization is due to the fact that they precipitate as fine carbides, nitrides, and carbonitrides that pin the austenite grain boundaries. In the case of niobium, the primary precipitate is niobium carbonitride, NbCN. These precipitates restrict both grain growth and recrystallization of the austenite. The carbides and carbonitrides of vanadium generally precipitate at lower temperatures, either during the austenite-to-ferrite transformation or as interphase precipitates in the ferrite itself. Thus, these precipitates contribute mainly to the strength of the steel by precipitation hardening. Precipitation of the vanadium precipitates occurs on a very fine scale (~5 nm) during cooling through 845 to 650 °C (1550 to 1200 °F). Solid-solution hardening is the result of manganese, silicon, and uncombined nitrogen.

High-strength low-alloy steels are primarily hot rolled into the usual wrought product forms (sheet, strip, bar, plate, and structural sections) and are commonly furnished in the hot rolled condition. In addition to hot rolled products, HSLA steels are also furnished as cold rolled sheet and forgings. The main advantage of HSLA forgings, like hot rolled HSLA products, is that yield strengths in the range of 275 to 550 MPa (40 to 80 ksi) can be achieved without heat treatment. Base compositions by weight of these microalloyed ferrite-pearlite forgings are typically 0.3 to 0.50% C and 1.4 to 1.6% Mn. Low-carbon bainitic HSLA steel forgings have also been developed.

The HSLA steels are covered under numerous SAE and ASTM International specifications. The HSLA steels are extensively used as structural beams for bridge construction, off-shore oil and natural gas platforms, ship hull and deck plate, and electrical transmission towers and poles. In automobiles, HSLA steels are used for safety applications such as ultrahigh-strength impact door beams and energy-absorbing bumper assemblies and for increasing fuel economy through thinner and lighter-weight chassis sections. The HSLA steels are also used for large-diameter gas transmission pipelines.

## 20.8 Dual-Phase Steels

A dual-phase steel is one that consists of islands of hard martensite embedded in a tougher continuous ferrite matrix. A mixture of fine ferrite and austenite grains is produced by heating into the two-phase $\alpha + \gamma$ field, followed by quenching to convert the austenite to martensite. Thus, the final microstructure consists of a mixture of ferrite and martensite (Fig. 20.16) with the possibility of some retained austenite within the martensite islands. The microstructural constituents that can be present in dual-phase steels after processing are a fine-grained ferrite matrix, some proeutectoid ferrite that formed from the austenite during cooling, and martensite or lower bainite, depending on the alloy content and the $M_s$ temperature. In addition, dual-phase steels also contain some retained austenite, and when vanadium or niobium is present, there can be carbonitride particles embedded within the matrix.

In a typical process, the sheet is passed through a continuous annealing furnace where it is heated into the $\alpha + \gamma$ range and then quenched. Typically, the objective is to produce approximately 15 to 20% austenite. Sometimes, it is reheated to temper the martensite, but normally this is not done because it tends to lower the tensile strength. In some processes, the sheet is cold rolled prior to annealing. In this case, the deformed ferrite recrystallizes during heating. With appropriate control of the rate of heating and the degree of cold work, a fine-grained mixture of ferrite and austenite can be obtained. However, the time at the high temperature has to be short to prevent grain growth. Another production method is to finish the hot rolling stage so that 80 to 90% ferrite forms and to allow the remaining austenite to transform to martensite later while the steel is being rolled into a coil. For this to be successful, the steel must have suitable continuous cooling transformation characteristics. Alloying to produce a bay in the TTT diagram creates a window so that metastable austenite can be coiled before martensite or bainite begins.

The unique characteristic of dual-phase steels is the continuous yielding behavior during deformation; that is, there is a lack of a yield point during deformation (Fig. 20.17). This provides increased uniform elongation and work hardening so that parts produced from a dual-phase steel actually gain strength during the forming operation. Dual-phase steels yield at relatively low stresses and then work harden

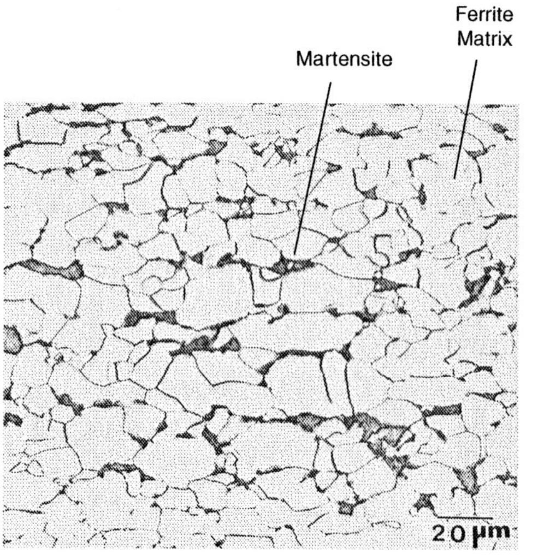


**Fig. 20.16** Microstructure of dual-phase steel. Source: Ref 17

rapidly. Any retained austenite usually transforms to martensite during deformation. The tensile strength varies approximately linearly with the percent martensite or lower bainite. Typical values are 550 MPa (80 ksi) for 10% by volume rising to 760 MPa (110 ksi) at 30%. Elongations of approximately 20% are typical.

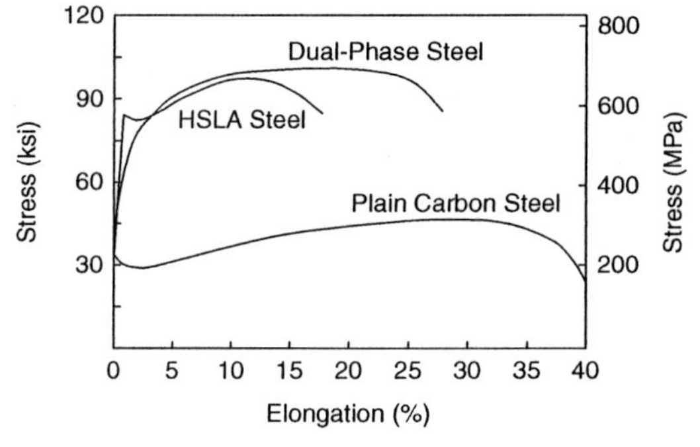


**Fig. 20.17** Stress-strain curves for plain carbon, high-strength low-alloy (HSLA), and dual-phase steels. Source: Ref 18

Dual-phase steels are used in applications such as automobile wheel rims and wheel disks. Because of their energy-absorbing characteristics, dual-phase steels are also used in critical locations of automobiles for safety to protect the occupants in the event of a crash.

## 20.9 TRIP Steels

The term TRIP is derived from the mechanism of transformation-induced plasticity.

The nominal composition by weight of these steels is 0.25% C, 2% Mn, 2% Si, 10% Cr, 9% Ni, and 5% Mo. They contain a high percentage of retained austenite (10 to 15%). The austenite transforms to martensite during forming of the part, thus providing enhanced formability, or it transforms on impact, as in an automotive crash. Similar to dual-phase steels, TRIP steels are annealed in the intercritical region, but instead of direct cooling to room temperature to form martensite, they are isothermally treated (Fig. 20.18). The products of transformation

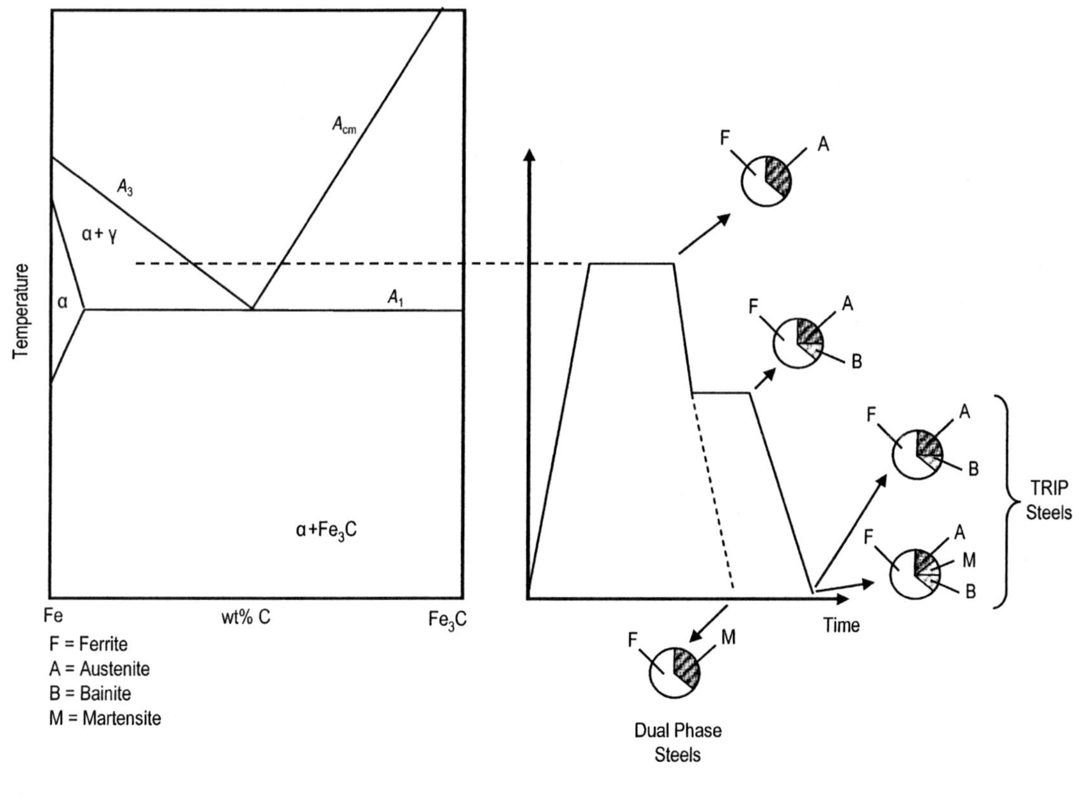


**Fig. 20.18** Intercritical annealing cycles. Source: Ref 19

therefore include bainite and retained austenite. The retained austenite later transforms to martensite under the action of stress, hence the term *transformation-induced plasticity*. The strain-induced martensitic transformation increases strain hardening and helps to delay necking instability during forming operations. The isothermal hold during heat treatment is designed to produce large, dispersed volume fractions of retained austenite in the ferrite matrix after intercritical annealing. In TRIP steels, alloying and isothermal treatments are designed to produce bainite and maximize retained austenite. The temperature of the isothermal hold is important; too high a temperature will produce excessive amounts of bainite, and too low a temperature can result in excessive amounts of martensite. Since silicon cannot dissolve in the crystal structure of cementite and therefore prevents cementite formation, TRIP steels may contain as much as 1.2 to 1.5 wt% Si. However, silicon can cause surface finish problems during forming.

## ACKNOWLEDGMENTS

Sections of this chapter were adapted from *Alloying: Understanding The Basics* edited by J.R. Davis, ASM International, 2001; "High Fracture Toughness Steels" by T.V. Phillip and T.J. McCafferty; and "Austenitic Manganese Steels" in *Properties and Selection: Irons, Steels, and High-Performance Alloys*, Volume 1, *ASM Handbook,* ASM International, 1990.

## REFERENCES

1. Austenitic Manganese Steels, *Properties and Selection: Irons, Steels, and High-Performance Alloys,* Vol 1, *ASM Handbook,* ASM International, 1990
2. J.R. Davis, *Alloying: Understanding The Basics,* ASM International, 2001
3. T.V. Philip and T.J. McCafferty, High Fracture Toughness Steels, *Properties and Selection: Irons, Steels, and High-Performance Alloys,* Vol 1, *ASM Handbook,* ASM International, 1990
4. E.C. Bain and H.W. Paxton, *Alloying Elements in Steel,* 2nd ed., American Society for Metals, 1961
5. High-Strength Structural and High-Strength Low-Alloy Steels, *Properties and Selection: Irons, Steels, and High-Performance Alloys,* Vol 1, *ASM Handbook,* ASM International, 1990
6. N.E. Dowling, *Mechanical Behavior of Materials,* 2nd ed., Prentice Hall, 1999
7. B.L. Bramfitt, Effects of Composition, Processing, and Structure on Properties of Irons and Steels, *Materials Selection and Design,* Vol 20, *ASM Handbook,* ASM International, 1997
8. Mechanical Properties of Carbon and Alloy Steels, *Metals Handbook Desk Edition,* 2nd ed., ASM International, 1998
9. W.D. Callister, *Fundamentals of Materials Science and Engineering,* 5th ed., John Wiley & Sons, Inc., 2001
10. G. Krauss, *Steels: Processing, Structure, and Performance,* 3rd ed., ASM International, 2005
11. E.R. Parker, *Metall. Trans. A,* Vol 8, 1977, p 1025
12. W.M. Imrie, *Philos. Trans. R. Soc. (London) A,* Vol 282, 1976, p 91
13. M. Schmidt and K. Rohrbach, Heat Treatment of Maraging Steels, *Heat Treating,* Vol 4, *ASM Handbook,* ASM International, 1991
14. B.L. Bramfitt and S.J. Lawrence, Metallography and Microstructures of Carbon and Low-Alloy Steels, *Metallography and Microstructures,* Vol 9, *ASM Handbook,* ASM International, 2004
15. I. Kozasu, Processing—Thermomechanical Controlled Processing *Constitution and Properties of Steels,* Vol 7, *Materials Science and Technology,* VCH, Weinheim, Germany, 1992, p 183–217
16. R.E. Reed-Hill and R. Abbaschian, *Physical Metallurgy Principles,* 3rd ed., PWS Publishing Company, 1994
17. A.O. Benscoter and B.L. Bramfitt, Metallography and Microstructures of Low-Carbon and Coated Steels, *Metallography and Microstructures,* Vol 9, *ASM Handbook,* ASM International, 2004
18. M.S. Rashid and B.V.N. Rao, Tempering Characteristics of a Vanadium-Containing Dual-Phase Steel, *Fundamentals of Dual-Phase Steels,* TMS-AIME, 1977, p 249–264
19. L. Laquerbe, J. Neutjens, P. Harlet, F. Caroff, and P. Cantinieaux, New Processing Route for the Production of Silicon-Free TRIP-Assisted Cold-Rolled and Galvanized Steels, *41st MWSP Conference Proceedings,* ISS, Vol XXXVII, 1999, p 89–99

## SELECTED REFERENCES


- "AerMet 100 Alloy" datasheet, Carpenter Technology Corporation, 1995
- B.L. Bramfitt, Carbon and Alloy Steels, *Handbook of Materials Selection,* John Wiley & Sons, Inc., 2002
- F.C. Campbell, *Manufacturing Technology for Aerospace Structural Materials,* Elsevier Scientific, 2006
- J.O. Morlett, H. Johnson, and A. Troiano, *J. Iron Steel Insti.,* Vol 189, 1958, p 37
- K. Rohrbach and M. Schmidt, Maraging Steels, *Properties and Selection: Irons, Steels, and High-Performance Alloys,* Vol 1, *ASM Handbook,* ASM International, 1990
- W.F. Smith, *Structure and Properties of Engineering Alloys,* 2nd ed., McGraw-Hill, Inc., 1993

# CHAPTER 21

# Surface Hardening of Steel

THERE ARE SOME APPLICATIONS where it is necessary to have a hard, wear-resistant surface but a tough, shock-resistant inner core. For example, cams, gears, and shafts require hard surfaces to resist wear but tough inner cores to resist shock. While a low-carbon steel containing 0.1 wt% C will have a tough core, its surface hardness will be low after hardening. On the other hand, a high-carbon steel containing 0.8 wt% C will have a high surface hardness after hardening, but the core will not be tough and shock resistant. There are two approaches to this problem. One is to use a medium-carbon steel and only harden the surface through heat treatment. The other approach is to diffuse carbon into the surface layers of a low-carbon steel. When heat treated, the high-carbon surface layers will attain a much higher hardness than the low-carbon core. This method of case hardening, called carburizing, is feasible if small, fast-diffusing elements, such as carbon or nitrogen, are used that will form hard carbides or nitrides.

## 21.1 Surface Hardening by Localized Heat Treatment

Since these processes require a high enough carbon content to obtain the required hardness, a medium-carbon steel with a carbon content of 0.35 to 0.50 wt% is usually selected. Initially, the part is hardened by conventional quenching and tempering to produce the desired core hardness. Sometimes, normalizing will produce the desired core hardness. The surface is then reheated into the austenitization range and immediately quenched to produce fresh martensite at the surface. The part is then retempered to produce the desired surface hardness. The hard surface layers and the soft core will generally be separated by a cushion layer of bainite that helps in reducing cracking and spalling. In addition to the hard surface, the surface layer is usually in a state of compression, which improves fatigue cracking resistance.

### *21.1.1 Flame Hardening*

The objective of flame hardening is to austenitize the steel at and near the surface and then to remove the flame and rapidly quench the work to produce martensite. The surface is heated by a gas flame created by burning acetylene, propane, or natural gas. The relatively low thermal conductivity of steel enables the surface regions to be austenitized using high rates of energy input without the interior being significantly affected. Flame hardening can be as simple as an operator with a torch or can consist of automated systems equipped with quench jets that follow right behind the torches. The torch can also be used to temper the martensite. Many variations are used, ranging from hand-held torches to automated ignition, burn, and quench assemblies, for example, a rotating shaft within a surrounding stationary array of burners.

Flame hardening is a very rapid and efficient method for producing cases as deep as 6.3 mm (1/4 in.), but the maximum hardness that can be obtained (50 to 60 HRC) is less than can be attained with through hardening. Unless the process is automated, it can be difficult to control the case depth, and prolonged heating can result in a case depth deeper than desired. Since only the surface is hardened, when the part is quenched, there is less chance of distortion or warpage. It is often used where small quantities of parts require hardening, the part is large and bulky, or the heat treating facilities are limited.

### *21.1.2 Induction Hardening*

In induction hardening, heat is supplied by surrounding the part with an inductor coil carrying a high-frequency current in the range of 2 to 500 kHz. Higher frequencies result in a shallower depth of heating and are therefore used for smaller-diameter workpieces. The coil acts like the primary winding of a transformer. The oscillating field produced by the induction coil induces electrical eddy currents in the steel within a certain depth of the outer surface, called the skin depth, which decreases as the frequency is increased. The eddy currents produce Joule resistance heating ($I^2R$) in the skin depth that rapidly raises the surface temperature. Additional heating is supplied by hysteresis losses, and the surface usually attains the austenization temperature within a few seconds. The part is then quenched to form martensite on the surface layers. Since the copper inductor coils are subject to radiation heating, they are made from hollow, water-cooled copper tubing. Where possible, the part is slowly rotated during heating to obtain more uniform heating.

Induction hardening is readily automated. It is more adaptable than flame heating because a wide variety of coil configurations are possible. A number of different coil configurations and their resultant magnetic fields are shown in Fig. 21.1. In many cases, when the needed depth of hardening is quite shallow, the induction heating time is completed within a few seconds. Irregular shapes can be handled quite readily with induction heating. Skin current can penetrate crevices and holes as well as exterior surfaces. To provide a uniform starting fine-grained microstructure, the steel is often normalized prior to induction hardening.

Induction hardening is generally used to produce relatively thin cases. Larger depths, such as 3.2 mm ($^1/_8$ in.), can be attained by leaving the current in contact with the surface for a longer period of time and by operating at lower frequencies. The case depth can be controlled more accurately in induction hardening than with other processes. The depth can be controlled by varying the frequency, the current, and the amount of time the current is in contact with the part. The higher the frequency, the more the current tends to flow over the outer surface only. Induction hardening provides outstanding resistance to warpage, distortion, oxidation, and scale formation due to the short heating time and to the fact that only a small portion of the part requires heating.

For induction heating processes of short duration, the depth to which the steel is austenitized is small. Shape changes due to thermal expansion and transformation of structure are accommodated by plastic flow in the hot metal. Then, when the rim transforms to martensite during cooling, it tries to expand, but the core is relatively cool and resists plastic flow. The result is that the surface can be forced into compression, which improves fatigue resistance. During treatments in which the heating depth is greater, the situation can be reversed, and the surface can be put into tension, which, of course, is detrimental to fatigue resistance.

The main disadvantage of induction hardening is the cost of the equipment and the requirement for a skilled technician to initially set up the process. However, once it is set up, a relatively unskilled technician can operate it.

## 21.2 Case Hardening

The surface hardness of low-carbon steels can be increased significantly by diffusing carbon into the surface layers at elevated temperatures, followed by quenching and tempering. Since the surface layers have higher carbon contents than the core, the surface layers attain a much

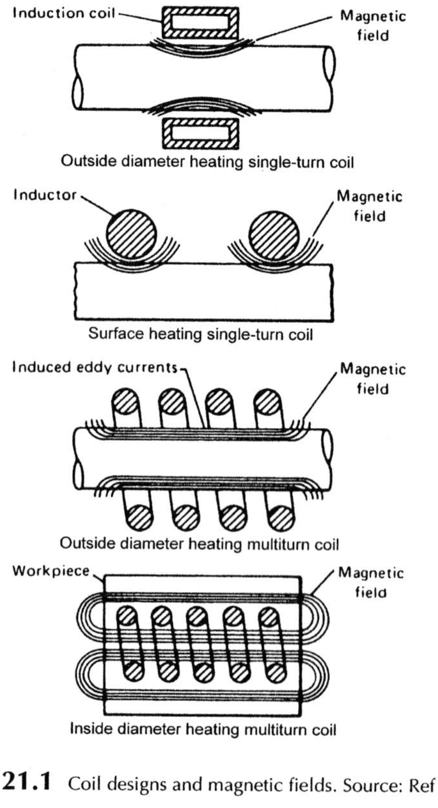


**Fig. 21.1** Coil designs and magnetic fields. Source: Ref 3

higher hardness than the inner core during heat treatment. The end result is a hard, wear-resistant surface with a tough inner core. This process, called carburizing, is one of several case-hardening processes. Another case-hardening process is nitriding. Since nitriding is done at a lower temperature, the part is quenched and tempered before nitriding. A third process, called carbonitriding, which is similar to carburizing, diffuses both carbon and nitrogen atoms into the surface.

## 21.3 Carburizing

Carburizing is conducted by heating a low-carbon steel into the single-phase austenitic field, generally between 845 and 955 °C (1550 and 1750 °F), where the steel has a high solubility for carbon. After holding for the appropriate time, the part is either quenched or cooled to room temperature. If it is cooled to room temperature, it then must be reheated for quenching. After quenching, the part is then tempered in the normal manner. Carburizing produces a wear-resistant high-carbon case on top of a tough low-carbon steel core. Steels used for carburizing usually have carbon contents of approximately 0.2 wt%, with carburized cases containing up to 0.8 to 1.0 wt% C.

A simple formula developed by Einstein can be used to predict the case depth:

$$\text{Case depth} = k\sqrt{t} \quad \text{(Eq 21.1)}$$

where

$$k = \sqrt{2D}$$

and $D$ is the diffusion coefficient. The case depth is a function of the surface concentration, temperature, and time. Carburizing can be done in a solid, liquid, or gaseous medium. A comparison of these different carburizing processes is shown in Table 21.1.

### 21.3.1 Pack Carburizing

Pack carburizing is a solid-state process in which the parts are placed in a heat-resistant steel box containing a solid carburizing medium such as charcoal. The parts are then heated to 875 to 925 °C (1600 to 1700 °F) and held for up to 8 h, depending on the desired case depth. Higher temperatures and longer times can be used to produce a greater case depth. The carbonaceous material combines with air present in the box to form carbon monoxide:

$$C + O_2 \rightarrow CO \quad \text{(Eq 21.2)}$$

Carbon monoxide then decomposes into carbon dioxide and carbon atoms, freeing the carbon atoms to diffuse into the steel surface:

$$2CO \rightarrow CO_2 + C \quad \text{(Eq 21.3)}$$

Barium carbonate (10 to 15 wt%) is usually added to facilitate the reaction. Barium carbonate decomposes to form:

$$BaCO_3 \rightarrow BaO + CO_2 \quad \text{(Eq 21.4)}$$

and the carbon dioxide reacts with charcoal to form carbon monoxide:

$$CO_2 + C \leftrightarrow 2CO \quad \text{(Eq 21.5)}$$

Note that this reaction is reversible, and as the temperature is increased, the rate of carburization increases. Since the rate of change in case

**Table 21.1 Comparison of carburizing processes**

| Process | Process temperature °C | Process temperature °F | Typical case depth μm | Typical case depth mils | Case hardness, HRC | Typical base metals | Process characteristics |
|---|---|---|---|---|---|---|---|
| Pack | 815–1095 | 1500–2000 | 125–1525 | 5–60 | 50–63 | Low-carbon steels, low-carbon alloy steels | Low equipment costs; difficult to control case depth accurately |
| Gas | 815–980 | 1500–1800 | 75–1525 | 3–60 | 50–63 | Low-carbon steels, low-carbon alloy steels | Good control of case depth; suitable for continuous operation; good gas controls required; can be dangerous |
| Liquid | 815–980 | 1500–1800 | 50–1525 | 2–60 | 50–65 | Low-carbon steels, low-carbon alloy steels | Faster than pack and gas processes; can pose salt disposal problem; salt baths require frequent maintenance |
| Vacuum | 815–1095 | 1500–2000 | 75–1525 | 3–60 | 50–63 | Low-carbon steels, low-carbon alloy steels | Excellent process control; bright parts; faster than gas carburizing; high equipment costs |

depth at a particular carburizing temperature is proportional to the square root of time, the rate of carburization is highest at the beginning of the cycle and gradually diminishes as the cycle is extended (Fig. 21.2). If it is necessary to prevent any areas from being carburized, they can be masked by electroplating with copper to a thickness of 0.075 to 0.10 mm (0.003 to 0.004 in.). At the carburizing temperature, carbon is insoluble in copper.

Pack carburizing has several advantages over other case-hardening processes. It involves minimal capital expense and is fairly foolproof. It is especially practical when only a few small parts require surface hardening at one time. The main disadvantages are that the process is rather slow and it is dirty.

### 21.3.2 Liquid Carburizing

Liquid carburizing is carried out in a bath of fused salt containing 20 to 50 wt% sodium cyanide, up to 40 wt% sodium carbonate, and varying amounts of either sodium or barium chloride. The bath temperature ranges from 870 to 955 °C (1600 to 1750 °F). Carburization time ranges from 5 min up to 1 h, depending on the desired case thickness. Salt bath carburizing is often conducted on small parts that are loaded in baskets that are then immersed in the bath. After carburizing, the parts can be immediately quenched to produce the martensitic case. The process is useful for producing shallow cases of 0.10 to 0.25 mm (0.004 to 0.010 in.), although case depths up to 5.0 mm (0.20 in.) can be attained. Since the bath contains sodium cyanide (NaCNO), some nitrogen is also released and diffuses into the surface, yielding additional hardness. Since some nitrogen is absorbed into the surface and immediately hardens the part, parts that are liquid carburized generally are not machined after carburizing. An advantage of salt baths is that they offer fast heating and accurate temperature control. On the downside, cyanides are extremely poisonous, efficient fume extraction systems are required, and waste disposal is problematic. In addition, the parts must be rinsed to remove the salts after carburizing to prevent rusting.

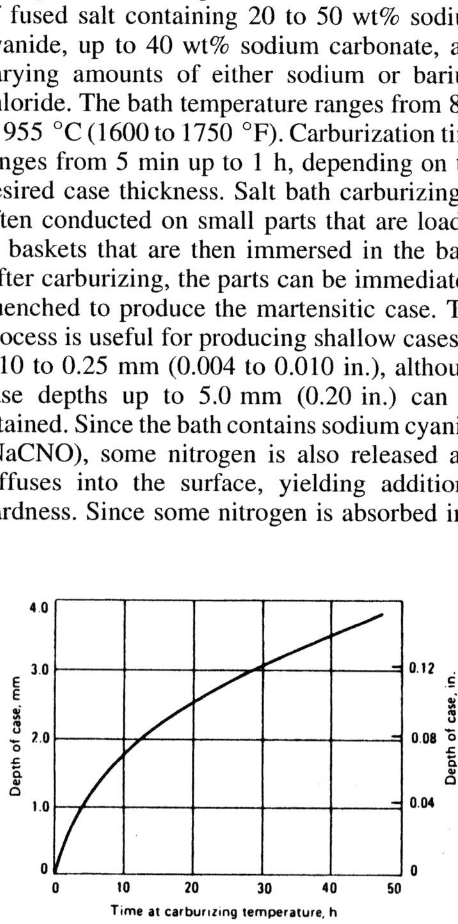


**Fig. 21.2** Effect of time on case depth during pack carburizing at 925 °C (1700 °F). Source: Ref 1

Different bath temperatures can be used to produce different case depths; low-temperature baths produce shallow cases, and high-temperature baths produce deep cases. Low-temperature cyanide baths (light-case baths) are usually operated in the temperature range of 845 to 900 °C (1550 to 1650 °F), although for certain specific effects this range is sometimes extended to 790 to 925 °C (1450 to 1700 °F). High-temperature cyanide baths (deep-case baths) are usually operated in the temperature range of 900 to 955 °C (1650 to 1750 °F). High-temperature baths are used for producing cases 0.50 to 3.0 mm (0.020 to 0.120 in.) deep. In some instances, even deeper cases are produced, up to approximately 6.4 mm (0.250 in.), but the most important use of these baths is for the rapid development of cases 1.0 to 2.0 mm (0.040 to 0.080 in.) deep.

Salt baths are usually contained in relatively small chambers or tanks, so it can be impractical to immerse large, odd-shaped parts into the liquid. For this reason, liquid carburizing is usually restricted to surface hardening of small parts.

### 21.3.3 Gas Carburizing

The vast majority of carburized parts are processed by gas carburizing, using either natural gas, propane, or butane. The part is heated to approximately 900 °C (1650 °F) for 3 to 4 h in a carbon-rich atmosphere, usually a mixture of carbon monoxide and water vapor. Typical case depths are less than 2.5 mm (0.1 in.), usually in the range of 0.5 to 1.5 mm (0.020 to 0.060 in.). Carburization temperatures are usually in the range of 845 to 955 °C (1550 to 1750 °F). Carburization times depend on the desired case depth. A common definition of the case depth is that depth below the surface at which the hardness is 50 HRC in the final quenched and tempered part.

In gas carburizing, carbon is transferred from the carburizing atmosphere to the part surface.

The carbon diffuses slowly into the bulk of the part, and a carbon concentration gradient below the surface is established. The driving force for the carburizing reaction is called the carbon potential. Within the steel part, the high-carbon surface has a higher carbon potential than the low-carbon interior; thus, carbon tends to diffuse from the surface toward the center. Similarly, the carburizing atmosphere has a higher carbon potential than does the surface of the steel. If, during processing, the atmosphere carbon potential should fall below the carbon potential at the steel surface, then carbon will be removed from the steel (decarburization). The carbon concentration gradient of carburized parts is a function of the carburizing temperature and time, type of cycle, carbon potential of the furnace atmosphere, and the original composition of the steel.

The source of carbon is a carbon-rich furnace atmosphere produced either from gaseous hydrocarbons, such as methane ($CH_4$), propane ($C_3H_3$), and butane ($C_4H_{10}$), or from vaporized hydrocarbon liquids. Controlled carburizing atmospheres are produced by blending a carrier gas with an enriching gas, which serves as the source of carbon. The usual carrier endothermic gas is not merely a diluent but also accelerates the carburizing reaction at the surface of the parts. The amount of enriching gas required depends primarily on the rate at which carbon is absorbed by the workpiece. The carrier gas is a blend of carbon monoxide, hydrogen, and nitrogen, with smaller amounts of carbon dioxide, water vapor, and methane produced by reacting a hydrocarbon gas, such as methane, propane, or butane, with air.

By assuming that the atmosphere consists of a carrier gas produced from methane that serves as the source of the carbon for the workpiece, the main constituents of the atmosphere are CO, $N_2$, $H_2$, $CO_2$, $H_2O$, and $CH_4$. Of these constituents, $N_2$ is inert, acting only as a diluent. The amounts of CO, $CO_2$, $H_2$, and $H_2O$ present are very nearly the proportions expected at equilibrium from the reversible reaction:

$$CO + H_2O \leftrightarrow CO_2 + H_2 \quad \text{(Eq 21.6)}$$

given the particular ratios of carbon, oxygen, and hydrogen in the atmosphere. Methane is present in amounts well in excess of the amount that would be expected if all the gaseous constituents were in equilibrium.

Although the sequence of reactions involved in carburizing is not known in detail, it is known that carbon can be added or removed rapidly from steel by the reversible reactions:

$$2CO \leftrightarrow C + CO_2 \quad \text{(Eq 21.7)}$$

and

$$CO + H_2 \leftrightarrow C + H_2O \quad \text{(Eq 21.8)}$$

However, a carburization process based solely on the decomposition of CO would require large flow rates of atmosphere gas to produce appreciable carburizing.

The methane enrichment of endothermic gas provides carbon for the process by slow reactions such as:

$$CH_4 + CO_2 \rightarrow 2CO + 2H_2 \quad \text{(Eq 21.9)}$$

and

$$CH_4 + H_2O \rightarrow CO + 3H_2 \quad \text{(Eq 21.10)}$$

which reduces the concentrations of $CO_2$ and $H_2O$, respectively. These reactions regenerate CO and $H_2$, thereby directing the reactions of Eq 21.7 and 21.8 to the right. Because the methane content of carburizing atmospheres is usually far above the content that is expected at equilibrium, given the $CO_2$ and $H_2O$ contents present, the reactions in Eq 21.9 and 21.10 do not approach equilibrium. The sum of the reactions in Eq 21.7 and 21.9 and in Eq 21.8 and 21.10 reduces to:

$$CH_4 \rightarrow C + 2H_2 \quad \text{(Eq 21.11)}$$

Therefore, with constant $CO_2$ content and constant dewpoint, the net atmosphere composition change during carburizing is a reduction in methane content and an increase in the hydrogen content. In most commercial operations, gas flow rates are high enough and the rate of methane decomposition is low enough to prevent a large buildup of hydrogen during a carburizing cycle. However, with carburizing loads having high surface area, there is a drop in the CO content of 1 to 3% at the beginning of the cycle, when the carbon demand is greatest. This is caused by the dilution of the furnace atmosphere with hydrogen. Carbon potential control during carburizing is achieved by varying the flow rate of the hydrocarbon enriching gas

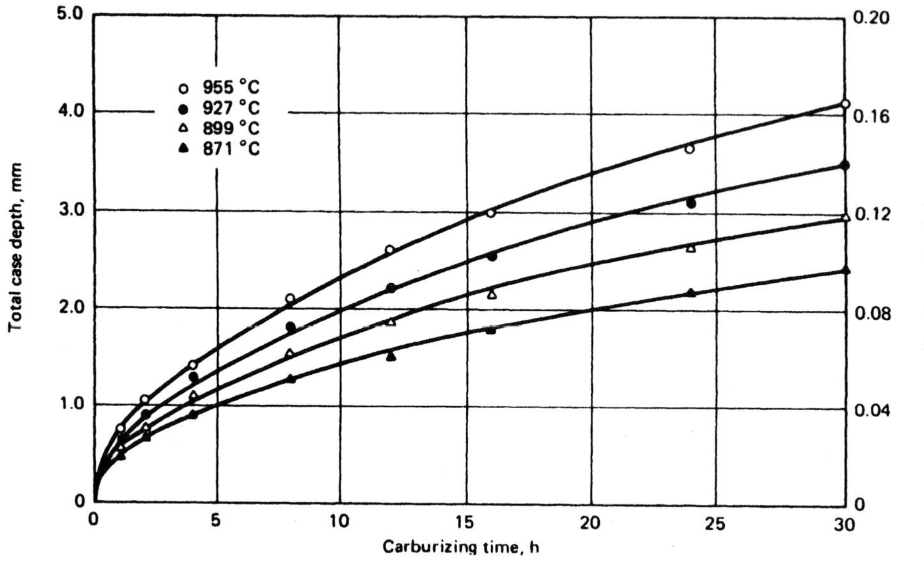


**Fig. 21.3** Effect of time and temperature on case depth during gas carburizing. Source: Ref 1

while maintaining a steady flow of endothermic carrier gas.

The maximum rate at which carbon can be added to steel is limited by the rate of diffusion of carbon in austenite, but the diffusion rate increases greatly with increasing temperature. For example the rate of carbon addition at 925 °C (1700 °F) is approximately 40% greater than at 870 °C (1600 °F). The temperature most commonly used for gas carburizing is 925 °C (1700 °F). This temperature permits a reasonably rapid carburizing rate without rapid deterioration of furnace equipment. The carburizing temperature is sometimes raised to 955 or 980 °C (1750 or 1800 °F) to shorten the time of carburizing for parts requiring deeper cases. For shallow-case carburizing, lower temperatures are used because case depth can be controlled more accurately with the slower rate of carburizing obtained at lower temperatures.

For consistent results, the temperature must be uniform throughout the work load, because if thinner parts reach the carburizing temperature first, they will begin to carburize well before thicker parts. This can produce a variability in the case depth from part to part and within a single part. In addition, soot can be deposited on cold parts exposed to a carburizing atmosphere. Therefore, the workpiece should be heated to the carburizing temperature in a near-neutral furnace atmosphere. In batch furnaces, parts can be heated in the endothermic carrier gas until

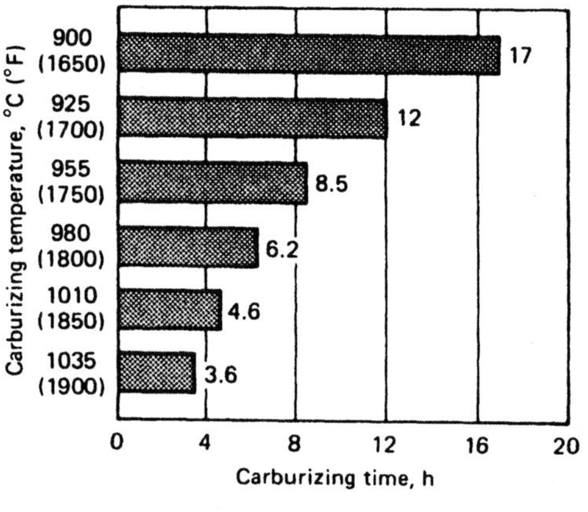


**Fig. 21.4** Effect of temperature on time during gas carburizing. Source: Ref 1

they reach the furnace temperature, then carburizing can commence with the addition of the enriching gas.

The effect of time and temperature on total case depth is shown in Fig. 21.3, which assumes carbon-saturated austenite at the surface of the workpieces. When the surface carbon content is less than the saturation value, case depths will be less than indicated in the figure. The carburizing time decreases with increasing temperature, as shown in Fig. 21.4, to produce a 1.5 mm (0.06 in.) case depth in 8620 steel. In addition to the time at the carburizing temperature, several hours may be required to bring large workpieces or heavy loads of smaller parts to the carburizing temperature. For work quenched directly from the carburizing furnace, the cycle may be further lengthened by allowing time for the work to cool from the carburizing temperature to approximately 845 °C (1550 °F) prior to quenching. If the work load is exposed to the carburizing atmosphere during heating, some carburizing will occur before the nominal start of carburizing. Similarly, additional diffusion and interchange of carbon with the atmosphere will occur during cooling prior to quenching. Thus, the actual case depth achieved may differ significantly from the values shown in Fig. 21.3 and 21.4.

Gas carburizing furnaces can be divided into two major categories: batch and continuous furnaces. In a batch-type furnace, the work load is charged and discharged as a single unit or batch. In a continuous furnace, the work enters and leaves the furnace in a continuous stream. Continuous furnaces are favored for the high-volume production of similar parts with total case depth requirements of less than 2.0 mm (0.08 in.).

The microstructure in the carburized case depends on its temperature-time history. A major difference exists in samples that are quenched directly from the carburizing furnace as compared to samples that are first cooled and then reaustenitized and quenched. In the latter case, the reaustenitization of the ferrite produced on cooling from the carburization furnace produces a fine-grained austenite in contrast to that which exists after prolonged carburization, because the reaustenitization temperature is lower, usually below the $A_{cm}$ line. In this case, the steel matrix is austenitized, and the proeutectoid cementite forms into globular particles that are dispersed in the martensite when the samples are quenched. This microstructure consists of fine crystals of very hard martensite and a dispersion of proeutectoid particles that are beneficial for wear resistance. In contrast, the austenitic grain size at the end of the carburization time is large. Often, the part is cooled below the $A_{cm}$ before it is quenched. This reduces the thermal shock by decreasing the temperature difference between the sample and the quenching medium. However, proeutectoid films can grow along the austenite grain boundaries during this stage, and these films are retained along the prior-austenite grain boundaries during quenching. Many of the martensitic plates are bigger because the grains in which they form are large, and also, the retained austenite is less finely dispersed. A comparison of the microstructures for gas-carburized and heat treated 9310 steel with different carbon contents is shown in Fig. 21.5.

### *21.3.4 Vacuum Carburizing (Adapted from Ref 1)*

Compared to conventional gas carburizing, vacuum carburizing offers several advantages: (1) excellent uniformity and repeatability resulting from the high degree of process control possible with vacuum furnaces, (2) better mechanical properties due to the lack of intergranular oxidation, and (3) potentially reduced cycle times, particularly when the higher process temperatures possible with vacuum furnaces are used. The disadvantages of vacuum carburizing are predominantly related to equipment costs and throughput.

Vacuum carburizing is a four-step process:

1. Heat and soak at carburizing temperature to ensure temperature uniformity throughout steel
2. Boost to increase the carbon content of austenite
3. Diffusion to provide gradual case/core transition
4. Oil quenching. In addition, a reheat step prior to quenching may also be necessary for grain refinement.

The first step is to heat the steel being carburized to the desired carburizing temperature, typically in the range of 845 to 1040 °C (1550 to 1900 °F), and then soak at the carburizing temperature only long enough to ensure that the steel is uniformly at temperature. Oversoaking can result in a reduction in toughness due to grain growth. A rough vacuum of 13 to 40 Pa (0.1 to 0.3 torr) is used to prevent surface oxidation. In the boost step, the vacuum chamber is backfilled to a partial pressure with either a pure hydrocarbon gas or a mixture of hydrocarbon gases. A minimum partial pressure of hydrocarbon gas is required to ensure rapid carburizing that varies with the carburizing temperature, the carburizing gas composition, and the furnace construction. Typical partial pressures vary from 1.3 to 6.7 kPa (10 to 50 torr) in furnaces of graphite construction to 13 to 27 kPa (100 to 200 torr) in furnaces of ceramic construction. The diffusion step enables the diffusion of carbon inward from the carburized

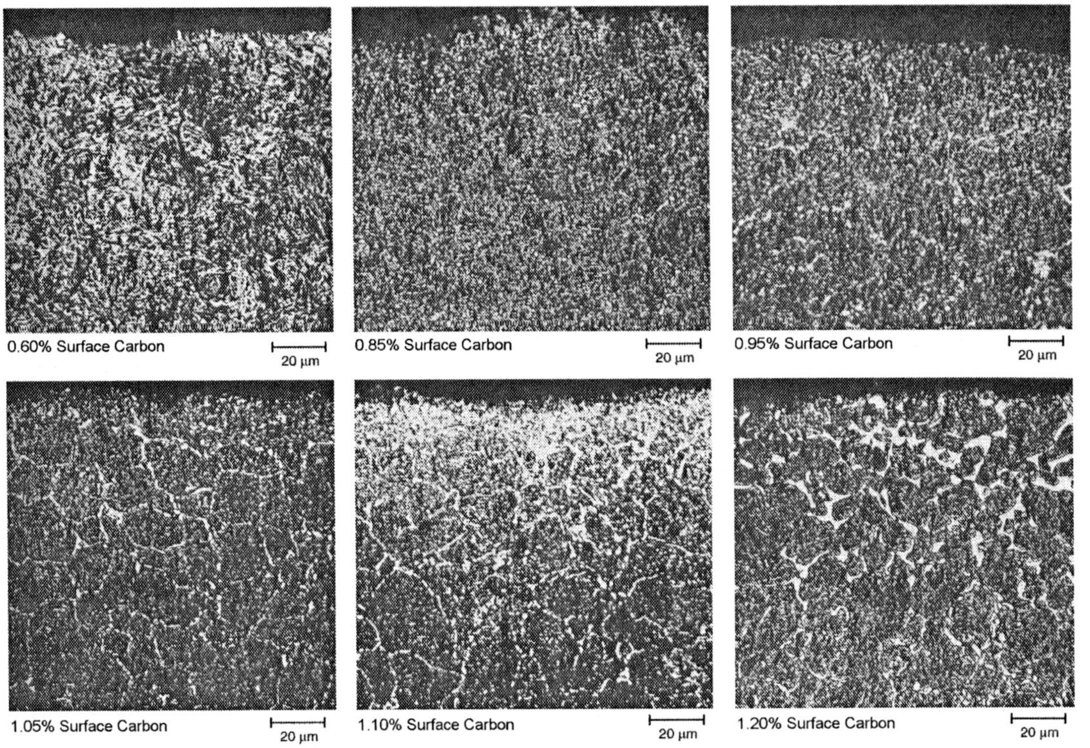


**Fig. 21.5** 9310 steel, gas carburized microstructures 4 h at 925–940 °C (1700–1725 °F), furnace cooled, austenitized at 815–830 °C (1500–1525 °F), oil quenched and tempered 4 h at 150 °C (300 °F). Source: Ref 4

surface, resulting in a lower surface carbon content and a more gradual case/core transition. The diffusion step is usually performed in a rough vacuum at the same temperature used for carburizing. If a reheat step is not going to be used, the steel is directly quenched in oil, usually under a partial pressure of nitrogen.

### *21.3.5 Plasma (Ion) Carburizing (Adapted from Ref 1)*

Plasma or ion carburizing is basically a vacuum process using glow discharge to introduce carbon-bearing ions to the steel surface for subsequent diffusion. This process is effective in increasing carburization rates because the process bypasses the several dissociation steps that are required to produce active soluble carbon. For example, because of the ionizing effect of the plasmas, active carbon for adsorption can be formed directly from methane gas. High temperatures can be used because the process takes place in a vacuum, thus producing a greater carburized case depth than either gas or vacuum carburizing (Fig. 21.6). Other advantages of plasma carburizing include improved case uniformity, insensitivity to steel composition, and environmental improvements.

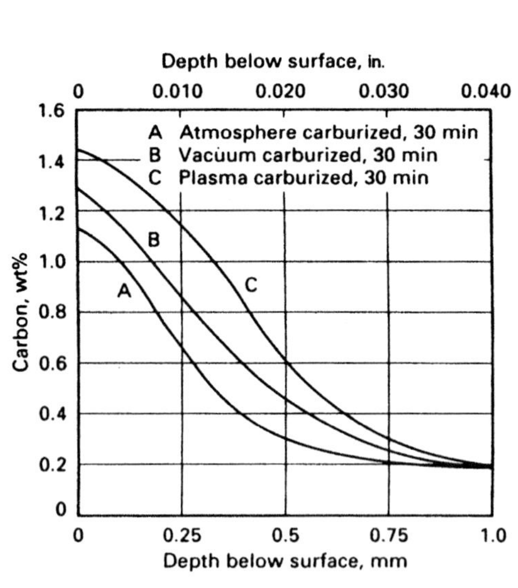


**Fig. 21.6** Comparison of carburization methods. 8620 steel carburized at 980 °C (1800 °F) for 30 min. Source: Ref 1

The microstructure of a steel gear tooth that was ion carburized is shown in Fig. 21.7. This gear was ion carburized at 920 °C (1690 °F), austenitized at 830 °C (1525 °F), oil quenched, and then tempered at 150 °C (300 °F). The microstructure is tempered martensite with some evidence of carbide or retained austenite. Plasma carburizing provides a much cleaner and safer environment than gas carburizing systems, and there is no fire hazard or toxic gases such as carbon monoxide.

A typical set-up for plasma carburizing is shown schematically in Fig. 21.8. The workpiece (cathode) is at ground potential, and the positive potential needed to establish and maintain the glow discharge is fed into the vacuum enclosure through a suitable insulated lead to a counter electrode (the anode). Auxiliary heating elements, either radiant or induction, surround the workpiece to heat it to the carburizing temperature, because the heat losses of the plasma are insufficient to heat the work load to the carburizing temperature of 900 to 1000 °C (1650 to 1830 °F).

Plasma (ion) nitriding is similar to plasma carburizing in that a plasma is formed in a vacuum using high-voltage electrical energy, and the nitrogen ions are accelerated toward the workpiece. The ion bombardment heats the part, cleans the surface, and provides active nitrogen. The process provides better control of case chemistry, case uniformity, and lower part distortion than gas nitriding. Ion nitriding can be performed at temperatures as low as 370 °C (700 °F), which minimizes residual stresses. Because loads are gas cooled, they do not experience distortion from temperature gradients or martensite formation.

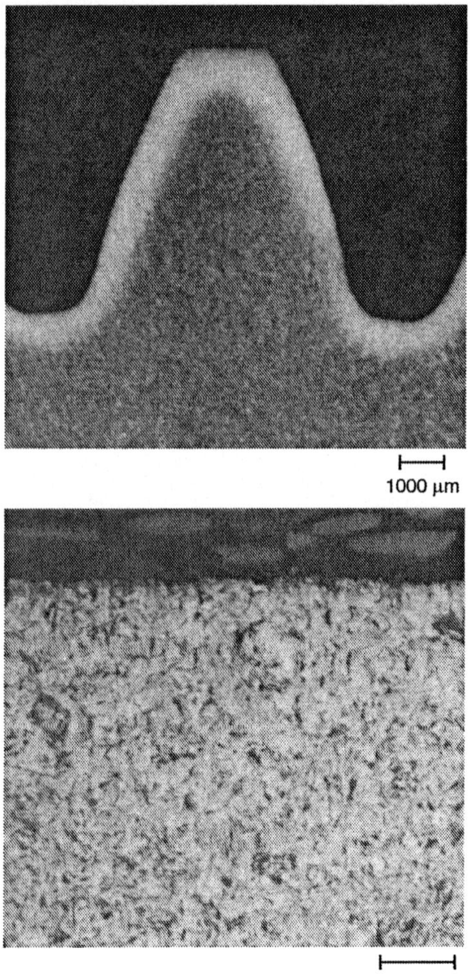


Fig. 21.7 Ion-carburized gear tooth, 2H2N4A steel, ion carburized at 920 °C (1690 °F), austenitized at 830 °C (1525 °F), oil quenched and tempered at 150 °C (300 °F). Source: Ref 4

## 21.4 Nitriding

Carburizing requires that the steel be quenched and then tempered. In contrast, nitriding is done at temperatures below the austenitization temperature. Since this removes the distortion inherent in the martensitic transformation during hardening, nitriding allows excellent dimensional control. Medium-carbon steels that have already been hardened by quenching and

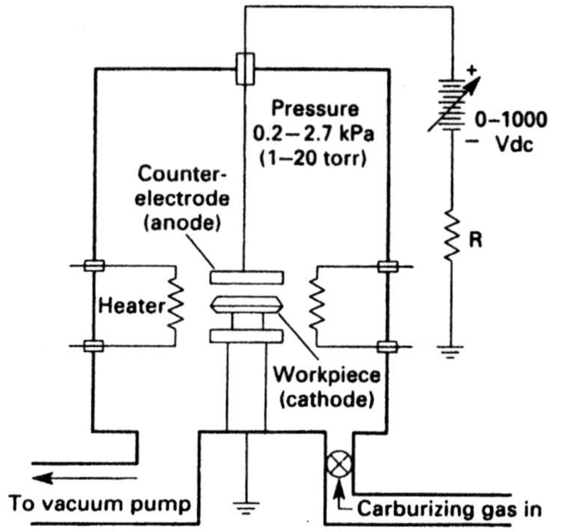


Fig. 21.8 Schematic of plasma (ion) carburizing. Source: Ref 1

tempering respond very well to nitriding. Similar to carburizing, nitriding can be done in several different media, as shown in the Table 21.2 comparison.

Nitriding is conducted in atmospheres that decompose ammonia to provide nitrogen to the surface. The nitrogen diffuses into the steel and also combines with the iron at the surface to form iron nitride according to the reaction:

$$NH_3 \rightarrow N + 3H \quad \text{(Eq 21.12)}$$

The nitriding case depth is shallow, usually less than 0.50 mm (0.020 in.) deep, even though nitriding times can exceed 100 h. During single-stage nitriding treatments, in which a single nitriding atmospheric composition is maintained, a white layer of iron nitride is formed. This iron nitride layer is hard but can crack and spall. When this is unacceptable, the layer is removed by surface grinding. Alternatively, a two-stage nitriding process can be used in which, after the first stage, the atmospheric conditions are changed so that iron nitride no longer forms at the surface, and the existing layer is removed as the nitrogen dissolves into the steel.

Unlike carburizing, nitriding is conducted between 500 and 550 °C (930 and 1020 °F). Since this is below the austenite formation temperature, the steel must be quenched and tempered before nitriding. Nitriding improves wear resistance, fatigue resistance, and corrosion resistance. The case of a nitrided steel contains hard $\gamma'$ ($Fe_4N$) and $\varepsilon$ ($F_{2\text{-}3}N$) intermetallic compounds. Steels containing aluminum, chromium, vanadium, tungsten, and molybdenum are suitable for nitriding because they readily form nitrides that are stable at the nitriding temperatures. For example, aluminum forms very fine AlN precipitates. Molybdenum, in addition to its contribution as a nitride former, also reduces the risk of embrittlement at nitriding temperatures.

Nitriding has a number of advantages. It produces the hardest cases; hardness values as high as 70 HRC are obtainable. No heat treatment is required after hardening, which, along with the low temperatures employed, minimizes warpage and distortion. The corrosion resistance of nitrided parts is better than that of carburized parts. Humidity, water, salt conditions, and other corrosive media are not as harmful to nitrided parts. Nitrided parts have good elevated-temperature resistance. Reheating parts to 540 to 595 °C (1000 to 1100 °F) for short periods does not affect their hardness, while long-term exposures to 315 to 425 °C (600 to 800 °F) will affect carburized but not nitrided parts. The disadvantages of nitriding are primarily the long cycle times and the inherent cost. A 0.75 mm (0.030 in.) case depth may take several days to produce, and ammonia gases are more expensive than the natural gases used for carburizing. Other disadvantages include some size growth that occurs during nitriding, and the extreme hardness produced precludes machining after nitriding.

### 21.4.1 Gas Nitriding

In gas nitriding, either a single-stage or a double-stage process can be used when nitriding with ammonia gas. Sources of nitrogen include $NH_3$, $NH_3$-$H_2$ mixtures, $NH_3$ mixed with an endothermic gas, and $NH_3$-$N_2$-$CO_2$ mixtures. The depths obtainable with nitriding are less than with carburizing, and the processing times are longer. A depth of 0.25 to 0.40 mm (0.01 to 0.015 in.) can be obtained in approximately 48 h.

**Table 21.2 Comparison of nitriding processes**

| | Process temperature | | Typical case depth | | | | |
|---|---|---|---|---|---|---|---|
| Process | °C | °F | μm | mils | Case hardness, HRC | Typical base metals | Process characteristics |
| Gas | 480–595 | 900–1100 | 125–760 | 5–30 | 50–70 | Alloy steels, nitriding steels, stainless steels | Hardest cases from nitriding steels; quenching not required; low distortion; process is slow; is usually a batch process |
| Salt | 510–565 | 950–1050 | 2.5–760 | 0.1–30 | 50–70 | Most ferrous metals, including cast irons | Usually used for thin, hard cases $<25$ μm (1 mil); no white layer; most are proprietary processes |
| Ion | 345–565 | 650–1050 | 75–760 | 3–30 | 50–70 | Alloy steels, nitriding steels, stainless steels | Faster than gas nitriding; no white layer; high equipment costs; close case control |

Either a single- or a double-stage process can be used when nitriding with anhydrous ammonia. In the single-stage process, a temperature in the range of approximately 500 to 525 °C (925 to 975 °F) is used, and the dissociation rate ranges from 15 to 30%. This process produces a brittle, nitrogen-rich white nitride layer at the surface, as shown in the micrograph of 4140 steel in Fig. 21.9. The double-stage process has the advantage of reducing the thickness of the white nitrided layer. The use of a higher-temperature second stage decreases the case hardness and increases the case depth. The first stage of the double-stage process is, except for time, a duplication of the single-stage process. The second stage may proceed at the nitriding temperature employed for the first stage, or the temperature may be increased to 550 to 565 °C (1025 to 1050 °F); however, at either temperature, the rate of dissociation in the second stage is increased to 65 to 85% (preferably 75 to 85%). Generally, an external ammonia dissociator is necessary for obtaining the required higher second-stage dissociation.

### 21.4.2 Liquid Nitriding

Liquid nitriding, conducted in a molten salt bath containing either cyanides or cyanates, is conducted at temperatures similar to gas nitriding. As opposed to salt bath carburizing, liquid nitriding can be used to treat finished parts because dimensional stability is maintained due to the subcritical temperatures used in the process. Furthermore, at the lower nitriding temperatures, liquid nitriding adds more nitrogen and less carbon to ferrous materials than is obtained with high-temperature treatments, because ferrite has a much greater solubility for nitrogen (0.4% maximum) than carbon (0.02% maximum).

## 21.5 Carbonitriding

Carbonitriding is a modified form of gas carburizing rather than a form of nitriding. The modification consists of introducing ammonia into the gas carburizing atmosphere to add nitrogen to the carburized case as it is being produced. Nascent nitrogen forms at the work surface by the dissociation of ammonia in the furnace atmosphere, and the nitrogen diffuses into the steel simultaneously with carbon. Typically, carbonitriding takes place at lower temperatures and shorter times than gas carburizing, producing a shallower case. Steels with carbon contents up to 0.2 wt% are commonly carbonitrided. It is usually conducted in a molten cyanide bath. Carbonitriding is similar to liquid cyaniding. Because of problems in disposing of cyanide-bearing wastes, carbonitriding is often preferred over liquid cyaniding. In terms of case characteristics, carbonitriding differs from carburizing and nitriding in that carburized cases normally do not contain nitrogen,

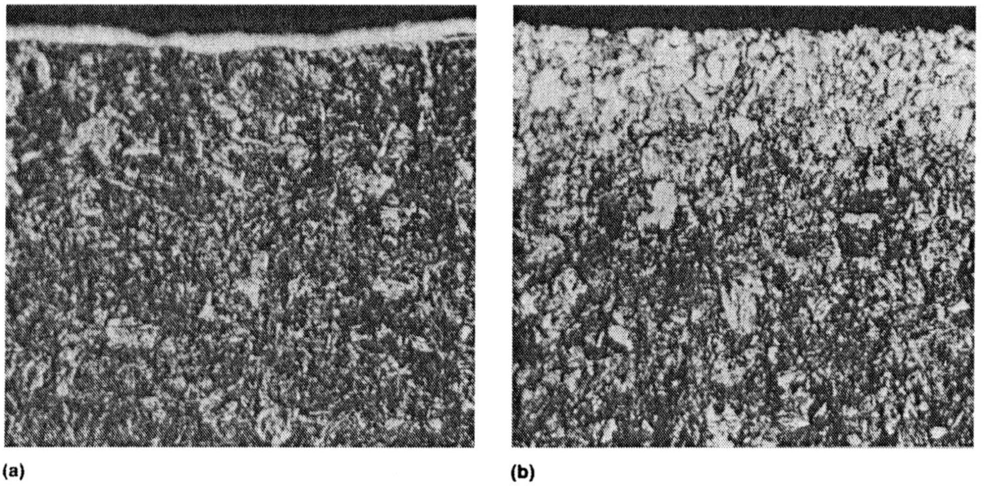

(a) (b)

**Fig. 21.9** Effect of one- and two-stage nitriding on white layer. (a) Single stage, gas nitrided for 24 h at 525 °C (975 °F). (b) Double stage, gas nitrided for 5 h at 525 °C (975 °F) followed by second stage at 565 °C (1050 °F) for 24 h. Original magnification: 400×. Source: Ref 1

**Table 21.3 Comparison of carbonitriding processes**

| Process | Process temperature °C | Process temperature °F | Typical case depth μm | Typical case depth mils | Case hardness, HRC | Typical base metals | Process characteristics |
|---|---|---|---|---|---|---|---|
| Gas | 760–870 | 1400–1600 | 75–760 | 3–30 | 50–65 | Low-carbon steels, low-carbon alloy steels, stainless steel | Lower temperature than carburizing (less distortion); slightly harder case than carburizing; gas control critical |
| Liquid (cyaniding) | 760–870 | 1400–1600 | 2.5–125 | 0.1–5 | 50–65 | Low-carbon steels | Good for thin cases on noncritical parts; batch process; salt disposal problems |
| Ferritic nitrocarburizing | 565–675 | 1050–1250 | 2.5–25 | 0.1–1 | 40–60 | Low carbon steels | Low distortion process for thin case on low-carbon steel; most processes are proprietary |

and nitrided cases contain nitrogen primarily, whereas carbonitrided cases contain both. A comparison of carbonitriding processes is given in Table 21.3.

Carbonitriding is used primarily to impart a hard, wear-resistant case, generally from 0.075 to 0.76 mm (0.003 to 0.030 in.) deep. Since nitrogen increases the hardenability of steel, a carbonitrided case has better hardenability than a carburized case. One major advantage of carbonitriding is that the nitrogen absorbed during processing lowers the critical cooling rate of the steel. Thus, the hardenability of the case is significantly greater when nitrogen is added by carbonitriding than when the same steel is only carburized (Fig. 21.10). Full hardness with less distortion can be achieved with oil quenching or, in some instances, even with gas quenching employing a protective atmosphere as the quenching medium.

Often, carburizing and carbonitriding are used together to achieve much deeper case depths and better engineering performance than could be obtained by using only the carbonitriding process. The process generally consists of carburizing at 900 to 955 °C (1650 to 1750 °F) to give the desired total case depth (up to 2.5 mm, or 0.100 in.), followed by carbonitriding for 2 to 6 h in the temperature range of 815 to 900 °C (1500 to 1650 °F) to add the additional carbonitrided case. The parts can then be oil quenched to obtain a deeper effective and thus harder case than would have resulted from the carburizing process alone. The addition of the carbonitrided surface increases the case residual-compressive stress level and thus improves contact fatigue resistance as well as increases the case strength gradient. When the carburizing/carbonitriding processes are used together, the ratio of effective case depth at 50 HRC to total case depth may vary from approximately 0.35 to 0.75, depending on the case hardenability, core hardenability, section size, and quenchant used.

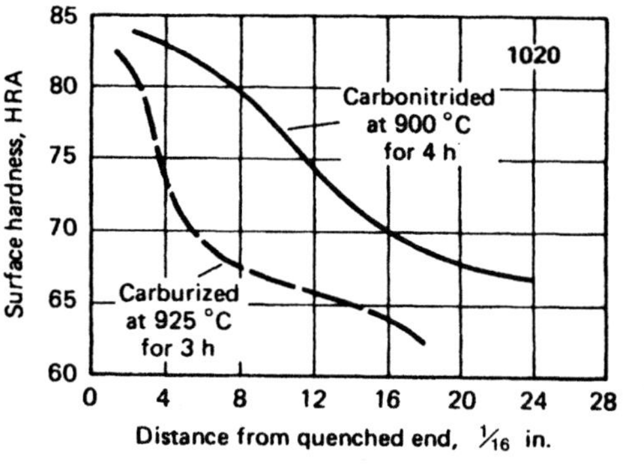


Fig. 21.10 Comparison between carburizing and carbonitriding. Source: Ref 1

## 21.6 Hardfacing

Hardfacing is the application of a coating or cladding to a substrate for the purpose of reducing surface damage. The principal hardfacing alloys are cobalt, nickel, and copper alloys and iron-base alloys with manganese or chromium. With the exception of the copper alloys and manganese steels, wear resistance is provided by the presence of hard chromium carbide particles. Some alloy compositions have tungsten carbide additions to further improve wear resistance. Hardfacing materials are applied by either welding or thermal spraying.

Weld overlay is a form of hardfacing that is applied by oxyacetylene or an arc welding process using hardfacing welding rods or electrodes (Fig. 21.11). Oxyacetylene welding overlays are used on steels where maximum hardness and minimum crack susceptibility are required. Although the rate of deposition is not as high as for other processes, oxyacetylene welding has the advantage of minimizing undesirable base metal dilution and loss of hardness of the hardfacing alloy. The absence of a steep thermal gradient in oxyacetylene welding reduces cracking or spalling because thermal stresses are reduced. Arc welding overlays are applied by gas tungsten arc welding (GTAW), shielded metal arc welding (SMAW), and gas metal arc welding (GMAW). GTAW yields very clean deposits with high rates of deposition. However, the high heat input results in steep thermal gradients, causing alloy dilution and loss of hardness in the overlay, along with increased cracking susceptibility from high thermal stresses. GTAW is often used where thin overlays are required. SMAW produces high deposition rates and intermediate dilution. Although some porosity and cracking are present, such discontinuities are usually acceptable in applications such as earthmoving and mining shovels that require thick overlays. It is generally necessary to apply several layers to achieve the desired thickness. GMAW is not as widely used for hardfacing as the other arc welding processes. A composite

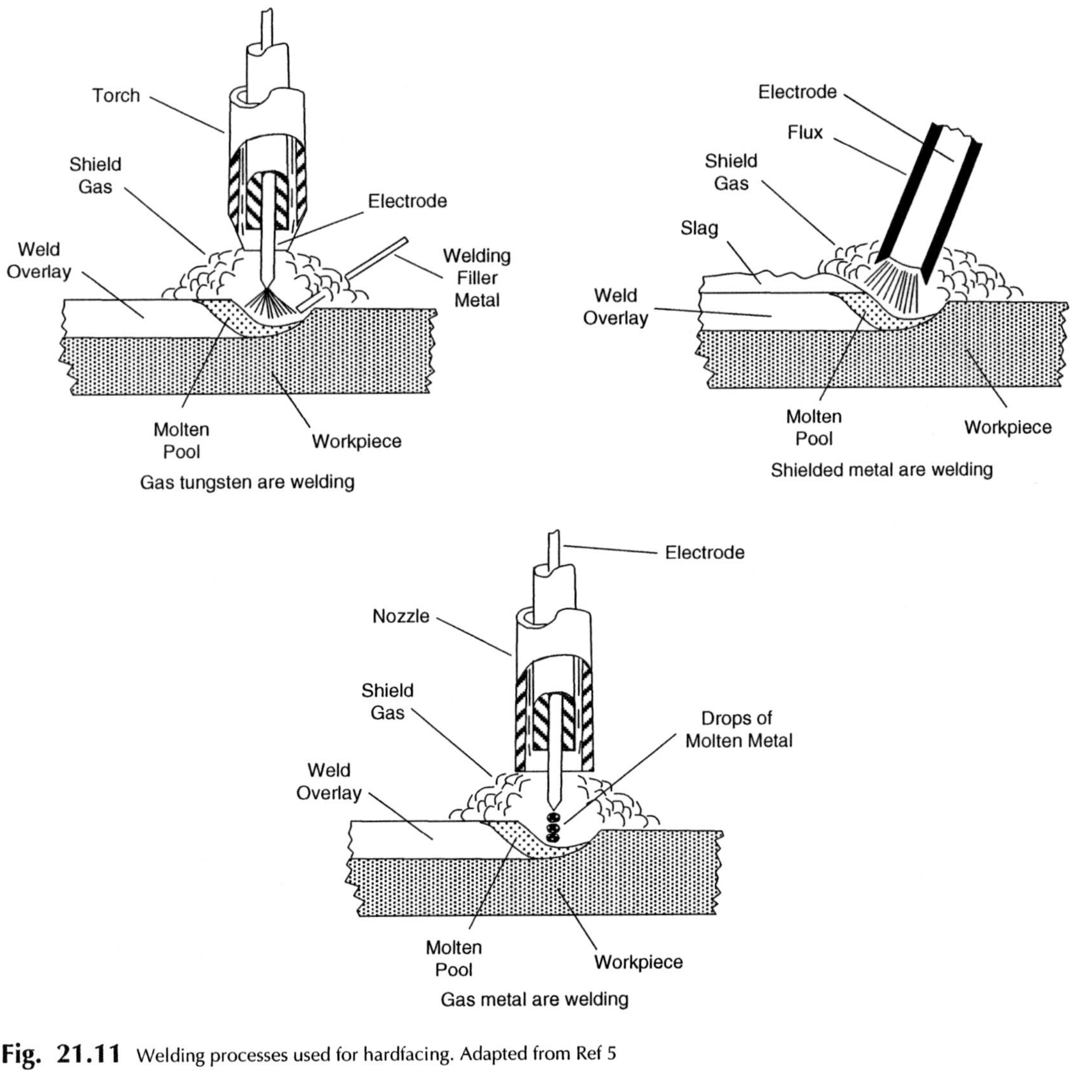


**Fig. 21.11** Welding processes used for hardfacing. Adapted from Ref 5

filler wire is normally used that consists of a tubular steel filled with metallic powder of the hardfacing alloy. GMAW allows for high deposition rates and low dilution by the base metal.

Thermal spraying is a group of processes in which finely divided metallic or nonmetallic materials are deposited in a molten or semimolten condition to form a coating. The material may be in the form of powder, ceramic, rod, or wire. Since the surface of the part does not heat up appreciably during spraying as it does in welding overlay deposition, distortion during thermal spraying is minimal. The four principal thermal spraying processes are flame spraying, spray and fuse, plasma spraying, and high-velocity oxyfuel thermal spraying (Fig. 21.12). In thermal spraying, an oxyfuel gas flame is used as the heat source. Compressed air is used for atomizing and propelling the droplets to the workpiece.

**Flame Spraying.** There are two variations of flame spraying; in one, sometimes referred to as metalizing, the coating source material is a metal wire, while the other uses a powder. In both variants, the material is fed through a gun and a nozzle and is melted by the oxyfuel gas flame. Flame spraying produces hard, thin coatings that are deposited quickly and uniformly, with deposits ranging from approximately 0.25 to 2.0 mm (0.01 to 0.08 in.) thick. However, flame-sprayed coatings are usually porous and brittle and do not resist excessive mechanical abuse.

**Spray and fuse** is a variation of flame spraying in which the part is flame sprayed and then fused to the substrate by using either a heating torch or by heating the part in a furnace. Spray and fuse coatings are usually made of nickel or cobalt self-fluxing alloys that contain silicon or boron that depress the melting point of the alloy to below that of steel. Fusing is performed at 1010 to 1315 °C (1850 to 2400 °F), depending on the particular alloy being used. The silicon and boron additions react with oxide films on the substrate and particles, allowing the particles to wet the surface and diffuse into the substrate.

**Plasma spraying** uses a plasma torch as the heat source for melting and propelling the particles to the workpiece. The temperature of the plasma arc is so much higher than that produced by flame spraying that coating materials with higher melting points can be applied. The powder is suspended in a carrier gas, which the plasma arc immediately melts and propels toward the workpiece. Since inert gas and high gas temperatures are used, the mechanical and metallurgical properties of the coatings are generally superior to flame spraying, and the bond strengths are higher.

**High-velocity oxyfuel** thermal spraying uses a mixture of oxygen and a combustible gas (e.g., acetylene) that is fed into the barrel of a gun along with a charge of surfacing powder. The mixture is ignited, and the detonation wave accelerates the molten or semimolten powder toward the workpiece. The cycle is repeated many times a second. The noise level is extremely high, and the process must be performed in a soundproof enclosure.

## 21.7 Other Surface-Hardening Processes

There are many other surface-hardening processes. A few are listed in Table 21.4.

**Aluminizing** applies a thin adherent coating on nickel- and cobalt-base superalloys to provide high-temperature oxidation resistance at temperatures up to approximately 1150 °C (2100 °F).

**Siliconizing,** the diffusion of silicon into steel, produces wear-resistant cases with hardnesses up to 50 HRC. The process is carried out in the temperature range of 925 to 1040 °C (1700 to 1900 °F). The workpiece is heated in contact with a silicon-bearing material such as silicon carbide, with chlorine gas used as a catalyst. The case depth ranges from 0.025 to 10 mm (0.001 to 0.4 in.), depending mainly on the carbon content of the base metal. The case produced contains approximately 14 wt% Si and is essentially an iron-silicon solid solution. The wear resistance is enhanced by the low coefficient of friction and antigalling properties.

**Chromizing** introduces chromium into the surface layers of the base metal. The process is not restricted to ferrous metals and may be applied to nickel, cobalt, molybdenum, and tungsten to improve corrosion and heat resistance. When it is applied to steel, it converts the surface layer into a stainless steel case. If the steel contains appreciable amounts of carbon (>0.60 wt%), chromium carbides will precipitate and increase wear resistance.

**Titanium carbide coatings** are used primarily for applications requiring good wear resistance.

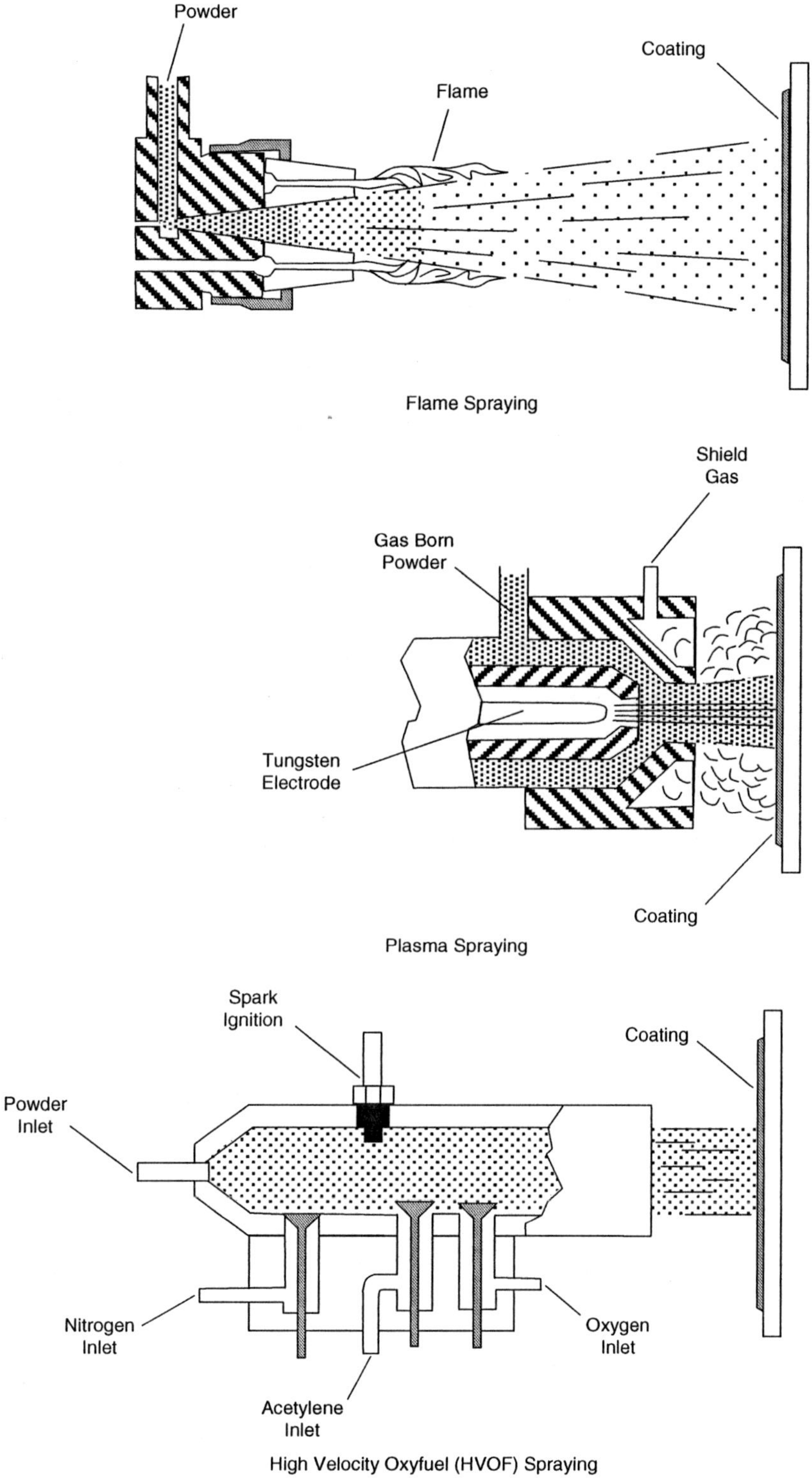


**Fig. 21.12** Thermal spraying processes used for hardfacing. Adapted from Ref 5

**Table 21.4 Comparison of other surface-hardening processes**

| Process | Process temperature | | Typical case depth | | Case hardness, HRC | Typical base metals | Process characteristics |
|---|---|---|---|---|---|---|---|
| | °C | °F | μm | mils | | | |
| Aluminizing (pack) | 870–980 | 1600–1800 | 25–1000 | 1–40 | <20 | Low-carbon steels | Diffused coating used for oxidation resistance at elevated temperatures |
| Siliconizing by chemical vapor deposition | 925–1040 | 1700–1900 | 25–1000 | 1–40 | 30–50 | Low-carbon steels | For corrosion and wear resistance; atmosphere control is critical |
| Chromizing by chemical vapor deposition | 980–1095 | 1800–2000 | 25–50 | 1–2 | Low-carbon steel, <30; high-carbon steel, 50–60 | High- and low-carbon steels | Chromized low-carbon steels yield a low-cost stainless steel; high-carbon steels develop a hard, corrosion-resistant case. |
| Titanium carbide | 900–1010 | 1650–1850 | 2.5–13 | 0.1–0.5 | >70 | Alloy steels, tool steels | Produces a thin carbide (TiC) case for resistance to wear; high temperature may cause distortion |
| Boriding | 400–1150 | 750–2100 | 13–50 | 0.5–2 | 40–70 | Alloy steels, tool steels, cobalt and nickel alloys | Produces a hard compound layer; mostly applied over hardened tool steels; high process temperature can cause distortion |

**Boronizing** is a thermochemical surface-hardening process that can be applied to a wide variety of ferrous, nonferrous, and cermet materials. It is frequently used for tooling applications where additional surface hardness is required.

## ACKNOWLEDGMENTS

Sections of this chapter were adapted from *Surface Engineering,* edited by J.R. Davis, ASM International, 2001, "Localized Heat Treating" and "Case Hardening of Steel" in *Metals Handbook Desk Edition,* 2nd ed., ASM International, 1998.

## REFERENCES

1. Case Hardening of Steel, *Metals Handbook Desk Edition,* 2nd ed., ASM International, 1998
2. J.R. Davis, *Surface Engineering,* ASM International, 2001
3. Localized Heat Treating, *Metals Handbook Desk Edition,* 2nd ed., ASM International, 1998
4. Metallography and Microstructures of Case-Hardening Steel, *Metallography and Microstructures,* Vol 9, *ASM Handbook,* ASM International, 2004
5. B.J. Moniz, *Metallurgy,* 2nd ed., American Technical Publishers, Inc., 1994

## SELECTED REFERENCES

- D.A. Brandt and J.C. Warner, *Metallurgy Fundamentals,* The Goodheart-Willcox Company, Inc., 2005
- R.A. Higgins, *Engineering Metallurgy—Applied Physical Metallurgy,* 6th ed., Arnold, 1993

# CHAPTER 22

# Tool Materials

A COMMON REQUIREMENT for tools is that the workpiece should have a minimal adverse effect on the tool in order for the tool to do its job successfully and for it to have a reasonable lifetime. The properties that are important when selecting a tool steel include:

- *Elastic strength:* The tool should resist deformation better than the workpiece.
- *Edge retention:* The tool material must be capable of being sharpened and then able to resist becoming blunt rapidly. This is a combination of machinability and hardness.
- *Wear resistance:* The abrasion resistance of a tool often determines its useful life.
- *Shock resistance:* In many operations, tools are loaded rapidly and must therefore be capable of sustaining stresses created during both mechanical and thermal shock loading. Toughness is a measure of the ability of the tool to absorb impact without breaking.
- *High-temperature stability:* During operations conducted at elevated temperatures, such as hot forming and high-speed cutting, it is important that the tool retain its properties rather than undergo rapid microstructural changes that degrade its properties. Hot hardness or red hardness is a measure of retained hardness.

A tool steel may be defined as any steel used to make tools for cutting, forming, or otherwise shaping a material into a part. Many tools are subjected to extremely high loads that are rapidly applied. Tools must withstand these loads a great number of times without breaking and without undergoing excessive wear or deformation. In many applications, tool steels must provide this capability under conditions that produce high temperatures and high temperature gradients in the tool. No single tool material combines maximum wear resistance, toughness, and resistance to softening at elevated temperatures. Thus, there is a fairly wide variety of different tool steels for different applications (Fig. 22.1).

Most tool steels are wrought products that are generally melted in relatively small-tonnage electric arc furnaces to economically achieve composition tolerances, cleanliness, and precise control of melting conditions. Special ladle refining and secondary remelting processes, such as electroslag remelting and vacuum arc remelting, are used to meet stringent demands regarding tool steel quality and performance. Powder metallurgy processing is also used in making tool steels. It provides a more uniform carbide size and distribution in large sections and can produce special compositions that are difficult or impossible to produce by wrought metallurgy.

One of the advantages of tool steels is that they can initially be easily formed and/or machined into the required shape and then heat treated to provide hardness and wear resistance. On the other hand, many extremely hard and wear-resistant materials, such as ceramics, are difficult to form into complex shapes. It is this machinability that accounts for the wide use of tool steels. There are many different kinds of tool steels. For any specific application, the steel should have adequate properties to perform properly and also be an economical choice. As Fig. 22.1 indicates, when one property is optimized, other properties are generally lessened, so a balance must be sought in materials selection.

Tool steels are usually supplied in the annealed or spheroidized condition. The tool producer is then responsible for machining or forming the metal and any subsequent heat treatments. Almost always, the tool will be heat treated to produce tempered martensite. When the desired properties can be obtained by using plain carbon steels, then they are generally the most economical choice. Otherwise alloy steels are required.

Hardenability is an important characteristic that is a measure of the depth below the surface that can be hardened (depth of case hardening). This controls the size of the tool that can be hardened throughout. Hardenability also dictates the necessary quench rates, which is important because the incidence of quench cracking, distortion, and dimensional changes can be minimized by reducing the cooling rate required to produce martensite. Many tool steels have molybdenum, chromium, and manganese as alloying additions to improve hardenability.

A high carbon content is required to obtain tempered martensite with a high hardness. In addition, wear resistance is enhanced by the presence of hard second-phase carbides. Stable alloy carbides coarsen more slowly than cementite and are therefore much more effective than cementite at higher temperatures. Therefore, carbide formers such as chromium, molybdenum, tungsten, and vanadium are used along with sufficient carbon to form alloy carbides while also providing the martensite matrix with a high hardness. To dissolve the carbides in austenite during heat treatment, high austenitization temperatures are often necessary. During tempering, the carbides then precipitate to form a fine uniform distribution. However, high austenitization temperatures create the risk of excessive grain growth during austenitization, so grain-boundary pinning is often achieved with vanadium carbide particles.

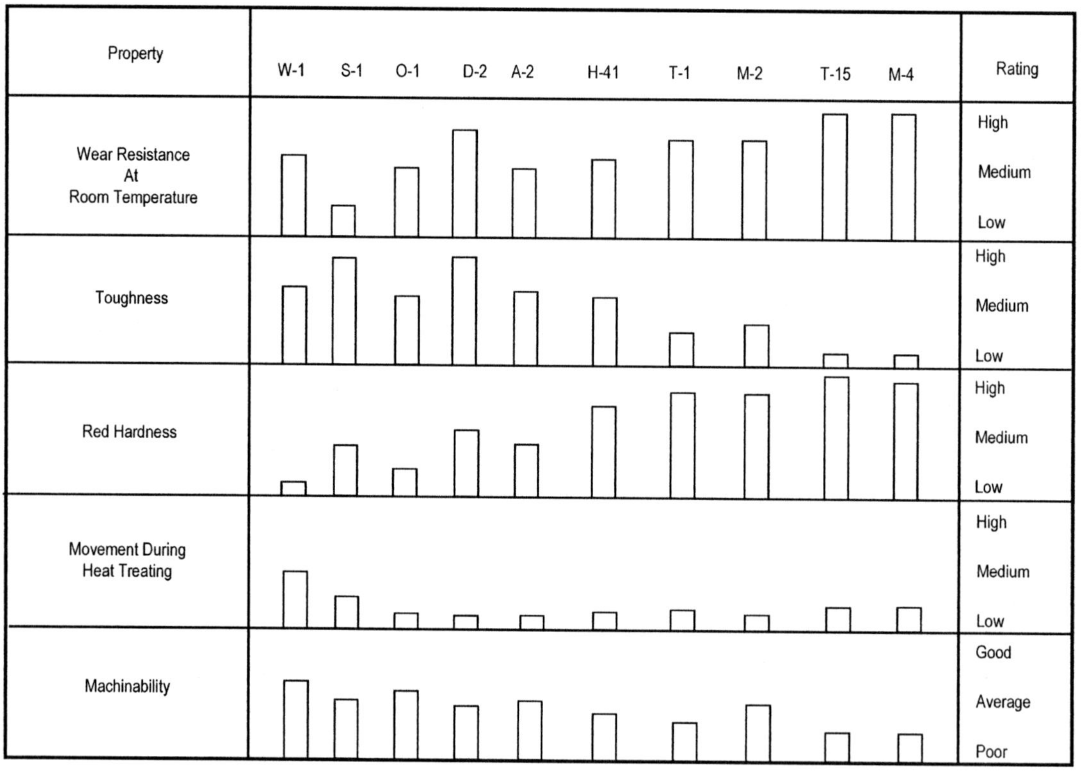


**Fig. 22.1** Comparison of tool steel properties

**Table 22.1 Compositions of representative group W water-hardening tool steels**

| Designation | | Composition, % | | | | | | | |
|---|---|---|---|---|---|---|---|---|---|
| AISI | UNS | C | Mn | Si | Cr | Ni | Mo | W | V |
| W1 | T72301 | 0.70–1.50(a) | 0.10–0.40 | 0.10–0.40 | 0.15 max | 0.20 max | 0.10 max | 0.15 max | 0.10 max |
| W2 | T72302 | 0.85–1.50(a) | 0.10–0.40 | 0.10–0.40 | 0.15 max | 0.20 max | 0.10 max | 0.15 max | 0.15–0.35 |
| W3 | T72305 | 1.05–1.15 | 0.10–0.40 | 0.10–0.40 | 0.40–0.60 | 0.20 max | 0.10 max | 0.15 max | 0.10 max |

(a) Specified carbon ranges are designated by suffix numbers. Source: Ref 1

In tools used at high temperatures, it is important that the steel resist oxidation and decarburization. Alloying additions of chromium are used to help prevent oxidation.

The AISI classification of tool steels includes seven major categories: water-hardening tool steels, shock-resisting tool steels, cold work tool steels, hot work tool steels, special-purpose tool steels, mold tool steels, and high-speed tool steels. The categories of tool steels are:

| Type | Classification |
|---|---|
| W | Water-hardening tool steels |
| S | Shock-resisting steels |
| O | Oil-hardening cold work steels |
| A | Air-hardening, medium-alloy, cold work steels |
| D | High-carbon, high-chromium, cold work steels |
| H10 to H19 | Chromium hot work steels |
| H21 to H26 | Tungsten hot work steels |
| H41 to H43 | Molybdenum hot work steels |
| L | Low-alloy special-purpose tool steels |
| P | Low-carbon mold steels |
| M | Molybdenum high-speed steels |
| T | Tungsten high-speed steels |

High-speed steels are used in high-speed cutting tools such as drill bits and end mills. Hot work tool steels are used in operations that utilize dies for punching, shearing, and forming materials at elevated temperatures, while cold work steels are used in similar operations at room temperature.

## 22.1 Water-Hardening Steels

Water-hardening steels, also called group W steels, contain carbon as the principal alloying element. They are basically plain carbon steels and are also referred to as carbon tool steels. They contain from 0.7 to 1.5 wt% C with a manganese content of approximately 0.25 wt%. For applications where toughness is the primary consideration, the carbon content is restricted to 0.7 to 0.75 wt%; for applications where toughness and hardness are equally important, the carbon content is increased to 0.75 to 0.90 wt%; and for applications requiring maximum wear resistance, the carbon content is 0.90 to 1.4 wt%. Some contain small amounts of chromium or vanadium, and a few contain both vanadium and chromium. Chromium increases hardenability and wear resistance, while vanadium is added to maintain a fine grain size and enhance toughness. The compositions of several group W steels are listed in Table 22.1. In general, the group W steels are less expensive than the other alloy tool steels.

An examination of a time-temperature transformation (TTT) diagram, as in the diagram for W1 in Fig. 22.2, shows that these steels are very shallow hardening. Sections more than approximately 13 mm (0.5 in.) thick generally have a hard case over a strong, tough, resilient

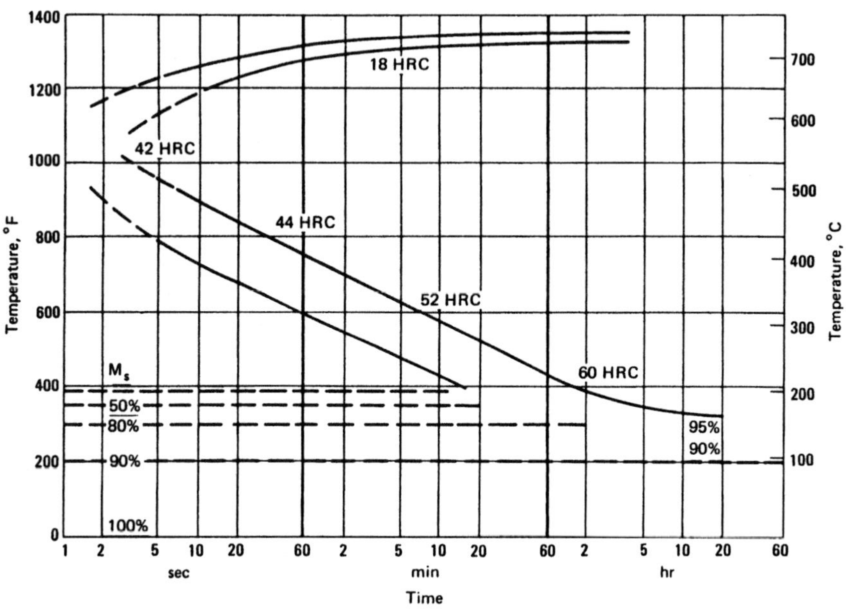


**Fig. 22.2** W1 isothermal transformation diagram. Composition: 1.14 C, 0.22 Mn, 0.61 Si. Austenitized at 790 °C (1455 °F). Source: Ref 4

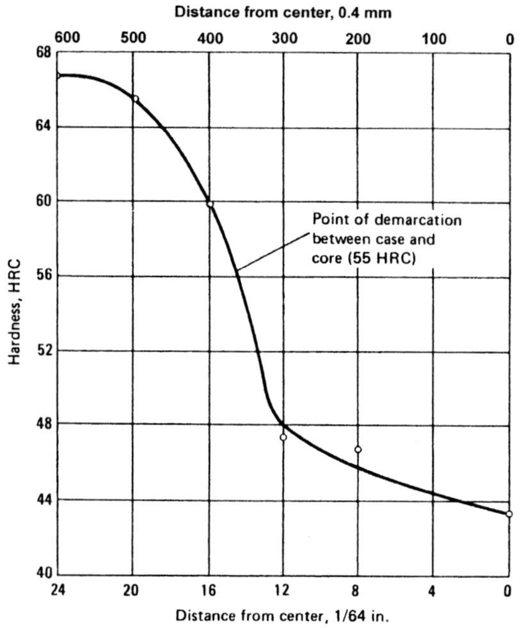


**Fig. 22.3** Hardness penetration curve for W1 tool steel. Composition: 1.06 C, 0.36 Mn, 0.27 Si, 0.01 S, 0.015 P, 0.05 Cr. 19 mm (3/4 in.) round bar, brine quenched from 815 °C (1500 °F). Pretreated by oil quenching after 40 min at 870 °C (1600 °F). Source: Ref 4

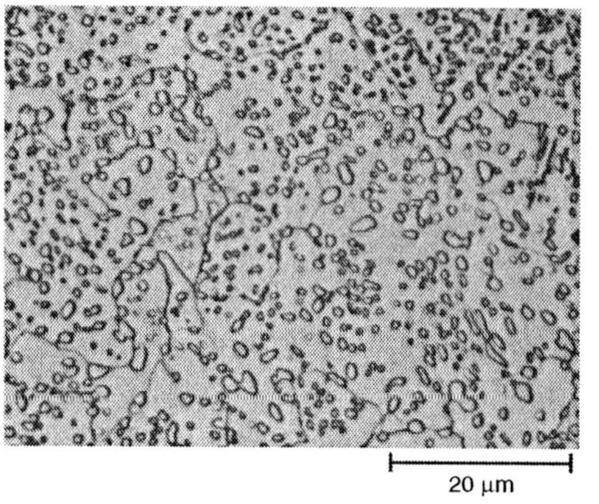


**Fig. 22.4** AISI W4 water-hardening tool steel (0.96C-0.66Mn-0.23Cr), as-received (full annealed). 170 HB. Structure consists of spheroidal cementite in a ferrite matrix. Original magnification: 1000×. Source: Ref 5

core. Except for small diameters that are through hardening, the microstructure is usually tempered martensite on the surface, with a softer, tougher core of fine pearlite. Typically, the surface hardness is in the range of HRC 58 to 65, with a core hardness of HRC 38 to 43. A hardness penetration curve for W1 is shown in Fig. 22.3. Since group W steels have only minimal amounts of alloying elements, they have low resistance to softening at elevated temperatures. They must be water quenched for high hardness and are thus subject to considerable distortion. Because of their low red hardness (i.e., the ability of the cutting edge to retain hardness at temperatures where the steel attains a dull red color), group W steels cannot be used when appreciable heat is developed at the cutting edge, such as during high-speed cutting. Their use as cutting tools is limited to conditions involving low speeds and relatively light depth of cuts of soft materials, such as wood, brass, aluminum, and unhardened low-carbon steels. They are suitable for cold heading, striking, coining, embossing tools, woodworking tools, hard metal cutting tools such as taps and reamers, wear-resistant machine tool components, and cutlery.

The manufacture of carbon tool steels begins with a forging process that produces the desired bar stock. Forging is carried out at temperatures fairly high in the austenite phase field but below those that would cause dissolution of the vanadium carbides. The steel is air cooled after forging. It is then annealed, sometimes after a normalization treatment, to produce a suitable microstructure. The purpose of annealing is to improve machinability, eliminate any residual stresses, and provide a microstructure that will respond uniformly to heat treatment. The desirable microstructure that achieves this condition is either a finely spheroidized structure (Fig. 22.4) or a mixture of fine spheroidite and pearlite.

Water-hardening tool steels are austenitized above the $A_1$ line but below the $A_{cm}$ line (Fig. 11.4 in Chapter 11, "Heat Treatment of Steel"). Therefore, the microstructure before quenching consists of austenite grains with proeutectoid cementite particles and possibly embedded vanadium carbides. Vanadium carbide restricts grain growth during austenitization and is retained within the martensite after quenching. The austenitization temperatures are high enough for chromium to be dissolved. An important contribution of dissolved chromium is that it retards the rate of tempering, which enables some residual stress relief without too much softening. Figure 22.5 shows the microstructure of an austenitized, quenched, and tempered sample. Chromium that becomes part of cementite improves wear resistance. The risk of quench cracking increases as the carbon content increases and as the severity of the quench increases. Quench cracking also becomes more likely as the austenitic grain size increases.

## 22.2 Shock-Resisting Steels

Group S shock-resisting steels were developed for applications where toughness—the ability to absorb repeated impacts—is paramount. The carbon content is approximately 0.50 wt% for all group S steels, which produces a combination of high elastic strength, high toughness, and low-to-medium wear resistance. The principal alloying elements are manganese, silicon, chromium, tungsten, and molybdenum in various combinations. Silicon strengthens the ferrite, while molybdenum and chromium increase hardenability and aid in wear resistance. Tungsten imparts some degree of red hardness. Group S steels are rated as fair in regard to red hardness and wear resistance, and the hardness is usually kept below 60 HRC for toughness. Group S steels are used primarily for chisels, rivet sets, punches, driver bits, and other applications requiring high toughness and resistance to shock loading. Types S1 and S7 are also used for hot punching and shearing, which requires some heat resistance. However, the older type S1 has decreased in popularity because it contains 2.5 wt% W, which makes it relatively more expensive than the other S types without any real advantage over the others. Group S steels exhibit excellent toughness at high strength levels, so they are often considered for nontooling or structural applications.

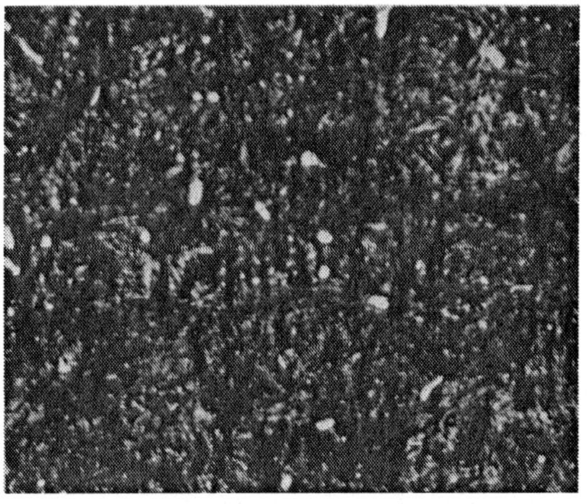

**Fig. 22.5** Hardened (HRC 64) W1 tool steel. Dark background is martensite; white dots are carbide. Austenitized at 790 °C (1450 °F), brine quenched, and tempered at 165 °C (325 °F). Source: Ref 6

Group S steels (Table 22.2) vary in hardenability from shallow hardening (S2) to deep hardening (S7). Compare the TTT diagram for S5 in Fig. 22.6 to the previous diagram for W1 in Fig. 22.2. The higher alloy content of S5 has shifted the nose of the TTT diagram to the right, allowing thicker sections to be through hardened. Group S steels require relatively high austenitizing temperatures to achieve optimal hardness. Type S2 is normally water quenched, while types S1, S5, and S6 are oil quenched, and type S7 is normally air cooled, except in thick sections that are oil quenched. Type S7 is deeper hardening due its higher molybdenum (1.4 wt%) and chromium (3.25 wt%) contents.

## 22.3 Cold Work Steels

Because they do not have the alloy content necessary to make them resistant to softening at elevated temperature, cold work steels are restricted to applications that do not involve prolonged or repeated heating above approximately 205 to 260 °C (400 to 500 °F). There are three categories of cold work steels: oil-hardening steels (group O), air-hardening steels (group A), and high-carbon, high-chromium steels (group D).

**Table 22.2 Compositions of representative group S shock-resisting tool steels**

| Designation | | Composition, % | | | | | | | |
|---|---|---|---|---|---|---|---|---|---|
| AISI | UNS | C | Mn | Si | Cr | Ni | Mo | W | V |
| S1 | T41901 | 0.40–0.55 | 0.10–0.40 | 0.15–1.20 | 1.00–1.80 | 0.30 max | 0.50 max | 1.50–3.00 | 0.15–0.30 |
| S2 | T41902 | 0.40–0.55 | 0.30–0.50 | 0.90–1.20 | ... | 0.30 max | 0.30–0.60 | ... | 0.50 max |
| S5 | T41905 | 0.50–0.65 | 0.60–1.00 | 1.75–2.25 | 0.50 max | ... | 0.20–1.35 | ... | 0.35 max |
| S6 | T41906 | 0.40–0.50 | 1.20–1.50 | 2.00–2.50 | 1.20–1.50 | ... | 0.30–0.50 | ... | 0.20–0.40 |
| S7 | T41907 | 0.45–0.55 | 0.20–0.90 | 0.20–1.00 | 3.00–3.50 | ... | 1.30–1.80 | ... | 0.20–0.30 |

Source: Ref 1

### 22.3.1 Oil-Hardening Cold Work Steels (Group O)

Group O steels are among the most widely used tool steels. Their properties include high as-quenched hardness, high hardenability from oil quenching, freedom from distortion on quenching fairly intricate sections, and the ability to hold a sharp cutting edge. The four group O tool steels vary in primary alloying elements, as well as in alloy content, even though they are similar in general characteristics and are used for similar applications. Type O1 contains manganese, chromium, and tungsten. Type O2 is alloyed primarily with manganese. Type O6 contains silicon, manganese, and molybdenum. Type O7 contains manganese and chromium and has a tungsten content higher than that of type O1. The compositions of the group O steels are shown in Table 22.3.

The most important property of group O steels is their high resistance to wear at normal temperatures, a result of their high carbon content. On the other hand, group O steels have a low resistance to softening at elevated temperatures. The ability of group O steels to fully harden by oil quenching results in lower distortion and greater safety (i.e., less tendency to crack) in hardening than for the water-hardening tool grades. Tools made from these steels can be successfully repaired or renovated by welding if proper procedures are followed.

Group O steels are used extensively for dies and punches for blanking, trimming, drawing, flanging, and forming. Surface hardness of 56 to 62 HRC, obtained through oil quenching followed by tempering at 175 to 315 °C (350 to 600 °F), provides a suitable combination of mechanical properties for most dies made from type O1, O2, or O6. Type O7 has lower

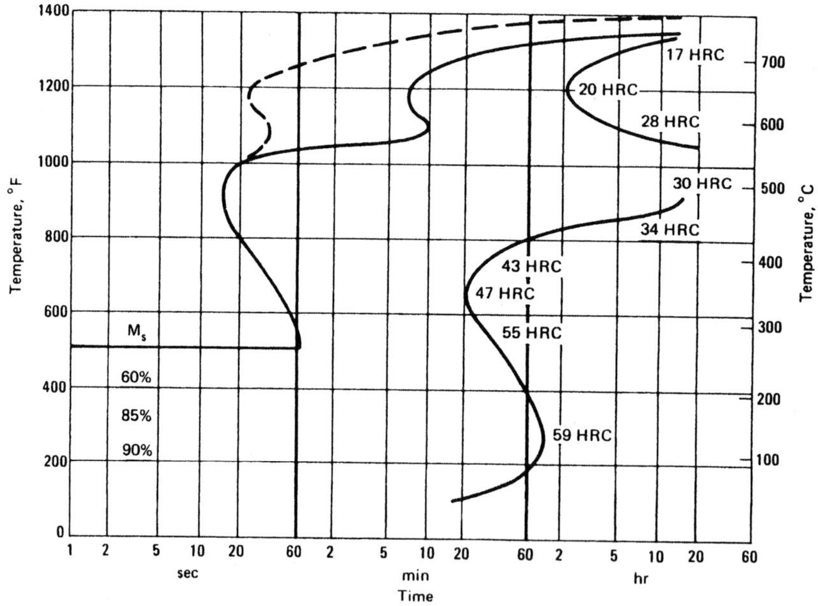


**Fig. 22.6** S5 isothermal transformation diagram. S5 containing 0.60 C, 0.75 Mn, 1.90 Si, 0.25 Cr, 0.30 Mo. Austenitized at 900 °C (1650 °F). Source: Ref 4

**Table 22.3 Compositions of representative group O oil-hardening tool steels**

| Designation | | Composition, % | | | | | | | |
|---|---|---|---|---|---|---|---|---|---|
| AISI | UNS | C | Mn | Si | Cr | Ni | Mo | W | V |
| O1 | T31501 | 0.85–1.00 | 1.00–1.40 | 0.50 max | 0.40–0.60 | 0.30 max | ... | 0.40–0.60 | 0.30 max |
| O2 | T31502 | 0.85–0.95 | 1.40–1.80 | 0.50 max | 0.50 max | 0.30 max | 0.30 max | ... | 0.30 max |
| O6 | T31506 | 1.25–1.55 | 0.30–1.10 | 0.55–1.50 | 0.30 max | 0.30 max | 0.20–0.30 | ... | ... |
| O7 | T31507 | 1.10–1.30 | 1.00 max | 0.60 max | 0.35–0.85 | 0.30 max | 0.30 max | 1.00–2.00 | 0.40 max |

Source: Ref 1

hardenability but better general wear resistance than any other group O steels, and it is more often used for tools requiring keen cutting edges. Oil-hardening tool steels are also used for machinery components (e.g., cams, bushings, and guides) and for gages, where good dimensional stability and wear resistance are needed.

### 22.3.2 Air-Hardening, Medium-Alloy, Cold Work Steels (Group A)

The group A steels contain enough alloying elements to enable them to achieve full hardening in sections of 10 cm (4 in.) diameter on air cooling from the austenitizing temperature. Because they are air hardening, group A tool steels exhibit minimum distortion and little tendency to crack during hardening. Manganese, chromium, and molybdenum are the principal alloying elements used to provide deep hardening. As a result of their relatively high alloy content, they exhibit some degree of secondary hardening during tempering (Fig. 22.7) and thus can be used at higher temperatures than the group W, O, and S steels.

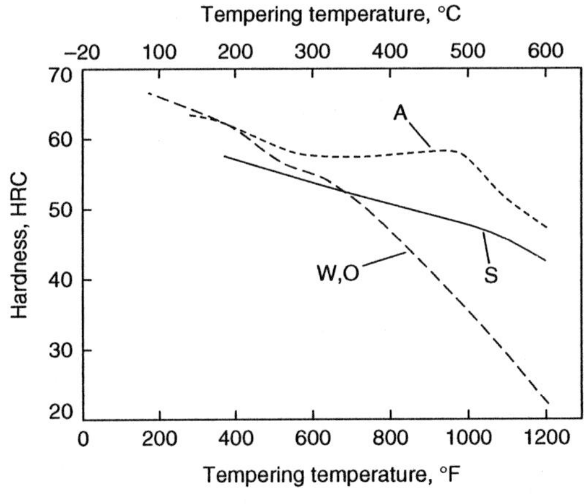


**Fig. 22.7** Tempering curves for several grades of tool steels

The compositions of a number of group A steels are given in Table 22.4. Types A2, A3, A7, A8, and A9 contain a high percentage of chromium (5 wt%), which provides moderate resistance to softening at elevated temperatures. Types A4, A6, and A10 are lower in chromium content (1 wt%) and higher in manganese content (2 wt%). They can be hardened from temperatures approximately 110 °C (200 °F) lower than those required for the high-chromium types, further reducing distortion and undesirable surface reactions during heat treatment. To improve toughness, more silicon is added to type A8 than A2 through A7, and both silicon and nickel are added to types A9 and A10. Because of the high carbon and silicon contents of type A10, graphite is formed in the microstructure. As a result, A10 has much better machinability in the annealed condition and somewhat better resistance to galling and seizing in the fully hardened condition than other group A steels.

Typical applications for group A steels include shear knives, punches, blanking and trimming dies, forming dies, and coining dies. The inherent dimensional stability of these steels makes them suitable for gages and precision measuring tools. In addition, the extreme abrasion resistance of type A7 makes it suitable for brick molds, ceramic molds, and other highly abrasive applications. The complex chromium or chromium-vanadium carbides in group A steels enhance their wear resistance. Therefore, these steels perform well under abrasive conditions at less than full hardness. Although cooling in still air is adequate for producing full hardness in most tools, massive sections can be hardened by cooling in an air blast or by interrupted quenching in hot oil.

**Table 22.4 Compositions of representative group A air-hardening, medium-alloy, cold work tool steels**

| Designation | | Composition, % | | | | | | | |
|---|---|---|---|---|---|---|---|---|---|
| AISI | UNS | C | Mn | Si | Cr | Ni | Mo | W | V |
| A2 | T30102 | 0.95–1.05 | 1.00 max | 0.50 max | 4.75–5.50 | 0.30 max | 0.90–1.40 | ... | 0.15–0.50 |
| A3 | T30103 | 1.20–1.30 | 0.40–0.60 | 0.50 max | 4.75–5.50 | 0.30 max | 0.90–1.40 | ... | 0.80–1.40 |
| A4 | T30104 | 0.95–1.05 | 1.80–2.20 | 0.50 max | 0.90–1.20 | 0.30 max | 0.90–1.40 | ... | ... |
| A6 | T30106 | 0.65–0.75 | 1.80–2.50 | 0.50 max | 0.90–1.20 | 0.30 max | 0.90–1.40 | ... | ... |
| A7 | T30107 | 2.00–2.85 | 0.80 max | 0.50 max | 5.00–5.75 | 0.30 max | 0.90–1.40 | 0.50–1.50 | 3.90–5.15 |
| A8 | T30108 | 0.50–0.60 | 0.50 max | 0.75–1.10 | 4.75–5.50 | 0.30 max | 1.15–1.65 | 1.00–1.50 | ... |
| A9 | T30109 | 0.45–0.55 | 0.50 max | 0.95–1.15 | 4.75–5.50 | 1.25–1.75 | 1.30–1.80 | ... | 0.80–1.40 |
| A10 | T30110 | 1.25–1.50 | 1.60–2.10 | 1.00–1.50 | 0.90–1.20 | 1.55–2.05 | 1.25–1.75 | ... | ... |

Source: Ref 1

Table 22.5 Compositions of representative group D high-carbon, high-chromium, cold work tool steels

| Designation | | Composition, % | | | | | | | | |
|---|---|---|---|---|---|---|---|---|---|---|
| AISI | UNS | C | Mn | Si | Cr | Ni | Mo | W | V | Co |
| D2 | T30402 | 1.40–1.60 | 0.60 max | 0.60 max | 11.00–13.00 | 0.30 max | 0.70–1.20 | ... | 1.10 max | ... |
| D3 | T30403 | 2.00–2.35 | 0.60 max | 0.60 max | 11.00–13.50 | 0.30 max | ... | 1.00 max | 1.00 max | ... |
| D4 | T30404 | 2.05–2.40 | 0.60 max | 0.60 max | 11.00–13.00 | 0.30 max | 0.70–1.20 | ... | 1.00 max | ... |
| D5 | T30405 | 1.40–1.60 | 0.60 max | 0.60 max | 11.00–13.00 | 0.30 max | 0.70–1.20 | ... | 1.00 max | 2.50–3.50 |
| D7 | T30407 | 2.15–2.50 | 0.60 max | 0.60 max | 11.50–13.50 | 0.30 max | 0.70–1.20 | ... | 3.80–4.40 | ... |

Source: Ref 1

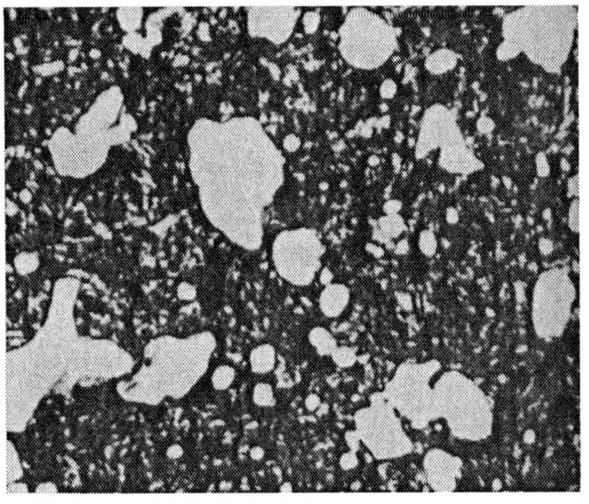

**Fig. 22.8** D7 tool steel, austenitized at 1040 °C (1900 °F), air cooled, tempered at 540 °C (1000 °F). Rockwell C 61. Structure consists of small and massive carbide particles (white) in a matrix of tempered martensite. Original magnification: 1000×. Source: Ref 6

### *22.3.3 High-Carbon, High-Chromium, Cold Work Steels (Group D)*

The excellent wear resistance of the group D steels is a result of their high chromium (12 wt%) and carbon (1.40 to 2.50 wt%) contents. Differences in their wear resistance are mainly due to differences in their carbon contents. For example, D2 (1.4 to 1.6 wt% C) contains 30 to 40 wt% fewer carbides than D3 (2.00 to 2.35 wt% C). With the exception of type D3, they also contain 0.70 to 1.20 wt% Mo. All group D tool steels except type D3 are air hardening and attain full hardness when air cooled. Type D3 is almost always quenched in oil; however, small parts can be austenitized in a vacuum and then gas quenched. Therefore, tools made of D3 are more susceptible to distortion and are more likely to crack during hardening. All of the group D steels have high resistance to high-temperature oxidation and good resistance to staining when hardened and polished. The compositions of a number of group D steels are given in Table 22.5.

Group D steels have high resistance to softening at elevated temperatures. These steels also exhibit excellent resistance to wear, especially type D7, which has the highest carbon and vanadium contents. All group D steels, particularly the higher-carbon types D3, D4, and D7, contain massive amounts of carbides, which makes them susceptible to edge brittleness. A number of large carbide particles can be seen in the micrograph of D7 steel (Fig. 22.8).

Typical applications of group D steels include long-run dies for blanking, forming, thread rolling, and deep drawing, dies for cutting laminations, brick molds, gages, burnishing tools, rolls, and shear and slitter knives.

## 22.4 Hot Work Steels

Hot work steels (group H) were developed to withstand the combinations of heat, pressure, and abrasion associated with operations involving punching, shearing, or forming of metals at high temperatures. Group H tool steels usually have medium carbon contents (0.35 to 0.45 wt%) and combined chromium, tungsten, molybdenum, and vanadium contents of 6 to 25 wt%. As a result of their high alloy contents, the group H steels require double tempering to convert all of the retained austenite to tempered martensite. Group H steels are divided into three subgroups: chromium-base hot work steels (types H10 to H19), tungsten-base hot work steels (types H21 to H26), and low-carbon, molybdenum-base hot work steels (types H42 and H43). The compositions of a number of hot work steels are given in Table 22.6.

### *22.4.1 Chromium Hot Work Steels*

Types H10 to H19 have good resistance to heat softening because of their medium chromium contents and the addition of carbide-forming elements such as molybdenum,

**Table 22.6 Compositions of representative group H tool steels**

| Designation | | Composition, % | | | | | | | | |
|---|---|---|---|---|---|---|---|---|---|---|
| AISI | UNS | C | Mn | Si | Cr | Ni | Mo | W | V | Co |
| **Chromium hot work steels** | | | | | | | | | | |
| H10 | T20810 | 0.35–0.45 | 0.25–0.70 | 0.80–1.20 | 3.00–3.75 | 0.30 max | 2.00–3.00 | ... | 0.25–0.75 | ... |
| H11 | T20811 | 0.33–0.43 | 0.20–0.50 | 0.80–1.20 | 4.75–5.50 | 0.30 max | 1.10–1.60 | ... | 0.30–0.60 | ... |
| H12 | T20812 | 0.30–0.40 | 0.20–0.50 | 0.80–1.20 | 4.75–5.50 | 0.30 max | 1.25–1.75 | 1.00–1.70 | 0.50 max | ... |
| H13 | T20813 | 0.32–0.45 | 0.20–0.50 | 0.80–1.20 | 4.75–5.50 | 0.30 max | 1.10–1.75 | ... | 0.80–1.20 | ... |
| H14 | T20814 | 0.35–0.45 | 0.20–0.50 | 0.80–1.20 | 4.75–5.50 | 0.30 max | ... | 4.00–5.25 | ... | ... |
| H19 | T20819 | 0.32–0.45 | 0.20–0.50 | 0.20–0.50 | 4.00–4.75 | 0.30 max | 0.30–0.55 | 3.75–4.50 | 1.75–2.20 | 4.00–4.50 |
| **Tungsten hot work steels** | | | | | | | | | | |
| H21 | T20821 | 0.28–0.36 | 0.15–0.40 | 0.15–0.50 | 3.00–3.75 | 0.30 max | ... | 8.50–10.00 | 0.30–0.60 | ... |
| H22 | T20822 | 0.30–0.40 | 0.15–0.40 | 0.15–0.40 | 1.75–3.75 | 0.30 max | ... | 10.00–11.75 | 0.25–0.50 | ... |
| H23 | T20823 | 0.25–0.35 | 0.15–0.40 | 0.15–0.60 | 11.00–12.75 | 0.30 max | ... | 11.00–12.75 | 0.75–1.25 | ... |
| H24 | T20824 | 0.42–0.53 | 0.15–0.40 | 0.15–0.40 | 2.50–3.50 | 0.30 max | ... | 14.00–16.00 | 0.40–0.60 | ... |
| H25 | T20825 | 0.22–0.32 | 0.15–0.40 | 0.15–0.40 | 3.75–4.50 | 0.30 max | ... | 14.00–16.00 | 0.40–0.60 | ... |
| H26 | T20826 | 0.45–0.55 | 0.15–0.40 | 0.15–0.40 | 3.75–4.50 | 0.30 max | ... | 17.25–19.00 | 0.75–1.25 | ... |
| **Molybdenum hot work steels** | | | | | | | | | | |
| H42 | T20842 | 0.55–0.70 | 0.15–0.40 | ... | 3.75–4.50 | 0.30 max | 4.50–5.50 | 5.50–6.75 | 1.75–2.20 | ... |

Source: Ref 1

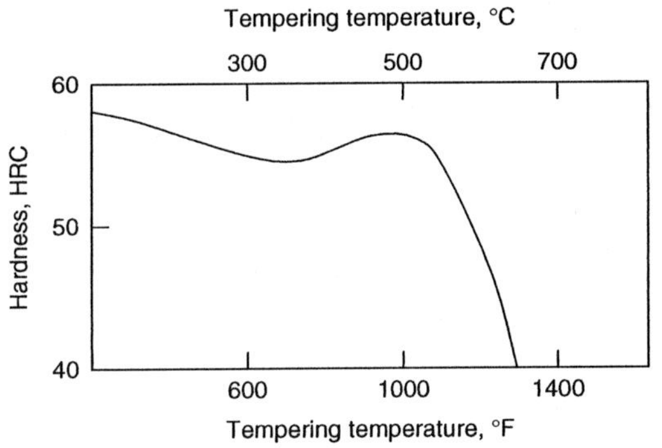


**Fig. 22.9** Secondary hardening of H11 tool steel. Source: Adapted from Ref 7

tungsten, and vanadium. The low carbon and low total alloy contents promote toughness at the normal working hardness of 40 to 55 HRC. Higher tungsten and molybdenum contents increase hot strength but slightly reduce toughness. Vanadium is added to increase resistance to erosive wear at high temperatures. Silicon additions improve oxidation resistance at temperatures up to 800 °C (1475 °F).

All of the chromium hot work steels are deep hardening; some can be air hardened in thicknesses up to 30 cm (12 in.) The air-hardening qualities of these steels result in low distortion during hardening. Chromium hot work steels are especially well adapted to hot die work of all kinds, particularly extrusion dies, die-casting dies, forging dies, mandrels, and hot shears. Most of these steels have carbon and alloy contents low enough that tools made from them can be water cooled in service without cracking.

Tool steel H11 and modifications of this alloy are used to make certain highly stressed structural parts, such as torsion bars and landing gear components. Material for such demanding applications is produced by vacuum arc remelting, which results in extremely low residual gas contents, excellent microcleanliness, and a high degree of structural homogeneity. The chief advantage of H11 over conventional high-strength steels is its ability to resist softening during continued exposure to temperatures up to 540 °C (1000 °F) (Fig. 22.9) and also to provide moderate toughness and ductility at room temperature, with 1725 to 2070 MPa (250 to 300 ksi) tensile strength. In addition, because of its secondary hardening characteristics, H11 can be tempered at high temperatures, resulting in nearly complete relief of residual stresses, which is necessary for maximum toughness at high strength levels. Other important advantages of chromium hot work steels for structural and hot work applications include ease of forming and working, good weldability, a relatively low coefficient of thermal expansion, acceptable thermal conductivity, and above-average resistance to oxidation and corrosion.

### *22.4.2 Tungsten Hot Work Steels*

The principal alloying elements in tungsten hot work steels (types H21 to H26) are carbon, tungsten, chromium, and vanadium. The higher alloy contents of these steels make them more resistant to high-temperature softening and erosive wear than the chromium hot work steels. However, their high alloy contents also make them more prone to brittleness at normal working hardness (45 to 55 HRC) and make it difficult for them to be safely water cooled in service.

Although tungsten hot work steels can be air hardened, they are usually quenched in oil or hot salt to minimize scaling. When air hardened, they exhibit low distortion. Tungsten hot work steels require higher hardening temperatures than chromium hot work steels, making them more likely to scale when heated in an oxidizing atmosphere.

These steels are used to make mandrels and extrusion dies for high-temperature applications, such as the extrusion of brass, nickel alloys, and steel, and they are also suitable for hot forging dies.

### *22.4.3 Molybdenum Hot Work Steels*

There are only two active molybdenum hot work steels: types H42 and H43. These alloys contain molybdenum, chromium, vanadium, carbon, and varying amounts of tungsten. They are similar to tungsten hot work steels, having almost identical characteristics and uses. Although their compositions resemble those of various molybdenum high-speed steels, they have a low carbon content and greater toughness. The principal advantage of types H42 and H43 over tungsten hot work steels is their lower cost. They are more resistant to heat checking (i.e., cracking) than are the tungsten hot work steels, but, in common with all high-molybdenum steels, they require greater care in heat treatment, particularly with regard to decarburization and control of austenitizing temperature.

There are also a number of proprietary low-alloy steel grades that are used for hot forging die holders and die blocks. The principal alloying elements in these steels are nickel, molybdenum, and chromium, with vanadium and silicon as smaller additions. The total alloying content is generally small enough that adequate machinability is retained in prehardened die blocks of these grades. Characterized by high toughness and, in some instances, good heat resistance, these steels have very good hardenability. As a trade-off to their good toughness, they are generally heat treated to relatively low hardness. Because of low hardness, their wear resistance is only moderate. However, they possess good resistance to shock loading (such as encountered in hammer forging), heat checking, and catastrophic failure. Because of the generally low tempering temperatures, these die steels are employed primarily in hammer operations where the contact times, during which heat transfer to the dies can occur, are short.

## 22.5 Low-Alloy Special-Purpose Steels

Low-alloy special-purpose steels, also called group L steels, contain small amounts of chromium, vanadium, nickel, and molybdenum. At one time, seven steels were listed in this group, but because of falling demand, only types L2 and L6 remain. Type L2 is available in several carbon contents, from 0.50 to 1.10 wt%. Its principal alloying elements are chromium and vanadium, which make it an oil-hardening steel of fine grain size. Type L6 contains small amounts of chromium and molybdenum, as well as 1.50 wt% Ni for increased toughness.

Although both L2 and L6 are considered oil-hardening steels, large sections of L2 are often quenched in water. Type L2 steel, containing 0.50 wt% C, is capable of attaining approximately 57 HRC as-oil-quenched, but it will not through harden in sections of more than approximately 13 mm (0.5 in.) thickness. Type L6, which contains 0.70 wt% C, has an as-quenched hardness of approximately 64 HRC. It can maintain hardness above 60 HRC in sections up to 76 mm (3 in.) thick. Group L steels are generally used for machine parts, such as arbors, cams, chucks, and collets, and for other special applications requiring good strength and toughness.

Typical compositions of L and P group steels ate found in Table 22.7.

## 22.6 Mold Steels

Mold steels, also called group P steels, contain chromium and nickel as the principal alloying elements. Types P2 and P6 are carburizing steels produced to tool steel quality standards. They have very low hardness and low resistance to work hardening in the annealed condition. These factors make it possible to produce a mold impression by cold hubbing. Hubbing is a technique for forming a mold cavity by forcing a hardened steel master hub, an exact replica of the article to be formed, into a softer die blank. After the impression is formed, the mold is carburized, hardened, and tempered to a surface hardness of approximately 58 HRC. Types P4 and P6 are deep hardening, and with type P4, full hardness in the carburized case can be achieved by air cooling.

Types P20 and P21 normally are supplied heat treated to 30 to 60 HRC, a condition in which they can be machined readily into large, intricate dies and molds. Because these steels are prehardened, no subsequent high-temperature heat treatment is required, and distortion and size changes are avoided. However, when used for plastic molds, type P20 is sometimes carburized and hardened after the impression has been machined. Type P21 is an aluminum-containing precipitation-hardening steel that is supplied prehardened to 32 to 36 HRC. This steel is preferred for critical finish molds because of its ability to take a high polish.

Group P steels have low resistance to softening at elevated temperatures, with the exception of P4 and P21, which have medium resistance. Group P steels are used almost exclusively in low-temperature die-casting dies and in molds for the injection or compression molding of plastics. Plastic molds can require massive steel blocks up to 76 cm (30 in.) thick and weighing as much as 9100 kg (10 tons). Because these large die blocks must meet stringent requirements for soundness, cleanliness, and hardenability, electric arc melting, vacuum degassing, and special deoxidation treatments have become standard practice in the production of group P tool steels. In addition, ingot casting and forging practices have been refined so that a high degree of homogeneity can be achieved.

## 22.7 High-Speed Steels

High-speed steels were developed primarily for cutting tools that machine metals. High-speed steels are characterized by high red hardness and wear resistance. Their outstanding attribute is red hardness, the ability to cut metal at temperatures where the tool itself gets hot enough to glow a dull red color (~650 °C, or 1200 °F). Typically, a HRC of 52 at 540 °C (1000 °F) and a HRC of 48 at 595 °C (1100 °F) can be achieved. There are two classifications of high-speed steels: molybdenum high-speed steels (group M) and tungsten high-speed steels (group T). Group M steels, having approximately 40% lower initial cost, constitute more than 95% of all high-speed steel produced in the United States. A listing of the compositions of select group M and T high-speed steels is given in Table 22.8.

Group M and T high-speed steels are essentially equivalent in performance. Typical applications for both include cutting tools, such as drills, reamers, end mills, milling cutters, taps, and hobs. Some grades are also satisfactory for cold work applications, such as cold header die inserts, thread rolling dies, punches, and blanking dies. While not as wear resistant as cemented carbides, their toughness allows them to outperform cemented carbides in delicate tools and interrupted cutting applications.

Typically, high-speed steels contain 0.25 to 0.3 volume fraction of carbides. The following carbides are found in high-speed steels:

- MC, which is vanadium rich
- $M_2C$, which is a tungsten- or molybdenum-rich carbide

**Table 22.7 Composition of representative low-alloy special-purpose and low-carbon mold steels**

| Designation | | Composition, % | | | | | | | |
|---|---|---|---|---|---|---|---|---|---|
| AISI | UNS | C | Mn | Si | Cr | Ni | Mo | V | Al |
| **Low-alloy special-purpose tool steels** | | | | | | | | | |
| L2 | T61202 | 0.45–1.00 | 0.10–0.90 | 0.50 max | 0.70–1.20 | ... | 0.25 max | 0.10–0.30 | ... |
| L6 | T61206 | 0.65–0.75 | 0.25–0.80 | 0.50 max | 0.60–1.20 | 1.25–2.00 | 0.50 max | 0.20–0.30(a) | ... |
| **Low-carbon mold steels** | | | | | | | | | |
| P2 | T51602 | 0.10 max | 0.10–0.40 | 0.10–0.40 | 0.75–1.25 | 0.10–0.50 | 0.15–0.40 | ... | ... |
| P3 | T51603 | 0.10 max | 0.20–0.60 | 0.40 max | 0.40–0.75 | 1.00–1.50 | ... | ... | ... |
| P4 | T51604 | 0.12 max | 0.20–0.60 | 0.10–0.40 | 4.00–5.25 | ... | 0.40–1.00 | ... | ... |
| P5 | T51605 | 0.10 max | 0.20–0.60 | 0.40 max | 2.00–2.50 | 0.35 max | ... | ... | ... |
| P6 | T51606 | 0.05–0.15 | 0.35–0.70 | 0.10–0.40 | 1.25–1.75 | 3.25–3.75 | ... | ... | ... |
| P20 | T51620 | 0.28–0.40 | 0.60–1.00 | 0.20–0.80 | 0.40–2.00 | ... | 0.30–0.55 | ... | ... |
| P21 | T51621 | 0.18–0.22 | 0.20–0.40 | 0.20–0.40 | 0.50 max | 3.90–4.25 | ... | 0.15–0.25 | 1.05–1.25 |

(a) Optional. Source: Adapted from Ref 1

**Table 22.8 Compositions of representative high-speed tool steels**

| Designation | | Composition, % | | | | | | | | |
|---|---|---|---|---|---|---|---|---|---|---|
| AISI | UNS | C | Mn | Si | Cr | Ni | Mo | W | V | Co |
| **Molybdenum high-speed steels** | | | | | | | | | | |
| M2 | T11302 | 0.78–0.88; 0.95–1.05 | 0.15–0.40 | 0.20–0.45 | 3.75–4.50 | 0.30 max | 4.50–5.50 | 5.50–6.75 | 1.75–2.20 | ... |
| M7 | T11307 | 0.97–1.05 | 0.15–0.40 | 0.20–0.55 | 3.50–4.00 | 0.30 max | 8.20–9.20 | 1.40–2.10 | 1.75–2.25 | ... |
| M10 | T11310 | 0.84–0.94; 0.95–1.05 | 0.10–0.40 | 0.20–0.45 | 3.75–4.50 | 0.30 max | 7.75–8.50 | ... | 1.80–2.20 | ... |
| M30 | T11330 | 0.75–0.85 | 0.15–0.40 | 0.20–0.45 | 3.50–4.25 | 0.30 max | 7.75–9.00 | 1.30–2.30 | 1.00–1.40 | 4.50–5.50 |
| M33 | T11333 | 0.85–0.92 | 0.15–0.40 | 0.15–0.50 | 3.50–4.00 | 0.30 max | 9.00–10.00 | 1.30–2.10 | 1.00–1.35 | 7.75–8.75 |
| M35 | T11335 | 0.82–0.88 | 0.15–0.40 | 0.20–0.45 | 3.75–4.50 | 0.30 max | 4.50–5.50 | 5.50–6.75 | 1.75–2.20 | 4.50–5.50 |
| M36 | T11336 | 0.80–0.90 | 0.15–0.40 | 0.20–0.45 | 3.75–4.50 | 0.30 max | 4.58–5.50 | 5.50–6.50 | 1.75–2.25 | 7.75–8.75 |
| M41 | T11341 | 1.05–1.15 | 0.20–0.60 | 0.15–0.50 | 3.75–4.50 | 0.30 max | 3.25–4.25 | 6.25–7.00 | 1.75–2.25 | 4.75–5.75 |
| M42 | T11342 | 1.05–1.15 | 0.15–0.40 | 0.15–0.65 | 3.75–4.25 | 0.30 max | 9.00–10.00 | 1.15–1.85 | 0.95–1.35 | 7.75–8.75 |
| M44 | T11344 | 1.10–1.20 | 0.20–0.40 | 0.30–0.55 | 4.00–4.75 | 0.30 max | 6.00–7.00 | 5.00–5.75 | 1.85–2.20 | 11.00–12.25 |
| M47 | T11347 | 1.05–1.15 | 0.15–0.40 | 0.20–0.45 | 3.50–4.00 | 0.30 max | 9.25–10.00 | 1.30–1.80 | 1.15–1.35 | 4.75–5.25 |
| **Tungsten high-speed steels** | | | | | | | | | | |
| T1 | T12001 | 0.65–0.80 | 0.10–0.40 | 0.20–0.40 | 3.75–4.50 | 0.30 max | ... | 17.25–18.75 | 0.90–1.30 | ... |
| T2 | T12002 | 0.80–0.90 | 0.20–0.40 | 0.20–0.40 | 3.75–4.50 | 0.30 max | 1.0 max | 17.50–19.00 | 1.80–2.40 | ... |
| T4 | T12004 | 0.70–0.80 | 0.10–0.40 | 0.20–0.40 | 3.75–4.50 | 0.30 max | 0.40–1.00 | 17.50–19.00 | 0.80–1.20 | 4.25–5.75 |
| T5 | T12005 | 0.75–0.85 | 0.20–0.40 | 0.20–0.40 | 3.75–5.00 | 0.30 max | 0.50–1.25 | 17.50–19.00 | 1.80–2.40 | 7.00–9.50 |
| T6 | T12006 | 0.75–0.85 | 0.20–0.40 | 0.20–0.40 | 4.00–4.75 | 0.30 max | 0.40–1.00 | 18.50–21.00 | 1.50–2.10 | 11.0–13.00 |
| T15 | T12015 | 1.50–1.60 | 0.15–0.40 | 0.15–0.40 | 3.75–5.00 | 0.30 max | 1.00 max | 11.75–13.00 | 4.50–5.25 | 4.75–5.25 |

Source: Ref 1

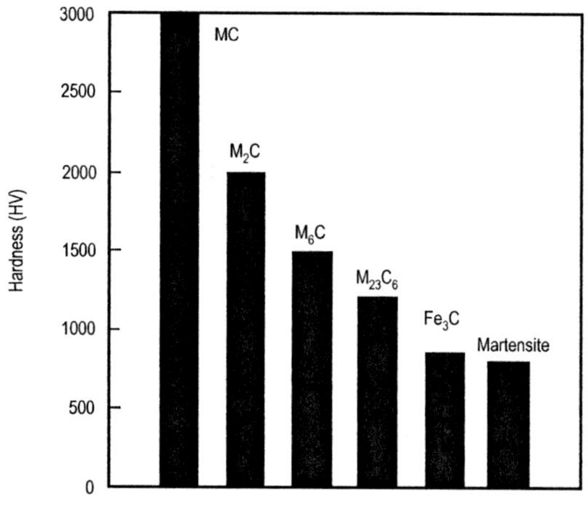


**Fig. 22.10** Relative hardness of alloy carbides in high-speed steels. Source: Ref 1

- $M_6C$, such as $Fe_3W_3C$ and $Fe_4W_2C$, where chromium, vanadium, and cobalt dissolve in $Fe_3W_3C$
- $M_{23}C_6$, which is chromium rich and dissolves iron, vanadium, molybdenum, and tungsten.

The hardness of these carbides (Fig. 22.10) is much greater than cementite or martensite.

Heat treatments used to harden high-speed steels first produce a mixture of austenite and alloy carbides. Initially, the microstructure is an array of ferrite grains and alloy carbides. When austenite begins to form, the microstructure contains austenite, ferrite, and carbides. Both the ferrite grains and the carbide particles restrict grain growth of the austenite. Then, $M_{23}C_6$ begins to dissolve, and dissolution is complete when 1095 °C (2000 °F) is reached on heating. Then, partial solution of the other carbides occurs until the austenite typically contains approximately 7 to 12 vol% alloy carbides of the types MC and $M_6C$. On cooling, there is an immediate tendency for the carbides to precipitate, especially on austenite grain boundaries. However, the diffusion rates of tungsten and molybdenum are very slow, and very slow cooling is required for significant carbide precipitation to occur on the grain boundaries and within the grains. The as-quenched steel contains 60 to 80 vol% martensite, 15 to 30 vol% retained austenite, and 5 to 12 vol% MC and $M_6C$. The as-quenched martensite is very hard, very brittle, has high residual stresses, and should be tempered immediately. The usual tempering range is 510 to 595 °C (950 to 1100 °F). During tempering, precipitation of alloy carbides also occurs in the retained austenite, which is also supersaturated. This reduces the concentration of dissolved alloying elements in the austenite, which raises the $M_s$ temperature. ($M_s$ is the temperature of an alloy system at which martensite starts to form on cooling.) This causes more martensite to form when the steel is cooled after being tempered. Double or triple tempering is necessary to eliminate the retained austenite.

### *22.7.1 Molybdenum High-Speed Steels*

Major alloying elements in group M high-speed steels include carbon, molybdenum, tungsten, chromium, vanadium, and sometimes cobalt. Group M steels have slightly greater toughness than group T steels at the same hardness. Otherwise, the mechanical properties of the two groups are similar. However, the cost of the group M steels is lower than the group T steels because of the use of molybdenum rather than tungsten.

The high carbon and alloy contents produce a large number of hard, wear-resistant carbides in the microstructure, particularly in the grades containing more than 1.5 wt% V and more than 1.0 wt% C. In particular, carbon and vanadium increase wear resistance by forming very hard vanadium carbide (VC). The addition of cobalt to some grades improves their ability to resist dulling at high cutting temperatures.

As a result of their high alloy contents, they are very deep hardening. Many can be fully hardened in thicknesses up to 76 mm (3 in.) by quenching in oil or molten salt. The TTT diagram for M2 steel (Fig. 22.11) illustrates that relative slow cooling of even thick sections will produce martensite. Because of the high content of ferrite stabilizers, the steel must be heated to within approximately 85 °C (150 °F) of its melting point to form austenite and dissolve sufficient carbon and other alloying elements so that it can then be hardened by air cooling. Secondary hardening (Fig. 22.12) due to the precipitation of fine alloy carbides occurs during tempering and during use.

A typical heat treatment for a group M high-speed steel (e.g., M42) would be:

1. Preheat in an air-circulating oven at 540 °C (1000 °F)
2. Preheat in a salt bath at 815 °C (1500 °F)
3. Austenitize in a salt bath at 1190 °C (2175 °F) for 5 min
4. Quench in a salt bath at 680 °C (1250 °F), followed by air cooling to room temperature
5. Triple temper at 540 °C (1000 °F) for 2 h (each time) in an air-circulating oven. (That is, temper for 2 h, cool to room temperature, retemper 2 h, cool again. A third tempering cycle is recommended.)

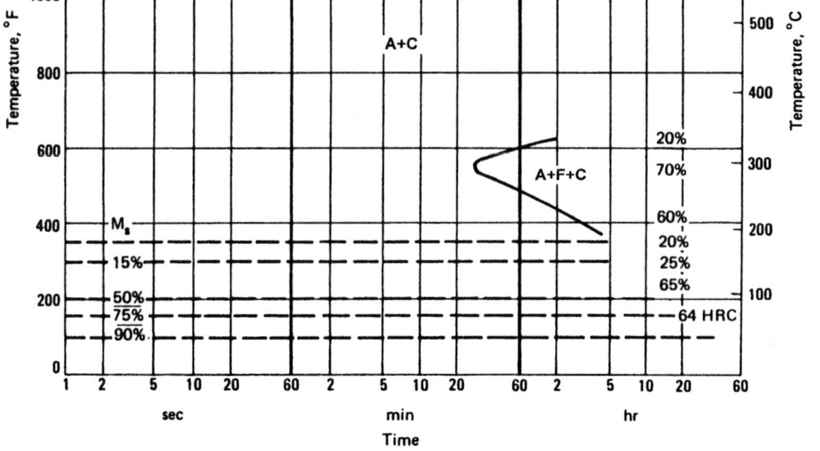


**Fig. 22.11** TTT diagram for M2 high-speed tool steel. Modified M2 containing 0.83 C, 0.32 Mn, 0.25 Si, 3.89 Cr, 4.30 Mo, 1.30 V, 5.79 W. Austenitized at 1220 °C (2225 °F). Prior condition, annealed. A, austenite; F, ferrite; C, carbide. Source: Adapted from Ref 4

Group M steels can be damaged by overheating during austenitizing. They are more sensitive than group T steels to hardening conditions, particularly austenitizing temperature and atmosphere. A normal microstructure for M4 and an overheated microstructure for M7 steel are shown in Fig. 22.13. The overheated microstructure shows evidence of incipient melting during austenitizing, as evidenced by the eutectic structure. Group M steels also have a tendency to decarburize during heat treatment, which results in a soft cutting edge that will immediately dull during metal cutting. Figure 22.14(a) shows a decarburized edge, and Fig. 22.14(b) compares a carburized M7 edge.

The maximum hardness obtainable in group M high-speed steels depends on the composition. For those with lower carbon contents, maximum hardness is usually approximately 65 HRC. Maximum hardness of the higher-carbon cobalt-containing steels is 69 to 70 HRC. However, few industrial applications exist for steels at the maximum hardness. Usually, the heat treatment is adjusted to provide a hardness of 66 to 68 HRC.

There are two grades, M50 and M52, with lower alloy content that are considered intermediate high-speed steels and are used for less demanding applications.

### 22.7.2 Tungsten High-Speed Steels

The principal alloying elements in the group T high-speed steels include carbon, tungsten, chromium, vanadium, and sometimes cobalt. Type T1 was the first high speed, developed partly as a result of the work of Taylor and White, who found in the early 1900s that certain tungsten alloy steels exhibited red hardness during metal-cutting operations.

Group T high-speed steels are all deep hardening when quenched from their recommended hardening temperature of 1205 to 1300 °C (2200 to 2375 °F). The maximum hardness of tungsten high-speed steels varies with carbon content and, to a lesser degree, with alloy content. A hardness of at least 64.5 HRC can be developed in any high-speed steel. Those types that have high carbon contents and high carbide contents, such as T15, can be hardened to 67 HRC.

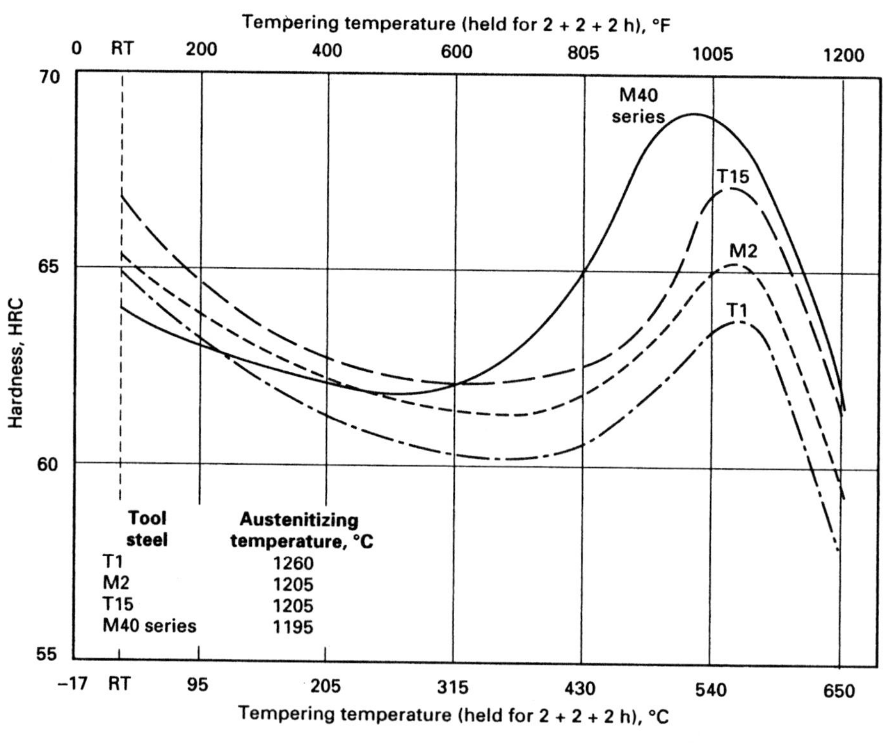


**Fig. 22.12** Plot of hardness vs. tempering temperature for selected high-speed tool steels. Source: Ref 8

## 22.8 Powder Metallurgy Tool Steels

As a result of pronounced ingot segregation, conventional tool steels often contain a coarse, nonuniform microstructure accompanied by low transverse properties and problems with size control and hardness uniformity during heat treatment. Carbide segregation in M2 steel annealed bars (Fig. 22.15) shows the longitudinal sections where segregation gets worse as the diameter increases. Rapid solidification of the atomized powders used in powder metallurgy (PM) tool steels eliminates this segregation and produces a very fine microstructure with a uniform distribution of carbides and nonmetallic inclusions. A distinguishing feature

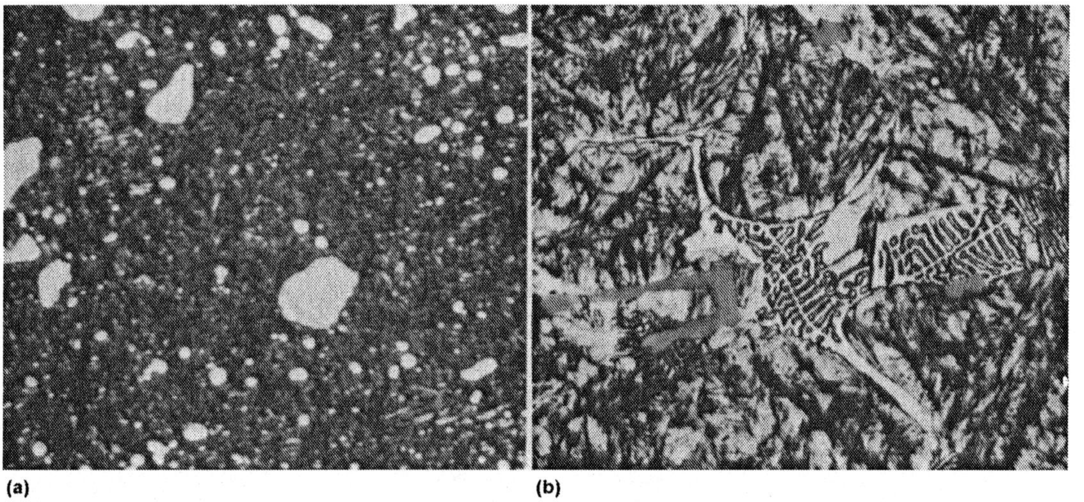

(a) (b)

Fig. 22.13 Hardened M4 and M7 tool steels. (a) M4 austenitized at 1218 °C (2225 °F), quenched in a salt bath, and double tempered (2 h plus 2 h) at 552 °C (1025 °F), air cooled. Normal microstructure seen, dark martensite with angular and spheroidal carbides (white). (b) M7 austenitized at 1260 °C (2300 °F) with the same quench and double temper. The overheated structure shows reprecipitated carbide eutectic and grain-boundary carbide in the coarse martensite that retains some austenite (white). Both have original magnification: 1000×. Source: Ref 6

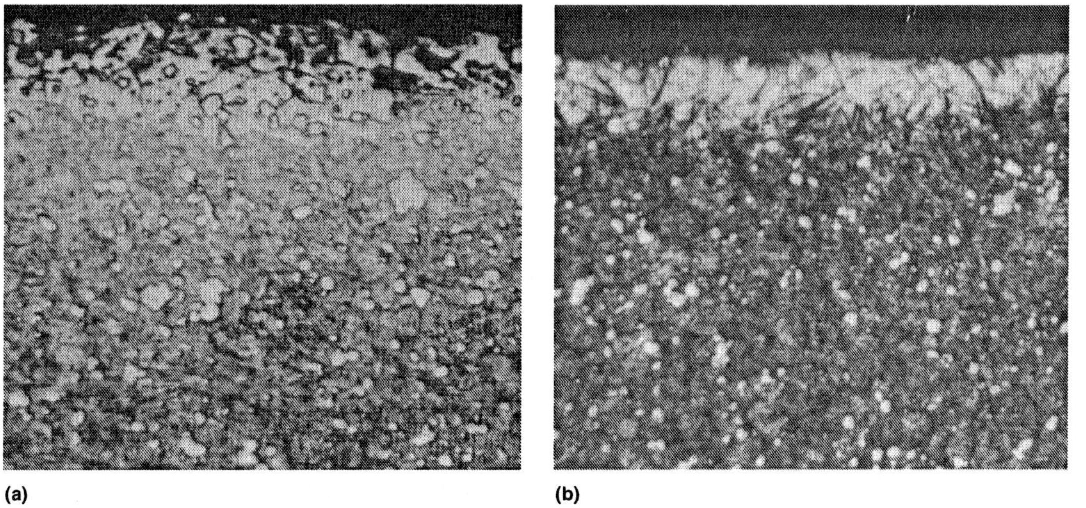

(a) (b)

Fig. 22.14 M7 high-speed tool steel. (a) Decarburized steel with white layer toward top that is ferrite at surface containing carbide spheroids and black oxide. (b) Carburized with white layer consistency predominantly of martensite and retained austenite. Original magnification: 750×. Source: Ref 6 of PM high-speed tool steels is the uniform distribution and small size of the primary carbides. A comparison of the microstructures of PM and wrought high-speed steels is shown in Fig. 22.16. The PM process has been used primarily for the production of advanced high-speed tool steels. However, it is now also being applied to the manufacture of improved cold work and hot work tool steels.

Advantages of PM high-speed steels compared to their wrought counterparts include better machinability, better grindability, better dimensional control during heat treatment, and superior cutting performance under difficult conditions where high edge toughness is essential. In addition, the alloying flexibility of the PM process allows the production of new tool steels that cannot be made by conventional ingot processes because of segregation-related hot workability problems. Examples are the highly alloyed superhigh-speed steels, such as CPM Rex 20, CPM Rex 76, and ASP 60, and the highly wear-resistant cold work tool steels, such as CPM 9V and CPM 10V, in Table 22.9.

The higher hardness attainable with PM high-speed tool steels, along with their greater amount of alloy carbides, constitutes a significant advantage over wrought high-speed steels. Temper resistance, or hot hardness, is largely determined by the composition and growth of the secondary hardening carbides and is promoted by vanadium, molybdenum, and cobalt. These elements can be used in larger amounts in PM high-speed steels than in wrought steels without degrading properties. The uniform distribution and small size of the carbides in PM high-speed steels represents an important toughness advantage that is important

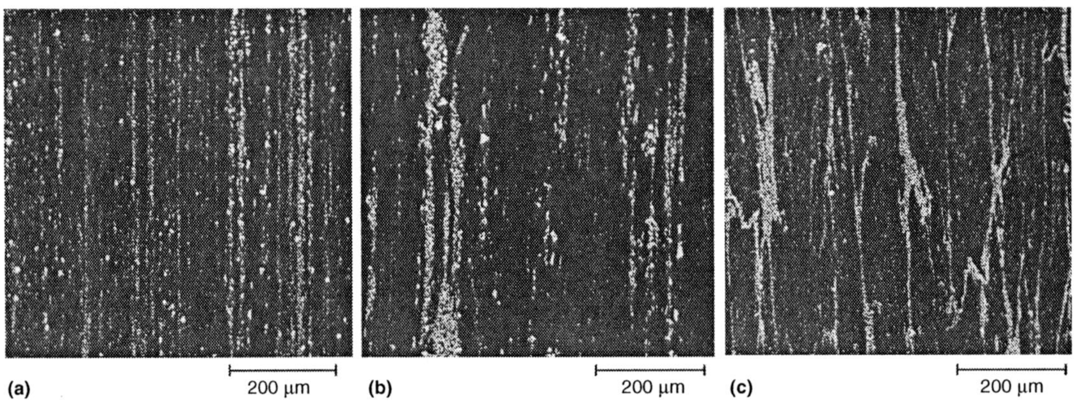


**Fig. 22.15** Carbide segregation at center of M2 round bars of various diameters. (a) 27 mm (1 1/6 in.). (b) 67 mm (2 5/8 in.). (c) 105 mm (4 1/8 in.). Original magnification: 100×. Source: Ref 5

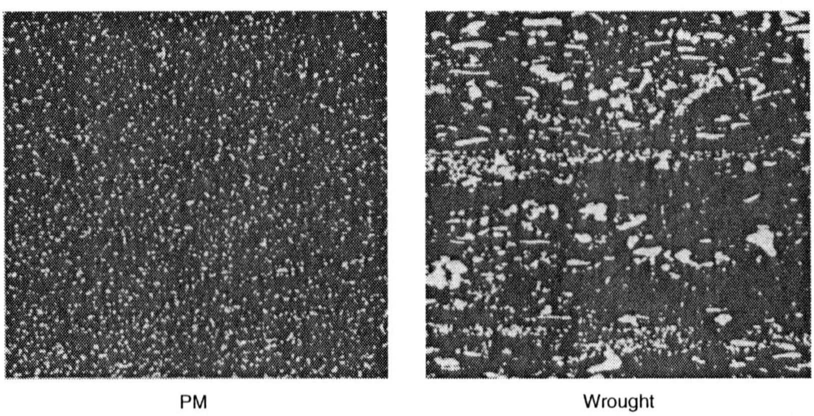


**Fig. 22.16** Powder metallurgy (PM) and wrought high-speed steels. Source: Ref 2

in interrupted cutting, where microchipping of the cutting edge can occur.

## 22.9 Cemented Carbides

In addition to high-speed steels, other important types of cutting tool materials include cast cobalt alloys, cemented carbides, and ceramics. As shown in the Fig. 22.17 comparison, ceramics retain their hardness best at elevated temperatures, followed closely by cemented carbides. The primary advantage of high-speed steels is their superior toughness compared to ceramics and cemented carbides. They can be used in heavy interrupted cuts, where ceramics or cemented carbides would immediately chip. Thus, ceramics and cemented carbides are used primarily for continuous cutting applications or ones where interrupted cutting is very limited. In these applications, they can have some productivity gains compared to high-speed steels (Fig. 22.18).

Cemented carbides are composites consisting of hard carbide particles in a metallic matrix

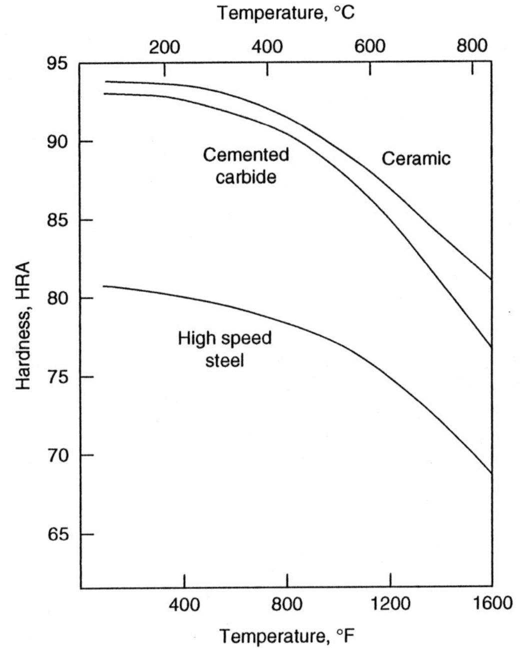


**Fig. 22.17** Elevated-temperature performance of cutting tool materials. Source: Ref 9

**Table 22.9 Nominal compositions of representative powder metallurgy tool steels**

| Trade names(a) | AISI | Nominal composition, % | | | | | | | Hardness, HRC |
|---|---|---|---|---|---|---|---|---|---|
| | | C | Cr | W | Mo | V | Co | S | |
| **Powder metallurgy high-speed tool steels** | | | | | | | | | |
| Micro-Melt 23<br>ASP 23/2023 | M3 | 1.28 | 4.20 | 6.40 | 5.00 | 3.10 | ... | ... | 65–67 |
| Micro-Melt 30<br>ASP 30/2030 | ... | 1.28 | 4.20 | 6.40 | 5.00 | 3.10 | 8.5 | ... | 66–68 |
| Micro-Melt 60<br>ASP 60/2060 | ... | 2.30 | 4.00 | 6.50 | 7.00 | 6.50 | 10.50 | ... | 67–69 |
| CPM Rex M4 | M4 | 1.35 | 4.25 | 5.75 | 4.50 | 4.00 | ... | 0.06 | 64–66 |
| CPM Rex M42 | M42 | 1.10 | 3.75 | 1.50 | 9.50 | 1.15 | 8.0 | ... | 66–68 |
| CPM Rex 45 | ... | 1.30 | 4.00 | 6.25 | 5.00 | 3.00 | 8.25 | 0.03 | 66–68 |
| CPM Rex 45HS | ... | 1.30 | 4.00 | 6.25 | 5.00 | 3.00 | 8.25 | 0.22 | 66–68 |
| CPM Rex 20 | M62 | 1.30 | 3.75 | 6.25 | 10.50 | 2.00 | ... | ... | 66–68 |
| CPM Rex 25 | M61 | 1.80 | 4.00 | 12.50 | 6.50 | 5.00 | ... | ... | 67–69 |
| CPM Rex T15 | T15 | 1.55 | 4.00 | 12.25 | ... | 5.00 | 5.0 | 0.06 | 65–67 |
| Micro-Melt M48<br>CPM Rex 76 | M48 | 1.50 | 3.75 | 10.0 | 5.25 | 3.10 | 9.00 | 0.06 | 67–69 |
| **Powder metallurgy cold work tool steels** | | | | | | | | | |
| Micro-Melt All-LVC<br>CPM 9V | ... | 1.78 | 5.25 | ... | 1.30 | 9.00 | ... | 0.03 | 53–55 |
| Micro-Melt All<br>CPM 10V | All | 2.45 | 5.25 | ... | 1.30 | 9.75 | ... | 0.07 | 60–62 |
| **Powder metallurgy hot work tool steels** | | | | | | | | | |
| CPM H13 | H13 | 0.40 | 5.00 | ... | 1.30 | 1.05 | ... | ... | 42–48 |
| CPM H19 | H19 | 0.40 | 4.25 | 4.25 | 0.40 | 2.10 | 4.25 | ... | 44–52 |
| CPM H19V | ... | 0.80 | 4.25 | 4.25 | 0.40 | 4.00 | 4.25 | ... | 44–56 |

(a) CPM, Crucible Powder Metallurgy (Crucible Materials Corporation). ASP, Anti-Segregation Process (Stora Kopparberg and ASEA). Micro-Melt, Carpenter Powder Products. Source: Adapted from Ref 2 binder, usually cobalt but sometimes a nickel or iron alloy. The microstructure of a typical tungsten carbide composite is shown in Fig. 22.19. The matrix phase, which imparts ductility, toughness, and thermal conductivity, typically comprises 10 to 15% of the volume. The original cemented carbide was tungsten carbide (WC) particles embedded in a cobalt matrix; however, many combinations exist today. Applications include metal cutting, mining, construction, rock drilling, metal forming, and structural wear components, with metal cutting being by far the most important, consuming approximately half of the annual production. Cemented carbide tools retain their hardness at very high temperatures and give longer tool lives during machining than conventional high-speed tool cutters.

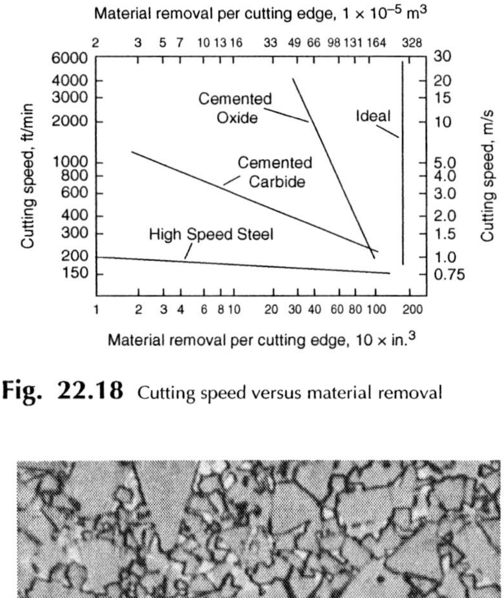


**Fig. 22.18** Cutting speed versus material removal

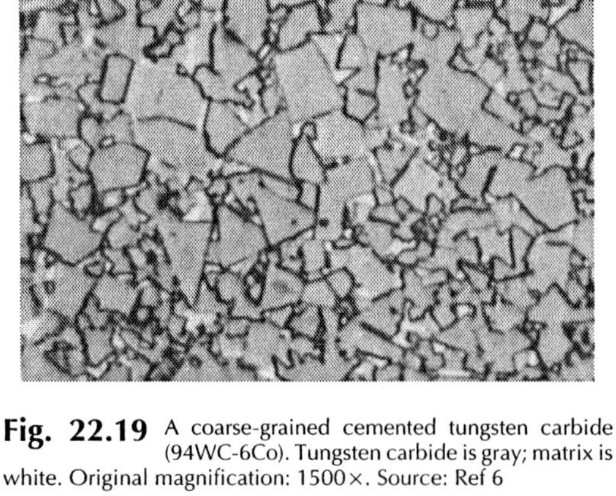

**Fig. 22.19** A coarse-grained cemented tungsten carbide (94WC-6Co). Tungsten carbide is gray; matrix is white. Original magnification: 1500×. Source: Ref 6

Cemented carbides are made by PM, in which a powder mixture is first prepared, then consolidated under high pressures and liquid-stage sintered at 1370 to 1480 °C (2500 to 2700 °F) to produce a blank that is finished by diamond grinding. During liquid-stage sintering, the binder melts and flows between and bonds to the carbide particles. The properties of sintered carbides are profoundly affected by their microstructure, including the compositions of the phases and the sizes, shapes, and distribution of the carbide particles. For example, if there is a deficiency of carbon, some of the WC particles can dissolve, and a mixed cobalt-tungsten carbide eta phase can form, which causes serious embrittlement. Carbides used in cemented hardmetals include WC, TiC, TaC, NbC, and mixed carbides of WC-TiC, WC-TiC-TaC, and WC-TiC-(Ta,Nb)C.

The basic WC-Co straight grades have excellent resistance to abrasive wear and are widely used in metal cutting. Although alloys with cobalt contents ranging between 3 and 30 wt% are available, the most widely used alloys for machining contain 10 to 12 wt% Co as the binder. The higher the WC content, the better the wear resistance but the more brittle the tool becomes.

## 22.10 Cutting Tool Coatings

Hard, wear-resistant surface coatings are often used as a means of improving tool life. These treatments increase surface hardness and wear resistance while reducing the coefficient of friction. Important considerations when selecting a surface treatment are substrate material, coating process temperature, coating thickness, coating hardness, and the material to be machined. There are many surface treatments and treatment processes available. It has been estimated that up to 60 to 70% of all carbide tools are now coated. The surface hardness produced by a number of different coatings is shown in Fig. 22.20. The material to be machined is an important consideration when selecting a coating. For example, titanium nitride (TiN) coatings work well on ferrous materials but immediately break down when machining titanium. Thin-film diamond coats work well when machining aluminum but are totally ineffective with ferrous alloys.

The primary methods of applying coatings to cutting tools are chemical vapor deposition

(CVD) and physical vapor deposition (PVD). Typical PVD coatings on carbide substrates are shown in Fig. 22.21.

Chemical vapor deposition, developed in the late 1960s-early 1970s, was the first method used for coating cemented carbide cutting tools. In the CVD process, the tools are heated in a sealed reactor to approximately 1000 °C (1830 °F), where gaseous hydrogen and volatile compounds supply the coating material constituents. Typical CVD coatings include titanium carbide (TiC), TiN, and titanium carbonitride (TiCN). More recently, CVD is being used to produce thin-film diamond coatings for machining graphite and nonferrous alloys. The typical thickness of CVD coatings ranges from ~5 to 20 μm. Because CVD coatings are applied at high temperatures and have a greater coefficient of thermal expansion than the substrate, they develop residual tension stresses on cooling to room temperature, which can cause cracking and spalling during interrupted cutting. However, the high processing temperatures ensure good bonding between the substrate and the coating. To reduce the residual-stress problem, the medium-temperature CVD process was developed in the 1980s to allow coating deposition at lower temperatures, between 705 and 900 °C (1300 and 1650 °F). The lower processing temperatures reduce thermally induced cracking in the coating.

Physical vapor deposition, the other major process used to produce cutting tool coatings, emerged in the 1980s as a viable process for applying hard coatings to both cemented carbide tools and high-speed steels. The PVD coatings are deposited in a vacuum using various processes, such as evaporation or sputtering. Electron beam evaporation of a titanium in a vacuum chamber and reaction with a nitrogen plasma to deposit TiN was the first successful application of PVD for cutting tools. Because PVD is a low-pressure process, PVD coatings are relatively thin and only cover areas within the line-of-sight of the coating source. The chief difference between PVD and CVD is the relatively low

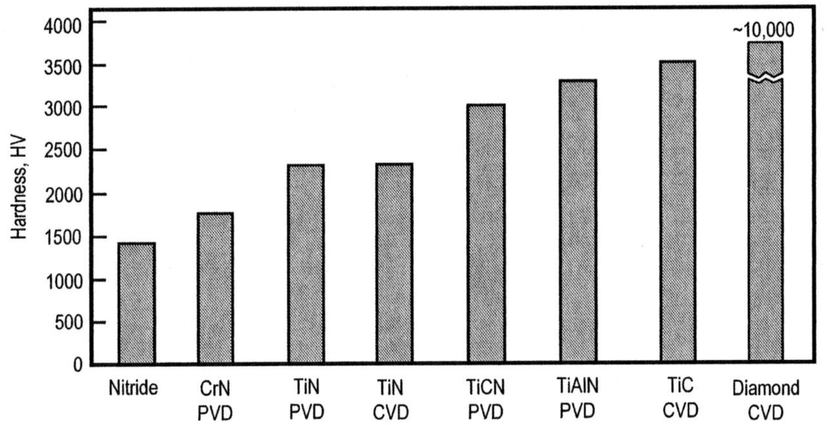


**Fig. 22.20** Hardness of coatings for tool materials. PVD, physical vapor deposition; CVD, chemical vapor deposition

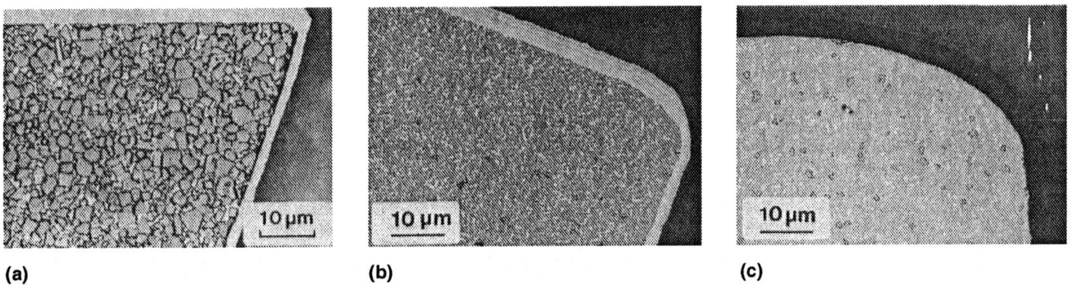


**Fig. 22.21** Physical vapor deposition coatings on cemented carbide substrates. (a) TiN. (b) TiCN. (c) TiAlN. Source: Ref 3

processing temperature for PVD (~500 °C, or 930 °F). The PVD coatings work well with high-speed and high-alloy tool steels. The PVD process temperatures are more than 28 °C (50 °F) below the tempering temperature of high-speed and high-alloy tool steels, nearly eliminating softening, distortion, and part growth. Other benefits of PVD coatings include finer coating grain sizes, smoother surface finishes, freedom from thermally induced cracking, and built-in compressive stresses that help resist crack initiation and propagation. Minimizing crack formation and propagation can help prevent premature tool failure.

A wide variety of hard coating materials are currently in use in machining applications. The most common is TiN, used on many high-speed steel and carbide tools. TiN coatings, which can be deposited by either CVD or PVD, have high hardness and low coefficients of friction that reduce wear, erosion, and abrasion. Titanium carbonitride coatings, which are harder than TiN, provide improved wear resistance when cutting carbon and alloy steels and cast irons. A newer coating, titanium-aluminum nitride (TiAlN), has improved hot hardness and oxidation resistance relative to TiN. A key to the performance of TiAlN is the addition of aluminum, which may oxidize during machining to form a very thin layer of $Al_2O_3$. With an increased emphasis on high speed and dry machining, aluminum oxide has become a major cutting tool coating material. Deposited using CVD, $Al_2O_3$ has excellent oxidation and wear resistance. It also becomes less thermally conductive as it gets hotter, acting as an effective heat barrier. The result is high hot hardness, wear resistance, and thermal protection at high cutting speeds, even in hardened work materials.

Multilayer coatings have become common place, some of them proprietary to their suppliers. The basic theory underlying the use of multiple coating layers is that each layer has its own function. One layer may have high hardness, another chemical wear resistance, and another oxidation resistance. Materials used to make up the various layers of a multilayer coating depend on the application. Layers may consist of several different materials or may be alternating layers of only a couple of materials. For example, a coating with alternating layers of $Al_2O_3$ and TiN is reported to be especially effective in high-speed machining of cast irons and steels. Such thin layers are considered nanofilms, and they can also be varied in microstructure to retard the growth of cracks.

Superhard coatings for cutting tools became a reality in the 1990s when research into making diamond adhere to cemented carbide tool substrates was commercialized. Diamond-coated carbide tools have since made inroads in applications traditionally reserved for polycrystalline diamond-brazed tools, such as high-speed machining of aluminum as well as nonmetallic materials such as graphite, ceramics, and advanced composites. Diamond films are produced using a modified CVD process. Diamond-coated cutting tools work in a way analogous to a grinding wheel. The cutting edge is composed of thousands of individual diamond crystals. Each crystal dulls and then fractures, presenting a new, sharp edge to the workpiece. The extreme hardness of diamond alone would make it a desirable cutting tool material, but its coefficient of friction is also very low. When machining abrasive aluminum alloys, the work material tends to flow very smoothly across a diamond-coated surface, minimizing adhesion and built-up edges and resulting in fine surface finishes.

## ACKNOWLEDGMENTS

Sections of this chapter were adapted from "Tool Steels" in *Metals Handbook Desk Edition*, 2nd ed., ASM International, 1998, "P/M Tool Steels" by K.E. Pinnow and W. Stasko in *Properties and Selection: Irons, Steels, and High-Performance Alloys,* Volume 1, *ASM Handbook,* ASM International, 1990, and "Surface Engineering of Carbide, Cermet, and Ceramic Cutting Tools" by A.T. Santhanan and D.T. Quinto in *Surface Engineering,* Volume 5, *ASM Handbook,* ASM International, 1994, p 900–908.

## REFERENCES

1. Tool Steels, *Metals Handbook Desk Edition,* 2nd ed., ASM International, 1998
2. K.E. Pinnow and W. Stasko, P/M Tool Steels, *Properties and Selection: Irons, Steels, and High-Performance Alloys,* Vol 1, *ASM Handbook,* ASM International, 1990
3. A.T. Santhanam and D.T. Quinto, Surface Engineering of Carbide, Cermet, and Ceramic Cutting Tools, *Surface Engineering,* Vol 5, *ASM Handbook,* ASM International, 1994, p 900–908

4. *Heat Treater's Guide: Practices and Procedures for Irons and Steels,* 2nd ed., ASM International, 1995
5. G.F. Vander Voort, Metallographic Techniques for Tool Steels, *Metallography and Microstructures,* Vol 9, *ASM Handbook,* ASM International, 2004, p 644–669
6. Microstructure of Tool Materials, *Atlas of Microstructures of Industrial Alloys,* Vol 7, *Metals Handbook,* 8th ed., American Society for Metals, 1972
7. Heat Treating of Specific Classes of Tool Steels, *Heat Treating,* Vol 4, *ASM Handbook,* ASM International, 1991
8. J.R. Davis, Ed., Introduction to Heat Treating of Tool Steels, *ASM Specialty Handbook: Tool Materialsm,* ASM International, 1995
9. E.W. Goblier, Advances in Cemented Carbide Tooling, *Met. Prog.,* Aug 1966, p 95–97


## SELECTED REFERENCES


- A.M. Bayer and L.R. Walton, Wrought Tool Steels, *Properties and Selection: Irons, Steels, and High-Performance Alloys,* Vol 1, *ASM Handbook,* ASM International, 1990
- J.R. Davis, *Alloying: Understanding The Basics,* ASM International, 2001
- Heat Treating of Specific Classes of Tool Steels, *Heat Treating,* Vol 4, *ASM Handbook,* ASM International, 1991
- G. Krauss, *Steels: Processing, Structure, and Performance,* 3rd ed., ASM International, 2005
- K. Narasimhan, How to Select the Right Carbide Insert, *Moldmak. Technol.,* Jan 2003
- P. Payton, *Metallurgy of Tool Steels,* John Wiley & Sons, Inc., 1962
- G.A. Roberts, J.C. Hamaker and A.R. Johnson, *Tool Steels,* 3rd ed., ASM International, 1962
- A.T. Santhanam, P. Tierney and J.L. Hunt, Cemented Carbides, *Properties and Selection: Nonferrous Alloys and Special-Purpose Materials,* Vol 2, *ASM Handbook,* ASM International, 1990
- W.F. Smith, *Structure and Properties of Engineering Alloys,* 2nd ed., McGraw-Hill, Inc., 1993

# CHAPTER 23

# Stainless Steels

IRON IS EXTRACTED from ores that are mainly iron oxides. After being extracted, it can be used to manufacture steels. Steels generally are degraded by the formation of oxides and hydroxides, either by oxidation when heated or by corrosion from environments that contain water and oxygen. Thus, the iron in steels will return to its former state under favorable conditions by forming rust or scale. In addition, the corrosion product is not effective protection, and the steel will continue to rust. Oxidation occurs when carbon steel is heated in oxidizing conditions. The oxide scale spalls off, and oxidation will continue as long as the steel is hot. This behavior, along with other corrosion processes, such as dissolution in dilute acids, imposes obvious limitations on the use of carbon and low-alloy steels.

The addition of chromium dramatically improves the corrosion and oxidation resistance of steel. Chromium oxidizes and forms a thin, tightly adherent layer of oxide ($Cr_2O_3$) on the surface that prevents or minimizes further corrosion. The oxide forms and heals itself in the presence of oxygen. Steels with sufficient chromium form the family of steels called stainless steels. Although the annual tonnage of stainless steels is only a small fraction of that of carbon and low-alloy steels, its range of uses is wide.

For a steel to be stainless, it must contain at least 11.5 wt% Cr, with at least 12 wt% Cr required for aqueous corrosion resistance. At 12 wt% Cr, they become passive in aqueous solutions; even higher chromium content is necessary for corrosion resistance in nonaqueous solutions. Few stainless steels contain more than 30 wt% Cr or less than 50 wt% Fe. Other important alloying elements include nickel, molybdenum, copper, titanium, aluminum, silicon, niobium, nitrogen, sulfur, and selenium. Most stainless steels have carbon present in amounts ranging from approximately 0.03 wt% to 1.0 wt%. A summary of some of the compositional and property linkages in the stainless steel family is shown in Fig. 23.1.

As the binary iron-chromium phase diagram in Fig. 23.2 shows, chromium stabilizes ferrite and forms a gamma ($\gamma$) loop in which austenite is the stable phase. When the chromium content exceeds 12 wt%, it is possible for ferrite to exist at all temperatures. The binary system also contains the intermetallic sigma ($\sigma$) phase. Sigma is an extremely brittle tetragonal structure that is usually avoided. The phase diagram shows that for some compositions, solid solutions can be transformed into austenite by heating them into the gamma loop, and, when cooled, austenite transforms into ferrite. This transformation forms the basis for the heat treatable martensitic stainless steels.

Stainless steels are used in a wide variety of applications. Most of the structural applications are in the chemical and power engineering industries, which account for more than a third of the market for stainless steel products. These applications include an extremely diversified range of uses, including nuclear reactor vessels, heat exchangers, oil industry tubes, components for chemical processing and pulp and paper industries, furnace parts, and boilers used in fossil-fuel-fired electric power plants.

There are five types of stainless steels: austenitic, ferritic, duplex, martensitic, and precipitation-hardening steels. These five types of stainless steel have a somewhat simplified classification system, as follows:

- Austenitic stainless steels with low nickel: 2*xx* series
- Austenitic stainless steels: 3*xx* series
- Ferritic stainless steels: 4*xx* series
- Duplex stainless steel: (Manufacturer's designation)
- Martensitic stainless steels: 4*xx* series
- Precipitation-strengthening stainless steels: *xx*-*x* PH

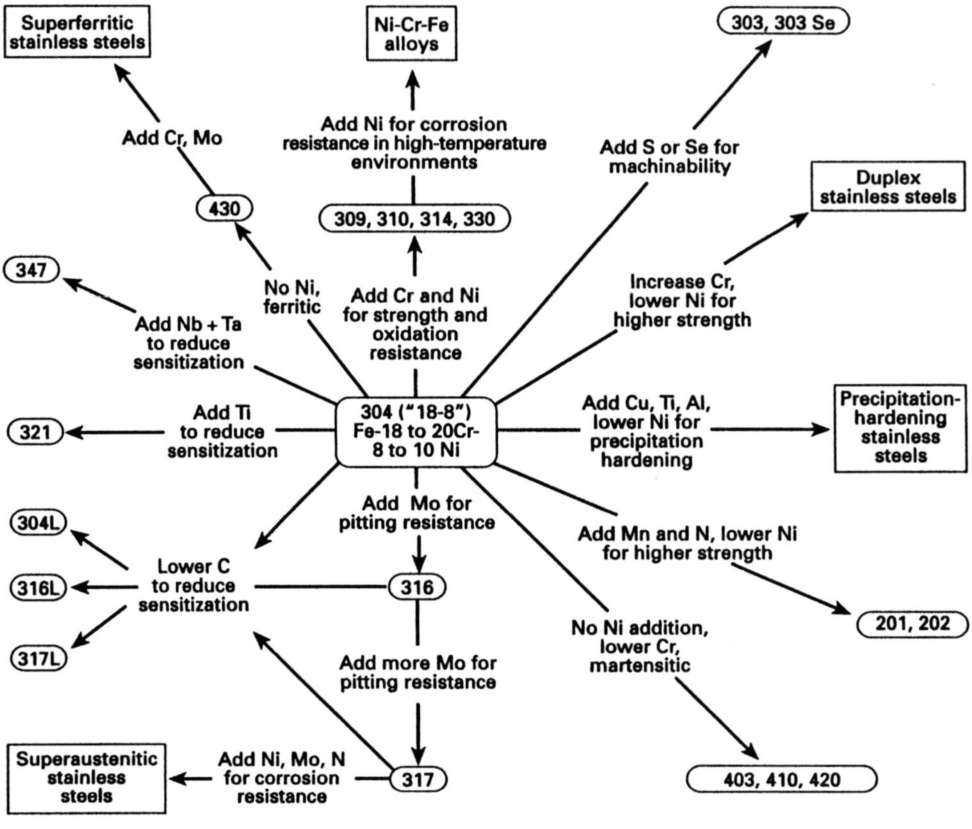


**Fig. 23.1** Compositional and property linkages in stainless steel alloys. Source: Ref 1

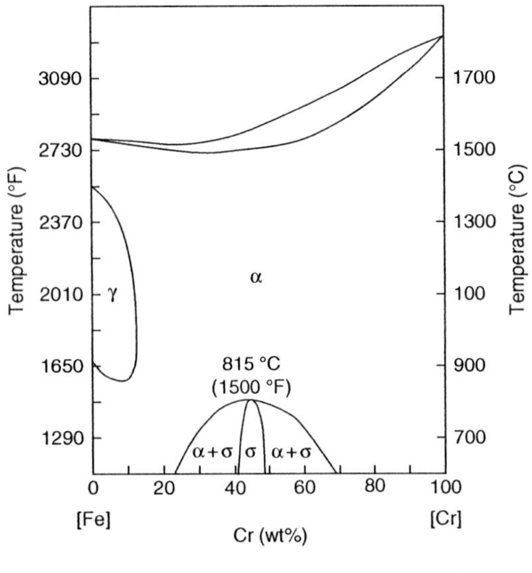


**Fig. 23.2** Iron-chromium phase diagram

The classification system for the stainless steels differs from the SAE/AISI system for low-alloy steels in that the last two digits (*xx*) do not represent the carbon content and have no particular compositional meaning. Unfortunately, the classification system is somewhat confusing, with the ferritic and martensitic stainless steels both being in the 4*xx* series. The 2*xx* series of austenitic stainless steels were developed during the 1950s, when nickel became scarce. In these steels, manganese and nitrogen are substituted for a portion of the nickel content in order to maintain strength.

## 23.1 Argon Oxygen Decarburization

Stainless steels are usually produced in an induction or electric arc furnace, sometimes under vacuum. To refine stainless steel, the argon oxygen decarburization (AOD) process is used. In AOD, an argon oxygen gas mixture is injected through the molten steel to remove carbon without a substantial loss of chromium, the main alloying element in stainless steel. The AOD process for refining stainless steels allows accurate control of the chemistry, in particular, control of carbon and sulfur. Prior to AOD, carbon could not be removed during refining without also removing chromium. Most stainless steels are available with different levels of carbon. For resistance to intergranular corrosion, a low carbon content is preferred, usually a maximum of 0.03 wt% C. Such a stainless steel is referred to as an "L" grade, for example, 304L and 316L. For aqueous corrosion resistance, the lower the carbon content the better the resistance. For high-temperature service, the opposite is true, and some minimum amount of carbon is required for both tensile and creep-rupture strength. Low-carbon grades could only be produced by starting with low-carbon raw materials, specifically low-carbon ferrochrome. The expense of low-carbon ferrochrome meant that the L grades were inherently more expensive. AOD enables reduction of carbon to very low levels, even if the starting stock contains higher carbon.

Using AOD, it is also possible to add a very small, precisely controlled amount of nitrogen. This does not harm intergranular corrosion resistance but increases the room-temperature tensile properties. Like carbon, sulfur can now readily be reduced to very low levels, typically less than 0.005 wt%. Usually, stainless steels intended for plate are refined to a low sulfur level to improve hot workability. Plate is generally formed and welded, usually with little machining by the user. Low sulfur is quite detrimental to machinability. Since bar products are usually machined, most stainless bar actually must be resulfurized to some level (~0.02 wt%) for improved machinability. Precise control of chemistry, in particular nitrogen, has permitted development of the superaustenitic 6% Mo grades. The ability to closely control nitrogen as an alloying addition has also tremendously improved the weldability of duplex stainless steels.

## 23.2 Ferritic Stainless Steels

Ferritic stainless steels (4*xx* series) are essentially iron-chromium alloys with body-centered cubic (bcc) crystalline structures. Ferritic stainless steels contain 11.5 to 30 wt% Cr, with most compositions containing 17 to 26 wt% Cr. The corrosion resistance increases with increasing chromium contents; that is, the range of environments in which the passive film ($Cr_2O_3$) is stable increases. Molybdenum additions also increase the stability of the protective film. Ferritic stainless steels may also contain small amounts of manganese, molybdenum, silicon, nickel, aluminum, titanium, and niobium. Molybdenum increases corrosion resistance, particularly to pitting. Niobium and titanium are used to stabilize against intergranular corrosion. Sulfur or selenium can be added to improve machinability. The compositions of some ferritic stainless steels are given in Table 23.1, and mechanical properties are shown in Table 23.2. Since expensive nickel is not used as an alloying element, ferritic stainless steels are less expensive than austenitic stainless steels. Note that ferritic stainless steels, in fact,

**Table 23.1 Composition of select ferritic stainless steels**

| | Composition, wt% | | | | | |
|---|---|---|---|---|---|---|
| Type | C | Cr | Mo | Ni | N | Other |
| 405 | 0.08 | 11.5–14.5 | ... | ... | ... | 0.10–0.30 Al |
| 406 | 0.06 | 12.0–14.0 | ... | 0.5 | ... | 2.75–4.25 Al; 0.6 Ti |
| 409 | 0.08 | 10.5–11.75 | ... | 0.5 | ... | Ti = 6 × C min to 0.75 max |
| 409Cb | 0.02 | 12.5 | ... | 0.2 | ... | 0.4 Nb |
| 429 | 0.12 | 14.0–16.0 | ... | ... | ... | ... |
| 430 | 0.12 | 16.0–18.0 | ... | ... | ... | ... |
| 430F | 0.12 | 16.0–18.0 | 0.6 | ... | ... | 0.06 P; 0.15 min S |
| 430Fse | 0.12 | 16.0–18.0 | ... | ... | ... | 0.15 min Se |
| 434 | 0.12 | 16.0–18.0 | 0.75–1.25 | ... | ... | ... |
| 436 | 0.12 | 16.0–18.0 | 0.75–1.25 | ... | ... | Nb + Ta = 5 × %C min |
| 439 | 0.07 | 17.00–19.00 | ... | 0.5 | ... | Ti = 0.20 + 4 (C + N) min to 1.0 max |
| 441 | 0.02 | 18.0 | ... | 0.3 | ... | 0.7 Nb, 0.3 Ti |
| 442 | 0.20 | 18.0–23.0 | ... | ... | ... | ... |
| 446 | 0.20 | 23.0–27.0 | ... | ... | ... | ... |

Source: Ref 1

most stainless steels except for the high-strength martensitic steels, contain rather low carbon contents.

Ferritic stainless steels have very good corrosion resistance in many environments. However, their mechanical properties are not very good. Since they all have a bcc structure, they exhibit a ductile-to-brittle transition temperature (DBTT) similar to alloy steels. However, if the carbon and nitrogen are held to low levels (<0.015 wt%), the transition temperature will be below room temperature. In the common alloys where the steel contains nitrogen, phosphorus, sulfur, and carbon, the DBTT can be higher than room temperature. For example, type 446 with 27 wt% Cr has a DBTT of approximately 150 °C (300 °F). In addition, since thicker sections have more constraint during plastic deformation, the DBTT increases with thickness. In contrast, thin sheets, in which yielding can occur through the thickness, remain ductile and are formable at temperatures well below room temperature. The grain size, the interstitial carbon and nitrogen contents, and the presence of various second phases also affect the DBTT. Fine grain sizes, low interstitial contents, and the elimination of second phases through proper heat treatment increase ductility and toughness. Melting practices, such as AOD and vacuum melting, and stabilization with titanium and niobium additions improve ductility and toughness. Since their grain size cannot be refined by heat treatment, large grains produced in the heat-affected zones at welds can reduce the local strength. In ferritic stainless steels with a high carbon content, the undesirable and brittle sigma ($\sigma$) phase may form during long exposures to high temperatures. For these reasons, ferritic stainless steels have not been very popular for structural applications.

The ferritic stainless steels are ferromagnetic. They can have good ductility and formability, but high-temperature strengths are relatively low compared to the austenitic grades. Toughness may be somewhat limited at low temperatures and in heavy sections. The ferritic stainless steels cannot be strengthened by heat treatment. As a result of their bcc crystalline structures, the ferritic stainless steels generally have higher yield strengths and lower ductilities than the face-centered cubic (fcc) austenitic stainless steels. Also, because the strain-hardening rates of ferrite are relatively low and cold work significantly lowers ductility, the ferritic stainless steels are not often strengthened by cold work. Since ferritic stainless steels are single phase and diffusion occurs fairly rapidly in the open bcc structure, grain growth during elevated-temperature processing can be a problem. Ferritic steels start coarsening rapidly at approximately 620 °C (1150 °F) as compared to 900 °C (1650 °F) for austenitic steels.

The typical microstructure of ferritic stainless steels consists of carbide particles embedded in a matrix of ferrite (Fig. 23.3). Typical annealed yield and tensile strengths for ferritic stainless steels are 240 to 380 and 415 to 585 MPa (35 to 55 and 60 to 85 ksi), respectively. Percent elongations tend to range between 20 and 35%. Higher strengths, up to 515 MPa (75 ksi) for yield strength and 655 MPa (95 ksi) for tensile strength, are obtained in the more highly alloyed superferritic stainless steels. Superferritic stainless steels contain high chromium contents (25 to 28%), with molybdenum usually at 3% or greater. Nickel may be added in amounts up to

**Table 23.2 Typical properties of select ferritic stainless steels**

| Steel | Condition | Tensile strength | | Yield strength | | Elongation, in 50 mm (2 in.), % | Reduction in area, % | Hardness, HB |
|---|---|---|---|---|---|---|---|---|
| | | MPa | ksi | MPa | ksi | | | |
| 405 | Annealed bar | 480 | 70 | 275 | 40 | 30 | 60 | 150 |
| | Cold drawn bar | 585 | 85 | 480 | 70 | 20 | 60 | 185 |
| 409 | Annealed bar | 450 | 65 | 240 | 35 | 25 | . . . | 75 HRB |
| 430 | Annealed bar | 515 | 75 | 310 | 45 | 30 | 65 | 155 |
| | Annealed and cold drawn | 585 | 85 | 480 | 70 | 20 | 65 | 185 |
| 442 | Annealed bar | 515 | 75 | 310 | 45 | 30 | 50 | 160 |
| | Annealed at 815 °C (1500 °F) and cold worked | 545 | 79 | 425 | 62 | 35.5 | 79 | 92 HRC |
| 446 | Annealed bar | 550 | 80 | 345 | 50 | 25 | 45 | 86 HRB |
| | Annealed at 815 °C (1500 °F) and cold drawn | 605 | 88 | 462 | 67 | 26 | 64 | 96 HRB |

Source: Ref 2

4%. This combination, combined with careful control of interstitials, results in a desirable combination of good mechanical properties and resistance to general corrosion, pitting, and stress-corrosion cracking (SCC). These properties make them attractive alternatives to the austenitic stainless steels commonly plagued by chloride SCC.

Annealing for recrystallization of cold-worked microstructure is conducted at 760 to 955 °C (1400 to 1750 °F), followed by rapid cooling. Rapid cooling is necessary to prevent the formation of phases that are detrimental to ductility and toughness. Ferritic stainless steels are susceptible to the formation of stretcher strains during severe drawing or forming operations. However, an even more serious problem is "ridging" or "roping," which is not a yield-point phenomenon but is due to a crystallographic textural effect. Rope marks are depressions on one side of a sheet that match elevations on the other side of the sheet, severely marring the appearance of the part. The sheet thickness remains constant, and the markings are parallel to the rolling direction.

There are three types of embrittlement that can occur in ferritic stainless steels.

**Sigma-Phase Embrittlement.** When ferritic stainless steels are aged for very long times between 540 and 760 °C (1000 and 1400 °F), the intermetallic phase sigma (σ) precipitates, causing embrittlement. Sigma is a brittle iron-chromium compound that forms when the chromium content and ferrite stabilizers are high. Type 446, with its high chromium content, is the most susceptible alloy. The embrittlement can be removed by reheating above 815 °C (1500 °F), where the sigma phase redissolves. However, reheating is often not feasible. Fortunately, the precipitation process is very slow, requiring several hundred hours for embrittlement to occur. Thus, it does not happen during the usual times of cooling or welding. However, it can be important if the component must operate for extended times in the 540 to 760 °C (1000 to 1400 °F) range.

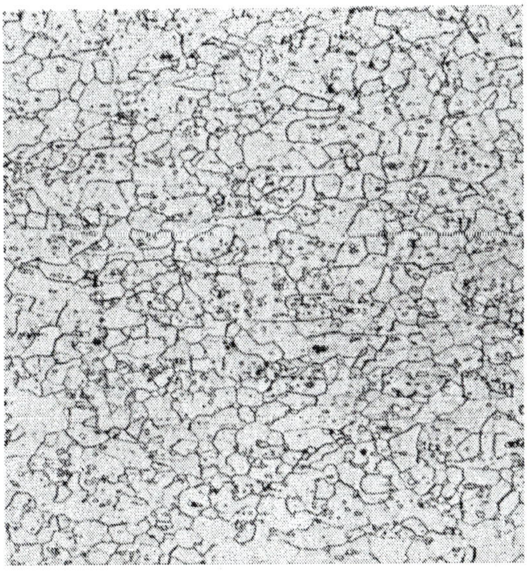

**Fig. 23.3** Microstructure of annealed 446 stainless steel strip. Original magnification: 100×. Source: Ref 3

**885 °F Embrittlement.** This embrittlement causes a loss of impact toughness and is caused by holding at temperatures between 400 and 540 °C (750 and 1000 °F). Embrittlement is associated with the precipitation of a coherent chromium-rich phase (α′) throughout the grains. It can be reversed by reheating the steel above 815 °C (1500 °F) and then cooling. Quenching is unnecessary because this form of embrittlement requires long times to form.

**High Temperature Embrittlement.** When ferritic steels containing normal amounts of carbon, nitrogen, phosphorus, and sulfur are heated above 980 °C (1800 °F) and then air cooled, as, for example, during welding, they can become severely embrittled and also lose their corrosion resistance near grain boundaries. The phenomenon is similar to sensitization in austenitic steels and is caused by precipitation of chromium carbides and nitrides on grain boundaries. Three methods are used to combat this problem: (1) ensure very low interstitial contents during the steel-making process, (2) add small amounts of very strong carbide and nitride formers (e.g., vanadium, niobium, and titanium) to scavenge these elements, and (3) add small amounts of aluminum, copper, platinum, palladium, silver, and vanadium. The first two methods can be successful; however, as the chromium content increases so does the minimum acceptable level of interstitial solutes. At 35 wt% Cr, the interstitial level cannot be reduced low enough by the first two methods. In these cases, the third method is used. It has been found that small additions of aluminum, copper, platinum, palladium, silver, and vanadium in the range of 0.1 to 1.3 wt% can improve the as-welded ductility and corrosion, even in the presence of relatively high quantities of interstitial solutes.

The corrosion resistance of ferritic stainless steels can range from moderate for the low-to-medium chromium-containing alloys to outstanding for the superferritics (e.g., type 444), which have high chromium and molybdenum contents. The low-chromium (11 wt%) alloys (e.g., types 405 and 409) have fair corrosion and oxidation resistance and good fabricability at low cost. Type 409 stainless steel, with the lowest chromium level (10.5 to 11.75 wt%), is the least expensive of the ferritic stainless steel series and is used for automotive exhaust systems because it far outlasts carbon steel. The intermediate-chromium (16 to 18 wt%) alloys, such as type 430, resist mild oxidizing acids and organic acids and are used in food-handling equipment. Type 434, which includes molybdenum for improved corrosion resistance, is used for automotive trim. The high-chromium (19 to 30 wt%) alloys, which include types 442 and 446 as well as the superferritics, are used for applications that require a high level of corrosion and oxidation resistance. By controlling interstitial element content through AOD processing, it is possible to produce grades with unusually high chromium and molybdenum contents and very low carbon contents (as low as 0.01 wt%). These highly alloyed superferritics offer exceptional resistance to localized corrosion due to aqueous chlorides. Localized corrosion, such as pitting, crevice corrosion, and SCC, are problems that plague many austenitic stainless steels. Therefore, the superferritics are often used in heat exchangers and piping systems for chloride-containing aqueous solutions and seawater.

## 23.3 Martensitic Stainless Steels

Martensitic stainless steels (4*xx* series) are Fe-Cr-C alloys that are ferromagnetic, hardenable by heat treatment, and generally resistant to corrosion only in relatively mild environments. Chromium content is generally in the range of 10.5 to 18 wt%, with carbon contents of 0.10 to as high as 1.2 wt%. They are sometimes classified as low-carbon and high-carbon martensitic stainless steels, with the higher-carbon grades attaining the highest strength levels. The chromium and carbon contents are balanced to ensure a martensitic structure. Because chromium content is kept as high as possible for corrosion resistance, austenite stabilizers, such as carbon, manganese, and nickel, are added to expand the austenite phase field. A high chromium content significantly shifts the nose of the time-temperature transformation diagram to the right, increasing the time for the start of the transformations to over 300 s. Thus, many sections are air hardenable, and martempering is also feasible. In addition, chromium retards tempering, and residual stresses can be removed during tempering before appreciable softening occurs. Other elements, such as niobium, silicon, tungsten, and vanadium, are added to modify the tempering response after hardening. Small amounts of nickel can be added to improve corrosion resistance and toughness. Sulfur or selenium is added to some grades to improve machinability. The compositions of a number of the martensitic types are listed in Table 23.3, and mechanical properties are given in Table 23.4.

The most commonly used alloy is type 410, which contains approximately 12 wt% Cr and

**Table 23.3 Composition of select martensitic stainless steels**

| Type | Composition, wt% | | | | | | | |
|---|---|---|---|---|---|---|---|---|
| | C | Mn | Si | Cr | Ni | P | S | Other |
| 403 | 0.15 | 1.00 | 0.50 | 11.5–13.0 | ... | 0.04 | 0.03 | ... |
| 410 | 0.15 | 1.00 | 1.00 | 11.5–13.5 | ... | 0.04 | 0.03 | ... |
| 414 | 0.15 | 1.00 | 1.00 | 11.5–13.5 | 1.25–2.50 | 0.04 | 0.03 | ... |
| 416 | 0.15 | 1.25 | 1.00 | 12.0–14.0 | ... | 0.06 | 0.15 min | 0.6 Mo |
| 416Se | 0.15 | 1.25 | 1.00 | 12.0–14.0 | ... | 0.06 | 0.06 | 0.15 min Se |
| 420 | 0.15 min | 1.00 | 1.00 | 12.0–14.0 | ... | 0.04 | 0.03 | ... |
| 420F | 0.15 min | 1.25 | 1.00 | 12.0–14.0 | ... | 0.06 | 0.15 min | 0.6 Mo |
| 422 | 0.20–0.25 | 1.00 | 0.75 | 11.5–13.5 | 0.5–1.0 | 0.04 | 0.03 | 0.75–1.25 Mo; 0.75–1.25 W; 0.15–0.3 V |
| 431 | 0.20 | 1.00 | 1.00 | 15.0–17.0 | 1.25–2.50 | 0.04 | 0.03 | ... |
| 40A | 0.60–0.75 | 1.00 | 1.00 | 16.0–18.0 | ... | 0.04 | 0.03 | 0.75 Mo |
| 440B | 0.75–0.95 | 1.00 | 1.00 | 16.0–18.0 | ... | 0.04 | 0.03 | 0.75 Mo |
| 440C | 0.95–1.20 | 1.00 | 1.00 | 16.0–18.0 | ... | 0.04 | 0.03 | 0.75 Mo |

Source: Ref 1

0.15 wt% C to provide strength. As the carbon content increases for types 420, 440A, 440B, and 440C, the strength also increases. The 440A, B, and C alloys have increased chromium contents to maintain corrosion resistance. Molybdenum is added to improve mechanical properties or corrosion resistance, as in type 422. Nickel is added to types 414 and 431 for the same reasons. When higher chromium levels are used to improve corrosion resistance, nickel additions are used to maintain the desired microstructure and to prevent excessive free ferrite. The limitations on alloy content required to maintain the desired fully martensitic structure restrict the corrosion resistance to only moderate levels.

To produce martensite in a stainless steel, the alloy must be transformed from the austenite phase field. The gamma loop shown in Fig. 23.2 is the region between 800 and 1400 °C (1470 and 2550 °F) and 0 to 12.7 wt% Cr. Since austenite only exists in this restricted region, the steel must be heated within this temperature range and quenched to room temperature to form martensite. Martensitic stainless steels contain added carbon, which expands the gamma loop to allow higher chromium contents to be used. Because they can be heat treated, the martensitic stainless steels generally have higher strength than the austenitic and ferritic stainless steels. The DBTTs of these steels are approximately, or barely below, room temperature, which means that they are not useful in cryogenic applications.

In the annealed condition, martensitic stainless steels have yield strengths of approximately 275 MPa (40 ksi), which can be moderately increased by cold working. The martensitic stainless steels can be annealed by either process annealing or full annealing to obtain maximum ductility and machinability. However, martensitic stainless steels are typically heat treated by quenching and tempering to yield strength levels as high as 1965 MPa (285 ksi), depending on the carbon content. The lower-carbon alloys have good ductility and toughness properties, which decrease as strength increases. Martensitic stainless steels are hardened by austenitizing at 925 to 1065 °C (1700 to 1950 °F), followed by oil quenching or air cooling. In general, higher austenitizing temperatures provide better strength and corrosion resistance. Since the thermal conductivity of these steels is low, preheating at 760 to 790 °C (1400 to 1450 °F) is used to equalize stresses and minimize the chances of distortion or cracking.

The low-carbon grades, being high-strength structural steels, contain approximately 0.15 wt% C and have good weldability, formability, and impact toughness. They are normally oil quenched and then tempered. When they are tempered at lower temperatures, they

**Table 23.4 Typical properties of select martensitic stainless steels**

| Steel | Condition | Tensile strength, MPa | Tensile strength, ksi | Yield strength, MPa | Yield strength, ksi | Elongation, in 50 mm (2 in.), % | Reduction in area, % | Hardness, HB |
|---|---|---|---|---|---|---|---|---|
| 403 | Annealed bar | 517 | 75 | 276 | 40 | 35 | 70 | 82 HRB |
| | Tempered bar | 765 | 111 | 586 | 85 | 23 | 67 | 97 HRB |
| 410 | Oil quenched from 980 °C (1800 °F), tempered at 540 °C (1000 °F) | 1089 | 158 | 1006 | 146 | 13 | 70 | ... |
| | Oil quenched from 980 °C (1800 °F); tempered at 40 °C (104 °F) | 1524 | 221 | 1227 | 178 | 15 | 64 | 45 HRB |
| 414 | Annealed bar | 793 | 115 | 621 | 90 | 20 | 60 | 235 |
| | Cold-drawn bar | 896 | 130 | 793 | 115 | 15 | 58 | 270 |
| | Oil quenched from 980 °C (1800 °F); tempered at 650 °C (1200 °F) | 1006 | 146 | 800 | 116 | 19 | 58 | ... |
| 420 | Annealed bar | 655 | 95 | 345 | 50 | 25 | 55 | 195 |
| | Annealed and cold drawn | 758 | 110 | 690 | 100 | 14 | 40 | 228 |
| 431 | Annealed bar | 862 | 125 | 655 | 95 | 20 | 55 | 260 |
| | Annealed and cold drawn | 896 | 130 | 758 | 110 | 15 | 35 | 270 |
| | Oil quenched from 980 °C (1800 °F); tempered at 650 °C (1200 °F) | 834 | 121 | 738 | 107 | 20 | 64 | ... |
| | Oil quenched from 980 °C (1800 °F); tempered at 40 °C (104 °F) | 1434 | 208 | 1145 | 166 | 17 | 59 | 45 HRC |
| 440C | Annealed bar | 758 | 110 | 448 | 65 | 14 | 25 | 97 HRB |
| | Annealed and cold-drawn bar | 862 | 125 | 690 | 100 | 7 | 20 | 260 |
| | Hardened and tempered at 315 °C (600 °F) | 1965 | 285 | 1896 | 275 | 2 | 10 | 580 |

Source: Ref 2

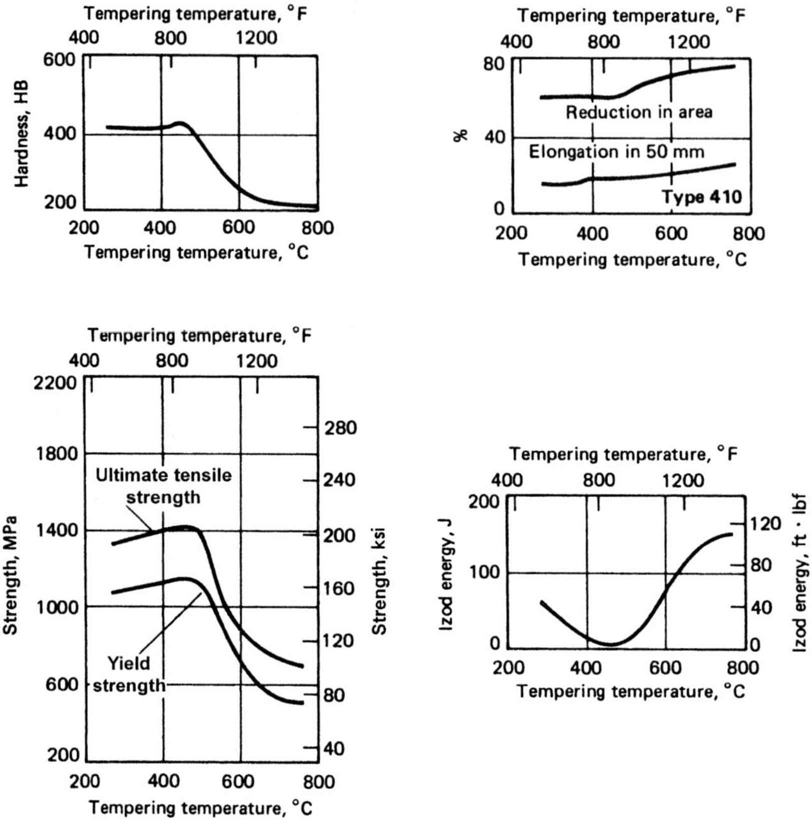


**Fig. 23.4** Effects of tempering temperature on type 410 martensitic stainless steel. Austenitized 30 min at 925 °C (1700 °F), oil quenched to room temperature, double stress relieved at 175 °C (350 °F) for 15 min, water quenched to room temperature, tempered as shown for 2 h. Source: Ref 4

have high strengths and lower toughness than those tempered at higher temperatures. The effects of tempering temperature on the properties of type 410 steel (0.15 wt% C) are shown in Fig. 23.4. Tempering in the range of 440 to 540 °C (825 to 1000 °F) is avoided because it adversely affects ductility and impact strength.

The microstructure of type 440C, a high-carbon grade that has been hardened and tempered, is shown in Fig. 23.5. The microstructure contains primary and secondary carbides in a tempered martensitic matrix. The effects of tempering temperature on the properties of type 440C stainless steel (0.95 to 1.20 wt% C) are shown in Fig. 23.6. Depending on the heat treatment, tensile strength values range from 760 MPa (110 ksi) for the annealed condition to 1965 MPa (285 ksi) for the fully hardened condition. However, the impact properties of type 440C are low at all tempering temperatures.

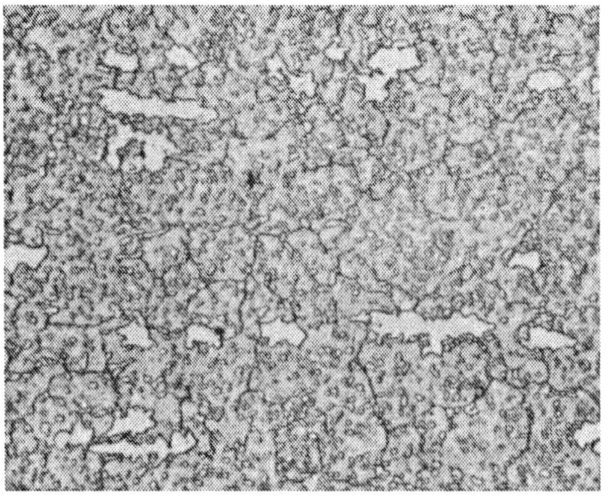

**Fig. 23.5** Microstructure of hardened 440C stainless steel. Preheated 1/2 h at 760 °C (1400 °F), austenitized 1/2 h at 1025 °C (1875 °F), air cooled to 65 °C (150 °F), double tempered 2 h at 425 °C (800 °F). Original magnification: 500×. Source: Ref 3

Martensitic stainless steels are used in applications requiring good tensile strength, creep, and fatigue strength properties, in combination with moderate corrosion resistance and heat resistance up to approximately 650 °C (1200 °F). The lower-carbon grades are used in the chemical, petrochemical, and power-generation industries. Type 420 and similar alloys are used in cutlery, valve parts, gears, shafts, and rollers. Applications for the high-carbon-level grades (i.e., type 440 grades) include cutlery, surgical and dental instruments, scissors, springs, valves, gears, shafts, cams, and ball bearings.

## 23.4 Austenitic Stainless Steels

Austenitic stainless steels (2*xx* and 3*xx* series) constitute the largest stainless family in terms of number of alloys and usage. The addition of nickel, which is a strong austenite stabilizer, overcomes the ferrite-stabilizing effect of chromium (Fig. 23.7), and all of these steels have the fcc structure. Since there is no phase change on cooling, they cannot be hardened by heat treatment. The austenitic stainless steels are essentially nonmagnetic in the annealed condition and can be hardened only by cold working. The microstructure of the annealed type 304 stainless steel strip (Fig. 23.8) consists of equiaxed grains of austenite. Austenitic stainless steels have excellent low-temperature toughness, weldability, and corrosion resistance. They usually possess excellent cryogenic properties and good high-temperature strength and oxidation resistance. Since austenitic stainless steels have the fcc structure, they do not have a DBTT and are often used in cryogenic applications. The compositions of some austenitic stainless steels are given in Table 23.5, and some mechanical properties are shown in Table 23.6.

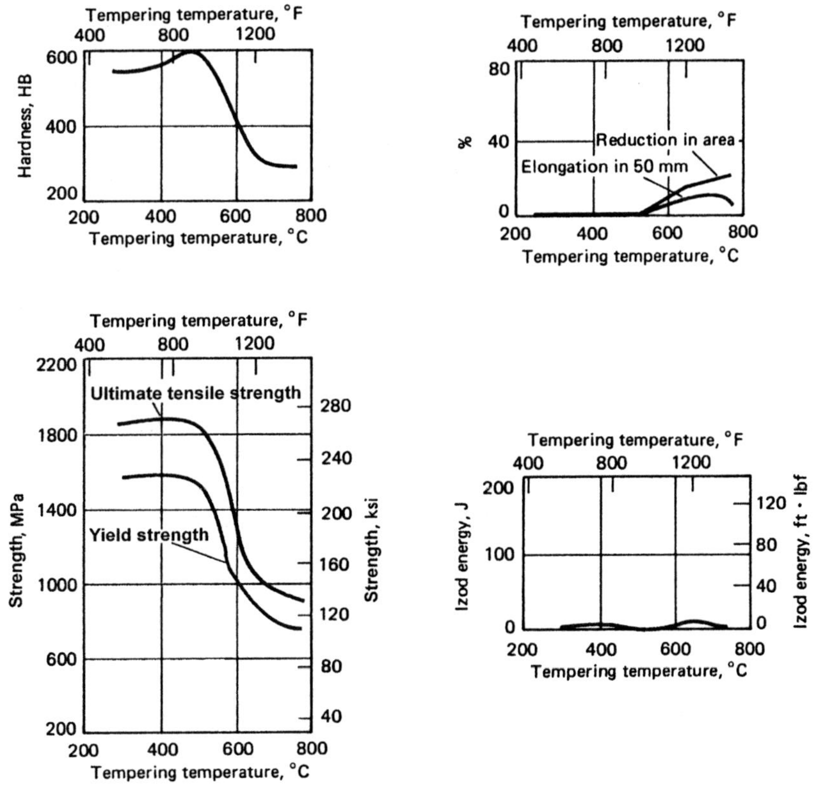


**Fig. 23.6** Effects of tempering temperature on type 440C martensitic stainless steel. Austenitized 1 h at 925 °C (1700 °F) and 2 h at 1040 °C (1900 °F), oil quenched to 65–95 °C (150–200 °F), double stress relieved at 175 °C (350 °F) for 15 min, water quenched to room temperature, tempered as shown for 2 h. Source: Ref 4

Austenitic stainless steels form the 2*xx* and 3*xx* series of alloys. Chromium content generally varies from 16 to 26 wt%, nickel content is less than or equal to approximately 35 wt%, and manganese content is less than or equal to 15 wt%. The 200-series steels contain nitrogen, 4 to 15 wt% Mn, and lower nickel contents with up to 7 wt% Ni; since they contain less nickel, they are less expensive than the 300-series. The 300-series steels contain larger amounts of nickel and up to 2 wt% Mn. Molybdenum, copper, silicon, aluminum, titanium, and niobium can be added to obtain certain desirable properties, such as halide pitting resistance or oxidation resistance. The original type 304 austenitic stainless steel contains 18 to 20 wt% Cr and 8 to 12 wt% Ni and is often referred to as 18-8 stainless steel for the chromium and nickel content.

The yield strengths of austenitic stainless steels are not high, comparable to those of mild steels. Typical minimum mechanical properties of annealed 300-series steels are yield strengths

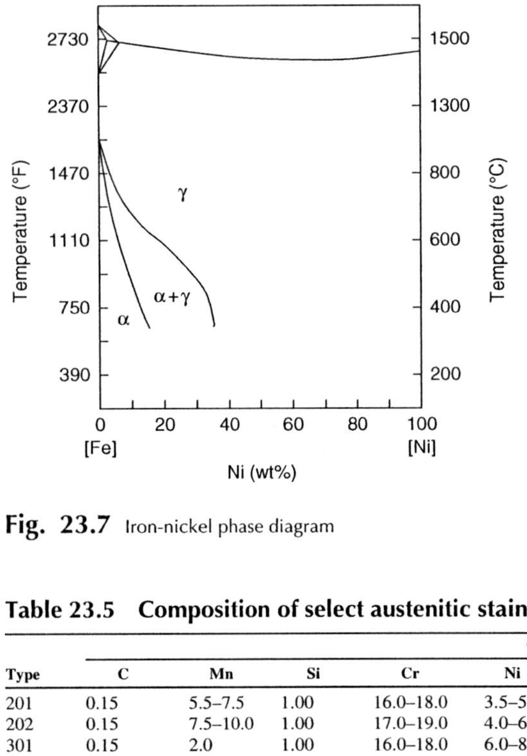


**Fig. 23.7** Iron-nickel phase diagram

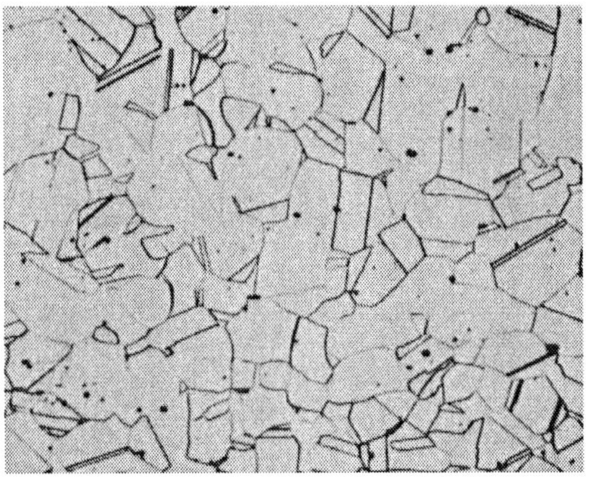

**Fig. 23.8** Microstructure of annealed 304 stainless steel strip. Original magnification: 250×. Source: Ref 3

**Table 23.5 Composition of select austenitic stainless steels**

| Type | Composition, % C | Mn | Si | Cr | Ni | P | S | Other |
|---|---|---|---|---|---|---|---|---|
| 201 | 0.15 | 5.5–7.5 | 1.00 | 16.0–18.0 | 3.5–5.5 | 0.06 | 0.03 | 0.25 N |
| 202 | 0.15 | 7.5–10.0 | 1.00 | 17.0–19.0 | 4.0–6.0 | 0.06 | 0.03 | 0.25 N |
| 301 | 0.15 | 2.0 | 1.00 | 16.0–18.0 | 6.0–8.0 | 0.045 | 0.03 | ... |
| 302 | 0.15 | 2.0 | 1.00 | 17.0–19.0 | 8.0–10.0 | 0.045 | 0.03 | ... |
| 303 | 0.15 | 2.0 | 1.00 | 17.0–19.0 | 8.0–10.0 | 0.20 | 0.15 min | 0.6 Mo |
| 304 | 0.08 | 2.0 | 1.00 | 18.0–20.0 | 8.0–10.5 | 0.045 | 0.03 | ... |
| 304N | 0.08 | 2.0 | 1.00 | 18.0–20.0 | 8.0–10.5 | 0.045 | 0.03 | 0.10–0.16 N |
| 308 | 0.08 | 2.0 | 1.00 | 19.0–21.0 | 10.0–12.0 | 0.045 | 0.03 | ... |
| 309 | 0.20 | 2.0 | 1.00 | 22.0–24.0 | 12.0–15.0 | 0.045 | 0.03 | ... |
| 309S | 0.08 | 2.0 | 1.00 | 22.0–24.0 | 12.0–15.0 | 0.045 | 0.03 | ... |
| 310 | 0.25 | 2.0 | 1.50 | 24.0–26.0 | 19.0–22.0 | 0.045 | 0.03 | ... |
| 316 | 0.08 | 2.0 | 1.00 | 16.0–18.0 | 10.0–14.0 | 0.045 | 0.03 | 2.0–3.0 Mo |
| 316F | 0.08 | 2.0 | 1.00 | 16.0–18.0 | 10.0–14.0 | 0.20 | 0.10 min | 1.75–2.5 Mo |
| 316H | 0.04–0.10 | 2.0 | 1.00 | 16.0–18.0 | 10.0–14.0 | 0.045 | 0.03 | 2.0–3.0 Mo |
| 316L | 0.03 | 2.0 | 1.00 | 16.0–18.0 | 10.0–14.0 | 0.045 | 0.03 | 2.0–3.0 Mo |
| 316LN | 0.03 | 2.0 | 1.00 | 16.0–18.0 | 10.0–14.0 | 0.045 | 0.03 | 2.0–3.0 Mo; 0.10–0.16 N |
| 316N | 0.08 | 2.0 | 1.00 | 16.0–18.0 | 10.0–14.0 | 0.045 | 0.03 | 2.0–3.0 Mo; 0.10–0.16 N |
| 321 | 0.08 | 2.0 | 1.00 | 17.0–19.0 | 9.0–12.0 | 0.045 | 0.03 | 5 × %C min Ti |
| 330 | 0.08 | 2.0 | 0.75–1.5 | 17.0–20.0 | 34.0–37.0 | 0.04 | 0.03 | ... |
| 347 | 0.08 | 2.0 | 1.00 | 17.0–19.0 | 9.0–13.0 | 0.045 | 0.03 | 10 × %C min Nb |
| 347H | 0.04–0.10 | 2.0 | 1.00 | 17.0–19.0 | 9.0–13.0 | 0.045 | 0.03 | 8 × %C min, 1.0 max Nb |
| 348 | 0.08 | 2.0 | 1.00 | 17.0–19.0 | 9.0–13.0 | 0.045 | 0.03 | 0.2 Co; 10 × %C min Nb; 0.10 Ta |
| 348H | 0.04–0.10 | 2.0 | 1.00 | 17.0–19.0 | 9.0–13.0 | 0.045 | 0.03 | 0.2 Co; 10 × %C min, 1.0 max NB; 0.10 Ta |

Source: Ref 1

of 205 to 275 MPa (30 to 40 ksi), ultimate tensile strengths between 515 and 760 MPa (75 and 110 ksi), and elongations of 40 to 60%. Annealed 200-series alloys have somewhat higher yield strengths, ranging from 345 to 485 MPa (50 to 70 ksi). Higher strengths are possible in cold-worked forms, especially in drawn wire, in which a tensile strength of 1205 MPa (175 ksi) or higher is possible.

Limiting the carbon content is important in austenitic stainless steels. When heated, carbon forms chromium carbide that precipitates on the austenite grain boundaries and produces sensitization. Because the chromium is tied up as carbide, the regions adjacent to the boundaries will be depleted in chromium, and corrosion can take place. Sensitization is reversible by heating the steel to temperatures between 1040 and 1150 °C (1900 and 2100 °F), followed by rapid cooling to room temperature. The high temperature dissolves the carbides, and the rapid cooling prevents reprecipitation of the carbides. Thus, austenitic stainless steels are quench annealed by heating to between 1040 and 1095 °C (1900 and 2000 °F) to dissolve all of the carbon in austenite and then rapidly cooled to room temperature to preserve a supersaturated solid solution of carbon in austenite. Since the $M_S$ temperature is normally just below room temperature, austenite may transform to martensite either by a subzero treatment or during cold working.

Austenitic stainless steels are sometimes classified as having stable austenite or metastable austenite. The microstructure of the stable alloys remains austenitic during cold working, while that of the metastable alloys is transformed to a mixture of austenite and martensite. That is, strain- or deformation-induced martensite formation occurs in response to cold working. The difference between the strain-hardening behavior of a metastable alloy (type 301) and a

**Table 23.6 Typical properties of select austenitic stainless steels**

| Steel | Condition | Tensile strength | | Yield strength | | Elongation, in 50 mm (2 in.), % | Reduction in area, % | Hardness, HB |
|---|---|---|---|---|---|---|---|---|
| | | MPa | ksi | MPa | ksi | | | |
| 201 | Annealed | 760 | 110 | 380 | 55 | 52 | ... | 87 HRB |
| | 50% hard | 1035 | 150 | 760 | 110 | 12 | ... | 32 HRC |
| | Full hard | 1275 | 185 | 965 | 140 | 8 | ... | 41 HRC |
| | Extra hard | 1550 | 225 | 1480 | 215 | 1 | ... | 43 HRC |
| 202 | Annealed bar | 515 | 75 | 275 | 40 | 40 | ... | ... |
| | Annealed sheet | 655 | 95 | 310 | 45 | 40 | ... | ... |
| | 50% hard sheet | 1030 | 150 | 760 | 110 | 10 | ... | ... |
| 301 | Annealed | 725 | 105 | 275 | 40 | 60 | 70 | ... |
| | 50% hard | 1035 | 150 | 655 | 95 | 54 | 61 | ... |
| | Full hard | 1415 | 205 | 1330 | 193 | 6 | ... | ... |
| 302 | Annealed strip | 620 | 90 | 275 | 40 | 55 | ... | 80 HRB |
| | 25% hard strip | 860 | 125 | 515 | 75 | 12 | ... | 25 HRC |
| | Annealed bar | 585 | 85 | 240 | 35 | 60 | 70 | 80 HRB |
| 303 | Annealed bar | 620 | 90 | 240 | 35 | 50 | 55 | 160 |
| | Cold drawn | 690 | 100 | 415 | 60 | 40 | 53 | 228 |
| 304 | Annealed bar | 585 | 85 | 235 | 34 | 60 | 70 | 149 |
| | Annealed and cold drawn | 690 | 100 | 415 | 60 | 45 | ... | 212 |
| | Cold-drawn high tensile | 860 | 125 | 655 | 95 | 25 | ... | 275 |
| 310 | Annealed sheet | 620 | 90 | 310 | 45 | 45 | ... | 85 HRB |
| | Annealed bar | 655 | 95 | 275 | 40 | 45 | 65 | 160 |
| 316 | Annealed sheet | 580 | 84 | 290 | 42 | 50 | ... | 79 HRB |
| | Annealed bar | 550 | 80 | 240 | 35 | 60 | 70 | 149 |
| | Annealed and cold-drawn bar | 620 | 90 | 415 | 60 | 45 | 65 | 190 |
| 321 | Annealed sheet | 620 | 90 | 240 | 35 | 45 | ... | 80 HRB |
| | Annealed bar | 585 | 85 | 240 | 35 | 55 | 65 | 150 |
| | Annealed and cold-drawn bar | 655 | 95 | 415 | 60 | 40 | 60 | 185 |
| 330 | Annealed sheet | 550 | 80 | 260 | 38 | 40 | ... | ... |
| | Annealed bar | 585 | 85 | 290 | 42 | 45 | ... | 80 HRB |
| 347 | Annealed sheet | 655 | 95 | 275 | 40 | 45 | ... | 85 HRB |
| | Annealed bar | 620 | 90 | 240 | 35 | 50 | 65 | 160 |
| | Annealed and cold-drawn bar | 690 | 100 | 450 | 65 | 40 | 60 | 212 |

Source: Ref 2

stable one (type 304) is shown in the stress-strain curves of Fig. 23.9. Type 304 exhibits a parabolic-shaped curve, typical for normal strain hardening. On the other hand, type 301 exhibits an accelerated strain-hardening effect after approximately 10 to 15% deformation. This accelerated work hardening is due to the formation of martensite from metastable austenite. When the metastable alloys are subjected to severe forming operations, such as deep drawing, the forming is often done at elevated temperatures (above the martensitic deformation, or $M_d$, temperature) to avoid the formation of martensite.

The lean austenitic stainless steels (e.g., types 302 and 304) offer general corrosion resistance in the atmosphere, in many aqueous media, in the presence of foods, and in oxidizing acids such as nitric acid. As shown in Fig. 23.1, type 321 is essentially type 304 with additions of titanium to stabilize carbides against sensitization. Similarly, type 347 has additions of niobium and tantalum. Molybdenum is added to types 316/316L to provide pitting resistance in phosphoric and acetic acids and dilute chloride solutions, as well as corrosion resistance in sulfurous acid. An even higher molybdenum content of 3 wt% in type 316L further enhances pitting resistance. Nitrogen is added to enhance strength at room temperature and to reduce the rate of chromium carbide precipitation and the susceptibility to sensitization. Nitrogen is also added to molybdenum-containing alloys to increase resistance to chloride-induced pitting and crevice corrosion. Higher amounts of chromium and/or nickel are used to enhance high-temperature oxidation resistance in types 309, 310, and 330. Copper and nickel can be added to improve resistance to reducing acids, such as sulfuric acid (type 320). The superaustenitic stainless steels are a special class of austenitic stainless steels that contain high levels of nickel (18 to 25 wt%), molybdenum (6 wt%), and, in some cases, nitrogen (0.15 to 0.25 wt%). These alloys are designed for severely corrosive environments. They provide improved resistance to SCC, pitting, and crevice corrosion relative to the standard 300 series of austenitic alloys. Higher nickel contents improve chloride SCC resistance, whereas molybdenum and nitrogen provide improved pitting and crevice corrosion resistance.

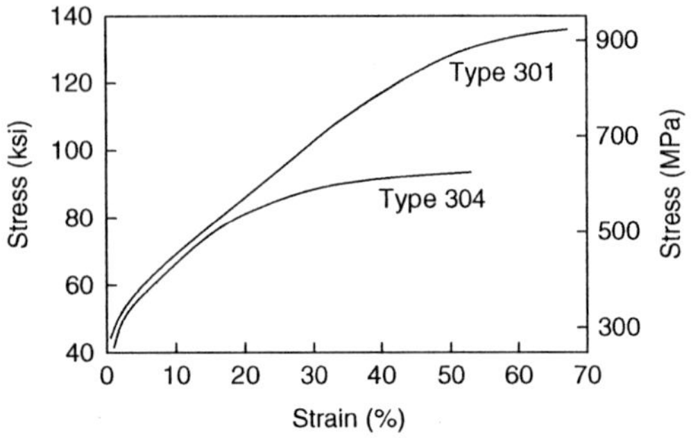


**Fig. 23.9** Stress-strain curves for types 301 and 304 stainless steels. Source: Ref 5

Although stainless steels provide resistance against general corrosion and pitting, austenitic stainless steels can be susceptible to intergranular corrosion by sensitization. The reader may want to refer to Section 18.2.8 in Chapter 18, "Corrosion," for a description of intergranular corrosion. Susceptible stainless steels are those that have normal carbon contents (generally $>0.04$ wt%) and do not contain titanium and niobium carbide stabilizing elements. Sensitization is caused by the precipitation of chromium carbides at grain boundaries during exposure to temperatures from 450 to 870 °C (840 to 1600 °F), with the maximum effect occurring near 675 °C (1250 °F). The resulting depletion in chromium adjacent to the chromium-rich carbides provides a selective path for intergranular corrosion. Precipitation commonly occurs from the heat of welding, but it may also result from slow cooling after annealing or from prolonged exposure to intermediate temperatures (~455 to 800 °C, or 850 to 1470 °F) in service. For exposures at very long times or at the high end of this temperature range, diffusion of chromium back into the depleted zone can restore the corrosion resistance.

An effective means of combating intergranular corrosion in stainless steels is to restrict the carbon content to the alloy. In the stainless L-grades, limiting the carbon content of a maximum of 0.03 wt% is often sufficient. High chromium and molybdenum additions also reduce the chance of intergranular attack. However, even better performance can be obtained from the stabilized types, which contain sufficient titanium and niobium that combine preferentially with carbon to form titanium and niobium carbides.

Since austenitic stainless steels have appreciable amounts of nickel, they are expensive steels, but they have the overall best corrosion resistance of the stainless steels. The stabilized and molybdenum grades are used in the chemical processing industries and for welded structures. High-chromium types have high oxidation and scaling resistance and are used for steam pipes, boiler tubes, and furnace parts. Alloys containing 25 wt% Cr and 20 wt% Ni with titanium or niobium additions have good creep resistance at temperatures up to 705 °C (1300 °F).

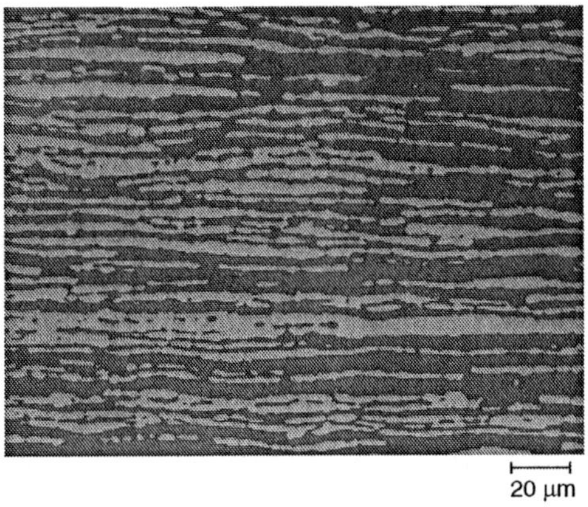


**Fig. 23.10** Duplex microstructure of alloy 255 (Ferralium). Ferrite-dark, austenite-light. Source: Ref 6

## 23.5 Duplex Stainless Steels

Duplex stainless steels are two-phase alloys based on the Fe-Cr-Ni system. These materials typically comprise approximately equal proportions of ferrite and austenite phases in their microstructures (Fig. 23.10). They have low carbon contents (<0.03 wt%) and additions of molybdenum, nitrogen, tungsten, and copper. Typical chromium contents are 20 to 30 wt%, with nickel in the range of 5 to 8 wt%. The compositions of some select duplex stainless steels are given in Table 23.7, and mechanical properties are shown in Table 23.8. The corrosion characteristics of duplex stainless steels are

**Table 23.7 Composition of duplex stainless steels**

| | Composition, wt% | | | | | | | | | |
|---|---|---|---|---|---|---|---|---|---|---|
| Type | C | Mn | S | P | Si | Cr | Ni | Mo | $N_2$ | Other |
| 44LN | 0.03 | 2.0 | 0.03 | 0.045 | 1.00 | 24.0–6.0 | 5.5–6.5 | 1.2–2.0 | 0.14–0.20 | ... |
| ... | 0.03 | 1.0 | 0.03 | 0.030 | 0.75 | 24.0–26.0 | 5.5–7.5 | 2.5–3.5 | 0.10–0.30 | 0.10–0.50 W, 0.20–0.80 Cu |
| ... | 0.03 | 1.2–2.0 | 0.03 | 0.03 | 1.4–2.0 | 18.0–19.0 | 4.25–5.25 | 2.5–3.0 | 0.05–0.10 | ... |
| UR45N | 0.03 | 2.0 | 0.02 | 0.03 | 1.00 | 21.0–23.0 | 4.5–6.5 | 2.5–3.5 | 0.08–0.20 | ... |
| 2304 | 0.03 | 2.5 | 0.04 | 0.04 | 1.0 | 21.5–24.5 | 3.0–5.5 | 0.05–0.60 | 0.05–0.20 | 0.05–0.60 Cu |
| ... | 0.03 | 1.5 | 0.03 | 0.04 | 1.0 | 24.0–24.0 | 4.5–6.5 | 2.9–3.9 | 0.10–0.25 | 1.5–2.5 Cu |
| 2507 | 0.03 | 1.2 | 0.02 | 0.035 | 1.0 | 24.0–26.0 | 6.0–8.0 | 3.0–5.0 | 0.24–0.32 | 0.5 Cu |
| Zeron 100 | 0.03 | 1.0 | 0.01 | 0.03 | 1.0 | 24.0–26.0 | 6.0–8.0 | 3.0–4.0 | 0.30 | 0.5–1.0 Cu, 0.5–1.0 W |
| Type 329 | 0.06 | 1.0 | 0.03 | 0.04 | 0.75 | 23.0–28.0 | 2.5–5.0 | 1.0–2.0 | ... | ... |
| 7 Mo Plus | 0.03 | 2.0 | 0.01 | 0.035 | 0.60 | 26.0–29.0 | 3.5–5.2 | 1.0–2.5 | 0.15–0.35 | ... |

Source: Ref 1

**Table 23.8 Minimum properties of duplex stainless steels**

| | | Ultimate tensile strength | | Yield strength | | | |
|---|---|---|---|---|---|---|---|
| Product form | Condition | MPa | ksi | MPa | ksi | Elongation, % | Maximum hardness, HRC |
| **44LN** | | | | | | | |
| Plate, sheet, strip | Annealed | 690 | 100 | 450 | 65 | 25 | 220 HB |
| **2205** | | | | | | | |
| Plate, sheet, strip | Annealed | 620 | 90 | 450 | 65 | 25 | 32 |
| **2304** | | | | | | | |
| Tubing | Annealed | 600 | 87 | 400 | 58 | 25 | 30.5 |
| **Ferralium 255** | | | | | | | |
| Plate, sheet, strip | Annealed | 760 | 110 | 550 | 80 | 15 | 32 |
| **Type 329** | | | | | | | |
| Plate, sheet, strip | Annealed | 620 | 90 | 485 | 70 | 15 | 28 |
| **7-Mo PLUS** | | | | | | | |
| Plate, sheet, strip | Annealed | 680 | 100 | 485 | 70 | 15 | 31 |

Source: Ref 7

similar to austenitic stainless steels. However, they have higher strength and better resistance to SCC than austenitic stainless steels. Specific advantages of duplex stainless steels compared to conventional 300-series stainless steels are approximately twice the strength of austenitic stainless steels, improved toughness and ductility compared to ferritic grades, and superior chloride SCC resistance and pitting resistance.

Duplex stainless steels have fairly high yield strengths, ranging from 550 to 690 MPa (80 to 100 ksi). The high alloy content and the presence of a bcc ferritic matrix make duplex stainless steels susceptible to embrittlement and loss of mechanical strength, particularly toughness, after prolonged exposure to elevated temperatures. For this reason, the upper use temperature is generally less than 300 °C (570 °F).

For severe-environment offshore applications, the availability of superduplex (25 wt% Cr) stainless steel alloys in a variety of forms is important. The 25 wt% Cr superduplex materials have a carefully controlled composition and balanced austenitic/ferritic structure, with substantial molybdenum and nitrogen contents. Bar, forgings, castings, sheet, plate, pipe/tube, welding consumables, flanges, fittings, and fasteners are available. They also offer excellent castability, weldability, and machinability. These features are complemented by excellent fatigue resistance and galvanic compatibility with other high-alloy stainless steels.

Duplex stainless steels are widely used in the oil and gas, petrochemical, and pulp and paper industries. They are commonly used in aqueous chloride-containing environments and as replacements for austenitic stainless steels that have suffered from either chloride SCC or pitting during service.

## 23.6 Precipitation-Hardening Stainless Steels

Precipitation-hardenable (PH) stainless steels are chromium-nickel grades that can be hardened by an aging treatment. The PH steels were developed to provide high strength and toughness while maintaining good corrosion resistance. They were developed to fill the gap imposed by the limits to strengthening austenitic and ferritic steels by solid solution and work hardening and by the limited ductility and toughness of the high-carbon martensitic types. The PH stainless steels develop high strength and toughness through additions of aluminum, titanium, niobium, vanadium, and/or nitrogen, which form precipitates during an aging heat treatment. Important properties of the PH stainless steels are ease of fabrication, high strength, good ductility, and excellent corrosion resistance. There are two main types of PH stainless steels: semiaustenitic and martensitic. The semiaustenitic grades are essentially austenitic in the solution-annealed condition. After fabrication operations are completed, they can be transformed to martensite by an austenite-conditioning heat treatment that converts the austenite to martensite, followed by precipitation hardening. The martensitic types are already martensitic in the solution-annealed condition and only require precipitation hardening after fabrication. The compositions of some PH stainless steels are listed in Table 23.9,

**Table 23.9 Composition of precipitation-hardening stainless steels**

| | Composition, wt% | | | | | | | | |
|---|---|---|---|---|---|---|---|---|---|
| Alloy | C | Mn | Si | Cr | Ni | Mo | P | S | Other |
| **Martensitic type** | | | | | | | | | |
| PH13-8 Mo | 0.05 | 0.10 | 0.10 | 12.25–13.25 | 7.5–8.5 | 2.0–2.5 | 0.01 | 0.008 | 0.90–1.35 Al; 0.01 N |
| 15-5PH | 0.07 | 1.00 | 1.00 | 14.0–15.5 | 3.5–5.5 | . . . | 0.04 | 0.03 | 2.5–4.5 Cu; 0.15–0.45 Nb |
| 17-4PH | 0.07 | 1.00 | 1.00 | 15.0–17.5 | 3.0–5.0 | . . . | 0.04 | 0.03 | 3.0–5.0 Cu; 0.15–0.45 Nb |
| Custom 450 | 0.05 | 1.00 | 1.00 | 14.0–16.0 | 5.0–7.0 | 0.5–1.0 | 0.03 | 0.03 | 1.25–1.75 Cu; 8 × %C min Nb |
| Custom 455 | 0.05 | 0.50 | 0.50 | 11.0–12.5 | 7.5–9.5 | 0.50 | 0.04 | 0.03 | 1.5–2.5 Cu; 0.8–1.4 Ti; 0.1–0.5 Nb |
| **Semiaustenitic type** | | | | | | | | | |
| PH15-7 Mo | 0.09 | 1.00 | 1.00 | 14.0–16.0 | 6.50–7.75 | 2.0–3.0 | 0.04 | 0.04 | 0.75–1.50 Al |
| 17-7PH | 0.09 | 1.00 | 1.00 | 16.0–18.0 | 6.50–7.75 | . . . | 0.04 | 0.04 | 0.75–1.50 Al |
| AM-350 | 0.07–0.11 | 0.50–1.25 | 0.50 | 16.0–17.0 | 4.0–5.0 | 2.50–3.25 | 0.04 | 0.03 | 0.07–0.13 N |
| AM-355 | 0.10–0.15 | 0.50–1.25 | 0.50 | 15.0–16.0 | 4.0–5.0 | 2.50–3.25 | 0.04 | 0.03 | 0.07–0.13 N |
| **Austenitic type** | | | | | | | | | |
| A-286 | 0.08 | 2.00 | 1.00 | 13.5–16.0 | 24.0–27.0 | 1.0–1.5 | 0.025 | 0.025 | 1.90–2.35 Ti; 0.35 max Al; 0.10–0.50 V; 0.0030–0.0100 B |

Source: Ref 1

and mechanical properties are given in Table 23.10.

The semiaustenitic alloys generally are supplied from the mill in the solution-annealed condition (condition A). In condition A, these alloys can be formed almost as easily as if they were true austenitic stainless steels. The alloy 17-7PH has approximately the same chromium and nickel contents as austenitic type 301 stainless but also contains 1.2 wt% Al for precipitation hardening. After fabrication in the soft condition, the austenite is conditioned to allow transformation to martensite. Because of their relatively high hardness in the solution-annealed condition, the martensitic types are used principally in the form of bar, rod, wire, and heavy forgings and only to a minimal extent in the form of sheet. The martensitic PH steels, before aging, are similar to the chromium martensitic stainless steels (e.g., 410 or 431) in their general fabrication characteristics.

The conditioning treatment for the semiaustenitic alloys consists of heating to a high enough temperature to remove carbon from solid solution and precipitate it as chromium carbide ($Cr_{23}C_6$). Removing carbon and some chromium from the austenite matrix makes the austenite unstable and, on cooling to the $M_s$ temperature, the austenite transforms to martensite. As shown in Fig. 23.11, 17-7PH is conditioned at 760 °C (1400 °F) and then cooled to 15 °C (60 °F) to produce the T-condition. If the conditioning is done at a higher temperature (955 °C, or 1750 °F), fewer carbides are precipitated, and the steel must be cooled to a lower temperature (−80 °C, or −110 °F) to transform the austenite to martensite, producing the R-100 condition. The final step is precipitation hardening, which is carried out in the 480 to 650 °C (900 to 1200 °F) range. During precipitation hardening, aluminum in the martensite combines with some of the nickel to produce precipitates of NiAl and $Ni_3Al$.

Since the martensitic PH grades are martensitic after solution annealing, they do not require conditioning but only a precipitation-hardening treatment. As shown in the right side of Fig. 23.11, 17-4PH is solution annealed at 1040 °C (1900 °F), followed by air cooling. Precipitation hardening during aging at 480 °C (900 °F) produces the H-900 condition. Typical microstructures are shown in Fig. 23.12. The tempering response of a number of PH stainless steels is shown in Fig. 23.13. If stress corrosion is a concern, the PH steels should be aged at the highest temperature that will maintain an adequate strength level.

Some dimensional changes are experienced during the heat treatment of the semiaustenitic steels. A dimensional expansion of approximately 0.114 mm/mm (0.0045 in./in.) occurs during the transformation from the austenitic to the martensitic condition, and during aging, a contraction of approximately 0.013 mm/mm (0.0005 in./in.) takes place. Vapor blasting of scaled parts after final heat treatment is recommended because of the hazards of intergranular corrosion in inadequately controlled acid pickling operations.

## 23.7 Cast Stainless Steels

Stainless steel castings are usually classified as either corrosion-resistant castings, which are

**Table 23.10 Properties of precipitation-hardening stainless steels**

| | Ultimate tensile strength | | Yield strength | | | |
|---|---|---|---|---|---|---|
| Alloy | MPa | ksi | MPa | ksi | Elongation, % | Hardness, HRB |
| **Martensitic types** | | | | | | |
| PH13-8Mo(a) | 1520 | 220 | 1410 | 205 | 6–10 | 45 HRC (min) |
| 15-5PH(b) | 1310 | 190 | 1170 | 170 | ... | ... |
| 17-4PH(b) | 1310 | 190 | 1170 | 170 | 5–10 | 40 HRC (min) |
| Custom 450(b) | 1240 | 180 | 1170 | 170 | 3–5 | 40 HRC (min) |
| Custom 455(a) | 1530 | 222 | 1450 | 205 | ≤4 | 44 HRC (min) |
| **Semiaustenitic types** | | | | | | |
| PH15-7Mo(b) | 1650 | 240 | 1590 | 230 | 1 | 46 HRC (min) |
| 17-7PH(a) | 1450 | 210 | 1310 | 190 | 1–6 | 43 HRC (min) |
| AM-350(c) | 1140 | 165 | 1000 | 145 | 2–8 | 36 HRC (min) |
| AM-355(c) | 1170 | 170 | 1030 | 150 | 12 | 37 HRC (min) |
| **Austenitic type** | | | | | | |
| A-286(d) | 860–965 | 125–140 | 655 | 95 | 4–15 | 24 HRC (min) |

(a) Aged at 510 °C (950 °F). (b) Aged at 480 °C (900 °F). (c) Aged at 540 °C (1000 °F). (d) Aged at 730 °C (1350 °F). Source: Ref 1 used in aqueous environments below 650 °C (1200 °F), or heat-resistant castings, which are suitable for service temperatures above 650 °C (1200 °F). However, this line of demarcation in terms of application is not always distinct, particularly for steel castings used in the range from 480 to 650 °C (900 to 1200 °F). The usual distinction between corrosion-resistant (C-type) and heat-resistant (H-type) cast steels is based on carbon content, with the heat-resistant grades normally having higher carbon contents.

The corrosion resistance of cast corrosion-resistant steels greatly depends on low carbon contents and the absence of precipitated carbides. Therefore, cast corrosion-resistant alloys are generally low in carbon, usually lower than 0.20 wt% and sometimes lower than 0.03 wt%.

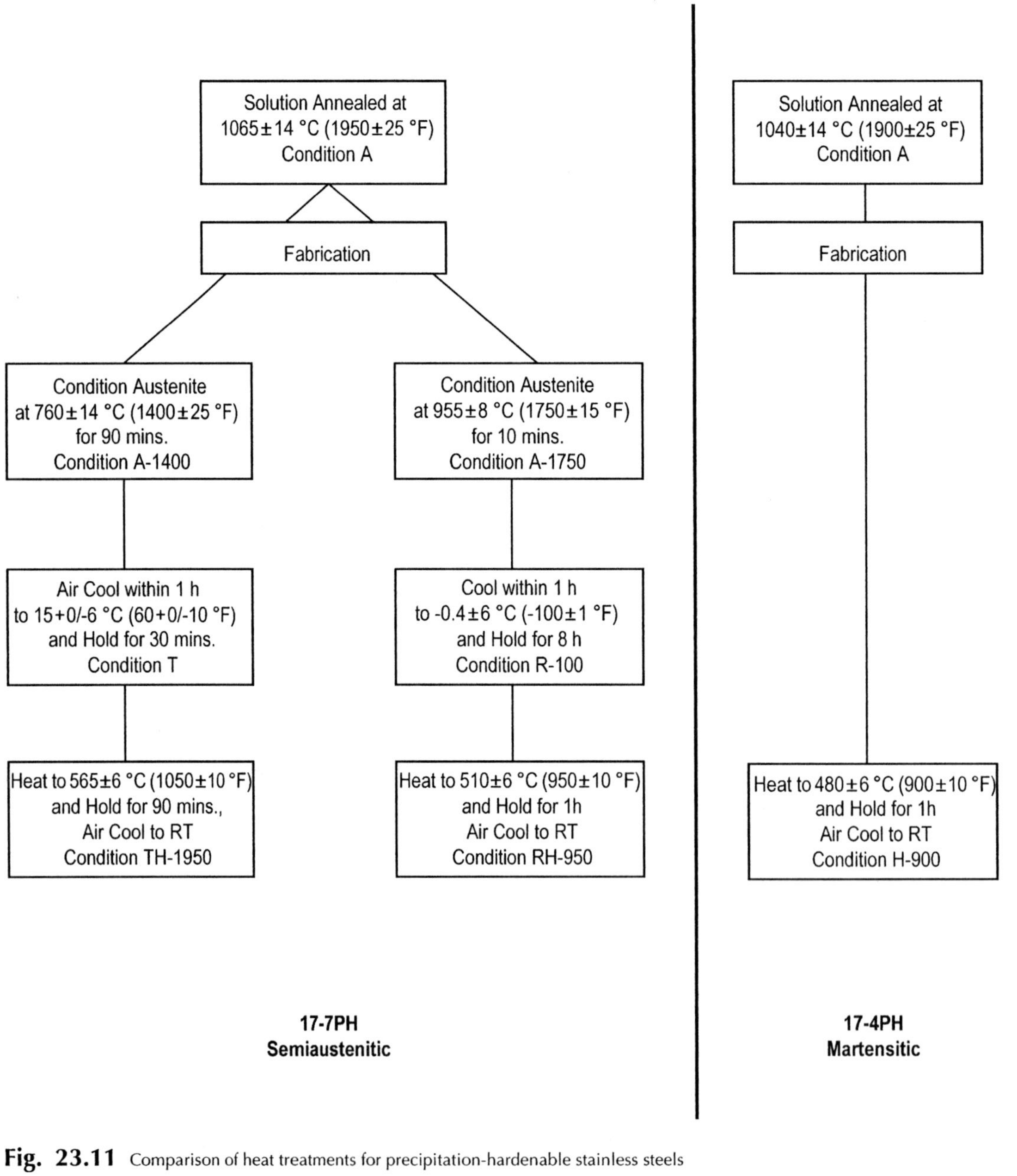


**Fig. 23.11** Comparison of heat treatments for precipitation-hardenable stainless steels

All cast corrosion-resistant steels contain more than 11 wt% Cr, and most contain from 1 to 30 wt% Ni, with the majority containing 18 to 22 wt% Cr and 8 to 12 wt% Ni. The addition of nickel to iron-chromium alloys improves ductility and impact strength. In addition, an increase in nickel content increases corrosion resistance in neutral chloride solutions and weakly oxidizing acids. The addition of molybdenum increases resistance to pitting attack by chloride solutions and extends the range of passivity in low-oxidizing solutions.

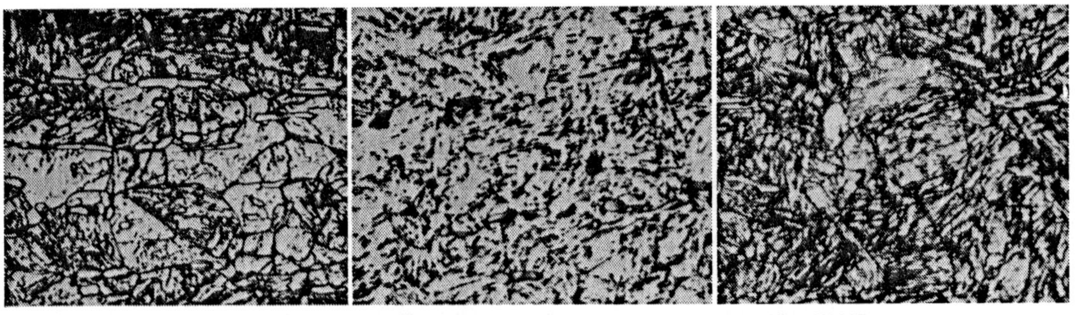

**Condition A**
**Solution treated at 1040 °C (1900 °F)**
**Air cooled**

**Condition H900**
**Solution treated at 1040 °C (1900 °F)**
**Air cooled to room temperature**
**Aged 1 h at 480 °C (900 °F)**

**Condition H1025**
**Solution treated at 1040 °C (1900 °F)**
**Air cooled to room temperature**
**Aged 1 h at 550 °C (1025 °F)**

**Fig. 23.12** Microstructures of heat treated 17-4PH steel. Original magnification: 1000 × . Source: Ref 3

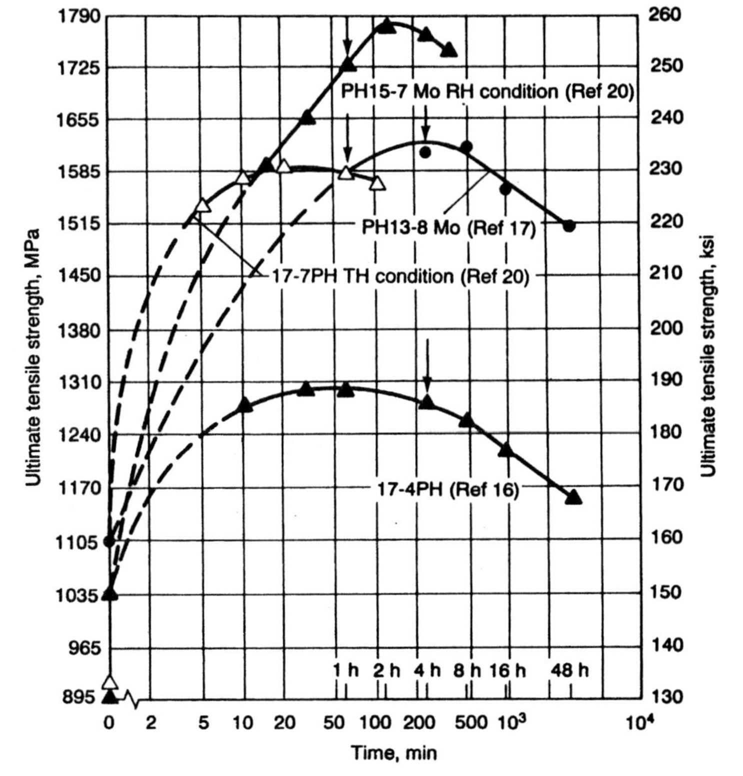


**Fig. 23.13** Tempering response of several precipitation-hardenable stainless steels. Source: Ref 8

The addition of copper to duplex (ferrite in austenite) nickel-chromium alloys produces alloys that can be precipitation hardened to higher strength and hardness. The addition of copper to single-phase austenitic alloys also greatly improves their corrosion resistance by sulfuric acid. In all Fe-Cr-Ni stainless alloys, corrosion resistance by environments that cause intergranular attack can be improved by lowering the carbon content.

Castings are classified as heat resistant if they are capable of sustained operation while exposed either continuously or intermittently to operating temperatures in excess of 650 °C (1200 °F). The major difference between heat-resistant alloys and their corrosion-resistant counterparts is the carbon content. With only a few exceptions, carbon in the cast heat-resistant alloys falls in the range from 0.3 to 0.6 wt%, compared with 0.01 to 0.25 wt% C normally used for the cast corrosion-resistant grades. This difference in carbon results in significant changes in properties; for example, the increased carbon content imparts higher creep-rupture strength in the cast heat-resistant grades. A wide range of mechanical properties is attainable, depending on the selection of alloy composition and heat treatment. Tensile strengths ranging from 475 to 1310 MPa (69 to 190 ksi) and hardness from 130 to 400 HB are available among the cast corrosion-resistant alloys. Similarly, wide ranges exist in yield strength, elongation, and impact toughness.

## 23.8 Schaeffler Constitution Diagram

A number of diagrams, originally developed for use during welding, show the effects of various combinations of austenite- and ferrite-stabilizing elements on the ferrite content in stainless steels. The ferrite-stabilizing elements, similar to chromium, are molybdenum, silicon, and niobium, while the austenite-stabilizing elements, similar to nickel, are manganese, carbon, and nitrogen. Nickel and chromium equivalents are calculated according to the various strengths of these elements in stabilizing austenite or ferrite. Plotting the chromium and nickel equivalents on opposing axes provides a graphic depiction of the relationship between composition and microstructure for stainless steel welds. The Schaeffler diagram (Fig. 23.14) has become known as the "roadmap" of stainless steels. The compositional ranges of the ferritic, martensitic, austenitic, and duplex alloys have been superimposed on this diagram, making it useful in predicting the type of stainless steel as a function of its alloy content.

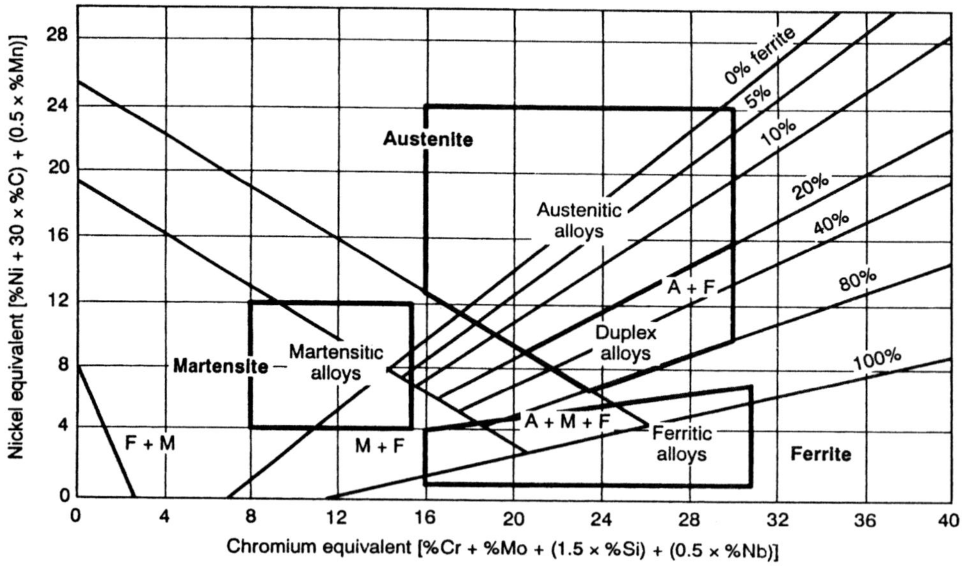


Fig. 23.14 Schaeffler constitution diagram for stainless steels. Compositions are by weight. Source: Ref 9

## ACKNOWLEDGMENTS

Sections of this chapter were adapted from "Selection and Application of Wrought Stainless Steels" in *Metals Handbook Desk Edition,* 2nd ed., ASM International, 1998.


## REFERENCES


1. Wrought Stainless Steels: Selection and Application, *Metals Handbook Desk Edition,* 2nd ed., ASM International, 1998
2. B.L. Branfitt, Structure/Property Relationships in Irons and Steels, *Metals Handbook Desk Edition,* 2nd ed., ASM International, 1998
3. Microstructure of Wrought Stainless Steels, *Atlas of Microstructures of Industrial Alloys,* Vol 7, *ASM Metals Handbook,* 8th ed., ASM International, 1972
4. J. Douthett, Heat Treating of Stainless Steels, *Heat Treating,* Vol 4, *ASM Handbook,* ASM International, 1991
5. *Making, Shaping and Treating of Steel,* 9th ed., United States Steel Company, 1971
6. G.F. Vander Voort, G.M. Lucas, and E.P. Manilova, Metallography and Microstructures of Stainless Steels and Maraging Steels, *Metallography and Microstructures,* Vol 9, *ASM Handbook,* ASM International, 2004
7. S.D. Washko and G. Aggen, Wrought Stainless Steels, *Properties and Selection: Irons, Steels, and High-Performance Alloys,* Vol 1, *ASM Handbook,* ASM International, 1990
8. B. Pollard, Selection of Wrought Precipitation-Hardening Stainless Steels, *Welding, Brazing, and Soldering,* Vol 6, *ASM Handbook,* ASM International, 1993
9. J.C. Lippold, Introduction to the Selection of Stainless Steels, *Welding, Brazing, and Soldering,* Vol 6, *ASM Handbook,* ASM International, 1993


## SELECTED REFERENCES


- M. Blair, Cast Stainless Steels, *Properties and Selection: Irons, Steels, and High-Performance Alloys,* Vol 1, *ASM Handbook,* ASM International, 1990
- F.C. Campbell, *Manufacturing Technology for Aerospace Structural Materials,* Elsevier Scientific, 2006
- Cast Stainless Steels, *Metals Handbook Desk Edition,* 2nd ed., ASM International, 1998
- J. Kelly, Stainless Steels, *Handbook of Materials Selection,* John Wiley & Sons, Inc., 2002
- G. Kraus, *Steels: Processing, Structure, and Performance,* 3rd ed., ASM International, 2005

# CHAPTER 24

# Cast Irons

IRON-CARBON ALLOYS with a carbon content of less than 2 wt% are normally classified as steels, while alloys with greater than 2 wt% C are classified as cast irons. The carbon content of the eutectic in the Fe-$Fe_3C$ system is 4.3 wt%, and since these compositions have microstructures that are too brittle to be of much use, cast irons are normally hypoeutectic compositions with less than 4.3 wt% C.

The ductility of cast irons is very low; they cannot be rolled, drawn, or worked at room temperature. In fact, most cast irons are not malleable at any temperature. However, they melt readily and can be cast into intricate shapes that are usually machined to final dimensions. Since casting is the only viable fabrication route for these alloys, they are known as cast irons. Although cast irons are brittle and have lower strength properties than most steels, they are cheap, more readily cast than steels, and have other useful properties. With proper alloying, good foundry practice, and appropriate heat treatment, the properties of any cast iron can be varied over a wide range. Large tonnages of high-quality cast irons are produced annually.

The form in which carbon is present in cast irons largely determines its properties, that is, whether it is cementite ($Fe_3C$) or graphite. Carbon is present as cementite in white cast irons. The name "white cast iron" derives from the white and lustrous appearance of the fracture surface. While white cast irons are extremely hard and have good wear resistance, they are brittle and cannot be used in applications with dynamic or impact loads. In gray cast irons, carbon is present as graphite, making it much more ductile than white cast iron. Gray cast iron is so called because of the gray, fibrous appearance of the fracture surface. The properties of the gray cast irons are largely influenced by the shape, size, and distribution of the graphite.

The form of carbon in cast irons is determined primarily by modifying the composition and controlling the cooling rate during casting. Certain alloying elements promote graphite formation, while others promote cementite formation. Those that promote graphite formation (graphitizing elements) include silicon, aluminum, nickel, cobalt, and copper. Silicon is the most powerful graphitizing element and is therefore the most important alloying element in the gray cast irons. Alloying elements that promote cementite formation (whitening elements) include sulfur, vanadium, chromium, tin, molybdenum, and manganese. The relative potencies of these alloying elements to affect microstructure are shown in Fig. 24.1.

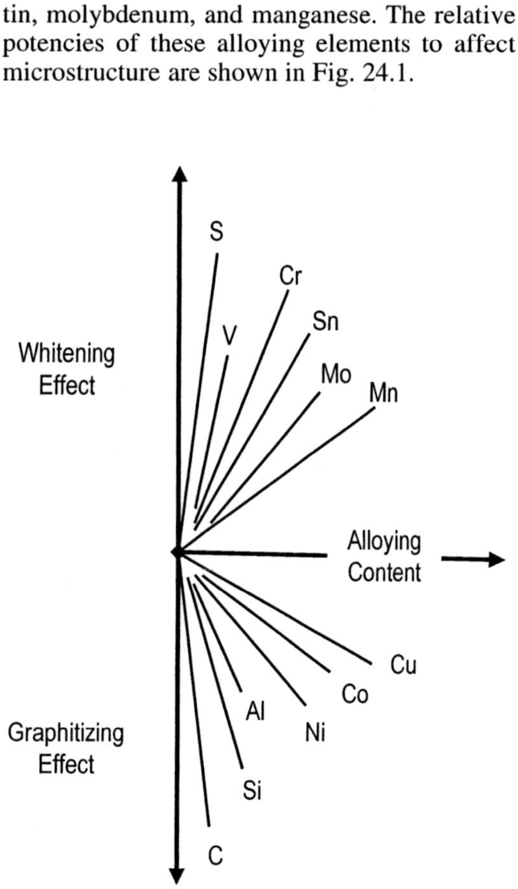


**Fig. 24.1** Effect of alloying elements on microstructure of cast iron. Source: Ref 3

The compositional range of carbon and silicon for common cast irons, as compared with steel, is shown in Fig. 24.2. The carbon content of cast irons is in excess of the maximum solubility of carbon in austenite, which is shown by the lower dashed line. A high carbon content increases the amount of graphite or $Fe_3C$. High carbon and silicon contents increase the graphitization potential of the iron as well as its castability. The effects of silicon additions to the iron-carbon phase diagram are shown in Fig. 24.3. Silicon additions of 2.4 and 4.8 wt% decrease both eutectic and eutectoid carbon contents while also decreasing the maximum solid solubility of carbon in austenite. Therefore, silicon additions reduce the carbon content of the pearlite and cause the eutectic and eutectoid reactions to take place over a range of temperatures and at higher temperatures than when no silicon is present. The temperature range increases as the amount of silicon increases.

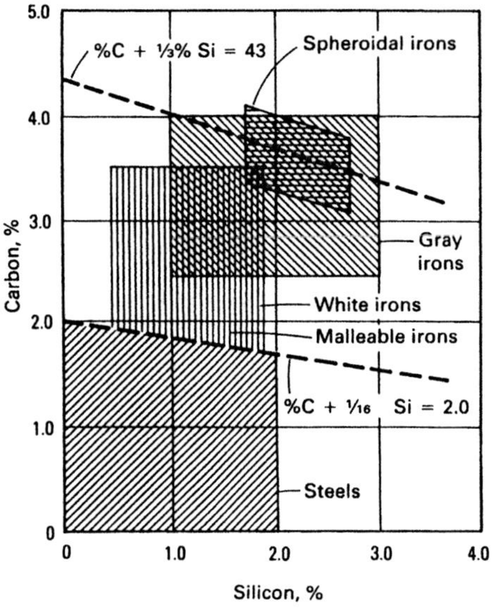


**Fig. 24.2** Approximate ranges of carbon and silicon for steel and cast irons. Source: Ref 2

Since both carbon and silicon influence the nature of iron castings, it is necessary to develop an approximation of their impact on solidification. The carbon equivalent (CE) is a parameter that accounts for the influence of composition on microstructure according to:

$$\begin{aligned}CE = \%C + 0.3(\%Si) + 0.33(\%P)\\ -0.027(\%Mn) + 0.4(\%S)\end{aligned} \quad \text{(Eq 24.1)}$$

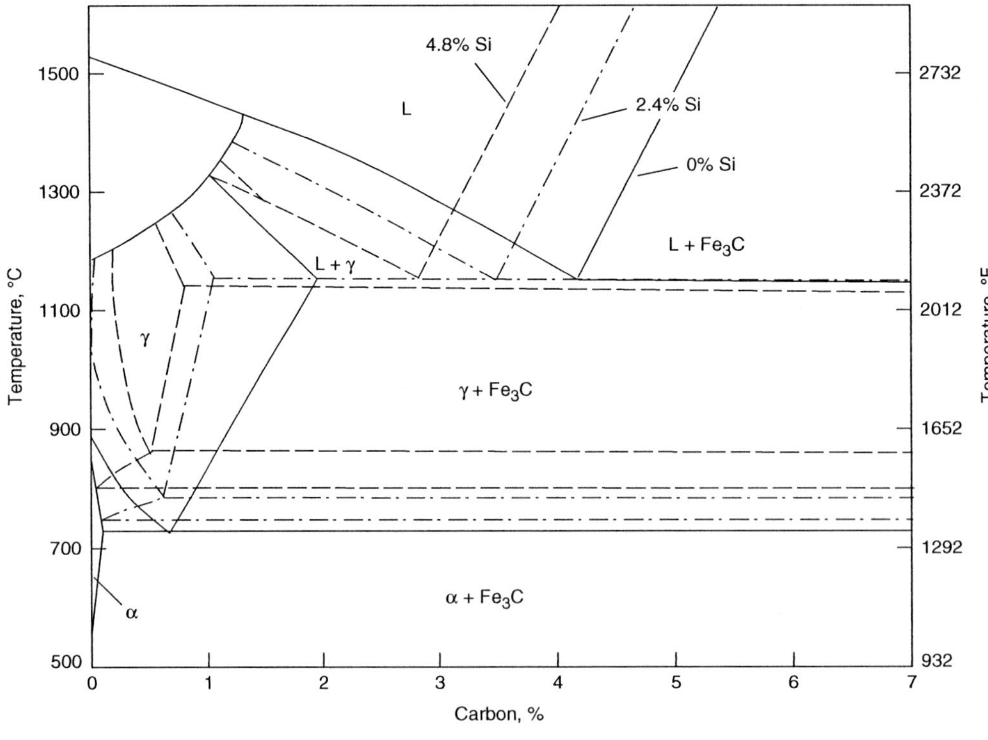


**Fig. 24.3** Effects of silicon on iron-carbon phase diagram. Source: Ref 4

Although increasing the carbon and silicon contents improves the graphitization potential and therefore decreases the chilling tendency, the tensile strength is adversely affected (Fig. 24.4) because of ferrite formation and coarsening of the pearlite. The manganese content varies as a function of the matrix, that is, ferrite or pearlite. Since manganese is a strong pearlite promoter, it can be as low as 0.1 wt% in ferritic irons and as high as 1.2 wt% in pearlitic irons.

Comparing the CE with the eutectic composition in the iron-carbon system (4.3% C) indicates whether a cast iron will behave as a hypoeutectic or hypereutectic alloy during solidification. When the CE is near the eutectic value, the liquid state persists to a relatively low temperature, and solidification takes place over a small temperature range. This latter characteristic is important in achieving uniform properties within a given casting. In hypoeutectic cast irons, a lower CE increases the tendency for white or mottled iron to form on solidification. In hypereutectic irons (CE greater than approximately 4.3%), there is a tendency for kish graphite (proeutectic graphite that forms and floats free in the molten iron) to precipitate on solidification under normal cooling conditions.

The strengths of cast irons can also be increased with alloying elements. Since the strengths of cast irons with a ferrite matrix are lower than those with a pearlitic matrix, alloying elements that suppress the formation of ferrite and increase the amount of pearlite will increase the strength. Alloying elements such as chromium, molybdenum, and tungsten are used for this purpose. The addition of small amounts of tin (~0.1 wt%) is also effective in preventing ferrite formation. The heat resistance of cast irons can be increased by alloying with 18 to 20 wt% Ni and 2 to 3 wt% Cr. There are three groupings of alloying elements.

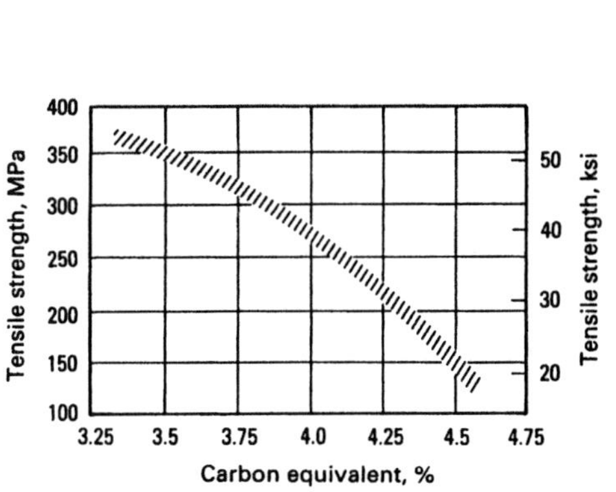


**Fig. 24.4** Effect of carbon equivalence on tensile strength of gray iron. Source: Ref 2

**Silicon and aluminum** increase the graphitization potential for both the eutectic and eutectoid transformations and increase the number of graphite particles. They form solid solutions in the matrix. Because they increase the ferrite/pearlite ratio, they decrease strength and hardness.

**Nickel, copper, and tin** increase the graphitization potential during the eutectic transformation but decrease it during the eutectoid transformation thus increasing the pearlite/ferrite ratio. This second effect is due to the retardation of carbon diffusion. Because they increase the amount of pearlite, they increase strength and hardness. These elements form solid solutions in the matrix.

**Chromium, molybdenum, tungsten, and vanadium** decrease the graphitization potential for both the eutectic and eutectoid transformations. Thus, they increase the amount of carbides and pearlite. The alloying elements concentrate principally in the carbides, forming $(FeX)_nC$-type carbides, but also alloy the α-Fe solid solution. As long as carbide formation does not occur, these elements increase strength and hardness. Above a certain level, any of these elements will determine the solidification of a structure with both graphite and $Fe_3C$ (mottled structure); they contribute to a lower strength but higher hardness.

The deleterious effect of sulfur is balanced by the effect of manganese. Without manganese, undesirable iron sulfide (FeS) will form at grain boundaries. If the sulfur content is balanced by manganese, harmless manganese sulfide (MnS) particles will be distributed within the grains. The optimal ratio between manganese and sulfur to achieve an FeS-free structure and a maximum amount of ferrite is:

$$\text{wt\% Mn} = 1.7(\text{wt\% S}) + 0.15 \qquad \text{(Eq 24.2)}$$

In addition to tying up sulfur, the small MnS particles act as nucleation sites for graphitization. Other minor elements, such as antimony, arsenic, bismuth, lead, magnesium, cerium, and calcium, can significantly alter both the graphite morphology and the microstructure of the matrix.

Slow cooling rates favor the formation of graphite, while faster rates promote more cementite. The formation of graphite rather than cementite is favored by:

- High carbon contents
- High silicon contents
- Slow cooling from solidification
- Absence of carbide formers, such as chromium and molybdenum

The interrelationship between composition and cooling rates on the microstructural development in cast irons is outlined in Fig. 24.5.

Cast irons can also be classified as either unalloyed cast irons or alloy cast irons. Unalloyed cast irons are essentially Fe-C-Si alloys containing small amounts of manganese, phosphorus, and sulfur. The five types of commercial cast irons are white, gray, ductile, malleable, and compacted graphite. Their compositional ranges are given in Table 24.1, and some of their distinguishing features are summarized in Table 24.2. With the exception of white cast iron, all cast irons have microstructures that consist of a graphite phase in a matrix that can be ferritic, pearlitic, bainitic, tempered martensitic, or combinations thereof. The four types of graphitic cast irons are roughly classified according to the morphology of the graphite phase. Gray iron has flake-shaped graphite, ductile iron has nodular or spherically shaped graphite, compacted graphite iron (also called vermicular graphite iron) is intermediate between these two, and malleable iron has irregular, globular-shaped graphite (temper carbon) that forms during tempering of white cast iron.

## 24.1 White Cast Iron

White cast irons usually have less carbon and silicon than other cast irons, typically approximately 2.5 wt% C and 0.5 wt% Si, so that graphite does not form. Therefore, in white cast iron, the carbon is present as cementite ($Fe_3C$) rather than graphite. This results in a hard but somewhat brittle structure. All white cast irons

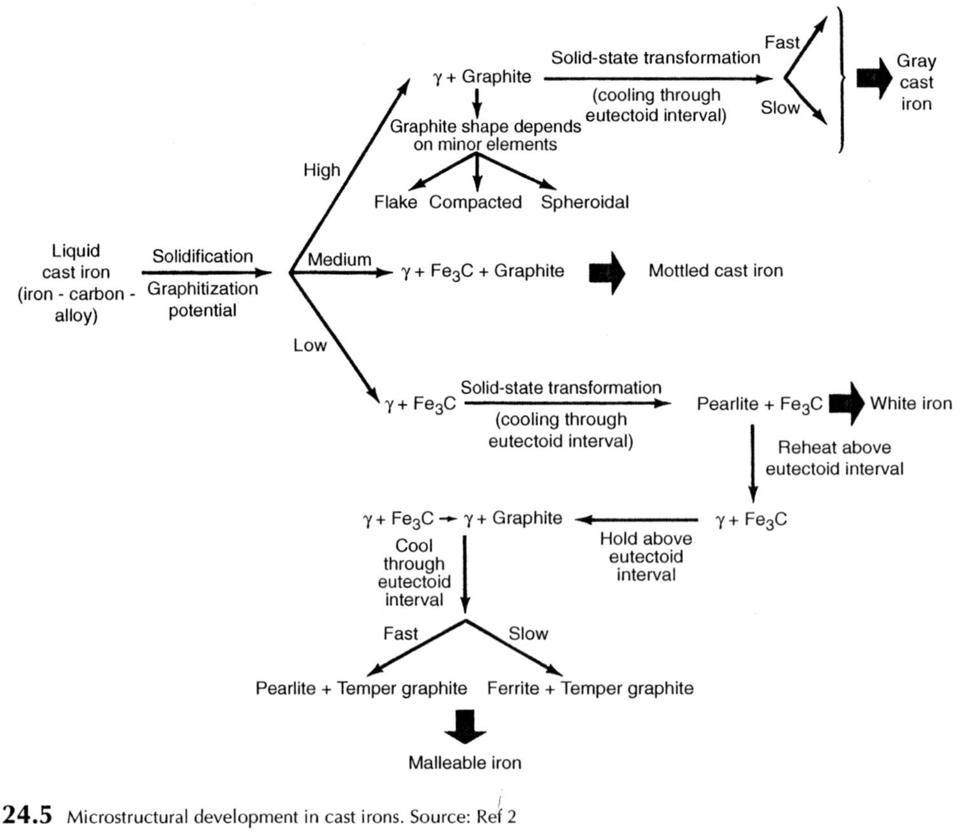


**Fig. 24.5** Microstructural development in cast irons. Source: Ref 2

are hypoeutectic alloys. The cooling of a 2.5% C alloy can be described using the iron-carbon phase diagram shown in Fig. 24.6. At point $x_1$, the alloy consists of a uniform solution of carbon dissolved in liquid iron. When the liquidus line is crossed at point $x_2$, solidification begins by the formation of austenite crystals containing approximately 1% C. As the temperature falls, primary austenite continues to solidify, with its composition moving down and to the right along the solidus line toward point $C$. At the same time, the liquid is becoming richer in carbon, and its composition also moves down and to the right along the liquidus line toward point $E$. At the eutectic point temperature (1148 °C, or 2098 °F), the alloy now consists of austenite dendrites containing approximately 2% C and a liquid solution containing 4.3% C. The liquid solidifies to form a eutectic mixture of austenite and cementite known as ledeburite. Since the eutectic reaction takes place at a high temperature, ledeburite tends to be a coarse mixture rather than the fine mixture typical of most eutectics. As the temperature continues to fall between points $x_3$ and $x_4$, the solubility of carbon in austenite decreases along the $A_{cm}$ line indicated as $CJ$. This results in precipitation of proeutectoid cementite, most of which is deposited on the cementite already present. At the eutectoid temperature of 738 °C (1360 °F), the remaining austenite, containing 0.68 wt% C, transforms to pearlite. As previously shown in Fig. 24.1, the elements manganese, molybdenum, tin, chromium, vanadium, and sulfur contribute to the formation of cementite rather than ferrite. Since white cast iron contains a relatively large amount of cementite as a continuous interdendritic network, the iron is hard and wear resistant but also extremely brittle and difficult to machine. Along with these alloying additions, silicon is limited to 0.5 to 1.2 wt% to avoid the formation of graphite flakes. Micrographs of typical white cast iron microstructures are shown in Fig. 24.7.

Applications that take advantage of this hard and abrasion-resistant structure include rolls for rolling mills, grinding plates, cement mixers, crushing balls, and extrusion dies. White cast iron is also a starting material for the production of malleable cast irons (Section 24.4 in this chapter).

**Table 24.1 Compositional ranges for unalloyed cast irons**

| | Composition, wt% | | | | | |
|---|---|---|---|---|---|---|
| Type of Iron | C | Si | Mn | P | S | Fe |
| Gray | 2.5–4.0 | 1.0–3.0 | 0.2–1.0 | 0.002–1.0 | 0.02–0.25 | bal |
| Compacted graphite | 2.5–4.0 | 1.0–3.0 | 0.2–1.0 | 0.01–0.1 | 0.01–0.03 | bal |
| Ductile | 3.0–4.0 | 1.8–2.8 | 0.1–1.0 | 0.01–0.1 | 0.01–0.03 | bal |
| White | 1.8–3.6 | 0.5–1.9 | 0.25–0.8 | 0.06–0.2 | 0.06–0.2 | bal |
| Malleable | 2.2–2.9 | 0.9–1.9 | 0.15–1.2 | 0.02–0.2 | 0.02–0.2 | bal |

Source: Ref 5

## 24.2 Gray Cast Iron

Gray cast irons are one of the most widely used types of cast irons. The microstructure of gray cast irons consists of flake graphite, typically in a pearlite matrix. Gray irons are Fe-C-Si alloys that usually contain 2.5 to 4 wt% C, 1 to 3 wt% Si, and, depending on the desired microstructure, manganese additions. Pearlite-matrix gray irons have manganese contents as high as 1.2 wt%, while ferrite-matrix gray irons have as little as 0.1 wt% Mn. Sulfur and phosphorus also are present in small amounts as residual impurities. A typical structure consists of graphite flakes embedded in a matrix of pearlite (Fig. 24.8). In a two-phase matrix consisting of ferrite and pearlite, the strength increases as the amount of pearlite increases.

Strength is also highly dependent on the graphite morphology and distribution. Coarse graphite flakes act as stress concentrations and decrease the tensile strength. Fracture occurs along the graphite flakes, so that the crack propagates almost entirely through the graphite.

**Table 24.2 Cast iron classifications**

| Commercial designation | Carbon-rich phase | Matrix(a) | Fracture | Final structure after |
|---|---|---|---|---|
| Gray iron | Lamellar graphite | P | Gray | Solidification |
| Ductile iron | Spheroidal graphite | F, P, A | Silver-gray | Solidification or heat treatment |
| Compacted graphite iron | Compacted (vermicular) graphite | F, P | Gray | Solidification |
| White iron | $Fe_3C$ | P, M | White | Solidification and heat treatment(b) |
| Malleable iron | Temper graphite | F, P | Silver-gray | Heat treatment |

(a) F, ferrite; P, pearlite; A, austenite; M, martensite. (b) White irons are not usually heat treated, except for stress relief and to continue austenite transformation. Source: Ref 5

This failure mode gives the fracture surface a dull gray appearance. In tensile tests, elongations of only 1% are typical. Tensile strength decreases with increasing quantity and size of the graphite flakes. Therefore, increasing carbon and silicon contents tend to reduce tensile strength.

Cooling that is too rapid may produce mottled iron, in which carbon is present in the form of both primary cementite ($Fe_3C$) and graphite. Very slow cooling of irons that contain large percentages of silicon and carbon is likely to produce a matrix predominantly of ferrite with coarse graphite flakes.

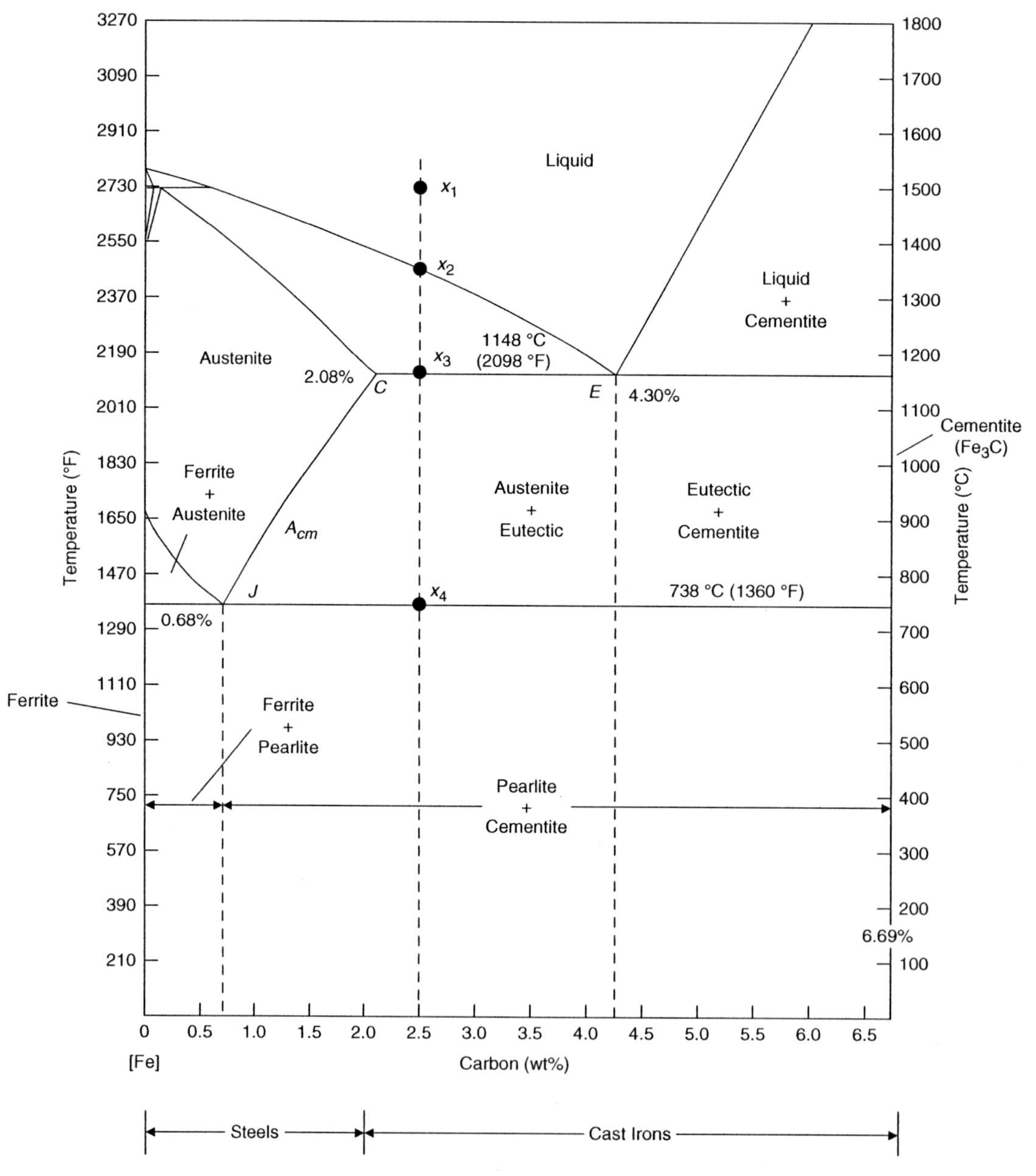


**Fig. 24.6** The iron-carbon diagram. Source: Ref 6

The mechanical and physical properties of gray iron are determined in large part by the shape, size, volume fraction, and distribution of the graphite flakes. The deeply etched specimen in Fig. 24.9 reveals the complex structure of graphite flakes. A method for evaluating graphite flake distribution and size has been jointly developed by the American Foundrymen's Society (AFS) and ASTM International and is given in ASTM A 247. There are five graphite flake morphologies classified as "A" to "E" and illustrated in Fig. 24.10.

**Type A** graphite flakes are uniformly distributed and randomly oriented throughout the iron matrix. Type A graphite is observed in inoculated irons with moderate cooling rates. The degree of undercooling during solidification is minimal for cast irons with this type of graphite. Type A graphite is associated with gray cast irons with the best mechanical properties.

**Type B** graphite (rosette pattern) is formed in gray irons of near-eutectic composition that solidify with a greater amount of undercooling than irons with type A graphite. Type B flake graphite is common with moderately thin castings (approximately 10 mm, or 0.375 in.) and along the surface of thicker castings. Large eutectic cell size and small undercoolings are common in cast irons exhibiting this type of graphite.

**Type C** graphite occurs in hypereutectic (high-carbon-content) irons as a result of solidification with minimum undercooling. Type C graphite precipitates during the primary freezing of the iron. Often called kish graphite, it appears as straight, coarse plates. Type C graphite greatly reduces the mechanical properties of the iron, and a rough surface finish results when machined. However, this graphite morphology results in attractive thermal properties, making the iron useful in applications requiring a high degree of heat transfer.

**Types D and E** graphite form when the amount of undercooling is high but is not sufficient to cause carbide formation. Both types are found in interdendritic regions. Type D graphite is randomly distributed, while type E flakes have a preferred orientation. Alloying elements, such as titanium and aluminum, promote undercooled graphite structures. The iron matrix associated with undercooled graphite is usually ferrite, because formation of the fine, highly branched flakes reduces carbon diffusion distances and

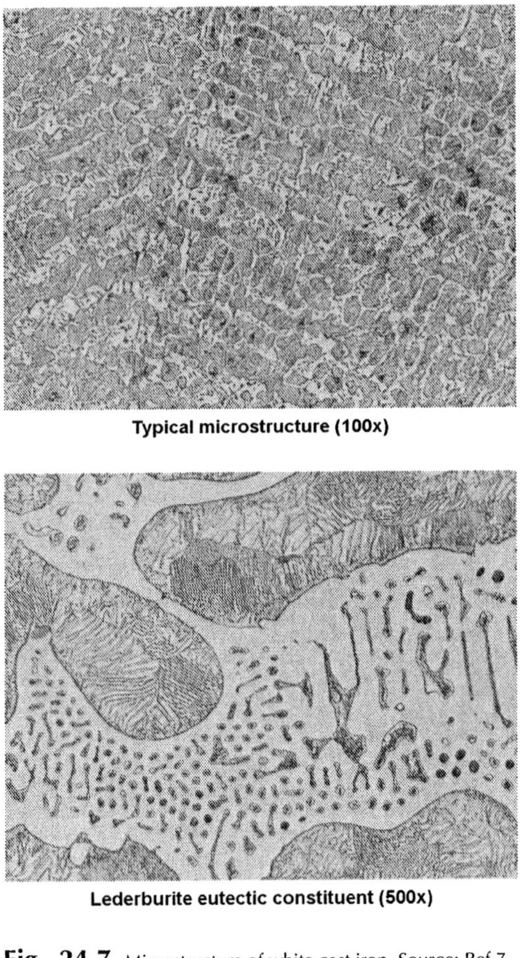


**Fig. 24.7** Microstructure of white cast iron. Source: Ref 7

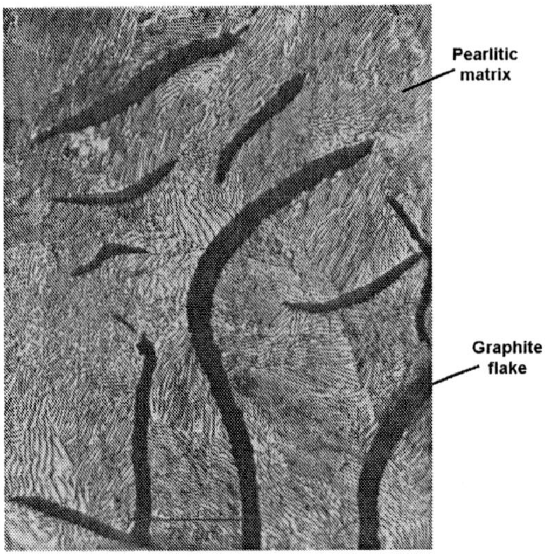


**Fig. 24.8** Class 30 gray cast iron. Source: Ref 8

results in a low-carbon matrix. Because ferrite has a lower tensile strength than pearlite, there is a reduction in the strength.

Not only is the graphite shape important but also the graphite size, because it is directly related to strength (Fig. 24.11). Similar to the standards for graphite flake distribution, there are also ASTM/AFS standards for graphite flake size, shown in Fig. 24.12. Large flakes are associated with irons having high CE values and slow cooling rates. Hypoeutectic irons and irons subjected to rapid solidification generally exhibit small, short flakes. Small flakes, because they disrupt the matrix to a lesser extent, lead to higher tensile properties and smoother surface finishes, while larger flakes are desirable in applications requiring high thermal conductivity and high damping capacity.

The cooling rate, like the chemical composition, can significantly influence the as-cast structure and therefore the mechanical properties. The cooling rate of a casting is primarily a function of its cross-sectional size. The dependence of structure and properties on cross section is termed section sensitivity. Casting of gray irons depends on the fluidity of the molten metal and on the cooling rate. Fluidity depends primarily on the amount of superheat above the freezing temperature (liquidus), with greater superheating leading to increased fluidity. As the total carbon content decreases, the liquidus temperature increases, and the fluidity at a given pouring temperature therefore decreases. However, the minimum thickness of section of gray iron that may be poured is more likely to depend on the cooling rate of the section than on the fluidity of the metal.

A high cooling rate during solidification tends to favor the formation of cementite rather than graphite. That is, the higher the cooling rate for any given cast iron composition, the "whiter" (that is, the larger the cementite fraction) and more brittle the casting is likely to be. The effect of cooling rate is important when choosing a suitable iron for thin-section castings. For example, consider a gray iron casting in which a fine microstructure is required. In thin sections, it would cool so rapidly that cementite would form in preference to graphite. The resulting completely white iron casting would be brittle.

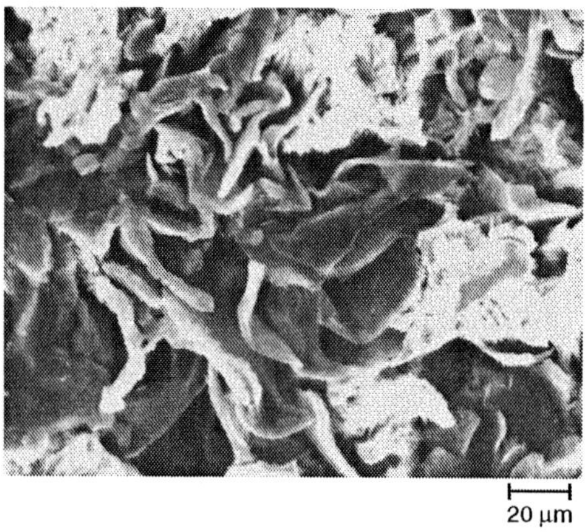


**Fig. 24.9** Flake graphite in gray iron specimen deep etched to reveal flakes. Original magnification: 500×. Source: Ref 9

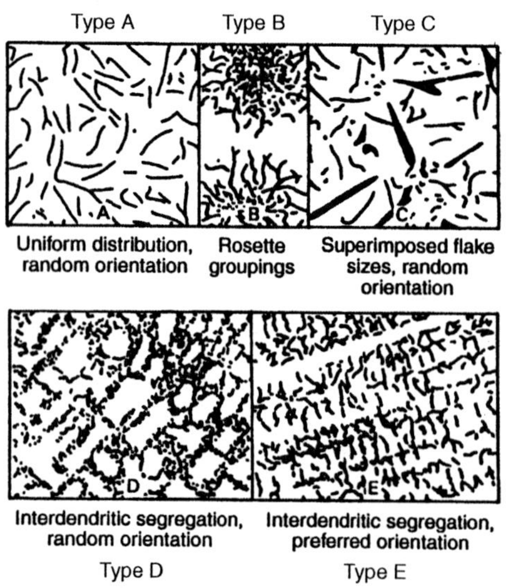


**Fig. 24.10** ASTM/AFS graphite flake types. Source: Ref 8

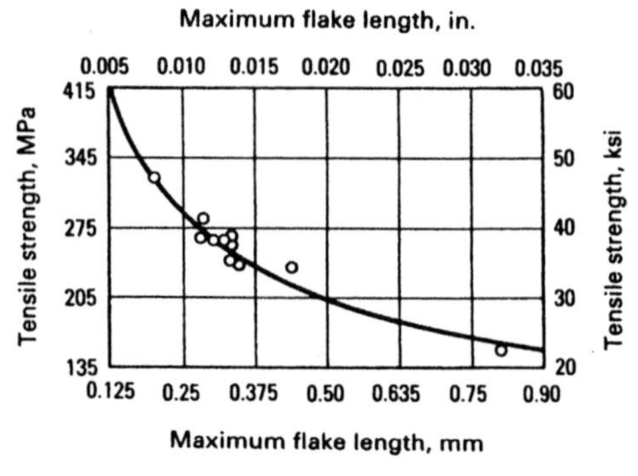


**Fig. 24.11** Effect of graphite flake length on tensile strength of gray iron. Source: Ref 2

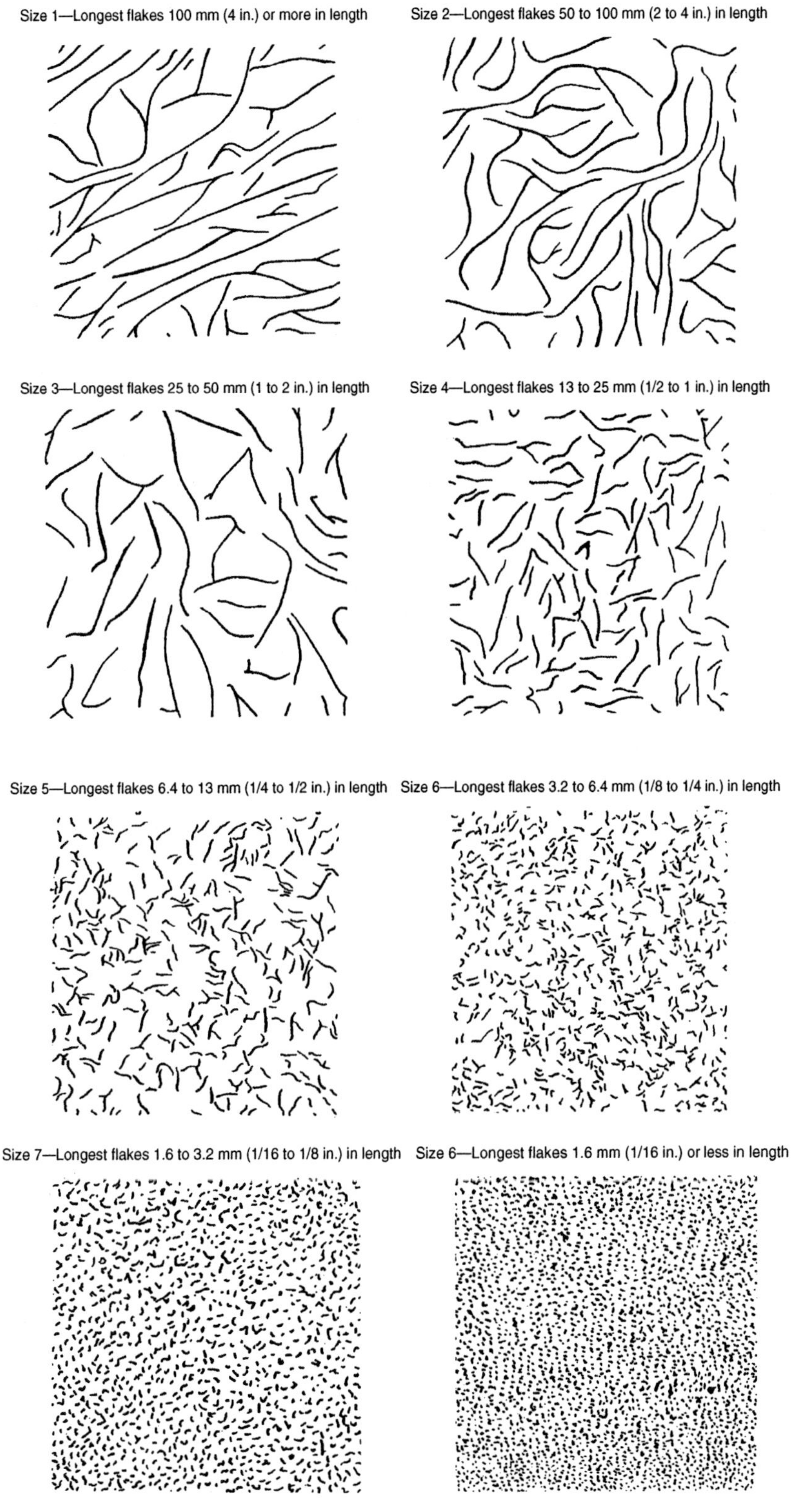


**Fig. 24.12** ASTM/AFS graphite flake size. Source: Ref 10

The effect on microstructure of section thickness and cooling rate is demonstrated by casting a stepped bar of iron of a suitable composition, shown in Fig. 24.13. The thin sections of the bar cooled so quickly that cementite solidified (chilled white zone on the graph), as indicated by a white fracture surface and high hardness values. The thicker sections, having cooled more slowly, are graphitic (represented by the gray-shaded regions of the graph) and therefore softer. The mold exerts a chilling effect, causing many castings to have a hard, white skin on the surface. This is often noticeable when taking the first cut during a machining operation. The cutting tool is immediately dulled by the hard, white skin.

The smallest section that can be cast gray without massive carbides depends on metal composition and also on foundry practices. For example, by adjusting the silicon content or by adding inoculants (i.e., graphitizing agents) to the ladle or, more effectively, to the stream of metal as it is poured, the foundryman can decrease the minimum section size that will contain cementite. The graphite structure can be refined by superheating the molten metal prior to casting. When the metal is heated approximately 110 °C (200 °F) above its melting point, all nuclei are dissolved in the melt. On cooling in the mold, crystallization starts at a lower temperature, but a larger number of nuclei are formed during significant undercooling, resulting in a finer grain size. Using this superheating method, the strength of gray cast iron can be increased by 20 to 25%.

The treatment of liquid cast iron is important because it can dramatically change the nucleation and growth conditions during solidification, affecting both the graphite morphology and the resultant properties. Good foundry practice for gray iron includes the use of inoculants, that is, the addition of minute amounts of minor elements before pouring. Inoculants include calcium, aluminum, titanium, zirconium, silicon carbide, calcium silicide, or combinations of these. The inoculants act as nucleation centers, leading to uniform, fine graphite structures, and aid in removing gas and deoxidizing the melt. As shown in Fig. 24.14, inoculation improves the tensile strength.

Gray cast irons are classified according to their tensile strengths (Table 24.3). For example, class 20 iron will have a nominal tensile strength of 138 MPa (20 ksi), while class 60 would have a tensile strength of approximately 415 MPa (60 ksi). However, strength is not a major criterion for selection. For example, for machine tools and other parts subject to vibrations, the superior damping capacity of a lower-strength grade would be advantageous. As the

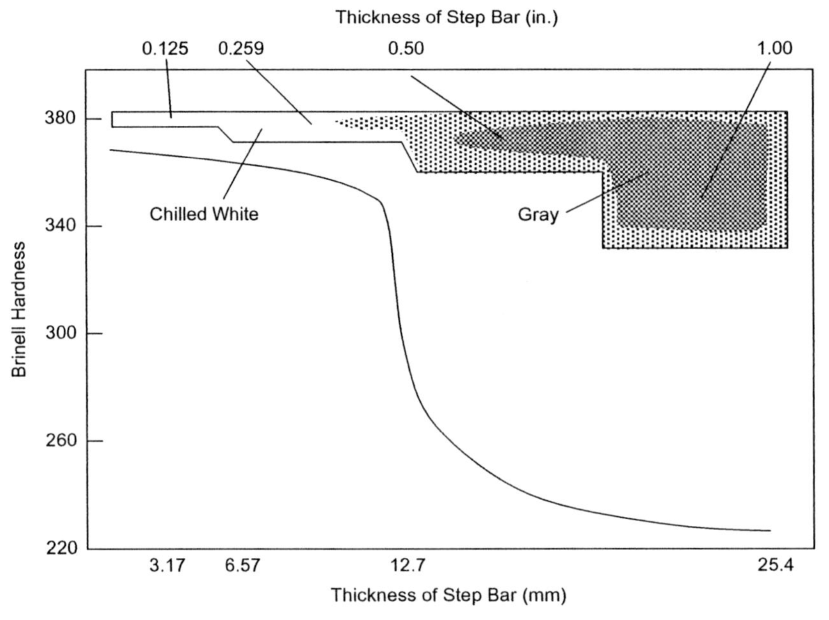


**Fig. 24.13** Effect of cooling rate on gray cast iron. Source: Ref 11

tensile strength increases from class 20 to 60, other mechanical properties increase: high-temperature strength, shear strength, modulus of elasticity, machinability (to a fine finish), and wear resistance. Lower-strength grades are easier to machine, have better resistance to thermal shock, have higher damping capacities, and are more amenable to casting in thin sections. In general, the combined carbon and silicon content decreases as the strength increases. Tensile strength decreases with increasing section size. For example, a gray iron with a tensile strength of 310 MPa (45 ksi) in a 0.04 mm (1 in.) section will have a tensile strength of only 207 MPa (30 ksi) in a 75 mm (3 in.) section. The decreased cooling rate with the larger section results in larger graphite flakes and a reduction in combined carbon.

Yield strength, elongation, and reduction of area are seldom determined for gray iron in standard tension tests. The elongation at fracture of gray iron is very small, on the order of only 0.6%. When gray iron is used for structural applications such as machinery foundations or supports, compressive strength is of greater importance than the tensile strength. The mechanical properties listed in Table 24.3 show the high compressive strength of gray irons; some typical stress-strain curves for gray iron are shown in Fig. 24.15.

Gray iron does not obey Hooke's law, and the modulus in tension is usually arbitrarily determined as the slope of the line connecting the origin of the stress-strain curve with the point corresponding to 1/4 of the tensile strength (secant modulus). The tangent modulus, or the

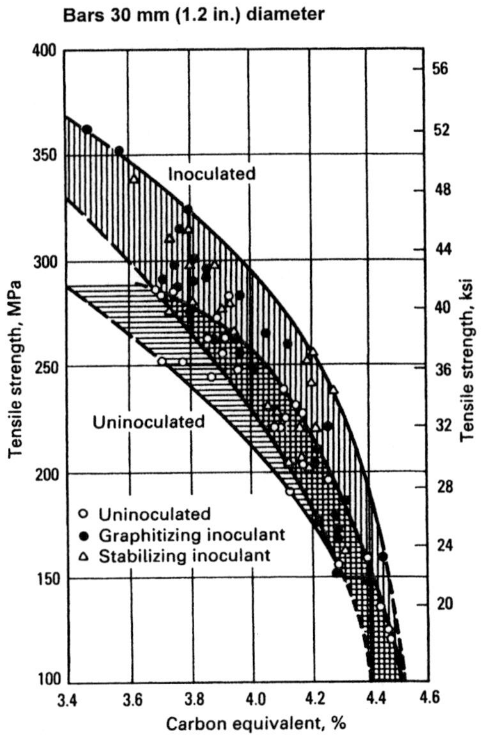


**Fig. 24.14** Effect of inoculating on tensile strength. Source: Ref 2

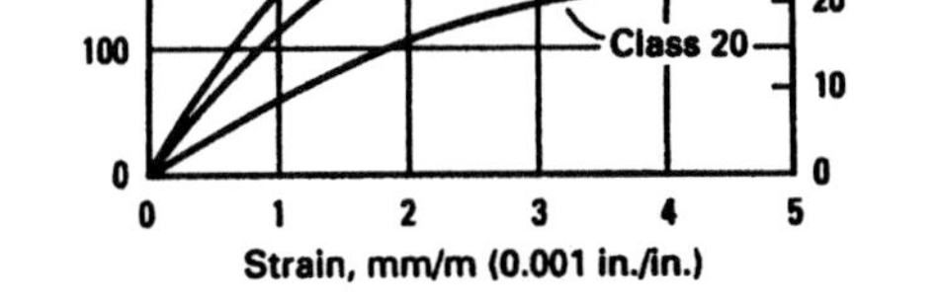


**Fig. 24.15** Typical tension stress-strain curves for gray cast iron. Source: Ref 8

**Table 24.3 Mechanical properties of gray cast irons**

| ASTM A48 class | Tensile strength MPa | Tensile strength ksi | Torsional shear strength MPa | Torsional shear strength ksi | Compressive strength MPa | Compressive strength ksi | Reversed bending fatigue limit MPa | Reversed bending fatigue limit ksi | Hardness, HB |
|---|---|---|---|---|---|---|---|---|---|
| 20 | 152 | 22 | 179 | 26 | 572 | 83 | 69 | 10 | 156 |
| 25 | 179 | 26 | 220 | 32 | 669 | 97 | 79 | 11.5 | 174 |
| 30 | 214 | 31 | 276 | 40 | 752 | 109 | 97 | 14 | 210 |
| 35 | 252 | 36.5 | 334 | 48.5 | 855 | 124 | 110 | 16 | 212 |
| 40 | 293 | 42.5 | 393 | 57 | 965 | 140 | 128 | 18.5 | 235 |
| 50 | 362 | 52.5 | 503 | 73 | 1130 | 164 | 148 | 21.5 | 262 |
| 60 | 431 | 62.5 | 610 | 88.5 | 1293 | 187.5 | 169 | 24.5 | 302 |

Source: Ref 8

slope of the stress-strain curve near the origin, is another method. The hardness of gray iron, as measured by Brinell or Rockwell testers, is an intermediate value between the hardness of the soft graphite and that of the harder metallic matrix. Variations in graphite size and distribution will cause wide variations in hardness, even though the hardness of the metallic matrix is constant.

Gray cast iron is a fairly cheap material. It is easy to cast because of its low melting point and because there is almost no liquid-to-solid shrinkage. Machining is easy because the soft graphite flakes cause the chips to break into many small chips. The soft graphite flakes give gray cast irons very good damping properties and are the irons frequently used for machine tool bases.

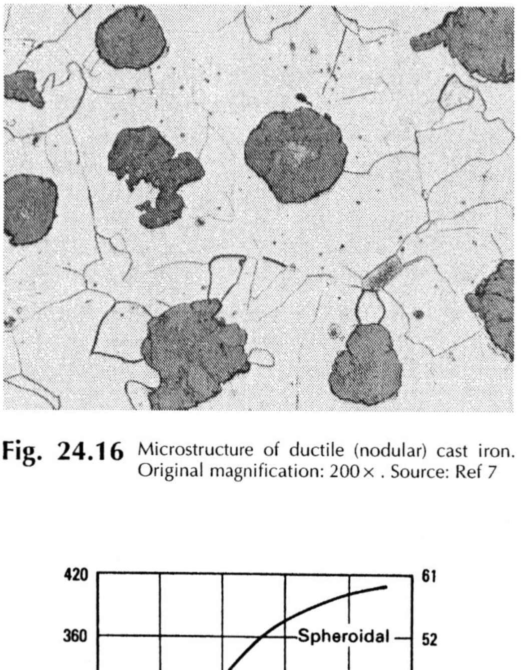

**Fig. 24.16** Microstructure of ductile (nodular) cast iron. Original magnification: 200×. Source: Ref 7

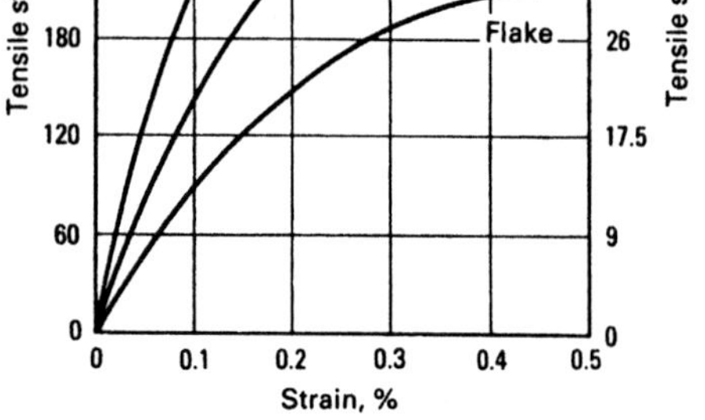


**Fig. 24.17** Influence of graphite morphology on stress-strain behavior. Source: Ref 2

## 24.3 Ductile Cast Iron

Ductile cast iron, also known as nodular iron, spheroidal iron, and spherulic iron, is cast iron in which the graphite is present as tiny nodules or spheroids. Spheroidal graphite particles form during solidification because of the presence of small amounts of specific alloying elements. The nodule-forming addition, usually magnesium or cerium, is added to the ladle just before pouring, causing the graphite to form as spheres or nodules (Fig. 24.16), which significantly improves the toughness and ductility. In fact, graphite morphology is the single most important factor affecting the mechanical properties of any cast iron (Fig. 24.17). The spheroidal shape minimizes stress concentrations. Therefore, ductile cast irons have higher strengths and toughness compared with a similar graphite size and distribution of a gray cast iron.

The most widely used additive element for the production of spheroidal graphite is magnesium. The amount of residual magnesium required to produce spheroidal graphite is generally 0.03 to 0.05 wt%. The precise level depends on the cooling rate, with higher cooling rates requiring less magnesium. However, the silicon and carbon contents are still important. The basic guidelines for determining silicon and carbon content are shown in Fig. 24.18. Alloying elements have the same influence on structure and properties as for gray iron. Because a spheroidal

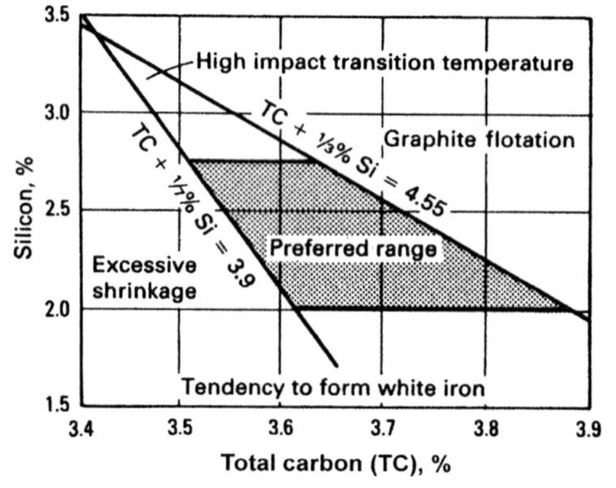


**Fig. 24.18** Carbon and silicon contents for ductile iron. Source: Ref 2

graphite morphology makes more effective use of the mechanical properties of the matrix, alloying is more common in ductile iron than in gray iron.

When changing the cooling rate, similar effects to those discussed for gray iron also occur in ductile iron, but the section sensitivity of ductile iron is lower because spheroidal graphite is less affected by cooling rate than flake graphite. The liquid treatment of ductile iron is more complex than that of gray iron. Liquid treatment of ductile iron occurs in two stages: (1) modification, which consists of magnesium or magnesium alloy treatment of the melt for the purpose of avoiding flake graphite and promoting spheroidal graphite, and (2) inoculation (actually postinoculation, i.e., after the magnesium treatment), to increase the nodule count. Increasing the nodule count is important, because a higher nodule count is associated with less chilling tendency (Fig. 24.19) and a higher as-cast ferrite/pearlite ratio.

Postcasting heat treatment is used extensively in the processing of ductile iron because better advantage can be taken of the matrix structure than for gray iron. The heat treatments usually applied are for stress relieving, annealing to produce a ferritic matrix, normalizing to produce a pearlitic matrix, hardening to produce temper martensitic structures, and austempering to produce a ferritic bainite. The advantage of austempering is that it results in ductile irons with approximately twice the tensile strength for the same toughness. A comparison between some mechanical properties of austempered ductile iron and standard ductile iron is shown in Fig. 24.20.

## 24.4 Malleable Cast Iron

The starting material for malleable cast iron is a hypoeutectic white cast iron with a low silicon content. Depending on the heat treatment, either a black or white malleable cast iron can be produced. When white irons are heated treated at 800 to 970 °C (1470 to 1780 °F), the $Fe_3C$ decomposes, and temper graphite is formed (Fig. 24.21). Most malleable iron is produced by this technique and is called blackheart malleable iron. Whiteheart malleable iron can be produced by decarburization of the as-cast white iron.

The composition of malleable irons must be selected to produce a white as-cast structure and to allow for fast annealing times. Although higher carbon and silicon contents reduce the heat treatment time, they must be limited to ensure a graphite-free structure on solidification. In addition, both tensile strength and elongation decrease with a higher carbon equivalent. However, it is not enough to control the carbon equivalent. The annealing time depends on the number of graphite nuclei available for graphitization, which in turn depends on several factors, including the carbon/silicon ratio. As shown in Fig. 24.22, a lower carbon/silicon ratio (i.e., a greater silicon content for a constant

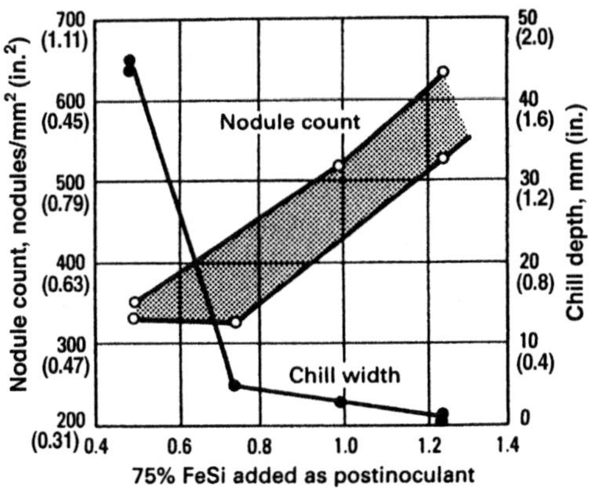


**Fig. 24.19** Postinoculation of ductile cast iron. Source: Ref 2

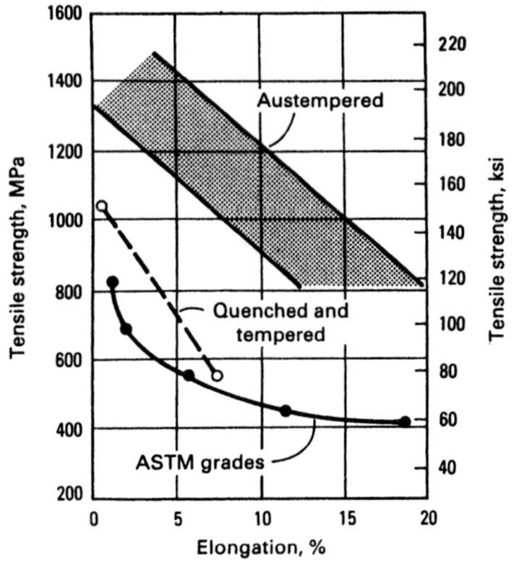


**Fig. 24.20** Strength and elongation of austempered ductile iron. Source: Ref 2

carbon equivalent) results in more temper graphite clusters. This in turn translates into shorter annealing times. In black tempering, cementite is converted to graphite by the reaction:

$$Fe_3C \rightarrow 3Fe + C \quad \text{(Eq 24.3)}$$

A two-stage heat treatment is used to convert white cast irons to malleable cast.

**Stage 1—Graphitization.** White iron castings are heated above the eutectoid temperature, usually between 900 and 955 °C (1650 and 1750 °F), and held for 3 to 20 h, depending on the composition, structure, and size of the castings. The iron carbide of the white iron is transformed to temper carbon and austenite. Temper carbon is formed at the interface between primary carbide and austenite. The growth around the nuclei occurs by a reaction involving diffusion and carbide decomposition.

**Stage 2—Cooling.** Austenite is transformed into ferrite, pearlite, or martensite, depending on the cooling cycle after graphitization.

*Ferritic Malleable Iron.* After the first-stage graphitization treatment, the casting is fast cooled to between 730 and 760 °C (1350 and 1400 °F) and then slowly cooled at a rate of approximately 3 to 16 °C/h (5 to 29 °F/h). During cooling, austenite is transformed to a mixture of ferrite and graphite, with the graphite depositing on the existing particles of temper carbon.

*Pearlitic Malleable Iron.* The castings are slowly cooled to approximately 870 °C (1600 °F) and then air cooled. The rapid cooling rate transforms austenite to pearlite, and temper carbon nodules form in a matrix of pearlite.

*Tempered Martensitic Malleable Iron.* The castings are cooled in the furnace to 845 to 870 °C (1550 to 1600 °F), held for 15 to 30 min to allow them to equalize, and then oil quenched to develop a martensitic matrix. The castings are then tempered at 595 to 730 °C (1100 to 1350 °F) to develop the desired hardness and strength. The final microstructure consists of temper carbon nodules in a tempered martensitic matrix.

While the objective of white tempering, like black tempering, is to decompose the cementite to graphite, a different approach is used. In white tempering, the decomposition of cementite is accomplished by burning out carbon in an oxidizing atmosphere according to the following reactions:

$$Fe_3C \rightarrow 3Fe + C \quad \text{(Eq 24.4)}$$

$$CO_2 + C \rightarrow 2CO \quad \text{(Eq 24.5)}$$

The heat treatment time to form malleable cast iron is long, normally 2 to 5 days, and is therefore expensive. For this reason, it has largely been replaced in applications by ductile cast iron.

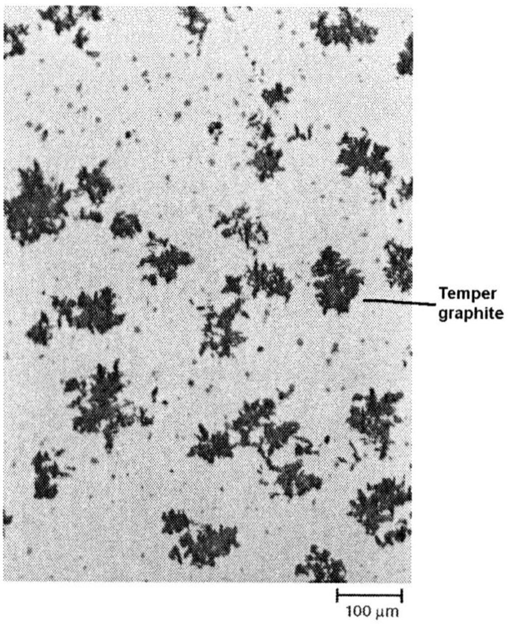


**Fig. 24.21** Microstructure of malleable cast iron. Original magnification: 100×. Source: Ref 9

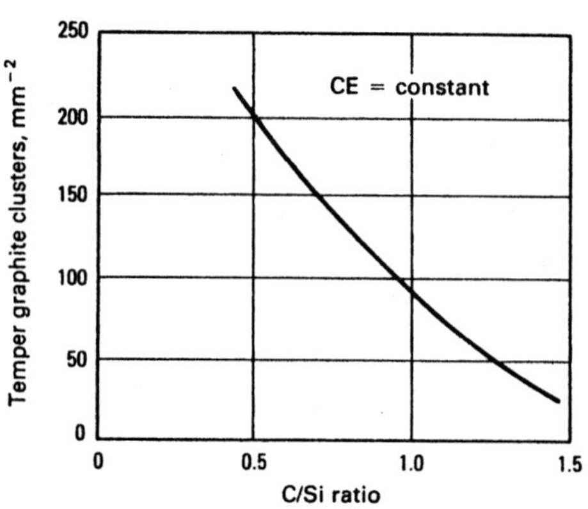


**Fig. 24.22** Influence of carbon/silicon (C/Si) ratio on temper graphite clusters. CE, carbon equivalent. Source: Ref 2

## 24.5 Compacted Graphite Iron

Compacted graphite irons have a graphite morphology that is neither flake (gray iron) nor spheroidal (ductile iron). In a compacted graphite iron, there is no flake graphite, and the spheroidal graphite content is less than 20%. At least 80% of the graphite is compacted (vermicular or wormlike). Typical compacted graphite iron microstructures are shown in Fig. 24.23. The optical micrograph on the left shows the normal appearance of compacted graphite, while the deeply etched sample on the right shows that the graphite consists of interconnected clusters that have a wormlike appearance.

This graphite morphology allows more effective use of the matrix, yielding higher strengths and ductilities than gray irons containing flake graphite. Compared to ductile iron, the castability of compacted graphite iron is better. In addition, the interconnected graphite provides better thermal conductivity and damping capacity than spheroidal graphite. As for ductile iron, the graphite shape is controlled through the addition of minor alloying elements; spheroidizing (magnesium, calcium, and/or rare earth) elements are combined with antispheroidizing (titanium and/or aluminum) elements to produce this intermediate structure. The first compacted graphite irons were the result of adding too little magnesium or cerium to spheroidal iron melts.

In general, the mechanical and physical properties of compacted graphite irons fall between those of gray irons and ductile irons. Compared to gray irons, compacted graphite irons are stronger and more ductile, while compared to ductile irons, compacted graphite irons have a lower coefficient of thermal expansion, higher thermal conductivity, better thermal shock resistance, and higher damping capacity. Compacted graphite iron can be substituted for gray iron in applications where the strength of gray iron is insufficient but in which a change to ductile iron is undesirable because of the less favorable casting properties. Because the thermal conductivity of compacted graphite iron is higher than that of ductile iron, compacted graphite iron is preferred for elevated-temperature applications and/or service under thermal fatigue conditions.

## 24.6 Alloy Cast Irons

Alloying elements are often added to cast irons to improve abrasion resistance, corrosion resistance, or heat resistance.

**Abrasion-Resistant Cast Irons.** Highly alloyed white irons are used for abrasion-resistant applications in crushing, grinding, and materials handling. They usually contain substantial amounts of chromium, which forms hard carbides. Maximum hardness is obtained with a fully martensitic matrix; hardness also increases

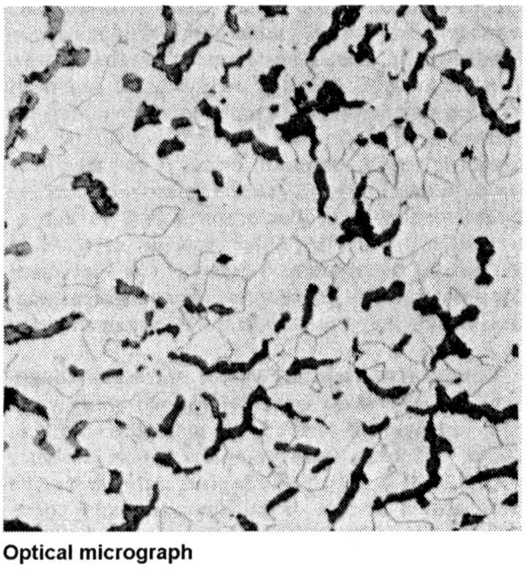

Optical micrograph

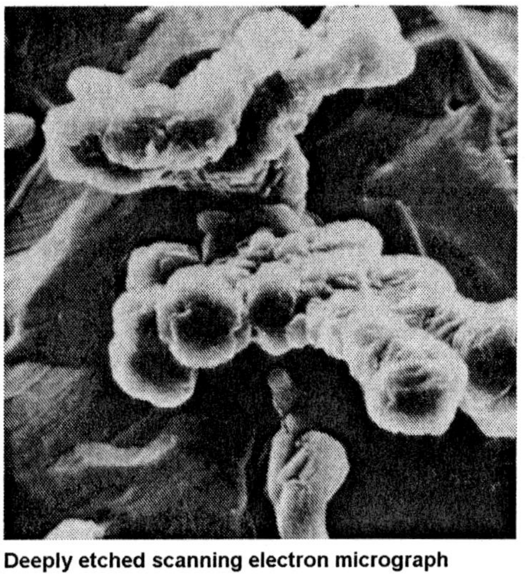

Deeply etched scanning electron micrograph

**Fig. 24.23** Microstructure of compacted graphite iron. Original magnification: 395×. Source: Ref 12

as the carbon content increases. Silicon increases the temperature at which martensite forms, making heat treatment for martensitic alloy white irons easier. However, since higher carbon and silicon contents tend to promote graphite formation during solidification, they must be used in conjunction with alloying elements such as chromium to stabilize the carbide. Depending on the heat treatment, hardness values ranging from 41 to 59 HRC can be achieved.

**Corrosion-Resistant Cast Irons.** The corrosion resistance of gray cast iron is enhanced by the addition of appreciable amounts of nickel, chromium, and copper, singly or in combination, or silicon in excess of approximately 3 wt%. Up to 3 wt% Si is normally present in all cast irons. In larger amounts, silicon is considered an alloying element. It promotes the formation of a strongly protective surface film under oxidizing conditions, such as exposure to oxidizing acids. Relatively small amounts of molybdenum and/or chromium can be added in combination with high silicon. The addition of nickel to gray iron improves its resistance to reducing acids and provides a high resistance to caustic alkalis. Chromium assists in forming a protective oxide that resists oxidizing acids, although it is of little benefit under reducing conditions. Copper has a smaller beneficial effect on resistance to sulfuric acid.

**Heat-resistant cast irons** are basically alloys of iron, carbon, and silicon having high-temperature properties improved by the addition of $>3$ wt% of chromium, nickel, molybdenum, aluminum, and silicon. Silicon and chromium increase the resistance to heavy scaling by forming a light surface oxide that is impervious to oxidizing atmospheres; however, both elements reduce the toughness at elevated temperatures. Molybdenum also increases high-temperature strength. Aluminum additions reduce both growth and scaling but adversely affect room-temperature mechanical properties.

## ACKNOWLEDGMENTS

Sections of this chapter were adapted from the section on Cast Irons in *Metals Handbook Desk Edition,* 2nd Edition, ASM International, 1998; and from Classification and Basic Metallurgy of Case Iron by D.M. Stefanescu in *Properties and Selection: Irons, Steels, and High-Performance Alloys,* Volume 1, *ASM Handbook,* ASM International, 1990.

## REFERENCES

1. Cast Irons, *Metals Handbook Desk Edition,* 2nd ed., ASM International, 1998
2. D.M. Stefanescu, Classification and Basic Metallurgy of Cast Iron, *Properties and Selection: Irons, Steels, and High-Performance Alloys,* Vol 1, *ASM Handbook,* ASM International, 1990
3. M. Tisza, *Physical Metallurgy for Engineers,* ASM International, 2001
4. K.G. Schmitt-Thomas, *Metallkunde fur das Maschinenwesen Band II,* (*Metallurgy for Mechanical Engineering,* Vol 2), Springer-Verlag, 1989
5. Basic Metallurgy of Cast Irons, *Metals Handbook Desk Edition,* 2nd ed., ASM International, 1998
6. G. Krauss, Microstructures, Processing, and Properties of Steels, *Properties and Selection: Irons, Steels, and High-Performance Alloys,* Vol 1, *ASM Handbook,* ASM International, 1990
7. B.L. Bramfitt, Effects of Composition, Processing, and Structure on Properties of Irons and Steels, *Materials Selection and Design,* Vol 20, *ASM Handbook,* ASM International, 1997
8. Gray Iron, *Metals Handbook Desk Edition,* 2nd ed., ASM International, 1998
9. J.M. Radzikowska, Metallography and Microstructures of Cast Iron, *Metallography and Microstructures,* Vol 9, *ASM Handbook,* ASM International, 2004
10. S.H. Avner, *Introduction to Physical Metallurgy,* 2nd ed., McGraw-Hill Book Co., 1974
11. R.A. Higgins, *Engineering Metallurgy—Applied Physical Metallurgy,* 6th ed., Arnold, 1993
12. Compacted Graphite Iron, *Metals Handbook Desk Edition,* 2nd ed., ASM International, 1998

## SELECTED REFERENCES

- Alloy Cast Irons, *Metals Handbook Desk Edition,* 2nd ed., ASM International, 1998
- W.F. Smith, *Principles of Materials Science and Engineering,* McGraw-Hill, 1986

# CHAPTER 25

# Copper

COPPER was probably the first metal used by man. It could be found in small quantities in the metallic state, and, being soft, it was readily shaped into ornaments and weapons during the Bronze Age. In addition, many copper ores can easily be reduced to the metal, and since these ores often contain other minerals, it is very probable that copper alloys were produced inadvertently during smelting. Until recently, copper ranked only behind iron as the most widely used commercial metal. However, due to its rather high cost, it has been replaced by lower-cost metals in some applications and now ranks third behind iron and aluminum. The attributes of copper include availability, excellent formability, good strength when alloyed, high thermal conductivity, good corrosion resistance, and a pleasing color. However, it is the very high electrical conductivity of pure copper that leads to its extensive use as electrical wiring.

Copper is often used in the unalloyed form because pure copper has better electrical conductivity than copper alloys. Pure copper is used extensively for wires and cables, electrical contacts, and a wide variety of other parts that are required to pass electrical current. Approximately 50% of U.S. production is used for wire and cable. It is also alloyed with zinc to form the brasses and with tin to form the bronzes. Coppers and certain brasses, bronzes, and cupronickels are used extensively for applications requiring good conduction of heat, such as automobile radiators, heat exchangers, and home heating systems. Because of their outstanding ability to resist corrosion in water and other aqueous solutions, coppers, brasses, bronzes, and cupronickels are used for pipes, valves, fittings, and coinage.

Copper has a density of 8.93 g/cm$^3$ (0.323 lb/in.$^3$), an elastic modulus of 128 GPa (19 msi), and a melting point of 1083 °C (1981 °F). The electrical conductivity of commercially pure copper is approximately 101% IACS (International Annealed Copper Standard), second only to that of commercially pure silver (~103% IACS). The IACS, which was established in 1913, specified that an annealed copper wire 1 m long with a cross-sectional area of 1 mm$^2$ should have a resistance no greater than 0.017241 Ω at 20 °C. Such a wire was defined as having a conductivity of 100%. Since 1913, higher-purity metals are now commonly produced, thus explaining numbers greater than 100% IACS. The thermal conductivity for copper is also high at 398 W/m · K (226 Btu/ft · h · °F). The outstanding electrical and thermal conductivity of copper as compared to other pure metals is shown in Table 25. 1.

Copper and its alloys are readily cast for subsequent hot or cold working into plate, sheet, rod, wire, or tube through standard rolling, drawing, extrusion, forging, machining, and joining methods. Copper and copper alloy tubing can be made by piercing and tube drawing as well as by continuous induction welding

**Table 25.1 Electrical resistivity and thermal conductivity of select pure metals**

| Metal | Electrical resistivity at 293 K, μΩ · cm | Thermal conductivity, W/m · K$^{-1}$ | Relative electrical conductivity (copper = 100) | Relative thermal conductivity (copper = 100) |
|---|---|---|---|---|
| Silver | 1.63 | 419 | 104 | 106 |
| Copper | 1.694 | 397 | 100 | 100 |
| Gold | 2.2 | 316 | 77 | 80 |
| Aluminum | 2.67 | 238 | 63 | 60 |
| Beryllium | 3.3 | 194 | 51 | 49 |
| Magnesium | 4.2 | 155 | 40 | 39 |
| Tungsten | 5.4 | 174 | 31 | 44 |
| Zinc | 5.96 | 120 | 28 | 30 |
| Nickel | 6.9 | 89 | 24 | 22 |
| Iron | 10.1 | 78 | 17 | 20 |
| Platinum | 10.58 | 73 | 16 | 18 |
| Tin | 12.6 | 73 | 13 | 18 |
| Lead | 20.6 | 35 | 8.2 | 8.8 |
| Titanium | 54 | 22 | 3.1 | 5.5 |
| Bismuth | 117 | 9 | 1.4 | 2.2 |

Source: Ref 2

of strip. Copper and its alloys owe their excellent fabricability to their face-centered cubic crystal structure. Copper is hot worked in the range of 760 to 870 °C (1400 to 1600 °F), annealed between cold working steps at 370 to 650 °C (700 to 1200 °F), and stress relieved at 205 to 345 °C (400 to 650 °F). Many of the applications of copper and its alloys take advantage of its work-hardening capability, with cold working during the final forming steps providing the desired strength-ductility combinations. Copper can be cold reduced almost limitlessly without anncaling, but hcavy dcformation ($>$ 80 to 90%) may result in a preferred crystal orientation or texturing. Since textured metal has different properties in different directions, it is undesirable for some applications. To avoid preferred orientation and textures, copper and many copper alloys are hot worked to nearly the finished size. Hot working reduces the as-cast grain size from approximately 1.0 to 10 mm (0.04 to 0.4 in.) to approximately 0.10 mm (0.004 in.) or less and yields a soft, texture-free structure suitable for finishing by cold working.

Cold working increases both tensile strength and yield strength, but it has a more pronounced effect on yield strength. For most coppers and copper alloys, the tensile strength of the hardest cold-worked temper is approximately two times the tensile strength of the annealed temper. However, the yield strength of the hardest cold-worked temper can be as much as five to six times that of the annealed temper. Work-hardened metal can be returned to a soft state by annealing. During annealing of simple single-phase alloys, deformed and highly stressed crystals are transformed into stress-free crystals by recovery, recrystallization, and grain growth. In severely deformed metal, recrystallization occurs at lower temperatures than in a lightly deformed metal. Also, when severely deformed metal is recrystallized, the grains are smaller and more uniform in size. Grain size can be controlled by proper selection of cold working and annealing schedules. Large amounts of cold work, fast heating rates to the annealing temperature, and short annealing times favor fine grain sizes. Larger grain sizes are normally a result of a combination of limited deformation and long annealing times.

Copper and copper alloys are readily joined by mechanical methods, such as crimping, staking, riveting, and bolting, although soldering, brazing, and welding are the most widely used processes for bonding copper.

Copper can be alloyed to improve its strength without degrading ductility or workability. However, additions of alloying elements degrade electrical and thermal conductivity. The choice of alloy and condition is often a trade-off between strength and conductivity. Alloying also changes the color from reddish-brown to yellow with zinc, as in brasses, and to metallic white or silver with nickel, as in U.S. cupronickel coinage.

## 25.1 Copper Production

Most copper comes from copper sulfide deposits that go through ore dressing procedures and various smelting operations (Fig. 25.1). Copper sulfide concentrates are smelted in a reverberatory furnace to produce a matte, which is a mixture of copper and iron sulfides. The copper sulfides in the matte are then converted

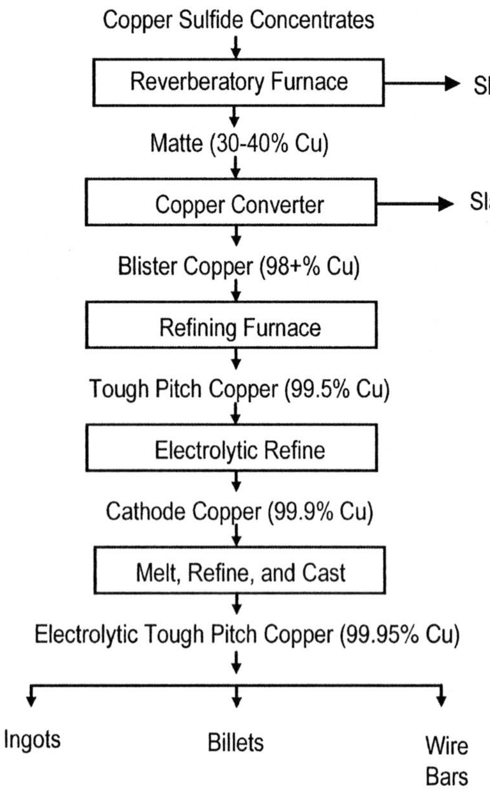


**Fig. 25.1** Processing steps in cooper refining

to blister copper by blowing air through the matte. Blister copper is elemental copper along with approximately 2% impurities. During the process of blowing air through the matte, some of the copper is oxidized to $Cu_2O$. Next, the blister copper is fire refined, which converts most of the $Cu_2O$ back into elemental copper, with approximately 0.5% $Cu_2O$ still remaining. The copper is now called tough pitch copper, which is used for some applications. The remainder of the tough pitch copper is electrolytically refined to produce 99.95% electrolytic tough pitch copper, which is used for electrical applications requiring the highest conductivity. In the electrolytic refining process shown in Fig. 25.2, a tough pitch copper anode is reduced, with the impurities settling to the bottom. The purified copper is deposited on a starting sheet of pure copper at the cathode. Oxygen-free high-conductivity copper is finally produced by depositing the cathodes under a reducing atmosphere that prevents oxidation.

## 25.2 Wrought Copper Alloys

Alloying elements are added to copper to optimize strength, ductility, and thermal stability, without unacceptable losses in fabricability, electrical/thermal conductivity, or corrosion resistance. Copper and copper alloys can be divided into the families shown in Table 25. 2. The coppers are essentially commercially pure copper, which are ordinarily soft and ductile and contain less than approximately 0.7% total impurities. The high-copper alloys, or dilute coppers, contain small amounts of various alloying elements, such as beryllium, cadmium, chromium, or iron. The brasses contain zinc, the phosphor bronzes contain phosphorus, the aluminum bronzes contain aluminum, the silicon bronzes contain silicon, and the copper-nickels (cupronickels) and nickel silvers contain nickel.

A list of selected wrought copper alloy compositions and their properties is given in Table 25. 3. Copper alloys have excellent hot and cold ductility, although usually not to the same degree as unalloyed copper. Even alloys with large amounts of solid-solution-hardening elements, such as zinc, aluminum, tin, and silicon, are readily processed by cold working beyond 50% before an intermediate anneal is required to permit additional processing. The temper designations for wrought copper and copper alloys are based primarily on the amount of cold work in the finished product. A partial

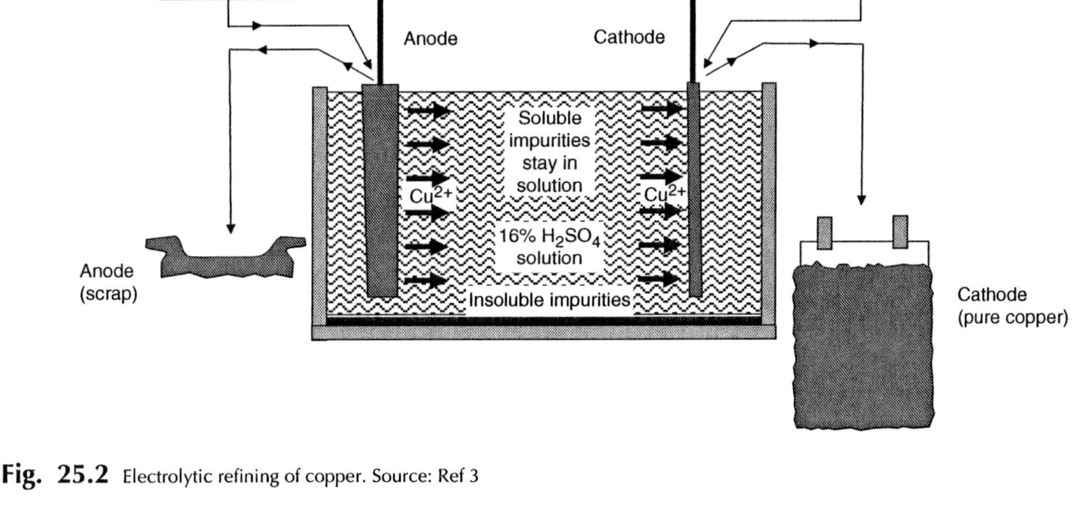


**Fig. 25.2** Electrolytic refining of copper. Source: Ref 3

listing of the rather extensive temper designations for copper and its alloys is shown in Table 25. 4.

## 25.3 Pure Coppers

Electrolytic tough pitch (ETP) copper contains a minimum of 99.9 wt% Cu and 0.02 to 0.05 wt% oxygen in the form of $Cu_2O$. It is the least expensive of the commercial coppers and is used for wire, rod, plate, and sheet. The presence of oxygen has both good and bad effects. Oxygen often ties up impurities that can adversely affect conductivity by forming harmless oxides. However, it can also combine with hydrogen at high temperatures (>400 °C, or 750 °F) to form water vapor (steam) that results in blistering. Therefore, these coppers should not be used for applications that require welding.

**Table 25.2 Classification of copper alloys**

| Alloy | UNS No. | Composition |
|---|---|---|
| **Wrought alloys** | | |
| Coppers | C10100-C15760 | >99% Cu |
| High-copper alloys | C16200-C19600 | >96% Cu |
| Brasses | C20500-C28580 | Cu-Zn |
| Leaded brasses | C31200-C38590 | Cu-Zn-Pb |
| Tin brasses | C40400-C49080 | Cu-Zn-Sn-Pb |
| Phosphor bronzes | C50100-C52400 | Cu-Sn-P |
| Leaded phosphor bronzes | C53200-C54800 | Cu-Sn-Pb-P |
| Copper-phosphorus and copper-silver-phosphorus alloys | C55180-C55284 | Cu-P-Ag |
| Aluminum bronzes | C60600-C64400 | Cu-Al-Ni-Fe-Si-Sn |
| Silicon bronzes | C64700-C66100 | Cu-Si-Sn |
| Other copper-zinc alloys | C66400-C69900 | ... |
| Copper-nickels | C70000-C79900 | Cu-Ni-Fe |
| Nickel silvers | C73200-C79900 | Cu-Ni-Zn |
| **Cast alloys** | | |
| Coppers | C80100-C81100 | >99% Cu |
| High-copper alloys | C81300-C82800 | >94% Cu |
| Red and leaded red brasses | C83300-C85800 | Cu-Zn-Sn-Pb (75–89% Cu) |
| Yellow and leaded yellow brasses | C85200-C85800 | Cu-Zn-Sn-Pb (57–74% Cu) |
| Manganese bronzes and leaded manganese bronzes | C86100-C86800 | Cu-Zn-Mn-Fe-Pb |
| Silicon bronzes, silicon brasses | C87300-C87900 | Cu-Zn-Si |
| Tin bronzes and leaded tin bronzes | C90200-C94500 | Cu-Sn-Zn-Pb |
| Nickel-tin bronzes | C94700-C94900 | Cu-Ni-Sn-Zn-Pb |
| Aluminum-bronzes | C95200-C95810 | Cu-Al-Fe-Ni |
| Copper-nickels | C96200-C96800 | Cu-Ni-Fe |
| Nickel silvers | C97300-C97800 | Cu-Ni-Zn-Pb-Sn |
| Leaded coppers | C98200-C98800 | Cu-Pb |
| Special alloys | C99300-C99750 | ... |

Oxygen-free high-conductivity (OFHC) copper is produced from electrorefined cathode copper by casting under a reducing atmosphere of carbon monoxide and nitrogen, so that oxygen is prevented from entering the copper. However, even though OFHC copper contains 99.95% Cu, the electrical conductivity (101% IACS) is approximately the same as that for ETP copper. The reason for the lack of improvement is the presence of iron impurities, which are tied up as oxides in ETP copper. However, OFHC copper does not have the blistering problem that ETP copper has, since there is no dissolved oxygen to form steam in the presence of hydrogen. A typical microstructure of hot rolled OFHC copper is shown in Fig. 25.3. An even higher purity of copper can be produced by remelting select copper cathodes to produce CDA 101 copper, which is 99.99% Cu. However, these extra processing steps increase the cost of oxygen-free coppers as compared to the ETP grades. Deoxidized coppers can also be produced by adding phosphorus, which converts the oxygen to phosphorus pentoxide ($P_2O_5$). Even though phosphorus retained in solution with copper reduces the electrical conductivity, excess phosphorus is sometimes desirable because it absorbs oxygen during hot working and allows the material to be welded.

As shown in Fig. 25.4, the presence of impurities reduces electrical conductivity; however, the reduction caused by the presence of certain elements in small amounts is not great. For example, up to 1 wt% Cd is added to telephone wires to provide greater strength. This alloy, when cold worked by drawing, has a tensile strength of approximately 462 MPa (67 ksi) compared to 338 MPa (49 ksi) for cold-worked pure copper, and the electrical conductivity is still over 90% of that for soft pure copper. Other elements have more deleterious effects on conductivity. For example, as little as 0.04 wt% P reduces the electrical conductivity to approximately 75% of that for pure copper.

Work hardening is the only strengthening mechanism used with pure copper, limited by the amount of ductility required for the application. Worked-hardened copper can be recrystallized by annealing at temperatures as low as 250 °C (480 °F), depending on the prior degree of cold work and time at temperature. While this facilitates processing, it also means that softening resistance during long-term exposures at moderately elevated temperatures can be a concern, especially in electrical and electronic applications where $I^2R$ heating is a factor. For applications above room temperature, thermal softening can occur over extended periods, and characteristics such as the half-softening temperature should be considered, that is, the temperature at which the worked metal softens to half its original hardness after a specific exposure time, usually 1 h.

## 25.4 Copper Alloys

Copper can be hardened by solid-solution alloying and by work hardening. In addition, a limited number of copper alloys can be strengthened by precipitation hardening. The commonly used solid-solution alloying elements listed in approximate order of increasing effectiveness are zinc, nickel, manganese, aluminum, tin, and silicon. Commercial alloys represent the entire range of available solid-solution compositions of each element, with up to 35% Zn, 50% Ni, 50% Mn, 9% Al, 11% Sn, and 4% Si by weight.

Work hardening is the principal hardening mechanism used for most copper alloys, the degree of which depends on the type and amount of alloying elements and whether the alloying elements remain in solid solution or form a dispersoid or precipitate phase. Even those alloys that are precipitation hardenable are often provided in the mill-hardened tempers; that is, they have been processed by cold working before or after precipitation hardening. The degree of work hardening achieved by cold working several single-phase copper alloys is shown in the curves in Fig. 25.5. Many copper alloys are used in wrought forms in a cold-worked temper, chosen to produce the desired combination of work-hardened strength and formability.

**Table 25.3 Properties of select copper alloys**

| Alloy | UNS No. | Nominal composition | Treatment | Ultimate tensile strength, MPa | Ultimate tensile strength, ksi | Yield strength, MPa | Yield strength, Ksi | Elongation, % | Rockwell hardness |
|---|---|---|---|---|---|---|---|---|---|
| **Pure copper** | | | | | | | | | |
| Oxygen-free high conductivity | C10200 | 99.95 Cu | ... | 228–455 | 33–66 | 69–365 | 10–53 | 55–4 | ... |
| **High-copper alloys** | | | | | | | | | |
| Beryllium-copper | C17200 | 97.9Cu-1.9Be-0.2Ni or Co | Annealed | 490 | 71 | ... | ... | 35 | 60 HRB |
| | | | Hardened | 1400 | 203 | 1048 | 152 | 2 | 42 HRC |
| **Brass** | | | | | | | | | |
| Gilding, 95% | C21000 | 95Cu-5Zn | Annealed | 248 | 36 | 76 | 11 | 45 | 52 HRF |
| | | | Hard | 393 | 57 | 352 | 51 | 5 | 64 HRB |
| Red brass, 85% | C23000 | 85Cu-15Zn | Annealed | 283 | 41 | 90 | 13 | 47 | 64 HRF |
| | | | Hard | 434 | 63 | 407 | 59 | 5 | 73 HRB |
| Cartridge brass, 70% | C26000 | 70Cu-30Zn | Annealed | 359 | 52 | 131 | 19 | 55 | 72 HRF |
| | | | Hard | 531 | 77 | 441 | 64 | 8 | 82 HRB |
| Muntz metal | C28000 | 60Cu-40Zn | Annealed | 379 | 55 | 117 | 17 | 45 | 80 HRF |
| | | | Half-hard | 490 | 71 | 352 | 51 | 15 | 75 HRB |
| High-lead brass | C35300 | 62Cu-36Zn-2Pb | Annealed | 352 | 51 | 117 | 17 | 52 | 68 HRF |
| | | | Hard | 421 | 61 | 317 | 46 | 7 | 80 HRB |
| **Bronze** | | | | | | | | | |
| Phosphor bronze, 5% | C51000 | 95Cu-5Sn | Annealed | 352 | 51 | 172 | 25 | 55 | 40 HRB |
| | | | Hard | 586 | 85 | 579 | 84 | 9 | 90 HRB |
| Phosphor bronze, 10% | C52400 | 90Cu-10Sn | Annealed | 483 | 70 | 248 | 36 | 63 | 62 HRB |
| | | | Hard | 710 | 103 | 655 | 95 | 16 | 96 HRB |
| Aluminum bronze | C60800 | 95Cu-5Al | Annealed | 421 | 61 | 172 | 25 | 66 | 49 HRB |
| | | | Cold rolled | 703 | 102 | 441 | 64 | 8 | 94 HRB |
| Aluminum bronze | C63000 | 81.5Cu-9.5Al-5Ni-2.5Fe-1Mn | Extruded | 689 | 100 | 414 | 60 | 15 | 96 HRB |
| | | | Half-hard | 814 | 118 | 517 | 75 | 15 | 98 HRB |
| High-silicon bronze | C65500 | 96Cu-3Si-1Mn | Annealed | 441 | 64 | 214 | 31 | 55 | 66 HRB |
| | | | Hard | 655 | 95 | 407 | 59 | 8 | 95 HRB |
| **Copper-nickel** | | | | | | | | | |
| Cupronickel, 30% | C71500 | 70Cu-30Ni | Annealed | 386 | 56 | 124 | 18 | 36 | 40 HRB |
| | | | Cold rolled | 586 | 85 | 552 | 80 | 3 | 86 HRB |
| **Nickel silver** | | | | | | | | | |
| Nickel silver | C75700 | 65Cu-23Zn-12Ni | Annealed | 427 | 62 | 193 | 28 | 35 | 55 HRB |
| | | | Hard | 593 | 86 | 524 | 76 | 4 | 89 HRB |

**Table 25.4 Select temper designations for copper alloys**

| Designation | Description |
|---|---|
| **Cold-worked tempers** | |
| H00 | 1/8 hard |
| H01 | 1/4 hard |
| H02 | 1/2 hard |
| H03 | 3/4 hard |
| H04 | Hard |
| H06 | Extra hard |
| H08 | Spring |
| H10 | Extra spring |
| H12 | Special spring |
| H13 | Ultra spring |
| H14 | Super spring |
| H50 | Extruded and drawn |
| H52 | Pierced and drawn |
| H55 | Light drawn; light cold rolled |
| **As-manufactured tempers** | |
| M07 | As-continuous cast |
| M10 | As-hot forged and air cooled |
| M11 | As-forged and quenched |
| M20 | As-hot rolled |
| M30 | As-hot extruded |
| M40 | As-hot pierced |
| M50 | As-hot pierced and rerolled |
| **Annealed tempers** | |
| O10 | Cast and annealed |
| O11 | As-cast and precipitation heat treated |
| O20 | Hot forged and annealed |
| O25 | Hot rolled and annealed |
| O30 | Hot extruded and annealed |
| O31 | Extruded and precipitation heat treated |
| O40 | Hot pierced and annealed |
| O50 | Light annealed |
| **Solution-treated and cold-worked tempers** | |
| TB00 | Solution heat treated only |
| TD00 | TB00 cold worked to 1/8 hard |
| TD01 | TB00 cold worked to 1/4 hard |
| TD02 | TB00 cold worked to 1/2 hard |
| TD03 | TB00 cold worked to 3/4 hard |
| TD04 | TB00 cold worked to full hard |
| **Cold-worked and precipitation-hardened tempers** | |
| TH01 | TD01 and precipitation hardened |
| TH02 | TD02 and precipitation hardened |
| TH03 | TD03 and precipitation hardened |
| TH04 | TD04 and precipitation hardened |
| **Precipitation-hardened and cold-worked tempers** | |
| TF00 | TB00 and precipitation hardened |
| TL00 | TF00 cold worked to 1/8 hard |
| TL01 | TF00 cold worked to 1/4 hard |
| TL02 | TF00 cold worked to 1/2 hard |
| TL03 | TT00 cold worked to 3/4 hard |
| TL04 | TF00 cold worked to full hard |
| TL08 | TF00 cold worked to spring hard |
| TL10 | TF00 cold worked to extra spring hard |

## 25.5 Brasses

Brasses, probably the most important of the copper alloys, are alloys of copper and zinc, with zinc contents as high as 45 wt%. Other alloying additions include tin, aluminum, silicon, manganese, nickel, and lead. Normally, the amount of these additional elements is approximately 4 wt% or less. Since zinc is less expensive than copper, there is an economic incentive to use high zinc contents. Also, the strength increases with higher zinc contents. However, the corrosion resistance of the brasses is generally inferior to that of pure copper. Zinc also reduces the melting temperature and the electrical conductivity. With increasing zinc content, the color of brasses changes from golden red (5 wt% Zn) to golden yellow (15 wt% Zn) to yellow (37 wt% Zn) to red-yellow (40 wt% Zn). Machinability can be improved by the addition of lead but with some sacrifice in cold working properties.

As shown in the copper-zinc phase diagram (Fig. 25.6), α copper will dissolve up to 32.5 wt% Zn at the solidus temperature of 902 °C (1656 °F), the proportion increasing to 39.0 wt% at 454 °C (849 °F). Although the amount of zinc decreases with further decreases in temperature, diffusion is very sluggish at temperatures below 454 °C (849 °F), and with ordinary industrial cooling rates, the amount of zinc that can remain in solid solution in

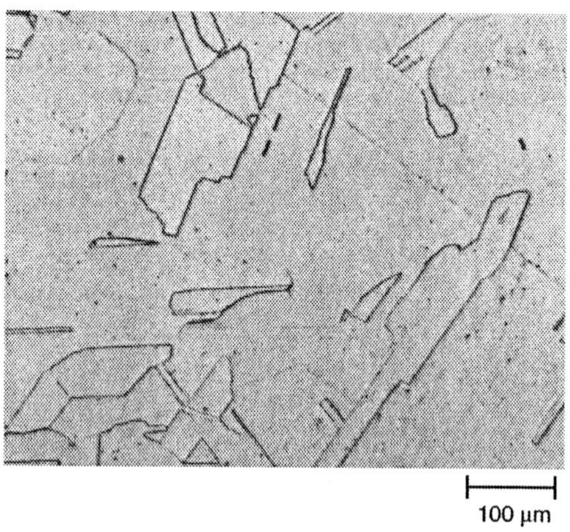


**Fig. 25.3** Oxygen-free copper hot rolled bar with large, equiaxed, twinned grains. Original magnification: 100×. Source: Ref 4

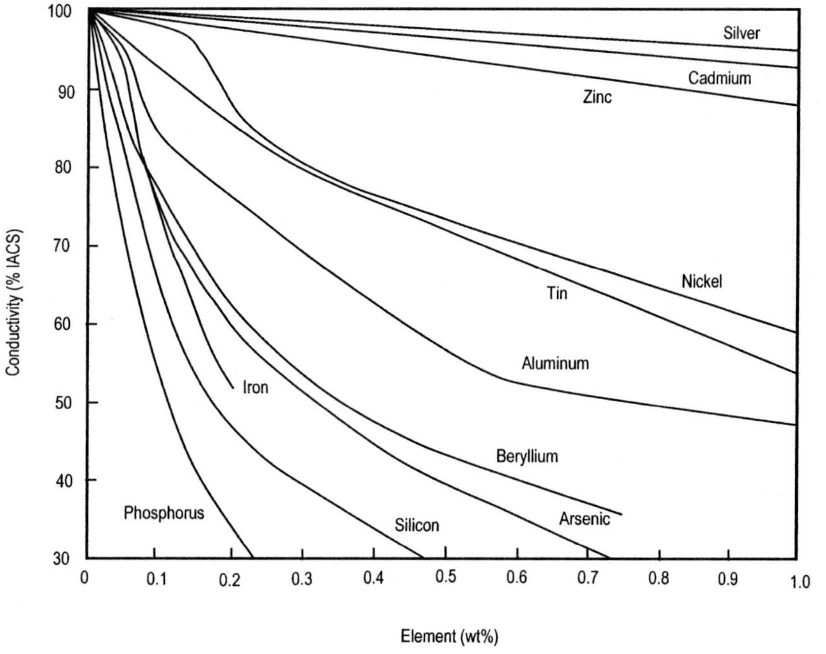


**Fig. 25.4** Effects of alloying elements and impurities on conductivity of copper. Source: Ref 5

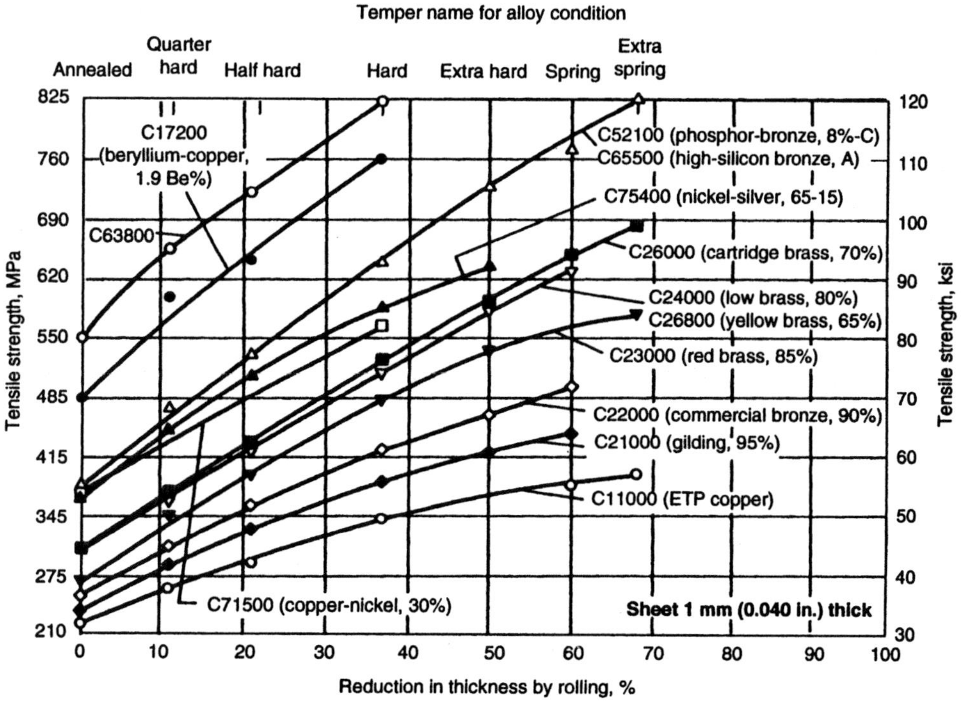


**Fig. 25.5** Tensile strength vs. reduction in thickness. Source: Ref 6

copper at room temperature is approximately 39 wt%. When the amount of zinc is increased beyond 39 wt%, the intermediate ordered β′ phase, equivalent to CuZn, will form on slow cooling. This phase is hard at room temperature and has limited ductility but becomes plastic when it transforms to the disordered β phase above 454 °C (849 °F). Unlike copper, which is face-centered cubic, and zinc, which is hexagonal close-packed, the beta phases are body-centered cubic. The high-temperature β phase is disordered, while the lower-temperature β′ phase is ordered (Fig. 25.7). These alloys are easy to machine and hot form but are not very amenable to cold forming, because the β′ phase is brittle at room temperature. There is also

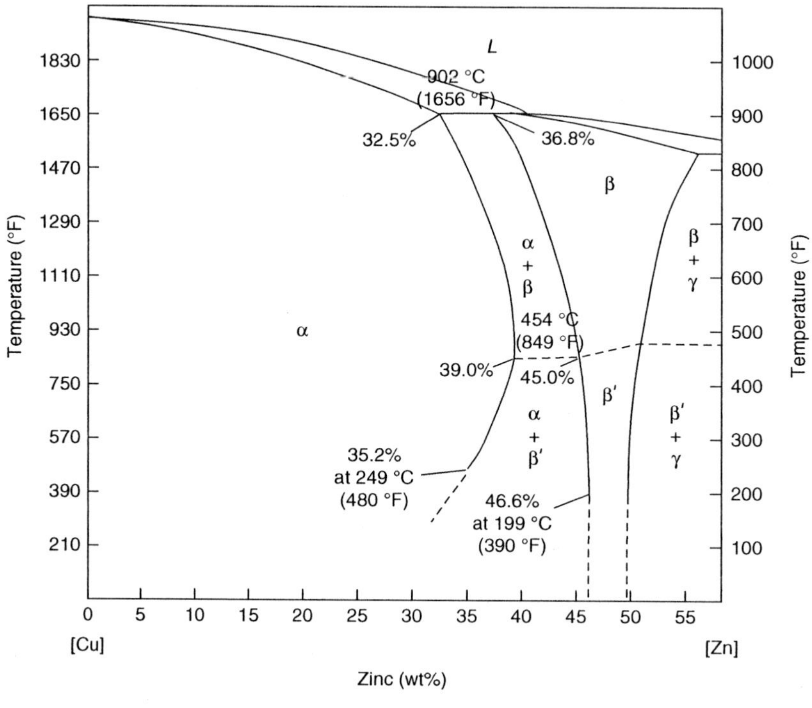


**Fig. 25.6** Copper-zinc phase diagram

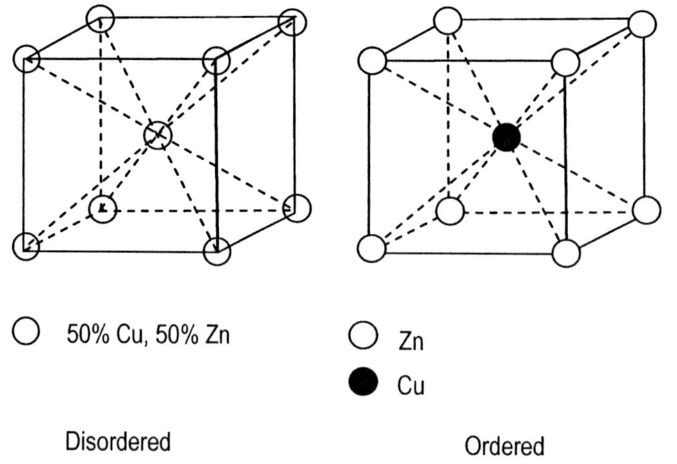


**Fig. 25.7** Disordered and ordered structures in 50%Cu-50%Zn β brasses

a rapid reduction in the impact strength, with a simultaneous increase in hardness and tensile strength. The maximum strength of 538 MPa (78 ksi) is attained at 44 wt% Zn. The microstructure of two-phase as-cast Muntz metal (60Cu-40Zn) is shown in Fig. 25.8. Naval brass, an alloy with improved corrosion resistance for use in marine environments, is obtained by replacing approximately 1 wt% of the zinc with tin. Further increases in the zinc content beyond 50 wt% cause the γ phase to form, which is too brittle for engineering alloys.

Since alloys that contain only the α phase are quite soft and ductile at room temperature, the α brasses are very amenable to cold working. However, as a result of the hard, ordered β′ phase, brasses containing both the α + β′ phases are rather hard, with a low capacity for cold work. The α + β′ brasses are hot worked in the temperature range where β′ transforms to β. Thus, the α brasses are often referred to as cold work brasses, and the α + β′ brasses are known as hot work brasses. The α brasses are generally cold worked, and a limited amount of cold work can be applied to the α + β′ brasses, which contain only small amounts of β′. In general, the α + β′ brasses are hot worked, and then, any cold working is used to finish the shape or to produce the desired amount of cold working. The ductility of the α brasses actually increases with increasing zinc contents up to a maximum of 30 wt% Zn (Fig. 25.9). The microstructure of annealed cartridge brass (70Cu-30Zn), shown in Fig. 25.10, contains extensive annealing twins. However, high-purity copper and zinc are required to reach maximum ductilities, which increases cost. Thus, mild steel with low interstitials has replaced α brasses in many applications. The α brasses are also sensitive to annealing temperature in that overheating can lead to rapid grain growth, which can produce rough "orange-peel" surfaces during forming operations. For example, cartridge brass, when

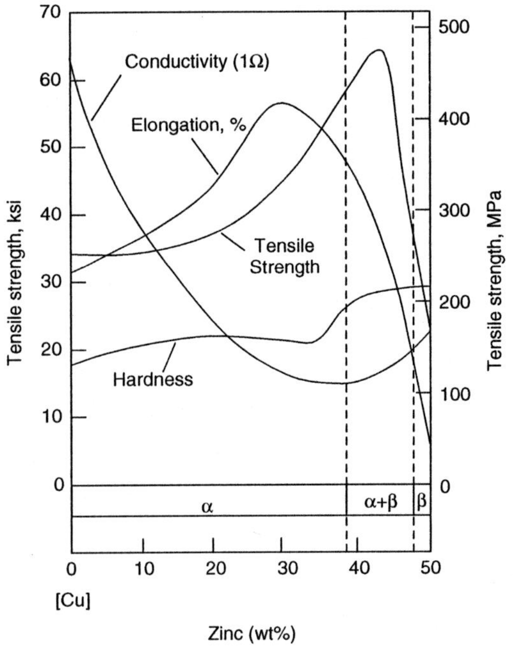


**Fig. 25.9** Effects of zinc on properties of copper. Source: Ref 5

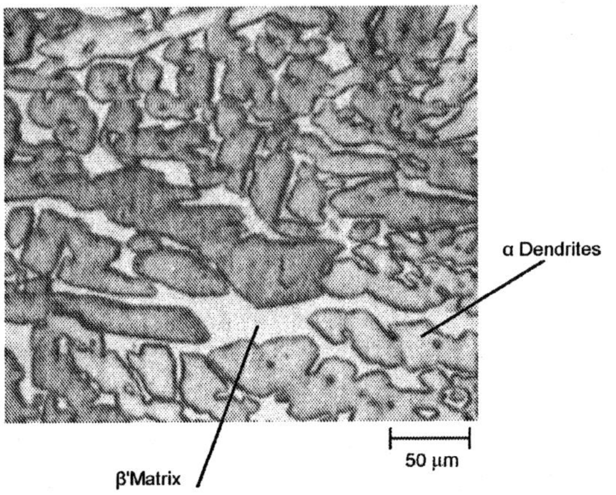


**Fig. 25.8** As-cast Muntz metal. Original magnification: 210×. Source: Ref 4

**Fig. 25.10** Annealed cartridge brass. Original magnification: 75×. Source: Ref 4

annealed at 760 °C (1400 °F), develops a rough texture during deep-drawing operations.

Cold-worked α brasses are also subject to stress-corrosion cracking, originally called season cracking. The term *season cracking* was used to describe the spontaneous cracking of stored cartridge cases in India during the monsoon season. It was particularly prevalent when the damp atmosphere contained ammonia emanating from nearby cavalry stables. Alloys with less than 15 wt% Zn have good corrosion and stress-corrosion resistance. Alloys with more than 15 wt% Zn need a stress-relieving heat treatment at approximately 260 °C (500 °F) to avoid stress-corrosion cracking.

Alloying elements added to α brasses include tin in amounts up to 1.0 wt% to improve corrosion resistance; small amounts of arsenic (0.01 to 0.05 wt%) to improve corrosion resistance and inhibit dezincification; up to 2.0 wt% Pb to improve machinability; up to 2.0 wt% Al to provide corrosion resistance, particularly to impingement attack; and nickel, which also improves corrosion resistance. Lead-containing brass or cast brass is often used for machined parts. Lead additions of 1 to 2 wt% do not alloy with the brass but form spherical particles that provide lubrication during machining. However, cast brasses are usually fairly brittle and machine well because the chips fragment easily.

Since the $\beta'$ phase is relatively hard at room temperature, $\alpha+\beta'$ brasses are stronger and have lower ductilities than the α brasses. When cooling from elevated temperatures in the $\alpha+\beta$ phase field, the α phase precipitates at $\beta'$ phase boundaries. Faster cooling results in larger amounts of transformed $\beta'$ phase, which increases the strength and hardness, while slower cooling allows more precipitation of the softer α phase.

The tin brasses contain 0.3 to 3.0 wt% Sn to enhance corrosion resistance and strength. Besides improving corrosion resistance in copper-zinc tube alloys, tin also provides good combinations of strength, formability, and electrical conductivity required for various electrical connectors.

Special brasses are those that contain additional alloying elements such as aluminum, iron, manganese, nickel, silicon, and tin in the range of 0.1 to 10 wt%. Additional alloying elements are used to improve corrosion resistance and increase strength, especially high-temperature strength. Aluminum increases strength and improves the resistance to corrosion and oxidation. Iron and manganese improve bearing properties and corrosion resistance. Nickel improves high-temperature strength and corrosion resistance. Tin and silicon improve bearing properties and corrosion resistance. The strength of special brasses ranges from 296 to 793 MPa (43 to 115 ksi), with elongations of 45 to 10%.

**Nickel silvers** are ternary alloys of copper, zinc, and nickel with compositions of 50 to 70 wt% Cu, 5 to 40 wt% Zn, and 5 to 30 wt% Ni. As a result of their silver color, they are called nickel silvers. Depending on the number of phases present, nickel silvers can be divided into single-phase and two-phase alloys. The single-phase alloys contain 60 to 63 wt% Cu and 7 to 30 wt% Ni, with the remainder being zinc. Although they exhibit only fair hot working properties, they are readily cold worked. They also serve as an excellent base for plating with chromium, silver, or nickel and provide a brilliant polish and good corrosion resistance. The two-phase alloys contain, by weight-approximately 45% Cu, 45% Zn, and 10% Ni and are easily hot worked into intricate shapes. However, their cold working properties are inferior to the single-phase alloys. Other alloying additions include manganese (0.4 wt%), iron (0.1 to 5 wt%), aluminum (0.5 to 2 wt%), and occasionally lead (1 to 3 wt%) for improved machinability. Nickel silvers have good corrosion resistance, high strength, and good elastic (spring) stability. They can also be used at low temperatures, where they maintain their ductility. Nickel silvers are white in color, which makes them suitable for the manufacture of spoons, forks, and other tableware.

## 25.6 Bronzes

Although the term *bronze* has been historically associated with copper-tin alloys, other alloying elements are now used to produce bronzes, such as aluminum bronzes and silicon bronzes.

**Tin Bronzes.** Although bronzes cost more than brasses, their superior corrosion resistance and strength can justify their cost in many applications. Tin is a potent solid-solution hardener. Solid-solution alloys nominally contain 0.8 to 8 wt% Sn, usually with a small addition of phosphorus for deoxidation. These alloys provide an excellent combination of strength, formability, softening resistance, electrical conductivity, and corrosion resistance. As the tin content increases, the wear and chemical resistance increases. Tin bronzes are produced as both wrought and cast alloys. Bronzes containing more than approximately 8 wt% Sn are too brittle to be cold worked; however, bronzes with as much as 20 wt% Sn are used for castings. The wrought alloys can be easily formed by cold working. The cast alloys are used mainly for bearings; that is, the cast structure contains hard particles that resist wear embedded in a matrix of ductile phase that resists shock. Cold-worked tin bronzes have tensile strengths of 345 to 690 MPa (50 to 100 ksi), yield strengths of 276 to 586 MPa (40 to 85 ksi), and elongations of 7 to 20%, while cast bronzes, with 10 to 12 wt% Sn, have tensile strengths of 241 to 296 MPa (35 to 43 ksi), yield strengths of 110 to 152 MPa (16 to 22 ksi), and elongations of 18 to 30%. In spite of these good properties, the relatively high cost of copper and tin has led to a decline in the use of bronzes.

The rate of diffusion of copper and tin in each other is much lower than it is with copper and zinc. In addition, structural changes below approximately 400 °C (750 °F) take place in copper-tin alloys with extreme sluggishness. As a result, cast bronze cooled to room temperature under normal industrial conditions will not exhibit the structure indicated by the phase diagram shown in Fig. 25.11. The eutectoid transformation ($\delta \rightarrow \alpha + \varepsilon$) at 350 °C (660 °F) occurs only under extremely slow cooling, slower than encountered in industrial processes. Thus, the phase ε ($Cu_3Sn$) is never seen in the structure of a cast bronze containing more than 11.0 wt% Sn. Further, due to the slow diffusion rate of copper and tin atoms below 345°C (650 °F), the precipitation of ε from α alloys containing less than 11.0 wt% Sn will not occur. For practical purposes, the equilibrium diagram below 400 °C (750 °F) can be ignored, and one can assume that whatever structure has been attained at 400 °C (750 °F) will be retained at room temperature under normal cooling rates.

Similar to brasses, the α phase, being a solid solution, is tough and ductile, so α-phase alloys can be cold worked. However, the δ phase is an intermetallic compound containing the composition $Cu_{31}Sn_8$, which is a hard, brittle phase that makes the $\alpha + \delta$ bronzes rather brittle. Therefore, the δ phase must be absent for alloys that

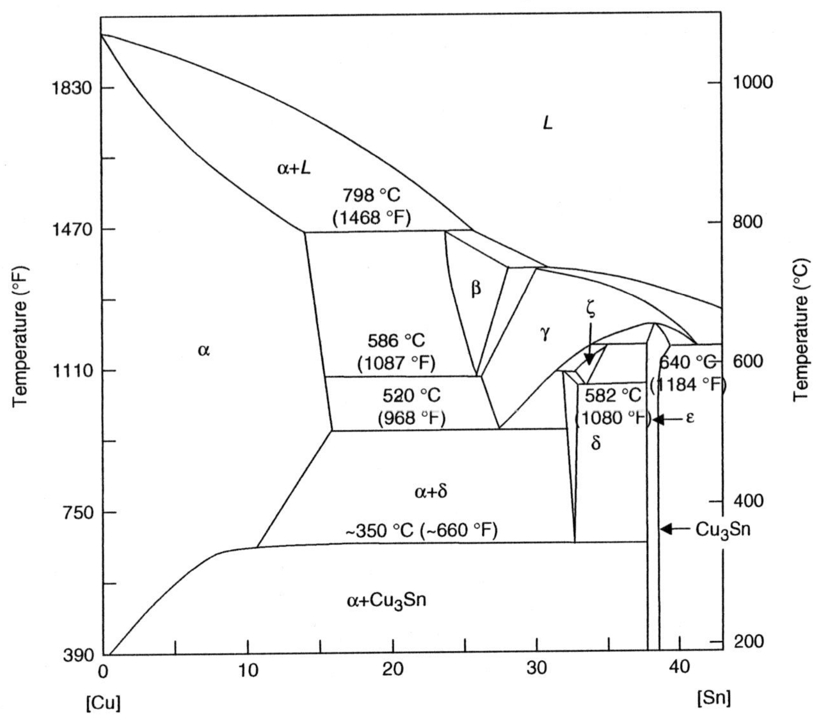


**Fig. 25.11** Copper-tin phase diagram

require cold working. Due to heavy coring, cast alloys with as little as 6.0 wt% Sn will have particles of δ at the boundaries of the cored α crystals. The inner cores of the α-phase crystals will be rich in copper, with the outer fringes correspondingly richer in tin to such an extent that the δ phase is formed. To make such an alloy amenable for cold work, the δ phase can be absorbed by prolonged homogenization annealing (e.g., 6 h at 705 °C, or 1300 °F), which promotes diffusion so that equilibrium is attained and a uniform α-phase structure is produced. Subsequent air cooling, or even furnace cooling, will be too rapid to permit precipitation of any ε phase, so the uniform α structure will be retained at room temperature. By using a homogenization heat treatment to produce a uniform α structure, it is then possible to cold work bronzes containing as much as 14 wt% Sn, although in practice, only alloys with up to 7 wt% Sn are produced in wrought form.

Most tin bronzes contain small amounts of phosphorus (0.05 wt%) as a result of deoxidization prior to casting. However, true phosphorus bronzes contain phosphorus as a deliberate alloying element, usually in the range of 0.1 to 1.0 wt%. Wrought phosphor bronzes contain up to 8.0 wt% Sn and up to 0.3 wt% P. Phosphorus not only increases the tensile strength but also improves corrosion resistance.

Bronzes containing zinc are produced in both wrought and cast product forms. The wrought alloys contain up to 3.0 wt% Sn and up to 2.5 wt% Zn and are mainly used for coinage. The replacement of tin with zinc reduces cost, since zinc is only approximately a tenth the cost of tin. Zinc is also a deoxidizer and improves casting fluidity. The best-known cast alloy is admiralty metal, which contains 10 wt% Sn and 2 wt% Zn. While it is no longer used for naval ordnance, it is still widely used where strong, corrosion-resistant castings are required.

Up to 2.0 wt% Pb is sometimes added to both bronzes and brasses to improve machinability. Larger lead additions are used for some bearings, because these bronzes permit 20% higher loading than do lead- or tin-base white metals. The thermal conductivity of these bronzes is also high, and since heat is dissipated more quickly, they can be used at high speeds. With normal lubrication, they have excellent wear resistance, but, equally important, seizure resistance is high since lead will function as a temporary lubricant should normal lubrication fail.

**Aluminum bronzes** have good strength and corrosion resistance. Aluminum provides solid-solution strengthening and work-hardening ability as well as improving corrosion resistance. Aluminum additions improve both strength and ductility (Fig. 25.12). Aluminum bronzes are similar to the brasses but have better oxidation resistance. Since they contain 3 to 14 wt% Al, a protective layer of alumina ($Al_2O_3$) contributes to their outstanding corrosion resistance. Other alloying elements include iron (grain refinement), manganese (deoxidation), nickel (corrosion resistance), arsenic (salt solution resistance), and silicon (elevated-temperature resistance). Therefore, the main industrial uses of aluminum bronzes depend on attributes such as the ability to retain strength at elevated temperatures, particularly when certain other elements are present; high resistance to oxidation at elevated temperatures; good corrosion resistance at ordinary temperatures; good wearing properties; and a pleasing color that makes some of these alloys useful for decorative purposes, for example, as a substitute for gold in imitation jewelry.

Similar to the copper-zinc alloys, when the aluminum bronzes are slowly cooled under normal cooling rates, transformations below approximately 400 °C (750 °F) are extremely sluggish, and the copper-aluminum phase diagram (Fig. 25.13) below 480 °C (900 °F) can be ignored. One can assume that the $\alpha + \gamma_2$ structure persists in alloys containing between 9.4 and 16.2 wt% Al that are slowly cooled to room temperature. Like the brasses, the

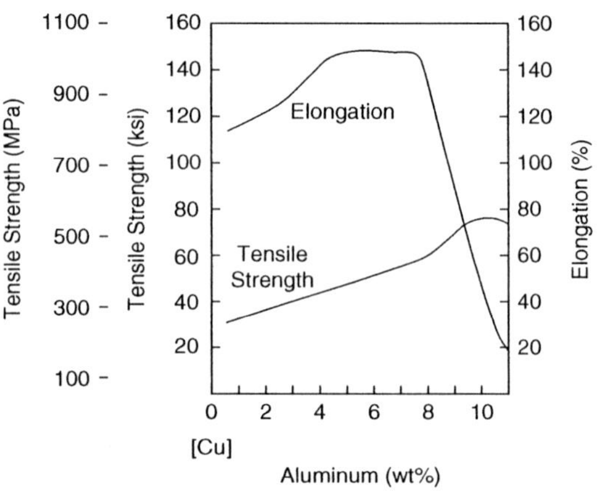


**Fig. 25.12** Effect of aluminum content on properties of aluminum bronzes. Source: Ref 7

aluminum bronzes can be divided into two main groups: the cold working and hot working wrought alloys or casting alloys. The phase diagram indicates that a solid-solution α containing up to 9.4 wt% Al at room temperature is formed. Like the other α solid solutions based on copper, it is quite ductile. With more than 9.4 wt% Al, the very hard and brittle $\gamma_2$ phase is formed. The $\gamma_2$ phase is an intermetallic compound with the formula $Cu_9Al_4$ and results in overall brittleness.

The α-phase aluminum bronze alloys are normally hardened by cold working. They contain between 4.0 and 7.0 wt% Al and occasionally up to 4.0 wt% Ni for improved corrosion resistance. Since the composition of these α-phase alloys can be adjusted to give a color similar to that of 18 karat gold, they are often used for decorative articles. However, many products originally made from aluminum bronze are now produced from lower-cost gold-colored anodized aluminum. The $\alpha + \gamma_2$ hot working and casting alloys contain 7.0 to 12.0 wt% Al, with other elements such as iron, nickel, or manganese. The hot working alloys contain from 7.0 to 10.0 wt% Al, with up to 5.0 wt% each of iron and nickel. The aluminum bronzes usually contain 1 to 5 wt% Fe, which provides elemental dispersions to promote dispersion strengthening and grain size control. These alloys are used for chemical engineering applications, especially for components exposed to high temperatures and where corrosion resistance is required.

Modifications of some aluminum bronze solid-solution alloys can be achieved by adding elements that react to form dispersions of intermetallic particles, which have a grain-refining and strengthening effect. As a result, higher strengths can be produced with less cold working, resulting in better formability at higher strength levels. Because these modifications do not require large amounts of costly elements, the gains are economical. For example, alloy C63800 (95Cu-2.8Al-1.8Si-0.4Co) is a high-strength alloy with an annealed tensile strength of 565 MPa (82 ksi) and tensile strengths of 662 to 896 MPa (96 to 130 ksi) for the standard rolled tempers. The cobalt addition provides intermetallic particles, resulting in dispersion strengthening.

**Silicon bronzes** can be used as lower-cost alternatives to tin bronzes. Silicon bronzes, containing 1 to 4 wt% Si, have good strength

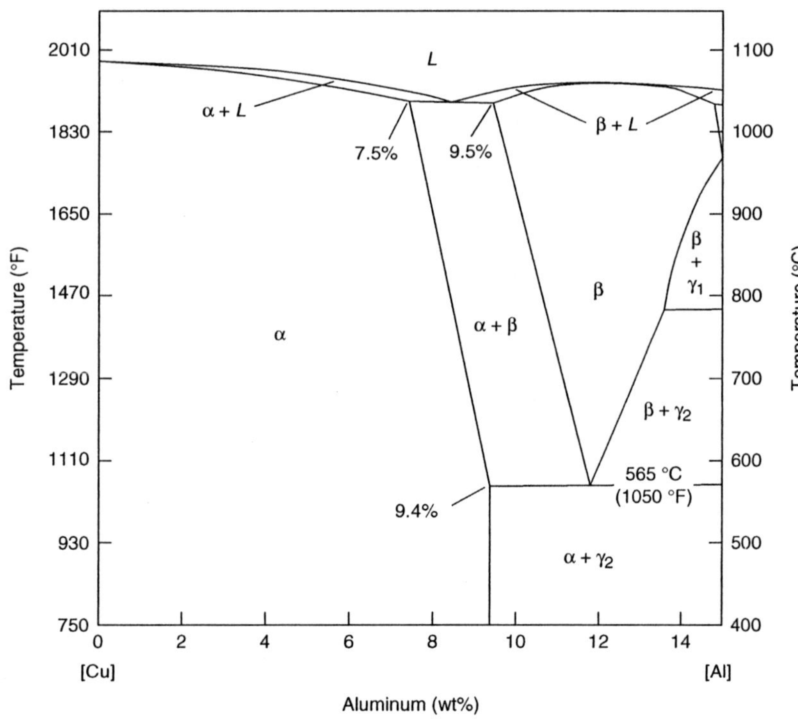


**Fig. 25.13** Portion of copper-aluminum phase diagram

and ductility. Normally, they also contain small additions of manganese, iron, and zinc. Silicon is a deoxidizer and increases the strength and corrosion resistance, while manganese improves high-temperature strength and wear resistance. These alloys have excellent corrosion resistance combined with high strength and toughness. They can easily be hot or cold worked, machined, and welded. They are often used in chemical processing plants. Two important silicon bronzes are Cu-3Si-1Mn and Cu-2Si-0.5Mn. Silicon is a deoxidizer and increases strength and corrosion resistance, while manganese improves high-temperature strength and wear resistance.

## 25.7 Copper-Nickel Alloys

Copper-nickel or cupronickel alloys usually contain 2 to 45 wt% Ni. Since copper and nickel form a complete series of solid solutions, many of their properties change continually with composition. As shown in Fig. 25.14, the addition of nickel to copper significantly improves strength, with the yield strength, tensile strength, and fatigue strength reaching a broad maximum at approximately 70 wt% Ni. Even a small amount of nickel (1.5%) doubles the impact toughness. The color of copper changes with increasing nickel content, and with approximately 15 wt% Ni, the copper-nickel alloys are nearly silver-white. Copper-nickel alloys are annealed at 540 to 690 °C (1000 to 1275 °F) and should not be heated above 800 °C (1475 °F). They are very sensitive to oxygen, lead, sulfur, and carbon, which embrittles them. Manganese is used as a deoxidizer, and iron increases corrosion resistance. They can be hot and cold worked and have excellent machinability. A Cu-25% Ni alloy is used for the clad coinage of the U.S. dime, quarter, and half-dollar. The coins contain a copper core that is clad on the surfaces with the copper-nickel alloy (Fig. 25.15). The same alloy is used for the U.S. nickel.

Copper-nickel alloys are also used because of their special electrical properties and as corrosion-resistant materials. The electrical resistance of copper increases with increasing nickel additions, reaching a maximum at approximately 50 wt% Ni (Fig. 25.14). Therefore, since the resistivity can be varied over rather wide ranges, copper-nickel alloys are used for electrical resistors. The alloy constantan (45Ni-55Cu) is widely used for thermocouples. The corrosion resistance of copper is improved considerably by additions of nickel. With alloys containing 5 to 30 wt% Ni, passive layers are

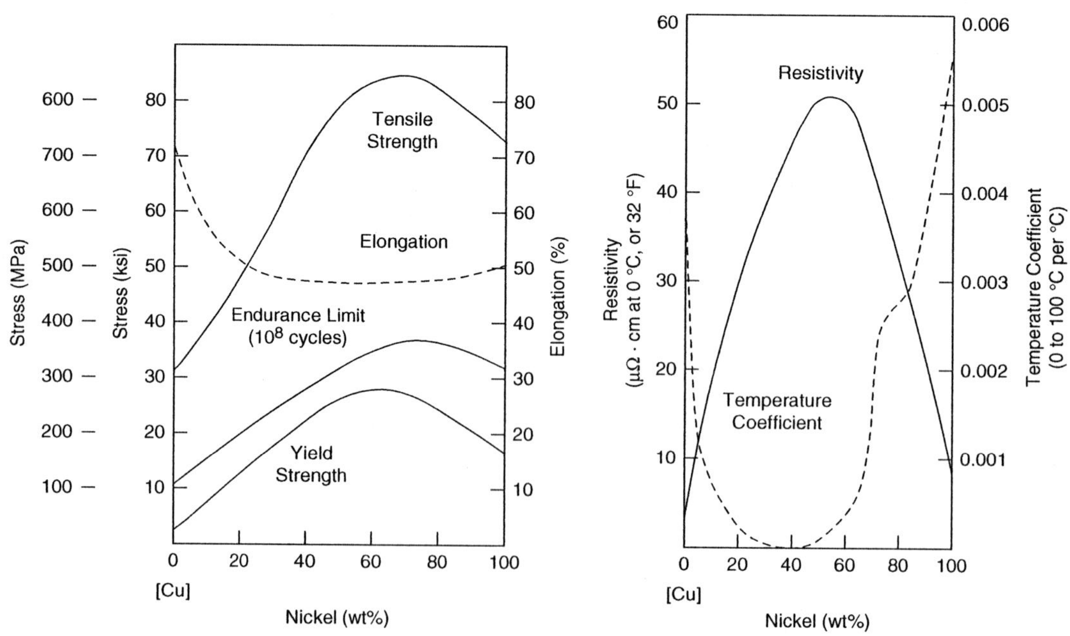


**Fig. 25.14** Effect of nickel on properties of copper. Source: Ref 8

formed in the presence of oxygen. Iron and manganese are also added for resistance to seawater.

## 25.8 Beryllium-Copper

Precipitation hardening can produce very high strengths but is limited to only a few copper alloys. Wrought beryllium-coppers can be precipitation hardened to the highest strength levels attainable in copper-base alloys. Even higher strengths can be obtained by work hardening the metal before or after aging. The biggest drawbacks to these alloys are the expense and toxicity of beryllium.

There are two commercially significant alloy families with two different ranges of beryllium. The red alloys contain 0.2 to 0.7 wt% Be along with additions of nickel or cobalt totaling 1.4 to 2.7 wt%, depending on the alloy. These low-beryllium alloys have relatively high conductivity (e.g., 50% IACS) and retain the pink luster of other low-alloy coppers. The red alloys have yield strengths ranging from approximately 172 to 552 MPa (25 to 80 ksi) without heat treatment to greater than 896 MPa (130 ksi) after precipitation hardening, depending on the degree of cold work. The more highly beryllium systems contain from 1.6 to 2.0 wt% Be and approximately 0.25 wt% Co. These alloys are frequently called the gold alloys because of the shiny luster imparted by the substantial amount of beryllium present. The gold alloys are the higher-strength beryllium-coppers; they can attain yield strengths ranging from approximately 207 to 690 MPa (30 to 100 ksi) in the solution-treated condition to above 1379 MPa (200 ksi) after aging. Due to the higher beryllium content, the conductivity of the gold alloys is lower than that of the red alloys. However, conductivity ranging from approximately 20% to greater than 30% IACS is obtained in wrought products, depending on the amount of cold work and the heat treatment schedule.

Beryllium-copper alloys are solution heat treated in the range of 760 to 955 °C (1400 to 1750 °F) and then aged at 260 to 565 °C (500 to 1050 °F) to produce a beryllium-rich, coherent precipitate. The specific temperatures are chosen for the particular alloy and desired property combination, as shown by the ranges in Fig. 25.16. The precipitation sequence during aging consists of the formation of solute-rich Guinier-Preston zones, followed in sequence by coherent platelets of the metastable intermediate phases $\gamma'$ and $\gamma''$. Overaging is marked by the appearance of the B2 ordered equilibrium $\gamma$ (BeCu) phase as particles within grains and along grain boundaries. Cobalt and nickel additions form dispersoids of equilibrium (Cu, Co, or Ni)Be that restrict grain growth during solution treating at elevated temperatures. A cold working step following solution treating is often used to increase the age-hardening response. For example, alloy C17200 (97.8Cu-1.8Be-0.4Co) can be processed to reach high tensile strengths by solution heat treating

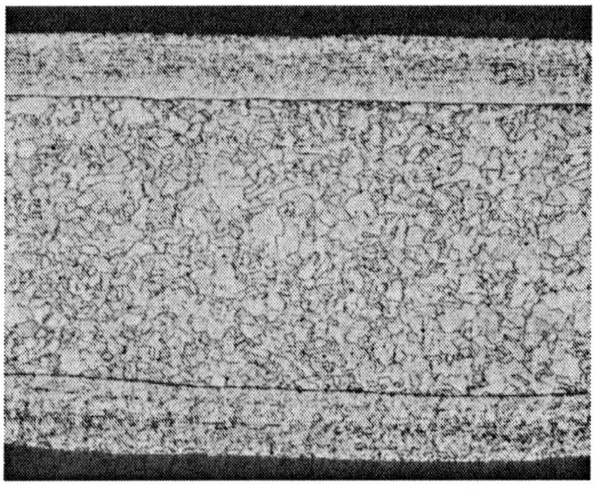

Fig. 25.15 Copper-nickel clad coinage alloy. Original magnification: 50×. Source: Ref 9

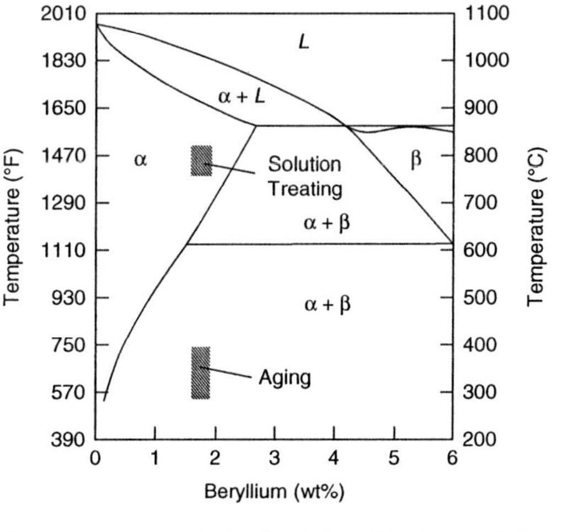


Fig. 25.16 Precipitation hardening of high-strength beryllium-copper alloys

(469 MPa, or 68 ksi), then cold rolling to the hard temper (758 MPa, or 110 ksi), and finally aging (1413 MPa, or 205 ksi). The microstructures of a solution-treated and quenched alloy and a fully aged alloy are shown in Fig. 25.17. While they are commercially available in the solution-treated condition, they are commonly provided in the mill-hardened temper with the optimal strength/ductility/conductivity combination suitable for the application.

Other precipitation-hardenable alloys include C15000; C15100 (zirconium-copper); C18200, C18400, and C18500 (chromium-coppers); C19000 and C19100 (copper-nickel-phosphorus alloys); and C64700 and C70250 (copper-nickel-silicon alloys). These precipitation-hardened alloys have somewhat lower strengths but better electrical conductivities than the beryllium-coppers.

## 25.9 Copper Casting Alloys

Copper casting alloys are available as sand, continuous, centrifugal, permanent mold, and some die castings. They are generally similar to their wrought counterparts but have their own unique composition/property characteristics. For example, the ability to add up to 25 wt% Pb, which would not be possible for a wrought alloy, provides compositions in which dispersions of lead particles help prevent galling in bearing applications. Copper casting alloys are used for their corrosion resistance and their high thermal and electrical conductivity. Since the liquidus and solidus freezing range is narrow (Fig. 25.6), cast brass alloys display only minimal segregation on freezing, while the freezing range is much wider for tin bronzes (Fig. 25.11), and they display much more segregation on freezing.

The most common brass alloys are the general-purpose cast red brass (85Cu-5Zn-5Sn-5Pb), used for valves and plumbing hardware, and cast yellow brass (60Cu-38Zn-1Sn-1Pb), which also is widely used for cast plumbing system components. A few weight percent of nickel, tin, and manganese are also used in certain alloys. Cast brasses are brittle and cannot be deformed; however, they have high strengths.

Cast phosphor bronzes contain up to 13.0 wt% Sn and up to 1.0 wt% P, and are used mainly for bearings and other components where a low friction coefficient is desirable, coupled with high strength and toughness. Phosphorus is usually present in cast alloys as copper phosphide ($Cu_3P$), which is a hard compound that forms a ternary eutectoid with the α and δ phases. The presence of a hard phase in a soft matrix makes these alloys good bearing materials with a low coefficient of friction. Arsenic and phosphorus also improve corrosion resistance, with phosphorus also improving the fluidity of casting alloys.

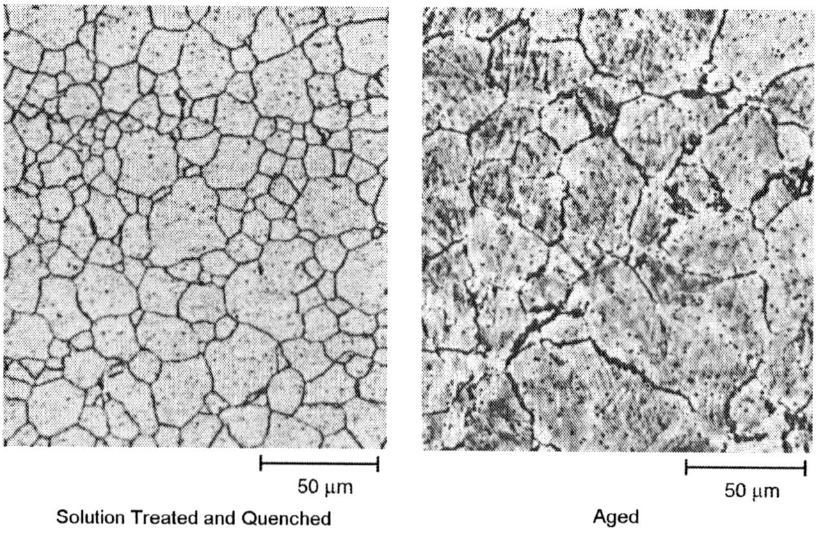


**Fig. 25.17** Microstructure of beryllium-copper alloy. Original magnification: 300×. Source: Ref 4

Cast manganese and aluminum bronzes have higher tensile strengths than cast brasses or tin bronzes, in the range of 448 to 896 MPa (65 to 130 ksi). As with wrought alloys, cast aluminum bronze alloys commonly contain an iron addition (0.8 to 5.0 wt%) to provide iron-rich particles for grain refinement and added strength. In addition, at aluminum levels in the range of 9.5 to 10.5 wt%, or 8.0 to 9.5 wt% Al along with nickel or manganese additions, the alloys are heat treatable for added strength. Depending on section thickness and the cooling rate of the casting, as well as the alloy composition and heat treatments, the microstructures can be rather complex. The aluminum bronzes can be completely or partially annealed in the $\beta$ field and quenched to form $\beta$ martensite with $\alpha$ needles. Aging tempers the martensite by the precipitation of fine $\alpha$ needles. One of the aluminum bronze alloys (79.5Cu-10.5Al-5Ni-5Pb) is used for its combination of high strength and good corrosion resistance. Through heat treatment, the intermetallic $\kappa$ phase, with its complex composition (Fe,Ni,Cu)Al and the CsCl-type crystalline structure, provides strengthening in any of its morphologies, that is, as globular particles, fine precipitates, or as a component of cellular eutectoid colonies.

Cast manganese and aluminum alloys are widely used in marine engineering for pump rods, valve fittings, propellers, propeller shafts, and bolts. They are also used for valve seats and spark plug bodies in internal combustion engines, for brush holders in generators, for heavy-duty bearings, for gear wheels, for pinions and worm wheels, and in the manufacture of nonsparking tools such as spanners, wrenches, shovels, and hammers in potentially dangerous gas, paint, oil, and explosives industries. However, they are softer and therefore inferior to beryllium-copper for some applications.

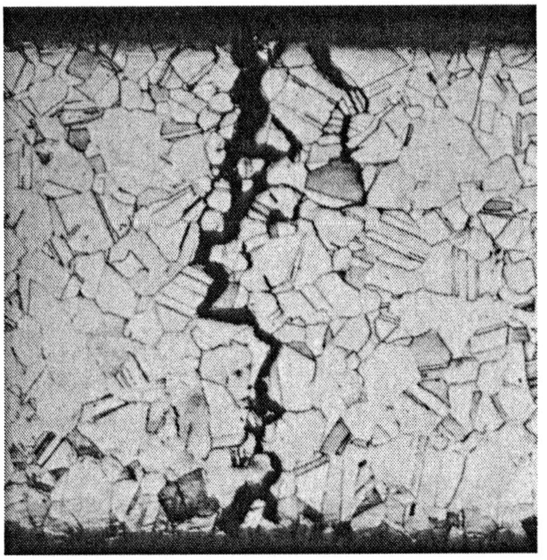

**Fig. 25.18** Stress-corrosion cracking of brass. Original magnification: 100×. Source: Ref 9

## 25.10 Corrosion

Pure copper resists corrosive attack quite well in most environments. However, some copper alloys are susceptible to hydrogen embrittlement and stress-corrosion cracking. Tough pitch coppers, which contain cuprous oxide, are susceptible to hydrogen embrittlement when exposed to a reducing atmosphere. As a consequence, most copper alloys are deoxidized and thus are not susceptible to hydrogen embrittlement.

Stress-corrosion cracking most commonly occurs in brasses that are exposed to ammonia or amines, exhibiting branched intergranular cracking (Fig. 25.18). Brasses containing more than 15 wt% Zn are the most susceptible. Copper and most copper alloys that either do not contain zinc or have a low zinc content generally are not susceptible to stress-corrosion cracking. To relieve residual stresses, brass alloys can be stress relieved or annealed after forming.

Dealloying or dezincification is another form of corrosion that affects zinc-containing copper alloys. Copper-zinc alloys containing more than 15 wt% Zn are susceptible to dezincification, where selective removal of zinc leaves a relatively porous and weak layer of copper and copper oxide. During dezincification, both $Cu^+$ and $Zn^{2+}$ ions go into solution. Copper, being more noble than zinc, is redeposited as a porous, spongy mass, as shown in Fig. 25.19. Corrosion continues beneath the primary corrosion layer, resulting in gradual replacement of sound brass by weak, porous copper. Dezincification eventually penetrates the metal, weakening it structurally and allowing liquids or gases to leak through the porous mass in the remaining structure. Additions of small amounts of phosphorus, arsenic, or tin are used to inhibit dezincification.

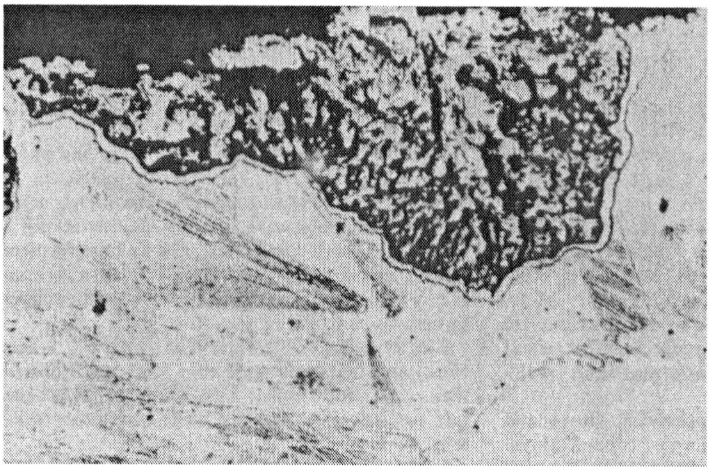

**Fig. 25.19** Plug-type dezincification in α brass. Original magnification: 160×. Source: Ref 10

## ACKNOWLEDGMENTS

Sections of this chapter were adapted from "Copper and Copper Alloys: Effects of Composition, Processing, and Structure on Properties of Nonferrous Alloys" by R.N. Caron and J.T. Staley in *Materials Selection and Design,* Volume 20, *ASM Handbook,* ASM International, 1997.

## REFERENCES

1. R.N. Caron and J.T. Staley, Copper and Copper Alloys: Effects of Composition, Processing, and Structure on Properties of Nonferrous Alloys, *Materials Selection and Design,* Vol 20, *ASM Handbook,* ASM International, 1997
2. E.A. Brandes, *Smithells Metal Reference Book,* 6th ed., Butterworths, 1983
3. A.G. Guy, *Elements of Physical Metallurgy,* 2nd ed., Addison-Wesley Publishing Company, 1959
4. R.N. Caron, R.G. Barth, and D.E. Tyler, Metallography and Microstructures of Copper and Its Alloys, *Metallography and Microstructures,* Vol 9, *ASM Handbook,* ASM International, 2004
5. J.H. Mendenall, "Understanding Copper Alloys," Olin Brass Corporation, 1977
6. Copper and Copper Alloys, Introduction and Overview, *Metals Handbook Desk Edition,* 2nd ed., ASM International, 1998
7. R.A. Higgins, *Engineering Metallurgy—Applied Physical Metallurgy,* 6th ed., Arnold, 1993
8. F.T. Sisco, *Modern Metallurgy for Engineers,* 2nd ed., Pitman Publishing Co.
9. Microstructure of Copper and Copper Alloys, *Atlas of Microstructures of Industrial Alloys,* Vol 7, *Metals Handbook,* 8th ed., American Society for Metals, 1972
10. Corrosion Characteristics of Copper and Copper Alloys, *Metals Handbook Desk Edition,* 2nd ed., ASM International, 1998

## SELECTED REFERENCES

- W. Heller, Copper-Based Alloys, *Structure and Properties of Nonferrous Alloys,* Vol 8, *Materials Science and Technology,* VCH, 1996
- P.J. Macken and A.A. Smith, "The Aluminum Bronzes," United Kingdom Copper Development Association, 1966
- D.E. Tyler and W.T. Black, Introduction to Copper and Copper Alloys, *Properties and Selection: Nonferrous Alloys and Special-Purpose Materials,* Vol 2, *ASM Handbook,* ASM International, 1990

# CHAPTER 26

# Aluminum

ALUMINUM has many outstanding attributes that lead to a wide range of applications, including:

- Good corrosion and oxidation resistance
- High electrical and thermal conductivities
- Low density
- High reflectivity
- High ductility and reasonably high strength
- Relatively low cost

Aluminum is a consumer metal of great importance. Aluminum and its alloys are used for foil, beverage cans, cooking and food-processing utensils, architectural and electrical applications, and structures for boats, aircraft, and other transportation vehicles. As a result of a naturally occurring tenacious surface oxide film ($Al_2O_3$), a great number of aluminum alloys have exceptional corrosion resistance in many atmospheric and chemical environments. Its corrosion and oxidation resistance is especially important in architectural and transportation applications. On an equal weight and cost basis, aluminum is a better electrical conductor than copper. Its high thermal conductivity leads to applications such as radiators and cooking utensils. Its low density is important for hand tools and all forms of transportation, especially aircraft. Wrought aluminum alloys display a good combination of strength and ductility. Aluminum alloys are among the easiest of all metals to form and machine. The precipitation-hardening alloys can be formed in a relatively soft state and then heat treated to much higher strength levels after forming operations are complete. In addition, aluminum and its alloys are not toxic and are among the easiest to recycle of any of the structural materials.

## 26.1 Aluminum Metallurgy

Aluminum is a lightweight metal with a density of 2.70 g/cm$^3$ (0.1 lb/in.$^3$) and a moderately low melting point of 655 °C (1215 °F). Since it has a face-centered cubic crystalline structure, the formability of aluminum and aluminum alloys is good. The good formability is further aided by its rather low work-hardening rate. Aluminum alloys are classified as either wrought or cast alloys. Some of the wrought alloys are hardened by work hardening, while others are precipitation hardenable. Likewise, some of the cast alloys can be hardened by precipitation hardening, while others cannot. Some of the important properties of each of the wrought alloy series are given in Table 26.1.

Microstructural control is extremely important in the production and processing of aluminum alloys. Important microstructural features include constituent particles and dispersoids. Constituent particles are coarse intermetallic compounds that form by eutectic decomposition during ingot solidification. Some are soluble, while others are virtually insoluble. The insoluble compounds usually contain the impurity elements iron or silicon and form compounds such as $Al_6(Fe,Mn)$, $Al_2Fe$, $Al_7FeCu_2$, and $\alpha Al(Fe,Mn,Si)$. The soluble compounds are equilibrium intermetallic compounds of one of the major alloying elements, such as $CuAl_2$ or $SiMg_2$. One of the major reasons for ingot homogenization before hot working is to dissolve these soluble compounds. During hot working, the large insoluble compounds are broken up and aligned as stringers in the working direction. Dispersoids are smaller submicron particles (typically 0.05 to 0.5 μm) that form during ingot homogenization by solid-state precipitation from elements that have only limited solubility and that diffuse slowly. Once they form, they resist dissolution and/or coarsening. They usually consist of one of the transition elements; examples are $Al_{20}Cu_2Mn_3$, $Al_{12}CrMg_2$, and $Al_3Zr$. Dispersoids are useful in retarding recrystallization and grain growth. Other microstructural features that can affect properties include oxide inclusions, porosity,

**Table 26.1 Major attributes of wrought aluminum alloys**

***1xxx*: Pure Al.** The major characteristics of the *1xxx* series are:

- Strain hardenable
- Exceptionally high formability, corrosion resistance, and electrical conductivity
- Typical ultimate tensile strength range: 69–186 MPa (10–27 ksi)
- Readily joined by welding, brazing, soldering

***2xxx*: Al-Cu Alloys.** The major characteristics of the *2xxx* series are:

- Heat treatable
- High strength at room and elevated temperatures
- Typical ultimate tensile strength range: 186–428 MPa (27–62 ksi)
- Usually joined mechanically, but some alloys are weldable
- Not as corrosion resistant as other alloys

***3xxx*: Al-Mn Alloys.** The major characteristics of the *3xxx* series are:

- High formability and corrosion resistance; medium strength
- Typical ultimate tensile strength range: 110–283 MPa (16–41 ksi)
- Readily joined by all commercial procedures
- Hardened by strain hardening

***4xxx*: Al-Si Alloys.** The major characteristics of the *4xxx* series are:

- Some heat treatable
- Good flow characteristics; medium strength
- Typical ultimate tensile strength range: 172–379 MPa (25–55 ksi)
- Easily joined, especially by brazing and soldering

***5xxx*: Al-Mg Alloys.** The major characteristics of the *5xxx* series are:

- Strain hardenable
- Excellent corrosion resistance, toughness, weldability; moderate strength
- Building and construction, automotive, cryogenic, marine applications
- Typical ultimate tensile strength range: 124–352 MPa (18–58 ksi)

***6xxx*: Al-Mg-Si Alloys.** The major characteristics of the *6xxx* series are:

- Heat treatable
- High corrosion resistance, excellent extrudability; moderate strength
- Typical ultimate tensile strength range: 124–400 MPa (18–58 ksi)
- Readily welded by gas metal arc welding and gas tungsten arc welding methods
- Outstanding extrudability

***7xxx*: Al-Zn Alloys.** The major characteristics of the *7xxx* series are:

- Heat treatable
- Very high strength; special high-toughness versions
- Typical ultimate tensile strength range: 221–607 MPa (32–88 ksi)
- Mechanically joined

***8xxx*: Alloys** with Al/other elements (not covered by other series). The major characteristics of the *8xxx* series are:

- Heat treatable
- High conductivity, strength, hardness
- Typical ultimate tensile strength range: 117–414 MPa (17–60 ksi)
- Common alloying elements include Fe, Ni, and Li

Source: Ref 1

grain size and shape, and crystallographic textures that can lead to anistropic properties.

Strengthening of non-heat-treatable alloys is a result of a combination of solid-solution strengthening, second-phase constituents, dispersoid precipitates, and work hardening. The alloys normally hardened by work or strain hardening include the commercially pure aluminums (1*xxx*), the aluminum-manganese alloys (3*xxx*), some of the aluminum-silicon alloys (4*xxx*), and the aluminum-magnesium alloys (5*xxx*). These can be work hardened to various strength levels with a concurrent reduction in ductility. Since these alloys will undergo recovery at moderate temperatures, they are used mainly for lower-temperature applications.

The highest strength levels are attained by the precipitation-hardenable alloys, which include the aluminum-copper alloys (2*xxx*), the aluminum-magnesium + silicon alloys (6*xxx*), the aluminum-zinc alloys (7*xxx*), and the aluminum-lithium alloys of the 8*xxx* series. For the cast alloys, this includes the aluminum-copper alloys (2*xx.x*), some of the aluminum-silicon + copper and/or magnesium alloys (3*xx.x*), and the aluminum-zinc alloys (7*xx.x*). One rather disappointing property of high-strength aluminum alloys is their fatigue performance. Increases in static tensile properties have not been accompanied by proportionate improvements in fatigue properties (Fig. 26.1). The precipitation-hardened alloys exhibit only minimal fatigue improvement because of two factors: (1) cracks initiating at precipitate-free zones adjacent to grain boundaries, and (2) the re-solution of precipitate particles when they are cut by dislocations. The cut precipitate particles become smaller than the critical size for thermodynamic stability and redissolve.

## 26.2 Aluminum Alloy Designation

A four-digit numerical designation system, developed by the Aluminum Association, is used to designate wrought aluminum and aluminum alloys. As shown in Table 26.2, the first digit defines the major alloying element of the series. The 1*xxx* series is handled a little differently than the 2*xxx* through 8*xxx* series. In the 1*xxx* series of commercially pure aluminums, the last two of the four digits in the designation indicate the minimum aluminum percentage. These digits are the same as the two digits to the right of the decimal point in the minimum aluminum percentage when expressed to the nearest 0.01%. When the second digit is a number other than zero, it indicates that special control has been used to control one or more of the naturally occurring impurities.

In the 2*xxx* through 8*xxx* alloy series, the second digit in the designation indicates the alloy modification. When the second digit is zero, it indicates the original alloy. The numbers 1 through 9, assigned consecutively, indicate modifications of the original alloy. Explicit rules, have been established for determining whether a proposed composition is merely a modification of a previously registered alloy or if it is an entirely new alloy. The last two of the four digits in the 2*xxx* through 8*xxx* series have no special significance but serve only to identify the different alloys in the series.

For cast alloys (Table 26.3), a four digit numerical designation incorporating a decimal point is used. The first digit indicates the major alloying element, while the second and third digits identify the specific alloy. For the 1*xx.x* series, the second two digits indicate purity. The last digit, which is separated from the others by a

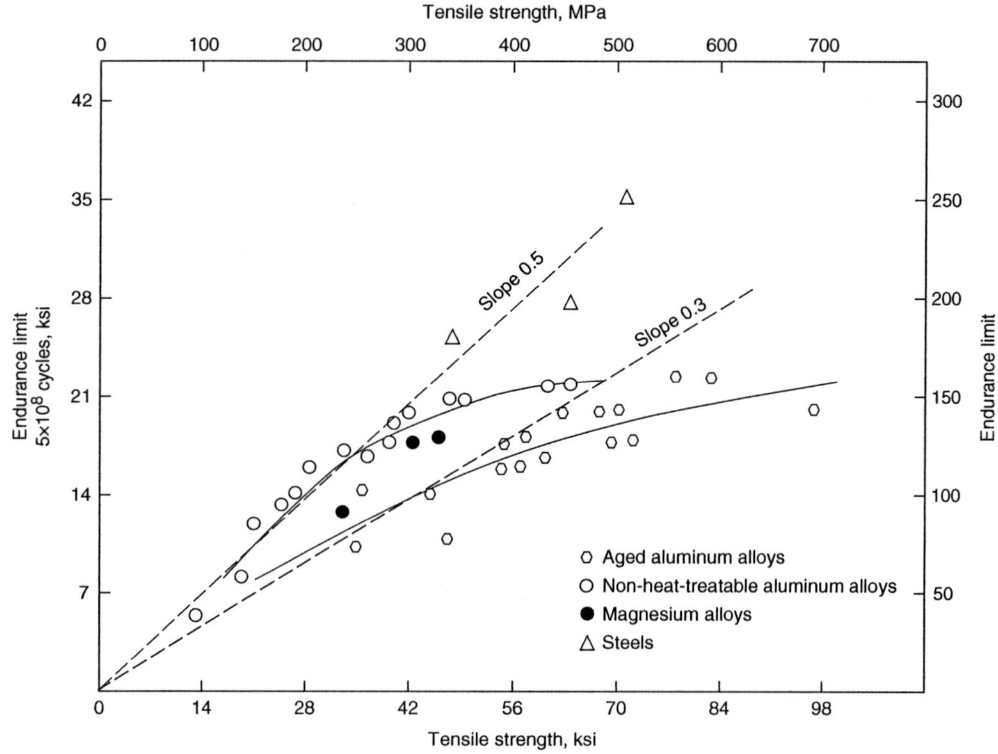


**Fig. 26.1** Fatigue strength comparison for aluminum. Source: Ref 2

**Table 26.2 Designations for aluminum wrought alloys**

| Series | Aluminum content or main alloying element |
|---|---|
| 1*xxx* | 99.00% minimum |
| 2*xxx* | Copper |
| 3*xxx* | Manganese |
| 4*xxx* | Silicon |
| 5*xxx* | Magnesium |
| 6*xxx* | Magnesium and silicon |
| 7*xxx* | Zinc |
| 8*xxx* | Others |
| 9*xxx* | Unused |

**Table 26.3 Designations for aluminum casting alloys**

| Series | Aluminum content or main alloying element |
|---|---|
| 1*xx*.0 | 99.00% minimum |
| 2*xx*.0 | Copper |
| 3*xx*.0 | Silicon with copper and/or magnesium |
| 4*xx*.0 | Silicon |
| 5*xx*.0 | Magnesium |
| 6*xx*.0 | Unused |
| 7*xx*.0 | Zinc |
| 8*xx*.0 | Tin |
| 9*xx*.0 | Other |

decimal point, indicates the product form, whether casting or ingot; the zero after a period identifies the alloy as a cast product. If the period is followed by the number 1, it indicates an ingot composition that would be supplied to a casting house. A modification of an original alloy, or of the impurity limits for unalloyed aluminum, is indicated by a serial letter preceding the numerical designation. The serial letters are assigned in alphabetical sequence starting with "A" but omitting "I," "O," "Q," and "X," the "X" being reserved for experimental alloys. For example, the designation A357.0 indicates a higher purity level than the original alloy 357.0.

The temper designations for aluminum alloys are shown in Table 26.4. The basic temper designations are as follows.

**As-fabricated (F)** applies to products in which there is no special control over the thermal conditions or work hardening (if applied), and there are no mechanical property limits.

**Annealed (O)** applies to wrought and cast products that are annealed to produce the lowest strength condition. Cast products are often annealed to improve ductility and dimensional stability.

**Work hardened (H)** applies only to wrought products that have been strengthened by work or strain hardening. A subsequent thermal treatment is sometimes used to produce some reduction in strength. The work-hardened condition H is always followed by one digit and sometimes two. The first digit signifies the specific work-hardening process, and the second digit gives the amount of residual hardening.

**Solution heat treated (W)** applies to products that are solution heat treated. The W condition is unstable since alloys will slowly age at room temperature. Wrought heat treatable alloys are often formed in the W condition since their formability is almost as good as the annealed condition. In this case, they are refrigerated after solution heat treating but before forming to retard natural aging. The refrigeration temperature must be in the range of −45 to −75 °C (−50 to −100 °F).

**Solution heated treated and aged (T)** applies to products that have been solution heat treated and either aged at room temperature (naturally aged) or aged at elevated temperature (artificially aged). The specific aging treatment is designated with a "T" followed by a number (1 through 10) for the specific aging treatment. Wrought products are also stress relieved after solution treating but before aging, designated as T*x*5*x*. This is conducted by stretching in tension (T*x*51), compression (T*x*52), or a combination of tension and compression (T*x*54). The small amount of stress relief (1 to 5%) reduces warpage during machining and improves the fatigue and stress-corrosion resistance. If the product is an extrusion, a third digit may be used. The number "1" for extruded products indicates the product was straightened by stretching, while

**Table 26.4 Temper designations for aluminum alloys**

| Suffix letter "F," "O," "H," "T," or "W" indicates basic treatment condition | First suffix digit indicates secondary treatment used to influence properties | Second suffix digit for condition H only indicates residual hardening |
|---|---|---|
| F—As-fabricated | | |
| O—Annealed-wrought products only | | |
| H—Cold worked, strain hardened | | |
| | 1—Cold worked only | 2—1/4 hard |
| | 2—Cold worked and partially annealed | 4—1/2 hard |
| | 3—Cold worked and stabilized | 6—3/4 hard |
| | | 8—Hard |
| | | 9—Extra hard |
| W—Solution heat treated | | |
| T—Heat treated, stable | | |
| T1—Cooled from an elevated-temperature shaping operation + natural aged | | |
| T2—Cooled from an elevated-temperature shaping operation + cold worked + natural aged | | |
| T3—Solution treated + cold worked + natural aged | | |
| T4—Solution treated + natural aged | | |
| T5—Cooled from an elevated-temperature shaping operation + artificial aged | | |
| T6—Solution treated + artificial aged | | |
| T7—Solution treated + overaged | | |
| T8—Solution treated + cold worked + artificial aged | | |
| T9—Solution treated + artificial aged + cold worked | | |
| T10—Cooled from an elevated-temperature shaping operation + cold worked + artificial aged | | |

Source: Ref 3

the number "0" indicates that it was not mechanically straightened.

## 26.3 Aluminum Alloys

The principal alloying elements in wrought aluminum alloys include copper, manganese, magnesium, silicon, and zinc. Alloys containing copper, magnesium + silicon, and zinc are precipitation hardenable to fairly high strength levels, while those containing manganese or magnesium are hardened primarily by cold working.

### *26.3.1 Wrought Non-Heat-Treatable Alloys*

The wrought non-heat-treatable alloys include the commercially pure aluminum alloys (1*xxx*), the aluminum-manganese alloys (3*xxx*), the aluminum-silicon alloys (4*xxx*), and the aluminum-magnesium alloys (5*xxx*). These alloys cannot be hardened by heat treatment and are therefore hardened by a combination of solid-solution strengthening (Fig. 26.2) and cold working (Fig. 26.3). The chemical compositions of a number of wrought non-heat-treatable alloys are shown in Table 26.5, and representative mechanical properties are given in Table 26.6.

**Commercially Pure Aluminum Alloys (1*xxx*).** The 1*xxx* alloys normally have tensile strengths in the range of 69 to 186 MPa (10 to 27 ksi). The 1*xxx* series of aluminum alloys include both the superpurity grades (99.99%) and the commercially pure grades containing up to 1 wt% impurities or minor additions. The last two digits of the alloy number denote the two digits to the right of the decimal point of the percentage of the material that is aluminum. For example, 1060 denotes an alloy that is 99.60% Al. The more prevalent commercially pure grades (99.0 wt% minimum aluminum) are available in most product forms and are used for applications such as electrical conductors, chemical processing equipment, aluminum foil, cooking utensils, and architectural products. Since these alloys are essentially free of alloying additions, they exhibit excellent corrosion resistance to atmospheric conditions. The most

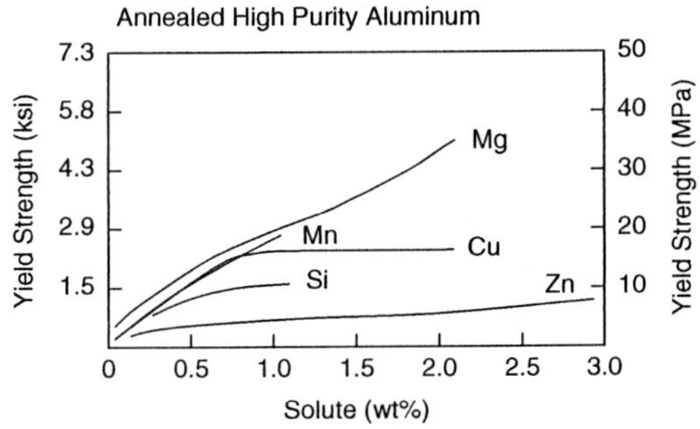


**Fig. 26.2** Solid-solution strengthening of aluminum. Source: Ref 4

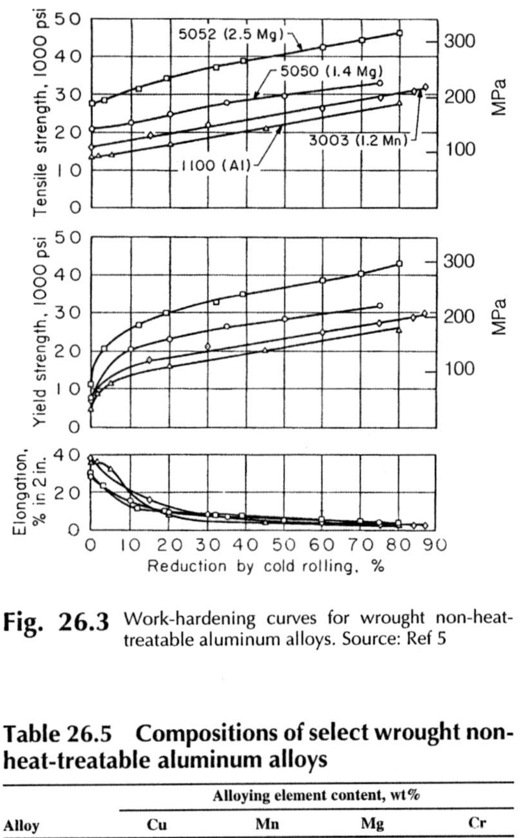


**Fig. 26.3** Work-hardening curves for wrought non-heat-treatable aluminum alloys. Source: Ref 5

**Table 26.5 Compositions of select wrought non-heat-treatable aluminum alloys**

| Alloy | Alloying element content, wt% | | | |
|---|---|---|---|---|
| | Cu | Mn | Mg | Cr |
| 3003 | 0.12 | 1.2 | ... | ... |
| 3004 | ... | 1.2 | 1.0 | ... |
| 3005 | ... | 1.2 | 0.4 | ... |
| 3105 | ... | 0.6 | 0.5 | ... |
| 5005 | ... | ... | 0.8 | ... |
| 5050 | ... | ... | 1.4 | ... |
| 5052 | ... | ... | 2.5 | 0.25 |
| 5252 | ... | ... | 2.5 | ... |
| 5154 | ... | ... | 3.5 | 0.25 |
| 5454 | ... | 0.8 | 2.7 | 0.12 |
| 5056 | ... | 0.12 | 5.0 | 0.12 |
| 5456 | ... | 0.8 | 5.1 | 0.12 |
| 5182 | ... | 0.35 | 4.5 | ... |
| 5083 | ... | 0.7 | 4.4 | 0.15 |
| 5086 | ... | 0.45 | 4.0 | 0.15 |

All contain iron and silicon as impurities. Source: Ref 6

popular 1*xxx* alloy is alloy 1100; it has a tensile strength of 90 MPa (13 ksi), which can be increased to 165 MPa (24 ksi) by work hardening. The 1*xxx* series are also used for electrical applications, primarily alloy 1350, which has relatively tight controls on impurities that would adversely affect electrical conductivity.

**Aluminum-Manganese Alloys (3*xxx*).** The 3*xxx* alloys are often used where higher strength levels are required along with good ductility and excellent corrosion resistance. The aluminum-manganese alloys contain up to 1.25 wt% Mn; higher amounts are avoided because the presence of iron impurities can result in the formation of large primary particles of $Al_6Mn$, which causes embrittlement. Additions of magnesium provide improved solid-solution hardening, as in the alloy 3004, which is used for beverage cans, the highest single usage of any aluminum alloys, accounting for approximately 1/4 of the total usage of aluminum. Their moderate strength (ultimate tensile strengths of 110 to 297 MPa, or 16 to 43 ksi) often eliminates their consideration for structural applications. These alloys are welded with 1*xxx*-, 4*xxx*-, and 5*xxx*-series filler alloys, dependening on the specific chemistry, specific application, and service requirements.

**Table 26.6 Mechanical properties of select wrought non-heat-treatable aluminum alloys**

| Alloy | Temper | Tensile strength, MPa | Tensile strength, ksi | Yield strength, MPa | Yield strength, ksi | Elongation in 50 mm (2 in.), % |
|---|---|---|---|---|---|---|
| 3003 | O | 100 | 16 | 40 | 6 | 30 |
| | H14 | 125 | 18 | 115 | 17 | 9 |
| | H18 | 165 | 24 | 150 | 22 | 5 |
| 3004 | O | 180 | 26 | 70 | 10 | 20 |
| | H34 | 240 | 35 | 200 | 29 | 9 |
| | H38 | 285 | 41 | 250 | 36 | 5 |
| | H19 | 295 | 43 | 285 | 41 | 2 |
| 3005 | O | 130 | 19 | 55 | 8 | 25 |
| | H14 | 180 | 26 | 165 | 24 | 7 |
| | H18 | 240 | 35 | 225 | 33 | 4 |
| 3105 | O | 115 | 17 | 55 | 8 | 24 |
| | H25 | 180 | 26 | 160 | 23 | 8 |
| | H18 | 215 | 31 | 195 | 28 | 3 |
| 5005 | O | 125 | 18 | 40 | 6 | 25 |
| | H34 | 160 | 23 | 140 | 20 | 8 |
| | H38 | 200 | 29 | 185 | 27 | 5 |
| 5050 | O | 145 | 21 | 55 | 8 | 24 |
| | H34 | 190 | 28 | 165 | 24 | 8 |
| | H38 | 220 | 32 | 200 | 29 | 6 |
| 5052 | O | 195 | 28 | 90 | 13 | 25 |
| | H34 | 260 | 38 | 265 | 31 | 10 |
| | H38 | 290 | 42 | 255 | 37 | 7 |
| 5252 | O | 180 | 26 | 85 | 12 | 23 |
| | H25 | 235 | 34 | 170 | 25 | 11 |
| | H28 | 285 | 41 | 240 | 35 | 5 |
| 5154 | O | 240 | 35 | 115 | 17 | 27 |
| | H34 | 290 | 42 | 230 | 33 | 13 |
| | H38 | 330 | 48 | 270 | 39 | 10 |
| | H12 | 240 | 35 | 115 | 17 | 25 |
| 5454 | O | 250 | 36 | 115 | 17 | 22 |
| | H34 | 305 | 44 | 240 | 35 | 10 |
| | H111 | 250 | 36 | 125 | 18 | 18 |
| | H112 | 250 | 36 | 125 | 18 | 18 |
| 5056 | O | 290 | 42 | 150 | 22 | 35 |
| | H18 | 435 | 63 | 405 | 59 | 10 |
| | H38 | 415 | 60 | 345 | 50 | 15 |
| 5456 | O | 310 | 45 | 160 | 23 | 24 |
| | H112 | 310 | 45 | 165 | 24 | 22 |
| | H116 | 350 | 51 | 255 | 37 | 16 |
| 5182 | O | 275 | 40 | 130 | 19 | 21 |

Source: Ref 6

**Aluminum-Silicon Alloys (4*xxx*).** The 4*xxx* series of alloys is not as widely used as the 3*xxx* and 5*xxx* alloys. Ultimate tensile strengths range from 172 to 379 MPa (25 to 55 ksi). Because of the relatively high silicon content, the 4*xxx* series has excellent flow characteristics, making them the alloys of choice for two major applications. Alloy 4032 is used for forged pistons; the high silicon content contributes to complete filling of complex dies and provides wear resistance in service. The 4*xxx* alloys are also used for weld and braze filler metals, where the silicon content promotes molten metal flow to fill grooves and joints during welding and brazing. Although aluminum-silicon alloys will not respond to heat treatment, some of the 4*xxx* alloys also contain magnesium or copper, which allows them to be hardened by precipitation heat treating.

**Aluminum-Magnesium Alloys (5*xxx*).** The 5*xxx* alloys have the highest strengths of the non-heat-treatable alloys, with tensile strengths ranging from 124 to 434 MPa (18 to 63 ksi). They develop moderate strengths when work hardened; have excellent corrosion resistance, even in saltwater; and have very high toughness, even at cryogenic temperatures to near absolute zero. They are readily weldable by a variety of techniques, at thicknesses up to 20 cm (8 in.). Since aluminum and magnesium form solid solutions over a wide range of compositions, alloys containing magnesium in amounts from 0.8 to approximately 5 wt% are widely used. The 5*xxx*-series alloys have relatively high ductility, usually in excess of 25%.

The 5*xxx* alloys, although still having very good overall corrosion resistance, can be subject to intergranular and stress-corrosion cracking attack. In alloys with more than 3 to 4 wt% Mg, there is a tendency for the β phase ($Mg_5Al_8$)

to precipitate at the grain boundaries, making the alloy susceptible to grain-boundary attack. The precipitation of β occurs slowly at room temperature but can accelerate at elevated temperatures or under highly work-hardened conditions. A second problem that can be encountered with the 5*xxx* alloys is one of age softening at room temperature; that is, over a period of time, there is some localized recovery within the work-hardened grains. To avoid this effect, a series of H3 tempers is used in which the alloy is work hardened to a slightly greater level and then subjected to a stabilization aging treatment at 120 to 150° C (250 to 300 °F). This treatment also helps reduce the tendency for β precipitation.

The 5*xxx* alloys are used extensively in the transportation industries for boat and ship hulls; dump truck bodies; large tanks for carrying gasoline, milk, and grain; and pressure vessels, especially where cryogenic storage is required. The weldability of these alloys is excellent, and they have excellent corrosion resistance.

### 26.3.2 Wrought Heat Treatable Alloys

The wrought heat treatable alloys include the aluminum-copper (2*xxx*) series, the aluminum-magnesium-silicon series (6*xxx*), the aluminum-zinc (7*xxx*) series, and the aluminum-lithium alloys of the 8*xxx* series. These alloys are strengthened by precipitation hardening, which is covered in more detail in Chapter 9, "Precipitation Hardening," in this book. The importance of precipitation hardening of aluminum alloys can be appreciated by examining the data presented in Fig. 26.4 for naturally aged 2024 and artificially aged 7075. Note the dramatic increase in strength of both due to precipitation hardening, with only moderate reductions in elongation. The chemical compositions of a number of the wrought heat treatable aluminum alloys are given in Table 26.7, and the mechanical properties of a number of alloys are shown in Table 26.8.

**Aluminum-Copper Alloys (2*xxx*).** The high-strength 2*xxx* and 7*xxx* alloys are competitive on a strength-to-weight ratio with the higher-strength but heavier titanium and steel alloys and thus have traditionally been the dominant structural material in both commercial and military aircraft. In addition, aluminum alloys are not embrittled at low temperatures and become even stronger as the temperature is decreased, without significant ductility losses, making them ideal for cryogenic fuel tanks for rockets and launch vehicles.

The wrought heat treatable 2*xxx* alloys generally contain magnesium in addition to copper as an alloying element. Other significant alloying additions include titanium to refine the grain

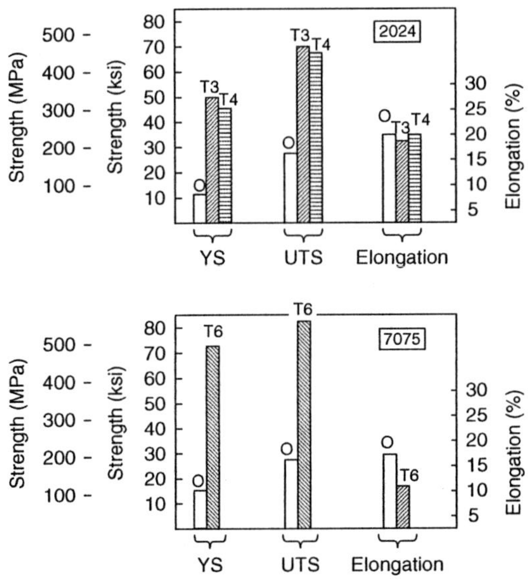


**Fig. 26.4** Effect of heat treatment on 2024 and 7075. YS, yield strength; UTS, ultimate tensile strength. Source: Ref 7

**Table 26.7 Compositions of select wrought heat treatable aluminum alloys**

| Alloy | Alloying element content, wt% | | | | | | | |
|---|---|---|---|---|---|---|---|---|
| | Fe | Si | Cu | Mn | Mg | Cr | Zn | Zr |
| 2008 | 0.40(a) | 0.65 | 0.9 | 0.3(a) | 0.4 | ... | ... | ... |
| 2219(b) | 0.30(a) | 0.20(a) | 6.3 | 0.3 | ... | ... | ... | 0.18 |
| 2519(b) | 0.39(a)(c) | 0.30(a)(c) | 5.8 | 0.3 | 0.25 | ... | ... | 0.18 |
| 2014 | 0.7(a) | 0.8 | 4.4 | 0.8 | 0.5 | ... | ... | ... |
| 2024 | 0.50(a) | 0.50(a) | 4.4 | 0.6 | 1.5 | ... | ... | ... |
| 2124 | 0.30(a) | 0.20(a) | 4.4 | 0.6 | 1.5 | ... | ... | ... |
| 2224 | 0.15(a) | 0.12(a) | 4.4 | 0.6 | 1.5 | ... | ... | ... |
| 2324 | 0.12(a) | 0.10(a) | 4.4 | 0.6 | 1.5 | ... | ... | ... |
| 2524 | 0.12(a) | 0.06(a) | 4.25 | 0.6 | 1.4 | ... | ... | ... |
| 2036 | 0.50(a) | 0.50(a) | 2.6 | 0.25 | 0.45 | ... | ... | ... |
| 6009 | 0.50(a) | 0.8 | 0.4 | 0.5 | 0.6 | ... | ... | ... |
| 6061 | 0.7(a) | 0.6 | 0.3 | ... | 1.0 | 0.2 | ... | ... |
| 6063 | 0.50(a) | 0.4 | ... | ... | 0.7 | ... | ... | ... |
| 6111 | 0.4(a) | 0.9 | 0.7 | 0.3 | 0.8 | ... | ... | ... |
| 7005 | 0.40(a) | 0.35(a) | ... | 0.45 | 1.4 | 0.13 | 4.5 | 0.14 |
| 7049 | 0.35(a) | 0.25(a) | 1.6 | ... | 2.4 | 0.16 | 7.7 | ... |
| 7050 | 0.15(a) | 0.12(a) | 2.3 | ... | 2.2 | ... | 6.2 | 0.12 |
| 7150 | 0.15(a) | 0.10(a) | 2.2 | ... | 2.4 | ... | 5.4 | 0.12 |
| 7055 | 0.15(a) | 0.10(a) | 2.3 | ... | 2.1 | ... | 8.0 | 0.12 |
| 7075 | 0.50(a) | 0.40(a) | 1.6 | ... | 2.5 | 0.25 | 5.6 | ... |
| 7475 | 0.12(a) | 0.10(a) | 1.6 | ... | 2.2 | 0.20 | 5.7 | ... |

(a) Maximum allowable amount.
(b) 2219 and 2519 also contain 0.10% V and 0.06% Ti.
(c) 0.40% max Fe plus Si. Source: Ref 6

structure during ingot casting, and transition element additions (manganese, chromium, and/or zirconium) that form dispersoid particles ($Al_{20}Cu_2Mn_3$, $Al_{18}Mg_3Cr_2$, and $Al_3Zr$), which help control the wrought grain structure. Iron and silicon are considered impurities and are held to an absolute minimum, because they form intermetallic compounds ($Al_7Cu_2Fe$ and $Mg_2Si$) that are detrimental to both fatigue and fracture toughness.

Due to their superior damage tolerance and good resistance to fatigue crack growth, the 2*xxx* alloys are used for aircraft fuselage skins for lower wing skins on commercial aircraft. The 7*xxx* alloys are used for upper wing skins, where strength is the primary design driver. The alloy 2024-T3 is normally selected for tension-tension applications because it has superior fatigue performance in the $10^5$ cycle range as compared to the 7*xxx* alloys.

Alloy 2024 has been the most widely used of the 2*xxx* series, although there are now newer alloys with better performance. Alloy 2024 is normally used in the solution-treated, cold-worked, and then naturally aged condition (T3 temper). Cold working is achieved at the mill by roller or stretcher rolling, which helps to produce flatness along with 1 to 4% strains. It has a moderate yield strength (448 MPa, or 65 ksi) but good resistance to fatigue crack growth and fairly high fracture toughness. Another common heat treatment for 2024 is the T8 temper (solution treated, cold worked, and artificially aged). Like the T3 temper, cold working prior to aging helps in nucleating fine precipitates and reduces the number and size of grain-boundary precipitates. In addition, the T8 temper reduces the susceptibility to stress-corrosion cracking.

One of the developments that led to improved properties in the high-strength 2*xxx* and 7*xxx* aluminum alloys is impurity control, specifically the reduction in the impurities iron and silicon. While 2024 has a combined iron and silicon

**Table 26.8 Mechanical properties of select wrought heat treatable aluminum alloys**

| | | | Tensile strength | | Yield strength | | Elongation in 50 mm |
|---|---|---|---|---|---|---|---|
| Alloy | Temper | Product form | MPa | ksi | MPa | ksi | (2 in.), % |
| 2008 | T4 | Sheet | 250 | 36 | 125 | 18 | 28 |
| | T6 | Sheet | 300 | 44 | 240 | 35 | 13 |
| 2014 | T6, T651 | Plate, forging | 485 | 70 | 415 | 60 | 13 |
| 2024 | T3, T351 | Sheet, plate | 450 | 65 | 310 | 45 | 18 |
| | T361 | Sheet, plate | 495 | 72 | 395 | 57 | 13 |
| | T81, T851 | Sheet, plate | 485 | 70 | 450 | 65 | 6 |
| | T861 | Sheet, plate | 515 | 75 | 490 | 71 | 6 |
| 2224 | T3511 | Extrusion | 530 | 77 | 400 | 58 | 16 |
| 2324 | T39 | Plate | 505 | 73 | 415 | 60 | 12 |
| 2524 | T3, T351 | Sheet, plate | 450 | 65 | 310 | 45 | 21 |
| 2036 | T4 | Sheet | 340 | 49 | 195 | 28 | 24 |
| 2219 | T81, T851 | Sheet, plate | 455 | 66 | 350 | 51 | 10 |
| | T87 | Sheet, plate | 475 | 69 | 395 | 57 | 10 |
| 2519 | T87 | Plate | 490 | 71 | 430 | 62 | 10 |
| 6009 | T4 | Sheet | 220 | 32 | 125 | 18 | 25 |
| | T62 | Sheet | 300 | 44 | 260 | 38 | 11 |
| 6111 | T4 | Sheet | 285 | 41 | 165 | 24 | 25 |
| | T6 | Sheet | 350 | 51 | 310 | 45 | 10 |
| 6061 | T6, T6511 | Sheet, plate, extrusion, forging | 310 | 45 | 275 | 40 | 12 |
| | T9 | Extruded rod | 405 | 59 | 395 | 57 | 12 |
| 6063 | T5 | Extrusion | 185 | 27 | 145 | 21 | 12 |
| | T6 | Extrusion | 240 | 35 | 215 | 31 | 12 |
| 7005 | T5 | Extrusion | 350 | 51 | 290 | 42 | 13 |
| 7049 | T73 | Forging | 540 | 78 | 475 | 69 | 10 |
| 7050 | T74, T745X | Plate, forging, extrusion | 520 | 74 | 450 | 65 | 13 |
| 7150 | T651, T6151 | Plate | 600 | 87 | 560 | 81 | 11 |
| | T77511 | Extrusion | 650 | 94 | 615 | 89 | 12 |
| 7055 | T7751 | Plate | 640 | 93 | 615 | 89 | 10 |
| | T77511 | Extrusion | 670 | 97 | 655 | 95 | 11 |
| 7075 | T6, T651 | Sheet, plate | 570 | 83 | 505 | 73 | 11 |
| | T73, T735X | Plate, forging | 505 | 73 | 435 | 63 | 13 |
| 7475 | T7351 | Plate | 505 | 73 | 435 | 63 | 15 |
| | T7651 | Plate | 455 | 66 | 390 | 57 | 15 |

Source: Ref 6

impurity level of 0.50 wt%, the newer alloy 2224 contains a maximum iron and silicon level of only 0.22 wt%. This lower fraction of impurities produces a better combination of strength and fracture toughness (Fig. 26.5). Other improvements for the plate materials include increasing the amount of cold work by stretching after quenching and the development of improved aging procedures.

The high-strength 2*xxx* alloys, which usually contain approximately 4 wt% Cu, are the least corrosion resistant of the aluminum alloys. Therefore, sheet products are usually clad on both surfaces with a thin layer of an aluminum alloy containing 1 wt% Zn. Since the clad is anodic to the underlying core alloy, it preferentially corrodes, leaving the core protected. The clad, which amounts to 1.5 to 10% of the thickness, is applied during hot rolling. Since the clad is weaker than the core alloy, there is a slight sacrifice in mechanical properties, especially fatigue cracking resistance.

**Aluminum-Magnesium-Silicon Alloys (6*xxx*).** The combination of magnesium (0.6 to 1.2 wt%) and silicon (0.4 to 1.3 wt%) in aluminum forms the basis of the 6*xxx* precipitation-hardenable alloys. During precipitation hardening, the intermetallic compound $Mg_2Si$ provides the strengthening. Manganese or chromium is added to most 6*xxx* alloys for increased strength and grain size control. Copper also increases the strength of these alloys, but if present in amounts over 0.5 wt%, it reduces the corrosion resistance. These alloys are widely used throughout the welding fabrication industry, are used predominantly in the form of extrusions, and are incorporated in many structural components.

The 6*xxx* alloys are heat treatable to moderately high strength levels, have better corrosion resistance than the 2*xxx* and 7*xxx* alloys, are weldable, and offer superior extrudability. With a yield strength comparable to that of mild steel, 6061 is one of the most widely used of all aluminum alloys. The highest strengths are obtained when artificial aging is started immediately after quenching. Losses of 21 to 28 MPa (3 to 4 ksi) in strength occur if these alloys are room-temperature aged for 1 to 7 days. Alloy 6063 is widely used for general-purpose structural extrusions because its chemistry allows it to be quenched directly from the extrusion press. Alloy 6061 is used where higher strength is required, and 6071 where the highest strength is required.

The 6*xxx* alloys can be welded, while most of the 2*xxx* and 7*xxx* alloys have very limited weldability. However, these alloys are solidification crack sensitive and should not be arc welded without filler material. The addition of adequate amounts of filler material during arc welding processes is essential to prevent base metal dilution, thereby preventing the hot cracking problem. They are welded with both

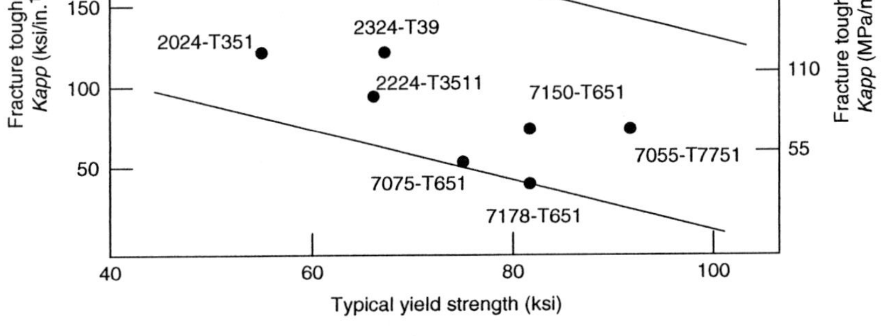


**Fig. 26.5** Fracture toughness versus yield strength for high-strength aluminum alloys. Source: Ref 8

4*xxx* and 5*xxx* filler materials, depending on the application and service requirements.

Although the 6*xxx* alloys have not traditionally been able to compete with the 2*xxx* and 7*xxx* alloys in applications requiring high strength, a relatively new alloy (6013-T6) has 12% higher strength than clad 2024-T3, with comparable fracture toughness and resistance to fatigue crack growth rate, and it does not have to be clad for corrosion protection.

**Aluminum-Zinc Alloys (7*xxx*).** The wrought heat treatable 7*xxx* alloys are even more responsive to precipitation hardening than the 2*xxx* alloys and can obtain higher strength levels, approaching tensile strengths of 690 MPa (100 ksi). These alloys are based on the Al-Zn-Mg(-Cu) system. The 7*xxx* alloys can be naturally aged but are not because they are not stable if aged at room temperature; that is, their strength will gradually increase with increasing time and can continue to do so for years. Therefore, all 7*xxx* alloys are artificially aged to produce a stable alloy.

Although the Al-Zn-Mg alloys cannot attain as high a strength level as those containing copper, they have the advantage of being weldable. In addition, the heat provided by the welding process can serve as the solution heat treatment, and they will age at room temperature to tensile strengths of approximately 310 MPa (45 ksi). The yield strengths may be as much as twice that of the commonly welded alloys of the 5*xxx* and 6*xxx* alloys. To reduce the chance of stress-corrosion cracking, these alloys are air quenched from the solution heat treating temperature and then overaged. Air quenching reduces residual stresses and reduces the electrode potential in the microstructure. The aging treatment is often a duplex aging treatment of the T73 type. The commonly welded alloys in this series, such as 7005, are predominantly welded with the 5*xxx*-series filler alloys.

The Al-Zn-Mg-Cu alloys attain the highest strength levels when precipitation hardened. Since these alloys contain up to 2 wt% Cu, they are the least corrosion resistant of the series. However, copper additions reduce the tendency for stress-corrosion cracking because they allow precipitation hardening at higher temperatures. As a class, these alloys are not weldable and are therefore joined with mechanical fasteners. The best known of these alloys is alloy 7075.

Some of the newer alloys have been produced to optimize their fracture toughness and resistance to corrosion, primarily stress-corrosion cracking and exfoliation corrosion. This has been accomplished through a combination of compositional control and processing, primarily through the development of new overaging heat treatments. As for some of the newer 2*xxx* alloys, reduced iron and silicon impurity levels are used to maximize fracture toughness. As an example, the older 7075 alloy contains a total iron and silicon content of 0.90 wt%, while it has been reduced to a maximum of only 0.22 wt% in 7475. The improvement in fracture toughness with reduced impurity levels is shown in Table 26.9, where the newer alloys 7149 and 7249 have higher fracture toughness values that the original 7049 composition.

One of the problems with 7075 and similar alloys when they are heat treated to the peak-aged T6 temper has been stress-corrosion cracking (SCC). Thick plate, forging, and extrusions of these alloys are particularly vulnerable when stressed in the through-the-thickness (short-transverse) direction. In response to these in-service failures, a number of overaged T7 tempers have been developed. Although there is some sacrifice in strength properties, they have dramatically reduced the occurrence of SCC failures. The T73 temper reduces the yield strength of 7075 by 15% but increases the SCC threshold stress by a factor of 6. Additional overaged tempers (T74, T75, and T77) have been developed that provide trade-offs in strength and SCC and exfoliation corrosion resistance. The T77 temper, developed by Alcoa, is of particular interest because

**Table 26.9 Effect of impurity content on high-strength aluminum extrusions**

| Alloy and temper | Composition, wt % | | | Fracture toughness yield strength | | Fracture toughness ultimate tensile strength | | Elongation, % | $K_{Ic}$ | |
|---|---|---|---|---|---|---|---|---|---|---|
| | Si max | Fe max | Mn max | MPa | ksi | MPa | ksi | | $MPa \cdot m^{1/2}$ | $ksi \cdot in.^{1/2}$ |
| 7049-T73511 | 0.25 | 0.35 | 0.35 | 503 | 73 | 552 | 80 | 11.6 | 26 | 24 |
| 7149-T73511 | 0.15 | 0.20 | 0.20 | 517 | 75 | 565 | 82 | 13.3 | 33 | 30 |
| 7249-T73511 | 0.20 | 0.12 | 0.12 | 531 | 77 | 579 | 84 | 13.3 | 37 | 34 |

Source: Ref 9

it maintains strength levels close to the T6 temper. This temper is a variation of a treatment called retrogression and reaging and produces the best combination of mechanical properties and corrosion resistance. Although the specific heat treatment parameters depend on the alloy composition, the part is first heat treated to the T6 temper. It is then reheated to approximately 205 °C (400 °F) for 1 h, water quenched, and reaged again for 24 h at 120 °C (250 °F).

**Other Aluminum Alloys (8*xxx*).** The 8*xxx* series is reserved for those alloys with less commonly used alloying elements, such as iron, nickel, and lithium. A series of electrical conductor alloys contains iron and nickel for strength, with only a minimal loss in conductivity. A number of aluminum-iron alloys have been studied for high-temperature applications. In addition, the 8*xxx* series, along with the 2*xxx* series, contains some of the high-strength aluminum-lithium alloys that have been evaluated for aerospace applications.

Aluminum-lithium alloys, part of both the 2*xxx* and 8*xxx* alloy series, are attractive for aerospace applications, owing to their reduced densities and higher elastic moduli. Lithium is an even lighter element than aluminum, and each 1 wt% of lithium alloyed with aluminum reduces the density by 3%. In addition, lithium additions increase the elastic modulus of aluminum alloys, with each 1 wt% Li producing an increase in the modulus by approximately 6%. However, lithium is a very active metal, making melting and casting difficult and expensive. Aluminum-lithium alloys cost approximately three times as much as conventional competitive high-strength aluminum alloys.

Aluminum-lithium alloys were initially produced as early as the 1950s and have progressed through several generations of improvements. The so-called second-generation alloys that were produced in the 1980s were designed as drop-in replacements for existing high-strength alloys. These alloys were classified as high strength (2090, 8091), medium strength (8090), and damage tolerant (2091, 8090). These alloys contain 1.9 to 2.7 wt% Li, which results in approximately a 10% lower density and 25% higher specific stiffness than equivalent 2*xxx* and 7*xxx* alloys. One of the major initial applications was for structural members on the C-17 transport aircraft. However, due to property and manufacturing problems, they were removed and replaced with conventional high-strength aluminum alloys. Technical problems included excessive anisotropy in the mechanical properties, lower-than-desired fracture toughness and ductility, hole cracking and delamination during drilling, and low SCC thresholds. The anisotropy experienced by these alloys is a result of the strong crystallographic textures that develop during processing, with the fracture toughness problem being one of primarily low strength in the short-transverse direction.

A third generation of alloys has been developed with lower lithium contents. One success story is alloy 2195, which has a lower copper content and has replaced 2219 for the cryogenic fuel tank on the space shuttle, where it provides a higher strength, higher modulus, and lower density than 2219.

## 26.4 Melting and Primary Fabrication

To produce pure aluminum, alumina ($Al_2O_3$) is first extracted from the mineral bauxite, which contains approximately 50% $Al_2O_3$. In the Bayer process, a sodium hydroxide solution is used to precipitate aluminum hydroxide, which is then calcined to form alumina. Alumina is then converted to pure aluminum by electrolysis, using the Hall-Héroult process illustrated in Fig. 26.6. The cell is lined with carbon cathodes, and consumable electrodes are gradually fed into the top of the cell. The electrolyte is cryolite ($Na_3AlF_6$) with 8 to 10 wt% $Al_2O_3$ dissolved in it. The cell operates at temperatures in the range of 955 to 1010 °C (1750 to 1850 °F), with a power rating of 10 to 12 kW·h/kg aluminum. Pure aluminum (99 wt%) is reduced at the cathode and forms a molten pool in the bottom of the cell, which is drained from the bottom and cast into aluminum ingots. Since impurities such as iron and silicon are reduced along with the aluminum, the raw materials used in the Bayer process are carefully controlled to minimize these metal oxides. Since the production of aluminum takes a lot of electrical energy, and recycling aluminum takes much less energy, a large portion of general-purpose aluminum currently is made from recycled material.

During casting of aluminum alloy ingots, oxides and gases must be controlled. Aluminum oxidizes rapidly in the liquid state and reverts back to alumina. Once the oxide becomes entrapped in the liquid metal, it is difficult to remove and remains dispersed in the liquid metal. The main source of oxygen is moisture from the furnace charge. Oxides are removed by fluxing with gases, solids, or molten salts. Various filtration methods are also used to remove oxides during pouring. Hydrogen is the only gas with appreciable solubility in molten aluminum and can cause blisters in heat treated parts and porosity in aluminum castings. Again, moisture from the furnace charge is the main source of hydrogen. Hydrogen can also be introduced from moisture in the air and products of combustion. Hydrogen is removed by bubbling chlorine, nitrogen, or argon gas through the melt.

The semicontinuous direct-chill process is the mostly widely used process for casting commercial ingots that will receive further processing, such as rolling, extrusion, or forging. It produces fine-grained ingots at high production rates. As shown in Fig. 26.7, the molten aluminum is poured into a shallow, water-cooled mold of the required shape. When the metal begins to freeze, the false bottom in the mold is lowered at a controlled rate, and water is sprayed on the freshly solidified metal as it exits the mold. A water box or spray rings are placed around the ingot to rapidly cool the ingot. Metals with low melting points, such as magnesium, copper, and zinc, are added directly to the molten charge, while high-melting-point elements (e.g., titanium, chromium, zirconium, and manganese) are added as master alloys. Inoculants, such as titanium and titanium-boron, are added to reduce hot cracking and refine grain size.

One of the main advantages of direct-chill continuous casting is that it helps to eliminate the segregation that occurs in high-strength alloys produced by the older tilt-casting procedures. These alloys, when produced by tilt casting, are highly segregated because of the broad solidification temperature ranges and the shape of the freezing front. Direct-chill casting eliminates most of this type of segregation because the liquid metal freezing front is almost horizontal, and the liquid metal freezes from the bottom to the top of the ingot. During direct-chill casting, the pouring temperature is maintained at only approximately 28 °C (50 °F) above the liquidus temperature; this helps reduce oxide formation and hydrogen pickup and produces a fine-grained structure. It can also produce fairly large ingots at slow speeds, a necessary requirement for the high-strength alloys to prevent cracking. Typical casting speeds are in the range of 2.5 to 13 cm/min (1 to 5 in./min).

### 26.4.1 Rolling Plate and Sheet

Rolled aluminum is the most common of the wrought aluminum product forms. Sheet is defined as rolled aluminum in the range of 0.15 to 6.3 mm (0.006 to 0.250 in.) thick. If the thickness is greater than 6.3 mm (0.250 in.), then it is called plate. Foil refers to aluminum product that is less than 0.15 mm (0.006 in.) thick. Aluminum foil, sheet, and plate are

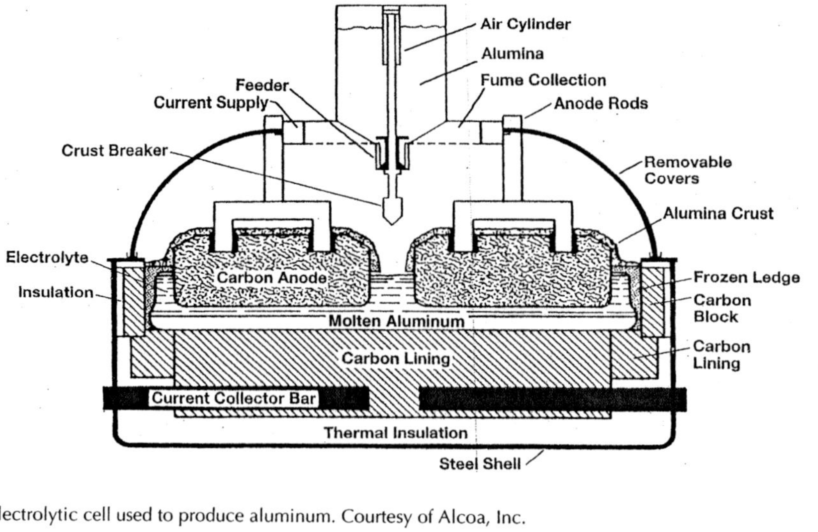


**Fig. 26.6** Electrolytic cell used to produce aluminum. Courtesy of Alcoa, Inc.

produced from aluminum ingots using the following steps:

1. Scalping of the ingot
2. Preheating and homogenizing the ingot
3. Reheating the ingot, if required, to the hot rolling temperature
4. Hot rolling to form a slab
5. Intermediate annealing
6. Cold rolling along with intermediate anneals to form foil and sheet product forms

**Scalping of Ingots.** To prevent surface defects on the cast ingot from being rolled into the surface, approximately 6.3 to 9.5 mm (0.250 to 0.375 in.) are removed from the surfaces to be rolled. Some alloys, such as 1100, 3003, and 3015, have fairly smooth as-cast surfaces and do not require scalping. Other, more highly alloyed ingots, such as magnesium-containing alloys and high-strength aircraft alloys, are always scalped.

**Preheating or homogenizing** puts into solid solution all the constituents that are soluble and reduces the coring that occurs during the casting process. It relieves stresses in the ingot and makes the cast structure more uniform and more readily hot worked. Soaking temperatures and times depend on the specific alloy. For example, 1100 aluminum is soaked for approximately 1 h at 455 to 510 °C (850 to 950 °F), while 7075 requires up to 24 h at 455 to 470 °C (850 to 875 °F). Ingots that are going to be clad during hot rolling are scalped after preheating to avoid the heavy oxidation that occurs during the long preheating cycle. This allows the cladding to form a better bond during hot rolling.

**Hot rolling** conducted above the recrystallization temperature produces a fine grain structure

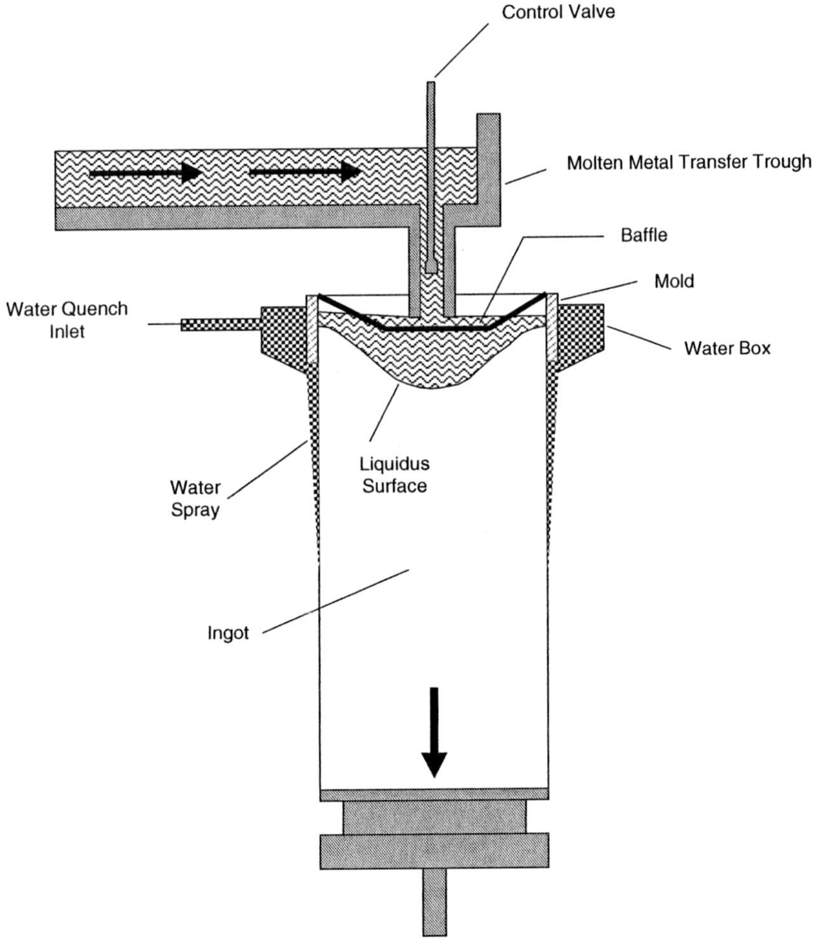


**Fig. 26.7** Semi-continuous direct-chill casting. Source: Ref 10

and a minimum of grain directionality. The upper limit is set by the lowest-melting-point eutectic present in the alloy, while the lower temperature is the temperature at which the metal is hot enough to be sufficiently reduced with each pass through the mill without cracking. The ingot is removed from the soaking pit and initially put through a series of four-high rolling mills to break down the ingot structure. Breakdown temperatures are in the range of 400 to 540 °C (750 to 1000 °F), with continuous rolling temperatures of approximately 290 to 455 °C (550 to 850 °F). Since the work rapidly lengthens in the direction of rolling, it is necessary to remove the slab from the mill, turn it around, and then cross roll it to produce wide sheet or plate. The grain structure becomes elongated in the rolling direction, as shown in Fig. 26.8. This results in anistropic mechanical properties in which the properties are lowest in the through-the-thickness (short-transverse) direction. Rolling into thinner plate and sheet is then conducted on five-stand four-high mills, with successive reductions at each mill. Aluminum sheet stock, which can exit the last mill at speeds approaching 485 km/h (300 mph), is coiled into large coils prior to cold rolling.

**Intermediate Annealing.** Since hot rolling produces some cold work, coiled aluminum sheet stock is given an intermediate anneal prior

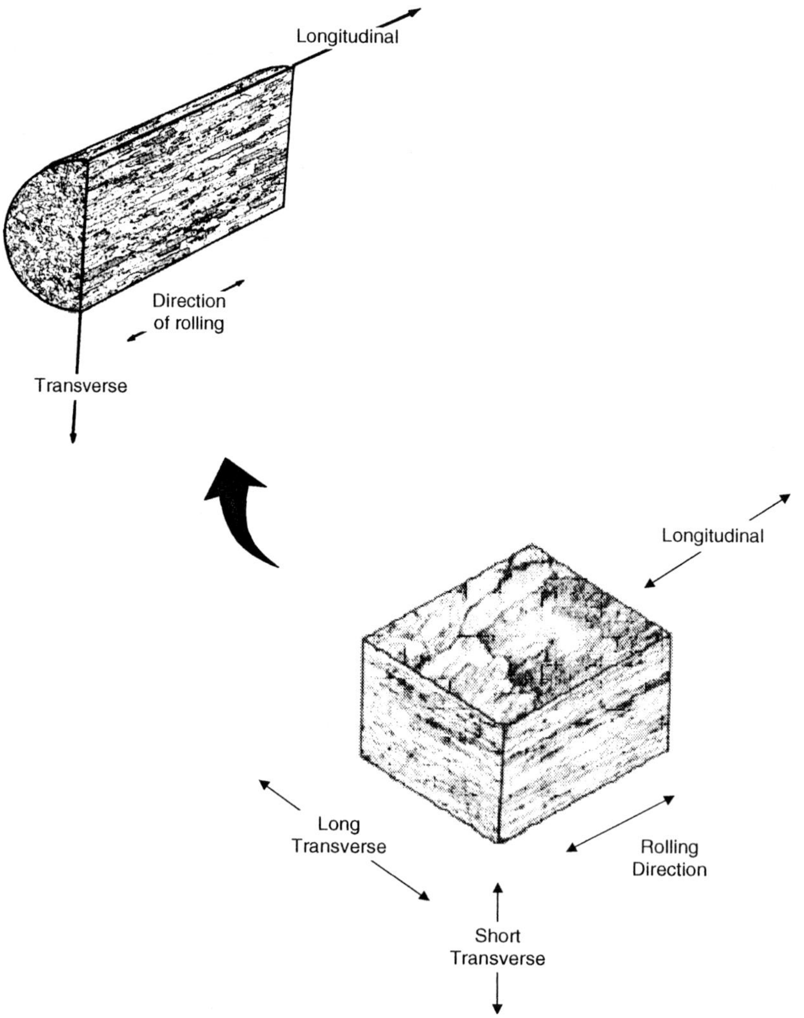


**Fig. 26.8** Grain directionality due to rolling. Original magnification at 40×. Source: Ref 11

to cold rolling. Since the amount of cold work introduced during hot rolling is sufficient to cause recrystallization, annealing not only lowers the strength and increases ductility, it also imparts a fine grain structure.

**Cold Rolling.** Sheet and foil are cold rolled after hot rolling to produce a much better surface finish and control the strength and ductility through work hardening. Again, four-high mills are used along with lubricants and, if a bright surface finish is required, polished rolls. Depending on the amount of reduction and the final strength required, intermediate anneals are conducted during the cold rolling process. Normally, reductions in the range of 45 to 85% are taken between anneals.

### *26.4.2 Extrusion*

Both cold and hot extrusion methods are used to produce extruded aluminum shapes. Cold or impact extrusions are made by a single sharp blow of a punch into a die cavity that contains a blank or slug of the correct size and shape. Almost all aluminum alloys can be formed by impact extrusion. The slugs are annealed and then generally impact extruded at room temperature, although the temperature may rise to as much as 230 °C (450 °F) due to the extensive plastic deformation. The slugs are also lubricated prior to extrusion to prevent excessive galling and die wear.

Direct hot extrusion is used to make structural shapes. In the direct extrusion process, the cylindrical ingot is preheated and then extruded in the temperature range of 340 to 510 °C (650 to 950 °F), depending on the specific alloy. The preheated ingot is placed in a hydraulic press and squeezed at high pressure through a steel die to produce the desired shape. During extrusion, the metal flows most rapidly at the center of the ingot. Since oxide and surface defects are left in the last 10 to 15% of the ingot, this part of the ingot (called the butt) is discarded. The 6*xxx*-series alloys, because of their easy extrudability, are the most popular alloys for producing shapes. The 2*xxx*- and 7*xxx*-series alloys are used in applications requiring higher strength; however, these alloys are more difficult to extrude.

## 26.5 Casting

Aluminum castings can offer significant cost savings by reducing the number of components and the associated assembly cost. Three types of casting processes are used extensively for aluminum alloys: sand casting for small numbers of large pieces, permanent mold casting for small and medium part sizes, and die castings for small parts where a large quantity can justify the cost of the die-casting tooling.

### *26.5.1 Aluminum Casting Alloys*

The major attributes of the aluminum casting series are shown in Table 26.10. Although all of the aluminum casting alloys are covered in this section, it should be emphasized that the

**Table 26.10 Major attributes of cast aluminum alloys**

**2*xx.x*: Al-Cu Alloys.** The major characteristics of the 2*xx.x* series are:

- Heat treatable; sand and permanent mold castings
- High strength at room and elevated temperatures; some high-toughness alloys
- Approximate ultimate tensile strength range: 131–448 MPa (19–65 ksi)

**3*xx.x*: Al-Si + Cu or Mg Alloys.** The major characteristics of the 3*xx.x* series are:

- Heat treatable; sand, permanent mold, and die castings
- Excellent fluidity; high-strength/some high-toughness alloys
- Approximate ultimate tensile strength range: 131–276 MPa (19–40 ksi)
- Readily welded

**4*xx.x*: Al-Si Alloys.** The major characteristics of the 4*xx.x* series are:

- Non-heat-treatable; sand, permanent mold, and die castings
- Excellent fluidity; good for intricate castings
- Approximate ultimate tensile strength range: 117–172 MPa (17–25 ksi)

**5*xx.x*: Al-Mg Alloys.** The major characteristics of the 5*xx.x* series are:

- Non-heat-treatable; sand, permanent mold, and die
- Tougher to cast; provides good finishing characteristics
- Excellent corrosion resistance, machinability, surface appearance
- Approximate ultimate tensile strength range: 117–172 MPa (17–25 ksi)

**7*xx.x*: Al-Zn Alloys.** The major characteristics of the 7*xx.x* series are:

- Heat treatable; sand and permanent mold cast (harder to cast)
- Excellent machinability and appearance
- Approximate ultimate tensile strength range: 207–379 MPa (30–55 ksi)

**8*xx.x*: Al-Sn Alloys.** The major characteristics of the 8*xx.x* series are:

- Heat treatable; sand and permanent mold castings (harder to cast)
- Excellent machinability
- Bearings and bushings of all types
- Approximate ultimate tensile strength range: 103–207 MPa (15–30 ksi)

Source: Ref 1

aluminum-silicon alloys with magnesium and/or copper, the 3*xx.x* alloys, are by far the most widely used of the aluminum casting alloys. The silicon addition greatly increases liquid metal fluidity and produces superior castings. The order of the alloy series in order of decreasing castability is 3*xx.x*, 4*xx.x*, 5*xx.x*, 2*xx.x*, and 7*xx.x*.

**Commercially pure aluminum alloys (1*xx.x*)** are used only for applications where high electrical conductivity is needed and strength is not very important.

**Aluminum-Copper Alloys (2*xx.x*).** The heat treatable aluminum-copper alloys that contain 4 to 5 wt% Cu include the highest-strength aluminum castings available and are often used for premium-quality aerospace products. The ductility can also be quite good, if prepared from ingot containing less than 0.15 wt% Fe. When these alloys are cast in permanent molds, special gating and risering techniques are required to relieve shrinkage stresses. However, when done correctly, the aluminum-copper alloys are capable of producing high-strength and ductile castings. Aluminum-copper alloys have somewhat marginal castability and are susceptible to SCC when heat treated to high strength levels. Manganese can be added to combine with iron and silicon and reduce the embrittling effect of insoluble phases. However, this reduces the castability. Manganese-containing alloys are mainly sand cast; when they are cast in permanent molds, silicon must be added to increase fluidity and reduce hot shortness, but silicon additions reduce ductility. Aluminum-copper alloys retain reasonable strength at elevated temperatures (up to 315 °C, or 600 °F). The high-temperature strength is a result of copper, nickel, and magnesium additions.

**Aluminum-Silicon+Copper and/or Magnesium Alloys (3*xx.x*).** The 3*xx.x* alloys are the workhorses of the aluminum casting industry, accounting for more than 95% of all die castings and 80% of all sand and permanent mold castings produced. Silicon is by far the most important alloying element in aluminum casting alloys. Silicon greatly improves the fluidity of molten aluminum, especially when the amount approaches the eutectic composition. Silicon increases fluidity, reduces cracking, and improves feeding to minimize shrinkage porosity.

The most widely used aluminum casting alloys are those containing between 9.0 and 13.0 wt% Si. These alloys are of approximately eutectic composition (Fig. 26.9), which makes them suitable as die-casting alloys, since their freezing range is small. However, they form a rather coarse eutectic structure (Fig. 26.9a) that is refined by a process known as modification, where small amounts of sodium (~0.01% by weight) are added to the melt just before casting. The sodium delays the precipitation of silicon when the normal eutectic temperature is reached and also causes a shift of the eutectic composition toward the right in the phase diagram. Therefore, as much as 14 wt% Si may be present in a modified alloy without any primary silicon crystals forming in the structure (Fig. 26.9b). It is thought that sodium collects in the liquid at its interface with the newly formed silicon crystals, inhibiting and delaying their growth. Thus, undercooling occurs and new silicon nuclei are formed in large numbers, resulting in a relatively fine-grained eutectic structure. Modification raises the tensile strength and the elongation in the manner shown in Fig. 26.10. The relatively high ductility of this cast eutectic alloy is due to the solid-solution phase in the eutectic constituting nearly 90% of the total structure. Therefore, the solid-solution phase is continuous in the microstructure and acts as a cushion against much of the brittleness arising from the hard silicon phase. More recently, modifying with strontium is replacing sodium, because there is less loss during melting due to oxidation or evaporation; overmodification (too much addition) is not a problem because it forms the innocuous compound $SrAl_3Si_3$, and strontium suppresses the formation of large primary silicon particles in hypereutectic alloys.

**Aluminum-Silicon alloys (4*xx.x*)** are used when good castability and good corrosion resistance are requirements. Alloys of the 4*xx.x* group, based on the binary aluminum-silicon system and containing from 5 to 12 wt% Si, are used in applications where combinations of moderate strength and high ductility and impact resistance are required.

**Aluminum-Magnesium Alloys (5*xx.x*).** The aluminum-magnesium casting alloys are single-phase binary alloys with moderate-to-high strength and toughness. The aluminum-magnesium alloys have excellent corrosion resistance. High corrosion resistance, especially to seawater and marine atmospheres, is the primary advantage of castings made of these alloys. The best corrosion resistance requires low impurity content (both solid and gases), and

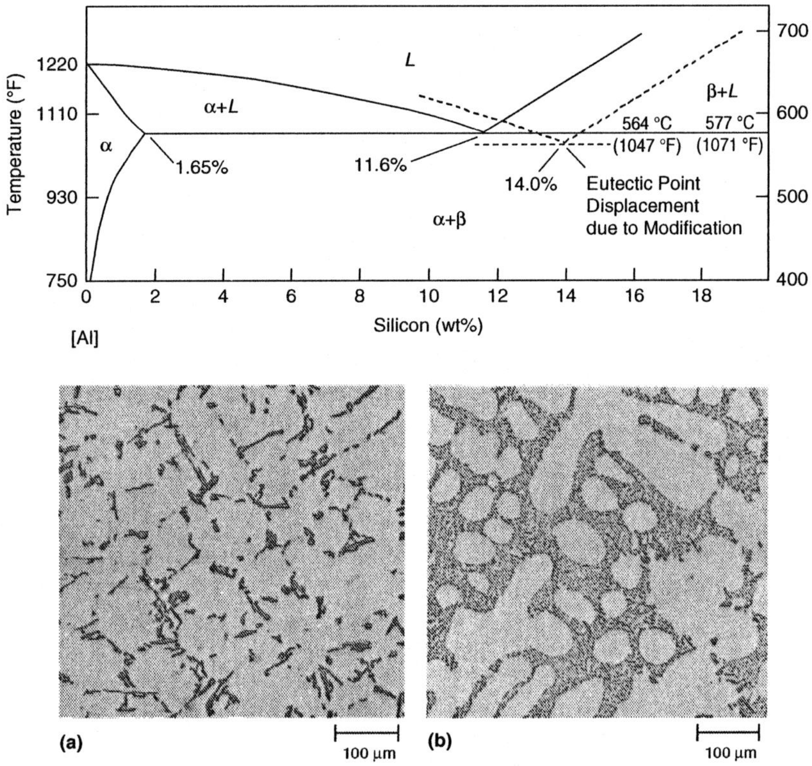


**Fig. 26.9** Modification of aluminum-silicon casting alloys. (a) Unmodified. (b) Modified. Original magnification: 100×. Source: Ref 12

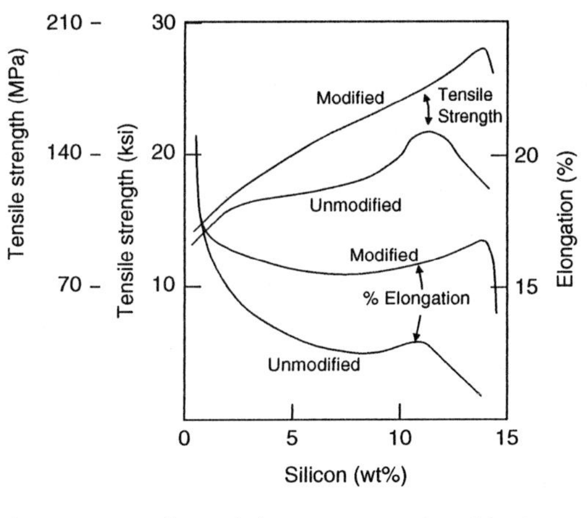


**Fig. 26.10** Effects of silicon content and modification on aluminum casting alloy

thus, alloys must be prepared from high-quality metals and handled with great care at the foundry. The aluminum-magnesium alloys lack fluidity, are prone to hot tearing, and have a greater tendency to oxidize when molten. The 5*xx.x* alloys are weldable, have good machinability, and have an attractive appearance when anodized.

**Aluminum-Zinc Alloys (7*xx.x*).** The aluminum-zinc casting alloys do not have the good fluidity or shrinkage-feeding characteristics of the silicon-containing alloys, and hot cracking can be a problem in large, complex shapes. However, the aluminum-zinc alloys are used where the castings are going to be brazed, because they have the highest melting points of all of the aluminum casting alloys. The 7*xx.x* alloys have moderate-to-good tensile properties in the as-cast condition. Annealing can be used to increase dimensional stability. They are also capable of self-aging at room temperature after casting, reaching quite high strengths after 20 to 30 days. Therefore, they are used in applications requiring moderate-to-high strength levels, where a full solution heating and quenching operation could cause severe warpage or cracking. They have good machinability and resistance to general corrosion but can be somewhat susceptible to SCC. They should not be used at elevated temperatures due to rapid overaging.

**Aluminum-Tin alloys (*8xx.x*)** are used for cast bearings and bushings where good compression strength is required. Heat treatment to the T5 condition improves their compression strength. Aluminum-tin alloys with approximately 6 wt% Sn, along with copper and nickel additions for strengthening, have excellent lubricity due to the tin content. These alloys can be cast by sand or permanent mold casting, but they are susceptible to hot cracking and therefore have rather poor castability. The best bearing properties are obtained when the casting has a small interdendritic spacing, produced by rapid cooling rates.

### *26.5.2 Aluminum Casting Control*

Molten aluminum alloys are extremely reactive and readily combine with other metals, gas, and sometimes with refractories. Molten aluminum dissolves iron from crucibles; therefore, aluminum is usually melted and handled in refractory-lined containers. For convenience in making up the charge, and to minimize the chance of error, most foundries use standard prealloyed ingot for melting rather than doing their own alloying. Most alloying elements used in aluminum castings, such as copper, silicon, manganese, zinc, nickel, chromium, and titanium, are not readily lost by oxidation, evaporation, or precipitation. Alloying elements that melt at temperatures higher than the melting temperature of aluminum, such as chromium, silicon, manganese, and nickel, are added to the molten metal as alloy-rich ingots or master alloys. Some elements, such as magnesium, sodium, and calcium, which are removed from the molten bath by oxidation and evaporation, are added in elemental form to the molten bath, as required, to compensate for loss.

Because of its reactivity, molten aluminum is easily contaminated. The principal contaminants are iron, oxides, and inclusions. When the iron content exceeds 0.9 wt%, an undesirable acicular grain structure develops in the thicker sections of the casting. When the iron content exceeds 1.2 wt% in the higher-silicon alloys, sludging is likely to occur, particularly if the temperature drops below 650 °C (1200 °F). To prevent sludging, the quantity wt%Fe + 2(wt%Mn) + 3(wt%Cr) should not exceed 1.9 wt%. When this quantity exceeds 1.9 wt%, the castings are likely to contain hard spots that impair machining and may initiate stress cracks in service. Iron also causes excessive shrinkage in aluminum castings and becomes more severe as the iron content increases beyond 1 wt%.

Oxides must be removed from the melt; if they remain in the molten metal, the castings will contain harmful inclusions. Magnesium is a strong oxide former, making magnesium-containing alloys difficult to control when melting and casting. Oxides of aluminum and magnesium form quickly on the surface of the molten bath, developing a thin, tenacious skin that prevents further oxidation as long as the surface is not disturbed. Molten aluminum also reacts with moisture to form aluminum oxide, releasing hydrogen. Oxidation is also caused by excessive stirring, overheating of the molten metal, pouring from too great a height, splashing of the metal, or disturbing the metal surface with the ladle before dipping. Oxides that form on the surface of the molten bath can be removed by surface-cleaning fluxes. These fluxes usually contain low-melting-point ingredients that react exothermically on the surface of the bath. The oxides separate from the metal to form a dry, powdery, floating dross that can be skimmed. Some denser oxides sink to the bottom and are removed by gaseous fluxing or through a drain hole located in the bottom of the furnace.

Hydrogen is the only gas that dissolves to any extent in molten aluminum alloys and, if not removed, will result in porosity in the castings. As shown in Fig. 26.11, the solubility of hydrogen is significantly higher in the liquid than in the solid state. During cooling and solidification, the solubility decreases, and hydrogen is precipitated as porosity. Hydrogen is introduced in the molten aluminum by moisture and dirt in the charge and by the products of combustion. Degassing fluxes to remove hydrogen are used after the surface of the bath has been fluxed to remove oxides. The degassing fluxes also help to lift fine oxides and particles to the top of the bath. Removal of hydrogen by degassing is a mechanical action; hydrogen gas does not combine with the fluxing gases. Degassing fluxes include chlorine gas, nitrogen-chlorine mixtures, and hexachloroethane.

The grain size of aluminum alloy castings can be as small as 0.13 mm (0.005 in.) in diameter to as large as 13 mm (0.5 in.) in diameter. Fine-grained castings are desired for several reasons. First, while any porosity is undesirable, coarse porosity is the most undesirable. Since the coarseness of porosity is proportional to grain size, porosity in fine-grained castings is finer and less harmful in fine-grained castings. Second, shrinkage and hot cracking are usually associated with coarse-grained structures. A finer grain size minimizes shrinkage, resulting in sounder castings. Third, the mechanical properties, such as tensile strength and ductility, are better for fine-grained castings than those of coarse-grained castings.

The grain size of aluminum alloy castings is influenced by pouring temperature, solidification rate, and the presence or absence of grain refiners. For all aluminum alloys, the grain size increases as the pouring temperature is increased. This is the main reason that aluminum alloy castings should be poured at the lowest temperature that will produce a sound casting. Rapid solidification rates produce finer grain sizes; therefore, castings done in steel dies will have finer grain sizes due to the faster solidification rates than those produced by sand casting. Grain-refining elements, such as titanium, boron, and zirconium, are helpful in producing finer-grained castings.

## 26.6 Heat Treating

The most common heat treatments of aluminum alloys are precipitation hardening and annealing. The details of the precipitation-hardening process have previously been covered in Chapter 9, "Precipitation Hardening," in this book. Annealing can be used for both the non-heat-treatable and the precipitation-hardenable grades of wrought and cast alloys. In this chapter, only the specifics of annealing of aluminum alloys are covered.

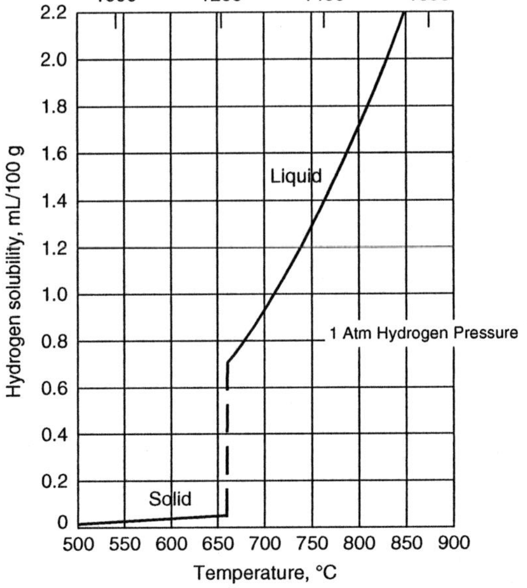


**Fig. 26.11** Solubility of hydrogen in aluminum. Source: Ref 13

### 26.6.1 Annealing

Full annealing (O temper) produces the lowest strength and highest ductility. Cold-worked products will normally undergo recrystallization, while hot-worked products may remain unrecrystallized, depending on the amount of cold work introduced during hot deformation. The recrystallization temperature is not a fixed value; it depends on alloy composition, amount of cold work, rate of heating, and time at temperature.

For both the non-heat-treatable and heat treatable alloys, reduction or elimination of the strengthening effects of cold working is accomplished by heating at temperatures in the range of 260 to 440 °C (500 to 825 °F), depending on the specific alloy. A 1 h soak at a temperature of $345 \pm 8$ °C ($650 \pm 15$ °F) is a satisfactory annealing treatment for the 1*xxx*- and 5*xxx*-series alloys. Longer times and higher temperatures are necessary for the 3*xxx* alloys. High heating rates to the annealing temperature are desirable to give finer grain structure. The time at temperature depends on the type of anneal, the thickness of the material, the method of furnace loading, and the temperature. The time at temperature for a full anneal is usually 1 h. For heat treatable alloys, the cooling rate should be slow enough so that precipitation reactions do not occur. A cooling rate not exceeding 25 °C/h (50 °F/h) is usually sufficient.

Annealing to remove the effects of cold work is conducted at approximately 345 °C (650 °F). If it is necessary to remove the hardening effects of heat treatment or of cooling from hot working temperatures, a treatment designed to produce a coarse, widely spaced precipitate is used, consisting of soaking at 415 to 440 °C (775 to 825 °F), followed by slow cooling (25 °C/h, or 50 °F/h max) to approximately 260 °C (500 °F). The high diffusion rates during soaking and slow cooling permit maximum coalescence of precipitate particles that result in minimum hardness. In the 7*xxx* alloys, partial precipitation occurs, and a second treatment of soaking at 230 ± 5 °C (450 ± 10 °F) for 2 h is required.

In annealing, it is important to ensure that the proper temperature is reached in all portions of the load; therefore, it is common to specify a soaking period of at least 1 h. The maximum annealing temperature is only moderately critical; however, temperatures exceeding 415 °C (775 °F) can result in oxidation and grain growth. Relatively slow cooling, in still air or in the furnace, is recommended for all alloys to minimize distortion.

Stress-relief anneals are used to reduce stresses without causing recrystallization. Temperatures in the range of 220 °C (425 °F) will produce a reasonable degree of stress relief. The 5*xxx* series of alloys tend to soften at room temperature after cold working. They are normally given stabilization anneals at the mill by heating to 120 to 150 °C (250 to 300 °F) to ensure the stability of mechanical properties after shipment.

## 26.7 Fabrication

Aluminum alloys are forged using hammers, mechanical presses, and hydraulic presses. Forging is conducted in the range of 360 to 470 °C (680 to 880 °F), depending on the specific alloy. Somewhat surprisingly, aluminum alloys generally require greater working forces for an equal amount of deformation than low-carbon steels, due to the difference in flow stress levels at their optimal hot working temperatures. Therefore, equipment used for forging aluminum must supply higher forces than that used for the low-carbon steels. For larger and more complex parts, hydraulic presses are preferred.

As a result of their face-centered cubic crystalline structure and their relatively low rates of work hardening, aluminum alloys are readily formable at room temperature. The choice of temper for forming depends on the severity of the forming operation and the alloy being formed. Aluminum alloys can be readily formed at room temperature in either the O or W temper. For alloys formed in the W (solution heat treated) temper, it is normal practice to refrigerate the solution heat treated material to prevent natural aging before forming.

Aluminum alloys are extremely easy to machine. Cutting speeds as high as 5 surface m/s (1000 surface ft/min) are common. The implementation of high-speed machining during the 1990s allowed even higher metal removal rates; three times greater metal removal rates are typical. In addition, since the cuts are light, most of the heat is removed with the chips. This allows extremely thin walls and webs to be machined without distortion. The significance is that parts that once had to be assembled from many formed pieces can now be machined from a single block of aluminum, resulting in a weight-competitive assembly at a large cost savings. A comparison of a sheet metal built-up assembly and a high speed machine integral assembly is shown in Fig. 26.12.

As a metal class, aluminum alloys are rather difficult to weld but can be welded by gas metal arc welding, gas tungsten arc welding, and resistance welding. The 2*xxx* and copper-containing 7*xxx* are either very difficult to weld or unweldable by conventional arc welding methods. However, a relatively new process, called friction stir welding, illustrated in Fig. 26.13, is capable of welding even the most difficult of the aluminum alloys. In this process, the weld joint never becomes a true liquid; it is a solid-state process.

## 26.8 Corrosion

The corrosion and oxidation resistance of aluminum is due to a very adherent oxide film ($Al_2O_3$) that immediately forms on exposure to air. Aluminum is corrosion resistant in neutral solutions but is attacked by both basic and acidic solutions. Most, but not all, aluminum alloys are less corrosion resistant than pure aluminum. General corrosion resistance of aluminum alloys is usually an inverse function of the amount of copper used in the alloy. Thus, the 2*xxx*-series alloys are the least corrosion-resistant alloys, since copper is their primary alloying element and all have appreciable (approximately 4 wt%) levels of copper.

Some 7*xxx* series alloys contain approximately 2 wt% Cu in combination with magnesium and zinc to develop strength. Such alloys are the strongest but least corrosion resistant of their series. Low-copper aluminum-zinc alloys, such as 7005, are also available and have become more popular recently. However, copper does have a beneficial effect on the SCC resistance of the 7*xxx* alloys by allowing them to be precipitated at higher temperatures without loss of strength in the T73 temper.

Among the 6*xxx*-series alloys, higher copper content (1 wt% in 6066) generally decreases

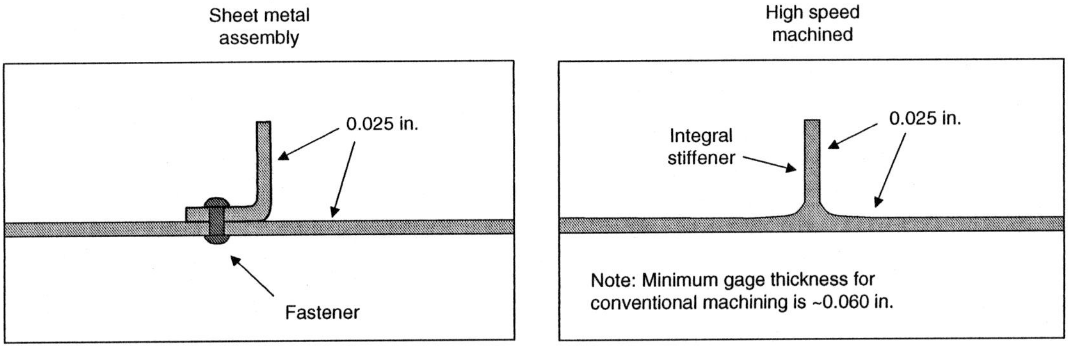


| Was | | Now | |
|---|---|---|---|
| Number of Pieces | 44 | Number of Pieces | 6 |
| Number of Tools | 53 | Number of Tools | 5 |
| Design and Fabrication hr (Tools) | 965 | Design and Fabrication hr (Tools) | 30 |
| Fabrication hr | 13.0 | Fabrication hr | 8.6 |
| Assembly Manhours | 50 | Assembly Manhours | 5.3 |
| Weight (lb) | 9.58 | Weight (lb) | 8.56 |

**Fig. 26.12** Comparison between assembled and high-speed machined assembly

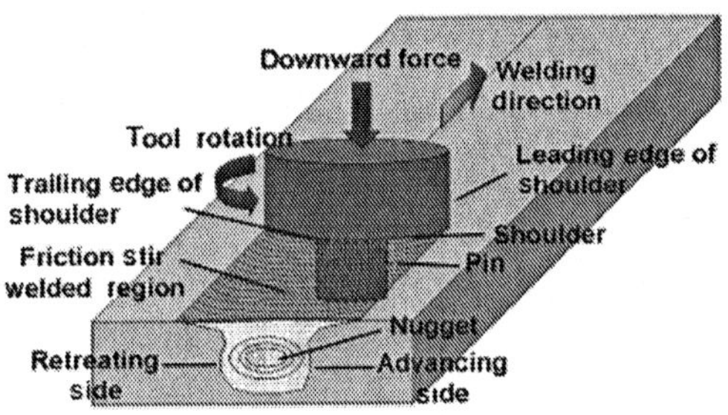


**Fig. 26.13** Friction stir welding process. Source: Ref 14

corrosion resistance, but most 6*xxx*-series alloys contain little copper. Some other alloying elements also decrease corrosion resistance. Lead (added to 2011 and 6262 for machinability), nickel (added to 2018, 2218, and 2618 for elevated-temperature service), and tin (used in 8*xx.x* castings) tend to decrease the corrosion resistance but not enough to matter in most applications. Many of the 5*xxx*-series alloys have corrosion resistance as good as commercially pure aluminum, are more resistant to saltwater, and thus are useful in marine applications.

Single-phase alloys tend to be more corrosion resistant than the two-phase precipitation-hardened alloys. With multiple phases, galvanic cells can arise between the phases. Therefore, the 3*xxx* and 5*xxx* alloys are more corrosion resistant than the 2*xxx* and 7*xxx* alloys. In the 2*xxx* alloys, precipitation of $CuAl_2$ at the grain boundaries can cause a depletion of copper adjacent to the boundaries, making these regions anodic relative to the centers of the grains, resulting in rapid intergranular corrosion. Of the precipitation-hardenable alloys, the 6*xxx* alloys have the best corrosion resistance but are not as strong as the heat treated 2*xxx* and 7*xxx* alloys.

The naturally forming alumina ($Al_2O_3$) coating is thin (0.005 to 0.015 mm, or 0.0002 to 0.0006 in. thick) and a poor base for paint. Two types of coatings, chemical conversion coatings and anodizing, are used to form a more uniform and thicker oxide for enhanced corrosion protection. Chemical conversion coatings produce a porous and absorptive oxide (0.05 to 0.076 mm, or 0.002 to 0.003 in. thick) that is very uniform and morphologically tailored to bond well with paint primers. The oxides are chromate- or phosphate-based, which further aids in corrosion protection.

To further enhance corrosion resistance, finished parts are frequently anodized before being placed in service to increase the thickness of the $Al_2O_3$ layer on the surface. Anodizing is an electrolytic process that produces thicker (0.05 to 0.13 mm, or 0.002 to 0.005 in.) and more durable oxides than those produced by conversion coatings; therefore, it provides better corrosion resistance. Both sulfuric and chromic acid baths are used along with an electrical current to deposit a porous oxide layer on the surfaces. The component is the anode in an electrolytic cell, while the acid bath serves as the cathode. Anodizing consists of degreasing, chemical cleaning, anodizing, and then sealing the anodized coating in a slightly acidified hot water bath. Depending on the temperature and time of the anodizing operation, the oxide layer can range from 5 to 13 μm (0.2 to 0.5 mil) to provide enhanced corrosion protection to the underlying metal. After anodizing, the pores in the oxide film are sealed by placing the part in slightly acidified 80 to 90 °C (180 to 200 °F) water. This sealing treatment converts the aluminum oxide to aluminum monohydrate, which expands to fill the pores.

## REFERENCES


1. J.G. Kaufman, Aluminum Alloys, *Handbook of Materials Selection,* John Wiley & Sons, Inc., 2002, p 89–134
2. P.C. Varley, *The Technology of Aluminum and Its Alloys,* Newnes-Butterworths, London, 1970
3. J.R. Kissell, Aluminum Alloys, *Handbook of Materials for Product Design,* McGraw-Hill, 2001, p 2.1–2.178
4. R.E. Sanders et al., *Proceedings of the International Conference on Aluminum Alloys—Physical and Mechanical Properties* (Charlottesville, VA), Engineering Materials Advisory Services, Warley, U.K., 1941, 1986
5. W.A. Anderson, *Aluminum,* Vol 1, American Society for Metals, 1967
6. R.N. Caron and J.T. Staley, Aluminum and Aluminum Alloys: Effects of Composition, Processing, and Structure on Properties of Nonferrous Alloys, *Materials Selection and Design,* Vol 20, *ASM Handbook,* ASM International, 1997
7. F.C. Campbell, *Manufacturing Technology for Aerospace Structural Materials,* Elsevier Scientific, 2006
8. B. Smith, The Boeing 777, *Adv. Mater. Process.,* Sept 2003, p 41–44
9. J.-P. Immarigeon et al., Lightweight Materials for Aircraft Applications, *Mater. Charact.,* Vol 35, 1995, p 41–67
10. Aluminum Casting Principles, Lesson 5, *Aluminum and Its Alloys,* ASM Course, American Society for Metals, 1979
11. K.R. Van Horn, Ed., *Aluminum,* Vol 1, *Properties, Physical Metallurgy and Phase Diagrams,* American Society for Metals, 1967
12. M. Warmuzek, Metallographic Techniques for Aluminum and Its Alloys, *Metallography and* Microstructures, Vol 9, *ASM Handbook,* ASM International, 2004
13. E.L. Rooy, Aluminum and Aluminum Alloys, *Casting,* Vol 15, *ASM Handbook,* ASM International, 1988
14. R.S. Mishra and M.W. Mahoney, *Friction Stir Welding and Processing,* ASM International, 2007


## SELECTED REFERENCES


- Alloy and Temper Designation Systems for Aluminum, *Metals Handbook Desk Edition,* 2nd ed., ASM International, 1998
- Aluminum Foundry Products, *Metals Handbook Desk Edition,* 2nd ed., ASM International, 1998
- Aluminum Wrought Products, *Metals Handbook Desk Edition,* 2nd ed., ASM International, 1998
- J.W. Bray, Aluminum Mill and Engineered Wrought Products, *Properties and Selection: Nonferrous Alloys and Special-Purpose, Materials,* Vol 2, *ASM Handbook,* ASM International, 1990
- R.F. Gaul, Hot and Cold Working Aluminum Alloys, Lesson 7, *Aluminum and Its Alloys,* ASM Course, American Society for Metals, 1979
- Heat Treating of Aluminum Alloys, *Heat Treating,* Vol 4, *ASM Handbook,* ASM International, 1991
- I.J. Polmear, *Light Alloys—Metallurgy of the Light Metals,* 3rd ed., Butterworth Heinemann, 1995
- W.F. Smith, Precipitation Hardening of Aluminum Alloys, Lesson 9, *Aluminum and Its Alloys,* ASM Course, American Society for Metals, 1979
- E.A. Starke and J.T. Staley, Application of Modern Aluminum Alloys to Aircraft, *Prog. Aerosp. Sci.,* Vol 32, 1996, p 131–172
- W.M. Thomas and E.D. Nicholas, Friction Stir Welding for the Transportation Industries, *Mater. Des.,* Vol 18 (No. 4/6), 1997, p 269–273
- J.C. Williams and E.A. Starke, Progress in Structural Materials for Aerospace Systems, *Acta Mater.,* Vol 51, 2003, p 5775–5799

# CHAPTER 27

# Magnesium and Zinc

MAGNESIUM has the lowest density (1.738 g/cm$^3$, or 0.063 lb/in.$^3$) of the structural metals, with a density of approximately 2/3 that of aluminum and 1/4 that of steel. Although magnesium alloys have only moderate tensile strengths, in the range of 140 to 345 MPa (20 to 50 ksi), and a modulus of elasticity of only 45 GPa (6.5 msi), due to their low densities, they exhibit favorable specific strengths (tensile strength/density) and specific moduli (modulus/density) comparable to other structural metals.

The majority of the annual production of magnesium is used for alloying elements in aluminum alloys, with only approximately 15% of the annual production being used for structural applications, with the majority of these being castings. Magnesium and its alloys are used in a wide variety of structural applications, such as automotive, industrial, materials handling equipment, kitchen appliances, hand-held tools, luggage frames, computer housings, cellular phones, and ladders. Magnesium is relatively inexpensive and easy to cast, machine, and weld; its electrical and thermal conductivity and heat capacity are relatively high. Magnesium alloys have very good damping capacity, and castings have found applications in high-vibration environments.

Over the years, one of the major drawbacks of magnesium alloys has been corrosion. Magnesium occupies the highest anodic position on the galvanic series and, as shown in Fig. 27.1, can be subject to rather severe corrosion. The corrosion problem is due to the impurity elements iron, nickel, and copper. The severe effect of iron on the corrosion susceptibility of pure magnesium is shown in Fig. 27.2. However, the use of higher-purity magnesium alloys, as shown for a number of casting alloys in Fig. 27.3, has led to corrosion resistance approaching that of some of the competing aluminum casting alloys.

## 27.1 Magnesium Metallurgy

Pure magnesium has a hexagonal close-packed crystalline structure that restricts slip at room temperature to the basal planes. Magnesium alloys work harden rapidly at room temperature, and their ductility is low. Since additional slip planes become operative at elevated temperature, wrought magnesium alloys are normally formed at temperatures greater than 205 °C (400 °F), normally in the range of 345 to 510 °C (650 to 950 °F), depending on the

**Fig. 27.1** Severely corroded magnesium part. Source: Ref 1

specific alloy. Wrought alloys also develop crystallographic texturing during mechanical deformation, leading to anisotropic mechanical properties. For example, a rolled sheet with a tensile strength of 220 MPa (32 ksi) and 2% elongation measured parallel to the rolling direction may display higher properties (e.g., a tensile strength of 262 MPa, or 38 ksi, and 8% elongation) when measured transverse to the rolling direction.

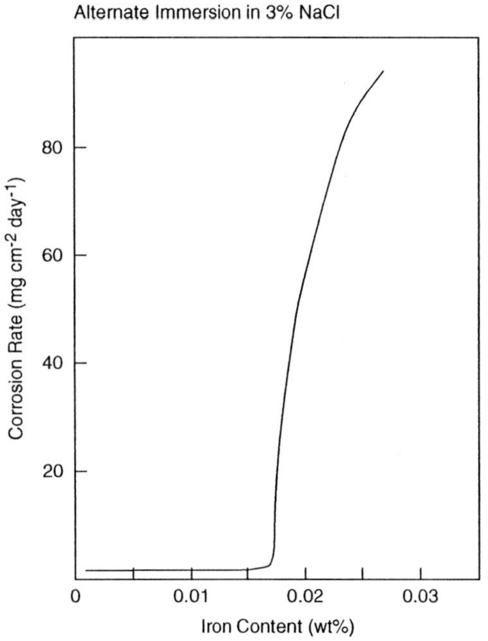


**Fig. 27.2** Effect of iron on corrosion of pure magnesium. Source: Ref 3

In addition, the yield strength in compression for wrought products is only approximately 40 to 70% of that in tension. During hot working, the individual crystals deform primarily by basal slip, and the basal planes rotate so that they tend to become oriented parallel to the working direction. This positions the grains so that they form twins when loaded in compression, thus lowering the compressive strength. Since castings do not develop texture, the compressive yield strength of castings is approximately equal to the tensile yield strength. As a result of these factors, combined with the fact that magnesium wrought products are more expensive than comparable aluminum wrought products, magnesium alloy castings are the much more widely used product forms.

Magnesium also has a rather low melting point (650 °C, or 1202 °F), which increases its susceptibility to elevated-temperature creep. However, through improved alloying techniques, the creep resistance of magnesium alloys has been significantly improved.

The most important alloying additions are aluminum, zinc, and zirconium. Aluminum provides solid-solution strengthening and widens the freezing range, making the alloy easier to cast. As aluminum is added to magnesium, the strength continuously increases as the aluminum content is increased up to 10 wt% Al, but the elongation peaks at approximately 3 wt% Al. Alloys with 3 wt% Al have the highest ductility, and those with 9% Al have the best strength, but those with approximately 6 wt% provide the best combination of strength and ductility. Zinc behaves in a manner similar to aluminum; the

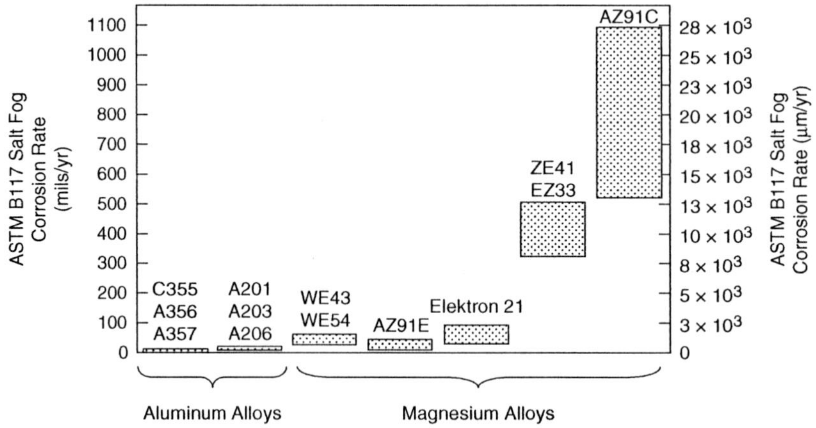


**Fig. 27.3** Corrosion comparison of aluminum and magnesium casting alloys. Source: Ref 4

ductility reaches a maximum at a 3 wt% addition, and a good combination of strength and ductility occurs with 5 wt% Zn. However, zinc causes hot shortness when present in amounts greater than 1 wt% in alloys containing 7 to 10 wt% Al. Zinc also improves the corrosion resistance by combining with the harmful impurities iron and nickel. Zinc is also used in conjunction with zirconium, rare earths, or thorium to produce precipitation-hardenable alloys.

Manganese is used to improve corrosion resistance of Mg-Al and Mg-Al-Zn alloys by removing iron due to the formation of harmless intermetallic compounds. The amount of manganese that can be added is restricted to approximately 1.5 wt% due to its low solubility in magnesium. Silicon greatly increases the fluidity of molten magnesium, thereby increasing its castability; however, silicon decreases corrosion resistance if iron is present. Silicon also provides increased creep resistance.

Zirconium is a powerful grain refiner, as shown in Fig. 27.4. However, zirconium cannot be used in combination with aluminum or manganese because it forms brittle intermetallic compounds that destroy ductility. The remarkable effectiveness of zirconium in refining the grains of cast magnesium has been explained by the similarities in crystal structure and lattice parameters of the two elements. Zirconium is such an important alloying addition that a whole series of aluminum-free magnesium-zirconium alloys have been developed. Zirconium additions are usually kept below 0.8 wt% since, at higher concentrations, it readily forms compounds with iron, aluminum, silicon, carbon, oxygen, and nitrogen and reacts with hydrogen to form a hydride that is insoluble in magnesium.

Rare earth (RE) elements are potent solid-solution strengtheners. The REs are usually added as natural mixtures of either mischmetal or as didymium. Mischmetal contains approximately 50 wt% Ce, with the remainder being mainly lanthanum and neodymium, while didymium contains approximately 85 wt% Nd and 15 wt% Pr. Rare earths and cerium also provide precipitation-hardening capability. As little as 1 wt% RE additions increase strength and reduce the tendency for weld cracking. Weld cracking and porosity are reduced because they narrow the freezing range. Silver alloying

Pure Magnesium

Pure Magnesium
+
Zirconium

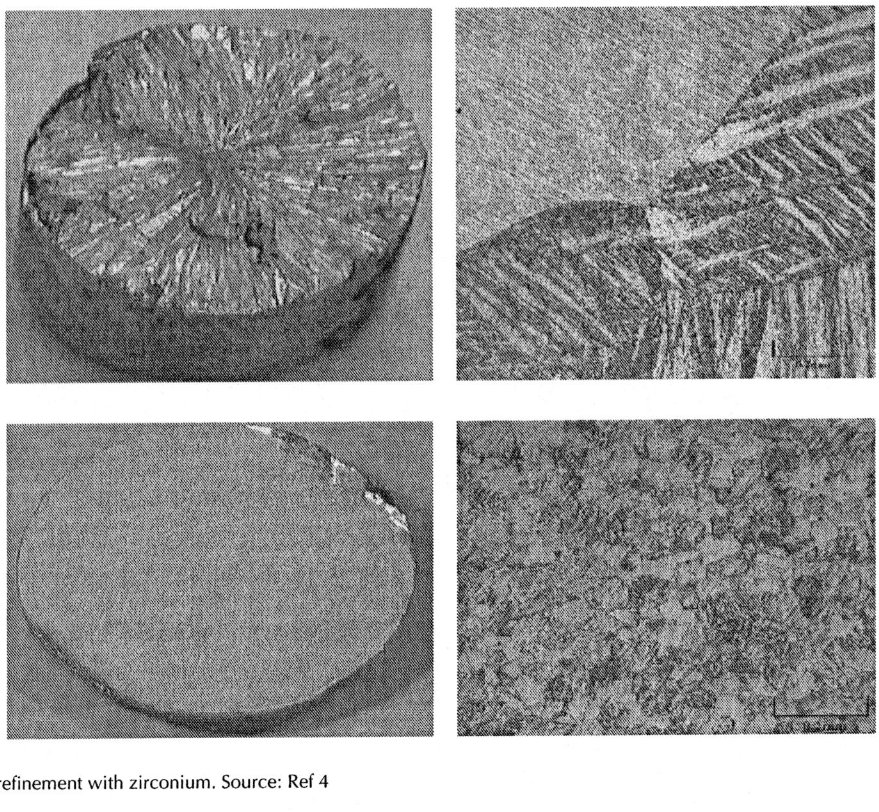

**Fig. 27.4** Grain refinement with zirconium. Source: Ref 4

additions, although expensive, greatly enhance the strength response during precipitation hardening. Both thorium (2 to 3 wt%) and yttrium (4 to 5 wt%) additions increase creep resistance. A number of thorium-containing alloys have been developed; however, even though thorium is only mildly radioactive, the use of these alloys has declined due to potential health concerns.

The elements iron and nickel are harmful impurities that greatly reduce corrosion resistance. Copper is also often considered, along with iron and nickel, as an impurity but is actually used as an alloying addition in some magnesium alloys. Iron is by far the most troublesome of the three, because nickel and copper are more readily controlled by selecting the purity of the starting materials. The iron that is present is controlled by additions of $MnCl_2$ to the melt during casting.

Although magnesium alloys are produced in both wrought and cast product forms, cast alloys are much more widely used. Some of the wrought alloys are strengthened by cold working, while others are precipitation hardenable. The cast alloys are used in the as-cast, annealed, or precipitation-hardened conditions. The alloys themselves are normally one of two classes: aluminum-containing alloys or zirconium-containing alloys. The tensile properties of magnesium alloys are moderate and generally range from 69 to 345 MPa (10 to 50 ksi) yield strength and 138 to 379 MPa (20 to 55 ksi) tensile strength, with elongations of 1 to 15%.

Since the main source of magnesium is seawater, it is a metal with an inexhaustible supply. Although magnesium can be produced by several extractive metallurgy processes, the most widely used process involves precipitating magnesium in dolomite [$CaMg(CO_3)_2$] and seawater as insoluble magnesium hydroxide [$Mg(OH)_2$], which is then treated with hydrochloric acid to produce magnesium chloride. Electrolytic cells are used to convert $MgCl_2$ to magnesium metal and chlorine gas.

## 27.2 Magnesium Alloy Designation

Magnesium alloys are designated by a combination of letters and numbers, which was established by ASTM International. The system, shown in Table 27.1, covers both the chemical compositions and the tempers. As an example, in the alloy EQ21A-T6, the first part of the designation, EQ, identifies rare earths (E) and silver (Q) as the main alloying elements. The second part, 21, gives the rounded-off percentages of both rare earths and silver, respectively; the rare earth addition is 2.25 wt%, which is rounded down to 2 wt%, while the silver content is 1.5 wt%, which is rounded down to 1 wt%. The third part, A, identifies that this is the original composition of the alloy EQ21. In the third part letter designation, A is the original composition, B is the second modification, C is the third modification, D indicates a high-purity version, and E is a high-purity, corrosion-resistant composition. Finally, the fourth part, T6, denotes that the alloy is solution treated and artificially aged. The cold working (H) and heat treat (T) temper designations are essentially the same as those used for aluminum alloys. Since cast alloys are the most prevalent product forms, the most widely used tempers are T4, T5, and T6. The compositions of magnesium casting alloys are shown in Table 27.2, and the minimum tensile properties of the cast alloys are given in Table 27.3.

## 27.3 Magnesium Casting Alloys

Magnesium castings are used in structural applications because of their low weight and good damping characteristics. Molten magnesium alloys have a very low viscosity, allowing the metal to flow long distances and fill narrow mold cavities. Their relatively low melting points allow the use of hot chamber die casting, and their minimal reactivity with steel below 705 °C (1300 °F) allows the use of inexpensive steel crucibles and molds. Since magnesium and its alloys readily react with air in the molten state, it is necessary to use a protective flux during all melting operations. Fluxes include mixtures of various proportions of $MgCl_2$, KCl, $CaF_2$, MgO, and $BaCl_2$. Since the fluxes have nearly the same density as molten magnesium, they float on the surface and form a scaly crust that protects the underlying molten magnesium.

During sand casting, molten magnesium will react with moisture in the sand to produce surface blackening and localized porosity, both of which reduce mechanical properties. Therefore, inhibitors are added to the sand to prevent the reaction.

As for aluminum casting alloys, hydrogen can cause porosity problems during solidification. Like aluminum, the main sources of hydrogen are moisture in the charge, moisture in the air, or hydrogen from the combustion products used for melting. Degassing is conducted by bubbling an insoluble gas, such as pure argon or nitrogen, throughout the melt. The formation of $MgCl_2$ allows chlorine to be added more rapidly and simultaneously removes MgO by fluxing.

### *27.3.1 Magnesium-Aluminum-Base Casting Alloys*

Magnesium-aluminum alloys were the original casting alloys. Most alloys contain 8 to 9 wt% Al with small amounts of zinc and manganese. Since the magnesium-aluminum alloys form a large and variable grain size when cast, they must be inoculated during casting. For those alloys that do not contain zirconium, the melt is refined by the addition of small briquettes of volatile carbon-containing compound. Grain refinement occurs to the inoculation of the melt with $Al_4C_3$ or $AlN \cdot Al_4C_4$. Die-cast alloys freeze so rapidly that inoculation is not required.

In the magnesium-aluminum phase diagram (Fig. 27.5), there is a eutectic between the terminal solid solution and the brittle intermetallic compound β ($Mg_{17}Al_{12}$). The eutectic structures contain more compound than a solid solution, and since the compound is brittle, alloys with a eutectic network are also brittle. When the aluminum content exceeds 8%, discontinuous precipitation of the β phase occurs at the grain boundaries, leading to a decrease in ductility. The phase diagram also shows that aluminum has a maximum solubility of 12.7 wt% at 436 °C (817 °F), which decreases down to approximately 2 wt% at room temperature. While this type of solubility would seem to indicate that these alloys would be amenable to precipitation hardening, unfortunately the resulting precipitate is coarse and results in only moderate hardening. While many of the alloying additions to magnesium form eutectics with decreasing solid solubility that lead to precipitation hardening, the strength that results from this hardening mechanism is much less than that observed with the precipitation-hardened aluminum alloys. Since the response to precipitation heat treatment is poor, most of these alloys are used in either the as-cast

**Table 27.1 ASTM International designation for magnesium alloys**

| First part | Second part | Third part | Fourth part |
|---|---|---|---|
| Indicates the two principal alloying elements | Indicates the amounts of the two principal alloying elements | Distinguishes between different alloys with the same percentages of the two principal alloying elements | Indicates condition (temper) |
| Consists of two code letters representing the two main alloying elements arranged in order of decreasing percentage (or alphabetically if percentages are equal) | Consists of two numbers corresponding to rounded-off percentages of the two main alloying elements and arranged in the same order as alloy designations in the first part | Consists of a letter of the alphabet assigned in order as compositions become standard | Consists of a letter followed by a number (separated from the third part of the designation by a hyphen) |
| A, aluminum<br>B, bismuth<br>C, copper<br>D, cadmium<br>E, rare earth<br>F, iron<br>G, magnesium<br>H, thorium<br>K, zirconium<br>L, lithium<br>M, manganese<br>N, nickel<br>P, lead<br>Q, silver<br>R, chromium<br>S, silicon<br>T, tin<br>W, yttrium<br>Y, antimony<br>Z, zinc | Whole numbers | A, first compositions, registered with ASTM<br>B, second compositions, registered with ASTM<br>C, third compositions, registered with ASTM<br>D, high purity, registered with ASTM<br>E, high corrosion resistant, registered with ASTM<br>X1, not registered with ASTM | F, as fabricated<br>O, annealed<br>H10 and H11, slightly strain hardened<br>H23, H24, and H26, strain hardened and partially annealed<br>T4, solution heat treated<br>T5, artificially aged only<br>T6, solution heat treated and artificially aged<br>T8, solution heat treated, cold worked, and artificially aged |

Source: Ref 5

or annealed condition. Annealing helps to redissolve any β phase, leading to a more homogeneous solid solution.

The addition of zinc also helps to strengthen the alloys by a combination of solid solution and precipitation hardening, but the additions are limited due to an increasing susceptibility to hot cracking during solidification. Zinc refines the precipitate and increases the strength by a combination of solid-solution strengthening and precipitation hardening (Fig. 27.6). The addition of zinc must be balanced by a concurrent decrease in the aluminum content to keep the total below approximately 11 wt%. For example, AZ63 (6Al-3Zn) is one alloy, while AZ91 (9Al-1Zn) is another alloy, with both containing less than 11%.

The early Mg-Al-Zn castings had severe corrosion problems when exposed to wet or moist environments. However, small additions of manganese (0.2 wt%) helped in improving the corrosion resistance; manganese reacts with iron impurities to form relatively harmless intermetallic compounds. AZ91 (Mg-9.5Al-0.5Zn-0.3Mn) is the most widely used of these casting alloys. The more recent version of this alloy, AZ91E, contains much less of the harmful impurities iron, nickel, and copper. Iron is limited to 0.015 wt%, nickel is limited to 0.0001 wt%, and copper to 0.015 wt%. This higher-purity grade leads to much improved corrosion resistance, as much as a 100 times improvement in salt fog tests compared to the earlier version, AZ91C. The corrosion resistance of AZ91E is comparable to some aluminum casting alloys. A comparison of the improved corrosion resistance of AZ91E compared to the earlier alloy AZ91C is shown in Fig. 27.7. A good dispersion of $Mg_{17}Al_{12}$ particles in die-cast AZ91D is illustrated in Fig. 27.8. With appropriate heat treatments, this phase can be dissolved in solution and then precipitated throughout the microstructure. When higher ductility is required, there are some alloys with lower aluminum contents, such as AM20 (Mg-2Al-0.5Mn), AM50 (Mg-5Al-0.3Mn), and AM60 (Mg-6Al-0.2Mn).

**Table 27.2 Nominal compositions of magnesium casting alloys**

| Alloy | | Composition, wt% | | | | | | |
|---|---|---|---|---|---|---|---|---|
| ASTM No. | UNS No. | Al | Mn(a) | Zn | Th | Zr | Rare earths | Other |
| **Sand and permanent mold castings** | | | | | | | | |
| AM100A | M10100 | 10.0 | 0.1 | 0.3 | ... | ... | ... | ... |
| AZ63A | M11630 | 6.0 | 0.15 | 3.0 | ... | ... | ... | ... |
| AZ81A | M11810 | 7.6 | 0.13 | 0.7 | ... | ... | ... | ... |
| AZ91C | M11914 | 8.7 | 0.13 | 0.7 | ... | ... | ... | ... |
| AZ91E | M11918 | 8.7 | 0.13 | 0.7 | ... | ... | ... | 0.005 Fe(b) |
| AZ92A | M11920 | 9.0 | 0.10 | 2.0 | ... | ... | ... | ... |
| EQ21A | M12210 | ... | ... | ... | ... | 0.7 | 2.25(c) | 1.5 Ag |
| EZ33A | M12330 | ... | ... | 2.55 | ... | 0.75 | 3.25 | ... |
| HK31A | M13310 | ... | ... | 0.3 | 3.25 | 0.7 | ... | ... |
| HZ32A | M13320 | ... | ... | 2.1 | 3.25 | 0.75 | 0.1 | ... |
| K1A | M18010 | ... | ... | ... | ... | 0.7 | ... | ... |
| QE22A | M18220 | ... | ... | ... | ... | 0.7 | 2.15(c) | 2.5 Ag |
| QH21A | M18210 | ... | ... | 0.2 | 1.1(d) | 0.7 | 1.05(c)(d) | 2.5 Ag |
| WE43A | M18430 | ... | 0.15 | 0.2 | ... | 0.7 | 3.4(e) | 4.0 Y |
| WE54A | M18410 | ... | 0.15 | 0.2 | ... | 0.7 | 2.75(e) | 5.0 Y |
| ZC63A | M16331 | ... | 0.25 | 6.0 | ... | ... | ... | 2.7 Cu |
| ZE41A | M16410 | ... | 0.15 | 4.25 | ... | 0.7 | 1.25 | ... |
| ZE63A | M16630 | ... | ... | 5.75 | ... | 0.7 | 2.55 | ... |
| ZH62A | M16620 | ... | ... | 5.7 | 1.8 | 0.75 | ... | ... |
| ZK51A | M16510 | ... | ... | 4.55 | ... | 0.75 | ... | ... |
| ZK61A | M16610 | ... | ... | 6.0 | ... | 0.8 | ... | ... |
| **Die casting** | | | | | | | | |
| AM60A | M10600 | 6.0 | 0.13 | 0.22 | ... | ... | ... | 0.5 Si; 0.35 Cu |
| AS41A | M10410 | 4.25 | 0.20 | 0.12 | ... | ... | ... | 1.0 Si |
| AS41B | M10412 | 4.25 | 0.35 | 0.12 | ... | ... | ... | 1.0 Si |
| AZ91A | M11910 | 9.0 | 0.13 | 0.7 | ... | ... | ... | 0.5 Si |
| AZ91B | M11912 | 9.0 | 0.13 | 0.7 | ... | ... | ... | 0.5 Si; 0.35 Cu |
| AZ91D | M11916 | 9.0 | 0.15 | 0.7 | ... | ... | ... | ... |
| AM60B | M10602 | 6.0 | 0.24 | 0.22 | ... | ... | ... | ... |
| AM50A | M10500 | 4.9 | 0.26 | 0.22 | ... | ... | ... | ... |

(a) Minimum. (b) If iron exceeds 0.005%, the iron-to-manganese ratio should not exceed 0.032. (c) Rare earth elements are in the form of didymium (a mixture of rare earth elements made chiefly of neodymium and praseodymium). (d) Thorium and didymium total is 1.5 to 2.4%. (e) Rare earths are 2.0 to 2.5% and 1.5 to 2.0% Nd for WE43A and WE54A, respectively, with the remainder being heavy rare earths. Source: Ref 5.

The improved ductility with reduced aluminum content is due to the reduced amount of β phase around the grain boundaries.

The Mg-Al and Mg-Al-Zn alloys have some susceptibility to the formation of microporosity, but, in general, their castability is good and their corrosion resistance is satisfactory for many environments. They can be used at temperatures up to 110 to 120 °C (230 to 250 °F); at higher temperatures, their creep resistance becomes a problem. Creep in magnesium alloys occurs primarily by grain-boundary sliding, and as the temperature in these alloys increases, the β phase at the grain boundaries softens, allowing grain-boundary sliding. Peak-aged (T6 temper) magnesium-aluminum alloys can be subject to stress-corrosion cracking when the part is stressed to values exceeding 50% of its yield strength.

For better creep resistance, silicon is added, which forms fine, hard particles of $Mg_2Si$ along the grain boundaries to help retard grain-boundary sliding. Examples are the alloys AS21 (Mg-2Al-1Si-0.4Mn) and AS41(Mg-4Al-1Si-0.3Mn). However, even these alloys do not compete with the creep properties of competing die-cast aluminum alloys.

**Table 27.3 Minimum mechanical properties for magnesium casting alloys**

| Alloy-temper | Tensile strength, Mpa | Tensile strength, ksi | Yield strength, Mpa | Yield strength, ksi | Elongation in 50 mm (2 in.), % | Hardness, HB |
|---|---|---|---|---|---|---|
| **Sand and permanent mold castings** | | | | | | |
| AM100A-T6 | 241 | 35 | 117 | 17 | ... | 69 |
| AZ63A-T6 | 234 | 34 | 110 | 16 | 3 | 73 |
| AZ81A-T4 | 234 | 34 | 76 | 11 | 7 | 55 |
| AZ91C-T6 | 234 | 34 | 110 | 16 | 3 | 70 |
| AZ91E-T6 | 234 | 34 | 110 | 16 | 3 | 70 |
| AZ92A-T6 | 234 | 34 | 124 | 18 | 1 | 81 |
| EQ21A-T6 | 234 | 34 | 172 | 25 | 2 | 78 |
| EZ33A-T5 | 138 | 20 | 96 | 14 | 2 | 50 |
| HK31A-T6 | 186 | 27 | 89 | 13 | 4 | 66 |
| HZ32A-T5 | 186 | 27 | 89 | 13 | 4 | 55 |
| K1A-F | 165 | 24 | 41 | 6 | 14 | ... |
| QE22A-T6 | 241 | 35 | 172 | 25 | 2 | 78 |
| QH21A-T6 | 241 | 35 | 186 | 27 | 2 | ... |
| WE43A-T6 | 221 | 32 | 172 | 25 | 2 | 85 |
| WE54A-T6 | 255 | 37 | 179 | 26 | 2 | 85 |
| ZC63A-T6 | 193 | 28 | 125 | 18 | 2 | 60 |
| ZE41A-T5 | 200 | 29 | 133 | 19.5 | 2.5 | 62 |
| ZE63A-T6 | 276 | 40 | 186 | 27 | 5 | ... |
| ZH62A-T5 | 241 | 35 | 152 | 22 | 5 | 70 |
| ZK51A-T5 | 234 | 34 | 138 | 20 | 5 | 65 |
| ZK61A-T6 | 276 | 40 | 179 | 26 | 5 | 70 |
| **Die castings** | | | | | | |
| AM50A | 200 | 29 | 110 | 16 | 10 | ... |
| AM60A and B | 220 | 32 | 130 | 19 | 8 | ... |
| AS41A and B | 210 | 31 | 140 | 20 | 6 | ... |
| AZ91A, B, and D | 230 | 34 | 160 | 32 | 3 | ... |

Source: Ref 5

The addition of copper to magnesium-zinc alloys produces a higher response to precipitation hardening and increases the ductility. The alloy ZC63 (Mg-6Zn-3Cu-0.5Mn) is a copper-containing sand casting alloy. The copper increases the eutectic temperature, which allows higher solution heat treating temperatures and thus more complete solution of zinc and copper. During aging, the precipitate is more evenly distributed throughout the matrix instead of being concentrated at the grain boundaries.

### *27.3.2 Magnesium-Zirconium-Base Casting Alloys*

Due to the remarkable grain-refining ability of zirconium, a series of magnesium casting alloys containing Mg-Zn-Zr were initially produced. Alloys in the Mg-Zn-Zr system develop high yield strengths with reasonably good ductility. However, they are more expensive than the Mg-Al-Zn alloys and more difficult to cast due to the occurrence of shrinkage, microporosity, and cracking. The alloys ZK51 (Mg-4.6Zn-0.75Zr) and ZK61 (Mg-6Zn-0.7Zr) were initially developed as sand casting alloys. They contain 5 to 6 wt% Zn for strengthening along with approximately 0.7 wt% Zr for grain refinement. Both ZK51 and ZK61 produce fairly high strength levels when heat treated to the T5 or T6 tempers, respectively. However, since these alloys are difficult to cast and are not repairable by welding, they are not widely used.

**Magnesium-Rare Earth-Zinc-Zirconium Casting Alloys.** The RE elements are lanthanum, mischmetal, cerium, and didymium, and

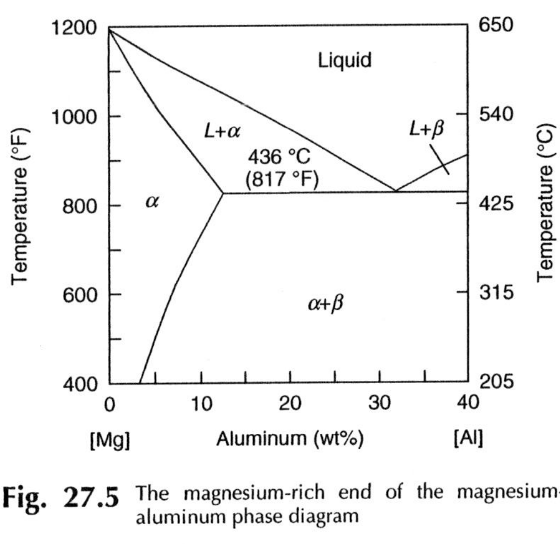


**Fig. 27.5** The magnesium-rich end of the magnesium-aluminum phase diagram

they have different effects on the magnesium alloys. The RE metals form eutectics with magnesium, which produces good castablility since the low-melting-point eutectics form at the grain boundaries and tend to suppress the formation of microporosity and hot cracking. These alloys are relatively free of microporosity but are more prone to shrinkage defects and dross inclusions than the Mg-Al-Zn alloys. On aging during precipitation heat treatments, precipitates form with the grains. This produces better creep resistance due to the combined strengthening effects of grain-boundary phases and precipitates within the grains. Zinc additions also improve the strength properties. One common alloy is ZE41 (Mg-4Zn-1.3RE-0.7Zr), which has moderate strength in the T5 condition that is maintained up to 150 °C (300 °F). Due to the removal of zinc from the solid solution during the formation of the Mg-Zn-RE phases at the grain boundaries, the room-temperature strengths tend to be rather low: 138 MPa

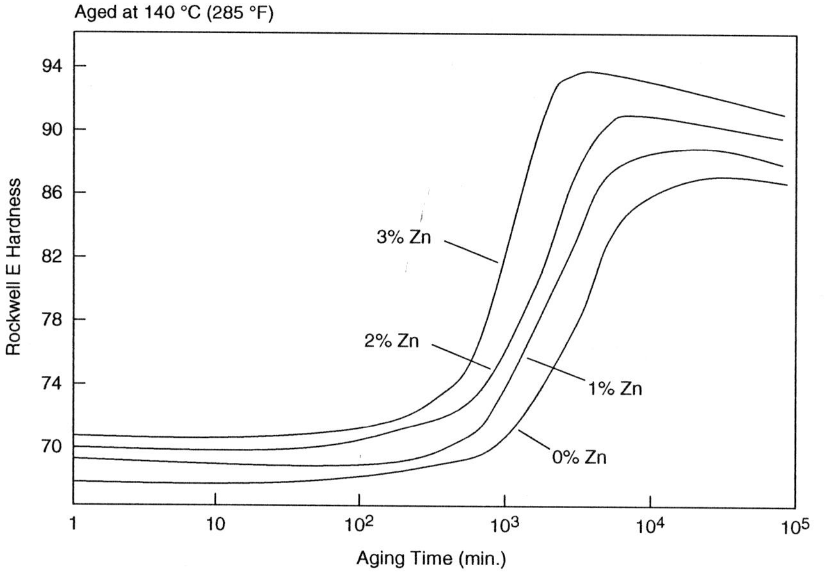


**Fig. 27.6** Aging curves for Mg-9wt%Al alloy with various zinc additions. (Zinc compositions are given in wt%).

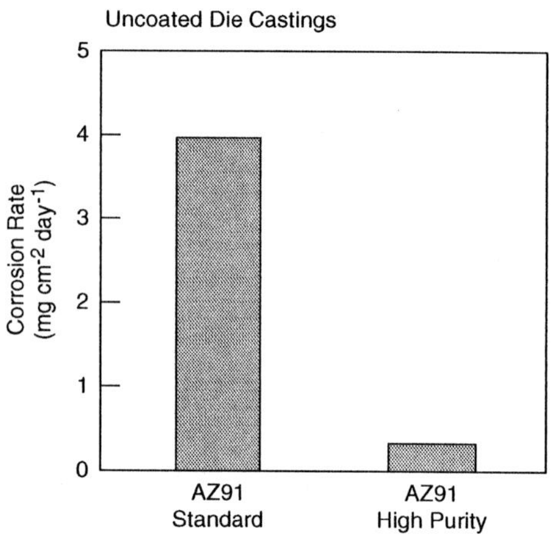


**Fig. 27.7** Effect of purity on corrosion resistance of AZ91 alloy. Source: Ref 6

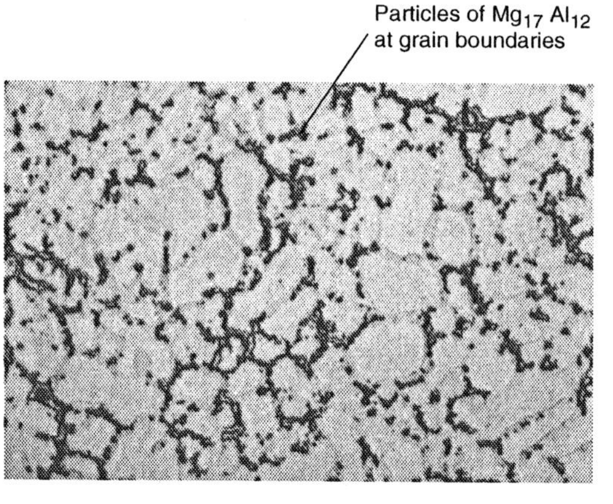


**Fig. 27.8** Microstructure of ZA91A-F die casting alloy. Original magnification 500×. Source: Ref 7

(20 ksi) for EZ33-T5 (Mg-2.7Zn-3.2RE-0.7Zr) and 200 MPa (29 ksi) for ZE41-T5.

**Magnesium-Thorium Casting Alloys.** Additions of thorium to magnesium increase creep resistance, and these alloys, such as HK31A-T6 (Mg-3.2Th-0.7Zr), can be used at temperatures approaching 175 °C (350 °F). Thorium also improves castability, and these alloys are weldable. Zinc additions in the alloys HZ32A (Mg-3.2Th-2.2Zn-0.7Zr) and HZ62A (Mg-3.2Th-6Zn-0.7Zr) further increase creep resistance. However, as previously mentioned, thorium is mildly radioactive, and the use of these alloys has decreased due to safety concerns.

**Magnesium-Silver Casting Alloys.** Silver greatly enhances the response to precipitation hardening. Silver is added to the Mg-RE-Zr alloys to improve their tensile properties. Several compositions exist that are comparable to the aluminum-base casting alloys. The major problem with the silver-containing alloys is their higher cost. The most widely used of the silver-containing alloys in QE22, which contains 2.5 wt% Ag and 2.5 wt% RE. When heat treated to the T6 condition, QE22A has a tensile strength of 241 MPa (35 ksi). It also retains good properties to approximately 260 °C (500 °F) but is generally restricted to 205 °C (400 °F) for creep-critical applications. The microstructure of sand-cast QE22A-T6 is shown in the Fig. 27.9 micrograph.

**Magnesium-Yttrium Casting Alloys.** The most recently developed high-temperature alloys contain approximately 4 to 5 wt% Y. These alloys have high strength with good creep resistance at temperatures up to 300 °C (570 °F). In addition, they have superior corrosion resistance compared to other high-temperature alloys, comparable to some aluminum-base casting alloys. However, pure yttrium is expensive and is also difficult to alloy with magnesium because of its high melting point and its strong affinity for oxygen. Therefore, a mischmetal containing 75 wt% Y in REs is used instead of pure yttrium. As an example of their high-temperature stability, the alloy WE43 (Mg-4Y-3.25RE-0.5Zr) has a room-temperature tensile strength of 248 MPa (36 ksi) when heat treated to the T6 condition, which it maintains after long-term aging (5000 h) at 205 °C (400 °F). The effect of 205 °C (400 °F) exposure on the room-temperature strength of WE43 is shown in Fig. 27.10. A relatively new alloy, Elektron 21 (Mg-4RE-0.3Zn-0.5Zr), offers many of the advantages of WE43; however, the cost is lower and the castability is better. Instead of using yttrium, neodymium and gadolinium are used along with zinc and zirconium.

## 27.4 Wrought Magnesium Alloys

Wrought magnesium alloys are much less widely used than castings, and a relatively limited number of alloys are available in wrought form. Since magnesium is somewhat more expensive than aluminum, and aluminum is much more easily cold formed, aluminum has a decided cost advantage over wrought alloys.

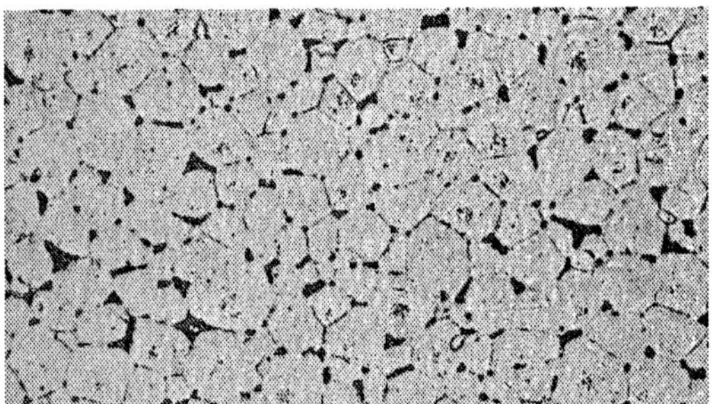

**Fig. 27.9** Microstructure of sand-cast QE22A-T6 magnesium alloy. Original magnification: 100×. Source: Ref 8

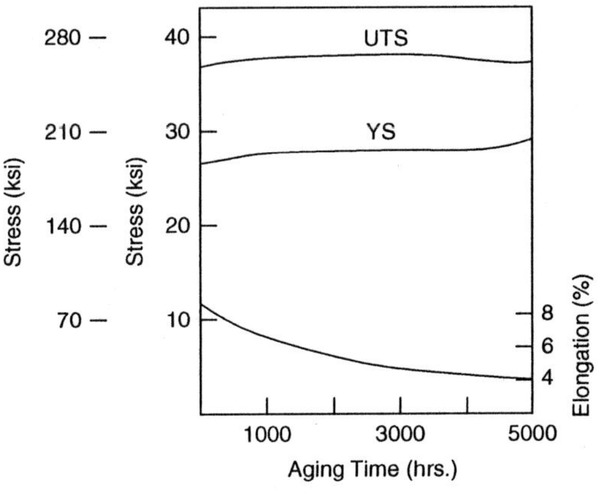


**Fig. 27.10** Effect of 205 °C (400 °F) aging on tensile properties of WE43A-T6. UTS, ultimate tensile strength; YS, yield strength. Courtesy of *Magnesium Electron, Ltd.*

Consequently, wrought magnesium alloys have rather limited use. Wrought magnesium alloys are produced as sheet and plate, extruded bars, billets, shapes, and, to some limited extent, forgings. The compositions of wrought magnesium alloys are shown in Table 27.4, and minimum mechanical properties are given in Table 27.5.

AZ31B (Mg-3Al-1Zn-0.3Mn) is the most widely used sheet and plate alloy and is available in several tempers. It is strengthened by a combination of solid-solution strengthening, grain size control, and cold working. It has a moderate strength with good ductility and formability. In the annealed (O) condition, the yield strength is 152 MPa (22 ksi), the tensile strength is 255 MPa (37 ksi), and the elongation is 21%. When it is worked to the H24 temper, the yield strength increases to 221 MPa (32 ksi), and the tensile strength rises to 290 MPa (42 ksi), while the elongation decreases to 15%. It can be used up to 95 °C (200 °F) in both tempers. The fine, elongated grain morphology in the warm-worked microstructure is illustrated in Fig. 27.11. Note the much more uniform microstructure as compared to the cast microstructures

**Table 27.4 Nominal compositions of wrought magnesium alloys**

| Alloy | | | Composition, wt% | | | | | |
|---|---|---|---|---|---|---|---|---|
| ASTM No. | UNS No. | Product form(a) | Al | Mn (min) | Zn | Th | Zr | Other |
| AZ31B | M11310 | F, S, E | 3.0 | 0.20 | 1.0 | ... | ... | ... |
| AZ31C | M11312 | S, E | 3.0 | 0.15 | 1.0 | ... | ... | ... |
| AZ61A | M11610 | F, E | 6.5 | 0.15 | 0.95 | ... | ... | ... |
| AZ80A | M11800 | F, E | 8.5 | 0.12 | 0.5 | ... | ... | ... |
| HK31A | M13310 | S | 3.0 | ... | 0.3 | 3.25 | 0.7 | ... |
| HM21A | M13210 | F, S | ... | 0.45 | ... | 2.0 | ... | ... |
| LA141A | M14141 | S | 1.25 | 0.15 | ... | ... | ... | 14 Li |
| M1A | M15100 | E | ... | 1.6 | ... | ... | ... | 0.3 Ca |
| ZE10A | M16100 | S | ... | 0.15 | 1.25 | ... | ... | 0.17 RE(b) |
| ZK40A | M16400 | E | ... | ... | 4.0 | ... | 0.45 | ... |
| ZK60A | M16600 | F, E | ... | ... | 5.5 | ... | 0.45 | ... |

(a) S, sheet and plate; F, forging; E, extruded bar, shape, tube, and wire. (b) RE, rare earths. Source: Ref 5

**Table 27.5 Minimum mechanical properties for wrought magnesium alloys**

| Alloy-temper | Tensile strength | | Yield strength | | Elongation in 50 mm (2 in.), % | Hardness, HB |
|---|---|---|---|---|---|---|
| | Mpa | ksi | Mpa | ksi | | |
| **Extruded bars, rods, and shapes** | | | | | | |
| AZ31B-F | 220–240 | 32–35 | 140–150 | 20–22 | 7 | ... |
| AZ61A-F | 260–275 | 38–40 | 145–165 | 21–24 | 7–9 | ... |
| AZ80A-F | 290–295 | 42–43 | 185–195 | 27–28 | 4–9 | ... |
| AZ80A-T5 | 310–325 | 45–47 | 205–230 | 30–33 | 2–4 | ... |
| M1A-F | 200–205 | 29–32 | ... | ... | 2–3 | ... |
| ZK40A-T5 | 275 | 40 | 255 | 37 | 4 | ... |
| ZK60A-F | 295 | 43 | 215 | 31 | 4–5 | ... |
| ZK60A-T5 | 295–310 | 43–45 | 215–250 | 31–36 | 4–6 | ... |
| **Forgings** | | | | | | |
| AZ31B-F | 234 | 34 | 131 | 19 | 6 | ... |
| AZ61A-F | 262 | 38 | 152 | 22 | 6 | ... |
| AZ80A-F | 290 | 42 | 179 | 26 | 5 | ... |
| AZ80A-T5 | 290 | 42 | 193 | 28 | 2 | ... |
| HM21A-T5 | 228 | 33 | 172 | 25 | 3 | ... |
| ZK60A-T5 | 290 | 42 | 179 | 26 | 7 | ... |
| ZK60A-T6 | 296 | 43 | 221 | 32 | 4 | ... |
| **Sheet and plate** | | | | | | |
| AZ31B-O | 221 | 32 | ... | ... | 9–12 | ... |
| AZ31B-H26 | 241–269 | 35–39 | 145–186 | 21–27 | 6 | ... |
| HK31A-O | 200–207 | 29–30 | 97–124 | 14–18 | 12 | ... |
| HK31A-H24 | 228–234 | 33–34 | 172–179 | 25–26 | 4 | ... |
| HM21A-T81 | 234 | 34 | 172 | 25 | 4 | ... |
| LA141A-T7 | 124–131 | 18–19 | 103 | 15 | 10 | ... |
| ZE10A-O | 200–207 | 29–30 | 83–124 | 12–18 | 12–15 | ... |
| ZE10A-H24 | 214–248 | 31–36 | 138–172 | 20–25 | 6 | ... |

Source: Ref 5

in Fig. 27.8 and 27.9. The alloys HK31A (Mg-0.7Th-3.2Cu) and HM21A (Mg-2Th-0.8Mn) are suitable for use at temperatures up to 315 and 345 °C (600 and 650 °F), respectively. Of the two, HM21A has better strength and creep resistance. For maximum creep resistance, alloy HK31 requires a T6 heat treatment, while HM21 is cold worked prior to aging (T8 temper). A special quality is used for photoengraving. Alloy PE, which is a modified version of AZ31B containing 3.3 wt% Al and 0.7 wt% Zn, has excellent flatness, corrosion resistance, and etchability.

Extrusions are made from several types of magnesium alloys. For normal strength requirements, one of the AZ (Mg-Al-Zn) alloys is usually selected, with higher aluminum contents producing higher strengths. Alloy AZ31B (Mg-3Al-1Zn-0.3Mn) is a widely used moderate-strength grade with good formability. Alloy AZ31C, which is a lower-purity variation of AZ31B, is used for lightweight structural applications that do not require maximum corrosion resistance. Alloys M1A (Mg-1.5Mn) and ZM21A (Mg-2Zn-1Mn) can be extruded at higher speeds than AZ31B but are more limited due to their lower strengths. Alloy AZ10A (Mg-1Al-0.2Mn), with its low aluminum content, is lower strength than AZ31B, but it can be welded without subsequent stress relief. Alloys AZ61A (Mg-6.5Al-1Zn) and AZ80A (Mg-8.5Al-0.5Zn) can be artificially aged for additional strength but with some sacrifice in ductility. Alloy ZK60A (Mg-5.5Zn-0.5Zr) is used where high strength and good toughness are required. The banded structure produced by hot extrusion of ZK60 is shown in Fig. 27.12. It is heat treatable and normally used in the artificially aged (T5) condition. Since the extrusion process is carried out at approximately the solution heat treating temperature and the extruded shape cools in air fairly rapidly, it is usually necessary to only age the alloy after extrusion.

Forgings are made of AZ31B, AZ61A, AZ80A, HM21A, and ZK60A. Alloys HM21A and AZ31B can be used for hammer forgings, while the other alloys are almost always press forged. For forging, magnesium alloys are normally heated to 345 to 510 °C (650 to 950 °F), depending on the composition of the specific alloy. Alloy AZ80A has greater strength than AZ61A but requires the slowest rate of deformation of the Mg-Al-Zn alloys. Alloy ZK60A has essentially the same strength as AZ80A but with greater ductility. To develop maximum properties, AZ80A, ZK60A, and HM21A are heat treated to the artificially aged (T5) condition. AZ80A can also be heat treated to the T6 temper to provide maximum creep stability. It should be noted that although forgings have the highest strengths of the various magnesium product forms, they are sometimes specified because of their pressure tightness, machinability, and lack of warpage rather than for their strength.

## 27.5 Magnesium Heat Treating

Depending on alloy composition, wrought magnesium alloys are annealed at 290 to 455 °C

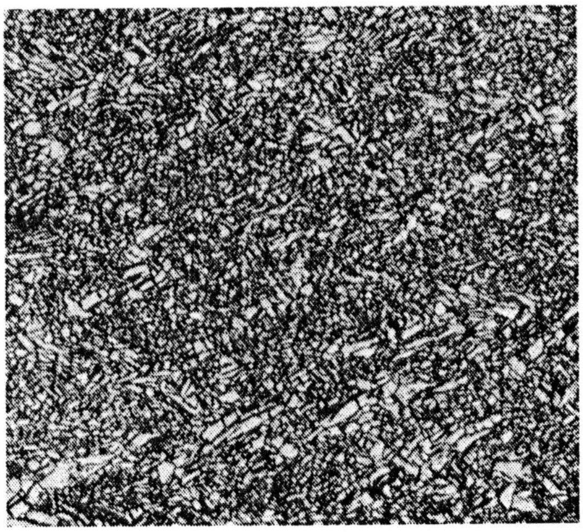

**Fig. 27.11** Microstructure of warm-worked AZ31B-H24 sheet. Original magnification: 250×. Source: Ref 7

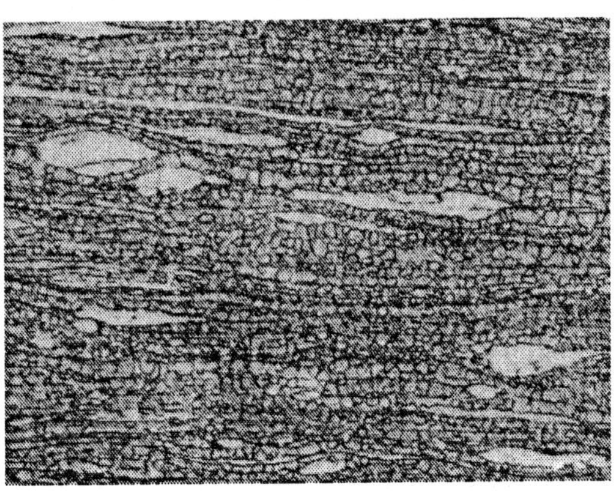

**Fig. 27.12** Section of hot-extruded ZK60 magnesium alloy. Original magnification: 250×. Source: Ref 6

(550 to 850 °F) for 1 to 4 h. Since most forming operations are conducted above 205 °C (400 °F), the need for full annealing is less than with other structural metals. Stress relieving, which is conducted in the range of 150 to 425 °C (300 to 800 °F), is used with wrought alloys to reduce stresses produced by cold working, forming, straightening, and welding. Important reasons for stress relieving castings include (1) prevention of stress corrosion in alloys containing more than 1.5 wt% Al, (2) reduction of stresses induced during weld repair, (3) allowance of precision machining to close dimensional tolerances, and (4) reduction in warpage and distortion. Due to the low modulus of elasticity (44 GPa, or 6.5 msi) of magnesium, even moderate residual stresses can cause appreciable elastic strains, leading to distortion. During casting, residual stresses result from nonuniform contractions during solidification, thermal gradients during heat treatment, machining-induced stresses, and weld repairs.

Solution heat treating is used with cast alloys to reduce the brittle eutectiferous networks that form during casting. Solution heat treated castings have a more uniform matrix that improves both strength and ductility. The aging treatment normally does not improve the tensile strength very much, but the yield strength is increased along with some reduction in ductility. However, the ductility decrease still leaves enough ductility for most casting applications. The most commonly used precipitation-hardening treatments for cast magnesium alloys are solution treating and naturally aging (T4), naturally aging after casting (T5), and solution treating and artificially aging (T6).

Solution heat treating is conducted at 390 to 525 °C (730 to 980 °F). The parts are placed in a furnace preheated to 260 °C (500 °F) and slowly heated to the solution heat treating temperature. Although there are exceptions, slow heating to the solution treating temperature is used to avoid melting of eutectic compounds and the formation of grain-boundary voids. During solution heat treating, burning can occur and appear as surface exudations, gray-black powder on the surface, and voids on the surface and part interior. Burning is caused by too high a solution heat treating temperature, too rapid heating to the solution temperature, and the presence of water vapor in the heat treating furnace. Protective furnace atmospheres containing sulfur hexafluoride, sulfur dioxide, or carbon dioxide are used to prevent excessive oxidation. Sulfur dioxide is particularly effective in removing moisture. When the solution heat treating temperature is reached, the components are held at the solution heat treating temperature for times ranging from 16 to 24 h. Long hold times are used because the solution treatment also homogenizes the cast structure. To prevent castings from sagging under their own weight, castings often require support fixtures during solution heat treating. Some magnesium alloys are subject to excessive grain growth during solution heat treating; however, there are special heat treatments that can minimize grain growth.

Air quenching is used to minimize distortion and cracking. Still air is normally sufficient; however, for dense loads or components that have thick sections, forced-air cooling can be used. Although it increases the danger of quench cracking, hot water must be used to develop the best strength properties for the silver-containing alloys QE22 and QH21. Glycol quenchants can also be used to help prevent distortion. Artificial aging consists of heating to 170 to 230 °C (335 to 450 °F) and holding for 5 to 25 h. Unfortunately, hardness cannot be used for verification of heat treatment. Tensile test specimens must either be cut from a portion of the casting or cast as separate tensile test bars.

## 27.6 Magnesium Fabrication

As previously discussed, wrought magnesium alloys are usually formed at elevated temperatures. Room-temperature forming is used for only the most mild contours. Minimum bend radii for annealed sheet formed at room temperature are 5 to 10*T* for annealed sheet and 10 to 20*T* for work-hardened sheet, where *T* is the sheet thickness. During cold forming, springback can be as much as 30° for a 90° bend. The AZ (magnesium-aluminum-zinc) alloys should be stress relieved after cold forming to prevent stress corrosion.

When magnesium alloys are formed at their recommended temperatures (345 to 510 °C, or 650 to 950 °F), they are highly formable and are routinely formed by a number of forming processes, including press brake forming, deep drawing, manual and power spinning, rubber pad forming, stretch forming, drop hammer forming, and impact extrusion. Forming magnesium alloys at elevated temperatures has several advantages: (1) forming operations can usually be conducted in one step without the need for intermediate anneals, (2) parts can be made to closer tolerances with less springback, and (3) hardened steel dies are not necessary for most forming operations.

Magnesium is among the easiest of all of the metals to machine. Machining is usually conducted dry, using large depths of cut and high feed rates. Very tight dimensional tolerances are easily achieved, with surface finishes as fine as 0.075 to 0.13 μm (3 to 5 μin.). Although machining is usually conducted dry, cutting fluids can be used to reduce the chances of distortion and minimize the danger of fire when chips are fine, as during finish machining. When magnesium chips ignite, they burn with a brilliant white light.

Magnesium alloys are welded by gas-shielded arc welding processes, primarily gas tungsten arc welding and gas metal arc welding. Magnesium alloy filler wires are used along with argon or helium inert gas. The weld beads normally have a rather fine grain size, averaging less than 0.25 mm (0.01 in.) in diameter. Residual stresses and the tendency for some alloys to crack are minimized by preheating, postweld heating, and stress relieving. In the Mg-Al-Zn alloys, higher aluminum contents aid weldability by refining the grain structure, while zinc contents of more than 1 wt% increase hot shortness and the tendency for weld cracking. Weld joints in the Mg-Al-Zn alloys and alloys containing more than 1 wt% Al require stress relieving, because they are subject to stress-corrosion cracking if not stress relieved. Alloys with high zinc contents (i.e., 5 to 6 wt% Zn) are very susceptible to weld cracking and have poor weldability. Weld-repaired castings are normally heat treated after welding to either the T4, T5, or T6 tempers. If the casting is not heat treated after weld repair, it is usually stress relieved.

## 27.7 Magnesium Corrosion Protection

For ordinary outdoor applications, substantial improvement over the earlier alloys has been attained by the use of high-purity alloys and by improvements in foundry practice, particularly with respect to fluxing treatments. Although commercial alloys are reasonably stable under inland atmospheres, it is desirable to paint the component unless it is definitely known that the exposure conditions are favorable. Seacoast locations involving direct contact with salts are definitely corrosive to magnesium alloys, and more extensive corrosion-prevention measures, such as chemical conversion coatings or anodizing treatments, are required.

The optimal corrosion protection is provided by an anodizing treatment, followed by the application of an organic paint system. Both treatments roughen the surface and chemically modify it for maximum paint adhesion. The most effective are the anodizing treatments, of which a number exist, such as the older Dow 17 and HAE and the newer and improved Tagnite, Keronite, and Magoxid treatments.

The process flow for the Dow 17 process is shown in Fig. 27.13.

The part is first alkaline cleaned and then anodized in a solution of $NH_4HF_2$, $Na_2Cr_2O_7 \cdot 2H_2O$, and $H_3PO_4$ heated to 70 to 80 °C (160 to 180 °F) using either an alternating or direct current (54 to 540 A/m$^2$, or 5 to 50 A/ft$^2$). A two-layer coating is produced; the first layer is a thin, light-green coating (6 μm, or 0.2 mil) that forms at lower voltages, followed by a thicker, dark-green coating (30 μm, or 1.2 mils) formed at higher voltages. The thicker coating enhances corrosion protection and forms an excellent base for paint but can be susceptible to spalling under impact, deformation, or flexing.

Prior to paint application, porous casting surfaces are normally filled with a penetrating resin. The primer bases contain zinc chromate or titanium dioxide pigments for improved corrosion resistance. Both air-drying and baked-on paints are used, with the baked-on paints being harder and more resistant to solvents. Finishes are chosen based on the application: vinyl alkyds for alkali resistance, acrylics for salt spray resistance, alkyd enamels for exterior durability, and epoxies for abrasion resistance. Vinyls can withstand temperatures up to approximately 150 °C (300 °F). Higher-temperature finishes include modified vinyls, epoxies, modified epoxies, epoxy-silicones, and silicones. The integrity of the paint system must be maintained when the component is placed in service, because the chemical conversion and anodized surfaces will readily corrode if exposed to the atmosphere.

## 27.8 Zinc

After iron, aluminum, and copper, zinc is the fourth most highly used industrial metal.

The main uses of zinc are galvanizing steel for corrosion protection; zinc-base die castings; and as an alloying element in copper, magnesium, and aluminum alloys. Die casting accounts for approximately 1/3 of the annual zinc consumption. Applications include small parts, such as carburetors and fuel pump housings. An advantage of zinc, compared to magnesium and aluminum die castings, is that zinc has a lower melting point, resulting in less die wear and better surface finishes. On the down side, zinc is considerably heavier (density of 7.133 g/cm$^3$, or 0.258 lb/in.$^3$) and is being replaced in applications where weight is important.

Commercially pure zinc has a tensile strength of only approximately 110 MPa (16 ksi), but it can be increased to approximately 415 MPa (60 ksi) by alloying with aluminum. Its low melting point (420 °C, or 788 °F) and high fluidity contribute to good castability, permitting the casting of thin-wall components. Zinc and its alloys are relatively low cost and do not pose any environmental problems. Zinc and its alloys are used extensively in the automotive, hardware, home appliance, and electrical parts industries.

Zinc is second only to magnesium at the anodic end of the electromotive series and as such is used as a sacrificial anode in galvanized coatings on iron and steels. Approximately 40 wt% of zinc consumed is used in the galvanizing of steel, where it can be applied electrolytically or by hot dipping. Since zinc is anodic compared to iron, small pinholes or breaks in the coating will not result in corrosion of the steel. This works well since the zinc itself corrodes rather slowly in neutral solutions. Zinc parts themselves have good corrosion resistance, provided that impurities are held to a minimum and they are not galvanically coupled to a more cathodic metal. A thin, adherent surface film or zinc "patina" [$ZnCO_3 \cdot 3Zn(OH)_2$] readily forms in air, giving zinc alloys their resistance to atmospheric corrosion.

Zinc recrystallizes and creeps at room temperature and cannot be significantly work hardened. As a result, applications for wrought alloys are limited, and most applications are for lowly-stressed zinc die castings. With a low melting temperature, it should not be surprising that high diffusion rates, natural aging reactions, and active creep mechanisms would be active at room temperature, since room temperature is approximately 0.43 of the absolute melting point. Commercially pure zinc does not have a stable modulus of elasticity, with values ranging from 70 to 140 GPa (10 to 20 msi). The addition of 4 wt% Al allows a definite modulus and yield strength to be measured. However, at 95 °C (200 °F), the tensile strength is only 65 to 75% of room-temperature values.

As a result of its hexagonal close-packed (hcp) crystalline structure, the room-temperature forming characteristics are poor. Zinc has a $c/a$ ratio greater than the ideal close-packed value of 1.633. Not only is slip difficult because of the limited number of available slip systems, but cleavage readily occurs along the basal plane. However, twinning mechanisms contribute to deformation straining. Zinc has a ductile-to-brittle transition temperature just below room temperature and is notch sensitive.

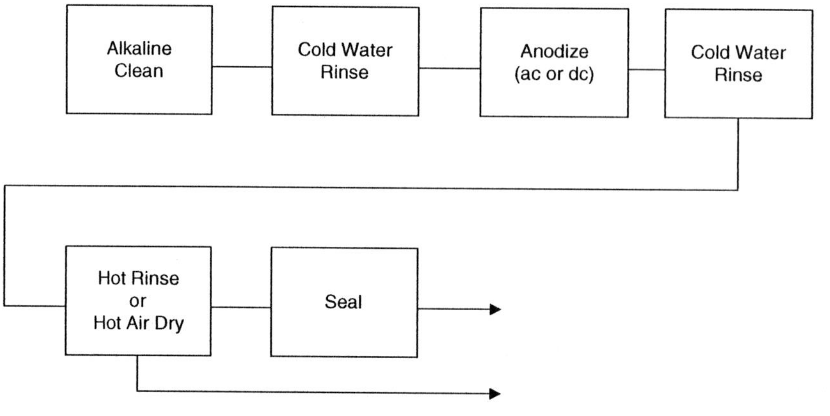


**Fig. 27.13** Anodizing treatment for magnesium alloys. Source: Ref 9

## 27.9 Zinc Casting Alloys

The relatively low solidification temperature range for zinc alloys permits a faster rate of die casting than with other competing alloys, such as those of aluminum and magnesium. Aluminum and copper, along with small amounts of magnesium (<0.08 wt%), are the main alloying elements in cast and wrought zinc alloys. These alloying elements provide second-phase particles, or precipitates, that increase strength while retarding recrystallization, grain growth, and creep processes. The compositions and properties of several cast and wrought zinc alloys are given in Table 27.6.

Aluminum is the principal alloying element in zinc casting alloys, providing good castability, easy finishing, higher strength, and freedom from intergranular corrosion. While molten zinc attacks iron-base alloys, the addition of aluminum greatly reduces the dissolution rate, allowing the use of ferrous die-casting equipment. The casting alloys can be divided into two principal classes based on aluminum content: the hypoeutectic alloys, with approximately 4 wt% Al for optimal mechanical properties, and those with greater levels of aluminum. The hypereutectic alloys, with >5 wt% Al, have higher strength, hardness, and stiffness but lower ductility. Copper and magnesium are the other common alloying elements. Copper is added for strength, hardness, and creep resistance. However, with increasing amounts of copper, ductility decreases, and the degree of dimensional growth increases with time.

Impurities, especially lead, cadmium, and tin, must be tightly controlled to avoid intergranular corrosion. The use of high-purity zinc, along with additions of 0.02 wt% Mg or 0.01 wt% Ni, greatly reduces the corrosion susceptibility. Magnesium also has the added benefit of lowering the ductile-to-brittle transition temperature. However, magnesium levels exceeding approximately 0.08 wt% lead to hot shortness (the tendency to separate along grain boundaries when stressed or deformed at temperatures near the melting point) during cooling of the casting. In general, exceeding the strict impurity limits will lead to intergranular corrosive attack and warping and cracking with time after casting.

The die-casting alloys (No. 2 to 7) are basically a family of Zn-4%Al alloys. This composition provides excellent casting characteristics and optimal strength and ductility. Copper additions provide strength, hardness, and creep resistance, but the amount is limited by the expansion or swelling it induces with time. The expansion with time by alloy No. 2 (2.5 wt% Cu) may be too excessive for some applications. On the other hand, alloy No. 3 (0 wt% Cu) has low creep resistance and should not be exposed to temperatures above 50 °C (120 °F) under load. The microstructure of die-cast AG40A (No. 3) is shown in Fig. 27.14.

Gravity-fed sand and permanent castings are made of alloys containing higher aluminum contents. Alloys ZA-8, ZA-12, and ZA-27 have higher strength, creep, and fatigue resistance than the 4 wt% Al alloys. Their improved creep resistance is directly related to a coarser cast grain structure. They are also more dimensionally stable and have lower densities than the 4 wt% Al alloys. Alloy ZA-27 is the lowest-density (5.00 g/cm$^3$, or 0.181 lb/in.$^3$), the

**Table 27.6 Nominal compositions and properties of select zinc alloys**

| UNS No. | Common name | Nominal composition, wt% | Tensile strength, Mpa | Tensile strength, ksi | Elongation, % | Hardness, HB |
|---|---|---|---|---|---|---|
| **Die-casting alloys** | | | | | | |
| Z35541 | No. 2, AC43A | Zn-4Al-2.5Cu-0.04Mg | 359 | 52 | 7 | 100 |
| Z33520 | No. 3, AG40A | An-4Al-0.04Mg | 283 | 41 | 10 | 82 |
| Z35531 | No. 5, AC41A | Zn-4Al-Cu-0.05Mg | 329 | 48 | 7 | 91 |
| Z33523 | No. 7, AG40B | Zn-4Al-0.015Mg-0.012Ni | 283 | 41 | 14 | 76 |
| Z35635 | ZA-8 | Zn-8Al-1Cu-0.025 Mg | 374 | 54 | 8 | 103 |
| Z35630 | ZA-12 | Zn-11Al-1Cu-0.025Mg | 404 | 58 | 5 | 100 |
| Z35840 | ZA-27 | Zn-27Al-2Cu-0.015Mg | 426 | 62 | 2 | 119 |
| **Wrought alloys (hot rolled condition)** | | | | | | |
| Z21220 | ... | Zn-0.06Pb-0.06Cd | 150–170 | 21–25 | 52–30 | 43 |
| Z44330 | ... | Zn-1Cu | 170–210 | 24–30 | 50–35 | 52 |
| Z41320 | ... | Zn-0.8Cu-0.15Ti | 221–290 | 32–42 | 38–21 | 61 |

Source: Ref 2

strongest, and the hardest zinc casting alloy but also has low ductility and impact strength.

## 27.10 Wrought Zinc Alloys

Wrought zinc alloys are restricted for structural applications because zinc alloys have a tendency to creep, even at room temperature, and rolled zinc alloys develop deformation anisotropy as a result of their hcp crystalline structure. One of the primary uses for the wrought alloys is for drawn battery cans.

Wrought zinc alloys are fabricated as flat rolled, wire drawn, extruded, and forged products. Zinc alloys are hot worked at 120 to 150 °C (250 to 300 °F), but even at room temperature, the heat generated during processing or forming is generally sufficient to cause recrystallization, eliminating the need for annealing. As a result of the crystallographic texture resulting from working this hcp material, the mechanical and thermal expansion properties are anisotropic (directional) in the rolled product, with the orientation transverse to the rolling direction exhibiting higher tensile strength and lower thermal expansion than the longitudinal orientation (rolling direction).

The eutectoid alloy of Zn-22%Al is superplastic, exhibiting tensile elongations exceeding 2500% when tested at the superplastic temperature of 250 °C (480 °F) and is easily formed into complex shapes at 250 to 270 °C (480 to 520 °F). When solution treated, quenched, and annealed, the alloy forms a fine-grain microstructure consisting of small, equiaxed grains that promote superplasticity. When heated above 275 °C (530 °F) and slowly cooled to room temperature, it loses its superplastic properties. Additions of 0.5 wt% Cu and 0.02 wt% Mg can be added to enhance creep strength. The superplastic forming process is used to make electronic enclosures and business machine parts.

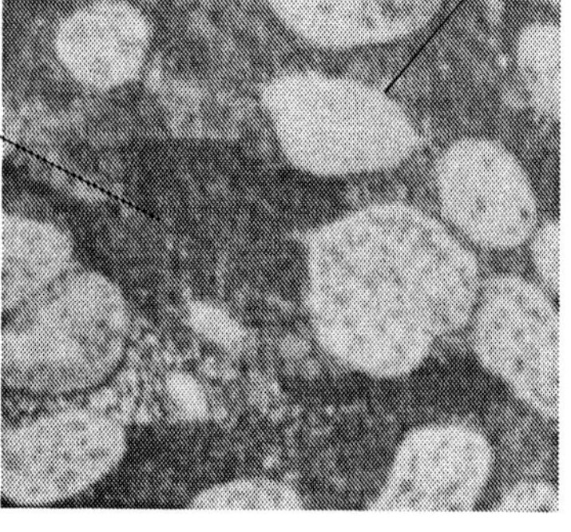


**Fig. 27.14** Microstructure of die-cast AG40A (No. 3) zinc alloy. Original magnification: 1000×. Source: Ref 10

### ACKNOWLEDGMENTS

Sections of this chapter were adapted from "Selection and Application of Magnesium and Magnesium Alloys" by S. Housch, B. Mikucki, and A. Stevenson in *Properties and Selection: Nonferrous Alloys and Special-Purpose Materials* in Volume 2, *ASM Handbook,* ASM International, 1990; and "Zinc and Zinc Alloys: Effects of Composition, Processing, and Structure on Properties of Nonferrous Alloys" by R.N. Caron and J.T. Staley in *Meterials Selection and Design,* Volume 20, *ASM Handbook,* ASM International, 1997.

### REFERENCES

1. S. Housh, B. Mikucki, and A. Stevenson, Selection and Application of Magnesium and Magnesium Alloys, *Properties and Selection: Nonferrous Alloys and Special-Purpose Materials,* Vol 2, *ASM Handbook,* ASM International, 1990
2. R.N. Caron and J.T. Staley, Zinc and Zinc Alloys: Effects of Composition, Processing, and Structure on Properties of Nonferrous Alloys, *Materials Selection and Design,* Vol 20, *ASM Handbook,* ASM International, 1997
3. J.D. Hanawalt, C.E. Nelson, and J.A. Peloubet, *Trans. AIME,* Vol 147, 1942, p 273
4. P. Lyon, "Electron 21 for Aerospace and Specialty Applications," AeroMet Conference and Exposition, June 7–10, 2004, Seattle, WA.
5. Selection and Applications of Magnesium and Magnesium Alloys, *Metals Handbook Desk Edition,* 2nd ed., ASM International, 1998
6. W. Unsworth, *Met. Mater.,* Feb 1988, p 82
7. R.N. Caron and J.T. Staley, Magnesium and Magnesium Alloys: Effects of Composition, Processing, and Structure on Properties of Nonferrous Alloys, *Materials Selection and Design,* Vol 20, *ASM Handbook,* ASM International, 1997

8. Microstructure of Magnesium Alloys, *Atlas of Microstructures of Industrial Alloys,* Vol 7, *Metals Handbook,* 8th ed., American Society for Metals, 1972
9. J.E. Hillis, Surface Engineering of Magnesium Alloys, *Surface Engineering,* Vol 5, *ASM Handbook,* ASM International, 1994
10. Microstructure of Zinc and Zinc Alloys, *Atlas of Microstructures of Industrial Alloys,* Vol 7, *Metals Handbook,* 8th ed., American Society for Metals, 1972

## SELECTED REFERENCES

- R.M. Brick, A.W. Pense, and R.B. Gordon, *Structure and Properties of Engineering Materials,* 4th ed., McGraw-Hill Book Company, 1977
- F.C. Campbell, *Manufacturing Technology for Aerospace Structural Materials,* Elsevier Scientific, 2006
- Heat Treating of Nonferrous Alloys, *Metals Handbook Desk Edition,* 2nd ed., ASM International, 1998
- T.V. Padfield, Metallography and Microstructures of Magnesium and Its Alloys, *Metallography and Microstructures,* Vol 9, *ASM Handbook,* ASM International, 2004
- H. Proffitt, Magnesium Alloys, *Casting,* Vol 15, *ASM Handbook,* ASM International, 1988
- W.F. Smith, *Structure and Properties of Engineering Alloys,* 2nd ed., McGraw-Hill Inc., 1993
- A. Stevenson, Heat Treating of Magnesium Alloys, *Heat Treating,* Vol 4, *ASM Handbook,* ASM International, 1991

# CHAPTER 28

# Titanium

THE PRIMARY ADVANTAGES of titanium alloys are their combination of relatively low densities, high strengths, and excellent corrosion resistance. With a density of 4.5 g/cm$^3$ (0.16 lb/in.$^3$), titanium is classified as a light metal and is only approximately half as heavy as steel and nickel-base superalloys. Yield strengths vary from 480 MPa (70 ksi) for some grades of commercial titanium to approximately 1100 MPa (160 ksi) for structural alloys. In addition to their static strength advantage, titanium alloys have much better fatigue strength than the other lightweight alloys, such as those of aluminum and magnesium. Titanium alloys can be used at moderately elevated temperatures, as high as 370 to 595 °C (700 to 1100 °F), depending on the specific alloy. In addition, some alpha titanium alloys, especially the low interstitial grades, can be used in cryogenic applications because they do not exhibit a ductile-to-brittle transition. As a result of their attractive combination of properties, titanium alloys are used extensively in aerospace for both airframe and engine components. For example, titanium alloys comprise approximately 42% of the airframe weight of the new F-22 fighter aircraft. In commercial passenger aircraft engines, titanium alloys are used for the fan, the low-pressure compressor, and approximately ⅔ of the high-pressure compressor.

Although titanium is a highly reactive metal, a very stable and highly adherent protective oxide film forms on its surface. This oxide film, which forms instantly when fresh metal surfaces are exposed to air and/or moisture, provides the excellent corrosion resistance of titanium. Titanium alloys are frequently used in chemical-processing equipment as a result of their excellent corrosion resistance. They also have outstanding compatibility with the human body and are used for prostheses and dental implants.

The biggest disadvantage of titanium alloys is their relatively high cost. Since titanium is a very reactive metal with a high melting point (1720 °C, or 3130 °F), ingot casting and primary fabrication procedures are complicated and expensive. Secondary fabrication processes, such as forming and machining, are also usually more costly than those for other competing metals.

## 28.1 Titanium Metallurgy

Pure titanium at room temperature has an alpha (α) hexagonal close-packed (hcp) crystal structure, which transforms to a beta (β) body-centered cubic (bcc) structure at a temperature of approximately 885 °C (1625 °F). This transformation temperature, known as the beta transus temperature, can be raised or lowered depending on the type and amount of impurities or alloying additions. As a result of the hcp crystalline structure, alloys with appreciable amounts of alpha must be formed at elevated temperatures, while those with predominantly bcc structures exhibit varying degrees of room-temperature formability.

At room temperature, commercially pure titanium is composed primarily of the alpha phase. As alloying elements are added to titanium, they tend to change the amount of each phase present and the beta transus temperature in the manners shown in Fig. 28.1.

Alpha stabilizers are those elements that increase the beta transus temperature by stabilizing the alpha phase and include aluminum, oxygen, nitrogen, and carbon. Aluminum, the principal alpha stabilizer, increases tensile strength, creep strength, and elastic modulus.

Beta stabilizers are elements that decrease the beta transus temperature. Beta stabilizers are classified into two groups: beta isomorphous and beta eutectoid. The isomorphous alpha phase results from the decomposition of the metastable beta in the first group, whereas in the second group, an intimate eutectoid mixture of alpha and a compound form. The isomorphous group consists of elements that are completely miscible in the beta phase; included in this group are molybdenum, vanadium, tantalum, and niobium. The eutectoid-forming group, which has eutectoid temperatures as much as 335 °C (600 °F) below the transformation temperature of unalloyed titanium, includes manganese, iron, chromium, cobalt, nickel, copper, and silicon. However, the eutectoid reactions in a number of these alloys are very sluggish, so that in reality, these alloys tend to behave as if the reaction does not exist. Molybdenum is an important beta stabilizer that promotes hardenability and short-time elevated-temperature strength. Vanadium is another important beta stabilizer, although it is less effective than molybdenum in providing elevated-temperature strength. Niobium improves oxidation resistance at high temperatures.

Tin and zirconium are considered neutral because they neither raise nor lower the beta transus temperature. Tin has extensive solid solubility in both the alpha and beta phases and is often used as a solid-solution strengthener in conjunction with aluminum to achieve higher strength without embrittlement. Zirconium forms a continuous solid solution with titanium and increases strength at low and intermediate temperatures.

Titanium has a great affinity for interstitial elements, such as oxygen and nitrogen, and readily absorbs them at elevated temperatures, which increases strength and reduces ductility (Fig. 28.2). Hydrogen is always minimized in titanium alloys because it causes hydrogen embrittlement by the precipitation of hydrides, so the maximum limit allowed is approximately

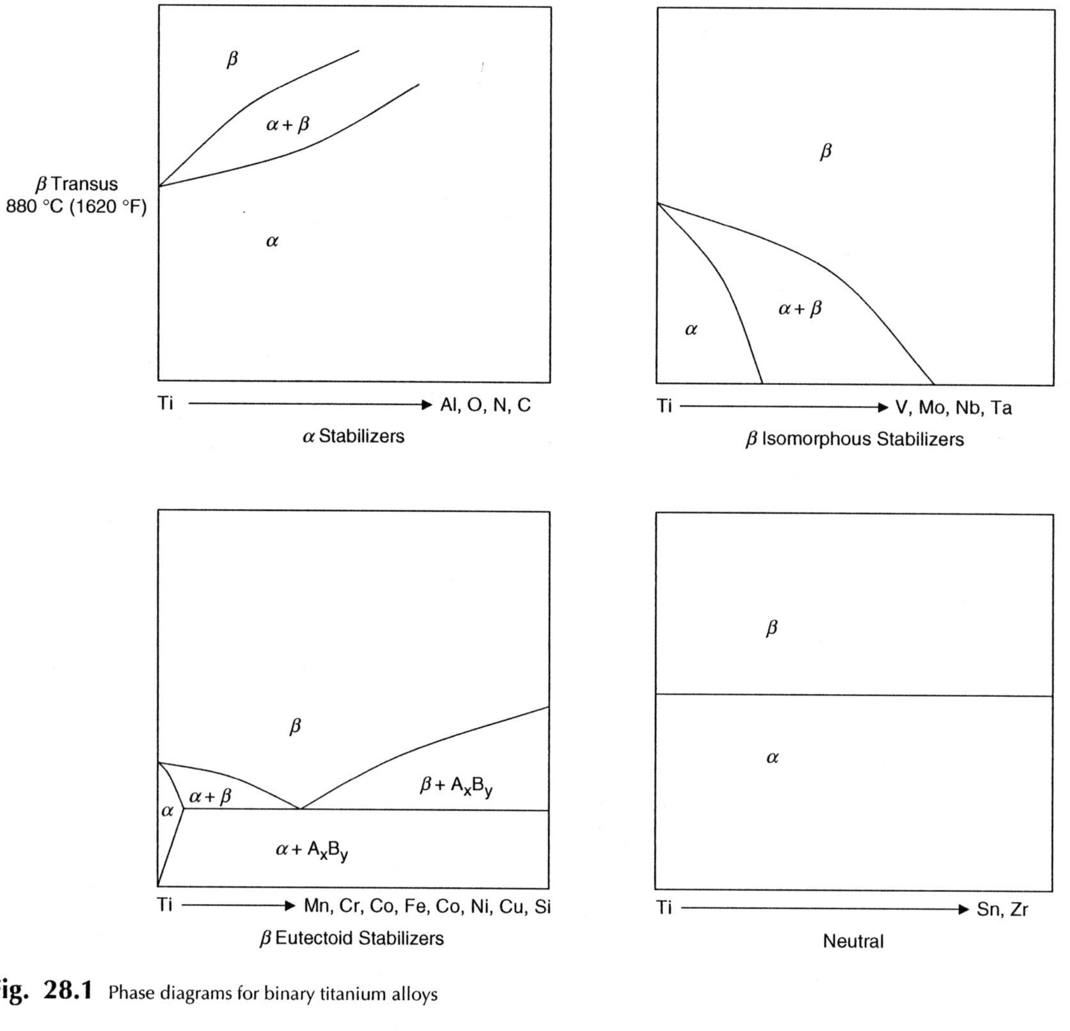


**Fig. 28.1** Phase diagrams for binary titanium alloys

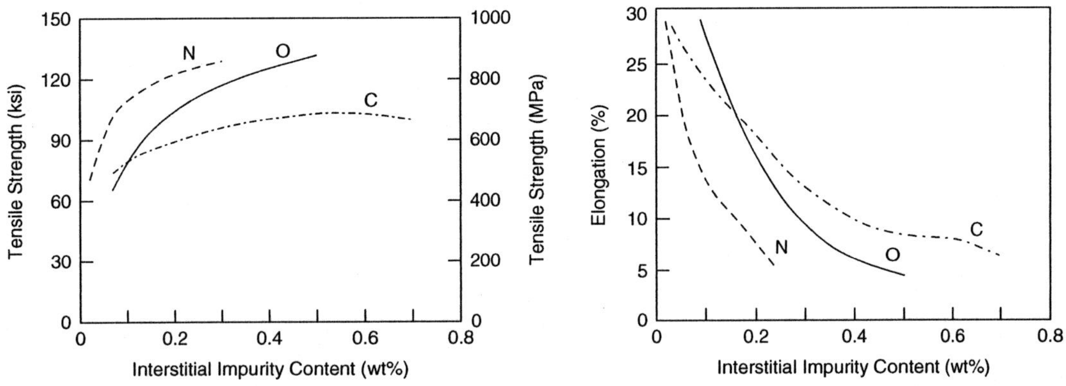


**Fig. 28.2** Effects of interstitial content on strength and ductility of titanium. Source: Ref 3

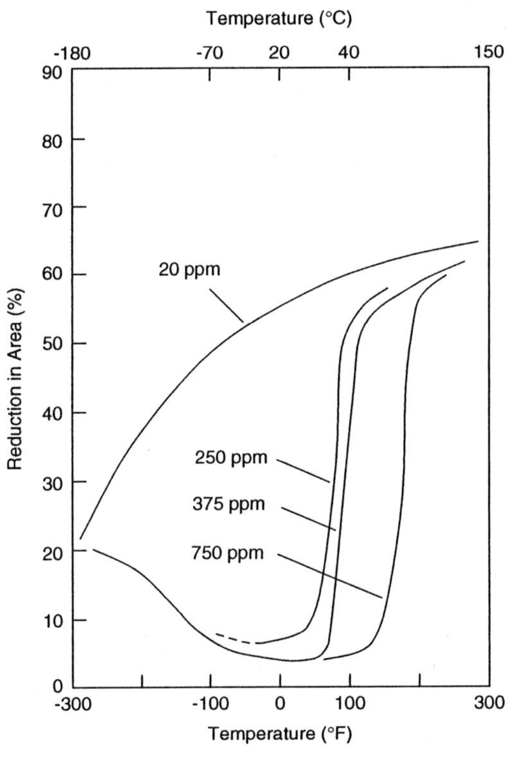


**Fig. 28.3** Effect of hydrogen content (ppm) on ductility of alpha titanium. Source: Ref 3

0.015 wt% (~100 ppm). When the solubility limit of hydrogen in titanium (~100 to 150 ppm for commercially pure titanium) is exceeded, hydrides begin to precipitate. Absorption of several hundred ppm of hydrogen results in embrittlement (Fig. 28.3) and the possibility of stress cracking. Note that the addition of 20 ppm does not cause embrittlement, but when the hydrogen content goes up to 250 ppm, the reduction in area is seriously impaired.

## 28.2 Titanium Alloys

Titanium alloys are classified according to the amount of alpha and beta retained in their structures at room temperature. Classifications include commercially pure, alpha and near-alpha, alpha-beta, and metastable beta. The commercially pure and alpha alloys have essentially all-alpha microstructures. Beta alloys have largely all-beta microstructures after air cooling from the solution-treating temperature above the beta transus. Alpha-beta alloys contain a mixture of alpha and beta phases at room temperature. Within the alpha-beta class, an alloy that contains much more alpha than beta is often called a near-alpha alloy. The names superalpha and lean-beta alpha are also used for this type of alpha-beta alloy. While these classifications are useful, many of them are actually very close to each other in the total amount of beta stabilizer present, as illustrated in the Fig. 28.4 phase diagram. For example, Ti-6Al-4V is classified as an alpha-beta alloy, and Ti-6Al-2Sn-4Zr-2Mo is classified as a near-alpha alloy, yet they differ little in the total amount of beta stabilizer concentration. The properties of a number of commercially important alloys are given in Table 28.1.

### *28.2.1 Commercially Pure Titanium*

Commercially pure titanium wrought products are used primarily for applications requiring corrosion resistance. They are also useful in applications requiring high ductility for fabrication but relatively low strength in service. Yield strengths range from 170 to 480 MPa (25 to 70 ksi). Basically, oxygen and iron contents determine the strength levels of commercially pure titanium. In the higher-strength grades, oxygen and iron are intentionally added to the residual amounts already in the sponge to provide extra strength. On the other hand, carbon and nitrogen usually are held to minimum residual levels to avoid embrittlement. When good ductility and toughness are desired, the extra-low interstitial (ELI) grades are used. In ELI grades, carbon, nitrogen, oxygen, and iron must be held to acceptably low levels, because they lower the ductility of the final product.

Several grades of commercially pure (CP) titanium are available, with the mechanical properties controlled by the oxygen (0.18 to 0.4%) and iron (0.2 to 0.5%) contents. Grades with higher oxygen and iron have higher strengths and cost less but have somewhat lower ductilities, fracture toughness, and corrosion resistance. The CP grades have good formability, are readily weldable, and have excellent corrosion resistance. The CP grades are supplied in the mill-annealed condition, which permits limited forming at room temperature; however, most forming operations are conducted at 150 to 480 °C (300 to 900 °F). Property degradation can be experienced after forming if the material is not stress relieved. Common applications of CP alloys are corrosion-resistant tubing, tanks, and fittings in the chemical-processing industry. A large titanium heat exchanger is shown in Fig. 28.5. Titanium-palladium alloys with nominal palladium contents of approximately 0.2 wt% are used in applications requiring excellent corrosion resistance in chemical-processing or storage applications where the media is mildly reducing or fluctuates between oxidizing and reducing.

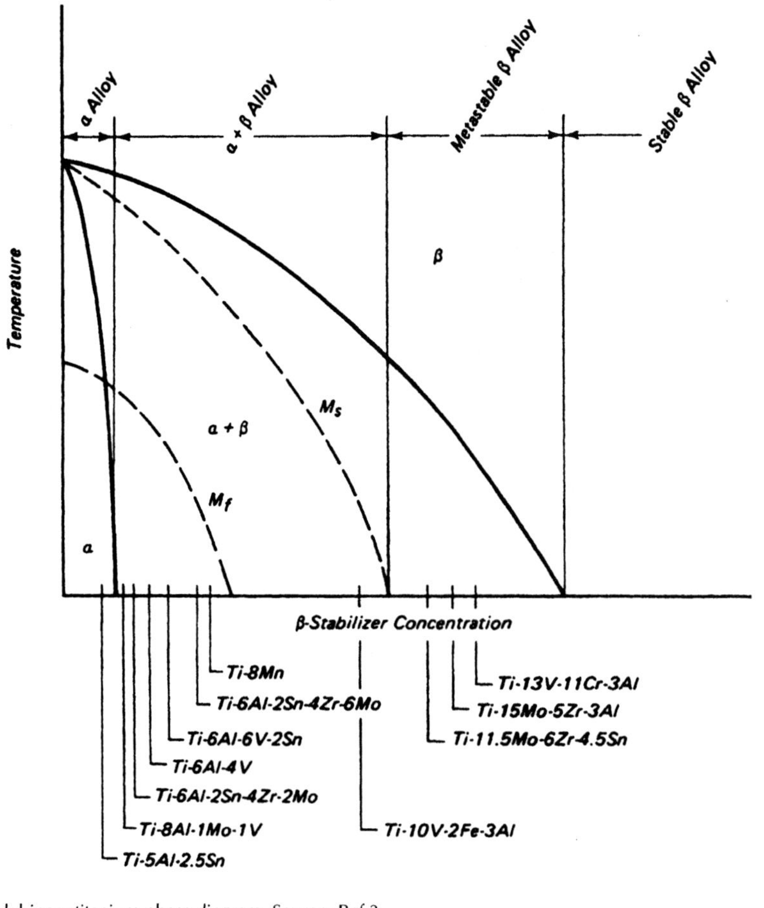


**Fig. 28.4** Pseudobinary titanium phase diagram. Source: Ref 3

### *28.2.2 Alpha and Near-Alpha Alloys*

Aluminum is the principal alloying element in the alpha and near-alpha alloys. Aluminum provides solid-solution strengthening and oxidation resistance and reduces density. Other additions include the neutral elements tin and zirconium, along with small amounts of beta stabilizers. Alpha and near-alpha alloys are slightly less corrosion resistant but higher in strength than unalloyed titanium. They develop moderate strengths and have good notch

**Table 28.1 Properties of selected titanium alloys**

| Alloy | Nominal composition | Condition(a) | Ultimate tensile strength, MPa | Ultimate tensile strength, ksi | Yield strength, MPa | Yield strength, ksi | Modulus of elasticity, GPa | Modulus of elasticity, msi | Elongation, % | Reduction in area, % |
|---|---|---|---|---|---|---|---|---|---|---|
| **Commercially pure** | | | | | | | | | | |
| Grade 1 | 0.03N, 0.20Fe, 0.18O | Annealed | 240 | 35 | 170 | 25 | 103 | 14.9 | 24 | 30 |
| Grade 2 | 0.03N, 0.30Fe, 0.25O | Annealed | 345 | 50 | 275 | 40 | 103 | 14.9 | 20 | 30 |
| Grade 4 | 0.05N, 0.50Fe, 0.40O | Annealed | 550 | 80 | 485 | 70 | 103 | 14.9 | 15 | 25 |
| **Alpha and near-alpha** | | | | | | | | | | |
| Ti-5-2.5 | Ti-5Al-2.5Sn | Annealed | 790 | 115 | 760 | 110 | 110 | 16.0 | 16 | 40 |
| Half 6-4 | Ti-3Al-2.5V | Annealed | 620 | 90 | 520 | 75 | 107 | 15.5 | 20 | ... |
| Ti-6242S | Ti-6Al-2Sn-4Zr-2Mo-0.25Si | Annealed | 900 | 130 | 830 | 120 | 114 | 16.5 | 15 | 35 |
| Ti-8-1-1 | Ti-8Al-1Mo-1V | Annealed | 900 | 130 | 830 | 120 | 124 | 18.0 | 15 | 28 |
| **Alpha-beta** | | | | | | | | | | |
| Ti-6-4 | Ti-6Al-4V | Annealed | 900 | 130 | 830 | 120 | 114 | 16.5 | 14 | 30 |
| | | STA | 1170 | 170 | 1100 | 160 | 114 | 16.5 | 10 | 25 |
| Ti-6-4 ELI(b) | Ti-6Al-4V | Annealed | 830 | 120 | 760 | 110 | 114 | 16.5 | 15 | 35 |
| Ti-6-6-2 | Ti-6Al-6Sn-2V | Annealed | 1035 | 150 | 1000 | 145 | 110 | 16.0 | 14 | 30 |
| | | STA | 1275 | 185 | 1170 | 170 | 110 | 16.0 | 10 | 20 |
| Ti-6246 | Ti-6Al-2Sn-4Zr-6Mo | STA | 1300 | 189 | 1170 | 170 | 114 | 16.5 | 10 | 23 |
| Ti-6-22-22S | Ti-6Al-2Sn-2Zr-2Mo-2Cr-0.25Si | Annealed | 1035 | 150 | 965 | 140 | 122 | 17.7 | ... | ... |
| | | STA | 1275 | 185 | 1140 | 165 | 122 | 17.7 | 11 | 33 |
| **Beta** | | | | | | | | | | |
| Ti-10-2-3 | Ti-10V-2Fe-3Al | STA | 1170 | 170 | 1100 | 160 | 112 | 16.2 | 10 | 19 |
| Ti-15-3 | Ti-15V-3Al-3Cr-3Sn | Annealed | 770 | 114 | 770 | 112 | ... | ... | 22 | ... |
| | | STA | 1100 | 159 | 985 | 143 | ... | ... | 12 | ... |
| Beta C | Ti-3Al-8V-6Cr-4Mo-4Zr | STA | 1275 | 185 | 1180 | 171 | ... | ... | 11 | 13 |

(a) STA, solution treated and aged. (b) ELI, extra-low interstitial

**Fig. 28.5** Large titanium heat exchanger. Source: Ref 1

toughness. They have medium formability and are weldable.

Ti-5Al-2.5Sn is the only true alpha alloy that is commercially produced. The remainder of the commercially available alloys are near-alpha alloys. Ti-5Al-2.5Sn is quite ductile, and the ELI grade retains ductility and toughness at cryogenic temperatures. Since Ti-5Al-2.5Sn is a single-phase alloy containing only alpha, it cannot be strengthened by heat treatment. Except for cryogenic applications, the use of Ti-5Al-2.5Sn has declined as alloys with better forming properties and higher creep resistance have been developed.

Near-alpha alloys contain small amounts beta phase dispersed in an otherwise all-alpha matrix. The near-alpha alloys generally contain 5 to 8 wt% Al. There is a practical limit to the amount of aluminum that can be added as an alloying element. If the equivalent aluminum content exceeds 9 wt%, the brittle intermetallic compound α-2 ($Ti_3Al$) forms, which adversely affects ductility. The aluminum equivalent is determined by summing the following weight percentages:

$$Al + \frac{1}{3}Sn + \frac{1}{6}Zr + 10(O + C + 2N)$$

The older alloy Ti-8Al-1Mo-1V has an aluminum equivalent that exceeds 9 wt% and has experienced instability problems and loss of ductility during long-term elevated-temperature exposures.

The near-alpha alloys retain their strength to high temperatures and have good creep resistance in the range of 315 to 595 °C (600 to 1100 °F). Silicon additions (0.10 to 0.25 wt%), which precipitate fine silicides that hinder dislocation climb, are used to enhance creep resistance. High-temperature near-alpha alloys that can be used up to 540 °C (1000 °F) include Ti-6242S (Ti-6Al-2Sn-4Zr-2Mo-0.25Si) and IMI 829 (Ti-5.5Al-3.5Sn-3Zr-1Nb-0.3Si). Alloys for service temperatures up to 595 °C (1100 °F) are IMI 834 (Ti-5.8Al-4Sn-3.5Zr-0.7Nb-0.5Mo-0.35Si) and Ti-1100 (Ti-6Al-2.8Sn-4Zr-0.4Mo-0.4Si), a modification to Ti-6242S. These alloys also perform well in cryogenic applications.

### *28.2.3 Alpha-Beta Alloys*

As their name implies, alpha-beta alloys contain both the alpha and beta phases. Again, aluminum is the principal alpha stabilizer that strengthens the alpha phase. Beta stabilizers, such as vanadium, also provide strengthening and allow these alloys to be hardened by solution heat treating and aging (STA). Alpha-beta alloys have a good combination of mechanical properties, rather wide processing windows, and can be used at temperatures up to approximately 315 to 400 °C (600 to 750 °F).

Although the metallurgy of titanium heat treatment is complex, the response to heat treatment is a result of the instability of the high-temperature beta phase at lower temperatures. Heating an alpha-beta alloy to the solution-treating temperature produces a higher ratio of beta phase. During quenching, the beta is transformed to beta and titanium martensite ($\alpha'$). During subsequent aging at an intermediate temperature, decomposition of the unstable martensite and the small amount of residual beta phase occurs to provide strengthening. Since the response to heat treatment is a function of cooling rate from the solution temperature, the section sizes that can be through hardened are limited. As the percentage of beta stabilizers increases, the strength increases during STA. However, the martensite formed in titanium alloys is not like the extremely hard and strong martensite formed during the heat treatment of steels. For example, the tensile strength of Ti-6Al-4V only increases from 900 to 1170 MPa (130 to 170 ksi) on STA, while the tensile strength of 4340 steel can be increased from 760 to 1930 MPa (110 to 280 ksi) by heat treatment.

The weldability of the alpha-beta alloys is not as good as the near-alpha alloys, but their formability is better. The alloys that contain smaller percentages of beta stabilizers, known as "lean" alloys, are more weldable. As the amount of beta stabilizers increases, the weldability decreases.

The alpha-beta alloys include Ti-6Al-4V, which is the workhorse of the aerospace industry. It accounts for approximately 60 wt% of the titanium used in aerospace and up to 80 to 90 wt% of that used for airframes. Ti-6Al-4V ELI is used for fracture-critical structures and for cryogenic applications. While the commercial grade of Ti-6Al-4V has an oxygen content of 0.16 to 0.18%, the ELI grade is limited to 0.10 to 0.13%. With its higher oxygen content, the strength of the commercial grade is slightly higher, but the ductility and fracture toughness of the ELI grade is higher. A comparison of the properties of commercial and ELI Ti-6Al-4V is given in Fig. 28.6.

The microstructure of alpha-beta alloys can take different forms, ranging from equiaxed to acicular or some combination of both. Equiaxed structures are formed by working the alloy in the alpha-beta range and annealing at lower temperatures (Fig. 28.7). Acicular structures (Fig. 28.7c) are formed by working or heat treating above the beta transus and rapid

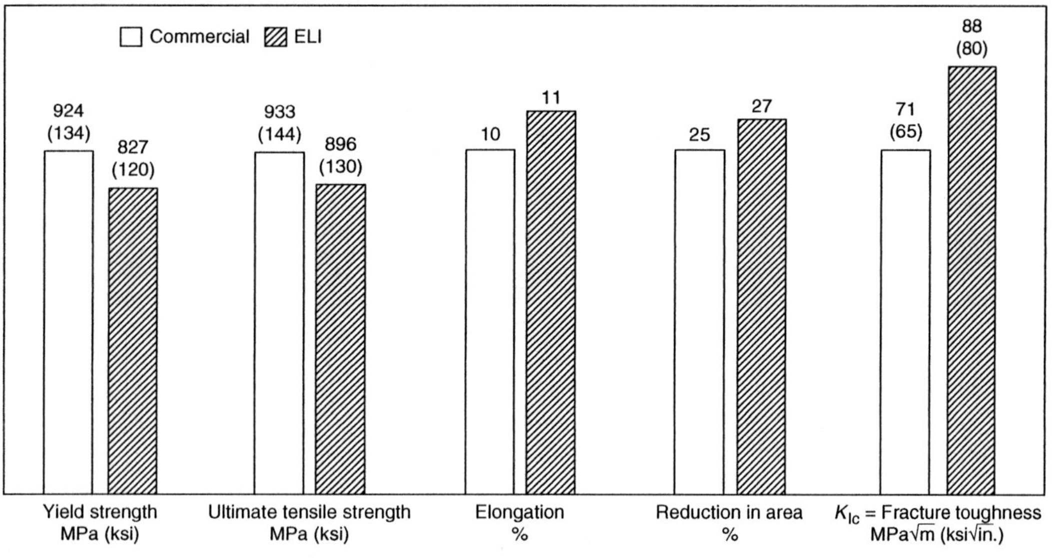


**Fig. 28.6** Room-temperature properties of commercial versus extra-low interstitial (ELI) Ti-6Al-4V. Source: Ref 4

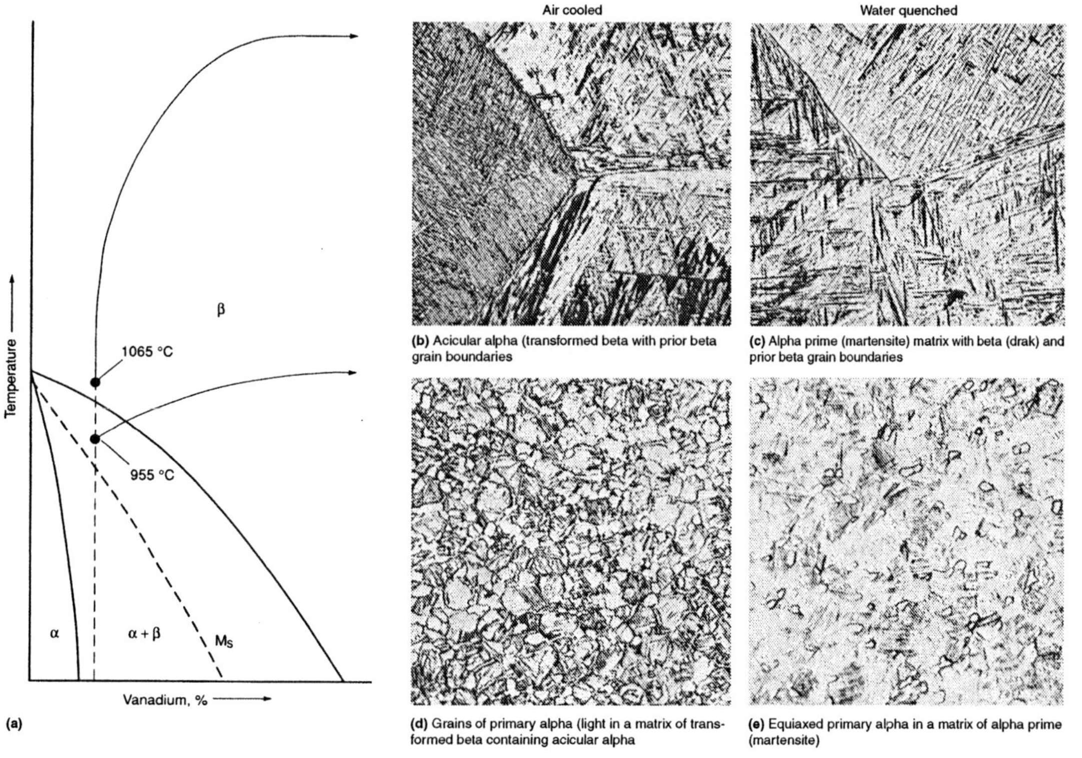


**Fig. 28.7** Ti-6Al-4V microstructures produced by cooling from different temperatures. Original magnification: 250×. Source: Ref 1

cooling. Rapid cooling from temperatures high in the alpha-beta range (Fig. 28.7d and e) will result in equiaxed primary (prior) alpha and acicular alpha from the transformation of beta structures. Generally, there are property advantages and disadvantages for each type of structure. Equiaxed structures have higher ductility and formability, higher threshold stresses for hot salt stress corrosion, higher strength for an equivalent heat treatment, and better low-cycle fatigue (crack initiation) properties. The advantages of the acicular structures are better creep properties, higher fracture toughness, superior stress-corrosion resistance, and lower fatigue crack propagation rates.

In the alpha-beta alloys, the presence of nonequilibrium phases, such as titanium martensite or metastable beta, results in substantial increases in tensile and yield strengths following the aging treatment. No response to aging occurs on furnace cooling from solution temperatures. Only a slight response occurs on air cooling, which produces the Fig. 28.7(b) and (d) microstructures. The greatest response is experienced with water quenching from the solution temperature, typical of microstructures shown in Fig. 28.7(c) and (e). A good response to aging takes place on water quenching from the beta field, as in Fig. 28.7(c); however, ductilities are quite low. The best combination of properties can be produced by solution treating and rapidly quenching from close to but below the beta transus temperature (Fig. 28.7e), followed by an aging treatment.

### *28.2.4 Beta Alloys*

Beta alloys are sufficiently rich in beta stabilizers and lean in alpha stabilizers that the beta phase can be completely retained with appropriate cooling rates. Beta alloys are metastable, and precipitation of alpha phase in the metastable beta is a method used to strengthen the alloys. Beta alloys contain small amounts of alpha stabilizing elements as strengthening agents.

As a class, beta and near-beta alloys offer increased fracture toughness over alpha-beta alloys at a given strength level. Beta alloys also exhibit better room-temperature forming and shaping characteristics than alpha-beta alloys; higher strength than alpha-beta alloys at temperatures where yield strength, instead of creep strength, is required; and better response to STA in heavier sections than the alpha-beta alloys. They are limited to approximately 370 °C (700 °F) due to creep. It should be noted that since beta alloys contain alloying additions of the heavy transition metals, their densities increase with increasing amounts of beta stabilizing elements. In addition, some have limited weldability.

Beta alloys, such as Ti-10V-2Fe-3Al, Ti-15V-3Cr-3Al-3Sn, and Beta 21S (Ti-15Mo-3Al-2.7Nb-0.25Si), are high-strength alloys that can be heat treated to tensile strength levels approaching 1380 MPa (200 ksi). In general, they are highly resistant to stress-corrosion cracking. Alloy Ti-10V-2Fe-3Al is used for forgings because it can be forged at relatively low temperatures, offering flexibility in die materials and forging advantages for some shapes. It is used extensively in the main landing gear of the Boeing 777. An advantage of Ti-15V-3Cr-3Al-3Sn is the ability to cold form the solution-treated material in thin gages and then age to high strengths.

Beta alloys can be categorized as either solute-rich or solute-lean alloys, depending on their alloy content. The more highly stabilized solute-rich alloys include Beta C (Ti-3Al-8V-6Cr-4Mo-4Zr) and Ti-15-3 (Ti-15V-3Cr-3Al-3Sn), while the solute-lean alloys, sometimes referred to as beta-rich alpha-beta alloys, include the important forging alloy Ti-10-2-3 (Ti-10V-2Fe-3Al).

For the solute-lean alloys, the thermomechanical processing window is quite critical in developing the desired microstructure and the resultant tensile properties and fracture toughness. In the case of Ti-10V-2Fe-3Al, thermomechanical processing can produce microstructures ranging from fully transformed, aged beta structures to controlled amounts of elongated primary alpha in an aged beta matrix. This latter microstructure is generally preferred for forgings.

Solute-rich beta titanium alloys are generally defined as metastable beta alloys that are too stable to decompose isothermally to a beta and omega (ω) phase mixture, as distinguished from solute-lean alloys that form an omega phase during aging. The omega phase is a metastable phase that forms in solute-lean beta alloys whenever the direct formation of alpha is difficult. For the solute-rich alloys, thermomechanical processing is less critical because they develop an extremely fine microstructure during processing. However, the aging sequence, including temperature and time, is important in producing a uniform precipitation without the occurrence of grain-boundary alpha. Excessive grain-boundary alpha precipitation is detrimental to alloy ductility, fatigue strength, and stress-corrosion cracking resistance. A low then high aging temperature sequence, or cold or warm working prior to aging, is used to provide a uniform precipitate. The introduction of a cold working step prior to aging, as is often done with Beta C, helps to produce a fine precipitate with a good combination of strength and ductility.

## 28.3 Melting and Primary Fabrication

Titanium for ingot production may be either titanium sponge or reclaimed scrap (revert). In both cases, stringent specifications must be met for control of ingot composition. Most important are the hard, brittle, and refractory titanium oxide, titanium nitride, or complex titanium oxynitride particles that, if retained through subsequent melting operations, could act as crack initiation sites, especially during fatigue loading. The product flow for various titanium product forms is shown in Fig. 28.8.

Titanium sponge is manufactured by first chlorinating rutile ore and reducing the resulting $TiCl_4$ with either sodium (Hunter process) or magnesium (Kroll process) metal. Sodium-reduced sponge is leached with acid to remove the NaCl by-product of reduction. Magnesium-reduced sponge may be leached, inert gas swept, or vacuum distilled to remove the excess $MgCl_2$ by-product. Vacuum distilling results in lower residual levels of magnesium, hydrogen, and

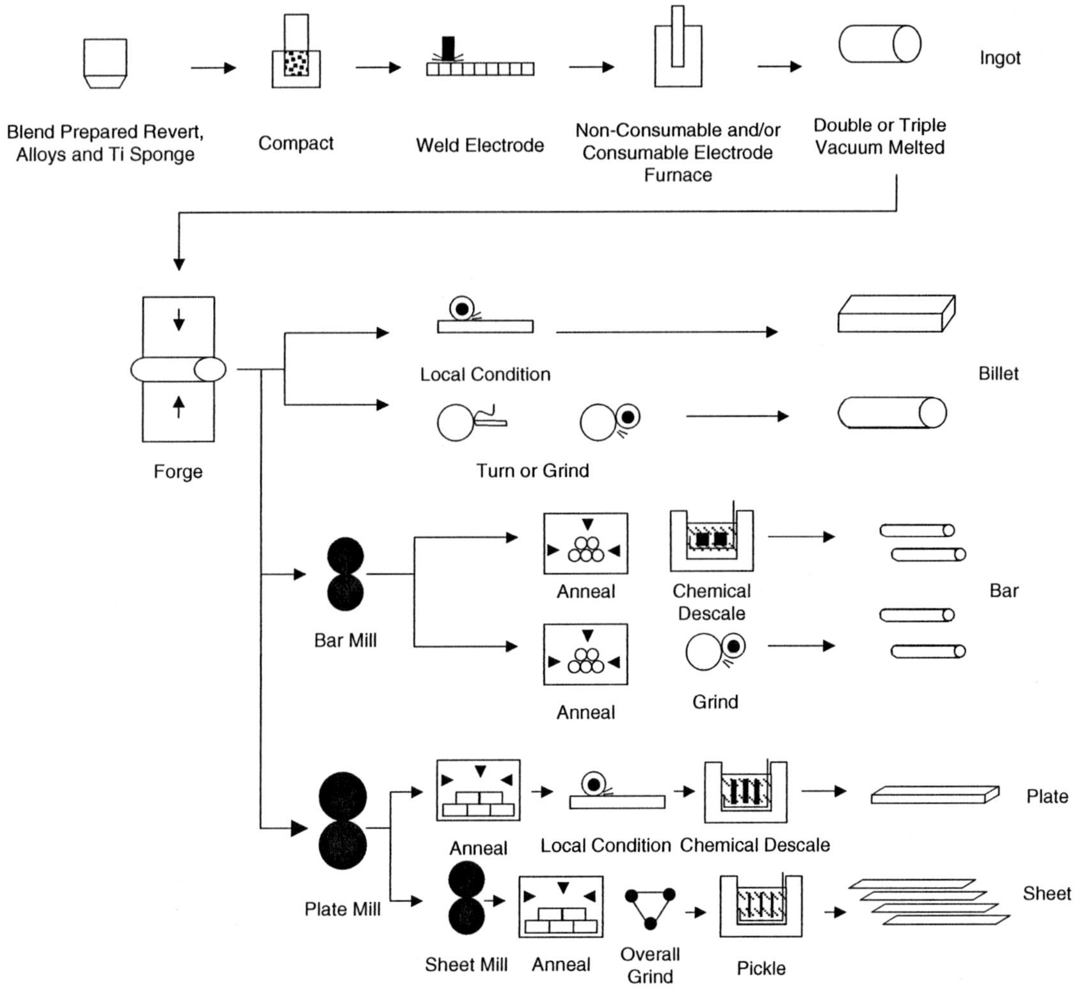


**Fig. 28.8** Titanium product flow

chlorine. Modern melting techniques remove volatile substances from sponge, so that high-quality ingot can be produced regardless of which method is used for production of sponge. Revert makes production of ingot titanium more economical than production solely from sponge. If properly controlled, revert is fully acceptable and can be used even in materials for critical structural applications.

### *28.3.1 Melting*

Titanium sponge, revert, and alloy additions are welded together to form an electrode and then vacuum arc melted, as shown schematically in Fig. 28.9. Even though this is the initial melting operation, it is actually called vacuum arc remelting. Ingots from the first melt are then used as the consumable electrodes for second-stage melting. Double melting is used for all applications to ensure an acceptable degree of homogeneity in the resulting product. Triple melting is used to achieve better uniformity. Triple melting also reduces oxygen-rich or nitrogen-rich inclusions in the microstructure to very low levels by providing an additional melting operation to dissolve them. All melting operations must be done under vacuum to eliminate the introduction of oxygen, nitrogen, and hydrogen. Newer hearth melting technologies using electron beams or plasma as a heat source are casting commercially pure slabs in ore-melting operations. Ingots are normally 650 to 900 mm (26 to 36 in.) in diameter and weigh 3600 to 9000 kg (8000 to 20,000 lb).

Segregation in titanium ingot must be controlled because it leads to several different types of imperfections that cannot be readily eliminated by homogenizing heat treatments or combinations of heat treatment and primary mill processing.

Type I imperfections, usually called high-interstitial defects, are regions of interstitially stabilized alpha phase that have substantially higher hardness and lower ductility than the surrounding material and that also exhibit a higher beta transus temperature. They arise from very high nitrogen or oxygen concentrations in sponge, master alloy, or revert. Type I imperfections frequently, but not always, are associated with voids or cracks. Although type I imperfections sometimes are referred to as low-density inclusions, they often are of higher density than is normal for the alloy.

Type II imperfections, sometimes called high-aluminum defects, are abnormally stabilized alpha-phase areas that may extend across several beta grains. Type II imperfections are caused by segregation of metallic alpha stabilizers, such as aluminum, and contain an excessively high proportion of primary alpha, having a microhardness only slightly higher than that of the adjacent matrix. Type II imperfections sometimes are accompanied by adjacent stringers of beta, areas of low aluminum content and low hardness. This condition is generally associated with closed solidification pipe into which alloy constituents of high vapor pressure migrate, only to be incorporated into the microstructure during primary mill fabrication. Stringers normally occur in the top portions of ingots and can be detected by macroetching or anodized blue etching. Material containing stringers usually must undergo metallographic review to ensure that the indications revealed by etching are not artifacts.

Beta flecks, another type of imperfection, are small regions of stabilized beta in material that has been alpha-beta processed and heat treated. In size, they may be less than 1 μm, or they may encompass several prior beta grains. Beta flecks are either devoid of primary alpha or contain less

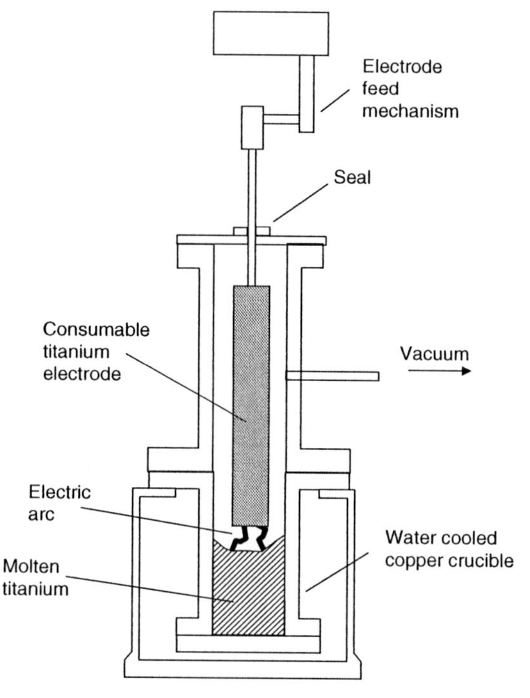


**Fig. 28.9** Vacuum arc melting of titanium ingots

than some specified minimum level of primary alpha. They are caused by localized regions either abnormally high in beta stabilizer content or abnormally low in alpha stabilizer content. Beta flecks are attributed to microsegregation during solidification of ingots of alloys that contain strong beta stabilizers. They are most often found in products made from large-diameter ingots. Beta flecks also may be found in beta-lean alloys, such as Ti-6Al-4V, that have been heated to a temperature near the beta transus during processing.

Type I and type II imperfections are not acceptable in aircraft-grade titanium because they degrade critical design properties. Beta flecks are not considered harmful in alloys lean in beta stabilizers if they are to be used in the annealed condition. On the other hand, they constitute regions that incompletely respond to heat treatment, and, for this reason, microstructural standards have been established for allowable limits on beta flecks in various alpha-beta alloys. Beta flecks are more objectionable in beta-rich alpha-beta alloys than in leaner alloys and are not acceptable in beta alloys.

### *28.3.2 Primary Fabrication*

Primary fabrication includes the operations performed at the mill to convert ingot into products. Besides producing these shapes, primary fabrication hot working is used to refine the grain size, produce a uniform microstructure, and reduce segregation. It has long been recognized that these initial hot working operations will significantly affect the properties of the final product. Titanium alloys are available in most mill product forms: billet, bar, plate, sheet, strip, foil, extrusions, wire, and tubing; however, not all alloys are available in all product forms. The wrought product forms of titanium and titanium alloys, which include forgings and the typical mill products, constitute more than 70 wt% of the market in titanium and titanium alloy production.

Generally, the first breakdown of production ingot is a press cogging operation done in the beta-temperature range. Modern processes use substantial amounts of working below the beta transus to produce billets with refined structures. These processes are carried out at temperatures high in the alpha-beta region to allow greater reduction and improved grain refinement with a minimum of surface rupturing. Where maximum fracture toughness is required, beta processing, or alpha-beta processing followed by beta heat treatment, is generally preferred. Final properties of titanium alloys are strongly influenced by the amount of processing both above and below the beta transus temperature and the extent of recrystallization. Such processing affects the strength of high-alpha grades in large section sizes. With modern processing techniques, billet and forged sections readily meet specified tensile properties prior to final forging.

## 28.4 Forging

Forging is a common method of producing wrought titanium alloy articles. Forging is more than just a shaping process. Forging sequences and subsequent heat treatment can be used to control the microstructure and resulting properties of the product. The key to successful forging and heat treatment is the beta transus temperature. The possible temperature regions for forging and/or heat treatment of a typical alpha-beta alloy such as Ti-6Al-4V are shown in Fig. 28.10. The higher the processing temperature in the alpha-beta region, the more beta that is available to transform on cooling. On quenching from above the beta transus, a completely transformed, acicular structure arises. The exact form of the globular (equiaxed) alpha and the transformed beta structures produced by processing depends on the forging temperature relative to the beta transus, which varies from heat to heat, and the degree and nature of

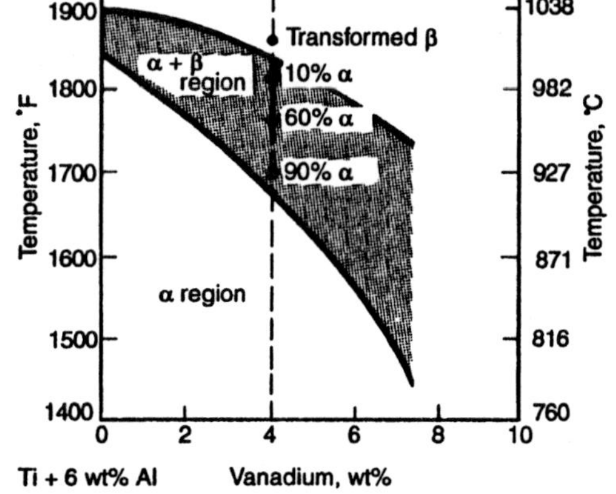


**Fig. 28.10** Forging temperatures for titanium alloys. Source: Ref 2

deformation produced. Section size is important, and the number of working operations can be significant. Conventional forging may require two or three operations, whereas isothermal forging may require only one. A schematic of a conventional alpha-beta forging and subsequent heat treatment sequence is shown in Fig. 28.11. The solution heat treatment offers a chance to modify the as-forged microstructure, while the aging cycle modifies the transformed beta structures to an optimal dispersion and increases strength.

Microstructural control is basic to successful processing of titanium alloys. Undesirable structures (grain-boundary alpha, beta flecks, "spaghetti," or elongated alpha) can interfere with optimal property development. Titanium ingot structures can carry over to affect the forged product. Beta processing, despite its adverse effects on some mechanical properties, can reduce forging costs, while isothermal forging offers a means of reducing forging pressures and/or improving die fill and part detail and provides better microstructural control. Isothermal beta forging is finding use in the production of more creep-resistant components of titanium alloys.

Typical microstructures representative of those most commonly found in alpha-beta alloys are shown in Fig. 28.12. Proceeding from Fig. 28.12(a) to (d) will generally result in progressively decreased tensile and fatigue strengths, with increasing improvements in damage-tolerance properties. The difference in microstructure between Fig. 28.12(a) and Fig. 28.12(b) is caused by the differences in working history. The temperature during sheet rolling decreases as rolling proceeds, and the final rolling temperature is significantly lower than the final forging temperature. Thus, there is less retained beta at the final working temperature for the sheet, and a predominantly "globular"-type alpha microstructure results (the black features in Fig. 28.12a are retained or transformed beta). The final forging temperature is significantly higher, with more retained beta accounting for the higher amount of lamellar-transformed beta microstructure. The slow cooling of the recrystallized annealed structure (Fig. 28.12c) permits the primary alpha to grow during cooling, consuming most of the beta. Retained beta is observed at some alpha-alpha boundaries and triple points. Solution treating and aging is not commonly used for Ti-6Al-4V (Fig. 28.12e) but is the standard heat treatment for aerospace fasteners.

The effects of beta forging, as compared to conventional alpha-beta processing, are summarized in Table 28.2. Although yield strengths after beta forging are not always as high as after alpha-beta forging, values of notched tensile strength and fracture toughness are consistently higher for beta-forged material. The beta-forged alloys tend to show a transformed beta or acicular microstructure, as in Fig. 28.12(d), whereas alpha-beta-forged alloys show a more equiaxed structure, as in Fig. 28.12(b). Trade-offs are required for each structural type (acicular versus equiaxed), because each structure has unique capabilities.

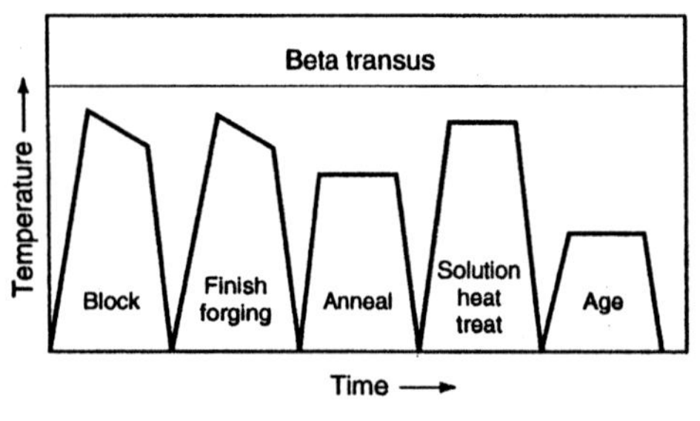


**Fig. 28.11** Typical thermomechanical processing sequence for alpha-beta titanium forgings. Typical temperatures during processing would be 955 °C (1750 °F) for the forging and solution treatment, 730 °C (1350 °F) for annealing, and 540 °C (1000 °F) for aging. Typical times during processing would be 30 min to 2 h for both annealing and solution treatment, and 8 h for aging. Source: Ref 2

## 28.5 Casting

Titanium can be cast in machined graphite molds, rammed graphite molds, and by investment casting. Investment casting is used to produce the largest and most complex castings. Since titanium castings can develop porosity during solidification, hot isostatic pressing (HIP) is used to close the internal porosity. Welding before HIP is used to repair any porosity that is exposed to the surface.

The HIP schedule that has become the industry standard is 2 h at 900 °C (1650 °F) under argon pressurized to 105 MPa (15 ksi). Hot isostatic pressing ensures that subsurface microporosity will be healed and therefore will not become exposed on a subsequently machined or polished surface to mar the finish or to act as a possible site for fatigue crack propagation.

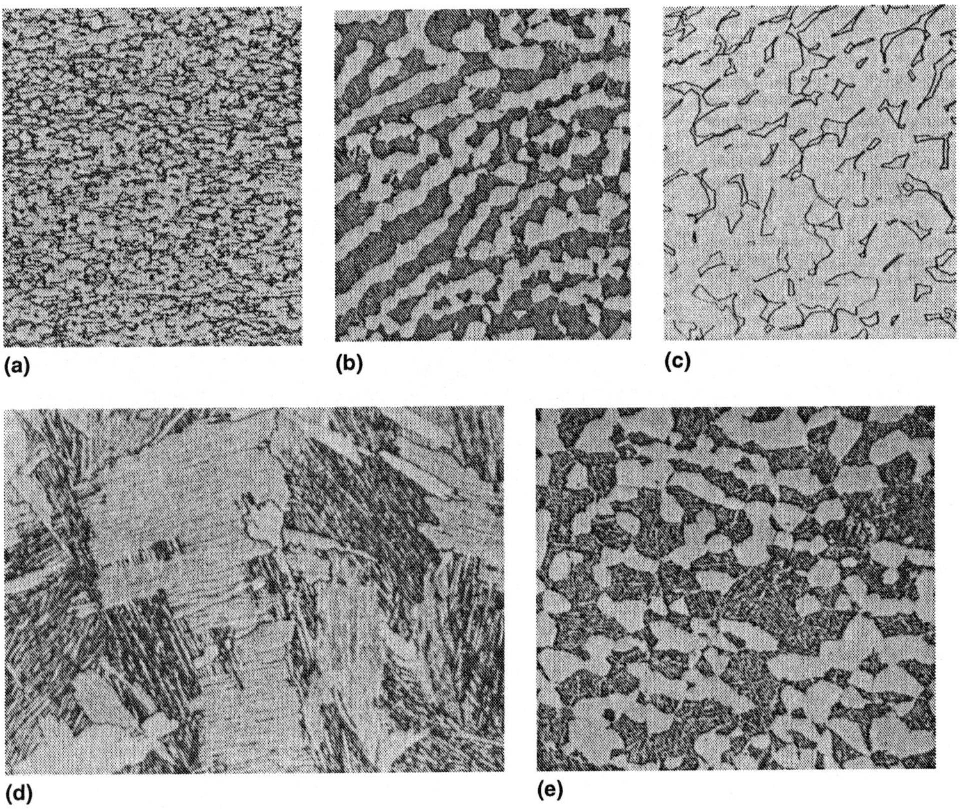

(a) (b) (c) (d) (e)

**Fig. 28.12** Effects of thermomechanical processing on microstructure of Ti-6Al-4V. (a) Sheet, rolled starting at 925 °C (1700 °F), annealed for 8 h at 730 °C (1350 °F), and furnace cooled. Structure consists of slightly elongated grains of alpha (light) and intergranular beta (gray). Original magnification: 250×. (b) As-forged at 955 °C (1750 °F), below the beta transus. Elongated alpha (light), caused by low reduction (20%) of a billet that had coarse, platelike alpha, in a matrix of transformed beta containing acicular alpha. Original magnification: 250×. (c) Plate, recrystallize-annealed at 925 °C (1700 °F) 1 h, cooled to 760 °C (1400 °F) at 50 to 55 °C/h (90 to 100 °F/h), then air cooled. Equiaxed alpha with intergranular beta. Original magnification: 500×. (d) Forging, beta annealed 2 h at 705 °C (1300 °F) exhibiting 92% basketweave structure. (e) Forging, solution treated 1 h at 955 °C (1750 °F), water quenched, and annealed 2 h at 705 °C (1300 °F). Equiaxed alpha grains (light) in transformed beta matrix (dark) containing fine acicular alpha. Original magnification: 500×. Source: Ref 2

**Table 28.2 Effect of beta forging on Ti-6Al-4V**

| Property | Beta forging |
|---|---|
| Strength | Lower |
| Ductility | Lower |
| Fracture toughness | Higher |
| Fatigue life | Lower |
| Fatigue crack growth rate | Lower |
| Creep strength | Higher |
| Aqueous stress-corrosion cracking resistance | Higher |
| Hot salt stress-corrosion cracking resistance | Lower |

Source: Ref 5

Cast titanium alloys are equal or nearly equal in strength to wrought alloys of the same compositions. However, typical ductilities are below the typical values for comparable wrought alloys but still above the guaranteed minimum values for the wrought metals. Because castings of Ti-6Al-4V have been used in aerospace applications, the most extensive data have been developed for this alloy. Generally, an improvement in fatigue properties and a reduction in the scatter of fatigue data is achieved through HIP (Fig. 28.13). Castings of alloys such as Ti-6Al-4V will generally have static and fatigue properties lower than wrought products. Fracture properties, such as fracture toughness, fatigue crack growth rate, and stress-corrosion resistance, are superior to those of mill-annealed wrought Ti-6Al-4V.

## 28.6 Heat Treating

Heat treatments for titanium alloys include stress relieving, annealing, and STA. Titanium and titanium alloys are heat treated to reduce residual stresses developed during fabrication (stress relieving); to produce an optimal combination of ductility, machinability, and dimensional and structural stability (annealing); to increase strength (STA); and to optimize special properties such as fracture toughness, fatigue strength, and high-temperature creep strength.

Not all heat treatments are applicable to all titanium alloys. The alpha and near-alpha titanium alloys can be stress relieved and annealed, but high strength cannot be developed in these alloys by any type of heat treatment. The beta alloys, on the other hand, contain metastable beta that allows strengthening during aging as the retained beta decomposes. The beta alloys offer great potential for age hardening and frequently use the stability of their beta phase to provide large-section hardenability. For beta alloys, stress relieving and aging treatments can be combined, and annealing and solution treating may be identical operations. Finally, the alpha-beta alloys exhibit heat treatment characteristics between that of the alpha class and the beta class. Alpha-beta alloys can exhibit hardening from the decomposition of beta, but these alloys do not exhibit the same section size hardenability as the beta alloys, due to the lesser amounts of retained beta. Nonetheless, the alpha-beta alloys are the most versatile in that certain microstructures can be enhanced by processing in either the alpha-beta region or the beta-phase region. Beta processing of near-alpha alloys for creep strength is useful because the near-alpha characteristic permits them to be worked or heat treated in the beta-phase field without risk of the loss of room-temperature ductility encountered with other titanium alloys processed in this way. The near-alpha alloys may also be worked high in the alpha-beta region to obtain an intermediate microstructure with a mixture of equiaxed and acicular alpha.

### 28.6.1 Stress Relief

Titanium and titanium alloys can be stress relieved without adversely affecting strength or ductility. Stress-relieving treatments decrease the undesirable residual stresses that result from nonuniform hot forging deformation from cold forming and straightening, asymmetric machining of plate (hogouts) or forgings, welding and cooling of castings, and residual thermal stresses generated during the cooling of parts with nonuniform cross sections. Stress relieving is normally conducted at 480 to 595 °C (900 to 1100 °F) for commercially pure titanium, 480 to 760 °C (900 to 1400 °F) for the alpha and near-alpha alloys, 480 to 705 °C (900 to 1300 °F) for the alpha-beta alloys, and 705 to 815 °C (1300 to 1500 °F) for the beta alloys.

Stress relief helps maintain shape stability and eliminates unfavorable conditions, such as the loss of compressive yield strength, commonly known as the Bauschinger effect. Stress-relieving treatments must be based on the metallurgical response of the alloy involved.

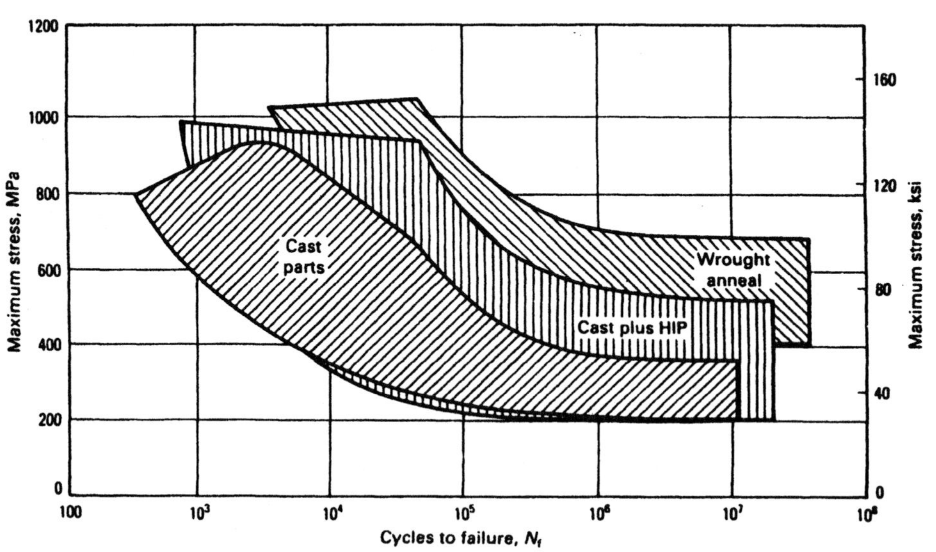


**Fig. 28.13** Effect of hot isostatic pressing (HIP) on fatigue properties of Ti-6Al-4V investment castings. Room temperature smooth bar, tension-tension fatigue, R = +0.1 Source: Ref 2

Generally, this requires holding at a temperature sufficiently high to relieve stresses without causing an undesirable amount of precipitation or strain aging in alpha-beta and beta alloys, nor producing undesirable recrystallization in single-phase alloys that rely on cold work for strength. During stress relief of STA titanium alloys, care should be taken to prevent overaging to lower strength. This usually involves selection of a time-temperature combination that provides only partial stress relief. Furnace or air cooling is acceptable. Uniformity of cooling is critical, particularly in the temperature range from 480 to 315 °C (900 to 600 °F). Oil or water quenching should not be used to accelerate cooling, because this can induce residual stresses by unequal cooling.

### 28.6.2 Annealing

Annealing is used to increase fracture toughness, room-temperature ductility, dimensional and thermal stability, and creep resistance. Many titanium alloys are placed in service in the annealed state. Because improvement in one or more properties generally is obtained at the expense of some other property, the annealing cycle should be selected according to the objective of the treatment. A comparison of common annealing treatments is shown in Table 28.3. Although the specific annealing treatments vary widely, annealing is normally conducted at 650 to 760 °C (1200 to 1400 °F) for commercially pure titanium, 705 to 900 °C (1300 to 1650 °F) for the alpha and near-alpha alloys, 650 to 815 °C (1200 to 1500 °F) for the alpha-beta alloys, and 705 to 815 °C (1300 to 1500 °F) for the beta alloys.

Mill annealing is a general-purpose treatment given to all mill products. It may not be a full anneal and may leave traces of cold or warm working in the microstructures of heavily worked products, particularly sheet. Duplex and beta annealing alter the shapes, sizes, and distributions of phases for improved creep resistance or fracture toughness. Both recrystallization and beta-annealing treatments are used to improve fracture toughness. Beta annealing is conducted at temperatures above the beta transus of the alloy being annealed.

In alpha-beta alloys, thermal stability is a function of beta-phase transformations. During cooling from the annealing temperature, beta may transform and, under certain conditions and in certain alloys, may form the brittle intermediate phase omega. A stabilization annealing treatment is designed to produce a stable beta phase capable of resisting further transformation when exposed to elevated temperatures in service. Alpha-beta alloys that are lean in beta, such as Ti-6Al-4V, can be air cooled from the annealing temperature without impairing their stability. However, furnace (slow) cooling may promote formation of the brittle $Ti_3Al$ intermetallic compound. Slight increases in strength (up to 34 MPa, or 5 ksi) can be achieved in Ti-6Al-4V and in Ti-6Al-6V-2Sn by cooling from the annealing temperature to 540 °C (1000 °F) at a rate of 56 °C/h (100 °F/h).

### 28.6.3 Solution Treating and Aging

A wide range of strength levels can be obtained in alpha-beta or beta alloys by solution treating and aging. To obtain high strength with adequate ductility, it is necessary to solution treat at a temperature high in the alpha-beta field, normally 28 to 83 °C (50 to 150 °F) below the beta transus of the alloy. If high fracture toughness or improved resistance to stress corrosion is required, beta annealing or beta solution treating may be desirable. A change in the solution-treating temperature of alpha-beta alloys alters the amount of beta phase and consequently changes the response to aging. Selection of solution-treating temperature

**Table 28.3 Effects of different anneal cycles on titanium properties**

| Property | Mill anneal(a) | Recrystallization anneal(b) | Duplex anneal(c) | Beta anneal(d) |
|---|---|---|---|---|
| Ultimate tensile strength | High | Low | Low | Low |
| Ductility | High | High | High | Lower |
| Fatigue strength | Intermediate | Intermediate | Lower | Lower |
| Fracture toughness | Lowest | High | Intermediate | Highest |
| Fatigue crack growth rate | Lowest | Intermediate | Intermediate | Highest |
| Creep resistance | Lowest | Lowest | Intermediate | Highest |

(a) Approximately 170–250 °C (300–450 °F) below beta transus, air cool. (b) Approximately 28–55 °C (50–100 °F) below beta transus, slow cool. (c) Approximately 28–55 °C (50–100 °F) below beta transus, air cool followed by mill anneal. (d) Usually 28–55 °C (50–100 °F) above beta transus, air cool usually is based on practical considerations, such as the desired level of tensile properties and the amount of ductility to be obtained after aging. The effects of solution-treating temperature on the strength and ductility of Ti-6Al-4V sheet are shown in Fig. 28.14.

Because solution treating involves heating to temperatures only slightly below the beta transus, proper control of temperature is essential. If the beta transus is exceeded, tensile properties, especially ductility, are reduced and cannot be fully restored by subsequent thermal treatment. Although the reduction in ductility is not drastic and may be acceptable, the near-alpha and alpha-beta alloys are usually solution treated below the beta transus to obtain an optimal balance of ductility, toughness, and creep strength.

Beta alloys may be obtained from producers in the solution-treated, solution-treated and aged, as-forged, or annealed conditions, depending on product form, gage, and if forming is to be done. If reheating is required, soak times should be only as long as necessary to obtain complete solutioning. Solution-treated temperatures for beta alloys are above the beta transus. However, since no second phase is present, grain growth can proceed rapidly.

The cooling rate after solution heat treating has an important effect on the strength of alpha-beta alloys. For most alpha-beta alloys, quenching in water or an equivalent quenchant is required to develop the desired strength levels. The time between removing from the furnace and the initiation of the quench is usually approximately 7 s for alpha-beta alloys and

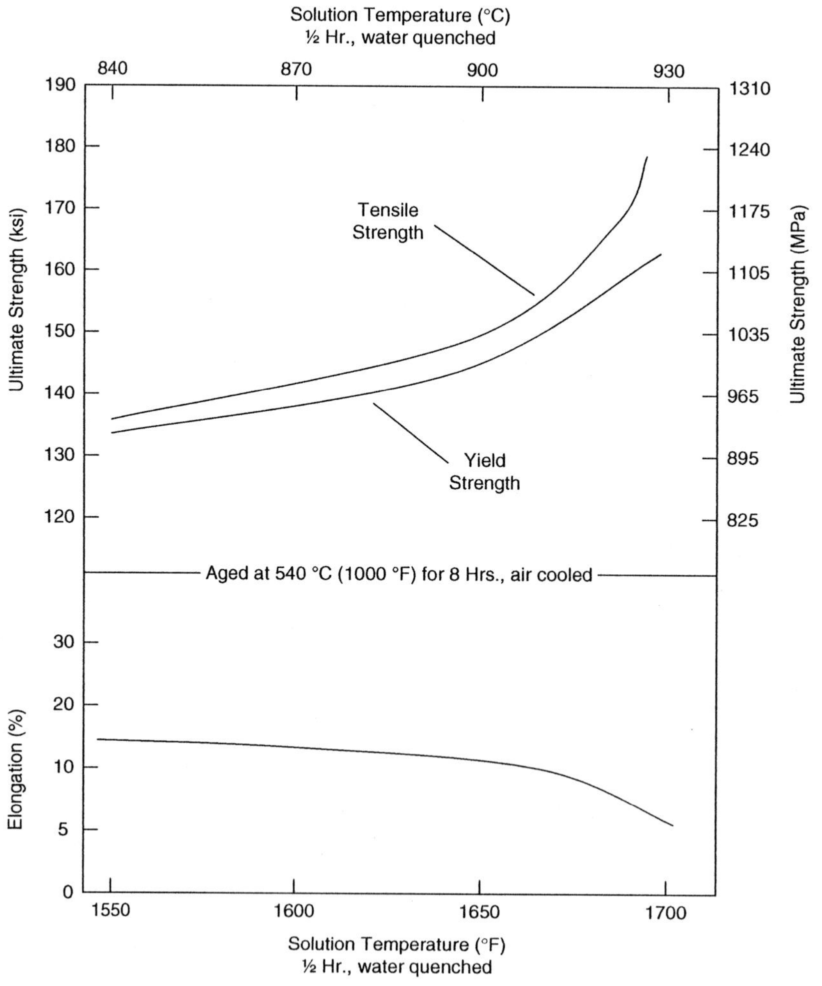


**Fig. 28.14** Effect of solution heat treating temperature on Ti-6Al-4V sheet

as long as 20 s for beta alloys. For alloys with appreciable beta stabilizing elements and moderate section thickness, air or fan cooling is usually adequate. Essentially, the amount and type of beta stabilizers in the alloy will determine the depth of hardening. Unless an alloy contains appreciable amounts of beta stabilizers, it will not harden through thick sections and will exhibit lower properties in the center, where the cooling rates are lower.

The final step in heat treating titanium alloys to high strength consists of reheating to an aging temperature between 425 and 650 °C (800 and 1200 °F). During aging of some highly beta-stabilized alpha-beta alloys, beta transforms first to a metastable transition phase referred to as omega phase. Retained omega phase, which produces unacceptable brittleness, can be avoided by severe quenching and rapid reheating to aging temperatures above 425 °C (800 °F). Aging above 425 °C (800 °F) is generally adequate to eliminate omega and precipitate alpha. However, because a coarse alpha phase forms, this treatment may not produce optimal strength properties. Aging at or near the annealing temperature will result in overaging. This condition, called solution treated and overaged, is sometimes used to obtain modest increases in strength while maintaining satisfactory toughness and dimensional stability.

### *28.6.4 Heat Treating Control*

Titanium reacts with the oxygen, water, and carbon dioxide found in oxidizing heat treating atmospheres and with hydrogen formed by decomposition of water vapor. Unless the heat treatment is performed in a vacuum furnace or in an inert atmosphere, oxygen will react with the titanium at the metal surface and produce an oxygen-enriched, hard, brittle layer called alpha case. Alpha case must be removed before the part is placed in service, because it can readily initiate fatigue cracking. However, the danger of hydrogen pickup is of even greater importance than that of oxidation. Current specifications limit hydrogen content to a maximum of 125 to 200 ppm, depending on alloy and mill form. Above these limits, hydrogen embrittles titanium alloys, reducing impact strength and notch tensile strength and causing delayed cracking.

Unlike most aluminum alloys and heat treatable ferrous alloys, hardness is not a good measure of the adequacy of the thermomechanical processes accomplished during the forging and heat treatment of titanium alloys and therefore is not used to verify the processing of titanium alloys. Instead, mechanical property tests, for example, tensile tests and fracture toughness, and metallographic/microstructural evaluation are used to verify the thermomechanical processing of titanium alloy forgings. Mechanical property and microstructural evaluations vary, ranging from the destruction of forgings to the testing of extensions and/or prolongations forged integrally with the parts.

## 28.7 Fabrication

Titanium is difficult to form at room temperature and exhibits a high degree of springback due to its yield-strength-to-modulus ratio. To compensate for the springback, titanium must be extensively overformed or, as is done most frequently, hot sized after cold forming. Hot forming, conducted at temperatures from 595 to 815 °C (1100 to 1500 °F), is normally used to form titanium alloys. Hot forming allows the material to deform more readily, simultaneously stress relieves the deformed material, and minimizes springback. Titanium also tends to creep at elevated temperature, and therefore, creep forming, performed by holding the part under load at the forming temperature, is another alternative for achieving the desired shape without having to compensate for extensive springback.

Titanium and its alloys are susceptible to the Bauschinger effect, where a plastic deformation in one direction causes a reduction in yield strength when stress is applied in the opposite direction. The Bauschinger effect is most pronounced at room temperature, where plastic deformation (1 to 5% tensile elongation) introduces a significant loss in compressive yield strength. For example, a 2% tensile strain applied to solution-treated Ti-6Al-4V causes the compressive yield strength to drop to less than half; however, a full anneal will restore the properties. As a result of the Bauschinger effect, all cold- and warm-formed structural parts must be annealed. Temperatures as low as the aging temperature remove most of the Bauschinger effect in solution-treated beta titanium alloys. However, the aging temperature is not sufficient for alpha and alpha-beta alloys. Heating or plastic deformation at temperatures above the normal aging temperature for solution-treated Ti-6Al-4V causes overaging to occur, and, as a result, mechanical properties decrease.

Titanium is difficult to machine because of its high reactivity, low thermal conductivity, relatively low modulus, and high strength at elevated temperatures. When machining titanium, it is important to use slow speeds, maintain high feed rates, use flood cooling, maintain sharp tools, and use rigid setups. Extreme caution should be used when using grinding as a final machining operation, because the fatigue strength can be reduced significantly.

Adhesive bonding, mechanical fastening, metallurgical bonding, and welding are used to join titanium and its alloys. The first three processes do not affect the properties of these metals as long as joints are properly designed. Metallurgical bonding includes all solid-state joining processes in which diffusion or deformation plays the major role in bonding. Because these processes are performed at temperatures below but close to the beta transus, metallurgical effects caused by the high process temperatures or from contamination can cause problems. However, properly processed joints have properties similar to the base metal, and, because diffusion bonding is carried out at a temperature high in the alpha-beta field, material properties are similar to those resulting from high-temperature annealing. With most alloys, a final low-temperature anneal produces properties characteristic of typical annealed material and provides thermal stability. The details of superplastic forming and superplastic forming-diffusion bonding are discussed in Chapter 16, "Deformation Processing," in this book.

Titanium alloys can be welded by gas tungsten arc welding in an inert atmosphere or electron beam or laser welded. Electron beam and laser welds are normally made without filler metal, and weld beads have high depth-to-width ratios. This combination allows excellent welds to be made in heavy sections, with properties very close to those of the base metal. All fusion welding must be done under strict environmental controls to avoid pickup of interstitials that can embrittle the weld. Small- and moderate-sized weldments can be enclosed within environmentally controlled chambers during welding. Larger weldments can be made with the aid of portable chambers that only partly enclose the components to maintain a protective atmosphere on both front and back sides of the weld until it has cooled below approximately 540 °C (1000 °F).

## ACKNOWLEDGMENTS

Sections of this chapter were adapted from "Wrought Titanium and Titanium Alloys" by S. Lampman in *Properties and Selection: Nonferrous Alloys and Special-Purpose Materials,* Volume 2, *ASM Handbook,* ASM International, 1990, and "Processing of Titanium and Titanium Alloys" by R.R. Boyer in *Metals Hand book Desk Edition,* 2nd ed., ASM International, 1998.

## REFERENCES

1. S. Lampman, Wrought Titanium and Titanium Alloys, *Properties and Selection: Nonferrous Alloys and Special-Purpose Materials,* Vol 2, *ASM Handbook,* ASM International, 1990
2. R.R. Boyer, Processing of Titanium and Titanium Alloys, *Metals Handbook Desk Edition,* 2nd ed., ASM International, 1998
3. M.J. Donachie, *Titanium: A Technical Guide,* 2nd ed., ASM International, 2000
4. Y.V.R.K. Prasad et al., Titanium Alloy Processing, *Adv. Mater Process.,* June 2000, p 85–89
5. J.C. Williams, Titanium Alloys: Production, Behavior and Application, *High Performance Materials in Aerospace,* Chapman & Hall, 1995, p 85–134

## SELECTED REFERENCES

- P. Allen, Titanium Alloy Development, *Adv. Mater. Process.,* Oct 1996, p 35–37
- R.R. Boyer, An Overview of the Use of Titanium in the Aerospace Industry, *Mater. Sci. Eng. A,* Vol 213, 1996, p 103–114
- F.C. Campbell, *Manufacturing Technology for Aerospace Structural Materials,* Elsevier Scientific, 2006
- J.D. Cotton, L.P. Clark, and H.R. Phelps, Titanium Alloys on the F-22 Fighter Aircraft, *Adv. Mater. Process.,* May 2002, p 25–28
- M.J. Donachie, Selection of Titanium Alloys for Design, *Handbook of Materials Selection,* John Wiley & Sons, Inc., 2002, p 201–234
- S.J. Gerdemann, Titanium Process Technologies, *Adv. Mater. Process.,* July 2001, p 41–43

- I.J. Polmear, *Light Alloys—Metallurgy of the Light Metals,* 3rd ed., Butterworth Heinemann, 1995
- R.W. Schutz and S.R. Seagle, Method for Improving Aging Response and Uniformity in Beta-Titanium Alloys, U.S. Patent 5,201,967, April 13, 1993
- J.C. Williams and E.A. Starke, Progress in Structural Materials for Aerospace Systems, *Acta Mater.,* Vol 51, 2003, p 5775–5799

# CHAPTER 29

# Nickel and Cobalt

NICKEL AND NICKEL ALLOYS have an excellent combination of corrosion, oxidation, and heat resistance, combined with good mechanical properties. Therefore, they are used extensively in aggressive environments, such as in the chemical processing, pollution control, power generation, electronic, and aerospace industries. Nickel is ductile and can be made by conventional processing methods into castings, powder metallurgy parts, and various hot- and cold-worked wrought products. Commercially pure nickel has a moderately high melting temperature (1468 °C, or 2647 °F), a density of 8.89 g/cm$^3$ (0.322 lb/in.$^3$), and an elastic modulus of 209 Pa (30 msi). Nickel is ferromagnetic, with a Curie temperature of 358 °C (676 °F), and possess good electrical (25% IACS, or International Annealed Copper Standards and thermal conductivity (7.0 W/m · K, or 48 Btu · in./h · ft$^2$ · °F). Nickel is used principally as an alloying element to increase the corrosion resistance of ferrous and copper alloys, with only approximately 13% of the annual production used for nickel-base alloys. Approximately 60% is used in stainless steel production, with another 10% in alloy steels and 2.5% in copper alloys. Aside from corrosion- and heat-resistant applications, nickel is also used in special-purpose alloys, such as electrical resistance, controlled expansion, magnetic, and shape memory alloys.

## 29.1 Melting of Nickel

There are four major types of nickel mineral deposits: Ni-Cu-Fe sulfides, nickel silicates, nickel laterites, and serpentines. The sulfide deposits, which are located in Canada, provide most of the Western world's supply. Nickel-copper-iron sulfide ore is crushed and ground, and the iron sulfide is separated magnetically. The remaining nickel and copper ores are then separated by flotation. There are three viable processes for refining nickel: pyrometallurgy, hydrometallurgy, and vapormetallurgy. In the pyrometallurgy process, the nickel concentrate is roasted, smelted in a reverberatory furnace, and converted to a bessemer matte that consists mainly of nickel and copper sulfides. The copper-nickel matte is cooled under controlled conditions to form discrete crystals of nickel and copper sulfides and a nickel-copper metallic alloy. After the cooled matte is crushed and ground, the metallic alloy is separated magnetically and treated. The remaining copper and nickel sulfides are separated by froth flotation. The nickel sulfide is roasted to produce various grades of nickel oxides that are then converted to pure nickel and nickel alloys.

Hydrometallurgy processes include electrorefining and electrowinning, both of which are electrolytic processes used to produce pure nickel cathode sheet. In the electrofining process, a nickel electrolytic refining cell, consisting of a slab of crude nickel to be refined (anode) and a thin nickel starting sheet (cathode), is immersed in an aqueous electrolyte, and direct current is passed through the cell, with pure nickel being deposited on the cathode. In the electrowinning process, soluble anodes of copper sulfide are electrolyzed to extract nickel from a leach liquor. In vapormetallurgy, a carbonyl process uses a gas-to-metal transformation to extract pure nickel from an impure nickel oxide. Oxide is reduced with hydrogen, and the nickel reacts selectively with carbon monoxide to form gaseous nickel carbonyl. The gas is then decomposed by heating to yield high-purity nickel as a powder or pellet.

## 29.2 Nickel Metallurgy

As a result of its face-centered cubic (fcc) crystal structure, nickel has excellent ductility and toughness. Alloying is used to further improve corrosion and heat resistance. The austenitic nickel matrix ($\gamma$) can be strengthened by solid-solution hardening, carbide precipitation, precipitation hardening, and/or work hardening.

Solid-solution hardening is provided by cobalt, iron, chromium, molybdenum, tungsten, vanadium, titanium, and aluminum. Nickel forms a complete series of solid solutions with copper and has nearly complete solubility with iron. In addition, it can dissolve approximately 35 wt% Cr, approximately 20 wt% each of molybdenum and tungsten, and approximately 5 to 10 wt% each of aluminum, titanium, manganese, and vanadium. The degree of solid-solution hardening is related to the atomic size difference between nickel and the alloying element, and the ability of the solute atoms to interfere with dislocation motion. Tungsten, molybdenum, niobium, tantalum, and aluminum, when in solution, are strong solution hardeners. The slower-diffusing elements, tungsten, niobium, tantalum, and molybdenum, are also effective alloying additives at temperatures above 0.6 $T_m$, where diffusion-controlled creep strength becomes important. Iron, cobalt, titanium, chromium, and vanadium are weaker solid-solution-hardening elements.

Although nickel itself is not a carbide former, the presence of carbon as an alloying element leads to the formation of carbides. The carbides most frequently found in nickel alloys include MC, $M_6C$, $M_7C_3$, and $M_{23}C_6$, where "M" is the carbide-forming element or elements, such as chromium, which forms $Cr_7C_3$. In general, carbides improve the elevated-temperature strength if they are located on the grain boundaries as discrete particles. However, if they form continuous grain-boundary films, they can cause embrittlement.

The precipitation of $\gamma'$ $Ni_3(Al,Ti)$ in the $\gamma$ nickel matrix provides significant precipitation hardening. The $\gamma$ precipitate has the same fcc structure as the austenitic matrix and a lattice constant mismatch of less than 1%, which provides stability at high temperatures. The volume percent of $\gamma'$ is important because high-temperature strength increases with increasing amounts of $\gamma'$. The amount of $\gamma'$ is a function of the hardening-element content in the alloy. Strong $\gamma'$ formers include aluminum, titanium, niobium, and tantalum.

Nickel alloys used in low-temperature applications are often hardened by cold working. The rate of work hardening for nickel alloys is greater than that for low-carbon steel and is usually less than that of the austenitic stainless steels. The work-hardening rate generally increases with the compositional complexity of the alloy, being moderately low for nickel and nickel-copper alloys to moderately high for nickel-chromium and the more highly alloyed high-temperature alloys. Precipitation-hardened alloys have higher work-hardening rates than their solid-solution-hardened counterparts. A comparison of the work-hardening characteristics of several nickel alloys, along with several other alloys, is shown in Fig. 29.1.

## 29.3 Nickel Alloys

Nickel alloys can be divided into alloys that combine corrosion and heat resistance, superalloys for high-temperature applications, and special nickel alloys. Special nickel alloys include electrical-resistance alloys, low-expansion alloys, magnetically soft alloys, and shape memory alloys. The important class of nickel-base superalloys used for jet turbines is discussed in Chapter 30, "Superalloys," in this book.

### *29.3.1 Corrosion- and Heat-Resistant Nickel Alloys*

A list of selected corrosion- and heat-resistant nickel alloys, along with their mechanical properties, is given in Table 29.1.

**Commercially Pure and Low-Alloy Nickels.** The commercially pure (CP) nickel grades, Nickel 200 to 205, are highly resistant to many corrosive media, especially reducing environments but also oxidizing environments due to the formation of a passive nickel oxide surface film. The CP nickel grades are used in the chemical processing and electronics industries. They are hot worked at 650 to 1230 °C (1200 to 2250 °F), annealed at 705 to 925 °C (1300 to 1700 °F), and hardened by cold working. The tensile properties of annealed sheet (462 MPa, or 67 ksi, tensile strength; 152 MPa, or 22 ksi, yield strength; and 47% elongation) can be increased by cold rolling up to 758 MPa (110 ksi) tensile strength, 634 MPa (92 ksi) yield strength, and 8% elongation. Because of its 0.08 wt% C content (0.15 wt% maximum), Nickel 200 is a CP (99.6 wt%) wrought nickel with a good balance of mechanical properties (Fig. 29.2) and resistance to a range of corrosive media. A typical microstructure for Nickel 200 is shown in Fig. 29.3. The microstructure of CP nickel consists of grains of gamma solid solution. Twinning may be present in cold-worked or annealed material. Carbon is present as finely distributed spheroidal graphite in cast materials. Nickel 200 is used for a variety of processing equipment, particularly to maintain product

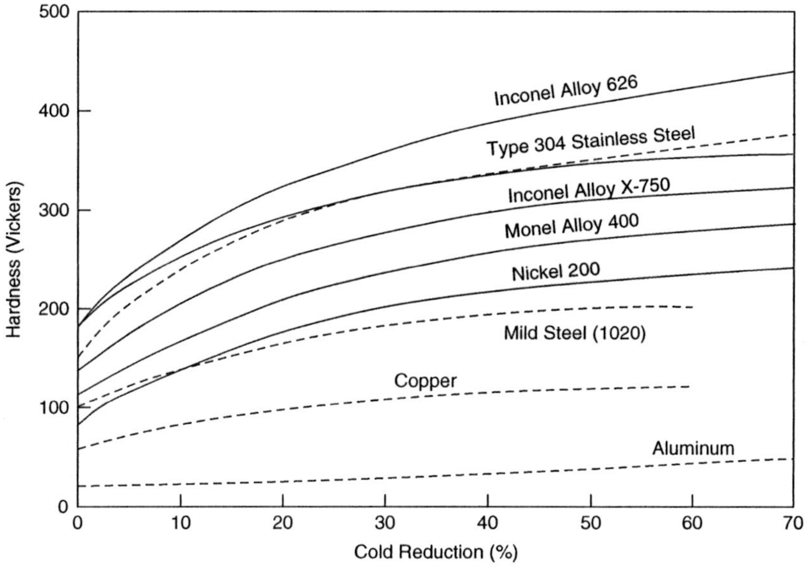


**Fig. 29.1** Effect of cold work on hardness. Source: Ref 1

**Table 29.1 Composition and properties of select nickel alloys**

| Alloy | Nominal composition, wt% | Tensile strength | | Yield strength | | Elongation in 50 mm (2 in.), % | Rockwell hardness |
|---|---|---|---|---|---|---|---|
| | | MPa | ksi | MPa | ksi | | |
| **Commercially pure and low-alloy nickels** | | | | | | | |
| Nickel 200 | 99.0% Ni | 462 | 67 | 148 | 21.5 | 47 | 109 HB |
| Nickel 201 | 99.0% Ni | 403 | 58.5 | 103 | 15 | 50 | 129 HB |
| Nickel 211 | Ni-4.75Mn-0.75Fe | 530 | 77 | 240 | 35 | 40 | ... |
| Duranickel 301 | Ni-4.5Al-0.5Ti | 1170 | 170 | 862 | 125 | 25 | 30–40 HRC |
| **Nickel-copper alloys** | | | | | | | |
| Alloy 400 | Ni-31Cu-2.5Fe | 550 | 80 | 240 | 35 | 40 | 110–150 HB |
| Alloy K-500 | Ni-30Cu-2Fe-1.5Mn-2.7Al-0.6Ti | 1100 | 160 | 790 | 115 | 20 | 300 HB |
| **Nickel-molybdenum and nickel-silicon alloys** | | | | | | | |
| Hastelloy B | Ni-28Mo-5.5Fe-2.5Co | | | | | | |
| Sheet | | 834 | 121 | 386 | 56 | 63 | 92 HB |
| Investment cast | | 586 | 85 | 345 | 50 | 10 | 93 HB |
| Hastelloy D | Ni-9.25Si-3Cu-1.5Co | 793 | 115 | ... | ... | ... | 30–39 HRC |
| **Nickel-chromium-iron alloys** | | | | | | | |
| Alloy 600 | Ni-15Cr-8Fe | 655 | 95 | 310 | 45 | 40 | 75 HRB |
| Alloy 601 | Ni-23Cr-14Fe | 655 | 95 | 276 | 40 | 50 | 90 HRB |
| Alloy 800 | Ni-21Cr-39.5Fe-0.4Ti-0.4Al | 600 | 87 | 295 | 43 | 44 | 138 HB |
| Alloy 690 | Ni-29Cr-9Fe | 725 | 105 | 348 | 50 | 41 | 88 HRB |
| Alloy 751 | Ni-15Cr-7Fe-1Nb-2Ti | 1310 | 190 | 976 | 142 | 22 | 352 HB |
| **Nickel-chromium-molybdenum alloys** | | | | | | | |
| Alloy C-276 | Ni-15.5Cr-16Mo-5.5Fe-3.75W-1.25Co + V | 785 | 114 | 372 | 54 | 62 | 209 HB |
| Alloy 625 | Ni-21.5Cr-9Mo-3.65Nb + Ta-2.5Fe | 930 | 135 | 517 | 75 | 42.5 | 190 HB |
| **Nickel-chromium-iron-molybdenum-copper alloys** | | | | | | | |
| Hastelloy G | Ni-22.25Cr-19.5Fe-6.5Mo-2Cu + Co,Nb,Ta | 690 | 100 | 320 | 47 | 50 | 79 HRB |
| Alloy 825 | Ni-21.5Cr-30Fe-3Mo-2.25Cu + Al | 690 | 100 | 310 | 45 | 45 | ... |

Source: Ref 2

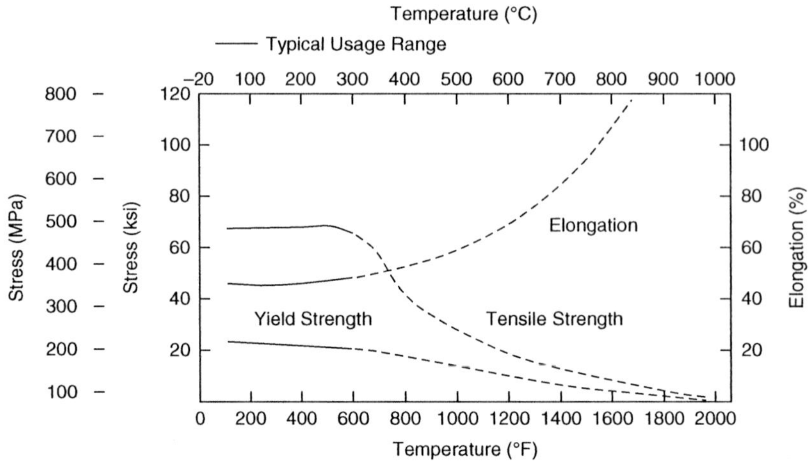


Fig. 29.2 Typical mechanical properties of annealed, commercially pure Nickel 200. Source: Ref 3

purity in handling foods, synthetic fibers, and alkalis. However, Nickel alloy 200 should not be used above 315 °C (600 °F), because embrittlement results from the precipitation of intergranular graphite in the range of 425 to 650 °C (800 to 1200 °F). Nickel 201 is also a CP (99.6 wt%) wrought nickel but with a lower carbon content than Nickel 200. It can be used at higher temperatures than Nickel 200 because it is not subject to graphite embrittlement. Typical properties up to 650 °C (1200 °F) are shown in Fig. 29.4. The lower carbon content also reduces the hardness, making Nickel 201 particularly useful for cold-formed parts.

The low-alloy nickels contain 94 wt% minimum nickel. Nickel 211 forms a solid solution with the 5 wt% Mn addition, which protects it against sulfur in service environments. As little as 0.005 wt% S can cause liquid embrittlement at grain boundaries in the range of 640 to 740 °C (1185 to 1365 °F). Duranickel alloy 301 (Ni-4.5Al-0.5Ti), with its aluminum and titanium additions, has the corrosion resistance of CP nickel along with strengthening provided by the precipitation of $\gamma'$. A typical heat treated microstructure is shown in Fig. 29.5. This sample was solution annealed for 30 min at 980 °C (1800 °F), water quenched to room temperature, and then aged for 20 h at 480 °C (900 °F), followed by water quenching. There are sufficient alloying additions in alloy 301 to lower the Curie temperature, making the alloy weakly ferromagnetic at room temperature. Duranickel alloy 301 is used for springs requiring high electrical conductivity, parts of equipment requiring good

Fig. 29.3 Cold-drawn and annealed Nickel 200. Source: Ref 4

thermal conductivity, and magnetostrictive units that are operated under stress conditions for which the fatigue strength of Nickel 200 is inadequate.

**Nickel-copper alloys** are strong and tough, offering corrosion resistance in various environments, including brine, sulfuric, and other acids, and are immune to chloride ion stress corrosion. They are used in chemical processing and pollution control equipment. Alloy 400 (Monel) is a nickel-copper alloy with 31 wt% Cu that has high strength and excellent corrosion resistance in a range of media, including

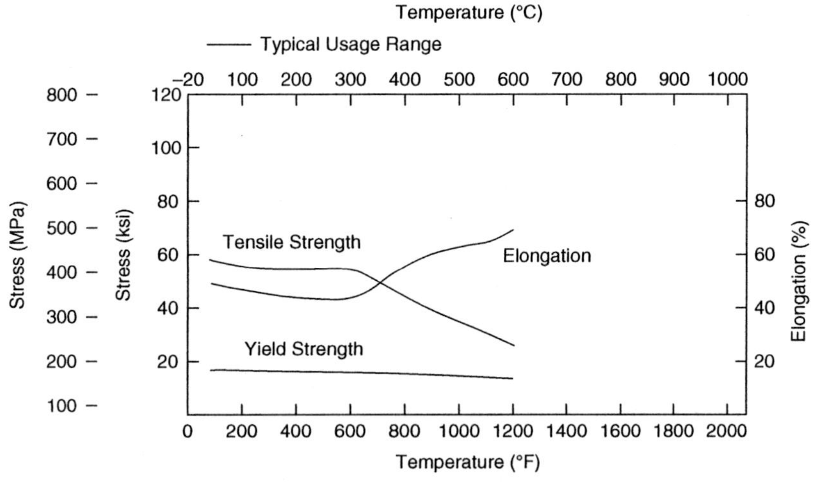


**Fig. 29.4** Typical mechanical properties of annealed nickel 201. Source: Ref 3

seawater, hydrofluoric acid, sulfuric acid, and alkalis. Typical mechanical properties of alloy 400 are shown in Fig. 29.6. It has excellent properties in seawater under high-velocity conditions, such as in propellers, propeller shafts, pump shafts, impellers, and condenser tubes, where resistance to cavitation and erosion are important. Corrosion rates in strongly agitated seawater usually do not exceed 25 μm/year (1 mil/year).

Alloy K-500 is a precipitation-hardenable nickel-copper alloy that combines the corrosion resistance of alloy 400 with greater strength and hardness. The addition of aluminum (2.30 to 3.15 wt%) and titanium (0.35 to 0.85 wt%) form the basis for precipitation of $\gamma'$ during aging, which increases the strength (Fig. 29.7). It also has low magnetic permeability and is nonmagnetic to temperatures as low as −100 °C (−150 °F). Alloy K-500 is used for pump shafts, oil well tools and instruments, doctor blades and scrapers, springs, fasteners, and marine propeller shafts.

**Nickel-Molybdenum and Nickel-Silicon Alloys.** Molybdenum improves the resistance to reducing acids, such as hydrochloric acid. Nickel-molybdenum alloys containing greater than 25% Mo develop exceptional resistance to hydrochloric acid. The high molybdenum and silicon additions to Hastelloy B and D promote good corrosion resistance in the presence of hydrochloric and sulfuric acids. The as-cast and annealed structures of Hastelloy B are shown in Fig. 29.8. The as-cast structure consists of

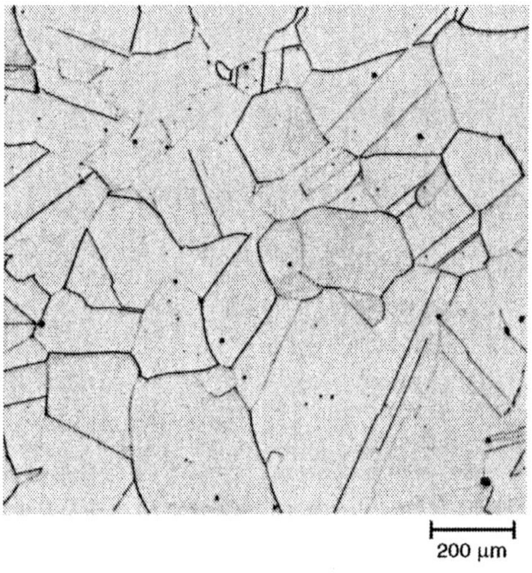


**Fig. 29.5** Precipitation heat treated Duranickel 301. Source: Ref 4

carbides at the grain boundaries that form islands in the matrix as a result of annealing. The chief weakness of these alloys is their relatively poor resistance to oxidizing agents, even if the oxidizer is present in only small amounts.

**Nickel-Chromium-Iron Alloys.** This family of alloys was developed for high-temperature oxidizing environments. They typically contain 50 to 80 wt% Ni, which permits the addition of other alloying elements to improve strength and corrosion resistance while maintaining toughness. The Ni-Cr-Fe alloys contain chromium ranging from 14 to 30 wt% and iron from 3 to 40 wt%. Since they form a protective surface film of $Cr_2O_3$, they have excellent corrosion resistance in many severe environments, including immunity to chloride ion stress-corrosion cracking. They also have good oxidation and sulfidation resistance along with good strength at elevated temperatures. They have a maximum operating temperature in the neighborhood of 1205 °C (2200 °F). Alloy 600 (Ni-15Cr-8Fe) is a single-phase alloy (Fig. 29.9) that can be used at temperatures from cryogenic to 1095 °C (2000 °F). The yield strength of strip in the annealed condition (207 to 310 MPa, or 30 to 45 ksi) can be work hardened by cold rolling to reach yield strengths of 830 to 1100 MPa (120 to 160 ksi) and then can retain most of the strength increase up to approximately 540 °C (1000 °F). Alloy 600 has good oxidation resistance at high temperatures and resistance to chloride ion stress-corrosion cracking, corrosion by high-purity water, and caustic corrosion. It is used for furnace components, in chemical and food processing, in nuclear engineering, and for sparking electrodes.

Alloy 601 is another alloy intended for high-temperature service. As shown in Fig. 29.10, it retains its strength up to 650 °C (1200 °F). Alloy 601 has exceptional resistance to oxidation at high temperatures, even better than alloy

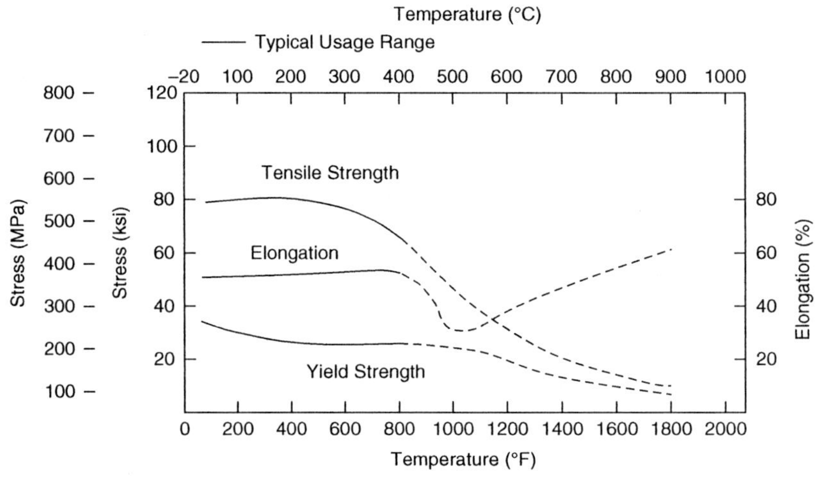


**Fig. 29.6** Typical mechanical properties of annealed Monel 400. Source: Ref 3

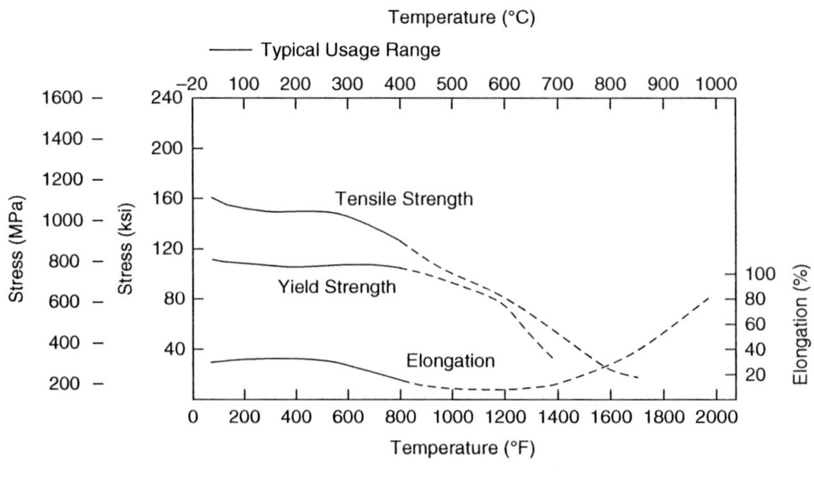


**Fig. 29.7** Typical mechanical properties of precipitation-hardened Monel K-500. Source: Ref 3

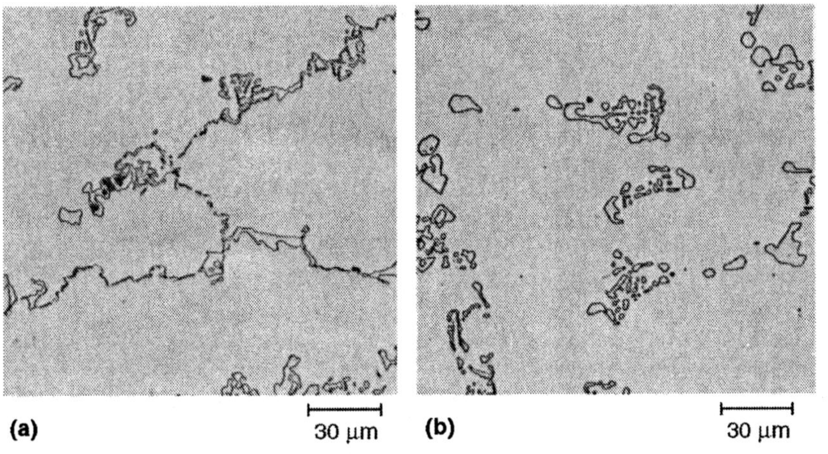


**Fig. 29.8** Microstructure of Hastelloy B. (a) As-cast structure. (b) After annealing. Source: Ref 5

600. This alloy forms a protective oxide coating that resists scaling even under the severe conditions of high-temperature cyclic exposure. The resistance of alloy 601, compared to alloys 600 and 800, in oxidation tests at 1150 °C (2100 °F) is shown in Fig. 29.11. In these tests, the specimens were exposed to 1150 °C (2100 °F) for 10 consecutive periods of 50 h each. After each exposure, the specimens were cooled to room temperature, brushed lightly to remove any loose oxide, and then weighed to determine weight change. The superior oxidation resistance of alloy 601 is due to the amounts of nickel, chromium, and aluminum. During high-temperature exposure, these elements form an extremely protective and adherent surface oxide film. In addition, a slight amount of internal oxidation occurs and increases the chromium content in the surface oxide. This protective oxide layer is shown in Fig. 29.12.

Alloy 800 is a nickel-chromium alloy with good strength and excellent resistance to oxidation and carburization in high-temperature atmospheres. It also resists corrosion in many aqueous environments. It maintains a stable, austenitic structure during prolonged exposure to high temperatures. Alloy 800 is used for process piping, heat exchangers, carburizing equipment, heating element sheathing, and nuclear steam generator tubing. Alloys 800H and 800HT have the same basic composition as alloy 800 but with significantly higher creep-rupture strength. The higher creep strengths result from close control of the carbon, aluminum, and titanium contents in conjunction with a high-temperature anneal

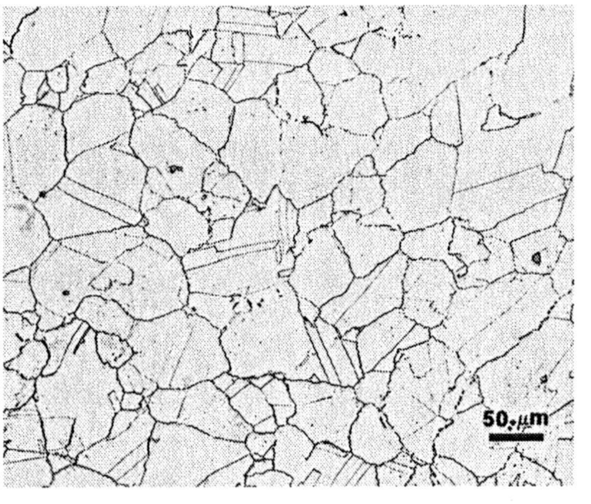


**Fig. 29.9** Forged section of alloy 600. Source: Ref 5

that produces a coarse grain size. Alloys 800H and 800HT are used in chemical and petrochemical processing, in power plants for superheater and reheater tubing, in industrial furnaces, and for heat treating equipment.

**Nickel-Chromium-Molybdenum Alloys.** The addition of chromium and molybdenum to nickel produces alloys with resistance to both oxidizing and reducing corrosives. Resistance to oxidizing environments is provided by the chromium, while molybdenum gives the resistance to reducing environments. Thus, these alloys are the most corrosion resistant of the nickel alloys. The Ni-Cr-Mo alloys are used in the chemical processing, pollution control, and waste treatment industries to utilize their excellent combination of heat and corrosion resistance. Alloys in this family, such as C-276 and alloy 625, are even more versatile as a result of their excellent welding characteristics and the corrosion resistance of their welded structures. The molybdenum additions to these alloys also improve resistance to pitting and crevice corrosion. Aluminum improves the protective surface oxide film, and the carbide formers titanium and niobium are used to stabilize the alloys against chromium carbide sensitization.

Alloy C-276 is a Ni-Cr-Mo alloy with a tungsten addition that has excellent corrosion resistance in a wide range of severe environments. The high molybdenum content makes it especially resistant to pitting and crevice corrosion. The low carbon content minimizes carbide precipitation during welding to maintain corrosion resistance in as-welded structures. Alloy C-276 is used in pollution control, chemical processing, pulp and paper production, and waste treatment plants.

Alloy 625 is a Ni-Cr-Mo alloy with a niobium addition that, along with molybdenum, stiffens the alloy matrix and provides high strength without a strengthening heat treatment. It resists a wide range of severely corrosive environments and is especially resistant to pitting and crevice corrosion. Alloy 625 is used in chemical processing, aerospace and marine engineering, pollution control equipment, and nuclear reactors.

Alloy 825 is a Ni-Cr-Fe-Mo alloy with a copper addition. It has excellent resistance to both reducing and oxidizing acids, to stress-corrosion cracking, and to localized attack such as pitting and crevice corrosion. The alloy is especially resistant to sulfuric and phosphoric acids. It is used for chemical processing, pollution control equipment, oil and gas well piping, nuclear fuel reprocessing, acid production, and pickling equipment.

### 29.3.2 Specialty Nickel Alloys

Unique combinations of properties are available with other nickel-base alloys designed

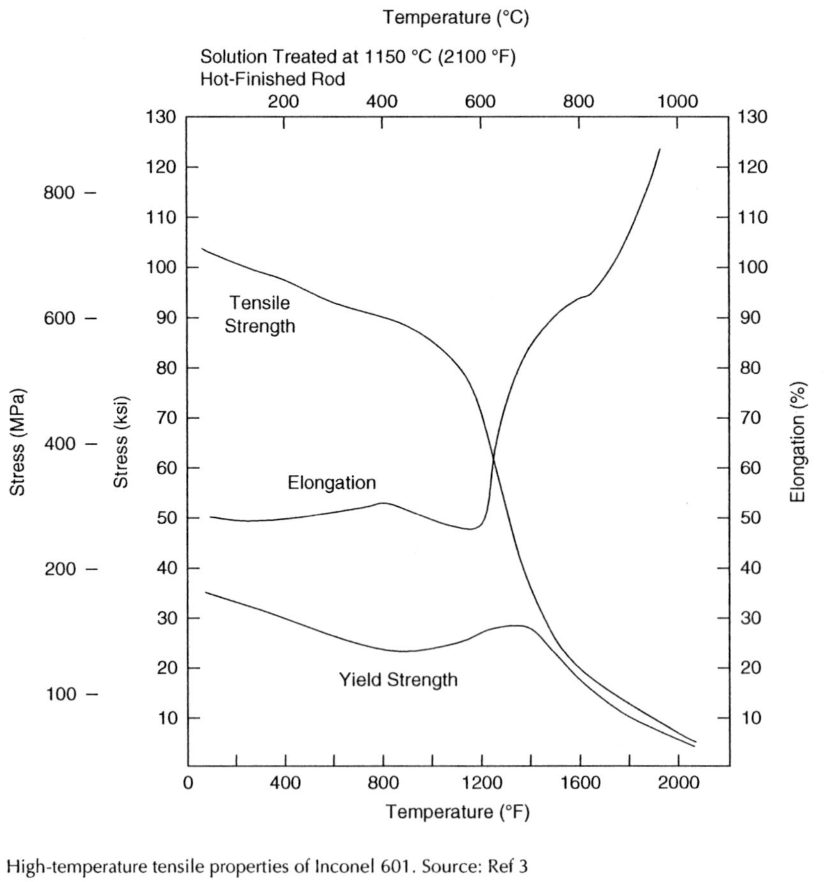


**Fig. 29.10** High-temperature tensile properties of Inconel 601. Source: Ref 3

for special applications. While some of these properties are also available to some extent with the alloys described previously, the alloys that follow were developed for these specific applications.

**Electrical resistance alloys** can be divided into two groups: (1) those that are used in instruments and control equipment to measure and regulate electrical characteristics (resistance alloys) and (2) those that are used in furnaces and appliances to generate heat (heating alloys).

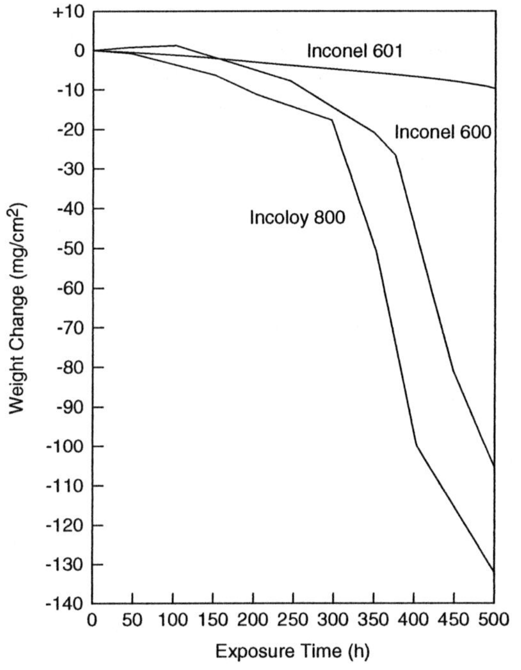


**Fig. 29.11** Relative oxidation resistance of alloys 601, 600, and 800. Oxidation tests at 1150 °C (2100 °F), 50 h cycles, cool to RT between cycles. Source: Ref 3

Examples of resistance alloys include copper-nickel alloys containing 2 to 45 wt% Ni, Ni-Cr-Al alloys containing 35 to 95 wt% Ni, Ni-Cr-Fe alloys containing 35 to 60 wt% Ni, and Ni-Cr-Si alloys containing 70 to 80% Ni. Heating alloys contain 35 to 95 wt% Ni with at least 15 wt% Cr to form an adherent surface oxide to protect against oxidation and carburization at temperatures up to 1010 to 1205 °C (1850 to 2200 °F) in air. Examples are Ni-20Cr, Ni-15Cr-25Fe, and Ni-20Cr-3Al-3Fe.

**Magnetic Alloys.** Nickel alloys are used for both soft and hard magnets. Soft magnet alloys are characterized by their high magnetic permeability and little or no retentivity, meaning they lose magnetism if the magnetic field is removed. These alloys contain 50 to 79 wt% Ni and have single-phase structures. Applications include magnets for rotors and generators. Permanent or hard magnetic alloys contain nickel, cobalt, and aluminum and have two-phase structures. Permanent nickel alloy magnets are used in telephone receivers, flow meters, and loudspeakers.

**Shape memory alloys** are a group of alloys that have the ability to return to some previously defined shape or size when subjected to an appropriate thermal cycle. Generally, these materials can be plastically deformed at relatively low temperature and, on exposure to some higher temperature, will return to their shape prior to the deformation. Materials that exhibit shape memory only on heating are referred to as having a one-way shape memory. Some materials also undergo a change in shape on recooling. These materials are said to have a two-way shape memory. Nitinol (50Ti-50Ni) is one of the best known and most useful of the shape memory alloys. Nitinol has been used in

**Fig. 29.12** Oxide layer on alloy 601 exposed to 1150 °C (2100 °F) for 500 h. Original magnification: 75×. Source: Ref 6

medical applications for blood clot filters, hydraulic couplings for aircraft, force actuators, proportional control devices, and superelastic eyeglass frames.

A shape memory alloy can be further defined as one that yields thermoelastic martensite. In this case, the alloy undergoes a martensitic transformation of a type that allows the alloy to be deformed by a twinning mechanism below the transformation temperature. The deformation is then reversed when the twinned structure reverts to the parent phase upon heating. The martensitic transformation that occurs in shape memory alloys yields a thermoelastic martensite that develops from a high-temperature austenite phase with long-range order. The transformation does not occur at a single temperature but over a range of temperatures that varies with each alloy system. The usual way of characterizing the transformation and naming each point in the cycle is shown in Fig. 29.13. Most of the transformation occurs over a relatively narrow temperature range, although the beginning and end of the transformation during heating or cooling actually extends over a much larger temperature range. The transformation also exhibits a hysteresis in that the transformation on heating and on cooling does not overlap. This transformation hysteresis (shown as *T* in Fig. 29.13) varies with the alloy system.

## 29.4 Iron-Nickel Alloys

Nickel has a pronounced effect on the thermal expansion of iron. Alloys can be designed to have a very low thermal expansion or display uniform and predictable expansion over certain temperature ranges. Low-expansion alloys include various binary iron-nickel alloys and several ternary alloys of iron combined with nickel-chromium, nickel-cobalt, or cobalt-chromium alloying. Many of the low-expansion alloys are identified by their trade names:

- Invar or Nilo is a 64Fe-36Ni alloy with the lowest thermal expansion coefficient of iron-nickel alloys.
- Kovar is a 54Fe-29Ni-17Co alloy with a coefficient of expansion closely matching those of standard types of borosilicate glass.
- Elinvar is a 52Fe-36Ni-12Cr alloy with a zero thermoelastic coefficient; that is, the modulus of elasticity is constant over a wide temperature range.
- Super Invar is a 63Fe-32Ni-5Co alloy with an expansion coefficient smaller than Invar but over a narrower temperature range.

Alloys of iron and nickel have coefficients of linear expansion ranging from a small negative value (−0.5 ppm/°C) to a large positive value (20 ppm/°C). The effect of nickel content on the linear expansion of iron-nickel alloys at room temperature is shown in Fig. 29.14 In the range of 30 to 60% Ni, it is possible to select alloys with tailored expansion characteristics. Iron-36Ni alloy has the lowest expansion of the iron-nickel alloys and maintains nearly constant dimensions during normal variations in temperature. Higher nickel content results in greater thermal expansion, which allows specific expansion rates to be selected by adjusting the nickel content. The addition of cobalt to the nickel-iron matrix produces alloys with a low coefficient of expansion, a constant modulus of

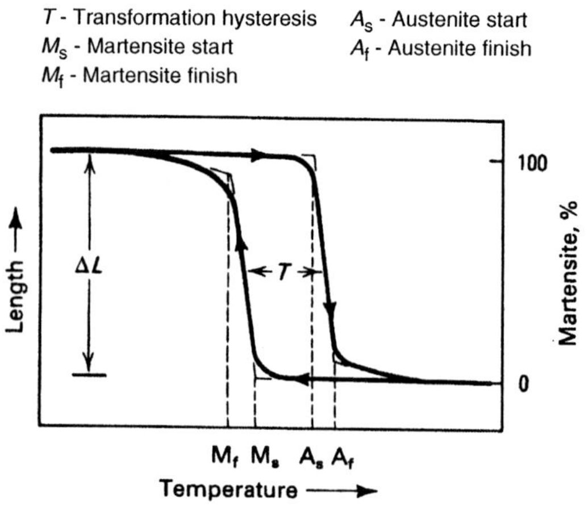


**Fig. 29.13** Typical transformation versus temperature curve for shape memory alloys. Source: Ref 7

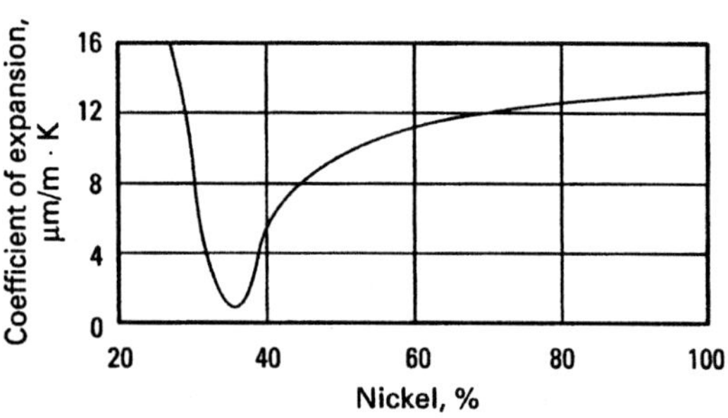


**Fig. 29.14** Coefficient of linear expansion versus nickel content for iron-nickel alloys measured at 68 °F (20 °C). Iron-nickel alloys contain 0.4% Mn and 0.1% C. Source: Ref 8

elasticity, and high strength. With niobium and aluminum additions, these alloys can also be strengthened by precipitation-hardening heat treatments.

## 29.5 Cobalt and Cobalt Alloys

Cobalt falls between iron and nickel on the periodic table and has many properties that are similar to those other two transition metals. Cobalt has a melting point of 1493 °C (2719 °F), a density of 8.8 g/cm$^3$ (0.322 lb/in.$^3$), a coefficient of thermal expansion of 13.8 μm/m·K (7.66 μin./in.·°F), a thermal conductivity of 69.0 W/m·K (478 Btu·in./h·Ft$^2$·°F), and an elastic modulus of 207 Pa (30 msi), all rather similar to the respective values of iron and nickel. All three are ferromagnetic, but the Curie temperature of cobalt (1120 °C, or 2050 °F) is significantly higher than that of iron (770 °C, or 1418 °F) or nickel (358 °C, or 676 °F).

Cobalt is not as abundant as its two companion metals, making cobalt approximately 100 times more costly than iron and approximately 8 times more costly than nickel. The largest use of cobalt is in superalloys that are used to make parts for gas turbine engines. Cobalt is also used in corrosion- and wear-resistant alloys, high-speed tool steels, cemented carbide tools, magnetic materials, and low-expansion alloys. Due to their excellent wear resistance, elevated-temperature hardness, and corrosion resistance, cobalt alloys are commercially used in gas turbine engines, earthmoving equipment, and as bearing materials. Cobalt alloys are often used for their wear resistance, either in solid forms or as welded or thermally sprayed overlays to hardface other structural materials. They are also available as powder metallurgy products and in wrought sheet, bar, and tube forms.

Cobalt has a hexagonal close-packed (hcp) crystal structure (ε-cobalt) at room temperature that transforms to fcc (α-cobalt) at 417 °C (783 °F). Chromium, tungsten, and molybdenum stabilize the hcp phase, while iron and nickel stabilize the fcc structure. However, the fcc-to-hcp transformation is very sluggish in cobalt alloys; therefore, alloys usually have the metastable fcc structure. The α-to-ε transformation usually occurs by a strain-induced martensitic reaction that contributes to the high work-hardening rates of cobalt alloys. Carbon, one of the principal alloying elements, forms carbide phases that lead to increased hardness, elevated temperature strength, creep resistance, and resistance to abrasive wear.

Cobalt-base alloys may be grouped as wear-resistant, corrosion-resistant, or heat-resistant alloys. The single largest application of cobalt-base alloys is for wear resistance. In heat-resistant applications, cobalt is more widely used as an alloying element in nickel-base alloys than as cobalt-base alloys. Alloying elements such as chromium, tungsten, and molybdenum provide solid-solution hardening and form hard, abrasion-resistant carbides; chromium also imparts corrosion resistance. Generally, the softer and tougher compositions are used for high-temperature applications, and the harder grades are used for wear resistance.

This chapter examines cobalt wear-resistant alloys and cobalt corrosion-resistant alloys. Cobalt-base superalloys are covered in Chapter 30, "Superalloys," and cemented carbide cutting tools are discussed in Chapter 22, "Tool Materials," in this book. A number of the wear-resistant and corrosion-resistant alloys are listed in Table 29.2.

### *29.5.1 Cobalt-Base Wear-Resistant Alloys*

The cobalt wear-resistant alloys, known as Stellites, were originally developed by Elwood Haynes in approximately 1900 for their wear and hardness properties. He named them Stellites after the Latin *stella,* for star, because of their lustrous appearance. The Stellite alloys contain large amounts of the carbide-forming elements chromium, tungsten, and molybdenum, along with 0.1 to 3.3 wt% C to allow the formation of hard carbide particles. Stellite is now a registered trade name of the Deloro Stellite Company, Inc.

The microstructures generally consist of chromium-rich $M_7C_3$ carbides and, in the higher-tungsten alloys, the tungsten-rich $M_6C$ carbide, with enough chromium and tungsten alloying element left in solid solution to provide a tough, solid-solution-hardened matrix. The chromium-rich $M_{23}C_6$ carbide is prevalent in the lower-carbon grades, such as Stellite 21. Chromium is restricted to no more than 30 wt% to avoid precipitation of the chromium-rich embrittling σ phase. The microstructures therefore consist of hard, complex alloy carbides in a tough, solid-solution alloy matrix. The microstructures of a number of cobalt-base wear-resistant alloys are shown for comparison in Fig. 29.15.

Stellite alloys 1, 6, and 12 are derivatives of the Co-Cr-W alloys originally developed by Haynes. The main difference between the Stellite alloys is their carbon contents and thus the carbide volume fraction of the alloy. For example, Stellite alloy 1, which contains 2.5 wt% C and 13 wt% W, has approximately 28.5 wt% hard carbides in the matrix, while Stellite alloy 6, which contains 1.2 wt% C and 4.5 wt% W, only has approximately 12 wt% hard carbides. Therefore, Stellite alloy 1 is harder and more abrasion resistant but also more brittle than Stellite alloy 6. In general, alloys with more than approximately 1.3 wt% C must be cast because of their limited ductility. The higher-carbon alloys are applied as hardfacings on other structural metals, using weld overlays or spray coatings. Stellite alloy 21 contains molybdenum, rather than tungsten, for solid-solution strengthening and considerably less carbon. As a result of the high molybdenum content, and the fact that most of the chromium is in solution rather than in $M_7C_3$ carbides, Stellite alloy 21 is more corrosion resistant than Stellite alloys 1, 6, and 12.

The Stellite alloys are generally used in the form of castings or weld overlays (hardfacing alloys). The alloys can also be applied by plasma and laser spraying techniques. Powder metallurgy (PM) versions of Stellite alloys, typically containing $\leq$1 wt% B to enhance sintering, are available for applications where the PM process is cost-effective for high volumes of relatively small components.

Laves-phase alloys include the Tribaloy family of wear-resistant materials, such as T-400 and T-800. The Tribaloy alloys contain excessive amounts of molybdenum and silicon, which cause the formation of a large volume fraction (approximately 50%) of the hard and corrosion-resistant intermetallic compound Laves phase (CoSiMo). The carbon content is held as low as possible to discourage the precipitation of carbides, which, if formed, would tie up chromium and molybdenum. Although the massive amounts of hard Laves-phase particles limit room-temperature ductility, they are readily formed as thermal sprayed or PM components to provide exceptionally good wear resistance.

The high-carbon alloys, such as alloy 6B and alloy 6K, are essentially wrought versions of the hardfacing Stellite alloys. Wrought processing improves the homogeneity, increases ductility, and substantially modifies the geometry of the carbide precipitates by forming abrasion-resistant blocky carbides. These alloys are Co-Cr-W-C quaternary alloys, with chromium providing strength and corrosion resistance in addition to being the chief carbide former. Tungsten provides additional solid-solution

**Table 29.2 Cobalt-base wear- and corrosion-resistant alloys**

| | Nominal composition, wt% | | | | | | | | | |
|---|---|---|---|---|---|---|---|---|---|---|
| Alloy trade name | Co | Cr | W | Mo | C | Fe | Ni | Si | Mn | Others |
| **Stellite wear-resistant alloys** | | | | | | | | | | |
| Stellite 1 | bal | 30 | 13 | 0.5 | 2.5 | 3 | 1.5 | 1.3 | 0.5 | ... |
| Stellite 4 | bal | 30 | 14 | 1 (max) | 0.57 | 3 (max) | 3 (max) | 2 (max) | 1 (max) | ... |
| Stellite 6 | bal | 29 | 4.5 | 1.5 (max) | 1.2 | 3 (max) | 3 (max) | 1.5 (max) | 1 (max) | ... |
| Stellite 12 | bal | 30 | 8.3 | ... | 1.4 | 3 (min) | 1.5 | 0.7 | 2.5 | ... |
| Stellite 21 | bal | 27 | ... | 5.5 | 0.25 | 3 (max) | 2.75 | 1 (max) | 1 (max) | 0.007 B (max) |
| Stellite 23 | bal | 24 | 5 | ... | 0.4 | 1 | 2 | 0.6 | 0.3 | ... |
| Stellite 27 | bal | 25 | ... | 5.5 | 0.4 | 1 | 32 | 0.6 | 0.3 | ... |
| Stellite 31 | bal | 25.5 | 7.5 | ... | 0.5 | 2 (max) | 10.5 | 1 (max) | 1 (max) | ... |
| Stellite 190 | bal | 26 | 14.5 | 1 (max) | 3.3 | 3 (max) | 3 (max) | 2 (max) | 1 (max) | ... |
| Stellite 306 | bal | 25 | 2 | ... | 0.4 | ... | 5 | ... | ... | 6 Nb |
| Stellite F | bal | 25 | 12.3 | 1 (max) | 1.75 | 3 (max) | 22 | 2 (max) | 1 (max) | ... |
| **Laves-phase wear-resistant alloys** | | | | | | | | | | |
| T-400 | bal | 9 | ... | 29 | ... | ... | ... | 2.5 | ... | ... |
| T-800 | bal | 18 | ... | 29 | ... | ... | ... | 3.5 | ... | ... |
| **Wrought wear-resistant alloys** | | | | | | | | | | |
| 6B | bal | 30 | 4 | 1.5 (max) | 1 | 3 (max) | 2.5 | 0.7 | 1.4 | ... |
| 6K | bal | 30 | 4.5 | 1.5 (max) | 1.6 | 3 (max) | 3 (max) | 2 (max) | 2 (max) | ... |
| **Corrosion-resistant alloys** | | | | | | | | | | |
| Ultimet (1233) | bal | 26 | 2 | 5 | 0.06 | 3 | 9 | 0.3 | 0.8 | 0.08 N |
| MP159 | bal | 19 | ... | 7 | ... | 9 | 25.5 | ... | ... | 3 Ti, 0.6 Nb, 0.2 Al |
| MP35N | bal | 20 | ... | 10 | ... | ... | 35 | ... | ... | ... |

Source: Ref 9

strengthening. Alloy 6B contains approximately 12.5 wt% carbides of the $M_7C_3$ and $M_{23}C_6$ types in a ratio of 9 to 1. Alloy 6K exhibits an even greater carbide volume fraction, again with $M_7C_3$ as the predominant carbide. Haynes alloy 6B is available in plate, sheet, and bar wrought forms in a worked or annealed matrix. In wrought form, alloy 6B has a yield strength of 620 MPa (90 ksi), a tensile strength of 1000 MPa (145 ksi), and an elongation of 11%.

The physical and mechanical properties of six commonly used cobalt wear-resistant alloys are shown in Table 29.3. These alloys display moderately high strengths and hardness, with the high-carbon-content alloys having the greatest hardness but lower ductilities. The enhanced ductility of alloy 6B is a result of wrought processing rather than casting. Abrasion data are presented for six wear-resistant alloy compositions in Fig. 29.16, along with data for 316L stainless steel and D2 tool steel (60 HRC) for comparison. Within the Stellite alloy group, abrasion resistance is a function of carbon and tungsten content. As the carbon content increases in the chromium-tungsten Stellite alloys, the tungsten content is also increased. This results in an increase in carbide content and higher hardness.

### 29.5.2 Corrosion-Resistant Cobalt Alloys

Although the cobalt-base wear-resistant alloys possess some resistance to aqueous corrosion, they are limited by grain-boundary carbide precipitation, a lack of alloying elements dissolved in the matrix, and, in the case of the cast and hardfacing materials, by chemical segregation in the microstructure. As a result of their homogeneous microstructures and lower carbon contents, the wrought cobalt-base high-temperature alloys, which typically contain tungsten rather than molybdenum, are even more resistant to aqueous corrosion; however, they still fall well short of the Ni-Cr-Mo alloys in corrosion performance.

Several low-carbon wrought Co-Ni-Cr-Mo alloys have been developed for providing excellent corrosion resistance along with high strength and toughness. Molybdenum additions instead of tungsten provide a greater degree of resistance to a variety of wet corrosive media. In addition, the carbon contents in these alloys

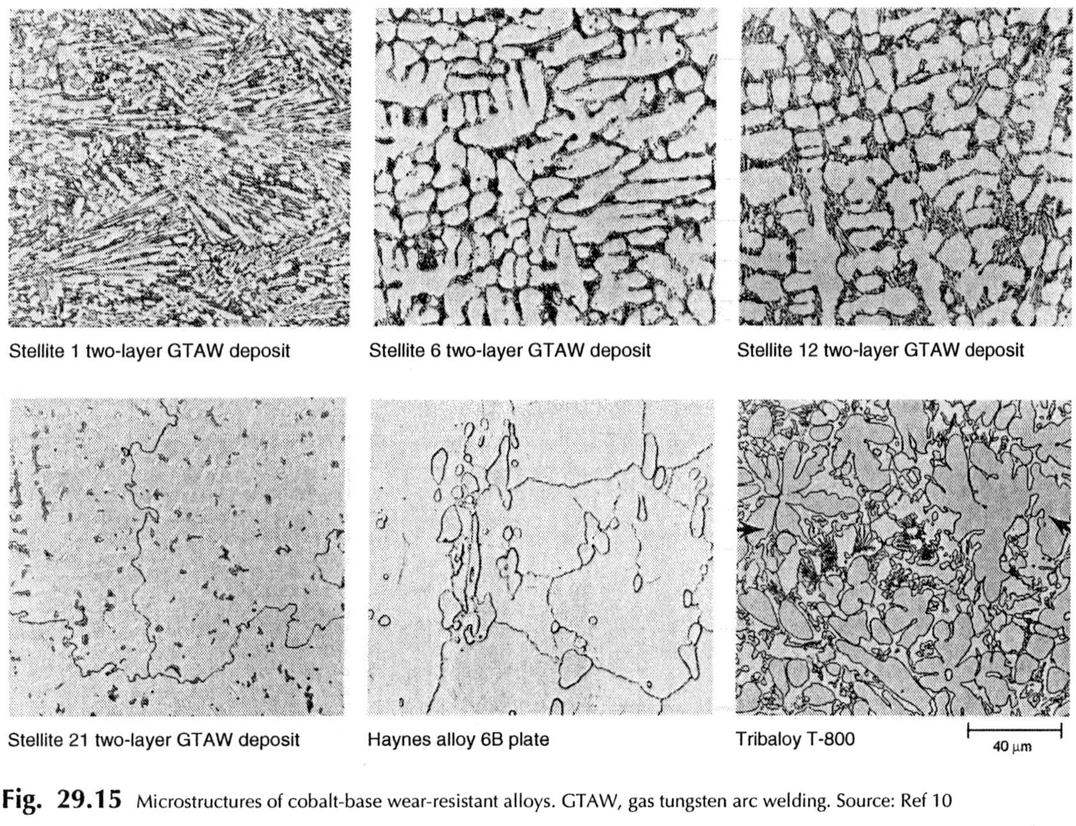


**Fig. 29.15** Microstructures of cobalt-base wear-resistant alloys. GTAW, gas tungsten arc welding. Source: Ref 10

are held within the soluble range to improve resistance to heat-affected zone sensitization during welding. The higher-nickel-content alloys (MP159 and MP35N) also exhibit improved resistance to stress-corrosion cracking. In these alloys, the carbon is minimized to avoid carbide precipitation along grain boundaries, maximizing corrosion resistance. The presence of chromium in cobalt alloys permits them to passivate by forming adherent films of $Cr_2O_3$, as in the stainless steels. These alloys are provided in various wrought forms, in the work-hardened or work-hardened plus age-hardened condition. They work harden rapidly due to a strain-induced transformation that provides a dispersion of fine hcp platelets. The tensile strength of the annealed condition (896 MPa, or 130 ksi) can be increased to 1586 MPa (230 ksi) with approximately 50% cold reduction and raised still further (2000 MPa, or 290 ksi) after a subsequent aging treatment at approximately 540 to 595 °C (1000 to 1100 °F) for 4 h. In this aged condition, the alloy retains favorable levels of ductility and toughness, exhibiting approximately 10% elongation and 46% reduction in area. The alloys exhibit good strength and ductility at elevated temperatures as well.

Because of their biocompatibility, nonmagnetic Co-Cr-Mo alloys, such as Co-28.5Cr-6Mo (ASTM F75 and F799), have been used for orthopedic implants. When made by investment casting, these alloys are strengthened by carbide particles. When made by forging or hot isostatic pressing, they are hardened by a combination of nitrogen and carbide strengthening mechanisms. In addition to the excellent corrosion resistance

**Table 29.3 Mechanical and physical properties of cobalt-base wear-resistant alloys**

| | Alloy | | | | | |
|---|---|---|---|---|---|---|
| **Property** | **1** | **6** | **12** | **21** | **6B** | **T-800** |
| Hardness, HRC | 55 | 40 | 48 | 32 | 37 | 58 |
| Yield strength; MPa (ksi) | ... | 545 (79) | 648 (94) | 496 (72) | 621 (90) | ... |
| Ultimate tensile strength; MPa (ksi) | 621 (90) | 896 (130) | 938 (136) | 689 (100) | 1000 (145) | ... |
| Elongation, % | <1 | 1 | <1 | 9 | 11 | ... |
| Thermal expansion coefficient, μm/m · °C | | | | | | |
| From 20 to 100 °C (68–212 °F) | 10.5 | 11.4 | 11.5 | 11.0 | 13.9 | ... |
| From 20 to 500 °C (68–930 °F) | 12.5 | 14.2 | 13.3 | 13.1 | 15.0 | 12.6 |
| From 20 to 1000 °C (68–1830 °F) | 14.8 | ... | 15.6 | ... | 17.4 | 15.1 |
| Thermal conductivity, W/m · K | ... | ... | ... | ... | 14.8 | 14.3 |
| Specific gravity | 8.69 | 8.46 | 8.56 | 8.34 | 8.39 | 8.64 |
| Electrical resistivity, μΩ · m | 0.94 | 0.84 | 0.88 | ... | 0.91 | ... |
| Melting range, °C (°F) | | | | | | |
| Solidus | 1255 (2291) | 1285 (2345) | 1280 (2336) | 1186 (2167) | 1265 (2309) | 1288 (2350) |
| Liquidus | 1290 (2354) | 1395 (2543) | 1315 (2400) | 1383 (2521) | 1354 (2470) | 1352 (2465) |

Source: Ref 11

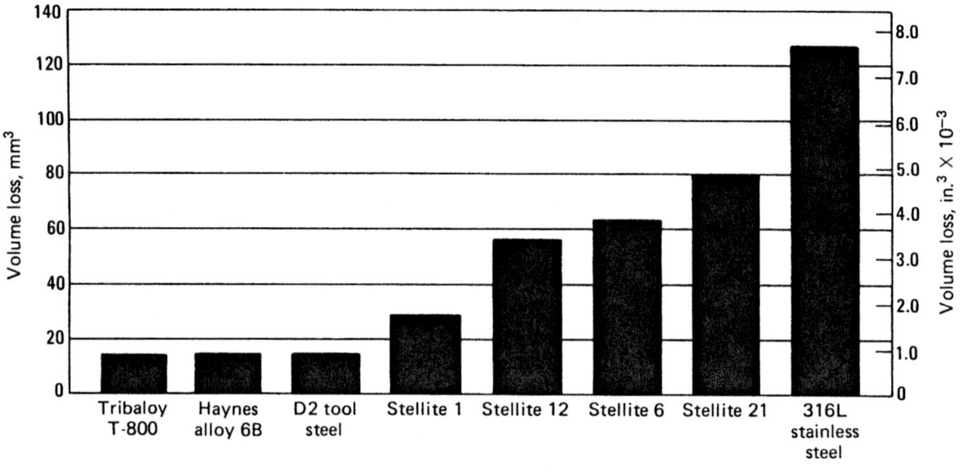


**Fig. 29.16** Abrasion data for various cobalt-base alloys. Source: Ref 11

for these applications, these alloys also have high-cycle fatigue resistance as well as strength, ductility, and wear resistance.

### ACKNOWLEDGMENTS

Sections of this chapter were adapted from "Nickel and Nickel Alloys: Effects of Composition, Processing, and Structure on Properties of Nonferrous Alloys;" and "Cobalt and Cobalt Alloys: Effects of Composition, Processing, and Structure on Properties of Nonferrous Alloys," both by R.N. Caron and J.T. Staley in *Materials Selection and Design,* Volume 20, *ASM Handbook,* ASM International, 1997.

### REFERENCES

1. T.H. Bassford and J. Hoiser, Nickel and Nickel Alloys, *Handbook of Materials Selection,* John Wiley & Sons, Inc., 2002
2. R.N. Caron and J.T. Staley, Nickel and Nickel Alloys: Effects of Composition, Processing, and Structure on Properties of Nonferrous Alloys, *Materials Selection and Design,* Vol 20, *ASM Handbook,* ASM International, 1997
3. "Product Handbook of High Performance Alloys," Special Metals Corporation, 2001
4. W.L. Mankins, Metallography and Microstructures of Nickel and Nickel-Copper Alloys, *Metallography and Microstructures,* Vol 9, *ASM Handbook,* ASM International, 2004
5. G.F. Vander Voort, G.M. Lucas, and E.P. Manilova, Metallography and Microstructures of Heat-Resistant Alloys, *Metallography and Microstructures,* Vol 9, *ASM Handbook,* ASM International, 2004
6. "Inconel Alloy 601 Data Sheet," Special Metals Corporation
7. D.E. Hodgson, M.H. Wu, and R.J. Biermann, Shape Memory Alloys, *Properties and Selection: Nonferrous Alloys and Special-Purpose Materials,* Vol 2, *ASM Handbook,* ASM International, 1990
8. W.L. Mankins and S. Lamb, Nickel and Nickel Alloys, *Properties and Selection: Nonferrous Alloys and Special-Purpose Materials,* Vol 2, *ASM Handbook,* ASM International, 1990
9. R.N. Caron and J.T. Staley, Cobalt and Cobalt Alloys: Effects of Composition, Processing, and Structure on Properties of Nonferrous Alloys, *Materials Selection and Design,* Vol 20, *ASM Handbook,* ASM International, 1997
10. D. Klarstrom, P. Crook, and J.W. Delore, Metallography and Microstructures of Cobalt and Cobalt Alloys, *Metallography and Microstructures,* Vol 9, *ASM Handbook,* ASM International, 2004
11. P. Crook, Cobalt and Cobalt Alloys, *Properties and Selection: Nonferrous Alloys and Special-Purpose Materials,* Vol 2, *ASM Handbook,* ASM International, 1990

### SELECTED REFERENCES

- J.R. Davis, Cobalt and Cobalt Alloys, *Alloying: Understanding The Basics,* ASM International, 2001
- B.J. Moniz, *Metallurgy,* American Technical Publishers, Inc., 1994

# CHAPTER 30

# Superalloys

SUPERALLOYS are nickel-, iron-nickel-, and cobalt-base alloys generally used at temperatures above approximately 540 °C (1000 °F). They exhibit a combination of mechanical strength and resistance to surface degradation.

Superalloys have been used in aircraft, industrial, and marine gas turbines, nuclear reactors, aircraft skins, spacecraft structures, petrochemical production, orthopedic and dental prostheses, and environmental protection applications. Today's modern, high-performance aircraft jet engine could not operate without the major advances made in superalloy development over the past 50 years. Typical applications for superalloys in a jet engine are shown in Fig. 30.1. The remarkable improvements in engine operating temperature as a result of superalloy development are shown in Fig. 30.2. Although developed for high-temperature use, some are used at cryogenic temperatures and others at room temperature. Applications continue to expand but at lower rates than in previous decades. Aerospace remains the predominant application on a volume basis.

Superalloys are used in cast, rolled, extruded, forged, and powder-processed forms. Sheet, bar, plate, tubing, shafts, airfoils, disks, and pressure

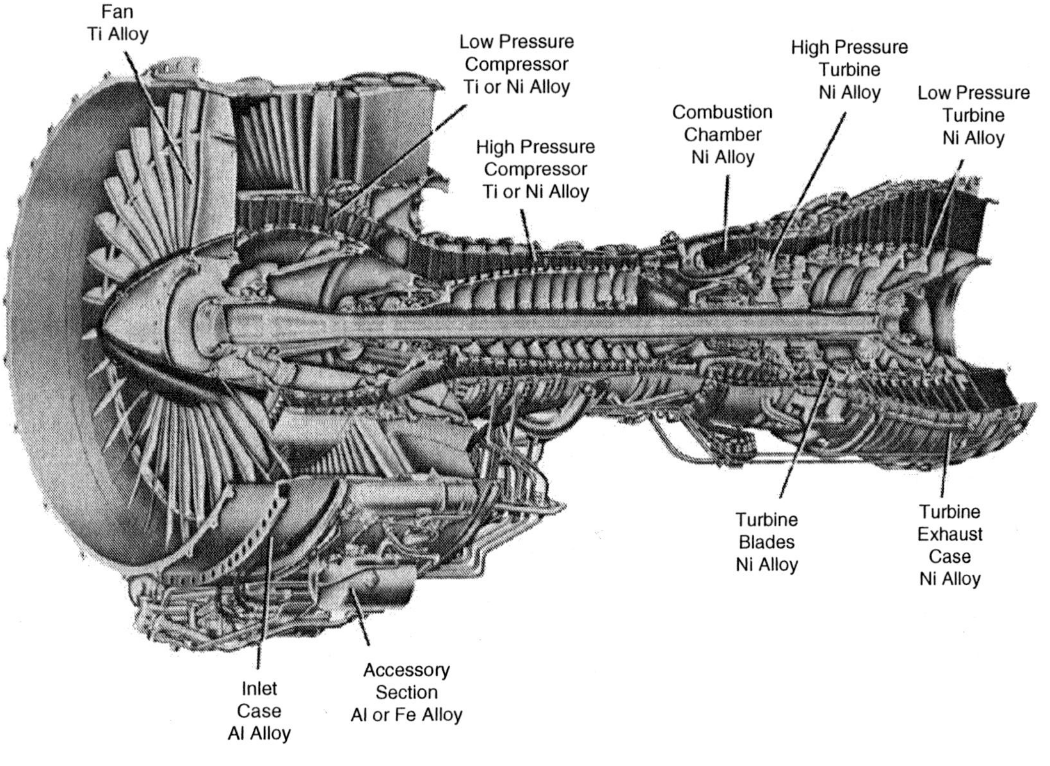


**Fig. 30.1** Typical materials selection for jet engine components

vessel cases are some of the shapes produced. The more highly alloyed compositions normally are processed as castings. Fabricated structures can be built up by welding or brazing, but many highly alloyed compositions containing a high amount of hardening phase are difficult to weld.

Properties can be controlled by adjustments to composition, by thermomechanical processing, and by heat treatment. Excellent elevated-temperature strengths are available in finished products. A comparison of the relative stress-rupture behavior of the three alloy classes is shown in Fig. 30.3. Nickel-base superalloys are used for the most stringent high-temperature applications. A noteworthy feature of nickel-base alloys is their use in load-bearing applications at temperatures in excess of 80% of their incipient melting temperatures, a percentage that is higher than for any other class of common engineering alloys.

## 30.1 Superalloy Metallurgy

All three types of superalloys contain the austenitic (γ) face-centered cubic (fcc) crystalline structure. Even though iron has a body-centered cubic structure and cobalt a hexagonal close-packed structure at room temperature, these alloys contain sufficient alloying additions that the fcc crystalline structure is stable at all temperatures. Because of the fcc crystalline structure, many of the wrought superalloys can be formed at room temperature. However, some of the compositions are so highly alloyed that they cannot even be produced in wrought forms; they must be produced by alternate routes, such as casting or powder metallurgy. Superalloys are relatively heavy, with densities in the range of 7.8 to 9.4 g/cm$^3$ (0.282 to 0.340 lb/in.$^3$). Density is influenced by alloying additions, with elements such as aluminum, titanium, and chromium reducing the density and elements such as tungsten, rhenium, and tantalum increasing it.

The main strengthening mechanisms for nickel and iron-nickel superalloys are solid-solution hardening and precipitation hardening. In addition, grain-boundary carbides are important in providing high-temperature stability. The austenitic matrix (γ) is capable of dissolving large percentages of solid-solution hardeners, such as cobalt, iron, chromium, molybdenum, and tungsten. Additions of aluminum and titanium not only provide solid-solution

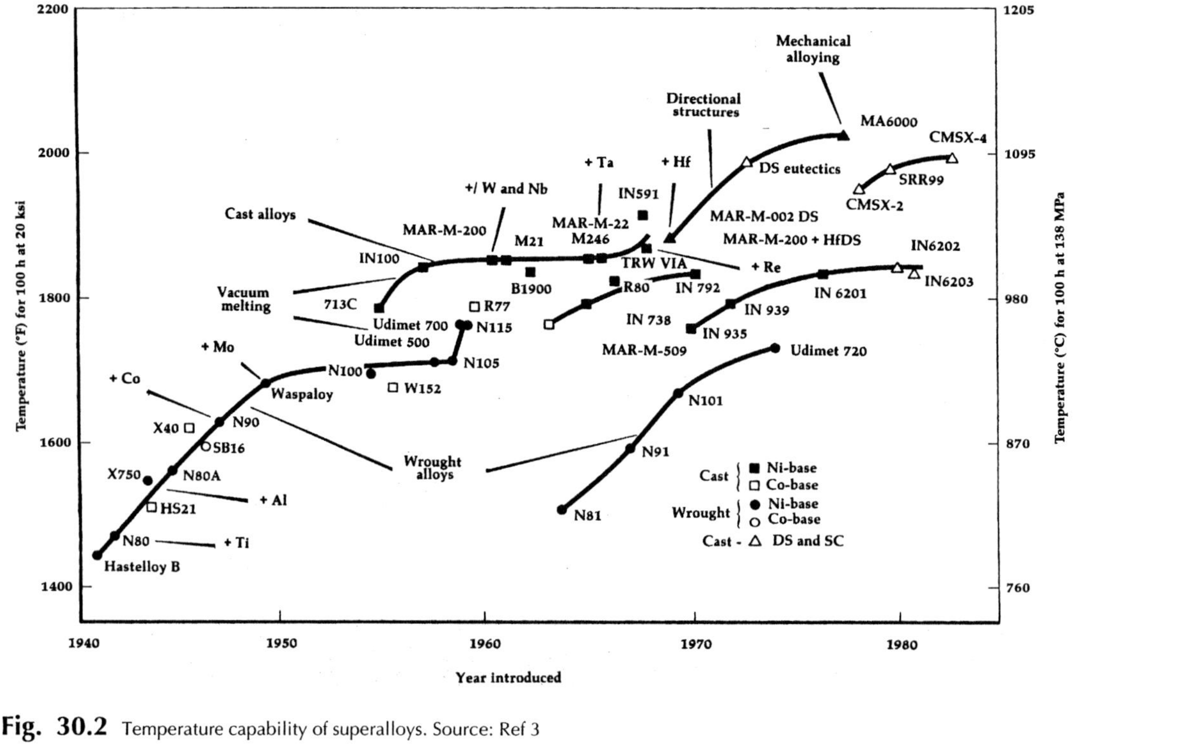


**Fig. 30.2** Temperature capability of superalloys. Source: Ref 3

strengthening, they also form the basis for precipitation hardening. Slowly diffusing elements, such as molybdenum and tungsten, help in reducing creep at high temperatures.

The most important precipitate in both nickel and iron-nickel superalloys is $\gamma'$, which is the fcc intermetallic compound $Ni_3(Al,Ti)$, either in the form of $Ni_3Al$ or $Ni_3Ti$. Other elements, notably niobium, tantalum, and chromium, also enter $\gamma'$. In nickel-base alloys, $\gamma'$ is typically $Ni_3(Al,Ti)$, but if cobalt is added, it can substitute for some nickel as $(Ni,Co)_3(Al,Ti)$. Since the $\gamma'$ precipitate has only approximately 0.1% lattice mismatch with the $\gamma$ matrix, $\gamma'$ precipitates are both homogeneous and have a low surface energy, leading to extraordinary high-temperature stability. The coherency between the precipitate and matrix is maintained to high temperatures, and the precipitate coarsens very slowly, so that the precipitate overages slowly, even at temperatures as high 0.7 $T_m$. As a result of the degree of ordering increasing with increases in temperature, alloys with a high volume of $\gamma'$ exhibit the rather anomalous behavior of becoming stronger up to approximately 700 °C (1300 °F). As the combined content of aluminum and/or titanium increases, the amount of $\gamma'$ precipitate increases, resulting in better high-temperature performance (Fig. 30.4). The degree of lattice mismatch between the $\gamma'$ precipitate and the $\gamma$ matrix determines precipitate morphology. The precipitate morphology is spherical for mismatches of 0 to 0.2%, cubical for mismatches of 0.5 to 1% (Fig. 30.5), and platelike at mismatches above approximately 1.25%. Grain-boundary $\gamma'$, a film of $\gamma'$ along the grain boundaries that forms during heat treatment and service exposure in the stronger alloys, is believed to further improve stress-rupture properties.

The important superalloy Inconel 718 relies on niobium additions to form the body-centered tetragonal ordered $\gamma''$ ($Ni_3Nb$) precipitate, which is coherent with the gamma matrix. However, a large mismatch produces strains on the order of 2.9%. This phase provides very high strength at low to intermediate temperatures but becomes unstable at temperatures above approximately 650 °C (1200 °F).

The roles of the various alloying elements are shown in Table 30.1. Compositions of commercial superalloys can become quite complex, with some alloys containing as many as a dozen

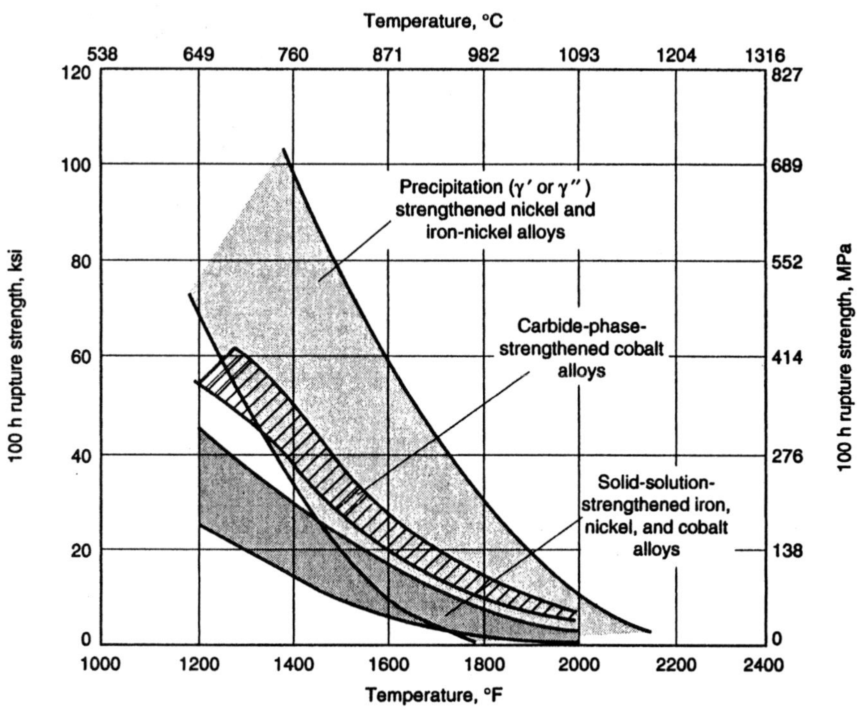


**Fig. 30.3** Stress-rupture comparison of wrought superalloys. Source: Ref 2

alloying elements. High-temperature oxidation resistance is provided by both chromium and aluminum, with chromium producing the protective surface oxide $Cr_2O_3$ and aluminum the even more stable $Al_2O_3$. Important solid-solution strengtheners include molybdenum, tantalum, tungsten, and rhenium. Rhenium is used in some of the latest cast nickel-base alloys because it also reduces the high-temperature coarsening rate of $\gamma'$. While aluminum and titanium additions are necessary to promote $\gamma'$ precipitation, as their contents are increased, the melting point becomes lower and the forging range becomes narrower. This has forced alternate fabrication processes, such as casting and/or powder metallurgy, to be used for some of the highest-service-temperature nickel-base superalloys. Cobalt is sometimes added to increase the volume percentage of $\gamma'$ and $\gamma''$ precipitates. Small additions of boron, zirconium, carbon, and hafnium are important additions to polycrystalline alloys because they reduce the tendency for grain-boundary cracking. However, since single-crystal alloys do not have grain boundaries, they are often reduced or eliminated in these alloys.

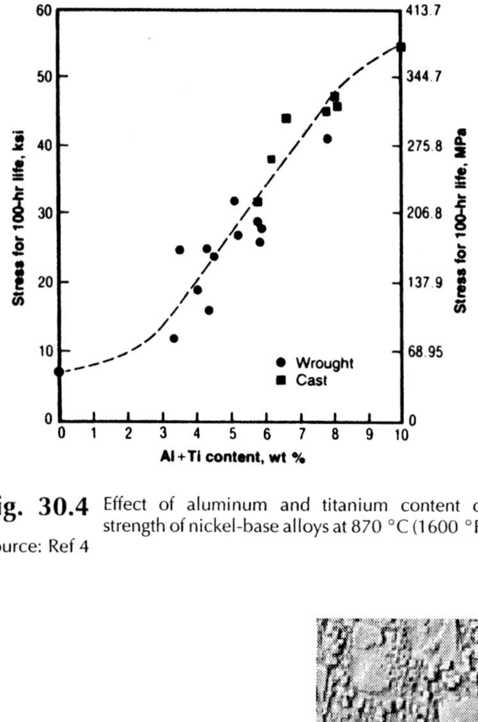


**Fig. 30.4** Effect of aluminum and titanium content on strength of nickel-base alloys at 870 °C (1600 °F). Source: Ref 4

While grain-boundary carbides strengthen the grain boundaries and reduce stress along the grain boundaries, their most important function is to retard the creep mechanism of grain-boundary sliding during high-temperature service. If there are no carbides along the grain boundaries, voids can form and cause excessive grain-boundary sliding. Carbon is added to form carbides in both the matrix and at the grain boundaries. In nickel-base alloys, wrought alloys contain approximately 0.02 to 0.2 wt% C, while cast alloys can contain carbon contents as high as 0.6 wt%. For carbides to be effective, they must form a discontinuous chain of carbides along the grain boundaries. Brittleness results when a continuous film of carbides is present along the grain boundaries.

The important carbides are MC, $M_{23}C_6$, $M_6C$, and, to a more limited extent, $M_7C_3$.

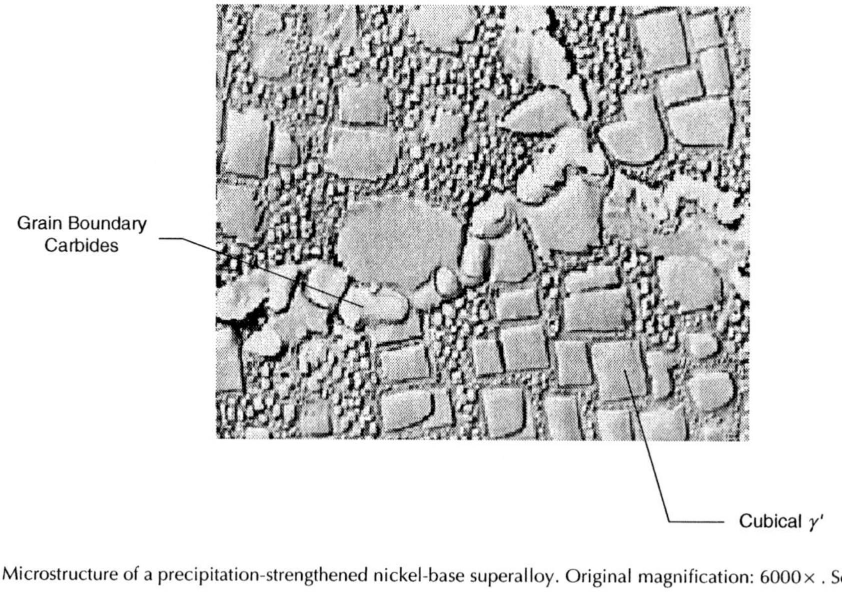


**Fig. 30.5** Microstructure of a precipitation-strengthened nickel-base superalloy. Original magnification: 6000×. Source: Ref 5

In nickel-base alloys, "M" represents titanium, tantalum, niobium, or tungsten. MC carbides, which form during ingot casting and are not as desirable as the other carbides, decompose during heat treatment to form the more stable $M_{23}C_6$ and $M_6C$ carbides. Chromium is the main metal in $M_{23}C_6$ carbides, which forms either during heat treatment or during service at temperatures between 760 and 980 °C (1400 and 1800 °F). Since $M_{23}C_6$ carbides tend to precipitate along the grain boundaries, they are useful in preventing creep. When the molybdenum and tungsten contents are high, $M_6C$ carbides can form at temperatures in the range of 815 to 980 °C (1500 to 1800 °F). $M_7C_3$ carbides are not nearly as prevalent as the other carbides. They also precipitate on the grain boundaries and are beneficial when they are discrete particles but detrimental if they form a grain-boundary film. Small additions of boron are also essential to the improved creep resistance of polycrystalline superalloys. However, because the level of boron added rarely exceeds 0.03 wt%, and because it is often substantially less than the solubility limit of 0.01 wt%, the volume fraction of borides tends to be quite small. Borides are hard particles that also form at grain boundaries. The boride is found in superalloys in the form of $M_3B_2$, with a tetragonal unit cell.

Embrittling topologically close-packed (tcp) phases can form during heat treatment or during elevated-temperature service. The most important tcp phases are σ, μ, and Laves. Since these phases usually form as thin plates or needles, they result in a loss of room-temperature ductility and lower elevated-temperature stress-rupture strengths. Alloys containing a high level of the transition metals tantalum, niobium, chromium, tungsten, and molybdenum are most susceptible to tcp formation. The hardness of σ and its platelike morphology cause premature cracking, leading to low-temperature brittle failure, although yield strength is largely unaffected. However, the major effect is on elevated-temperature rupture strength. Sigma formation depletes refractory metals in the matrix, causing a loss of matrix strength. Also, high-temperature fracture can occur along σ plates rather than along the normal intergranular path, resulting in sharply reduced rupture life. The tcp phases are usually minimized by compositional control. As with most high-performance alloys, hydrogen, oxygen, and nitrogen are considered detrimental and are held to very low levels.

**Table 30.1 Role of alloying elements in superalloys**

| Alloy additions | Solid-solution strengtheners | γ′ formers | Carbide formers | Grain-boundary strengtheners | Oxide scale formers |
|---|---|---|---|---|---|
| Chromium | X | ... | X | ... | X |
| Aluminum | ... | X | ... | ... | X |
| Titanium | ... | X | X | ... | ... |
| Molybdenum | X | ... | X | ... | ... |
| Tungsten | X | ... | X | ... | ... |
| Boron | ... | ... | ... | X | ... |
| Zirconium | ... | ... | ... | X | ... |
| Carbon | ... | ... | ... | X | ... |
| Niobium | ... | X | X | ... | ... |
| Hafnium | ... | ... | X | X | ... |
| Tantalum | ... | X | X | X | ... |

Source: Ref 6

## 30.2 Commercial Superalloys

The three types of superalloys may be further subdivided as either wrought or cast. Over the last 50 years, a large number of alloys have been invented and studied; however, only a few are extensively used. The compositions of a number of wrought alloys are shown in Table 30.2, and some cast alloy compositions are given in Table 30.3. Note that the wrought alloy listing shows that nickel and iron-nickel alloys are available as both solid-solution- and precipitation-hardening alloys, while the cobalt-base alloys are only available in the solid-solution-hardened condition. In other words, cobalt-base superalloys are not precipitation hardenable.

### *30.2.1 Nickel-Base Superalloys*

Nickel-base superalloys are the most complex and the most widely used for the hottest parts of turbine engines. They currently constitute over 50% of the weight of advanced aircraft engines. Nickel-base superalloys are available as either solid-solution-strengthened alloys for lower-temperature applications or as precipitation-hardened alloys for higher-temperature use. When the most highly alloyed compositions are cast, they can be used to the highest homologous temperature ($T/T_m$) of any common alloy system.

Many nickel-base superalloys contain 10 to 20 wt% Cr, a combined aluminum and titanium content up to 8 wt%, 5 to 15 wt% Co, and small amounts of boron, zirconium, magnesium, and carbon. Other common additions are molybdenum, niobium, and tungsten, all of which provide solid-solution strengthening and form carbides. Chromium and aluminum are also necessary to improve surface stability through the formation of $Cr_2O_3$ and $Al_2O_3$, respectively.

The most important class of nickel-base superalloys are those strengthened by intermetallic compound precipitation in the fcc austenitic matrix. For nickel-titanium/aluminum alloys, the strengthening precipitate is $\gamma'$. Typical alloys strengthened by the precipitation of $\gamma'$ are the wrought alloys Waspaloy and Udimet 720 and the cast alloy René 80. Most wrought nickel-base alloys contain between 20 and 45 vol% $\gamma'$, while cast alloys can contain as much as 60 vol% $\gamma'$. As the amount of $\gamma'$ increases, the elevated-temperature resistance increases; however, the workability of wrought alloys decreases and eventually reaches a point where other fabrication methods, such as casting or powder metallurgy, must be used. Inconel 718 is the single most important wrought superalloy that is sometimes classified as a nickel-base alloy and sometimes as an iron-nickel-base alloy. It contains niobium and is therefore strengthened primarily by $\gamma''$ ($Ni_3Nb$). Alloys that contain niobium with titanium and/or aluminum, such as Incoloy 909, are strengthened by the combination of $\gamma''$ and $\gamma'$.

Inconel 718 is one of the strongest and most widely used superalloys, but it rapidly loses strength in the range of 650 to 815 °C (1200 to 1500 °F) because of the instability of the $\gamma''$ precipitate. The widespread use of Inconel 718 is due to its excellent combination of mechanical properties, moderate price, and ease of processing, including weldability. The total use of Inconel 718 may account for as much as 60 wt% of the total consumption of superalloys. Recently, the upper use temperature of 650 °C (1200 °F) has been extended up to 700 °C (1300 °F), making it a candidate to replace the more expensive alloy Waspaloy in some applications. The principal differences in the new 718Plus, compared with the original composition, are an increase in the aluminum and titanium contents; additions of cobalt and tungsten, which replace some iron; and the addition of a small amount of phosphorus.

Solid-solution-strengthened nickel-base superalloys, such as the Hastelloy series, are used primarily for low-temperature service, often in corrosive environments. The solid-solution-strengthened nickel-base alloys also derive some additional strengthening from carbide

**Table 30.2 Nominal compositions of select wrought superalloys**

| Alloy | Composition, wt% | | | | | | | | | | |
|---|---|---|---|---|---|---|---|---|---|---|---|
| | Cr | Ni | Co | Mo | W | Nb | Ti | Al | Fe | C | Other |
| **Nickel base** | | | | | | | | | | | |
| Solid-solution hardened | | | | | | | | | | | |
| Hasteloy X | 22.0 | 49.0 | <1.5 | 9.0 | 0.6 | ... | ... | 2.0 | 15.8 | 0.15 | ... |
| Inconel 625 | 21.5 | 61.0 | ... | 9.0 | ... | 3.6 | ... | 0.2 | 2.5 | 0.05 | ... |
| Nimonic 75 | 19.5 | 75.0 | ... | ... | ... | ... | 0.4 | 0.15 | 2.5 | 0.12 | <0.25 Cu |
| Precipitation hardened | | | | | | | | | | | |
| Astroloy | 15.0 | 56.5 | 15.0 | 5.25 | ... | ... | 3.5 | 4.4 | <0.3 | 0.06 | 0.03 B, 0.06 Zr |
| Inconel 100 | 10.0 | 60.0 | 15.0 | 3.0 | ... | ... | 4.7 | 5.5 | <0.6 | 0.15 | 1.0 V, 0.06 Zr, 0.015 B |
| Inconel 706 | 16.0 | 41.5 | ... | ... | ... | ... | 1.75 | 0.2 | 37.5 | <0.08 | 2.9 (Nb + Ta), <0.15 Cu |
| Nimonic 95 | 19.5 | 53.5 | 18.0 | ... | ... | ... | 2.9 | 2.0 | <5.0 | <0.15 | +B+Zr |
| René 95 | 14.0 | 61.0 | 8.0 | 3.5 | 3.5 | 3.5 | 2.5 | 3.5 | <0.3 | 0.16 | 0.01 B, 0.5 Zr |
| Waspaloy | 19.5 | 57.0 | 13.5 | 4.3 | ... | ... | 3.0 | 1.4 | <2.0 | 0.07 | 0.006 B, 0.09 Zr |
| **Iron-nickel base** | | | | | | | | | | | |
| Solid-solution hardened | | | | | | | | | | | |
| 19-9 DL | 19.0 | 9.0 | ... | 1.25 | 1.25 | 0.4 | 0.3 | ... | 66.8 | 0.30 | 1.10 Mn, 0.60 Si |
| Haynes 556 | 22.0 | 21.0 | 20.0 | 3.0 | 2.5 | 0.1 | 0.1 | 0.3 | 29.0 | 0.10 | 0.50 Ta, 0.02 La, 0.002 Zr |
| Incoloy 802 | 21.0 | 32.5 | ... | ... | ... | ... | ... | 0.58 | 44.8 | 0.36 | ... |
| Precipitation hardened | | | | | | | | | | | |
| A-286 | 15.0 | 26.0 | ... | 1.25 | ... | ... | 2.0 | 0.2 | 55.2 | 0.04 | 0.005 B, 0.3 V |
| Inconel 718 | 19.0 | 52.5 | ... | 3.0 | ... | 5.1 | 0.91 | 0.5 | 18.5 | <0.08 | <0.15 Cu |
| Incoloy 903 | <0.1 | 38.0 | 15.0 | 0.1 | ... | 3.0 | 1.4 | 0.7 | 41.0 | 0.04 | ... |
| **Cobalt base** | | | | | | | | | | | |
| Solid-solution hardened | | | | | | | | | | | |
| Haynes 25 (L605) | 20.0 | 10.0 | 50.0 | ... | 15.0 | ... | ... | ... | 3.0 | 0.10 | 1.5 Mn |
| Haynes 188 | 22.0 | 22.0 | 37.0 | ... | 14.5 | ... | ... | ... | <3.0 | 0.10 | 0.90 La |
| MP35-N | 20.0 | 35.0 | 35.0 | 10.0 | ... | ... | ... | ... | ... | ... | ... |

precipitation. They typically contain chromium, molybdenum, iron, or tungsten dissolved in solid solution.

Another class of nickel-base superalloys includes oxide-dispersion-strengthened alloys, such as the mechanically alloyed MA-754 and MA-6000E, which are strengthened by dispersion of inert particles, primarily yttria ($Y_2O_3$), coupled in some cases with $\gamma'$ precipitation, as in MA-6000E.

### 30.2.2 Iron-Nickel-Base Superalloys

Iron-nickel-base superalloys evolved from austenitic stainless steels and are based on the principle of combining both solid-solution-hardening and precipitate-forming elements. As a class, the iron-nickel superalloys have useful strengths to approximately 650 °C (1200 °F). The austenitic matrix is based on nickel and iron, with at least 25 wt% Ni needed to stabilize the fcc phase. Other alloying elements, such as chromium, partition primarily to austenite to provide solid-solution hardening. Most alloys contain 25 to 45 wt% Ni. Chromium in the range of 15 to 28 wt% is added for oxidation resistance at elevated temperature, while 1 to 6 wt% Mo provides solid-solution strengthening. The main elements that facilitate precipitation hardening are titanium, aluminum, and niobium.

The strengthening precipitates are primarily $\gamma'$ ($Ni_3Al$), $\eta$ ($Ni_3Ti$), and $\gamma''$ ($Ni_3Nb$). Elements that partition to grain boundaries, such as boron and zirconium, suppress grain-boundary creep, resulting in significant increases in rupture life. Boron in quantities of 0.003 to 0.03 wt% and, less frequently, small additions of zirconium are added to improve stress-rupture properties and hot workability. Zirconium also forms the MC carbide ZrC. Another MC carbide (NbC) is found in alloys that contain niobium, such as Inconel 706. Vanadium also is added in small quantities to iron-nickel superalloys to improve both notch ductility at service temperatures and hot workability. Based on their composition and strengthening mechanisms, there are several groupings of iron-nickel superalloys.

Alloys that contain 25 to 35 wt% Ni and are strengthened by $\gamma'$ include alloys such as V-57 and A-286. The $\gamma'$ phase is titanium rich in these alloys, and care must be taken to avoid an excessively high titanium-to-aluminum ratio, resulting in the replacement of $\gamma'$ by the less effective strengthening precipitate $\eta$. A second subgroup, of which Inconel X750 and Incoloy 901 are examples, contains at least 40 wt% Ni, as well as higher levels of solid-solution-strengthening and precipitate-forming elements. A third subgroup, based on the Fe-Ni-Co system strengthened by $\gamma'$, combines low thermal expansion coefficients and relatively high strength to a temperature of 650 °C (1200 °F). Examples include Incoloy 903 and 909. Their unusually low thermal expansion coefficients result from the elimination of ferrite stabilizing

**Table 30.3 Nominal compositions of select cast superalloys**

| | Composition, wt% | | | | | | | | | | | | |
|---|---|---|---|---|---|---|---|---|---|---|---|---|---|
| Alloy | C | Ni | Cr | Co | Mo | Fe | Al | B | Ti | Ta | W | Zr | Other |
| **Nickel base** | | | | | | | | | | | | | |
| CMS30-2 | ... | 66.2 | 8 | 4.6 | 0.6 | ... | 56 | ... | 1 | 6 | 8 | 6 | ... |
| Inconel 713C | 0.12 | 74 | 12.5 | ... | 4.2 | ... | 6 | 0.012 | 0.8 | 1.75 | ... | 0.1 | 0.9 Nb |
| Inconel 738 | 0.17 | 61.5 | 16 | 8.5 | 1.75 | ... | 3.4 | 0.01 | 3.4 | ... | 2.6 | 0.1 | 2 Nb |
| MAR-M-247 | 0.15 | 59 | 8.25 | 10 | 0.7 | 0.5 | 5.5 | 0.015 | 1 | 3 | 10 | 0.05 | 1.5 Hf |
| PWA 1480 | ... | bal | 10 | 5.0 | ... | ... | 5.0 | ... | 1.5 | 12 | 4.0 | ... | ... |
| René 41 | 0.09 | 55 | 19 | 11.0 | 10.0 | ... | 1.5 | 0.01 | 3.1 | ... | ... | ... | ... |
| René 80 | 0.17 | 60 | 14 | 9.5 | 4 | ... | 3 | 0.015 | 5 | ... | 4 | 0.03 | ... |
| René 80 Hf | 0.08 | 60 | 14 | 9.5 | 4 | ... | 3 | 0.015 | 4.8 | ... | 4 | 0.02 | 0.75 Hf |
| René N4 | 0.06 | 62 | 9.8 | 7.5 | 1.5 | ... | 4.2 | 0.004 | 3.5 | 4.8 | 6 | ... | 0.5 Nb, 0.15 Hf |
| Udimet 700 | 0.1 | 53.5 | 15 | 18.5 | 5.25 | ... | 4.25 | 0.03 | 3.5 | ... | ... | ... | ... |
| Waspaloy | 0.07 | 57.5 | 19.5 | 13.5 | 4.2 | 1 | 1.2 | 0.005 | 3 | ... | ... | 0.09 | ... |
| **Iron-nickel base** | | | | | | | | | | | | | |
| Inconel 718 | 0.04 | 53 | 19 | ... | 3 | 18 | 0.5 | ... | 0.9 | ... | ... | ... | 0.1 Cu, 5 Nb |
| **Cobalt base** | | | | | | | | | | | | | |
| AirResist 215 | 0.35 | 0.5 | 19 | 63 | ... | 0.5 | 4.3 | ... | ... | 7.5 | 4.5 | 0.1 | 0.1 Y |
| FS30-414 | 0.25 | 10 | 29 | 52.5 | ... | 1 | ... | 0.010 | ... | ... | 7.5 | ... | ... |
| Haynes 25 | 0.1 | 10 | 20 | 54 | ... | 1 | ... | ... | ... | ... | 15 | ... | ... |
| MAR-M 918 | 0.05 | 20 | 20 | 52 | ... | ... | ... | ... | 0.2 | 7.5 | ... | 0.5 | ... |
| 30–40 | 0.50 | 10 | 22 | 57.5 | ... | 1.5 | ... | ... | ... | ... | 7.5 | ... | 0.5 Mn, 0.5 Si |

elements. In each of these alloys, a small quantity of titanium is present to allow the formation of $\gamma'$ during aging. These alloys are used for shafts, rings, and casings to permit the reduction of clearances between rotating and static components.

### 30.2.3 Cobalt-Base Superalloys

Cobalt-base superalloys are strengthened by a combination of solid-solution strengthening and carbides. Unfortunately, cobalt-base superalloys cannot be strengthened by precipitation hardening; no coherent, ordered precipitate, such as $\gamma'$ in nickel-base alloys, has been found for the cobalt-base alloys. Cast cobalt alloys contain approximately 50 to 60 wt% Co, 20 to 30 wt% Cr, 5 to 10 wt% W, and 0.1 to 1.0 wt% C. Wrought alloys contain approximately 40 wt% Co and high nickel contents (~20 wt%) for increased workability.

Cast alloys rely on carbide strengthening to a much greater extent than the wrought alloys. A variety of carbides have been found in cobalt alloys, including MC, $M_{23}C_6$, and $M_6C$. In both L-605 and Haynes 188, $M_6C$ transforms into $M_{23}C_6$ after long-time (3000 h) exposures at temperatures in the range of 816 to 927 °C (1500 to 1700 °F). The MC carbides are only found in alloys containing tantalum, niobium, zirconium, titanium, or hafnium.

Haynes 25 is the best known of the wrought alloys and has been widely used for hot sections of gas turbines, components for nuclear reactors, devices for surgical implants, and for cold-worked fasteners and wear pads. Haynes 188 is an alloy that was specially designed for sheet metal components, such as combustors and transition ducts in gas turbines. The composition of Haynes 188 provides excellent qualities, such as oxidation resistance at temperatures up to 1100 °C (2000 °F), hot corrosion resistance, creep resistance, room-temperature formability, and ductility after long-term aging at service temperatures.

Cobalt-base superalloys have flatter stress-rupture curves than other superalloys above approximately 1000 °C (1830 °F). Thus, they are often used for low-stress, high-temperature applications. Also, cobalt-base alloys have superior hot corrosion resistance at high temperatures, probably a result of their considerably higher chromium contents. Cobalt-base alloys generally exhibit better weldability and thermal fatigue resistance than nickel-base alloys. Another advantage of cobalt-base alloys is the capability to be melted in air or argon, in contrast to the vacuum melting required for nickel- and iron-nickel-base alloys that contain the reactive metals aluminum and titanium.

Unlike nickel-base alloys, which have a high tolerance for alloying elements in solid solution, cobalt-base alloys are more likely to precipitate undesirable platelike $\sigma$, Laves, and similar tcp phases. Increased susceptibility to Laves phase precipitation occurs in the presence of silicon in tungsten-containing alloys such as L-605. Haynes 188 was developed with reduced tungsten, increased nickel, and controlled silicon contents in order to avoid this problem.

## 30.3 Melting and Primary Fabrication

Superalloys are used in the cast, rolled, extruded, forged, and powder-produced forms, as shown in the overall process flow in Fig. 30.6. Since wrought alloys are generally more uniform, with finer grain sizes, and have superior tensile and fatigue properties, they are used in applications where tensile strength and fatigue resistance are paramount, such as disks. On the other hand, cast alloys have more alloy segregation, with coarser grain sizes that produce better creep and stress-rupture properties. They are used where creep and stress rupture are the main design criteria, such as for turbine blades. The alloy content of some advanced heat-resistant superalloys has become so high that they can only be produced by casting or powder metallurgy.

The process flow for a forged superalloy component is shown in Fig. 30.7. Production starts with ingot melting, which can involve as many as three separate melting processes. Ingot melting is followed by homogenization and then cogging to break down the as-cast structure. Final forging is then used to produce the desired shape and refine the microstructure.

### 30.3.1 Melting

The three primary melting processes are (1) vacuum induction melting (VIM) to produce the desired alloy composition, (2) electroslag refining (ESR) to remove oxygen-containing inclusions, and (3) vacuum arc remelting (VAR) to reduce compositional segregation that occurs during ESR solidification. In the VIM process, liquid metal is produced under vacuum in an induction-heated crucible. Double VIM may be employed for some critical applications. In general, vacuum melting improves cleanliness and trace element-sensitive properties, such as fatigue strength, ductility, and impact strength. Secondary remelting using ESR and/or VAR further refines the metal by eliminating gases, nonmetallic and metallic impurities, and inclusions. Both the ESR and VAR processes allow directional solidification of an ingot from bottom to top, yielding high density and macrostructural homogeneity as well as producing an absence of segregation and shrinkage cavities. The precipitation-hardening nickel-base alloys are generally double vacuum induction melted and then vacuum arc or electroslag remelted.

The structure of superalloy ingots consists of primary dendrites, which are solute lean, and interdendritic regions, which are solute rich. For those alloys with sufficient aluminum, titanium, or niobium to exhibit useful age-hardening response, it is necessary to homogenize the ingot prior to deformation processing. Homogenization consists of extended exposures (up to 48 h) at temperatures approaching the incipient melting temperature of the alloy. The degree of segregation that must be removed by homogenization is a function of both alloy composition and melting practice. While homogenization treatments are generally effective in producing a great leveling of microscale concentration differences in superalloys, some residual minor differences may remain in areas containing large primary dendrites.

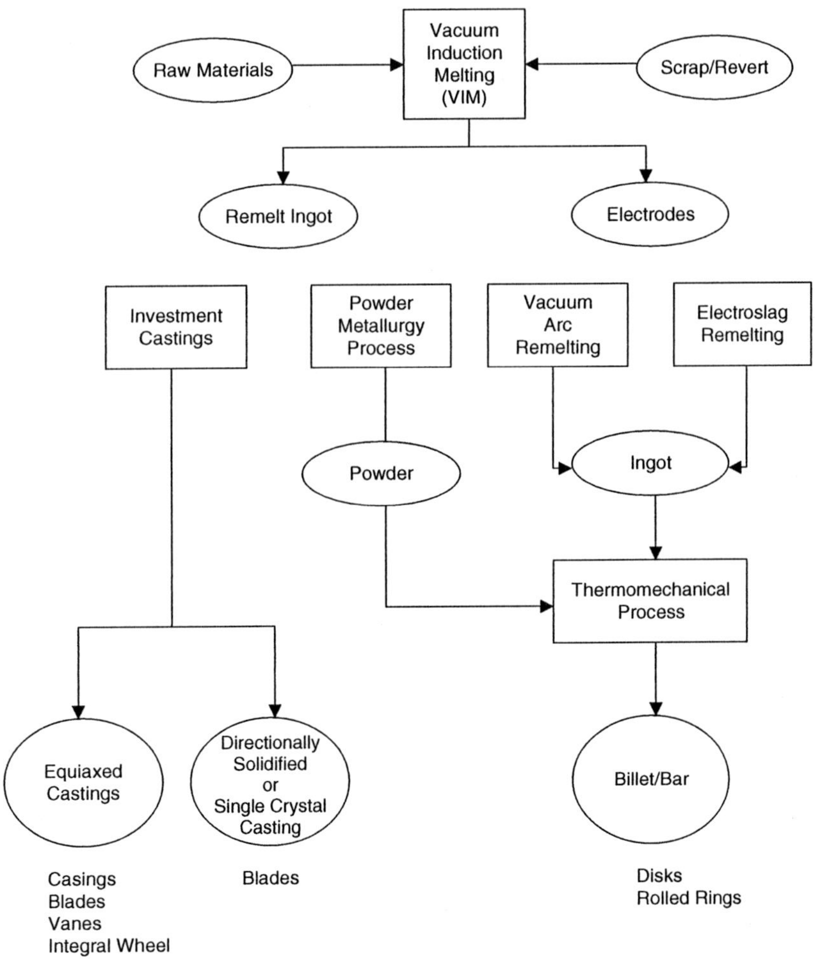


**Fig. 30.6** Process flow for superalloy components. Source: Ref 7

### *30.3.2 Wrought Alloy Primary Fabrication*

Wrought superalloys are manufactured in all mill forms common to the metal industry. Iron-nickel-, nickel-, and cobalt-base superalloys are produced conventionally as bar, billet, extrusions, plate, sheet, strip, wire, and forgings by primary mills.

Cast structure is first refined by cogging, a process of reducing ingot diameter and increasing length by multiple open-die forging and reheating operations. Initial ingot breakdown is conducted well above the $\gamma'$ solvus to allow chemical homogenization and microstructural refinement; however, care must be taken to make sure that the temperature does not exceed the incipient melting temperature. Uniform billet structure and improved surface finish are additional objectives of this process. Initial ingot breakdown is followed by methods such as rolling, forging, or extrusion. Sheet, bar, and ring rolling are commonly employed for secondary hot working of superalloys.

Forgings are used in both the turbine and compressor sections of gas turbines. Open-die forging is carried out in flat or swaging dies. Closed-die forging is used to produce shapes that match the impressions of dies attached to the ram and anvil. Both hammers and presses are used. For hammer forging, the stresses required for deformation are higher because of the higher strain rates employed. Billets for conventional superalloy forgings are preheated to 1040 to 1230 °C (1900 to 2250 °F). While most forging is carried out with steel tooling heated in the range of 200 to 430 °C (390 to 805 °F), isothermal forging is now widely used for near-net shape processing. Supcralloy or molybdenum alloy dies are heated to the same temperature as the forging, in the range of 650 to 980 °C (1200 to 1800 °F). Isothermal forging produces a uniform microstructure while requiring less material, thus lowering machining costs. For temperatures at which superplastic behavior of superalloys occurs, less powerful presses or hammers and slower die closure rates can be used. The benefits outweigh the increased costs of hot die tooling for expensive input materials.

Extrusion is used for the conversion of ingot to billet, especially for the stronger,

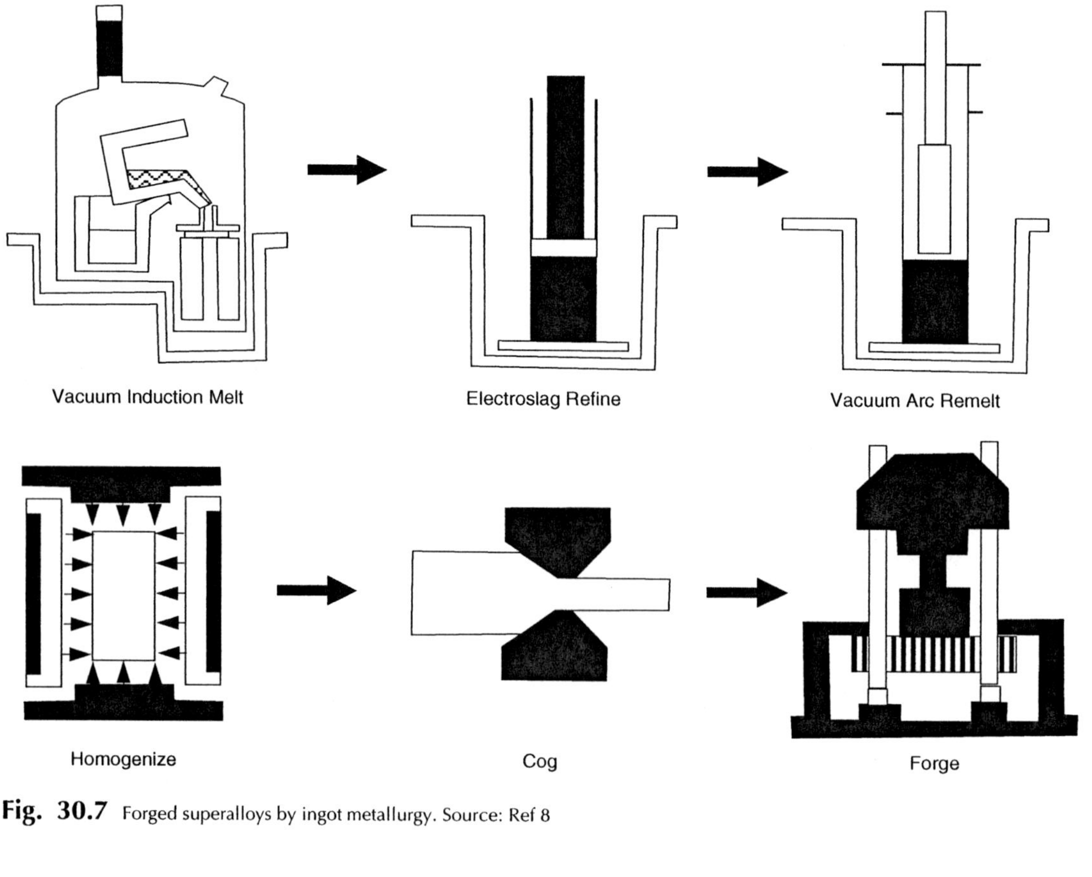


**Fig. 30.7** Forged superalloys by ingot metallurgy. Source: Ref 8

crack-prone alloys as well as for powder-processed alloys. Canning in mild steel or stainless steel is required to avoid chilling and surface cracking. Sheet and other semifinished products are produced by rolling, often with many reheats, frequent conditioning, and possibly encasing of the alloy in a can.

Grain structure can be controlled by thermomechanical processing in several iron-nickel-base alloys that have two precipitates present. Inconel 718 is the most important of these, with the primary precipitate being $\gamma''$ ($Ni_3Nb$) and the secondary precipitate $\delta$ ($Ni_3Nb$). The secondary precipitate is produced first by heat treating at 900 °C (1650 °F) for 8 h followed by working at approximatly 950 °C (1740 °F), below the $\eta$ solvus. Final working is carried out below the recrystallization temperature, and the alloy is subsequently recrystallized below the $\eta$ solvus. Finally, the alloy is direct aged by standard procedures. The result is a fine-grained alloy with higher tensile strength and improved fatigue resistance. Other thermomechanical working schedules are used to produce a double-necklace structure of fine grains surrounding the large grains formed during high-temperature recrystallization. Reductions of 25 to 50% are needed in the final working operations at 1080 to 1110 °C (1975 to 2030 °F) to produce the small, recrystallized grains in cast/wrought René 95.

### *30.3.3 Powder Metallurgy Fabrication*

Powder metallurgy (PM) processing is used for two different types of nickel-base superalloys. For some very highly alloyed compositions, PM is used to produce parts that would crack during forging if made by conventional ingot metallurgy. Powder Metallurgy has also been used to make a special class of oxide-dispersion-strengthened (ODS) alloys for improved high-temperature creep-resistant applications.

Certain alloys, such as René 95 and Inconel 100, contain such high volume fractions of alloying elements that they would crack during forging if made by conventional ingot metallurgy. The PM process for these alloys is shown schematically in Fig. 30.8. Vacuum induction is again used to produce the desired alloy composition. The VIM ingot is then remelted and atomized to produce a powder. This is followed by sieving to remove large particles and inclusions that are ~50 to 100 μm in diameter. The powder is then placed in a metal container (canning) that is vacuum degassed and welded shut. Hot isostatic pressing or extrusion is used to consolidate the powder. The PM billets are then forged to produce the final shape. A comparison between conventional ingot metallurgy and forging with the PM approach is shown in Fig. 30.9.

The process flow for production of ODS alloy parts is shown in Fig. 30.10. Powders of the desired composition are first blended and then mechanically alloyed using a high-energy dry ball milling process. In the ball milling process, the powders are repeatedly fractured and then rewelded together. Ball milling is conducted until a homogeneous particle composition is produced. Approximately 1 wt% of very fine yttria (25 nm) powder, which produces the dispersion strengthening, is added to the powder mix during blending. During the mechanical alloying process, the small yttria particles become entrapped and welded into the metal fragments. After mechanical alloying, the powder is packed into steel tubes and extruded. The extruded billet is then hot rolled to produce product forms such as bar, plate, sheet, and wire. Since most applications are for creep-resistant microstructures, a grain-coarsening anneal at 1315 °C (2400 °F) is conducted after fabrication.

## 30.4 Heat Treatment

Full annealing and precipitation hardening are the two main heat treatments used for superalloys. Stress relieving is usually avoided since it is conducted in the critical 650 to 870 °C (1200 to 1600 °F) temperature range that causes excessive carbide precipitation. A low-temperature stress equalization treatment, conducted at less than 650 °C (1200 °F), that avoids excessive carbide precipitation is sometimes used, but the amount of stress relief at the lower temperatures is, at best, only partial.

### *30.4.1 Annealing*

Annealing of solid-solution-strengthened superalloys means full annealing, that is, complete recrystallization and the attainment of maximum softness. Typical annealing heat treatments are conducted at minimum temperatures of 950 to 1120 °C (1750 to 2050 °F) for 5 to 20 min, followed by rapid cooling through the range of 650 to 870 °C (1200 to 1600 °F). Full annealing must be followed by fast cooling in the range of 650 to 870 °C (1200 to 1600 °F) to prevent carbide precipitation.

Annealing is used mainly to reduce hardness and increase ductility to facilitate forming or machining, prepare for welding, relieve stresses after welding, produce specific microstructures, or soften age-hardened structures by resolution of second phases. Most wrought superalloys can be cold formed but are more difficult to form than austenitic stainless steels. Severe cold forming operations may require several intermediate annealing operations. Annealing should be conducted after welding the precipitation-hardenable alloys if highly restrained joints are involved. For alloys sensitive to strain-age cracking, heating rates must be rapid. If the configuration of the weldment does not permit annealing, aging may be used for stress relieving in alloys not prone to strain-age cracking. Reheating during hot working operations is similar to annealing in that the aim is to promote adequate formability of the metal being deformed. Control of temperature can be critical to resultant properties, because varying degrees of recrystallization and control of grain growth may be desired. In most standard operations, heating or reheating for hot working is a full annealing step, with recrystallization and dissolution of all or most secondary phases.

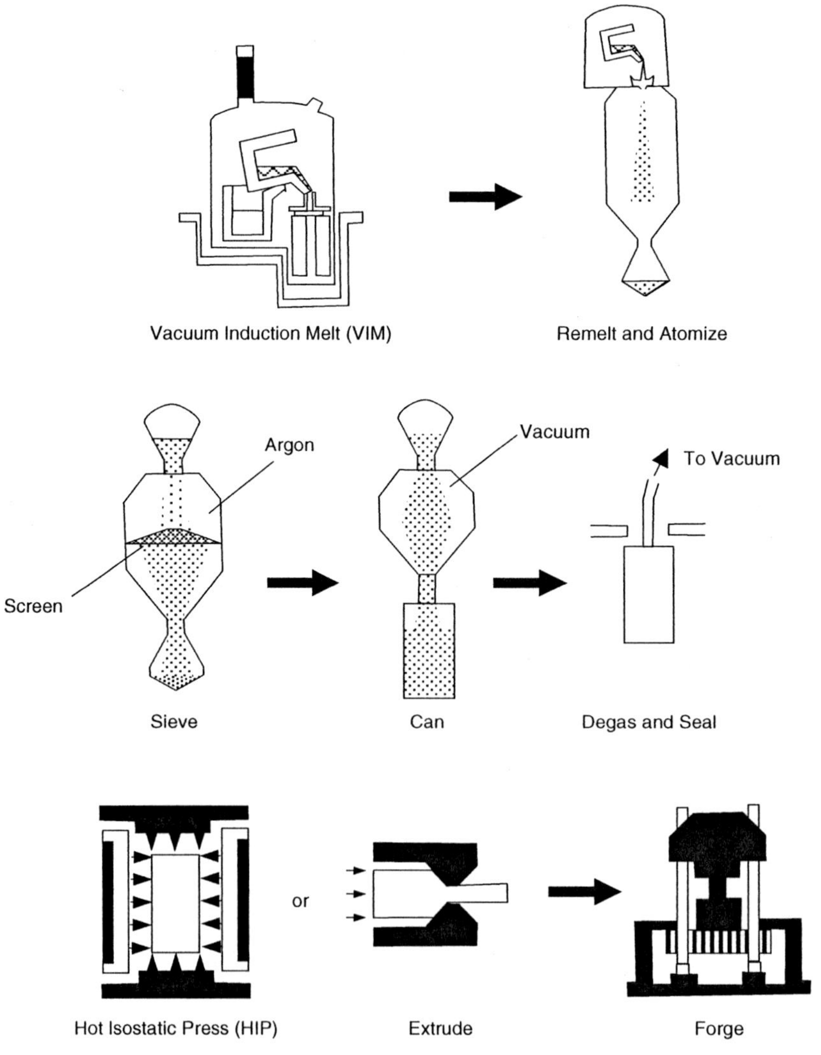


**Fig. 30.8** Forged superalloys by powder metallurgy. Source: Ref 8

### 30.4.2 Precipitation Hardening

Precipitation hardening of nickel and iron-nickel superalloys is conducted using standard precipitation-hardening procedures, that is, solution heat treating, quenching, and then artificially aging at an intermediate temperature. While this sounds rather straightforward, the actual precipitation-hardening treatments are quite varied and sometimes rather complex, as a

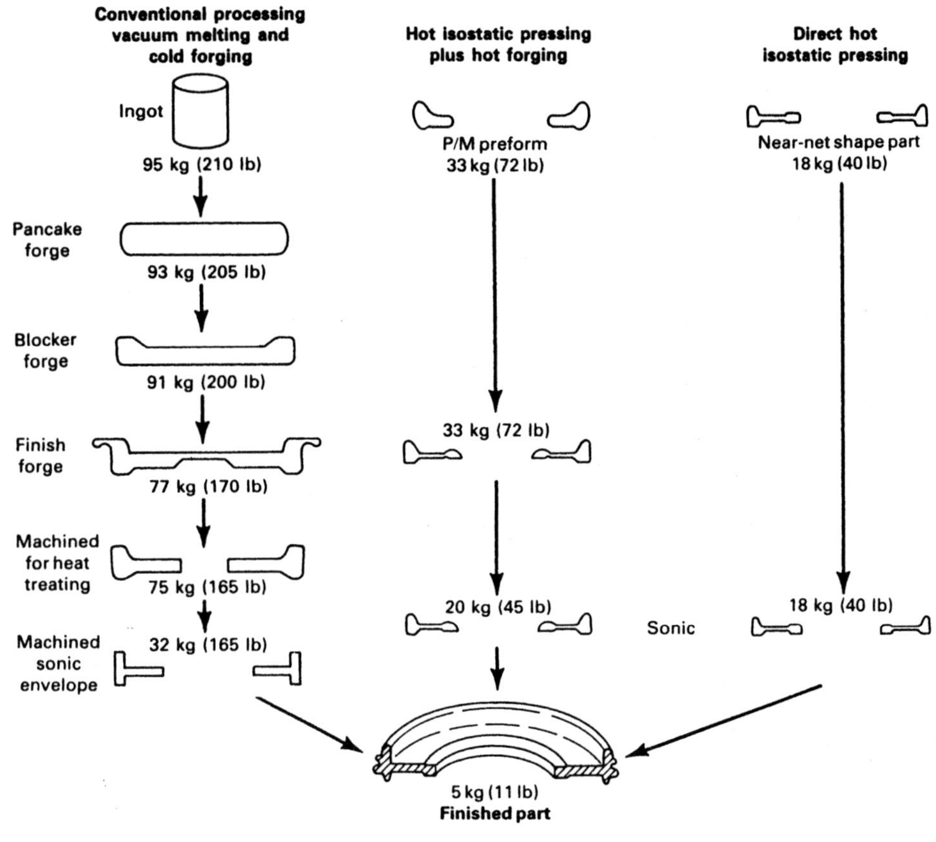


**Fig. 30.9** Material and fabrication savings with powder metallurgy (PM). HIP, hot isostatic pressing. Source: Ref 5

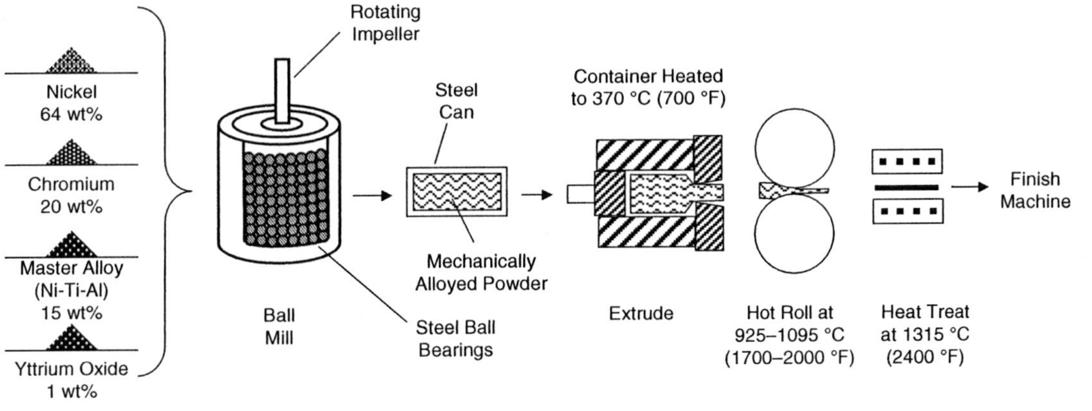


**Fig. 30.10** Mechanical alloying powder metallurgy process. Source: Ref 5

result of the many compositions used and the varied applications for the end products. Thus, only general principles can be given here.

Solution-treating temperatures range from approximately 980 to 1230 °C (1800 to 2250 °F) or even up to 1315 °C (2400 °F) for some single-crystal alloys. Nickel and iron-nickel superalloys are solution treated to dissolve nearly all γ′ and/or γ″ and carbides other than the very stable MC carbides. In some alloys, the solution-treating temperature will depend on the properties desired. A higher solution-treatment temperature will produce a larger grain size for optimal creep-rupture properties, while a lower solutioning temperature retains a finer grain for optimal short-time tensile properties at elevated temperature, improved fatigue resistance, or improved resistance to notch-rupture sensitivity. A higher solution-treating temperature will also result in some grain growth and more extensive carbide dissolution. The principal objective of a high-temperature solution treatment is to put the hardening phases into solution and dissolve all but the most stable carbides. After aging, the resulting microstructure of these wrought alloys consists of large grains that contain the principal aging phases (γ′ and/or γ″) and a heavy concentration of carbides along the grain boundaries. A lower solution-treating temperature dissolves the principal aging phases without grain growth or significant carbide solution. A comparison of the mechanical properties of these two different heat treatments (fine grain versus coarse grain) for Waspaloy is shown in Fig. 30.11. For some wrought superalloys (e.g., Nimonic 80A and Nimonic 90), an intermediate solution-treating temperature is selected to produce a mechanical property compromise. For other alloys (e.g., Udimet 500 and Udimet 700), an intermediate-temperature aging treatment is used to tailor the grain boundaries for improved creep-rupture properties.

The purpose of quenching after solution treating is to maintain at room temperature the supersaturated solid solution obtained during solution treating. Quenching permits a finer age-hardening precipitate size. Cooling methods commonly used include oil and water quenching as well as various forms of air or inert gas cooling. Aging treatments strengthen precipitation-hardenable alloys by causing the

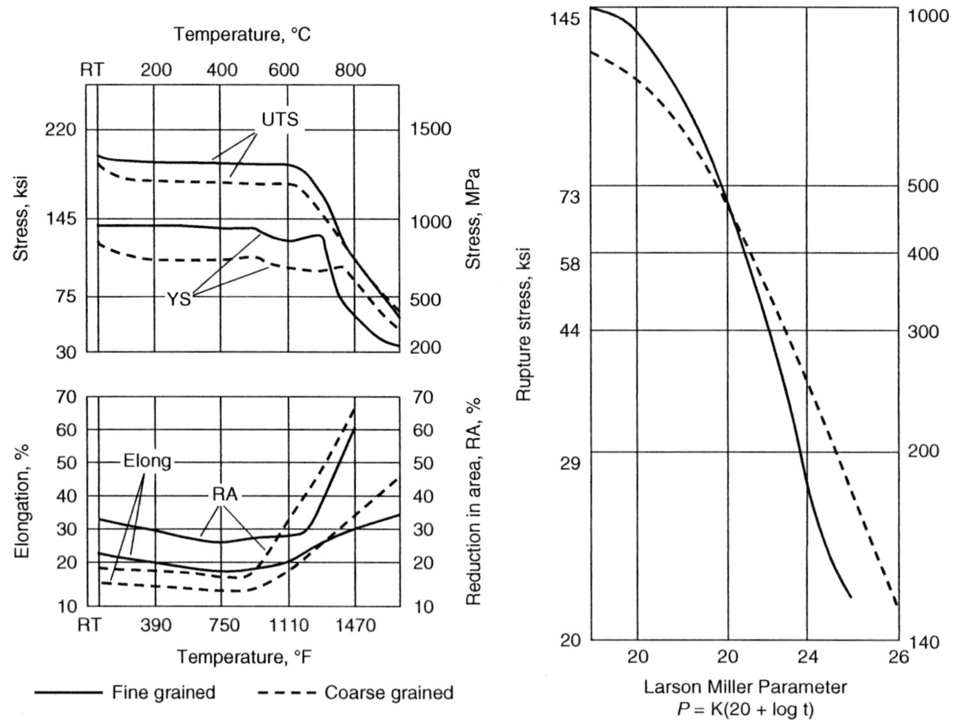


**Fig. 30.11** Effect of heat treatment on mechanical properties of Waspaloy. UTS, ultimate tensile strength; YS, yield strength. Source: Ref 9

precipitation of additional quantities of one or more phases from the supersaturated matrix that is developed by solution treating. Aging heat treatments usually range from 870 to 980 °C (1600 to 1800 °F) for times of approximately 4 to 32 h, followed by air cooling. Factors that influence the selection and number of aging steps and aging temperature include the type and number of precipitating phases available, anticipated service temperature, precipitate size, and the combination of strength and ductility desired. Because dimensional changes can occur during aging, finish machining is conducted after aging.

A two-step aging treatment is commonly used, with the first treatment in the range of 850 to 1100 °C (1560 to 2010 °F) for up to 24 h. Aging at one or more lower temperatures, for example, at 760 °C (1400 °F) for 16 h, completes the precipitation of $\gamma'$. The finer $\gamma'$ produced in the second aging treatment is advantageous for tensile strength as well as for rupture life.

Carbide distribution also is controlled by the heat treatment schedule. Modifications to the heat treatment procedure often are required to avoid problems with carbide films at grain boundaries. For example, a solution treatment of René 41 at 1175 °C (2150 °F) leads to subsequent precipitation of a grain-boundary film of $M_{23}C_6$, with deleterious effects on mechanical properties. Therefore, a lower solution-treatment temperature of approximately 1075 °C (1970 °F) is used to preserve the fine-grained, as-worked structure with well-dispersed $M_6C$. The application of diffusion or overlay environmental coatings is also sometimes conducted during solution treating, although, in some cases, the coating treatment coincides with an aging treatment.

### *30.4.3 Cast Superalloy Heat Treatment*

Heat treatment of cast superalloys includes stress relieving, homogenization, solution heat treatment, and aging heat treatments. Some polycrystalline superalloy castings are used in the as-cast condition. However, stress-relief treatments are commonly used if complex castings develop residual cooling stresses or if stringent dimensional tolerances are required. Castings requiring weld repair are always stress relieved after welding. When heating to high temperatures, as in annealing or solution treating, slow heat-up rates are used to promote homogenization and reduce the danger of incipient melting.

Although many conventionally cast polycrystalline nickel-base alloys are not solution heat treated, all directionally solidified (DS) and single-crystal (SX) alloys are solution heat treated. Since the more highly alloyed cast nickel alloys must be heated to a temperature between the solvus and solidus to dissolve all alloying elements in solution, a soak time of 2 to 6 h is required to produce a homogeneous distribution of alloying elements. Since precipitation occurs during cooling from the solution-treating temperatures, fast cooling rates (e.g., gas furnace quenching) are used to prevent coarsening of the $\gamma'$ precipitate that can degrade the mechanical properties. Hot isostatic pressing (HIP) can be used to improve the quality by healing internal casting porosity. A typical HIP cycle would be 1190 °C (2175 °F) for 4 h at 103 MPa (15 ksi) argon pressure.

## 30.5 Fabrication

Investment casting is an extremely important method of producing superalloys for gas turbine blades and vanes. The fundamentals of investment casting superalloys are covered in Chapter 7, "Solidification and Casting," in this book. The importance of the DS and SX processes, along with improved alloy development, can be appreciated by comparing their creep performance with polycrystalline castings (Fig. 30.12). The introduction of DS allowed approximately a 28 °C (50 °F) increase in operating temperature, while the SX process produced another 28 °C (50 °F) increase.

Forming processes include drawing, spinning, stretch forming, and press brake forming. Since the hot forming range (925 to 1260 °C, or 1700 to 2300 °F) is high and rather narrow, cold forming with intermediate anneals is preferred for thin sheets. During cold forming operations, superalloys work harden rapidly. The rate of work hardening for four superalloys, compared to type 302 stainless steel, is shown in Fig. 30.13. The iron-nickel-base alloy A-286 work hardens similarly to type 302, while the others work harden at higher rates. The cobalt-base alloy S-816 displays the highest work-hardening rate, which means that greater forming forces will be required and more intermediate anneals needed to prevent cracking. As the percentage of

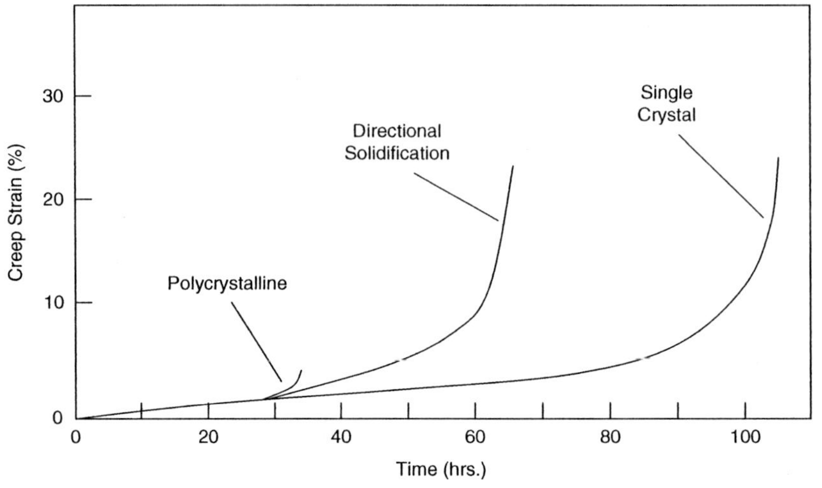


**Fig. 30.12** Creep comparison of a nickel-base superalloy for different casting procedures. Alloy: Mar-M200, 205 MPa (30 ksi), 980 °C (1800 °F). Source: Ref 4

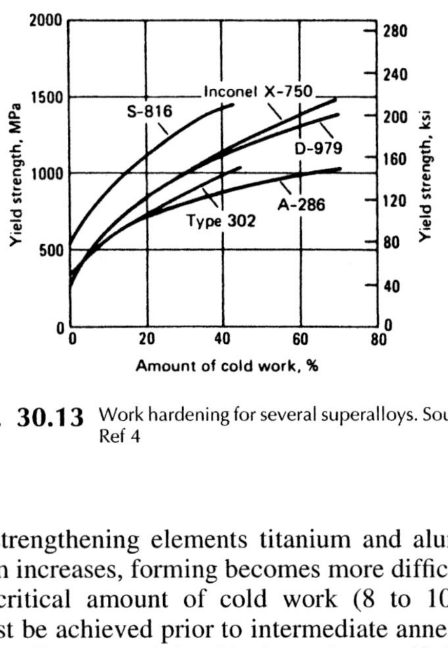


**Fig. 30.13** Work hardening for several superalloys. Source: Ref 4

γ′-strengthening elements titanium and aluminum increases, forming becomes more difficult. A critical amount of cold work (8 to 10%) must be achieved prior to intermediate anneals. Annealing material with less than sufficient amounts of cold work can result in abnormal grain growth, which can cause undesirable "orange-peel" effects during subsequent forming operations. Severely cold-formed parts should be fully annealed after cold forming. If annealing would cause distortion, the part can be formed to within 10% of its final shape, annealed, pickled, and then given the final forming. When Inconel 718 is produced as a fine-grained sheet, it can be superplastically formed at 980 °C (1800 °F) using a slow strain rate to strains up to 250%.

Superalloys, especially the more highly alloyed precipitation-hardened nickel-base alloys, are some of the most difficult of all alloys to machine. However, since superalloys cover a wide range of alloys, they will exhibit different degrees of difficulty in machining. For example, the iron-nickel-base alloys, which resemble stainless steels, are easier to machine than the nickel- and cobalt-base alloys. In general, as the amount of alloying elements increases for higher-temperature service, the alloy becomes more difficult to machine.

The solid-solution-strengthened iron-nickel-, nickel-, and cobalt-base superalloys are readily welded by arc welding processes such as gas tungsten arc welding and gas metal arc welding. However, the precipitation-hardened iron-nickel- and nickel-base alloys are susceptible to hot cracking during welding or may crack after welding (delayed cracking). The susceptibility to hot cracking, a function of the aluminum and titanium contents that form γ′, is shown in Fig. 30.14. Cracking usually occurs in the heat-affected zone, and welding is usually restricted to wrought alloys with approximately 35 vol% γ′. Casting alloys with high aluminum and titanium contents are considered unweldable because they will usually hot crack during the welding operation.

## 30.6 Coating Technology

Superalloys react with oxygen, and oxidation is the prime environmental effect on these alloys. At moderate temperatures, approximately 870 °C (1600 °F) and below, general uniform oxidation is not a major problem. At higher temperatures, nickel- and cobalt-base superalloys are attacked by oxygen. The level of oxidation resistance at temperatures below approximately 980 °C (1800 °F) is a function of chromium content, because $Cr_2O_3$ forms as a protective oxide. At temperatures above approximately 980 °C (1800 °F), aluminum content becomes more important in oxidation resistance, because $Al_2O_3$ forms as a protective oxide. Chromium and aluminum can contribute in an interactive fashion to oxidation protection. Thus, the higher the chromium level, the lower the aluminum content that may be required. However, the aluminum contents of many superalloys are insufficient to provide long-term $Al_2O_3$ protection.

The development of higher-strength nickel-base superalloys led to reductions in chromium content and to greater oxidation attack and susceptibility to hot corrosion. To protect against local oxidation and hot corrosion, two different types of coatings have been developed: aluminide diffusion coatings and overlay coatings. In recent years, thermal barrier coatings have also been developed to provide substantial reductions in temperatures on superalloy surfaces. Coating selection is based on knowledge of oxidation/corrosion behavior in laboratory, pilot-plant, and field tests. Attributes that probably will be required for successful coating selection include:

- High resistance to oxidation and/or hot corrosion
- Ductility sufficient to provide adequate resistance to thermomechanical fatigue
- Compatibility with the base alloy
- Low rate of interdiffusion with the base alloy
- Ease of application and low cost relative to improvement in component life
- Ability to be stripped and reapplied without significant reduction of base metal dimensions or degradation of base metal properties

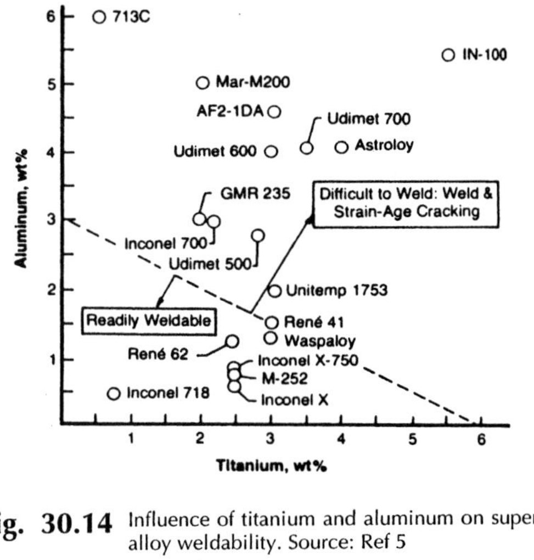


**Fig. 30.14** Influence of titanium and aluminum on superalloy weldability. Source: Ref 5

**Aluminide Diffusion Coatings.** The most common type of coating for environmental protection of superalloys is the aluminide diffusion coating, which develops an aluminide (CoAl or NiAl) outer layer with enhanced oxidation resistance. This outer layer is developed by the reaction of aluminum with the nickel or cobalt in the base metal. Aluminide diffusion coatings generally are applied by pack processes, but slurry, electrophoretic, and other techniques have been used. Customarily, deposition is done at an intermediate temperature, followed by diffusion in a controlled atmosphere furnace at approximately 1040 to 1120 °C (1900 to 2050 °F). Some use has been made of aluminides containing chromium or silicon, and, in recent years, noble metals such as platinum have been used to enhance the oxidation resistance of aluminides. The oxidation resistance of aluminide coatings is derived from formation of protective $Al_2O_3$ scales. Aluminide diffusion coatings are generally thin, approximately 50 to 75 μm (2 to 3 mils). They consume some base metal in their formation and, although deposited at lower temperatures, are invariably diffused at temperatures from approximately 1040 to 1120 °C (1900 to 2050 °F) prior to being placed in service.

**Overlay coatings** are generally MCrAl or MCrAlY coatings and are derived directly from the deposition process. They do not require diffusion for their formation. The constituent denoted "M" in these designations has, at various times, been iron, cobalt, nickel, or combinations of nickel and cobalt. Overlay coatings are applied by physical vapor deposition (PVD) in vacuum chambers. They also may be applied by plasma spray techniques. Low-pressure plasma spray techniques produce coatings with properties comparable to, or better than, those of PVD vacuum-produced coatings. A high-temperature heat treatment at 1040 to 1120 °C (1900 to 2050 °F) is performed to homogenize the coating and to ensure its adherence to the substrate. MCrAl coatings are

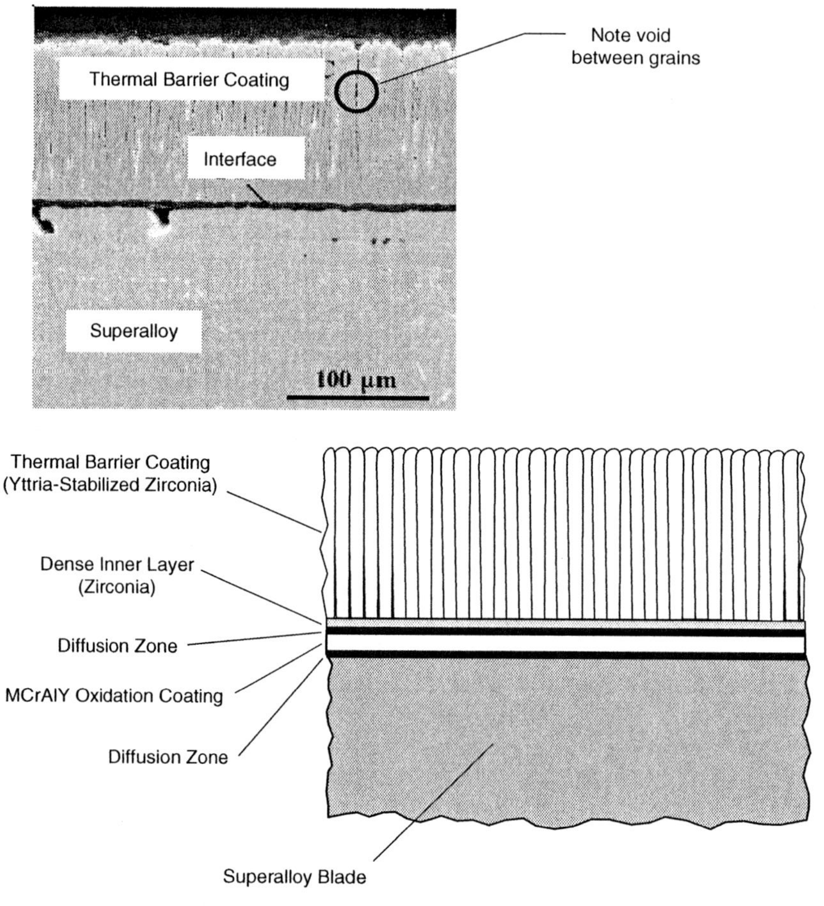


**Fig. 30.15** Thermal barrier coating. Source: Ref 5

approximately twice as thick as aluminide coatings, and, through incorporation of yttrium, overlay coatings can be made to have improved corrosion resistance. An advantage of MCrAlY coatings is that their compositions can be tailored to produce greater or lesser amounts of chromium or aluminum within the coating, and thus, the protectivity and mechanical properties of the coating can be balanced for optimal performance.

**Thermal barrier coatings** (TBCs) have provided enough insulation for superalloys to operate in temperature environments as much as 165 °C (300 °F) above their customary operating range. The TBCs are ceramics, most notably plasma-sprayed partially stabilized zirconia, such as the one shown in Fig. 30.15. These ceramic coatings use an underlay of a corrosion-protective layer, typically an overlay coating, such as an MCrAlY, that provides for oxidation resistance and the necessary roughness for topcoat adherence. Thermal barrier coatings do not provide oxidation protection for superalloys.

## ACKNOWLEDGMENTS

Sections of this chapter were adapted from "Wrought and P/M Superalloys" by N.S. Stoloff in *Properties and Selection: Nonferrous Alloys and Special-Purpose Materials,* Volume 2, *ASM Handbook,* ASM International, 1990; and "Superalloys" by M.J. Donachie and S.J. Donachie in *Metals Handbook Desk Edition,* 2nd Edition, ASM International, 1998.

## REFERENCES

1. N.S. Stoloff, Wrought and P/M Superalloys, *Properties and Selection: Irons, Steels, and High-Performance Alloys,* Vol 1, *ASM Handbook,* ASM International, 1990

2. M.J. Donachie and S.J. Donachie, Superalloys, *Metals Handbook Desk Edition,* 2nd ed., ASM International, 1998
3. W.J. Molloy, Investment-Cast Superalloys—A Good Investment, *Adv. Mater. Process.,* Oct 1990, p 23–30
4. E.F. Bradley, *Superalloys: A Technical Guide,* ASM International, 1988
5. J.R. Davis, *Heat-Resistant Materials,* ASM International, 1997
6. R.E. Smallman and R.J. Bishop, *Modern Physical Metallurgy and Materials Engineering,* 6th ed., Butterworth Heinemann, 1999
7. G.E. Maurer, Primary and Secondary Melt Processing—Superalloys, *Superalloys, Supercomposites, and Superceramcis,* Academic Press, 1989, p 64–96
8. M.G. Benz, "Preparation of Clean Superalloys," GE Technical Information Series 98CRD128, General Electric Research and Development Center, 1998
9. F. Schubert, Temperature- and Time-Dependent Transformation: Application to Heat Treatment of High-Temperature Alloys, *Superalloy Source Book,* ASM International, 1989, p 88

## SELECTED REFERENCES

- F.C. Campbell, *Manufacturing Technology for Aerospace Structural Materials,* Elsevier Scientific, 2006
- I.A. Choudhury and M.A. El-Baradie, Machinability of Nickel Based Super Alloys: A General Review, *J. Mater. Process. Technol.,* Vol 77, 1998, p 278–284
- D.R. Clarke and S.R. Phillpot, Thermal Barrier Coating Materials, *Mater. Today,* June 2005, p 23
- J.R. Davis, *Alloying: Understanding The Basics,* ASM International, 2001
- D.A. DeAntonio et al., Heat Treating of Superalloys, *Heat Treating,* Vol 4, *ASM Handbook,* ASM International, 1991
- M.J. Donachie and S.J. Donachie, *Superalloys: A Technical Guide,* 2nd ed., ASM International, 2002
- R.L. Kennedy, W.D. Cao, T.D. Bayha, and R. Jeniski, "Developments in Wrought Nb Containing Superalloys (718 + 100 °F)," TMS, 2003
- R. Schafrik and R. Sprague, Gas Turbine Materials, *Adv. Mater. Process.,* March–June 2004

CHAPTER 31

# Refractory Metals

THE REFRACTORY METALS include niobium, tantalum, molybdenum, tungsten, and rhenium. Refractory metals have one characteristic in common: their exceptionally high melting points. For example, tungsten melts at 3410 °C (6170 °F), which is more than double that of iron and ten times that of lead. These metals are considered refractory because of their high melting points, high-temperature mechanical stability, and resistance to softening at elevated temperatures. A comparison of some of the properties of the refractory metals compared to iron, copper, and aluminum is given in Table 31.1. Their good strength and stiffness retention at elevated temperatures is shown in Fig. 31.1 and 31.2, respectively. They all have high densities, low vapor pressures, and relatively low thermal expansions compared to other metals. Unfortunately, despite their high-temperature resistance, all of them oxidize severely at relatively low temperatures and therefore require oxidation-resistant coatings for elevated-temperature use. Protective coating systems have been developed, primarily for niobium alloys in aerospace applications, to enable their use in high-temperature oxidizing environments.

The refractory metals all have body-centered cubic crystalline structures except for rhenium, which has a hexagonal close-packed (hcp) structure. Rhenium has some very attractive properties but is extremely scarce and therefore very expensive. Excluding rhenium, the remaining four refractory metals can be classified roughly into two groups. Molybdenum and tungsten both have very high elastic moduli and high ductile-to-brittle transition temperatures (DBTTs). On the other hand, niobium and tantalum have lower moduli, and the DBTTs are below room temperature. In all cases, their DBTTs (Fig. 31.3) are sensitive to impurity elements that segregate to the grain boundaries.

Because of their high melting points and ease of oxidation, refractory metals are usually produced in powder form. Refractory metals are extracted from ore concentrates, processed into chemicals, and then into powders. The powders are consolidated into finished products or mill shapes and ingots for further processing. All processing must be conducted in either a vacuum or other protective atmosphere to prevent catastrophic oxidation.

## 31.1 Niobium

Niobium, formerly called columbium in the United States, has a lower density (8.57 g/cm$^3$, or 0.31 lb/in.$^3$), a lower elastic modulus (103 GPa, or 15 msi), and a lower melting point

**Table 31.1 Comparative properties of refractory metals**

| Metal | Crystal structure(a) | Specific gravity | Melting point | | Modulus of elasticity | | Coefficient of thermal expansion | |
|---|---|---|---|---|---|---|---|---|
| | | | °C | °F | GPa | msi | ppm/°C | ppm/°F |
| **Refractory metals** | | | | | | | | |
| Niobium | bcc | 8.57 | 2468 | 4474 | 103 | 15 | 7.3 | 4.1 |
| Tantalum | bcc | 16.6 | 2996 | 5425 | 186 | 27 | 6.5 | 3.6 |
| Molybdenum | bcc | 10.2 | 2610 | 4730 | 324 | 47 | 4.9 | 2.7 |
| Tungsten | bcc | 19.2 | 3410 | 6170 | 400 | 58 | 4.6 | 2.6 |
| Rhenium | hcp | 21.0 | 3180 | 5755 | 462 | 67 | 6.7 | 3.7 |
| **Metals for comparison** | | | | | | | | |
| Iron | bcc | 7.87 | 1538 | 2798 | 207 | 30 | 11.8 | 6.5 |
| Copper | fcc | 8.93 | 1085 | 1981 | 117 | 17 | 16.5 | 9.2 |
| Aluminum | fcc | 2.70 | 660 | 1220 | 62 | 9 | 23.6 | 13.1 |

(a) bcc, body-centered cubic; hcp, hexagonal close-packed; fcc, face-centered cubic. Source: Ref 1

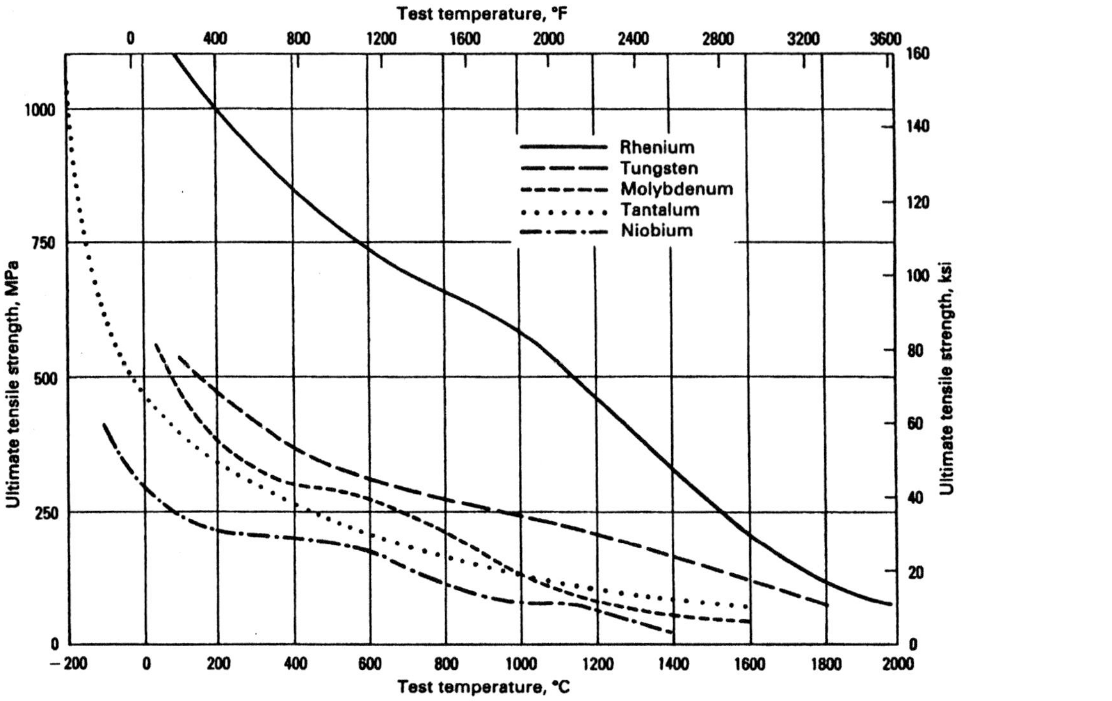


Fig. 31.1 Temperature versus tensile strength for refractory metals. Source: Ref 2

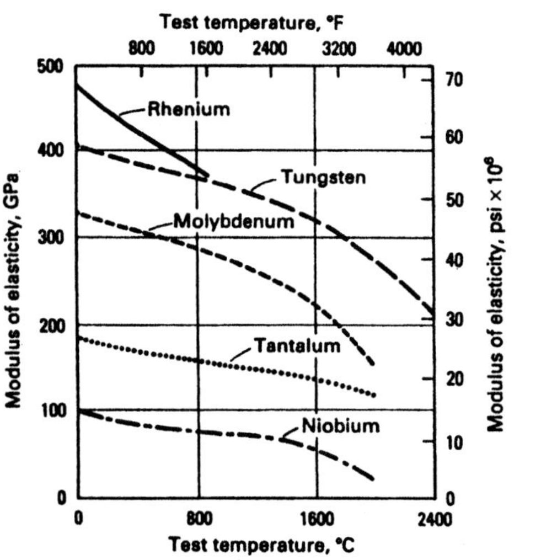


Fig. 31.2 Temperature versus elastic modulus for refractory metals. Source: Ref 2

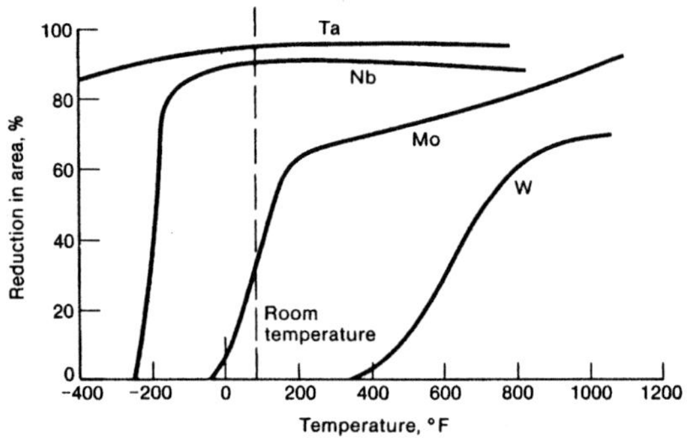


Fig. 31.3 Ductile-to-brittle transition for recrystallized refractory metals. Source: Ref 3

(2468 °C, or 4474 °F) than the other refractory metals. The DBTT of −145 to −200 °C (−225 to −330 °F) is lower than the other refractory metals except for tantalum. Since the DBTT rises with impurity content, strict control of impurities is necessary during alloy processing.

Niobium is used as an alloying element in steels, superalloys, and nonferrous alloys, accounting for approximately 95% of its production. Since niobium has approximately half the density of tantalum, the other ductile refractory metal, its alloys have been more widely used for aerospace applications. Niobium alloys are being considered for the next generation of high-temperature turbine engine alloys, replacing nickel-and cobalt-base superalloys. To date, a totally reliable coating system to prevent oxidation of niobium alloys has yet to be developed.

Oxidation-resistant coatings are susceptible to spalling during thermal cycling due to differences in the coefficients of thermal expansion between the base metal and the coating.

Niobium and its alloys can be made by powder metallurgy (PM) techniques but are usually made by consumable electrode vacuum arc melting or by electron beam melting. Ingots are hot worked at 790 to 1205 °C (1450 to 2200 °F), followed by cold working and forming at 205 to 315 °C (400 to 600 °F). Niobium alloys can be recrystallized at 1010 to 1300 °C (1850 to 2375 °F) and stress relieved at 900 to 1010 °C (1650 to 1850 °F) in vacuum or inert atmospheres. Niobium can be hot worked and cold worked at room temperature in either the stress-relieved or the recrystallized conditions. Unalloyed niobium is basically soft and malleable, able to withstand extremely high amounts of cold working (70 to 80%) between anneals. Because of its high fabricability, niobium is commercially available in plate, sheet, wire, and tube forms.

Niobium powder is fabricated by standard PM methods. Powder is pressed into bars and presintered in a vacuum induction furnace until the bars have sufficient strength to be clamped into pairs between the electrodes of the sintering furnace. Under high vacuum, a high current is then passed through the bars until a temperature just below the melting point is reached. When cooled, the bars are rolled to consolidate the pores, and the bars are returned to the sintering furnace for resintering at 2300 °C (4170 °F). The metal then can be forged, rolled, or swaged to almost theoretical density.

To facilitate processing, tight control of impurities must be maintained during manufacturing. Contamination by interstitial impurity elements impairs room-temperature ductility. Niobium must be protected against oxygen, nitrogen, and hydrogen absorption during manufacturing and during service. The oxides, nitrides, and hydrides formed by these gases are detrimental to ductility. A complex series of oxides is formed on exposure to air at elevated temperatures, including $NbO_x$, NbO, $NbO_2$, and three types of $Nb_2O_5$. The oxidation rate fluctuates between linear and parabolic, depending on the type of oxide forming and the sequence in which the oxides build up on the surface. Niobium also has a high solubility for oxygen, which it absorbs internally from the surface oxide that forms at temperatures above approximately 400 °C (750 °F). Oxygen diffuses into the structure at a rapid rate, producing a hard but embrittled oxygen-affected zone that can cause as much as a 70% loss in ductility when only 15% of the cross section is contaminated. The diffusion occurs so rapidly that this degree of contamination can occur in just 1 min in air at 590 °C (1100 °F). Niobium absorbs hydrogen in the temperature range of 245 to 955 °C (475 to 1750 °F) and forms an embrittling hydride at 480 to 600 °C (900 to 1110 °F). The limits for the impurity elements are 100 ppm carbon, 200 ppm oxygen, 100 ppm nitrogen, and only 15 ppm hydrogen. Niobium and its alloys are protected from these impurities by melting and casting in vacuum or by processing in sacrificial molybdenum containers. Small alloying additions of zirconium and yttrium are added to tie up oxygen as internal oxide particles to promote better ductility. Before placing niobium products in service, they can be protectively coated with aluminide or silicide compounds, such as hafnium silicide or iron-chromium silicide applied as a mixture of Si-20Cr-20Fe.

The properties of a number of niobium alloys are given in Table 31.2. Solid-solution

**Table 31.2 Niobium alloys**

| | | | | | Typical high-temperature strength | | | | | | | |
|---|---|---|---|---|---|---|---|---|---|---|---|---|
| | | | | | Temperature | | Tensile strength | | Temperature | | 10 h rupture strength | |
| Nominal alloy additions, wt% | Common designation | Most common product form | Usual condition(a) | Low-temperature ductility class(b) | °C | °F | MPa | ksi | °C | °F | MPa | ksi |
| **Niobium and niobium alloys** | | | | | | | | | | | | |
| None | Unalloyed Nb | All | Rx | A | 1095 | 2000 | 69 | 10 | 1095 | 2000 | 34 | 5 |
| 1Zr | NB-1Zr | All | Rx | A | 1095 | 2000 | 158 | 23 | 1095 | 2000 | 96 | 14 |
| 1Zr, 0.1C | PWC-11 | All | Rx | A | 1095 | 2000 | 130 | 19 | ... | ... | ... | ... |
| 10Hf, 1Ti, 0.7Zr | C-103 (KB13) | All | Rx | A | 1095 | 2000 | 186 | 27 | ... | ... | ... | ... |
| 10Ta, 10W | SCb-291 | Bar, sheet | Rx | A | 1095 | 2000 | 220 | 32 | 1095 | 2000 | 62 | 9 |
| 10W, 10 Hf, 0.1Y | C-129Y | Sheet | Rx | A | 1315 | 2400 | 179 | 26 | 1095 | 2000 | 103 | 15 |
| 28Ta, 11W, 0.8Zr | FS-85 | Sheet | Rx | A | 1315 | 2400 | 158 | 23 | 1315 | 2400 | 83 | 12 |

(a) Rx, recrystallized. (b) A, excellent cryogenic ductility. Source: Ref 2 strengthening is greatest for alloying elements that have a large atomic size misfit, a low diffusivity, and a high melting point (for thermal stability). Substitutional alloying additions include tungsten, molybdenum, tantalum, and vanadium. Tungsten and molybdenum are effective strengtheners but increase density and significantly degrade low-temperature ductility. However, the addition of as little as 6 at.% W doubles the creep strength of niobium. Tantalum provides moderate strengthening and does not degrade ductility but increases density. Vanadium is an effective strengthener but impairs creep strength. Additions that form carbides include zirconium, hafnium, and titanium, while carbon and nitrogen form interstitials. Zirconium and hafnium form carbides and carbonitrides for the dispersion-strengthened alloys. These carbides and carbonitrides are key contributors to the higher-strength alloys, such as Cb-1 (Nb-30W-1Zr-0.06C) and B-88 (Nb-28W-2Hf-0.07C).

Alloy C-103 can easily be fabricated into sheet and rod products and is weldable with proper precautions. It has been used for rocket components requiring moderate strength at temperatures in the range of 1095 to 1315 °C (2000 to 2400 °F). The microstructure contains alloying-element oxide particles for grain size control and dispersion strengthening. The microstructure of annealed sheet is shown in Fig. 31.4. Its service temperature range is 980 to 1315 °C (1800 to 2400 °F). Alloys C-129Y, FS-8, and Cb-752 have higher elevated-temperature tensile and creep strengths than C-103 while maintaining good fabricability and thermal stability. Alloy C-129Y is solution annealed at 1850 °C (3360 °F) and aged at 1200 °C (2190 °F) for 1 h to form dispersoids of hafnium and yttrium oxides that extend the service temperature up to 1650 °C (3000 °F). Alloy Nb-1Zr is used in nuclear applications because it has a low thermal neutron absorption cross section, good corrosion resistance, and good resistance to radiation damage. It is also used extensively for liquid metal systems operating at 980 to 1205 °C (1800 to 2200 °F). Superconducting niobium alloys are also used in the construction of high-field-strength magnets for research purposes.

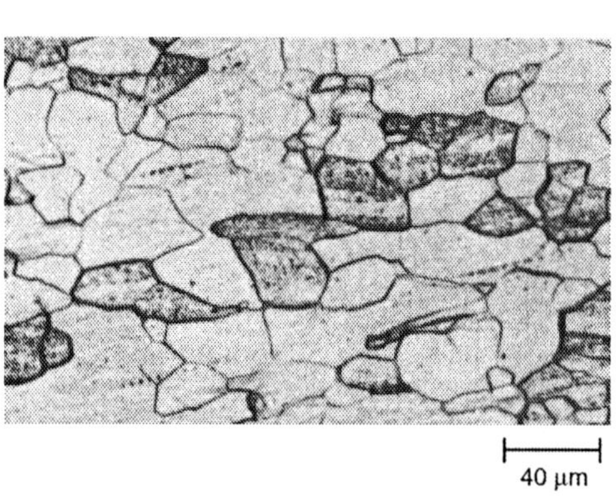


**Fig. 31.4** Niobium alloy C-103 annealed sheet. Arc melted, hot extruded, warm rolled, and annealed. Cold rolled to finished size. Final annealed in vacuum at 1290 °C (2350 °F) for 1 h. Source: Ref 4

Niobium has good corrosion resistance to many environments, a result of its naturally forming protective oxide. It is resistant to strong nitric acid and hydrochloric acid solutions, even when the hydrochloric acid is present under oxidizing conditions. Niobium offers nearly the corrosion resistance of tantalum and nearly the melting temperature of molybdenum. Its cost is approximately 1/6th that of tantalum and 25% more than molybdenum. Niobium mill products are used in the fabrication of corrosion-resistant process equipment, including reaction vessels, columns, bayonet heaters, shell and tube heat exchangers, U-tubes, rupture diaphragms, and orifices. At elevated temperatures, niobium reacts with the halogens, oxygen, nitrogen, hydrogen, carbon, and sulfur. Niobium resists attack by most gases below 205 °C (400 °F). However, at 345 °C (650 °F), it will begin to oxidize. It will react with nitrogen above 300 °C (570 °F) and is embrittled by hydrogen above 250 °C (480 °F).

## 31.2 Tantalum

Tantalum has a high melting point (3000 °C, or 5425 °F), a high density (16.6 g/cm$^3$, or 0.60 lb/in.$^3$), some tolerance for interstitial elements, a reasonable elastic modulus, and excellent room-temperature ductility with greater than 20% elongation. It is readily weldable and has a DBTT of −270 °C (−450 °F). Tantalum and its alloys also have excellent corrosion resistance. At temperatures up to approximately 150 °C (300 °F), tantalum forms a thin stable oxide film ($Ta_2O_5$) that has corrosion resistance similar to glass, being attacked by only fluorides, fuming sulfuric acid (oleum), free sulfur trioxides, and strong caustics. It is used in heat exchangers where heat must be transferred to or from corrosive fluids and vapors. Tantalum reaction vessels have been exposed to boiling hydrochloric or nitric acid for as long as 30 years without appreciable corrosion. Approximately half of the tantalum production is used for tantalum oxide capacitors. Tantalum is used as an alloying element in superalloys, and tantalum carbide is an important constituent in carbide cutting tools. Mill products constitute approximately 25% of tantalum consumption.

Tantalum powder is compacted, sintered into bars, and then arc or electron beam melted to produce ingots. The ingots are of high purity and have excellent ductility and good weldability. Tantalum and its alloys can be cold rolled into sheet and foil and can be formed and drawn into many shapes. Its excellent ductility and low work-hardening coefficient make tantalum and its alloys readily cold workable, permitting cold reductions of 75% or higher with intermediate anneals. Tantalum and its alloys are rolled at 260 to 370 °C (500 to 700 °F). Recrystallization annealing is conducted at 1205 to 1400 °C (2200 to 2550 °F) in vacuum or inert atmospheres. The room-temperature annealed tensile strength of 207 to 379 MPa (30 to 55 ksi) can be increased to 1380 MPa (200 ksi) by cold working. While it has good elevated-temperature stability, tantalum must be alloyed to improve its low strength at elevated temperatures.

A number of tantalum alloys are listed in Table 31.3. Tantalum has complete solid solubility with tungsten, niobium, and molybdenum and extensive solubility with hafnium, zirconium, rhenium, and vanadium. Tungsten and molybdenum exert the greatest solid-solution strengthening effect on tantalum. For example, by adding tungsten, the elevated-temperature strength of Ta-10W is increased from 62 MPa (9 ksi) for unalloyed tantalum to 345 MPa (50 ksi) at 1315 °C (2400 °F). Ta-10W is used at temperatures up to 2480 °C (4500 °F), where high strength in a corrosive environment is required. It has approximately twice the tensile strength of pure tantalum and yet retains the corrosion resistance and a good portion of the ductility of pure tantalum. Ta-2.5W is useful in applications where high corrosion resistance and good formability are required but lower strength can be tolerated. It has higher strength than pure tantalum while maintaining the fabricability characteristics of pure tantalum. It is available in basically the same sizes and shapes as pure tantalum at a comparable cost. The alloy Ti-111 (Ta-8W-2Hf) has even higher strength and creep resistance than Ta-10W alloy, with comparable formability and room-temperature properties. The elevated-temperature properties of tantalum alloys are generally superior to those of niobium but lower than those of molybdenum or tungsten alloys.

Carbide dispersions provide superior creep and high-temperature strength. High-temperature annealing treatments that increase the grain size also increase the creep resistance. Alloys strengthened from a dispersed second phase by the addition of carbon exhibit superior creep and high-temperature yield properties. Some of these alloys (e.g., Astar-811C) have creep properties similar to those of molybdenum and tungsten alloys (Fig. 31.5). The doped tantalum alloys contain small additions of yttria, thoria, silicon, or yttrium that provide fine particles to resist grain growth at elevated

**Table 31.3 Tantalum alloys**

| Nominal alloy additions, wt% | Common designation | Most common product form | Usual condition(a) | Low-temperature ductility class(b) | Typical high-temperature strength: Tensile, Temperature °C | °F | Tensile strength MPa | ksi | 10 h rupture, Temperature °C | °F | 10 h rupture strength MPa | ksi |
|---|---|---|---|---|---|---|---|---|---|---|---|---|
| **Tantalum and tantalum alloys** | | | | | | | | | | | | |
| None | Unalloyed Ta | All | Rx | A | 1315 | 2400 | 62 | 9 | 1315 | 2400 | 7 | 1 |
| 7.5W (PM alloy)(c) | FS-61 | Wire, strip | CW | A | 25 | 77 | 1140 | 165 | ... | ... | ... | ... |
| 2.5W, 0.15Nb | FS-63 | All | Rx | A | 95 | 200 | 315 | 46 | ... | ... | ... | ... |
| 25W | KBI-6 | All | Rx | A | 95 | 200 | 315 | 46 | ... | ... | ... | ... |
| 10W | Ta-10W | All | Rx | A | 1315 | 2400 | 345 | 50 | 1315 | 2400 | 140 | 20 |
| 8W, 2Hf | T-111 | All | Rx | A | 1315 | 2400 | 255 | 37 | ... | ... | ... | ... |
| 8W, 1Re, 1Hf, 0.025C | Astar 811C | All | Rx | A | 1315 | 2400 | 275 | 40 | ... | ... | ... | ... |
| 40Nb | KBI-40 | All | Rx | A | 260 | 500 | 290 | 42 | ... | ... | ... | ... |
| 37.5Nb, 2.5W, 2Mo | KBI-41 | All | Rx | A | 260 | 500 | 515 | 75 | ... | ... | ... | ... |

(a) CW, cold worked; Rx, recrystallized. (b) A, excellent cryogenic ductility. (c) PM, powder metallurgy. Source: Ref 2 temperature. The microstructure of yttria-stabilized tantalum is shown in Fig. 31.6.

The electronics industry is a major market for high-purity tantalum oxide capacitors, and the development of higher-capacitance powders has expanded its applications. With these powders, less material is needed to make a component of a given capacitance rating; that is, high capacitance permits miniaturization. The dielectric constant of the oxide ($Ta_2O_5$) is stable over the temperature range of −55 to 120 °C (−70 to 250 °F). Tantalum oxide capacitors are used in computers, military hardware, color television sets, radios, and cell phones.

Tantalum is the refractory metal that is most resistant to corrosion. The only media that can affect it are fluorine, hydrofluoric acid, sulfur trioxide, fuming sulfuric acid, concentrated strong alkalis, and certain molten salts. The corrosion resistance of tantalum can be compared to that of glass. However, tantalum withstands higher temperatures and offers the fabrication advantages of a metal. Tantalum equipment is frequently used in conjunction with glass, glass-lined steel, and other nonmetallic construction materials. The natural oxide film on tantalum is responsible for its excellent corrosion resistance to nitric, hydrochloric, bromic, and sulfuric acids. It is also resistant to the liquid metals bismuth, lead, lithium, magnesium, mercury, potassium, sodium, and sodium-potassium alloys. However, when exposed to air or an oxidizing atmosphere above 260 °C (500 °F), tantalum readily oxidizes and must be protected during processing and while in service.

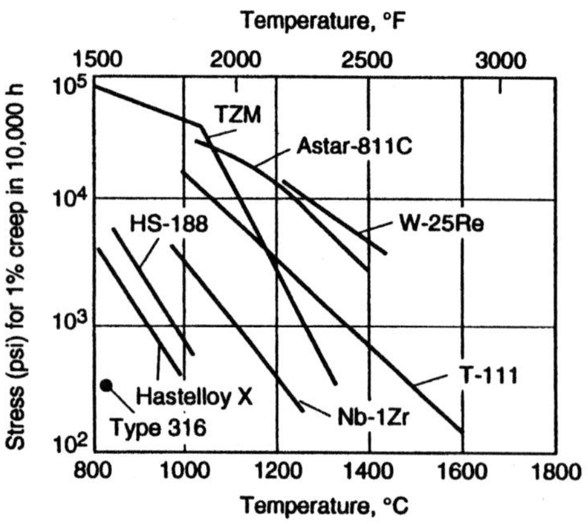


**Fig. 31.5** Long-term creep property comparison. Source: Ref 2

## 31.3 Molybdenum

More molybdenum is consumed annually than any other refractory metal. Although molybdenum is the most readily available and the least expensive of the refractory metals, it is brittle at room temperature, which greatly restricts its uses. Molybdenum has a high melting temperature (2610 °C, or 4730 °F) and retains its strength to high temperatures. It also has a high specific elastic modulus, high thermal conductivity, low coefficient of thermal

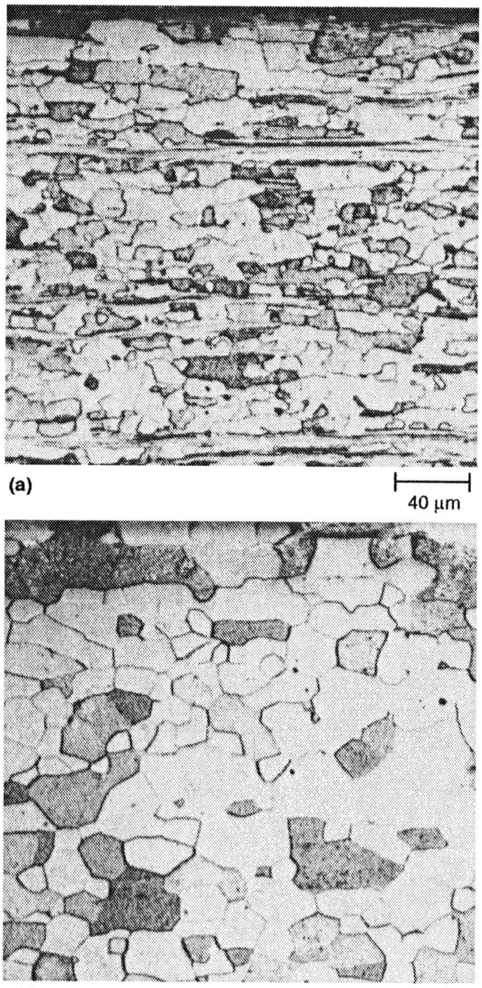


**Fig. 31.6** Ta-250 ppm $Y_2O_3$ powder metallurgy sheet. (a) Longitudinal cross section showing banding from residual cold work resulting from yttria stabilization. ASTM grain size 9. (b) Same processing, with added 30 min final anneal at 2000 °C (3630 °F). Longitudinal cross section. ASTM grain size 8. Source: Ref 4

expansion, and good resistance to thermal shock and fatigue. Molybdenum is used primarily as an alloying element in steels, cast irons, and heat- and corrosion-resistant alloys. Molybdenum additions improve hardenability, toughness, abrasion resistance, corrosion resistance, strength, and creep resistance. Approximately 90% of molybdenum is used for alloying elements, with only approximately 5% for molybdenum mill products.

Molybdenum powder is consolidated or vacuum melted to produce billets that are used for further working or refining. Vacuum-melted billets generally have lower oxygen and nitrogen contents than those produced from powder billets. Molybdenum and its alloys have a very low solubility for oxygen, nitrogen, and carbon at room temperature. When cooled from the molten state or from temperatures near the melting point, the interstitial elements are rejected as oxides, nitrides, and carbides. If the impurity content is high enough, a continuous brittle grain-boundary film will form that severely limits plastic flow. Warm working below the recrystallization temperature breaks up the grain-boundary film and produces a fibrous structure that has good strength parallel to the working direction. Molybdenum sheet can be readily bent in the direction of rolling, but it is difficult to make a bend at 90° to the rolling direction. Sheet metal suitable for deep drawing is produced by cross rolling to reduce sheet anisotropy. Molybdenum is supplied in the worked, stress-relieved, and annealed condition, which has a DBTT of −20 to 40 °C (−5 to 105 °F). Molybdenum is extremely sensitive to impurities at grain boundaries and exhibits intergranular fractures when in the fully recrystallized condition. If the wrought, stress-relieved and, annealed product is be heated above its recrystallization temperature (1150 °C, or 2100 °F) in service, the DBTT increases to 40 to 80 °C (100 to 175 °F), making the metal brittle when cooled to room temperature.

Mechanical properties of unalloyed molybdenum are a function of the degree of work done below the recrystallization temperature. For the best ductility, parts should be given at least a 50% reduction in area. The metal has unusual strength and hardness at elevated temperatures. However, when hot strength is necessary, an alloy rather than pure molybdenum is generally chosen. Alloying additions of titanium and zirconium produce alloys having hot strength and recrystallization temperatures well above those of unalloyed molybdenum. A comparison of partially and fully recrystallized molybdenum microstructures is shown in Fig. 31.7. The microstructure of wrought molybdenum consists of elongated, interlocked cold-worked grains with directional mechanical properties.

The properties of a number of molybdenum alloys are given in Table 31.4. There are several classes of molybdenum alloys.

**Carbide-strengthened alloys** are produced in both vacuum-melted and PM grades. Carbide-strengthened alloys have fine dispersions of carbides that strengthen the material and

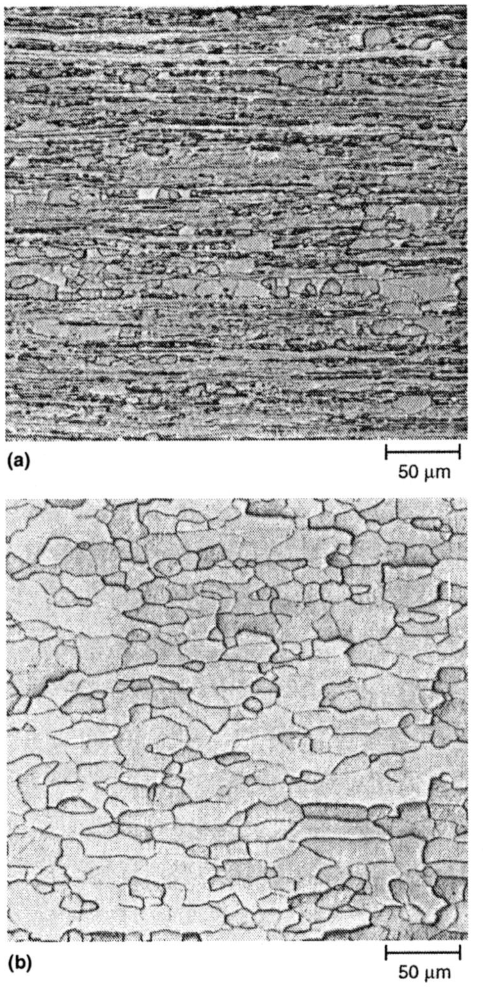


**Fig. 31.7** (a) Commercially pure molybdenum, rolled to 0.040 in. thick sheet, annealed at 900 °C (1650 °F) for 1 h. Longitudinal section. Partly crystallized. (b) Sheet annealed 15 min at 1350 °C (2460 °F). Completely recrystallized. Source: Ref 4

increase the recrystallization temperature. The most widely carbide-strengthened alloy is TZM, which contains 0.5% Ti and 0.08% Zr. The finer grain structure of TZM and the formation of TiC and ZrC on the grain-boundaries inhibit grain growth and grain-boundary failures. Because of its high recrystallization temperature (1370 °C, or 2500 °F), TZM has good stability at high temperatures. At approximately 1095 °C (2000 °F), the strength of TZM is about twice that of unalloyed molybdenum. The recrystallization temperature of TZM is approximately 280 °C (500 °F) higher than pure molybdenum, and it offers better weldability. It can be used for structural applications operating at over 980 °C (1800 °F) under conditions where unalloyed molybdenum is normally used. It has proved to be one of the best materials for hot work dies because of its combination of high hot hardness, high thermal conductivity, low thermal expansion, and high resistance to heat checking as compared to hot work steels. The primary uses of TZM are for high-temperature tooling and glassmaking equipment. TZM costs approximately 25% more than pure molybdenum and costs approximately 5 to 10% more to machine. It is available in sheet and rod form in basically the same size range as pure molybdenum.

**Solid-solution alloys** are alloyed primarily with tungsten or rhenium. The molybdenum-tungsten alloys have good chemical resistance and are used in equipment for handling molten zinc. Mo-30W was developed for the zinc industry because it resists the corrosive effects of molten zinc. It has also proved effective in rocket nozzles and has the potential of offering enhanced performance in applications where any erosive effects are a factor. The importance of molybdenum-rhenium alloys is due to the "rhenium effect," which results in a significantly lower DBTT. The most common alloys contain 5, 41, and 47.5 wt% Re. The 5Re and 41Re alloys are used for thermocouple wire and for structural applications, while the Mo-50Re (47.5 wt% Re) alloy is typically specified for high-temperature structural components. Mo-50Re offers the strength of molybdenum with the ductility and weldability of rhenium. However, it is a costly alloy and is only available in a very limited size range. It offers significant advantages in thin-foil applications for high-temperature delicate parts, especially those that must be welded. Mo-41Re does not develop sigma phase, making the alloy even more ductile after exposure to high temperatures. The alloy HWM-25 (Mo-25W-1Hf-0.07C) contains both carbide-forming and substitutional elements to provide improved high-temperature strength.

**Dispersion-strengthened PM alloys** rely on second-phase particles, introduced or produced during powder processing, to increase the resistance to recrystallization and stabilize the recrystallized grain structure. These alloys have enhanced high-temperature strength and

**Table 31.4 Molybdenum alloys**

| Nominal alloy additions, wt% | Common designation | Most common product form | Usual condition(a) | Low-temperature ductility class(b) | Typical high-temperature strength: Temperature, °C | °F | Tensile sterngth, MPa | ksi | Temperature, °C | °F | 10 h rupture strength, MPa | ksi |
|---|---|---|---|---|---|---|---|---|---|---|---|---|
| **Molybdenum and molybdenum alloys** | | | | | | | | | | | | |
| None | Unalloyed Mo(c) | All | SRA | B-C | 980 | 1800 | 50 | 7 | 980 | 1800 | 175 | 25 |
| 0.5Ti, 0.08Zr, 0.03C | TZM (MT-104)(c) | All | SRA | B-C | 980 | 1800 | 600 | 87 | 1315 | 2400 | 140 | 20 |
| 1.2Ti, 0.3Zr, 0.1C | TZC | All | SRA | B-C | 980 | 1800 | 800 | 116 | 1315 | 2400 | 190 | 28 |
| 1.2Hf, 0.05C | MHC(HCM) | All | SRA | B-C | 980 | 1800 | 800 | 116 | 1315 | 2400 | 210 | 30 |
| 0.5Zr, 1.5Hf, 0.2C | ZHM | All | SRA | B-C | 980 | 1800 | 800 | 116 | 1400 | 2550 | 200 | 29 |
| 25W | 25W | All | SRA | B-C | 980 | 1800 | 330 | 48 | ... | ... | ... | ... |
| 30W | 30W(c) | All | SRA | B-C | 980 | 1800 | 350 | 51 | 1095 | 2000 | 140 | 20 |
| 1Hf, 0.07C, 25W | HWM-25 (Mo25WH) | All | SRA | C | 980 | 1800 | 900 | 131 | 1300 | 2370 | 200 | 29 |
| 5Re | 5Re | All | SRA or Rx | B | 980 | 1800 | 400 | 58 | 1650 | 3000 | 7 | 1 |
| 41Re | 41Re | All | SRA or Rx | B | 980 | 1800 | 600 | 87 | ... | ... | ... | ... |
| 47.5Re | 50Re | All | SRA or Rx | B | 980 | 1800 | 580 | 84 | 1595 | 2900 | 27 | 4 |
| 0.5$ZrO_2$ | Z-6 | All | SRA | B-C | 980 | 1800 | 280 | 41 | ... | ... | ... | ... |
| 150K, 300Si (ppm) | MH(HD) | Wire, sheet | CW, SRA | B | 980 | 1800 | 300 | 44 | ... | ... | ... | ... |
| 200K, 300Si, 100Al (ppm) | KW | Wire, sheet | CW, SRA | B | 980 | 1800 | 300 | 44 | ... | ... | ... | ... |

(a) CW, cold worked; Rx, recrystallized; SRA, stress-relief annealed. (b) B, excellent room-temperature ductility; C, may have marginal ductility at room temperature. (c) Available in both powder metallurgy and arc-cast forms. Source: Ref 2 improved low-temperature ductility. Dispersion-strengthened alloys do not have particularly good strength at low temperatures, but the dispersions are very effective at increasing their resistance to recrystallization and provide good creep properties. Doped molybdenum alloys were the first dispersion-strengthened materials. (A dopant is essentially an impurity added to a metal in small amounts to alter its properties.) They are similar to the doped tungsten alloys developed for lamp filaments and were initially designed to satisfy lighting industry requirements for creep-resistant molybdenum parts. Dispersions of reactive metal oxides also are used to improve the elevated-temperature strength and creep resistance. An example is the zirconia dispersion-strengthened alloy Z-6 (0.5% $ZrO_2$).

The corrosion resistance of molybdenum is similar to tungsten. Molybdenum particularly resists nonoxidizing mineral acids. It is relatively inert to carbon dioxide, ammonia, and nitrogen to 1095 °C (2000 °F) and also in reducing atmospheres containing hydrogen sulfide. Molybdenum offers excellent resistance to corrosion by iodine vapor, bromine, and chlorine up to clearly defined temperature limits. It is resistant to molten bismuth, lithium, magnesium, potassium, and sodium but is attacked by molten tin, aluminum, iron, nickel, and cobalt. When heated in atmospheres containing oxygen above 425 °C (800 °F), molybdenum forms a volatile oxide ($MoO_3$) that evolves as a white, odorless vapor, necessitating the use of protective coatings for applications above 500 °C (930 °F) to avoid rapid thinning of the metal. Coatings that form a thin surface layer of $MoSi_2$ have outstanding oxidation resistance up to approximately 1650 °C (3000 °F). In a vacuum, uncoated molybdenum has a virtually unlimited life at high temperature.

## 31.4 Tungsten

Tungsten has the highest melting point of all metals (3410 °C, or 6170 °F), a high elastic modulus (400 GPa, or 58 msi), high density (19.254 g/cm$^3$, or 0.70 lb/in.$^3$), and good stability at elevated temperatures. However, it is very brittle at room temperature, with a DBTT of 150 to 300 °C (300 to 570 °F), and is also notch sensitive. The low coefficient of thermal expansion and good thermal conductivity of tungsten provide good dimensional stability over a wide temperature range. Approximately 25% of the annual production of tungsten is made into mill products, with another 60% used in cutting tools in the form of tungsten carbide. Tungsten alloys are used in radiation shields, as counterweights and inertial weights, kinetic energy penetrators, heavy-duty electrical contacts, and targets for x-ray tubes.

Tungsten can be used at temperatures as high as 2500 °C (4530 °F). Although the metal is refractory, its oxide films become volatile above approximately 540 °C (1000 °F). Thus, for use at elevated temperatures, tungsten parts must be coated, used in a vacuum, or surrounded by an inert atmosphere. Typical uses include incandescent lamp filaments, electron tube electrodes, and various types of heating elements. Silicide coatings and noble metal claddings, such as platinum-gold alloys, are effective oxidation-resistant coatings.

Powder metallurgy is used to produce most tungsten parts. Whether formed into ingots or bars for further processing or into final shapes, powders are pressed and sintered in hydrogen atmospheres at temperatures of 1800 °C (3270 °F) or higher. Less common methods for producing tungsten products from powders include vacuum arc melting, electron beam melting, zone melting and refining, slip casting and sintering, gas pressure bonding, and plasma spraying.

Processing involves extrusion and forging, with rolling and drawing used as followup for making the finish forms. The alloys are "warm" worked at approximately 1200 °C (2190 °F), at temperatures above the DBTT and below the recrystallization temperature of 1370 °C (2500 °F), in order to avoid embrittlement caused by the crack sensitivity of recrystallized grain boundaries. Tungsten is embrittled by impurity elements that segregate to recrystallized grain boundaries, especially the interstitial elements oxygen, carbon, and phosphorus. Processing by warm working above 650 °C (1200 °F), along with intermediate anneals below the recrystallization temperature, develops a microstructure with elongated, cold-worked grains. In this condition, the material can be further worked or used in applications requiring hardness and strength at moderate temperatures below the recrystallization temperature. For higher-temperature use, the more stable recrystallized grain structure needed for creep resistance and mechanical stability is obtained by alloy doping techniques.

The properties of a number of tungsten alloys are shown in Table 31.5. Classes of tungsten alloys include the following.

**Unalloyed tungsten** mill products are PM products that are strongly dependent on deformation history, purity, and testing orientation. Wire made of unalloyed tungsten can be woven into a mesh and used as electrical resistance heating elements in high-temperature vacuum furnaces. Tungsten sheet is used for heat shields in high-temperature furnaces, while sheet and foil are used for radiation shields in medical x-ray equipment.

**Solid-solution alloys** contain various amounts of molybdenum (2 to 20 wt%) or rhenium (1 to 25 wt%). Molybdenum helps refine the grain size of arc-cast tungsten, lowers the melting point, and reduces density. Molybdenum forms a continuous solid solution with tungsten. The tungsten-molybdenum alloys are used primarily for applications requiring improved machinability, where lower strength is acceptable. Rhenium improves the ductility of tungsten, improves strength, and decreases the DBTT and recrystallization temperatures. Rhenium is the only alloying element known to lower the DBTT of tungsten. Tungsten-rhenium alloys also have good room-temperature ductility and are much easier to machine and weld than pure tungsten.

**Doped tungsten alloys** contain minute quantities of aluminum (15 ppm), silicon (50 ppm), potassium (90 ppm), and oxygen (35 ppm). The AKS alloys contain aluminum, potassium, and silicon compounds that form small bubbles during sintering in hydrogen. The bubbles form along the grain boundaries, providing enhanced creep properties. Instead of equiaxed grains in the recrystallized structures, very long, interlocking grains are produced in tungsten wire that reduce sagging during service and increase the life of the wire. These alloys are used for electric lamp filaments that can reach temperatures as high as 2400 °C (4350 °F).

**Dispersion-strengthened tungsten alloys** are stronger than any other metal at temperatures above 2000 °C (3630 °F). They contain dispersions of 1 to 2 wt% thoria ($ThO_2$) or hafnium carbide (HfC) and may also include additions of rhenium. Tungsten-thoria alloys have traditionally been used for gas tungsten arc welding electrodes but have fallen from favor somewhat due to their radioactive properties. However, arc welding electrodes of tungsten alloyed with ceria (W-2$CeO_2$), zirconia (W-0.25$ZrO_2$), and lanthanum oxide (W-1$La_2O_3$) are available. Of all the tungsten alloys, those in the W-Re-HfC system have the highest strength.

**Tungsten-Heavy Alloys.** If high-temperature strength is not a requirement, more economical liquid-phase sintered alloys can be used. Tungsten-heavy alloys typically consist of 90 to 98 wt% W in combination with a binder phase, such as nickel, iron, and/or cobalt. For these alloys, tungsten powder is alloyed with 6 to 7 wt% Ni or 4 to 7 wt% Cu and 3 to 5 wt% Fe and then liquid-phase sintered. The microstructure of the as-sintered condition of the tungsten-heavy metal alloy 90W-6Ni-4Cu is shown in Fig. 31.8. Spheroidal particles of tungsten (white and gray) are embedded in a

**Table 31.5 Tungsten alloys**

| | | | | | Typical high-temperature strength | | | | | | | |
|---|---|---|---|---|---|---|---|---|---|---|---|---|
| | | | | Low-temperature ductility class(b) | Temperature | | Tensile strength | | Temperature | | 10 h rupture strength | |
| Nominal alloy additions, wt% | Common designation | Most common product form | Usual condition(a) | | °C | °F | MPa | ksi | °C | °F | MPa | ksi |
| **Tungsten and tungsten alloys** | | | | | | | | | | | | |
| None | Unalloyed W | Wrought bar | SRA | D | 1000 | 1830 | 620 | 90 | ... | ... | ... | ... |
| | | Sheet, wire | SRA | D | 1650 | 3000 | 125 | 18 | 1650 | 3000 | 50 | 7 |
| None | Unalloyed W | CVD sheet | Rx | D | 1000 | 1830 | 565 | 82 | ... | ... | ... | ... |
| Al, K, Si (ppm levels) | Doped AKS W | Wire | CW | C | 1650 | 3000 | 650 | 94 | ... | ... | ... | ... |
| 1$ThO_2$ | W-1%$ThO_2$ | Bar, sheet, wire | SRA | D | 1650 | 3000 | 255 | 37 | ... | ... | ... | ... |
| 2$ThO_2$ | W-2%$ThO_2$ | Bar, sheet, wire | SRA | D | 1650 | 3000 | 205 | 30 | 1650 | 3000 | 125 | 18 |
| 15Mo | W-15Mo | Bar, sheet | SRA | D | 1650 | 3000 | 250 | 36 | 1650 | 3000 | 85 | 12 |
| 4Re | W-4Re | Wire, ribbon | SRA or Rx | C | 1650 | 3000 | 150 | 22 | ... | ... | ... | ... |
| 25Re | W-25Re | All | SRA or Rx | B | 1650 | 3000 | 275 | 40 | 1650 | 3000 | 85 | 12 |
| 4Re, HfC | W-4Re-HfC | Wire | SRA | C | 1650 | 3000 | 620 | 90 | ... | ... | ... | ... |

(a) CW, cold worked; Rx, recrystallized; SRA, stress-relief annealed. (b) B, excellent room-temperature ductility; C, may have marginal ductility at room temperature; D, normally brittle at room temperature. Source: Ref 2 matrix of copper-nickel solid solution (white). These high-density alloys are used for counterweights, balances, radiation shielding, and kinetic energy warheads.

Tungsten has a low resistance to oxidation at temperatures greater than 1120 °C (2050 °F) because of cracking in the oxide layer that allows further oxide penetration. Protective oxidation-resistant coatings are applied in the form of metals, oxides (zirconia), or other compounds (carbides, nitrides, borides, and silicides). Metal coating systems include hafnium-tantalum, nickel-chromium (good to 1000 to 1400 °C, or 1830 to 2550 °F), platinum, and rhodium (good to 1650 °C, or 3000 °F) and usually are applied over a diffusion barrier layer between the tungsten alloy and the coating. Tungsten has good resistance to water and atmospheric corrosion at ambient temperatures and is resistant to many severe environments that readily attack other metals. Tungsten also has good resistance to several liquid metals, including sodium, mercury, gallium, and magnesium, and to oxide ceramics, such as alumina, magnesia, zirconia, and thoria. It is often used for crucibles to melt these materials in an inert atmosphere. Although resistant to hydrogen, tungsten reacts at high temperatures with carbon dioxide, carbon monoxide, nitric oxide, nitrogen dioxide, and sulfur.

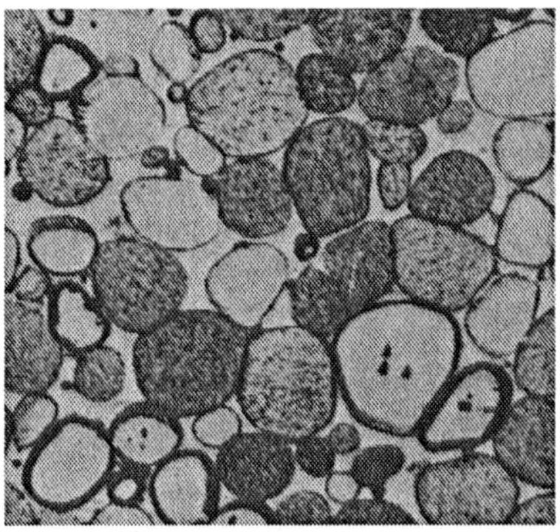

**Fig. 31.8** As-sintered tungsten-heavy alloy. Original magnification: 200×. Source: Ref 1

## 31.5 Rhenium

The scarcity and high cost of rhenium have limited its use. For example, the addition of 3 wt% Re to tungsten wire doubles the cost of the wire. Nevertheless, the unique properties of rhenium make it attractive to use in specialized applications. Rhenium is a strong, ductile, refractory metal with an hcp crystal structure. It has a very high density (21.02 g/cm$^3$, or 0.760 lb/in.$^3$), elastic modulus (460 GPa, or 67 msi), and melting point (3180 °C, or 5755 °F). It has good mechanical stability at elevated temperatures, offering good resistance to thermal shock and wear, and higher creep resistance and strength than the other refractory elements. As the temperature increases from room temperature to 725 °C (1340 °F), the modulus decreases only approximately 20%. In the annealed condition, it has a room-temperature tensile strength of 1170 MPa (170 ksi). It still has a tensile strength of 48 MPa (7 ksi) at a temperature as high as 2710 °C (4910 °F) (Table 31.6).

Compared with other refractory metals, rhenium has superior tensile and creep-rupture strength over a wide temperature range. At temperatures up to 2800 °C (5070 °F) and at high stresses, the rupture life of rhenium is longer than that of tungsten. The high-temperature stress-rupture properties of several refractory metals are shown for comparison in

**Table 31.6 Unalloyed rhenium**

| Nominal alloy additions, wt% | Common designation | Most common product form | Usual condition(a) | Low-temperature ductility class(b) | Typical high-temperature strength: Temperature °C | °F | Tensile strength(c) MPa | ksi | Temperature °C | °F | 10 h rupture strength MPa | ksi |
|---|---|---|---|---|---|---|---|---|---|---|---|---|
| **Rhenium** | | | | | | | | | | | | |
| None | Unalloyed Re(d) | All | Rx | A | 20 | 70 | 1172/2324 | 170/337 | 1300 | 2400 | 55 | 8 |
| | | | | | 500 | 900 | 786/1200 | 114/174 | 2200 | 4000 | 21 | 3 |
| | | | | | 1000 | 1800 | 586/855 | 85/124 | 2800 | 5070 | 4.1 | 0.6 |
| | | | | | 1500 | 2700 | 262/276 | 38/40 | ... | ... | ... | ... |
| | | | | | 2000 | 3600 | .../103 | .../15 | ... | ... | ... | ... |
| | | | | | 2300 | 4200 | 53/... | 7.7/... | ... | ... | ... | ... |

(a) Rx, recrystallized. (b) A, excellent cryogenic ductility. (c) First value, annealed material; second value, wrought material cold worked 15%. (d) Available in both powder metallurgy and arc-cast forms. Source: Ref 2

Fig. 31.9. Of the refractory metals, the melting point of rhenium (3180 °C, or 5755 °F) is second only to tungsten. Because it does not have a DBTT, rhenium retains its ductility from subzero to high temperatures.

Rhenium and its alloys are processed by PM technology. Powders are compacted, vacuum presintered, and then sintered in hydrogen at high temperatures. Fabrication is accomplished by cold working with frequent annealing steps. Rhenium has high cold ductility, but because of its very high work-hardening rate, it requires the use of light deformation passes with frequent intermediate anneals, either stress relief or recrystallization at 1230 to 1630 °C (2240 to 2960 °F). Hot deformation must be carried out in vacuum or hydrogen to prevent hot cracking caused by formation of a low-melting-point oxide that penetrates grain boundaries during hot working in air. The metal catastrophically oxidizes in air at moderately elevated temperatures, forming $Re_2O_7$, which melts at 297 °C (567 °F) and boils at 363 °C (685 °F), evolving as a white cloud. Thus, the metal must be protected from oxidation during processing and during service.

The high work-hardening rate of rhenium at room temperature translates into rapid increases in strength with cold work. For an annealed condition starting with a tensile strength of 1035 MPa (150 ksi), cold working only 10% increases the tensile strength to 1895 MPa (275 ksi); at 20% cold work, the tensile strength increases to 2000 MPa (290 ksi); and with 40% cold work, it increases still further to 2690 MPa (390 ksi). The metal must be given frequent intermediate anneals during working. Crystallographic texturing of the hcp crystal structure and the high work-hardening rate lead to mechanical property anisotropy in processed sheet.

Platinum-rhenium catalysts are the major type of rhenium products and account for approximately 85% of rhenium consumption. Rhenium is also used as an alloying element in tungsten-rhenium and molybdenum-rhenium alloys, which derive their properties from the "rhenium effect." A rhenium addition simultaneously improves strength, plasticity, and weldability, lowers the DBTT of wrought products, and reduces the degree of recrystallization embrittlement. The greatest improvements in properties are obtained with additions of 10 to 26 wt% Re to tungsten and 11 to 50 wt% Re to molybdenum. Rhenium is also a solid-solution-strengthening alloying element in superalloys. A number of directionally solidified and single-crystal nickel-base superalloys contain up to 3 wt% Re. Among the elements, rhenium has a melting point exceeded only by those of tungsten and carbon and a density exceeded only by those of osmium, iridium, and platinum. A DBTT does not exist for the pure metal. It is unique among refractory metals in that it does not form carbides.

A rather unique application for rhenium is for exhaust nozzles for the small, low-thrust engines that boost communication satellites

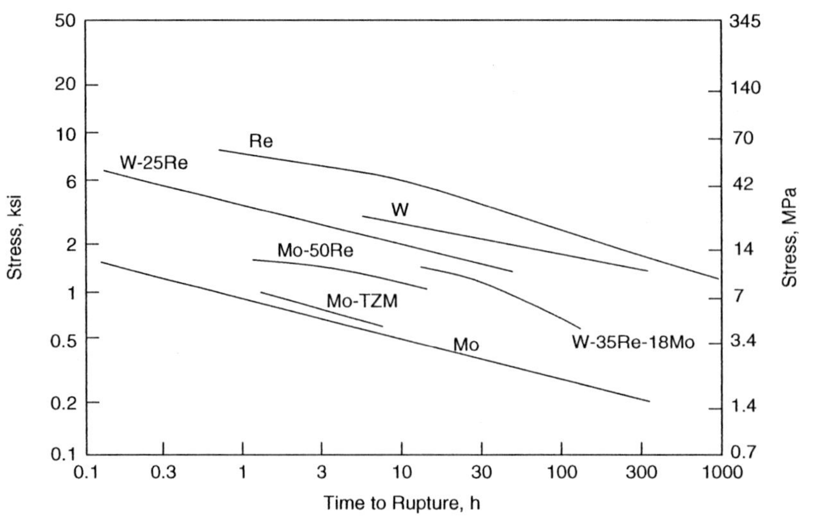


**Fig. 31.9** Stress rupture at 2200 °C (4000 °F) for several refractory metals

from low-Earth orbit into their final geosynchronous orbit. The engines are required to fire for times ranging from 4 to 6 h at exhaust temperatures approaching 1980 °C (3600 °F). These nozzles are first produced by chemically vapor depositing a 50 to 75 μm thick coating of oxidation-resistant iridium onto a graphite mandrel. Then, the 7.6 mm (0.3 in.) thick rhenium nozzle is applied by chemical vapor deposition on the iridium layer. Finally, the graphite mandrel is removed by chemical leaching.

Rhenium is not attacked by molten copper, silver, tin, and zinc. It dissolves readily in molten iron and nickel. It is resistant to hydrochloric acid and shows good resistance to saltwater corrosion and the mechanical effects of erosion. At elevated temperatures, rhenium stands up well in hydrogen and inert atmospheres.

## 31.6 Fabrication

Selection of a specific refractory alloy is often based on fabricability rather than on strength or corrosion resistance. Niobium, tantalum, and their alloys are the most easily fabricated refractory metals. They can be formed, machined, and joined by conventional methods. They are ductile in the pure state and have higher interstitial solubilities for carbon, nitrogen, oxygen, and hydrogen. Because of the higher solubilities in niobium and tantalum, these embrittling contaminants normally do not present problems during fabrication. However, tantalum and niobium dissolve sufficient amounts of oxygen at elevated temperatures to destroy ductility at normal operating temperatures. Therefore, elevated-temperature fabrication of these metals is used only when necessary. Protective coatings or atmospheres are mandatory unless some contamination can be tolerated. The allowable level of contamination, in turn, determines the maximum permissible exposure time in air at elevated temperature.

The cold working properties of niobium are excellent. Niobium is a very ductile metal that can undergo cold reductions of more than 95% without failure. The metal can be easily forged, rolled, or swaged directly from ingot at room temperature. Annealing is necessary after the cross-sectional area has been reduced by approximately 90%. Heat treating at 1205 °C (2200 °F) for 1 h causes complete recrystallization of material cold worked over 50%. The annealing process must be performed either in an inert gas or in a high vacuum. Niobium is well suited to deep drawing. The metal may be cupped and drawn to tube, but special care must be taken with lubrication. Sheet metal can also be easily formed by general sheet metal working techniques.

Tantalum can also be fabricated by bending, roll forming, and welding. It can be cold worked with standard equipment and can undergo cold reductions of more than 95% without failure. It can be rolled, forged, blanked, formed, and drawn. Tantalum can also be resistance welded, electron beam or gas tungsten inert gas welded, brazed, and riveted. Tantalum does have a tendency to stick to tooling during metal forming operations. To avoid this, specific lubricant and die material combinations are required in high-pressure forming operations. Most sheet metal work in tantalum is performed on metal with a thickness ranging from 0.10 to 1.5 mm (0.004 to 0.060 in.). Annealing tantalum is accomplished by heating in a high vacuum to temperatures above 1095 °C (2000 °F).

Molybdenum, molybdenum alloys, tungsten, and tungsten alloys require special fabrication techniques. Fabrication involving mechanical working should be performed below the recrystallization temperature. These materials have very limited solubilities for carbon, nitrogen, oxygen, and hydrogen. Because the residual levels of these elements required to avoid embrittlement are impractically low, the microstructure must be controlled to ensure a sufficiently low DBTT.

Molybdenum can be formed using conventional processes and tooling, but fabrication must be done above the DBTT and below the recrystallization temperature (1150 °C, or 2100 °F). The rate of working and stress concentrations are important because the DBTT is shifted upward with an increase in the strain rate. Welding must be performed in an inert atmosphere or a vacuum. Weldments must be immediately stress relieved after welding to prevent cracking in the heat-affected zone.

Tungsten is very difficult to machine and fabricate. With experience, it can be turned. Milling can only be done with great difficulty and high cost by those most experienced with it. Forming must be done at very high temperatures and with careful stress relieving. Welding is not recommended, and riveting is difficult at best.

The best method for machining is electrical discharge machining.

## 31.7 Refractory Metal Protective Coatings

Despite their high melting temperatures, the refractory metals react with oxidizing atmospheres at moderate temperatures, and this restricts their use in high-temperature applications. Tantalum oxidizes in air above approximately 300 °C (570 °F), niobium above 425 °C (800 °F), molybdenum and tungsten above 500 °C (930 °F), and rhenium above 600 °C (1110 °F). Oxidation-resistant protective coatings include intermetallic compounds, such as silicides and aluminides that form compact or glassy oxide layers, alloys that form compact oxide layers, noble metals that resist oxidation, and stable oxides that provide a physical barrier to the penetration of oxygen.

Although coating systems have been developed for all of the refractory metals and their alloys, most of the research has been directed toward the lower-density niobium alloys used for aerospace. Niobium and its alloys are the only refractory metals for which large parts have been coated to prevent oxidation in high-temperature service. During the Apollo space program, processes were developed for applying slurry coatings of complex aluminides and subsequently silicides. The most successful coating is Si-20Cr-20Fe, made using elemental powder suspended in a lacquer along with a gelling agent. Other variants contain hafnium silicide, which gives the final coating a higher remelt temperature. Methods for applying the slurry include dipping, spraying, and touch-up painting. Following application of approximately 0.076 mm (0.003 in.) of slurry per side, the coating is heated to 1300 to 1400 °C (2370 to 2550 °F) for reaction bonding and diffusion. Silicide coatings enable niobium to be used at temperatures up to 1650 °C (3000 °F). Compared with the base metal, the coated metal has lower strength and ductility, higher emissivity, and increased weight.

### ACKNOWLEDGMENTS

Sections of this chapter were adapted from "Refractory Metals and Alloys" in *Metals Handbook Desk Edition,* 2nd ed., ASM International, 1998.

### REFERENCES

1. R.N. Caron and J.T. Staley, Effects of Composition, Processing, and Structure on Properties of Nonferrous Alloys, *Materials Selection and Design,* Vol 20, *ASM Handbook,* ASM International, 1997
2. Refractory Metals and Alloys, *Metals Handbook Desk Edition,* 2nd ed., ASM International, 1998
3. J.L. Walter, *Alloying,* ASM International, 1988, p 422
4. Metallography and Microstructures of Refractory Metals and Alloys, *Metallography and Microstructures,* Vol 9, *ASM Handbook,* ASM International, 2004

### SELECTED REFERENCES

- J.R. Davis, *Alloying: Understanding The Basics,* ASM International, 2001
- J.B. Lambert, Refractory Metals and Alloys, *Properties and Selection: Nonferrous Alloys and Special-Purpose Materials,* Vol 2, *ASM Handbook,* ASM International, 1990
- B.J. Moniz, *Metallurgy,* American Technical Publishers, Inc., 1994
- A.M. Russell and K.L. Lee, *Structure-Property Relationships in Nonferrous Alloys,* Wiley-Interscience, 2005
- W.F. Smith, *Structure and Properties of Engineering Alloys,* 2nd ed., McGraw-Hill, Inc., 1993

## CHAPTER 32

# Miscellaneous Nonferrous Metals

A NUMBER of miscellaneous nonferrous metals and their alloys are covered in this chapter. Zirconium and hafnium are reactive metals that are used in nuclear reactors. Beryllium is an extremely lightweight metal that, as a result of its high specific modulus, is used in high-value aerospace structures. Lead and tin form the most widely used solders, and lead is the most important of the battery materials. Gold, silver, and the platinum group are all precious metals that, in addition to being measures of wealth, have important industrial applications. Finally, the fusible alloys are a group of over 100 alloys that have melting points lower than that of lead-tin eutectic solder.

## 32.1 Zirconium

Zirconium is a reactive metal similar to titanium, but it has a 50% higher density. Zirconium has excellent resistance to many corrosive media, including superheated water, and is almost transparent to thermal energy neutrons. These properties, combined with its good mechanical properties, make zirconium very useful in nuclear power applications, especially as cladding for uranium fuel and for other reactor components.

Zirconium and most of its alloys exhibit strong anisotropy because of its room-temperature, hexagonal close-packed (hcp) $\alpha$ phase which undergoes an allotropic transformation to a body-centered cubic (bcc) $\beta$ phase structure at approximately 870 °C (1600 °F). The most common alloys are rather dilute $\alpha$ alloys with characteristics generally similar to those of unalloyed zirconium. Zirconium work hardens rapidly and reaches its maximum strength and hardness after only 20% cold working. Zirconium is highly anistropic in the cold-worked condition and moderately anistropic in the fully recrystallized condition. The yield strength is higher in the transverse direction, while the tensile strength is higher in the longitudinal direction. Recrystallization begins at approximately 510 °C (950 °F).

The most common zirconium alloys are Zircaloy-2 and Zircaloy-4. Except for being somewhat stronger and less ductile than unalloyed grades, the Zircaloys are quite similar to unalloyed zirconium in metallurgical behavior. Zircaloy-2 (Zr-1.5Sn-0.15Fe-0.10Cr-0.05Ni) is a zirconium-tin alloy with small amounts of iron, chrome, and nickel that is highly resistant to hot water and steam. Zircaloy-4 (Zr-1.5Sn-0.20Fe-0.10Cr) is a variation of Zircaloy-2, but it contains no nickel and has a greater but more closely controlled iron content. It absorbs less hydrogen than Zircaloy-2 when exposed to corrosion in water and steam. Both alloys are widely used in nuclear reactors. The Zircaloys are forged in the $\beta$ region at 1065 °C (1950 °F), water quenched, and then hot worked below 845 °C (1550 °F) in the $\alpha$ region. Hot working in this region produces a thin, uniform distribution of intermetallics in the matrix microstructure that is completely $\alpha$.

The only other zirconium alloy that has significant commercial importance is Zr-2.5Nb. The mechanical and physical properties of Zr-2.5Nb are very similar to those of the Zircaloys. Zr-2.5Nb has high corrosion resistance to superheated water and is heat treatable to a higher level of strength than that which can be achieved with the Zircaloy alloys. It was developed for use in the pressure tubes of heavy-water reactors. Nonnuclear grades, also referred to as industrial alloys, are used primarily in chemical processing equipment.

Zirconium is a reactive metal and has a high affinity for oxygen. When zirconium is exposed to an oxygen-containing environment, an adherent protective oxide film forms on its surface. This film is formed spontaneously in air or water at room temperature and below. As a result, zirconium is very resistant to corrosive attack in most mineral and organic acids, strong alkalis, saline solutions, and some molten salts. Zirconium is not attacked by oxidizing media unless halides are present. Like titanium, it is susceptible to the formation of embrittling hydrides when exposed to hydrogen at elevated temperatures.

## 32.2 Hafnium

Hafnium, like titanium and zirconium, is a reactive metal but has a higher melting point (1760 °C or 3200 °F), a higher hcp-to-bcc transformation temperature, and a much higher density (13.09 g/cm$^3$, or 0.48 lb/in.$^3$). It is also much more scarce and thus much more expensive. Hafnium has a high nuclear absorption cross section for thermal energy neutrons; thus, it is highly opaque to thermal neutrons, in contrast to the very low absorption cross section of zirconium, and is used as nuclear reactor control rods. Previously, the principal application of hafnium was as a neutron absorption material. However, more hafnium is currently used as an alloying element in superalloys than in reactor applications. Hafnium is a grain-boundary strengthener in superalloys where 1.5 to 2 wt% Hf is added to directionally solidified superalloys to prevent grain-boundary cracking.

Like other reactive metals, hafnium resists attack by many chemicals because of the thin, tenacious layer of oxide that forms naturally on the surface. The corrosion properties are unaffected as long as this thin layer is not penetrated by reactants at increasing temperatures. Hafnium is superior to zirconium and Zircaloy alloys in corrosion resistance in water and steam, molten alkali metals, and air at temperatures below 400 °C (750 °F). However, at higher temperatures, hafnium begins to react appreciably with oxygen, nitrogen, and hydrogen. The oxides and nitrides formed at these elevated temperatures tend to remain at the surfaces; however, hydrogen diffuses rapidly and forms hydrides throughout the metal. Oxides, nitrides, and hydrides all lead to impaired ductility.

## 32.3 Beryllium

Beryllium has a very low density (1.848 g/cm$^3$, or 0.067 lb/in.$^3$), a moderately high melting point (1283 °C, or 2341 °F), and a high elastic modulus (303 GPa, or 44 msi). It has good electrical conductivity (41% IACS, or International Annealed Copper Standard) and thermal conductivity (210 W/m·K, or 125 Btu/ft·h·°F). The specific modulus (elastic modulus/density) is higher than that of aluminum, magnesium, or titanium. The high specific modulus, combined with low thermal expansion, translates into good physical stability and low distortion during inertial or thermal stressing, making beryllium useful for critical aerospace components. Its exceptionally good dimensional stability and ability to reflect infrared radiation makes beryllium well suited for mirrors, either as polished or plated substrate. The near transparency of beryllium to x-rays and other high-energy electromagnetic radiation is responsible for its widespread use in foil gage as windows in x-ray tubes and radiation-detection devices.

While beryllium can be melted and cast, the resulting casting has coarse grains that are difficult to process, and attempts to refine them by alloying have been largely unsuccessful. A fine grain size is produced primarily by powder metallurgy (PM) techniques, from which it inherits the fine grain size (5 to 15 μm). Powder consolidation is done by vacuum hot pressing, hot isostatic pressing, or cold isostatic pressing followed by vacuum sintering. Beryllium powder and dust are toxic, which further increases its cost, since the manufacturing environment must be carefully controlled.

Consolidated beryllium shows only modest ductility (~3% tensile elongation) at room temperature, which is attributed to a large covalent bonding component in its atomic bonding in the *c*-axis direction and to its hcp crystal structure, which is limited at room temperature to only one slip direction on two crystal planes, the basal and prism. This low ductility is not improved even when impurities are reduced to low levels. However, beryllium does show improved ductility at temperatures above 205 to 260 °C (400 to 500 °F), where more crystal deformation modes become available. For example, at 400 °C (750 °F), it has a tensile elongation as high as 50%. Beryllium transforms to a bcc phase at 1270 °C (2318 °F), only 13 °C (23 °F) below its melting point.

The properties of a number of pure commercial grades are given in Table 32.1. Commercially available grades of pure beryllium are distinguished by impurity levels and BeO content. The metal normally contains 0.7 to

**Table 32.1 Properties of several pure beryllium grades**

| Property | Grade | | | | | | |
|---|---|---|---|---|---|---|---|
| | S-200F | S-200FH | S-200FC | S-65 | I-70H | I-220H, Gr. 1 | O-30 |
| Ultimate tensile strength, MPa (ksi) (spec. min.) | 324 (47) | 414 (60) | 261 (38) | 289 (42) | 345 (50) | 448 (65) | 400 (58) |
| Yield strength, MPa (ksi) (spec. min.) | 241 (35) | 296 (43) | 171 (25) | 206 (30) | 207 (30) | 345 (50) | 297 (43) |
| Modulus, MPa (ksi) (spec. min.) | 303 (44) | 303 (44) | 303 (44) | 303 (44) | 303 (44) | 303 (44) | 303 (44) |
| Elongation at failure, % (spec. min.) | 2 | 3 | 2 | 3 | 2 | 2 | 3 |
| Precision elastic limit (PEL), MPa (ksi) | ... | 27 (4) typical | ... | ... | 3 (0.4) | 6 (0.1) | 2.5 (0.36) |
| Beryllium assay, % (min) | 98.5 | 98.5 | 98.5 | 99 | 99 | 98 | 99 |
| BeO, % (max) | 1.5 | 1.5 | 1.5 | 1.0 | 0.7 | 2.2 | 0.5 |
| Coefficient of thermal expansion at 25 °C, ppm/ °C (ppm/ °F) | 11.4 (6.3) | 11.4 (6.3) | 11.4 (6.3) | 11.4 (6.3) | 11.4 (6.3) | 11.4 (6.3) | 11.4 (6.3) |
| Thermal conductivity at 25 °C, W/m · K (Btu · ft/ft$^2$ · hr · °F) | 216 (125) | 216 (125) | 216 (125) | 216 (125) | 216 (125) | 216 (125) | 216 (125) |
| Electrical conductivity at 20°C, %IACS | 40.7 | 40.7 | 40.7 | 40.7 | 40.7 | 40.7 | 40.7 |
| Specific heat at 20 °C, J/kg · K (Btu/lb · °F) | 1925 (0.46) | 1925 (0.46) | 1925 (0.46) | 1925 (0.46) | 1925 (0.46) | 1925 (0.46) | 1925 (0.46) |
| Density, g/cm$^3$ (lb/in.$^3$) | 1.85 (0.067) | 1.85 (0.067) | 1.85 (0.067) | 1.85 (0.067) | 1.85 (0.067) | 1.85 (0.067) | 1.85 (0.067) |
| Latent heat of fusion, Btu/lb | 560 | 560 | 560 | 560 | 560 | 560 | 560 |
| Melting point, °C (°F) | 1285 (2345) | 1285 (2345) | 1285 (2345) | 1285 (2345) | 1285 (2345) | 1285 (2345) | 1285 (2345) |
| Sonic velocity, M/s (ft/s) | 12,588 (41,300) | 12,588 (41,300) | 12,588 (41,300) | 12,588 (41,300) | 12,588 (41,300) | 12,588 (41,300) | 12,588 (41,300) |
| Fracture toughness, $K_{IC}$, MPa $\sqrt{in.}$ (ksi$\sqrt{in.}$) | 10.6–12.3 (9–12) | 10.6–12.3 (9–12) | ... | 10.6–12.3 (9–12) | 10.6–12.3 (9–12) | 10.6–12.3 (9–12) | 10.6–12.3 (9–12) |
| Powder | Impact ground | Impact ground | Impact ground | Impact ground | Impact ground | Impact ground | Gas-atomized, spherical |
| Consolidation process | VHP(a) | HIP(b) | CIP(c) | VHP(a) | HIP(b) | HIP(b) | HIP(b) |
| Isotropy | Slight anisotropy | Isotropic | Slight anisotropy | Slight anisotropy | Isotropic | Isotropic | Superior isotropy |
| Typical uses | Structural | Structural high isotropy | Scanning mirrors | Fusion reactors, nuclear | Optical | High PEL/high strength | Optical |

(a)VHP, vacuum hot pressed. (b)HIP, hot isostatically pressed. (c)CIP, cold isostatically pressed and sintered. Courtesy of Brush Wellman Engineered Materials

4.25 wt% BeO, which is unavoidable and increases with the fineness of the beryllium powder used in the consolidation process. It has extremely low solubility for oxygen and readily forms oxide particles that are situated at grain boundaries, which helps to control grain growth without being harmful to ductility below 1.2% BeO. Powder compaction is normally conducted either by vacuum hot pressing (VHP) or hot isostatic pressing (HIP). The VHP process involves loading powder into a graphite die and vibrating. The loaded die is placed into a vacuum hot press, a vacuum is established, and the powder is consolidated under 3.4 to 14 MPa (500 to 2000 psi) at temperatures between 1000 and 1100 °C (1830 and 2020 °F). Densities in excess of 99% of theoretical are obtained in diameters of 20 to 183 mm (8 to 72 in.). The HIP process is similar except that the powder must be enclosed in a mild steel can prior to consolidation. After canning, the can is degassed under vacuum at 595 to 705 °C (1100 to 1300 °F) to remove all air and gases absorbed on the particle surfaces. The can is then sealed and put through the HIP consolidation cycle at 103 MPa (15 ksi) and 760 to 1100 °C (1400–2010 °F). Although HIP is generally more expensive than VHP, it allows better control of grain size because it allows greater latitude in temperature selection. Also, HIP is capable of attaining 100% of theoretical density. The improvement in anisotropy for the HIP-processed material is primarily a result of being able to use lower processing temperatures to keep the grain size fine and apply uniform and high pressure in all directions during consolidation. After powder consolidation, beryllium can be processed by conventional metalworking technologies, such as rolling and extrusion. The mechanical properties are improved by metalworking in the predominant direction of metal flow. Biaxial and/or triaxial deformation can be used to obtain more balanced properties.

Beryllium is too reactive to alloy with most common metals because it reacts to form stable compounds. Most of the research into beryllium-base alloys has been conducted on the beryllium-aluminum system. An example is the alloy AlBeMet AM162, which contains 62% Be and 38% Al by weight. The alloy is made by PM followed by cold isostatic pressing to reach approximately 80% of its final density. The material is then rolled or extruded to its final shape and density. With its low density (2.10 g/cm$^3$, or 0.076 lb/in.$^3$) and high elastic modulus (200 GPa, or 29 msi), AlBeMet AM162 is 3.9 times stiffer than aluminum alloy 7075-T73 and titanium alloy Ti-6Al-4V in tension tie applications. When compared to unalloyed beryllium, AlBeMet AM162 offers improved ductility (10% elongation) with a yield strength of 283 MPa (41 ksi) and a tensile strength of 338 MPa (49 ksi). The disadvantages of AlBeMet AM162 are the limitations on extrusion and rolled sheet sizes available and the high cost (approximately 400 times the cost of aluminum). While this alloy is somewhat more difficult to machine than aluminum (20% lower speeds and feeds), it is easier to machine than titanium. Like all beryllium-containing alloys, the machining process needs to be controlled due to the hazardous nature of beryllium. Physical properties of AlBeMet AM162 and several other beryllium alloys are summarized in Table 32.2.

**Table 32.2 Properties of selected beryllium alloys**

| Property | Beryllium S200F/AMS7906 | AlBeMet AM162H/AMS7911 | AlBeMet AM140 (Sheet/Ext) | E-Material E-60 |
|---|---|---|---|---|
| Density, g/cm$^3$ (lb/in.$^3$) | 1.85 (0.067) | 2.10 (0.076) | 2.28 (0.082) | 2.51 (0.091) |
| Modulus, GPa (msi) | 303 (44) | 193 (28) | 150 (22) | 331 (48) |
| Ultimate tensile strength, GPa (ksi) | 324 (47) | 262 (38) | 276 (40) | 273 (39.3) |
| Yield strength, GPa (ksi) | 241 (35) | 193 (28) | 186 (27) | NA |
| Elongation, % | 2 | 2 | 14/16 | <0.05 |
| Fatigue strength, MPa (ksi) | 261 (39) | 97 (14) | 145 (2.1)/97 (14) | NA |
| Thermal conductivity, W/m · K (Btu/h · ft · °F) | 216 (125) | 210 (121) | 204 (119) | 240 (138) |
| Heat capacity, J/g · °C (Btu/lb · °F) | 1.95 (0.46) | 1.56 (0.373) | 1.416 (0.334) Est. | 1.26 (0.310) |
| Coefficient of thermal expansion, ppm/ °C (ppm/ °F) | 11.3 (6.3) | 13.9 (7.7) | 16.5 (9.1) | 6.1 (3.4) |
| Electrical resistivity, ohm · cm | $4.2 \times 10^{-6}$ | $3.5 \times 10^{-6}$ | To be determined | NA |
| Chemical composition | Commercially pure | 62 wt% Be, 38 wt% Al | 40 wt% Be, 60 wt% Al | Metal-matrix composite; 27–31 at % Be, 69–73 at % BeO |

NA, not applicable

Beryllium forms a thin protective oxide (BeO) coating that provides good atmospheric corrosion resistance at room temperature and excellent resistance to pure water up to 300 °C (570 °F). However, corrosion resistance is impaired by impurities. The presence of carbides or chlorides can cause a reaction with moist air at room temperature. Above 760 °C (1400 °F), beryllium reacts with oxygen and nitrogen but not at all with hydrogen at any temperature. It is attacked by many dilute common acids at room temperature. It can be anodized or chromate coated for improved corrosion resistance.

The main concern associated with handling beryllium is the effect on the lungs when excessive amounts of respirable beryllium powder or dust are inhaled. Sensitive individuals exposed to airborne beryllium may develop chronic beryllium disease (CBD), an affliction of the lungs formerly known as berylliosis. Users should comply with occupational safety and health standards applicable to beryllium in Title 29, Part 1910, Code of Federal Regulations. The current Occupational Safety and Health Administration (OSHA) permissible exposure limit (PEL) for beryllium is 2 $\mu g/m^3$ during an 8 h shift. There is an effort underway to reduce the PEL to 0.2 $\mu g/m^3$ over an 8 h shift. Individual susceptibilities to CBD differ, so the level constituting excessive exposure varies.

## 32.4 Lead

Lead is characterized by a number of high properties—density (11.32 $g/cm^3$, or 0.41 $lb/in.^3$), malleability, lubricity, electrical conductivity, and coefficient of thermal expansion—and some low properties—strength, creep resistance, elastic modulus, elastic limit, hardness, and melting point (327 °C, or 621 °F). The toxicity of lead has long been known, but it was not until the 20th century that the full extent of its danger was appreciated. Lead poisoning can result from drinking contaminated water, inhalation of dust, skin contact with lead or lead compounds, and ingestion of contaminated food or paint chips. Lead can severely damage the brain and kidneys, causing permanent neurological damage, weakness and tremor in the extremities, personality disorders, paralysis, coma, and even death. A particular danger is to children who are attracted to the sweet taste of lead-base paint chips; therefore, lead should be removed from homes/structures to avoid ingestion by humans, particularly young children.

The most important applications of lead and lead alloys are lead acid storage batteries, solders, cable sheathing, and building construction materials. Other important applications include x-ray shielding, counterweights, bearings, ammunition, type metal, terneplate, and foil. In addition, lead is used as an alloying element in steel and copper alloys to improve machinability, and it is used in low-melting-point (fusible) alloys. Increasingly, selenium is replacing lead content in free-machining steels.

The principal use of lead is in the manufacture of lead acid storage batteries. Nearly 85% of the lead consumed in the United States is used for battery applications. Battery grids consist of a series of plates made from either cast or wrought calcium-lead or antimony-lead that are pasted with a mixture of lead oxides and immersed in a concentrated electrolyte solution of sulfuric acid.

Lead-tin solders are among the most widely used of all joining materials. When used as a solder, it is normally alloyed with varying amounts of tin to produce solders with different melting points. For environmental and health reasons, there is a considerable amount of effort being expended in developing lead-free solders. Lead-tin solders can be obtained with melting temperatures as low as 180 °C (360 °F) and as high as 315 °C (600 °F). Industrial solder alloys include a wide variety of material combinations, from 100% Pb to 100% Sn. The reader may want to refer to Section 6.3.2 in Chapter 6, "Phase Diagrams," for a discussion of solders in relation to the lead-tin eutectic phase diagram. Silver can be added to lead-tin solders, yielding solders with good tensile, creep, and shear strengths.

Lead is very soft and is often alloyed with antimony, tin, calcium, and arsenic to give it greater hardness. Terneplate, a roofing and weather-sealing material, is fabricated by hot dipping steel strip into lead with 3 to 15 wt% Sn. Babbitts are alloys of lead, antimony, and tin that are used for sliding bearing applications. Arsenic additions provide higher strength.

## 32.5 Tin

Like lead, tin has a low melting point (230 °C, or 450 °F). Tin has a bcc structure at room temperature but transforms to a brittle diamond cubic structure when cooled below 13 °C (56 °F). While the transformation is so sluggish that it normally is not a problem, if it does occur, the volume expansion that occurs with the transformation is so great that the alloys can disintegrate into powder. This very thing happened to Napoleon's army in 1812 when they were retreating from Russia during a particularly cold winter. The cast tin buttons on the soldiers' uniforms disintegrated, adding to their misery. In all, over 500,000 of Napoleon's troops perished during their retreat in the harsh Russian winter.

The three primary uses of tin are for tinplating of steel for corrosion resistance, as a main alloying element in solders, and for bearing alloys. The largest single use of tin is for tinplating of steel, accounting for approximately 40% of the worldwide consumption. Tinplating is applied by electrodeposition on continuous strips of rolled steel. The largest use in the United States is for solders. Tin is often alloyed with lead, which produces a series of solders that melt at lower temperatures. Additions of antimony and silver are used to increase the strength of these solders.

Pewter is a tin-base white metal containing antimony and copper. Originally, pewter was an alloy of tin and lead, but to avoid toxicity and dullness of finish, lead is not used in modern pewter. Modern compositions contain 1 to 8 wt% Sb and 0.25 to 3.0 wt% Cu. Pewter casting alloys usually have less copper than wrought alloys because copper reduces fluidity. Pewter is malleable and ductile and does not require annealing during fabrication. Typical pewter products include coffee and tea services, trays, steins, mugs, bowls, plates, vases, and candlesticks.

Tin has a low coefficient of friction and is useful as a bearing material. Since tin is not very strong, it is alloyed with copper and antimony for increased hardness, tensile strength, and fatigue resistance. Normally, the quantity of lead in these alloys, called tin-base babbitts, is limited to 0.35 to 0.5 wt% to avoid formation of the tin-lead eutectic that significantly reduces strength properties. Lead-base babbitts, which contain up to 10 wt% Sn and 12 to 18 wt% Sb, are not as strong as tin-base babbitts but cost less. In addition to the tin-base and lead-base babbitts, there is a series of intermediate lead-tin bearing alloys. These alloys have tin and lead contents between 20 and 65 wt% plus various amounts of antimony and copper. Increasing the tin content provides higher hardness and greater ease of casting. Bearing alloys must maintain a balance between softness and strength. Aluminum-tin bearing alloys represent a compromise between the requirements for high fatigue strength and the need for good surface properties, such as softness, seizure resistance, and embeddability. Aluminum-tin bearing alloys will support significantly higher loads than the tin- or lead-base bearing alloys.

## 32.6 Gold

Gold is a bright yellow metal that is very dense, very malleable, highly ductile, and has high thermal and electrical conductivity. Gold is not attacked by common acids and is not oxidized by air at any temperature. Thus, gold does not tarnish. Gold is scarce; the world's annual production of gold could be melted into a single cube approximately 5 m (15 ft) on a side. The primary uses of gold are for coinage, jewelry, dentistry, and semiconductor leads. For jewelry, it is commonly alloyed with copper, zinc, nickel, and silver. Nickel additions are used to make white gold.

Gold and its alloys are the most popular of the jewelry metals, as they have been for thousands of years. Jewelry alloys are classified by their karat ratings, generally 10k gold (42 wt% Au), 14k gold (58 wt% Au), 18k gold (75 wt% Au), and 24k gold (100 wt% Au). The elements that make up the balance of these alloys are silver and the base metals copper, zinc, and nickel. White gold consists of a group of Au-Ni-Cu alloys containing approximately 80 wt% Au.

Gold is used in electronic devices, particularly in printed circuit boards, connectors, keyboard contactors, and miniaturized circuitry. Because electronic devices employ low voltages and currents, it is important that the coated components remain completely free from tarnish films and that they remain chemically and metallurgically stable for the life of the equipment.

Gold brazing alloys are used for joining jet and rocket engine components made from nickel- or cobalt-base superalloys or stainless steel. Joint ductility is high because weld metal reactions with the base metal are minimized. They are also used for joining hermetically sealed vacuum tube components, where their low vapor pressure is advantageous. The common brazing alloys include gold-copper alloys, gold-nickel alloys, and gold-nickel-palladium alloys.

## 32.7 Silver

Silver is a bright white metal that is very soft and malleable. Pure silver has the lowest electrical resistivity of any metal and is an excellent conductor of heat. It is the least expensive of the precious metals. Silver is also widely used for jewelry. Until recently, it was a coinage metal, but the rising cost of silver has led to it being replaced with lower-cost metals. Like gold, silver is used in the electronics industry where high electrical conductivity is required. Another important use for silver is silver solders, which have higher melting points and greater strengths than the lead-base solders. It is resistant to corrosion in many organic acids and in sodium and potassium hydroxide but is susceptible to corrosion in mineral acids. Silver resists oxidation in dry and moist air environments at room temperature; a thin oxide film may form at slightly elevated temperatures. Above approximately 455 °C (850 °F), the silver oxide dissociates, and the surface is clean once again. Silver is susceptible to attack by sulfur and ozone.

The high reflectivity of silver makes it particularly attractive for applications in jewelry and tableware. Strength, hardness, and wear resistance are obtained through the use of alloys, in particular sterling silver, which contains 92.5 wt% Ag and 7.5 wt% Cu. Silver alloys containing copper (e.g., Ag-10Cu) are traditional coinage alloys.

Silver-base brazing filler metal alloys are characterized by low melting temperatures and their ability to wet solid base metals; they are used for brazing steel, cast iron, stainless steel, and copper alloys. They are also used to braze some reactive and refractory metals, although their use restricts the upper service temperature of the joint.

Silver is used for medium- and heavy-duty electrical contacts, particularly when alloyed with a dispersion of cadmium oxide. Here, advantage is taken of the high thermal and electrical conductivity and low surface-contact resistance of silver. The cadmium oxide particles prevent sticking and welding and minimize arc erosion. Dental amalgam is a silver-mercury alloy for restoring lost tooth structure. In this application, it is commonly referred to as a silver filling.

## 32.8 Platinum Group

The platinum group metals include ruthenium, rhodium, palladium, osmium, iridium, and platinum. All are by-products of copper, gold, and silver refining. The platinum group metals are dense, refractory, and oxidation resistant. They are among the least abundant elements in the Earth's crust, which makes them expensive. They have close-packed structures (face-centered cubic or hexagonal close-packed), relatively good thermal and electrical conductivities, and good oxidation and corrosion resistance. Platinum and palladium are ductile and easily fabricated; rhodium and iridium have limited ductility; and ruthenium and osmium are hard, brittle materials with high elastic moduli.

**Platinum** is the most abundant of the platinum group metals and is also the most widely used. Applications are based on its corrosion resistance, high melting point, ductility, electronic properties, and overall appearance. Platinum remains bright when heated in air at all temperatures up to the melting point and is soluble only in acids that generate free chlorine. The most common application for platinum and its alloys is as a catalyst. Platinum is an outstanding catalyst for oxidation, as in the production of $H_2SO_4$ and $HNO_3$; for hydrogenation, as in the production of vitamins and other chemicals; in the petroleum reforming process, as in the production of high-octane gasoline; and automobile exhaust converter systems. A more visible use of platinum, which is even more precious than gold, is for jewelry.

**Palladium** is similar to platinum in its appearance, ductility, and strength but has a lower melting point, lower corrosion/tarnish resistance, and a much lower density. Overall, it is the second most widely used of the platinum group elements, with applications for electrical contacts. In the automotive industry, palladium is employed as a component of catalyst alloys; palladium alloy catalysts also have applications in the chemical and pharmaceutical industries. Other uses of palladium alloys are in medical and dental materials, jewelry, and brazing alloys. It can be electroplated, electroformed, and deposited by electroless methods.

**Rhodium, iridium, ruthenium, and osmium,** the remaining platinum-group metals, are all hard white metals that have little ductility at room temperature. They are used primarily as alloying elements in other precious metals.

## 32.9 Fusible Alloys

Fusible, or low-melting-point, alloys include binary, ternary, quaternary, and quinary alloys containing bismuth, lead, tin, cadmium, and indium. The term *fusible alloy* refers to any of the more than 100 white metal alloys that melt at relatively low temperatures, that is, below the melting point of tin-lead eutectic solder (180 °C, or 360 °F). The melting points of these alloys are as low as 47 °C (116 °F).

Many of the fusible alloys used in industrial applications are based on eutectic compositions (Table 32.3). At room temperature, such an alloy has sufficient strength to hold parts together at a specific elevated temperature; however, the fusible alloy link will melt, thus disconnecting the parts. In fire sprinklers, the links melt when dangerous temperatures are reached, releasing water from piping systems and extinguishing the fire. Boiler plugs and furnace controls operate similarly; an increase in temperature beyond the safety limits of the furnace or boiler operation will melt the plug. When the fusible alloy link melts, pressure (or heat) in the boiler is dissipated or delivery of fuel is interrupted, allowing the temperature to return to a safe level. In addition to eutectic alloys, each of which melts at a specific temperature, there are numerous noneutectic fusible alloys that melt over a range of temperatures (Table 32.4).

In certain alloys, normal thermal contraction due to cooling after solidification can be partly, completely, or more than compensated for by thermal expansion during aging. For example, bismuth alloys containing 33 to 66 wt% Pb exhibit a net expansion after solidification and subsequent aging. Some fusible alloys show no contraction (shrinkage) and expand rapidly while still warm; others show slight shrinkage during the first few minutes after solidification and then begin to expand; in still others, expansion does not commence until some time after the fusible alloy casting has cooled to room temperature.

The expansion characteristics (net expansion, net contraction, and little or no volume change) have consequences in the casting and machining of the alloys. For example, a wood pattern used for making molds must have somewhat greater dimensions than those desired in the casting to compensate for shrinkage of the casting on solidification and during cooling to room temperature. Where metal patterns are cast from a master wood pattern, these two allowances must be made unless the alloy used for the metal

**Table 32.3 Compositions and melting temperatures of selected eutectic fusible alloys**

| Composition, % | | | | | Melting point | |
|---|---|---|---|---|---|---|
| Bi | Pb | Sn | Cd | Other | °C | °F |
| 44.70 | 22.60 | 8.30 | 5.30 | 19.10 In | 47 | 117 |
| 49.00 | 18.00 | 12.00 | ... | 21.00 In | 58 | 136 |
| 50.00 | 26.70 | 13.30 | 10.00 | ... | 70 | 158 |
| 51.60 | 40.20 | ... | 8.20 | ... | 91.5 | 197 |
| 52.50 | 32.00 | 15.50 | ... | ... | 95 | 203 |
| 54.00 | ... | 26.00 | 20.00 | ... | 102.5 | 217 |
| 55.50 | 44.50 | ... | ... | ... | 127 | 255 |
| 58.00 | ... | 42.00 | ... | ... | 138.5 | 281 |
| ... | 30.60 | 51.20 | 18.20 | ... | 142 | 288 |
| 60.00 | ... | ... | 40.00 | ... | 144 | 291 |
| ... | ... | 67.75 | 32.25 | ... | 177 | 351 |
| ... | 38.14 | 61.86 | ... | ... | 183 | 362 |
| ... | ... | 91.00 | ... | 9.00 Zn | 199 | 390 |
| ... | ... | 96.00 | ... | 3.50 Ag | 221 | 430 |
| ... | 79.7 | ... | 17.7 | 2.60 Sb | 236 | 457 |
| ... | 87.0 | ... | ... | 13.00 Sb | 247 | 477 |

Source: Ref 1

**Table 32.4 Compositions and melting points of selected noneutectic fusible alloys**

| Composition, % | | | | | Yield temperature | | Melting range | |
|---|---|---|---|---|---|---|---|---|
| Bi | Pb | Sn | Cd | Other | °C | °F | °C | °F |
| 50.50 | 27.8 | 12.40 | 9.30 | ... | 70.5 | 159 | 70–73 | 158–163 |
| 50.00 | 34.5 | 9.30 | 6.20 | ... | 72 | 162 | 70–79 | 158–174 |
| 50.72 | 30.91 | 14.97 | 3.40 | ... | 72.5 | 163 | 70–84 | 158–183 |
| 42.50 | 37.70 | 11.30 | 8.50 | ... | 72.5 | 163 | 70–90 | 158–194 |
| 35.10 | 36.40 | 19.06 | 9.44 | ... | 75 | 167 | 70–101 | 158–214 |
| 56.00 | 22.00 | 22.00 | ... | ... | 96 | 205 | 95–104 | 203–219 |
| 67.00 | 16.00 | 17.00 | ... | ... | 96 | 205 | 95–149 | 203–300 |
| 33.33 | 33.34 | 33.33 | ... | ... | 101 | 214 | 101–143 | 214–289 |
| 48.00 | 28.50 | 14.50 | ... | 9.00 Sb | 116 | 241 | 103–227 | 217–440 |
| 40.00 | ... | 60.00 | ... | ... | 150 | 302 | 138–170 | 281–338 |
| 48.0 | 25.63 | 12.77 | 9.60 | ... | ... | ... | 61–65 | 142–149 |

Source: Ref 1

patterns possesses zero shrinkage. Fusible alloys with eutectic compositions are often used for casting metal patterns from wood masters because they undergo enough expansion to allow cleaning of production castings without reducing the final dimensions below tolerances. The expansion characteristics of fusible alloys are often used when a metal part is to be firmly anchored in a lathe chuck. After the part is machined, the extra fusible alloy is melted away.

In general, the load-bearing capacity of fusible alloys is good, although some deformation will occur under prolonged stress. In addition, hardness and other mechanical properties of many fusible alloys change gradually with time, probably because of the same microstructural changes caused by aging that affects expansion or shrinkage.

## ACKNOWLEDGMENTS

Sections of this chapter were adapted from "Zirconium and Hafnium," "Lead and Lead Alloys," "Tin and Tin Alloys," "Precious Metals and Alloys," and "Fusible Alloys," all in *Metals Handbook Desk Edition*, 2nd Edition, ASM International, 1998.

## REFERENCES

1. Fusible Alloys, *Metals Handbook Desk Edition,* 2nd ed., ASM International, 1998

## SELECTED REFERENCES

- S.H. Avner, *Introduction to Physical Metallurgy,* 2nd ed., McGraw-Hill Book Co., 1974
- Lead and Lead Alloys, *Metals Handbook Desk Edition,* 2nd ed., ASM International, 1998
- J.M. Marder, Powder Metallurgy Beryllium, *Powder Metal Technologies and Applications,* Vol 7, *ASM Handbook,* ASM International, 1998
- B.J. Moniz, *Metallurgy,* American Technical Publishers, Inc., 1994
- Precious Metals and Alloys, *Metals Handbook Desk Edition,* 2nd ed., ASM International, 1998
- A.M. Russell and K.L. Lee, *Structure-Property Relationships in Nonferrous Alloys,* Wiley-Interscience, 2005
- W.F. Smith, *Structure and Properties of Engineering Alloys,* 2nd ed., McGraw-Hill, Inc., 1993
- G.D. Smith and E. Zysk, Corrosion of the Noble Metals, *Corrosion,* Vol 13, *ASM Handbook,* ASM International, 1987
- W. Speer and O.S. Es-Said, Applications of an Aluminum-Beryllium Composite for Structural Aerospace Components, *Eng. Fail. Anal.,* Vol 11, 2004, p 895–902
- A.J. Stonehouse and J.M. Marder, Beryllium Grades and Their Designations, *Properties and Selection: Nonferrous Alloys and Special-Purpose Materials,* Vol 2, *ASM Handbook,* ASM International, 1990
- A.J. Stonehouse and J.M. Marder, Beryllium: Health and Safety Considerations, *Properties and Selection: Nonferrous Alloys and Special-Purpose Materials,* Vol 2, *ASM Handbook,* ASM International, 1990
- A.J. Stonehouse and J.M. Marder, Beryllium: Powder Consolidation Methods, *Properties and Selection: Nonferrous Alloys and Special-Purpose Materials,* Vol 2, *ASM Handbook,* ASM International, 1990
- Tin and Tin Alloys, *Metals Handbook Desk Edition,* 2nd ed., ASM International, 1998
- Zirconium and Hafnium, *Metals Handbook Desk Edition,* 2nd ed., ASM International, 1998

CHAPTER 33

# Metal-Matrix Composites

METAL-MATRIX COMPOSITES (MMCs) are capable of providing higher-temperature operating limits than their base metal counterparts, and they can be engineered for improved strength, stiffness, thermal conductivity, abrasion resistance, creep resistance, or dimensional stability. However, due to their high cost, commercial applications for MMCs are sparse, especially for the extremely expensive continuous fiber-reinforced MMCs. There are some limited uses for discontinuously reinforced MMCs but almost no current applications for continuously reinforced MMCs.

The matrix phase of an MMC is a monolithic alloy, usually a low-density nonferrous alloy, and the reinforcement consists of high-performance carbon, metallic, or ceramic additions, as depicted in Fig. 33.1. Reinforcements, characterized as either continuous or discontinuous, may constitute from 10 to 70 vol% of the composite. Continuous fiber reinforcements include graphite, silicon carbide (SiC), boron, aluminum oxide ($Al_2O_3$), and refractory metal wires. Continuous reinforcements usually consist of small multifilaments or much larger monofilaments. Small multifilaments, typically less than 20 μm (0.8 mil) in diameter, usually consist of tows, or bundles, with thousands of fibers and are derived from textile materials. Monofilaments are much larger, ranging from 100 to 140 μm (4.0 to 5.5 mils) in diameter, and are made one filament at a time by chemical vapor deposition on a small substrate, normally tungsten wire or carbon fiber. Discontinuous reinforcements are mainly silicon carbide whiskers ($SiC_w$); particles (p) of silicon carbide ($SiC_p$), alumina ($Al_2O_3$), and titanium diboride ($TiB_2$); and short or chopped fibers (c) of alumina or graphite.

The primary MMC fabrication processes are either liquid-phase or solid-state processes. Liquid-phase processing is generally considerably less expensive than solid-state processing. Characteristics of liquid-phase-processed discontinuous MMCs include low-cost reinforcements, such as silicon carbide particles; low-temperature-melting matrices, such as aluminum and magnesium; and near-net-shaped parts. Liquid-phase processing results in intimate interfacial contact and strong reinforcement-to-matrix bonds but also can result in the formation of brittle interfacial layers as a result of interactions with the high-temperature liquid matrix. Liquid-phase processes include various casting processes, liquid metal infiltration, and spray deposition.

Solid-state processes, in which no liquid phase is present, are usually associated with some type of diffusion bonding process, whether the matrix is in a thin sheet or powder form. Although the processing temperatures are lower for solid-state diffusion bonding, they are often still high enough to cause significant reinforcement degradation. In addition, the pressures are almost always higher than for the liquid-state processes.

The choice of a fabrication process for any MMC is dictated by many factors, the most important being: preservation of reinforcement strength, preservation of reinforcement spacing and orientation, promotion of wetting and bonding between the matrix and reinforcement, and minimization of reinforcement damage, primarily due to chemical reactions between the matrix and reinforcement. The interface between the reinforcement and matrix is important in determining mechanical properties. While a good bond is needed to effectively transfer stresses between the reinforcement and the matrix, chemical reactions between the two during elevated-temperature processing or during in-service exposures can cause embrittling reactions.

## 33.1 Aluminum-Matrix Composites

Most commercial work on MMCs has focused on aluminum as the matrix metal. The combination of light weight, environmental resistance, and useful mechanical properties makes aluminum well suited for use as a matrix material. The melting point of aluminum is high enough for many applications, yet low enough to make composite processing reasonably convenient. Also, aluminum can accommodate a variety of reinforcements. Although much of the early work on aluminum MMCs concentrated on continuous fibers, most of the present work is focused on discontinuously reinforced particle or whisker aluminum composites, also known as discontinuously reinforced aluminum (DRA). The DRA composites are easier to manufacture at lower production costs and have relatively isotropic properties. However, as shown in Fig. 33.2, higher-performance composites are produced by the more expensive continuous fiber reinforcements. At the opposite end of the cost/performance spectrum are the particle-reinforced cast metal composites. In between lie medium-priced composites, including those produced by preform infiltration and powder metallurgy techniques.

Because more aluminum MMCs are produced than all other MMCs combined, the Aluminum Association developed a standard designation system for aluminum MMCs that has since been adopted by the American National Standards Institute (ANSI) in ANSI 35.5-1992, using the following format: matrix/reinforcement/vol% form.

For example, 2124/SiC/25w describes alloy 2124 reinforced with 25 vol% of silicon carbide whiskers; 7075/$Al_2O_3$/10p is the alloy 7075

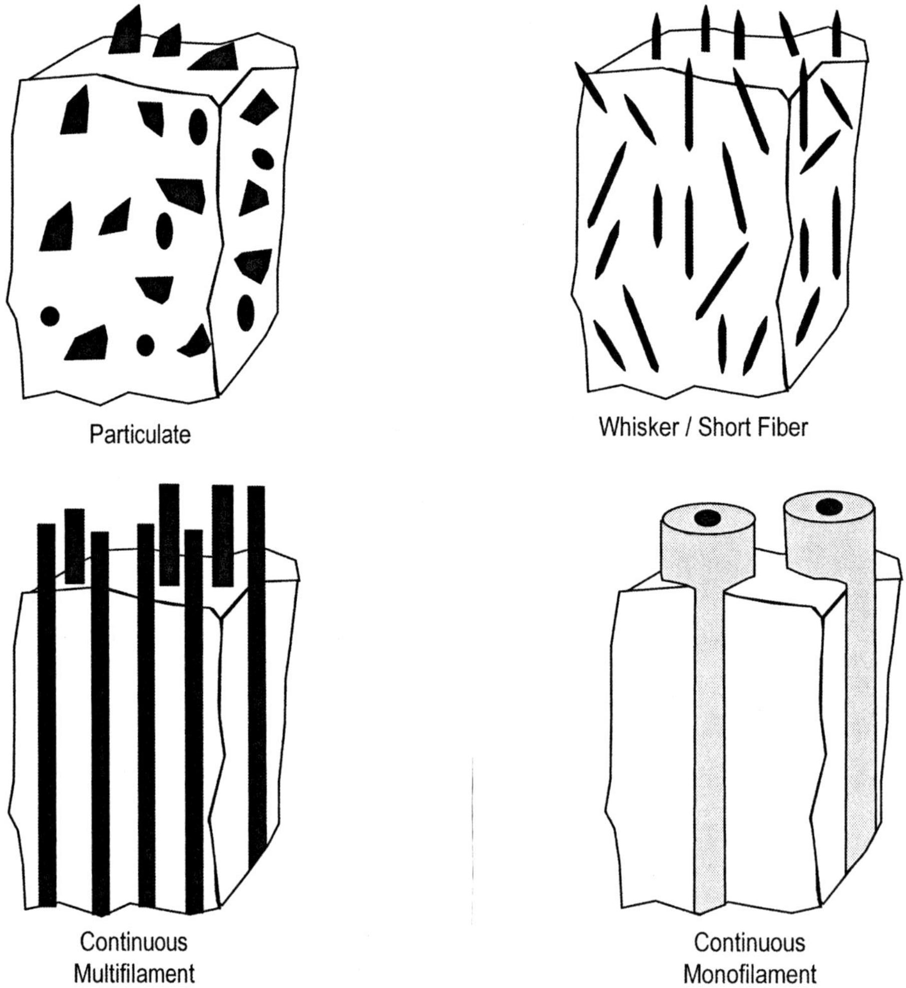


**Fig. 33.1** Metal-matrix composite reinforcements. Source: Ref 3

reinforced with 10 vol% of alumina particles; 6061/SiC/47f is the alloy 6061 reinforced with 47 vol% of continuous SiC fibers; and A356/C/05c is the casting alloy A356 reinforced with 5 vol% of chopped graphite fibers.

### 33.1.1 Discontinuously Reinforced Aluminum Composites

The most commonly used reinforcement materials in DRA composites are SiC and $Al_2O_3$, although silicon nitride ($Si_3N_4$), $TiB_2$, and graphite have also been used in some specialized applications. For example, graphite/aluminum composites have been developed for tribological applications because of their excellent antifriction properties, wear resistance, and nonseizing characteristics. The properties of discontinuous reinforcements used to reinforce aluminum alloys are given in Table 33.1.

Mechanical properties of MMCs depend on the amount, size, shape, and distribution of the dispersed reinforcement phase; the mechanical properties of the matrix material; and the nature of the interface. By definition, a composite material requires an amount of dispersed phase (>1 vol%) of a size (>1 μm) that allows the component to be load bearing and not merely control dislocation movement, as in dispersion-strengthened materials. The shape of the reinforcement is so important in determining its load-bearing capacity that composites are classified on this basis: (1) fiber-reinforced composites with both continuous and discontinuous fibers and (2) particle- or whisker-reinforced composites. The reinforcement aspect ratio (length/diameter) generally characterizes the shape. In continuous fiber composites, the load is applied directly to both the matrix and the fiber, with the fibers being the major load-bearing constituent. In discontinuous fiber- or particle-reinforced composites, the load is transmitted to the dispersoid through the matrix.

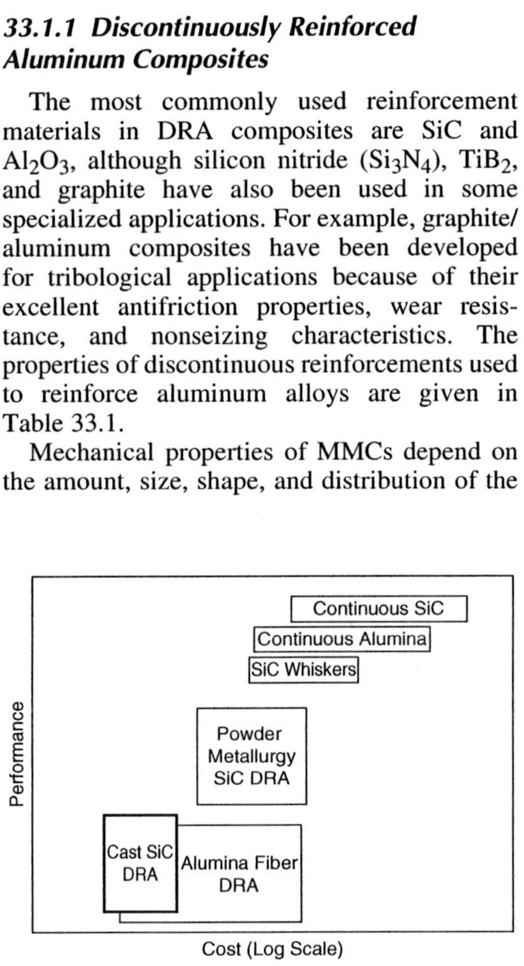


**Fig. 33.2** Performance/cost trade-offs for metal-matrix composites. DRA, discontinuously reinforced aluminum. Source: Ref 2

The influence of hard particle reinforcement (e.g., SiC) on the relevant mechanical and physical properties of DRAs can be summarized as follows:

- Both the ultimate tensile strength and yield strength increase with an increase in reinforcement volume fraction.
- The fracture toughness, ductility, percent elongation, and strain to failure decrease with increasing reinforcement volume fraction.
- The modulus of elasticity increases with increasing reinforcement volume fraction.
- The thermal and electrical conductivities and the coefficient of thermal expansion decrease with increasing reinforcement volume fraction.

The effects of particulate SiC volume fraction on the properties of DRAs are shown in

**Table 33.1 Common reinforcements for discontinuous aluminum composites**

| Property | Reinforcement | | | | |
|---|---|---|---|---|---|
| | $SiC_p$ | $Al_2O_{3p}$ | $TiB_{2p}$ | $Al_2O_{3c}$(a) | $SiC_w$(b) |
| Density, g/cm$^3$ | 3.21 | 3.97 | 4.5 | 3.3 | 3.19 |
| Diameter, μm | ... | ... | ... | 3–4 | 0.1–1.0 |
| Coefficient of thermal expansion, $10^{-6} \cdot K^{-1}$ | 4.3–5.6 | 7.2–8.6 | 8.1 | 9 | 4.8 |
| Tensile strength, MPa (ksi) | 100–800(c) (14.5–116)(c) | 70–1000(c) (10–145)(c) | 700–100(c) (101.5–145)(c) | >2000 (>290) | 3000–4000 (435–2030) |
| Young's modulus, GPa ($10^6$ psi) | 200–480 (29–70) | 380 (55) | 514–574 (75–83) | 300 (44) | 400–700 (58–101.5) |
| Elongation, % | ... | ... | ... | 0.67 | 1.23 |

(a) Saffil (Saffil Ltd.) (96%$Al_2O_3$-4%$SiO_2$). (b) >98% SiC. (c) Transverse rupture strength of bulk material. Source: Ref 1

Fig. 33.3. Typical fiber volumes of DRAs used for structural applications are usually limited to 15 to 25%, because higher volumes result in low ductility and fracture toughness.

### *33.1.2 Processing DRA Composites*

Processing methods for MMCs include various casting processes, liquid metal infiltration, spray deposition, and powder metallurgy. Casting is currently the most inexpensive method of producing MMCs and lends itself to the production of large ingots, which can be further worked by extrusion, hot rolling, or forging. In any casting process, wetting and distribution of reinforcements are important processing parameters. In addition, any mechanical process, such as blending or stirring, that severely abrades short fibers will tend to break them up and reduce their aspect ratios.

**Stir Casting.** In the stir casting process, the reinforcement filler is introduced in a continuously stirred molten matrix and then cast by sand, permanent mold, or pressure die casting. As shown in Fig. 33.4, an impeller mixer generates a vortex that draws the reinforcement from the top down into the melt. An inert atmosphere or a vacuum is essential to avoid gas entrapment.

Difficulties can include segregation and settling of the reinforcements in the melt, reinforcement agglomeration, reinforcement fracture during stirring, and, since the time in the melt is long, extensive interfacial reactions. Good reinforcement wetting can be a problem, and reinforcements are often pretreated prior to casting to promote wetting. Particle agglomeration and sedimentation in the melt can cause inhomogeneities. Reinforcement redistribution can be caused by gravity effects and from the reinforcement being pushed by an advancing solidification front. Proper stirring helps to prevent many of these problems.

Matrix alloys include aluminum-silicon casting compositions specially designed for MMC processing. Reinforcements include 10 to 20 μm- (0.4 to 0.80 mil)-sized SiC or $Al_2O_3$ particles with volume fractions ranging from 10 to 20%. A typical microstructure is shown in Fig. 33.5. The automotive industry is the main proponent of cast DRAs. Current or potential applications include brake rotors and drums, brake calipers, brake pad backing plates, and cylinder liners.

**Rheocasting,** also referred to as compocasting, is similar to stir casting, but instead of the particulates being stirred into a fully liquid melt, they are stirred into a lower-temperature two-phase liquid + solid melt. The particulate reinforcement becomes trapped in the two-phase liquid + solid melt, forming a slurry, which helps to minimize segregation. Stirring reduces

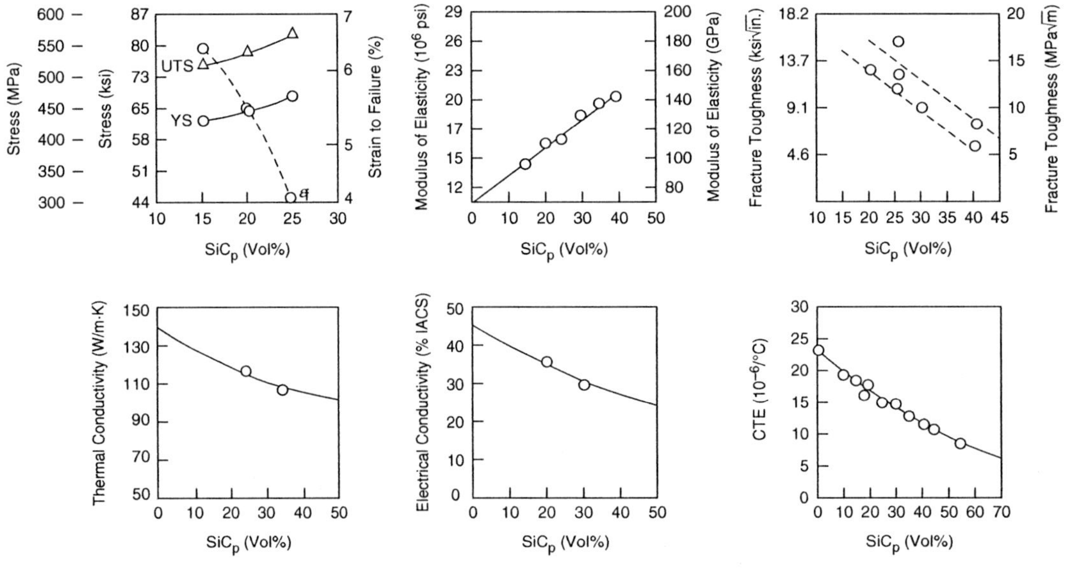


**Fig. 33.3** Properties of SiCp discontinuously reinforced aluminum composites. UTS, ultimate tensile strength; YS, yield strength; CTE, coefficient of thermal expansion. Source: Ref 1

the melt viscosity, resulting in a mutual interaction between the matrix melt and the reinforcement, which enhances wetting and bonding between the two. Slurry viscosity can be reduced by increasing the mixing shear rate and/or by increasing the slurry temperature. Since compocasting can be done at lower temperatures than conventional casting techniques, there is less chance of reactions occurring between the reinforcement and the liquid matrix. Particles and discontinuous fibers of SiC, $Al_2O_3$, TiC, $Si_3N_4$, graphite, mica, glass, slag, magnesium oxide, and boron carbide have been incorporated into vigorously agitated, partially solidified aluminum alloy slurries by compocasting.

**Squeeze Casting.** In squeeze casting, a hydraulic press is used to slowly apply high pressure, usually in the range of 10 to 100 MPa (1.5 to 15 ksi), to the melt and force it into a reinforcement bed, typically a short-fiber preform, as illustrated in Fig. 33.6. Since solidification occurs rapidly under high pressure, shrinkage porosity is eliminated. Squeeze-cast parts are usually fine grained with excellent surface finishes.

Preforms for infiltration are usually prepared by sedimentation of short fibers or particles from a suspension. A Saffil (Saffil Ltd.) alumina short-fiber preform is shown in Fig. 33.7. Binders are frequently required to maintain preform integrity for handling. Binders (5 to 10 wt%) can either be fugitive ones that are burned off during casting or high-temperature silica or alumina compounds that require firing before casting.

Squeeze casting is attractive because it minimizes material and energy use, produces net shape components, and offers the capability for selective reinforcement. Both discontinuous and continuous aluminum-copper, aluminum-silicon, and aluminum-magnesium casting alloys reinforced with up to 45 vol% SiC have been produced.

**Liquid Metal Infiltration.** Pressure infiltration casting (PIC) is similar to squeeze casting except that gas, rather than mechanical pressure, is used to promote consolidation. In PIC, an evacuated particulate or fiber preform with molten metal is subjected to isostatically applied

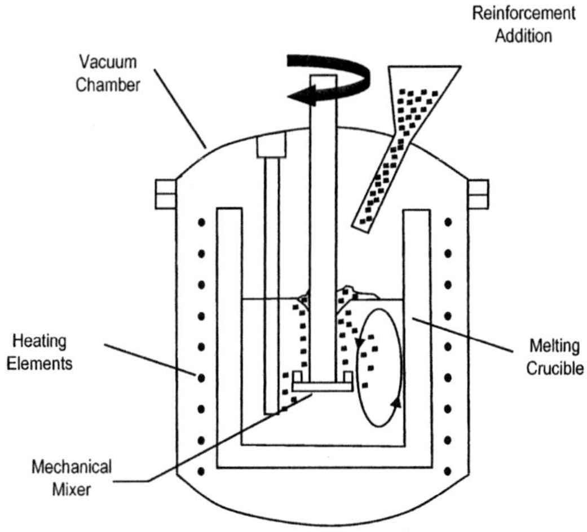


**Fig. 33.4** Stir casting. Source: Ref 7

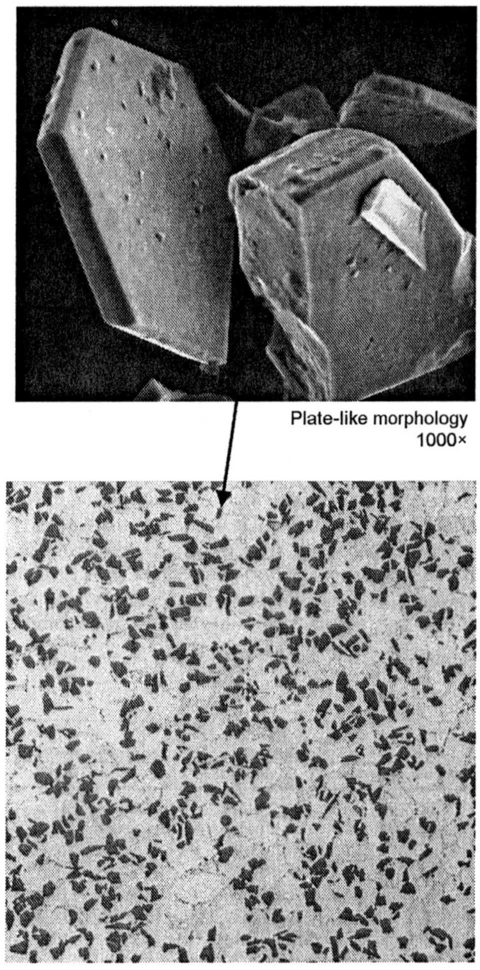


**Fig. 33.5** Microstructure of 20 vol% SiCp discontinuously reinforced aluminum. Source: Ref 1, 5

gas pressure. After the mold is preheated and the aluminum melt has reached the desired superheat temperature, inert gas pressure in the range of 1 to 10 MPa (150 to 1500 psi) is applied. As liquid aluminum infiltrates the preform, the pressure acting on the mold quickly dissipates as it approaches the isostatic state; therefore, the mold only supports the pressure difference for a very short period of time, so that large and expensive molds are not needed.

Important applications for PIC $SiC_p/Al$ composites are packages, substrates, and support structures for electronic components. Typical requirements include a low coefficient of thermal expansion (CTE) to reduce mechanical stresses imposed on the electronic device during attachment and operation, high thermal conductivity for heat dissipation, high stiffness to minimize distortion, and low density for minimum weight. Compared with conventional aluminum alloys, composites with high loadings of SiC particles have much lower CTEs and significantly higher elastic moduli, with little or no thermal conductivity or density penalty.

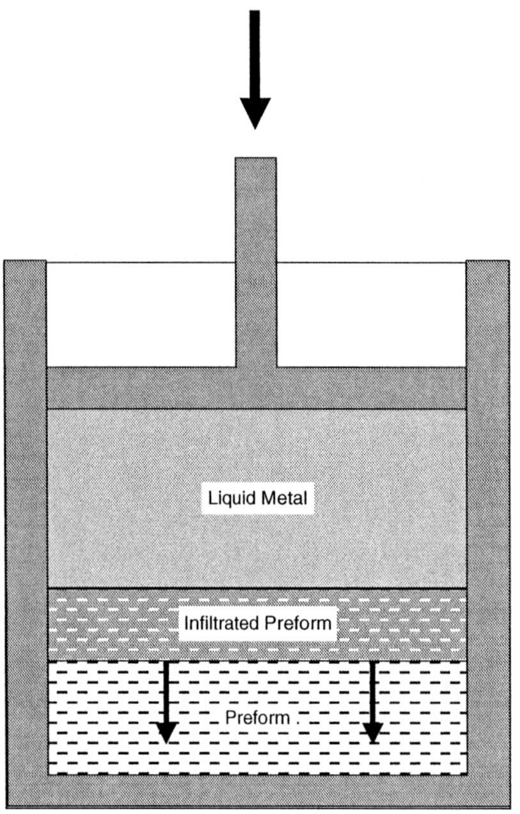


**Fig. 33.6** Squeeze casting. Source: Ref 6

The Primex (Lanxide) process is a pressureless metal infiltration process. Key process ingredients include the aluminum alloy, a nitrogen atmosphere, and the presence of magnesium. During heating to the infiltration temperature (~750 °C, or 1380 °F), the magnesium reacts with the nitrogen atmosphere to form magnesium nitride ($Mg_3N_2$), which is the infiltration enhancer that allows the molten aluminum alloy to infiltrate the reinforcing phase without the necessity of applied pressure or vacuum. During infiltration, the $Mg_3N_2$ is reduced by the aluminum to form a small amount of aluminum nitride (AlN). The AlN is found as small precipitates and as a thin film on the surface of the reinforcing phase. Magnesium is released into the alloy by this reaction. The pressureless infiltration process can produce a wide array of engineered composites by tailoring the alloy chemistry, particle type, shape, size, and loading. Particulate loading in cast composites can be as high as 75 vol%, given the right combination of particle shape and size.

The most widely used cast composite produced by liquid metal infiltration is an Al-10Si-1Mg alloy reinforced with 30 vol% SiC. The 1 wt% Mg present is obtained during infiltration by the reduction of the $Mg_3N_2$. This system can be used for all casting processes except die casting. The composite most used for die casting is based on this system, with the addition of 1 wt% Fe. Alloy modifications can be made to the alloy prior to infiltration or in the crucible prior to casting. The only universal alloy restriction for this composite system is the presence of magnesium to allow the formation of the $Mg_3N_2$. For SiC-containing systems, silicon must also be present in sufficient quantity to

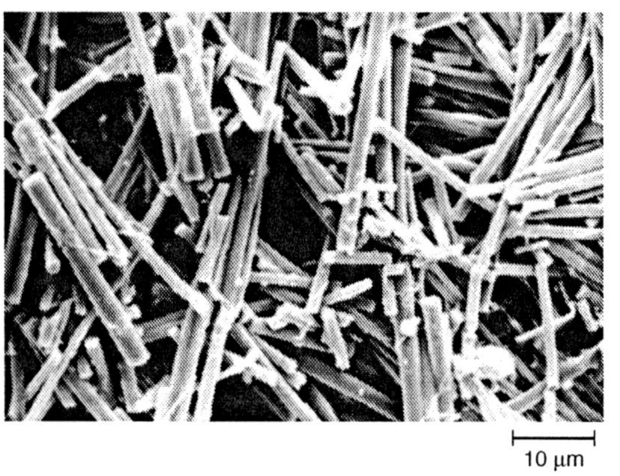


**Fig. 33.7** Saffil alumina fiber preform

suppress the formation of aluminum carbide ($Al_4C_3$).

**Spray deposition** involves atomizing a melt and, rather than allowing the droplets to solidify totally as for metal powder manufacture, collecting the semisolid droplets on a substrate. Spray deposition is a hybrid rapid solidification process, because the metal experiences a rapid transition through the liquidus to the solidus, followed by slow cooling from the solidus to room temperature, resulting in a fine-grained structure. The production of MMCs by spray deposition can be conducted by introducing particulates into the standard spray deposition metal spray, leading to codeposition with the atomized metal on the substrate, as in the Osprey process shown in Fig. 33.8. Inherent in spray processes are composites with minimal reinforcement degradation, little segregation, and fine-grained matrices. A uniform particulate distribution is produced with a typically 95 to 98% dense aluminum matrix. The advantages of spray deposition processes include fine grain sizes and minimal reinforcement degradation. Disadvantages include high porosity levels and the resulting need to further process these materials to achieve full consolidation.

A number of aluminum alloys containing SiC particulate have been produced by spray deposition. These include aluminum-silicon casting alloys and wrought alloys of the 2*xxx*, 6*xxx*, 7*xxx*, and the aluminum-lithium alloys of the 8*xxx* series. Significant increases in specific modulus can be obtained with SiC-reinforced 8090 alloy. Products that have been produced by spray deposition include solid and hollow extrusions, forgings, sheet, and remelted pressure die castings.

**Powder Metallurgy (PM).** When higher-strength discontinuous MMCs are required, PM

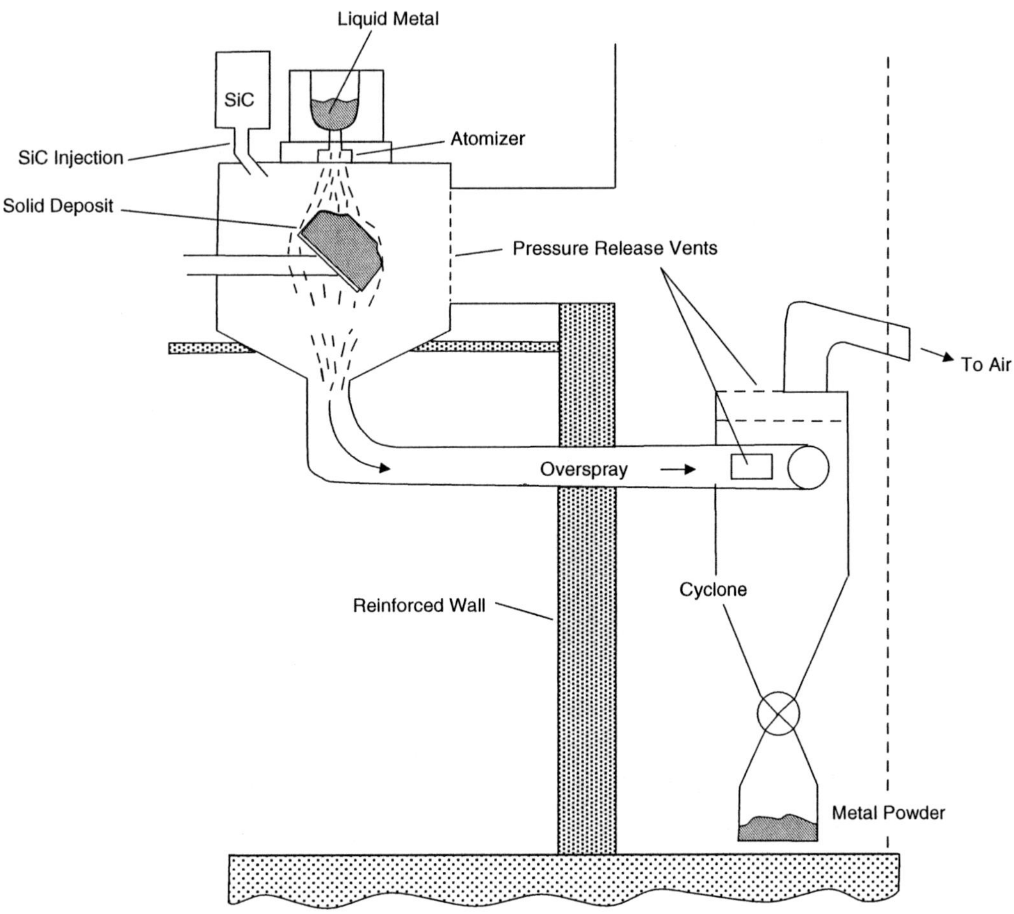


**Fig. 33.8** Osprey spray process. Source: Ref 3

processes are often used, because the PM process minimizes segregation, brittle reaction products, and residual stresses from solidification shrinkage. The PM technique generally produces properties superior to those obtained by casting and liquid metal infiltration techniques. Powder metallurgy processing of MMCs is possible for reinforcements of both SiC particulates and whiskers, although $Al_2O_3$ particles and $Si_3N_4$ whiskers have also been employed. It should be noted that PM processes also can be used to make continuous fiber MMCs, in particular, continuous fiber-reinforced titanium matrix composites.

Processing, shown in Fig. 33.9, involves (1) blending of the gas-atomized matrix alloy and reinforcement in powder form; (2) compacting by cold pressing the homogeneous blend to roughly 80% density; (3) degassing the open pore preform to remove volatile contaminants such as lubricants, mixing and blending additives, water vapor, and gases; and (4) consolidation by vacuum hot pressing or hot isostatic pressing. The consolidated cylindrical billets can then be subsequently extruded, rolled, or forged.

Whisker-reinforced MMCs may experience some whisker alignment during extrusion or rolling. Control of whisker alignment enables production of MMC product forms with directional properties needed for some applications. Cross rolling of sheet establishes a more planar whisker alignment, producing a two-dimensional isotropy. The mechanical properties of whisker-reinforced aluminum MMCs are superior to particle-reinforced composites at any common volume fraction. However, whiskers are more expensive than particles, they are more difficult to disperse uniformly in the matrix, and there are health hazard concerns for whiskers.

## 33.2 Continuous Fiber Aluminum MMCs

As previously shown in Fig. 33.2, aluminum MMCs reinforced with continuous fibers provide the highest performance/strength. Some mechanical properties of several continuous fiber-reinforced MMCs are given in Table 33.2. Of the reinforcements shown, boron and silicon carbide (SCS-2) are monofilaments, and graphite and Nextel 610 (3M) are multifilaments. The smaller and more numerous multifilament tows are difficult to impregnate using solid-state processing techniques, such as diffusion

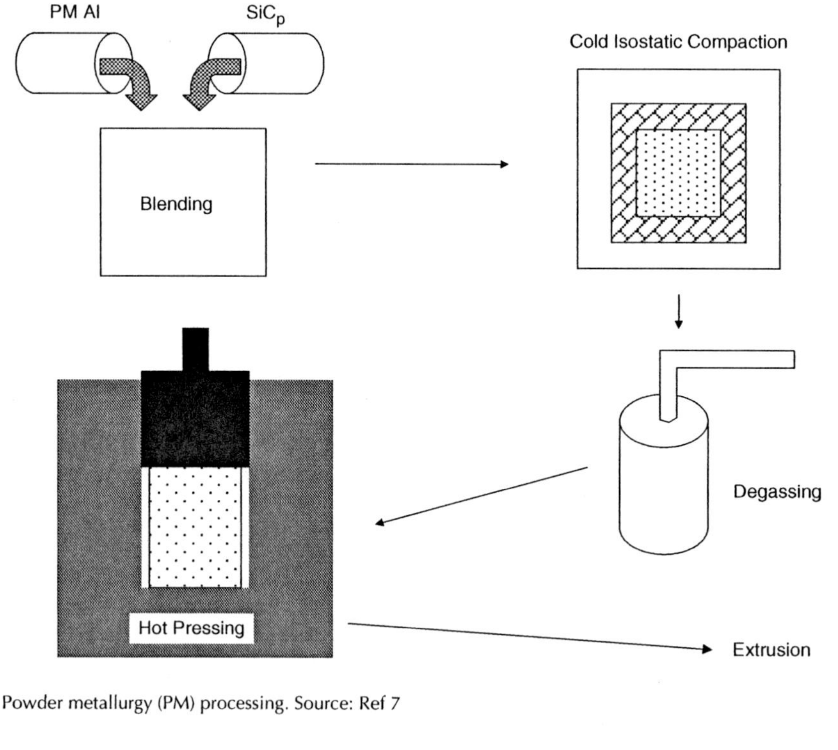


**Fig. 33.9** Powder metallurgy (PM) processing. Source: Ref 7

bonding, because of their small size and tightness of their tow construction. One of the main advantages for both continuous fiber aluminum and titanium-matrix composites is their ability to be used at higher temperatures than their base metals (Fig. 33.10). However, because of their high cost, most applications have been limited, even in the aerospace industry.

Boron/aluminum is the oldest of the continuous fiber MMC systems. The microstructure of a boron/aluminum composite is shown in Fig. 33.11. Applications include tubular truss members in the midfuselage structure of the Space Shuttle orbiter and cold plates in electronic microchip carrier multilayer boards. Fabrication processes for boron/aluminum composites are based on hot press diffusion bonding of alternating layers of aluminum foil and boron fiber mats (foil-fiber-foil processing) or plasma spraying methods.

After continuous silicon carbide monofilaments were developed, they largely replaced boron fibers because they have similar properties, and silicon carbide fibers are not degraded by molten aluminum. One such silicon carbide fiber is SCS, which can be manufactured with any of several surface chemistries to enhance bonding with a particular matrix, such as aluminum or titanium. The SCS-2 fiber, tailored for aluminum, has a 1 μm (0.04 mil) thick carbon-rich coating that increases in silicon content toward its outer surface.

Hot molding is a low-pressure, hot pressing process designed to fabricate SiC/Al parts at significantly lower cost than is possible with the solid-state diffusion bonding process. Because SCS-2 fibers can withstand molten aluminum for long periods, the bonding temperature can be raised into the liquid + solid region to ensure good aluminum flow and consolidation at low pressure, eliminating the need for high-pressure diffusion bonding. The mold is an integrally heated, slip-cast ceramic that contains the profile of the finished part. A plasma-sprayed aluminum

**Table 33.2 Mechanical properties of continuous fiber aluminum-matrix composites**

| Property | B/6061 Al | SCS-2/6061 Al(a) | P100 Gr/6061 Al | Nextel/610/Al(b) |
|---|---|---|---|---|
| Fiber content, vol% | 48 | 47 | 43.5 | 60 |
| Longitudinal modulus, GPa ($10^6$ psi) | 214 (31) | 204 (29.6) | 301 (43.6) | 240 (35) |
| Transverse modulus, GPa ($10^6$ psi) | ... | 118 (17.1) | 48 (7.0) | 160 (23.2) |
| Longitudinal strength, MPa (ksi) | 1520 (220) | 1462 (212) | 543 (79) | 1600 (232) |
| Transverse strength, MPa (ksi) | ... | 86 (12.5) | 13 (2) | 120 (17.4) |

(a) SCS-2 is a silicon carbide fiber.
(b) Nextel 610 (3M) is an alumina fiber. Source: Ref 1

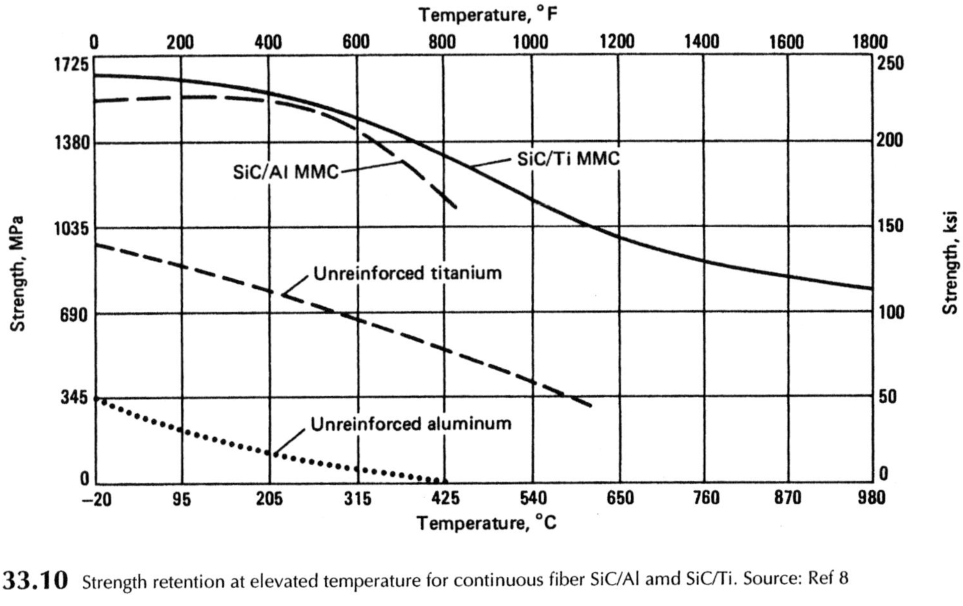


**Fig. 33.10** Strength retention at elevated temperature for continuous fiber SiC/Al amd SiC/Ti. Source: Ref 8

preform is laid into the mold, heated to near the aluminum melting temperature, and pressure consolidated in an autoclave by a metallic vacuum bag. SiC/Al MMCs exhibit increased strength and stiffness compared to unreinforced aluminum, with no weight penalty. In contrast to the base aluminum alloy, the composite retains its room-temperature tensile strength at temperatures up to 260 °C (500 °F).

Graphite/aluminum MMC development was initially prompted by the commercial appearance of strong and stiff graphite fibers in the 1960s. Graphite fibers offer a range of properties, including an elastic modulus up to 966 GPa (140 msi) and a negative CTE down to $-1.62 \times 10^{-6}$/°C ($-0.9 \times 10^{-6}$/°F). However, graphite/aluminum composites are extremely difficult to process. A deleterious reaction between graphite and aluminum, poor wetting of graphite by molten aluminum, and oxidation of the graphite fiber are significant technical barriers to the production of these composites. In addition, graphite fibers in a matrix of either aluminum or magnesium can cause galvanic corrosion of the matrix. Two processes are currently used for making commercial aluminum MMCs: liquid metal infiltration of the matrix on spread tows and hot press bonding of spread tows sandwiched between sheets of aluminum. With both processes, secondary processing, such as diffusion bonding, is needed to make structural elements.

Precision space structures with strict tolerances on dimensional stability need stiff, lightweight materials that exhibit low thermal distortion. Graphite/aluminum MMCs have the potential to meet these requirements. A unidirectional P100 Gr/6061 aluminum tube has an elastic modulus in the fiber direction (0° direction) significantly greater than that of steel, and it has a density approximately one-third that of steel.

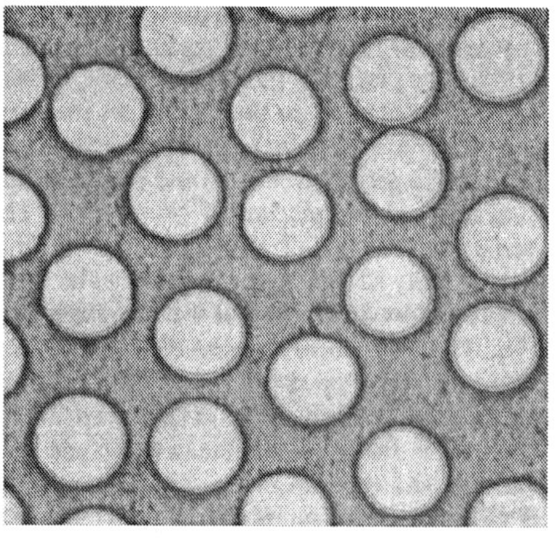

**Fig. 33.11** Monofilament boron/aluminum composite. Source: Ref 1

Aluminum oxide fiber/aluminum MMCs can be fabricated by a number of methods, but liquid or semisolid-state processing techniques are commonly used. Aluminum oxide fibers, which include continuous Nextel 610 and short-fiber Saffil, provide improved properties as compared with unreinforced aluminum alloys. For example, the composites have improved resistance to wear and thermal fatigue deformation and a reduced CTE. Continuous fiber aluminum oxide/aluminum MMCs are fabricated by constructing a preform by arranging aluminum oxide fiber tapes in the desired orientation, inserting the preform into the mold, and infiltrating molten aluminum into the preform with a vacuum assist.

## 33.3 Titanium-Matrix Composites

Titanium is attractive as a matrix metal because of its good specific strength at both room and moderately elevated temperature and its excellent corrosion resistance. Although continuous fiber titanium-matrix composites (TMCs) are even more expensive than continuous fiber aluminum-matrix composites, they are more useful from a structural standpoint because they can be used at much higher temperatures. Potential applications include replacements for much heavier superalloys in some portions of jet turbine engines and as skin materials for hypersonic aircraft. Much of the recent work on continuous fiber TMCs was conducted in the early 1990s in support of the hypersonic National Aerospace Plane (NASP) project.

### *33.3.1 Continuous Fiber TMCs*

Silicon carbide monofilaments are the reinforcement of choice for TMCs. There are several producers of silicon carbide monofilaments, but since the fibers are expensive and the demand is very limited, they are manufactured in limited quantities. All silicon carbide monofilaments are produced by chemically vapor depositing SiC on a small substrate of either tungsten or carbon. They are then coated with a thin layer, usually carbon or titanium diboride ($TiB_2$), to help prevent them from reacting with the titanium matrix during consolidation or when placed in service. Since these fibers are large (100 to 140 μm, or 4.0 to 5.5 mils) and stiff (400 to 430 GPa, or 58 to 62 msi), they are extremely difficult to hold in position during processing, and their bend radii are limited. The best known of the silicon carbide monofilaments is SCS-6, which is coated on a carbon core and has a diameter of 142 μm. A micrograph of a consolidated SCS-6 TMC laminate is shown in Fig. 33.12. In the inset, the protective carbon coating on the fiber is visible, along with a small reaction layer. Typically, fiber contents range from 30 to 40 vol%, with mechanical properties similar to those shown in Table 33.3.

Conventional-matrix alloys include Ti-6Al-4V for low-temperature applications and Ti-6Al-2Sn-4Zr-2Mo (Ti-6242) when higher creep resistance is required or when the temperature is higher than the maximum use temperature for Ti-6Al-4V. Although beta titanium alloys are not noted for their high-temperature creep resistance, they have the advantage that they are much easier to cold roll into foil needed for the foil-fiber-foil processing method. The beta alloy Ti-15V-3Cr-3Sn-3Al has been used for this method but does not have sufficient oxidation resistance. During the NASP program, Beta 21S (Ti-15Mo-2.8Nb-3Al-0.2Si) was developed with improved oxidation resistance in the 705 to 815 °C (1300 to 1500 °F) range. Titanium aluminide ordered intermetallics, such as Ti-22Al-23Nb and Ti-22Al-26Nb, have also been evaluated as high-temperature-matrix materials.

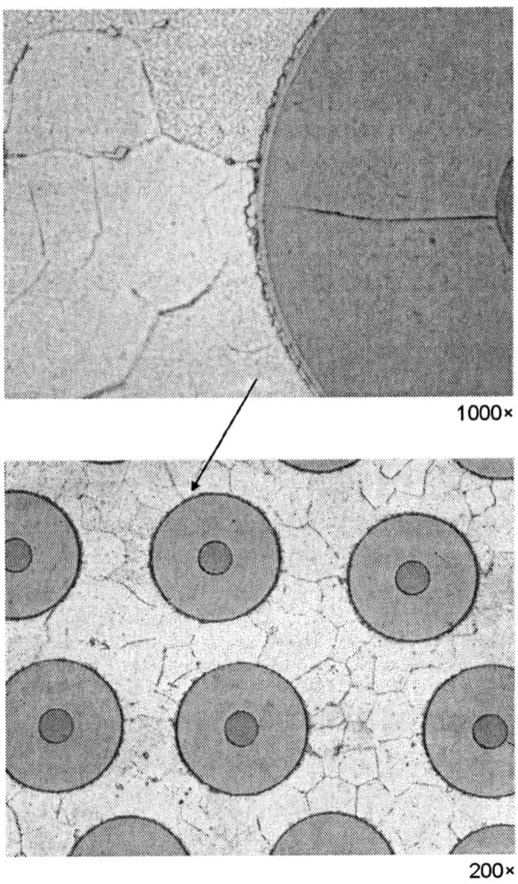


**Fig. 33.12** Silicon carbide monofilament/titanium-matrix composite

### *33.3.2 TMC Processing Techniques*

Processing techniques for continuous fiber TMCs include foil-fiber-foil processing, tape casting, plasma spraying, powder metallurgy, and electron beam deposition of the titanium matrix directly onto the fibers. Of these, the foil-fiber-foil method is probably the most widely studied.

**Foil-Fiber-Foil Process.** In this method, a fiber mat is sandwiched between thin layers of titanium foil, as depicted in Fig. 33.13. The fiber mat is a uniweave system in which the relatively large-diameter silicon carbide monofilaments are straight and parallel and held together by a cross weave, usually molybdenum, titanium, or titanium-niobium wire or ribbon. The cross weave of metallic ribbon is visible in the Fig. 33.14 mat. Since thin titanium foil is required for this method, beta titanium alloys, such as Ti-15V-3Cr-3Sn-3Al, are normally cold rolled down to a foil with a thickness of 0.11 mm (0.0045 in.). The plies are cut, laid-up on a consolidation tool, and consolidated by either vacuum hot pressing or hot isostatic pressing (HIP).

**Green Tape Process.** Another method for making continuous fiber TMCs involves placing a layer of titanium foil on a mandrel and filament

**Table 33.3 Mechanical properties of continuous fiber SCS-6/titanium-matrix composites**

| Property | SCS-6/Ti-6Al-4V |
|---|---|
| Fiber content, vol% | 37 |
| Longitudinal modulus, GPa ($10^6$ psi) | 221 (32) |
| Transverse modulus, GPa ($10^6$ psi) | 165 (24) |
| Longitudinal strength, MPa (ksi) | 1447 (210) |
| Transverse strength, MPa (ksi) | 413 (60) |

Source: Ref 1

winding the silicon carbide fiber over the foil in a collimated manner to produce a unidirectional single ply. An organic fugitive binder, such as an acrylic adhesive, is used to maintain the fiber spacing and alignment once the preform is cut from the mandrel. In this method, shown in Fig. 33.15, the fibers are wound onto a foil-covered rotating drum, oversprayed with resin, and then the layer is cut from the drum to provide a flat sheet of monotape. The organic binder is burned off prior to the HIP cycle or during the early portions of the vacuum hot pressing cycle.

**Plasma Tape Spraying.** In plasma spraying, the resin binder is replaced with a plasma-sprayed matrix. Plasma spraying eliminates potential contamination problems from organic residue during the consolidation cycle and speeds the process by not having to outgas an organic binder. One potential disadvantage of plasma spraying is that titanium, being an extremely reactive metal, can absorb oxygen from the atmosphere, potentially leading to embrittlement problems. This method has been evaluated primarily for titanium aluminide-matrix composites since it is difficult to roll these materials into thin foil.

### *33.3.3 TMC Consolidation Procedures*

The two primary consolidation procedures are vacuum hot pressing and HIP. High-temperature/short-time roll bonding was used some years ago but only to a very limited extent.

**Vacuum Hot Pressing.** This technique begins with sealing the lay-up in a stainless steel envelope that is placed in a vacuum hot press. After evacuation, a small positive pressure is applied via the press platens to hold the filaments in place during the initial 450 to 550 °C (800 to 1000 °F) soak used to decompose and remove the binder under the action of a dynamic vacuum. The temperature is then gradually increased to a level where the titanium flows around the fibers under increased pressure, and the foil interfaces are diffusion bonded together. Each fabricator uses a specific set of consolidation parameters, although typical ranges are 900 to 950 °C (1650 to 1750 °F) at 40 to 70 MPa (6 to 10 ksi) pressure for 60 to 90 min.

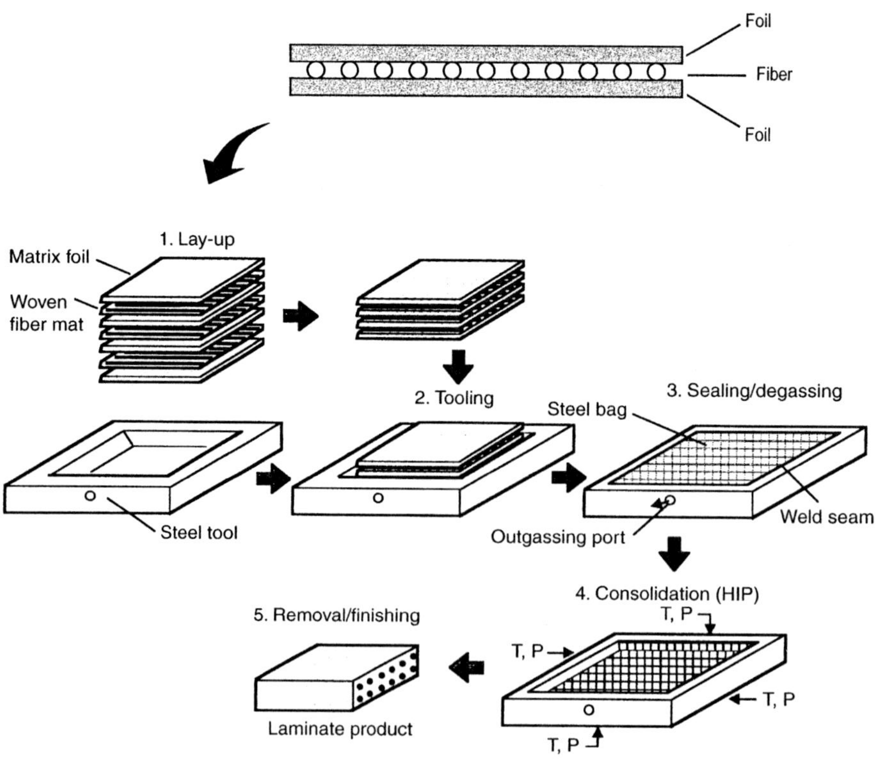


**Fig. 33.13** Foil-fiber-foil fabrication process. Top photo, courtesy of The Boeing Company; bottom image, HIP, hot isostatic pressing; T, temperature; P, pressure. Source: Ref 2

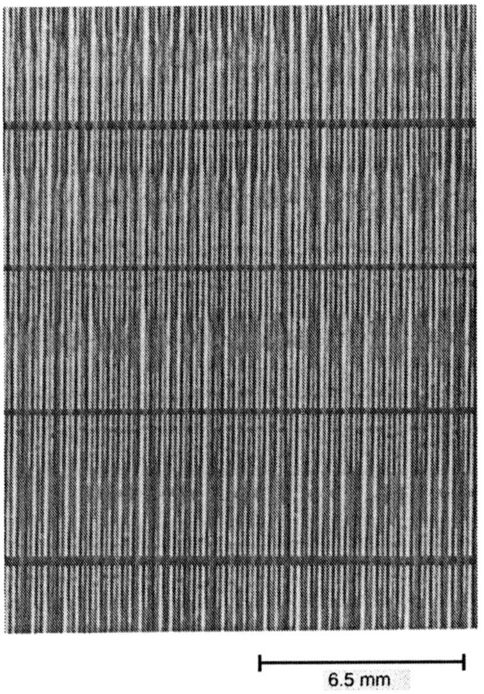


**Fig. 33.14** SiC uniweave fabric showing cross weave of metallic ribbon. Source: Ref 2

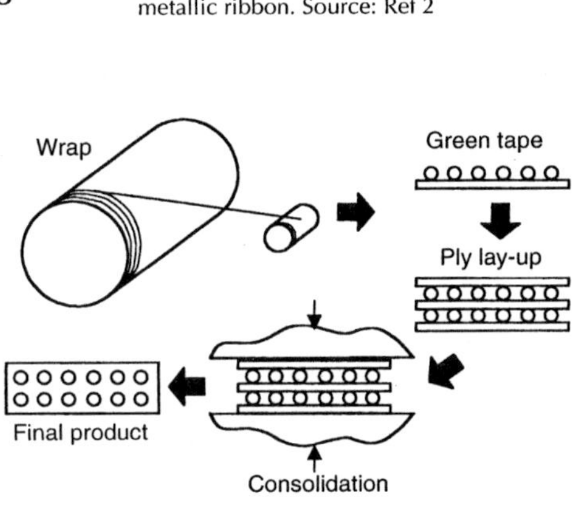


**Fig. 33.15** Green tape method for titanium-matrix composite fabrication. Source: Ref 2

**Hot isostatic pressing** has largely replaced vacuum hot pressing as the preferred consolidation method. The primary advantages of HIP consolidation are that the gas pressure is applied isostatically, alleviating the concern about uneven platen pressure, and the HIP process is much more amenable to making complex structural shapes. Typically, the part to be hot isostatically pressed is canned (or a steel bag is welded to a tool), evacuated, and then placed in the HIP chamber. For titanium, typical HIP parameters are 850 to 950 °C (1600 to 1700 °F) at 103 MPa (15 ksi) gas pressure for 2 to 4 h. Since HIP processing is a fairly expensive batch processing procedure, it is normal practice to load a number of parts into the HIP chamber for a single run.

### *33.3.4 Particle-Reinforced TMCs*

Particle-reinforced TMCs are processed by PM methods. Although a variety of materials have been studied, the most common combination is Ti-6Al-4V reinforced with 10 to 20 wt% TiC. These composites offer increased hardness and wear resistance over conventional titanium alloys. Properties of unreinforced and reinforced Ti-6Al-4V are compared in Table 33.4.

Unlike traditional composites where the reinforcements are distinct and separate constituents, in situ composites are systems in which the reinforcement forms within the metallic matrix during solidification. Directional solidification of certain eutectic systems can produce fibers or rodlike structures, as shown for the niobium-niobium carbide system in Fig. 33.16. While these structures can be produced, the growth rates are slow (~1 to 5 cm/h, or 0.4 to 2.0 in./h) due to the need to maintain a stable growth front, which requires a large temperature gradient. In addition, all of the reinforcement is aligned in the growth

**Table 33.4 Mechanical and physical properties of discontinuous $TiC_p$/Ti-matrix composites**

| Property | Ti-6Al-4V | 10 wt% TiC/Ti-6Al-4V | 20 wt% TiC/Ti-6Al-4V |
|---|---|---|---|
| Density, g/cm$^3$ (lb/in.$^3$) | 4.43 (0.160) | 4.45 (0.16) | 4.52 (0.162) |
| Tensile strength, MPa (ksi), at: | | | |
| RT | 896 (130) | 999 (145) | 1055 (153) |
| 540 °C (1000 °F) | 448 (65) | 551 (80) | 620 (90) |
| Modulus, GPa (10$^6$ psi), at: | | | |
| RT | 113 (16.5) | 133 (19.3) | 144 (21) |
| 540 °C (1000 °F) | 89 (13) | 105 (15.3) | 110 (16) |
| Fatigue limit (10$^6$ cycles), MPa (ksi) | 517 (75) | 275 (40) | ... |
| Fracture toughness MPa$\sqrt{m}$ (ksi$\sqrt{in.}$) | 55 (50) | 44 (40) | 32 (29) |
| Hardness, HRC | 34 | 40 | 44 |

RT, room temperature. Source: Ref 1

or solidification direction. There are also limitations in the nature and volume fraction of the reinforcement.

Another in situ method is to exploit exothermic reactions, such as the exothermic dispersion (XD) process. In the XD process, an exothermic reaction between two components produces a third component. Generally, a master alloy is produced that contains a high volume fraction of a ceramic reinforcing phase. The master alloy is then mixed and remelted with a base alloy to achieve the desired amount of particle reinforcement. In one process, aluminum, titanium, and boron are heated together to produce an exothermic reaction between the aluminum and titanium, producing a mixture of $TiB_2$ particles distributed in a titanium aluminide matrix. Strengths greater than 690 MPa (100 ksi) were measured at room temperature and at 800 °C (1470 °F) for a gamma titanium aluminide (Ti-45at.%Al) alloy reinforced with $TiB_2$ particulates.

Compared to the conventional MMCs, in situ MMCs exhibit several possible advantages: the in-situ-formed reinforcements are thermodynamically stable, leading to less degradation at elevated temperature; the reinforcement-to-matrix interfaces are clean, resulting in a strong interfacial bonding; and the in-situ-formed reinforcements are finer in size, and their distribution in the matrix is more uniform.

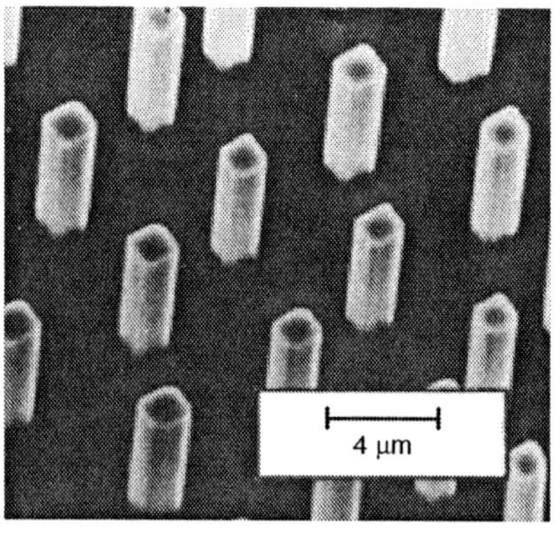


**Fig. 33.16** Rod-reinforced NbC/Nb eutectic. Source: Ref 9

## 33.4 Fiber-Metal Laminates

Fiber-metal laminates are a relatively new class of materials, possessing some of the characteristics of polymer-matrix composites and some characteristics of MMCs. The original fiber-metal laminate material was aramid-reinforced aluminum laminate, which consisted of high-strength aramid fibers sandwiched between thin sheets of aluminum alloy. More recently, the aramid fibers have been replaced with glass fibers. The new material, glass laminate aluminum reinforced (GLARE), is used for some of the fuselage skins on the Airbus A380 aircraft.

The details of a GLARE monolayer are shown in Fig. 33.17. The aluminum alloy sheet, typically 2024-T3 or 7475-T7651, is chemically cleaned and anodized for bonding using chromic acid or phosphoric acid anodizing. The aluminum sheet is primed with an epoxy primer that enhances the adhesive bond to the aluminum and helps in preventing interfacial corrosion of the aluminum. The S-2 glass-fiber fabric interlayer, which accounts for 59 vol% of the layer, is then impregnated and bonded to the aluminum sheet with two layers of an epoxy film adhesive. A number of these monolayers are repeated to produce the required final laminate thickness, which is autoclave cured.

The primary advantages of GLARE are better fatigue crack propagation resistance than aluminum, superior damage tolerance compared to aluminum, higher bearing strengths than carbon/epoxy, 10% lighter weight than aluminum, and lower cost than carbon/fiber composite, although higher cost than aluminum. In addition to GLARE, the introduction of titanium foils in carbon/epoxy laminates (TiGr) has been evaluated, primarily for improved bearing strengths.

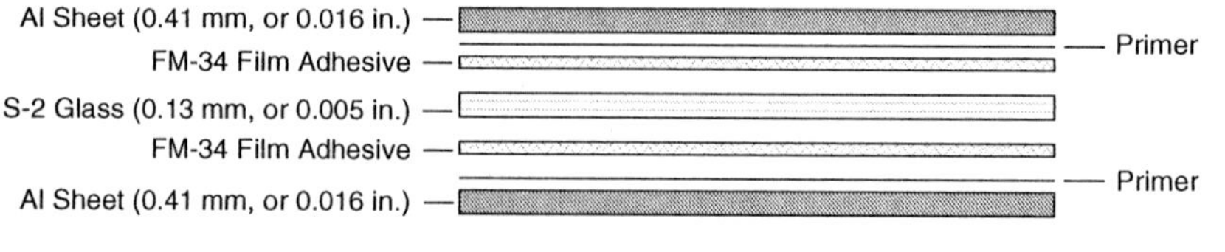


**Fig. 33.17** Typical glass laminate aluminum-reniforced ply construction

## ACKNOWLEDGMENTS

Sections of this chapter were adapted from "Metal-Matrix Composites" in *Metals Handbook Desk Edition*, 2nd Edition, ASM International, 1998; and "Processing of Metal-Matrix Composites," in *Composites*, Volume 21, *ASM Handbook*, ASM International, 2001.


## REFERENCES


1. Metal-Matrix Composites, *Metals Handbook Desk Edition,* 2nd ed., ASM International, 1998
2. Processing of Metal-Matrix Composites, *Composites,* Vol 21, *ASM Handbook,* ASM International, 2001, p 579–588
3. T.W. Clyne and P.J. Withers, *An Introduction to Metal Matrix Composites,* Cambridge University Press, 1993
4. D.R. Herling, G.J. Grant, and W. Hunt, Low-Cost Aluminum Metal Matrix Composites, *Adv. Mater. Process.,* July 2001, p 37–40
5. C.A. Smith, Discontinuous Reinforcements for Metal-Matrix Composites, *Composites,* Vol 21, *ASM Handbook,* ASM International, 2001
6. V.J. Michaud, Liquid State Processing, *Fundamentals of Metal-Matrix Composites,* Butterworth-Heinemann, 1993, p 3–22
7. F.C. Campbell, *Manufacturing Technology for Aerospace Structural Materials,* Elsevier Scientific, 2006
8. J.V. Foltz and C.M. Blackman, Metal-Matrix Composites, *Properties and Selection: Nonferrous Alloys and Special-Purpose Materials,* Vol 2, *ASM Handbook,* ASM International, 1990
9. D.M. Stefanescu and R. Ruxanda, Fundamentals of Solidification, *Metallography and Microstructures,* Vol 9, *ASM Handbook,* ASM International, 2004


## SELECTED REFERENCES


- M.R. Ghomashchi and A. Vikhrov, Squeeze Casting: An Overview, *J. Mater. Process. Technol.*, Vol 101, 2000, p 1–9
- A.K. Ghosh, Solid State Processing, *Fundamentals of Metal-Matrix Composites,* Butterworth-Heinemann, 1993, p 2–32 to 2–41
- J.A. McElman, Continuous Silicon Carbide Fiber MMCs, *Composites,* Vol 1, *Engineered Materials Handbook,* ASM International, 1987, p 858–866
- M.A. Mittnick, Continuous SiC Fiber Reinforced Materials, *21st International SAMPE Technical Conference,* Sept 1989, p 647–658
- T.S. Srivatsan, T.S. Sudarshan and E.J. Lavernia, Processing of Discontinuously-Reinforced Metal Matrix Composites by Rapid Solidification, *Prog. Mater. Sci.*, Vol 39, 1995, p 317–409
- A. Vlot and J.W. Gunnick, *Fibre Metal Laminates: An Introduction,* Kluwer Academic Publishers, 2001

# APPENDIX A

# Metric Conversions

| To convert from | To | Multiply by |
|---|---|---|
| **Area** | | |
| in.$^2$ | mm$^2$ | 6.451 600 E+02 |
| in.$^2$ | cm$^2$ | 6.451 600 E+00 |
| in.$^2$ | m$^2$ | 6.451 600 E−04 |
| ft$^2$ | m$^2$ | 9.290 304 E−02 |
| **Force** | | |
| lbf | N | 4.448 222 E+00 |
| kip (1000 lbf) | N | 4.448 222 E+03 |
| **Fracture toughness** | | |
| ksi (in.)$^{1/2}$ | MPa (m)$^{1/2}$ | 1.098 800 E+00 |
| **Length** | | |
| mil | μm | 2.540 000 E+01 |
| in. | mm | 2.540 000 E+01 |
| in. | cm | 2.540 000 E+00 |
| ft | m | 3.048 000 E−01 |
| yd | m | 9.144 000 E−01 |
| **Mass** | | |
| oz | kg | 2.834 952 E−02 |
| lb | kg | 4.535 924 E−01 |
| **Mass per unit area (areal weight)** | | |
| oz/in.$^2$ | kg/m$^2$ | 4.395 000 E+01 |
| oz/ft$^2$ | kg/m$^2$ | 3.051 517 E−01 |
| oz/yd$^2$ | kg/m$^2$ | 3.390 575 E−02 |
| lb/ft$^2$ | kg/m$^2$ | 4.882 428 E+00 |
| **Mass per unit volume (density)** | | |
| lb/in.$^3$ | g/m$^3$ | 2.767 990 E+01 |
| lb/in.$^3$ | kg/m$^3$ | 2.767 990 E+04 |
| lb/ft$^3$ | g/m$^3$ | 1.601 846 E−02 |
| lb/ft$^3$ | kg/m$^3$ | 1.601 846 E+01 |

| To convert from | To | Multiply by |
|---|---|---|
| **Pressure (fluid)** | | |
| lb/in.$^2$ (psi) | Pa | 6.894 757 E+03 |
| in. of Hg (60 °F) | Pa | 3.386 850 E+03 |
| atm (standard) | Pa | 1.013 250 E+05 |
| **Stress (force per unit area)** | | |
| lbf/in.$^2$ (psi) | MPa | 6.894 757 E−03 |
| ksi (1000 psi) | MPa | 6.894 757 E+00 |
| msi (1,000,000 psi) | MPa | 6.894 757 E+03 |
| **Temperature** | | |
| °F | °C | 5/9K(°F−32) |
| K | °C | K−273.15 |
| **Thermal conductivity** | | |
| Btu/ft · h · °F | W/m · K | 1.730 735 E+00 |
| **Thermal expansion** | | |
| in./in. · °F | m/m · K | 1.800 000 E+00 |
| in./in. · °C | m/m · K | 1.000 000 E+00 |
| **Velocity** | | |
| in./s | m/s | 2.540 000 E−02 |
| ft/s | m/s | 3.048 000 E−01 |
| ft/min | m/s | 5.080 000 E−03 |
| ft/h | m/s | 8.466 667 E−05 |
| **Viscosity** | | |
| poise | Pa · s | 1.000 000 E−01 |
| **Volume** | | |
| in.$^3$ | m$^3$ | 1.638 706 E−05 |
| ft$^3$ | m$^3$ | 2.831 685 E−02 |

# APPENDIX B

# Crystalline System Calculations

## B.1 Cubic Systems

The cubic crystal systems are regular cubes with a lattice parameter $a$ such that $a=b=c$ and $\alpha=\beta=\gamma=90°$. Therefore, only one lattice parameter ($a$) is required to define the cubic lattice. Within the cubic family of systems, there are three important variations: (1) the simple or primitive cubic in which there are atoms only at the corner points of the lattice; (2) the body-centered cubic (bcc) structure, which has an additional atom located at the center of the structure; and (3) the face-centered cubic (fcc) structure, which has an extra atom located on each of the six faces. The bcc and fcc structures are extremely important in metallurgy, with approximately 90% of industrially important metals crystallizing into one of these two structures.

### *B.1.1 Simple Cubic System*

Two types of graphical depictions of a simple cubic system are shown in Fig. B.1, namely the point model and the hard ball model. Although the hard ball model, with the outermost electron shells in contact, is the more realistic of the two, the point model is more commonly used since it is easier to visualize all of the atoms.

The relationship between the atomic radius ($r$) and the lattice constant ($a$) is shown in Fig. B.1 and is equal to:

$$a=2r \qquad \text{(Eq B.1)}$$

Another important value is the number of atoms ($N$) belonging to a unit cell. Examination of the hard ball model in Fig. B.1 reveals that each atom located at the corner point of the unit

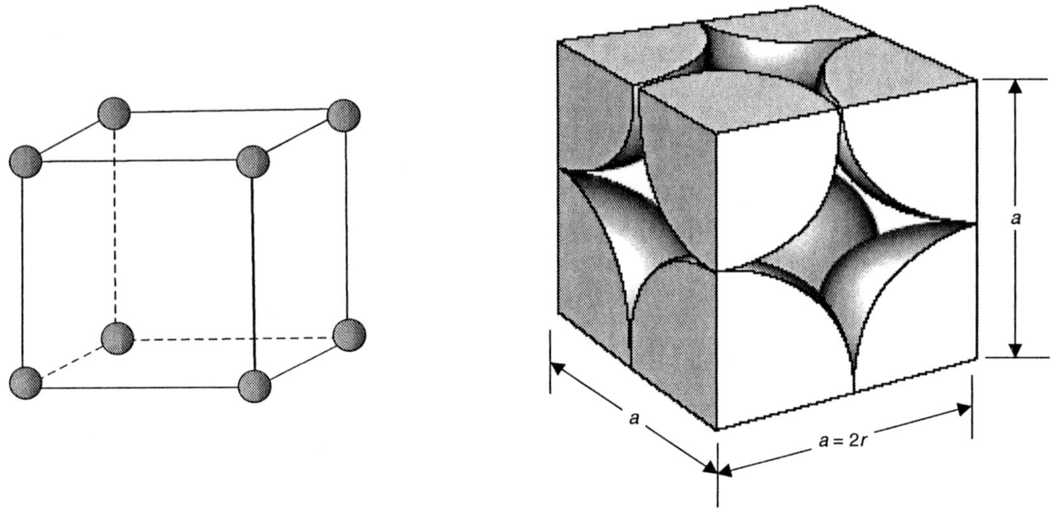


**Fig. B.1** Simple cubic structure

cell belongs to eight unit cells. Therefore, only 1/8 of each atom can be considered to belong to one unit cell. Since there are eight corners, or atomic positions, in the unit cell, the number of atoms belonging to the unit cell is:

$$N = 8\left(\frac{1}{8}\right) = 1 \quad \text{(Eq B.2)}$$

Thus, in a simple cubic crystal system, one full atom belongs to each unit cell.

In metals, the atoms tend to pack or occupy positions as close as possible to each other and to form the most dense lattice structure possible. The atomic packing factor (APF), which indicates the part of the volume of the unit cell that is actually occupied by atoms, is used to describe the denseness of the unit cell:

$$\text{APF} = \frac{V_a}{V_c} = \frac{\text{Volume of atoms belonging to unit cell}}{\text{Volume of unit cell}} \quad \text{(Eq B.3)}$$

The volume of atoms belonging to the simple unit cell can be calculated as:

$$V_a = N\frac{4\pi r^3}{3} \quad \text{(Eq B.4)}$$

Likewise, the volume of the simple cubic system is:

$$V_c = a^3 = 8r^3 \quad \text{(Eq B.5)}$$

Therefore, the APF is:

$$\text{APF} = \frac{N\frac{4\pi r^3}{3}}{8r^3} = \frac{N\pi}{6} \quad \text{(Eq B.6)}$$

Since $N = 1$ for the simple cubic system, the APF reduces to:

$$\text{APF} = \frac{\pi}{6} = 0.52 \quad \text{(Eq B.7)}$$

This means that only approximately half of the simple cubic system is occupied with atoms. Since metals tend to pack as closely as possible during crystallization, this rather loose packing explains why very few metals crystallize into the simple cubic system.

Another parameter used to define crystal systems is the coordination number (CN), or number of nearest neighboring atoms each atom in the structure possesses. In the hard ball model, it is the number of atoms that touch a given atom. Referring again to Fig. B.1, it can be seen that any atom is in contact with six neighboring atoms, thus CN = 6.

### B.1.2 Body-Centered Cubic System

The bcc system is shown in Fig. B.2. The bcc system is similar to the simple cubic system except that it has an additional atom located in the center of the structure. Since the center atom belongs completely to the unit cell in question,

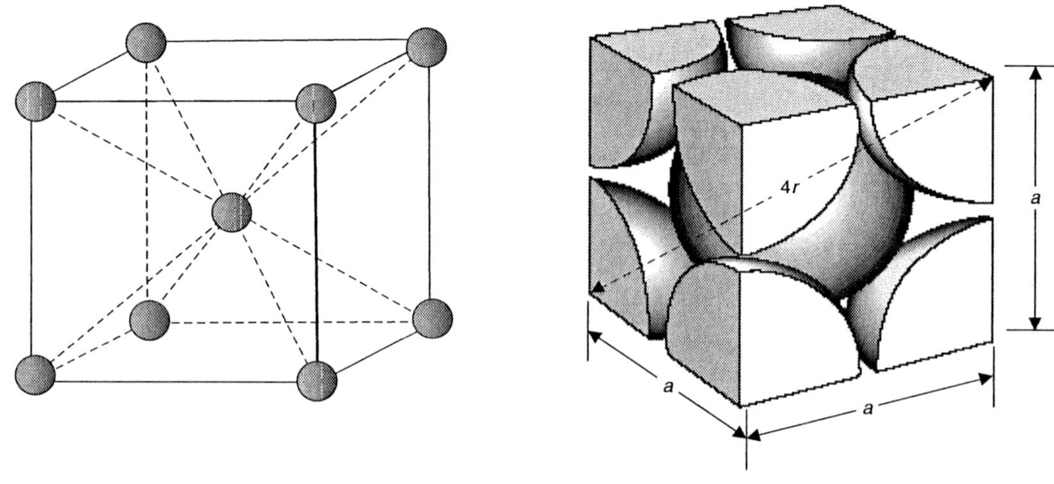


**Fig. B.2** Body-centered cubic structure

the number of atoms belonging to the bcc unit cell is:

$$N = 8 \times \frac{1}{8} + 1 = 2 \quad \text{(Eq B.8)}$$

The relationship between the lattice parameter $a$ and the atomic radius $r$ is:

$$a = \frac{4r}{\sqrt{3}} \quad \text{(Eq B.9)}$$

The APF can then be calculated to be:

$$\text{APF} = \frac{\frac{4N\pi r^3}{3}}{a^3} = \frac{\frac{4N\pi r^3}{3}}{\frac{64r^3}{3\sqrt{3}}} \quad \text{(Eq B.10)}$$

$$\text{APF} = \frac{\pi\sqrt{3}}{8} = 0.68 \quad \text{(Eq B.11)}$$

Since the APF for the bcc structure is significantly higher (0.68) than that for the simple cubic structure (0.52), the bcc structure is significantly denser, and many important metals, α-iron, for example, crystallize into the bcc structure.

The coordination number, or number of nearest neighbor atoms, for the bcc structure is eight (CN = 8), since the full center atom is in contact with eight neighboring atoms located at the corner points of the lattice.

### *B.1.3 Face-Centered Cubic System*

The fcc system is shown in Fig. B.3. As the name implies, in addition to the corner atoms, there is an atom centrally located on each face. Since each of the atoms located on the faces belongs to two unit cells and the eight corner atoms each belongs to eight unit cells, the number of atoms belonging to a unit cell is:

$$N = 8 \times \frac{1}{8} + 6 \times \frac{1}{2} = 4 \quad \text{(Eq B.12)}$$

Since, in the fcc structure, atoms are in contact along the face diagonals, the relationship between the lattice parameter and atomic radius is:

$$a = \frac{4r}{\sqrt{2}} \quad \text{(Eq B.13)}$$

The APF can then be calculated as:

$$\text{APF} = \frac{\frac{4N\pi r^3}{3}}{a^3} = \frac{\frac{4N\pi r^3}{3}}{\frac{64r^3}{2\sqrt{2}}} \quad \text{(Eq B.14)}$$

Substituting $N = 2$ into the equation, the APF becomes:

$$\text{APF} = \frac{\pi\sqrt{2}}{6} = 0.74 \quad \text{(Eq B.15)}$$

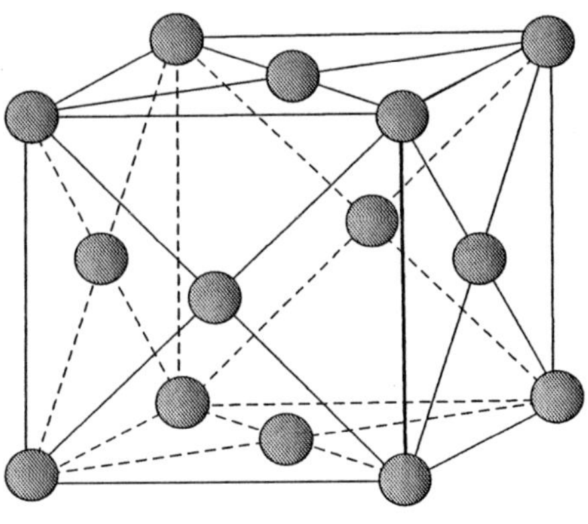

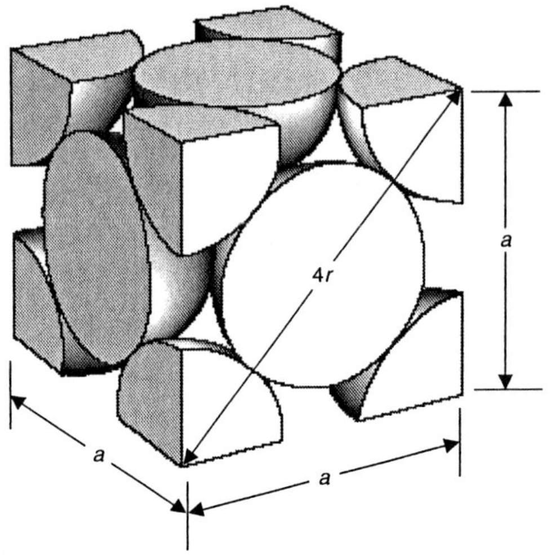


**Fig. B.3** Face-centered cubic structure

Since 74% of the fcc lattice is occupied by atoms, this is an even denser packing than that for the bcc structure. It occurs in many important metals, such as aluminum, copper, and nickel.

The fcc structure has a CN = 12. With 12 nearest atom neighbors, the fcc structure is the most efficient of the cubic structures.

## B.2 Hexagonal System

Another important crystal system is the hexagonal close-packed (hcp) system, shown in Fig. B.4. The lattice parameters are:

$$a_1 = a_2 = a_3 \neq c$$

$$\alpha_1 = \alpha_2 = \alpha_3 = 120°; \ \gamma = 90°$$

To determine the number of atoms belonging to a unit cell, note that atoms at the corner points belong to the six neighboring lattices, and they contribute 1/6 of an atom to the unit cell. Atoms at the face centers belong to two adjacent lattices and contribute 1/2 of an atom to the unit cell. Finally, the three atoms in the interior belong to the unit cell. Therefore, the number of atoms belonging to the unit cell is:

$$N = 12 \times \frac{1}{6} + 2 \times \frac{1}{2} + 3 \times 1 = 6 \quad \text{(Eq B.16)}$$

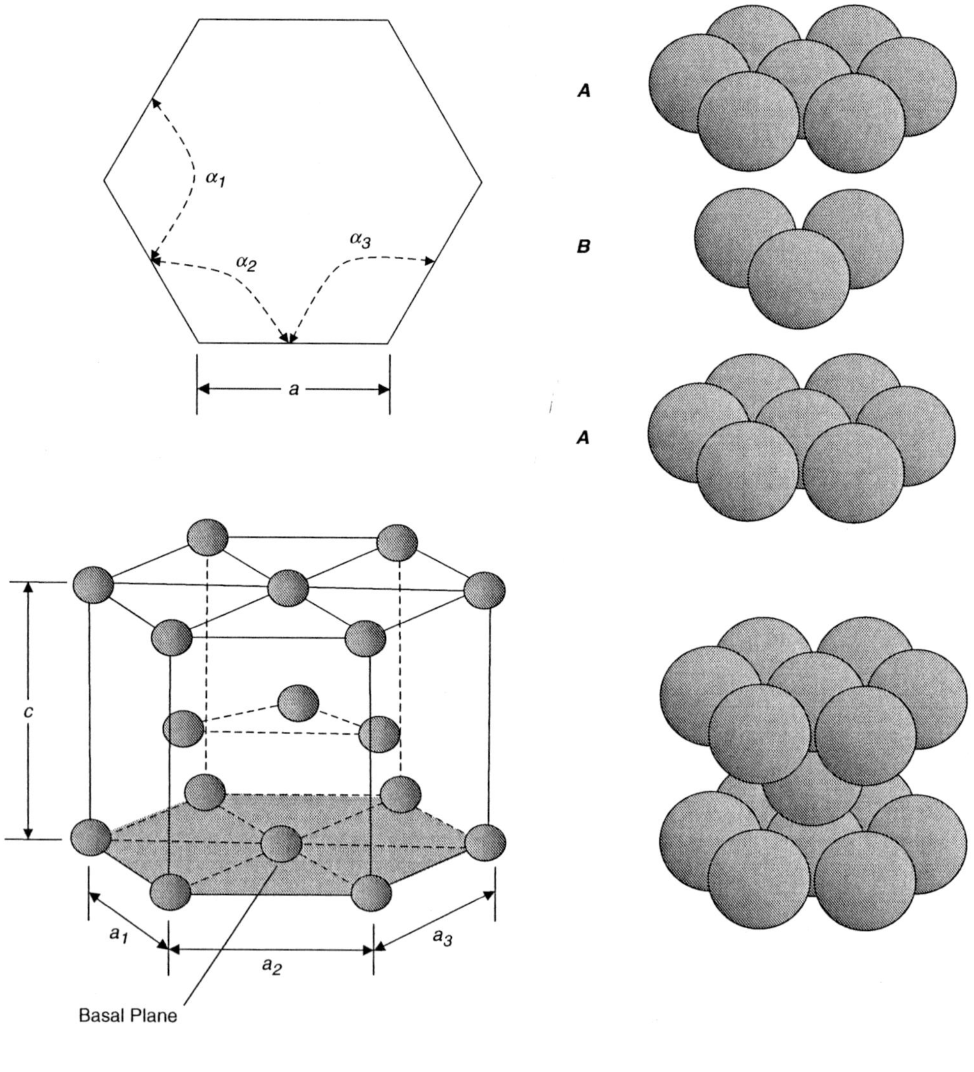


**Fig. B.4** Hexagonal closed-packed structure

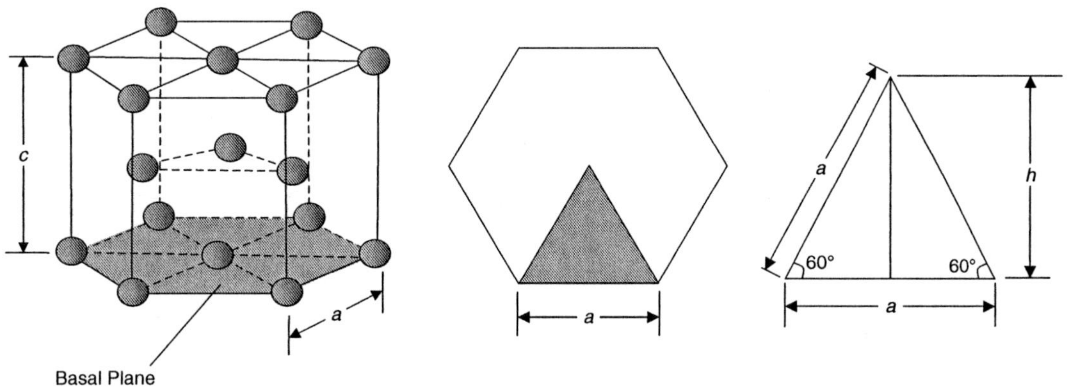


**Fig. B.5** Calculation of volume of hexagonal lattice

In an ideal hcp structure, the ratio of the lattice constants $c/a$ is 1.633. In this ideal packing arrangement, the layer between the two basal planes in the center of the structure is located close to the atoms on the upper and lower basal planes. Therefore, any atom in the lattice is in contact with 12 neighboring atoms, and thus CN = 12. It should be noted that there is often some deviation from the ideal ratio of $c/a = 1.633$. If the ratio is less than 1.633, it means that the atoms are compressed in the $c$-axis direction, and if the ratio is greater than 1.633, the atoms are elongated along the $c$-axis.

The relationships between the lattice parameters $a$ and $c$ and the atomic radius $r$ for an ideal hcp structure ($c/a = 1.633$) are:

$$a = 2r$$
$$c = 1.6333 \times 2r = 3.266r \quad \text{(Eq B.17)}$$

To determine the APF, first determine the volume of the hexagonal lattice using Fig. B.5. The area of the base hexagon is:

$$A = 3a^2 \sin 60^\circ = \frac{3\sqrt{3}}{2}a^2 \quad \text{(Eq B.18)}$$

The volume of the lattice is:

$$V_c = \frac{3\sqrt{3}}{2} 1.633a^3 \quad \text{(Eq B.19)}$$

The APF is then:

$$\text{APF} = \frac{V_a}{V_c} = \frac{\frac{4N\pi r^3}{3}}{\frac{3\sqrt{3}}{2} 1.633a^3} \quad \text{(Eq B.20)}$$

Substituting $a = 2r$ in Eq B.20 gives:

$$\text{APF} = \frac{N\pi}{(1.633)9\sqrt{3}} \quad \text{(Eq B.21)}$$

Finally, substituting the number of atoms belonging to the unit cell, $N = 6$, gives:

$$APF = 0.74 \quad \text{(Eq B.22)}$$

Note that this is the same value as was obtained for the fcc structure. Also, the coordination number obtained for the fcc structure, CN = 12, is the same as that for the hcp structure. A basic rule of crystallography is that if the coordination numbers of two different unit cells are the same, then they will both have the same packing factors. It should also be noted that both the fcc and hcp structures are what is known as close-packed structures with crystallographic planes having the same arrangement of atoms; however, the order of stacking the planes is different.

## ACKNOWLEDGMENTS

This appendix was adapted from *Physical Metallurgy for Engineers* by M. Tisza, ASM International, 2001.

## REFERENCE

1. M. Tisza, *Physical Metallurgy for Engineers,* ASM International, 2001

# APPENDIX C

# Crystallographic Planes and Directions

## C.1 Miller Indices for Cubic Systems

Special planes and directions within metal crystal structures play an important part in plastic deformation, hardening reactions, and other aspects of metal behavior. Crystallographic planes are identified by their Miller indices. Consider the general plane shown in Fig. C.1. The Miller indices for a given plane can be determined as follows:

1. The plane should be displaced, if necessary, with a parallel displacement to a position where it does not pass through the origin of the coordinate system. This is permissible because the coordinate system chosen is an arbitrary one that could be chosen anywhere on the vast expanse of a space lattice, and, as will be seen, parallel planes are equivalent.
2. The *a, b,* and *c* intercepts of the plane with the *x, y,* and *z* coordinates are determined.
3. The reciprocals of these numbers are determined. The reciprocals are denoted by the letters *h, k,* and *l,* respectively:

$$\frac{1}{a} \rightarrow h; \quad \frac{1}{b} \rightarrow k; \quad \frac{1}{c} \rightarrow l$$

4. The reciprocals normally result in fractions. Apply either multiplication or division to determine the smallest set of *h, k,* and *l* values that yield integer numbers. These numbers, when enclosed in brackets, such as $(h\ k\ l)$, give the Miller indices for the plane.

Several examples can be used to best show the calculation method. The Miller indices for the plane shown in Fig. C.2(a) would be:

1. Intercepts $x = 1, y = 1, z = 1$
2. Reciprocals $\frac{1}{x} = 1,\ \frac{1}{y} = 1,\ \frac{1}{z} = 1$
3. No fractions to clear
4. (111)

The Miller indices for the plane shown in Fig. C.2(b) would be:

1. The plane never intercepts the *z*-axis, so $x = 1, y = 2, z = \infty$.
2. Reciprocals $\frac{1}{x} = 1,\ \frac{1}{y} = \frac{1}{2},\ \frac{1}{z} = \frac{1}{\infty} = 0$
3. Clear fractions $\frac{1}{x} = 2,\ \frac{1}{y} = 1,\ \frac{1}{z} = \frac{1}{\infty} = 0$
4. (210)

The Miller indices for the plane shown in Fig. C.2(c) would be:

1. Since the plane passes through the origin, the plane must be moved. By moving it one lattice parameter on the *y*-axis, $x = \infty$, $y = -1, z = \infty$.
2. Reciprocals $\frac{1}{x} = 0,\ \frac{1}{y} = -1,\ \frac{1}{z} = 0$
3. No fractions to clear
4. $(0\bar{1}0)$ Note that the negative intercept of $-1$ on the *y*-axis is notated with a bar over the 1.

Finally, examine Fig. C.2(d) in which all of the side faces of the cube are indicated with planes. Remembering that the space lattice, for all purposes, is infinite in all three coordinate directions, and that the designation of the coordinate system *x, y,* and *z* is also arbitrary, it can be concluded that all of the face planes in the

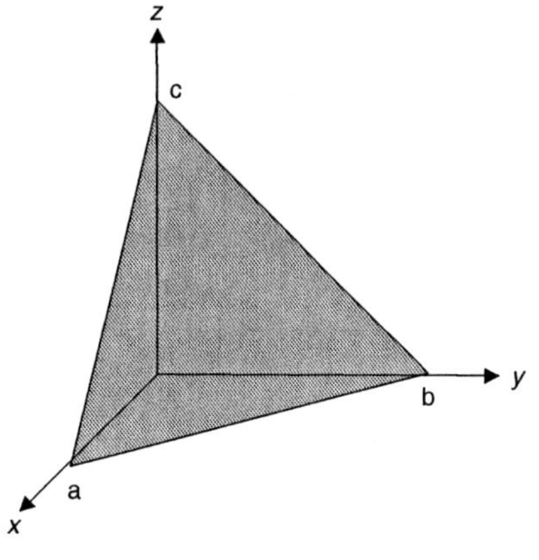


**Fig. C.1** Crystallographic plane indices for cubic systems

cube shown have the same atomic arrangement. These are known as crystallographic equivalent planes. A little analysis will show that all of these planes consist of the same three numbers, namely 1,0,0. Crystallographic equivalent planes are known as family planes and denoted by putting the integers in braces. In this case, the notation {100} represents all side faces of the cube collectively as a family plane.

## C.2 Miller-Bravais Indices for Hexagonal Crystal Systems

The crystalline planes in hexagonal crystal systems are identified in a slightly different manner. Besides the *h, k,* and *l* indices, there is a fourth index, *i,* that is used for hexagonal systems. In the Miller-Bravais system, the three axes in the basal plane are denoted by $a_1$, $a_2$, and $a_3$ indices, while *l* is used to denote the intercept with the *z*-axis. Thus, a plane will have the designation $(h\ k\ i\ l)$. As mentioned earlier, there are two different lattice parameters. The lattice parameter *a* is measured along the three axes of the basal plane, while the *c*-axis is measured in the direction of the *z*-axis perpendicular to the basal plane. Thus, $a_1 = a_2 = a_3 = a \neq c$.

The Miller indices for the basal plane shown in Fig. C.3(a) would be:

1. Intercepts $a_1 = a_2 = a_3 = \infty$; $c = 1$
2. Reciprocals $\frac{1}{a_1} = \frac{1}{a_2} = \frac{1}{a_3} = 0$; $\frac{1}{c} = 1$
3. No fractions to clear
4. (0001)

The Miller indices for the plane shown in Fig. C.3(b) would be:

1. Intercepts $a_1 = 1$, $a_2 = 1$, $a_3 = -\frac{1}{2}$; $c = 1$
2. Reciprocals $\frac{1}{a_1} = 1$, $\frac{1}{a_2} = 1$, $\frac{1}{a_3} = -2$; $\frac{1}{c} = 1$
3. No fractions to clear
4. $(11\bar{2}1)$

Since the four-axes representation contains redundancy, the equation $h + k = -i$ is valid for

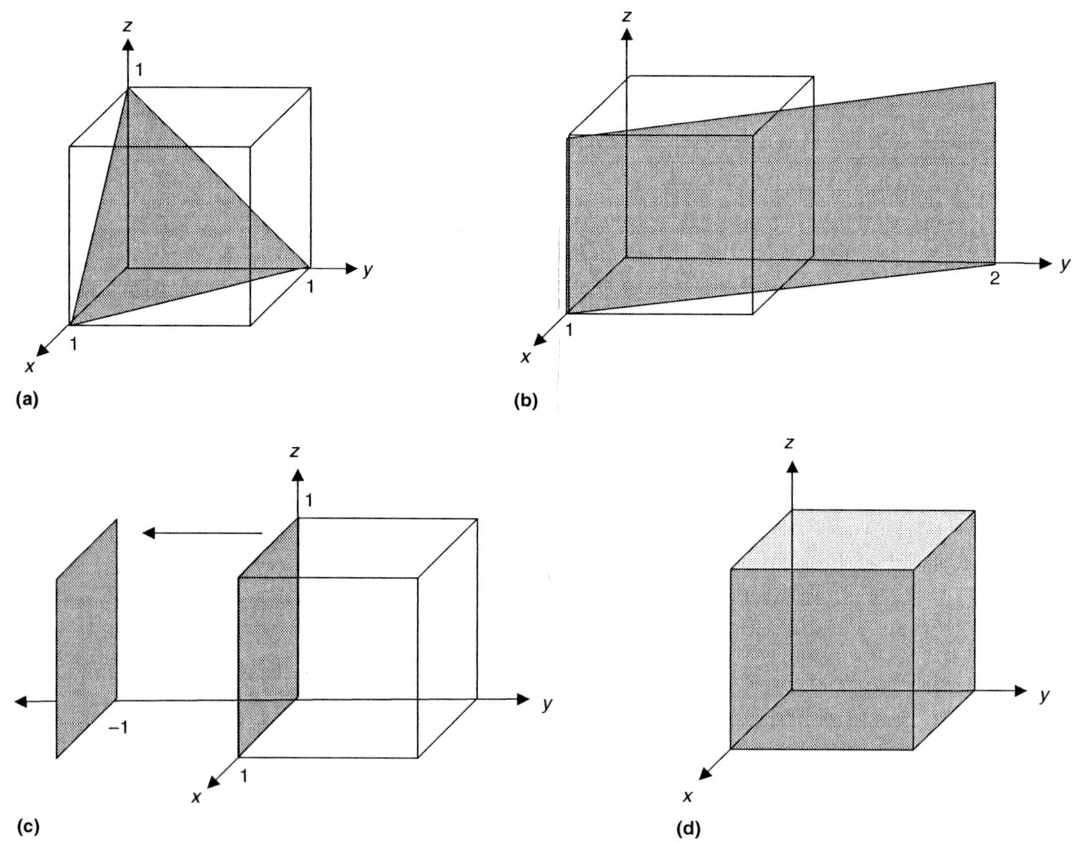


**Fig. C.2** Examples of Miller indices for planes

the indices. That is, the third index in the basal plane can be calculated from the other two.

## C.3 Crystallographic Directions in Cubic Crystal Structures

Directions in crystals are indicated by bracketed indices, for example, [110], and in the cubic system, the direction is always perpendicular to the plane having the same indices (Fig. C.4). More generally, a line in a given direction, such as [110], can be constructed in the following manner. Draw a line from the origin through the point having the coordinates $x = 1$, $y = 1$, $z = 0$, in terms of axial lengths. This line and all lines parallel to it are then in the given direction. Note that

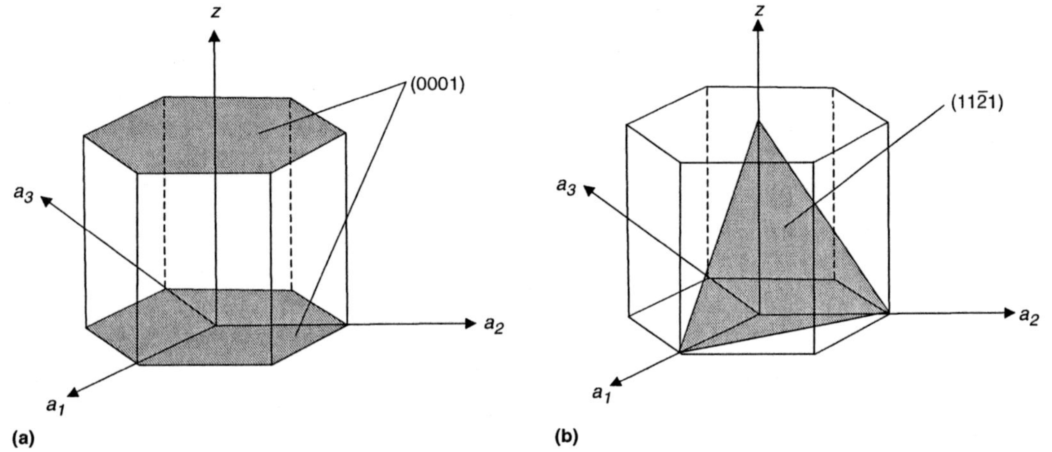


**Fig. C.3** Examples of Miller-Bravais indices for hexagonal planes

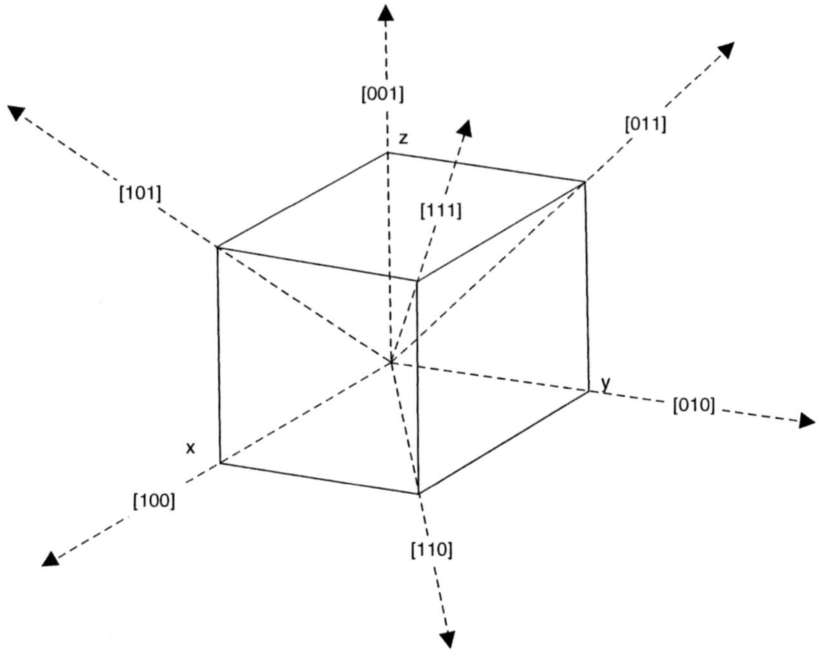


**Fig. C.4** Important direct o ıs in cubic cell

reciprocals are not involved in obtaining the Miller indices of directions. A family of equivalent directions, such as [100], [$\bar{1}$00], [010], [0$\bar{1}$0], [001], and [00$\bar{1}$], is designated as {100}.

## C.4 Crystallographic Directions in Hexagonal Crystal Structures

The direction system for the Miller-Bravais system for hexagonal crystal structures is extremely cumbersome and confusing. Miller-Bravais indices of direction are also expressed in terms of four digits. Like the indices for planes, the third digit must always equal the negative sum of the first two digits (i.e., $h+k=-i$). Thus, if the first two digits are 2 and $-1$, then the third digit must be $-1$; that is $(2)+(-1)=1 \rightarrow -1$. The method for finding the direction of the $a_1$ axis is shown in Fig. C.5. This axis has the same direction as the vector sum of the three vectors, one of them of length $+2$ along the $a_1$ axis, another of length $-1$ along the $a_2$ axis, and the third of length $-1$ along the $a_3$ axis. This yields direction indices of [2$\bar{1}\bar{1}$0]. This cumbersome method is required to satisfy the relationship $h+k=-i$. The corresponding indices of the $a_2$ and $a_3$ axes are [$\bar{1}$2$\bar{1}$0] and [$\bar{1}\bar{1}$20], respectively.

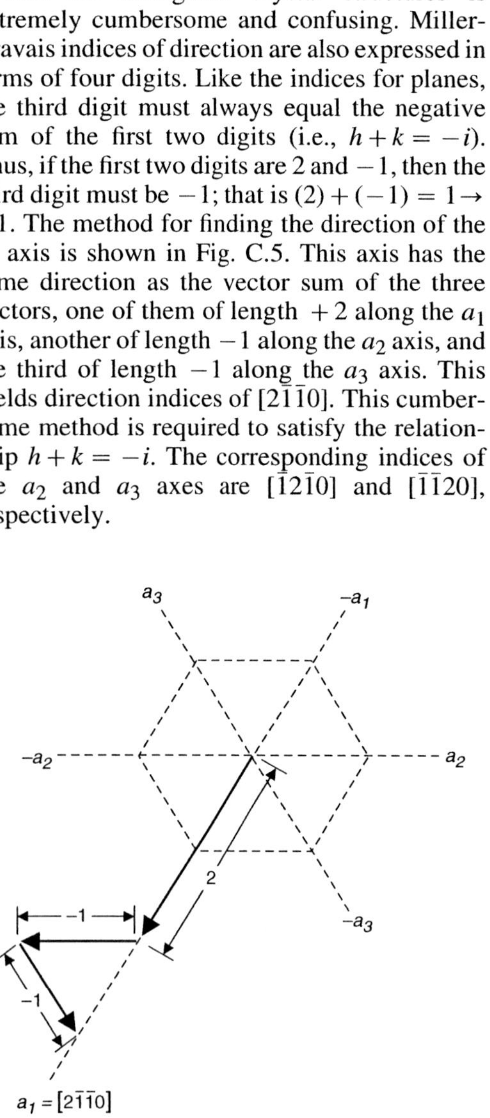


**Fig. C.5** Example of Miller-Bravais directional indices for hexagonal planes. Source: Ref 1

## C.5 X-Ray Diffraction for Determining Crystalline Structure

X-rays are a form of high-energy electromagnetic radiation with short wavelengths. Since the wavelengths of x-rays are approximately the same as the atomic spacings of metallic lattices, several x-ray techniques have been developed over the years to determine crystalline structure. Diffraction occurs when a wave encounters a series of regularly spaced obstacles that are capable of scattering the wave and have spacings that are comparable in magnitude to the wavelength. A diffracted beam is one in which a large number of scattered waves mutually reinforce one another.

When a beam of x-rays with a wavelength λ strikes a set of crystalline planes at some arbitrary angle, there will usually be no reflected beam because the rays reflected from the various crystal planes must travel paths of different lengths. In other words, although the incident rays are in phase, the reflected rays are out of phase and thus cancel one another. However, if each ray is out of phase with the preceding one by exactly one wavelength, or any whole integer wavelength ($n=1, 2, 3, \ldots$), then the reflected beam will consist of rays that are in phase again. The angle at which reflection occurs is known as the Bragg angle, θ.

Two parallel planes of atoms identified as $A$-$A'$ and $B$-$B'$, which have the same $h$, $k$, and $l$ Miller indices separated by the interplanar spacing $d_{hkl}$, are shown in Fig. C.6. Two in-phase rays, labeled 1 and 2, having a wavelength λ strike the surface at an angle θ. Upon impact, rays 1 and 2 are scattered by atoms $P$ and $Q$. Constructive interference of the scattered rays 1′ and 2′ occurs at an angle to the planes, if the path length difference between 1-$P$-1′ and 2-$Q$-2′ (i.e., $\overline{SQ}+\overline{QT}$) is equal to a whole number, $n$, of wavelengths. This condition for diffraction is:

$$n\lambda = \overline{SQ}+\overline{QT}$$

$$n\lambda = d_{hkl}\sin\theta + d_{hkl}\sin\theta$$

$$n\lambda = 2d_{hkl}\sin\theta \qquad \text{(Eq C.1)}$$

Equation C.1 is known as Bragg's law. Also, $n$ is the order of reflection, which may be any integer (1, 2, 3, ...) consistent with $\sin\theta$ not exceeding unity. Bragg's law is a simple expression relating the x-ray wavelength and interatomic spacing to the angle of the diffracted

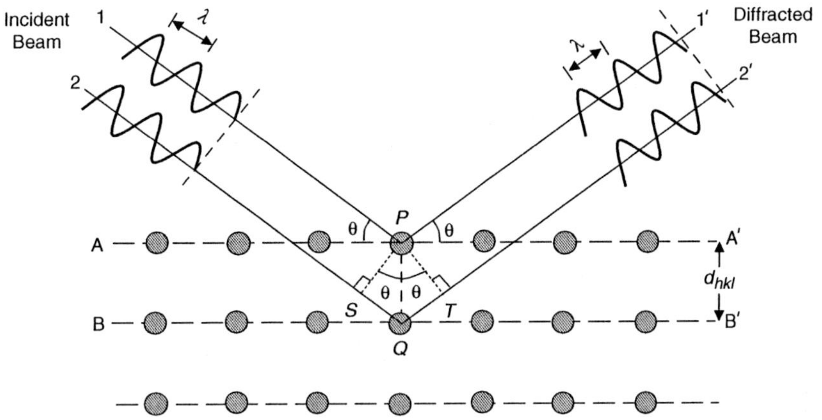


**Fig. C.6** Diffraction of x-rays by planes of atoms. Source: Ref 2

beam. If Bragg's law is not satisfied, then the reflected rays will be out of phase and cancel each other. Although for a given metallic substance, only a few values for $n$ and $d_{hkl}$ will satisfy Bragg's law, the values for λ and θ can be varied continuously over a wide range.

The magnitude of the distance between two adjacent and parallel planes of atoms (i.e., the interplanar spacing $d_{hkl}$) is a function of the Miller indices ($h$, $k$, and $l$) as well as the lattice parameter(s). For example, for crystal structures with cubic symmetry:

$$d_{hkl} = \frac{a}{\sqrt{h^2 + k^2 + l^2}} \qquad \text{(Eq C.2)}$$

where $a$ is the lattice parameter. There are also relationships similar to this one for the cubic system for the other crystalline systems.

There are a number of x-ray techniques that have been developed for studying crystalline structures. The Laue method uses a narrow collimated beam of white x-rays that strike a stationary single crystal. It is not suitable for the determination of lattice parameters but is used for the determination of crystal orientation and the detection of lattice imperfections. The rotating single-crystal method uses a constant wavelength beam while the diffracting crystal is rotated. This method is used mostly for determining the structure of single crystals. The Debye-Scherrer powder method uses an x-ray beam of constant wavelength and a specimen consisting of thousands of tiny crystals. Since there are a large number of powder particles with many different orientations, the diffracted beam produces a cone of radiation. Different reflection

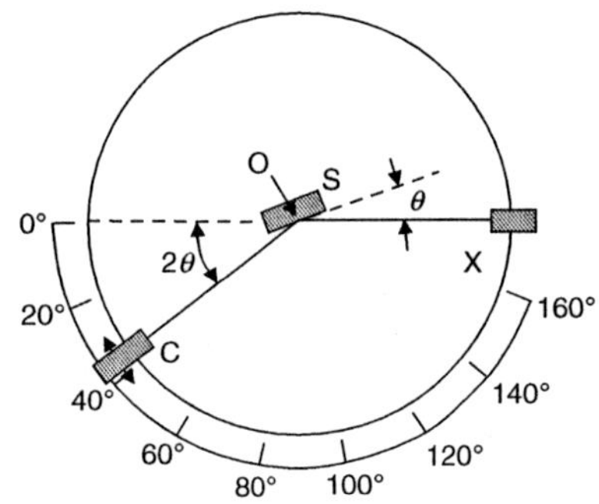


**Fig. C.7** Schematic of x-ray diffractometer. Source: Ref 2

cones are recorded on a film strip that allows the diffraction angles to be measured. The distance $d_{hkl}$ between the reflecting atomic planes is then calculated using Bragg's law.

In addition to the photographic methods, there are diffractometer methods that measure the intensity of the beam in counts per second diffracted from the specimen over a range of angles. A powdered or polycrystalline specimen consisting of many fine, randomly oriented particles is exposed to monochromatic x-radiation. Each powder particle (or grain) is a crystal, and having a large number of them with random orientations ensures that some particles are properly oriented such that every possible set of crystallographic planes will be available for diffraction. In a typical diffractometer (Fig. C.7), a specimen "S" is supported so that it

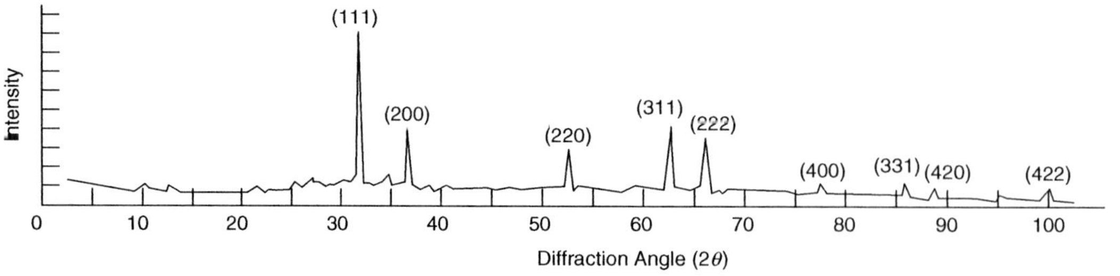


**Fig. C.8** Typical diffraction pattern. Source: Ref 2

rotates about the axis labeled "O." The monochromatic x-ray beam is generated at point "X," and the intensities of diffracted beams are detected with a counter labeled "C." The counter is mounted on a movable carriage that may also be rotated about the O-axis, with its angular position in terms of 2θ marked on a graduated scale. The carriage and specimen are mechanically coupled such that a rotation of the specimen through θ is accompanied by a 2θ rotation of the counter, which assures that the incident and reflection angles are maintained equal to one another. Collimators are incorporated within the beam path to produce a well-defined and focused beam. Filters are used to provide a near-monochromatic beam. As the counter moves at constant angular velocity, a recorder automatically plots the diffracted beam intensity monitored by the counter as a function of 2θ, where 2θ is the diffraction angle and measured experimentally. A diffraction pattern for a powdered specimen of lead is shown in Fig. C.8. The high-intensity peaks result when the Bragg diffraction condition is satisfied by some set of crystallographic planes. One of the primary uses of x-ray diffractometry is for the determination of crystal structure. The unit cell size and geometry may be resolved from the angular positions of the diffraction peaks, while the arrangement of atoms within the unit cell is associated with the relative intensities of these peaks.

## REFERENCES

1. R.E. Reed-Hill and R. Abbaschian, *Physical Metallurgy Principles,* 3rd ed., PWS Publishing Company, 1991
2. W.D. Callister, *Fundamentals of Materials Science and Engineering,* 5th ed., John Wiley & Sons, Inc., 2001

## SELECTED REFERENCES

- A.G. Guy, *Elements of Physical Metallurgy,* 2nd ed., Addison-Wesley Publishing Company, 1959
- M. Tisza, *Physical Metallurgy for Engineers,* ASM International, 2001

# Index

## A

# B

## D

## E

## G

# H

## N

# O



# P

## T

## Z